ASTRONOMY

Astronomy

The Human Quest for Understanding

Dale A. Ostlie

Weber State University

OXFORD
UNIVERSITY PRESS

OXFORD
UNIVERSITY PRESS

Great Clarendon Street, Oxford, OX2 6DP,
United Kingdom

Oxford University Press is a department of the University of Oxford.
It furthers the University's objective of excellence in research, scholarship,
and education by publishing worldwide. Oxford is a registered trade mark of
Oxford University Press in the UK and in certain other countries

Published in the United States of America by Oxford University Press
198 Madison Avenue, New York, NY 10016, United States of America

British Library Cataloguing in Publication Data
Data available

Library of Congress Control Number: 2022933866

ISBN 978–0–19–882582–1 (hbk)
ISBN 978–0–19–882583–8 (pbk)

DOI: 10.1093/oso/9780198825821.001.0001

Printed and bound by
CPI Group (UK) Ltd, Croydon, CR0 4YY

Dedicated to my parents

Dean Arthur Ostlie (1931–2015)
and
Dorothy Marie Carlson Ostlie (1931–2021)

My father served as a loving parent and my personal role model throughout my life. His quiet strength and leadership was an inspiration to me.

My mother's gifts of love and ceaseless commitment to her five children were truly remarkable.

No one could have asked for two more wonderful parents.

Acknowledgments

The writing of *Astronomy: The Human Quest for Understanding* took much longer than I had ever imagined. With so many new developments in astronomy and astrophysics over the past decade, simply keeping up-to-date with all of the scientific discoveries has practically been a full-time job, but then that is what makes the study of astronomy and astrophysics so rewarding. On too many occasions to count, new data and revelations required going back to previously completed sections of the book and making revisions. No doubt other significant discoveries will occur before the book ultimately appears in students' hands, or on their computer screens.

I must thank everyone at Oxford University Press for their endless patience when it must have seemed like this nearly nine-year project would never see the light of day. In particular, I extend my deepest thanks to Harriet Konishi for her early and enthusiastic support, and for her understanding with too many missed deadlines. I also greatly appreciate the guidance of Sonke Adlung and Francesca McMahon, who were willing to provide the time and flexibility to make this book the best this author can hope to produce. Of course, I would be remiss if I didn't thank my project editor, Giulia Lipparini, for her extremely helpful advice and for seeing the book to publication. In addition to Oxford University Press, the team at Integra Software Services was critical in finalizing the LaTeX 2_ε project and producing the print and ebook versions of this text. Specifically I would like to thank Sharmila Radha and Sam Augustin Durai Ebenazer, Project Managers, along with copy-editor Gwynneth Drabble for her very meticulous review of the text. This book is immensely improved through the many suggestions of everyone who participated in its production.

Colleagues and reviewers have also been critical in strengthening the book and ensuring its accuracy. Of course, despite every effort to make this textbook as error-free as possible, any mistakes or flaws in arguments are the complete responsibility of the author.

Finally, I reserve my greatest thanks to my wife, Candy, and to our wonderful adult children, Michael and Megan, who have supported me throughout. After transitioning to the status of "empty nesters," Candy and I made the decision to sell our home and move into our truck camper, touring the United States and parts of Canada from coast to coast while visiting as many national parks and monuments as possible. Remarkably, with the growing level of wireless connectivity across the nation, we have also been able to continue working remotely and full-time. This adventure in a new work-at-"home"/travel/recreation model has been truly rewarding, fun, and productive. At a time when Covid-19 was redefining the work environment for many millions of people, our already-nomadic lifestyle proved to be both timely and effective. Without my traveling companion's ongoing love, support, and remarkable patience with me in our tiny home with an amazing and constantly changing backyard, this project would have never been completed.

I hope that you enjoy *Astronomy: The Human Quest for Understanding*, and that it conveys to you the amazing history, current knowledge, and the excitement of discovery and awe that is astronomy and astrophysics. Always remember to keep looking up!

Dale A. Ostlie, Emeritus Professor
Weber State University
Ogden, Utah, USA

Contents

List of Tables   xi

Introduction   xiii
    About This Textbook   xiii
    Aids and Suggestions for Students   xvi

I The Process of Science through the Lens of Astronomy   1

1 The Nature of Science   3
    1.1 Science as a Human Enterprise   4
    1.2 Characteristics of a Scientific Theory   5
    1.3 Scientific Theories Must Be Self-Consistent   6
    1.4 Scientific Revolutions Are Rare and Alter our Paradigms   6
    1.5 Peer Review and the Self-Correcting Nature of Science   7
    1.6 What Is Pseudoscience?   8
    1.7 Science and Society: The Issue of Terminology   9
    Exercises   10

2 The Heavens: Realm of Mystery   12
    2.1 Observing the Changing Sky with the Naked Eye   13
    2.2 The Skywatchers   26
    Exercises   51

3 On the Path toward Modern Science   55
    3.1 Sumerian and Babylonian Astronomy   56
    3.2 The Ancient Greeks and the Dawn of Science   59
    3.3 An Opposing View: The Greek Heliocentric Model   70
    3.4 Medieval Astronomy in India and the Middle East   72
    3.5 The Celestial Sphere   74
    Exercises   86

4 The Copernican Revolution   91
    4.1 The European Renaissance   92
    4.2 The Heliocentric Universe of Nicolaus Copernicus 93
    4.3 The Motions of the Stars in the Heliocentric Universe   97
    4.4 The Changing Seasons   100
    4.5 The Source of Hipparchus's Precession   102
    4.6 The Moon's Motions: Phases and Eclipses   104
    4.7 Finally, a Natural Solution to the "Wandering Stars"   112
    4.8 The Quest for Understanding Continues   115
    4.9 Tycho Brahe and His Passion for Precision   119
    4.10 Johannes Kepler and Planetary Orbits   121
    4.11 Kepler's Three Laws in Their Original Form   125
    4.12 Galileo Galilei, the First True Scientist   131
    Exercises   141

5 Sir Isaac Newton's Universe   **145**
  5.1   The Life of Sir Isaac Newton   146
  5.2   Newton's Three Laws of Motion   154
  5.3   The Universal Law of Gravitation   165
  5.4   Newton Completes Kepler's Laws   173
  5.5   The Discovery of Uranus, Neptune, and Pluto   176
  5.6   The Enlightenment (The Age of Reason)   178
        Exercises   181

6 The Universality of Physical Law   **185**
  6.1   Determining the Size of Our Solar System   186
  6.2   Measuring the Speed of Light   189
  6.3   Weighing Earth with a Laboratory Balance: Measuring Big G 191
  6.4   The Conservation Laws   192
  6.5   Light is an Electromagnetic Wave   201
  6.6   Blackbody Radiation   213
  6.7   The Inverse Square Law of Light   220
  6.8   Heat as a Form of Energy   223
        Exercises   231

7 Revealing Secrets Hidden in Light and Matter   **237**
  7.1   Prisms, Scattering, Diffraction, and Spectral Lines   238
  7.2   The Periodic Table of the Elements   247
  7.3   Detecting Helium and Other Elements in the Sun   251
  7.4   Particles Smaller than Atoms   253
        Exercises   260

8 Modern Physics: New Science to Study the Universe   **263**
  8.1   Blackbody Radiation and Planck's Energy Quanta   264
  8.2   The Life of Albert Einstein   266
  8.3   The Special Theory of Relativity   274
  8.4   The Doppler Effect for Light   281
  8.5   Einstein's Explanation of Planck's Energy Quanta   285
  8.6   Bohr's Atom   288
  8.7   Quantum Mechanics: What Are the Chances?   293
  8.8   The Nucleus and Isotopes   298
        Exercises   301

II The Sun, Our Solar System, Exoplanets, and Life   307

9 The Sun, Our Parent Star   **309**
  9.1   What Is Our Sun Like?   310
  9.2   The Sun's Dynamic Atmosphere   317
  9.3   The Solar Cycle   328
  9.4   Using Computers for Exploration   334
  9.5   How to Build a Star   336
  9.6   The Sun's Nuclear Furnace   342
  9.7   Test, Test, and Test Some More   349
        Exercises   354

10 An Overview of the Solar System   **358**

10.1   Taking Inventory: A Taxonomy of the Solar System   359

10.2   Putting It in Perspective: A Scale Model   367

10.3   The Importance of Tides   370

        Exercises   376

11 The Rocky Planets and Our Moon   **379**

11.1   Mercury   380

11.2   Venus   386

11.3   Earth   394

11.4   The Moon   405

11.5   Mars   408

11.6   Their Interiors   421

11.7   Population Growth and the Exponential Function   431

11.8   Global Warming: Humans Impacting Environment   437

        Exercises   454

12 The Giant Planet Systems   **460**

12.1   Comparing the Giant Planets   462

12.2   The Moons of the Giant Planets   476

12.3   Rings around the Planets   491

        Exercises   497

13 Dwarf Planets and Small Bodies   **501**

13.1   Dwarf Planets   502

13.2   Asteroids and Meteorites   510

13.3   Comets   520

13.4   Impacts throughout the Solar System   530

        Exercises   536

14 Planets Everywhere and the Search for Extraterrestrial Life   **541**

14.1   Exoplanets in Abundance   542

14.2   The Development of Planetary Systems   551

14.3   In Search of Extraterrestrial Life   566

        Exercises   582

III   Stars and the Universe Beyond   **587**

15 Measuring the Stars   **589**

15.1   Observing the Cosmos across Wavelengths   590

15.2   The Brightness of Stars   599

15.3   How Far Away are they?   604

15.4   Determining Stellar Masses in Binary Star Systems   611

15.5   The Harvard Spectral Classification Scheme   616

15.6   Calculating the Radii of Stars   623

15.7   The Universality of Physical Law Revisited   624

15.8   The Birth of the Hertzsprung–Russell Diagram   627

15.9   Finding Patterns and Asking Why   633

        Exercises   634

16 The Lives of Stars   **638**
    16.1   The Interstellar Medium: The Realm of Gas and Dust   640
    16.2   A Star is Born   646
    16.3   The Life of an Aging Star   653
    16.4   The Pulsating Stars   665
    16.5   Clusters as Tests of Stellar Evolution Theory   673
            Exercises   678

17 The End of a Stellar Life   **684**
    17.1   Winds and White Dwarfs   685
    17.2   Supernovas: Going out with a Bang   691
    17.3   Neutron Stars   700
    17.4   Black Holes: Gravity's Ultimate Victory   709
            Exercises   721

18 Galaxies Galore   **726**
    18.1   Historical Studies of the "Universe"   727
    18.2   The Milky Way Galaxy   733
    18.3   A Taxonomy of Galaxies   749
    18.4   Galaxies Everywhere   754
    18.5   Supermassive Black Holes and Active Galaxies   760
    18.6   How to Build a Galaxy   765
            Exercises   769

19 The Once and Future Universe   **776**
    19.1   The Expanding Universe   777
    19.2   The Big Bang and Spacetime   779
    19.3   The Accelerating Universe   786
    19.4   The Early Universe   790
    19.5   The Λ-Cold Dark Matter Model   806
    19.6   The Meaning of "The Universe"   810
    19.7   The Scientist's Crystal Ball: Looking to the Future   815
            Exercises   821

Appendices
    A   The 88 constellations   829
    B   The 25 brightest stars   833
    C   The Messier catalog   834

Selected Answers   837

Credits   842

Glossary   843

Index   865

List of Tables

2.1 Day names based on naked-eye astronomical objects   22
2.2 The first five Long Count calendar units used in Mesoamerica   37
2.3 Named months of the Haab´ calendar and named days of the Tzolk´in calendar   38

4.1 The Greek alphabet   113
4.2 Kepler's third law for the naked-eye planets   128
4.3 Orbital data for the Galilean moons   143

5.1 Prefixes in the International System of Units   155

6.1 Wavelength and frequency ranges of the electromagnetic spectrum   213

7.1 The periodic table of the elements   250
7.2 Selected visible wavelengths of various elements (nm)   252

9.1 Characteristics of the Sun   312
9.2 Leptons   346

10.1 Comparing the rocky and giant planets   360
10.2 Dwarf planets and dwarf planet candidates   364
10.3 Scale model of the Solar System   368

11.1 Characteristics of the rocky planets   381
11.2 Historical data and future projections of global population, temperature change, and sea-level rise   451

12.1 Characteristics of the giant planets   463
12.2 Characteristics of the Solar System's largest moons   478
12.3 Freezing and boiling points of some substances   486

13.1 Meteor showers throughout the year   527
13.2 Blackbody temperatures, T_{bb}, across the Solar System as a function of semimajor axes, a 540

14.1 Orbits of TRAPPIST-1 planets   585

15.1 The relationship between magnitude differences and brightness ratios   602
15.2 Properties of main-sequence stars in the Harvard spectral classification scheme   631
15.3 Spectral classes or surface temperatures and absolute magnitudes or luminosities for selected stars   637

18.1 Properties of the Milky Way and its components   735
18.2 Data for an idealized Keplerian orbit and an idealized flat rotation curve   773

19.1 The components of the Λ-CDM model   807

Appendix A The 88 constellations   829
Appendix B The 25 brightest stars   833
Appendix C The Messier catalog   834

Introduction

About This Textbook   xiii

Aids and Suggestions for Students   xvi

About This Textbook

Purpose

Courses in introductory astronomy are offered at colleges and universities for two primary reasons: to provide an exciting general education course in the physical sciences, and to get students interested in astronomy, physics, and planetary sciences with the goal of enticing them to take additional coursework in those disciplines. The second purpose is certainly appropriate for introductory astronomy courses as they are often considered gateway courses to further scientific study, but it is really the general education aspect of introductory astronomy courses that motivates most students. However, most texts seem to be focused on the latter goal by introducing a large amount of detail without motivating the material in ways that resonate with most non-science majors. Unfortunately, this approach often fails to present the inherent excitement and fascination of the discipline. The general education student wants to know how the study of a subject outside of their major is going to be relevant to them personally, how the subject will enrich their lives, and why the subject should be generally important to them. Are there ties to what they are otherwise interested in?

Introductory astronomy courses tend to be very popular at institutions where they are offered. This is largely because astronomy inspires awe and wonder across such a wide range of the population. The beautiful and inspiring images that are provided by the great Earth-based and space-based observatories, together with Solar System exploration missions, constantly capture the human imagination. Astronomy also naturally inspires individuals to ask profound, age-old questions, such as how the universe began, how vast and how old the universe is, how did humanity come into existence, what is humanity's place in the universe, what is the fate of the universe, how old is Earth, are we alone in the universe, what are black holes, and what was the Big Bang?

By its very nature, astronomy is the ideal subject for a physical science general education course. Not only does it naturally inspire fascinating questions, but being the oldest of the sciences and at the same time one of the most rapidly advancing, astronomy readily lends itself to telling the continuing story of science as a human and creative enterprise. Astronomy is replete with inspiring stories of discovery, but also of wrong paths taken, and of revolutions that are not only scientific, but that have also directly impacted human culture, sometimes in very profound ways.

The Human Endeavor That is Science

An integrating theme throughout this text is an ongoing story of discovery, creativity, self-consistent explanations, and the requirement of evidence. Students who take introductory astronomy courses come from every discipline represented at the university. As such, it should be a fundamental goal of a general education science course to help the student understand that science is tightly integrated with our lives in very important and profound ways, even in ways that are not necessarily readily discernible. Said another way, it is important that science general education courses demonstrate their relevancy to students of all academic persuasions. It is also of utmost importance that a general education science course helps students understand that science is a way of knowing, unique in its reliance on continuous empirical testing, and in its ability to be self-correcting. This aspect of a general education science course is perhaps more critical today than it has ever been, given the rise in anti-science rhetoric and the denial of basic facts and well-established knowledge such as global climate change, biological evolution, the age of the universe, and the age of Earth. It is also important for students to understand that science is an integrated whole, and that the discipline of astronomy does not exist in isolation from other disciplines such as mathematics, geology, chemistry, and biology. At their best, all of the scientific disciplines inform one another in the continual progress of scientific understanding.

Student Learning Outcomes

This text is anchored in general education student learning outcomes designed for courses in the physical sciences. Although the specific wording of general education learning outcomes differs from one institution to another, those of one institution, Weber State University, are presented as characteristic examples:

General Education Mission of the Natural Sciences

The mission of the natural sciences general education program is to provide students with an understanding and appreciation of the natural world from a scientific perspective. Science is a way of knowing. Its purpose is to describe and explain the natural world, to investigate the mechanisms that govern nature, and to identify ways in which all natural phenomena are interrelated. Science produces knowledge that is based on evidence and that knowledge is repeatedly tested against observations of nature. The strength of science is that ideas and explanations that are inconsistent with evidence are refined or discarded and replaced by those that are more consistent. Science provides personal fulfillment that comes from understanding the natural world. In addition, experience with the process of science develops skills that are increasingly important in the modern world. These include creativity, critical thinking, problem solving, and communication of ideas. A person who is scientifically literate is able to evaluate and propose explanations appropriately. The scientifically literate individual can assess whether or not a claim is scientific, and distinguish scientific explanations from those that are not scientific.

Natural Sciences Learning Outcomes

After completing the natural sciences general education requirements, students will demonstrate their understanding of general principles of science:

1. **Nature of Science:** Scientific knowledge is based on evidence that is repeatedly examined, and can change with new information. Scientific explanations differ fundamentally from those that are not scientific.
2. **Integration of Science:** All natural phenomena are interrelated and share basic organizational principles. Scientific explanations obtained from different disciplines should be cohesive and integrated.
3. **Science and Society:** The study of science provides explanations that have significant impact on society, including technological advancements, improvement of human life, and better understanding of human and other influences on Earth's environment.
4. **Problem Solving and Data Analysis.** Science relies on empirical data, and such data must be analyzed, interpreted, and generalized in a rigorous manner.

Physical Sciences Learning Outcomes

Students will demonstrate their understanding of the following features of the physical world:

1. **Organization of Systems:** The universe is scientifically understandable in terms of interconnected systems. The systems evolve over time according to basic physical laws.
2. **Matter:** Matter comprises an important component of the universe, and has physical properties that can be described over a range of scales.
3. **Energy:** Interactions within the universe can be described in terms of energy exchange and conservation.
4. **Forces:** Equilibrium and change are determined by forces acting at all organizational levels.

The narrative of *Astronomy: The Human Quest for Understanding* clearly integrates the presentation of empirical data, models, and the theoretical underpinnings of science, and implicitly strives to continually address the mission statement and learning outcomes expressed above.

The Presentation of Data and the Use of Terminology

Introductory astronomy texts typically contain a tremendous amount of data and a barrage of jargon. Although it is certainly the case that data are a fundamental aspect of the sciences, and the means by which we evaluate our understanding of physical systems, an overwhelming amount of data simply confuses the student and leaves them wondering what is important. Is the learning of astronomy at an introductory level and the goal of addressing the stated learning outcomes contingent upon a great deal of memorization of basic data? *Astronomy: The Human Quest for Understanding* has been written with the belief that the answer to that question is no. Given virtually instantaneous access to enormous amounts of data today, how much detailed factual information should students really need to be responsible for in a general education course? This text attempts to keep the amount of data to a minimum, only presenting data within the context of important discoveries, and as justification for basic models and theories. Furthermore, the text is explicit regarding which data should be committed to memory and why, and which data are simply being presented as examples or for comparison purposes. Tabular data are presented for easy look-up as needed.

The same argument applies to the use of terminology in an introductory astronomy text. Texts are often filled with terms for specific features on the surfaces of planets and moons, classifications of solar flares, various layers of the solar corona, or specific classes of active galactic nuclei. While clearly important to the professional, or to students preparing for careers in astronomy, such jargon only confuses the general education student, leaving them to wonder if memorizing terms is the important point in science. *Astronomy: The Human Quest for Understanding* helps students understand that the simple memorization of terms is not equivalent to understanding. In courses designed for students who may only take one course in the physical sciences, memorization should not reflect the nature of the discipline; rather, it is important to always come back to the student learning outcomes as the anchor for what is important in an introductory science course. For quick reference to terms that are used in the text, an extensive glossary is included for reference, in addition to a complete index.

Astronomy is inherently an empirical discipline; however, most introductory texts tend to minimize the use of mathematics or eliminate it entirely in deference to the generally weak mathematical preparation of today's college student. In many cases, the necessary mathematics is relegated to optional text boxes that are too often not read by the student and may be ignored entirely within the course. Even worse, these boxes can lead the student to believe that he or she is simply not intelligent enough to understand the mathematical language of science, even at its most basic level. This is certainly not the impression we want to leave students with who already feel insecure about their mathematical preparation.

A different approach is taken in *Astronomy: The Human Quest for Understanding*. Although the mathematics is kept to a minimum, important key equations are not eliminated, but integrated directly into the narrative as a basic component of the text. This treatment of the mathematics emphasizes its importance in modern science and strengthens the explanations of fundamental concepts. To assist students in understanding the mathematical concepts, relevant equations are given descriptive names and are explicitly interpreted. Examples of how equations are implemented and enhance understanding are also presented. An important secondary outcome of this approach is to further demonstrate the importance of quantitative literacy requirements and abstract thought in a student's general education.

Aids and Suggestions for Students

Aids for Students

This icon indicates that more information or assistance is available on the textbook's website
AstronomyHumanQuest.com

Astronomy: The Human Quest for Understanding has been designed as an e-text that can also be printed in traditional form. To that end careful consideration has been given to the many features of the e-text that can benefit from embedded hyperlinks to other portions of the book or to external Internet resources, including a dedicated website at https://www.AstronomyHumanQuest.com that contains support materials. When only a portion of a URL (universal resource locator) is shown in the text, the link implicitly assumes that it refers to the text's website; as an example, Math_Review can be accessed through the website's homepage.

The following learning aids are contained throughout the text. In the electronic version of the text, embedded internal and external hyperlinks are also available.

- Any time you see a dark red highlighted word or phrase, it signals that additional web-based information is available, including online help through the textbook's website. The online help is typically in the form of tutorials and practice problems associated with the mathematics or graphics included in the text, or videos or animations designed to assist in visualizing concepts. The highlighted words or phrases are hperlinks in the e-text.

- Many of the figures in the text are also available on the text's website for enlargement or further study. Those that aren't available are withheld due to copyright issues. In those cases, links to the original figure may be provided.

- Figure credits are given in figure captions. Further information about various types of permissions are provided in the Figure Credits section of the text, including links to the source documents.

- In the e-text, internal hyperlinks to other locations within the text are included and highlighted in light blue; for example, page 309. These internal links direct you to previous discussions pertaining to the topic for review purposes, or to related topics. Internal hyperlinks are also embedded in the table of Contents and the List of Tables.

- References to parts, chapters, sections, subsections, pages, figures, or tables also contain implicit internal hyperlinks in the e-text, such as Chapter 1 or Table 2.1. These links allow you to quickly navigate to the referenced material.

- Terms or names highlighted in blue refer to information in the text's glossary, such as Sir Isaac Newton (1642–1727). Glossary entries are designed to give you a quick reminder of the meaning of a term or a very brief discussion of an individual that you have previously encountered. The blue highlighted terms or names are internal hyperlinks in the e-text.

- Critical information and important terms that you must understand in order to successfully complete the course are indicated within the body of the text or in the margins by highlighting them in green. Margin notes are further labeled as *Key Point(s)*. These key points also aid in reviewing the material. In this way, this critical information is immediately available within the context of the associated material, rather than collecting that information in a brief review section at the end of the chapter as is often done in textbooks.

  Key Point
  Key Points highlight especially important information that you should **fully understand**, and *not just memorize!*

- Astronomy texts often confuse students with the use of specialized units associated with quantities such as mass, distance, or time. This is particularly true when equations have been reduced to using specialized units in order to simplify their presentation. Textbooks typically use the same symbol if the variable is contained in an equation that uses the standard International System of Units (*Systéme international d'unités*, or SI units), or specialized units such as solar units or parsecs. To avoid this confusion, subscripts are provided to remind you when special units are required for the form of equation provided. For example, when mass is contained in an equation the symbol m is used, indicating the need for the standard SI mass unit of kilograms (kg), but if the equation requires solar units, the symbol m_{Sun} is used. Similarly, if distance should be in the standard unit of meters (m), d is used, but if parsecs are required, d_{pc} is used.

- Beyond science, SI units are the standard system of units used almost world-wide. United States customary units are often provided parenthetically in this textbook to assist readers who are otherwise unfamiliar with SI units. It is a goal of the author that you will become more "fluent" in the SI standard.
- Solutions to some of the end-of-chapter exercises are available following the appendices and can be quickly accessed through an *Answer* link in the e-text. Returning to the original exercise is accomplished by clicking on the corresponding exercise number in the Answers section of the text.

Support for the Mathematics Contained in the Textbook

Math tutorials are available at `Math_Review`

Mathematics is a truly universal language and the language of nature. As a result, it is necessary for scientists, and students of science, to become comfortable in that language. Although mathematics is kept to a minimum in this textbook, it would be a handicap to understanding astronomy and how the discipline of astronomy functions and advances if mathematics was completely eliminated in the pages that follow. To provide assistance in working with equations:

- Key equations are placed inside highlighted boxes and provided with descriptive titles; for example,

$$E = mc^2$$

Einstein's mass–energy relationship

- Care is taken to clearly interpret what the equations are saying.
- Equations are reduced to dealing with ratios whenever possible, thus avoiding the need to apply constants and obtain values that are generally meaningless to the majority of general education students.
- The behavior of each equation is emphasized, pointing out the dependence on, and sensitivity to, various physical quantities such as mass, distance, or temperature.
- Examples are included describing the applications of the equations and reinforcing their meanings within the appropriate context.
- Opportunities to practice using the equations are provided in end-of-chapter exercises.
- Extensive just-in-time, easily accessible online tutorials are provided for further explanation and practice.

Neuromythologies

You should never think of yourself as being unable to "do math and science." There is a common myth of brain function, referred to as a neuromyth, claiming that certain students are right-brained and others are left-brained. The implication is that right-brained individuals are creative types while those who are left-brained are more analytical and therefore more capable of being proficient in mathematics. This view traces its roots back to brain hemisphere separation surgeries in the 1970s that attempted to halt severe epileptic episodes in patients. Those cases were extremely unusual and impossible to generalize to all humans. In reality, the brain is highly interconnected across hemispheres, meaning that various types of thought processes involve both sides of the brain. Unfortunately, students (and some educators) have attached themselves to the idea that students tend to be more right- or left-brain dominant, subtly and erroneously handicapping students by leading them to believe that they are incapable of being analytical or creative, when in fact no such distinction exists. This outdated argument leads to self-fulfilling prophecies: I've been told I can't do it, therefore I can't do it. You are certainly able to "do math and science"; the challenge is, that like learning to ski, play a musical instrument, or write well, developing your analytic skills requires hard work, a great deal of practice, and commitment to a goal.[1]

A related neuromyth has to do with the belief in learning styles (sometimes referred to by VAK for visual, auditory, kinaesthetic). Multiple studies have also debunked this educational myth. The danger of this neuromyth is that it, too, handicaps student learning. We all learn through all of our available senses and we shouldn't limit ourselves by falsely believing that one type of learning is superior to others in our individual case.[2]

Recommendations for Strengthening your Study Skills

For many of you, this may be your first physical sciences course, and for some of you this may even be your first term in college. If you fall into either or both categories, it is critically important that you develop and/or enhance your study skills. It is typically the case that college-level courses are significantly more challenging than the high school courses you took in the past. In order to be successful in this course (or any college course for that matter) you need to be systematic about your approach to studying. A number of suggestions are included here that may aid you in earning good grades during your collegiate career.

- *Always* read the course syllabus. The syllabus contains critical information about the course, such as instructor contact information, office hours, material to be covered, grading policy, and perhaps assignment due dates, exam days, and the lecture schedule. Be sure to check the syllabus first before asking your instructor about information that is already contained in it.
- Be sure that you are setting aside enough time each week for studying. The general rule of thumb is that you should commit to studying two to three hours outside of class for every hour in class (and no, that is not hyperbole). This means that for a class that meets three hours per week, you should plan on spending an *additional* six to nine hours per week outside of class fully focused on the course.
- Don't bite off more than you can chew. Too often it is the case that part-time or full-time students take a heavier course load than their personal situation can accommodate. Students are sometimes so focused on taking enough credits to graduate in the fewest number of semesters or quarters possible that they simply cannot set aside enough time to be successful, resulting in the need to retake courses later at additional cost. If you must take a full load (and you are certainly encouraged to do so when feasible), ask yourself the critical question of whether or not it is possible to take all of the classes you

[1]Geake, John (2008) Neuromythologies in education, *Educational Research*, 50:2, 123–133.
[2]Kaufman, Scott Barry (2018), Enough With the "Learning Styles" Already!, *Scientific American*, December.

are enrolled in while also working twenty to forty hours per week and perhaps simultaneously caring for family members. It may be better for you to take longer to complete your education with fewer credit hours per term than it is to earn poor grades or even fail courses you simply didn't have time for in the first place. Failing courses or receiving poor grades will cause you to lose that scholarship anyway!

- **Study in an area that is free from distractions.** If you are a single student who is not responsible for taking care of one or more family members, you may find it convenient to study in the campus library, or another designated study area. If you are taking courses remotely through online education, attempt to find a distraction-free study location in your home or apartment. If you have other life commitments, such as caring for family members and/or working a part-time or full-time job, it may be necessary to study after all of the day's other duties have been completed or before they begin, so that you can have uninterrupted study time. Research has demonstrated that multitasking results in less efficiency and poorer performance in each of the activities involved.

- **Always read the assigned material** *before* **coming to class** (again, not hyperbole). Doing so means that you can focus more on the information being presented as a way to clarify what you have already read. **You will also be more prepared to ask questions in class** when your instructor covers a topic that you are struggling to understand. Besides, if you have a question and are confident enough to ask that question in class, *you will be a hero* to the other students present, because if you are confused about a topic chances are that the majority of the other students in the class are as well!

- Having read the material ahead of class time also means that you can be more intentional about your note-taking rather than trying to write down everything your instructor says in class. Simply being a stenographer doesn't constitute **active listening.** You need to be thinking about the presented material rather than being too busy just writing down any words that fly by. With the prevalence of smartphones, some students simply take pictures of boards after the instructor has filled them with information. Just taking a picture certainly does not constitute active learning. *Worse yet is not taking notes of any kind.*

- If possible, **find one or more study partners from your class to work with.** This proves to be a win–win for everyone involved. Even if you are doing better in class than members of your study group, you will find (and virtually any instructor will confirm this from their own experience in teaching) that only when you can clearly explain an idea to another person do you truly understand it. If you happen to be a weaker student than others in the group, you have a great opportunity to ask questions about topics you don't understand. This study approach is sometimes referred to as *peer-to-peer tutoring.* To quote a reasonably well-known scientist: "If you can't explain it simply, you don't understand it well enough" (Albert Einstein).

- **Your instructor is there to help you, including outside of class.** Make sure that you know when your instructor's office hours are (they are probably in the syllabus), and take the time to visit your instructor when you have questions that you haven't been able to find the answers to on your own or through your study group.

- Be an active reader. Reading a science textbook is nothing like reading a novel for relaxation! The content of a scientific text tends to be very information dense, meaning that you cannot expect to breeze through the reading material. When done correctly you will take significantly more time reading than you are accustomed to. You should always make sure that you truly understand the material before moving on. This may mean writing notes to yourself in the margins, working through examples presented in the text (not merely reading through them!), and going back to clarify and review prior material.
- You should also be aware that simply highlighting sentences in the text cannot be considered as being an active reader. What often happens with this common technique is that the student highlights the sentence without actually thinking about the content and context of the sentence at the time. Usually the student, with the best of intentions, plans to return to the sentence later on and then study it more carefully. What typically happens instead is that when it comes time to prepare for an exam, the student simply rereads the highlighted sentence in the same fashion as before, without careful analysis. It also frequently turns out to be the case when highlighting scientific texts that most of the sentences end up being highlighted because the sentences are so information dense. As a result, all you have in the end is a book with virtually every sentence highlighted and with no real direction for further study later on. You also wasted money on a highlighter! You can use the inline green highlighted notes and the *key points* in the margins of this textbook as an initial substitute for using your highlighter, but don't rely on those points alone; they simply point out, or summarize, especially critical information or concepts. You still need to read *and understand* the rest of the text as well.
- Students who first take a physical sciences course typically find themselves reading a section of the text and moving on, convinced that they understand what their eyes just moved across. In fact it is very easy for students to convince themselves that they understand the topic when in fact they don't. This is where active learning again comes into play. You should be able to explain what you have read in a clear and succinct way, and in your own words. If you encounter a worked example in the text, you should be able to solve the problem *and similar ones* on your own before moving on. Again, studying with peers greatly helps in this process.
- *Memorization does not equal understanding!* Too often students believe that they simply need to memorize terms and the concepts associated with them. Memorization alone does not mean that you have actually internalized the topic with deep understanding. It is certainly possible to memorize a sentence, for example, and recite it without having any idea what you just said. You will be encountering many terms that you are unfamiliar with while reading this text, and you will need to remember them, but that should not happen without understanding what the terms refer to and how that material is interconnected with other material that is presented. To aid in this process, terms in this text are included in a glossary (that are hyperlinked in the e-text) where you can have immediate access to brief summaries of the terms that are used.
- There is a method of teaching and learning in higher education that is sometimes employed in various disciplines, including in physics and astronomy

courses. The technique is often referred to as the flipped classroom. The general concept is simply an extension of what has already been mentioned in many of the previous study recommendations. Students are required to read and study the material themselves *prior to coming to class*, and the class time is then spent working in groups to solve problems, asking questions among your peers, and getting guidance from your instructor. Rather than your instructor being "the sage on the stage," your instructor serves as a "guide on the side." In some implementations of this pedagogical approach, a quiz may be given at the beginning of class as a check on how effectively you are learning the material ahead of class time (active learning). The quiz also provides information to your instructor about specific topics that the class as a whole may be struggling with and therefore should receive more focused attention during class time. When some students encounter this pedagogical method for the first time the reaction is often: "I didn't pay to have to learn the material on my own, I payed 'big bucks' for the professor to teach me!" In reality your tuition dollars have gone to get you the best education possible, and that includes you putting in a major effort through active participation in that education rather than simply being a passive consumer, an empty vessel to have knowledge poured into. The flipped classroom is a natural extension of active learning with the instructor being there to assist you in that process.

- Many students find it helpful to summarize the material presented in class as soon as possible after the class session (even if you read the material prior to class). Along with explaining a topic to one of your peers, writing is a great tool for solidifying ideas and exposing gaps in your understanding of the material.

- Students will often tell their instructors that *"I understand the material, but I just can't do the problems."* Such a statement is an immediate red flag to the instructor that the student has been fooling themself regarding the material. In reality, if you can't do the problems then you don't actually understand the material. Don't let yourself fall into this trap!

- Be aware that your instructor is not Amazon.com® so don't expect electronic response times like placing an order at Amazon.com®. Students often find it convenient to email (or even text) their instructors as soon as they get stuck on a problem. First, make sure that you know your instructor's policies regarding email, texting, and other forms of electronic communication (there is a good chance that *it is in the syllabus*). After all, your instructor has a life as well. If it is permissible to email your instructor with questions, don't expect your instructor to respond immediately. It will likely be several hours, or even a day or two, before your instructor can get back to you.

- Don't let yourself fall behind. Unlike some other courses that you may have had, science, engineering, and mathematics courses are usually structured so that new material builds on previously-covered material. Obviously, if you don't understand the prior material upon which the new material is based, you won't be able to fully understand the new material either.

- Always come to class whenever possible. That should be enough said; however, if you do need to miss class for some reason, *never ask your instructor if you missed something important!!!* Clearly, from your instructor's point of view, everything they discuss in class is considered important, otherwise your instructor wouldn't have covered that material.

- **You don't want to ask "Will this be on the test?"** Learning should never be just about the grade, it should be about taking advantage of an exciting and perhaps one-time opportunity to study a topic that will enrich your life and prepare you to be a citizen informed about how science works and how science is intimately intertwined with your life. This astronomy course, and other courses that you take, will also help to prepare you to be a life-long learner who can better understand scientific issues that face society, and for which you make decisions every day, either consciously or unconsciously, including evaluating candidates for elected public office. In any case, if you study appropriately you will be ready for the exam anyway.

- Make sure that you understand what plagiarism is and what its implications are (plagiarism is a form of stealing intellectual property and is most likely illegal). You should always strive to hold yourself to the highest ethical standards, including academic honesty. If you are unclear if a particular activity might not satisfy this high standard, be sure to ask your instructor. *Plagiarism and cheating are never acceptable and should not be allowed to occur.* If you know that another student is crossing the line, call them on it; don't let it pass unchallenged. Stand up for what is right.

- **Make sure that you leave time for yourself** so that you can recharge your batteries. A higher-education degree can be a long haul. Although completing a baccalaureate degree in four years is often considered typical, and it still is for many traditional-aged students living on residential campuses, for other students the time to graduation is often significantly longer. The average time to completion of baccalaureate degree in the United States is about six years, while the time for an associate degree is about three years.

- Finally, have fun! Whether you are a traditional-age or non-traditional student, you live on campus or off campus, you are taking the course face-to-face or online, you are working part-time, full-time, or not at all, you have family responsibilities or not, learning should be exciting and fun. This is an opportunity to open your mind to new ideas that can be phenomenal, mind-boggling realities of which you were previously unaware, or ideas that may challenge you to think differently about who you are, what you think you know, or what you believe you understand.

The Process of Science through the Lens of Astronomy

1 The Nature of Science 3
2 The Heavens: Realm of Mystery 12
3 On the Path toward Modern Science 55
4 The Copernican Revolution 91
5 Sir Isaac Newton's Universe 145
6 The Universality of Physical Law 185
7 Revealing Secrets Hidden in Light and Matter 237
8 Modern Physics: New Science to Study the Universe 263

A colorized version based on a woodcarving by an unknown artist that first appeared in Flammarion, Camille (1888), *L'Atmosphère: Météorologie Populaire*. (Raven, CC BY-SA 4.0)

The Nature of Science

1

*We can judge our progress by the courage of our questions
and the depth of our answers, our willingness to embrace
what is true rather than what feels good.*

Carl Sagan (1934–1996)

1.1   Science as a Human Enterprise   4
1.2   Characteristics of a Scientific Theory   5
1.3   Scientific Theories Must Be Self-Consistent   6
1.4   Scientific Revolutions Are Rare and Alter our Paradigms   6
1.5   Peer Review and the Self-Correcting Nature of Science   7
1.6   What Is Pseudoscience?   8
1.7   Science and Society: The Issue of Terminology   9
      Exercises   10

Fig. 1.1 Montage images are public domain except *Principia* cover page. (© Andrew Dun, CC BY-SA 2.0)

Introduction

Key Point
Science is a way of
knowing.

The word "science" is derived from the Latin word, *scientia*, which means "knowledge." Modern science is a way of knowing; an intellectual approach to studying nature and uncovering the processes by which nature operates. However, there are certainly other ways of knowing as well. When a composer writes a symphony, she is creating a work that explores our human experience in the form of music. When a historian studies the past, he is uncovering and interpreting portions of the long march of humanity toward where we find ourselves today. Law is a body of knowledge built on underlying guiding principles and past history that articulates how we live together as a complex, rich tapestry of cultures intertwined with traditional expectations, evolving worldviews, and individual needs and desires. Although all of these ways of knowing are equally important and enrich our lives by teaching us about ourselves and the world in which we live, this text focuses on science in particular, as a unique way of knowing within the human experience.

1.1   Science as a Human Enterprise

Key Point
The common description of
the scientific method.

There is a rich literature regarding the philosophical underpinnings of science and how science proceeds toward unveiling nature's secrets. You may have heard of the scientific method, for example. The scientific method as traditionally described implies that science proceeds by a well-defined series of steps: (a) making observations and/or conducting experiments, (b) developing a hypothesis for explaining those observations or experimental results, (c) using the hypothesis to make predictions of as-yet unobserved phenomena, (d) validating or falsifying those predictions, thereby supporting or discrediting the hypothesis, and (e) if a hypothesis is determined to be invalid then another hypothesis must be developed to explain the new results together with all relevant data from previous experiments or observations.

Key Point
Science is rarely as direct
and sanitized as the
common description of the
scientific method would
suggest.

While the scientific method as described above depicts what appears to be an almost formulaic and straightforward approach toward advancing our understanding of nature, the actual process of science is rarely that direct and sanitized. Instead, the history of science shows us that there are periods of rapid discovery and profound jumps in our understanding, but at other times science has only small, incremental successes. Just as importantly, science often deviates onto paths that result in dead ends, sometimes requiring us to backtrack and re-evaluate our underlying assumptions. As a human enterprise, scientists can also make mistakes that may derail further progress for a time. You may have an image of science as a long, straight trail through the wilderness of undiscovered truths about nature with scientists marching laboriously along the path, systematically unveiling previously unknown tidbits of knowledge. In reality, science is more like following a trail through a forest that contains many undocumented intersecting paths. Without knowledge about any of the optional paths, science may take an unproductive fork only to reach an intellectual dead end, forcing us to go back and attempt to locate a more productive path.

1.2   Characteristics of a Scientific Theory

Although in practice science is not the linear process you may have imagined, there still remain important characteristics of science that undergird its great success:

- Science is always based on an underlying set of assumptions about how nature works. Collectively, that set of assumptions is the theory on which future developments are based.
- A valid scientific theory must be able to make testable predictions that have the potential to be *falsified.* A testable prediction is often referred to as a hypothesis that is built upon the underlying theory.
- Results of experiments or observations could either validate or falsify the pre- dictions.
- If the experimental test or observation agrees with the prediction, then the underlying theory is further supported, but the theory has not been proven to be true. In fact, it is impossible to ever prove a scientific theory, it is only possible to disprove it. The reason a theory cannot be proven is that it is impossible to be 100% confident that some future prediction of the theory won't be shown to be incorrect based on additional experimentation or observation.
- If a prediction of the theory is falsified, then either the theory must be modified in an appropriately consistent manner, or the theory must be replaced by a new one that is able to *incorporate all of the previous relevant experimental and observational data and then lead to additional testable hypotheses.* The new theory doesn't build upon the previously rejected theory, the new theory is most likely an entirely new way of perceiving nature.
- A successful scientific theory must incorporate *all* relevant data, even if some of those data come from other scientific disciplines.

For example, as you will learn, our current understanding of the universe indicates that it is 13.8 billion years old. If that value for the age of the universe is correct, then clearly everything within the universe must be less than the age of the universe itself. Obviously, this includes the galaxies, stars, and planets, but it also includes the rocks on Earth and the Moon. Earth itself is estimated to be 4.54 billion years old, and the Moon is 4.51 billion years old. In turn, the Moon and Earth cannot be older than our Sun, which has existed for 4.57 billion years. In addition, any life on Earth could not have existed before the planet itself. To date, the oldest estimate of life on Earth dates back 4.41 billion years, although other estimates are slightly more recent at 3.77 billion years.

According to all available evidence, the chronology of the age of the universe is consistent with stellar astrophysics, geology, and biology. Within geology, the fossil record must be consistent with the ages of the organisms, plants, and animals recorded in the rock. Changes in the fossil record or variations within a given species must also be consistent with changes in genetic material. In other words, the evolution of the universe must be consistent with stellar evolution, which in turn must be consistent with geological evolution, biological evolution, and anthropology. To date, all available evidence supports those predictions and observations.

1.3 Scientific Theories Must Be Self-Consistent

For a scientific theory to be considered as valid (although never truly provable), it is not appropriate to accept the argument that, "well, most of the data fit, therefore the theory is good enough." Even a single unresolved bit of data may be enough to invalidate the theory. History is replete with such examples: Johannes Kepler (1571–1630) was able to match all of the orbital data of Mars using perfect circles except for two data points; the discrepancy forced Kepler to abandon a nearly two-millennia-old assumption about the motions of the planets in the heavens. Later, all of the planets in the Solar System except for Mercury followed orbits directly derivable from Sir Isaac Newton's (1642–1727) universal law of gravity and his three laws of motion, with Mercury's abnormal orbit ultimately resolved by Albert Einstein's (1879–1955) revolutionary general theory of relativity.

Key Point
A scientific theory must be self-consistent.

When predictions of the theory are shown to be incorrect, a simple ad hoc "patching" of the theory is not only intellectually unsatisfying, it also violates the inherent *mandate of self-consistency*. For example, there are those who believe in a so-called "young Earth" where our planet (and the entire universe) is only six- to ten-thousand years old. However, observational evidence clearly demonstrates that many (in fact the vast majority) of objects in the vastness of space are so far away that the light they produce would not have had sufficient time to reach Earth, begging the question of how then can these objects be seen today. One solution proposed by some young-Earth supporters is that the speed of light was once infinitely fast, meaning that light from the distant universe took no time at all to traverse the vast distances involved. It was only at some specified instant in time (claimed to be the moment of original human sin) that light's speed instantaneously changed to the finite value that is directly measurable today. Although the proposed solution to the problem does resolve the conflict, it is an ad hoc add-on to the laws of physics as currently understood. Moreover, the proposed solution does not address the many other threads of evidence that point toward an Earth that is 4.54 billion years old. The proposed solution is also inconsistent with the fossil record, the ages of stars, the ages of galaxies, and the age of the expanding universe itself. In addition, the proposal of an infinite speed of light is not testable and therefore not falsifiable, meaning that it is fundamentally not scientific. Favored personal ideas of how nature operates that are inconsistent with the available data cannot resort to processes that are supernatural or that violate the laws of physics. Any valid scientific theory must be entirely self-consistent and must be able to make testable predictions.

1.4 Scientific Revolutions Are Rare and Alter our Paradigms

Key Point
A well-established scientific theory becomes a paradigm.

When a prevailing scientific theory has been in place for a significant period of time (sometimes for centuries), that theory can become incorporated not only into the mindset of scientists, but also into society and our collective worldview. Such an embedded way of thinking is known as a paradigm. Science does not operate apart from society, but is an important component of society.

The classic example of a scientific revolution, known as the Copernican revolution, will be discussed in detail in Chapter 4 (page 91*ff*). In brief, the Copernican revolution overturned nearly 2000 years of believing that Earth is the center of the universe and that all celestial objects revolve around Earth. This old geocentric scientific paradigm had become fully integrated into the European view of religion, humanity, and our place in the universe. When Nicolaus Copernicus (1473–1543) proposed that the Sun, rather than Earth, is the center of the universe, it was nearly universally dismissed as being clearly impossible. However, over the course of more than 100 years, the view finally moved science, and society with it, toward seeing the heliocentric universe as being more consistent with observations and fully supported by Newtonian physics. The Copernican revolution also radically changed how we see ourselves in the universe.

While this example may seem as though science and society were very primitive and unenlightened prior to Copernicus, and that such a situation could never happen in the twenty-first century, society is full of similar examples today, including the Big Bang theory and the biological foundation of evolution. It has now been more than 100 years since Einstein first proposed that space and time are intimately linked as a continuous four-dimensional spacetime, and it has been 90 years since the development of quantum mechanics and its seemingly bizarre description of the atomic and subatomic world, and yet the general population has great difficulty fathoming these new ways of seeing nature. In fact, the Big Bang theory is a natural outgrowth of these ideas and current observational evidence. Similarly, biological evolution was first developed by Charles Darwin (1809–1882) more than 150 years ago, as a way of making sense of the tremendous variety of plant and animal species on our planet. Today, the theory of evolution underlies all of biology much like general relativity and quantum mechanics form the foundation of modern physics, and yet it is dismissed out-of-hand as simply being impossible by many members of our modern society.

1.5   Peer Review and the Self-Correcting Nature of Science

A fundamental component of modern science is the process of peer review. Rather than simply accepting scientific claims as truths about nature, modern science requires that the results of experimentation and observations, or proposed new hypotheses, be carefully reviewed by a group of the scientist's peers. In scientific journals, such peer review is generally anonymous, allowing the reviewer the opportunity to honestly critique the researcher's work without fear of repercussions if the review is negative. Typically, multiple reviewers are assigned independently to evaluate a journal submission. This process of peer review is not only designed to validate or reject submissions for publication, but the peer review process also provides the researcher with important information and suggestions to make the paper stronger. Peer review is also used in making decisions about grant funding through agencies such as the National Science Foundation (NSF), the National Aerodynamics and Space Administration (NASA), or the National Institutes of Health (NIH). Peer review is used by most private foundations, universities, and other sources of funding as well.

Key Point
Science is inherently
self-correcting.

It is this process of peer review, together with the requirement that a scientific theory must be able to make predictions that are potentially falsifiable, that makes science inherently self-correcting. Should errors or false claims survive the initial peer review process, the published results will become available for greater scrutiny by the entire scientific community, which amounts to a much-expanded form of peer review. If an experiment cannot be replicated by other scientists, for example, then the published results will be called into question and possibly disproved by further research. This is an example of the analogy presented earlier of science traveling down a path in a forest with many undocumented intersecting paths that may lead scientists temporarily along a route that ultimately proves to be unproductive. On those very rare occasions when a scientific theory is demonstrated to be invalid in its underlying assumptions, science may be forced to abandon the original path completely, go back to the trailhead, and look for an alternate route forward.

1.6   What Is Pseudoscience?

Key Point
Astrology is not scientific;
it is referred to as a
pseudoscience.

There do exist instances where lines of thinking can take on many of the appearances of science but are in fact missing one or more critical elements of a true scientific theory. One such example that is important in the present context is astrology, which claims to be able to predict one's future based on the positions of the planets and the Moon relative to the stars at the time of one's birth. Historically, astronomy and astrology were close cousins; in fact, some well-known early astronomers, such as Kepler, made some extra income by providing astrological predictions for patrons.

It is certainly understandable that astronomy and astrology were essentially one and the same prior to the development of modern science. Without a scientific foundation upon which societies could make sense of the world around them, superstition and mythology were the pillars that controlled human lives. Deities were represented by individual planets, stars, or constellations, and the presence of those deities at the time of birth were believed to affect how a person would develop and the characteristics that the person would possess.

However, modern science began to challenge those notions. For example, if astrology makes predictions about a person's future based on celestial positions, then those predictions must be testable. It is the impact of that testing that distinguishes astronomy from astrology. Although astrology can make predictions, those predictions have time and time again been shown statistically to be no better than random guess in predicting the patterns of people's lives. Since astrology fails the validity test it must be rejected. Unfortunately, astrology is still present in the twenty-first century despite continual falsification.

Another question arises about astrology: What is the mechanism by which celestial objects can affect our behavior? There must be some way in which the information about the location of celestial objects is imprinted on human beings. One obvious possibility is through the force of gravity, but it is easy to show that the equipment present in the delivery room at the time of birth has a greater gravitational influence on the baby than any of the celestial objects do. Perhaps it is light that is the information transmitter? In that case, all deliveries should be performed outside, under a Sun-lit or starry sky. To date, there is no known way in which celestial objects can directly influence a child at birth.

What about a philosophical argument? Most people choose to believe that they have free will, meaning that they are ultimately free to make the decisions that affect their own lives. What should I wear to class today? Where should I go skiing this weekend? Who should I marry? However, the notion of free will is clearly in conflict with the idea that the positions of the stars and planets at the time of my birth deterministically establish my future.

1.7 Science and Society: The Issue of Terminology

Finally, a word about terminology in science and in society. It is often the case that the scientific usage of a term and the common societal usage of the same term are inconsistent, thus leading to unfortunate misunderstanding. For example, in science the term *power* refers specifically to the amount of energy produced over a specified period of time, while in mathematics it refers to an exponent. In politics the term means something entirely different; someone in power can significantly impact a debate or political outcome!

Another very important example is the term we have already been using: *theory*. It is routinely heard in conversation that "it is just a theory," implying that the idea is easily dismissed. The phrase has often been invoked to reject biological evolution, the Big Bang, and even the nearly universally accepted explanation for human-driven global warming and climate change within the scientific community. By now, hopefully you have come to understand that the scientific usage of the term is dramatically different from, and in fact virtually in opposition to, the common interpretation generally found in society. As mentioned repeatedly, a scientific theory can never be proven, it can only be disproved, but that does not mean that a scientific theory is without merit. In reality, a mature scientific theory, such as the general theory of relativity, quantum mechanics, or biological evolution, is carefully and constantly tested. A scientific theory is also the underlying map that helps guide us down the path of discovery and understanding. If, at some point in the future, predictions based on the theory are shown to be invalid then the theory must be improved in a fully self-consistent way, or it must be rejected in favor of a more complete theory. But until that point is reached a scientific theory has the support and validation of an almost endless list of attempts to prove it wrong. A scientific theory can never be tossed aside simply because you don't like what it says about nature, or about our place in nature.

There are numerous examples of conflicts between common societal usage of a term and its scientific meaning. When those cases arise in this text they will be emphasized and discussed.

Science first and foremost teaches us to always question ideas and demand evidence. Science requires that we critically evaluate hypotheses and search for flaws in the logic or conflicts with any of the available data. We should never be satisfied with an explanation just because someone, including the author of this text, your instructor, or someone else in authority, simply stated that something is true. Indeed, this process of critical thinking should not be limited just to science, but it should

Key Point
Being able to **think critically** is the foundation of understanding and true learning.

also be the foundation upon which you evaluate any information being presented to you, whether it is found in news reports, articles in newspapers, in statements from politicians or others in leadership roles, or from friends. You should always be demanding:

Show me the evidence!

The value of a college education is not the learning of many facts but the training of the mind to think.

Albert Einstein (1879–1955)

The remainder of Part I of this text takes you on an intellectual exploration of our developing understanding of physical science as it applies to astronomy. At every point, you should be keeping the ideas regarding the nature of science in mind. Do the concepts presented satisfy the criteria for being a component of a true scientific theory? If not, ask yourself what is missing, is the concept consistent with all known data, or is the idea simply ad hoc? At times in this text, you will be witness to dramatic changes in our understanding of the universe, while at other times the science will develop at a slower pace, building on what was already previously known. If the puzzle pieces don't all fit, you should think about what is amiss.

What you learn in Part I will be applied more thoroughly in Parts II and III as we investigate our Solar System, other planetary systems, the stars, galaxies, and the universe as a whole.

Enjoy the journey!

 Exercises

All of the exercises are designed to further develop *your understanding* of the material by thinking carefully and critically about what you have read in this chapter. Answers to selected exercises are available in the back of the book.

True/False

1. Science always proceeds in a very linear way, with new knowledge always building on information and ideas from the past.
2. (*Answer*) A scientific theory must be able to make predictions that are potentially falsifiable.
3. A scientific paradigm is a theory that has been proven to be true.
4. (*Answer*) A truly scientific theory never makes assumptions about nature.
5. A scientific theory that addresses problems in physics has no relationship to science conducted in biology.
6. (*Answer*) A scientific theory must be able to explain the most important relevant data, but cannot be expected to explain every relevant piece of information.
7. If a prediction made by a scientific theory is proven to be false, scientists may decide that the error is not sufficiently important to reject or modify the theory.
8. (*Answer*) A scientific theory never invokes supernatural causes to explain currently unexplained phenomena.

9. A fundamental scientific theory is overturned by new evidence on a fairly regular basis, indicating that scientific knowledge is fleeting.

Multiple Choice

10. (*Answer*) Astrology is classified as a pseudoscience rather than a valid scientific process because
    (a) it does not make testable, and potentially falsifiable predictions.
    (b) tests of astrological predictions have repeatedly been shown to be successful only at the level of random guess.
    (c) there is no viable explanation for how the locations of celestial objects at the time of one's birth can be transmitted to and imprinted on the newborn child.
    (d) all of the above
    (e) (b) and (c) only
11. A scientific theory
    (a) serves as a map to discovering more about nature.
    (b) must be proven in order to be accepted by the scientific community.
    (c) has no influence on society.
    (d) all of the above.

12. (*Answer*) Which of the following statements is (are) *not* considered to be a feature(s) of a valid scientific theory?
    (a) Science must always build upon all previous models or theories about nature.
    (b) If a prediction of a scientific theory is shown to be incorrect, then the theory may be adjusted in any way that will explain why the prediction failed.
    (c) Science must be able to predict previously unknown information about nature.
    (d) A scientific theory must always have the ability to be falsified.
    (e) (a) and (b)
    (f) (c) and (d)

Short Answer

13. Summarize the key points of a scientific theory.
14. Explain why ideas such as creationism and intelligent design (the idea that a supernatural being guided biological evolution) are not considered to be truly scientific theories.
15. Describe the peer review process that is normally used before a scientific paper is accepted for publication.
16. Explain what is meant by saying that science is inherently self-correcting.
17. Visit a news service website such as The New York Times, The Washington Post, the Associated Press, or your local or regional paper and identify at least one scientific issue or statistical analysis that is associated with a current story. The article need not be entirely scientific, but science or statistics may be intertwined with, and inform, a current event. Explain how science or statistics is related to the story.
18. It has been argued that at least within the United States of America we now live in an anti-intellectual culture where the views of experts who may have spent lifetimes studying specific issues or classes of problems are considered to be no more valid than anyone else's thoughts on a topic. As an extension of that mindset, it has also been suggested that we live in an anti-science culture that is also a post-factual or post-truth culture where evidence no longer matters in decision-making for a significant portion of the population.

    Go to an Internet news site and identify an article that appears to make a controversial claim that seems to contradict, or is otherwise not reported by, a mainstream news organization. Topics to consider include denials of global warming and climate change, biological evolution, or statements made by politicians of others that need not be about scientific subjects. Sites you may want to explore can be found on this textbook's Chapter 1 resources webpage, or you can conduct an Internet search for "fake news" or another topic of interest.
    (a) What is the link to the article you are reviewing?
    (b) Give a brief summary of the claim or claims made in the article?
    (c) Does the article cite factual information that can be verified? Simply citing other Internet sources does not constitute being verifiable.
    (d) Have the claims made in the article been reviewed by one or more fact-checking organizations? If so, what are the fact checker's conclusions regarding the truth of the claims being made? Fact checking sites can also be found on this textbook's Chapter 1 resources webpage.
    (e) If the article makes claims that are disputed by others, what is the nature of the dispute? Do the claims of the article provide a strong argument in contradiction to the opposing view? How might the conflict be resolved?

19. In your own words, explain what Carl Sagan meant in his statement quoted at the beginning of Chapter 1.

2 The Heavens: Realm of Mystery

*Astronomy compels the soul to look upward and leads us
from this world to another.*

Plato (c. 428–348 BCE)

2.1	Observing the Changing Sky with the Naked Eye	13
2.2	The Skywatchers	26
	Exercises	51

Fig. 2.1 Delicate Arch, Utah, and the Milky Way. (public domain, CC0 1.0 Universal)

Introduction

The history of astronomy is a history of awe and wonder. It is also a history that is tied to mythologies, superstition, deities, religions, and the search for meaning. But beyond that, it is a history of natural human curiosity and our ongoing quest for understanding.

Just as a two-year-old child never seems to stop asking the question "Why?", humans have been asking that same question for millennia. Today, our understanding of the cosmos, and indeed our understanding of all fields of science, is the result of asking that fundamental question. But the question is not just asked in science, it is also asked in our literature as we try to make sense of human tragedy, or in our art and our music as we try to express human relationships with nature and with one another.

The science of astronomy is one manifestation of the natural human process of searching for understanding about the universe around us. Before the birth of modern science, astronomy and its pseudoscientific cousin, astrology, were also about finding meaning in our human experience and our place in the universe. As we have worked to understand the cosmos there have been advances, often slow and incremental, and occasionally dramatic and revolutionary. But there have also been many false starts and dead ends; after all, science is rarely, if ever, a linear process from the beginning of a puzzle to a deeper understanding of natural processes. Regardless of whether an avenue of exploration is successful or not, science is the very human intellectual enterprise of asking why, and looking for creative and self-consistent solutions to previously unexplained natural phenomena.

After a moon-lit climb of a 430 m (14,000 ft) peak in the San Juan mountains of southern Colorado one January many years ago, members of the summiting team sat on the mountaintop and entered into that age-old conversation about who we are and the meaning of our existence in the universe. One of the members of the team from New York City asked your author, who was still in college at the time, what he wanted to do with his life. When your author responded that he wanted to be an astronomer, the man from New York City replied that he thought it was a shame that we know so much about the universe because it takes all of the mystery and wonder out of gazing up at the night sky. As you explore astronomy in this course and begin to develop an understanding of the process of science and how we have come to understand something of the planets, stars, galaxies, and the universe as a whole, your author hopes that this journey of discovery upon which you are embarking will provide you with your own personal answer for the climber from New York City.

2.1　Observing the Changing Sky with the Naked Eye

Even a cursory look at the sky during the day or at night immediately tells us that objects in the sky are not fixed in place relative to Earth. There are some motions that are quickly discernible even to the untrained eye, while others may take days, months, years, or even centuries to detect. Some features associated with the changing sky may even require making observations from different locations on Earth.

Fig. 2.2 The Big Dipper (a portion of the constellation, Ursa Major) seen over Grand Teton National Park, Wyoming. [Photo by Robert C. Hoyle, National Park Ranger and former Professor of Astronomy (Copyright Robert C. Hoyle)]

Before we develop a complete, modern explanation of what we can see in the sky with the unaided eye, we will begin more simply by describing the various types of phenomena that pre-modern societies would have observed and for which they would have sought explanations meant to make sense of it all. Here we use the term "pre-modern" to describe a society that existed prior to the introduction of science as we practice it today.

Constellations and the Motions of the Stars

Humans have a particular talent for recognizing patterns. Even if you haven't spent much time looking up at the stars at night, assuming that you live in the northern hemisphere you can probably pick out the Big Dipper in the northern sky (Fig. 2.2). The Big Dipper is actually a portion of a larger set of stars known as Ursa Major, the Big Bear, seen in Fig. 2.3.

The Big Dipper within Ursa Major provides a convenient starting point for finding many of the other constellations in the northern sky. The Big Dipper can be seen in the middle of Fig. 2.3 where the handle of the dipper corresponds to the tail of Ursa Major and is comprised of the stars Alkaid, Mizar, and Alioth. The cup of the dipper is formed by Megrez, Phecda (also known as Phad), Merak, and Dubhe. The blue-colored lines depict the "connect-the-dots" patterns that the constellation figures are based on. By drawing a line from Merak to Dubhe and extending it, the line roughly points toward the "North Star," Polaris, in the constellation of Ursa Minor (Merak and Dubhe are sometimes referred to as the "pointing stars"). The star

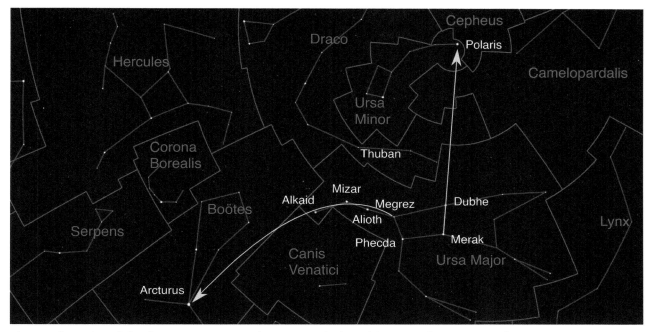

Fig. 2.3 The region of the sky near Ursa Major, the Big Bear. (Adapted from a *Stellarium* screenshot, GNU GPLv2)

Arcturus in the constellation of Boötes can be found by drawing a smooth arc from Megrez, that passes near Alioth, Mizar, and Alkaid, and then extending the curve ("arc over to Arcturus").

Many of the constellations are based primarily on the imaginary figures that ancient observers associated with their shared cultural experiences, representing gods, wild animals, or characters from mythology (of course, not all cultures of the past associated the same stars with the same imagery). In the northern hemisphere sky, today's constellations are due primarily to ancient Sumerian and Greek astronomy (Sections 3.1 and 3.2, respectively). In the southern hemisphere sky, constellations were proposed more recently by European and American astronomers after European explorers began to visit that part of the world. A small number of constellations were also created to fill in gaps in the constellation map.

In 1922 the international body governing modern professional astronomy, the International Astronomical Union (IAU), adopted 88 figures as the official set of constellations. Then, in 1930, the IAU drew boundaries around the constellations that included the stars of the imaginary figures as well as other stars within and around those figures. These 88 regions collectively contain every star in the sky, associating each star with one of the 88 figures adopted by the IAU. The red lines in Fig. 2.3 are the borders that define the regions of the sky named for the constellations that they contain. Today, when astronomers refer to a star in a particular region of the sky, such as R Vir within the region bounding Virgo, that star may not be a defining star of the imaginary figure, but it does lie within the official IAU boundaries defined in 1930. (Technically, a pattern of stars, such as the Big Dipper, is known as an asterism, and the IAU-defined region of the sky that contains the asterism is a constellation.)

Key Point
There are a total of 88 constellations, that cover the entire sky.

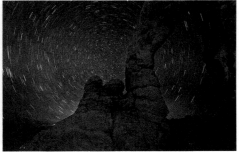

Fig. 2.4 *Left:* A 32 minute exposure of Arch Rock in Joshua Tree National Park, California. The streaks are the time-elapsed paths of stars as they rise in the east. Those streaks are referred to as star trails. (Elia Scudiero, CC BY 2.0) *Right:* A time-elapsed photo of Turret Arch in Arches National Park, Utah. The photo shows counterclockwise star trails around the north celestial pole (NCP). Polaris, the "North Star", is close to the NCP. (Arches National Park, public domain)

Video Tutorial:
Motions of the Stars, Part 1
is available on the Chapter
2 resources page.

Video Tutorial:
Motions of the Stars, Part 2
is available on the Chapter
2 resources page.

Just as the Sun appears to move across the sky from east to west during the day, stars travel from east to west (or northeast to northwest, or southeast to southwest) as well [Fig. 2.4 (*Left*)]. The exceptions are stars that are sufficiently close to two special points in the sky known as the north celestial pole and the south celestial pole. For those stars, their daily motions appear to make circles in the sky that are centered on the north celestial pole [Fig. 2.4 (*Right*)] or the south celestial pole. Stars that trace out complete circles in the sky are known as circumpolar stars; never rising or setting, they are always visible.

Although the Big Dipper is circumpolar at northern latitudes above 41° N, Ursa Major in its entirety is not, as can be seen by comparing Fig. 2.2 with Fig. 2.3. In order for all of Ursa Major to be circumpolar (including its rear left paw), the observer must be north of about 61.5° N, roughly the latitude of Anchorage, Alaska. Conversely, in order for all of Ursa Major to be below the horizon for at least a short period of time means that the observer would need to be farther south than 17° N, approximately the latitude of Acapulco, Mexico. Anywhere between those two extremes in latitude, portions of Ursa Major would not be circumpolar, but would instead rise in the northeast and set in the northwest. You can get some sense of that effect by studying Fig. 2.4. Stars sufficiently close to the north celestial pole are able to make complete circles over a 24-hour period, but those that are too far away will rise up above the horizon in the northeast and dive down below the horizon in the northwest, meaning that they trace only partial circles above the horizon. Because the heights of constellations in the sky depend on the observer's latitude, careful pre-modern observers of the night sky who traveled north or south a sufficient distance would certainly have been able to detect shifts in the vertical location of Ursa Major and other constellations in the sky and wondered why that happened.

Changes in the location of Ursa Major at sunset would undoubtedly have been noticed as well. At a latitude of 41° N, the Big Dipper is just above the northern horizon at sunset on January 1, but on July 1 the sunset location is high in the northern sky and upside down. To understand this, draw a circle centered on the north celestial pole that passes through Phecda (the north celestial pole is just barely to the left of, and very slightly below, Polaris in Fig. 2.3). Now rotate the circle around the north celestial pole until Ursa Major is laying on its back. Making note of the fact that these same orientations happen every winter and every summer, respectively, cultures would have associated the changing locations of stars in the sky with the changing seasons.

The situation is even more dramatic for an observer at mid-northern latitudes viewing constellations in the southern sky. The constellation of Orion, the Hunter (Fig. 2.5), is completely absent in the night sky during the summer, but is high and dominant in the southern sky at night in the winter months. As the seasons transition from fall to winter, Orion appears progressively higher in the sky shortly before sunrise on each successive day.

A studious observer of the stars at night will find that the positions of the stars at sunrise change over time, because the stars only take about 23 hours, 56 minutes, and 4 seconds to rise on two successive days. This is referred to as a sidereal day ("star day"); just under four minutes less time than it takes the Sun to rise two days in a row (a solar day). Over the course of one year, just under four minutes per day turns out to be 24 hours per year. In other words, after one year the same star will rise at the same time it did the previous year. This translates into different constellations being seen in the night sky at different times of the year.

Fig. 2.5 The winter constellation of Orion, the Hunter. (Adapted from a *Stellarium* screenshot, GNU GPLv2, and Johan Meuris, Free Art License)

The Sun's Annual Motion and the Changing Seasons

As the dominant object in the sky during the daytime, the motion of the Sun in the sky would have been easily discernible to pre-modern (and hopefully modern!) observers. The most obvious observation is that the Sun rises in the east and sets in the west, repeating the pattern every 24 hours. Not as obvious from one day to the next, but very noticeable over just a few days or weeks, is that the locations where the Sun rises and sets along the eastern and western horizons, respectively, change throughout the year in a systematic way. As depicted in Fig. 2.6, starting from their northernmost positions along the eastern and western horizons, respectively, the Sun's rising and setting positions move southward over a six-month period of time. After six months the Sun's rising and setting positions have reached their southernmost positions. The Sun's rising and setting positions then reverse course and move northward for the next six months, returning to their northernmost positions.

Video Tutorial: Motions of the Sun is available on the `Chapter 2` resources page.

Fig. 2.6 The setting Sun over the Great Salt Lake as seen from Ogden, Utah (41.2° N, 112° W). From left to right: First day of summer at 9:04 p.m. MDT on June 21, 2015 (northwest). First day of fall at 7:23 p.m. MDT on September 23 (due west). First full day of winter at 5:02 p.m. MST on December 21, 2015 (southwest). The location of the setting Sun on the first day of spring on March 19, 2016, 7:40 p.m. MDT occurred in the same location as for the first day of fall. (Dale A. Ostlie)

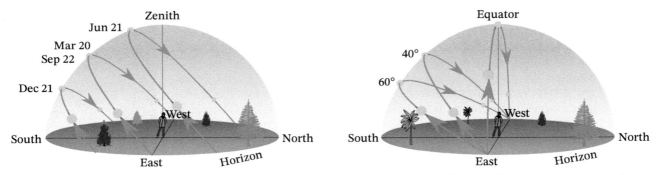

Fig. 2.7 *Left:* In the northern hemisphere the Sun is highest in the southern sky at midday on the first day of summer (June 21) and lowest in the southern sky at midday on the first day of winter (December 21). In this diagram the observer is assumed to be at a latitude of 40° N (actual tracks of the Sun over eight years are displayed in Fig. 2.8) *Right:* The path of the Sun on the first days of spring and fall at three different latitudes in the northern hemisphere; the equator (0°), 40° N, and 60° N.

If you could see the stars during the daytime you would also note that the Sun's position relative to the stars changes in a systematic way over the course of one year. This repeating, annual path of the Sun across the stars is known as the ecliptic.

Pre-modern observers of the sky would have also noted that the changing seasons and the lengths of the days are linked to the positions of the Sun. Referring to Fig. 2.7 (*Left*), starting on about December 21 at mid latitudes in the northern hemisphere, as the Sun moves northward, the days grow longer and warmer, and the Sun travels higher in the sky at midday before coming back down again. As the Sun moves high in the sky it is time to plant crops or perhaps move farther north or to higher elevations in search of animals to hunt, as some tribes of the Native Americans of North America would have done. After the Sun has completed its northern journey, it starts to head back south again. The days begin to get shorter, the weather cools, and the Sun's midday elevation is lower in the southern sky. It is time to prepare for the long winter that lies ahead.

For observers who travel sufficiently far north or south, they should notice other effects as well. If a tribe of Native Americans moved northward from mid latitudes, not only would they have noticed that the weather became colder, but the extreme rising and setting points of the Sun would shift farther north at the start of summer and farther south at the start of winter. At the same time, the Sun would rise and set at progressively shallower angles relative to the horizon and the Sun would be lower in the southern sky, even at its highest point during midday. The sunlight of summer days would last longer at more northerly latitudes but it would be less intense, and the winter days would be shorter while the nights would be longer than they are farther south. Figure 2.7 (*Right*) shows the Sun's path through the sky on the first days of spring and fall when the Sun rises directly east and sets directly west regardless of latitude. Comparing the two diagrams in Fig. 2.7 allows you to picture how the Sun moves across the sky at different times of the year for observers at different latitudes.

For those who live far enough north, like the Inuit of Alaska, northern Canada, and Greenland, or the inhabitants of northern Scandinavia or Russia, they see that the Sun's approach to the western horizon is so shallow that it doesn't actually cross the horizon for days or months at a time during the summer months (hence the reference as "Land of the Midnight Sun"). Similarly, when winter arrives, the precious light and heat of the Sun never makes an appearance, again for days or months at a time. As a result, the inhabitants find themselves in perpetual darkness during that time. In these far northern regions, the Sun circles the horizon during the summer months, moving higher in the southern sky at midday, and lower in the northern sky at midnight while never actually setting. Figure 2.9 shows that on the first day of summer at a latitude of 66 1/2° N the Sun just grazes the northern horizon at midnight, but on the first day of winter the Sun barely makes an appearance on the southern horizon at noon. This special latitude, known as the arctic circle, is nearly 200 km (120 mi) north of Fairbanks, Alaska and just north of Iceland. The arctic circle is 23 1/2° south of the North Pole (90° N).

If travelers move closer to the equator (0°) from mid northern latitudes, they will notice that the Sun rises more directly upward from the eastern horizon and sets more directly downward into the western horizon. Careful observers will also notice that the lengths of the days and nights vary much less than they do farther north or south. In fact, for those who may be directly on the equator, the lengths of the day and night are always 12 hours long all year, and the Sun rises straight up, perpendicular to the eastern horizon and sets straight down, perpendicular to the western horizon [recall Fig. 2.7 (*Right*)]. The excursions of the Sun north or south along the horizon are also much less dramatic and average daytime temperatures hardly change throughout the year. During the hottest part of the day the Sun is found very high in the sky. For those who study the motion of the Sun at latitudes between 23 1/2° N (referred to as the tropic of Cancer) and the equator they will note that the Sun actually passes directly overhead twice during the year, once moving into the northern sky and once headed back to the southern sky. For observers located right on the tropic of Cancer, the Sun reaches the point directly overhead at midday just once each year, on the first day of summer, and then heads south again.

With the tropic of Cancer cutting across central Mexico, the Maya, the inhabitants of Teotihuacán, and the Olmec before them, would have certainly noted the motions of the Sun just described. The tropic of Cancer also passes 18 km (11 mi) south of Muscat, Oman, 40 km (25 mi) south of Dhaka, Bangladesh, and through central Taiwan. The equator runs through the northern end of the Galapagos Islands where Charles Darwin (1809–1882) made his fundamental observations concerning biological evolution. The equator also runs just north of Quito, Ecuador,

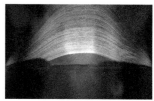

Fig. 2.8 "Perpetuity" by Regina Valkenborgh. A pin-hole camera attached to a dome at the Bayfordbury Observatory, University of Hertfordshire, captured daily arcs of the Sun from 2012 to 2020 at a latitude of 51.8° N. The highest arcs occurred on June 21 each year and the lowest on December 21. The camera was oriented directly south, similar to the observer in Fig. 2.7 (*Left*). The camera was constructed by the artist from a cider can with the interior lined with photographic paper that likely degraded somewhat over the record-setting eight-year exposure. The thousands of people who may have walked in front of the camera during that time became invisible over those eight years. [Regina Valkenborgh (obscura-photography.co.uk)]

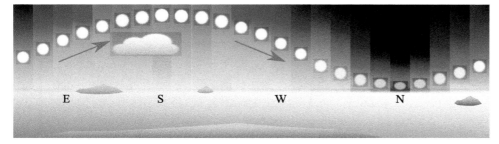

Fig. 2.9 A 360° artistic panorama of the path of the Sun over a twenty-four-hour period on the first day of summer as seen from the arctic circle.

through the northern edge of Lake Victoria in Uganda, and 150 km (90 mi) south of Singapore.

The behaviors of the Sun during the day and throughout the year described in the preceding paragraphs also occur in the southern hemisphere, but shifted by six months relative to the northern hemisphere. The most southerly latitude for which the Sun can be found directly overhead at midday for one day each year occurs on the first day of northern hemisphere winter (about December 21) at a latitude of 23 1/2° S (the tropic of Capricorn). The tropic of Capricorn passes just north of São Paulo, Brazil, about 300 km (200 mi) north of Johannesburg, South Africa, and through central Australia. At a latitude of 66 1/2° S, the antarctic circle, the Sun never sets for one day each year on the first day of northern hemisphere winter and never rises on the first day of northern hemisphere summer. The antarctic circle skirts much of the eastern-hemisphere coast of Antarctica and encompasses almost all of the continent.

Key Points
Definitions of solstices and equinoxes.

There are four unique locations in the sky defined by the Sun's annual motion. The point among the stars where the center of the Sun has traveled farthest south and begins its northward progression is referred to as the winter solstice. The vernal equinox is that location among the stars where the center of the Sun crosses directly above the equator moving northward (the term "equinox" refers to *equal* amounts of sunlight and darkness). The point among the stars where the center of the Sun has reached its northernmost extent is the summer solstice. Finally, when the center of the Sun crosses directly above the equator moving southward, the center of the Sun is passing the point among the stars known as the autumnal equinox. The astronomical definitions for the beginning of winter, spring, summer, and fall, respectively, occur when the center of the Sun crosses these points in the sky.

Phases of the Moon

Key Point
The phases of the Moon go through a constantly repeating cycle of approximately 29 1/2 days.

Observing the sky over a period of one month reveals another readily-apparent fact; the Moon goes through phases. As illustrated in Fig. 2.10, the Moon changes from being new moon, through waxing crescent, first quarter, waxing gibbous, full moon, waning gibbous, third quarter, waning crescent, and back to new moon with a repeating cycle of about 29 1/2 days. Even while going through its phases on a nearly monthly schedule, the Moon always presents the same "face" to Earth-bound observers.

New moon Oct 30	Waxing crescent Nov 3	First quarter Nov 7	Waxing gibbous Nov 11	Full moon Nov 14	Waning crescent Nov 16	Third quarter Nov 21	Waning crescent Nov 24	New moon Nov 29

Fig. 2.10 The phases of the Moon over a thirty-day period in 2016. (NASA's Scientific Visualization Studio)

Fig. 2.11 *Left*: A total lunar eclipse; September 27, 2015. (Alfredo Garcia, Jr, CC BY-SA 2.0) *Right*: A total solar eclipse (August 21, 2017). (NASA/Aubrey Gemignani)

Not only does the Moon's phase change noticeably on a daily basis, but so do its times of rising and setting, and its location in the sky. When the Moon is in waxing crescent it sets in the west shortly after sundown, having trailed the Sun across the sky. Conversely, when the Moon is in waning crescent it rises shortly before the Sun does, leading the Sun across the sky. Careful observations from one night to the next will quickly demonstrate that the Moon actually rises about 50 minutes later on each successive night. When the Moon is first quarter, it is seen to rise when the Sun is highest in the sky; when it is full the Moon rises at sunset; when the Moon is third quarter it rises near midnight; and when the Moon is new it rises when the Sun does. Since the lit portion of the Moon must be facing the Sun, it is illuminating (pun intended!) to study Fig. 2.10 and draw pictures of the location of the Sun and Moon as they travel across the sky during each phase, and correlate them with the descriptions in this paragraph. (Don't forget the importance of being an active reader.)

Another, more subtle aspect of the Moon's motion is that it can travel across the sky, either above or below the path that the Sun takes as it travels from horizon to horizon. This variation can be as much as 5° higher or lower in the sky relative to the Sun, implying that the Moon may be as much as 5° above or below the ecliptic. This is a fairly large range of positions in the sky, given that the full moon itself covers only 1/2° of the sky. It is also the case that if the Moon is above the Sun's path at one point in the lunar phase cycle then it will be equally far below the Sun's path one-half phase cycle later. This means that the Moon's position relative to the ecliptic can vary by as much as 20 lunar diameters. What can possibly cause such an apparently inexplicable behavior?

In what would surely have been one of the most bizarre and frightening phenomena seen in the sky for pre-modern observers, the Moon can on occasion turn a reddish color (sometimes described as blood red) during a full moon [Fig. 2.11 (*Left*)]. On extremely rare occasions, the Sun can completely disappear for a time during the day with a ghostly halo or ring of fire suddenly appearing around the blackened Sun. This is just what happened during the Great American Eclipse of August 21, 2017, seen in Fig. 2.11 (*Right*). There would certainly have

Key Point
The Moon rises approximately 50 minutes later every day.

Key Point
The Moon can be as much as 5° above or below the ecliptic.

Table 2.1 Day names based on naked-eye astronomical objects

Object	Latin	Spanish	English
Sun	diēs Sōlis	domingo	Sunday
Moon	diēs Lūnae	lunes	Monday
Mars	diēs Mārtis	martes	Tuesday
Mercury	diēs Mercuriī	miércoles	Wednesday
Jupiter	diēs Iovis	jueves	Thursday
Venus	diēs Veneris	viernes	Friday
Saturn	diēs Sāturnī	sábado	Saturday

been tremendous rejoicing and relief from a very grateful people when the Moon returned to normal or when the Sun reappeared! Even today, these **eclipses** garner a great deal of public attention, including special watch parties when they occur. Remarkably, there are still those in our supposedly modern, scientifically literate society who forecast apocalyptic events with the coming of eclipses, as occurred in the lead-up to what the media termed the "blood supermoon" lunar eclipse of September 27, 2015. Imagine what it must have been like in a pre-modern culture to see these events happen when you didn't know what was going on. Surely a great deal of effort could have gone into trying to explain and predict these amazing events.

Motions of the Wandering Stars

Although seeing a lunar eclipse or an extremely rare solar eclipse would have been the most spectacular and terrifying celestial events witnessed by pre-modern societies, tracking and trying to predict the behavior of the "wandering stars" were certainly the most perplexing. Even though all of the other stars in the sky remain in the same positions relative to each other, the five naked-eye "wandering stars" don't. These strange objects are known today as the planets Mercury, Venus, Mars, Jupiter, and Saturn. Combined with the other two naked-eye objects that move relative to the stars, the Sun and the Moon, these seven objects led to the names of the seven days of the week (Table 2.1).

Key Point
The planets usually move from west to east *relative* to the background stars although they occasionally undergo retrograde motion, temporarily moving from east to west.

Most of the time, the "wandering stars" appear to drift slowly from west to east relative to the fixed stars while still moving from east to west relative to the observer. An example of this behavior, the motion of Mars between August 1, 2022 and March 31, 2023, is depicted in Fig. 2.12. Another way to think about this motion is to imagine that the "wandering stars" appear to slip slightly or very slowly lag behind the fixed stars as they move across the sky. What is particularly odd about their behavior is that occasionally "wandering stars" will slow to a stop relative to the fixed stars and reverse direction to move slightly faster than the stars from east to west, as is the case for Mars between October 30, 2022 and January 14, 2023. The "wandering stars" then reverse direction again, returning to their original motion, as Mars does after January 14, 2023. The "wandering stars" don't even follow the same path among the stars each time they undergo this reverse, or **retrograde**, motion, nor do

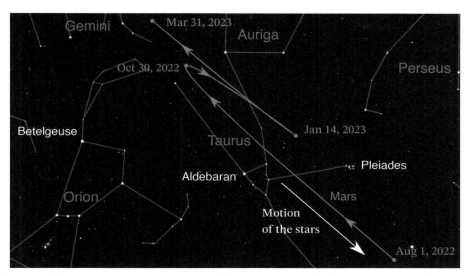

Fig. 2.12 The slow, retrograde motion of Mars relative to the background stars between October 30, 2022 and January 14, 2023. The planet is at its brightest on December 8. (Adapted from a *Stellarium* screenshot, GNU GPLv2)

they do it in the same place among the stars. Sometimes a planet's path appears as a loop relative to the stars while at other times it may be described as a *Z*-like zigzag pattern.

At least all of the planets do follow a fairly narrow band centered on the ecliptic as they move among the stars. The ecliptic passes through the twelve constellations of the astrological zodiac, as well as through a thirteenth constellation, Ophiuchus. The constellations (*signs*) of the zodiac are: Aries, Taurus, Gemini, Cancer, Leo, Virgo, Libra, Scorpio, Sagittarius, Capricornus, Aquarius, and Pisces. The existence of the pseudoscience of astrology and the concept of the zodiac bear witness to the importance that the Sun, the Moon, and the mysterious "wandering stars" played among at least some cultures, and inexplicably still do for some modern-day individuals as well.

Still another oddity of the "wandering stars" is the fact that there are two classes of behavior. Mars, Jupiter, and Saturn, referred to as superior planets, can be seen anywhere along the ecliptic at any time, including on the observer's meridian[1] at midnight. (The meridian is the imaginary line that runs across the sky from the horizon point straight north of the observer, up through the point in the sky directly over the observer's head, known as the observer's zenith, and back down again to the point on the observer's horizon that is straight south.) Mercury and Venus, known as inferior planets, are never visible on the observer's meridian at midnight. In fact, Mercury and Venus are never found more than 28° and 47° from the Sun, respectively, and are usually closer to the Sun in the sky. There are times when Mercury and/or Venus lead the Sun across the sky and times when one or both planets trail the Sun across the sky. This means that Mercury and/or Venus

Key Points
• The naked-eye superior planets (Mars, Jupiter, and Saturn) can be seen on the meridian at midnight.
• The inferior planets (Mercury and Venus) are never seen on the meridian at midnight; they always remain near the Sun.

[1]The time abbreviations a.m. and p.m. come from the Latin phrases *ante meridiem* and *post meridiem*; before midday and after midday.

may rise shortly before the Sun on the eastern horizon or set shortly after the Sun on the western horizon. Because Venus is typically the brightest object in the sky other than the Sun and the Moon, when it rises before the Sun it is frequently referred to as the "morning star" and when it sets after the Sun does, it is often known as the "evening star."

Finally, the visual appearance of the "wandering stars" is not constant over time. For example, pre-modern observers would have realized that the superior planets are always at their brightest during retrograde motion. In addition, Mars is noticeably red in color. To societies trying to make sense of these "wandering stars" there are certainly many phenomena that need explaining.

On the winter solstice of 2020 (December 21), the superior planets Jupiter and Saturn put on an extremely rare show. They were only 1/10 degree apart, an event that was visible around the world. This extraordinary alignment of the two "wandering stars" was popularly referred to as the "Christmas star." The last time they had approached one another that close together at night was in the Middle Ages, 800 years ago. Even today, the "wandering stars" provide treats for the observer.

The Feathery Stars and Streaks in the Sky

Although fairly rare events to the unaided eye, sky watchers may witness a strange sort of "star" that moves relative to normal stars. Comets may appear as fuzzy "stars" that have long, feathery or hazy "tails" attached to them (Fig. 2.13). Sometimes the "tails" may appear yellowish, white, or even bluish in color, with the possibility that some comets may exhibit two "tails," one of which will appear bluish. Unfortunately, the imagery associated with the term "tail" is a bit misleading because it suggests that the "tail" follows the comet across the sky, when in fact it is also possible that the "tail" can lead the comet as though, from our Earth-bound experiences, the comet is somehow moving backward. Comets may appear anywhere in the sky, not just near the ecliptic, and they move relatively quickly against the backdrop of the true stars, especially when they are seen near the eastern horizon before sunrise

Fig. 2.13 Comet McNaught on January 23, 2007. (Soerfm, CC BY 3.0)

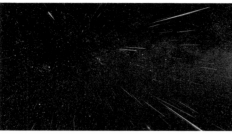

Fig. 2.14 *Left*: A time-elapsed photo of the 2009 Perseids meteor shower. A portion of the Milky Way is also visible. (NASA/JPL) *Center*: A 17 m (56 ft) diameter bolide exploded 19 to 24 km (12 to 15 mi) above Chelyabinsk, Russia on February 15, 2013 with an energy of roughly 30 times that of the first atomic bombs. (Aleksandr Ivanov, CC BY 3.0) *Right*: Barringer's Meteor Crater 30 km (18 mi) west of Winslow, Arizona and 60 km (37 mi) east-southeast of Flagstaff. The crater is 1.2 km (4100 ft) wide and 173 m (570 ft) deep. (Shane.torgerson, CC BY 3.0)

or near the western horizon after sunset. Are these strange apparitions close to us, perhaps in Earth's atmosphere, or are they at great distances from us? What could they possibly mean for a superstitious, pre-modern society? Are they messengers of some kind, foretelling of dramatic events to come, or that have just occurred? The appearance of a comet in the sky would have certainly raised questions of great import.

Much more routine are meteors. On a normal, dark night a patient observer can witness streaks in the sky coming from a variety of directions. But on some nights, the usually slow trickle can turn into a storm of streaks all seemingly coming from the same general region of the sky, like the meteor shower shown in Fig. 2.14 (*Left*). Could these strange events seemingly coming from the heavens be some kind of omen for a pre-modern society?

In some cases these fascinating streaks appear as blazing fireballs. Normally they simply put on a spectacular show, but occasionally fireballs can explode in the sky with tremendous energy, even causing significant destruction on the ground (exploding fireballs are termed bolides). Fireballs may even hit the ground with the potential for enormous, possibly even catastrophic, damage [Fig. 2.14 (*Center* and *Right*)]. What would it have been like for pre-modern observers to witness such phenomenal events? Based on archaeological evidence and minerals that instantly crystallized at high temperature, it appears that an airburst event occurred 3700 years ago above what is now Jordan, near the Dead Sea. The bolide may have resulted in the immediate destruction of communities in a circular area 25 km (16 mi) across where an estimated 40,000 to 65,000 people lived at the time. The devastating event also left the area uninhabitable for 600–700 years. It has been suggested that the event could have been the basis for the biblical story of the destruction of the cities, Sodom and Gomorrah.

Key Point
Fireballs can cause significant damage on the ground by exploding in the sky or striking the ground.

Stars that Suddenly Appear

Although exceedingly rare, occasionally a "new star" (nova in Latin) appears in the sky where one was not previously visible. Much like the unexpected appearance of a comet in the sky, the sudden presence of a previously unseen "star" must have been a very dramatic event for members of a pre-modern society. Figure 2.15 (*Center*)

Key Point
A previously unseen "new star" can suddenly appear.

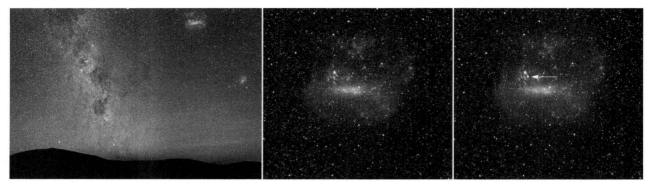

Fig. 2.15 *Left*: A portion of the Milky Way (nearly vertical, left of center) together with the Large Magellanic Cloud (LMC; top right) and the Small Magellanic Cloud (SMC; far right) as seen from La Silla Observatory, Chile, in the southern hemisphere. (ESO/H. Stockebrand) *Center*: The LMC before the appearance of the "new star," Supernova 1987A. (ESO) *Right*: The LMC just after the appearance of SN1987A (arrow). SN1987A was bright enough to be seen with the naked eye. (ESO)

and (*Right*) shows the appearance of the first naked-eye supernova in modern times, SN1987A. Naked-eye supernovas over the past one thousand years occurred in 1006, 1054, 1181, 1572, and 1604. These events are exceedingly rare indeed.

The Trail in the Sky

Key Point
The Milky Way appears as a diffuse band of light across the sky.

One last feature that is readily apparent to any observer with a dark sky is the diffuse band of light that stretches from horizon to horizon. The ubiquitous Milky Way, visible in Figs. 2.14 (*Left*) and 2.15 (*Left*), moves with the stars and yet appears to the naked eye to be something entirely different. The presence of the Milky Way would have been deeply meaningful to a pre-modern society that embraced mythology and superstition. Perhaps it is a trail the dead are destined to follow on their way to the afterlife, or perhaps it is a pathway of the gods.

2.2   The Skywatchers

In Section 2.1 we discussed the many ways in which celestial objects move, change, or make unexpected appearances in the sky, and how more down-to-Earth events, such as the varying seasons of the year, appear to be tied to the motions of some of those objects. Pre-modern, superstitious societies that would naturally place many of their deities in the heavens, and that might depend on fore-knowledge of seasonal weather variations, would certainly have spent significant time, and perhaps resources, observing the heavens for signs predicting the future. Societies that had also developed writing would have been able to record important events and search for historical patterns. Societies without sophisticated written language may have left clues as to their interest in astronomical phenomena through rock art, paintings, or even in their architecture.

Examples of pre-modern societies' observations of the sky are found all over the world and throughout human history. Entire books could be, and have been, written about these compelling archaeological finds and their relationships to cultural norms. This intersection between archaeology and astronomy is a fascinating field of study in its own right, known as archaeoastronomy. Within the broader discipline of archaeoastronomy is the study of origin mythologies; how did the universe begin and where did we as a people come from? The term cosmology is used in archaeoastronomy to refer to a culture's creation mythology and the origin of its people. Here, we can only hope to provide a very superficial overview of a few of the many examples of astronomy and cosmology that have existed throughout human history.

Observers in the Americas

Perhaps as early as 32 millennia before Christopher Columbus (1451–1506), sailing under the auspices of Spain (1492 CE), or Leif Erikson and the Norse Vikings before him (c. 1000 CE), set foot on the Americas, humans from Siberia had already made their way to Alaska and Canada. In multiple migrations during the last ice age, when sea levels were much lower than they are today, hunters of large animals followed their prey across the exposed Beringia land bridge, in what is now the Bering Sea. Once in northern North America, the First Peoples of the Americas spread out eastward and southward to populate both continents of the western hemisphere. It has also been suggested that some humans may have traveled by boat along the Pacific Coast rather than inland, perhaps as far south as Monte Verde in southern Chile. Evidence for these migrations can be found in radiometric carbon dating, archaeology, anthropology, linguistics, genetics, and DNA analysis.

Key Point
The First Peoples of the Americas arrived over the Beringia land bridge that connected present-day Siberia with Alaska across the Bering Sea.

The Skidi Pawnee

Prior to, and during the early days of the westward migration of European immigrants (c. 1500–1876 CE), the Pawnee were one of the many First Peoples that lived in the Great Plains of the North American continent. The Pawnee generally occupied the regions surrounding the Loup and Platte Rivers in what is now east-central Nebraska, and in the central border region of Nebraska and Kansas surrounding the Republican River. Pawnee sites are also known to have existed along the Big Blue River of southeast Nebraska.

The Pawnee's first encounter with Europeans may have been as early as 1541, although significant impact to their way of life didn't occur until about 1800. Between 1811 and 1857 the Pawnee signed four treaties with the United States government that first promised friendship, peace, and protection, but finally required that they give up claims to their ancestral lands in exchange for money, implements, and other assistance from the government. Unfortunately, increasing population from settlers and other plains tribes that were moved onto reservations in the same general region diminished the availability of bison (American buffalo) to hunt which, along with corn, was historically the primary food source of the Pawnee. The increasing population density also led to increased intertribal warfare. Combined with diseases previously unknown to the Pawnee and other tribes, including smallpox, cholera, and venereal disease, the total Pawnee population decreased from a high of perhaps 8000–12,000 individuals in 1800 to approximately 600 one century later. It is fortunate that much of the cultural heritage of the Pawnee has been preserved through anthropological research conducted in the late 1800s and early 1900s.

Fig. 2.16 La-Roo-Chuck-A-La-Shar (Sun Chief) wearing a bison robe painted with stars, and wearing a peace medal (1868). (Photographed by William H. Jackson, 1868, National Anthropological Archives, Smithsonian Institution)

Fig. 2.17 A chief's headdress, possibly inspired by the tail of a comet (recall Fig. 2.13). (Photographed by William H. Jackson, 1868, National Museum of the American Indian)

The Pawnee First Peoples consisted of four individual bands, with the Skidi ("Wolf") band being particularly interested in the sky as the realm of their gods, as evidenced in Figs. 2.16, 2.17, and 2.18. According to the Skidi cosmology, each of the stars in the sky was considered to be a god, but certain stars, along with the Sun, the Moon, and the five naked-eye planets, held particular significance in their culture.

Skidi creation mythology held that, in the beginning, all that existed was Tirawahut (the universe), Tirawa (its chief), Atira (Tirawa's wife), and the other gods. After informing the gods that he was about to create people who would be like himself, Tirawa instructed the gods where to stand in Tiawahut, and what their roles were to be. Sakuru (the Sun) was to stand in the east and he would provide light and warmth. Pah (the Moon) was to stand in the west and she would provide light when darkness comes. Tcuperekata (Bright Star or Evening Star; Venus) was ordered to stand in the west and she would become the mother of all things. Operikata (Great Star or Morning Star; probably Mars) was to be a great warrior and was told to stand in the east. Tirawa selected Karariwari (the Star-That-Does-Not-Move; Polaris, the North Star) to stand in the north and remain motionless, and he was also chosen to be chief of all of the gods of the heavens. Another star (Red Star; possibly Antares) was assigned to stand in the south and would only be seen occasionally during a certain season of the year. In addition, Tirawa told Black Star that he was to be the source of darkness and night. (It has been suggested that Black Star may have been a nova that made a short appearance in the constellation of Cassiopeia in 1572. It has also been suggested that Black Star could have been Vega, so chosen because a bolide may have been seen originating in the direction of Vega. Legend has it that two meteorites were collected the day after the bolide and kept in the sacred Big Black Meteoritic Star Bundle. An example of one sacred bundle is shown in Fig. 2.19.)

Tirawa also assigned four mid-quarter gods to hold up the heavens: Opiritakata (Yellow Star; which may have been Capella) was to stand in the northwest to be with the Sun when he sets, White Star (Sirius?) was to stand with the Moon in the southwest, the Big Black Meteoritic Star (Vega?) was told to stand in the northeast, and Red Star (Antares?) was assigned to the southeast. These four stars were to give the people sacred bundles which would hold prominent places among the Skidi Pawnee. They were also given additional powers: Big Black Meteoritic Star was given the power to create animal gods as well as people, and that the animals should also have the power to communicate with the people so that the people could understand their mysteries; Yellow Star was given the power to send bison to the people by means of the Wind; and White Star and Red Star were also given the power to create people.

Tirawa gave Evening Star the gods Cloud, Wind, Lightning, and Thunder, and told her that these four gods were to stand between her and the Garden. Tirawa then dropped a pebble in Cloud, Wind, Lightning, and Thunder, thus producing the world. But the world was covered with water so Tirawa gave the four mid-quarter gods war-clubs with which they touched the waters, separating water from dry land.

When the Sun and the Moon were given their assignments, Tirawa also gave the Sun permission to overtake the Moon and mate with her, at which time the Moon would disappear (new moon?). From their union a boy was to be born. Tirawa also told Morning Star that he could stay with Evening Star in his travels from east to west and they would produce a girl.

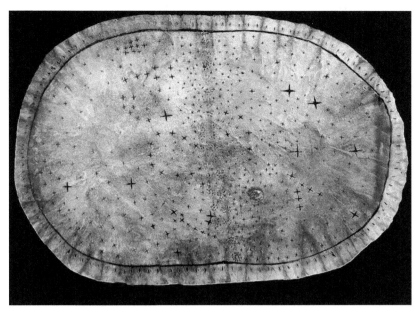

Fig. 2.18 A star map of the Skidi Pawnee measuring about 56 cm by 38 cm (22 in by 15 in), acquired by the Field Museum of Natural History in Chicago in 1906. Estimates of its age are between 200 and 400 years old. Numerous astronomical features are evident, including the Milky Way, Corona Borealis, Ursa Major, Ursa Minor, Polaris, and the clusters of the Pleiades and Hyades. The map may have wrapped meteorites in the Big Black Meteoritic Star bundle. (The Field Museum, CSA1623/c)

Fig. 2.19 An example of a sacred bundle of the Skidi: the sacred Skull Bundle, which was said to have contained the skull of the first Man. The bundle resides at the American Museum of Natural History, courtesy of the National Anthropological Archives, Smithsonian Institution. (Chamberlain, 1982,[2] Fig. 11; Courtesy of Malki Museum.)

The girl was born first and delivered to Earth by a funnel of Cloud's (a tornado?). Later, the boy was delivered to Earth in the same way. The gods also provided the girl and boy with the wisdom and tools they needed to survive. To recognize that a girl was delivered to Earth first, the boy was told to find a yellow bison (albino?), kill it, and construct a sacred Yellow Bundle from its remains.

---

[2]Chamberlain, Von Del (1982). *When stars came down to Earth: Cosmology of the Skidi Pawnee Indians of North America* (Ballena Press).

The girl and boy were joined together and a child was born who was the first fully human-born Man and Chief. Over time, the population of Pawnee grew and other villages were also discovered, these other villages presumably the result of the powers of the mid-quarter gods to create people. Several years after the death of the first Man, his skull was used to make the sacred Skull Bundle (Fig. 2.19). At one point the pole of a tipi was dropped on the bundle, crushing the skull; the skull of another chief was then exhumed to replace the skull of the first Man! Among other artifacts contained within the Skull Bundle was a bow and arrow, believed to have been given to the first Man by the gods who taught him how to use them for hunting.

The lodges (Fig. 2.20) of the Skidi reflected the great importance that the culture placed on the objects in the heavens. Figure 2.21 depicts the floor plan of an idealized, perfectly symmetric Skidi Pawnee earth lodge, where it is assumed that the horizon is flat and unobscured. The lodges were very similar in design to igloos used by the Inuit, but were much larger in size and constructed of earth rather than ice. The typical dimensions of a lodge measured 12 meters (40 ft) in diameter and 4 1/2 meters (15 ft) high, having an entryway that exceeded 3 meters (10 ft) in length and was about 2 meters (6 ft) high and 2 meters wide. The entryway was oriented directly east to honor Morning Star (Great Star). Above the fire pit there was a smoke hole with a diameter that would have been roughly 2/3 meter (2 ft) wide. The dome of the lodge was designed to represent the apparent dome of the sky stretching from horizon to horizon.

Within the lodge there was an altar along the west wall meant for a bison skull and a sacred bundle (the bundle may have included corn). Posts supporting the ceiling of the lodge represented Yellow Star (northwest), White Star (southwest), Big Black Meteoritic Star (northeast), and Red Star (southeast). Two additional posts were for the messengers of Big Black Meteoritic Star and Red Star. The fire pit was meant to represent the open mouth of the supreme god, Tirawa, the fire within the pit represented the Sun, and an earthen mound directly east of the entrance that was made from the excavated dirt of the fire pit represented the words that Tirawa utters.

Fig. 2.20 A Pawnee earth lodge, with a family standing in front of the entrance and other tribe members sitting on the roof. A full-size replica of an earth lodge can be seen at The Field Museum in Chicago, IL. (Photograph by William Henry Jackson, n.d. National Anthropological Archives, Smithsonian Institution)

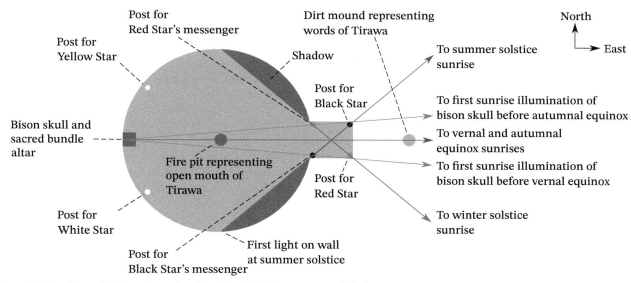

Fig. 2.21 A schematic floor plan of an idealized Skidi Pawnee earth lodge.

From the layout of the lodge it is clear that the lodge was useful as an astronomical calendar. For example, at the summer solstice the first light of the Sun at sunrise would be able to make it into the interior of the lodge along a path from Big Black Meteoritic Star's post to the post representing Big Black Meteoritic Star's messenger, striking the south wall of the lodge. Since the Sun's light cannot reach farther southeast of that point on the wall, the southeast portion of the wall would remain in shadow. As the Sun moved southward following the summer solstice on its way to the winter solstice, the position of first light in the interior at sunrise would move clockwise along the circular wall until it reached its other extreme extent on the opposite side of the hut from the location of the first light at the summer solstice. The position of the first light of day would then move counterclockwise as the days grew longer again. When the Sun is at either the vernal equinox or the autumnal equinox, the first light of sunrise would pass directly along the line from the dirt mound to the fire pit, striking the altar in the west-most position of the lodge. Two other alignments associated with the lodge were likely to have been of special importance to the Skidi as well; the first time light could reach the altar prior to the vernal equinox (which also corresponds to the last time light will reach the altar after the autumnal equinox), and the first time light could reach the altar prior to the autumnal equinox (which also corresponds to the last time light can reach the altar following the vernal equinox).

The alignments appear to have been important in signaling time for some of the many ceremonies that were conducted by the Skidi. For example, the last two alignments could have alerted the band that it was time to leave their villages of earthen lodges to begin the spring and autumn bison hunts. During those times of absence from their villages, the Skidi would become nomadic, following the bison while living in tipis.

Other alignments were also possible. Although not shown in Fig. 2.21, making observations through the smoke hole directly above the fire pit would have allowed the Skidi to see objects like the Pleiades star cluster in the constellation of Taurus as it passed through their field of view just before sunrise in July and just after sunset in February. The Skidi could have also observed Corona Borealis, believed to represent a semicircle of chiefs in council. Being nearly opposite in the sky from the Pleiades, Corona Borealis would have been visible just after sunset in July and just before sunrise in February. In addition, in some lodges there could have been a shrine to the Star-That-Does-Not-Move, Polaris. It would have been possible to place a shrine on the south side of the lodge so that the Star-That-Does-Not-Move would always shine down on it. Recall that the Star-That-Does-Not-Move was given the charge by Tirawa to be chief of all the other gods.

Finally, what about a ceremony for Morning Star/Great Star (Mars), who fathered the first girl with Evening Star (Venus)? Morning Star was considered to be the most powerful of all the gods, and was the warrior god. Apparently, such a ceremony only occurred once every few years and was triggered by a warrior having a vision in a dream that Morning Star was talking to him. The warrior would report his vision to the Morning Star priest, keeper of the Morning Star bundle, and the priest would confirm whether or not Morning Star was in the sky. If Morning Star was indeed present, an elaborate series of events would be initiated that included a replication of Morning Star overtaking Evening Star by having the vision warrior and a group of others traveling west to capture a girl from another tribe. The girl was returned to the Skidi and ultimately sacrificed by killing her with the bow and arrow contained in the Skull bundle. This would all take place only when Morning Star could witness the events. Appeasing the Great Star was believed to bring power and success to the warriors.

A great deal more could be said of the Skidi Pawnee and their focus on the Sun, Moon, planets, stars, the Milky Way, meteorites, and perhaps comets, but the discussion above should give some sense of the great importance that they placed on the sky. It is interesting to note that the stories of the Skidi differed significantly from most Native American tribes in North America in that the Skidi believed themselves to be descended from the sky rather than originating from Mother Earth. It is also worth pointing out that the Skidi placed more importance on the inter-cardinal positions of northeast, southeast, southwest, and northwest than did most other pre-modern societies, which typically referenced north, east, south, and west.

Key Point
In science it is important to ask whether or not a particular observation is truly significant or is it simply due to random chance. This is true whether the question has to do with pre-modern architectural alignments or some other entirely different sort of observation.

A cautionary note is important to mention here. Although the alignments of the earthen lodge discussed above are almost certainly correct, there are a great many objects in the sky, and other alignments may or may not have been intentionally designed. There are many pre-modern structures throughout the world where various astronomical alignments have been proposed. When considering suggested alignments it is important to ask whether or not the proposed alignment would have had significant meaning to the culture that built the structure, or is a proposed alignment simply a matter of random chance. For example, consider again Fig. 2.21; it is certainly possible to envision an alignment from the post for Yellow Star past the post for Red Star on to the horizon. Of course, there are many stars that rise in that direction throughout the year, some of which might even be fairly bright, but are those stars important to the culture? Archaeoastronomy has often suggested alignments only to withdraw those claims later because of the statistical likelihood that the alignments were simply accidental or had no meaning for the culture that constructed the structure being investigated.

The Ancestral Puebloan Peoples of Chaco Canyon

The culture of the Ancestral Puebloan Peoples[3] thrived for hundreds of years in what is now the Four-Corners region of the United States (southern Utah, southwestern Colorado, northern New Mexico, and northeastern Arizona). Chaco Canyon, a national historical site and an United Nations Educational, Scientific, and Cultural Organization (UNESCO) World Heritage site in northwest New Mexico, approximately 185 km (115 mi) west-northwest of Santa Fe, was a major center of Ancestral Puebloan life from 850 CE to 1250 CE. The canyon is in the high desert at an elevation of approximately 1.9 km (6200 ft) above sea level and is a major archaeological site containing the largest collection of Ancestral Puebloan ruins in the region.

Fig. 2.22 The great kiva at Chetro Ketl in Chaco Canyon. (National Park Service)

Not much is known of the details of the Chacoans beyond their construction skills (e.g., Fig. 2.22) and the likelihood that they served as a commercial center for distant outlying villages. However, we do know that they were careful and ingenuous observers of the sky. For example, it appears that the Chacoans recorded a bright, naked-eye supernova in a petroglyph (Fig. 2.23). The depiction of the "new star" to the left and slightly below the crescent moon corresponds to the relative positions of these objects in the sky on July 4, 1054 CE, when the supernova would have been visible.

Fig. 2.23 A Chacoan depiction of the supernova of 1054 CE, seen to the left of the crescent moon. (Alex Marentes, CC BY-SA 2.0)

More impressive is the feature that has been named the Sun Dagger. Located near the top of Fajada Butte some 135 m (440 ft) above the canyon floor are three nearly vertical rock slabs leaning against a cliff wall, which are believed to have been the result of a natural rock fall (see Fig. 2.24). In 1977, Anna Sofaer, a volunteer artist who was recording petroglyphs in Chaco Canyon, came across two spirals located behind the three slabs. On a return visit to the slabs she happened to be at the site on the day of the summer solstice. Not only was she there on the right day, but she was also there at the right time to witness an amazing event. At 11:15 a.m. a dagger of light, as depicted in Fig. 2.25, descended vertically through the larger of the two spirals, passing directly through the middle of that petroglyph. Later observations also indicated very precise "sun daggers" appearing on the petroglyphs signaling the vernal and autumnal equinoxes (smaller spiral) and the winter

Fig. 2.24 *Left:* Fajada Butte in Chaco Canyon, NM. (National Park Service) *Right:* The three rock slabs of the Ancestral Puebloan Sun Dagger. (Public domain)

---

[3]The Ancestral Puebloan Peoples had previously been referred to as the Anasazi, a Navajo term meaning "the Ancient Ones" or "enemy ancestors."

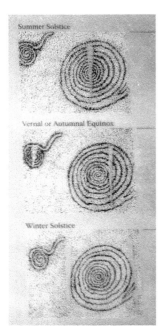

Fig. 2.25 A rendering of Fajada Butte "sun daggers" at the summer solstice, the vernal and autumnal equinoxes, and the winter solstice. (National Park Service)

solstice (bracketing the larger spiral). It has even been suggested that the spirals of the larger petroglyph could have also been used to track a complex 18.6-year eclipse cycle.

Due to the delicate nature of the alignments, the National Park Service closed Fajada Butte to visitors in the 1980s. Sadly, the action was too late in coming; in 1989 the rock slabs shifted, destroying the precise alignments required for the Sun Dagger to function properly. It is thought that traffic at the site may have eroded and compacted the soil, affecting the stability of the slabs. The site had functioned as an astronomical calendar for a millennium, only to succumb to modern human impact in the course of just one decade.

Mesoamerica

Mesoamerica is the region of Central America that is bounded on the north by the tropic of Cancer and on the south by northern Costa Rica. This area is just one of six locations on Earth where ancient civilization developed independently and just one of five regions where writing developed. While much focus is given to the intellectual advancements made in Europe, the Middle East, and Asia, in pre-Columbian times (before the arrival of Christopher Columbus) the cultures of Mesoamerica were in many ways equally sophisticated, and in some ways more advanced, than in other parts of the world. It is unfortunate that we cannot spend the amount of time studying these cultures that they deserve, but we can at least illustrate their amazing levels of progress in astronomy, mathematics, calendars, and architecture that were tied to their religious belief systems.

The initial human occupation of Mesoamerica began before 10,500 BCE, and by 7000 BCE inhabitants of the region made the transition from being hunter-gatherers to learning how to cultivate maize (corn), beans, squash, and chili, along with the domestication of turkeys and dogs. Once the ability to provide food became place-bound rather than having to follow animals to hunt, communities started to spring up with the ability to work collectively to the advantage of the whole. Communication between villages led to some commonality in religions, art, and architecture. The first evidence of pottery dates back to at least 2500 BCE, which corresponds loosely with the period of construction of Stonehenge in England and the Egyptian pyramids.

The first truly organized society was that of the Olmec, who appeared sometime between 1600 BCE and 1500 BCE in what are the present-day states of Veracruz and Tabasco, Mexico, and they survived until 400 BCE. For reference with Europe and the Middle East, the Olmec culture existed at the time of the Trojan war, Homer, Socrates, and Alexander the Great. Although relatively little is known of this culture, their influence on subsequent civilizations is clearly evident through their artwork and writing (e.g., Fig. 2.26). There is also evidence that they had an extensive long-distance trade network with other cultures.

The Mesoamerican Base-20 Number System

As early as 900 BCE, or perhaps even earlier, a profound intellectual development that likely originated with the Olmec was the creation of the base-20 number system, including the invention of zero. It may seem odd to point out the explicit invention of zero, but consider its modern-day meaning: a number and symbol designed to represent nothing at all. The Mesoamerican interpretation of zero was meant to indicate closure or completeness, rather than a lack of something. It is important to note that only the ancient Egyptians had developed a symbol for zero by that time; for example, the Roman numeral system (I, V, X, L, C, D, and M) has no representation for zero. The Mesoamerican zero was also used to develop a positional number system similar to our modern system with a "ones" place, a "tens" place, a "hundreds" place and so on, but even the Egyptians did not reach that level of sophistication. The Babylonians did have a positional numerical system at about the same time but simply represented a zero as an empty space.

It is so common-place for us to use a positional number system that we usually don't even think about what it is representing. Today, our system is based on ten symbols (0, 1, 2, 3, 4, 5, 6, 7, 8, 9) and the places those numbers occupy, meaning that our system is referred to as base-10. The origin of the base-10 system rests in the fact that we humans have ten fingers on our hands. However, consider the conditions in Mesoamerica where footwear was not required for environmental reasons other than to protect the bottoms of the feet while walking. In that case the ten toes on human feet are also available for counting, leading to twenty human digits that can be used.

Thinking back to how our base-10 positional system works, consider the number 2815. Without even thinking about it we read the number as "two thousand eight hundred fifteen," but what the number is actually representing is the sum of $2 \times 1000 + 8 \times 100 + 1 \times 10 + 5 \times 1$. Each place corresponds to a power of 10 where $1 = 10^0$, $10 = 10^1$, $100 = 10 \times 10 = 10^2$, and $1000 = 10 \times 10 \times 10 = 10^3$, where the integer exponents simply represent the number of times the base number (in this case 10) is multiplied by itself.

The Mesoamerican base-20 number system that was used to count money or objects (the *merchant system*) is similar to our base-10 system except that only three symbols were used: a dot (●) representing 1, a bar (———) representing 5, and a zero glyph (⬬) often symbolized as a sea shell although many other symbols have been used for zero as well (we will use the sea shell exclusively in this text). To construct numbers between 1 and 19, a stacked combination of dots and bars is used. For example, the number 17 requires 2 dots and 3 bars so that $2 \times 1 + 3 \times 5 = 17$:

For numbers greater than 19, digits are stacked vertically with the smallest place (the 1s place; 20^0) on the bottom of the stack, the 20s place (20^1) is next, then the 400s place (20^2) is above the 20s place, the 8000s place (20^3) is above the 400s place, and so on. This is very similar to how we write our Arabic numbers larger than 9 horizontally, with the smallest place on the right.

Fig. 2.26 San Lorenzo Monument 3 is an example of the characteristic colossal heads created by Olmec artisans. This head stands 1.78 m (5.84 ft) tall and is housed in the Museo de Antopologia de Xalapa, Universidad Veracruzana, Mexico. The largest known surviving head is the Rancho La Cobata head at 3.4 m (11 ft) high. [Maribel Ponce Ixba (frida27ponce), CC BY 2.0]

An exponents tutorial is available at
`Math_Review/Exponents`

Example 2.1

How did the Olmec write the number 2815 that we discussed several paragraphs earlier?

The first question we must ask ourselves is: what is the largest factor of 20 that remains less than 2815? That number will then occupy the top digit in the Olmec system. Since 2815 is less than 8000 (20^3), the highest place we will use is the 400s place (20^2). So, how many 400s go into 2815? $2815 \div 400 = 7.0375$, meaning that we will need 7 in the 400s place. After subtracting $7 \times 400 = 2800$ from 2815 we are left with 15. Repeating the process, we ask ourselves how many 20s go into 15, and the answer is 0. Finally, we have 15 left over for the 1s place. Putting it all together with number glyphs we have:

$$
\begin{array}{rcrcr}
7 \times 400 & = & 7 \times 20^2 & = & 2800 \\
0 \times 20 & = & 0 \times 20^1 & = & +0 \\
15 \times 1 & = & 15 \times 20^0 & = & +15 \\
\hline
& & & & 2815
\end{array}
$$

In order to avoid the cumbersome glyph notion, Mesoamerican numbers are sometimes written today in a more compact form that borrows elements from our base-10 system. The number used in the example, 2815, may be written more concisely in the Mesoamerican base-20 system as (7.0.15), remembering that the appropriate glyphs go in each position ( ), and should be stacked vertically.

The Maya

The Maya first appeared as early as 2600 BCE. By about 1800 BCE they had established village-based societies throughout the Yucatán peninsula, in what are today the southern Mexico states of Chiapas and Tabasco, and the region composed of Guatemala, Belize, western Honduras, and northern El Salvador. Over the first two millennia of their existence, the Maya were an unorganized people focused on survival, but they were certainly aware of, and influenced by, their Olmec neighbors to the west.

From about 250 CE to 900 CE, the Maya civilization was at its peak, with large and sophisticated city-based societies that utilized extensive agricultural efforts to support those cities. Some 20 or more city-states existed with their own independent political systems. The cities boasted magnificent architectural achievements, including the building of tiered pyramids, circular towers, ball courts, temples, and monuments. During this Classic Period, the Maya population numbered in the millions and were scattered throughout their vast region, with as many as 70 different dialects developing over time. Today there are perhaps six million Mayans, with most still living in their ancient homelands.

The Classic Period was also a time of impressive artistic development and intellectual advancements, including writing, incorporating the Olmec number system, calendars, and the study of astronomy. It is unfortunate that only three authenticated folding books of the Maya, known as codexes, remain today for careful study, although much can also be gleaned by studying the many sites scattered throughout Mesoamerica. Those books are named for the museums in which they are now kept: the Dresden Codex (Saxon State and University Library, Dresden, Germany; Fig. 2.27), the Madrid Codex (Museo de América, Madrid, Spain), and the Paris Codex (Bibliothéque Nationale de France, Paris, France). A fourth codex, the Grolier Codex (Museo Nacional de Antropología, Mexico City) is considered by many scholars to be a forgery, although debate on this point continues.

Fig. 2.27 Page 49 of the Dresden Codex, showing a portion of the Venus almanac. The figures depict the plumed serpent and god of war, Kukulkán (Quetzalcóatl). Note that some numbers are shown horizontally when associated with calendar glyphs. (Public domain)

Table 2.2 The first five Long Count calendar units used in Mesoamerica

Time period	Length of time in k´ins (days)		
1 day	= 1 k´in		
1 winal	= 20 k´ins		
1 tun	= 18 winals	= 18 × 20 k´ins	= 360 k´ins
1 k´atun	= 20 tuns	= 20 × 18 × 20 k´ins	= 7200 k´ins
1 b´ak´tun	= 20 k´atuns	= 20 × 20 × 18 × 20 k´ins	= 144,000 k´ins

The Mayan calendars

It is believed that the complex calendar systems of the Maya were first conceived by the Olmec but the Maya likely refined the calendars and certainly integrated them fully into their culture. The most straightforward of these calendars is the Long Count calendar. The starting date for the calendar is thought to be August 11, 3114 BCE, according to the most commonly accepted transformation from our modern Gregorian calendar. The Long Count calendar uses the same base-20 counting system we just discussed except for one important modification: in the Long Count calendar the value of the second digit ranges from 0 to 17 rather than from 0 to 19. This means that the value of the third digit is $1 \times 18 \times 20 = 360$ rather than 400, the fourth digit is $1 \times 18 \times 20 \times 20 = 7200$, and so on. The reason for this confusing modification likely arose from the desire to correspond loosely with the approximately 365 1/4 days in one year.

The naming convention invented by scholars of the Maya culture refers to one day as one *k´in*, which also represents the value of the first place in the Long Count calendar. The names of the first five places in the counting system and their associated values are listed in Table 2.2.

The ancient Maya philosophy regarding numbers and calendars is deeply embedded in their traditional belief system. The Mayan religion holds that all time is cyclic. Even the number zero is a part of that fundamental philosophy. As mentioned before, rather than representing emptiness or the absence of something, the Maya zero represents completion. The zero glyph hints at that sense of completeness or closure. The concept of a cycle also applies to the Long Count calendar; the calendar does not continue uninterrupted forever. From mythology, the last day of 13 b´ak´tuns occurred on the Long Count day of (12.19.19.17.19), which is believed to have corresponded to August 10, 3114 BCE. The next day, August 11, 3114 BCE [(13.0.0.0.0)], also represented by (0.0.0.0.0), was the start of a new Long Count cycle. Just like the old mechanical odometer of a car rolling over from 99,999 to 00,000, the Maya Long Count calendar restarts as well. After another 13 b´ak´tuns the same thing happens. In Maya mythology, the end of 13 b´ak´tuns represents the start of another natural cycle. The date of August 11, 3114 BCE was seen by the Maya as the creation of the current world; the world in which humans came to be.

In 2012 CE a social phenomenon occurred. There were those who were claiming that the world would come to an end on December 21, 2012 as foretold by the Maya Long Count calendar. The reason for this specific date is that December 20, 2012 was the last day of another 13 b´ak´tuns and therefore the next day would start a new Long Count series and a new world, requiring the end of the previous world.

Table 2.3 Named months of the Haab´ calendar and named days of the Tzolk´in calendar

Haab´ Month	Glyph Meaning	Tzolk´in Day	Association or Meaning
Pop	mat	Imix´	crocodile
Wo´	black conjunction	Ik´	wind
Sip	red conjunction	Ak´b´al	darkness
Sotz´	bat	K´an	sacrifice
Sek	death	Chikchan	snake
Xul	dog	Kimi	death
Yaxk´in´	new sun	Manik´	deer
Mol	water	Lamat	Venus
Ch´en	black storm	Muluk	jade
Yax	green storm	Ok	dog
Sak´	white storm	Chuwen	howler monkey
Keh	red storm	Eb´	rain
Mak	enclosed	B´en	young maize
K´ank´in	yellow sun	Ix	jaguar
Muwan´	owl	Men	eagle
Pax	planting time	Kib´	wax
K´ayab	turtle	Kab´an	earth
Kumk´u	granary	Etz´nab´	flint
Wayeb´	five unlucky days	Kawak	rain storm
		Ajaw	ruler (Sun)

Fig. 2.28 The Time Keeper and the Haab´. The glyphs on the outside circle represent the 19 months of the 365 day "vague year" calendar. (theilr, CC BY-SA 2.0)

Obviously, since you are reading this book, the end of the world didn't happen (again). In reality, nowhere in known Maya writings is there a mention of the end of the world at the end of the last cycle of 13 b´ak´tuns. Instead, it was really an opportunity to have a very big "New First B´ak´tun Eve" party!

A second calendar system, referred to as Haab´ or the "vague year," is a 365-day solar calendar. As depicted in the carving shown in Fig. 2.28, the Haab´ is composed of 18 named "months" of 20 days each, plus an additional nineteenth month comprised of only 5 days ($18 \times 20 + 5 = 365$). Since no attempt was made to adjust the Haab´ to the year of approximately 365 1/4 days as we do with our Leap Year every fourth year, the starting date of the Haab´ calendar would drift through the seasons by about one day every four years. To refer to a specific day in the Haab´ calendar, the day of the month would be followed by the month name, beginning with zero (or "Seating" day) as the first day of the month. For example, 0 Pop corresponds to the first day of the Haab´ calendar and 5 Sip corresponds to the *sixth* day in the third month. This system is very much like our reference to Ground Hog Day as being 2 February. The 19 named months are listed in Table 2.3, with Wayeb´ being the month with the "five unlucky days." After 365 days, the calendar starts over again, just as our calendar restarts on January 1, with the process continuing forever.

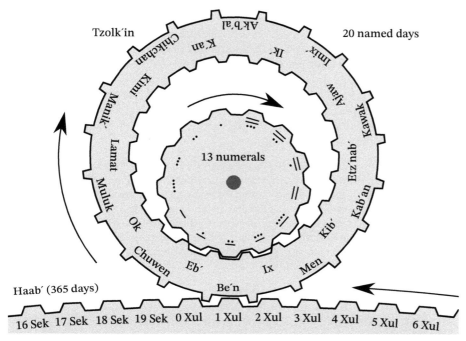

Fig. 2.29 The Tzolk´in and the Haab´ combine to create the Calendar Round. The Tzolk´in is composed of the top two gears and the Haab´ is shown as the bottom gear.

An additional (third) calendar of significance in Mesoamerican cultures is the 260-day Tzolk´in calendar. The Tzolk´in is composed of two cycles, the numbers 1 through 13 and 20 day names (the day names are also listed in Table 2.3). As we have already seen, the number 20 comes up frequently as a reflection of the base-20 number system. The number 13 is less obvious, but also shows up over and over again. It has been suggested that 13 corresponds to the number of major joints in the human body: 2 ankles, 2 knees, 2 hips, 2 shoulders, 2 elbows, 2 wrists, and 1 neck. Other researchers have suggested that 13 may be the number of heavens in ancient Mesoamerican religions.

In any case, the Tzolk´in calendar operates like two connected gears, one with 13 cogs and the other with 20 cogs. Starting the calendar at 1 Imix´, the next day would be 2 Ik´, then 3 Ak´b´al and so on. When 13 is reached the day would be 13 B´en. At this point the numbers start over again while the names continue with the next day being 1 Ix. At 7 Ajaw the day names start again, making the next day 8 Imix´. Every 260 days (13 × 20) the calendar is back to where it started; 1 Imix´. This cyclic process also continues forever.

Finally, when the Tzolk´in and Haab´ calendars are combined, the result is the Calendar Round. As depicted in Fig. 2.29, both calendars continue to cycle endlessly. Eventually the same numbered day name of the Tzolk´in calendar and the day and month from the Haab´ calendar will return to an initial combination. For example, 1 Imix´ 0 Pop is followed by 2 Ik´ 1 Pop, then 3 Ak´b´al 2 Pop, et cetera, until 1 Imix´ 0 Pop pops up again. This will occur when an integer number of Tzolk´ins have been completed at the same time that an integer number of Haab´s have been completed [mathematically we are looking for what is termed the least common multiple (LCM) of 260 and 365]. As a result, the two calendars will again align after 18,980 kin's, when 52 Haab's and 73 Tzolk'ins have been completed. In our calendar with close to 365 1/4 days per year, one complete Calendar Round would occur about every 51 years, 352 days.

The completion of a Calendar Round was recognized throughout Mesoamerica. In central Mexico for example, the day was associated with a New Fire celebration. According to Fray Bernardion de Sahagùn, a sixteenth-century Spanish Franciscan who extensively documented the culture:

> Behold what was done when the years were bound — when was reached the time when they were to draw the new fire, when now its count was accomplished. First they put out fires everywhere in the country round. And the statues, hewn in either wood or stone, kept in each man's home and regarded as gods, were all cast into the water. Also (were) these (cast away) — the pestles and the (three) hearth stones (upon which the cooking pots rested); and everywhere there was much sweeping — there was sweeping very clear. Rubbish was thrown out; none lay in any of the houses.

Now that was some serious "spring cleaning!"

It is worth noting that 52 Haab's can be divided evenly into four equal intervals of 13 Haab's (there's that critical number again). Each quarter of a Calendar Round was associated with one of the four cardinal directions (north, east, south, west).

Mayan Astronomy

Now that we have some understanding of the Mayan philosophy and sophistication regarding numbers, cyclic calendars, and a repeating universe, we can take a look at their remarkable achievements as observational astronomers. Like so many other pre-modern societies, the Maya were highly motivated to study the heavens given that the heavens were the realm of their deities.

Patterns are very common in astronomy. To list just a few of the many patterns that can be identified:

- slightly more than 365.256 days in one year (roughly 365 1/4 days),
- the repeating pattern of the rising and setting locations of the Sun along the horizon during the course of one year,
- just over 29.53 days between full moons (roughly 29 1/2 days), known as a synodic month,
- a little more than 27 1/3 days needed for the Moon to return to the same region among the stars (approximately 27 1/3 days), referred to as a sidereal month,
- the amount of time between successive lunar and/or solar eclipses, and
- the number of days between consecutive first appearances of a rising planet on the eastern horizon after it emerges from the glow of the Sun, known as heliacal risings of the planet.

With the development of writing, and hence the ability to record information over long periods of time, together with their prowess with numbers and calendars, the Maya were able to identify all of these patterns and even use them to predict future events, often with remarkable precision.

Without getting into the details concerning when eclipses may or may not occur (that subject will be explored in Section 4.6), let's investigate some of the underlying patterns that long-term observations and careful analysis can reveal.

The first such pattern might be the time between a lunar eclipse and a subsequent solar eclipse, or between a solar eclipse and a subsequent lunar eclipse. Since the Maya only dealt with integers (they didn't seem interested in, or perhaps hadn't developed, the capacity to work with, fractions) the fractional synodic month wasn't of importance. Instead, they considered the time between full and new moons to be either 14 or 15 days, numbers that show up frequently in their written records. Since eclipses can also appear approximately one-half year after a previous eclipse, or set of eclipses, intervals of 5, 5 1/2, or 6 "moons" (lunar phase cycles) were commonly recorded as well, corresponding to 148 ± 1, 163 ± 1, and 177 ± 1 days, respectively, where ± 1 stands for "plus or minus 1," meaning that 148 ± 1 is somewhere between 147 and 149. (The reason that 6 1/2 or 7 moons don't lead to eclipses has to do with a shifting of the orientation of the Moon's orbit relative to Earth's orbit.)

Example 2.2

Since the Moon's synodic period is 29.53 days, 5 moons would correspond to $5 \times 29.53 = 147.65$ days. Rounded off to the nearest integer, the Moon's synodic period is about 148 days. Since the Moon can be "close enough" to the correct position to cause at least a partial eclipse within about one day on either side of the exact value, eclipses could occur 147, 148, or 149 days apart (148 ± 1 days).

By observing eclipses over many years the Mayan astronomers realized that lunar eclipses tend to come in groupings of three, four, or five, with the interval between each eclipse being slightly less than six months. After a group of lunar eclipses ends, the astronomers could then look forward several years to anticipate the next eclipse cycle.

Key Point
The Maya were able to predict eclipses by looking for patterns.

The remarkable ability of the Maya to uncover celestial patterns can also be found in their knowledge of an 18.03 year eclipse cycle (very close to 6585 1/3 days), known as the **saros cycle** (saros means repetition). When an eclipse (either lunar or solar) occurs, another similar eclipse will occur again 18 years later. However, given the timing of the saros cycle, that eclipse 18 years later won't be observable from the same place on Earth, but it will be observable one-third of the way around Earth, which means that the Maya wouldn't see it (this is due to the additional 1/3 day in the cycle). After another 18 years an eclipse will happen two-thirds of the way around Earth relative to the home of the Maya, and after yet another 18 years the eclipse would finally return to the Mayan realm. In this way, the cycle observed by the Maya was actually three saros cycles, equaling $18 \times 3 = 54$ years.

From the perspective of the Maya and other cultures throughout Mesoamerica, the planet Venus was of particular importance due to its association with the mythical god, Quetzalcóatl (known as Kukulkán to the Maya), the feathered or plumed serpent in Fig. 2.30. The legend of Quetzalcóatl may date back as far as the Olmec culture, but was certainly present at Teotihuacán (Fig. 2.31), a city-state independent of the Maya (and now a UNESCO World Heritage site) that was established around 100 BCE and may have been occupied until sometime between 800

Fig. 2.30 A feathered serpent head depicting Quetzalcóatl, decorating the Temple of the Feathered Serpent in Teotihuacán. (Jami Dwyer, CC BY-SA 2.0)

Fig. 2.31 Teotihuacán. A view looking down the Avenue of the Dead with the Pyramid of the Sun seen in the distance off to the left. The photograph was taken from the Pyramid of the Moon. (Public domain)

and 900 CE. The remains of the city are located about 48 km (30 mi) northeast of modern-day Mexico City. At its peak, this huge city may have housed more than 125,000 persons, making it among the largest cities in the world at the time. Remarkably little is known about the Teotihuacán civilization, although it was certainly a center with significant commerce and influence throughout Mesoamerica.

One of the many variants of the Quetzalcóatl legend suggests that he was one of the gods associated with the creation of the world on five separate occasions, each of the first four ending at the conclusion of a cycle of 13 b'ak'tuns. Apparently Quetzalcóatl took on human form and first appeared in Teotihuacán. He then went to the city of Tollan, the center of the Toltec empire, as the famous leader Ce Acatl Topiltzin Quetzalcóatl to provide people with the calendar and instruct them in raising maize and other necessities of life. Some versions of the legend also suggest that Quetzalcóatl took his people to Chichén Itzá (rhymes with "chicken pizza"), in the north-central region of the Yucatán peninsula to protect them from invaders. As the legend goes, Quetzalcóatl was tricked by his brother, Tezcatlipoca, another of the creation gods, into getting drunk. When he awoke the next morning Quetzalcóatl was no longer a virgin and he was so upset with himself for violating his own strict rules of conduct that he traveled east to present-day Veracruz on the Gulf of Mexico. In one version of the story he threw himself into a funeral fire, at which point his heart rose up into the sky and became Venus. Another version suggests that Quetzalcóatl sailed off to the east on a ship made of serpents, vowing his return to rule again one day. In the east the planet Venus then appeared. This took place in the year One Reed.

Regardless of which version of the legend is adopted, it is evident that Quetzalcóatl was an important god in the pantheon of gods in the various Mesoamerican cultures over thousands of years, and that Quetzalcóatl was closely associated with Venus. Hence the motivation of the Maya to track Venus carefully over millennia.

Fig. 2.32 Two of the structures at Chichén Itzá, in the Yucatán peninsula, Mexico. *Left:* The pyramid known as the Castillo. *Right:* The Caracol tower, believed to be an observatory designed to observe the Sun and Venus. (Dale A. Ostlie)

Two of the structures at Chichén Itzá appear to be designed intentionally to observe both the Sun and Venus. Figure 2.32 shows photographs of the huge tiered pyramid known as Castillo (*Left*) and the cylindrical Caracol tower atop a two-tiered base (*Right*) The Castillo has nine tiers with steep stairs rising up each of the four sides. Each of the stairs have 91 steps, giving a total of $4 \times 91 = 364$ steps in total. Including the final step at the top of the pyramid, there are 365 stairs, as many as there are days in the Haab′. Ninety-one also happens to be the approximate number of days between the winter solstice, vernal equinox, summer solstice, and autumnal equinox. The alignment of the pyramid itself is such that the east–west orientation of its base points directly toward the setting position of the Sun on the days when the Sun passes through the local zenith. The most fascinating aspect of the Castillo is that on the vernal and autumnal equinoxes, shadows cast by the corner of the pyramid on the side of one of the staircases result in the image of an undulating snake near sunset. As the Sun sets, the snake appears to move down the side of the staircase and descends into the underground. At the base of the staircase is a carving of a large snake's head, presumably depicting the plumed serpent of the Mayan god, Kukulkán.

The Caracol also exhibits evidence of a design meant to observe both the Sun and Venus. The perpendicular to the base of the structure points in the direction of the setting position of Venus when it is as far north as it ever travels. The perpendicular to the upper platform points toward the setting position of the Sun on the day when it passes through the local zenith. The diagonal from the southwest corner of the upper platform to the northeast corner is aligned with the summer solstice sunrise. Reversing the direction points toward the winter solstice sunset.

The top cylindrical feature of the Caracol contains alignments through the "windows" that can be seen at the very top of the structure. Unfortunately, some time between the late nineteenth century and the early twentieth century, one-half of the cylindrical tower was destroyed, along with three of the six sighting windows. The three remaining windows do suggest additional observational alignments with the maximum northern and maximum southern settings of Venus along the horizon, as well as the setting of the Sun at the equinoxes. It has been proposed that perhaps other observations could have been made as well, such as observing the Pleiades, the Hyades, and Aldebaran, all in the constellation of Taurus. However, recall the caveat mentioned on page 32.

However, recall the caveat mentioned on page 32.

The portion of the Dresden Codex shown in Fig. 2.27 is one page of a series of pages that is a table of the heliacal rising and setting dates of Venus. It is believed that the Codex, written sometime during the eleventh or twelfth century CE at Chichén Itzá, is a copy of original texts that may have been written as early as the seventh or eighth century CE. Apparent on the page are depictions of Kukulkán/Quetzalcóatl together with numbers and specific dates. Combined with a correction table, also contained within the Codex, it would have been possible to predict the rising of Venus to within two hours over a period of 481 years; a remarkable accomplishment!

Key Point
The Maya were able to predict the rising of Venus to within two hours over 481 years.

The Arrival of the Spanish in Mesoamerica

In the first years of the sixteenth century, a bright comet appeared. Moctezuma II, king of the great Aztecs empire in central Mexico, believed that the comet was an omen forecasting his downfall. It was in 1519 (the year One Reed) when the Spaniard, Hernán Cortés (1485–1547), arrived from the east with 11 ships, 508 men, 16 horses, and a few cannons. Cortés came ashore in Veracruz on a complete Calendar Round anniversary of the departure of Quetzalcóatl from the same region of the Gulf of Mexico coastline. It had been understood for centuries that Moctezuma II assumed Cortés to be Quetzalcóatl, returning to rule as promised long ago. The association of Cortés with Quetzalcóatl then set the stage for the downfall of that great empire and the rapid and nearly complete transformation of all of Mesoamerica. In recent years that long-held understanding has been challenged, with the argument that the Franciscan fryers who had arrived with Cortés may have generated the story themselves. Whatever the truth may be, the impact of the arrival of the Spanish was devastating and irreversible. Through both battle and diseases brought to Mesoamerica for which the indigenous populations had no immunity, the great cultures nearly vanished.

The first archbishop of Yucatán, Diego de Landa (1524–1579), wrote shortly after the Spanish conquest:

> We found a large number of books in these characters [hieroglyphs] and, as they contained nothing in which there were not to be seen superstition and lies of the devil, we burned them all, which they [the Maya in the Yucatán city of Mani] regretted to an amazing degree, and which caused them much affliction.

Today, descendants of the Maya remain but much of the rich culture of the ancient peoples of Mesoamerica has been lost to history.

The Stonehenge Monument

Certainly the most famous of the ancient astronomical sites in Europe is Stonehenge (Fig. 2.33) on the Salisbury Plain of southern England. Another UNESCO World Heritage site, Stonehenge is located 130 km (80 mi) southwest of London near Amesbury.

Fig. 2.33 The Stonehenge monument. The Slaughter Stone is visible in the foreground and the Heel Stone (Fig. 2.34) is behind the photographer. (garethwiscombe, CC BY 2.0)

Legend has it that a monument once stood on storied Mount Killaraus in Ireland that had been erected by giants who brought huge stones from Africa with magical healing powers. Merlin, the wizard and teacher of King Arthur, dismantled the monument, and with the aide of 15,000 men, transported the stones to the present-day location of Stonehenge where the monument was reconstructed to honor those Britons who had died during a battle with the Saxons.

The Stonehenge monument we see today is, in part, a reconstruction of what the site may have looked like. Between 1901 and 1964 some stones were set in concrete to stabilize them, and other stones were even moved before being re-erected using cranes. Unfortunately, the official guidebooks of the monument left out those twentieth-century alterations, leading its millions of visitors with the impression that what they see is how the monument has looked for thousands of years. Those past changes also impact present-day archaeological research at the site. In the twenty-first century, careful conservation work has also been done at the site to help stabilize stones and limit deterioration over time.

Fig. 2.34 The Heel Stone at Stonehenge. (Heikki Immonen, CC BY 3.0)

Despite the changes in the twentieth-century, modern research indicates that some of the bluestones that comprised the Stonehenge monument had been used in an earlier monument at the Waun Mawn site in the Preseli Hills in Wales, approximately 240 km (150 mi) west of Stonehenge. Each bluestone weighs up to 3 tonnes[4] (up to 3.3 tons) and stands as high as 2 m (6 ft). Around 3100 BCE the people at the Waun Mawn site decided to migrate to the Salisbury Plain to join others who had long been holding ceremonial gatherings at the Stonehenge site which may have held spiritual significance as a place of cosmic harmony between Earth and sky (evidence indicates some construction by hunter-gatherers as early as 8000 BCE).

[4]One tonne, also referred to as a metric ton, equals 1000 kg. In North America, ton (or short ton) is more commonly used, and is defined as 2000 lb_m. A mass of 1 tonne is equivalent to 2204.62 lb_m = 1.10231 ton. [The distinction between pound-mass (lb_m) and pound (lb; or pound-force, lb_f) is discussed on page 154.]

In their relocation the people of Waun Mawn brought their bluestones with them, perhaps as symbols of their own ancestral identities. In all, about 80 bluestones from the Preseli Hills were used in the construction of Stonehenge, but only a fraction of them were used in the original Waun Mawn site.

Why that particular location for Stonehenge? Two of the stones appear to have been there for millions of years, and unlike the bluestones, are sarsen stones composed of native sandstone. The Heel stone (Fig. 2.34) and Stone 16 are among the largest of the sarsen stones and unlike other stones at the site they have not been re-shaped. As indicated in the schematic diagram of the monument in Fig. 2.35, those two stones also have a natural alignment with the summer solstice sunrise and winter solstice sunset. It may be that the unique alignment led to that site being chosen as a ceremonial gathering place, worthy of further development.

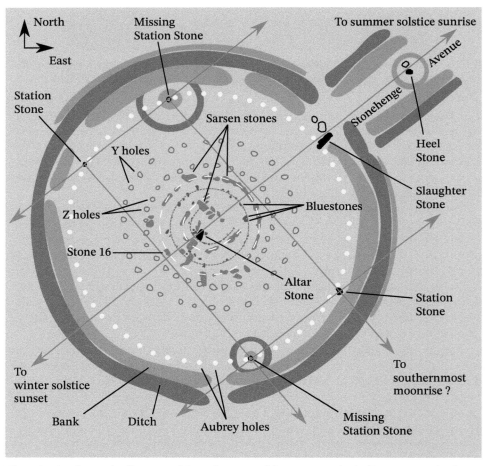

Fig. 2.35 A schematic diagram of Stonehenge and its astronomical alignments. The inner bluestone circle and horseshoe are outlined in gray dots, the sarsen stone circle and horseshoe are outlined with white dashes, and the missing stones referenced in the text are marked with black outlines. The white circles inside the ditch and bank are the Aubrey holes. The circular ditch and bank measures 110 m (360 ft) across. (Adapted from Adamsan, CC BY-SA 3.0)

The first period of construction around 3100 BCE involved the creation of a circular ditch and bank having a diameter approximately equal to the length of an American football field, 110 m (360 ft), and similar in size to the Waun Mawn site. The ditch and bank were built with a large opening to the northeast and a smaller one on the south side. A circle of 56 erect bluestones was also constructed, but the stones that once formed the circle were later moved to a new configuration. The holes that were left behind were used to bury the cremated remains of at least 63 individuals and are known as Aubrey holes, named for John Aubrey (1626–1697), the seventeenth-century scholar who first identified them.

Some 500 years later (2600 BCE), builders at Stonehenge attempted to enhance the monument by constructing two circles of bluestones. What is referred to today as the Altar Stone, at the center of the monument, might have been put in place at that time as well. Although it is no longer present, it appears that the Heel Stone had a companion, indicated by the black outline in Fig. 2.35. The Stonehenge Avenue, which extended more than 3 km (2 mi) to the River Avon, was also built during that period. Finally, a portal into the monument, composed of the unfortunately-named Slaughter Stone (now fallen over) and at least one other stone (now missing) were erected. From the center of the monument, the sight line past the Slaughter Stone to the Heel Stone and along the Stonehenge Avenue corresponds to the opening created by the ditch and bank builders five centuries earlier. Four Station Stones, two of which are now missing, were also likely erected during that intense period of construction.

Between 2600 BCE and 2400 BCE, more enormous sarsen stones were brought to the site from some 25 km (16 mi) north of Stonehenge. The stones were shaped through hammering that included creating points on the tops of some stones that were stood on end. Other sarsen stones had indentations carved into them so that they could be seated horizontally atop the points of the vertical stones. The vertical sarsen stones were erected in a circle, with the horizontal stones forming a cap for the circle, portions of which can be seen in Fig. 2.33. Each vertical sarsen stone stands 4.1 m (13 ft) high, is 2.1 m (7 ft) wide, and has a mass of 23 tonnes (25 tons). The cap stones are each about 3.2 m (10 ft) long, 1 m (3 ft) wide, and 0.8 m (2.5 ft) thick. Inside the sarsen stone circle a horseshoe of upright sarsen stones with caps was erected with the opening of the horseshoe pointing northeast toward the Heel Stone. The horseshoe stones were even larger than those that formed the circle, with masses up to 45 tonnes (50 tons).

Over the course of the next 800 years, it seems that there was some rearrangement of the bluestones. At one point, bluestones formed a circle between the sarsen stone circle and the sarsen stone horseshoe, together with an oval shape inside the horseshoe. The Altar Stone may have also been moved to the center of the bluestone oval and re-erected. Eventually, the northeastern portion of the bluestone circle was taken down, creating a horseshoe that was similar to the sarsen stone horseshoe. Construction at the site finally culminated around 1600 BCE with the digging of two rings of holes that formed concentric circles around the sarsen stone circle.

Scholars have worked to understand what purpose(s) Stonehenge served over the 1500 years that it was being used since the Waun Mawn brought their bluestones with them from Wales and perhaps for 5000 years or more before that. For example, along with marking summer solstice sunrise and winter solstice sunset it has been suggested that, in addition to being a burial site, Stonehenge may have served as a ceremonial healing center. Setting aside the amazing engineering feats of the builders, the mere fact that it was a center of activity for so long is truly impressive.

In 1963, Gerald Hawkins (1928–2003), a British-born astronomer, ran computer simulations to look for possible alignments within Stonehenge that correlate with celestial events along the horizon, of which he found many. He also proposed that the 56 Aubrey holes could have been used to calculate eclipses. Again recalling the cautionary note on page 32, just because an alignment exists does not mean that it was intended; today most of the alignments Hawkins proposed have been rejected. The suggestion that the Aubrey holes could have been used as an eclipse calculator also appears to be highly unlikely.

What clearly supports celestial alignments is the northeast orientation of the Stonehenge Avenue, the Heel Stone and its missing partner, the Slaughter Stone and its missing partner(s), and Stone 16. When viewed from the center of the monument, the Sun rises at the summer solstice along the line defined by the gap between the Heel Stone and its partner, and the gap between where the Slaughter Stone used to stand and its partner. The northeast–southwest alignment of the Station Stones also supports an alignment with the summer solstice sunrise, with an additional alignment in the opposite direction toward the winter solstice sunset to the southwest. More questionable, but certainly possible, is the northwest–southeast alignment of the Station Stones with the southernmost position of moonrise.

Based on estimates of slaughtered animals, it has been suggested that as many as 4000 individuals may have gathered at Stonehenge during its peak to celebrate the summer and winter solstices. DNA evidence even suggests that some visitors may have come from as far away as the Mediterranean and Scotland.

It is unfortunate that many of the original stones are missing today. The ready availability of stones made for a convenient quarry from which to obtain the materials necessary to construct the stone walls so prevalent in that part of rural England.

We may never know all of the significance that was attributed to this incredible monument, but it does point once again to the importance that ancient societies placed on tracking the seasons via motions in the heavens.

The Great Pyramid of Giza

As with many of the pyramids of the Maya and other Mesoamerican cultures, the Great Pyramid at Giza, Egypt [Fig. 2.36 (*Left*)] is carefully aligned astronomically. Built over a period of about 20 years beginning in 2580 BCE, the Great Pyramid was the first, and the only remaining, structure of the "Seven Wonders of the World." With a height of 146 m (481 ft) the Great Pyramid was the tallest structure in the world for 3800 years. Perhaps not surprising to you at this point, the base of the Great Pyramid is aligned with the four cardinal positions, north, east, south, and west. However, it is the precision of the alignment that is remarkable; no side is misaligned by more than 5 1/2 minutes of arc or 0.092°. This means that if a marker was placed 10 km (6.2 mi) from the Great Pyramid along the line defined by the most misaligned side, that marker would miss the alignment with the true cardinal position by only 16 m (52 ft). Even more amazing is the fact that each side of the square base averages 230.4 m (756 ft) in length, with no side being off by more than 20 cm (8 in). Impressive for a monument that was built 4600 years ago and constructed with some 2.3 million blocks of stone.

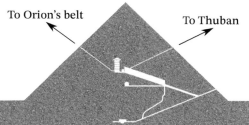

Fig. 2.36 *Left:* The Great Pyramid at Giza, Egypt. One of the pyramids in the region that are a part of a UNESCO World Heritage site. (Diego Delso, delso.photo, License CC-BY-SA) *Right:* The alignments of the "air shafts" from the King's chamber within the Great Pyramid at Giza. The north face is on the right side of the drawing.

The most perplexing features in the interior of the Great Pyramid are the "air shafts," small tubes on the north and south walls of the burial chamber of Pharaoh Khufu (the King's Chamber) and the Queen's Chamber [Fig. 2.36 (*Right*)]. Thought at one time to be ventilation shafts, that suggestion is highly unlikely given the length and narrow design of the shafts [typically about 30 cm (1 ft) on a side, although these dimensions do vary]. There are also multiple changes in direction near the chambers before the "air shafts" point upward and due north and south. In addition, the shafts from the Queen's Chamber never reach the surface of the pyramid and the ends of the King's Chamber were likely sealed after construction.

As with many other sites around the world, astronomical alignments have been proposed. In the King's Chamber it was proposed in the 1960s that the northern shaft was aligned with Thuban, the brightest star in the constellation of Draco. At the time of the construction of the pyramid, Thuban would have been closer to the north celestial pole than Polaris was. It was also suggested that the southern shaft was pointed toward Orion's belt when it crossed the meridian, Orion having been associated with Osiris, god of the dead (Fig. 2.37). The northern shaft of the Queen's Chamber would have pointed toward Mizar in Ursa Major when it crossed the meridian. Finally, the southern shaft of the Queen's Chamber was directed toward Sirius in Canis Major during its meridian crossing. Sirius was associated with the goddess Isis, who was considered to be the sister and wife of Osiris, and was goddess of health, marriage, and wisdom. Sirius was also the "Nile star," with the annual flooding of the Nile being due to the tears of Isis over the slaying of her husband. Interestingly, the mythology of Isis found its way into ancient Greek and Roman mythology as well.

Fig. 2.37 Osiris and winged Isis in the Isis Temple, Philae Island, Egypt. (I, Rémih, CC BY-SA 3.0)

Ancient Chinese Astronomy

The oldest known recorded astronomical observations were made in China. Inscribed bones and shells unearthed in a tomb dating back to about 4000 BCE indicate awareness of regions of the sky.

Ancient Chinese astronomers created multiple star catalogs and were successful in identifying the equinoxes and solstices. Careful record-keeping also indicates that astronomers recorded features on the surface of the Sun (sunspots and "eruptions" known as prominences). In addition, the Chinese made note of meteor showers, comets (perhaps even Comet Halley), and "new stars" that suddenly appeared in the sky, including the same supernova of 1054 CE that was recorded on Ancestral Puebloan rock art; recall Fig. 2.23.

The *Book of Documents*, also known as *Classic of History* or *Shangshu*, contains a record of the solar eclipse of October 22, 2137 BCE. For the hired astrologers of the kings and emperors, failure to predict eclipses could result in poor performance reviews. Legend has it that two drunken brothers, Hsi and Ho, who were both royal astrologers, unfortunately failed to predict the solar eclipse of 2137 BCE and were subsequently beheaded.

As was the case with Mayan astronomy, the long records of lunar and solar eclipses allowed the Chinese to identify the 5, 5 1/2, and 6-month recurring eclipse patterns, along with the 18-year saros cycle. By 8 BCE, Chinese astronomers had also identified the 135-month (11-year) tritos eclipse cycle that could be used in conjunction with the saros cycle to predict solar eclipses, a task that is much more difficult to do than predicting lunar eclipses. Then, by 206 CE, the Chinese astronomers could analyze the motion of the Moon itself, to assist them in making solar eclipse predictions.

The meticulous record-keeping of those Chinese astronomers has even proven to be useful to modern-day astronomical research when historical information has been needed. But what motivated such a dedication to observational astronomy? There were likely several reasons; the ability to predict dramatic events, thus validating the rule of various kings and emperors, the development of calendars to identify periodic events such as rituals and planting seasons, and simple human curiosity.

Not unlike the Mayan Haab' calendar (page 38), the Chinese developed a lunar–solar calendar that marked out 12 lunar months (phase cycles) of 354 days, plus an additional month every two to three years to bring the calendar back into alignment with the changing seasons of the solar year. However, when a new emperor came into power, one of the acts to solidify the emperor's rule was to wipe away the previous calendar and start anew. Although the nature of the calendar didn't necessarily change, the starting date of the calendar was reset. The calendar played a particularly important role in Chinese politics because it was believed that the heavens and Earth were interconnected, with events taking place in the sky reflecting events on Earth. The emperor, as the "Son of Heaven," was of course closely associated with that dynamic Heaven–Earth relationship.

It is worth noting that the cosmology of the ancient Chinese wasn't unchangeable; instead, Chinese astronomy was open to new ideas and models of the universe. Over time, at least three primary models were developed. The oldest dates back to the third century BCE. In this concept, the heavens were a hemispherical dome with Earth as an inverted, square-based bowl located within the dome. The space between the two bases was an ocean produced when the water from rains moved down the inverted Earth bowl. The hemispherical dome rotated, causing it to imitate the motions of the stars about the north celestial pole.

A second model came from the Hun Tiam school of thought, perhaps as early as the fourth century BCE, in which the heavens were a celestial sphere that encircled Earth. According to Zhang Heng (78–139 CE):

> The heavens are like a hen's egg and as round as a crossbow bullet; the Earth is like the yolk of the egg, and lies alone in the center. Heaven is large and Earth

small. Inside the lower part of the heavens there is water. The heavens are supported by vapor (*qi*), the Earth floats on the waters ... [The rotation of the heavens] goes on like that around the axle of a chariot.

The third model, from the Xuan Ye school, was remarkable in its conceptual thinking; the idea of an infinite universe. It has been attributed to Qi Meng, who lived sometime in the first two centuries CE. As described by Ge Hong (284–364) early in the fourth century CE:

The Sun, Moon and company of stars float freely in empty space, moving or standing still, and all of them are nothing but condensed vapor. The seven luminaries [the Sun, Moon, and the five naked-eye planets] sometimes appear and sometimes disappear, sometimes move forward and sometimes retrograde, seeming to follow each a different series of regularities. Their advances and recessions are not the same. It is because they are not rooted to any basis that their movements can vary so much; they are not in any way tied together. Among the heavenly bodies the Pole Star alone keeps its place, and the Great Bear never sinks below the horizon in the west as do the other stars. The speed of the luminaries depends on their individual natures, which shows they are not attached to anything, for if they were fastened to the body of heaven, this could not be so.

Although very progressive, the Xuan Ye school concept of the heavens and Earth did not seem to dominate Chinese astronomy. After 520 CE, it was the Hun Tiam school of thought that began to be considered as the correct picture of the universe.

It is unfortunate that even though there was some interaction between Europe and China prior to the mid-1300s CE, China's tendency toward isolationism began to prevail and interaction with the West ended for several centuries, halting any meaningful exchange of ideas during that time.

 ## Exercises

All of the exercises are designed to further develop *your understanding* of the material by thinking carefully and critically about what you have read in this chapter. Answers to selected exercises are available in the back of the book.

True/False

1. There are 95 official constellation as defined by the International Astronomical Union.
2. (*Answer*) The entire constellation of Ursa Major is circumpolar at a latitude of 45°.
3. All stars are visible from Earth's equator at some time during the course of one year.
4. (*Answer*) Orion is visible in the northern hemisphere at night in June.
5. Between the arctic circle and antarctic circle, the rising and setting positions of the Sun move along the horizon throughout the year, with the southernmost rising and setting positions occurring on about December 21.
6. (*Answer*) On the arctic circle, the Sun never sets for 24 hours at the summer solstice and on the antarctic circle it never sets for 24 hours at the winter solstice.
7. The Sun is at the zenith exactly one day each year forobservers between the tropic of Cancer and the tropic of Capricorn.

8. (*Answer*) The Moon is only visible at night.
9. The full moon can be seen rising at midnight.
10. (*Answer*) The waning crescent moon rises shortly before the Sun does.
11. A total solar eclipse can only occur when the Moon is new.
12. (*Answer*) A total lunar eclipse can occur during any lunar phase.
13. Predicting the appearance of comets in the sky was easily accomplished by all pre-modern societies.
14. (*Answer*) Venus can be seen at the observer's zenith soon after the Sun sets.
15. The earthen lodges of the Skidi Pawnee could be used to mark the seasons of the year.
16. (*Answer*) The star map of the Skidi Pawnee includes numerous identifiable features in the night sky.
17. Tracking the movements of Venus and Mars was of religious importance to the Skidi Pawnee.
18. (*Answer*) Chaco Canyon's Ancestral Puebloans were able to accurately mark the time of the summer solstice.

19. The Olmec developed a written base-20 number system.
20. (*Answer*) The Maya could predict lunar eclipses.
21. Venus was largely ignored in Mayan astronomy.
22. (*Answer*) The Mayan vague year (the Haab´) used a sophisticated process to adjust for the fact that Earth orbits the Sun in just over 365 1/4 days.
23. The Castillo in Chichén Itzá is a Mesoamerican pyramid that appears to have been of some astronomical significance.
24. (*Answer*) Stonehenge was used to predict eclipses.
25. The Great Pyramid at Giza has a base that is very nearly a perfect square, and therefore could not possibly have been built 4600 years ago.

Multiple Choice

26. (*Answer*) Polaris is
    (a) very close to the vernal equinox.
    (b) very close to the summer solstice.
    (c) in the constellation of Ursa Major.
    (d) about 1/2° from the north celestial pole.
27. As viewed from the northern hemisphere, the south celestial pole
    (a) is always visible.
    (b) rises in the east and sets in the west.
    (c) is located in the constellation of Virgo.
    (d) none of the above
28. (*Answer*) There is an equal amount of daylight and darkness when the Sun is at the
    (a) winter solstice   (b) vernal equinox
    (c) summer solstice   (d) autumnal equinox
    (e) both (a) and (c)   (f) both (b) and (d)
29. In the northern hemisphere, as an observer moves farther north, she would be able to note that
    (a) the amount of daylight progressively increases at the summer solstice and progressively decreases at the winter solstice.
    (b) the angle at which the Sun rises above the eastern horizon and sets into the western horizon becomes shallower.
    (c) both (a) and (b).
    (d) none of the above.
30. (*Answer*) The Sun will be directly overhead at least once each year
    (a) at the winter solstice.
    (b) between the tropic of Cancer and the tropic of Capricorn.
    (c) north of the arctic circle.
    (d) south of the antarctic circle.
31. The new moon rises at approximately
    (a) midnight   (b) 6:00 a.m.   (c) noon   (d) 6:00 p.m.
32. (*Answer*) The third quarter moon rises at approximately
    (a) midnight   (b) 6:00 a.m.   (c) noon   (d) 6:00 p.m.
33. Wednesday is named for the planet
    (a) Vulcan   (b) Mercury   (c) Venus
    (d) Mars   (e) Jupiter   (f) Saturn
34. (*Answer*) Which of the following statements is *not true*: The naked-eye superior planets
    (a) are Mars, Jupiter, and Saturn.
    (b) may be seen near the observer's zenith from anywhere on Earth.
    (c) can cross the observer's meridian at midnight.
    (d) can undergo retrograde motion.

35. The motions of the planets across the sky remain close to the
    (a) zenith   (b) south celestial pole
    (c) meridian   (d) ecliptic
36. (*Answer*) Mercury and Venus are always
    (a) near the observer's zenith.
    (b) circumpolar.
    (c) within approximately 28° and 47° of the Sun, respectively.
    (d) near the north celestial pole.
37. Prior to 1800 CE, the Skidi Pawnee
    (a) numbered between perhaps 8000 and 12,000 persons.
    (b) lived in what is today east-central Nebraska and the central border region of Nebraska and Kansas.
    (c) were purely an agrarian society.
    (d) all of the above.
    (e) (a) and (b) only.
38. (*Answer*) The Skidi Pawnee believed that
    (a) Venus was Morning Star and Mars was Evening Star.
    (b) Morning Star and Evening Star were the parents of the first girl.
    (c) The Sun and the Moon were the parents of the first boy.
    (d) Morning Star was the great warrior god.
    (e) all of the above.
39. Earthen lodges of the Skidi Pawnee could be used to
    (a) locate the position of the setting Sun along the horizon.
    (b) observe the Sun at the observer's zenith on the summer solstice.
    (c) see the Star-That-Does-Not-Move on any night of the year (assuming clear skies).
    (d) all of the above.
40. (*Answer*) The Ancestral Puebloan Peoples appear to have
    (a) observed a supernova in the eleventh century.
    (b) recorded rising and setting positions of Mars over a period of centuries.
    (c) been able to determine when the equinoxes and solstices would occur.
    (d) all of the above.
    (e) (a) and (c) only.
    (f) (b) and (c) only.
41. The Olmec were likely the Mesoamerican culture that developed the positional number system that used
    (a) base-2   (b) base-10   (c) base-16
    (d) base-20   (e) base-60
42. (*Answer*) An important development in the Mesoamerican number system was
    (a) the creation of a symbol for zero.
    (b) the ability to multiply large numbers.
    (c) understanding how exponents work.
    (d) negative numbers.
43. Our modern Arabic representation of the number, 4412, is represented in the Mesoamerican merchant number system as
    (a) ⋮⋮   (b) ▔▔   (c) ⋮   (d) ═
44. (*Answer*) Express this Mesoamerican merchant number in modern Arabic base-10 notation:

    (a) 40   (b) 6139   (c) 156,150   (d) 154,700

45. Calculate the Mesoamerican merchant sum, (3.18.4.7) + (2.15.13). Express your answer in condensed form.
    (a) (3.20.19.20)   (b) (5.33.17.7)   (c) (6.13.17.7)
    (d) (4.1.0.0)

46. (*Answer*) In the Tzolk´in calendar, starting with 1 Imix´, what day would it be 32 days later?
    (a) 32 Eb   (b) 6 Eb   (c) 11 Ik   (d) 12 Ik

47. In the text the point is made that 18,980 is the least common multiple (LCM) of 260 and 365. The LCM is defined as the smallest number such that each number in a set of numbers can be divided into the LCM with the results always being an integer number (no remaining fractions). Quick calculations will show that $18,980/260 = 73$ and $18,980/365 = 52$. Find the LCM of 20 and 70.
    (a) 1400   (b) 700   (c) 140   (d) 70   (e) 20

48. (*Answer*) Along with observing the horizon positions of the Sun during important astronomical days of the year, the Caracol tower in Chichén Itzá was also used to observe
    (a) Mercury   (b) Venus   (c) Mars   (d) Jupiter
    (e) Saturn

49. The saros cycle is
    (a) a two-wheeled transportation device used by the ancient Maya.
    (b) the length of time between successive heliacal risings of Venus.
    (c) an approximately 18-year eclipse cycle
    (d) the period of time between appearances of Sauron.

50. (*Answer*) Stonehenge was apparently used for millennia to note the first day of
    (a) summer   (b) spring   (c) autumn
    (d) winter   (e) more than one of the above

51. The Great Pyramid at Giza is remarkable in many ways, including
    (a) being the tallest man-made structure in the world for almost 4 millennia.
    (b) its very precise alignments with the cardinal directions, north, east, south, and west.
    (c) the use of nearly 2.3 million blocks in its construction.
    (d) the apparent alignment of so-called "air shafts" that pointed toward Thuban and Orion's belt from the King's Chamber and toward Sirius from the Queen's Chamber.
    (e) all of the above.

52. (*Answer*) As was true of ancient astronomers in the Americas, ancient Chinese astronomers
    (a) were able to predict eclipses.
    (b) recorded the supernova of 1054 CE.
    (c) constructed a calendar to note important events, both terrestrial and astronomical.
    (d) all of the above

Short Answer

53. Show that if stars rise nearly 4 minutes earlier on each successive day relative to the Sun, those same stars will rise about 24 hours ahead of the Sun after one year. To make the calculation easier, assume that the stars rise *exactly* 4 minutes earlier each day and that the year is only 360 days long. This means that the same stars will be in the same positions in the sky on the same day from one year to the next. For example,

Orion will be high in the southern sky at 10:00 p.m. in the northern hemisphere every year in mid January.

54. Referring to Fig. 2.7, draw a sketch similar to the figures that shows the paths of the Sun across the sky on June 21 and December 21 for an observer at the equator.

55. Referring to Fig. 2.7, draw a sketch similar to the figures that shows the paths of the Sun across the sky on June 21 and December 21 for an observer at 60° N.

56. Explain how the large spiral on Fajada Butte could be associated with the number 18.6, the length in years of an eclipse cycle.

57. (*Answer*) Referring to Table 2.2 calculate the number of days and years in 13 b´ak´tuns. Use the fact that there are approximately 365 1/4 = 365.25 days in one year.

58. Identify at least one city, nation, or prominent feature that lies within 300 km of the tropic of Cancer, the equator, and the tropic of Capricorn that are not mentioned in the text. You may find LatLong.net or Google Earth™ helpful.

59. (*Answer*) What does this merchant number correspond to in our Arabic base-10 notation:

60. Expand the concise form of the Mesoamerican merchant number (3.12.0) in glyph notation and convert the number to our base-10 system.

61. Write our base-10 number 18,352 using the Mesoamerican merchant base-20 glyph notation.

62. Using the Mesoamerican merchant base-20 number system, what is (1.12.19.5) + (13.5.15)? Express your answer in
    (a) the Mesoamerican concise notation,
    (b) glyph notation, and
    (c) in our base-10 system.

63. (a) What day is represented in Fig. 2.29? Include both the Tzolk´in and Haab´ calendars.
    (b) What day will it be 28 days later?

64. Two mentions were given in the text regarding predictions, just since 2012, that the world was about to come to an end; the end of a cycle of 13 b´ak´tuns in 2012 (page 37) and the blood supermoon in 2015 (page 22). Conduct an Internet search to identify at least one other end-of-the-world prophesy based on an anticipated astronomical event that was made since 1970 and that (obviously) didn't occur. Describe the nature of the prophesy and its societal impact for at least some groups of people.

Activities

65. Figure 2.18 depicts objects in the sky that would have been important to the Skidi Pawnee. The map is not an accurate representation of the relative locations of objects, but instead was likely used for storytelling, teaching, and recording important aspects of Skidi mythology. Use the free, open-source planetarium software package *Stellarium*, or other star maps or software tools, to find the following objects in the night sky, and also identify them on the Skidi star map. You should draw an elliptical outline of the map and then place and label the appropriate objects on the drawing.
    (a) the Milky Way (the Pawnee "spirit path")
    (b) The Hyades star cluster

Exercise 65 continues on the next page

65. *Continued*
   (c) The Pleiades star cluster
   (d) Corona Borealis (thought of as a semicircle of chiefs in council; more naked-eye stars are shown on the map than appear in the actual sky)
   (e) Ursa Minor (known as the "little stretcher" to the Pawnee, meant to carry the sick or dead; the stretcher is accompanied by a medicine man, a woman, and a number of chiefs)
   (f) Polaris ("Star-That-Does-Not-Move")
   (g) Ursa Major (the "big stretcher," also meant to carry the sick or dead, accompanied by a medicine man, a woman, and chiefs)
   (h) the crescent moon

On the Path toward Modern Science

3

All men by nature desire knowledge.

Aristotle (384–322 BCE)

3.1	Sumerian and Babylonian Astronomy	56
3.2	The Ancient Greeks and the Dawn of Science	59
3.3	An Opposing View: The Greek Heliocentric Model	70
3.4	Medieval Astronomy in India and the Middle East	72
3.5	The Celestial Sphere	74
	Exercises	86

Fig. 3.1 Star trails around the north celestial pole; Fayyoum, Egypt. (AhmedMosaad, CC BY-SA 4.0)

Introduction

In Section 2.2 you learned about numerous cultures and sites where astronomical knowledge was being gathered. In most cases, some of the information was useful for anticipating the changing of the seasons or establishing the times of important rituals. Often the data were also gathered in order to follow the activities of, and hopefully appease, the celestial deities that were a part of the culture's pantheon of supernatural gods. Thanks to the development of writing and basic mathematics in Mesoamerica and China, long-term patterns were also identified, leading to the ability to predict the appearance of specific planets or the occurrence of lunar and solar eclipses. These predictions also served the purpose of forecasting events that may befall leaders.

People in each of the cultures you studied were satisfied with simply making observations of what was taking place, and perhaps predicting when and where similar events might occur again. Other than invoking mystical deities, those cultures never asked the more fundamental questions: Are there underlying processes that drive the phenomena they were observing? Is it possible to go beyond predictions based on patterns in order to understand the source(s) of the patterns themselves?

3.1   Sumerian and Babylonian Astronomy

Modern science traces its roots back to Mesopotamia between the Tigris and Euphrates rivers in what is present-day Iraq (see Figs. 3.2 and 3.3). Mesopotamia is a portion of the Fertile Crescent, a relatively moist and fertile region in the otherwise arid and semi-arid regions of western Asia and northeastern Africa. The earliest cultures in the region trace back to around 6000 BCE (the beginning of the sixth millennium BCE) with evidence of houses and pottery. The Sumer civilization (the Sumerians) likely developed permanent settlements along the Euphrates sometime between 5500 BCE and 4000 BCE, and the oldest known form of writing, Sumerian cuneiform script, began to develop from pictographs sometime in the second half of the fourth millennium BCE (e.g., Fig. 3.4). It is estimated that modern excavations of the region have produced between one-half million to two million clay tablets containing cuneiform, of which perhaps 30,000 to 100,000 have actually been studied so far. The Sumerians also developed a sexagesimal number system (base 60), a portion of which is still in common use today in the form of 60 minutes in one hour, 60 seconds in one minute, and 60 arcminutes (60′) in one degree (1°), and 60 arcseconds (60″) in one arcminute.

Early in the history of Mesopotamia, separate city-states existed that interacted with one another through the sharing of culture and trade, with the dominant language being Sumerian. When Sargon of Akkad staged a palace coup in the late twenty-fourth century BCE, the city-states were absorbed into a unified Mesopotamia. At that point the Semitic language of Akkadian, which is related to Hebrew and Aramaic, became the principle language of the region.

As was the case in China, the Mesopotamians created a 354-day calendar based on 12 months, each month equaling one lunar phase cycle. The slippage of the calendar by 11 days each solar year was resolved, just as the Chinese did, by inserting an extra month in the calendar every few years. The problem of when to insert these

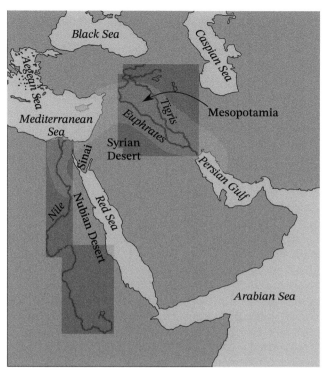

Fig. 3.2 Mesopotamia and surrounding regions. Mesopotamia was a part of the Fertile Crescent, shown in green. (Adapted from Nafsadh, CC BY-SA 4.0)

Fig. 3.3 A 2013 CE map of the region in and around the Fertile Crescent. You should compare this map with Fig. 3.2. (Central Intelligence Agency, 2014[1])

"intercalary" months was eventually systematized by establishing seven years out of every nineteen years as intercalary years. The start of each month was defined to be when the thin crescent moon first appeared after a new moon. Starting in

[1]Central Intelligence Agency (2014). *The CIA World Factbook 2015* (Skyhorse Publishing).

Fig. 3.4 An example of a Sumerian cuneiform tablet dating from the end of the third millennium BCE. The tablet contains the Sumerian creation myth and is currently housed in the Louvre Museum in Paris, France. (Louvre Museum, CC BY-SA 3.0)

the third millennium BCE, a second, 360-day "ideal" calendar was used for administrative and astronomical purposes, defined simply as 12 months of 30 days each, meaning that it slipped by 5 1/4 days for each solar year of 365 1/4 days.

In addition to developing calendars, astronomical observations in early Mesopotamian history were carried out for astrological purposes, much like they were in other parts of the world. The gathering and recording of astronomical data allowed astrologers to interpret omens and possibly predict future astronomical events. However, some advances did occur; for example, it was understood early in the third millennium BCE that the planet Venus, which was associated with the evening and morning god, Inanna, was in fact the same object. That reality wasn't realized in Europe for another 2000 years.

A two-tablet collection of data, the MUL.APIN, was written in 686 BCE, which is a copy of a much older document that was probably written about 1000 BCE. The MUL.APIN contains tables of stars and constellations, including heliacal rising dates of 34 stars and constellations using the "ideal" calendar. The tablets also include a mathematical method of determining the rising and setting times of the Moon each month, mathematical procedures for calculating the lengths of days and nights, and information about astrological omens.

As we have seen multiple times with other ancient, superstitious cultures, the Mesopotamians were particularly concerned with eclipses that might foretell some upcoming tumultuous event, such as the start of a war or the death of a king. In the seventh century BCE, when a lunar eclipse occurred and an astrologer forecast the king's death, it was not uncommon for a substitute king to be assigned to the throne, usually a criminal or a prisoner of war. During the period of potential danger, the substitute king would enjoy virtually all of the privileges of the king, including the run of the palace. In the meantime the actual king would still remain in the palace without its perks, to rule in anonymity. After it was determined that any threat to the real king had passed, the substitute king would be removed from the throne and executed, and the real king would return to visible rule.

As the story of the paranoid kings of the seventh century BCE illustrates, the desire to accurately predict eclipses (both lunar and solar) was of great importance. As with the Maya and the Chinese, astronomers were able to use their meticulous recordings of past eclipses to identify the necessary eclipse patterns and make eclipse predictions. According to cuneiform tablets, it seems that so-called "astronomical diaries" were produced annually from at least 652 BCE (and perhaps as much as one century earlier) until the first century CE, giving roughly 800 years of valuable observational data.

In 612 BCE, after the fall of Assyria as the dominant political force in Mesopotamia, the city of Babylon became the center of power in the region. Beginning sometime in the fifth century BCE, Babylonian astronomers developed a more accurate method of specifying the locations of stars, planets, the Sun, the Moon, and other celestial phenomena. By realizing that the Sun, the Moon, and the planets follow a narrow band through the stars, and taking inspiration from their "ideal" 360-day calendar, the Babylonians divided the path up into twelve segments of thirty angular intervals each, giving what are today the twelve signs of the zodiac and 360° in a circle.

Unification by Alexander the Great

It was in 331 BCE when Alexander the Great (Fig. 3.5), king of the ancient Greek kingdom of Macedonia, conquered Babylon. Before his death, Alexander, having never been defeated in battle, had succeeded in creating an empire that stretched from Greece around the eastern Mediterranean to Egypt, across much of what is today known as the Middle East, and as far east as northwest India, bounded on the north by the Himalayan mountains. During his reign he created 20 cities, all of them called Alexandria, the most famous of which is Alexandria, Egypt. The city of Kandahar, in Afghanistan, was also originally named Alexandria in honor of the great king. Knowing of his military successes, many of his potential targets simply surrendered to him, including the city of Jerusalem.

Despite his many conquests, Alexander was only 32 years of age when he died in the palace of Nebuchadnezzar II in Babylon, following a two-week illness. It seems that the illness was related to drinking a large quantity of wine. One of the Babylonian astronomical diaries that was based on the "ideal" calendar noted his death rather succinctly along with the weather for the day: "the 29th, the king died, clouds."

The legacy of Alexander is vast, making him one of the most influential people in world history. Of importance for this text was the spread of the Greek culture in the East which increased the communication of astronomical information between the ancient Greek philosophers and the Babylonian, Egyptian, and Indian astronomers.

3.2   The Ancient Greeks and the Dawn of Science

Thales of Miletus (c. 624–546 BCE)

Fig. 3.6 Thales of Miletus (c. 624–546 BCE). Illustration from Wallis (1875).[2] (Public domain)

Fig. 3.5 Statue of Alexander the Great (356–323 BCE) in the Istanbul Archaeological Museum. (Giovanni Dall'Orto; attribution, via Wikimedia Commons)

Several centuries before the conquests of Alexander the Great, a new way of thinking was starting to develop in ancient Greece, which included the modern country of Greece and regions around the Aegean Sea (you may want to again refer to the maps in Figs. 3.2 and 3.3). Thales of Miletus (Fig. 3.6), whom Aristotle (384–322 BCE) more than two centuries later would call the first philosopher in the Greek tradition, was born on the western coast of Anatolia near the present-day village of Balat, Turkey. Thales rejected mythological (supernatural) explanations for natural events, including those observed in the heavens. For this reason he is sometimes considered to be the "father of science."

Knowing principles of geometry, Thales was able to determine the distance of ships from shore. He also used geometry to calculate the heights of the Egyptian pyramids. This was accomplished by measuring the length of a pyramid's shadow at the same moment that

[2]Wallis, Ernst (1875). *Illustrerad verlds historia: Egyptens, Vestra Asiens Och Greeklands Historia Till Romerska Herraväldets Tid*, volume 1 (Central Printing Company's Publishers: Stockholm).

his own shadow equaled his height. By understanding the nature of similar triangles (triangles that have the same three internal angles but proportionately longer or shorter sides) he realized that when the length of his shadow equaled his height, then the length of the pyramid's shadow must equal its height.

Pythagoras of Samos (c. 570–495 BCE)

Known today for the mathematical theorem regarding right triangles that bears his name, Pythagoras of Samos (Fig. 3.7) was well traveled. He went to Egypt to study geometry, he studied arithmetic with the Phoenicians, and astronomy with the Chaldeans of Mesopotamia. We don't know a great deal about Pythagoras except through the writings of others, but we do know that he had a great love of numbers. We also know that Pythagoras believed that numbers were the representation of nature. The small integer ratios of frequencies in musical intervals, such as 2:1 for an octave or 3:2 for a perfect fifth, had a profound impact on Pythagoras and the religious Pythagorean cult that he started. Pythagoras also believed strongly in the importance of abstract thinking when considering the natural world, meaning that he thought that one should seek to uncover natural influences beyond what is immediately observable.

Fig. 3.7 Bust of Pythagoras of Samos (c. 570–495 BCE). (Public domain)

Plato (c. 428–348 BCE)

In ancient Greece, astronomy was considered to be a branch of mathematics, and the philosophy of the Pythagoreans had a significant impact on the natural philosophy of Plato (Fig. 3.8), who was a student of Socrates (c. 470–399 BCE). In order to provide an environment for intellectual challenge, Plato established a school in Athens that some consider to be the world's first university. Plato wrote prolifically on many subjects, but of importance for our story is a question he posed for the Greek mathematicians. According to an account written by Simplicius of Cilicia (c. 490–560 CE) in the sixth century, nearly one thousand years after the fact, Plato is said to have asked "what are the hypotheses that, by smooth, circular, and ordered motions, can save the planetary phenomena?"[3]

Fig. 3.8 Plato (c. 428–348 BCE). Marble copy of a portrait made by Silanion c. 370 BCE. (Public domain)

Within Plato's question rests a set of assumptions that would guide the thinking of natural philosophers, astronomers, and society collectively for nearly two thousand years! It seemed apparent to Plato and others that the cosmos is centered on Earth; the universe is geocentric. We watch the Sun, the Moon, the planets, and the stars move about their heavenly paths as though they are going around Earth. In Plato's time there

[3] In recent years the claim, as stated by Simplicius, that it was Plato who required "smooth, circular, and ordered motions" for the heavens has been called into question. Perhaps the initial concept rests with Socrates, Aristotle, or Eudoxus. However, we won't concern ourselves with that debate here; what is important is that these requirements were placed on viable models in ancient Greece.

was no evidence that Earth itself is traveling through space, but Plato was willing to consider that Earth could be spinning on an axis passing through the North and South Poles. Is it not natural that the pageantry of the heavens should unfold around us?

Plato's question took the assumptions about planetary motions another step. He was stating, as though without debate, that planetary motions are "smooth, circular, and ordered" (commonly referred to as uniform and circular motion). In other words, any models for planetary motions must only rely on perfect circles (or spheres) and constant speeds. However, recalling Fig. 2.12 (page 23) and the discussion of retrograde motion, the motions of the planets certainly do not exhibit the "smooth, circular, and ordered motions" that Plato apparently assumed to be true. Plato's question created a difficult challenge meant to "save the planetary phenomena."

Plato was deeply committed to the development of the intellect and abstract thinking. In particular, Plato's philosophy was built on the belief that only ideas are true; observations on the other hand are transient and can deceive. The observations of planetary retrograde motions are simply the consequences of more fundamental, permanent, ideas of a geocentric universe founded on perfect circles and constant speeds.

Eudoxus of Cnidus (408–355 BCE)

Like Pythagoras before him, Eudoxus traveled extensively in order to gain knowledge. Along with studying under Plato in Athens, Eudoxus also went to Italy to learn mathematics and medicine, and to Egypt to study astronomy and additional mathematics. He also developed his own school in Cyzicus (near Erdek, in modern-day Turkey). For a time he returned to Athens and took over from Plato while Plato traveled to Syracuse.

Eudoxus is considered by many to be among the greatest mathematicians of ancient Greece and perhaps second only to Archimedes of Syracuse (c. 287–212 BCE) in all of antiquity. He had begun to think of numbers as a continuum rather than as simple integers or integer ratios. He also began to think of nature in a geometric way instead of just as numbers and arithmetic.

Taking up Plato's challenge, Eudoxus developed a series of nested geocentric crystalline spheres around a fixed, non-rotating Earth to explain motions in the heavens. As depicted in Fig. 3.9, using a sphere to explain the paths that the stars follow in the sky is relatively straightforward; simply place the stars in their relative positions on a geocentric sphere (the celestial sphere). The celestial sphere then rotates on an axis that connects the north celestial pole (NCP) and the south celestial pole (SCP) with a period of about 23^h56^m (one sidereal day; recall the brief discussion on page 17). This simple model nicely produces motions of the stars across the sky that are seen at various latitudes, as shown for example back in Fig. 2.4. (We won't go into the details of the celestial sphere now, but this important and still useful, although incorrect, model is the subject of Section 3.5.)

Modeling the motion of the Sun using geocentric spheres was slightly more complicated; Eudoxus placed another sphere, called the ecliptic sphere, inside the celestial sphere, but tilted the new sphere by 23 1/2° (you should recall that the latitudes of the tropic of Cancer, the tropic of Capricorn, the arctic circle, and the

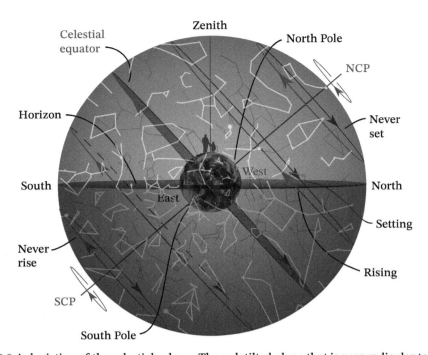

Fig. 3.9 A depiction of the celestial sphere. The red, tilted plane that is perpendicular to the line connecting the north celestial pole (NCP) and the south celestial pole (SCP) represents the extended plane of Earth's equator. The intersection of the plane with the celestial sphere is a great circle known as the celestial equator. The dark blue plane represents the local horizon of the observers with their zenith directly above them. Note that the observers are unable to see what is below the horizon because Earth blocks their view. You should compare the star paths with the seasonal paths of the Sun in Fig. 2.7.

antarctic circle are all related to this important 23 1/2° angle). The axis of the ecliptic sphere was attached to the celestial sphere so that its poles would be carried around by the celestial sphere with the celestial sphere's rotation period. The ecliptic sphere would then rotate slowly about its own axis in the opposite direction with a period of one year. In this way the Sun would appear to *slip backward* by just under 1° per day (exactly 360° per year) relative to the stars, giving a day of 24 hours. The time required for the celestial sphere to rotate that 1° is close to four minutes; this is the additional four minutes needed for the celestial sphere's $23^h 56^m$ period plus the motion of the ecliptic sphere to give the Sun's 24-hour daily motion. The Sun would also return to the same place among the stars after one year. It is this shifting of the Sun on the ecliptic sphere relative to the celestial sphere that also explains why different constellations are seen at different times of the year. The combined motion of the celestial sphere and the ecliptic sphere, together with the ecliptic sphere's 23 1/2° tilt, is also able to explain the shifting positions of the Sun along the horizon at sunrise and sunset as well as its height at midday during its daily crossing of the meridian throughout the year. The "equator" of the ecliptic sphere is what produces the ecliptic path of the Sun among the stars. Eudoxus also used a third sphere to fine-tune the model.

Eudoxus's modeling of the Moon was accomplished in a similar way, but with its three required spheres nested inside the required spheres for the Sun, with each one attached to the sphere immediately outside of it. The outermost of the Moon's spheres had its axis tilted about 5° relative to the axis of the ecliptic sphere in order to produce the behavior that the Moon is sometimes above and sometimes below the ecliptic by as much as 5°. The combination of spheres also had to rotate in such a way as to mimic the 29 1/2 day period of the Moon's phases.

Modeling the motions of the naked-eye planets (the "wandering stars") was even more complex, with four nested spheres being required for each of them. Remember that the complicated retrograde motion of the planets exhibited in Fig. 2.12 had to be replicated, along with their motions close to the ecliptic.

By the time Eudoxus had fine-tuned his geocentric, geometric model of the motions in the heavens he required a total of 27 nested spheres, successively attached to one another; one for the celestial sphere, three for the Sun, three for the Moon, and four each for the five planets. In each case the orientation of their axes, their rotation periods, and their rotation directions had to be carefully adjusted. All of this complexity was necessary in order meet Plato's call to "save the planetary phenomena."

Despite all of his effort, Eudoxus's model of the heavens using geocentric spheres was fundamentally flawed in numerous major ways:

- Recalling the discussion on page 24, the superior planets are brightest in the middle of their retrograde motions.
- Careful observations show that the superior planets also appear larger at mid retrograde.
- The Moon appears larger during some full moons than it does during other full moons (giving rise to the term "supermoon"; page 22).
- The speed of the Moon against the background stars is not constant.
- The lengths of each of the seasons are not exactly equal; approximately 89 days of winter, 93 days of spring, 94 days of summer, and 90 days of fall.
- The Sun doesn't travel along the ecliptic at a constant speed throughout the year.
- Finally, the Sun appears very slightly larger at certain times of the year than it does at other times.

Aristotle (384–322 BCE)

Certainly the greatest of Plato's and Eudoxus's students was Aristotle (Fig. 3.10), whose writings had tremendous impact on western science for two millennia. Aristotle's intellectual interests weren't just limited to science; he also wrote extensively about logic, ethics, poetry, music, politics, and government. After Plato's death, Aristotle was invited by King Philip II of Macedonia to serve as the tutor of his son, Alexander (soon to be the Great), during which time Aristotle encouraged Alexander toward eastern conquest. Apparently Aristotle, who was unquestionably ethnocentric, was not much of a fan of the Persians.

After eight years as the tutor of Alexander and two other future kings, Ptolemy and Cassander, Aristotle returned to Athens. It was after his return that Aristotle did much of his greatest work. Along with his many other contributions, Aristotle essentially defined the ground rules for the development of science through the sixteenth century.

Fig. 3.10 Bust of Aristotle (384–322 BCE). Roman copy in marble after a Greek bronze original by Lysippos (c. 390–c. 300 BCE). (Public domain)

Although he was Plato's student, Aristotle had a very different view of nature. Rather than believing that ideas are the only true form of reality, Aristotle believed that reality must be observable and empirical.

In what is known today as Aristotelian physics, Aristotle believed all objects on Earth involved four elements, earth, water, air, and fire, and that the motions of objects will ultimately cease when each element's natural location is reached. He also believed that the elements could transform from one to another. However, Aristotle also believed that a fifth, perfect and unalterable element known as the ether occupies the heavens, beginning with the sphere of the Moon. He proposed that the ether formed all of the heavenly bodies. As with those who came before him, Aristotle believed that the heavens must be perfect because the pantheon of the gods resided in the heavenly realm. Therefore a unique, perfect, element must compose the heavenly bodies and all of the space between them. Aristotle also emphatically dismissed the idea of voids, empty regions where nothing at all existed. He is famously quoted as saying that "Nature abhors a vacuum." (As we will see, variations on the concept of the ether persisted into the early part of the twentieth century.)

Aristotle did not believe that the universe could be infinite in extent. Arguing in terms of the geocentric spheres of Eudoxus, Aristotle pointed out that if the universe was infinite then, while the celestial sphere rotated, the stars would need to move an infinite distance in a finite amount of time, something which is clearly impossible.

How did Aristotle explain the motions in the heavens if there was only one element that could not seek another, different location? In an odd sort of theology, he proposed the concept of "an immortal, unchanging being, ultimately responsible for all wholeness and orderliness in the sensible world." This immortal being was known as the prime mover, and is the ultimate cause of all motion in the heavens, starting with the rotation of the celestial sphere. Aristotle even extended the concept to include unmoved movers, invisible deities for each geocentric sphere of Eudoxus. He also included counter-rotating spheres to correct for undesired interactions between planets. In the end, depending on the particular revision, Aristotle incorporated either 47 or 55 spheres in his extension of Eudoxus's model in order to "save the planetary phenomena."

Aristotle's view of the celestial realm clearly differed significantly from the view of Thales, who rejected supernatural explanations for physical phenomena. However, Aristotle's explanation of terrestrial motions did maintain Thales's view. [An interesting aside: The concept of the unmoved mover was discussed and elaborated upon by the Catholic theologian, Saint Thomas Aquinas (1226–1274).]

There is some evidence to suggest that toward the end of his life Aristotle began to doubt Eudoxus's model of employing geocentric spheres. Aristotle was concerned that the model simply didn't explain all of the phenomena adequately; a possible foreshadowing of the modern scientific requirement that a theory must explain all known observations and experimental results.

Eratosthenes of Cyrene (c. 276–195 BCE)

If you had learned in prior schooling that western civilizations in the Middle Ages and earlier believed that Earth is flat, you were misinformed. Scholars since at least the time of the ancient Greek philosophers have known that Earth is not flat, but essentially spherical in shape. This is easily seen during a lunar eclipse, such as the one shown in Fig. 3.11. As the Moon enters or leaves Earth's shadow, a curved, darkened region moves across the surface of the Moon. This is simply a consequence of a round Earth casting the shadow. As a second example, as a ship sails away from a harbor and over the horizon, the ship doesn't just drop off the end of Earth; instead it is able to turn around and return to the safety of the harbor. With careful observations it is also possible to notice that the bottom of the ship begins to disappear before the mast does. No, the ship is not sinking (hopefully!), it is just that the bottom of the ship passes below the horizon before the top of the mast does.

Fig. 3.11 A partial lunar eclipse; December 10, 2011. (Tom Ruen, public domain)

  Eratosthenes estimated the circumference of Earth by using shadows cast on the ground by the Sun (Fig. 3.12). He knew that in Egypt's Nile-River city of Syene (known today as Aswan), it was possible for sunlight to reach the bottom of a well when the Sun was almost at the zenith. In other words, the Sun must be nearly directly over Syene when the bottom of the well is illuminated. It is also the case that the latitude of Syene (24°05′) is situated very near the tropic of Cancer. Recall (page 19) that on the tropic of Cancer the Sun is directly overhead for only one day each year, the first day of summer (the summer solstice). By placing a vertical stick in the ground in Alexandria, which is north of Syene, Eratosthenes was able to measure the shadow cast there on the summer solstice and determine that the Sun was not directly overhead, but was instead at an angle of about 1/50 of a complete circle south of the zenith. Assuming that Alexandria is straight north of Syene (which it

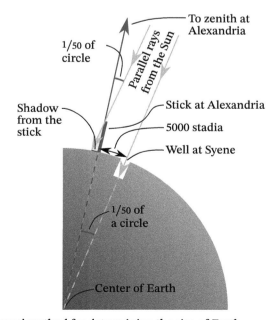

Fig. 3.12 Eratosthenes' method for determining the size of Earth.

A units conversion tutorial is available at
`Math_Review/Unit_`
`Conversions`

Fig. 3.13 An armillary sphere. Städtisches Museum Göttingen. (Public domain)

is not) then Alexandria must be 1/50 of the circumference of Earth north of Syene. Finally, according to geographers of the pharaohs, Alexandria is 5000 stadia from Syene (one stade being the length of a particular stadium). This implies that the circumference of Earth must be about 5000 stadia × 50 = 250,000 stadia. There is some debate about exactly what one stade equals in today's length measurements, but the most commonly accepted value is that 1 stade = 185 m. If that is the case then Eratosthenes would have arrived at a circumference of

$$250,000 \; \text{stadia} \times 185 \text{ m} / \text{stade} = 4.6 \times 10^7 \text{ m} = 46{,}000 \text{ km} \quad (28{,}900 \text{ mi}).$$

This value is too large by about 15%, but not bad given the assumptions and crude measurements involved.[4]

Besides obtaining a reasonable estimate for the size of Earth, Eratosthenes' accomplishments included measuring the tilt of Earth's axis, creating the first map of the world that included meridians and lines parallel to the equator, attempting to determine the distances to the Moon and the Sun, and inventing the armillary sphere (Fig. 3.13), an astronomical device composed of rings that represent celestial latitude and longitude, the celestial sphere, and the ecliptic. When he was 30 years of age, Eratosthenes was recruited by pharaoh Ptolemy III Euergetes to be a librarian at the legendary Library of Alexandria. He became head librarian in five years.

Apollonius of Perga (c. 262–190 BCE)

Fig. 3.14 A bust of Apollonius of Perga (c. 262–190 BCE), the "Great Geometer." (Public domain)

As we have already discussed on page 63, using geocentric spheres to model the motions of the Sun, the Moon, and the planets was fundamentally flawed in a number ways, perhaps the most obvious being that the superior planets become brighter and larger during retrograde, peaking in the middle of the retrograde path. It is also readily apparent that the Moon varies significantly in size over time. The Sun appears larger at various times as well, but this is a more subtle observation. Each of these effects strongly suggests that the planets, the Moon, and the Sun are closer to Earth at some times than they are at other times. However, modeling these celestial objects on crystalline spheres centered on Earth prohibits changes in distance.

Apollonius of Perga (Fig. 3.14), envisioned a model of the universe based on circles as depicted in Fig. 3.16. His model featured a small circle, called an epicycle, on which a planet is placed. The epicycle was centered on a larger circle (a deferent), with the deferent centered on Earth. The deferent was also oriented so that it passed through the zodiac. The epicycle would rotate at a constant speed while the center of the epicycle moved at a constant speed around the deferent. As the epicycle carried the planet inside the deferent, retrograde motion occurred (again refer back to Fig. 2.12). The speed that the planet took along its path (the green line) is illustrated by the spacing of the dots along the path; when the dots are spread out the planet would be moving along the path more rapidly than when the dots are close together. Note how slowly the planet moves in the middle of the retrograde loop relative to excursions outside of the deferent.

Apollonius's model was able to maintain the spirit of Plato's requirement regarding uniform and circular motion for a geocentric model if one ignores the issues that, according to the model, the planet is not exactly going around Earth at a

---

[4]The term percent (symbolized by %) is actually a concatenated word meaning "per cent," or per one hundred; for example, there are 100 cents in one dollar, and 100 years in one century. When you see the symbol % following a number, the meaning is that you multiply the number by 1/100 = 0.01.

constant distance from it, nor is the speed of the planet a constant as observed from Earth. In fact, the epicycle was designed explicitly to adjust those behaviors.

In addition to his models of planetary motions, Apollonius also made valuable contributions to the mathematics of geometry; to his contemporaries Apollonius was known as the "Great Geometer."

Hipparchus of Nicaea (c. 190–120 BCE)

Hipparchus of Nicaea (Fig. 3.15) added another level of sophistication several decades after Apollonius developed his epicycle–deferent model. As shown in Fig. 3.16, instead of keeping the deferent centered on Earth, Hipparchus shifted Earth away from the center of the deferent circle, making it eccentric. As the epicycle moves around the deferent at constant speed, the off-center Earth meant that the planet would be even closer to Earth during retrograde, causing the planet to look that much bigger and brighter.

The motions of the Sun and Moon were explained in a similar way, although an epicycle wasn't required for the Sun. What was required was shifting the center of the Sun's deferent slightly away from Earth. The orientation of the Sun's deferent aligned perfectly with the ecliptic, while the orientations of the other deferents were tilted slightly relative to the Sun's deferent. In addition, further fine-tuning of positions could be accomplished by tilting the epicycles slightly relative to their specific deferents. By combinations of sizes of epicycles relative to deferents, the distances of the deferent centers from Earth, the speeds of the planets around the epicycles relative to the speeds of the epicycles around the deferents, and the tilts of the individual deferents and epicycles, Hipparchus was able to replicate the

Fig. 3.15 A portrait of Hipparchus of Nicaea (c. 190–120 BCE) by Raphael. (Public domain)

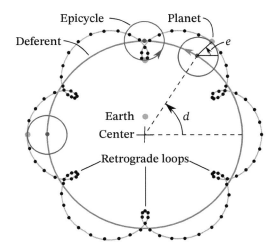

Fig. 3.16 Apollonius first suggested modeling retrograde motion using deferents and epicycles with Earth fixed at the center of the deferent. Hipparchus introduced the idea of shifting Earth away from the center of the deferent to an eccentric point. The green line traces out the path of the planet as it moves counterclockwise around the epicycle while the center of the epicycle moves counterclockwise around the deferent. Both the deferent angle, d and the epicycle angle, e, change at constant rates. The dots are evenly spaced in time, indicating that the planet moves very slowly across the sky during retrograde, but quickly between loops.

observed planetary motions reasonably well. Of course, this required an amazing number of precise adjustments to the model. Remember that all of this was based solely on careful naked-eye observations of the planets moving among the stars on the celestial sphere.

Hipparchus also made numerous other significant contributions to astronomy, in many ways more important than his modifications to the epicycle–deferent model of Apollonius. For example, he created an important catalog of stars that included their brightnesses and positions. It is uncertain how many stars were included in his catalog (no copy survived to the modern age) but estimates vary from 850 and 1025. In order to record brightnesses, Hipparchus created a numerical scale, marking the brightest stars in the sky as magnitude 1, somewhat dimmer stars were labeled as magnitude 2, and so on, with the very dimmest stars that could be seen with the naked eye being classified as magnitude 6.

Although we don't know exactly how he recorded the positions of stars in the sky, his scheme was based on, or at least influenced by, a system developed much earlier by the Babylonians. Hipparchus also had the coordinates of stars that were included in a Babylonian catalog. By comparing his own observations with those ancient observations, Hipparchus noticed a systematic change in recorded positions. Hipparchus had inadvertently discovered a constant and very slow shift in the locations of the vernal and autumnal equinoxes. He attempted to estimate how long it would take for the equinoxes to travel across the stars and return to the same places again, although the value he arrived at was much shorter than the true value of about 26,000 years. So far the equinoxes have only traveled about one-twelfth of their lap among the stars, or 30° of a complete circle since Hipparchus first discovered the motion. Today this important motion is known as precession, which we will revisit in Section 3.5.

Given his accomplishments, it is obvious that Hipparchus's mathematical skills were very advanced for the time. Hipparchus is even recognized as the first to develop the new mathematics of trigonometry.

Ptolemy (c. 90–168 CE)

The final act of our nearly 800-year-long Greek geocentrist play, with a much longer Mesopotamian prelude, brings us to another resident of Alexandria, Egypt, Ptolemy (Fig. 3.17), whose writing would strongly influence western thought in areas of mathematics, geography, astrology, and astronomy for the next 1400 years! (Ptolemy was also known by his Roman name; Claudius Ptolemæus.) Ptolemy's world maps in his treatise *Geography* were marked with well-defined latitude and longitude lines, north was placed at the top of the map, and major known land masses were included (you can blame Ptolemy for north being "up," as in going "up to Alaska"). Of course, at that time North and South America were unknown to Europe and Asia, as was Antarctica. Much of Ptolemy's work in geography was based on prior work of Eratosthenes and Hipparchus, although he used a significantly smaller estimate of the size of the world than Eratosthenes had arrived at. In addition, he stretched Asia far beyond its actual extent. Because of Ptolemy's errors, when Christopher Columbus went to Queen Isabella and King Ferdinand of Spain in search of funding for his 1492 expedition to find a new trade route from Europe to Asia by sailing

Key Point
Hipparchus created the magnitude scale for quantifying the brightness of stars.

Key Point
Hipparchus also discovered the precession of the equinoxes.

Fig. 3.17 Ptolemy (c. 90–168 CE). (Unknown artist, in Thevet, 1584;[5] public domain)

[5]Thevet, André (1584). *Vrais pourtraits et vies des homes illustres grecz, latins et payens* (Paris).

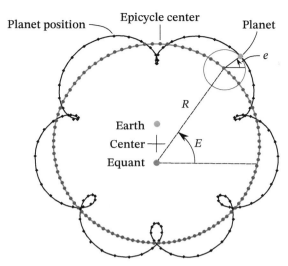

Fig. 3.18 Ptolemy introduced an equant to the model of Hipparchus (recall Fig. 3.16). For a superior planet, as the angle E changes at a constant rate, the distance R from the equant to the center of the epicycle increases when the epicycle approaches Earth, and it decreases when the epicycle moves to the opposite side of the deferent. As a result, the center of the epicycle moves fastest along the deferent when the epicycle is closest to Earth, as indicated by the blue dots on the deferent. The blue dots on the deferent and the black squares on the planet's path are evenly spaced in time. The angle e also changes at a constant rate.

west, he incorrectly believed that Asia was much closer to Europe than it turned out to be via a western route. Of course, he hadn't anticipated running into the Americas first. Columbus left Spain more confident of finding a new route than he had a right to be.

Ptolemy's *Almagest* represented virtually the entirety of all that was known of astronomy at the time, much of which was based on prior work of others, particularly Hipparchus. He adopted Hipparchus's catalog of the stars, but adjusted the coordinates of the stars to account for more than two hundred years of precession. In order to do so, he used the trigonometry of Hipparchus and may have extended it so that it applied to the surface of the celestial sphere rather than just flat surfaces.

Based on Hipparchus's epicycle–deferent–eccentric model, Ptolemy introduced solar, lunar, and planetary motions that included even more features. As seen in Fig. 3.18, Ptolemy introduced a new point inside of the deferent, known as an equant. The equant was situated on the opposite side of the deferent center from Earth and the same distance away from the center. Now, rather than the epicycle moving at a constant speed around the deferent (in km/s for example), the line between the equant and the center of the epicycle changes at a constant *angular speed* (degrees/s). The effect of this change is to cause the epicycle to actually move fastest along the deferent when it is farthest from the equant and closest to Earth, and slowest when it is closest to the equant and farthest from Earth. By carefully looking at the spacing of dots in Fig. 3.18 you can see that they are closer together at the bottom of the figure than they are at the top. This is despite the fact that the time interval between each dot remains the same. You should also note that the retrograde loops are thinner near Earth and wider on the opposite side of the deferent.

Key Point
Ptolemy introduced additional adjustments to the epicycle–deferent– eccentric model, including an equant and constant *angular* motion.

The spacing between retrograde loops is also affected, meaning that spacings along the zodiac are no longer even.

To give you a sense of the extremes to which Ptolemy went in order to model celestial motions as accurately as possible, consider the following: In order to mimic the motions of the inferior planets, Mercury and Venus, Ptolemy tied the centers of their epicycles to a line from Earth to the Sun as the Sun follows its own deferent. In this way, Mercury and Venus could never be more than 28° and 47°, respectively, from the Sun at any time (page 23). Mercury's equant also moved around yet another circle in order to replicate all of the subtleties of its motion. Ptolemy's final model for the motion of the Moon included *two* deferents and an epicycle; the Moon would travel around an epicycle which traveled around the outer deferent while the center of the outer deferent was linked to a point that traveled around the inner deferent.

Needless to say, from Apollonius to Hipparchus to Ptolemy, the models became more and more complex with circles no longer centered on Earth. In addition, speeds didn't remain constant, only angular speeds needed to be constant. Does all of this complexity "save the planetary phenomena" (page 60) in the way originally proposed by Plato five centuries earlier?

Ptolemy never envisioned actual circles with connecting lines; the diagrams simply illustrated how the celestial objects moved. Instead, he envisioned nested crystalline spheres within the starry celestial sphere, much like Eudoxus except that the centers were all different. The epicycles would be their own smaller crystalline spheres moving in spaces between the larger, supposedly "Earth-centered" ones. The centers of the epicycles traced out the representative deferents, while the epicycle crystalline spheres rotated in such a way as to represent the motions of the Sun, Moon, and planets around the circular epicycles of his diagrams. In keeping with Aristotle's concept that "nature abhors a vacuum," each of the crystalline spheres was filled with the perfect fifth element, the ether.

Key Point
Are ancient Greek geocentric models truly scientific?

At this point it is important to point out that it is theoretically possible to mimic any pattern of motion for a heavenly body going around Earth given a sufficient number of circles, centers, tilts, speeds, and other tweaks. As such, important questions must be raised at this point. Does this ever-increasing level of complexity seem satisfying in some intellectually or aesthetically pleasing way? Can these models fundamentally explain *why* the celestial motions occur without invoking supernatural influences? Are these models truly scientific in the modern sense of the term? You may want to go back and review the discussion of the nature of science found in Chapter 1.

3.3   An Opposing View: The Greek Heliocentric Model

Not all ancient thinkers were sold on the idea that Earth is the center of the universe. The concept of a heliocentric (Sun-centered) universe appears to date back at least as far as the religious cult known as the Orphics, according to their poetic hymns from perhaps as early as 1800 BCE:

> Hear golden Titan! Glowing like gold, you who strides above, oh heavenly light
> ... you who combines the epochs ... You are the world ruler ... With your golden

lyre, draw on the *harmonious path of the world* ... [you] who wanders through fire and *moves around in a circle* [emphasis added].

A second hymn contains the line "You who occupies the center of the home of the greatest and eternal fire," and a third hymn mentions a round Earth that spins on its axis. The Orphics are believed to have been followers of the teachings of the mythical poet and musician, Orpheus.

Some followers of Pythagoras, especially Philolaus of Croton (c. 470–385 BCE), had proposed a central fire (not the Sun), with the Earth, the Sun, the Moon, the planets, and the stars on crystalline spheres around the fire. Earth circled the fire in 24 hours and the Sun circled it in one year. In this model, another planet, essentially a counter-Earth called Anticthon, also orbited the central fire with a period of 24 hours, but on the other side of the central fire so that it was never visible from Earth. Given the Pythagoreans great love of the mystery of numbers, perhaps Anticthon was added to make the number of celestial bodies and associated crystalline spheres equal 10 (a holy number to the Pythagoreans). So much for a testable, falsifiable, scientific theory. Interestingly, according to the biographer, Plutarch (c. 46–120 CE), after carefully studying Philolaus's work, "Plato, near the end of his days, had regrets for his older opinion by which he unfittingly placed Earth at the center of the Universe."

Although based on the work of those who came before him, it is Aristarchus of Samos (Fig. 3.19) who is considered to have first proposed a truly heliocentric model of the universe with the Sun at the center, the planets, including Earth, in circular orbits around the Sun, and the Moon in a circular orbit around Earth; very much like our modern understanding of our Solar System and the distant stars. Although original writings of Aristarchus have not survived history, Archimedes of Syracuse (c. 287–212 BCE) gave us a recounting of Aristarchus's model of the heavens. According to Archimedes, in his book, *The Sand Reckoner*:

Fig. 3.19 Bust of Aristarchus of Samos (c. 310–230 BCE). (Eliseevmn, CC BY-SA 4.0)

> You [King Gelon] are aware the "universe" is the name given by most astronomers to the sphere, the center of which is the center of the Earth, while its radius is equal to the straight line between the center of the Sun and the center of the Earth. This is the common account as you have heard from astronomers. But Aristarchus has brought out a book consisting of certain hypotheses, wherein it appears, as a consequence of the assumptions made, that the universe is many times greater than the 'universe' just mentioned. His hypotheses are that the fixed stars and the Sun remain unmoved, that the Earth revolves about the Sun on the circumference of a circle, the Sun lying in the middle of the floor, and that the sphere of the fixed stars, situated about the same center as the Sun, is so great that the circle in which he supposes the Earth to revolve bears such a proportion to the distance of the fixed stars as the center of the sphere bears to its surface.

Apparently Aristarchus was even able to determine the correct order of the planets. To explain the apparent motion of the "fixed stars," Aristarchus required that Earth itself is spinning as it moves in its circular path around the Sun.

Aristarchus also developed an argument as to why no noticeable shift in the positions of the stars is visible as Earth goes around the Sun. This shift, known as parallax, arises from looking at an object from two different directions. As an example, conduct the following simple experiment shown in Fig. 3.20: Hold up one

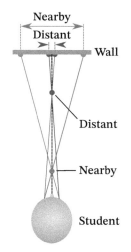

Fig. 3.20 Seen from above, when you observe an object that is nearby (the red dot) with one eye and then with the other, it appears to shift back and forth by a greater amount than an object farther away (the blue dot).

of your fingers just a few inches from your face. If you first look at your finger with one eye and then the other while keeping your finger's location fixed, you will see that your finger seems to move in relation to background objects. If you now hold your finger out at arm's length you will also see your finger apparently move, but not nearly as much as when it was only a few inches away. Aristarchus's argument for the lack of parallax was simply that the stars are too far away; the shifts are too small to be detected. This argument is essentially contained within Archimedes' account in the last part of the final sentence from the quote above: "the sphere of the fixed stars, situated about the same center as the Sun, is so great that the circle in which he supposes the Earth to revolve bears such a proportion to the distance of the fixed stars as the center of the sphere bears to its surface."

Aristarchus's heliocentric model was not rejected out of hand. It seems that Seleucus of Seleucia (c. 190–c. 150 BCE), a Hellenistic Babylonian astronomer who lived one century after Aristarchus and who was a contemporary of Hipparchus of Nicaea (c. 190–120 BCE), referred to retrograde motion as an *apparent phenomenon*, suggesting that heliocentrism was accepted by him. According to the model, retrograde motion arises due to *relative* motions of Earth and the planets as they travel around the Sun (something we will discuss in greater detail in Section 4.7). Apparently Hipparchus had explored the possibility of a heliocentric model, but ultimately rejected it on the grounds that he couldn't produce sufficiently satisfactory motions for the planets using perfect circles alone.

The heliocentric model did survive until at least the fourth century CE. Emperor Julian (336–363 CE), who briefly led the Roman Empire from 360 CE until his death three years later, was a strong supporter of Aristarchus's model.

3.4   Medieval Astronomy in India and the Middle East

At its apex in 117 CE, when Ptolemy was a young man, the Roman Empire occupied a vast region that included all of the lands around the Mediterranean Sea, the portion of Europe west of the German Rhine River, Britain, the northernmost portions of Africa, including Egypt, and parts of the Middle East, including Mesopotamia, Assyria, and the Fertile Crescent (recall Figs. 3.2 and 3.3). Within the borders of the Roman Empire were the historic centers of ancient Greek philosophy; the Greek peninsula, the eastern Mediterranean, and Alexandria, Egypt.

Key Point
In the Roman Empire, Latin was the common language of the west while Greek was the common language of the east.

Due to its size, in 285 CE Emperor Diocletian (245–311) divided the Roman Empire into two halves so that it could be more easily governed. It was bisected by a mostly north–south line with the present-day Italian peninsula to the west and the Greek peninsula to the east. The western half of the Empire continued to be governed from Rome while the eastern portion was governed from Byzantium, which later became Constantinople under the rule of Constantine the Great (c. 272–337), and is now Istanbul, Turkey. Latin became the formal language in the west and Greek in the east.

Constantine was the first emperor to declare conversion to Christianity and he played a major role in the Edict of Milan in 313, in which tolerance for Christianity became an official position of the Empire, thus ending a generally sporadic and localized prosecution of adherents to the religion. Constantine also called the First Council of Nicaea in 325, a pivotal moment in early Christianity. In 380, Theodosius the Great (347–395), the last emperor to rule over a united Eastern and Western

Roman Empire, issued the Edict of Thessalonica proclaiming Nicene Trinitarian Christianity to be the official religion of the Empire, while also ceasing support for traditional Roman polytheism.

When Emperor Romulus Augustus (c. 461–c. 507), ruler of the Western Roman Empire, was overthrown in 476 by the Germanic King, Flavius Odoacer (433–493), the west began to break into smaller governments. Meanwhile, the Eastern Roman Empire became the Byzantine Empire. Following the death of the Prophet Muhammad ibn Abdullāh (c. 570–632), the religion of Islam, that originated in Mecca and Medina in present-day Saudi Arabia, began to spread throughout the Middle East, northern Africa, and the Iberian Peninsula (present-day Spain and Portugal). It wasn't until 1453, over one thousand years later, that the Byzantine Empire fell to the Ottoman Turks.

Key Point
The Western Roman Empire fell in 476 CE.

The reason for this quick history lesson has to do with the influence these events had on intellectual advancement in the centuries that followed the culminating work of Ptolemy. Science and mathematics never live in a vacuum independent of other human activities.

Following the work of Ptolemy, the west entered a period of relative quiet in terms of scientific and mathematical development; so much so that they lost connection with the works of the ancient Greeks. The east, on the other hand, continued to study the great Greek works, and in many cases translated them into local languages. Ptolemy's own work, *Almagest* (meaning "greatest"), derives its name not from the original title that Ptolemy gave it, but from the Arabic name, *al-majisṭī*. Today, some of the ancient texts survive, not in their original Greek, but from the translations of them.

The Roman numeral system that you likely learned in grade school makes use of capital letters to represent values; I = 1, V = 5, X = 10, L = 50, and so on. If you have ever tried to do arithmetic with that system, you know how awkward it is. There are a lot of reasons for this, not the least of which is that there is no representation of zero (recall the success of the Mesoamerican base-20 system, with its representation of zero, discussed on page 35). Our modern ten-digit decimal number system actually originated in Hindu India between the first and fourth centuries CE and made its way through Persia to the Arab world. The base-10 Hindu–Arabic numeral system was only introduced to western Europe in 1202 by Leonardo Bonacci (c. 1170–1250), a gifted Italian mathematician who is better known by his nickname, Fibonacci, and the famous mathematical sequence that he introduced to the world: 0, 1, 1, 2, 3, 5, 8, 13, 21, 34, … , where each new number is the sum of the previous two numbers.

Major advances in algebra, geometry, and trigonometry occurred in the Middle East during the time of relative intellectual hiatus in Europe. The term "algebra" comes from the title of a book, *Hidab al-jabr wal-muqubala*, written by the mathematician Moḥammed ibn-Mūsā al-Khwārizmī (c. 780–850), in Baghdad in about 825; al-jabr was a set of mathematical operations he discussed. "Algorithm" comes from the Latinized version of his name, Algoritmi.

Astronomy plays a particularly important role in Islam due to three requirements of the faith. The first involves determining the start of the Islamic month, defined to be the appearance of the thin crescent for the first time following a new moon. While this would seem to be a straightforward observation project, imagine the issues that arise due to cloudy skies, views along the horizon when obstructions may be in the way (hills, trees, buildings, et cetera), and even people's eyes playing

Key Point
The medieval Islamic world produced important advances in mathematics and astronomy.

tricks on them. The importance of getting the date right is also tied to Ramadan, the Islamic month of fasting, which is based on twelve lunar months (a total of 354 or 355 days) rather than on a solar year of 365 1/4 days. As a result, the start of Ramadan drifts through the solar calendar, starting about eleven days earlier every year.

The second reason for the importance of astronomy has to do with the requirement that prayers should always be performed facing toward the religious shrine, the Ka'ba, in Mecca. Determining the location on the surface of a nearly spherical Earth without the aid of GPS, or even modern time-keeping devices, is a non-trivial proposition, but one that can be solved in part through astronomical observations and analysis. (You will learn more about this challenge in Section 3.5.)

Finally, the third need for astronomy in Islam comes from determining when the five daily prayers are to occur, which are specified in terms of the height of the Sun in the sky.

Beyond the practical needs of astronomy, as with all cultures there were purely intellectual interests as well. During the course of over one thousand years of observations of the night sky since the publication of the *Almagest*, and concerns over some of Ptolemy's solutions to planetary motions, Arab astronomers made modifications to his models that improved their predictive capabilities, and, one could argue, were also improvements from a philosophical standpoint. Ibn al-Shātir (1304–1375), for example, was able to replace Ptolemy's eccentric circles and the equant by placing a planet on an epicycle that traveled on an epicycle on an epicycle on a deference circle! By doing so, he was successful in fulfilling Plato's original requirement, based on the available data, of a truly Earth-centered motion using perfect circles at constant speeds (al-Shātir was able to "save the phenomena").

Al-Shātir's work, and the work of other Arab astronomers and mathematicians, made its way to Europe during the European Renaissance, largely by way of the Iberian Peninsula. It seems that Johannes Müller (1436–1476), a German astronomer and mathematician better known as Regiomontanus, may have been informed by the work of the Islamic astronomers. His text, *Epitome of the Almagest*, was used extensively in the years after its publication, including by the astronomer Nicolaus Copernicus (1473–1543).

3.5   The Celestial Sphere

Key Point
The stars and constellations appear as though they are attached to an illusionary celestial sphere that is centered on a fixed Earth.

There is a common theme that exists across most of the models of the heavens that have been discussed thus far; the idea that the stars are fixed on a sphere that is either rotating around a fixed Earth, or is fixed while Earth rotates on its own axis. Either way, the observational effect is the same; the stars appear to be spinning around us *slightly faster than once each day* (approximately once every 23^h56^m; see page 17). In the Greek tradition, the sphere was assumed to be crystalline with the stars attached to it (think about the extreme cases of being at the North or South Pole and being at the equator).

Going back to what you learned in Section 2.1 about the apparent motions of the stars in the sky, it is possible to develop a model that does a very good job of explaining those motions. Although certainly not valid as an accurate modern representation of stellar motions, the model of a celestial sphere spinning around a fixed Earth is still helpful in many ways, including:

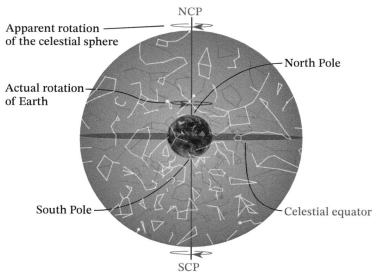

Fig. 3.21 The celestial sphere.

- describing the apparent positions of stars relative to each other,
- predicting which stars will be up in the sky and where they will be at a specific time of night during a given time of year,
- determining which stars will be circumpolar at a given latitude on Earth,
- calculating the angles at which stars rise in the east and set in the west,
- creating a coordinate system to catalog the locations of specific stars, and
- with one addition to the model, describing approximately where the Sun is in the sky relative to the background stars.

(We know today that the stars are all at different distances from Earth, but that fact is completely undetectable to the unaided eye. As a result, for millennia observers of the night sky believed that the stars were in fact located on a celestial sphere.)

The Observer's Latitude

As illustrated in Fig. 3.21, imagine that Earth is at rest at the center of the universe while the celestial sphere is rotating about an axis that passes through Earth's North Pole, its center, and its South Pole. The nroth celestial pole and south celestial pole are located on the celestial sphere directly above Earth's North and South Poles, respectively, as first discussed in Section 2.1.

Extending the celestial sphere model a bit further, imagine a circle around the sphere that is directly over Earth's equator everywhere. This imaginary circle is referred to as the celestial equator. Just like the equator on a globe of Earth, the circle that is the celestial equator is exactly halfway between the north celestial pole and the south celestial pole on the surface of the celestial sphere.

If you have ever been in a planetarium like the one shown in Fig. 3.22, the dome of the planetarium is just a representation of the celestial sphere with the stars and constellations projected onto the dome rather than being permanently attached to it. The center of the room, where the star projector is typically located, is where Earth is located in Fig. 3.21. Although there are stars below your horizon when you

Fig. 3.22 A non-digital Spitz planetarium projector. The star projector is the ball in the upper right corner of the picture. The dome is shown in the background. The bottom of the dome represents the observer's horizon. (Astorrs, CC BY-SA 2.5)

are outside, Earth blocks your view of those stars. Technically, the planetarium horizon is at the base of the curved dome, which may or may not correspond to the planetarium floor because the planetarium may have vertical walls between the dome and the floor.

The other major difference between the model of the celestial sphere and the planetarium is that the dome itself doesn't actually rotate, but the rotation is simulated by having the projections of the stars move instead. Older, non-digital planetariums accomplish this by having a globe with holes in it located in the middle of the planetarium. The holes are positioned in the arrangement of the actual stars in the sky, like those that form the constellation of Orion. A bright bulb in the center of the globe causes light to shine through the holes and onto the dome.

To simulate the spinning of the celestial sphere, the globe in the center of a non-digital planetarium spins along an axis that represents the rotation axis of the celestial sphere passing through the north celestial pole and south celestial pole. Tilting the projector simulates being at different latitudes on Earth. Comparing Fig. 3.22 to the various orientations in Fig. 3.23, you can imagine that, if the projector is pointed straight up, that orientation would correspond to being at the North Pole with the north celestial pole at your zenith and the celestial equator on your horizon (the orientation of the celestial sphere in this case is also the same as the one depicted in Fig. 3.21). If the projector is set with its rotation axis being horizontal, the north celestial pole would be straight north and on your horizon with the celestial equator passing through your zenith while intersecting your horizon directly east and west of your position, just as if you were observing the sky from Earth's equator. It is important to realize that in its North Pole position the projector's axis is perpendicular to the horizon (90°), which corresponds to the latitude of the North Pole (90° N). In its horizontal position, the projector's axis is parallel to the horizon (0°), corresponding to the equator's latitude of 0°. To simulate an observer at 41° N, you would simply tilt the projector's axis 41° above horizontal, which would put the north celestial pole 41° above the northern horizon, similar to the orientation of the projector in Fig. 3.22. You can imagine tilting the celestial sphere in just the same way to produce the same effects, with you as the observer sitting in the middle. (By the way, hopefully you noticed the relationship between the two words "horizontal" and "horizon.")

It is important to note that, as was mentioned in Fig. 2.4, Polaris, the "North Star," is less than 1° from the north celestial pole. As a result, as long as you are in the northern hemisphere, measuring the angle up from your northern horizon to Polaris gives a reasonable estimate of your latitude. Unfortunately, no similar bright "South Star" exists in the southern hemisphere.

The Daily Motions of the Stars

A video tutorial on the effects of latitude on the motions of stars is available on the Chapter 3 resources page.

Now that you have a mental image of how a non-digital planetarium projector and dome can replicate the celestial sphere, think for a moment about what you would see as the projector (or celestial sphere) rotates. When the projector is oriented as if you are at the North Pole, the north celestial pole would be at your zenith and all of the stars north of the celestial equator would simply make circles around the horizon, never rising or setting [Fig. 3.23 (*Top left*)]. In other words, all of the stars that you would ever see at the North Pole are circumpolar stars. This also means that you

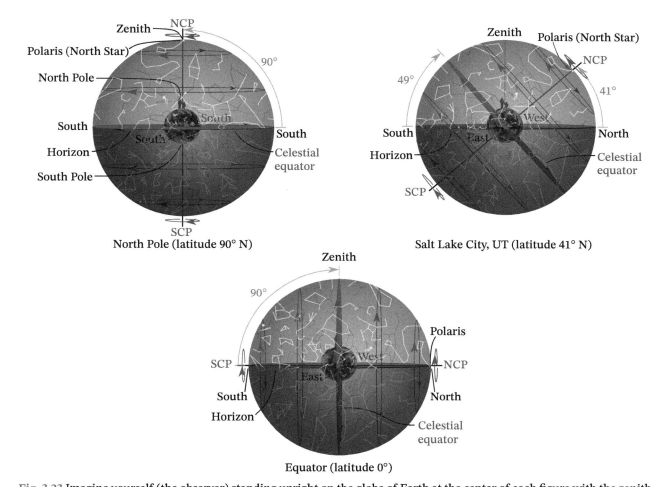

Fig. 3.23 Imagine yourself (the observer) standing upright on the globe of Earth at the center of each figure with the zenith position directly over your head. The tilt of the celestial sphere for observers at the North Pole (*Top left*; latitude = 90° N), at a latitude of 41° N (*Top right*; Salt Lake City, Utah), and at the equator (*Bottom*; latitude = 0°). North is to the right for the second and third drawings and east is in the front. There is no direction of north (or east or west) if you are at the North Pole; every direction you look is south!

would never see any stars that are south of the celestial equator simply because they always remain below your horizon. On the other hand, if the projector is configured as though you are observing the sky from the equator [Fig. 3.23 (*Bottom*)], as the projector (or celestial sphere) rotates you would see every star in the sky rise straight up from the eastern (or northeastern or southeastern) horizon and set straight down below the western (or northwestern or southwestern) horizon. In addition, if you wait for a complete rotation you would be able to see every star in the heavens even though only one-half of them would be up at any given moment (assuming that the Sun doesn't light up the daytime sky!).

What about at a latitude of $\ell = 41°$ N? As depicted in Fig. 3.23 (*Top right*), you would see some stars up all of the time, so long as they are within 41° of the north celestial pole (or more than 49° north of the celestial equator; 90° − 41° = 49°).

You would see other stars rise at an angle of 49° above the eastern horizon and then set at the same angle below the western horizon. You would also never see some stars, those that are within 41° of the south celestial pole. Which stars are circumpolar depends on the latitude of the observer. More generally, at any latitude, ℓ, between the North Pole and the equator, stars rise at angles with respect to the horizon of $90° − \ell$. Stars within the angle ℓ of the north celestial pole are circumpolar, while stars within ℓ of the south celestial pole are never visible.

These are the effects we talked about earlier and that are shown in the actual star trails recorded in Fig. 2.4; you should compare those time-elapsed photographs with the discussion above. It is important that you can sketch these figures for any specified latitude and you should be able to determine which stars are circumpolar as well as how to calculate the angle at which stars rise above the horizon and set below it.

It should be pointed out that there is a very slight error present in every figure depicting the observer's horizon. As you may have noticed, the horizon is always drawn through the center of Earth rather than immediately under the observer's feet, as it should be. Although Earth seems large by our personal standards, it is but a speck amid the cosmos. This means that the extremely tiny shift from the surface of Earth to its center is essentially meaningless for routine observations with the naked eye.

The Daily and Annual Motions of the Sun, and the Earth's Seasons

How do we modify the model of the celestial sphere to include the motions of the Sun? It is simply a matter of adding a circle that represents the ecliptic centered on Earth that goes around the celestial sphere. This is identical to using one of Eudoxus's tilted crystalline spheres or one of Ptolemy's deferents. In either case, the plane of the circle must be tilted by 23 1/2° with respect to the celestial equator as shown in both parts of Fig. 3.24. The Sun then travels once per year around the ecliptic but in the opposite direction from the spin of the celestial sphere. Overall, the Sun still rises in the east and sets in the west due to the much faster rotation rate of the celestial sphere, but the Sun lags behind the stars by about four minutes per day. In this way, the 23^h56^m rotation period of the celestial sphere plus the 4-minute lag of the Sun means that the Sun rises once every 24 hours. Again, as pointed out on page 17, 4 minutes per day works out to be 24 hours over 360 days. Lagging behind by 4 minutes per day also corresponds to moving along the ecliptic at a rate of 1° per day, or 360° (one complete circle) in 360 days.

Key Point
The Sun's daily path is always essentially parallel to the celestial equator, regardless of latitude.

To better understand the motion of the Sun in the sky, imagine yourself at the North Pole with the north celestial pole at your zenith; see Fig. 3.23 (*Top left*) and Fig. 3.24 (*Left*). At this latitude you would see the Sun making circles around the sky, parallel to your horizon. More generally, as illustrated in both drawings in Fig. 3.24, regardless of the latitude of the observer, the Sun's daily motion is essentially a circular path that is approximately parallel to the celestial equator. (It is because the Sun moves slowly north or south during its annual trip around the ecliptic that the daily motion of the Sun is not exactly parallel to the celestial equator.) Only for observers at the North and South Poles do the celestial equator and the horizon coincide, causing the Sun to trace circles parallel to the horizon at those latitudes.

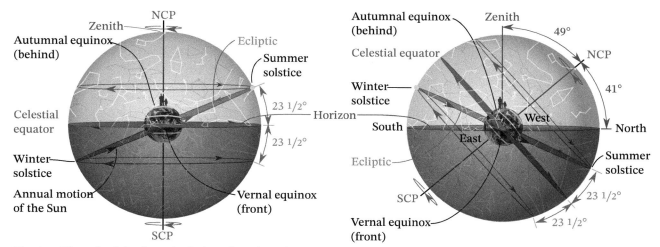

Fig. 3.24 The celestial sphere, including the celestial equator (red), the horizon (blue), and the ecliptic (green), seen from the perspective of observers standing at the North Pole (*Left*) and at a latitude of 41° N (*Right*). In the right-hand diagram, north is to the right and east is out of the page. The Sun's motion combines the daily rotation of the celestial sphere with the much slower annual motion of the Sun along the ecliptic in the opposite direction (slightly less than 1° per day). The approximate daily motions of the Sun on the first days of spring, summer, autumn, and winter are displayed with the red circles. The Sun is shown at its position on the celestial sphere on the first day of summer in the left-hand diagram and (six months later) on the first day of winter in the right-hand diagram. You may want to compare the daily paths of the Sun in the right-hand diagram with those in Fig. 2.7 (*Left*). The positions of the vernal equinox (in front of Earth in the diagrams), the summer solstice, the autumnal equinox (behind Earth in the diagrams), and the winter solstice are marked. The yellow lines on Earth are, from north to south, the arctic circle, the tropic of Cancer, the equator, the tropic of Capricorn, and the antarctic circle.

Referring again to Fig. 3.24 (*Left*), when spring first arrives at the North Pole on about March 20, the Sun is crossing the celestial equator from south to north; it is this intersection that defines the location of the vernal equinox on the celestial sphere (page 20). When the Sun is at the vernal equinox, its daily path hugs the horizon, traveling all the way around the horizon in 24 hours. On each succeeding day the circle that the Sun makes around the sky moves progressively higher above the horizon until about June 21, the first day of summer. The Sun is now as high in the sky as it will ever get, with the Sun's center at 23 1/2° above the horizon and the celestial equator; the center of the Sun is now at the summer solstice. As the summer season continues, the Sun moves closer to the horizon while it follows its annual route around the ecliptic. On approximately September 22, when the Sun returns to the celestial equator, it again makes a circle around the horizon, having reached the intersection of the ecliptic with the celestial equator traveling from north to south, which is the defining point of the autumnal equinox. After another three months of traveling progressively farther south along the ecliptic, the Sun arrives at its most southerly point on the celestial sphere on about December 21, when it is 23 1/2° below the celestial equator. The center of the Sun is now located on the winter solstice. After the Sun passes the winter solstice it heads back north along the ecliptic toward the vernal equinox and a new spring.

It is important to understand that the vernal equinox, the summer solstice, the autumnal equinox, and the winter solstice are not points in time; they are points on the celestial sphere that move with the celestial sphere as it spins (or more accurately, as Earth rotates). Common usage has confused those points in the heavens with dates on a calendar. When spring started on March 20, 2022 at 11:33 AM EDT (4:33 PM in Greenwich, England), this precise time is when the center of the Sun moves across the vernal equinox.

At the North Pole, when the Sun is between the vernal equinox and the autumnal equinox, the Sun is north of the celestial equator and above your horizon, meaning that there are six months of perpetual daylight. For the other six months the Sun is south of the celestial equator and below your horizon, resulting in six months of continual darkness. For someone observing from the South Pole [imagine yourself standing on the "bottom" of Earth as depicted in Fig. 3.24 (*Left*)], the six months of daylight occurs from the autumnal equinox to the vernal equinox.

We now turn our attention to what these motions look like if you are not observing from either the North Pole or the South Pole. In particular, imagine yourself as an observer located somewhere south of the North Pole but still in the northern hemisphere (on or north of Earth's equator). The descriptions for observers south of the equator are the same, except that the seasons are reversed. In other words, if you want to go skiing during the northern hemisphere's summer and don't want to ski on a melting glacier, head to Chile or New Zealand.

Suppose that you are located at a mid-northern latitude such as 41° N [refer again to Fig. 3.23 (*Top right*) and Fig. 3.24 (*Right*)]. As before, you can imagine that you are standing on Earth at the top of the diagrams. In this case, the entire celestial sphere is tilted so that the north celestial pole is 41° above the point on your horizon that is due north of your location. In both diagrams your horizon is shaded and horizontal. (To visualize the directions to the cardinal points it may help to picture yourself laying a flat map on the horizon with north to the right.)

At 41° N, when the Sun's annual path along the ecliptic carries it northward across the vernal equinox, its daily motion would cause it to rise due east of your location and set due west (please review the composite of sunset pictures in Fig. 2.6). This is because the celestial equator always intersects the horizon directly east and directly west of an observer at any latitude. Since exactly one-half of the celestial equator is always above the horizon, on the first day of spring the Sun will be above the horizon for one-half of the day and below the horizon for the other half; there will be twelve hours of daylight and twelve hours of night. This is actually true everywhere on Earth, except precisely at the North and South Poles.

When the Sun reaches the summer solstice it will rise as far to the northeast at it ever gets, and will set at its northernmost point in the northwest. This is also the day when the Sun spends the greatest fraction of the day above the horizon. For observers north of the tropic of Cancer, at noon the Sun reaches its highest elevation of the year when it's at the summer solstice. To see this, study Fig. 3.24 (*Right*) carefully, noting that even if the Sun hasn't yet reached the summer solstice in the diagram, over one-half of the circle that the Sun travels along during the day is above the horizon. You may also recall Fig. 2.7 illustrating the motions of the Sun on the first days of spring, summer, fall, and winter.

How long daylight lasts depends heavily on latitude. If you are on the equator, the amount of daylight equals the amount of darkness (12 hours each) all year long. You can see how this is the case by referring back to Fig. 3.23 (*Bottom*) and noting that exactly one-half of each circle is above the horizon. You can also see

Key Point
Six months of daylight and six months of darkness at the North and South Poles.

Key Points
The amount of daylight, the height of the Sun as it crosses the observer's meridian at local noon, and the locations of the rising and setting Sun depend on latitude and the location of the Sun along the ecliptic throughout the year.

Key Point
Don't forget to be an active reader! Draw a sketch to help you visualize what is being discussed.

this by visualizing what Fig. 3.24 (*Right*) would look like at the equator (latitude, $\ell = 0°$). Certainly on the first day of spring or the first day of fall, when the Sun is on the vernal equinox or the autumnal equinox, respectively, the daily rotation of the celestial sphere will carry the Sun straight up from the horizon point due east of the observer, the Sun will then pass directly through the observer's zenith at noon six hours later, and then set straight below the horizon due west of the observer. At midnight the Sun would be directly below the observer. But what happens at other times of the year? On the first day of summer the Sun will be 23 1/2° north of the celestial equator along the ecliptic and so will rise 23 1/2° north of east, but still straight up out of the horizon. The Sun will then cross the observer's meridian 23 1/2° north of the zenith and set 23 1/2° north of west. On the first day of winter, the Sun will be 23 1/2° south of the celestial equator on the ecliptic and so will rise 23 1/2° south of east, again straight up out of the horizon. The Sun will then cross the observer's meridian 23 1/2° south of the zenith and set 23 1/2° south of west. (Describing the motions of the Sun at the arctic circle and the tropic of Cancer are left as exercises.)

At this point, by using the model of the celestial sphere and the ecliptic we have been able to describe all of the solar and stellar motions we had previously studied in Section 2.1. These are the same motions and seasons that non-modern cultures were trying to understand largely through mythology until the time of the ancient Greek philosopher-scientists. However, even in the case of the ancient Greeks they still invoked deities such as the prime mover to explain physically why the heavens, and the objects in them, moved.

Video Tutorial: Motions of the Sun with latitude and the seasons is available on the Chapter 3 resources page.

Finding Your Location on Earth

In order to identify a specific point on Earth requires two numbers: the latitude and longitude of that location. However, before those numbers can be determined, reference locations must be defined. Terrestrial latitude is based on measuring the number of degrees north or south of the equator along a hypothetical meridian that runs from the North Pole to the South Pole while passing through the location of interest. (This is the same meridian we have been referring to for someone observing the sky, except that instead of envisioning an imaginary line passing over the observer's head from north to south, we picture the line as running under the observer's feet.) Terrestrial longitude is based on the prime meridian, where longitude is set to 0°. The prime meridian was chosen in 1884, during an international conference of 22 countries, to be a half-circle that runs from the North Pole, through a point at the Royal Observatory in Greenwich, England, to the South Pole (Fig. 3.25). Today's modern definition of the prime meridian is based on a detailed process that takes into account motions of Earth's tectonic plates (see Section 11.3). The modern prime meridian is currently about 102 m east of the Royal Observatory's original line. Terrestrial longitude is the number of degrees east or west of the prime meridian, measured along a line of constant latitude (a parallel) to the local meridian of the point of interest. The eastern hemisphere is that half of Earth east of the prime meridian between the prime meridian and 180° (on the opposite side of Earth from the prime meridian). The western hemisphere is that part of Earth west of the prime meridian up to 180° longitude. By straddling the prime meridian you can stand with one foot in the eastern hemisphere and one foot in the

Fig. 3.25 The prime meridian at the Royal Observatory, Greenwich, London, United Kingdom. (Andres Rueda, CC BY 2.0)

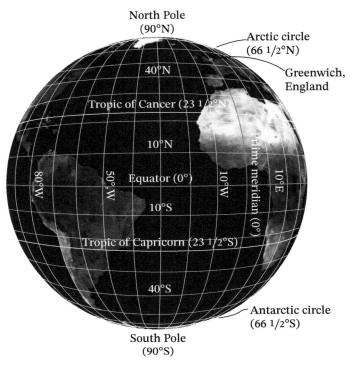

Fig. 3.26 Earth's coordinate system is based on latitude (degrees north or south of the equator) and longitude (degrees east or west of the prime meridian). (Image of Earth by NASA/Goddard Space Flight Center Scientific Visualization Studio)

western hemisphere. Figure 3.26 illustrates Earth's latitude–longitude coordinate system.

We have already seen that it is fairly straightforward to get a reasonable estimate of your local latitude (assuming that you are in the northern hemisphere) by measuring the angle from your ideal northern horizon (no mountains or other features allowed) up to Polaris, the "North Star." To be more precise, you must locate the center of the motions of the circumpolar stars (Polaris being one of them), which is the north celestial pole, and measure the angle from your horizon to that position in the sky. Although there is no equivalent to the "North Star" in the southern hemisphere, if you can identify the position of the south celestial pole above your southern horizon, you can determine your latitude south of the equator in precisely the same way.

Historically, it proved to be much more difficult to determine longitude with any level of precision based on the positions of the stars. The reason is simple; the stars are constantly moving from east to west, or around the north celestial pole and the south celestial pole. If the stars were fixed in the sky and you knew which ones were directly over the prime meridian it would be possible to think in terms of following a hypothetical path in the sky parallel to the celestial equator from an identified star to a star directly above the prime meridian, but of course that doesn't work. However, since the stars do move at a constant rate, *if* you could know when a star passed over the prime meridian and then passed over your head (your local meridian) you could deduce your longitude based on the elapsed time. For example, if a star passed over the prime meridian six hours before it crossed your local meridian then you must be one-quarter of the way around the planet from the prime meridian (6/24 = 1/4), implying that you are at a longitude of 90° W. To accomplish this task requires two things, knowledge of when stars pass over the prime meridian on any given day (remember that stars cross the meridian about four minutes earlier every day relative to the Sun), and knowledge of your local time, without artificial time zones, relative to the time of the prime meridian. The time at the prime meridian is a universal standard. Once known as Greenwich mean time (GMT), based on the average elapsed time between transits of the Sun over the prime meridian, the standard is now referred to as coordinated universal time (UTC) and is based on the elapsed time measured by atomic clocks.[6]

Determining when stars cross the prime meridian on any given day is easily accomplished by making observations from the prime meridian over the course of one year and recording that information in a table. It may seem surprising today, but the process of determining your local time was the hard part, especially for a sailor on the open ocean. This was because, until shortly before the American Revolutionary War, the best available time-keeping devices were pendulum clocks (like grandfather clocks). Obviously, trying to keep time on a rolling deck with a device that relied on a swinging pendulum was problematic.

The lack of knowledge of one's longitude could be, and sometimes was, a matter of life or death. Without knowing your longitude on the open sea, it becomes essentially impossible to locate islands for resupply without sailing along a constant latitude. In a tragic case in 1707, a fleet of British ships sank just 65 km (40 mi) off Land's End in southwest England after foundering on the rocks of the Isles of Scilly because their admiral, Sir Clowdesley Shovell (1650–1707), a highly regarded captain, didn't know his position with sufficient accuracy.

In 1714, in an effort to find a solution to the critical "longitude problem," the British Parliament offered grants of £2000 for promising experiments in developing an accurate seafaring clock, and £20,000 for clearly solving the problem (a large sum of money in those days). It took until 1761 before John Harrison (1693–1776), a carpenter, developed a spring-driven device that passed the stringent test set by Parliament 47 years earlier. Over the course of a nine-week voyage to Jamaica his clock lost only five seconds, amounting to an accuracy of about 1/60 of a degree of longitude.

Remarkably, today you can immediately determine your current latitude and longitude from the United States' Global Positioning System (GPS) or other global navigation satellite systems (GNSS) using your smartphone or some other electronic device dedicated to mapping. Among other applications, the precision of these systems helps first responders find you in an emergency, and allows companies to develop self-driving cars. GPS and GNSS also provide elevation information relative to sea level.

---

[6]Technically, there are different versions of universal time depending on how measurements are made. Other versions are based on precise measurements of the positions of distant celestial objects.

Celestial Coordinates

In order for astronomers to study specific stars in the sky, they must first be able to find them. The brightest stars, such as Sirius or Vega, are well known, but what about the billions of other stars, galaxies, and other objects, sometimes extremely faint, that are out there?

By now you should have noticed a very strong resemblance between the celestial sphere and a globe of Earth. The equatorial coordinate system (Fig. 3.27), one of the coordinate systems commonly used by astronomers, completely replicates Earth's coordinate system of latitude and longitude (Fig. 3.26). It too requires two references, the first being the celestial equator, and the second is the vernal equinox. Just like latitude on Earth, declination (dec) on the celestial sphere is the number of degrees north or south of the celestial equator along the meridian that contains the celestial object of interest. The analogy to longitude is right ascension (RA), which is always measured east from the vernal equinox along the celestial equator to the intersection of the object's meridian with the celestial equator.

An important difference does exist between longitude and right ascension in terms of the units that are used: while longitude uses degrees east or west of the prime meridian, right ascension measures angles in "hours," where 1 h equals 15°,

Key Point
The equatorial coordinate system is based on declination (dec, analogous to latitude) and right ascension (RA, analogous to longitude).

A time and degrees arithmetic tutorial is available at Math_Review/Time_Degrees

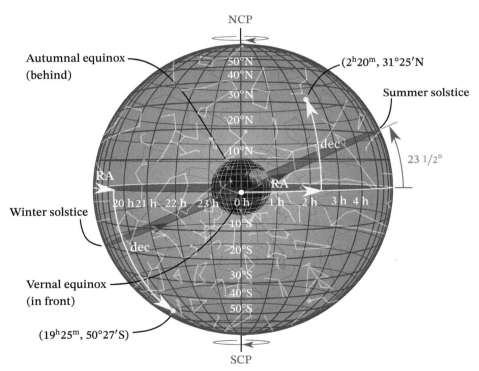

Fig. 3.27 Recalling Fig. 3.26, the equatorial coordinate system for the celestial sphere is based on declination (dec, analogous to Earth's latitude) and right ascension (RA, analogous to Earth's longitude). The reference point on the celestial sphere is the vernal equinox (analogous to the intersection of Earth's prime meridian with its equator). Right ascension is always measured eastward from the vernal equinox in units of sidereal hours.

or 24 h = 360°. The reason for using a time unit as an angle is because time lets the observer know how long it will take for the object of interest to pass the observer's meridian after the vernal equinox has done so.

The "hour" used in right ascension isn't the same period of time as the hour on your wall clock. Because the Sun "slips" backward along the ecliptic by 4 minutes per rotation, it is more appropriate to think about how long it takes for the vernal equinox to cross the observer's meridian on two successive passages, rather than how long it takes the Sun to do so. As a result, the hours used as coordinate units of right ascension are associated with sidereal time (page 17). It is common for astronomers to keep track of both solar and sidereal time at an observatory. Solar time essentially tracks daylight and darkness, while sidereal time keeps track of the passage of the vernal equinox across the observer's meridian, and therefore where the stars are in the sky at any given moment.

Key Point
One solar day is the average amount of time required for the Sun to cross the observer's meridian on two consecutive days. One sidereal day is the amount of time required for the vernal equinox (or a star) to cross the observer's meridian on two consecutive days. One sidereal day is approximately four minutes shorter than one solar day.

Example 3.1

The right ascension of Sirius, in the constellation of Canis Major, is about 06^h45^m and its declination is close to $-16°42'$. Betelgeuse, in Orion, has a right ascension of 05^h55^m and a declination of $+07°24'$. For an observer at a latitude of 41° N, where is Sirius relative to Betelgeuse and the vernal equinox? At what time of year are these two stars near the observer's meridian at midnight?

At a latitude of 41° N, the north celestial pole is 41° above the observer's northern horizon and the celestial equator crosses the observer's local meridian 41° south of the zenith or $90° - 41° = 49°$ above the observer's southern horizon (again refer to Fig. 3.24). Since Sirius is south of the celestial equator (its declination is negative), it will be $49° - 16°42' = 32°18'$ above the southern horizon at the time it crosses the observer's meridian. Sirius will also cross the observer's meridian 06^h45^m sidereal time after the vernal equinox does.

Using the same arguments, Betelgeuse will cross the observer's meridian at an altitude of $+56°24'$ above the southern horizon and 05^h55^m sidereal time after the vernal equinox does.

Since the right ascension of Betelgeuse is smaller than Sirius, it will cross the observer's local meridian slightly less than 1 hour before Sirius does ($06^h45^m - 05^h55^m = 50^m$). Given that the declination of Betelgeuse puts it in the northern hemisphere of the celestial sphere while Sirius is in the celestial sphere's southern hemisphere, Betelgeuse will also be higher in the sky than Sirius by about 24° when each of them crosses the meridian [$+07°24' - (-16°42') = 24°06'$].

To determine the best time of year to observer Betelgeuse and Sirius, think about where they are relative to the vernal equinox. Since Betelgeuse is about 6 hours behind the vernal equinox (one-quarter of a celestial sphere rotation cycle), this means that the vernal equinox will be setting when Betelgeuse is at its highest in the sky. If you want the best viewing of Betelgeuse and Sirius, then they should be high in the sky at midnight, meaning that the Sun must be on the other side of the celestial sphere. Since the vernal equinox is on the western horizon when the two stars are near the meridian, the Sun must be one-quarter of the way around the celestial sphere from the vernal equinox. With the Sun "slipping" along the ecliptic toward the vernal equinox, if the Sun is one-quarter of the way around the ecliptic from the vernal equinox, it must be near the winter solstice. Sirius (in Canis Major) and Betelgeuse (in Orion) are highest in the sky in early winter.

Precession of the Equinoxes

When you learned about the work of Hipparchus of Nicaea it was pointed out that he discovered that the equinoxes move very slowly along the ecliptic, making a complete lap around the ecliptic roughly once every 26,000 years (25,772 years to be more precise). There are multiple important consequences of precession, including changes in the locations of the north and south celestial poles, and the vernal equinox, summer solstice, autumnal equinox, and winter solstice on the celestial sphere. Today, the north celestial pole points to within 1° of Polaris (the North Star) in Ursa Minor, but at the time the Egyptian pyramids were built, the north celestial pole was pointing close to Thuban in Draco (this is why Thuban was considered as a possible alignment for one of the "air shafts" in Fig. 2.36; Thuban's position on the celestial sphere is labeled in Fig. 2.3). In ancient times the vernal equinox was in the constellation of Aries, while the summer solstice and the winter solstice were in Cancer and Capricorn, respectively. It is for these reasons that the vernal equinox is sometimes referred to as the "First Point in Aries," 23 1/2° N latitude is known as the tropic of Cancer, and 23 1/2° S latitude is the tropic of Capricorn. Today, the vernal equinox is located in Pisces and is headed toward Aquarius ("the dawning of the Age of Aquarius" for those of you who grew up in the 1960s and 1970s, or more likely who had parents or grandparents who did). The summer solstice is currently in Taurus, the autumnal equinox is in Virgo, and the winter solstice is in Sagittarius.

Precession also impacts how the equatorial coordinate system is used to locate specific stars in the sky. This is because the stars' coordinates are constantly changing as the vernal equinox moves along the ecliptic; but at least the coordinates are changing in a way that is understood and can be accounted for. It is, therefore, necessary to be specific about when the coordinates are recorded in tables and then use formulas that allow astronomers to "precess" the coordinates from those in the tables to their actual values at the time that observations are to be made.

Exercises

All of the exercises are designed to further develop *your understanding* of the material by thinking carefully and critically about what you have read in this chapter. Answers to selected exercises are available in the back of the book.

True/False

1. Modern science traces its roots back to medieval Europe.
2. (*Answer*) The Sumerians developed the oldest known form of writing sometime in the mid-3000s BCE.
3. The Mesopotamians developed a calendar that was based on the phase cycle of the Moon.
4. (*Answer*) An "ideal" calendar created in the third millennium BCE contained 360 days, using 12 months of 30 days each.
5. We trace our concept of 360° in a circle back to the ancient Greeks.
6. (*Answer*) It is appropriate to defer to supernatural influences when science is unable to explain a natural phenomenon. This is because if the phenomenon hasn't already been explained the phenomenon is fundamentally unexplainable by science.

7. Thales of Miletus is sometimes considered to be the "father of science."
8. (*Answer*) Plato challenged the Greek mathematicians to explain planetary motion by assuming that Earth is the center of all heavenly motions, and that planetary motions must be based on perfect circles and constant speeds.
9. The superior planets are moving most slowly against the stars, and are at their dimmest during retrograde motion.
10. (*Answer*) Eudoxus of Cnidus believed, contrary to Plato, that Earth goes around the Sun.
11. The stars rise approximately four minutes earlier every day relative to the Sun.
12. (*Answer*) The path that the Sun follows annually against the background stars is called the ecliptic.
13. The Moon is never seen more than 5′ above or below the ecliptic.

14. (*Answer*) Aristotle believed in the fundamental importance of observational and empirical evidence.
15. According to Aristotle there exist four terrestrial elements (earth, water, air, and fire) and one perfect celestial element (ether).
16. (*Answer*) Aristotle did not believe in empty space that was devoid of any elements.
17. The discovery that the vernal equinox moves slowly along the ecliptic was made by Hipparchus of Nicaea.
18. (*Answer*) Ptolemy's *Almagest* was the definitive work in astronomy from the second century CE until the 1500s.
19. None of the ancient Greeks thought that Earth went around a central Sun.
20. (*Answer*) From an observational point of view, a spinning Earth under fixed stars looks very different to an observer on Earth than a rotating celestial sphere around a fixed Earth.
21. If Polaris is seen very close to your zenith, you must be located near the equator.
22. (*Answer*) If the celestial equator is on your horizon then you must be at the equator.
23. For an observer at a latitude of 55° N, the north celestial pole will be 55° above your northern horizon.
24. (*Answer*) For an observer at a latitude of 55° N, stars with declinations of less than 35° N will be circumpolar.
25. If you live for a year at a latitude of 20° S, you will be able to observe the Sun at your zenith on two separate days.
26. (*Answer*) At a latitude of 25° N, the Sun will come within 25° of the observer's zenith on the first day of spring.

Multiple Choice

27. A legacy of the Sumerian sexagesimal number system still exists in modern society in the form of
    (a) the number of seconds in one minute.
    (b) the number of minutes in one hour.
    (c) the number of arcminutes in one degree.
    (d) the number of degrees in a complete circle.
    (e) all of the above.
28. (*Answer*) The standard Mesopotamian calendar of 354 days was based on _____ months, with each month having an average of _____ days. The months were based on _____ .
    (a) 4; 88.5; the average lengths of the seasons
    (b) 5.9; 60; the base-60 system of mathematics
    (c) 10; 35.4; the average number of days the Sun spends in each constellation of the zodiac
    (d) 12; 29.5; the phase cycle of the Moon
29. The zodiac, on which astrology is based, contains 12 constellations, rather than the thirteen that actually have the ecliptic passing through them. This is because
    (a) the Babylonians divided the zodiac up into 12 zones of 30° each, in a fashion similar to the ideal calendar of 12 months containing 30 days each.
    (b) there are exactly 12 cycles of lunar phases in the time it takes the Sun to complete one circle around the ecliptic.
    (c) 12 was a sacred number to the Sumerians.
    (d) a regular dozen is better than a baker's dozen.
30. (*Answer*) Alexander the Great
    (a) lived in the fourth century BCE.
    (b) was a student of Aristotle's.

    (c) unified through conquest the region of the world from the Mediterranean to India; one consequence of which was facilitating communication between intellectual centers.
    (d) established or otherwise renamed 20 cities, all of which were named Alexandria.
    (e) all of the above.
31. The "father of science" is often considered to be
    (a) Thales of Miletus.   (b) Pythagoras of Samos.
    (c) Plato.   (d) Aristotle.
    (e) Ptolemy.   (f) Hipparchus of Nicaea.
32. (*Answer*) _____ is usually credited with proposing that all celestial motions must be explained by being Earth-centered, perfectly circular, and moving uniformly.
    (a) Thales of Miletus   (b) Pythagoras of Samos
    (c) Plato   (d) Aristotle
    (e) Ptolemy   (f) Hipparchus of Nicaea
33. _____ believed that the outcome of motion was to stop when the natural level of the material was reached.
    (a) Thales of Miletus   (b) Pythagoras of Samos
    (c) Plato   (d) Aristotle
    (e) Ptolemy   (f) Hipparchus of Nicaea
34. (*Answer*) _____, which was discovered by Hipparchus of Nicaea, means that _____ is very slowly moving around the ecliptic roughly once every 26,000 years.
    (a) Precession, the north celestial pole
    (b) The lunar phase cycle, the Moon
    (c) Precession, the vernal equinox
    (d) The north celestial pole, the winter solstice
35. _____ is recognized for his work, _____, which became the definitive work on astronomy until the sixteenth century, CE.
    (a) Pythagoras of Samos, *The Music of the Spheres*
    (b) Plato, *Timaeus*
    (c) Aristotle, *Physics*
    (d) Hipparchus of Nicaea, *Trigonometry*
    (e) Ptolemy, *The Almagest*
36. (*Answer*) _____ argued that _____ occupied the center of the universe.
    (a) Pythagoras of Samos, numbers
    (b) Plato, the ether
    (c) Aristotle, Sun
    (d) Aristarchus of Samos, the Sun
37. At the equator, _____ are circumpolar.
    (a) all stars
    (b) no stars
    (c) no stars in the celestial sphere's western hemisphere
    (d) all stars in the celestial sphere's northern hemisphere
    (e) none of the above
38. (*Answer*) At the North Pole, _____ are circumpolar.
    (a) all stars
    (b) no stars
    (c) no stars in the celestial sphere's western hemisphere
    (d) all stars in the celestial sphere's northern hemisphere
    (e) none of the above
39. For an observer at 30° N, Polaris will be close to _____ above the observer's _____ horizon.
    (a) 30°, northern   (b) 60°, southern
    (c) 70°, northern   (d) 90°, southern

40. (*Answer*) On the first day of summer for an observer at a latitude of 50° N, how high will the Sun get above the southern horizon at local noon?
    (a) 73 1/2°   (b) 63 1/2°   (c) 50°   (d) 40°   (e) 16 1/2°

41. On the first day of fall for an observer at a latitude of 50° N, how high will the Sun get above the southern horizon at local noon?
    (a) 73 1/2°   (b) 63 1/2°   (c) 50°   (d) 40°   (e) 16 1/2°

42. (*Answer*) On the first day of winter for an observer at a latitude of 50° N, how high will the Sun get above the southern horizon at local noon?
    (a) 73 1/2°   (b) 63 1/2°   (c) 50°   (d) 40°   (e) 16 1/2°

43. The equatorial coordinates of the center of the Sun's disk on the first day of spring are _____ and _____ for its right ascension and declination, respectively.
    (a) $0^h, 0°$ (b) $6^h, +23\ 1/2°$N
    (c) $12^h, 0°$ (d) $18^h, -23\ 1/2°$N

44. (*Answer*) The equatorial coordinates of the winter solstice are _____ and _____ for its right ascension and declination, respectively.
    (a) $0^h, 0°$ (b) $6^h, +23\ 1/2°$N
    (c) $12^h, 0°$ (d) $18^h, -23\ 1/2°$N

Short Answer

45. Explain why many civilizations first developed calendars based on the phases of the Moon rather than the length of the year.

46. The Babylonian and Greek sexagesimal number system can be written in a concise form, in the same way that the Mesoamerican base-20 system can. For example, the concise form of the Babylonian number (2.58.45) is equal to $2 \times 60^2 + 58 \times 60^1 + 45 \times 10^0 = 2 \times 3600 + 58 \times 60 + 45 \times 1 = 107{,}250$ in our more familiar base-10 system.
    (a) What is the base-10 value of (26.5.52.59)?
    (b) Write the base-10 number, 360, in the sexagesimal number system.
    (c) Calculate the sexagesimal sum of (42.5.16.59) and (3.40.1), expressing your answer in the concise notation. (Hint: The numbers can be added from right to left, carrying forward a "1" just like adding numbers in base-10, except that the maximum value of each place is 59. For example, (3.57) + (2.8) = (6.5). Since 57 + 8 in base-10 is 65, it is necessary to carry a "1" that is worth 60.)
    (d) Convert both of the numbers in the sum of part (c) to base-10 digits and compute the sum.
    (e) Verify that your results in part (c) and part (d) agree.

47. The latitude of the Great Pyramid of Giza, Egypt is very nearly 30° N. How high does the Sun get in the southern sky at Giza during the first day of summer?

48. Explain why everyone standing on the surface of Earth has their own personal horizon.

49. Redraw the diagram in Fig. 3.24 (*Right*), except show the locations of the vernal equinox, summer solstice, autumnal equinox, winter solstice, and ecliptic six hours later.

50. (a) Construct a diagram similar to those in Fig. 3.24 (*Right*) for an observer located on the arctic circle. Be sure to label the observer's zenith and horizon, as well as the locations of the north and south celestial poles, the celestial equator, the ecliptic, and the angle between the observer's northern horizon and the north celestial pole. Also identify the summer solstice and the winter solstice.
    (b) Sketch and describe the path of the Sun in the sky on the first day of summer. Where is the Sun at its lowest point in the sky on that day, and at what time of day does that happen?
    (c) Sketch and describe the path of the Sun in the sky on the first day of winter. Where is the Sun at its highest point in the sky on that day, and at what time of day does that happen?

51. (a) Construct a diagram similar to those in Fig. 3.24 (*Right*) for an observer located on the tropic of Cancer. Be sure to label the observer's zenith and horizon, as well as the locations of the north and south celestial poles, the celestial equator, the ecliptic, and the angle between the observer's northern horizon and the north celestial pole. Also identify the summer solstice and the winter solstice.
    (b) Sketch and describe the path of the Sun in the sky on the first day of summer. Where is the Sun at its highest point in the sky on that day?
    (c) Based on your answer to part (b), how can the tropic of Cancer be defined?

52. What is your current latitude and longitude? You may find the GPS on your smartphone, LatLong.net, Google Earth™, or some other online tool helpful.

53. During the discussion on page 80 regarding the height of the Sun above the southern horizon for an observer in the northern hemisphere on the first day of summer, the restriction was made that the observer must be on or north of the tropic of Cancer. Explain what is unique about the situation when the observer is between the tropic of Cancer and the equator. What happens to the position of the Sun at noon at those latitudes?

54. It is the first day of fall in the northern hemisphere and an astronomer at a latitude of 30° N wants to study objects near her zenith at midnight.
    (a) Approximately what right ascension should her observing targets have?
    (b) Is her answer to part (a) true if she was in the southern hemisphere at a latitude of 30° S on that same day?
    (c) Would she see the same objects high in the sky at both latitudes (30° N and 30° S)? Why or why not?

55. In the previous exercise, why was it not necessary to specify the observer's longitude?

56. (a) What is the range in declination for stars that will be visible to you if you are at a latitude of 30° N?
    (b) Are there stars that you have no hope of observing at 30° N? If so, what equatorial coordinates do they correspond to?

57. The Gemini Observatory is a pair of telescopes, one on Mauna Kea in Hawai'i, and the other on Cerro Pachón, a mountain in the Chilean Andes. Why would astronomers build telescopes in both places?

58. Your author states on page 71 that the Pythagoreans' proposal of the "counter-Earth," Antichthon, is not consistent with "a testable, falsifiable, scientific theory." Explain why that was the case at the time when the counter-Earth was proposed.

59. (a) Explain why Eudoxus's model of geocentric spheres could not possibly be correct despite being able to explain (at least to some extent) the retrograde motions of the planets. What could it not explain?

(b) What aspect(s) of a modern scientific theory make Eudoxus's model fundamentally unscientific? Explain. (You may want to review the characteristics of a scientific theory in Chapter 1.)

60. Although Aristotle is sometimes considered the first true scientist, suggest a counter-argument to that claim based on his model of celestial motion.

61. Compare what you learned about Mesoamerican astronomy to the astronomy of the Babylonians. Can either or both be considered as truly scientific? Why or why not?

62. Minneapolis, Minnesota has a latitude of 45° 00′ N and Des Moines, Iowa has a latitude of 41° 35′ N. Minneapolis is almost exactly 370 km (234 mi) straight north of Des Moines.
    (a) Calculate the number of degrees that Minneapolis is north of Des Moines.
    (b) Estimate the circumference of Earth in both kilometers and miles by thinking about a full circle that goes around Earth and includes (very nearly) the North Pole, the two cities, and the South Pole. Consider the fraction of the full 360° circle that corresponds to your answer in part (a) and compare it to the fraction of the complete distance around the circle.

    You have essentially replicated the approach used by Eratosthenes of Cyrene around the start of the second century BCE to determine the circumference of Earth.

63. A popular American television show (2009–2016) was based in the Hamptons on Long Island, north of New York City. Toward the end of the series the characters would enjoy the beauty of the sunset while reclining on their porch overlooking the Atlantic ocean. What was wrong with those scenes?

Activities

64. For this exercise you will want to use an online search engine to locate a website that provides information about times of sunrise and sunset at your location, as well as dates and times of the equinoxes and solstices.
    (a) List the days and times of the upcoming solar crossings of the (i) winter solstice, (ii) vernal equinox, (iii) summer solstice, and (iv) autumnal equinox. Be sure to cover an entire year; in other words, if it is currently spring 2022, the next event will be the summer solstice 2022 so you will also want to include summer solstice 2023.
    (b) What are the rising and setting times of the Sun on the days of the solstices and equinoxes you identified in part (a)?
    (c) Based on your results in part (b), determine how much time the Sun spends above the horizon on each of the solstice and equinox days.
    (d) Determine the amount of increase or decrease (the difference) in the time the Sun is above the horizon at your location between
        i. winter solstice and the next vernal equinox,
        ii. vernal equinox and the next summer solstice,
        iii. summer solstice and the next autumnal equinox,
        iv. autumnal equinox and the next winter solstice.
    (e) How do the four sets of sunlight periods compare? Are they nearly equal? Do one or more stand out as being noticeably longer or shorter? A complete theory of planetary motion would need to account for any discrepancies that exist.

65. For this exercise you will want to use an online search engine to locate a website that provides information about times of sunrise and sunset at your location. A labeled sheet of graph paper can be downloaded for this problem from the `Chapter 3` resources webpage. A review tutorial on graphing is also available in `Math_Review`.
    (a) Create a table with columns for:

Date	Rise time	Set time	Amt daylight

    for every Sunday over one calendar year. If some of the times are in daylight savings time, be sure to convert them to standard time so you are comparing "apples to apples."
    (b) Using the downloaded graph paper, plot your data with the amount of daylight on the vertical axis and date on the horizontal axis, and draw a smooth curve through the data points.
    (c) On which dates do the lengths of the days change most rapidly? These dates will be when the curve is steepest, either increasing lengths of days or decreasing lengths of days.
    (d) On your graph, mark the dates of the next winter solstice, vernal equinox, summer solstice, and autumnal equinox.
    (e) Are the equinox and solstice days located at any special places on your graph? If so, describe what is happening on your graph at or very near those points.
    (f) Summarize what you have learned about the changing lengths of days throughout the year based on the results of this exercise.

66. (*You will want to start this activity early enough so that you can make your first measurement as the Sun crosses your local meridian.*)
    In this activity you will use the technique that Thales of Miletus used for determining the heights of pyramids (page 59) to make measurements of the heights of two objects; you and an inanimate object (such as a tree, a telephone pole, a scoreboard, a building, a pyramid, or something else that will cast a clear, uninterrupted, shadow). This exercise will need to be done outdoors on a clear day with a clear line of sight to the path of the Sun (but never look directly at the Sun!). You will also need to be able to see your entire shadow and the shadow of the other object during this exercise. You may also want to team up with one or two another students in your class (unless you can be in multiple places at once). Be sure to read the entire exercise before starting so you know when and where you need to make measurements.
    (a) Record your latitude on Earth and the date when you do this exercise. If you don't know your latitude there are a myriad of tools that you can use, such as LatLong.net or the GPS on your smartphone.
    (b) Find a straight stick that is at least 60 cm long, perhaps a meter stick, a yard stick, or a broom handle and set it in the ground vertically. Record the height of the stick *that is above ground*.
    (c) Draw a circle on the ground that is centered on the stick and that has a radius equal to the height of the stick above the ground. As the Sun moves across the sky the length of the shadow cast by the stick will change.
        i. Record the time of day and the length of the shadow cast by the stick at the moment when the shadow is at its *shortest* (this will be sometime near, but not

necessary exactly at, noon at your location). The Sun is crossing your local meridian at this moment.

ii. Is the end of the shadow inside or outside of the circle you drew on the ground? In other words, is the shadow shorter or longer than the height of your stick above the ground?

iii. Carefully sketch this situation by drawing a line on a sheet of paper with a scaled length corresponding to the length of the stick. For every 5 cm of actual length, draw a length of 1 cm on your paper. If your stick is 60 cm above the ground, the length of the stick on the paper should be $60/5 = 12$ cm high.

iv. Sketch the shadow from the base of the stick on the paper so that the shadow's length is perpendicular to the sketch of your stick. Be sure to use the same scale for the shadow as you did for the stick.

v. On your sketch, place and label a mark along the line of the shadow that equals the height of the stick. This mark corresponds to the circle that you drew on the ground.

vi. Complete the triangle on your sheet by drawing a dashed line from the end of the shadow to the top of the stick.

vii. Using a protractor, measure the angle that is made by the shadow and the dashed line. This angle corresponds to the angular height of the Sun above the horizon when the shadow is at its shortest.

viii. Will you be able to measure your height and the height of the inanimate object later on in the day by using the same technique that Thales used for the pyramid? Recall that Thales conducted his measurements when the length of his shadow equaled his own height. Is there any period during the year when you would be able to carry out this exercise successfully at your latitude? Why or why not?

If it is impossible to carry out this measurement due to the Sun not rising high enough in the sky at your latitude during the period of time the exercise is assigned, stop here; otherwise continue on to the next part.

(d) As the day goes on, the length of the stick's shadow will grow. At the moment when the length of the stick touches the circle, record the time when this occurs.

(e) Here is where your partner(s) come in handy. Standing straight with your feet together, at the same moment when the shadow of the stick touches the circle, have one of your partners mark the end of your own shadow. How far is your shadow from the point directly below the top of your head? This should be your height.

(f) Again at the moment that the stick's shadow touches the circle, your other partner should mark the position of the end of the shadow of the inanimate object you are determining the height of. Measure the length of the object's shadow from directly below the highest point of the object. This equals the height of the object. What value did you get for the height of the object?

(g) On a sheet of paper, draw a diagram of the stick in the ground and the length of its shadow from the stick using a length of 1 cm on the sheet of paper for every 10 cm of length of the stick and the shadow. Do the same thing for you standing vertically and your shadow with the same 1:10 ratio. Although the lengths aren't identical for the two triangles, the two triangles should have similar shapes (the same set of three interior angles). Indicate the values of the three angles on each triangle (you may use a protractor for this step). Verify that two angles are 45° and the third is 90°. (An aside: Recall from basic geometry that the three angles of any flat triangle, regardless of its shape, always add up to 180°.)

(h) It is because the triangles are similar that this technique works. Explain.

(i) Estimate the error in the measurement of your height as follows: Calculate the difference between the measured value of your height and your actual height (in cm). Divide that difference by your actual height (again in cm). Multiply by 100 to convert from a decimal fraction to a percentage.

The Copernican Revolution

4

God, who founded everything in the world according to the norm of quantity, also has endowed man with a mind which can comprehend these norms.

Johannes Kepler (1571–1630)

4.1	The European Renaissance	92
4.2	The Heliocentric Universe of Nicolaus Copernicus	93
4.3	The Motions of the Stars in the Heliocentric Universe	97
4.4	The Changing Seasons	100
4.5	The Source of Hipparchus's Precession	102
4.6	The Moon's Motions: Phases and Eclipses	104
4.7	Finally, a Natural Solution to the "Wandering Stars"	112
4.8	The Quest for Understanding Continues	115
4.9	Tycho Brahe and his Passion for Precision	119
4.10	Johannes Kepler and Planetary Orbits	121
4.11	Kepler's Three Laws in their Original Form	125
4.12	Galileo Galilei, the First True Scientist	131
	Exercises	141

Fig. 4.1 Copernican planosphere. (Cellarius, 1660;[1] public domain)

---

[1]Cellarius, Andreas (1660). *Harmonia Macrocosmica* (J. Janssonius; Amsterdam).

Introduction

Thus far we have explored the astronomy of some non-western, pre-modern societies in Chapter 2, and we have studied the contributions to our evolving astronomical knowledge provided by the Sumerians, the Babylonians, the ancient Greeks, and Hindu and Islamic astronomers and mathematicians in Chapter 3.

In this chapter we will begin exploring what is usually considered to be the start of a critical shift in how people formulated questions about nature and how it is manifested in our observations and our experiments. Are theories capable of being either verified or falsified? Do they make predictions of future observations or experimental results that haven't yet been performed? Can what we observe be explained without invoking supernatural influences or deities?

These are the questions for which we begin to seek answers throughout the remainder of our intellectual quest to understand the cosmos.

4.1   The European Renaissance

Rediscovering Ancient Greek Philosophy

Fig. 4.2 Leonardo da Vinci's *Vitruvian Man*, circa 1490, in the Gallerie dell' Accademia in Venice, Italy. (Public domain)

As pointed out in Section 3.4, knowledge of the great ancient Greek philosophers largely disappeared in Europe for nearly 1000 years, but began to reappear in the twelfth century. This rediscovery of the ancient Greek texts occurred following the fall of the Islamic medieval kingdom of Toledo (Spain) to Christian conquerors in 1085 and the capture of Arab-held Sicily by the Normans in 1091. The result of these conquests was a mix of Arabic-speaking scholars together with Latin and Greek scholars.

Gerard of Cremona (c. 1114–1187), an Italian, became part of the Toledo School of Translators so that he could translate Ptolemy's *Almagest* from Arabic to Latin, which he completed c. 1175. Unbeknownst to Gerard, the *Almagest* had already been translated from the original Greek to Latin c. 1160 in Sicily, but Gerard's translation turned out to be the most widely used. In all, Gerard translated 87 books in science and mathematics from Arabic to Latin.

During this period the first European universities were forming; the University of Bologna (Italy) in 1088, the University of Paris (France) c. 1150, and the University of Oxford (England) in 1167. Based on the educational model of teaching the trivium (critical thinking through grammar, logic, and rhetoric), followed by the quadrivium (arithmetic, geometry, music, and astronomy), the new universities were eager to become reacquainted with the works of the Greek masters, especially in mathematics and astronomy.

The capture of Constantinople (modern-day Istanbul) during the Fourth Christian Crusade in 1204 set the stage for the eventual overthrow of the Byzantine Empire by the Ottoman Turks in 1453, which in turn resulted in an influx of Byzantine scholars to western Europe, adding impetus to Greek and Roman studies. Along with scientists and mathematicians came poets, artists, writers, musicians, architects, philosophers, politicians, and theologians; the scientific and mathematical renaissance of the twelfth century was expanding into a much broader European Renaissance[2] that would continue into the seventeenth century.

The Renaissance period was rich with innovation, discovery, and new ways of thinking. The history of that period is replete with famous figures, including Leonardo da Vinci (1452–1519; his famous *Vitruviam Man* is shown in Fig. 4.2), Christopher Columbus (1451–1506), Niccolò Machiavelli (1469–1527), Sir Thomas More (1478–1535), Martin Luther (1483–1546), John Calvin (1509–1564), Michelangelo (1475–1564; his sculpture, *David*, is pictured in Fig. 4.3), William Shakespeare (1562–1616), and many others. Johannes Gutenberg's (1398–1468) printing press with movable type, first used in 1439, played a crucial role in the recording and dissemination of knowledge that helped to fuel the Renaissance, the Protestant Reformation, and the scientific revolution to come (a Gutenberg Bible is seen in Fig. 4.4).

Fig. 4.3 Michelangelo's *David* in the Accademic Gallery in Florence, Italy. (Rico Heil, CC BY-SA 3.0)

Fig. 4.4 A Gutenberg Bible (c. 1455) in the New York Public Library. [NYC Wanderer (Kevin Eng), CC BY-SA 2.0]

Within science, the philosophical movement that came out of the Renaissance resulted in a greater emphasis on observation and evidence in reaching plausible, rather than definitive, conclusions. This method of inductive reasoning leaves open the possibility that conclusions based on currently available evidence could be wrong. Recall that in 1492, when Columbus relied on Ptolemy's *Geography* in an attempt to sail west to Asia, the absence of North and South America on the great Greek master's maps suggested that Ptolemy's other works, including the *Almagest*, could be fallible as well.

4.2   The Heliocentric Universe of Nicolaus Copernicus

Who Was Copernicus?

Nicolaus Copernicus (Fig. 4.5), was born in Toruń in the Kingdom of Poland during the European Renaissance, to a merchant father and a mother who was the daughter of a rich merchant. Recalling the list of players from Section 4.1 during the late 15th and early 16th centuries, you will see that Copernicus was a contemporary of many of the great explorers, artists, philosophers, and theologians of that era. It was a time ripe with change.

Copernicus was not a professional astronomer or mathematician, but was in many ways a "Renaissance man" (a polymath) in much the same way that other famous Renaissance players were. Perhaps the classic example of a polymath is Leonardo da Vinci, who was not only a remarkable painter (e.g., *The Last Supper*, *Mona Lisa*), but was also a scientist and inventor.

When his father died in about 1483, Copernicus's uncle, Lucas Watzenrode the Younger (1447–1512), helped guide his education and career. Copernicus matriculated at the Jagiellonian University of Kraków in 1491, where he studied the traditional trivium and quadrivium. Copernicus also had the opportunity to study Aristotelian philosophy, together with arithmetic, geometry, and astronomy (subjects of the quadrivium), and he likely attended numerous special lectures on astronomy as well, although he never actually obtained a degree there.

Fig. 4.5 Nocolaus Copernicus (1473–1543). (Public domain)

---

[2]Renaissance comes from the French word for rebirth.

In 1496, Copernicus was sent to study canon and civil law at the University of Bologna by his uncle (he eventually received his doctorate in law). During his time at Bologna, Copernicus met a famous astronomer, Domenico Maria Novara da Ferrara (1454–1504), and became his assistant for a time. It was Novara who likely introduced Copernicus to the writings of Regiomontanus regarding Ptolemaic astronomy. In 1497, while at Bologna, Copernicus made his first known astronomical observation, that of the passing of the Moon in front of the bright star Aldebaran, verifying some inconsistencies in Ptolemy's model of the Moon's motion. Copernicus also made observations of a partial eclipse of the Moon in 1500 when he was in Rome to give a lecture on astronomy.

Copernicus's uncle, now Prince-Bishop of Warmia (a semi-independent state of the Roman Catholic Church) provided him with a position in the office of the canon at the Warmia chapter in 1501. One of the opportunities of the position was to receive funding for additional education which he took advantage of almost immediately, enrolling in the University of Padua to study medicine for the next two years. Upon his return from Padua, Copernicus served as his uncle's secretary and physician until at least 1510 or perhaps until the Prince-Bishop's death in 1512.

Sometime between 1510 and 1512, Copernicus moved to Frombork, a small town on the Baltic Sea, although he remained employed as the canon lawyer and was at the center of the complex politics of the Prince-Bishopric of Warmia throughout much of the remainder of his life. Copernicus was even nominated to be Bishop at one point, but was not elected. He also continued to practice some medicine, was involved in economics and monetary theory, and even published a bit of poetry. In addition, at his own leisure Copernicus conducted at least half of his 60 recorded astronomical observations from Frombork.

The Heliocentric Model Renewed

Although his occupation was in law, with some work as a physician, Copernicus possessed a strong education in both mathematics and astronomy. He was also well-versed in the philosophies of the ancient Greek masters (Section 3.2). It seems that Copernicus had a particular interest in the mathematical philosophy of the Pythagoreans regarding the primacy of numbers in the physical world. In addition, he was apparently heavily influenced by Plato's view that observations can deceive while the realm of ideas holds the key to truth. Copernicus also shared Plato's belief that only uniform and circular motion must apply to the motions of the heavens. Of course, Copernicus was fully knowledgeable in Aristotelian physics, including the work of Aristarchus of Samos and his heliocentric model of the universe, proposed some 1800 years earlier (Section 3.3).

Copernicus had some reservations about Ptolemy's geocentric models however, especially his use of the equant (Fig. 3.18), much as Ibn al-Shātir (1304–1375) did (Section 3.4). Sometime between 1510 and 1514, Copernicus wrote a draft of his early thoughts about planetary motions and the structure of the universe in a work that would only be published in its entirety in 1878, more than three centuries after his death. In what became called *Nicolai Copernici de hypothesibus motuum coelestium a se constitutis commentariolus* in publication, or *Commentariolus* (*Little Commentary*) for short, Copernicus wrote:

[Ptolemy proposed] certain equant circles, because of which the planet does not appear to move always with a uniform speed, neither on its deferent orb, nor around the proper center [of the universe] ... [Ptolemy's theory] neither sufficiently achieved nor [was] sufficiently in accord with reason ... [I sought to develop] a more rational system of circles, from which all apparent irregularity would result, while everything would move uniformly around its proper center, as the principle of perfect motion requires.

Within *Commentariolus*, Copernicus specified seven postulates from which he would develop his heliocentric model:

Key Points
Copernicus's heliocentric postulates

1. There is no one center of all the celestial spheres (*orbium caelestium*) or spheres (*sphaerae*).
2. The center of the Earth is not the center of the universe, but only the center towards which heavy things move and the center of the lunar sphere.
3. All spheres surround the Sun as though it were in the middle of all of them, and therefore the center of the universe is near the Sun.
4. The ratio of the distance between the Sun and the Earth to the height of the sphere of the fixed stars is so much smaller than the ratio of the semidiameter of the Earth to the distance of the Sun that the distance between the Sun and Earth is imperceptible compared to the great height of the sphere of the fixed stars.
5. Whatever motion appears in the sphere of the fixed stars belongs not to it but to the Earth. Thus the entire Earth along with the nearby elements rotates with a daily motion on its fixed poles while the sphere of the fixed stars remains immovable and the outermost heaven.
6. Whatever motions appear to us to belong to the Sun are not due to [motion] of the Sun but [to the motion] of the Earth and our sphere with which we revolve around the Sun just as any other planet. And thus the Earth is carried by more than one motion.
7. The retrograde and direct motion that appears in the planets belongs not to them but to the [motion] of the Earth. Thus, the motion of the Earth by itself accounts for a considerable number of apparently irregular motions of the heavens.

Initially, Copernicus only shared his *Commentariolus* with a few close friends, not intending to have his thoughts on a model of the universe published. However, his ardent supporter, Georg Joachim (1514–1574), encouraged him to publish his results, as did Copernicus's good friend, the Bishop of Chelmno, Tiedemann Giese (1480–1550). (Joachim changed his name to Rheticus to avoid association with his father, who was beheaded for sorcery.) The result of that encouragement was the historic text, *De revolutionibus orbium coelestium*, published in 1543 (Fig. 4.6).

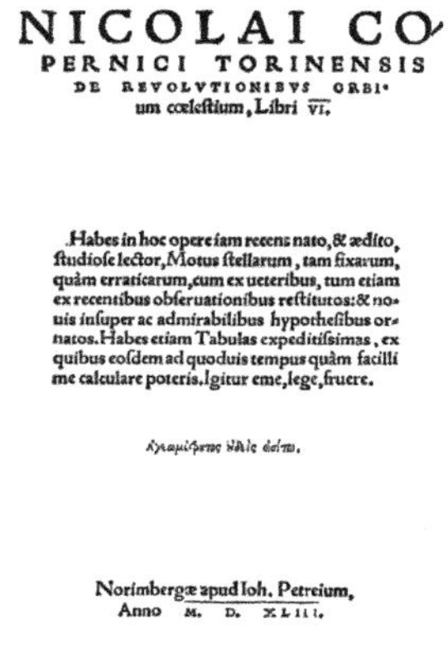

Copernicus included a preface to *De revolutionibus* that was written directly "To His Holiness, Pope Paul III" (1468–1549). A portion of the preface reads:

I have no doubt that acute and learned astronomers will agree with me if, as this discipline especially requires, they are willing to examine and consider, not superficially but thoroughly, what I adduce in this volume in proof of these matters. However, in order that educated and uneducated alike may see that I do not run away from the judgment of anybody at all, I have preferred dedicating my studies to Your Holiness rather than anyone else. For even in this very remote [corner] of the Earth where I live you are considered the highest

NICOLAI CO
PERNICI TORINENSIS
DE REVOLVTIONIBVS ORBI-
um coelestium, Libri vi.

Habes in hoc opere iam recens nato, & ædito, studiose lector, Motus stellarum, tam fixarum, quàm erraticarum, cum ex ueteribus, tum etiam ex recentibus obseruationibus restitutos: & no-uis insuper ac admirabilibus hypothesibus or-natos. Habes etiam Tabulas expeditissimas, ex quibus eosdem ad quoduis tempus quàm facilli me calculare poteris. Igitur eme, lege, fruere.

Ἀγεωμέτρητος ἤδει εἰσίτω.

Norimbergæ apud Ioh. Petreium,
Anno M. D. XLIII.

Fig. 4.6 *De revolutionibus orbium coelestium* by Nicolaus Copernicus, 1543. (Public domain)

authority by virtue of the loftiness of your office and your love for all literature and astronomy too …

Perhaps there will be babblers who claim to be judges of astronomy although completely ignorant of the subject and, badly distorting some passage of Scripture to their purpose, will dare to find fault with my undertaking and censure it. I disregard them even to the extent of despising their criticism as unfounded. For it is not unknown that Lactantius, otherwise an illustrious writer but hardly an astronomer, speaks quite childishly about the Earth's shape, when he mocks those who declared that the Earth has the form of a globe. Hence scholars need not be surprised if any such persons will likewise ridicule me. Astronomy is written for astronomers.

Clearly, Copernicus believed that his proposed heliocentric model of the universe, depicted in Fig. 4.7, was an improvement over the geocentric Ptolemaic model and he was willing to stand by what he had written. Legend has it that Copernicus saw the first printed copy of *De revolutionibus* on the last day of his life, after which he died peacefully.

Copernicus gave oversight of the printing of *De revolutionibus* to Rheticus, who unfortunately could not complete that project. Rheticus in turn handed the project over to Andreas Osiander (1498–1552), a Lutheran theologian who believed only in divine revelation of truth. Osiander added an unauthorized, anonymous preface to the book that reads in part:

Fig. 4.7 The heliocentric universe of Copernicus from *De revolutionibus orbium coelestium*. (Public domain)

There have already been widespread reports about the novel hypotheses of this work, which declares that the Earth moves whereas the Sun is at rest in the center of the universe. Hence certain scholars, I have no doubt, are deeply offended and believe that the liberal arts, which were established long ago on a sound basis, should not be thrown into confusion. But if these men are willing to examine the matter closely, they will find that the author of this work has done nothing blameworthy. For it is the duty of an astronomer to compose the history of the celestial motions through careful and expert study. Then he must conceive and devise the causes of these motions or hypotheses about them. *Since he cannot in any way attain to the true causes* [emphasis added], he will adopt whatever suppositions enable the motions to be computed correctly from the principles of geometry for the future as well as for the past. The present author has performed both these duties excellently. For these hypotheses need not be true nor even probable. On the contrary, if they provide a calculus consistent with the observations, that alone is enough …

So far as hypotheses are concerned, let no one expect anything certain from astronomy, which cannot furnish it, lest he accept as the truth ideas conceived for another purpose, and depart from this study a greater fool than when he entered it.

In his preface, Osiander clearly wants to point out that *De revolutionibus* should only be considered as a tool for computation rather than any statement of reality regarding the fixed centrality of the Sun and the motions of Earth. This is unquestionably inconsistent with Copernicus's own intentions for his work.

De revolutionibus is generally considered to be the first great work of what has become known as the Copernican revolution, the scientific revolution that began in the early sixteenth century and culminated with the work of Sir Isaac Newton (1642–1727) in the eighteenth century. However, given that Copernicus never relinquished Plato's idea that the motions of the heavens must be uniform and circular, some have argued that, rather than being the first scientist of the new scientific revolution, Copernicus could be thought of as the last great astronomer in the long line that began with the ancient Greeks two millennia earlier.

Regardless of where we consider Copernicus to be in the historical development of science, placing the Sun at the center of our Solar System and relegating Earth to the position of a planet in that system:

Key Points
The successes and a failure of the heliocentric universe of Nicolaus Copernicus.

- provided a simple description of the daily motions of the stars (page 16),
- clearly described why there is an annual variation in which constellations are visible at night (page 16),
- explained the annual northern and southern motions of the Sun and the variations in the seasons (page 17),
- naturally resulted in the order and relative sizes of the planetary orbits,
- tremendously simplified the description of retrograde motion that was illustrated in Fig. 2.12,
- resulted in easier calculations of planetary positions, and
- allowed for a straight-forward model of precession (pages 68 and 86).

On the other hand, given his inability to imagine a universe that did not conform to Plato's uniform and circular motion ultimately meant that the heliocentric model of Copernicus required the philosophically unsatisfying addition of epicycles. The scientific revolution (or perhaps we can argue, the *evolution* of science) had started but was far from complete, and as we will see throughout this text, it is certainly not complete today. But the opportunity to contribute to humanity's long quest for understanding is what makes science so exciting and rewarding.

4.3   The Motions of the Stars in the Heliocentric Universe

The Daily Motions of the Stars

As described in the fourth postulate of his *Commentariolus*, Copernicus envisioned a universe where the distance to the celestial sphere was immense in relation to the distance between the Sun and Earth. Previously, some ancient Greek philosophers had made the same argument, but struggled to understand how an enormous celestial sphere could possibly spin at the incomprehensible speed required to complete a rotation in a single day (recall Fig. 3.21). If the celestial sphere was just 1000 times larger in diameter than Earth's orbit, for example, the sphere would need to move at nearly 11,000 km/s (24,000,000 mi/h) at its equator. The solution proposed by Copernicus was simple, the celestial sphere doesn't move at all (Postulate 5); instead, Earth rotates around an axis that passes through the North Pole, the center of the planet, and the South Pole (Fig. 4.8). This is completely analogous to you standing in the middle of a room and spinning. In doing so, you see different parts of the room go by as you rotate. You would see the same thing if you stood still and the room rotated around you in the opposite direction; but that hardly makes sense now, does it?

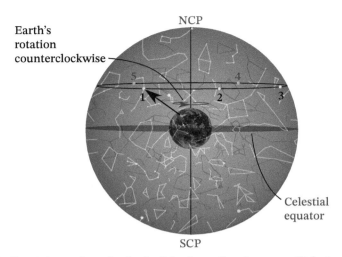

Fig. 4.8 As Earth rotates under a *fixed* celestial sphere, the observer will first see star 1 near her zenith, followed by star 2, then stars 3, 4, and 5. The view would be identical if the celestial sphere rotated in the opposite direction around a fixed Earth; see Fig. 3.21. (Image from NASA)

The Annual Changes in the Night Sky

On page 16 we discussed the fact that the constellations visible in the night sky change continually throughout the year. With Earth orbiting a fixed Sun, this well-known phenomenon is the simple consequence of us looking in different directions toward the celestial sphere during the course of our annual trip around the Sun, as depicted in Fig. 4.9. Since the Sun is visible during the daytime, if you look in the direction opposite to that of the Sun you see the stars and constellations that are visible at night. For example, at the beginning of August, the zodiac constellation of Capricornus is close to your meridian at midnight. On the same day, if you could see the stars while the Sun is up, you would see that the Sun *appears* to be located in the constellation of Cancer. Six months later, after completing one-half of our orbit, the situation is reversed; Cancer is near your meridian at midnight while the Sun lies in the direction of the constellation of Capricornus. The ecliptic path of the Sun through the zodiac is just the result of us looking at the Sun from different locations in our orbit. In reality, the ecliptic is the plane of Earth's orbit. (By the way, when you celebrate a birthday, anniversary, or the New Year, you are essentially celebrating the fact that you completed another trip around the Sun since the last such celebration.)

Key Point
The ecliptic is the plane of Earth's orbit. It is also the apparent path of the Sun among the background stars.

When we described the annual motion of the Sun along the ecliptic from the viewpoint of a geocentric universe on page 78 it was pointed out that the Sun slips backward nearly four minutes every day relative to the stars that are fixed on the celestial sphere. This apparent motion is explained in the Copernican model by combining the daily rotation of Earth on its own axis and Earth's annual orbital motion around the Sun, both of which are counterclockwise when viewed from above Earth's North Pole. Referring to Fig. 4.10, since there are 360° in a complete circle, in order for Earth to complete one revolution around the Sun in one year (365 1/4 days) it must move nearly 1° around its orbit each day. At the start of a

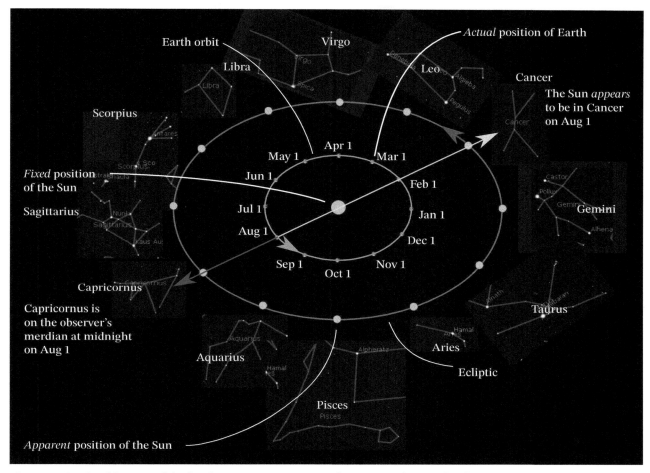

Fig. 4.9 As Earth orbits around the fixed Sun, our vantage point changes, resulting in the changing constellations throughout the year. (Don't forget that the celestial sphere is enormously large relative to Earth's orbit, unlike the depiction here.) The apparent oblong shapes of Earth's orbit and the ecliptic in the diagram are due to a viewpoint above and outside of Earth's orbit, rather than looking straight down on it. Earth's October 1 position is out of the page and its April 1 position is behind the page. (Constellations from *Stellarium* screenshots, GNU GPLv2; images from NASA)

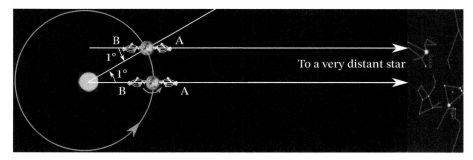

Fig. 4.10 Earth rotates 360° to bring the stars overhead two nights in a row (sidereal day), but Earth must rotate nearly 361° to bring the Sun back to the meridian two days in a row (solar day). One sidereal day is approximately 23^{h}56^m long, while one solar day is 24^h long. The angle is exaggerated to help visualize the effect. (Constellations from *Stellarium* screenshots, GNU GPLv2; images from NASA)

single day's motion note that the Sun is on the meridian for observer B (local noon), while on the opposite side of the planet, a specific distant star lies on observer A's meridian. One day later, because the stars are extremely far from Earth, after Earth has rotated on its axis 360° and it has moved nearly 1° in its orbit, observer A again has the same star on his meridian (the two lines in Fig. 4.10 on either side of the text, "to a very distant star," are essentially parallel and the spacing between the lines relative to the distance to the star is very small). However, the Sun is not yet on observer B's meridian after a 360° rotation of Earth. In order for the Sun to return to the meridian it is necessary for Earth to rotate an additional 1°. In other words, in order to bring the same star directly overhead on two consecutive days Earth must rotate 360°, but in order to bring the Sun back to the observer's meridian on two consecutive days Earth must rotate almost 361°. Rotating that extra (nearly) 1° requires almost 4 minutes. This is the difference between a sidereal day (star-based time) and a solar day (based on successive passages of the Sun across the observer's meridian).

4.4 The Changing Seasons

In Section 4.3 we discussed two of Earth's motions within the fixed celestial sphere, the rotation of Earth about its own north–south axis and the orbit of Earth around a fixed Sun. However, neither of these by itself explains the changes in the seasons, or the Sun's apparent annual excursions north and south of the celestial equator. Those observational effects were first discussed beginning on page 17.

Key to each of these effects is the ecliptic's 23 1/2° tilt relative to the celestial equator (Fig. 3.24). Recall that this tilt is directly related to the latitudes of the tropics of Cancer and Capricorn, as well as the arctic and antarctic circles.

In the heliocentric model of the universe, the 23 1/2° tilt can also be thought of as being between Earth's rotation axis and the direction perpendicular to the ecliptic (see Fig. 4.11). It is important to note that the direction that Earth's axis points remains very nearly constant throughout its orbit around the Sun. In the north the axis always points toward the north celestial pole (NCP), and toward the south celestial pole (SCP) in the south. Even though the orbit causes the center of Earth to move, and therefore the axis itself moves as well, Earth's orbit is tiny in comparison to the distance to Polaris or any other star. As a result, it is impossible to detect the orbital motion of the axis with the naked eye over the course of one year. This is just the point Copernicus made with his fourth postulate in the *Commentariolus*. This is the same argument that was made about the very nearly parallel sight lines in Fig. 4.10.

Because Earth's axis tilt is almost[3] constant throughout its orbit, in a sense Earth's northern hemisphere "leans" toward the Sun during a portion of the orbit, and "leans" away from the Sun later in the orbit. It is on June 21, when the center of the Sun's apparent position on the celestial sphere crosses the summer solstice and the Sun is as far north of the celestial equator as it ever gets, that the northern hemisphere is tilted most directly toward the Sun. Figure 4.12 shows an enlarged view of Earth from Fig. 4.11 with key latitudes labeled when Earth is at its June 21 position. Given that the Sun is very much larger than Earth and far from us, the rays of the Sun are traveling parallel to one another when they strike Earth.

---

[3]The "almost" caveat is a consequence of the 26,000 year precession period.

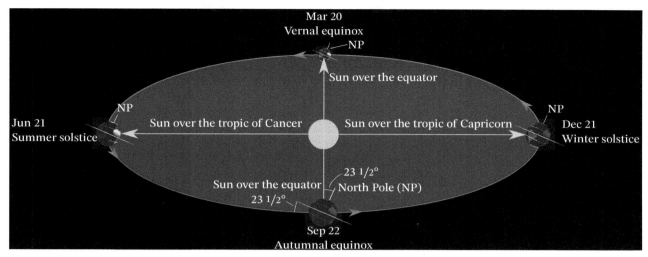

Fig. 4.11 Earth's rotation axis is tilted 23 1/2° with respect to the axis of its orbit around the Sun, and Earth's equatorial plane (yellow) is tilted 23 1/2° with respect to the its orbital plane (blue). As a result, the northern hemisphere leans most toward the Sun on about June 21 (when the center of the Sun is at the summer solstice) and farthest from the Sun on about December 21 (winter solstice). The Sun is directly above Earth's equator on about March 20 (vernal equinox) and September 22 (autumnal equinox). The orbit appears elongated because the observer is above and outside of Earth's orbit. Earth's March 20 location is behind the page, the September 22 position is in front of the page, and the June 21 and December 21 positions are in the plane of the page. (Images of Earth and the Sun courtesy of NASA)

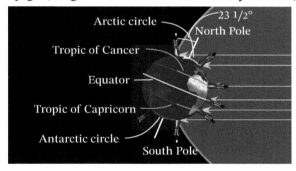

Fig. 4.12 An enlarged image of the June 21 position of Earth from Fig. 4.11. (Image of Earth courtesy of NASA's Scientific Visualization Studio)

Referring again to Fig. 4.12, for an observer on the arctic circle (66 1/2° N) on June 21, the Sun will be low in the southern sky at noon. Twelve hours later (at midnight) the observer on the arctic circle will be standing along the line that is perpendicular to the ecliptic (on the "top" of Earth in the diagram). As a result, she would need to look toward her northern horizon (toward the North Pole) to see the Sun's light just grazing Earth's surface; recall the 360° panorama of the Sun over a 24-hour period shown in Fig. 2.9. For an observer on the tropic of Cancer (23 1/2° N) on June 21, the rays from the Sun come straight down, meaning that the Sun is at the observer's zenith at noon. Viewing the Sun from the equator on that day would require that the observer look northward. For an observer on the tropic of Capricorn (23 1/2° S), the Sun would be even farther north. On the antarctic circle (66 1/2° S), where the observer would be standing on the "bottom" of Earth, the Sun's rays would just graze the surface looking north at noon. Additionally, on the

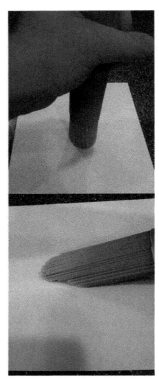

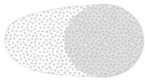

Fig. 4.13 Spaghetti in a paper tube provides an analogy for the intensity of light striking Earth's surface from (*top*) straight above the observer (the Sun at the observer's zenith) and at an angle (*center*). The drawing (*bottom*) comes from overlaying the patterns made in the two cases.

antarctic circle the Sun would never rise above the horizon because Earth's rotation would carry the observer into shadow and facing away from the Sun throughout the rest of the day.

On about December 21 (the winter solstice), the situation is the reverse of that shown in Fig. 4.12. On that date, the tilt of the axis is still the same, but instead of the rays coming from the right, they would now be coming from the left. The result is that it would now be the tropic of Capricorn that would get the sunlight coming straight down at noon. On the equator the observer would need to look south to see the Sun, while on the antarctic circle viewing the Sun over a 24-hour period would be very similar to Fig. 2.9 except that it would be highest in the sky in the north and graze the horizon in the south. Meanwhile, on the arctic circle the Sun would barely make an appearance at noon and then sink below the horizon.

Half way between the June 21 and December 21 positions of Earth in its orbit (on about September 22, the autumnal equinox) the Sun is directly over the equator at noon, moving farther south on each successive day. Six months later (approximately March 20, the vernal equinox), the Sun is again at the observer's zenith at noon on the equator, moving northward on succeeding days.

At this point it would be useful for you to compare what you see from Earth as illustrated in Fig. 2.7 with the geocentric model shown in Fig. 3.24 and the motions associated with the Copernican heliocentric universe just discussed.

So how is all of this related to the summer warmth and winter cold? The answer lies in the direction of the Sun's rays striking the ground. Figure 4.13 provides a useful analogy. Imagine a bundle of uncooked spaghetti noodles wrapped in a toilet paper tube. When the bundle is held vertically, perpendicular to a sheet of paper, the spaghetti makes a tight disk on the paper. However, when the bundle is tilted the spaghetti spreads out over the paper, covering a larger, more elongated, area. Think of each individual noodle as a single ray of sunlight. If the Sun is directly overhead (at the observer's zenith) the "bundle" of light rays forms a tight disk, but if the sunlight is arriving at an angle the same amount of light spreads out over a larger surface area. The result is that sunlight becomes progressively less intense as it strikes the ground at shallower angles. This also means that the energy the light delivers becomes more spread out, causing less warming of the ground. On June 21, when the northern hemisphere is "leaning" toward the Sun, the summer is warmer, but six months later, when the northern hemisphere is "leaning" away from the Sun, the winter is colder. The situation is reversed in the southern hemisphere. Contrary to a common misconception, the northern hemisphere's summer heating is not due to Earth being closer to the Sun in June; in fact, Earth is slighter farther from the Sun during the northern hemisphere summer than it is during the northern hemisphere winter.

4.5   The Source of Hipparchus's Precession

Beginning on page 86 we discussed the subtle effects of precession, first discovered by Hipparchus. This nearly 26,000-year cycle is due to the very slow shift in the direction of Earth's rotation axis, which is completely analogous to the familiar pivot of a toy top that leans over as it rotates on its own axis [see Fig. 4.14 (*Left*)]. The cause of the precession of a top is an interaction between its rotation and the gravitational force exerted by Earth on the top. In the same way, the precession of Earth's

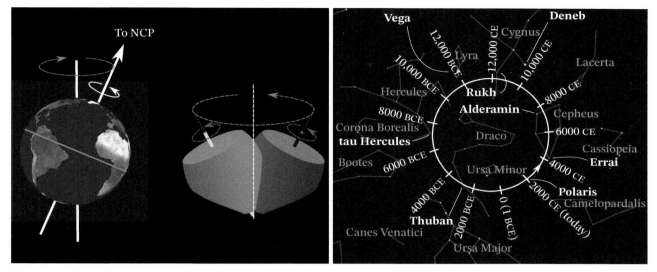

Fig. 4.14 *Left:* Earth precesses on its axis with a period of 26,000 years, similar to how a toy top precesses. (Image of Earth courtesy of NASA) *Right:* As a result of the precession, the location of the north celestial pole (NCP) moves through the heavens in a circle with a radius of 23 1/2°. Polaris is currently near the NCP, but will be 47° from the NCP in about 13,000 years, at which time the NCP will be very near the boundary between the constellations of Lyra and Hercules. (Star field and constellation background screenshot from *Stellarium*, GNU GPLv2)

axis is an interaction between the rotation of Earth and the gravitational forces from our Moon and the Sun (together with all other gravitational forces in the universe).

As Earth's axis slowly pivots in a circle, the location of the NCP among the stars changes as well. This is because the NCP is just the point in the sky directly over Earth's North Pole (the SCP moves in a similar fashion). Today the NCP is very close to Polaris [Fig. 4.14 (*Right*)], but 2000 years from now it will be fairly close to Errai in the constellation of Cepheus, and in 14,000 CE the NCP will be in the area of the star Vega, in Lyra. You may recall the discussion of the alignment of the "air shafts" in the Great Pyramid at Giza (Fig. 2.36) when it was mentioned that one of them was pointed toward Thuban, in the constellation of Draco; Thuban was effectively the North Star at the time the pyramids were built (the NCP was very close to Thuban in 2800 BCE). Because Earth's axis tilt is 23 1/2°, the diameter of the circle that the NCP makes among the stars is twice that, or 47°.

The NCP and SCP are not the only important locations in the sky that are affected by precession. Since the celestial equator is simply the extension of the plane of Earth's equator into the heavens, as Earth's rotation axis pivots, so does the celestial equator. As a result, the two intersections of the celestial equator with the ecliptic (the vernal equinox and the autumnal equinox) also shift systematically with a 26,000 year period. With the pivoting of Earth's rotation axis, the vernal equinox, the autumnal equinox, the winter solstice, and the summer solstice) move through the stars.

4.6   The Moon's Motions: Phases and Eclipses

The Moon's Phases

If you refer back to Copernicus's diagram from *De revolutionibus* (Fig. 4.7) you will see that Earth [the third orbit out from the Sun (Sol)] is shown with the Moon in orbit around it. Although the ancient Ptolemaic model did indicate the Moon going around Earth, that geocentric model required an arbitrary coupling of the Sun with the Earth–Moon system in order to produce the appropriate phases of the Moon. The Copernican model requires no such ad hoc coupling.

Figure 4.15 shows the true relative sizes of Earth, the Moon, and their orbital separation. Given the Moon's size and distance from Earth, it only subtends 1/2° of sky, or just 1/720 of a circle.

Figure 4.16 illustrates the Moon's phases at eight different locations in its orbit around Earth (not to scale), with the Sun off the left side of the diagram. The images that are labeled in Fig. 4.16 show what an observer on Earth would see when the Moon is at each position in its orbit. You should note that when looking down on Earth's North Pole and the Moon's orbit, both the rotation of Earth and the motion of the Moon in its orbit are in the counterclockwise direction, as is the rotation of the Moon about its own axis. As we discussed on page 20, the sequence of phases begins with the new moon, followed by waxing crescent, first quarter, waxing gibbous, full moon, waning gibbous, third quarter, waning crescent, and then back to new moon. This phase cycle (the synodic month) takes about 29 1/2 days, and is the same period we discussed repeatedly with regard to pre-modern societies in Section 2.2.

If you study Fig. 4.16 carefully, you will immediately see that the diagrams of the Moon displayed on its orbital path always have one hemisphere lit, the one directed toward the Sun. However, the only time we can see that entire hemisphere is when the Moon's phase is full. When the phase of the Moon is new, we can't see any of the Moon's lit hemisphere. In reality, the Moon's dark hemisphere does receive some sunlight that is reflected from Earth, but we still can't see it despite the fact that the new moon is high in the sky at local noon. This is because the reflected "earthshine" is overwhelmed by the brightness of daylight.

Starting with new moon, as the Moon moves in its orbit the portion of the lit hemisphere that we can see constantly increases and is said to be waxing. Initially, just a sliver of the lit hemisphere is visible (waxing crescent), building up to our being able to see one-half of the lit hemisphere, or one-quarter of the entire surface (first quarter). Beyond first quarter, an increasing percentage of the lit hemisphere becomes visible (waxing gibbous) until full moon occurs. During the second half of the Moon's orbit, progressively less and less of the lit hemisphere is visible and the phases are said to be waning, moving through waning gibbous, to third quarter (once again one-quarter of the entire surface is visible to us), and waning crescent, until the entire lit side is once again directed away from Earth during new moon.

Key Points
The Moon is slightly larger than 1/4 of the diameter of Earth, and the Moon is about 60 Earth radii distant from us, meaning that the angular size of the Moon is about 1/2°.

A NASA simulation of lunar phases is available on the Chapter resources page.

Fig. 4.15 The relative sizes of Earth, the Moon, and their orbital separation. The tiny angle that the Moon subtends as seen from the surface of Earth is 1/2°. (Images courtesy of NASA and NASA's Scientific Visualization Studio)

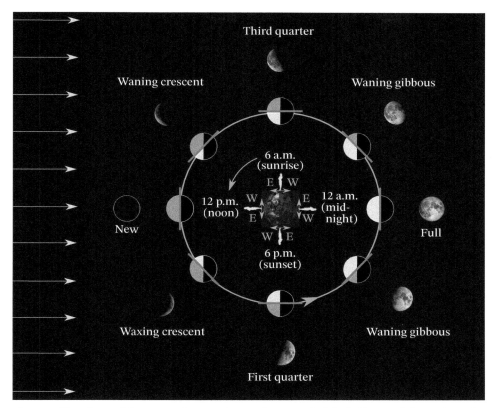

Fig. 4.16 The phase of the Moon that we observe from Earth (labeled images) depends on the location of the Moon in its orbit relative to the position of the Sun (sunlight is coming from the left). Where we see the Moon in the sky depends not only on the location of the Moon in its orbit, but also on the local time of the observer. E and W represent east and west directions, respectively. The straight red lines through the drawings of the Moon show the boundaries of what can be seen from Earth. The side of the Moon that is illuminated by the Sun is shown in white and light gray. The light gray portion is not visible from Earth. Note that when looking down on Earth's North Pole, Earth's rotation and the Moon's orbit and rotation are counterclockwise. (Images courtesy of NASA and NASA's Scientific Visualization Studio)

During the waxing half of the Moon's orbit it may be confusing to see the lit hemispheres in the sketches directed toward the Sun, but the images seem to have their lit sides turned away from the Sun. The reason for the discrepancy is simply the direction that we are looking in the sky; in Fig. 4.16, east is to the left at 6 a.m. while west is to the right, but at 6 p.m. the directions are reversed.

In order to make sense of the directions, think about where the Sun is, relative to the observer and which direction Earth is rotating. The observer's horizon is tangent to the surface of Earth directly under her feet. At about 6 a.m. at the equator the Sun is rising, meaning that the observer is being carried around by Earth toward the Sun. At noon the Sun is directly overhead, at 6 p.m. the observer is being carried away from the Sun and at midnight the Sun is on the other side of Earth, essentially below her feet. To establish the directions along the horizon, note that at 6 a.m. the eastern direction is pointed toward the rising Sun, while at 6 p.m. the western horizon is pointed toward the setting Sun. By carefully studying Fig. 4.16 you will see that the directions associated with the observer's horizon rotate around Earth along with the observer.

Example 4.1

Using Fig. 4.16 it is possible to determine the approximate time of day or night by observing the Moon, assuming of course that it is visible. In general, the Moon can be seen in the sky during daylight as long as it is not too close to new moon (such as a very thin waxing or waning crescent).

One obvious example is when you see a full moon near your meridian. Since the Moon must be on the opposite side of Earth from the Sun, in that case the local time must be approximately midnight.

However, suppose that you see the full moon on your eastern horizon. Given that it is on your eastern horizon the Moon must be rising, but a full moon is *always* on the other side of Earth from the Sun, which means that the Sun must be setting, implying a local time of about 6 p.m..

As another example, suppose that you see a waxing crescent moon near your eastern horizon. What would your approximate local time be? On initial inspection there may seem to be two options here: the observer is seeing the Moon at either 9 a.m. or 9 p.m.. To determine which, draw a line through the center of Earth that passes through the waxing crescent Moon. Since the Moon is on your eastern horizon it must be rising, which means that the observer must be getting carried toward the Moon rather than away from it. This implies that the local time must be 9 a.m.

Note that drawing an artificial horizon through the center of Earth rather than under the observer's feet is simply a convenience, but certainly doesn't introduce any significant error for our purposes; recall the relative sizes of Earth and the Moon, and their distance of separation shown in Fig. 4.15. You may have also realized that instead of simply estimating the local time, it is possible to determine either the approximate local time, the phase of the Moon, or the location of the Moon in the sky, given the other two pieces of information.

There is a bit of a deception related to Fig. 4.16. From the diagram it appears that the Moon completes an orbit at the same time that it completes its phase cycle. That conclusion is incorrect for the same reason that Earth's rotation period (its sidereal day) is slightly less than four minutes shorter than its solar day. Referring back to Fig. 4.10, recall that since Earth moves just under 1° in its orbit during the time it takes to rotate 360°, it must actually rotate almost 361° to bring the Sun overhead on two successive days. A similar situation applies for the Moon in its orbit. As depicted in Fig. 4.17, as the Moon orbits Earth, Earth is also moving along its own orbit. The lunar sidereal month corresponds to the true orbital period of the Moon relative to the distant stars, with the Moon traveling 360° in its orbit; this motion requires just 27 1/3 days. However, in order for the Moon to reach a point in its orbit when it is again on the opposite side of Earth from the Sun (from full moon to full moon, a lunar synodic month), it must travel in its orbit for an additional 2.2 days (an additional 29° beyond the 360° orbit). During one synodic month of about 29 1/2 days, Earth has completed just under 1/12 of its orbit. In order for the full moon to return to the observer's meridian Earth must rotate that additional 29°, which requires an additional 116 minutes of rotation over those 2.2 days, or

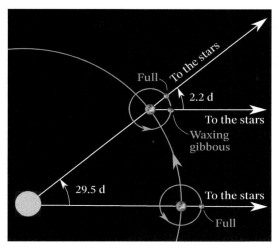

Fig. 4.17 Because Earth moves in its orbit around the Sun at the same time that the Moon orbits Earth, the synodic month from full moon to full moon is longer than the sidereal month of 27 1/3 days by 2.2 days, giving a synodic month of 29 1/2 days. This is also why the Moon rises 53 minutes later every day. Sizes and distances not to scale. (Images courtesy of NASA and NASA's Scientific Visualization Studio.)

53 minutes per day. One way to think about this is to realize that the Moon rises about 53 minutes later every day.

Therein lies the mismatch between using the Moon as the basis for a calendar and using the Sun for that purpose. When you also include the fact that Earth's rotation doesn't fit nicely into one complete Earth orbit, it becomes easier to understand why pre-modern cultures struggled to establish a reliable calendar. *If* Earth moved exactly 1/12 of an orbit in one synodic month, and *if* there were an integer number of days in each synodic month, say 30, then constructing a consistent calendar would be much simpler.

Lunar and Solar Eclipses

The occurrence of lunar and solar eclipses, like those shown in Fig. 2.11, can now be understood as nothing more than shadows being cast by Earth on the Moon during lunar eclipses, and the Moon's shadow being cast on Earth during solar eclipses. Let's begin our study of eclipses by looking at how lunar eclipses occur.

Although it may appear otherwise when you see the Moon in the sky, it is actually rather small and far from Earth (recall Fig. 4.15). With a radius that is 27% of Earth's radius and a distance from us that is just over 60 times the radius of Earth, the full moon's angular diameter is only about 1/2° wide. This means that 720 full moons set side by side would be required to form a complete circle around Earth ($2 \times 360°$). For comparison, if you hold your thumb up at arm's length, it subtends about 2°; in other words, your thumb appears to be four times wider than the full moon!

Given that Earth's diameter is almost four times larger than the Moon's diameter, as long as the Moon is directly on the opposite side of Earth from the Sun, the Moon is guaranteed to fall in Earth's shadow, as shown in Fig. 4.18 (*Left*). Of course,

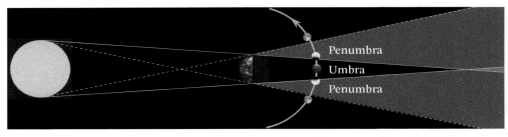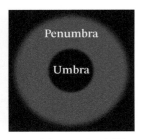

Fig. 4.18 *Left:* The geometry of lunar eclipses. The umbra is the portion of Earth's shadow where no part of the Sun is visible. Within the penumbra a portion of the Sun is blocked but another portion is visible. Note: The images of the Moon are as they would be seen from Earth, not from far above Earth's North Pole, as the diagram seems to suggest. If the images were correct from the diagram's vantage point, the hemispheres directed away from the Sun should be dark, as is the case for Earth's image. Sizes and distances not to scale. (Total lunar eclipse by Supportstorm, public domain; partial lunar eclipse by Tom Ruen, public domain; Sun, Earth, and full moon images courtesy of NASA) *Right:* The umbra and penumbra are essentially circular in cross-section because the Sun and Earth are nearly spherical.

this arrangement results in a full moon. Figure 4.18 (*Right*) shows what the shadows look like in cross-section at the location of the Moon's orbit.

There are actually two parts to the shadow cast by Earth. In the umbra no part of the Sun is visible from the Moon, but in the penumbra, Earth is only blocking a portion of the Sun's disk. Consequently, as the Moon moves in its orbit, it first passes into the penumbra which causes less light to be reflected back toward Earth, making the Moon appear a bit dimmer. From the penumbra the Moon moves into the umbra. During the transition from the penumbra to the umbra, a curved shadow is visible moving across the face of the Moon. It is this curved shadow that offers conclusive proof, even to pre-modern societies, that Earth is round.

When the Moon is fully within the umbra, observers on Earth can still see it, although it is much dimmer and appears reddish in color (Fig. 4.19). This is because some of the red component of the Sun's light can still reach the Moon and be reflected back toward Earth (sunlight is composed of all of the colors of a rainbow). In the same way that sunsets are red and the clear sky is blue on Earth (to be discussed in Section 7.1), light passing through Earth's atmosphere is scattered and bent, with the red light being scattered less than other colors. As illustrated in Fig. 4.20, the result is that it is red light passing through Earth's atmosphere that illuminates the Moon's surface.

Fig. 4.19 The total lunar eclipse of December 21, 2010. (Supportstorm, public domain)

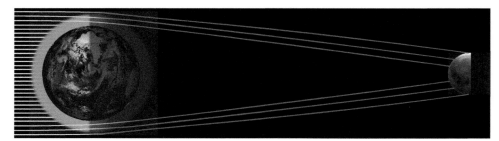

Fig. 4.20 Light from the Sun is scattered and bent as it passes through the thin layer of Earth's atmosphere. Blue light is scattered the most, yellow light less so, and red light the least. This allows red light to successfully pass through the atmosphere and be bent toward the Moon. (Total lunar eclipse by Supportstorm, public domain; Earth image courtesy of NASA)

Your everyday experience tells you that not every full moon is reddish in color, nor do you always see a curved shadow cast on the full moon by Earth before or after an eclipse, so what is going on? In reality, lunar eclipses are relatively uncommon; they typically happen 2 to 3 times per year rather than the 12 to 13 times per year when full moons occur. Even when lunar eclipses do occur, they may not become total eclipses, which occur only when the Moon completely enters the umbra; instead, as shown in Fig. 4.21, they may be partial eclipses, when only a portion of the Moon enters the umbra, or they may be penumbral eclipses, when the Moon passes through the penumbra but never actually enters the umbra. However, during the time leading up to and after a total lunar eclipse the Moon first enters the penumbra, resulting in a penumbral eclipse, then before entering the umbra entirely, the Moon is in a partial eclipse phase, becoming a total eclipse when it is fully inside the umbra. The situation is reversed when the Moon exits Earth's shadow.

The Moon's orbital plane is not aligned with the orbital plane of Earth (the ecliptic) but instead is tilted by about 5° with respect to the ecliptic, as shown in Fig. 4.22. This means that the Moon is often too far above or below the ecliptic for an eclipse to occur (remember that the Moon is only about 1/2° wide). Geometry tells us that the intersection of two planes forms a line; in the case of the ecliptic plane and the Moon's orbital plane, that line of intersection is the line of nodes. The two "nodes" are the two points in the Moon's orbit where it crosses the ecliptic. This geometry means that eclipses can only occur when the Moon is near the line of nodes.

In a situation that is similar to that of Earth's extremely slow 26,000-year precession of its rotation axis (Fig. 4.14), the plane of the Moon's orbit, together with the line of nodes, precesses with a period of 18.6 years. Even though the plane of

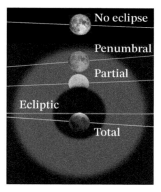

Fig. 4.21 The three types of lunar eclipses depend on the orientation of the Moon's orbit relative to the center of Earth's shadow (the plane of the ecliptic). (Total lunar eclipse by Supportstorm, public domain; partial lunar eclipse by Tom Ruen, public domain; full moon image courtesy of NASA)

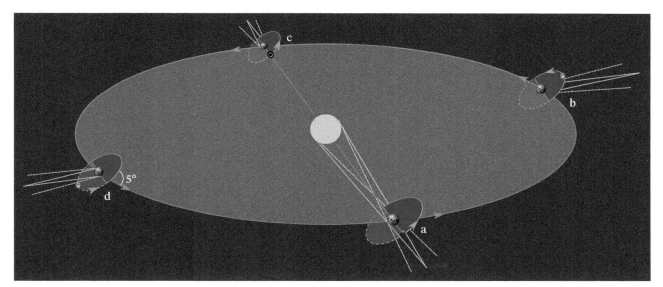

Fig. 4.22 The Moon's orbital plane is tilted approximately 5° with respect to the ecliptic. The line of nodes is indicated in red. (a) Lunar or (c) solar eclipses only occur when the full moon or new moon, respectively, are near the line of nodes. No eclipses occur in (b) or (d) when full and new moons occur above and below the plane of the ecliptic. Diagram components not to scale. (Total lunar eclipse by Supportstorm, public domain; total solar eclipse by Damien Deltenre, CC BY-SA 3.0; Sun, Earth, and full moon images courtesy of NASA)

the Moon's orbit is *almost constant* in its orientation throughout Earth's annual trip around the Sun, that orientation does pivot slowly in the same direction as Earth travels in its orbit. As a result, the Moon crosses the same node twice in a row, say from south to north, slightly faster than it completes one orbit around Earth. More specifically, the time required to cross the same node twice, known as a lunar draconic month, is 27.212 days while one sidereal month is 27.321 days long. At the same time, the phase cycle for the Moon (the synodic month) is 29.531 days long. (The draconic month gets its name from the Chinese myth of a dragon eating the Sun during an eclipse.)

The reason for specifying these three periods to that level of accuracy goes back to the amazing abilities of pre-modern astronomers to discover long-period patterns in the heavens. For example, as you learned on page 41, Mayan astronomers were aware of an 18.03 year eclipse cycle, known as the saros cycle. That cycle turns out to be an integer number of synodic months (233) and an integer number of draconic months (242), equaling very nearly 6585 1/3 days. Three saros cycles is almost exactly 19,756 days, or just over 54 years. Three saros cycles also turns out to be almost exactly 669 synodic months. That integer number of years and days means that once every 54 years the Moon is crossing the same node at the same location in its orbit with the same phase above the same location on Earth during the same season of the year. In other words, eclipse patterns repeat in the same region of the planet every 54 years. Uncovering such a pattern requires very careful observations, diligently recorded, for hundreds of years, combined with the ability to analyze those records to find the pattern. That is an impressive feat for societies that knew nothing of the actual motions of the Moon around Earth, Earth around the Sun, the tilt of the Moon's orbit relative to the ecliptic, or the precession of the line of nodes.

Solar eclipses are much less commonly seen than lunar eclipses. Whereas lunar eclipses (at least partial eclipses) can be seen by anyone on the nighttime hemisphere of Earth when they occur, total solar eclipses are only visible from a small swath of the planet, and only for a few minutes or so. Figure 4.23 shows a total eclipse (*Top left*), a partial eclipse (*Top center*), and an annular eclipse (*Top right*), the latter of which occurs when the Moon is too far from Earth to completely cover the disk of the Sun. Figure 4.23 (*Bottom*) shows a time-elapsed progression of the Sun passing behind the Moon, leading from a partial solar eclipse to totality, and then to a partial eclipse again, before completely reemerging from behind the Moon. The ghostly glow around the Moon during a total solar eclipse is caused by the extremely hot gases in the outermost part of the Sun's atmosphere becoming visible, which is normally overwhelmed by the Sun's much brighter disk.

As can be seen in the photograph in Fig. 4.24, when the Sun moves behind the Moon, the Moon casts a shadow on Earth. During the brief period of totality the sky changes to night, the atmosphere becomes noticeably cooler, animals (including humans!) have been known to exhibit unusual behavior, and the stars become visible. From a high vantage point or one with a clear view of the horizon you may even be able to see the approaching shadow of totality. If you ever have the opportunity to travel to the center path of totality during a solar eclipse, you should not pass it up; it will certainly be worth the effort. Figure 4.25 shows the dates and locations of total and annular solar eclipses from 2021 through 2040.

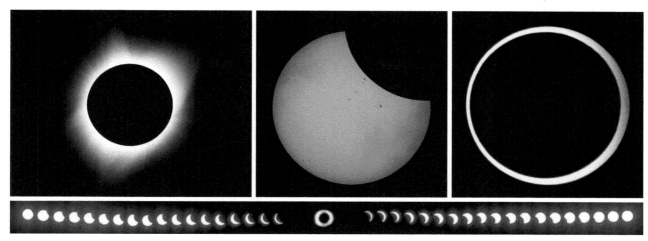

Fig. 4.23 *Top left:* The total solar eclipse of August 21, 2017 above Madras, Oregon. (NASA/Aubrey Gemignani) *Top center:* A partial solar eclipse following totality on August 21, 2017 as seen near Banner, Wyoming. Careful examination of the image shows the International Space Station transiting the disk of the Sun just below the Moon and to the right of the center of the Sun. (NASA/Joel Kowsky) *Top right:* An annular solar eclipse occurs when the Moon is too far from Earth to completely cover the Sun's disk. The photograph was taken from Middlegate, Nevada on May 20, 2012. (Smrgeog, CC BY-SA 3.0) *Bottom:* The progression of the total solar eclipse of August 1, 2008, as seen from Novosibirsk, Russia. As the Moon passes across the face of the Sun, more and more of the Sun becomes hidden from view until the disk of the Sun is completely obscured, allowing us to see the hot, glowing gases that surround the Sun. The sequence is reversed as the Moon moves away from the Sun. The image has been rotated; the actual path in the sky lies along the ecliptic. (Kalan, CC BY 3.0)

Note that on April 8, 2024 a total solar eclipse will pass across the west central coast of Mexico, across the eastern United States, and across southeastern Canada. If you will be traveling to the eclipse, plan far ahead of time as housing options will be filled months or even years before the eclipse, as was the case for the Great American Eclipse of August 21, 2017 that passed across the United States from the west coast to the east coast. Here's hoping for clear skies! *An important word of caution:* NEVER LOOK DIRECTLY AT THE SUN, DURING AN ECLIPSE OR OTHERWISE: WITHOUT PROPER EYE PROTECTION YOU CAN PERMANENTLY DAMAGE YOUR EYES!

Fig. 4.24 The shadow of the Moon on Earth during the total solar eclipse of March 9, 2016. (NASA image courtesy of the DSCOVR EPIC team)

The apparent rarity of total solar eclipses has to do with a bit of surprising geometry, as shown in Fig. 4.26: the Moon subtends 1/2° of sky, but coincidentally so does the Sun. This coincidence is only temporary in the history of the Earth–Moon system because the Moon is slowly moving farther from Earth, causing its apparent size in the sky to get progressively smaller. We are fortunate to have the opportunity to witness total solar eclipses during this period in human history. Because both the Sun and the Moon subtend 1/2°, only a very narrow umbral shadow is cast by the Moon on the surface of Earth; the extended shadow seen in Fig. 4.24 also includes the penumbra. As Earth rotates and the Moon moves in its orbit, the narrow path of totality within the umbra [typically about 150 km (90 mi) wide] and the wider shadow of the penumbra move across the planet. For those observers in the penumbral region of the eclipse, they will be able to see a partial solar eclipse with the curved surface of the Moon imprinted on the solar disk as in Fig. 4.23 (*Top center*).

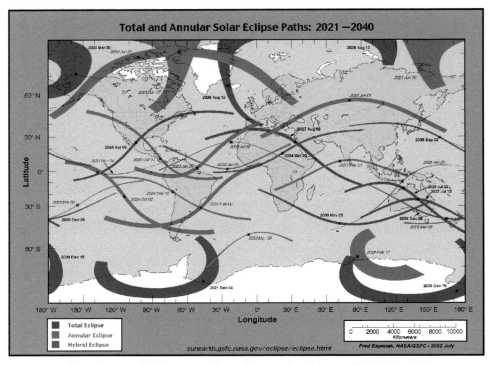

Fig. 4.25 Solar eclipse paths from 2021 through 2040. (Fred Espenak, NASA/GSFC)

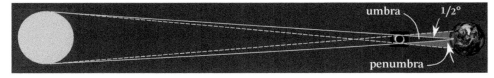

Fig. 4.26 The geometry of a total solar eclipse. Not to scale. (Total solar eclipse by Damien Deltenre, CC BY-SA 3.0; Sun and Earth images courtesy of NASA)

4.7   Finally, a Natural Solution to the "Wandering Stars"

The Relative Distances of the Planets

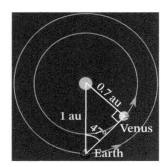

Fig. 4.27 The geometry of Venus, Earth, and the Sun when Venus is at its greatest angular distance from the Sun as seen from Earth. (Images courtesy of NASA)

Thus far, the motions and observational effects discussed in Sections 4.3 through 4.6 can all be replicated using Ptolemy's geocentric model of the universe. The motions of the planets proved to be much more challenging, however.

First among those challenges was determining the locations of the five known, naked-eye planets: Mercury, Venus, Mars, Jupiter, and Saturn. In fact, for the two millennia that the geocentric model reigned, it proved impossible to do so. Astronomers prior to Copernicus did realize that Mercury and Venus must always be near the Sun (see page 23), but in order for that to occur those inferior planets had to be coupled in an ad hoc way to the deferent on which the Sun's epicycle moved. On the other hand, within the Copernican heliocentric model the relative distances of all of the planets from the Sun can be determined by simple geometry, based on the distance between the Sun and Earth. That Sun–Earth distance is referred to as one astronomical unit (au).[4] It is important to note, though, that getting reasonably accurate true distances (as opposed to relative distances) would have to wait for more than a century after Copernicus died (see Section 6.1).

To see how geometry leads to relative planetary distances, consider the case of Venus shown in Fig. 4.27. As the brightest object in the sky after the Sun and the Moon, it is very easy to track the movements of Venus. You should recall that Venus is sometimes referred to as the morning star, when it is shining brightly in the east for at most about three hours before sunrise, and it is known as the evening star when it is shining in the western sky for up to three hours after sunset, but at no time will you ever see Venus more than 47° from the Sun. When Venus is at its maximum angular distance from the Sun, a right triangle is formed between the Sun, Earth, and Venus, with the 90° angle (a right angle) at the location of Venus. With the triangle's hypotenuse being the imaginary 1 au-long line connecting Earth and the Sun, the leg across from the 47° angle must be approximately 0.7 au, which is the distance between Venus and the Sun. [If you are familiar with trigonometry, 1 au × sin (47°) = 0.7 au.]

As inferior planets, Mercury (0.4 au from the Sun with a maximum angle of 28°) and Venus are never seen crossing your meridian at midnight, but the superior planets of Mars, Jupiter, and Saturn can cross your meridian at any time, including when the Sun is on the opposite side of Earth. The only way that this can occur is if the superior planets have orbits that are larger than Earth's orbit, as shown for Mars in Fig. 4.28.

When Mars, Earth, and the Sun are positioned in a straight line with Mars on the opposite side of Earth from the Sun, it is in this configuration when you are able to see Mars crossing your meridian at midnight. In order to determine the distance from Mars to the Sun, Copernicus realized that when Mars and Earth have moved in their orbits to the point where a right angle forms between Mars, Earth, and the Sun, Mars is crossing the meridian when the Sun is setting. This right triangle makes it possible to determine the angle between Earth and Mars at the Sun's location, which is accomplished by noting what fraction of one year, and therefore what fraction of one complete Earth orbit, occurred since opposition; this gives angle α in Fig. 4.28 [α is the Greek letter alpha (see Table 4.1 for the entire Greek alphabet); not all of astronomy and physics is Greek to you, but some of it certainly is]. Then, by knowing the sidereal orbital period of Mars it is also possible to determine the angle β (beta), based on what fraction of Mars' orbit was completed in the same interval of time. Subtracting β from α gives the desired angle as 49°, which, through the properties of a right triangle, immediately gave the Sun–Mars distance to be about 1.5 au. Exactly the same process could also be carried out for Jupiter (5.2 au) and Saturn (9.5 au).

Figure 4.29 shows the relative distances of the naked-eye planets from the Sun (but note that the sizes of the Sun and the planets are not to scale).

The Inclinations of Planetary Orbits

Copernicus was able to discern another important aspect of planetary motions as well, namely the inclinations of their orbits, also illustrated in Fig. 4.29. As we have already discussed, Earth's equatorial plane is inclined 23 1/2° with respect

[4]The astronomical unit was defined precisely in 2012; see page 189.

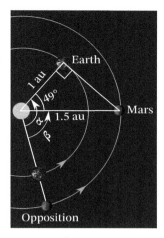

Fig. 4.28 The geometry of Mars, Earth, and the Sun when Mars is in the opposite direction from the Sun, and when Mars, Earth, and the Sun form a right triangle. (Images courtesy of NASA)

Table 4.1 The Greek alphabet

Name	Upper	Lower
Alpha	A	α
Beta	B	β
Gamma	Γ	γ
Delta	Δ	δ
Epsilon	E	ϵ
Zeta	Z	ζ
Eta	H	η
Theta	Θ	θ
Iota	I	ι
Kappa	K	κ
Lambda	Λ	λ
Mu	M	μ
Nu	N	ν
Xi	Ξ	ξ
Omicron	O	o
Pi	Π	π
Rho	P	ρ
Sigma	Σ	σ
Tau	T	τ
Upsilon	Y	υ
Phi	Φ	ϕ
Chi	X	χ
Psi	Ψ	ψ
Omega	Ω	ω

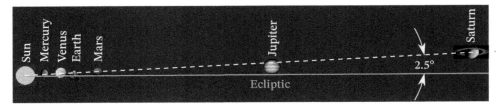

Fig. 4.29 The relative distances of the naked-eye planets from the Sun and their maximum excursions above and below the ecliptic (the inclination of Saturn's orbit is 2.5°). Placing all of the planets on the same side of the ecliptic is simply a convenience; they can be found on either side, depending on their locations in their orbits. The sizes of the planets and the Sun are not to scale. (Images courtesy of NASA)

to the ecliptic and our Moon's orbit is inclined about 5° to the ecliptic. Each of the planet's orbits are also inclined relative to the ecliptic by varying amounts, the largest being Mercury (7°) and Venus (3.4°). Although Saturn's orbital inclination (2.5°) is not as large as either Mercury's or Venus's, it does reach a greater distance from the ecliptic because it is so much farther from the Sun than either of the inferior planets. These inclinations are manifested observationally by noting that the planets do not follow exactly the same paths across the background stars that the Sun appears to follow; instead, they may be found slightly above or below the ecliptic. An example of this can be seen in the simulation of the motion of Mars in Fig. 2.12.

In an impressive bit of analysis, Copernicus also realized that the line of nodes of each planet's orbital plane intersecting the plane of Earth's orbit (the ecliptic plane) is oriented differently from those of every other planet. This means that the orbit of each planet tilts in a different direction from every other planet. In addition, he was even able to calculate their orientations. Although Copernicus didn't discover this, the orbital planes of the planets also precess, similar to the precession of our Moon's orbit, causing their lines of nodes to slowly change direction in the ecliptic plane.

The Answer for Retrograde Motion

The single greatest challenge for the geocentric model was the phenomenon of retrograde motion as the planets otherwise traveled slowly west to east relative to the background stars (again refer to Fig. 2.12). It was Plato's belief that uniform and circular motion should govern all motions in the heavens that led him to ask "what are the hypotheses that ... can save the planetary phenomena?" Epicycles were introduced into ancient Greek astronomy in order to resolve the retrograde behavior. Then as time went on, culminating with Ptolemy's *Almagest*, the model of planetary motion in an Earth-centered universe became progressively more complex and ad hoc.

Copernicus never deviated from Plato's claim that the motions in the heavens must be both uniform and circular, he only altered the object that was at the center of those perfect motions. However, by doing so, a natural solution to retrograde motion became apparent. Imagine two cars traveling in the same direction down a highway with the faster car in the left lane as shown in Fig. 4.30. As the faster car overtakes the slower one in the right lane, the line of sight from the faster car to the slower one changes direction, with the result that the slower one appears to move backward relative to a distant landscape. The same effect applies to the *apparent* motions of the planets.

Just like the rotation and orbital directions of Earth and our Moon, when viewed from a vantage point above Earth's North Pole all of the planets orbit the Sun in the same counterclockwise direction. In addition, the farther a planet is from the Sun the slower it moves and the farther it has to travel to complete an orbit. In the case of the superior planets, Earth is in the fast lane, overtaking and passing the slower moving outer planet. The situation is reversed for the inferior planets; they are the ones in the fast lane, overtaking a slower Earth.

As illustrated in Fig. 4.31, just as with the analogy of the faster car passing a slower one, when Earth overtakes Mars, Mars appears to slow down, stop, back up, and then resume its west-to-east motion. Over the course of six months, Earth travels one-half of an entire orbit while Mars travels just over one-quarter of an orbit (its sidereal orbital period is 1.9 years). Because the inclination and the orientation of Mars' orbit causes the line-of-sight to Mars to sometimes be above and sometimes be below the ecliptic plane, depending on the particular situation, this produces either an apparent loop or a Z-shaped path during retrograde.

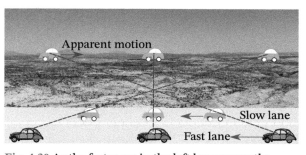

Fig. 4.30 As the faster car in the left lane passes the slower car in the right lane, the slower car appears to move backward against the more distant landscape. (Landscape image by Dale A. Ostlie)

Key Point
The true reason for retrograde motion of the planets.

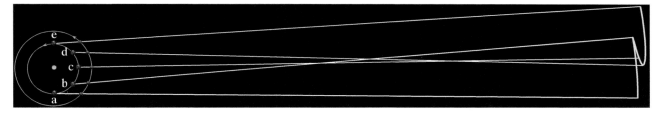

Fig. 4.31 The retrograde of Mars relative to the background stars in late 2022 and early 2023 shown in Fig. 2.12 is due to the more rapidly moving Earth overtaking Mars. The positions of the planets are shown for (a) August 1, 2022, (b) October 30, 2022, (c) December 8, 2022, (d) January 14, 2023, and (e) March 31, 2023. (Retrograde path from *Stellarium*, GNU GPLv2; images courtesy of NASA)

4.8   The Quest for Understanding Continues

Occam's Razor and the Ideal of Simplicity

In his *Commentariolus* Nicolaus Copernicus wrote: "[I sought to develop] a more rational system of circles, from which all apparent irregularity would result, while everything would move uniformly around its proper center, as the principle of perfect motion requires." Copernicus did not set out to revolutionize our understanding of the heavens; rather, his goal was to uncover a simpler, less convoluted way to explain the apparently complex set of motions that are observed from Earth. His great insight was that, by abandoning one fundamental assumption, namely that Earth must be unmoving, and replacing it with a fixed Sun about which Earth revolved, he could explain in a natural way the array of motions that had perplexed humanity throughout history.

Copernicus's quest for greater simplicity in our understanding of celestial motions is a philosophical tenet of science. More specifically, the belief that nature's behaviors should be reducible to as few basic assumptions as possible without needing to resort to inconsistent, ad hoc fixes, is a core component of modern science. One could argue that perhaps that aspect of his heliocentric model was Copernicus's greatest contribution. As we will see in the remainder of this text, science has progressively moved toward theories that are increasingly more encompassing and built on fewer assumptions, although assumptions never will be completely eliminated. Once again, please refer back to the opening chapter in this text, *The Nature of Science*, to review what is meant by the activity of "doing science."

Occam's razor is the philosophical premise that states if two explanations can explain the same set of phenomena, then the one with the least number of assumptions is generally the right one. The notion of the *razor* alludes to cutting away as many underlying assumptions possible. Occam's razor is attributed to an English Franciscan friar, philosopher, and theologian, William of Ockham (c. 1287–1347), although others before him certainly made similar arguments, including Pythagoras, Aristotle, and, ironically, even Ptolemy, who wrote "We consider it a good principle to explain the phenomena by the simplest hypothesis possible."

Key Point
The quest for the simplest possible explanation of phenomena in nature is a philosophical tenet of science known as Occam's razor.

Perfect Circular Motion Remains

As already mentioned, even though Copernicus abandoned an Earth-centered universe in favor of a Sun-centered one, a change that went counter to two millennia of established belief, he was unable to imagine a universe that didn't conform to the pre-conceived norms of perfection in the heavens: uniform and circular motion. There were certainly signs that perfectly circular motion required some adjustment. For example, not all solar eclipses are total, partial, or even penumbral; there are eclipses that are annular, as in Fig. 4.23 (*Right*). Since an annular eclipse occurs when the disk of the Moon is too small to cover the entire disk of the Sun, an annular eclipse requires that there must be a changing relationship between the distances to the Sun and the Moon: either the Moon has moved farther from Earth or the Earth–Moon system has moved farther from the Sun relative to the configuration required for a total solar eclipse. Such a situation is impossible if Earth's orbit around the Sun is perfectly circular with the Sun at the geometric center, and if the Moon's orbit around Earth is also perfectly circular with Earth at the center. Even without viewing an annular eclipse it is readily apparent that the angular size of the Moon changes noticeably over time. It requires more careful observation, but the same situation applies to the varying angular size of Mars. In order to remedy these obvious problems, Copernicus was required to resort to the addition of epicycles and other tweaks in order to make the fine adjustments needed for his model to agree with observations.

How Was the Copernican Model Received?

Despite the numerous successes of the Copernican model, they were fundamentally mathematical in nature, with no clear evidence that heliocentrism is correct. In that sense, Osiander was right at the time when he wrote in his unauthorized preface to *De revolutionibus* that "So far as hypotheses are concerned, let no one expect anything certain from astronomy, which cannot furnish it." At the time, there was no observational evidence or theoretical underpinnings that could unambiguously support the heliocentric model over the 2000-year-old geocentric model. Such evidence would only come much later.

Martin Luther (Fig. 4.32) was a professor of theology at the University of Wittenberg, in Germany, beginning in 1508 (the university had just been established six years earlier). While at Wittenberg, Luther gave birth to the Protestant Reformation on October 31, 1517, by nailing his ninety-five theses on the door of All Saint's Church in Wittenberg, thus challenging the doctrine of the Roman Catholic Church.

As a professor, it was common for Luther to host students and colleagues in his home for dinners, camaraderie, and conversation. During one of those dinners an assistant of his, Anton Lauterbach (1502–1569), recorded the following comment made by Luther on June 4, 1539, one of many such conversations recorded by Lauterbach and published twenty years later (well after Luther's death) in *Table Talk*:

Fig. 4.32 Martin Luther (1483–1546). [Painting by Lucas Cranach the Elder (c. 1472–1553), public domain]

> [Lauterbach prefaces Luther's comment:] There is mention of a certain new astrologer who wanted to prove that the Earth moves and not the sky, the Sun, and the Moon. This would be as if somebody were riding on a cart or in a ship and imagined that he was standing still while the Earth and the trees were moving. [According to Luther] "So it goes now. Whoever wants to be clever must agree with nothing that others esteem. He must do something of his own. This is what that fellow does who wishes to turn the whole of astronomy upside down. Even in these things that are thrown into disorder I believe the Holy Scriptures, for Joshua commanded the Sun to stand still and not the Earth."

Luther was very much a literalist in his understanding of the Christian Bible. Of course, it is again important to remember that the Aristotelian–Ptolemaic worldview had now been in place for two millennia. Furthermore, a moving Earth (spinning and orbiting the Sun) seemed to contradict common sense. It is also worth noting that the comment was made in casual conversation based on relatively little information, perhaps largely hearsay, four years prior to the publication of *De revolutionibus*.

Apparently Luther made no other reference to Copernicus or his theory, either in quotes attributed to him or in his own writings, but he did write about the relationship between science and the Bible:

> [In biblical studies] one must accustom oneself to the Holy Spirit's way of expression. With the other sciences, too, no one is successful unless he has first duly learned their technical language ... Now no science should stand in the way of another science, but each should continue to have its own mode of procedure and its own terms.

Interestingly, despite the use of the term "astrologer" to refer to Copernicus in *Table Talk*, Luther was quite clear in his thoughts about astrology versus astronomy, saying "astrology is entirely without proof ... All the astrological experiences are purely individual cases ... I do not believe that from such partial observations a science can be established." Even today, such a response to claims of meaning in astrological predictions would be very appropriate.

Fig. 4.33 Philipp Melanchthon (1497–1560). [Painting by Lucas Cranach the Elder (c. 1472–1553), public domain]

Luther's principal lieutenant, Philipp Melanchthon (Fig. 4.33), a professor of Greek at the University of Wittenberg beginning in 1518, who became a powerful educational reformer, giving particular emphasis to mathematics and astronomy. Just as Luther did, Melanchthon fully rejected the idea of a heliocentric universe because it appeared to be contrary to scripture, but he did support the mathematical simplicity of the Copernican model for computational purposes only. It was Melanchthon who supported Rheticus in his trip to study under Copernicus, thus ensuring the publication of *De revolutionibus*.

Melanchthon also supported the work of another Wittenberg professor, Erasmus Reinhold (1511–1553), who used *De revolutionibus* as the basis for computing new, more accurate, planetary tables, which were published in his *Prutenic Tables*. Reinhold's tables and Copernicus's work were used as the foundation for the Gregorian calendar, commissioned by Pope Gregory XIII (1502–1585) in 1582. It is the Gregorian calendar that is universally used today. (Sadly, Reinhold was a victim of the plague.)

Within Lutheran circles, the acceptance of Copernicus's work was clearly bifurcated; the core concept of heliocentrism was rejected, in part because of all lack of observational or theoretical confirmation, combined with an apparent contradiction to biblical dogma, but the mathematical simplicity was fully embraced. It is worth noting, however, that biblical stories were often taken to be somewhat allegorical within more traditional Catholic circles, and for many decades Copernicus and his ideas were not strictly rejected; instead, the response was largely ambivalence.

The Case of Giordano Bruno

Fig. 4.34 Bronze statue of Giordano Bruno (1548–1600) in Rome by Ettore Ferrari (1845–1929). (Jastrow, public domain)

In 1576, Thomas Diggs (1546–1595), an English astronomer, proposed that the fixed celestial sphere of the Copernican model should be replaced by an infinite number of stars scattered throughout an infinite universe. His proposal was contained within one of several appendices that he wrote for his late father's almanac. Written in English, it helped to popularize Copernicus's *De revolutionibus*.

Giordano Bruno (Fig. 4.34), an Italian-born Dominican friar, was known for his remarkable memory. He devised an elaborate mnemonics system to aid his memory, but there were those who simply believed it to be magic. Bruno, also known for his rather unorthodox thinking, took Diggs's proposal a step further. If an infinite universe is filled with an infinite number of stars, couldn't those stars have planets orbiting them, just like the planets orbiting our Sun? Going even further, if there are planets orbiting other stars, couldn't those planets be inhabited, just like Earth?

Bruno's infinite universe with an infinite number of populated worlds was certainly not his only radical idea. He traveled widely, including spending two years in England, and he tended toward reading books that were forbidden by the Roman Catholic Church at the time. His studies and non-conformist thinking led him to develop and teach theological views that conflicted with church doctrine. As a result of his refusal to recant his non-traditional beliefs, he was imprisoned in 1593 for heresy by the Roman Inquisition and underwent a seven-year trial, at the end of which he was sentenced and publicly burned at the stake in Rome. (A statue of Bruno now stands at the site of his execution; Fig. 4.34.) Although historians generally don't believe that he was executed because of his concept that we live in an infinite universe filled with an infinite number of other inhabited worlds, his view was very contrary to the Aristotelian–Ptolemaic worldview that still held sway at the time and almost certainly didn't help his case.

4.9   Tycho Brahe and His Passion for Precision

Tycho Brahe (Fig. 4.35) born three years after the death of Copernicus and the publication of *De revolutionibus*, and two years before the birth of Bruno. Traditionally referred to by his first name, Tycho was born into Danish nobility and abducted by his uncle, Jørgen Brahe (1515–1565), sometime before the age of two. His uncle argued that since Tycho's father [Otto Brahe (1518–1571)] already had a son, he too should be able to share in the "riches." Ultimately, Jørgen was given permission to raise Tycho, a decision that had a dramatic impact on both Tycho and the future of astronomy. While his brothers were raised to govern, Tycho was able to travel extensively with a tutor beginning when he was fifteen years old, visiting cities and universities. History suggests that Tycho developed an interest in astronomy at an early age which led him, while being tutored, to secretly read books on the subject because astronomy was not considered an important part of his education.

Fig. 4.35 Tycho Brahe (1546–1601). (The Museum of National History, public domain)

Tycho's uncle Jørgen died of pneumonia in 1565 following an accident when he and King Frederick II (1534–1588) both fell into a river after they had been drinking. Apparently the king fell in first and Jørgen tried to rescue him; the king survived.

After his uncle's death Tycho's father still wanted him to enter government but he refused, choosing instead to continue his education at the University of Rostock. While at the university, he attended a dance hosted by a professor where he got into an argument with another student over who was the better mathematician. Apparently Tycho and his rival were quite emotional about the issue because it resulted in a sword dual. The dual wasn't fatal for either combatant but Tycho did lose the bridge of his nose on that night in late December, 1566. To remedy the situation, for dress occasions Tycho had a nosepiece made out of gold and silver which he glued on. He had at least one lighter model made for everyday wear. (Tycho's tomb was opened on the three-hundredth anniversary of his death in 1901, when it was discovered that his skull was stained green around his nasal cavity. In 2012, Danish and Czech researchers analyzed bone fragments and discovered that one of his noses was made of bronze, an alloy that contained copper.) Following his father's death in 1571, Tycho inherited substantial wealth which allowed him to live and do as he pleased, including living with his mate whom he was unable to marry, given that she was not from nobility.

New Stars, Comets, and Planets

On November 11, 1572, while out walking one evening, Tycho noticed a new star in the constellation of Cassiopeia. [We know today that the "new star" was actually a supernova, now referred to as Tycho's Supernova (SN 1572), which is a star that exploded. (*Nova* is Latin for "new.")] He was able to show through the lack of any apparent parallax that his "new star" was farther from Earth than the Moon, and by observing it for several months realized that it moved with the stars, rather than traveling through them as planets do. Tycho was, and remained throughout his

Fig. 4.36 An engraving from 1587 of Tycho's mural quadrant in his Uraniborg observatory on the island of Hveen in Denmark. (Public domain)

Fig. 4.37 A woodcut of the Great Comet of 1577 by Georgium Jacobum von Datschitz. (Public domain)

Key Point
Tycho made precise naked-eye observations of the positions of the planets and stars.

life, a firm believer in the Aristotelian–Ptolemaic view of a geocentric universe in which the heavenly realm above Earth's atmosphere was perfect and unchanging, and that the motions were circular and uniform. Although his "new star" did not fit Tycho's strict view of the heavens, his decision to publish his discovery in a small book, *De nova stella* (1574), became another pivotal moment in history, both for solidifying his career and for the direction of astronomy.

Following the publication of his book, Tycho considered moving to Basel in present-day Switzerland, but King Fredrick II didn't want Denmark to lose the distinguished astronomer. To entice him to stay, Fredrick gave him the island of Hveen, located between Denmark and southern Sweden (Sweden later took over the island in 1658). Fredrick also gave him a substantial annual grant and rents from the forty or so farms on the island. With the funding, Tycho built a research institution and observatory that he named Uraniborg, meaning the "Castle of Urania," Urania being the Muse of Astronomy from Greek mythology. Within Uraniborg, Tycho constructed the most precise instruments ever built for astronomy, as depicted in Fig. 4.36. Later, due to concerns over stability and the effects of wind on his instruments, Tycho also built a second observatory below ground level, which he named Stjerneborg, the "Castle of the Stars." During its operation between 1576 and 1597 it is estimated that nearly 100 students and and other assistants worked at the observatories. He even had a printing facility built on the site for the publication of his results.

In 1577 another major astronomical event occurred, that proved to be a thorn in the side of the Aristotelian worldview: the appearance of the Great Comet of 1577, depicted in the woodcut in Fig. 4.37. As he did with the "new star," Tycho was able to demonstrate that the comet must be located well beyond the distance of the Moon because of the lack of any observable parallax. Although comets had been seen throughout history, there was a general belief that they were atmospheric phenomena and therefore not a challenge to the concept of an unchanging heavens. But the distance of the comet from Earth also raised another problem. Because comets moved across the stars, the implication was that the crystalline spheres upon which the Sun, the Moon, and the five planets were believed to travel, and upon which the stars existed, could not exist; certainly comets cannot pass through those solids.

As we will see in Section 4.10, the most important class of observations that Tycho made, and for which he built his observatories, were those of the positions of the stars and the five planets. In the mural depicted in Fig. 4.36 you can see Tycho seated in a giant quadrant carefully marked out in angles. The mural shows him sighting through a slit in the wall while aides record the angle and the time of each observation. Tycho's goal was to record positions with better than 1′ accuracy, and he claimed to have reached that goal, but later analysis of his published tables indicate that they were, on average, only accurate to about 2′, still a remarkable feat for the time. (1′, or 1 arcminute, is 1/30 of the angular width of a full moon.)

Following the death of Frederick II, Tycho's financial and political standing deteriorated. After arguments with Frederick's successor, Christian IV, Tycho left Hveen with his equipment in 1597 and moved to Prague in the present-day Czech Republic at the invitation of the Bohemian king and Holy Roman emperor, Rudolf II (1552–1612), where he became the official imperial astronomer. Tycho built a new observatory and remained in Prague until his death in 1601.

Tycho's Pseudo-Geocentric Universe

Despite being the most famous astronomer of his day, Tycho was never willing to accept the Copernican heliocentric model. He did believe in the mathematical simplicity of the model, but could not agree with its fundamental premise that Earth was a mere planet in motion around an immutable Sun, given the model's conflict with common sense and biblical doctrine. He also could not believe that the stars were enormously far away, leaving the universe with an incomprehensible amount of wasted space. He substantiated his argument regarding the nearness of the stars with errant estimates of the sizes that the stars would need to have if they were very distant.

In order to rectify the situation, Tycho developed his own planetary system. As seen in Fig. 4.38 he envisioned Earth at the center with the Moon and the Sun in orbit around it. He then placed the five planets in orbit around the Sun, with the orbit of Mars cutting across the Sun's orbit. The eighth sphere (the celestial sphere) is shown just beyond the orbit of Saturn. Tycho struggled with the idea of Mars' orbit cutting across the orbit of the Sun, impossible for orbits contained on crystalline spheres, but he eventually concluded that the space must be filled with a fluid rather than solid spheres.

In reality, Tycho's model is identical to the one developed by Copernicus, save the choice of perspective. Imagine riding in a car: from your perspective the world seems to be moving past you. Of course, if someone else is standing along the side of the road that person would see you move past him while he would perceive that the world is not moving. The same is true for Tycho's model versus the model of Copernicus. From the point of view of Tycho, the Sun is moving around Earth, but from Copernicus's point of view Earth is going around the Sun; all other aspects of the two perspectives are identical, with the exception of the distance to the stars. In the long term, the Tychonian model had little impact on astronomy's development, but it does illustrate the extent to which theologians, philosophers, and even astronomers would go in order to hold on to the 2000-year-old Aristotelian–Ptolemaic view of the universe that was so ingrained in the thinking of the Middle Ages.

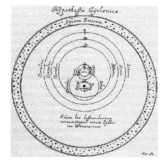

Fig. 4.38 The Tychonic pseudo-geocentric model of the universe. The Sun and the Moon orbit Earth (the black dot at the center) while the planets orbit the Sun. The stars are located on the crystalline celestial sphere. (Public domain)

4.10   Johannes Kepler and Planetary Orbits

Johannes Kepler (Fig. 4.39) born prematurely into a devote Lutheran family that was struggling financially. Young Johannes had a challenging childhood, having contracted smallpox, that affected his hands and left him with poor vision. Making his childhood more difficult, Johannes's father left the family when the boy was five years old and disgraced them by working as a mercenary, fighting against the Protestants in an uprising in the Netherlands. Johannes's father apparently died during that war.

Kepler's mathematical aptitude became apparent rather early on, when he entertained customers at his grandfather's inn, showing off his talent for numbers. Kepler was also introduced to astronomy at a young age. When he was six years old he witnessed the Great Comet of 1577, the same comet that Tycho had studied (Fig. 4.37). Kepler recalled that he "was taken [by his mother] … to a high place to

Fig. 4.39 A portrait of Johannes Kepler (1571–1630). (Smithsonian Libraries and Archives, public domain)

look at it." Then, when he was nine, his mother took him outside to observe a lunar eclipse, recalling that the Moon "appeared quite red."

Despite his early interest in astronomy and his mathematical prowess, Kepler's real passion was to become a philosopher and theologian, hoping to obtain a position in the new Lutheran Church that grew out of the Protestant Reformation. In 1589 he enrolled at the highly-regarded Lutheran University of Tübingen. Along with his other studies at Tübingen, Kepler happened to study astronomy under Michael Maestlin (1550–1631), perhaps the only professor in Europe at the time who was a committed Copernican. It was at this time that Kepler became a dedicated Copernican himself. [Maestlin had studied under Jacob Heerbrand (1521–1600), who was a former student of Luther and Melanchthon. In essence, Kepler was an academic "great-grandson" of Luther and his first lieutenant.]

When Kepler was nearing completion of his degree in 1594, he was offered a teaching position in mathematics at a Protestant school in Graz, in southern Austria. Despite his desire to become a minister, Kepler accepted the position, which was again one of those decisions that history would remember as having a major impact on the future of astronomy, at least in the short term.

Kepler was very much a believer in the philosophy of Plato regarding the precedence of ideas over observations that can deceive. He was also convinced, in the tradition of the Pythagoreans, of the divine importance of geometry and numbers. Given his deep Lutheran convictions, spirituality was a core aspect of his being, leading to a sort of mysticism in his writings. Kepler felt that there must be mathematical beauty in the positions and sizes of the planets, and that the Sun occupies the center of the universe because it is the realm of God.

Shortly after starting his teaching post in Graz, Kepler was lecturing to his students, trying to "inscribe within a circle many triangles, or quasi-triangles, such that the end of one was the beginning of the next. In this manner a smaller triangle was outlined by the points where the lines of the triangles crossed each other." Then he suddenly remembered that there are exactly five perfect platonic solids; solid objects constructed with straight lines on every edge, and that have exactly the same shape and surface area on each flat face, with the same number of faces meeting at each vertex of the solid (a cube is one of the platonic solids). Could it be that the reason there are six planets (including Earth) is because there are five platonic solids?

After making the "discovery" that the five platonic solids, with six spheres nested inside and outside of them, seemed to magically explain the existence of Earth and the five other planets, the young Kepler commented in a letter to his mentor, Maestlin, at the University of Tübingen, "I wanted to become a theologian, for a long time I was restless. Now, however, behold how through my effort God is being celebrated in astronomy." The construction of Kepler's model is shown in Fig. 4.40.

Without yet having access to the much more accurate data being obtained by Tycho and his research team at that time, Kepler recalls in his first book, *Mysterium Cosmographicum* (1596):

Fig. 4.40 Kepler's attempt at a geometric Solar System using five perfect solids and six spheres. The bottom drawing is a more detailed look inside the model. (Public domain)

> And then ... it struck me: why have plane [two-dimensional] figures among three-dimensional orbits? Behold reader, the invention and whole substance of this little book! In memory of that event, I am writing down for you the sentence in the words from the moment of conception: The Earth's orbit is the measure of all things; circumscribe around it a dodecahedron [twelve faces], and the

circle containing this will be Mars; circumscribe around Mars a tetrahedron [four faces], and the circle containing this will be Jupiter; circumscribe around Jupiter a cube [six faces], and the circle containing this will be Saturn. Now inscribe within the Earth an icosahedron [twenty faces], and the circle continued in it will be Venus; inscribe within Venus an octahedron [eight faces], and the circle contained in it will be Mercury. You now have the reason for the number of planets ...

This was the occasion and success of my labors. And how intense was my pleasure from this discovery can never be expressed in words. I no longer regretted the time wasted. Day and night I was consumed by the computing, to see whether this idea would agree with the Copernican orbits, or if my joy would be carried away in the wind. Within a few days everything worked, and I watched as one body after another fitted precisely into its place among the planets.

Kepler didn't believe that the platonic solids actually existed in the space between the planets; instead, he saw them as establishing the spacings within the celestial design of the Creator. As strange as his initial ideas about planetary motion seem to us today, they set the stage for his future work, which became profoundly important for both the future of astronomy and physics, and for science in general.

Tycho became aware of the mathematical talent of Kepler through his *Mysterium Cosmographicum*, and in December 1599 invited Kepler to join him in Prague to help analyze the data on planetary positions that he had amassed while at Hveen. In particular, Tycho was hoping that Kepler would be able to prove the validity of his Tychonic pseudo-geocentric model. As it turned out, an anti-Protestant movement was growing at the time and, when Kepler refused to join the Catholic Church, he and his family were forced to leave Graz. Kepler moved to Prague in August 1600 and began working for Tycho. In addition to having Kepler work on analyzing his planetary data, Tycho also convinced Rudolf II of the need to create new, more accurate astronomical tables based on his more precise observations, to replace the *Prutenic Tables* that had been developed by Erasmus Reinhold from the Copernican models. Kepler was given the assignment and began the long, tedious process of computing the new *Rudolphine Tables*, which were ultimately published in 1627, more than a quarter-century after the work began.

Two days after Tycho's unexpected death in October 1601, Kepler was appointed to succeed him as the imperial astronomer. In addition to the computation of the *Rudolphine Tables*, Kepler was to produce astrological forecasts for the emperor, something he had been doing for family, friends, and others since attending the University of Tübingen; this despite believing that trying to predict specific events was folly. To paraphrase in more modern parlance: "It paid the bills." Fortunately, Kepler was provided full access to all of Tycho's data as well, despite some legal challenges from the Brahe family. Thus began the most productive part of his career.

The Fall of Perfect Circular Motion

Kepler came at his study of astronomy from a fundamentally different perspective than anyone previous to him; he didn't treat the Moon and beyond differently from what occurs on Earth. Even Copernicus believed that events that occur in the heavenly realm are the result of mechanisms that don't apply to processes on Earth.

Key Point
Kepler was the first to reject the idea that the motions in the heavens are fundamentally different from terrestrial phenomena.

Fig. 4.41 William Gilbert (1544–1603). (Wellcome Library, public domain)

In 1600 William Gilbert (Fig. 4.41) published his treatise on magnetism, *De Magnete, Magneticisque Corporibus, et de Magno Magnete Tellure* (*On the Magnet and Magnetic Bodies, and on the Great Magnet the Earth*), leading Kepler to think that perhaps magnetism emanating from the Sun held the key to the motions of the planets, rather than angels or other supernatural causes keeping them in motion as had been believed for millennia. Writing in 1605, Kepler explained his view:

> I am much occupied with the investigation of the physical causes. My aim in this is to show that the celestial machine is to be likened not to a divine organism but rather to a clockwork ... insofar as nearly all the manifold movements are carried out by means of a single, quite simple magnetic force, as in the case of a clockwork all motions [are caused] by a simple weight. Moreover, I show how this physical conception is to be presented though calculation and geometry.

As Kepler analyzed Tycho's data he focused his attention on Mars, in particular. In applying the new, more precise data to his mystic model of platonic solids, Kepler came to realize that the fit was less satisfactory than he had previously believed. Many of the data points deviated from expected values by as much as 8′, which was unacceptable to Kepler given his reliance on the 2′ accuracy of Tycho's observations (8′ corresponds to about 1/4 the angular width of the full moon). Kepler simply wouldn't allow himself the luxury of believing that Tycho's data could be in error by an amount sufficient to explain the discrepancies. Instead, he realized that an oval shape fit the data more accurately, at which point he found the need to abandon another deeply held conviction of astronomy that dated back to the time of the great Greek philosophers, namely perfectly circular motion. In a letter to a colleague in 1607, Kepler wrote:

> When you say it is not to be doubted that all motions occur on a perfect circle, then this is false for the composite, i.e., the real motions. According to Copernicus, as explained, they occur on an orbit distended at the sides, whereas according to Ptolemy and Brahe on spirals. But if you speak of components of motion, then you speak of something existing in thought; i.e., something that is not there in reality. For nothing courses on the heavens except the planetary bodies themselves — no orbs, no epicycles.

Johannes Kepler no longer saw the need to restrict celestial motions to circles, or to a combination of circles, since they weren't real anyway. He was finally moving human understanding beyond that rigid architecture and opening up the possibility that the cosmos could be more varied and creative than that. Today, Kepler's "ovals" are known as ellipses, of which a circle is simply a special case.

In 1604, another naked-eye supernova appeared, the second in less than thirty years (the next such event clearly visible to the naked eye wouldn't appear until 1987). Although Kepler didn't discover it himself, he did show that it too was beyond the distance of the Moon, just as Tycho had done with the supernova of 1572. The 1604 supernova now bears Kepler's name.

In *Astronomia Nova*, published in 1609, Kepler describes the supernova of 1604 as well as what are known today as his first two laws of planetary motion. However, Kepler was still anchored in his Pythagorean philosophy of numerical harmonies, believing that there may be a mystical relationship between musical intervals and planetary motions. In his *Harmonices Mundi* (*The Harmony of the World*), which

Key Point
Kepler's first two laws were published in 1609 and the third was published in 1619.

was published a decade later (1619), he described what he believed to be his sought-after "music of the spheres." Within *Harmonices Mundi* he describes what is today known as Kepler's third law, an equation that describes the orbital period of a planet in terms of that planet's average distance from the Sun.

As Kepler would later state, "without [Tycho's] observation books everything that I have brought into the clearest light would have remained in darkness."

A statue honoring Tycho and Kepler is located in Prague (Fig. 4.42).

Fig. 4.42 Monument of Tycho Brahe (1546–1601) and Johannes Kepler (1571–1630) in Prague, Czech Republic. (Photograph by Josef Vajce, public domain)

4.11 Kepler's Three Laws in their Original Form

Kepler likely did not know the enormous impact that his three laws of planetary motion would have on the future of science. Although incomplete in their original form, Sir Isaac Newton would later use Kepler's laws to guide the development of his universal law of gravitation. Newton would also derive Kepler's laws from first principles, and in the process uncover a critical missing ingredient. We will explore Kepler's laws as expanded by Newton in Section 5.4, but for now it is instructive to study them as Kepler himself identified them.

Kepler's First Law: The Law of Ellipses

In order to better understand what the laws are saying, we must first look at what an ellipse actually is. As shown in Fig. 4.43, an ellipse can be drawn by attaching a string to two focal points (F and F'), and while keeping the string taut, sketch a closed path. The long axis of the ellipse is known as the *major axis* while one-half of the long axis is referred to as the semimajor axis. The length of the semimajor axis is traditionally labeled by the variable a. The short axis of the ellipse is the *minor axis*, and one-half of the minor axis is the semiminor axis, the length of which is labeled by the variable b. Mathematically, the idea of keeping the string taut as you draw the ellipse on a sheet of paper is equivalent to saying that $r + r' = 2a$, where

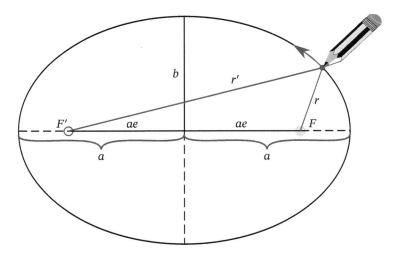

Fig. 4.43 The geometry of an ellipse.

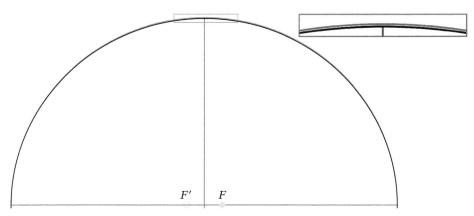

Fig. **4.44** One-half of the elliptical orbit of Mars (black) together with a perfect circle (red). The circle has a radius equal to the semimajor axis of Mars' orbit, and the center of the circle is located half way between the orbit's focal points. The location of the Sun is shown at F. The inset to the right is an enlarged view of the top portion of the orbit and circle showing where the deviation of the orbit from a perfect circle is the largest.

r and r' are the distances from the two focal points to any point on the ellipse (the length of the string and the length of the major axis of the ellipse are both $2a$).

Kepler's law of ellipses states that a planet orbits the Sun in an ellipse, with the Sun located at one focus of the ellipse. Figure 4.43 shows the Sun at the "principal" focal point, F. The secondary focal point, F', is indicated by an open circle because there is nothing physical actually located at that position in space; it is simply a mathematical point.

Both focal points are the same distance away from the center of the ellipse (the point where the major and minor axes cross). That distance is given by ae (meaning $a \times e$), where the variable e is called the **eccentricity** of the ellipse; e can have any value between 0 and 1, but not including 1 (the mathematical shorthand is $0 \leq e < 1$).[5] The eccentricity of the ellipse in Fig. 4.43 is $e = 0.7$. Since a is the length of the semimajor axis and ae is the distance from the center of the ellipse to the principal focus F, the closest point of the ellipse to the principal focus must be $a - ae = a(1 - e)$. In the case of the orbit of a planet, that point of closest approach to the Sun is known as **perihelion**. The farthest distance from the principal focus is given by $a + ae = a(1 + e)$. That greatest distance from the Sun is referred to as **aphelion**. For a perfect circle, $e = 0$, meaning that there is not separation between either focal point and the center of the ellipse, or between each other. This is why a circle can be referred to as simply a special form of an ellipse, with the radius of the circle being $r = a$.

It was fortuitous that Kepler focused his energy on studying the orbit of Mars. It turns out that, with the exception of Mercury, which is close to the Sun and therefore difficult to observe, Mars has the most non-circular orbit of all of the naked-eye planets, with an eccentricity of $e = 0.0934$. On the left-hand-side of Fig. 4.44 a perfect circle (red) is superimposed on the orbit of Mars (black), demonstrating just how nearly circular that planet's elliptical orbit is.

[5]Eccentricities can also be greater than or equal to 1 ($e \geq 1$), but those orbits are open-ended, meaning that an object makes a closest approach to the **principal focus** but then leaves, to never return again.

Recalling the third postulate of *Commentariolus* (on page 95), Copernicus didn't actually place the center of the universe exactly at the center of the Sun, but rather *near* the Sun. By looking at Fig. 4.44 you can see why that was necessary. Figure 4.44 also illustrates why the use of perfectly circular deferents with their small correcting epicycles worked so well for millennia. Even for the case of Mars, its orbit is just slightly inside a perfect circle on the minor axis when the circle is centered between the focal points of the orbit. All of the efforts made to tweak perfectly circular motion were in part to account for those minor deviations. The offset position of the Sun at the principal focus of an elliptical orbit translates into an offset position of Earth in a geocentric view of the universe. It is that shift that required Hipparchus, Ptolemy, and others to introduce the eccentric offset of Earth from the center of the deferent (recall Fig. 3.18). Obviously, all of those epicycle–deferent diagrams in Chapter 3 had highly exaggerated epicycles, eccentrics, and equants. Copernicus was also forced to use epicycles to preserve his belief in the ideal of uniform and circular motion.

Kepler's Second Law: The Law of Equal Areas

Kepler's law of equal areas states that an imaginary line connecting a planet to the Sun sweeps out equal amounts of area in equal time intervals, regardless of the location of the planet in its orbit. Figure 4.45 illustrates this for four equal time intervals. You should note that when the planet is closest to the Sun in its elliptical orbit it travels a greater distance during that time interval than it does when it is farthest from the Sun. This means that the planet is traveling fastest when it is closest to the Sun and slowest when it is farthest from the Sun. Imagine that the orbit represents a strangely shaped, elliptical pizza. If you were served the blue-colored slice you would be receiving exactly the same amount of pizza as someone else who was served the green, red, or aqua-colored pieces. It is the change

Key Point
Kepler's second law (law of equal areas)

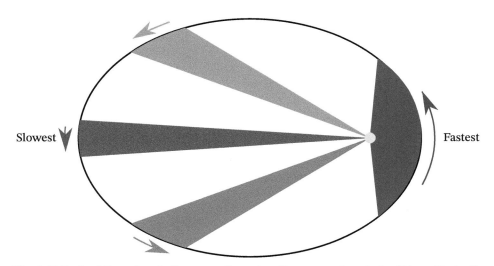

Fig. 4.45 Each of the colored slices were "swept out" over equal periods of time. Each slice has exactly the same amount of area. The Sun is shown at the principal focus.

Table 4.2 Kepler's third law for the naked-eye planets

Planet	a (au)	P (y)	a_{au}^3	P_y^2	a_{au}^3 / P_y^2
Mercury	0.387 10	0.240 85	0.058 01	0.058 01	0.999 95
Venus	0.723 34	0.615 19	0.378 45	0.378 46	0.999 98
Earth	1.000 00	1.000 02	1.000 00	1.000 00	0.999 96
Mars	1.523 66	1.880 82	3.537 24	3.537 48	0.999 93
Jupiter	5.203 36	11.861 78	140.881	140.701 8	1.001 24
Saturn	9.537 07	29.457 14	867.451	867.723 1	0.999 69

in speed of a planet orbiting the Sun that forced Ptolemy to introduce the equant to his geocentric model (again recall Fig. 3.18). It took Kepler to clearly eliminate any need for the equant by simply replacing perfect circular motion with elliptical orbits and variable speeds described by a straight-forward prescription.

Kepler's Third Law: The Harmonic Law

An exponents tutorial is available at
Math_Review/Exponents

In his attempt to attach some mystical meaning to the motions of the planets across the heavens, Kepler uncovered his harmonic law. Believing that the periods of the orbits were somehow related to musical intervals, he explored ways to find exponential relationships similar to the exponential intervals that exist in musical scales. Within music the ratio of frequencies (pitch) between adjacent half-steps is given by $2^{1/12}$ (there are 12 separate notes between octaves, such as between middle C and high C). The ratio in frequencies for two notes in an octave is exactly 2:1; for a perfect fifth, the ratio is 3:2, for a major third the ratio is 5:4, and so on. It is this seemingly cosmic relationship between pleasant sounding musical intervals and mathematics that meant so much to the Pythagoreans.

A variables tutorial is available at
Math_Review/Variables

Kepler found his "harmonies" in the relationship between the semimajor axes and orbital periods of the planets given by the ratio, $a_{au}^3/P_y^2 = 1$, where the semimajor axis lengths of the planets must be given in astronomical units and their orbital periods must be expressed in Earth years. Table 4.2 lists the semimajor axes (a), periods (P), semimajor axes cubed (a^3), and periods squared (P^2) for the six naked-eye planets. (The deviations from $a_{au}^3 / P_y^2 = 1$ for the planets in Table 4.2 are due to round-off errors and, as we will see in Chapter 5, various gravitational interactions throughout the Solar System.)

A graphing tutorial is available at
Math_Review/Graphing

The data in Table 4.2 are plotted in Figs. 4.46 and 4.47. In the first graph the orbital periods in years (P_y) are plotted as a function of the semimajor axes of the planets' orbits in astronomical units (a_{au}). The graph clearly shows that there is a well-defined relationship between period and average distance from the Sun for all of the naked-eye planets. The second graph is a plot of P_y^2 ($P_y \times P_y$) as a function of a_{au}^3 ($a_{au} \times a_{au} \times a_{au}$). Because the four inner planets are all bunched up together near the origin of the graph, an inset better demonstrates that the straight line continues all the way from Mercury through Mars and then to Saturn. The straight line visually illustrates the meaning of Kepler's harmonic law, as he first discovered it; the orbital period of a planet squared (expressed in years) equals the semimajor axis of its elliptical orbit (expressed in astronomical units).

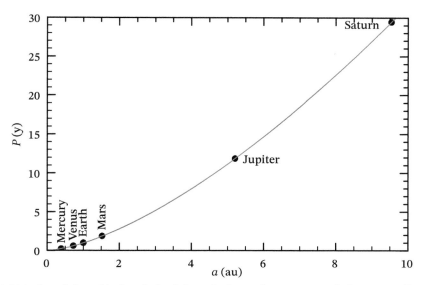

Fig. 4.46 A plot of the orbital periods of the naked-eye planets versus their average distances from the Sun. (Data from Table 4.2)

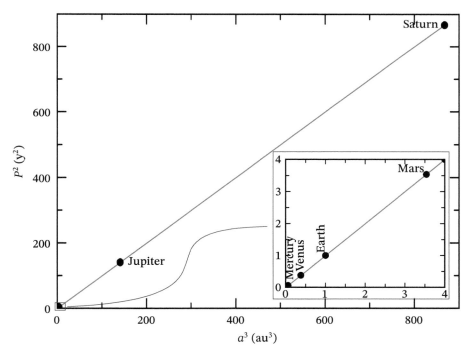

Fig. 4.47 A plot of the orbital periods *squared* of the naked-eye planets versus their average distances *cubed* from the Sun (data from Table 4.2). Plotting the data this way gives a straight line with a slope of 1. The inset shows that the straight line also passes through the data points for the four inner planets tucked down in the lower-left corner of the graph.

Example 4.2

Kepler's laws don't just apply to the naked-eye planets in our Solar System, but they also apply to any object orbiting the Sun. The asteroid Ceres (also classified as a dwarf planet) is located between Mars and Jupiter and has an orbital semimajor axis of 2.768 au. What is Ceres's orbital period?

There are a couple of ways you can determine an answer. From Fig. 4.46 locate Ceres's semimajor axis along the horizontal axis of the graph, draw a vertical line from the axis to the curve, and mark the point of intersection. Next, draw a horizontal line from the point of intersection with the curve over to the vertical axis on the left. You can then read off the answer as 4.6 years. Unfortunately, this approach is limited by how accurately you can read between the tick marks on both the horizontal and vertical axes.

A more direct approach uses the mathematical form of Kepler's harmonic law directly. Since $a_{au} = 2.768$ au, a_{au}^3 is given by

$$a_{au}^3 = a_{au} \times a_{au} \times a_{au} = 2.768 \text{ au} \times 2.768 \text{ au} \times 2.768 \text{ au} = 21.2079 \text{ au}^3.$$

But, according to Kepler's harmonic law, this is exactly equal to P_y^2, or $P_y \times P_y$. To find P_y itself you need to take the *square root* of P_y^2. Therefore

$$P_y = \sqrt{21.2079} \text{ years} = 4.605 \text{ years}.$$

In addition to finding a mathematical relationship between the orbits of the planets and their semimajor axes, Kepler also used the newly-discovered Galilean moons of Jupiter (see Section 4.12) to show that a similar relationship exists in that system as well, although the ratio a_{au}^3 / P_y^2 equals a constant other than one. You will have the opportunity to explore that relationship, and others as well, in the exercises at the end of the chapter.

In summary, Kepler's laws as he originally developed them are:

Key Point
Kepler's laws of planetary motion as Kepler originally deduced them.

1. (The law of ellipses) A planet orbits the Sun in an ellipse with the Sun at one focus of the ellipse.
2. (The law of equal areas) An imaginary line connecting a planet to the Sun sweeps out equal areas in equal time intervals.
3. (The harmonic law) The orbital period of a planet (P_y) measured in Earth years, when multiplied by itself (i.e., "P_y squared") equals the semimajor axis of the planet's orbit (a_{au}) measured in astronomical units multiplied by itself three times (i.e., "a_{au} cubed"):

$$P_y^2 = a_{au}^3. \tag{4.1}$$

Kepler's original third law (the harmonic law)

(Note that the subscripts "y" and "au" are there to remind you of the need to use years and astronomical units in the formula, rather than some other units, such as seconds or meters.)

4.12   Galileo Galilei, the First True Scientist

Galileo Galilei (Fig. 4.48) was born in Pisa, Italy, seven years before Kepler's birth in Germany. Raised by a father who was an accomplished lutenist and composer, Galileo also became a talented lutenist himself. When Galileo was about eight years old his family moved to Florence, where he was exposed to the great artists of the Renaissance and developed an appreciation for the aesthetic qualities of life and nature.

Encouraged by his father to prepare for the priesthood, Galileo chose instead to enter the University of Pisa to study medicine. But in yet another serendipitous event that affected life (and science) trajectories, Galileo attended a lecture on geometry that so fascinated him he decided to become a mathematician instead.

In 1588, while still early in his career, Galileo was hired as an instructor in the Accademia delle Arti del Disegno (Academy of the Arts of Drawing) in Florence, where he taught perspective and the use of strong light and dark contrast in art. One year later, Galileo became the chair of mathematics at the University of Pisa, then moved to the University of Padua in 1592, where he taught mathematics, mechanics, and astronomy. For reasons to be discussed below, in June 1610 Galileo was appointed "Chief Mathematician of the University of Pisa and Philosopher of the Grand Duke, without obligation to teach and reside at the University or in the city of Pisa, and with a salary of one thousand Florentine scudi per annum." Instead of living in Pisa, Galileo would reside in Florence for the rest of his life.

Fig. 4.48 Portrait of Galileo Galilei (1564–1642) by Justus Sustermans (1597–1681). (Public domain)

The First Telescopic Observations of the Heavens

People had been making use of lenses for centuries in devices such as eyeglasses (Kepler used a pair), but it wasn't until about 1600 that Hans Lippershey (1570–1619) was likely the first person to combine two lenses and notice that they made distant objects appear larger. (One version of the story has it that children, playing with lenses in his shop first put the combination together.) The German–Dutch lens maker, living in Middelburg, Netherlands, had invented the telescope. Hoping to profit from his invention, he petitioned the State General on October 2, 1608 to keep his invention secret while allowing him to make the device for his country only, and hoping also to receive a grant that would allow him to develop a similar device that could be looked through with both eyes. Unfortunately, knowledge of the device was already spreading; King Henry IV received one from the French ambassador at the Hague in 1608 and telescopes were being sold in Germany and Italy, and they were even being made in London by 1609.

It was in 1609 that news of the invention made its way to Galileo in Padua. Based on a description of the device, Galileo designed his own and began to improve upon it. A replica of one of his telescopes is shown in Fig. 4.49. What today

Fig. 4.49 A replica of one of Galileo's telescopes, on display at the Griffith Observatory in Los Angeles. (Photograph by Jim and Rhoda Morris, public domain)

seems like an obvious thing to do (at least for those reading this text, as well as astronomers and amateurs worldwide), Galileo not only looked at Earth-based objects with his telescope, he also pointed it up at the heavens. What this mathematician saw changed his career, and science, forever. [Galileo was not the first to study the heavens with a telescope; Thomas Harriot (1560–1621) had sketched the surface of the Moon four months earlier.]

Galileo was not initially a Copernican; in fact he was supporting the Ptolemaic view during lectures in Padua in 1597. When he received Kepler's *Mysterium Cosmographicum*, arguing in support of the heliocentric model, Galileo acknowledged it but refused to endorse the idea. However, his thoughts on the subject were evolving based in part on his own long-since discredited theory that ocean tides are produced by a sloshing of water back and forth due to Earth's motions. Any uncertainty that Galileo may have still had about the heliocentric model vanished forever when he made his telescopic observations and then reported them in a small, twenty-four page booklet titled *Sidereus Nuncius* (*The Starry Messenger*) in March 1610; Fig. 4.50.

As Galileo states at the beginning of *Sidereus Nuncius*:

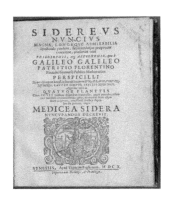

Fig. 4.50 The cover of Galileo's *Sidereus Nuncius*. (Public domain)

> In the present small treatise I set forth some matters of great interest for all observers of natural phenomena to look at and consider. They are of great interest, I think, both from their intrinsic excellence, and from their absolute novelty, and also on account of the instrument by the aide of which they have been presented to my apprehension.
>
> The number of the fixed stars which observers have been able to see without artificial powers of sight up to this day can be counted. It is therefore decidedly a great feat to add to their number, and to set distinctly before the eyes other stars in myriads, which have never been seen before, and which surpass the old, previously known, stars in number more than ten times.
>
> Again, it is a most beautiful and delightful sight to behold the body of the Moon, which is distant from us nearly sixty [semi-]diameters of the Earth, as near as if it was at a distance of only two of the same measures; so that the diameter of this same Moon appears about thirty times larger, its surface about nine hundred times, and its solid mass nearly 27,000 times larger than when it is viewed only with the naked eye; and consequently any one may know, with the certainty that is due to the use of our senses, that the Moon certainly does not possess a smooth and polished surface, but one rough and uneven, and, just like the face of the Earth itself, is everywhere full of vast protuberances, deep chasms, and sinuosities.
>
> Then to have got rid of disputes about the Galaxy or Milky Way, and to have made its nature clear to the very senses, not to say to the understanding, seems by no means a matter which ought to be considered of slight importance. In addition to this, to point out, as with one's finger, the nature of these stars which every one of the astronomers up to this time has called cloud-like, and to demonstrate that it is very different from what has hitherto been believed, will be pleasant, and very fine.
>
> But that which will excite the greatest astonishment by far, and which indeed especially moved me to call the attention of all astronomers and philosophers, is this, namely, that I have discovered Four Erratic Stars, neither known nor observed by any one of the astronomers before my time, which have their revolutions round a certain bright star, one of those previously known, like Venus and Mercury round the Sun, and are sometimes in front of it, sometimes behind it, though they never depart from it beyond certain limits. all which facts were

discovered and observed a few days ago by the help of a telescope devised by me, through God's grace first enlightening my mind.

Galileo's observations severely challenged Aristotelian science. In fact he would go on to make other important observations that were not first recorded in *Sidereus Nuncius*. Here is a list of his observations, as well as their importance in contradicting the Aristotelian–Ptolemaic worldview:

- *There are countless more stars in the heavens that can be seen with the naked eye.* Why would God create stars that are not directly observable by Man? Is the universe possibly infinite after all, as suggested by Thomas Diggs and Giordano Bruno? Tycho had argued that the stars could not be much farther away that the orbit of Saturn.
- *The Milky Way is not cloud-like, but is instead composed of a an enormous number of stars, sometimes clustered.* Again, why is a device needed to see objects in the heavens?
- *The Moon contains mountains and craters.* The objects in the heavens had always been believed to be perfect, meaning perfectly spherical and unblemished. How is it possible that the heavens are corruptible?
- *Jupiter has four moons in orbit around it* as evidenced by changes in where they are located relative to the planet, sometimes passing behind, and sometimes in front of it. The orbits of each of the moons also have easily measured periods. There should only be one center of the universe, specifically Earth, although the Tychonian pseudo-geocentric model, quite popular at the time, did suggest that the Sun and the Moon orbit Earth while the other planets orbit the Sun. [The four Galilean moons were discovered independently by Simon Marius (1573–1625), perhaps before Galileo, but he didn't start keeping notes until one day after Galileo documented his discovery.]
- *Venus is observed to have phases ranging from a thin crescent to full.* This observation strongly suggests that Venus must orbit the Sun. If so, then a crescent would be seen when Venus is between the Sun and Earth while a gibbous or full phase would be seen when Venus is on the other side of the Sun relative to Earth. Although this is possible from the Tychonian model, Ptolemy argued that Venus must orbit Earth either entirely on Earth's side of the Sun or the far side of the Sun. In the former case an observer would only be able to see crescent phases, while on the far side only gibbous or full phases would be visible.
- *The Sun has spots.* Surely if the heavens are perfect, then the giver of light, warmth, and life must be spotless. Galileo didn't actually discover sunspots first since they can be seen with the naked eye. The Chinese had known about them for centuries; in addition, John of Worcester (died c. 1140) sketched a sunspot in 1128. Telescopically, David Fabricius (1564–1617) and his son Johannes (1587–1615) observed sunspots in 1611 before Galileo observed them in 1612. (As mentioned before, DON'T LOOK FOR THEM YOURSELF WITHOUT PROPER EYE PROTECTION!!!)
- *Galileo observed Saturn's rings* (Fig. 4.52) although when he first saw them in 1610 he thought they might be two "planets" orbiting Saturn, similar to

Fig. 4.51 Sketches from Galileo's *Sidereus Nuncius* (English translation), from top to bottom: the Moon at waning gibbous and third quarter, the Pleiades star cluster in Taurus (only the large stars in the sketch are visible to the naked eye), and the four Galilean moons of Jupiter on different days, illustrating their motions. (Public domain. Based on the version by Carlos,[6] 1880)

[6]Carlos, Edward Stafford (1880). Galileo's *Sidereus nuncius* (English translation), edited and corrected by Peter Barker (2004, Byzantium Press: Oklahoma City).

Fig. 4.52 Galileo's sketch of Saturn's rings from 1616, published in his treatise, *Assayer* of 1623. (Public domain)

Jupiter's moons. In 1612 he was surprised to find that the two "planets" had disappeared, only to reappear in 1616, at which time he noted that instead of the fuzzy appendages being "planets," he observed and sketched "two half-ellipses" around Saturn. At the time he didn't realize that his "half-ellipses" were actually portions of that planet's extensive ring system. The vanishing of the "planets" in 1612 was the result of the plane of the very thin ring system being seen edge on.

- *It seems that Galileo may have seen Neptune for the first time*, although he didn't realize it was a planet. He did note that the "star" moved slowly relative to other nearby stars. Are not the stars all fixed on the celestial sphere?

Ironically, although his observations and his tidal theory made him a dedicated Copernican for the rest of his life, Galileo would never agree with Kepler that the correct orbital motions of the planets were elliptical. Like so many before him, even Galileo could not give up on the idea that the motions in the heavens must be circular, the epitome of Godly perfection.

Other Important Contributions to Science Made by Galileo

It was not only Galileo's telescopic observations that greatly impacted science, his work in engineering and natural philosophy (physics) did as well. For example, Galileo improved upon a geometric and military compass that could be used by both artillery gunners and surveyors. He also constructed a crude thermometer. And, in addition to his telescope, he invented a compound microscope as well.

While in the cathedral in Pisa, Galileo happened to notice that the chandelier seemed to swing back and forth with the same period regardless of the amplitude and the direction it was swinging (he used his own pulse to time the swings). In reality, the period of oscillation is only approximately independent of amplitude, but it is nearly true if the amplitude is small. Galileo also discovered that the square of the period of oscillation is related to the length of the pendulum. That principle has now been used in grandfather clocks for centuries.

Other than his telescopic observations, perhaps the most famous story of Galileo is that he dropped balls of different masses from the Leaning Tower of Pisa and found that they all fell at the same constantly accelerating rate (the acceleration due to gravity), regardless of mass. The validity of the story has been challenged by many historians, but others have argued that it could have occurred. Whether that experiment was actually performed or not, he did do similar experiments by using inclined planes that allowed for greater control of air resistance and friction. In any case, he reached the correct conclusion that, in the absence of other resistive forces, objects fall under the influence of gravity at precisely the same rate, independent of how much they weigh.

Also based on his experiments with objects sliding down inclined planes, Galileo came to the conclusion that if an object is moving on a flat, horizontal surface without any resistance then that object would continue traveling forever in a straight line at a constant speed (inertia). This result directly flies in the face of Aristotelian physics, which had argued for two thousand years that the natural motion of an object is to come to rest at its appropriate level (air, fire, water, or earth).

Key Point
All falling objects accelerate at the same rate in a vacuum, independent of their masses.

Key Point
An object in motion continues in motion in a straight line at a constant speed unless a force acts to alter that motion.

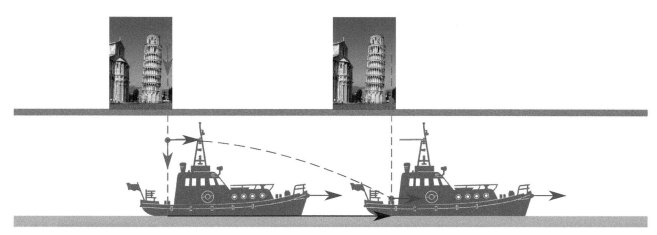

Fig. 4.53 Galilean relativity postulates that any experiment, such as dropping a ball, will result in the same outcome even if the experiment is done on a steadily moving ship. Here, a ball dropped from the Leaning Tower of Pisa drops straight down to the ground following the red dashed path. On the ship, the ball follows the red dashed path and lands on the deck of the ship at a point directly below where it was dropped. The green arrows represent the downward motion of the ball relative to the Leaning Tower of Pisa and relative to the ship. The blue arrows symbolize the forward speeds of the ball and the ship. The black arrow on the surface of the water indicates how far forward the ship and the ball have moved while the ball was falling. To someone watching from a boat not moving relative to the water, the ball will appear to fall along the blue dashed trajectory (it is a very fast ship!). (Photograph of the Leaning Tower of Pisa by Alkarex Malin äger, CC-BY-SA-3.0)

Yet another profound conclusion reached by Galileo was his concept of relative motion, known today as Galilean relativity. As depicted in Fig. 4.53, Galileo postulated that if an experiment were to be performed in a stationary laboratory, or on a steadily moving ship, the results of the experiments would be identical. For example, if a ball is dropped vertically in a laboratory, dropping the same ball on a ship that is moving in a straight line at a constant speed would result in the ball falling directly downward on the ship as well, *relative to the ship*. The reason that this happens is that the ball on the ship is already moving in the same direction as the ship and with the same speed when it is dropped. If you have never done so, it is easy to perform the experiment in a moving car (just don't be the one driving when you do the experiment). From outside the ship (or the car), an observer will see the ball move downward *and forward* as it falls.

Galileo's idea of relativity provided an answer to one of the major critiques from opponents of a moving Earth. If Earth is spinning on an axis, shouldn't a dropped ball be left behind while Earth moves under it? Galileo's response: No, because the ball is already moving in the direction that Earth is rotating, and it keeps moving in that direction as it is falling. In Fig. 4.53 the Leaning Tower of Pisa is actually moving at nearly 1600 km/h (1000 mi/h) in the direction of Earth's rotation, and the ball has that speed too.

It's worth mentioning that Galileo also tried to measure the speed of light, but his method was woefully inadequate for the task. He and an assistant stood about one mile apart with shuttered lanterns. First, Galileo uncovered his lantern, then when his assistant saw Galileo's lantern he would uncover the lantern he was holding. At that point, Galileo would note now long it had been between the time he uncovered his lantern and when he saw the second one. All Galileo could conclude

Key Point
Identical experiments conducted in a laboratory or on a ship that is moving in a straight line at a constant speed with respect to the laboratory give identical results.

from the experiment is that light is either infinitely fast (it isn't by the way), or simply too fast to be measured by his approach.

In his 1623 book, *The Assayer*, Galileo described his view of the role of mathematics in science:

> Philosophy [physics] is written in this grand book — I mean the universe — which stands continually open to our gaze, but it cannot be understood unless one first learns to comprehend the language and interpret the characters in which it is written. It is written in the language of mathematics, and its characters are triangles, circles, and other geometrical figures, without which it is humanly impossible to understand a single word of it; without these, one is wandering around in a dark labyrinth.

Key Point
Mathematics is the language of nature.

Not only did Galileo's natural mathematical orientation serve him well (as it did Kepler and Copernicus before him), but he also saw that science cannot advance without both a theoretical underpinning and experimental testing. Before Galileo, the process of carefully constructed, repeatable experiments that explore phenomena and test ideas was largely foreign in natural philosophy. Even with his telescope, Galileo conducted extensive experiments, including looking at an Earthbound object through his telescope and then going over to the object to make sure that he had not been deceived. In 1611 he stated: "Over a period of two years now, I have tested my instrument (or rather dozens of my instruments) by hundreds and thousands of experiments involving thousands and thousands of objects, near and far, large and small, bright and dark."

Key Point
Experimentation and empirical testing are fundamental to the nature of science.

Regarding Galileo's contributions, Albert Einstein said:

> Pure logical thinking cannot yield us any knowledge of the empirical world; all knowledge of reality starts from experience and ends in it. Propositions arrived at by purely logical means are completely empty as regards reality. Because Galileo saw this, and particularly because he drummed it into the scientific world, he is the father of modern physics — indeed, of modern science altogether.

Key Point
Galileo is considered to be the father of modern science.

The Galileo Affair

By 1610, when Galileo wrote *Sidereus Nuncius*, Copernicus's *De revolutionibus* was now more than sixty years old and had escaped any serious controversy. Although Copernicus's claim of the reality of heliocentrism was largely ignored, the simplicity of his model made calculations much easier, including in the creation of the Gregorian calendar. Conflict between Galileo and the Church would change the situation dramatically.

Galileo and Kepler lived at a time of growing upheaval in Europe. Following the Protestant Reformation came the Roman Catholic Counter-Reformation. In 1542 Pope Paul III created the permanent Congregation of the Holy Office of the Inquisition (commonly called the Roman Inquisition) as the Church's judicial arm. In the Council of Trent, which was convened intermittently from 1545 to 1563, one of the outcomes was the decree that the Church is the ultimate authority on interpretation of Scripture. A commission was also established to rule on whether or not certain books should be placed on the *Index Librorum Prohibitorium*, a list of prohibited books.

The Thirty Years' War would also begin in 1618, when the slowly disintegrating Holy Roman Empire tried to impose a consistent religion across its regions. The war started out as a fight between Protestant and Catholic states, but later evolved into a prolonged struggle between two great powers, the Kingdom of France and the House of Habsburg, the powerful house of the Holy Roman Empire. The Thirty Years' War would prove to be devastating, resulting in wide-spread famine and disease, dramatically reducing populations (especially in the German and Italian states), and bankrupting most of the states involved.

Within this religious and political climate, the preface to Galileo's *Sidereus Nuncius* contains the following formal review:

> The most excellent gentlemen, the Heads of the Ecclesiastical Council of Ten, designated below, having received and assurance of the Governing Council of the University of Padua, based on a report of the two persons deputized to investigate, that is from the reverend Father Inquisitor, and from the Secretary of the Senate, Giorgio Maraviglia, with an oath, that the book entitled SYDEREUS NUNCIUS, etc., by Lord Galileo Galilei, contains nothing contrary to the Holy Catholic Faith, principles, and good customs, and that it is fit for printing, grant a license, so that it may be printed in this city.
>
> Given on the first day of March 1610.
>
> Heads of the Ecclesiastical Council of Ten: Lord Marco Antonio Valaresso, Lord Nicolo Bon, Lord Lunardo Marcello
>
> Secretary to the most illustrious Council of Ten: Bartolomeo Comino
>
> 1610, 8 March, Registered in the records, on page 39.
>
> Office of Ioannes Baptista Breatto,
> Coadjutor of the Council of Blasphemy

When Galileo discovered the four moons orbiting Jupiter he proposed to call them the "Medicean Stars" in honor of "The Most Serene Cosimo De' Medici the Second, Fourth Grand-Duke of Tuscany." It was that dedication in *Sidereus Nuncius*, together with the gift of one of his telescopes, that caused the Grand Duke to appoint Galileo to his lifetime appointment to the University of Pisa without obligation to reside or teach there (page 131). Unfortunately, that generous appointment did not endear Galileo to the other academics, particularly the philosophers, who were the ones who taught physics at the time. Moreover, in a bit of irony, it was the philosophers who strongly opposed Galileo's attack on Aristotelian philosophy, rather than the Church authorities.

In 1611 Galileo was invited to Rome where he was immediately granted an audience with Pope Paul V (1552–1621) on April 1. In an unusual show of respect, Pope Paul V allowed Galileo to stand in his presence, rather than remain kneeling. Two weeks later Galileo was honored with an outdoor banquet where he demonstrated his telescope to "several theologians, philosophers, mathematicians, and others." A fresco painted more than two centuries later depicts the event; Fig. 4.54.

Even so, there was sufficient disbelief in what Galileo was reporting that some Aristotelians simply refused to look through his telescopes. According to one colleague who did look through a Galilean telescope,

> I tested this instrument of Galileo's in a thousand ways, both on things here below and on those above. Below, it works wonderfully; in the sky it deceives

Fig. 4.54 Fresco done in 1858: *Galileo Galilei showing the Doge of Venice how to use the telescope.* [Giuseppe Bertini (1825–1898); public domain]

one, as some fixed stars are seen double. I have as witnesses most excellent men and noble doctors ... and all have admitted the instrument to deceive.

In fact a great many stars that appear to be single stars to the naked eye are actually double star systems; Galileo's colleague was not being deceived, he just didn't believe what he was seeing.

In December 1613, a philosopher from the University of Pisa and a former student of Galileo's, Benedetto Castelli (1578–1643), was asked by the Grand Duchess Christina of Tuscany[7] (1565–1637) to comment on theological grounds regarding the idea that Earth moves. Castelli, who was also a Benedictine abbot, indicated that he did not believe that a moving Earth could be possible. Castelli also told Galileo of the conversation. In response, Galileo wrote a lengthy letter to the Grand Duchess outlining his own ideas about the relationship between theology and science. In it he mentions a statement first believed to have been made by Cardinal Cesare Baronius (1538–1607), "That the intention of the Holy Ghost is to teach us how one goes to heaven, not how heaven goes." By 1615 the *Letter to the Grand Duchess Christina of Tuscany* had been expanded and copied multiple times. In the first two paragraphs he wrote:

> Some years ago, as Your Serene Highness well knows, I discovered in the heavens many things that had not been seen before our own age. The novelty of these things, as well as some consequences which followed from them in contradiction to the physical notions commonly held among academic philosophers, stirred up against me no small number of professors — as if I had placed these things in the sky with my own hands in order to upset nature and overturn the sciences. They seemed to forget that the increase of known truths stimulates the investigation, establishment, and growth of the arts; not their diminution or destruction.
>
> Showing a greater fondness for their own opinions than for truth they sought to deny and disprove the new things which, if they had cared to look for themselves, their own senses would have demonstrated to them. To this end they hurled various charges and published numerous writings filled with vain arguments, and they made the grave mistake of sprinkling these with passages taken from places in the Bible which they had failed to understand properly, and which were ill-suited to their purposes.

In his passion to share his exciting new discoveries with the world, Galileo was not always diplomatic.

Being a devote Catholic, Galileo also believed that the Bible, when interpreted properly, was inerrant. In the *Letter* he wrote:

> I think in the first place that it is very pious to say and prudent to affirm that the holy Bible can never speak untruth — whenever its true meaning is understood. But I believe nobody will deny that it is often very abstruse, and may say things which are quite different from what its bare words signify. Hence in expounding the Bible if one were always to confine oneself to the unadorned grammatical meaning, one might fall into error...
>
> But I do not feel obliged to believe that the same God who has endowed us with senses, reason and intellect has intended us to forgo their use and by some

---

[7]Christina was the granddaughter of Christina De' Medici, Queen of France, and mother of Cosimo De' Medici the Second for whom Galileo had proposed naming Jupiter's moons.

other means to give us knowledge which we can attain by them. He would not require us to deny sense and reason in physical matters which are set before our eyes and minds by direct experience or necessary demonstrations.

Galileo knew that the philosophers were working to have Copernican books placed on the list of banned books by the Roman Inquisition's Commission, so he traveled to Rome in late 1615 (probably a mistake) to defend his position. After the first draft of the *Letter* the Inquisition had found no fault in Galileo's Catholic theology, but the philosophers had pressed Pope Paul V to review the case. Cardinal Robert Bellarmine (1542–1621), who was one of the judges in Giordano Bruno's trial, and some of the Jesuit fathers were already starting to argue heresy, given the edict from the Council of Trent that the Church alone was allowed to interpret Scripture. Disliking academic disputes, in February 1616 Pope Paul V had Cardinal Bellarmine inform Galileo that he was not to teach heliocentrism and the motion of Earth any longer. In addition, the Commission of the Roman Inquisition was told to forbid any Copernican book from arguing for reconciliation between heliocentrism and the Bible, or claim that heliocentrism is literally true. Use of Copernicus's *De revolutionibus* was suspended until two small "corrections" could be made.

Returning to Florence defeated, Galileo refrained from teaching heliocentrism for many years, but then Paul V died in 1621 and his successor, Pope Gregory XV (1554–1623), died two years later. Gregory XV's successor turned out to be Maffeo Barberini (1568–1644), who took the name Pope Urban VIII. Urban VIII had been a long-time supporter and friend of Galileo's, so Galileo naturally thought that he had a new ally in Rome. Traveling once again to Rome in 1624 to pay his respects to the new pope, Galileo was granted six audiences with Urban VIII. The Pope was himself an intellectual and wanted to build support for an intellectual climate in Italy, while arguing that the edict from the Council of Trent was only meant to limit interpretation of scripture, not inhibit science. For his part Galileo proposed to write a book supporting Urban VIII's goal.

For the next six years Galileo sporadically worked on the book. He intended to call it *Dialogue on the Tides*, in which he discussed his theory along with Copernicanism and Aristotelianism. When he completed the work, the book was submitted for review by the Commission's censors. After some editing to remove any conflicts with the edict, the book was approved for publication. Urban VIII, however, asked for additional changes; the name of the book should not contain mention of tides, nor should the text itself discuss tides. Although desiring to ensure that Italy was seen as supporting science, Urban didn't want the book to reflect a particular position on heliocentrism and the movement of Earth.

The book, with the required changes, was published in Italian in 1632 under the title *Dialogue*. Beginning in 1744, it was published as *Dialogue Concerning the Two Chief World Systems, Ptolemaic and Copernican*, which is the title by which it is most commonly referred to today. *Dialogue* quickly became very popular, but its printing was halted almost immediately. As with other books that had been written over time arguing for and against particular viewpoints, *Dialogue* was constructed as a three-person conversation. The players were Salviati, who argues for the Copernican system, Simplicio, who held to the Aristotelian worldview, and Segredo, who was an intelligent layman listening intently to the arguments while inevitably being swayed by the Copernican argument. Both Salviati and Segredo were named after good friends of Galileo, while Simplicio was supposedly named for Simplicius of Cilicia (c. 490–560 CE) who commented on Aristotle. It is generally believed however, that there was a hidden meaning in the name of Simplicio as the simpleton.

Urban VIII was outraged by the book, in part because he perceived that some of his own comments appeared in the failed arguments of Simplicio. Galileo was summoned back to Rome in February 1633 to stand trial for heresy, despite being gravely ill at the time. Kindly, the Grand Duke of Florence provided a stretcher and he was carried all the way from Florence to Rome, during the winter. During his trial, Galileo was threatened with torture to supposedly insure that he told the truth in all questioning. Thankfully, no torture was ever carried out.

In the end, on June 16, 1633, the Pope, not wanting to show weakness in a religious and politically tense environment, had *Dialogue* placed on the *Index Librorum Prohibitorium*. In addition, Galileo was found guilty of heresy. On June 22 Galileo knelt before those who had sat in judgment over him and read a long statement that stated in part:

> I, Galileo, son of the late Vincenzio Galilei, Florentine, aged seventy years, arraigned personally before this tribunal and kneeling before you, Most eminent and Lord Cardinals Inquisitors-General against heretical pravity throughout the entire Christian Commonwealth, having before my eyes and touching with my hands the Holy Gospels, swear that I have always believed, do believe, and with God's help will in the future believe all that is held, preached and taught by the Holy Catholic and Apostolic Church. But, whereas, after an injunction had been lawfully intimated to me by this Holy Office to the effect that I must altogether abandon the false opinion that the Sun is the center of the world and immobile, and that the Earth is not the center of the world and moves, and that I must not hold, defend, or teach, in any way, verbally or in writing, the said false doctrine, and after it had been notified to me that the said doctrine was contrary to Holy Scripture, I wrote and printed a book in which I treated this new doctrine already condemned and brought forth arguments in its favor without presenting any solution for them, I have been judged to be vehemently suspected of heresy, that is, of having held and believed that the Sun is the center of the world and immobile and that the Earth is not the center and moves.
>
> Therefore, desiring to remove from the minds of Your Eminences, and of all faithful Christians, this vehement suspicion rightly conceived against me, with sincere heart and unpretended faith I abjure, curse, and detest the aforesaid errors and heresies and also every other error, error and sect whatever, contrary to the Holy Church, and I swear that in the future I will never again say or assert verbally or in writing, anything that might cause a similar suspicion toward me.

Galileo was sentenced to life in prison, but the next day the sentence was commuted to house arrest in Florence for the remainder of his life. While initially to be spent in solitary confinement, the Inquisition later let him have visitors, including the poet John Milton (1608–1674). Galileo did write another very important work while under house arrest, *The Two Sciences*, in which his work in physics, and some work on the nature of materials, was presented, again by the characters Salviati, Segredo, and Simplicio. In order to be published, the work was smuggled to Leyden where it was printed by the Elzevirs. Galileo went blind during the final four years of his life, possibly due to his observations of the Sun through his telescopes.

On January 8, 1642, Vincenzo Viviani (1622–1703), a young scholar whom the Pope allowed to serve as a companion to Galileo, reported his death:

> With philosophic and Christian firmness he rendered up his soul to its Creator, sending it, as he liked to believe, to enjoy and to watch from a closer vantage point those eternal and immutable marvels which he, by means of a fragile device, had brought closer to our mortal eyes with such eagerness and impatience.

It wasn't until 1744, when the *Dialogue* was reprinted with its extended title, *Dialogue of Two Chief World Systems, Ptolemaic and Copernican*, and with an approved preface by a Catholic theologian, that Galileo's book was removed from the *Index Librorum Prohibitorium*. In 1758 Pope Benedict XIV (1740–1758) removed Copernicus's never-corrected *De revolutionibus orbium coelestium* from the *Index Librorum Prohibitorium*. Pope John Paul II (1920–2005) stated in 1978 that Galileo's Catholic theology in the *Letter to the Grand Duchess Christina of Tuscany* was better than those who condemned him. On October 31, 1992, after a 13-year review by Vatican experts, Pope John Paul II declared that the Church had erred in its condemnation of Galileo because of a "tragic mutual incomprehension."

As any historian will tell you, the past offers valuable lessons for the present and the future. The histories of science and religion are no exceptions.

 ## Exercises

All of the exercises are designed to further develop *your understanding* of the material by thinking carefully and critically about what you have read in this chapter. Answers to selected exercises are available in the back of the book.

True/False

1. The medieval educational model involved teaching the trivium followed by the quadrivium.
2. (*Answer*) The trivium focused on the development of critical thinking through the instruction of grammar, logic, and rhetoric.
3. The quadrivium involved the teaching of theology and astrology.
4. (*Answer*) The development and societal acceptance of scientific understanding is completely independent of social, philosophical, political, and religious influences.
5. Inductive reasoning is a philosophical view that any conclusion reached by a logical argument, and based on currently available evidence, is assumed to remain valid for all time.
6. (*Answer*) Copernicus was a young man at the time when Christopher Columbus inadvertently demonstrated flaws in Ptolemy's *Geography*.
7. Copernicus believed that the philosophical tenants of uniform and circular motion ascribed to Plato were fundamentally flawed.
8. (*Answer*) Copernicus's book, *De revolutionibus orbium coelestium*, was placed on the Index of forbidden books immediately after it was published.
9. One sidereal day elapses when Earth rotates 360° on its axis.
10. (*Answer*) One solar day elapses when Earth rotates 360° on its axis.
11. The 24 hours we measure everyday for the length of one day is one sidereal day.
12. (*Answer*) One sidereal day is approximately four minutes shorter than one solar day.
13. When viewed from above Earth's North Pole, Earth's rotation direction and its orbital direction are both counterclockwise.
14. (*Answer*) On January 1, Gemini is near your meridian at midnight.
15. Sunlight is most intense at noon on the tropic of Capricorn on about December 21.
16. (*Answer*) At the North Pole, the Sun is found at the observer's zenith at noon.
17. The orbital direction of the Moon is clockwise when viewed from above Earth's North Pole.
18. (*Answer*) The Moon is always within 1/2° of the ecliptic.
19. The Moon must be close to the line of nodes at the same time that the phase of the Moon is new in order for a solar eclipse to occur.
20. (*Answer*) It is possible for you to observe Venus crossing your meridian at sunset.
21. It is possible for you to see Jupiter crossing your meridian at sunset.
22. (*Answer*) Tycho was a strong proponent of the Copernican heliocentric system.
23. Tycho was considered to be the greatest naked-eye observational astronomer of his time.
24. (*Answer*) Kepler was influenced by the philosophies of Pythagoras and Plato.

25. Kepler never believed in perfectly circular motion.
26. (*Answer*) Kepler's first work on planetary orbits was based on spheres and nested platonic solids.
27. Kepler didn't believe that his three laws actually represented reality, only that they provided an excellent way to compute the motions of the planets through the heavens.
28. (*Answer*) Kepler's law of equal areas (his second law) relates the orbital speed of a planet around the Sun to its distance from the Sun at any point in its orbit.
29. The naked-eye planet with the largest orbital eccentricity is Saturn because it is farthest from the Sun.
30. (*Answer*) Galileo was the first person to build a telescope.
31. Galileo demonstrated that the Milky Way Galaxy is composed of giant clouds that stretch across the sky.
32. (*Answer*) In the absence of air, a heavy cannonball and a much-lighter golf ball will fall at the same rate.
33. If you are riding a bicycle and drop a ball, the ball will hit the ground behind the bicycle at the place where it was dropped.

Multiple Choice

34. (*Answer*) The quadrivium was composed of which of the following:
    (a) arithmetic   (b) geometry   (c) music
    (d) astronomy   (e) all of the above   (f) (a) and (d) only
35. Which of the following people lived during the European Renaissance?
    (a) Claudius Ptolemy   (b) Leonardo da Vinci
    (c) Martin Luther   (d) Nicolaus Copernicus
    (e) Christopher Columbus   (f) all of the above
    (g) (b) through (e) only   (h) (b) and (d) only
36. (*Answer*) Which of the following *was not* one of Copernicus's postulates in his *Commentariolus*:
    (a) Earth is not the center of the universe.
    (b) The celestial sphere is located just beyond the orbit of Saturn.
    (c) The Sun is located near the center of the universe.
    (d) The apparent motions of the fixed stars are due to the motions of Earth.
    (e) The apparent motions of the planets are due in part to motions of Earth.
    (f) All of the above were postulates contained in the *Commentariolus*.
37. When the Sun is located in Libra, _____ will be near your meridian at midnight.
    (a) Scorpius   (b) Aquarius   (c) Aries   (d) Gemini
38. (*Answer*) During the period between when the Sun crosses the vernal equinox and then the autumnal equinox, an observer at the North Pole experiences _____ of sunshine every day.
    (a) 0 hours   (b) 6 hours   (c) 12 hours   (d) 24 hours
39. The Sun just grazes the northern horizon at midnight on the winter solstice at what latitude?
    (a) arctic circle   (b) tropic of Cancer
    (c) equator   (d) tropic of Capricorn
    (e) antarctic circle   (f) none of the above
40. (*Answer*) The star tau Hercules will be the North Star again in about _____.
    (a) 6000 CE   (b) 15,000 CE   (c) 19,000 CE
    (d) 24,000 CE   (e) Never, Polaris is always the North Star.

41. Given that we always see the same "face" of the Moon from Earth, what is the rotation period of the Moon? The Moon's sidereal orbital period is 27.32 days and its synodic phase period is 29.53 days.
    (a) 0 days; the Moon doesn't rotate
    (b) 27.32 days
    (c) 29.53 days
    (d) There is not enough information provided to determine the Moon's rotation period.
42. (*Answer*) When observing our Solar System from above Earth's North Pole, which motions are counterclockwise?
    (a) the orbits of the inferior planets
    (b) the orbit of Earth
    (c) the orbit of the Moon
    (d) the orbits of the superior planets
    (e) the rotation of Earth about its axis
    (f) the rotation of the Moon about its axis
    (g) all of the above
    (h) (b) through (f) only
43. During a total lunar eclipse the Moon appears reddish in color because
    (a) lunar dust is ejected from the surface.
    (b) red light is scattered more than blue light when sunlight passes through Earth's atmosphere.
    (c) blue light is scattered more than red light and the red light is bent toward the Moon while passing through Earth's atmosphere.
    (d) the Moon is always reddish in color, but the color can only be seen with the naked eye when the Moon passes through Earth's shadow, cutting off the bright light from the Sun.
44. (*Answer*) In order for a total lunar eclipse to occur, the following condition(s) must be met:
    (a) the phase of the Moon must be full.
    (b) the Moon must be near the line of nodes.
    (c) the Moon must be almost 5° above or below the plane of the ecliptic.
    (d) all of the above.
    (e) (a) and (b) only.
45. Which of the following planets can never be seen crossing the meridian late at night?
    (a) Mercury   (b) Venus
    (c) Mars   (d) Jupiter
    (e) Saturn   (f) all of the above
    (g) (a) and (b) only   (h) (c), (d), and (e) only
46. (*Answer*) The following observations were attributed to Tycho Brahe:
    (a) planetary positions with typically better than $2'$ accuracy
    (b) the distance to the supernova of 1572, indicating that it was beyond the orbit of the Moon
    (c) the distance to the Great Comet of 1577, indicating that it was farther away than the Moon
    (d) all of the above
47. Kepler's contributions to astronomy included
    (a) the philosophical view that phenomena in the heavens are governed by the same laws as exist on Earth.
    (b) the beginning of the downfall of the belief in uniform and circular motion in the heavens.
    (c) support for the Tychonian planetary system.
    (d) all of the above.
    (e) (a) and (b) only.

48. (*Answer*) Which of the following represent Kepler's harmonic law (his third law)?
    (a) $P_y^2 = a_{au}^3$ (b) $P_y^2/a_{au}^3 = 1$ (c) $a_{au}^3/P_y^2 = 1$
    (d) For planets orbiting the Sun, the orbital period (in years) squared equals the semimajor axis (in astronomical units) cubed.
    (e) all of the above

49. A new, mysterious Planet X is discovered orbiting the Sun in an elliptical orbit with a semimajor axis of 25 au. Which of the following is closest to the orbital period that it must have?
    (a) 8.5 years   (b) 25 years   (c) 125 years
    (d) 625 years   (e) 15,625 years

50. (*Answer*) Galileo discovered the following with his telescopes:
    (a) The Moon has mountains and craters on its surface.
    (b) Jupiter has moons in orbit around it.
    (c) Venus goes through phases that include thin crescents, gibbous, and full.
    (d) The Milky Way is composed of a myriad of stars.
    (e) all of the above.
    (f) (a) and (b) only.

51. Which of the following will fall most rapidly in the absence of air?
    (a) a cannonball
    (b) a baseball
    (c) a golf ball
    (d) a ping pong (table tennis) ball
    (e) a feather
    (f) they will all fall at the same rate

52. (*Answer*) According to Galileo, if an object is set in motion without any resistance, it will
    (a) eventually stop when it obtains its natural level.
    (b) travel in a straight line at a constant speed.
    (c) continue on forever.
    (d) all of the above.
    (e) (a) and (b) only.
    (f) (b) and (c) only.

53. Which of the following statements is not true?
    (a) Mathematics is considered to be the language of nature.
    (b) Logic alone is sufficient to understand nature.
    (c) Experimentation and observation are fundamental components of true science.
    (d) all of the above.
    (e) (a) and (b) only.
    (f) (a) and (c) only.

Short Answer

54. Comment on the relationship between the four topics of the quadrivium from the viewpoint of Pythagorean philosophy.
55. What role did the invention of Gutenberg's printing press play in the development of knowledge during the early European Renaissance?
56. Explain how Osiander's preface to *De revolutionibus* is clearly inconsistent with Copernicus's views as expressed in his own preface, which was directed specifically to Pope Paul III.
57. In your own words explain the principal reason why the southern hemisphere is warmer in January than it is in July.
58. Why is it important to take precession into consideration when investigating potential astronomical alignments of archaeological sites?

59. (a) List all of the total and partial lunar eclipses and all of the total, partial, and annular solar eclipses that will be visible somewhere in the world between 2026 and 2030. You may wish to consult an internet source such as https://eclipse.gsfc.nasa.gov/eclipse.html.
    (b) How many lunar eclipses occur each year on average between 2026 and 2030?
    (c) How many solar eclipses occur each year on average during that same period?
    (d) Explain why solar eclipses are rare events in a particular location on Earth?

60. In the orbital configuration shown in Fig. 4.27, would Venus be considered as the morning star or the evening star? Explain your reasoning.

61. Draw a diagram that illustrates how Jupiter temporarily appears to travel in a retrograde direction relative to the background stars. Your diagram should accurately take into consideration that Earth's sidereal period is 1 year while Jupiter's sidereal period is 11.9 years. Your diagram should also reflect the fact that Jupiter's semimajor axis is 5.2 au compared to Earth's 1 au. Don't concern yourself with the slight eccentricities of the planets' orbits.

62. Did Kepler believe that his strong religious views were contrary to his astronomical discoveries? Explain your reasoning.

63. (*Answer*) Calculate a_{au}^3/P_y^2 for the orbit of our Moon around Earth where $a_{au} = 0.00256$ au and $P_y = 0.0747$ y. How does this value compare with the value obtained for the orbit of Earth around the Sun?

Table 4.3 Orbital data for the Galilean moons

Moon	a (au)	P (y)
Io	0.002 819	0.004 844
Europa	0.004 485	0.009 722
Ganymede	0.007 155	0.019 588
Callisto	0.012 585	0.045 691

A scientific notation tutorial is available at Math_Review/Scientific_Notation

64. The four largest moons of Jupiter, the Galilean moons, have orbital periods and semimajor axes given in Table 4.3.
    (a) Create a complete table similar to Table 4.2 and fill in each column. Note that you might find it easier to express your results for a_{au}^3 and P_y^2 in scientific notation.
    (b) Plot your data on a graph of P_y versus a_{au}. You can download prepared graph paper from the Chapter 4 resources webpage.
    (c) Plot your data on a graph of P_y^2 versus a_{au}^3. You can download prepared graph paper from the Chapter 4 resources webpage.
    (d) What is the average of a_{au}^3/P_y^2 from your table? Speculate on the significance of your result.

65. The Moon's orbit has an eccentricity of $e = 0.0549$. Calculate the ratio of the Moon's closest approach to Earth (called perigee; "gee" specifies Earth, similar to "geo") and its greatest distance from Earth (apogee). The formulas are identical to those for calculating perihelion and aphelion, respectively, as discussed on page 126. Note that the value of the semimajor axis cancels out in the ratio of perigee to apogee, giving $(1 - e)/(1 + e)$. What does this tell you about the change in the apparent size of the Moon at different locations in its orbit?

66. Venus ranges in distance from Earth between about 0.3 au and 1.7 au (a factor of almost 6), and yet its brightness in the sky doesn't vary dramatically. Considering the phases that Venus goes through in its orbit as seen from Earth, explain why its changes in brightness are fairly minor.

67. (a) Sketch a diagram of the orbits of Venus and Earth around the Sun.
    (b) On the diagram place Earth at one location in its orbit.
    (c) Relative to the position of Earth in part (b), indicate where Venus would need to be located in its orbit in order for us to observe a nearly full phase for the planet.
    (d) Repeat part (c), but this time locate Venus where it would be seen in a crescent phase.
    (e) In your own words, explain why the phases of Venus rule out the Ptolemaic model of the Solar System, but not the Tychonian model.

68. If you throw a ball straight up into the air while you are running, and continue to run along under it at a constant speed while it is in flight, you will be able to catch the ball when it falls.
    (a) Describe what someone else would see if she were standing still when you ran past her chasing your ball in the air.
    (b) How does this situation challenge the Aristotelian argument that Earth cannot be moving?
    (c) If you stand still relative to Earth when you throw the ball straight up into the air, are you truly motionless? Explain. How is it that the ball falls straight back down to you?

69. Galileo is generally considered to be the first true scientist. Given what you learned about the nature of science in Chapter 1 and Galileo's approach to studying nature, construct an argument that either supports or rejects that assertion.

70. In light of what you have learned about the Galileo affair, consider the present-day controversy that exists between the scientific model of biological evolution and conservative religious doctrine. What similarities exist between the two cases? What differences do you see between them?

71. In 2005 the United States District Court of the Middle District of Pennsylvania ruling in Kitzmiller v. Dover Area School District found the teaching of creationism and intelligent design in public schools to be unconstitutional based on the First Amendment separation of church and state. Before that ruling a number of state legislatures, local school districts, and individual teachers had attempted to mandate the inclusion of creationism and intelligent design in curricula as valid alternative theories to biological evolution. Compare and contrast this situation to the medieval Roman Catholic Church's rulings regarding Copernican books and Galileo's *Dialog*.

Activities

72. People often think that the Moon looks much larger when it is near the horizon than when it is high in the sky. During the night of a full moon, hold your camera or smartphone close to your face and take a picture of the Moon just above the eastern horizon with your outstretched arm and thumb extended just below the Moon. Instead of your thumb you can use a pen, ruler, or some other object for comparison with the Moon. Take a second, similar, picture of the Moon and your comparison object when the Moon is high in the sky near your meridian.
    (a) Record the date and time of each photograph.
    (b) Compare the two photographs. What can you conclude about the apparent sizes of the Moon near the horizon and near the meridian?
    (c) What appears larger, your thumb (or other object that you used), or the Moon?
    (d) Comment on the role of perception versus measurement in science.

73. If a lunar eclipse (either total or partial) is going to occur during the term you are enrolled in this course, take a series of photographs of the event and carefully record the time of each photo. You should start a couple of hours before eclipse maximum if possible, or continue for a couple of hours after eclipse maximum. You will be able to find detailed information about the eclipse, including when the eclipse will begin, end, and the time of maximum, from newspapers or at NASA's eclipse website:
    https://eclipse.gsfc.nasa.gov/eclipse.html.
    (a) Could you see the round shape of Earth's shadow on the Moon? If a total lunar eclipse occurred, at what times before or after totality did you see the curvature of Earth's shadow?
    (b) How long did totality last if it occurred during the eclipse?
    (c) Record your impressions of the event, including any changes in the coloration of the Moon.

Sir Isaac Newton's Universe

5

If I have seen farther than others, it is by standing upon the shoulders of giants.

Sir Isaac Newton (1642–1727)

5.1	The Life of Sir Isaac Newton	146
5.2	Newton's Three Laws of Motion	154
5.3	The Universal Law of Gravitation	165
5.4	Newton Completes Kepler's Laws	173
5.5	The Discovery of Uranus, Neptune, and Pluto	176
5.6	The Enlightenment (The Age of Reason)	178
	Exercises	181

Fig. 5.1 Third Latin edition of Sir Isaac Newton's *Philosophiæ Naturalis Principia Mathematica*. (Paul Hermans, CC BY-SA 3.0)

Introduction

Up to this point we have witnessed a slow evolution in the enterprise of science and its successes. The advances have come in fits and starts, especially prior to the work of Nicolaus Copernicus. In many ways it can be argued that what took place prior to the reintroduction of the heliocentric model by Copernicus led largely to dead ends in advancing our understanding of nature. Even with the contribution of Copernicus, key ingredients to making progress were still missing: (a) the reluctance to abandon long-held views that the heavens were fundamentally different from what occurs below the orbit of the Moon; (b) the role of testing ideas through critical observations and experimentation; and (c) the goal of not only explaining *what* was happening, but *why* it was happening, without invoking supernatural deities by way of explanation. Johannes Kepler is principally responsible for helping us to appreciate the first missing ingredient in our changing worldview. Galileo Galilei, with his telescope aimed at the heavens and his testing of motions on Earth, offered guidance on the second ingredient. But it was largely Sir Isaac Newton (1642–1727) who played the pivotal role in taking us from simply observing and documenting what was occurring to putting us on the path of seeking fundamental explanations for what happens throughout the universe, of which we occupy a very tiny corner indeed.

5.1   The Life of Sir Isaac Newton

Woolsthorpe Manor

Fig. 5.2 Sir Isaac Newton (1642–1727). [Portrait from 1689 by Sir Godfrey Kneller (1646–1723); public domain]

Isaac Newton (Fig. 5.2) was born on December 25, 1642 in Woolsthorpe Manor, near Colsterworth, England, about 145 km (90 miles) north-northwest of London. Newton's birth came two weeks before the one-year anniversary of Galileo's death, and five months short of the centennial of the death of Copernicus and the publication of his *De revolutionibus*. Kepler died twelve years before Newton's birth.

Isaac Newton's father, also named Isaac Newton (1606–1642), was a yeoman farmer who died three months prior to the premature birth of his son, leaving the newborn's mother, Hannah (c. 1610–1679) to raise him. Baby Isaac was so small at birth that it was said that he could fit inside a quart mug. Fortunately, the family was quite well off financially given the success of the Newton clan over previous generations. Hannah had brought financial resources to the marriage as well.

When he was three years old Isaac's mother remarried and moved in with her new husband, the Reverend Barnabas Smith (1582–1653), while Isaac was left in the care of his maternal grandparents, the Ayscoughs, at Woolsthorpe (the Smiths' residence was only a little over 2 km, or 1 ½ miles, from Woolsthorpe). It seems that Newton wasn't exactly fond of his stepfather; he admitted in a list of sins recorded at age 19 to "threatening my father and mother Smith to burn them and the house over them." Newton was also less than enamored with his grandfather, and apparently the feeling was mutual given that his grandfather left Isaac out of his will.

Although Newton did not get along with the Ayscoughs, they did insist that he become educated. This was unlike the pattern on his father's side of the family;

having always been farmers, his father and uncles could not even sign their own names. When the Reverend Smith died, Newton's mother returned to Woolsthorpe with three children from her second marriage, but fortunately Newton's education was able to continue.

Two years after his mother's return, Newton was sent to a grammar school in the nearby town of Grantham, where apparently the principal focus of his education was the study of Latin. However, at the age of 17 Newton was called home in order to manage the farm, which turned out to be a bit of a disaster. Newton was clearly not interested in the task, and rather than watching over the sheep he would let them wander into neighboring properties while he built models of water wheels, windmills, and sundials. It seems that he also frightened the townspeople by flying a lantern from a kite. As a result of his neglect, Newton's mother had to pay for damages done by the sheep. She decided it best to let him return to the grammar school and prepare for further education at a university.

Newton's somewhat traumatic childhood likely had a deep impact on him. Newton tended toward introversion and he waged personal attacks on his critics that would be characteristic of his schooling and his mathematical and scientific career in the years ahead.

A Student of Trinity College in Cambridge

Newton matriculated in Trinity College in June, 1661. Despite family wealth, Newton apparently received only a meager allowance, forcing him to have to work as a personal assistant to cover his costs.

It seems that Newton made very few friends while a student, but instead focused all of his time and energy on his studies. Initially, Newton followed a very traditional course of study in Aristotelian physics, ethics, and rhetoric, the works of Ptolemy (c. 90–168 CE), and the mathematics of Euclid of Alexandria (c. mid-4th–mid-3rd century BCE).[1] As became common for Newton, he set up notebooks with many headings in which he would record information, pose questions, and develop answers. However, sometime in early 1664, he abandoned these paths to learning that had been in place since the creation of the medieval universities and began to focus on more contemporary sources of knowledge, including Galileo's *Dialog*, the writings of the chemist and physicist Robert Boyle (1627–1691), and the mechanical models of the Solar System proposed by René Descartes (Fig. 5.3), in which the planets were viewed as being dragged around the Sun by swirling vortices in an ether fluid filling all of space.

It was in 1664 that Newton received a scholarship that would fund his studies at Cambridge for the next four years. Newton's attention was drawn to mathematics, in which he was entirely self-taught. He fully immersed himself in Descartes's *Geometry* and numerous other works. In less than one year he had become a master in the mathematics of the day, and began to move beyond what was then known. His first discovery in 1665 involved an expansion of a mathematical formula in an infinite sum of terms[2] that allowed him to determine the area under a curve on a graph: the process that is today known as integration (one of the two major

Fig. 5.3 René Descartes (1596–1650). [Portrait by Frans Hals (c. 1583–1666); public domain]

Key Point
In 1665, Newton developed the earliest form of the mathematics of the calculus, that involves quantities that change over location and/or time.

[1] Plane geometry, the common form of geometry studied by all students today, is known as Euclidean geometry.

[2] The binomial expansion theorem.

components of the calculus). He then went on to develop, based on prior work of Descartes, a method for determining the tangents of curves that he termed *flux-ions* (the beginning of differential calculus); not bad for an undergraduate research project. Newton was awarded his Bachelor of Arts (B.A.) degree in August 1665 but the university closed temporarily soon thereafter as a precaution due to the devastating Great Plague that swept through Europe. As a result, Newton returned home to Woolsthorpe for the next 18 months.

Newton's *annus mirabilis*, 1665–1666

Much later in life, Newton wrote of his time away from Cambridge:

> In the beginning of the year 1665 I found the Method of approximating series ... The same year in May I found the method of Tangents ... & in November had the first direct method of fluxions & the next year in January had the Theory of Colours & in May following I had entrance into yᵉ inverse method of fluxions ... the same year I began to think of gravity extending to yᵉ orb of the Moon & (having found out how to estimate the force with wᶜʰ [a] globe revolving within a sphere presses the surface of the sphere) from Keplers rule of the periodical times of the Planets being in sesquialterate proportion of their distances from the center of their Orbs, I deduced that the forces wᶜʰ keep the Planets in their Orbs must [be] reciprocally as the squares of their distances from the centers about wᶜʰ they revolve; & thereby compared the force requisite to keep the Moon in her Orb with the force of gravity at the surface of the earth, & found them answer pretty nearly. All this was in the two plague years of 1665–1666. *For in those days I was in the prime of my age for invention & minded Mathematics & Philosophy* [meaning physics] *more than at any time since.* [emphasis added]

Key Point
Newton spent two years away from Cambridge during the plague years of 1665–1666 intensely focused on mathematics and physics, during which time he conducted much of his initial work on mathematics, gravitation, orbital motion, and optics.

There is an almost mythical aspect to the legend surrounding the "miraculous year" (*annus mirabilis*) of 1665–1666. It may be the case that Newton developed the core of his ideas regarding the calculus, optics, gravitation, and the orbital motions of the planets during his 18 months of seclusion from academic life in Cambridge, but it is clear that he didn't fully develop those ideas during that time, the possible exception being his progress in mathematics. Regardless of the state of his work when Trinity College reopened and he returned to Cambridge, there is no doubt that his future phenomenal mathematical and scientific output was well on its way. The human race had never seen a scientific mind the like of Isaac Newton before, and perhaps since.

Lucasian Professor

Shortly after returning to Cambridge, Isaac Newton was selected for a fellowship that allowed him to stay indefinitely at Trinity College as a minor fellow at the university. One year later he received his Masters of Arts (M.A.) degree and was now one of 60 major fellows. Being a fellow meant that he received about £60 annually, a portion of which was for room and board. It also meant that he had no required duties, although he could tutor if he so chose (he tutored a total of three students over the next 28 years). In other words, he was free to do whatever he wanted to do.

In 1669 Newton became only the second Lucasian professor at Cambridge, and at the time one of only eight to hold an endowed chair at the university. While not a particularly prestigious chair at the time, the Lucasian Chair provided an additional £100 to his income, which, when combined with funding from his estate, provided him with a comfortable living. The Lucasian Chair did come with the requirement of giving an annual lecture series. Apparently Newton wasn't the most spellbinding lecturer, because it wasn't uncommon for him to give his required lectures to an empty hall! What any scientist or mathematician wouldn't give today to have the opportunity to sit in on a lecture by Isaac Newton. [Holding the Lucasian Chair today is a great honor, and many of the recipients are highly recognized mathematicians, astronomers, and physicists, including Stephen Hawking (1942–2018)].

Newton's Theory of Light and Color, and His Reflecting Telescope

Newton's first major scientific effort was directed toward understanding the nature of light. Although we know today that his underlying concept of light being composed of particles was fundamentally flawed (our modern understanding of the nature of light will be discussed in detail in Sections 6.5 and 8.5) he did advance our understanding of color through a series of experiments involving prisms.

It had been known for some time that when white light passes through a prism a band of varying colors is produced, from red to violet, just like the colors in a rainbow [Fig. 5.4 (*Left*)]. However, it was believed that some characteristic of the glass in the prism caused the colors to appear. Newton devised an experiment where he allowed light from the Sun to enter his laboratory, pass through a prism to create the colors, and then send the colors through a second prism, thus recombining them to produce white light again [Fig. 5.4 (*Right*)]. Based on his study of light, Newton concluded that white light is actually a combination of all of the colors and that the eye simply interprets that combination as white.

In 1669, based on his study of light and colors, Newton built the first telescope that was constructed with mirrors rather than lenses, a replica of which is shown in Fig. 5.5. He was motivated to do so by realizing that when light passes through glass lenses it is susceptible to the dispersion of light that is bent at slightly different angles into various colors. The result is a distortion of the image. If mirrors

Key Point
Newton designed and built the first reflecting telescope in 1669. The design is now referred to as a Newtonian telescope.

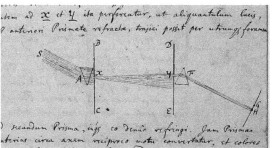

Fig. 5.4 *Left:* White light passing through a prism causes colors to separate. Red light is bent least and blue/violet light is bent the most. (Prism-rainbow.svg: Suidroot derivative work: Sceptre, CC BY-SA 3.0) *Right:* Newton's illustration of his dual prism experiment from his letter to the Royal Society on January 1, 1671. Sunlight (*S*) enters prism *A* and spreads out into its component colors. The colors enter the second prism *F* and recombine to produce sunlight again at *H*. (Public domain)

Fig. 5.5 A replica of Newton's telescope that was presented to the Royal Society in 1671 is housed in the Whipple Museum of the History of Science in Cambridge. (Photograph ©Andrew Dunn, CC BY-SA 2.0)

could be used instead of lenses then the light is reflected from the mirrored surface without such distortion. This is one of the reasons why all major telescopes today are constructed with mirrors rather than lenses.

The Royal Society

When the Royal Society, which had formed just nine years earlier as one of the first learned societies in the world, learned of the invention of Newton's telescope in 1671, they invited him to London to demonstrate his device. As a result, in 1672 he was elected as a fellow of the Society. Unfortunately, the election was not without controversy. At the time Robert Hooke (1635–1703), the curator of experiments for the Royal Society, wasn't convinced of the merits of Newton's theory of light and issued a condescending challenge to his conclusions. Given his inherently introverted nature, Newton chose to withdraw from the public eye and even asked to be removed from the rolls of the Royal Society. Thankfully, his petition was not acted upon at the time. When Newton did attend a meeting of the Royal Society, in 1674, he found that he was highly respected despite Hooke's criticism. Hooke would later argue that an aspect of the form of Newton's law of gravity was first formulated by him and that Newton essentially plagiarized his work, when in reality Newton had developed his ideas 20 years earlier but had not yet published them. This would not be the last time that Newton got embroiled in controversy because of his tendency to avoid the limelight and not publish his findings in a timely fashion.

When his nemesis, Hooke, died in 1703 Newton was elected as the Royal Society's president, a position he held for the remainder of his life. It turned out that Newton brought the same intensity he demonstrated in mathematics and physics to his position as president and he proved to be a very able administrator. He led the Society from near financial ruin to become a stable organization that remains a preeminent academic society today.

The *Principia*

In August 1684 Newton was visited by Sir Edmond Halley (1656–1742), for whom the famous comet is named. The visit was motivated by the observance of two great comets, in November and December 1680; the first was seen before sunrise and the second after sunset. The Astronomer Royal, John Flamsteed (1646–1719), had proposed that the two comets were actually one and the same, first moving toward the Sun and then moving away from it. Halley asked Newton, the now-famous mathematician, if he had ever calculated orbits based on an attractive force from the Sun. Newton immediately replied that he had but he was unable to locate his previous work. He then promised to redo his calculations and send them to Halley. In November, Halley received a nine-page document known as *De motu* (*Concerning motion*), in which Newton sketched out the orbital dynamics that Halley had wondered about. Highly impressed, Halley strongly encouraged Newton to publish his work through the Royal Society. Unfortunately, the Royal Society was unable to finance the project and so Halley took it upon himself to pay for it and to see to its publication. On July 5, 1687 the *Philosophiæ Naturalis Principia Mathematica* (*Mathematical Principles of Natural Philosophy*), shown in Fig. 5.6, was published in three books and became without question the greatest single work ever written in

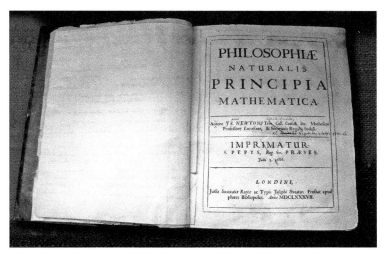

Fig. 5.6 Newton's own copy of the first edition of the *Principia*, with his edit marks in preparation for a future printing. On display at the Wren Library, Trinity College, Cambridge. (Photograph ©Andrew Dunn, CC BY-SA 2.0)

the physical sciences. The second edition, significantly enhanced over the first and even more significant in its impact on the future of science, appeared in 1713. A third edition, with minor changes, was published shortly before Newton's death in 1727.

The *Principia*, as it is commonly known, contains Newton's famous three laws of motion and the concept of mass, his universal law of gravitation and the derivation of Kepler's laws of planetary motion, the first true explanation of ocean tides, and his ideas about how science should be conducted. Newton also tied together the motions in the heavens and terrestrial motions as an integrated whole. In his original notes for the *Principia* Newton had acknowledged contributions by Hooke, but when Hooke made his charges of plagiarism, Newton deleted most of the references, leaving only one minor mention of Hooke in the treatise.

The Calculus Controversy

Newton used his method of fluxions (early calculus) throughout the *Principia* despite not having yet published that great advance in mathematics separately. However, in a manuscript dated October 1666, *De analysi per æquationes numero terminorum infinitas*, that was shared with a few close friends, Newton did discuss his work on the calculus. In June 1669 Isaac Barrow (1630–1677), the first Lucasian Professor and Newton's tutor, sent a copy of the manuscript to a colleague, John Collins (1625–1683), with a letter that stated in part: "Mr Newton, a fellow of our College, and very young ... but of an extraordinary genius and proficiency in these things."

Meanwhile, another young and brilliant mathematician from Germany, Gottfried Leibniz (Fig. 5.7), began work on a somewhat different approach to the calculus in 1674, publishing his results in 1684. When Newton's *Principia* appeared in 1687, Leibniz cried foul, arguing that Newton had plagiarized his work on the new mathematics. Newton had in fact communicated briefly with Leibniz in two

Fig. 5.7 A portrait of Gottfried Leibniz (1646–1716) by Johann Fredrich Wentzel (1670–1729), circa 1700. (Public domain)

detailed letters in 1676 regarding Leibniz's request for information pertaining to infinite series. Within the letters Newton also included his fluxions method. In addition, Leibniz had seen papers written by Newton that Collins had obtained. Regardless, Leibniz had not referenced Newton's fluxions method in his 1684 publication, and further, he challenged Newton's priority concerning the development of the calculus in a letter to the Royal Society in 1711, at the time Newton was its President. Furious, Newton organized a committee to study the matter. In the end, the historical record that Newton had amassed supported his earlier development of his fluxions method. The committee also confirmed that there was really only one calculus rather than two (Newton's and Leibniz's). Newton, of course, wrote the committee report. Unfortunately, the argument between Newton and Leibniz didn't really end until Leibniz's death in 1716.

In an interesting attempt to show that Leibniz's approach to the calculus was superior to Newton's, a friend of Leibniz, Johann Bernoulli (1667–1748), an outstanding mathematician in his own right, sent Newton two challenge problems in 1697. According to the documented story, Newton came home from a long, exhausting day, read the letter containing the challenge problems, solved them before going to bed, and sent the solutions back to Bernoulli anonymously. When Bernoulli received the correspondence he stated that "as the lion is recognized from his claw," meaning that the brilliant Newton is immediately recognized through his solutions.

In reality, it seems that Newton and Leibniz developed the calculus largely independently of one another and today it is Leibniz's mathematical notation that is widely used by mathematicians and scientists. As an example from history, the calculus debate illustrates one of the many reasons why publication is so important in research. Not only does it clearly establish priority, but the modern form of a scientific paper also contains an introductory section that clearly spells out previous work on the topic, who made those contributions, and when. A scientific paper is also subjected to anonymous peer review, where others knowledgeable about the topic are asked to look for flaws and gaps in the study and request corrections or additional changes prior to publication. Once a paper is published, it then becomes a public record of the researcher's work and can be further examined for errors or fallacies if they exist. The process of scientific publication is designed to be extremely rigorous, so as to ensure that publications do not plagiarize and that they contribute to the advancement of human knowledge. Of course, no human process is perfect and sometimes, although rarely, papers are retracted after publication because of some fundamental flaw or flaws in the paper that are considered severe and occasionally dishonest. Fortunately, the scientific publication process, with its checks and balances, is very robust, although sometimes frustratingly tedious. Today, with the immediate access to information through the internet, preprint servers have been established that allow researchers to read drafts of each other's work ahead of publication with the understanding that the papers are not yet finalized. One such server for physics (including astronomy and astrophysics), mathematics, and other academic disciplines is `https://arXiv.org/`.

Master of the Royal Mint

In 1695, the English economy was in trouble. England had been in a costly war with France, and criminals were both manufacturing fake coinage and cutting the edges off irregularly shaped coins and then melting down the scraps for profit. Without professional economists in government at the time, the advice of eight intelligent men was sought to recommend an approach to thwart the criminals. Newton was one of the men chosen for the task: evidence of the respect he was held in because of the *Principia*. The majority recommendation of the group, including Newton, was that new, round coins should be minted that would not have the protruding irregular points that could be cut off. Although an obvious solution, it required an enormous investment of time and resources to accomplish, including the establishment of five mints around the country. Newton was asked to undertake the task, which he accepted in 1696, causing him to leave Trinity College, Cambridge where he had been living for 35 years, and move to London. Although he was originally offered the post of Warden of the Mint, which was supposed to involve relatively little work, Newton invested himself in the job with his usual high level of intensity and commitment, successfully overseeing the recoinage in less than three years. He was later promoted to Master of the Mint, receiving one of the highest salaries in government. Although he didn't live extravagantly, Newton became a wealthy man. Because of his secure financial standing, Newton resigned his Lucasian Chair in 1701.

Alchemy

Along with his work in mathematics, physics, and astronomy, Isaac Newton was also fascinated with, and spent a great deal of time and energy studying, alchemy: the discipline of trying to change ordinary metals into precious metals, most notably gold. It may seem odd today that perhaps the greatest scientist who ever lived would spend time on such a pursuit, but our knowledge of chemistry and the nature of the atom was completely absent in Newton's day. As with everything else he did, Newton studied alchemy with great passion, collecting an impressive library of alchemical texts, and conducting many original experiments in his private laboratory.

Theology

Another topic that consumed Newton throughout his lifetime was theology. He was originally driven to the subject by the requirement that all fellows of Trinity College be ordained. Again, he read and studied extensively, burying himself in the Bible to the point that few people knew its contents as well. He also developed an enormous library of theological works. Newton became particularly interested in the development of the concept of the Christian Trinity in the fourth century (page 72), for which Trinity College was named. Through his study, Newton developed a rather unorthodox view of Christianity and could not accept the concept of the Trinity, although he was devoutly Christian. In part because of his tendency toward great personal secrecy and dislike of conflict, he kept his views private. His views did pose a serious problem for his future as an academic, however, given the requirement of ordination. Fortunately for him, and perhaps for the future of science, there were two positions at the College that were exempt from the ordination requirement by order of parliament, one of which was the Lucasian Chair, to which he had been appointed.

Recognition

With the publication of the *Principia*, Newton became the most respected scientist in Europe, despite his clashes with Hooke, Leibniz, and others. He also became well known outside scientific circles, through his leadership of the Royal Mint. In recognition of his contributions, Queen Anne knighted Newton in 1705, making Sir Isaac Newton the first scientist to be so honored.

When Newton died, on March 20, 1727, he was laid in state in the Jerusalem Chamber of Westminster Abbey in London, with the pall bearers being the Lord Chancellor, the Dukes of Montrose and Roxburgh, and the Earls of Pembroke, Sussex, and Macclesfield, all of whom were members of the Royal Society. He was interred on March 28, 1727, in a prominent location in the nave of Westminster Abbey. Four years after his death, Newton's heirs had a monument erected (Fig. 5.8) at his interment site that bore the inscription "Let Mortals rejoice That there has existed such and so great an Ornament to the Human Race."

In what may have been a fitting epitaph, of his own life Newton wrote:

> I do not know what I may appear to the world, but to myself I seem to have been only like a boy playing on the sea-shore, and diverting myself in now and then finding a smoother pebble or a prettier shell than ordinary, whilst the great ocean of truth lay all undiscovered before me.

Fig. 5.8 A monument to Sir Isaac Newton in Westminster Abbey. (Photograph by Klaus-Dieter Keller; public domain)

5.2   Newton's Three Laws of Motion

Among Sir Isaac Newton's many accomplishments, surely the most important were his development of the mathematics of the calculus, his three general laws describing the motions of objects under the influence of forces, and his specific description of the force due to gravitation. Although the calculus will not be considered here, his three laws of motion will be presented in this section, leaving the discussion of his gravitational force equation and its fundamental importance to astronomy for Section 5.3. However, before we can discuss each of the laws in turn, we must first define several quantities, some physical and some mathematical.

Distinguishing Mass and Weight

Key Point
Mass is an intrinsic quantity of a material, independent of anything else.

There is a common misconception that mass and weight are the same thing. Although they are related, mass is an intrinsic quantity of an object, while weight is a measure of how hard gravity pulls on that object. To add to the confusion, in the Imperial system of units and in the United States (US) customary units there is a unit of force called the pound-force (lb_f), usually simply referred to as the pound (lb), and a pound-mass (lb_m). When you get on a weight scale, your weight is determined by how much you compress springs in the scale due to Earth pulling you down. Your author weighs about 205 lb, but if he were to travel to the Moon where the gravitational pull is about one-sixth of what it is on Earth, his weight would be closer to 34 lb (205 lb / 6). Although this could be considered an instant weight reduction program, it would not be a mass reduction program since there would be no less of your author, only the pull on him due to gravity would be less. Because of

Table 5.1 Prefixes in the International System of Units

Factor	Prefix	Name		Factor	Prefix	Name	
10^{24}	Y	yotta	(septillion)	10^{-1}	d	deci	(tenth)
10^{21}	Z	zetta	(sextillion)	10^{-2}	c	centi	(hundredth)
10^{18}	E	exa	(quintillion)	10^{-3}	m	milli	(thousandth)
10^{15}	P	peta	(quadrillion)	10^{-6}	μ	micro	(millionth)
10^{12}	T	tera	(trillion)	10^{-9}	n	nano	(billionth)
10^{9}	G	giga	(billion)	10^{-12}	p	pico	(trillionth)
10^{6}	M	mega	(million)	10^{-15}	f	femto	(quadrillionth)
10^{3}	k	kilo	(thousand)	10^{-18}	a	atto	(quintillionth)
10^{2}	h	hecto	(hundred)	10^{-21}	z	zepto	(sextillionth)
10^{1}	da	deka	(ten)	10^{-24}	y	yocto	(septillionth)

the confusion that the term "pound" can cause, throughout the remainder of this text we will always refer to mass in its International System of Units (SI)[3] unit of kilograms. The SI system of units is also the internationally adopted standard for physical units in the vast majority of countries around the world, and relies heavily on prefixes like k for thousands (1 km = 1000 m); see Table 5.1. (Newton first described the concept of mass in the *Principia*.)

The Difference Between Scalars and Vectors

A scalar is any quantity that is described by one number *and an appropriate unit*. For example, mass is specified in SI units in kilograms; your author's mass is approximately 93 kg. That single number and its associated unit is sufficient to describe mass. Another commonly used scalar quantity is temperature; to conserve energy in the winter in colder climates, we are asked to set our thermostats to 20°C (68°F).

Key Point
A scalar describing a physical quantity has a magnitude (the number) and a unit.

To carry out calculations with scalars you simply use the common processes that you are familiar with: addition, subtraction, multiplication, and division. Just remember to always include units in every answer and in any intermediate steps. Only reporting that your answer is 42 is meaningless; 42 what? Including the proper unit clarifies what you mean, such as the building is 42 meters tall. Even if the reader knows that you meant height, 42 could have been 42 centimeters, 42 feet, 42 furlongs, 42 kilometers, or 42 light-years!

Example 5.1

An elevator carries you from the 2nd floor up to the 12th floor, and the distance between floors is 4 meters. How many floors did you go up and what was the change in your elevation?

---

[3]SI comes from the French name, Le Système international d'unités.

A unit conversion tutorial is available at
Math_Review/Unit_
Conversions

Since the number of floors is a scalar you are simply asked to calculate

12th floor − 2nd floor = 10 floors.

To determine your change in elevation you want to multiply the number of floors you traveled by the distance traveled per floor, which is also a scalar. (At this point you should start getting used to using units throughout your calculations, noting that *units cancel in fractions just like numbers do.* 4 m / floor reads "4 meters per floor.")

10 floors × (4 m / floor) = 40 m.

A **vector** is a quantity that contains an additional piece of information, namely direction. Velocity is one such example: velocity not only describes how fast something is moving (its speed, v, which is a scalar), but it also describes the direction in which the object is moving. If you tell someone that you are driving a car on interstate I-15 through Utah at 130 km/h (about 80 mi/h, which is legal along some sections of freeways in the western United States), that information only tells the individual how fast you are going (a scalar) but it doesn't tell that person which direction you are going. For a more complete description of your motion you would tell that person that you are traveling due north at 130 km/h. To clearly distinguish a vector from a scalar an arrow is placed over the mathematical symbol for the quantity and/or the symbol is set in boldface type (an arrow and boldface type are both used in this text; for example, velocity is represented by $\vec{v}$). When you hand-write a vector, simulating boldface is not necessary but the arrow definitely is. *A vector has both magnitude with the appropriate unit (the scalar part) and direction.*[4]

Along with placing an arrow over a mathematical symbol, vectors are represented graphically with an arrow on a *vector diagram*. The length of the arrow corresponds to the magnitude of the vector and the direction that the arrow points indicates the vector's direction. The length of the vector on the diagram is proportional to its magnitude relative to other vectors with the same units, but the length is not necessarily an absolute physical length such as centimeters, simply because vector quantities that don't measure distances don't have distance units; even if they did, the lengths might not be convenient, such as kilometers.

Example 5.2

On a graphical vector diagram where north is at the top of the graph, show two cars, A and B, where Car A is traveling north at 20 meters per second (20 m/s is equivalent to 72 km/h or 45 mi/h) and Car B is traveling west at 10 m/s (36 km/h, 22 mi/h).

As depicted in Fig. 5.9 (*Left*), since Car A is traveling at twice the speed of Car B, Car A's arrow needs to be twice as long to illustrate the two different magnitudes.

---

[4]Point of trivia: In the animated film, *Despicable Me* (2010), directed by Chris Renaud and Pierre Coffin [Feature film], Universal City, CA: Universal Pictures, the villain's name was Vector. He explained his name as "a mathematical term; a quantity represented by an arrow with both direction and magnitude."

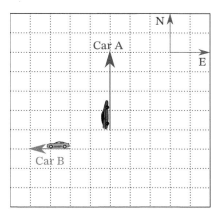 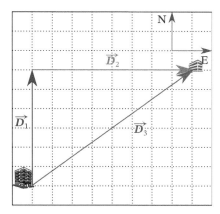

Fig. 5.9 *Left:* Car A is traveling north with twice the speed of Car B, which is traveling west. *Right:* A student walks to class first along displacement vector $\vec{D_1}$ and then along displacement vector $\vec{D_2}$. The total distance walked, $D_1 + D_2$ (adding the magnitudes of the two vectors), is greater than if the student had walked along vector $\vec{D_3}$, even though $\vec{D_3} = \vec{D_1} + \vec{D_2}$.

Also, Car A's arrow must be pointed north while Car B's arrow must be pointed west.

Note that there is nothing special about *where* the vectors were drawn on the diagram, since you were asked only to draw velocity vectors, not locations. Car A could be driving north from an intersection that Car B just passed, or Car A could be driving north in California while Car B could be driving west in Pennsylvania. The lengths and the directions of the vector arrows do matter, however.

Adding and subtracting vectors is more complicated than adding and subtracting scalars, because the direction must also be included. Adding and subtracting vectors can be done graphically by a process called "tip-to-tail." The idea is that if there are two arbitrary vectors $\vec{a}$ and $\vec{b}$, and you need to find $\vec{c} = \vec{a} + \vec{b}$, first draw vector $\vec{a}$ and then draw vector $\vec{b}$ by placing the tail of $\vec{b}$ at the tip of $\vec{a}$ (remember that positions of the vectors on a vector diagram don't matter, only their magnitudes and directions). The resultant vector $\vec{c}$ is the vector you get by starting at the tail of $\vec{a}$ and ending at the tip of $\vec{b}$. The process is best illustrated by an example:

Example 5.3

You are going from your residence hall to your astronomy classroom building by first walking north 300 m then east for 400 m. (a) Draw three vectors; the first vector $\vec{D_1}$ is your first leg of the walk, the second vector $\vec{D_2}$ is the second leg, and $\vec{D_3}$ is the vector that corresponds to where your classroom building is relative to where your residence hall is (in other words, the direction and distance you would have walked if you had been able to walk straight to the classroom building). As a mathematical vector equation you are finding $\vec{D_3} = \vec{D_1} + \vec{D_2}$. (b) Determine the total distance you walked. (c) Determine the direct distance from your residence hall to your classroom building.

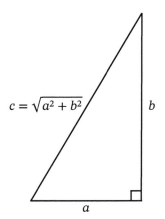

$$c = \sqrt{a^2 + b^2}$$

Fig. 5.10 A right triangle with legs of length a and b, and hypotenuse of length c, where $a^2 + b^2 = c^2$. The square in the lower right corner represents the right angle (90°).

(a) The vectors are sketched in Fig. 5.9 (*Right*).

(b) The total distance you walked is simply the sum of the magnitudes (lengths) of the two vectors, or $D_1 + D_2 = 300$ m $+ 400$ m $= 700$ m (scalar arithmetic).

(c) You may recall that the Pythagorean Theorem, $a^2 + b^2 = c^2$ for a right triangle (Fig. 5.10) gives the length of the hypotenuse, $c = \sqrt{a^2 + b^2}$, if the lengths of the legs a and b are known. From the theorem, the distance from your residence hall to the classroom building (the length or magnitude of vector $\vec{D_3}$) is

$$D_3 = \sqrt{D_1^2 + D_2^2} = \sqrt{(300 \text{ m})^2 + (400 \text{ m})^2} = \sqrt{(300^2 + 400^2) \text{ m}^2} = 500 \text{ m}.$$

The *magnitude* (length) of a vector is indicated by leaving the arrow off the vector's mathematical symbol and not using boldface type; this is why it is important to use the arrow when the direction is indicated. Note that although $\vec{D_3} = \vec{D_1} + \vec{D_2}$, it *is not true that* $D_3 = D_1 + D_2$. You walked 700 m even though the distance from your residence hall to the classroom building is only 500 m. Adding vectors requires paying attention to the geometry involved.

Multiplying a vector by a scalar is straightforward because the only thing the scalar does is change the magnitude of the vector, including changing the units if the scalar has units. If you multiply a vector by 2 with no units, the magnitude of the vector simply increases by a factor of two and a graphical representation of the vector is now twice as long. If you multiply a vector by −2, the magnitude of the vector is again twice as great, but it now points *in the opposite direction* from the original vector; all the minus sign does is reverse the direction of the vector.

We won't need to "multiply" one vector by another vector, although doing so is very useful in other contexts.

Velocity and Acceleration

The reason for our quick diversion into vectors has to do with the definitions of velocity and acceleration. Velocity, $\vec{v}$, is the change in the position vector, $\vec{r}$, divided by the elapsed time, and is represented by the equation,

$$\vec{v} = \frac{\vec{r_f} - \vec{r_i}}{t}, \tag{5.1}$$

The definition of velocity

where $\vec{r_f}$ is the final position of the object that is moving, $\vec{r_i}$ is the object's initial position, and t (a scalar) is the time that elapsed while changing positions. [The $\vec{D}$ vectors in Fig. 5.9 (*Right*) are position vectors starting from the student's dormitory; instead of $\vec{D_1}$, $\vec{D_2}$, and $\vec{D_3}$, they could have been designated $\vec{r_1}$, $\vec{r_2}$, and $\vec{r_3}$, respectively.]

Similarly, acceleration, $\vec{a}$, is the change in the velocity vector divided by the elapsed time:

$$\vec{a} = \frac{\vec{v_f} - \vec{v_i}}{t}, \tag{5.2}$$

The definition of acceleration

where $\vec{v}_f$ is the final velocity of the object that is moving, $\vec{v}_i$ is the object's initial velocity, and t is again the elapsed time.

When driving in a straight line you are undoubtedly comfortable with thinking about acceleration as speeding up; in other words, increasing your speed by pushing down on the accelerator pedal. When you speed up, your acceleration is considered to be positive. When you want to slow down you push on the brake pedal which is also a form of acceleration, only in this case the acceleration is negative (often referred to as deceleration).

Example 5.4

(a) Suppose you are driving straight north on a highway at a constant speed. You notice that it takes you 50 seconds to travel from one mile marker to the next. How fast are you going? Express your answer in miles per hour (mi/h) and meters per second (m/s). (There are 1609 meters in one mile.)

(b) You notice that you are traveling faster than the allowed speed limit of 60 mi/h (about 27 m/s) and you decide to slow down. It takes you 10 seconds to reach the speed limit. What was your acceleration in m/s?

Since this problem is simply along a straight line from south to north, there is no need to explicitly use vectors as long as we agree that going north is in the positive (+) direction and going south is in the negative (−) direction. The plus and minus signs indicate the directions for straight-line motion.

(a) Assuming that the starting position is arbitrarily set to be zero ($r_i = 0$ mi), then after $t = 50$ s you will be at a position $r_f = 1$ mi. Since 50 seconds is 5/6 minute or 0.83 minute, your speed (the scalar term for velocity) is

$$v = \frac{r_f - r_i}{t} = \frac{1 \text{ mile}}{0.83 \text{ min}} = \frac{1.2 \text{ miles}}{\text{minute}} = 1.2 \text{ mi/min}.$$

To convert your answer to miles per hour you must multiply by a conversion factor. Since 60 minutes = 1 hour, dividing both sides by 1 hour gives 1 = 60 minutes / 1 hour, which is the conversion factor we are looking for. When you change units you are really only multiplying by 1, which of course doesn't change the quantity, only the units used to express it:

$$\frac{1.2 \text{ miles}}{1 \text{ minute}} \times \frac{60 \text{ minutes}}{1 \text{ hour}} = \frac{72 \text{ miles}}{1 \text{ hour}} = 72 \text{ mi/h};$$

in other words, 1.2 mi/min = 72 mi/h, which is often written as 72 mph.

Next, to convert from mi/h to m/s you need to use two conversion factors, 1 mi = 1609 m and 1 h = 3600 s, or 1 = 1609 m / 1 mi and 1 = 1 hour / 3600 seconds:

$$\frac{72 \text{ miles}}{1 \text{ hour}} \times \frac{1609 \text{ meters}}{1 \text{ mile}} \times \frac{1 \text{ hour}}{3600 \text{ seconds}} = \frac{115{,}848}{3600} \frac{\text{meters}}{\text{second}} = 32 \text{ m/s}.$$

Note that, since it is also true that 1 = 1 mi / 1609 m, which way you write the conversion fraction depends on what it is you are trying to cancel and the units you want to end up with. For example, in the last equation, if we had written that conversion fraction the other way, the result would have been

$$\frac{72 \text{ mi}}{1 \text{ h}} \times \frac{1 \text{ mi}}{1609 \text{ m}} \times \frac{1 \text{ h}}{3600 \text{ s}} = \frac{72}{5{,}792{,}400} \frac{\text{mi} \times \text{mi}}{\text{m} \times \text{s}} = 0.000\,012 \frac{\text{mi} \times \text{mi}}{\text{m} \times \text{s}},$$

which, although technically not wrong, certainly isn't the kind of answer that we were looking for! This is just one example of why it is always important to include units in all calculations, not just the final answer.

(b) Since acceleration is the change in velocity over a specified period of time, and since you are traveling in a straight line, when you slow down your acceleration will be negative. With $v_i = 32$ m/s as your initial velocity, $v_f = 27$ m/s as your final velocity, and $t = 10$ s as the amount of time you spend slowing down,

$$a = \frac{v_f - v_i}{t} = \frac{27 \text{ m/s} - 32 \text{ m/s}}{10 \text{ s}} = \frac{-5 \text{ m/s}}{10 \text{ s}} = -0.5 \text{ m/s/s}.$$

The odd combination of units in your answer is read as "negative 0.5 meters per second per second," meaning that your velocity is decreasing by 0.5 meters per second for each second that you are slowing down (your speed was reduced by 5 meters per second over the 10 second interval). Remember, the minus sign means that you are slowing down since your final velocity is less than your initial velocity.

Rather than write the acceleration unit as m/s/s, it is more concise to combine the common units and indicate how many times they are present, just like multiplying the same number by itself some number of times. For example, you know that $2 \times 2 = 4$, but by using exponents it is also possible to write the same equation as $2^2 = 4$ where 2^2 is just short-hand for 2×2. Notice that your answer can be written as

$$-0.5 \frac{\text{m/s}}{\text{s}} = -0.5 \frac{\text{m}}{\text{s} \times \text{s}} = -0.5 \frac{\text{m}}{\text{s}^2} = -0.5 \text{ m/s}^2.$$

The combined unit of acceleration is usually read as "meters per second squared."

An exponents tutorial
is available at
Math_Review/Exponents

Key Point
Acceleration occurs with a change in speed, direction, or both.

The last example assumed that the traveler was moving only in a straight line and that velocity could be treated as a scalar, but this is not always, or even usually, the case. We, and most other moving objects in the universe, are frequently, and sometimes constantly, changing direction as we move, meaning that our velocity vectors are changing the direction they are pointing. In fact, even if your speed remains a constant (the magnitude of your velocity vector), your velocity can still be changing by virtue of the fact that you may be changing direction (driving around a curve for example). An object can experience an acceleration if that object (a) speeds up or slows down (change in magnitude), (b) changes direction, or (c) both.

Example 5.5

Imagine that our Moon is traveling in a perfectly circular orbit around Earth at a constant speed (this is only approximately true). By drawing vectors, determine in which direction the Moon is accelerating.

Since the time t in the definition of acceleration is always positive (unless someone figures out how to travel backward in time that is), the direction that $\vec{a}$ points must be the same direction as the vector $\Delta \vec{v} = \vec{v}_f - \vec{v}_i$, where $\Delta \vec{v}$ is read "delta v" (Δ is commonly used to represent the difference between two quantities; recall the

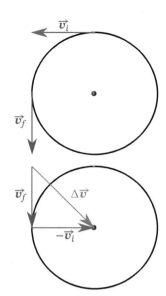

Greek alphabet in Table 4.1). However, subtraction is the same as adding the negative of the second quantity, or $\Delta\vec{v} = \vec{v}_f + (-\vec{v}_i)$. This means that we can just add $\vec{v}_f$ to a vector that is identical to $\vec{v}_i$ except that the vector would be pointed in the opposite direction and still have the same length as $\vec{v}_i$ (you are actually multiplying $\vec{v}_i$ by the unitless scalar, -1).

As shown in the top diagram in Fig. 5.11, consider two velocity vectors, the first velocity vector, $\vec{v}_i$, is the direction that the Moon is moving at the moment it is at the top of the circular orbit, and the second velocity vector, $\vec{v}_f$, is the direction the Moon is moving one-quarter of an orbit later. Given that velocity vectors are simply describing the direction and speed that an object is moving, they are not anchored to a specific location (recall the discussion in Example 5.2) and can be moved around for convenience. Since we are looking for a change in velocity between two points in an orbit, it makes sense to locate that change in velocity at the midpoint of the two velocities in question. As can be seen from the bottom diagram, adding $\vec{v}_f$ and $-\vec{v}_i$ "tip to tail" shows that the direction of the resultant vector, $\Delta\vec{v} = \vec{v}_f + (-\vec{v}_i)$, is toward the center of the idealized circular orbit, namely the location of Earth about which the Moon orbits.

Fig. 5.11 *Top:* Velocity vectors for the Moon in an idealized circular orbit around Earth at two locations: top and one-quarter of an orbit later. *Bottom:* The first velocity vector, $\vec{v}_i$, is subtracted from the second vector, $\vec{v}_f$. The resultant difference in velocity vectors, $\Delta\vec{v}$, is directed toward Earth (the center of the orbit).

There was nothing special about the two locations of the Moon in its idealized circular orbit around Earth in Example 5.5. In fact, using any two positions in the orbit and then subtracting the earlier velocity vector, $\vec{v}_i$, from the later one, $\vec{v}_f$, will result in a $\Delta\vec{v}$ vector that is directed toward the center of the circle at the point on the orbit half-way between the two positions. Even if the points are separated by a very, very tiny fraction of the circumference of the circular orbit, the $\Delta\vec{v}$ vector will still be pointed toward the center of the orbit. Of course if the two points are extremely close together their individual velocity vectors will become almost identical; the lengths will always be the same since the speed is constant in this idealized example, but the directions will only differ by a tiny amount. As a result, $\Delta\vec{v}$ will continue to point toward the center, but its length will get very short. This also means though that the time interval t will become very small because the Moon has only traveled for a short time over the short distance, with the result that dividing the magnitude of the $\Delta\vec{v}$ by t will give the same answer for the magnitude of the acceleration. No matter where the Moon is in its orbit, it always experiences an acceleration that is pointed directly toward Earth, and the acceleration always has the same magnitude (assuming a perfectly circular orbit).

Newton called the acceleration experienced by an object moving at a constant speed around a circle, centripetal acceleration, $\vec{a}_c$ (Fig. 5.12). He also showed that the magnitude of the centripetal acceleration vector is related to the orbital speed, v, and the radius of the orbit, r, by

$$a_c = \frac{v^2}{r}. \tag{5.3}$$

The magnitude of centripetal acceleration equation

The equation can be read as "the centripetal acceleration equals the square of the orbital speed divided by the radius of the orbit." Notice that the velocity vector is always tangent to the circular orbit and the centripetal acceleration vector is perpendicular to the velocity vector when the orbit is perfectly circular.

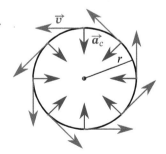

Fig. 5.12 Velocity vectors, $\vec{v}$, and centripetal acceleration vectors, $\vec{a}_c$, for eight locations in the Moon's orbit around Earth. $\vec{v}$ and $\vec{a}_c$ are perpendicular to each other at every point in a circular orbit. r is the radius of the orbit.

A scientific notation
tutorial is available at
Math_Review/Scientific_
Notation

Example 5.6

Use the equation for the magnitude of centripetal acceleration to calculate the centripetal acceleration of the Moon toward the center of Earth. The orbital speed of the Moon is 1023 m/s (2288 mi/h) and the radius of the Moon's orbit is 3.844×10^8 m or 384,400 km (238,900 mi).

From the centripetal acceleration equation

$$a_c = \frac{(1023 \text{ m/s})^2}{3.844 \times 10^8 \text{ m}} = \frac{1.054 \times 10^6 \text{ m}^2/\text{s}^2}{3.844 \times 10^8 \text{ m}} = 0.0027 \text{ m/s}^2.$$

Now that we have the preliminaries out of the way, we can present and discuss each of Newton's three laws of motion in turn. Newton's laws represent a major turning point in understanding motion, both on Earth and in the heavens. This is because his three laws present a framework for describing *why* motions occur, rather than simply *how* they happen. It is quite likely that you are already familiar with one or more of these laws, given that they are a part of the popular lexicon, especially his third law, which many do not realize originated with Newton.

Newton's First Law (The Law of Inertia)

Newton's first law states that:

> An object at rest remains at rest, and an object in motion remains in motion in a straight line at a constant speed unless the object is acted upon by an unbalanced force.

The general concept of the first law had been in existence long before Newton's work. It is generally believed that Galileo was the first person to argue that once an object is moving, it will keep moving unless there is some sort of push, pull, or resistance exerted. In other words, as Galileo described it (see page 134), an object possesses a quantity known as inertia. The first law is often referred to as the law of inertia.

This may seem obvious when you see the first law written down, but it wasn't at all obvious before Galileo's time. Prior to his experiments with balls rolling down inclines, Aristotelian physics argued that the natural motion of something was to come to rest at its proper level. This was because there was no clear understanding of friction or air resistance as impediments to motion. Galileo realized that if those resistive forces could be eliminated, objects would in fact just keep going in a straight line with a constant speed.

Key Point
The net force is the vector sum of all forces acting on an object.

The portion of the first law that refers to "an unbalanced force" means that an object that is motionless or moving in a straight line at a constant speed will only be affected if there is a force, or a series of forces added together, that are not counteracted by one or more additional forces. Since forces are vector quantities, another way to say this is that by adding together all of the force vectors acting on the object, if the resultant force vector has length zero, then the forces are balanced and there won't be any change in the object's motion, but if the sum total of all of the force vectors is not zero (i.e., unbalanced) then the direction and/or speed of the object will change. The sum of all forces acting on the object is known as the net force.

The first law as stated above can be shortened by thinking in terms of an object's velocity vector since a velocity vector specifies speed and direction. Rewriting the statement gives

> (Law of inertia) An object's velocity vector will remain unchanged unless the object is acted upon by a non-zero net force, $\overrightarrow{F}$.

Key Point
Newton's first law.

Newton's Second Law

Newton's second law is a mathematical relationship tying together what happens when a force is applied to an object and how much change occurs in its motion (its acceleration). How much the object responds depends on the object's mass (with mass corresponding to Galileo's inertia). Newton's second law states that

$$\overrightarrow{F} = m\overrightarrow{a}, \tag{5.4}$$

Newton's second law of motion

Key Point
Newton's second law.

the net force acting on an object equals the mass of the object times its acceleration (read "F equals m a").

As with all mathematical statements, it is important to be able to interpret Newton's second law. Simply memorizing $\overrightarrow{F} = m\overrightarrow{a}$ is not the same thing as understanding it. Remember that mathematics is the language of nature; it is very concise, very precise, and it has its own grammar (such as the rules of algebra), but ultimately it is important to be able to read and understand what the language is telling us about nature.

The second law is really describing how an object will respond to any and all forces that may act on it. If a ball is dropped it will accelerate downward because gravity is acting on it. If a pitcher in a baseball game throws a ball and a batter hits the ball, its direction of motion and its speed will change because of the force exerted on the ball by the bat.

How much acceleration occurs depends directly on the force applied to an object and how massive the object is. As Fig. 5.13 (*Top*) illustrates, for a given mass, m, the acceleration of an object increases in direct proportion to the force applied: if the force doubles, the acceleration doubles: if the force increases by a factor of three then so does the acceleration. Figure 5.13 (*Bottom*) shows what happens if the amount of force remains constant but the mass changes: if the mass doubles then the acceleration decreases by a factor of two; and if the mass increases by a factor of three, the acceleration decreases by a factor of three. If a major league batter hits a 0.145 kg baseball, the ball may travel well over 100 m, but if the same batter tries to hit a 7.0 kg bowling ball coming at him with the same speed as the baseball, the outcome won't be quite the same! The same applied force will result in a much smaller acceleration for the bowling ball relative to the baseball's acceleration. Mathematically, this can be seen by dividing both sides of the second law by mass and solving for acceleration, resulting in another form of the second law that describes what happens to an object based on the total force exerted on that object:

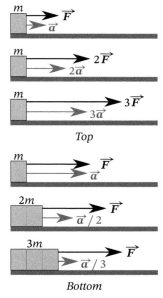

Top

Bottom

Fig. 5.13 *Top:* For a specified mass, m, as the force increases on the object, the object's acceleration increases in the same proportion. *Bottom:* For a specified force, $\overrightarrow{F}$, as the mass increases the acceleration decreases by the same factor.

$\vec{a} = \vec{F}/m$. Newton's second law in this form can be read as "the acceleration of an object is given by the net force applied to the object divided by that object's mass."

For a given applied net force, $\vec{F}$, the greater the mass, m, is in the denominator of the equation, the smaller the acceleration, $\vec{a}$, will be. If the applied net force is zero ($\vec{F} = 0$) then the resulting acceleration will be zero ($\vec{a} = 0$) and the velocity will be unchanged [$\vec{v}_f = \vec{v}_i$ according to the definition of acceleration [Equation (5.2)]. This is actually an even more concise and mathematical version of the first law: If $\vec{F} = 0$, then $\vec{a} = 0$ and $\vec{v}_f = \vec{v}_i$.

The second law in its first form [Equation (5.4)] tells us what the unit of force is; it is equal to mass (kg) times acceleration (m/s²), or kg × m/s². This combination of units in SI units is appropriately called the **newton (N)**; $1\text{ N} = 1\text{ kg m/s}^2$. To get a better feel for the magnitude of a 1 N force, note that $1\text{ N} = 0.2248\text{ lb}_f$, or $1\text{ lb}_f = 4.448\text{ N}$. Your author's weight of about 205 lb$_f$ corresponds to about 912 N.

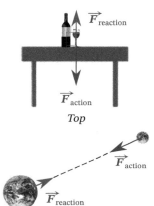

Newton's Third Law

Newton's third law is almost universally recognized, and almost universally misunderstood:

For every action there is an equal and opposite reaction.

The misunderstanding comes from the terms "action" and "reaction." Over time they have come to be recognized as almost anything, including someone's response to another person's behavior.

The real meaning of Newton's third law is much more exacting, where the terms "action" and "reaction" represent two different forces, equal in magnitude and opposite in direction. Mathematically, the third law can be written as

$$\vec{F}_{\text{reaction}} = -\vec{F}_{\text{action}}, \qquad (5.5)$$

Newton's third law of motion

and illustrated by two examples in Fig. 5.14.

The third law is telling us that for every force that exists, there is always another force that has exactly the same magnitude, but is pointed in the opposite direction. For example, if you push down on a table, the table pushes back on your hand with the same amount of force, but in the opposite direction. You can certainly feel that reaction force from the table. As another example, Earth exerts a gravitational pull on the Moon that keeps it in its orbit, but the Moon also exerts a force of the same magnitude on Earth, pulling on Earth in the opposite direction. If you drop your pen, Earth pulls on the pen gravitationally, causing the pen to accelerate toward Earth, but the pen is also pulling on Earth, causing Earth to accelerate toward the pen! You don't feel the acceleration of Earth toward the pen simply because Earth is somewhat more massive than the pen and therefore the second law tells us that Earth's acceleration toward the pen will be a lot less than the acceleration of the pen toward Earth. [If you're curious, Earth's mass is roughly 10^{26} (one hundred trillion trillion) times greater than the mass of your pen and so the acceleration of Earth is 10^{26} times smaller!]

Fig. 5.14 *Top:* The reaction force on the wine glass due to the table is equal in magnitude and opposite in direction to the force due to its weight. *Bottom:* The gravitational force exerted on the Moon by Earth is equal in magnitude and opposite in direction to the gravitational force exerted on Earth by the Moon. In either example, which force is considered the action force and which force is considered the reaction force is purely a matter of perspective. (Earth and Moon images courtesy of NASA)

Newton's Laws Summarized

Newton's laws of motion are given together for easy reference:

1. (Law of inertia) An object's velocity vector will remain unchanged unless the object is acted upon by a non-zero force, $\vec{F}$.
2. $\vec{F} = m\vec{a}$, or $\vec{a} = \vec{F}/m$; the acceleration an object experiences equals the force exerted on it divided by the object's mass.
3. $\vec{F}_{\text{reaction}} = -\vec{F}_{\text{action}}$; for every force there is an equal and opposite reaction force.

5.3   The Universal Law of Gravitation

Apples and Orbits

There is a famous story of Newton sitting under a tree one day when he was home at Woolsthorpe Manor during his *annus mirabilis* of 1665–1666, and an apple fell and struck him on the head. He then looked up, saw the Moon, and instantly realized that there existed a universal law of gravitation that caused both the apple to fall and the Moon to continue in its orbit around Earth. While the story is certainly allegorical, in his later years Newton repeatedly told a story of being inspired by the falling of an apple. A good friend of his, William Stukeley (1687–1765), reported a conversation he had had with Newton in 1726 regarding the story:

> we went into the garden, & drank tea under the shade of some appletrees; only he, & my self. amidst other discourse, he told me, he was just in the same situation, as when formerly, the notion of gravitation came into his mind. "why should that apple always descend perpendicularly to the ground," thought he to himself; occasion'd by the fall of an apple, as he sat in a contemplative mood. "why should it not go sideways, or upwards? but constantly to the earths center? assuredly, the reason is, that the earth draws it. there must be a drawing power in matter. & the sum of the drawing power in the matter of the earth must be in the earths center, not in any side of the earth. therefore dos this apple fall perpendicularly, or toward the center. if matter thus draws matter; it must be in proportion of its quantity. therefore the apple draws the earth, as well as the earth draws the apple."

Fig. 5.15 A sapling apple tree, located at Trinity College, Cambridge, that is believed to be descended from the original tree that Newton claimed inspired his search for a universal law of gravitation. (Loodog, CC BY-SA 3.0)

Newton certainly didn't have a moment when the universal law of gravitation entered his mind fully formed. Rather, the incident he referred to started him on a path of investigation that, through a great deal of effort and intuition, led him to one of the most famous scientific results of his career. Over the next few pages we will think through how his laws of motion and the empirical data that existed at the time led Newton to develop an equation representing the universal law of gravitation. While studying the reasoning for the first, time try to focus on the general arguments without worrying too much about the details of the mathematics. (Remember though that mathematics is simply a language that concisely and precisely describes the physical phenomena; you should always read the mathematics as you would a paragraph, trying to understand what it is telling you about nature.)

Fig. 5.16 *Left:* An apple falling toward Earth with the acceleration due to gravity, $g = 9.8 \, \text{m/s}^2$. (Adapted from public domain clip art) *Right:* The Moon experiences a centripetal acceleration toward the center of Earth of magnitude $a_c = 0.0027 \, \text{m/s}^2$; see Example 5.6. Note that the two vector lengths are not proportional. (Adapted from Adrian Michael, CC BY-SA 3.0)

Many decades before Newton began his search for the universal law of gravitation, Galileo had determined that a freely falling object (such as an apple) speeds up with a constant acceleration [see Fig. 5.16 (*Left*)]. The value of the acceleration due to gravity *near the surface of Earth* is:

Key Point
The acceleration due to gravity *near the surface of Earth.*

$$g_{\text{Earth}} = 9.8 \, \text{m/s}^2 = 32 \, \text{ft/s}^2, \tag{5.6}$$

Acceleration due to gravity *near Earth's surface*

where g_{Earth} (or simply g) is commonly used to represent the acceleration due to gravity near Earth's surface. Since force is $\vec{F} = m\vec{a}$, the force of gravity on an object can be found by replacing the general expression for acceleration, $\vec{a}$, with the acceleration due to gravity, g, or $\vec{F}_{\text{gravity}}$.

Newton's insight was that the same influence that caused the apple to accelerate downward "toward the center" might also be the cause of the centripetal acceleration of the Moon (recall that centripetal acceleration for circular motion always points toward the center of the circle; Fig. 5.12). To test his hypothesis, Newton invoked the relationship Kepler had found between orbital period and the semi-major axis of an ellipse [Kepler's original harmonic law, Equation (4.1)]. Although the relationship was based on the orbits of the planets, Newton believed that the same general relation might apply to the Moon as well. He was able to combine the centripetal acceleration equation with Kepler's harmonic law to find a relationship between acceleration due to gravity toward the center of Earth and the distance between the center of Earth and an object, such as the apple or the Moon. What he discovered is known as an inverse square law,

$$a = \frac{C}{r^2},$$

where a represents the acceleration, r is the distance of separation between their two centers, and C is some, yet to be determined, constant. The inverse square law states that the acceleration of a falling object toward Earth decreases with the square of the distance between their centers; C divided by distance (r) and then divided by distance again, or alternatively C divided by distance times distance ($r \times r = r^2$).

Example 5.7

Since the falling apple is one Earth radius from the center of Earth (1 R_{Earth}) and the center of the Moon is slightly more than 60 times farther from the center of Earth (60.3 R_{Earth}), Newton's inverse square law suggests that the acceleration of the Moon toward the center of Earth [its centripetal acceleration; see Fig. 5.16 (*Right*)] should be $60.3^2 = 3640$ times smaller than the acceleration of the apple toward the center of Earth, or

$$a_{c,\,Moon} = \frac{g}{60.3^2} = \frac{9.8 \text{ m/s}^2}{3640} = 0.0027 \text{ m/s}^2,$$

which agrees with the value we found in Example 5.6.

As Newton stated (page 148),

> I deduced that the forces ... must [be] reciprocally as the squares of their distances from the centers about wch they revolve; & thereby compared the force requisite to keep the Moon in her Orb with the force of gravity at the surface of the earth, & found them answer pretty nearly.

Having discovered a relationship between the acceleration toward the center of Earth of a falling object near its surface and the centripetal acceleration of the Moon, also toward the center of Earth, Newton's laws dictated that there must also be a relationship between the forces acting on them. From Newton's second law, the magnitude of the force acting on the apple must be $F_{apple} = m_{apple}\, g_{Earth}$, the mass of the apple times the acceleration due to gravity near Earth's surface. Similarly, the magnitude of the force acting on the Moon must be $F_{Moon} = m_{Moon}\, a_{c,\,Moon}$, the mass of the Moon multiplied by the Moon's centripetal acceleration. Since there is a general relationship between acceleration and distance from the center of Earth, the forces acting on the apple and the Moon must both be described by the inverse square law,

$$F_{gravity} = m\left(\frac{C_{Earth}}{r^2}\right),$$

where m is the mass of either the apple or the Moon, and C_{Earth} represents an unknown constant associated with the acceleration of mass m toward Earth.

Newton's third law provides another important clue to developing a universal law of gravitation: if there is a force acting on the apple, there must be an equal and opposite force acting on Earth, and similarly between the Moon and Earth [recall

Fig. 5.14 (*Bottom*)]. In Stukeley's recounting of the apple story on page 165, the last portion states that "if matter thus draws matter, it must be in proportion of its quantity, therefore the apple draws the earth, as well as the earth draws the apple." This suggests that the force of gravity acting on Earth must be

$$F_{\text{gravity}} = M_{\text{Earth}}\, a_{\text{Earth}} = M_{\text{Earth}} \left(\frac{C_m}{r^2} \right),$$

where M_{Earth} is the mass of Earth, the a_{Earth} is the acceleration of Earth toward the apple (or the Moon), and C_m represents a different unknown constant associated with the acceleration of Earth toward mass m (either the apple or the Moon).

If we only consider the mutual attraction between Earth and the apple, then according to Newton's third law the magnitude of the force that Earth exerts on the apple and the magnitude of the force that the apple exerts on Earth must be equal. Mathematically, this means that

$$F_{\text{gravity}} = m_{\text{apple}} \left(\frac{C_{\text{Earth}}}{r^2} \right) = M_{\text{Earth}} \left(\frac{C_{\text{apple}}}{r^2} \right).$$

It is common in science for symmetry to point us toward the solution to a problem, including within equations, and Newton's search for a universal law of gravitation is an excellent example of that aspect of nature. If the gravitational force that Earth exerts on the apple is equal and opposite to the gravitational force that the apple exerts on Earth, then both the mass of Earth and the mass of the apple should be explicit in the universal law of gravitation. This can occur if the two unknown constants, C_{Earth} and C_{apple} have their own masses in them. In other words, suppose $C_{\text{Earth}} = GM_{\text{Earth}}$ since Earth is causing the acceleration toward its center, and $C_{\text{apple}} = Gm_{\text{apple}}$, since the apple is causing the acceleration of Earth toward it, and G is a new constant that does not depend on the mass of either object. What we now have in the equations above is

$$F_{\text{gravity}} = m_{\text{apple}} \left(\frac{GM_{\text{Earth}}}{r^2} \right) = M_{\text{Earth}} \left(\frac{Gm_{\text{apple}}}{r^2} \right).$$

Notice that the same two masses are in both equations. Of course, G and $1/r^2$ are in both equations as well.

If we now no longer specify the apple and Earth, but simply use any two masses, identified by M and m, and we rearrange either equation, we finally arrive at Newton's universal law of gravitation,

$$F_{\text{gravity}} = G \frac{Mm}{r^2}, \tag{5.7}$$

Newton's universal law of gravitation

A universal law of gravitation tutorial is available through the Chapter 5 resources page

where G is a constant to be determined (the universal gravitational constant), M and m are the two masses being attracted toward one another (one mass could be the apple and the other mass could be Earth, for example; it doesn't matter which one is M or m), and r is the distance separating the centers of the two masses. In this equation the variables must all be expressed in SI units (kilograms and meters), and F_{gravity} is in the force unit of newtons. The G in Newton's universal law of gravitation is described as "Big G" so as not to confuse it with "little g," the acceleration due to gravity near the surface of Earth; 9.8 m/s².

As the name implies, Newton's universal law of gravitation applies to any two masses, anywhere in the universe. They could be an apple and Earth, the Sun and Jupiter, the Milky Way Galaxy and the Andromeda Galaxy, or you and someone else. Two people may be attracted to one another for other reasons, but every person is attracted to every other person on Earth, as well as to everything else in the universe, by the force of gravity.

The Mass of Earth and the Mystery of Big G

The magnitude of the gravitational force acting on a falling apple by Earth near Earth's surface is given by $F_{\text{gravity}} = m_{\text{apple}}\, g_{\text{Earth}}$, but it must also be given by Newton's universal law of gravitation. This means that

$$F_{\text{gravity}} = m_{\text{apple}}\, g_{\text{Earth}} \quad \text{and} \quad F_{\text{gravity}} = G\frac{M_{\text{Earth}}\, m_{\text{apple}}}{R_{\text{Earth}}^2}$$

must both give the same result, where R_{Earth} is the radius of Earth expressed in meters. By comparing the two equations, as depicted in Fig. 5.17, it must be true that

$$g_{\text{Earth}} = G\frac{M_{\text{Earth}}}{R_{\text{Earth}}^2} = 9.8 \text{ m/s}^2.$$

More generally, the acceleration due to gravity toward the center of a mass, M, a distance, r, from its center is given by

$$g = G\frac{M}{r^2}. \tag{5.8}$$

The acceleration due to gravity

Since weight is just the force of gravity acting on a mass, and the force of gravity on an object of mass m can be written as $F_{\text{gravity}} = mg$,

$$\text{Weight} = mg. \tag{5.9}$$

The relationship between weight and mass

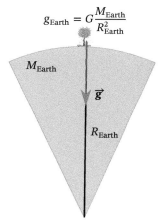

$$g_{\text{Earth}} = G\frac{M_{\text{Earth}}}{R_{\text{Earth}}^2}$$

Fig. 5.17 The acceleration due to gravity on Newton's apple near Earth's surface depends on the mass and radius of Earth. G is the universal gravitational constant.

As it turned out, in Newton's time it was impossible to actually determine the value of Big G. The value of little g had been known since Galileo's time, and even the ancient Greek philosophers had a reasonable estimate of the radius of Earth (Fig. 3.12), but the mass of Earth was unknown; only the product of Big G and the mass of Earth, M_{Earth}, could be calculated (GM_{Earth}). Without knowing Big G, the mass of Earth could not be determined, and without the mass of Earth, Big G wasn't knowable. As we will learn in Section 6.3, this dilemma wasn't resolved until more than 70 years after Newton's death. Even today, the combination of the two constants, GM_{Earth} is known with greater precision than either G or M_{Earth} separately.

Applying the Universal Law of Gravitation

It is gravity that controls the motions of the planets, moons, asteroids, comets, and ring systems in our Solar System, drives the collapse of giant gas and dust clouds to form stars and is the ultimate reason that stars eventually die. It explains the spiral structure of majestic galaxies, binds thousands of galaxies together to form immense clusters and superclusters of galaxies, and plays an important role in the evolution of the entire universe.

Fortunately, even without knowing the value of G it is possible to learn a great deal about the behavior of the force of gravity. In fact, we won't worry about using the universal law of gravitation to calculate forces in newtons, but it is important to understand how the force of gravity depends on masses and the distance of separation between the centers of the masses.

Example 5.8

What would happen to your weight if the mass of Earth were to be miraculously replaced by the mass of the Moon, but without changing the planet's radius?

The mass of the Moon is slightly less than one-eightieth of the mass of Earth, $m_{Moon} = 0.0123\,M_{Earth}$, where M_{Earth} is a unit representing a mass equal to the mass of Earth just as kg is a unit representing a mass equal to one kilogram. Since your weight is just the force of gravity exerted on you, and your own mass doesn't change in this example, your weight would become just 0.0123 times your current weight on Earth. Your current Earth weight is given by

$$\text{Earth weight} = F_{gravity} = G\frac{M_{Earth}\,m_{you}}{R_{Earth}^2}.$$

If Earth's mass magically decreased but its radius remains unchanged, then your weight would become

$$\text{New weight} = F_{gravity} = G\frac{(0.0123\,M_{Earth}) \times m_{you}}{R_{Earth}^2} = 0.0123 \times \text{Earth weight}.$$

As mentioned in the first subsection of Section 5.2, there wouldn't be any less of you, only the pull of gravity on your body would decrease. If your weight on Earth is 667 N (150 lb), your weight would now be $0.0123 \times 667\,N = 8.2\,N$ (1.8 lb).

Example 5.9

What would happen to your weight if you lived on a planet with Earth's mass, but it had a radius that is twice the radius of Earth?

Since the distance r in the denominator of the universal law of gravitation equation would double, and since r is squared, this means that the the denominator would increase by a factor of $2^2 = 4$. As a result, your weight on the planet would be

New weight = Earth weight /4.

If your weight on Earth is 667 N, then your new weight would be 667 N/4 = 167 N.

Example 5.10

In the first subsection of Section 5.2 it was pointed out that the weight of your author on the Moon would be about one-sixth of his weight on Earth, but in Example 5.8 we determined that by replacing the mass of Earth with the mass of the Moon, your weight would be about one-eightieth of its value on the surface of Earth (m_{Moon} = 0.0123 M_{Earth}). The discrepancy is because the Moon's radius is also smaller, being about one-quarter of the radius of Earth (R_{Moon} = 0.272 R_{Earth}), thus making the force of gravity greater because of r^2 in the denominator of the universal law of gravitation. Show that your weight on the Moon would be about one-sixth of its value on Earth.

In this case we need to replace both the mass of Earth and the radius of Earth in the universal law of gravitation with the values for the Moon. This gives

$$\text{New weight} = G\frac{(0.0123\ M_{Earth}) \times m_{you}}{(0.272\ R_{Earth})^2} = \left(\frac{0.0123}{0.272^2}\right) \times \left(G\frac{M_{Earth}\ m_{you}}{R_{Earth}^2}\right).$$

But the expression inside the last pair of parentheses is just the force of gravity exerted on you when you are on the surface of Earth. Furthermore, $0.0123/0.272^2 = 0.0123/(0.272 \times 0.272) = 0.116 = 1/6.03$. Therefore,

New weight = $0.116 \times$ Earth weight = Earth weight /6.03.

If your weight on Earth is 667 N, then your new weight would be 667 N/6.03 = 111 N.

Example 5.11

An exotic type of star, known as a neutron star, can have a mass of two times the mass of our Sun and a radius of just 10 km. In terms of the mass of Earth, the mass of a neutron star is roughly 700,000 M_{Earth} and its radius is approximately 0.002 R_{Earth}. How many times greater would your weight be on the surface of a neutron star than it is on the surface of Earth? (Note: Because of the immense force of gravity in this case, Newton's universal law of gravitation only approximates the true answer.)

Following the same procedure as in Example 5.10,

$$\text{New weight} = G\frac{700,000\ M_{Earth} \times m_{you}}{(0.002\ R_{Earth})^2} = \left(\frac{700,000}{0.002^2}\right) \times \text{Earth weight.}$$

This means your weight on the surface of the neutron star would be

New weight = $(1.75 \times 10^{11}) \times$ Earth weight;

your weight would be 175 billion times greater on the surface of the neutron star than it is on the surface of Earth. That's some serious weight gain!

Example 5.12

The mass of Mars is 11% of the mass of Earth ($M_{\text{Mars}} = 0.11\ M_{\text{Earth}}$) and its radius is slightly greater than 1/2 of Earth's radius ($R_{\text{Mars}} = 0.53\ R_{\text{Earth}}$).

(a) What is the acceleration due to gravity on Mars?
(b) If your mass on Earth is 70 kg, what is your mass on Mars?
(c) What would your weight be on Earth (in newtons)?
(d) What would your weight be on Mars (again in newtons)?

(a) Substituting the values into Equation (5.8) (the acceleration due to gravity equation),

$$g_{\text{Mars}} = G\frac{0.11\ M_{\text{Earth}}}{(0.53\ R_{\text{Earth}})^2} = \frac{0.11}{(0.53)^2}\frac{GM_{\text{Earth}}}{R_{\text{Earth}}^2} = 0.39 \times g_{\text{Earth}} = 3.8\ \text{m/s}^2.$$

(b) Trick question! Mass is an intrinsic quantity of the object and is independent of gravity or any source of acceleration. Your mass on Mars would be 70 kg.
(c) On Earth, your weight is $mg_{\text{Earth}} = 70\ \text{kg} \times 9.8\ \text{m/s}^2 = 686\ \text{kg m/s}^2 = 686\ \text{N}$.
(d) On Mars, your weight is $mg_{\text{Mars}} = 70\ \text{kg} \times 3.8\ \text{m/s}^2 = 266\ \text{N}$.

But What Causes Gravity?

Although Newton developed a general mathematical description of the effects of gravity on an object having mass, he wasn't able to explain *how* gravity acted on that object. In the *Principia* he wrote:

> But hitherto I have not been able to discover the cause of those properties of gravity from phenomena, and I feign no hypotheses ... And to us it is enough that gravity does really exist, and act according to the laws which we have explained, and abundantly serves to account for all the motions of the celestial bodies, and of our sea [the tides].

We would have to wait for more than two centuries before Albert Einstein gave us a theory of how gravity exerts its influence (Section 8.2).

5.4 Newton Completes Kepler's Laws

In the *Principia*, Newton described a "thought experiment" designed to illustrate the connection between the gravity acting on objects near the surface of Earth (e.g., an apple) and orbital motion, such as the Moon around Earth. As illustrated in his text (see Fig. 5.18), he asked us to consider a giant cannon atop a tall mountain, V, on an otherwise featureless Earth. If the cannon was given a small amount of gunpowder and then fired, the cannonball would naturally travel a short distance, striking the ground at location D on the sketch. A bit more gunpowder would result in a faster cannonball leaving the barrel and the cannonball would reach location E on Earth. With even more gunpowder and greater speed, the cannonball would be able to travel one-quarter of the way around the planet, reaching point F. With yet more gunpowder and speed, the result would be a flight half-way around the world, causing the cannonball to strike Earth's surface at point G. Note that in each of these cases the force of gravity on the cannonball is always directed toward the center of Earth just like the acceleration vector for centripetal force is directed toward the center of circular motion as illustrated in Fig. 5.12. Of course, given that Earth is roughly spherical in shape, during the flight of the cannonball Earth's surface is curving under the cannonball's flight.

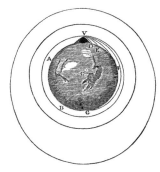

Fig. 5.18 Newton's mountain and cannon thought experiment as illustrated in the *Principia*. (Public domain)

Newton's sketch illustrates what happens next, should the cannon have enough gunpowder, and the cannonball sufficient speed. When fired from the mountaintop, the cannonball continues to curve due to the centrally directed force of gravity, while at the same time Earth curves continuously under the cannonball, with the result that the cannonball never strikes the surface, but instead completes an orbit, returning to the top of the mountain at V. Assuming that the cannon is moved out of the way (!) and there is no atmosphere, there is nothing to stop the cannonball, which means that it will continue to orbit forever. The other paths shown in the drawing indicate what would happen if a taller mountain and even more gunpowder was used. In fact, the shape (circular or elliptical) and size of the orbit depends on just what the initial speed and elevation of the cannonball is. Newton also argued that if given a sufficiently great initial speed "it might never fall to earth but go forward into the celestial spaces, and proceed in its spaces infinitum." In modern parlance, Newton's cannon is today's rocket, and the cannonball is one of the many thousands of satellites now in orbit around Earth (one of which is our Moon) or spacecraft that are headed out into interstellar space, including Voyagers 1, 2, and New Horizons.

Newton used Kepler's third law [the harmonic law, Equation (4.1)] to uncover the inverse square nature of the universal law of gravitation, and his own three laws of motion to introduce the masses of two attracting bodies into the law. Newton was then able to use his prodigious mathematical talent to derive each of Kepler's three laws directly from his gravitational force law rather than finding them buried within Tycho's data of the motions of the planets. In so doing, Newton discovered important missing pieces in Kepler's original formulation of the laws that had been previously undetected or not entirely understood.

As was described in Section 4.11, Kepler made the critical discovery that orbital motion can be elliptical rather than perfectly circular, but he missed the important fact that the center of the Sun is not what lies at one focus of a planet's elliptical orbit, it is the center of mass that resides at the principal focus, *F*, in Fig. 4.43.

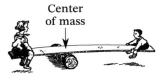

Center of mass

Fig. 5.19 Two children playing on a teeter-totter. (Adapted from public domain clip art)

For now you can think of the center of mass as simply the balance point between two masses. For example, consider two children playing on a teeter-totter (a seesaw) as shown in Fig. 5.19. For the game to be fair, the plank on which the children sit must be balanced. In order for this to occur, the heavier child will need to sit at the end of the shorter portion of the plank while the lighter child will sit at the end of the longer portion. The point on the plank where balance occurs is the center of mass. The reason this wasn't apparent to Kepler from Tycho's data is because the Sun is overwhelmingly more massive than any of the other planets. Even Jupiter, the most massive of the planets, is only 1/1000 of the mass of the Sun. This places the center of mass of the Sun–Jupiter system just outside the surface of the Sun rather than at its center. With Jupiter being 5.2 au away from the center of mass, the Sun's center is 5.2 au /1000 = 0.0052 au from the center of mass. Tycho's data simply weren't precise enough to see that the principal focus of Jupiter's elliptical orbit wasn't at the Sun's center, because the difference was just too small. However, if two stars are orbiting one another and their masses are equal, the principal focus for each star's orbit would be located in empty space half-way between the two stars. This is just what would happen if two children of identical mass were to sit on a teeter totter.

The other major difference that Newton discovered was that Kepler's third law [Equation (4.1)] also must include the sum of the masses of the two objects in orbit around one another. As Newton was able to show, the complete version of the law becomes

Key Point
Kepler's third law

$$(M_{\text{Sun}} + m_{\text{Sun}})\, P_{\text{y}}^2 = a_{\text{au}}^3, \tag{5.10}$$

Kepler's third law (the harmonic law)

where, in this form, M_{Sun} and m_{Sun} must be in terms of the mass of the Sun (solar units), P_{y} must be given in Earth years, and a_{au} is given in astronomical units (au). Kepler's third law states that for two objects orbiting one another, adding the two masses together (expressed in terms of the mass of the Sun) and then multiplying that sum by the orbital period of the two stars (expressed in Earth years) squared must equal their average separation (expressed in astronomical units) cubed. (Although Kepler's third law can be expressed in the fundamental basic SI units, that form is unnecessarily complicated for our needs here.)

Again, Tycho's data were simply not precise enough to allow Kepler to realize that a number other than 1 was actually multiplying P_{y}^2 for the planets orbiting the Sun. Considering the orbit of Jupiter, the sum of the masses (written in terms of the mass of the Sun) becomes $M_{\text{Sun, solar}} + m_{\text{Jupiter, solar}} = 1 + 0.001 = 1.001$. In other words, rather than Kepler's third law applied to Jupiter being $P_{\text{y}}^2 = a_{\text{au}}^2$, it is really $1.001 \times P_{\text{y}}^2 = a_{\text{au}}^3$; a very tiny adjustment indeed (an error of only one-tenth of one percent, 0.1%).

There was some hint of this missing ingredient (the sum of the masses) during Kepler's time. Recall at the end of Section 4.11 that Kepler realized that a form of his third law also works for the Galilean moons in orbit around Jupiter except that the ratio $a_{\text{au}}^3 / P_{\text{y}}^2$ had to take on a value other than 1. You may have also explored that reality in Exercises 4.63 and 4.64. Return to Table 4.2 and look at the value of Jupiter for $a_{\text{au}}^3 / P_{\text{y}}^2$.

Example 5.13

The planet Kepler 427b is one of thousands of planets known to orbit stars other than our Sun. Kepler 427b has an orbital period of 10.3 days and a semimajor axis of 0.0914 au. Determine the mass of its parent star assuming that the planet's mass is much smaller than the star's mass, and can therefore be neglected.

We must first convert the planet's orbital period to years. Since Earth's orbit around the Sun takes 365.256 days, the orbital period of Kepler 427b is

$$10.3 \text{ days} \times 1 \text{ year} / 365.256 \text{ days} = 0.0282 \text{ years}.$$

This means that the sum of the masses of the planet and its parent star is

$$M_{\text{Sun}} + m_{\text{Sun}} = 0.0914^3 / 0.0282^2 = 0.960 \text{ M}_{\text{Sun}},$$

where M_{Sun} is a commonly used unit that represents the mass of the Sun. (Notice that the italicized M_{Sun} is an equation variable and upright M_{Sun} is a unit equal to the mass of the Sun.) Since planets are generally much less massive than their parent stars, this result implies that Kepler 427b's parent star (Kepler 427) has a mass of about 0.960 M_{Sun}, making it only about 4% less massive than our own Sun.

The importance of Kepler's laws in astronomy, as derived by Newton, cannot be overstated. The laws provide the only direct way to determine the masses of objects orbiting one another, whether those objects are planets around stars, moons around planets, spacecraft around asteroids, stars around stars, or even galaxies around other galaxies. Newton's derivation of Kepler's laws was the beginning of a truly quantitative approach to understanding the heavens. Going forward, it is Newton's formulation that we will refer to as *Kepler's laws*.

Kepler's Laws of Orbital Motion Summarized

Kepler's laws, as modified by Newton, are listed together for easy reference (Newton's changes are emphasized):

1. (The law of ellipses) The shape of an object's orbit in a binary system is an ellipse with *the center of mass* of the system at one focus of the ellipse.
2. (The law of equal areas) An imaginary line connecting the orbiting object to *the center of mass* sweeps out equal amounts of area in equal time intervals.
3. (The harmonic law) *The total mass of a binary system,* their orbital period, and the semimajor axis (a_{au}) of their combined orbits are related by Kepler's third law.

$$(M_{\text{Sun}} + m_{\text{Sun}}) P_{\text{y}}^2 = a_{\text{au}}^3.$$

5.5   The Discovery of Uranus, Neptune, and Pluto

There were six known planets prior to the invention of the telescope: Mercury, Venus, Earth, Mars, Jupiter, and Saturn. There is also a seventh planet, Uranus, that is just barely visible to the naked eye, although it was never recognized as a planet before the age of telescopes. It may be that Hipparchus recorded it as a star in his catalog that then became a part of Ptolemy's *Almagest*. Others were also known to have observed Uranus in the late 1600s and in the 1700s, but it was Sir William Herschel (Fig. 5.20), a British astronomer using a telescope of his own design, who first realized in 1781 that Uranus was not a star. Initially he identified the object as a comet, but as other astronomers subsequently observed the object moving among the background stars and discovered that its orbit was nearly circular, Uranus became widely accepted to be a planet. (Along with being a telescope builder and an astronomer with many discoveries to his name, William Herschel was also an accomplished musician and composer, having written 24 symphonies and numerous concertos, sonatas, fugues, and other works. Herschel's sister, Caroline Lucretia Herschel (1750–1848), frequently aided him in his astronomical research throughout his career, as did his son, Sir John Frederick William Herschel (1792–1871), later in William's life.

Fig. 5.20 Sir Frederick William Herschel (1738–1822) painting by Lemuel Francis Abbott (1760–1802). (Public domain)

In Section 4.12 it was mentioned that Galileo may have observed Neptune with his small telescope (page 134), although he didn't realize that it was a planet. Neptune may well have been seen telescopically by others as well, but no one identified it as a planet. The true discovery that Neptune is a planet proved to be a powerful verification of Newton's new way of looking at the universe.

Almost immediately after the discovery of Uranus, Anders Johan Lexell (1740–1784) computed its orbit from the observational data and noticed some irregularities that could not be explained by the gravitational pull of the Sun and the other six known planets. He then proposed that perhaps there are other planets in the Solar System, as yet undiscovered, that could account for the discrepancies. While an undergraduate at Cambridge University, John Couch Adams (Fig. 5.21) stated that he believed it was possible to calculate the mass, position, and orbit of an as-yet unobserved planet by using the orbital data for Uranus and Newton's universal law of gravitation. After graduating in 1843, Adams began his calculations during his summer vacation, completing at least some of his work by September 1845. Meanwhile, a French astronomer, Urbain Le Verrier (Fig. 5.22), not knowing of Adams's research, presented results in June 1846 that gave the position of the mysterious planet, but not its mass or complete orbit. Two months later Le Verrier was also able to calculate the planet's mass and orbit, and then presented the new results. He sent his calculation of the planet's location in the sky to Johann Gottfried Galle (1812–1910) at the Berlin Observatory, which Galle received on September 23. That very night, Galle looked in the region suggested by Le Verrier and found Neptune just after midnight.

Fig. 5.21 John Couch Adams (1819–1892), circa 1870. (Public domain)

Since the two men had predicted the location of Neptune independently and almost simultaneously, there was some dispute between England and France over who should be given the credit. It was Adams in November 1846 who publicly acknowledged:

> I mention these dates merely to show that my results were arrived at independently, and previously to the publication of those of M. Le Verrier, and not

Fig. 5.22 Urbain Le Verrier (1811–1877). (Public domain)

with the intention of interfering with his just claims to the honours of the discovery; for there is no doubt that his researches were first published to the world, and led to the actual discovery of the planet by Dr. Galle, so that the facts stated above cannot detract, in the slightest degree, from the credit due to M. Le Verrier.

Today, Adams and Le Verrier are recognized as having made the mathematical discovery independently.

If a mathematical prediction of the location of a planet worked once, perhaps it could work again. Based on perceived remaining perturbations to the orbit of Uranus, Percival Lowell (Fig. 5.23), a wealthy businessman and astronomer from Boston, calculated the possible location of yet another planet in 1906. However, after a decade of searching, he was unable to locate the hypothetical planet. Then in 1930 Clyde Tombaugh (Fig. 5.24), working at Lowell Observatory, found a very faint object drifting extremely slowly against the background stars. Tombaugh had discovered Pluto in the general region predicted by Lowell 24 years earlier. But, as it turned out, the mathematical prediction was fallacious because the perceived remaining discrepancies in the orbit of Uranus after taking Neptune into account were simply too small to be meaningful; they were smaller than the observational errors inherent in the data that had been used in the prediction. In the end, Pluto was discovered by accident through a tedious study of a random region of the sky! We know today that there are many more objects out there that are relatively small worlds of rock and ices similar to Pluto. In 2006 the International Astronomical Union (IAU) "demoted" Pluto, classifying it and other similar objects as dwarf planets.

Fig. 5.23 Percival Lowell (1855–1916). **Photograph taken by James E. Purdy (1858–1933). (Public domain)**

Fig. 5.24 Clyde Tombaugh (1906–1997) **carrying a photographic plate holder at Lowell Observatory, Flagstaff, Arizona, in 1931. (Courtesy of New Mexico State University Library, Archives and Special Collections)**

Measurement Errors and Error Bars

The discovery of Pluto is an important lesson about not reading more into the data than is actually there. Every measurement in science contains a certain level of uncertainty, and scientists should always report the level of uncertainty that exists. Specifying a range of possible values around a measured value is referred to as assigning an error bar to the measurement. For example, suppose you try to measure the length of a book with a ruler that is marked out in centimeters, with each centimeter subdivided into ten millimeters as shown in Fig. 5.25. Without a more precise measuring tool you probably can't measure the length of the book to an accuracy of better than about 0.2 mm = 0.02 cm. As a result, you should really report the length of your book as 24.16 cm ± 0.02 cm, which is read as "24.16 cm plus

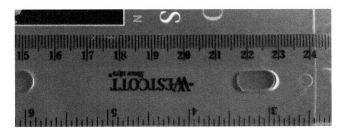

Fig. 5.25 Measuring the length of a book.

or minus 0.02 cm." What this means is that the actual length of the book is *probably* somewhere between 24.14 cm and 24.18 cm. In fact, in science, probabilities are also assigned to the likelihood that the actual measurement lies between the specified error bars.

You may be familiar with a similar type of error-reporting in political polls, where the pollster will report a "margin of error." If a pollster says that one candidate has 46% support in the polls and another has 43% support, with a margin of error of 3%, what the pollster is actually saying is that the poll indicates a tie between the two candidates given that the data are no more accurate than 3%. The poll indicates that one candidate's support is somewhere between 46% − 3% = 43% and 46% + 3% = 49%, while the other candidate's support is between 43% − 3% = 40% and 43% + 3% = 46%, meaning that the error bars for 46% and 43% overlap.

5.6   The Enlightenment (The Age of Reason)

Du Châtelet and the Acceptance of the *Principia* in France

Fig. 5.26 Émilie du Châtelet (1706–1749) portrait by Maurice Quentin de La Tour (1704–1788). (Public domain)

Despite the great success of Newton's *Principia*, it was not immediately accepted in all parts of Europe, especially not in France. In part due to strong nationalism, René Descartes and his mechanical model of a Solar System in which the planets and moons are dragged around in an ether full of swirling vortices remained the preferred theory. All of this changed when a translation of the original Latin version of the *Principia* was published posthumously in French, together with supporting commentary, by the mathematician, physicist, and author, Émilie du Châtelet (Fig. 5.26).

As a child, du Châtelet was encouraged by her father to study Latin, Greek, Italian, and German. She was also tutored in mathematics, science, and literature. Although apparently supportive of her education initially, Émilie's mother came to oppose her further education when the girl demonstrated significant gifts in the subjects. This was likely because of her mother's education in a convent, typical for girls at the time, and because encouragement of education for girls was not commonly accepted in French society.

Du Châtelet wed the Marquis Florent-Claude du Chastellet-Lomont in 1725 in an arranged marriage. By 1744, Émilie's husband was in the high position of Lieutenant General of France. They had three children together, although one died as a toddler.

During that period it was standard for both the man and the woman to take a lover and indeed du Châtelet did so on several occasions. One of her lovers was the highly regarded French philosopher, Voltaire (1694–1778), who was invited to live with her for a time with the approval of her husband. It has been suggested that the Marquis du Chastellet-Lomont agreed to the arrangement because Voltaire supported du Châtelet's extravagant lifestyle.

In any case, the famous Voltaire worked with, and encouraged, du Châtelet's growing interests in mathematics and science. He even had a laboratory constructed for their mutual use. At one point, Voltaire wrote in a letter to his friend, King Frederick II of Prussia, that du Châtelet was "a great man whose only fault was being a woman." Voltaire and du Châtelet did have a friendly rivalry in arguing their positions on philosophy and science. In 1738 they both entered a competition in the Paris Academy on the nature of fire, disagreeing with each other's ideas on the subject. Although neither of them won the first prize, their papers were both published, making du Châtelet the first woman ever to have had a paper published by that Academy.

In some of her own original work, du Châtelet conducted experiments to investigate the idea that energy could be transferred from one form to another, leading to her concept of a conservation of energy. But her greatest contribution came in her very ambitious project of translating the *Principia* into French. However, in 1749 she learned that she was pregnant from a new lover at the age of 42. Given her age, she feared that she would not survive the pregnancy, so she worked frantically to finish the translation, which she completed shortly before giving birth. Sadly, du Châtelet died a few days later and her baby died soon thereafter. To this day, her work remains the standard French translation of the *Principia*.

A Clockwork Universe and Determinism

When Newton penned his three laws of motion, not only did he describe a fundamentally new way of thinking about how nature worked, but he also opened a Pandora's Box of profound philosophical and theological questions. Newton's laws clearly describe how objects move once all of the forces acting on them are identified. Although certainly impossible in practice, theoretically speaking, if all of the particles in the universe can be identified, and if all of the forces between every particle can be calculated, then, according to Newtonian physics, for some initial description of the positions and velocities of the particles, it would be possible to determine how every particle will respond to the influence of every other particle for all eternity.

How does all of this pose philosophical and theological challenges? There is nothing in Newton's laws that limits "particles" to being planets, cars, or baseballs. Those "particles" can just as easily be you and me, or even the cells and individual atoms and neurons in our bodies. Newton's laws of motion raise the possibility that when the universe began, the initial configuration of the universe and its forces forever determined what would happen in the future. This view of the universe is sometimes referred to as a "clockwork universe;" the universe was wound up and set in motion, and future events and outcome are pre-determined. Do humans have any free will in such a universe? Are our everyday perceived decisions just the product of a universe set in motion 13.8 billion years ago?

The Dawning of a New Age of Reason and Critical Thought

Newton's *Principia* helped usher in what is today known as the Enlightenment, sometimes referred to as the Age of Reason. Some historians place the beginning of the Enlightenment with Francis Bacon (1561–1626), who espoused the importance of empiricism in scientific knowledge, while others argue that Descartes may be considered as the founder of the movement. However, it is Newton with his demonstration that motions could be understood by empirical means that firmly established the movement.

Along with Bacon, Descartes, and Newton, the philosophers of the Enlightenment included the Englishman John Locke (1632–1704), the Dutchman Baruch Spinoza (1632–1777), the Frenchman Voltaire, the American Benjamin Franklin (1706–1790), the Scotsman David Hume (1711–1776), the German Immanuel Kant (1724–1804), and the American Thomas Jefferson (1743–1826).

In essence, the Enlightenment, which existed through the end of the 18th century, embraced the idea that human reason can lead to understanding without appealing to prior forms of authority, including monarchs or the Church. The spread of this profound new way of thinking was aided by an explosion of printed materials. For example, although Newton's *Principia* was originally published in Latin and was generally inaccessible to most people anyway, given its level of complexity, authors began to publish general audience explanations of the contents of the book, including du Châtelet's French translation.

As a natural extension of the strength of human reason, and supported by the success of Newton, philosophers began to challenge the literal interpretation of the Bible. The appeal to supernatural events, including miracles, visitations by angels, and the resurrection of Jesus, were seen by some, including Jefferson, as unnecessary when interpreting the messages in the Christian New Testament. Jefferson went so far as to create his own version of the New Testament, commonly referred to as the *Jefferson Bible*, that he titled *The Life and Morals of Jesus of Nazareth*, where he cut and pasted (the old-fashioned way, with a razor and glue!) doctrines of Jesus without including any miracles or mention of the Resurrection.

Locke argued for religious tolerance as being an individual right that should not be affected by government, something that Jefferson called for as a "wall of separation between church and state" at the federal level. As a result, the philosophy of the Enlightenment made its way into the United States' Declaration of Independence and the Constitution.

Several forms of theology grew out of the Enlightenment and its challenge of authority, the two major theologies being deism and atheism. Deism is the notion of a clockwork universe: that God created the universe and then became an inactive observer, not interfering with what had been created. Atheism argues that there is no need for God at all if there are no miracles or interactions with angels.

The view that human reason can lead to understanding also led naturally to the notion that the governed should have a voice in government. It has been argued that this view contributed to the American and French Revolutions.

Certainly, the *Principia* alone wasn't solely responsible for revolutions or the theological views of deism and atheism (Newton was himself deeply religious, albeit in a rather unorthodox way), but skepticism of authority, empiricism, and the power of human reason forever changed how humanity sees itself and its place in the universe.

Looking toward your future from a deterministic perspective:

May the net force be with you!

Exercises

All of the exercises are designed to further develop *your understanding* of the material by thinking carefully and critically about what you have read in this chapter. Answers to selected exercises are available in the back of the book.

True/False

1. Isaac Newton was born on December 25, 1642.
2. (*Answer*) Newton began his university education by studying the ancient Greek philosophers of science and mathematics.
3. In his *annus mirabilis*, Newton completely developed his theory of light, his three laws of motion, the calculus, and his universal law of gravitation.
4. (*Answer*) After completing his degree at Trinity College, Newton spent the remainder of his life in Cambridge, Massachusetts.
5. Newton was first denied membership in the Royal Society because of his feud with Robert Hooke.
6. (*Answer*) The *Principia* is generally believed to be the most important book ever written in the physical sciences.
7. Newton's discoveries caused him to become an atheist.
8. (*Answer*) Newton never engaged in alchemy, the goal of trying to transform one type of metal into another by means of chemistry.
9. Newton helped rescue the English economy.
10. (*Answer*) Weight is an intrinsic quantity of matter.
11. When two vectors are added together, the length of the resultant vector is always equal to the lengths of the two vectors added together.
12. (*Answer*) If you increase the distance between two billiard balls by a factor of 3, the force of gravity between them decreases by a factor of 6.
13. You land on a world that has the same mass as Earth, but its radius is just one-half the radius of Earth. Your weight on this new world would be twice as much as it is on Earth.
14. (*Answer*) You land on a world that has the same mass as Earth, but its radius is twice the radius of Earth. Your weight on this new world would be four times as much as it is on Earth.
15. You land on a world that has one-half the mass of Earth but the same radius. Your weight on this new world would be one-half as much as it is on Earth.
16. (*Answer*) Newton realized that it was possible for a projectile to have enough speed that it could escape Earth and not fall back onto it or go in orbit around it.
17. The center of mass of the balanced rock shown in Fig. 5.27 is located over the middle of the spot where it is attached to the rest of the rock formation below it.
18. (*Answer*) It is fair to say that Neptune was discovered mathematically before it was ever observed to be a planet.
19. Percival Lowell successfully predicted where to find Pluto in the sky based on an analysis of the orbits of the other planets.
20. (*Answer*) The mass of a rock is measured in a laboratory and reported to be 3.2 ± 0.1 kg. This means that the rock's mass is likely between 3.1 kg and 3.3 kg.

Fig. 5.27 A balanced rock in Arches National Park, Utah (NPS/Debra Miller)

Multiple Choice

21. Newton's education at Cambridge University included studying the works of
    (a) Galileo   (b) Boyle   (c) Descartes
    (d) all of the above
22. (*Answer*) Newton is responsible for
    (a) the earliest development of the calculus.
    (b) demonstrating that sunlight is composed of a range of colors from red to violet.
    (c) inventing the reflecting telescope.
    (d) discovering that the natural motion of an object is to continue in a straight. line at a constant speed.
    (e) all of the above
    (f) (a), (b), and (c) only
23. The *Principia* contains
    (a) Newton's three laws of motion.
    (b) the universal law of gravitation.
    (c) the derivations of Kepler's laws of planetary motion.
    (d) an explanation of the tides found in Earth's oceans.
    (e) all of the above
    (f) (a), (b), and (c) only
24. (*Answer*) The nucleus of an atom has a diameter of about 1×10^{-15} m. Express the size using SI prefixes.
    (a) 1 Pm   (b) 1 km   (c) 1 mm   (d) 1 nm
    (e) 1 pm   (f) 1 fm   (g) 1 am   (h) 1 ym
25. A modern personal computer hard drive has a storage capacity of 5 TB. When the first personal computers with internal hard drives came out in the early 1980s, hard drive capacities were typically 10 MB. How many times more data can the modern hard drive hold compared to the ones in the 1980s?
    (a) 50   (b) 500   (c) 5000
    (d) 50,000   (e) 500,000   (f) 5,000,000

26. (*Answer*) A person walks 5 km straight south and then 12 km straight west. How far is the person from their starting point?
    (a) 5 km   (b) 12 km   (c) 13 km   (d) 169 km
    (e) none of the above

27. You are driving on a straight road at 115 km/h and come to a construction zone with a speed limit of 80 km/h. It takes you 10 seconds to slow down. What is your rate of acceleration in m/s²? Note that you will need to convert kilometers to meters and hours to seconds.
    (a) $3.5 \, \text{m/s}^2$ (b) $-0.97 \, \text{m/s}^2$ (c) $0.97 \, \text{m/s}^2$
    (d) $-9.7 \, \text{m/s}^2$ (e) $-3500 \, \text{m/s}^2$

28. (*Answer*) You are riding a merry-go-round that has a radius of 2 meters. At the point where you are standing, the merry-go-round's speed is 4 m/s. What is the magnitude of your centripetal acceleration? (Hang on tight!)
    (a) $2 \, \text{m/s}^2$ (b) $4 \, \text{m/s}^2$ (c) $8 \, \text{m/s}^2$ (d) $16 \, \text{m/s}^2$

29. A communications satellite in geosynchronous orbit around Earth has an orbital radius of 4.2×10^7 m and a speed of 3075 m/s. It's centripetal acceleration is
    (a) $0.0027 \, \text{m/s}^2$ (b) $0.225 \, \text{m/s}^2$ (c) $9.8 \, \text{m/s}^2$
    (d) $42 \, \text{m/s}^2$

30. (*Answer*) Two 2 kg blocks are separated by 5 m. One of the two blocks is replaced by a 6 kg block but the distance of separation remains unchanged. The gravitational force between the 2 kg block and the 6 kg block is _____ times _____ than it had been between the two 2 kg blocks.
    (a) 3, greater   (b) 3, less   (c) 4, greater
    (d) 4, less   (e) 6, greater   (f) 6, less
    (g) 12, greater   (h) 12, less

31. You land on a world that has the same mass as Earth, but its radius is one-third (1/3) the radius of Earth. Your weight on this planet would be _____ your weight is on Earth.
    (a) one-third (1/3)   (b) three times (3)
    (c) one-sixth (1/6)   (d) six times (6)
    (e) one-ninth (1/9)   (f) nine times (9)

32. (*Answer*) You land on a world that has twice Earth's mass, but it's radius is the same as Earth's. Your weight on this planet would be _____ what your weight is on Earth.
    (a) one-half of (1/2)   (b) twice (2)
    (c) one-fourth of (1/4)   (d) four times (6)

33. You land on a world that has four times Earth's mass and it's radius is twice (2 times) the radius of Earth. Your weight on this planet would be _____ what your weight is on Earth.
    (a) the same as
    (b) one-half of (1/2)   (c) two times (2)
    (d) one-fourth of (1/4)   (e) four times (4)
    (f) one-eight of (1/8)   (g) eight times (8)

34. (*Answer*) A planet is discovered to be orbiting a star in a circular orbit with a period of four years (4 y) at a distance of four astronomical units (4 au). What is the approximate mass of the star? (Don't reach for a calculator for this one!)
    (a) $0.5 \, \text{M}_{\text{Sun}}$ (b) $1 \, \text{M}_{\text{Sun}}$ (c) $2 \, \text{M}_{\text{Sun}}$
    (d) $4 \, \text{M}_{\text{Sun}}$ (e) $8 \, \text{M}_{\text{Sun}}$

35. A planet is discovered to be orbiting a star in a circular orbit with a period of five years (5 y) at a distance of ten astronomical units (10 au). What is the approximate mass of the star?
    (a) $0.5 \, \text{M}_{\text{Sun}}$ (b) $1.25 \, \text{M}_{\text{Sun}}$ (c) $3 \, \text{M}_{\text{Sun}}$
    (d) $4 \, \text{M}_{\text{Sun}}$ (e) $5 \, \text{M}_{\text{Sun}}$ (f) $40 \, \text{M}_{\text{Sun}}$

Short Answer

36. List Newton's three laws of motion.

37. When you are in a car that goes around a turn, you feel like you are being pulled toward the outside, away from the center of the turn. Explain this sensation in terms of Newton's first law, the law of inertia.

38. In the movie *Interstellar* (2014),[5] the robot TARS stated: "Newton's third law. The only way humans have ever figured out how to get somewhere is to leave something behind." Explain what is meant by that statement in the context of rockets operating in the vacuum of space.

39. Describe the difference between mass and weight.

40. (a) Explain the difference between a scalar quantity and a vector quantity.
    (b) List three examples of scalar quantities and three examples of vector quantities.

41. (*Answer*) One velocity vector ($\vec{A}$) points straight east with a magnitude of 10 km/s and a second velocity vector ($\vec{B}$ points straight north with a magnitude of 20 km/s. Using a ruler, draw vector $\vec{A}$ with a length of 2 cm. Next draw vector $\vec{B}$ starting at the tip (the end of the arrow) of vector $\vec{A}$, making sure that vector $\vec{B}$'s length is in the right proportion relative to vector $\vec{A}$. Finally, draw vector $\vec{C}$ such that $\vec{C} = \vec{A} + \vec{B}$. Again using a ruler, measure and record the length (the magnitude) of vector $\vec{C}$. Don't forget to include the proper units with the magnitude of vector $\vec{C}$.

42. In a process similar to Fig. 5.11, show that if the final velocity vector, $\vec{v}_f$, is positioned one-eighth (1/8) of the way around the circle, rather than one-fourth of the way around, the vector $\Delta\vec{v} = \vec{v}_f - \vec{v}_i$ would still point toward the center of the circle.

43. You are running slowly at 3 m/s and decide to speed up to 6 m/s, doing so in 3 seconds. What is your rate of acceleration? Remember to include the correct units in your answer.

44. The radius of the orbit of a geosynchronous communications satellite is 6.6 R_{Earth}. Following the procedure involving the inverse square law, that was used on page 167 to calculate the centripetal acceleration experienced by the Moon in its orbit around Earth, determine the centripetal acceleration experienced by the communications satellite. Compare your answer to the one you determined by another means in Exercise 29.

45. The acceleration due to gravity near Earth's surface is always about 9.8 m/s², regardless of what the object is that is falling. Why then does a feather fall so much more slowly than an apple if they are dropped from the same height?

46. The force of gravity acting on an apple near the surface of Earth is about 1 N while the force of gravity exerted on the Moon by Earth is 2×10^{20} N. It is also true that the centripetal acceleration experienced by the Moon is only 0.0027 m/s² but the acceleration of a freely falling apple is 9.8 m/s² near Earth's surface. Explain how this can be the case.

47. Describe Newton's universal law of gravitation in words. How does the force of gravity depend on the masses of two objects and on their distance of separation?

48. (*Answer*) You are in a classroom with no windows at school in Kansas and are suddenly "beamed" to another, identical room on another planet that has the same mass as Earth, but has a radius that is 20% smaller. How would you be able to

---

[5]*Interstellar* (2014). Directed by Christopher Nolan. [Feature film]. Hollywood, CA: Paramount Pictures.

explain to a friend who was beamed there with you that you are definitely not in Kansas anymore?

49. Refer back to Exercise 63 of Chapter 4. What is the physical meaning of the value that was calculated in that exercise?

50. You move to a new planet in a different planetary system and you are asked to determine the mass of the planet's parent star. Explain how you would do that, assuming that you can determine the distance from your new home to the star's center.

51. You read on a website that scientists have just discovered an Earth-killing asteroid orbiting the Sun in a very elliptical orbit with an average distance from the Sun of slightly less than one astronomical unit, but that the asteroid does get farther from the Sun than Earth for a short period once every two years, allowing it to intersect Earth's orbit. Is this possible? Why or why not?

52. (*Answer*) Our Sun orbits the center of the Milky Way Galaxy. The Sun's orbital period is 211 million years (2.11×10^8 y) and the Sun's average distance from the center of the Galaxy is 1.68 billion astronomical units (1.68×10^9 au). Estimate the mass of the Milky Way Galaxy that is inside the Sun's orbit.

53. On February 6, 1971 Alan Shepard (1923–1998), while on the surface of the Moon during the Apollo 14 mission, hit a one-handed golf shot with the head of a six-iron attached to rock-mining equipment (see Fig. 5.28). He estimated that the shot went about 200 yards despite not having much club speed; he was wearing a bulky spacesuit and backpack at the time. How was the ball able to travel so far? (The club is now in the United States Golf Association Museum with a replica in the Smithsonian Air and Space Museum in Washington, D.C.)

Fig. 5.28 Alan Shepard (1923–1998) hitting a golf ball on the surface of the Moon during the Apollo 14 mission on February 6, 1971; a true "Moon shot." (NASA)

54. Explain how John Couch Adams and Urbain Le Verrier were able to deduce the existence of Neptune without ever actually observing it.

55. Explain the importance of always including error bars in any measurement.

56. Discuss the relationship between Isaac Newton's work on motion and gravitation with the philosophical concept of determinism and the theological views of deism and atheism.

57. Describe the implications of the statement, "May the net force be with you."

Activities

58. It is often the case in physics and astronomy that experimental or observational data are gathered with the goal of verifying or refuting the prediction of an underlying theoretical model.

In this activity you will be producing graphs that include error bars. Specifically you will plot each data point with a dot and the ± errors around the data point using the symbol ⬤. The top bar of the symbol is set at the maximum value of the error range and the bottom bar is set at the bottom of the error range. For example, for 5 ± 3 the top bar would be at 8, the dot would be at 5 and the lower bar would be at 2.

(a) Plot the data for Case 1 from the table, including error bars. You can download a standard (Cartesian) sheet of graph paper from math_review/graphing/graph_templates, or alternatively use a graphing program such as Excel®. The horizontal axis should be labeled as x with a minimum value of 0 on the left end and a maximum value of 10 at the right end. Every other grid line should be labeled with increasing integers: 1, 2, … The vertical axis should be labeled as y and should also range from 0 to 10, with every other grid line labeled with increasing integers. The title of the graph should be "Case 1," written above the graph.

	Case 1		Case 2
x	y	x	y
1	3.4 ± 0.3	1	-1.0 ± 5.2
2	4.0 ± 0.4	2	2.5 ± 2.8
3	4.8 ± 0.3	3	0.3 ± 5.4
4	4.9 ± 0.8	4	6.7 ± 2.7
5	5.9 ± 0.3	5	4.2 ± 2.5
6	7.0 ± 0.5	6	9.5 ± 3.5
7	7.1 ± 0.4	7	6.0 ± 4.4
8	8.2 ± 0.6	8	13.5 ± 6.2
9	8.5 ± 0.3	9	10.2 ± 2.2

i. Suppose that a theory predicts that the data should follow a straight line with the equation $y = 0.6x + 3$. Calculate y from the equation for $x = 0$ and $x = 10$, plot those points on the graph, and draw a straight line between them using a straight edge, such as a ruler.

ii. Can other straight lines be drawn that stay within the error bars of the data? If so, can it be said that these other lines differ significantly from the theoretical line? Explain.

iii. Do the data appear to support or refute the theoretical line?

(b) Create a second plot using the data for Case 2. Again, you can download a sheet of graph paper from math_review/graphing/graph_templates or use a graphing program. Label the horizontal axis the same way you did before, but this time label the y axis from −20 to 20. Every fifth grid mark is an increase of 10 (every grid mark is an increase to 2). Label the $y = -10$, 0, and 10 grid marks. The title of the graph should be "Case 2".

Exercise 58 continues on the next page

58. *Continued*

(b)   i. The theoretical line for this data set is given by $y = 1.5x - 1.5$. Again calculate the theoretical values for $x = 0$ and $x = 10$, plot the points, and draw a straight line between them.

ii. Do the experimental data tightly constrain the theoretical line or are other lines also possible that have significantly different slopes (tilts)?

(c) Comment on the importance of always reporting error bars with experimental or observational data.

The Universality of Physical Law

6

Science is bound, by the everlasting vow of honour, to face fearlessly every problem which can be fairly presented to it.

Sir William Thomson (Lord Kelvin; 1824–1907)

6.1	Determining the Size of Our Solar System	186
6.2	Measuring the Speed of Light	189
6.3	Weighing Earth with a Laboratory Balance: Measuring Big G	191
6.4	The Conservation Laws	192
6.5	Light is an Electromagnetic Wave	201
6.6	Blackbody Radiation	213
6.7	The Inverse Square Law of Light	220
6.8	Heat as a Form of Energy	223
	Exercises	231

Fig. 6.1 M83. [NASA, ESA, and the Hubble Heritage Team (STScI/AURA); Acknowledgement: W. Blair (STScI/Johns Hopkins University) and R. O'Connell (University of Virginia)]

Introduction

The work of Sir Isaac Newton brought together the study of motions in the heavens with the study of motions on Earth. This revolutionary integration of astronomy and terrestrial science has become a fundamental tenent of modern astrophysics, that the same laws of nature apply throughout the universe, and for all time, without exception. Not only has Newton's universal gravitational constant, G, been the same everywhere in the universe since the beginning of time, but so have other fundamental physical constants that we will encounter throughout this text. The same laws that can be investigated in laboratories on Earth also apply to distant planetary systems and even in the most remote galaxies. The chemical elements that we find within our own bodies and on Earth are the same elements that we can identify in stars based on their tell-tale light signatures. Many of the molecules that exist naturally on Earth can even be identified drifting in interstellar space.

The universality of physical law undergirds our entire understanding of the cosmos. Without this constancy of nature we would be left only to marvel at the specks of light in the sky without comprehension, while the magnificent variety of phenomena on display throughout the universe, built upon a remarkable, foundational simplicity, would remain forever hidden from us.

6.1   Determining the Size of Our Solar System

The Distance to the Moon

The quest to determine the size of our own Solar System dates back to the time of the ancient Greek philosopher scientists. The only way in which this could be done at the time was through clever applications of geometry. Determining distances to the Moon, the Sun, and the planets started with first calculating the Moon's distance from Earth in terms of Earth's own radius. This can be accomplished by measuring the Moon's parallax; recall Fig. 3.20. Imagine making observations from two different locations on Earth with one location replacing your left eye in Fig. 3.20 and the other location replacing your right eye. The farther apart you can separate your two "eyes," the greater the apparent shift of the object will be. Since astronomical objects are very far away you want your "eyes" to be as far apart as possible so as to make the shift as large as possible.

The representation of the measurement technique shown in Fig. 6.2 uses an idealized right triangle, but it illustrates in principle the method that was employed for millennia to determine the Moon's distance from Earth. The approach is to measure the angle α (recall Table 4.1 listing the Greek alphabet) for an observer at position O at the same moment that a second observer at M has the Moon crossing her meridian. Knowing α then allows the determination of the distance from the center of Earth to the center of the Moon, d_{Moon}, in terms of the radius of Earth, r_{Earth}. In reality, it is impossible for an observer at O to be able to see the center of the Moon's disk while being exactly one-quarter of the way around Earth relative to another observer who has the center of the Moon's disk on his or her meridian.

Fig. 6.2 An idealized process for determining the distance to the Moon using parallax. r_{Earth} represents the radius of Earth, d_{Moon} is the distance from the center of Earth to the center of the Moon, and α is an angle measured from Earth's surface. Not drawn to scale. (Images courtesy of NASA)

Consequently, rather than the geometry being a true right triangle, a more complex situation would exist, but the basic approach is still the same.

An additional complication has to do with the timing of when observations from two different locations on Earth are made, given that Earth is rotating and the Moon is moving in its orbit. The more the Moon moves across the background of stars between the observations of the two observers, the more error gets introduced. One way to help overcome this problem is to record the position of the Moon relative to the stars when the observations are made. An even more accurate approach would be to record measurements when the Moon occults (passes in front of) a star. In any case, even the ancient Greek astronomers were able to get a fairly good estimate of the distance from the center of Earth to the center of the Moon. Recall that Newton used the knowledge that $d_{Moon} = 60.3\ R_{Earth}$ to help him develop his universal law of gravitation (Example 5.7).

Early Attempts to Determine the Distance to the Sun

Aristarchus, who had proposed a heliocentric Solar System, attempted to determine the distance to the Sun relative to the distance to the Moon by using a right triangle involving Earth, the Moon, and the Sun, as depicted in Fig. 6.3. When the phase of the Moon is either first quarter or third quarter (recall Fig. 4.16) the angle formed at the position of the Moon must be 90°. By measuring the angle between the Moon and the Sun from Earth, β (beta), it becomes theoretically possible to determine the distance to the Sun in terms of the distance to the Moon. However, Aristarchus's method is problematic for two reasons; in order for it to work the observer must be able to determine exactly when the phase of the Moon is either first quarter or third quarter, and measuring the angle β is very difficult with any precision, using the naked eye alone. Apparently Aristarchus determined β to be about 87°, when in fact it is about 89.9°. His value suggested that the Sun is about 19 times farther from Earth than the Moon is, when it is actually about 390 times farther away. Noting that during a solar eclipse the Moon just covers the disk of the Sun (see Fig. 4.23), he also concluded that the Sun's diameter must be about 19 times larger than that of the Moon, while the actual value is closer to 400 times larger. Eratosthenes, Hipparchus, and Ptolemy also tried to determine the Sun's distance relative to the distance to the Moon by using various geometrical techniques.

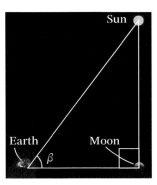

Fig. 6.3 Aristarchus's method of determining the distance to the Sun. Not drawn to scale. (Images courtesy of NASA)

Measuring the Astronomical Unit

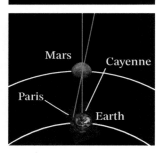

Fig. 6.4 Mars during its closest approach to Earth. The bottom figure is an expanded view of the top-central portion of the top figure. Not drawn to scale. (Images courtesy of NASA)

When Copernicus developed his heliocentric model of the Solar System he was able to use geometry to determine the *relative* distances of the planets from the Sun based on the distance of Earth from the Sun [the astronomical unit (au)], but his view of the actual size of the Solar System was based on the estimates of the Earth–Sun distance from the Greeks. In Kepler's original version of the harmonic law [Equation (4.1)] he related the orbital periods of the planets to their distances from the Sun in astronomical units. It was also Kepler who first realized that the determinations of the astronomical unit made by the Greeks were significantly smaller than the actual distance. With the invention of the telescope it became possible to make more accurate measurement of the angle β in Fig. 6.3, resulting in an improved estimate.

A significant improvement in the determination of the astronomical unit came in 1672, by making parallax measurements of the planet Mars. Giovanni Domenico Cassini (1625–1712) urged the French Academy of Sciences to put together an expedition to measure the position of Mars relative to the background stars from two different locations on Earth, at the time when Mars was at its closest approach to Earth (Fig. 6.4). The French Academy sent Jean Richer (1630–1696) to Cayenne in what is now Guiana in South America, just 5° north of the equator, while Cassini stayed in Paris to make the same observation. The measurement of the parallax as seen from the two locations was made easier because at the time Mars was only about 0.37 au from Earth. The measurement by Cassini and Richer gave a value for the astronomical unit of 21,600 times the radius of Earth, still smaller than the actual value of 23,456 R_{Earth}, but staggering large compared to any previous estimates.

Sir Edmond Halley argued that a more accurate measurement could be made by observing the rare transits of Venus across the face of the Sun. (Pairs of transits only occur once every century or so because the orbit of Venus is tilted relative to Earth's orbit.) When Venus is between Earth and the Sun it is only about 0.28 au from Earth, meaning that the parallax is even larger and therefore easier to measure than the parallax for Mars. Furthermore, it should be easier to accurately time when to make observations by watching to see just when the disk of Venus is completely inside the disk of the Sun (Fig. 6.5). Halley pointed out in 1679 that the next transits would occur in 1761 and 1769, so unfortunately he wouldn't live to seen the results of his method.

An international effort was organized to observe the Venus transit of 1761 by sending more than 100 observers to various points around the world. One of the researchers was the hapless Guillaume Le Gentil (1725–1792). In 1760 Le Gentil set sail for Pondicherry, a French colony in India, only to have war break out between France and Britain, which confused his travel plans. He was finally able to get passage on a frigate that was to arrive in time for the June 6, 1761, event but the ship was blown off course for a time. Still believing that they could arrive in port for the transit the captain continued on, only to learn that the British had captured Pondicherry and the ship was unable to dock. Although the skies were clear on June 6, Le Gentil could not make his observations with the necessary precision from the rolling deck of a ship. Instead of returning to France, he spent some time exploring the coast of Madagascar, after which he traveled to Manila in the Philippines to observe the June 3, 1769 transit. Unfortunately, the French Academy ordered

Fig. 6.5 The egress of Venus during the June 6, 2012, transit of the Sun. (NASA/SDO)

Le Gentil to return to Pondicherry to make the observations there. After having constructed a small observatory in Pondicherry from which to observe the transit, the sky was cloudy on that day! Returning to Paris twelve years after he left, Le Gentil found that his relatives believed him dead and they had divided up his estate. His wife had also remarried. After recovering some, but not all, of his money through legal proceedings (he had to pay expensive legal fees of course), Le Gentil decided to change profession. Ultimately, the international effort did yield an improved value for the astronomical unit. (Le Gentil became the subject of a Canadian play, *The Transit of Venus*, by Maureen Hunter, that was staged at the Royal Manitoba Theatre Centre in 1992. It was also made into an opera and performed by the Manitoba Opera Company in 2007.)

Originally the astronomical unit was defined to be the average distance between Earth and the Sun, taking into consideration that Earth's orbit is slightly elliptical. However, in 2012, based on increasingly precise determinations of the astronomical unit through the timing of light travel times from planetary spacecraft (Section 6.2), the International Astronomical Union simply *defined the astronomical unit* to be exactly 1 au $\equiv$ 149,597,870,700 m = $1.495\,978\,707\,00 \times 10^{11}$ m, which is approximately 150 million kilometers, or 93 million miles (the symbol $\equiv$ can be read as "is defined as"). Today, the astronomical unit isn't actually the true average distance between Earth and the Sun, but it is very nearly so.

6.2 Measuring the Speed of Light

Fig. 6.6 Portrait of Ole Christensen Rømer (1644–1710) c. 1700 by Jacob Conig (c. 1647–1724). (Public domain)

Galileo first tried to measure the speed of light by using widely separated lanterns (page 135), only to conclude that either the speed of light is infinitely fast, or it is too fast to be measured by his method. The problem with Galileo's method was that he and his assistant were too close together for slow human reaction times to be effective over the distance involved.

It was about forty years later that a Danish astronomer, Ole Christensen Rømer, (Fig. 6.6), noticed that when Earth is moving toward Jupiter, the passage of Io, one of Jupiter's four Galilean moons, seemed to move behind the planet earlier than expected, and when Earth was moving away from Jupiter, the planet eclipsed Io later than expected. Treating the orbit of Io like a clock with an orbital period of 1.77 days, Rømer concluded (correctly) that the changing eclipse times were due to differing distances between Jupiter and Earth as depicted in Fig. 6.7. From his analysis, Rømer estimated that light took about 22 minutes to traverse the diameter of Earth's orbit, indicating for the first time that the speed of light is in fact finite. Christiaan Huygens (Fig. 6.8) then combined Rømer's timing measurements with estimates of the astronomical unit to get

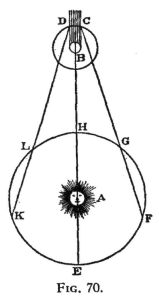

Fig. 70.

Fig. 6.7 Rømer's diagram of the eclipses of Io by Jupiter as seen from different locations in Earth's orbit. Jupiter is at B and Io passes behind the planet from C to D. The diagram is from Rømer, 1676.[1] (Public domain)

---

[1]Rømer never formally published his results, but presented them to the French Academy of Sciences. An anonymous news report, "Démonstration touchant le mouvement de la lumière trouvé par M. Roemer de l'Académie des sciences," was published on December 7, 1676, in the *Journal des sçavans*.

Fig. 6.8 Portrait of Christiaan Huygens (1629–1695) by Caspar Netscher (c. 1639–1684). (Public domain)

a value for the speed of light of about 220,000 km/s, approximately 80,000 km/s slower than the true speed of light. Sadly, most of Rømer's papers were destroyed in the Copenhagen Fire of 1728 that burned more than one-quarter of the city.

Over time, other methods were developed that improved on that first determination by Rømer and Huygens. In 1849, nearly two centuries after that initial effort, Hippolyte Fizeau (1819–1896) performed an experiment that was essentially the same as the one first carried out by Galileo, but with a much longer baseline (8 km), a mirror replacing the assistant with the second lantern, and a rapidly spinning toothed wheel (a gear) that replaced Galileo. In the experiment, a beam of light illuminated the spinning wheel, allowing a pulse of light to pass through a gap between two of the teeth. The pulse of light then traveled to the mirror and was reflected back to the wheel. By adjusting the spin rate of the wheel, the reflected light pulse could either be blocked by a tooth or allowed to pass through a gap. In this way, the spinning wheel acted as a timer to determine the time required for the light to travel the 16 km round-trip. Fizeau's method resulted in a value that was a bit too fast, but later modifications of the experiment by others, including using rotating mirrors instead of toothed wheels, began to converge on the actual speed of light. With greatly increased precision, scientists reached the point where measuring the speed of light was actually more precise than the way in which the length of one meter was defined. As a result, the speed of light became a defined quantity in 1983, and the meter became defined in terms of the speed of light rather than the other way around. Today the speed of light, universally represented by the symbol, c, *is defined* to be $c \equiv 299{,}792{,}458$ m/s $= 2.997\,924\,58 \times 10^8$ m/s. For our purposes, you should know that the speed of light is approximately

$$c = 300{,}000 \text{ km/s}, \tag{6.1}$$

The approximate speed of light

which in scientific notation is 3.0×10^8 m/s (about 186,000 mi/h). At that speed, the time required for light to travel from the Sun to Earth (1 au) is nearly

$$t = \frac{\text{distance}}{\text{speed}} = \frac{1 \text{ au}}{c} = \frac{150{,}000{,}000 \text{ km}}{300{,}000 \text{ km} / \text{s}} = 500 \text{ s} = 8.3 \text{ min.}$$

One way to think about this result is that if a major eruption were to happen on the surface of the Sun, we wouldn't see it occur for 500 seconds.

Since the speed of light is a constant in the vacuum of space, the time required for light to travel a distance can also be thought of as a measure of the distance itself; distance $= c \times$ time, or $d = ct$. In this context, the distance to the Sun is 8.3 light-minutes or 500 light-seconds.

The speed of light is another one of the fundamental constants of nature, and one that we will refer to frequently throughout this text. Although your author pointed out on page xxi of the study skills section of the Introduction's Aids and Suggestions For Students that "memorization does not equal understanding" there are some pieces of information that justify memorization; the approximate value of the speed of light, $c = 300{,}000$ km/s, is definitely one of those!

6.3 Weighing Earth with a Laboratory Balance: Measuring Big *G*

As we learned earlier (page 168), when Newton developed his universal law of gravitation,

$$F_{\text{gravity}} = G\frac{Mm}{r^2},$$

he was unable to determine the universal gravitational constant, G ("Big G"). This was because knowledge of the local acceleration due to gravity, g ("little g") was tied to both G and the mass of Earth, M_{Earth}, by the equation

$$g = G\frac{M_{\text{Earth}}}{R^2_{\text{Earth}}}.$$

A reasonably good estimate of the radius of Earth (R_{Earth}) was known in Newton's time, but it wasn't until 1798 that Henry Cavendish (Fig. 6.9), conducted a clever experiment in a laboratory that allowed for a direct determination of G.[3]

Cavendish was a very shy individual who some have suggested may have had a mental disability somewhere along the autism spectrum. However, despite that aspect of his private life he was a highly respected theoretical and experimental chemist and physicist. In the realm of chemistry he is credited with the discovery in 1766 of the lightest element in nature, hydrogen, which he referred to as "inflammable air." (We will be discussing hydrogen extensively throughout this text, given that it is by far the most abundant element in the universe.)

In all fairness it was Reverend John Michell (Fig. 6.10), a well-respected scientist, who first designed and built an apparatus that was meant to measure the density, and therefore the mass, of Earth, but unfortunately he died before the necessary experiments could be performed. The apparatus was eventually passed on to Cavendish, who rebuilt it to conduct the experiments.

Figure 6.11 shows a simplified diagram of the apparatus that was used. Two small lead balls were attached to the ends of a long wooden rod that was suspended from a thin wire. Two much larger and more massive lead balls were situated close to the smaller balls. The tiny force of gravity that was exerted on the smaller balls by the larger ones caused the wire to twist slightly. By measuring the amount of twist that occurred, the magnitude of the gravitational forces on the balls could be determined. Knowing all of the masses involved, as well as the separation distances between the large and small masses, means that the only unknown quantity in the universal law of gravitation is G. Once G is calculated, the mass of Earth can be computed from the equation for g. Because of the extreme sensitivity required for the experiment, Cavendish needed to eliminate any air currents, which he did by enclosing the large experiment inside a 10 ft×10 ft×2 ft (about 3 m×3 m×2/3 m) box with small windows that he could look through with a telescope to see the roughly 4 mm motion of the small balls.

Fig. 6.9 A sketch of Henry Cavendish (1731–1810) with his signature. (From the frontispiece of Wilson, 1851;[2] public domain)

Fig. 6.10 Reverend John Michell (1724–1793) (Public domain)

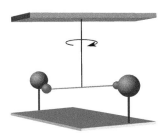

Fig. 6.11 The Cavendish experimental design.

[2]Wilson, George (1851). *The Life of the Hon. Henry Cavendish: Abstracts of his more important scientific papers*, Cavendish Society: London.

[3]As a point of trivia, Newton didn't originally write his equation in the form of Equation (5.7), he wrote it instead in terms of the weight density of Earth, which implicitly contained both G and M_{Earth}.

Big G is a tiny number, $G = 6.674\,30 \times 10^{-11}\,\mathrm{N\,m^2/kg^2}$. One way to put this number into perspective is to see just how weak the force of gravity actually is. Suppose that two 1-kg masses are placed 1 m apart and are allowed to move without any resistance of any kind. The mutual gravitational force exerted on the masses (don't forget Newton's third law) would initially cause them to accelerate toward one another at the rate of 66.7 *trillionths* of a meter per second per second. At that rate, it would take 1.5 hours for either mass to move 1 mm. It seems amazing that the force that controls the motions of the moons, planets, comets, asteroids, stars, galaxies, and even clusters of clusters of galaxies, is so weak. The reason that gravity has such a dominant influence is simply because every bit of mass in the universe attracts every other bit of mass, and objects in the universe can be very massive.

As suggested by the name, G is one of the fundamental constants of the universe. Although you can certainly memorize the number, simply realizing that its value is very small and that gravity is surprisingly weak is all that is really necessary for our purposes in this text. As we did in Section 5.3, it is most important that you understand how the force of gravity depends on the two masses involved (they are multiplied together) and their distance of separation, which is squared and in the denominator (the inverse square law).

6.4 The Conservation Laws

We learned in Chapter 5 that Newton's laws allow us to determine the motions of objects based on forces that are applied to them. Although true, sometimes the direct application of Newton's laws isn't always the most straightforward way to study what is happening. In fact, in many situations using Newton's laws directly can be so complicated that actually arriving at an answer is computationally impossible. This is certainly the case in a great many of the physical systems that are encountered in astronomy and astrophysics.

Rather than using Newton's laws, we will often want to look at several physical quantities in nature that remain constant no matter what happens within the system we are looking at. The term system can be loosely thought of as being an object or objects of interest while ignoring everything else in the universe. The Solar System is an example, or the Earth–Moon system. In some cases, we will want to consider a more restrictive type of system, known as a closed system, in which nothing, including light and heat, can enter or leave. In reality a truly closed system is never achieved, except in the case of the entire universe. After all, by definition, the entire universe is just that, everything there is; nothing can enter or leave the universe from outside of the universe. However, idealized approximations to a closed system can be imagined, such as everything we want to study being contained inside an opaque, sealed, well-insulated box where no light or heat can enter or leave.

A quantity of a system that never changes, no matter what is going on inside of system is known as a conserved quantity. In this section we will look at three conservation laws that will come up over and over again as we try to make sense of the natural processes that affect everything that happens in the universe, from collisions between two objects, to global warming on Earth, to the formation and lives of stars and planetary systems, and even to the history and future development of the universe as a whole.

Conservation of Linear Momentum

Newton's first law of motion requires that a non-zero net force must act on an object to alter its motion. Newton's third law requires that if two objects interact, their forces must be equal and opposite, meaning that the sum of their forces is exactly zero. Even though each object experiences a force, there is no net force acting on the two of them together due to their equal and opposite forces. Collectively, Newton's laws imply that if two objects are interacting, then either the two of them must be moving together with an overall common velocity vector, or their common center must be motionless. The same argument holds for any number of interacting particles; the common center of the group must have a constant velocity despite the individual motions of the particles that make up the group.

You are familiar with what this means in practice. If two children inside a car push each other away, the two of them may move as a result but their actions don't make the car itself move. Even if the car happened to be coasting with a constant velocity, their pushing each other still won't affect the car's motion (Fig. 6.12). As another example, if one American or Canadian football player tackles another player, the direction that the two of them move together immediately after the collision is the result of their individual masses and velocities before they collide.[4] Similarly, just because the planets in our Solar System are orbiting the Sun doesn't mean that the entire Solar System jumps around chaotically in space as a result; instead, the Solar System collectively moves through the Milky Way Galaxy as if it were a single object. Internal forces don't change external motions.

The linear momentum vector of a single object is equal to its mass times its velocity vector, and is commonly symbolized by the vector $\vec{p}$, or

$$\vec{p} = m\vec{v}. \tag{6.2}$$

The definition of linear momentum

In order to change an object's linear momentum vector, a non-zero net force must be exerted on that object. If object 1 interacts with object 2, Newton's third law tells us that the force on 1 due to 2 is equal and opposite to the force on 2 due to 1: $\vec{F_1} = -\vec{F_2}$. This means that the change in the linear momentum vector of object 1 is equal and opposite to the change in the linear momentum vector of object 2, which can be written mathematically as $\Delta\vec{p_1} = -\Delta\vec{p_2}$, where again Δ is read as "the change in"; for example, $\Delta\vec{p_1} = \vec{p}_{1,f} - \vec{p}_{1,i}$, where the subscripts i and f stand for "initial" and "final" (before an interaction and after the interaction, respectully). As a result, the *sum* of the linear momentum vectors of objects 1 and 2 doesn't change:

$$\vec{p}_{1,i} + \vec{p}_{2,i} = \vec{p}_{1,f} + \vec{p}_{2,f} = \text{constant};$$

the sum of the linear momentum vectors of objects 1 and 2 before an interaction equals the sum of the linear momentum vectors of the two objects after an interaction [or, said another way, the sum of the linear momentum vectors is a constant even though their individual linear momentum vectors can change ($\vec{p}_{1,i}$ *does not equal* $\vec{p}_{1,f}$ and $\vec{p}_{2,i}$ *does not equal* $\vec{p}_{2,f}$)]. More generally, the conservation of linear momentum requires that the *sum* of all of the linear momentum vectors in

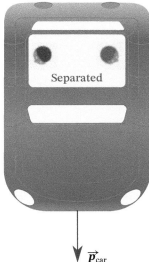

Fig. 6.12 Two children push each other inside a car, but the car's linear momentum is unaffected by interactions confined to the interior of the car. The car can be thought of as a system containing interacting objects.

---

[4]Soccer is referred to as football in much of the world.

Fig. 6.13 The Giotto space-craft that rendezvoused with Halley's Comet in 1986. (ESA)

the system remains unchanged, even though the individual linear momentum vectors of objects inside the system can change because of interactions with other objects in the system.

The only way the total linear momentum of the system of objects can change is if a force or forces from outside the system is applied to the group. This is just the situation for the two children pushing each other in the car: the total linear momentum of the car and its occupants is unaffected by the changes in linear momenta of the children when they push each other away.

Since mass is a part of the expression for linear momentum, an object has more linear momentum at a given velocity if it has more mass. In other words, it is a lot harder to stop a truck traveling at 100 km/h than a baseball traveling at 100 km/h. If a massive American football player tackles a much smaller player when running straight toward one another at the same speed, you know who "wins" the collision.

But mass isn't the only contributor to linear momentum: velocity can play an important role as well, even when masses are very different. If two football players with the same mass run straight at one another during a tackle, their direction after the collision will be the direction that the faster player was running before they hit. A fly won't do much to a truck except make a mess, but a speck of dust moving at 245,000 km/h (68 km/s) can negatively affect a big spacecraft. When the 960 kg Giotto spacecraft (Fig. 6.13) intercepted Halley's Comet in 1986, it carried a 50 kg aluminum and Kevlar shield to protect it from getting damaged by a speck of dust having that relative speed. Despite the protection, a collision with a speck of dust caused Giotto to start spinning just after the spacecraft's closest approach to the comet.

Conservation of Angular Momentum

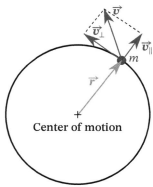

Fig. 6.14 The magnitude of the angular momentum is the product of mass, m, the perpendicular component of the velocity vector, $\vec{v}$, and the length of the vector, $\vec{r}$, from the center of motion to the mass ($L = mv_\perp r$). Recall that vectors can be thought of as being the addition of perpendicular vectors, in this case the component $\vec{v}_\perp$, that is perpendicular to $\vec{r}$, and the component $\vec{v}_\parallel$, that is parallel to $\vec{r}$; $\vec{v} = \vec{v}_\perp + \vec{v}_\parallel$.

Giotto started to spin because the dust speck didn't collide with the spacecraft at its exact center. The off-center collision imparted to the spacecraft a certain amount of angular momentum, $\vec{L}$. Like linear momentum, angular momentum is a vector and a conserved quantity (meaning that it doesn't change). We won't worry about the direction of the angular momentum vector for now, but its magnitude is given by the part of the linear momentum vector $\vec{p}$ that is perpendicular ($\perp$) to the vector $\vec{r}$ pointing from the center of the motion to the mass m, multiplied by the length of the vector $\vec{r}$; see Fig. 6.14. Mathematically,

$$L = p_\perp r = mv_\perp r. \tag{6.3}$$

The magnitude of angular momentum

Any object that moves around a center possesses angular momentum, whether it is a spinning top, a rotating ball, Earth spinning on its axis, or Earth orbiting the Sun.

The effects of angular momentum not only involve mass and velocity, but also how the mass is distributed relative to the rotation axis. Referring to Fig. 6.15, consider what happens to a figure skater during a spin. In order to increase his rate of spin, he pulls his arms in tightly and crosses his legs. Given that his overall mass doesn't change, making himself as narrow as possible means that the average distance of his mass from his rotation axis, r, becomes as small as possible.

Conservation of angular momentum requires that total angular momentum, $\vec{L}$, must stay constant, with the result that his rotational velocity, $v_\perp$, increases. To slow himself down he can spread his arms out wide and uncross his legs. Conservation of angular momentum is also used by gymnasts and divers for the same effect, to spin fast when tucked in a ball or along their lengths, and opening up to slow down.

How does the figure skater, gymnast, or diver initiate a spin in the first place? He applies a torque (τ, the Greek letter tau). As with angular momentum, torque is a vector quantity, but again we will only concern ourselves with its magnitude, which equals the magnitude of a perpendicular force ($\vec{F}_\perp$) multiplied by the distance r from the axis of rotation where the perpendicular force is applied:

$$\tau = F_\perp r. \tag{6.4}$$

The magnitude of torque

You are certainly familiar with the concept of torque. When you use a wrench to tighten or remove a nut on a bolt, you push or pull on the wrench in the direction you want the nut to rotate, and you apply the force at a location that is some distance away from the nut. The farther from the nut you apply the force, the more torque you exert. It is for that reason that it is much easier to tighten or remove a nut with a wrench than with your fingers alone. Just as a force alters linear momentum, torque alters angular momentum. It was the off-axis force of the collision between the dust speck and Giotto that resulted in a torque being applied to the spacecraft, causing it to spin.

Fig. 6.15 Tomáš Verner at the 2011 World Figure Skating Championships. (CC0 1.0, public domain)

Example 6.1

Our Moon is currently orbiting about 60.27 R_{Earth} from the center of Earth. At one point in the history of the Earth–Moon system, our Moon may have been orbiting Earth at about 25 R_{Earth} from Earth's center. Today, the Moon's orbit takes about 27.32 days to complete one orbit, but when the Moon was 25 R_{Earth} away, it would have only taken about 7.30 days to complete an orbit. As a result, while the Moon's orbital speed today is about 1023 m/s, back then it would have been 1585 m/s. Compare the amount of orbital angular momentum that the Moon has today to the amount of orbital angular momentum it had in the past.

The equation for angular momentum is $L = mv_\perp r$. It is possible to compare angular momenta by writing the equation as a ratio of now and then. To do so, divide the left-hand sides and the right-hand sides, giving

$$\frac{L_{\text{now}}}{L_{\text{then}}} = \frac{\cancel{m_{\text{Moon}}} \, v_{\perp,\,\text{now}} \, r_{\text{now}}}{\cancel{m_{\text{Moon}}} \, v_{\perp,\,\text{then}} \, r_{\text{then}}} = \left(\frac{v_{\perp,\,\text{now}}}{v_{\perp,\,\text{then}}}\right) \times \left(\frac{r_{\text{now}}}{r_{\text{then}}}\right).$$

The reason this works is because you are dividing both sides of the L_{now} equation by the same quantity, L_{then}. It is just that on the right-hand side of the equation, L_{then} is written in its equivalent algebraic form of $m_{\text{Moon}} v_{\perp,\,\text{then}} r_{\text{then}}$. Since the mass of the Moon is the same in the numerator and the denominator, those numbers simply cancel.

Evaluating the ratio, we find

$$\frac{L_{now}}{L_{then}} = \left(\frac{1023 \; \cancel{m/s}}{1585 \; \cancel{m/s}}\right) \times \left(\frac{60.27 \; \cancel{R_{Earth}}}{25 \; \cancel{R_{Earth}}}\right) = \left(\frac{1023}{1585}\right) \times \left(\frac{60.27}{25}\right)$$

$$= 0.6454 \times 2.411 = 1.556.$$

Moon's orbital angular momentum today is 1.56 times greater (56% greater) than it used to be in the distant past.

Notice that, by making ratios out of the equations, the units and the mass of the Moon all cancel. This is how Kepler's third law was written in the form shown in Equation (5.10). By dividing both sides of the more complex equation that involves "Big G," (which wasn't presented) by the same equation describing Earth's orbit (the Sun's mass, an orbital period of 1 year, and a semimajor axis of 1 au), several constants canceled out. The trade-off is that Equation (5.10) requires that the masses be given in terms of the mass of the Sun (1 M_{Sun}), the orbital period in terms of the orbital period of Earth (1 y), and the length of the semimajor axis in astronomical units (1 au).

The last example poses a couple of interesting questions: Since the Moon's orbital angular momentum has increased over time, what was the source of the torque required to cause that to happen? And since angular momentum of the Earth–Moon system must be conserved, where did that extra angular momentum of the Moon's orbit come from? The answer to the first question has to do with tides on Earth and the force of gravity between Earth and the Moon. Newton first explained Earth's ocean tides due to the gravitational influence of the Moon in the *Principia*. Similarly, Earth actually produces tides on the Moon (the shape of the Moon deforms slightly). The answer to the second question is that both Earth and the Moon rotated more rapidly in the past and have been slowing down ever since. Although the Moon's orbital angular momentum has increased, the rotational angular momenta of Earth and the Moon have decreased in such a way that the total angular momentum of the Earth–Moon system has remained constant.

Tides important in many areas of astronomy, from the Earth–Moon system to the evolution of the Solar System, stars orbiting one another, and interacting galaxies. Tides are discussed more fully in Section 10.3.

Conservation of Energy

Linear and angular momentum aren't the only pieces of the puzzle, though. People have known forever that pushing something on a very smooth surface is much easier than pushing something on a rough surface. Galileo investigated this phenomenon by rolling balls down inclines. He realized that he could make them move farther if he made the surfaces as smooth as possible (this idea actually led to Newton's first law, the law of inertia.) It is force due to friction that is clearly the culprit, but where did the motion go when the balls in Galileo's experiments finally stopped moving? Technically, both linear and angular momentum are still conserved, as they must be, with both forms of momentum getting transferred to the entire laboratory and the planet on which it is anchored! But there is also another important subtlety involved.

Researchers, including Émilie du Châtelet (page 178), began to consider the possibility that there is another quantity that is conserved. We know that pushing a heavy filing cabinet across a carpet is hard work, as is lifting a heavy object straight up, or hiking to the top of a mountain. Why is it that doing those things is so much work?]

In the context of motion, work has a formal definition: Work is defined as the magnitude of an applied force, $\vec{F}$, exerted on an object multiplied by how far that object is moved *in the direction of the applied force vector*, $d_{\parallel}$ (parallel to the force vector), or

$$\text{Work} = Fd_{\parallel}. \tag{6.5}$$

The definition of work

In the case of moving the filing cabinet, if you push with a force of 200 newtons over a distance of 5 meters, then the work that you have done is $(200 \text{ N}) \times (5 \text{ m}) = 1000 \text{ N m}$. Because this combination of units (N m) comes up so often in science, it is given its own name, the joule (J), named for James Prescott Joule (1818–1889), a physicist and beer maker. Moving the cabinet required 1000 J of work.

Fig. 6.16 Weightlifter Óscar Figueroa of Colombia lifting 177 kg at the 2012 Summer Olympics. (Mono693, CC BY-SA 3.0)

Example 6.2

If a 177 kg barbell (Fig. 6.16) is lifted from the ground to a height of 2 m, how much work is done?

The force needed to lift the 177 kg barbell is equal to the weight of that mass, which is given by its mass multiplied by the acceleration due to gravity near Earth's surface (page 166),

$$\text{force} = \text{weight} = mg_{\text{Earth}} = (177 \text{ kg}) \times (9.8 \text{ m/s}^2) = 1735 \text{ N}.$$

Applying that force over a height of 2 m, means that the work done is

$$\text{work} = Fd_{\parallel} = (1735 \text{ N}) \times (2 \text{ m}) = 3470 \text{ J}.$$

 All potential energy

 Less potential energy

Some kinetic kinetic energy

Du Châtelet conducted experiments (Fig. 6.17) into whether work that is done in moving an object is simply lost or if it shows up in some other form. To test her idea she lifted balls and then dropped them into clay. She noticed that the higher she lifted the balls before dropping them, the greater the deformation in the clay. The amount of deformation in the clay was not simply proportional to how fast the balls were going when they hit, however; instead, the deformation was related to the speed squared, v^2. The deformation was also related to the mass of the balls that were dropped, with more massive balls causing a greater degree of deformation. Du Châtelet had come to realize that the work done in lifting the balls up was proportional to the speed squared when they hit the clay multiplied by the mass of the balls.

Today we think of the work done in lifting the balls up as being stored in gravitational potential energy (PE) that is released as the balls fall. That potential energy is converted into an energy of motion, known as kinetic energy (KE), which is proportional to the mass of the moving object and its speed squared:

 Clay

Fig. 6.17 Du Châtelet's experiment. As a ball falls, its velocity increases as the ball's potential energy is converted to kinetic energy. When the ball strikes the clay, it ejects some of the clay in proportion to the ball's kinetic energy at the moment of impact.

$$KE = \frac{1}{2}mv^2. \qquad\qquad (6.6)$$

Kinetic energy

Du Châtelet's work was a first step in developing a new law of conservation of energy: energy is never created or destroyed, but it can change forms, such as potential energy, kinetic energy, and work (there are also many other forms of energy that will be discussed throughout this text).

Example 6.3

When the 177 kg barbell in the previous example is dropped from 2 m, how fast is it going when it reaches the ground?

Conservation of energy dictates that the potential energy that the barbell has when it is 2 m above the ground must be converted into kinetic energy as it falls. This means that the kinetic energy of the barbell when it reaches the ground is 3470 J. Consequently,

$$KE = \frac{1}{2}mv^2 = 3470 \text{ J.}$$

This can be solved for the velocity squared (v^2) by multiplying both sides of the equation by 2 and dividing both sides of the equation by m so that

$$\left(\frac{2}{m}\right) \times \left(\frac{1}{2}mv^2\right) = \left(\frac{2}{m}\right) \times 3470 \text{ J,}$$

or

$$v^2 = \frac{2 \times 3470 \text{ J}}{177 \text{ kg}} = 39.2 \text{ m}^2/\text{s}^2.$$

In order to calculate v itself, it is necessary to take the square root of 39.2 m²/s², so that

$$v = \sqrt{v^2} = \sqrt{39.2 \text{ m}^2/\text{s}^2} = 6.3 \text{ m/s.}$$

Note that if you look carefully at both examples you will find that the mass was included in the first example but it was then divided out again in the second one. In other words, it didn't matter what the mass of the barbell was, the final speed would have been exactly the same. But this shouldn't surprise you because that is just what Galileo and Newton told us long ago: the acceleration due to gravity doesn't depend on mass, and so every mass will speed up at the same rate and hit the ground at the same time (neglecting air resistance of course). But a more massive object will have more kinetic energy.

Escape Speed

If you were to take that same 177 kg barbell from the last two examples, or any mass for that matter, and carry it far above our planet's surface so that it is essentially infinitely far away and then let it fall (ignore everything else in the universe at this point!), it is possible to figure out how fast it will hit the ground. This requires some more advanced mathematics (calculus) because, according to Newton's universal law of gravitation [Equation (5.7)], the force of gravity decreases as the inverse square of the distance from the center of Earth [$1/r^2$; recall that the acceleration due to gravity at the surface is 9.8 m/s^2 and 0.0027 m/s^2 at the orbit of the Moon (Fig. 5.16)], but the basic idea is the same as in the previous two examples. The equation for the speed turns out to be

$$v_{\text{escape}} = \sqrt{\frac{2GM}{R}}, \tag{6.7}$$

Escape speed

where M is the mass of the planet and R is the planet's radius. The speed is labeled as the escape speed because that is also how fast something needs to be traveling in order to leave the surface and never fall back down again. If you could throw a baseball straight up with that speed it would escape the planet! For Earth, the escape speed works out to be 11.2 km/s (25,000 mi/h). The speed can also be calculated for any other object, such as from the surface of the Moon, from the surface of Mars, or even from the Solar System or our Galaxy.

Example 6.4

Suppose a 30-km-diameter (19 mi) asteroid similar to Ida in Fig. 6.18 came crashing down on Earth, striking with a speed equal to that of Earth's escape speed. How much kinetic energy would the asteroid have when it hit?

As a rough estimate, we can assume that the asteroid has a mass equal to that of an Earth rock of the same size. This would be perhaps 40,000 trillion kilograms, or $4 \times 10^{16} \text{ kg}$ (over $8 \times 10^{16} \text{ lb}_m$). The associated kinetic energy would then be roughly

$$\text{KE} = \frac{1}{2}mv^2 = \frac{1}{2} \times (4 \times 10^{16} \text{ kg}) \times (11{,}200 \text{ m/s})^2 = 2.4 \times 10^{24} \text{ J}.$$

So how much energy is 2.4×10^{24} J (2.4 trillion trillion joules)? The atomic bomb that was dropped on Nagasaki, Japan at the end of World War II yielded about 8.8×10^{13} J. By this measure the asteroid's kinetic energy would be roughly the same as *27 billion* Nagasaki bombs.

Fig. 6.18 The asteroid Ida and its moon, Dactyl. Ida's mass is 4×10^{16} kg. (NASA/JPL)

Could such a catastrophic event as described in the last example actually occur? It already did, about 66 million years ago when an asteroid (or perhaps a comet) crashed into Earth in what is now the Gulf of Mexico, just off the Yucatán peninsula (see Fig. 6.19). The impact formed the Chicxulub crater that is partially in the Gulf of Mexico and partially buried beneath the peninsula. That collision is believed to have been responsible for the extinction of the dinosaurs. Since an asteroid crashing into Earth would only have a minimum speed equal to Earth's escape speed, the calculation above is probably low. Assuming that the asteroid was traveling at closer to 25 km/s, the energy of the impact would have been about five times greater

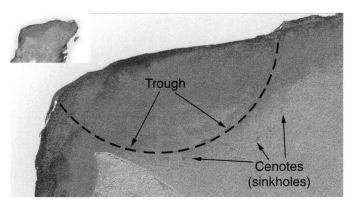

Fig. 6.19 Outline of the Chicxulub crater created 66 million years ago in the Gulf of Mexico and the Yucatán peninsula (inset). (NASA/JPL)

(1×10^{25} J). The crash of the asteroid that formed the Chicxulub crater was essentially an enormous version of one of du Châtelet's clay deformation experiments.

It is possible to use the same "trick" that we employed in Example 6.1 to write the escape speed equation [Equation (6.7)] in a form that doesn't require G. By dividing both sides of the equation by the value for the escape speed from the surface of Earth on the left-hand side and by the equation for Earth's escape speed on the right-hand side, and then solving for v_{escape}, we get

$$v_{escape} = 11.2 \text{ km/s} \times \sqrt{\frac{M_{Earth}}{R_{Earth}}},$$

where M_{Earth} and R_{Earth} are the mass and radius of the object that you want to know the escape speed of, written in terms of the mass and radius of Earth, respectively.

Example 6.5

The Moon's mass is 1.2% of the mass of Earth (0.012 M_{Earth}), and the Moon's radius is 27% of the radius of Earth (0.27 R_{Earth}). What is the speed of escape from the surface of the Moon?

Substituting into the escape speed equation written in terms the escape speed of Earth,

$$v_{escape} = 11.2 \text{ km/s} \times \sqrt{\frac{0.012}{0.27}} = 11.2 \text{ km/s} \times \sqrt{0.044} = 11.2 \text{ km/s} \times 0.21$$

$$= 2.36 \text{ km/s}.$$

The escape speed from the Moon is about 5 times slower than it is from the surface of Earth.

Example 6.6

The asteroid Ida (Fig. 6.18) has a mass that is only seven one-billionths (7×10^{-9}) of the mass of Earth, with an effective radius (it is not spherical) of about 15 km, approximately 0.2% (0.002) of the radius of Earth. What is the escape speed from Ida?

Following the same procedure as the last example,

$$v_{\text{escape}} = 11.2 \text{ km/s} \times \sqrt{\frac{0.000\,000\,007}{0.002}} = 11.2 \text{ km/s} \times 0.0019 = 0.02 \text{ km/s}.$$

0.02 km/s is the same as 20 m/s (45 mi/h). It wouldn't be difficult at all to throw a baseball fast enough to have it escape Ida.

6.5   Light is an Electromagnetic Wave

The Characteristics of a Wave

You should recall that it was Newton who first demonstrated that sunlight is composed of all of the colors of the rainbow (page 149). Newton also proposed that light is composed of small particles that emanated from a source to strike the eye. His hypothesis was a curious choice between the two competing ideas of the day: that light is made up of particles, or that light behaves like waves. What makes his hypothesis odd is that one of the experiments he conducted was to investigate a phenomenon known today as Newton's rings, a pattern of concentric bright and dark rings formed by looking through a lens resting on a reflecting surface (Fig. 6.20). Newton's rings are a classic example of a process known as wave interference; Newton's rings cannot be produced if light is composed of the particles he envisioned.

You are certainly familiar with wave phenomena, such as waves moving across the surface of a lake, a pond, or even a bathtub. Waves are also visible in flags on a breezy day. But to understand how wave interference occurs, we must first learn a bit about the characteristics of waves in general before we discuss light waves in greater detail.

As shown in Fig. 6.21, a wave is an oscillatory pattern with regularly repeating maximums and minimums. The maximums are known as *peaks* and the minimums are called *troughs*. The length of the repeating pattern, either from one peak to the next or from one trough to the next, is the wavelength of the wave, and is symbolized by the Greek letter λ (lambda). The amplitude of the wave is A, which is measured from the midpoint between a peak and a trough, and the speed of the wave is represented by v. Another important characteristic of a wave is its frequency, f, which can be thought of as simply counting the number of wave peaks that pass a fixed point in one second. The unit used to represent the frequency is the hertz (Hz), or the number of "cycles" per second (a cycle is the repeating pattern, say from peak to peak). Since the number of cycles is just a count without units,

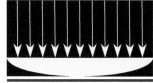
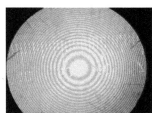

Fig. 6.20 *Top:* A lens resting on a mirror with light rays striking the top surface of the lens. *Bottom:* An example of Newton's rings as seen via a microscope looking down through the lens. (Warrencarpani, CC0 1.0, public domain)

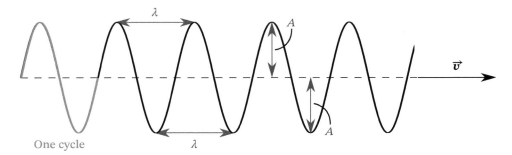

Fig. 6.21 A typical wave shape. λ is the wavelength, A is the wave amplitude, and v, the magnitude of the velocity vector, is the speed of the wave (in this case moving to the right).

1 Hz is the same as 1/s. You have probably encountered this frequency unit before with the AC (alternating current) electricity in your home wiring, which is typically 60 Hz in the United States. You are also probably aware that "over the air" radio stations broadcast at kHz (kilohertz, or thousands of hertz) and MHz (megahertz, or millions of hertz) frequencies. (Note: kHz for 10^3 Hz and MHz for 10^6 Hz are examples of prefixes commonly used in the SI system of units in place of powers of 10 in scientific notation, just as km and cm are shorthand for 10^3 m and 10^{-2} m, respectively; recall Table 5.1.)

The frequency, wavelength, and speed of a wave are related by the relationship

$$v = f\lambda, \tag{6.8}$$

The general relationship between speed, frequency, and wavelength

the speed of the wave equals the wave's frequency multiplied by its wavelength. Since wavelength is just a distance and frequency is the inverse of time, this speed–frequency–wavelength relationship is the same as the more familiar equation, $v = d/t$, speed equals distance traveled divided by the time required to travel that distance, which is actually nothing more than the definition of speed.

To understand the relationship, think about watching a wave move across a lake. If the speed of the wave is 1 m/s and the distance between peaks (the wavelength) is 1 m, then you would expect to see one peak move past a fixed point every second since it takes one second for a wave traveling at a speed of 1 m/s to move a distance of 1 m. If the length of the wave is 1 m but the wave is moving at a speed of 2 m/s, then you would expect to see two waves move past a fixed point in one second. As a result, you would have witnessed two cycles per second, or a frequency of 2 Hz.

Example 6.7

A wave has a speed of 4.4 m/s on the surface of the ocean, with a wavelength of 7 m. What is the wave's frequency?

The relationship between speed, frequency, and wavelength can be solved algebraically for frequency by dividing both sides of the equation by the wavelength (don't forget that you must always perform the same operation on both sides of the equality). This gives

$$\frac{v}{\lambda} = \frac{f\lambda}{\lambda},$$

or

$$f = \frac{v}{\lambda} = \frac{4.4 \text{ m/s}}{7 \text{ m}} = \frac{4.4}{7 \text{ s}} = 0.63 \text{ Hz}.$$

Since 0.63 is close to the fraction $2/3 = 0.67$, this result means that approximately two-thirds of a wave would move past a fixed location in one second, or two waves would move past that point every three seconds.

Example 6.8

The D string on a guitar oscillates with a frequency of 146.83 Hz and the wavelength is 1.3 m (twice the length of the string). How fast does a wave propagate along the string?

$$v = f\lambda = 146.83 \text{ Hz} \times 1.3 \text{ m} = 146.83/\text{s} \times 1.3 \text{ m} = 191 \text{ m/s}.$$

The speed of a wave along a tightly stretched metal string is quite fast, 191 m/s is equivalent to 427 mi/h.

Interference Between Two Waves

When two waves on the surface of an otherwise calm lake or puddle run into one another, you might have noticed that although they are a bit jumbled for a moment, they eventually pass right through each other and continue on their way like the ripples in Fig. 6.22. At the points where they collide, their heights combine. If both waves are at their peaks when they hit, the temporary result is a peak that is the sum of the amplitudes of the two wave peaks, but if one wave is at a peak while the other is at a trough, the two partially cancel each other out (they exactly cancel if the height of one peak is identical to the depth of the other's trough). The combining of waves is what is meant by the term *interference* mentioned earlier. The case of peaks combining with peaks and troughs combining with troughs is known as constructive interference, while destructive interference occurs when a peak and a trough combine. The top portion of Fig. 6.23 illustrates two waves adding to produce constructive interference and the bottom illustration shows two waves adding to produce destructive interference.

Newton's rings, shown in Fig. 6.20 (*Bottom*), are the result of one light wave reflecting from the curved bottom of the lens combining with a light wave reflecting from the flat mirror on which the lens rests. When the two waves arrive at the microscope, in some places the waves combine to produce constructive interference, which shows up as bright rings, while in other places the waves cancel each other

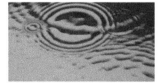

Fig. 6.22 Raindrops striking the surface of water and producing colliding ripples (waves). (Joka2000, CC BY 2.0)

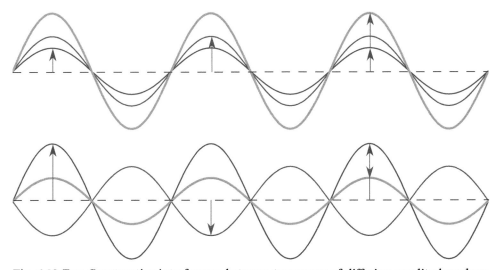

Fig. 6.23 *Top:* Constructive interference between two waves of differing amplitudes where peaks align. *Bottom:* Destructive interference between two waves of differing amplitudes where the peak of one wave aligns with the trough of the other wave. The blue and red arrows show the amplitudes of the waves, and the sums of the two arrows show the amplitudes of the resulting green waves. In the bottom illustration the amplitude of the red wave is actually subtracted from the amplitude of the blue wave, as indicated by the direction of the arrow.

out through destructive interference, producing dark rings. This is one of those rare cases where Newton's intuition failed him; light behaves like a wave rather than being composed of particles.

It turns out that the phenomenon of interference for light waves provides a very powerful tool that is utilized extensively in astronomy, physics, and other fields. To see why, consider what happens when two waves of identical amplitude and wavelength are added together, but with one wave shifted ahead of the other as illustrated in Fig. 6.24. In the top panel the wave shown in red is shifted slightly in front of the wave shown in blue, with the result that the sum of the two waves (the green wave) is much taller than either of the original waves. For light waves, the amplitude squared, $A^2 = A \times A$, of the wave is proportional to brightness, meaning that in this case the total amount of light would be brighter than either of the two original waves alone (note that the colors of the waves in the diagram are only meant to distinguish the waves, the colors do not imply the color of the light). In the middle panel the wave marked in red is shifted farther ahead of the blue wave, and now the sum of the two waves (again indicated by the green wave) is lower than either of the two original waves, meaning that the light is actually dimmer than if just one of the waves was present. Finally, in the bottom panel, the wave marked in red is shifted one-half of a wavelength ahead of the wave marked in blue, causing the sum of the two waves to be zero everywhere, a result referred to as *total destructive interference*. The result in the bottom panel demonstrates that the light from the two waves completely cancel out, so that no light is produced from the combination of the two waves. If the red-colored wave shifts even farther ahead, the light begins to return again, reaching maximum brightness when the red-colored wave is completely aligned with the blue one, producing *total constructive interference*. Because

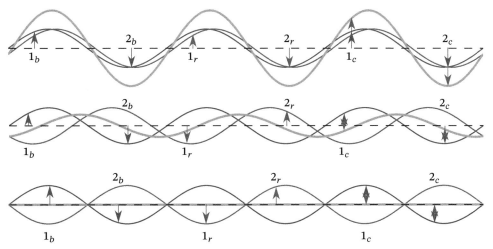

Fig. 6.24 Two waves of identical amplitude. From top to bottom: The red wave slightly ahead of the blue wave; the red wave two-fifths of a wavelength ahead of the blue wave; the red wave one-half wavelength ahead of the blue wave. The sum of the blue and red waves is the green wave. The bottom frame produces perfect destructive interference (the green wave has zero amplitude). 1_b and 1_r mark the same points in a cycle for the blue and red waves, respectively; similarly for points 2. 1_c and 2_c are the combined amplitudes of the blue and red waves at those locations.

a wave is exactly repeating every wavelength, by shifting the red wave 1 1/2 wavelengths forward, total destructive interference occurs again, and if the red wave is shifted 2 wavelengths forward, total constructive interference occurs.

This behavior is exactly what happens in Newton's rings experiment; the dark rings are produced when two rays of light are shifted by one-half wavelength (or 1 1/2 wavelengths, or 2 1/2 wavelengths, etc.) with respect to one another, thus canceling each other out. The centers of the bright rings occur when the waves are shifted by an integer number of wavelengths with respect to one another.

Experimentally Determining the Wavelengths of Visible Light

Thomas Young (Fig. 6.25) was the first person to measure the wavelengths of light by using the "trick" of shifting wavelengths, thereby producing interference patterns of alternating bright and dark areas on a screen,. As shown on the left side of Fig. 6.26, he accomplished this by sending light of a single color through two tiny slits that were spaced close together. If the light waves from the two slits both travel the same distance to the midpoint between the slits on a distant screen, the waves from each slit arrive with their peaks and troughs exactly aligned, resulting in constructive interference and a bright spot. Everywhere else on the screen the light from the two slits must travel different distances, where the extra distance traveled is shown in the blue inset box on the diagram. In the situation shown for the two light rays, the light from slit S_1 travels a shorter distance (d_1) to the location on the screen than does the light from slit S_2 ($d_1 < d_2$), and therefore slit S_2's wave peaks and troughs arrive behind those from slit S_1. If the shift is an integer number of wavelengths, constructive interference occurs and a bright spot forms,

Fig. 6.25 Thomas Young (1733–1829). [Henry Adlard, after Thomas Lawrence (1769–1830), public domain]

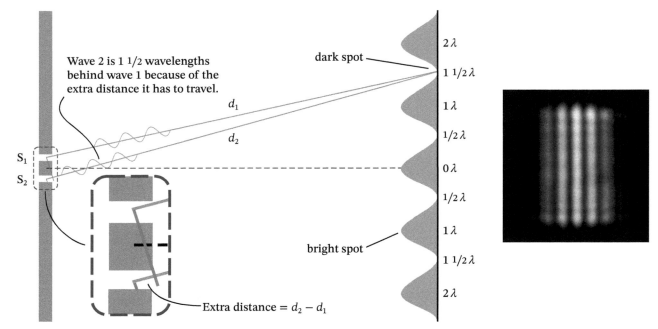

Fig. 6.26 *Left:* Young's double-slit experiment to measure the wavelengths of various colors of light. A location of destructive interference on the screen to the right is shown, where the distances traveled by the two light rays differ by 1 1/2 wavelengths, or $d_2 - d_1 = 1.5\lambda$. The height of the undulating pattern on the screen at a specific location corresponds to how bright the light is at that location. *Right:* Sunlight with all of its colors passed through a double-slit apparatus. The maximums of different colors occur at different locations because each color has a unique wavelength. The exception is at the center, where all colors are shifted by 0λ. (Aleksandr Berdnikov, CC BY-SA 4.0)

but if the shift is 1/2 wavelength, 1 1/2 wavelengths, 2 1/2 wavelengths, and so on, destructive interference occurs and a dark area results. The process of determining the wavelength of the light becomes a problem in geometry to determine the extra length traveled by the second wave. For the two waves shown in the diagram, the extra distance traveled from S_2 is $d_2 - d_1 = 1.5\lambda$. The wavelength, λ, can then be determined from the equation. Determining the extra distance traveled requires knowing the spacing between the slits, the distance to the screen from the slits, and how far the relevant bright or dark spot is from the center of the screen.

The Wavelengths and Frequencies of Visible Light

The wavelengths of visible light are very short, ranging from 400 nm for blue/violet light to 700 nm for red light (see Figs. 6.27 and 6.28), where 1 nm (one nanometer) is one-billionth of a meter. You probably have a ruler that has mm (millimeter) marks on it; 400 nm is the same as 0.0004 mm (four ten-thousandths of a millimeter). The wavelengths of visible light are well worth remembering.

Because light behaves like a wave, the relationship between frequency, wavelength, and speed applies to light just as it does for any other type of wave. In Section 6.2 we learned that the speed of light is a universal constant. As a result, for the special case of light waves we can replace the general representation for speed, v, with the speed of light, c, giving

Fig. 6.27 A rainbow's visible light ranges in wavelengths from 400 nm (blue/violet) to 700 nm (red). (Adapted from United States Department of the Interior)

$$f\lambda = c. \tag{6.9}$$

The frequency–wavelength–speed relationship for light

The frequency of a light wave multiplied by its wavelength equals the speed of light. Because the speed of light is 300,000 km/s (3.0×10^8 m/s), and the wavelengths of visible light are very short, the frequencies of visible light are extremely high.

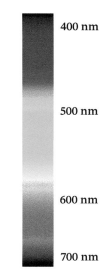

400 nm

500 nm

600 nm

700 nm

Fig. 6.28 The visible spectrum of light. (Public domain)

Example 6.9

Determine the frequencies of blue/violet light and red light.

Solving the frequency–wavelength–speed equation of light for frequency gives

$$f = \frac{c}{\lambda}.$$

Given that the speed of light is a constant, the shorter the wavelength of the light, the higher the frequency of the light. For the case of blue/violet light,

$$f_{blue} = \left(\frac{3 \times 10^8 \text{ m/s}}{400 \text{ nm}}\right) \times \left(\frac{1 \text{ nm}}{1 \times 10^{-9} \text{ m}}\right) = \frac{3 \times 10^8}{4 \times 10^{-7} \text{ s}} = 7.5 \times 10^{14} \text{ Hz},$$

or 750 trillion hertz. If you were actually able to see light waves go by, you would see 750 trillion peaks pass you every second.

Red light has a longer wavelength than blue/violet light, implying that the frequency is a little lower:

$$f_{red} = \frac{3 \times 10^8 \text{ m/s}}{700 \text{ nm}} = 4.3 \times 10^{14} \text{ Hz}.$$

Having established that light behaves like a wave, other questions naturally arise. Water waves propagate across water (obviously), and sound waves, which are pulses of increased pressure, move through air, water, and other substances, but what are light waves and what do they "wave" through?

Electricity and Magnetism

Electric and magnetic phenomena have been known since the dawn of civilization. Thales of Miletus (page 59) was aware that rubbing fur on amber would cause those materials to attract small particles like specks of dust. The Greeks were also aware that by rubbing long and hard enough, a spark could be generated. Of course, lightning was well-known but its properties were not understood to be related to static electricity.

Roman and Arab naturalists reported on the shocks that would occur by touching certain aquatic animals such as electric eels, or the electric catfish found in the Nile River (Fig. 6.29). As early as 2750 BCE, the Egyptians referred to electric catfish as "thunderers of the Nile" and protectors of other fish. Ancient physicians would recommend touching electric fish to cure headache.

Fig. 6.29 An electric catfish at the Steinhart Aquarium in San Francisco. (Stan Shebs, CC BY-SA 3.0)

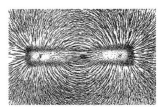

Fig. 6.30 Iron filings around a bar magnet. (Newton Henry Black, public domain)

Aristotle tells us that Thales also explored the peculiar behavior of a naturally occurring mineral called lodestone that was known to attract iron. It is possible that as early as 1000 BCE, hundreds of years before Thales, the Olmec of Central America (page 34) may have discovered that, when suspended by a filament, lodestone would point north. By the 12th century, the Chinese certainly used this property of the magnetic material as an aid to navigation. Because lodestone was known to attract iron, one reason proposed for why it could be used in navigation was that there was a gigantic, undiscovered iron island at the North Pole. Others suggested that lodestone was attracted to Polaris, the North Star.

Given that both lodestone and rubbed amber attracted materials, albeit of different types, early ideas of electricity and magnetism suggested that they somehow represented manifestations of the same phenomenon. It was in 1600 that William Gilbert published his *De Magnete* (page 124) proposing that Earth itself acted like a giant magnet similar to the bar magnet shown in Fig. 6.30, causing lodestone and magnetized needles to point north. Based on his experiments Gilbert also demonstrated that electricity and magnetism were in fact two separate phenomena. (Recall that Kepler used Gilbert's work to suggest that magnetism is what attracted the planets to the Sun and kept them in their orbits; he was wrong of course.)

Along with being a politician and statesman, Benjamin Franklin (Fig. 6.31) was also a respected scientist during the Enlightenment (Section 5.6). In June 1752 Franklin, with the help of his son William, purportedly flew a kite in a thunderstorm (Fig. 6.32) in order to demonstrate that lightning was in fact a form of static electricity. The experiment involved placing a sharp metal needle on the kite to attract electricity. The kite was attached to a silk string, and a metal key was tied near the base of the string close to where Franklin held the string. Attached to the key was a metal wire that extended down into a Leyden jar (a device for holding electrical charge). Franklin also tied a silk ribbon to the key that he could hold on to. Normally, silk does not conduct electricity, but when wet (as in a rain storm) electricity can travel along the silk. When the storm arrived sparks were seen striking the kite, filaments in the kite string stood up, and the jar became electrically charged. The legend has it that when Franklin moved the back of his hand near the key he received an electrical shock. There have been challenges to the story but it is at least theoretically possible, although it seems unlikely that Franklin could have lived to report the event. Others who tried to replicate the experiment weren't so fortunate! In any case, he was correct in showing that lightning is an electrical discharge. Based on his ideas about lightning, Franklin invented the lightning rod, which he first tried on his own house before it was installed on public buildings around Philadelphia.

Fig. 6.31 Benjamin Franklin (1706–1790). [After Joseph Duplessis (1725–1802); public domain]

Fig. 6.32 An engraving of Benjamin Franklin and his son William flying a kite in a thunderstorm. (From Gallaudet, 1832;[5] public domain)

Franklin also proposed that electricity is a fluid and that the overabundance or absence of the fluid gave materials their attractive properties. For example, rubbing amber with fur transferred this electrical fluid from one material to the other. Franklin is credited with assigning positive, +, and negative, −, to the overabundance or absence of the fluid, a notation that persists to this day. Two objects with net positive charges or two objects with net negative charges repel one another, while an object with a positive charge and an object with a negative charge attract one another; "opposites attract."

---

[5]Gallaudet, Thomas Hopkins (1832). *The Youth's Book on Natural Theology: Illustrated in Familiar Dialogues, with Numerous Engravings*, American Tract Society: New York.

It was Charles-Augustin de Coulomb (Fig. 6.33) who demonstrated in 1784 that the force between two spherical charges behaves in a way that is very similar to Newton's universal law of gravitation [Equation (5.7)]. Mathematically, Coulomb's law looks virtually identical:

$$F_{\text{electric}} = k_E \frac{q_1 q_2}{r^2}, \tag{6.10}$$

Coulomb's law of electrostatic force

Fig. 6.33 Charles-Augustin de Coulomb (1736–1806). (Public domain)

where q_1 and q_2 are the quantities of charge on each object, r is the distance between their centers, and k_E is another universal constant. Notice that the law is an inverse square law, just like Newton's universal law of gravitation. If q_1 and q_2 are both either positive or negative, then multiplying them together gives a positive result and the force is repulsive, but if q_1 is positive and q_2 is negative, or vice versa, then their product is negative and the force is attractive. The electrical force is much stronger than the gravitational force, but only shows up in relatively small to extremely small size scales. This is because, on large scales, like moons, planets, and stars, those objects tend to be electrically neutral, having the same amount of positive and negative charge so that the total charge is zero, resulting in a zero net force; in Coulomb's law for electric charges, if either q_1 or q_2 is zero, then F_{electric} must be zero.

Other experiments in electricity and magnetism began to demonstrate that even though they were not the same phenomena, they did at least have some relationship to each other. In 1820, Hans Christian Ørsted (1777–1851), a Danish physicist and chemist, noticed that when an electric current was passed through a wire it caused a deflection in a nearby magnetic compass. A little more than one decade later, the Englishman, Michael Faraday (Fig. 6.34), was able to demonstrate that by changing the amount of electrical current in a coil of wire, another nearby coil of wire would suddenly carry current through it without using a battery. The same result can be produced by moving a magnet through space near a coil of wire. Faraday also created the first prototype of a modern electric motor.

Electromagnetic Waves

Fig. 6.34 Michael Faraday (1791–1867). (Thomas Phillips, 1842; public domain)

It was in 1865, near the end of the American Civil War, that James Clerk Maxwell (Fig. 6.35), a brilliant mathematical physicist, was able to realize what is sometimes called "the second great unification in physics," the first being the unification of terrestrial and celestial motions by Newton. Taking the knowledge provided by a host of scientists before him, and guided by the symmetry inherent in the mathematical description of their discoveries, Maxwell was able to develop a set of four truly elegant equations that describe everything then known about electricity and magnetism, including the eerie effects that electric currents can affect magnets and that moving a magnet in the vicinity of a coil of wire produces an electric current in the wire.

Maxwell's work and the theoretical developments that came before him involve the concepts of electric and magnetic fields. A field may be described in terms of a set of imaginary vectors that trace out the various directions of force that one object could exert on another object. As an example, if you look back at Fig. 6.30,

Fig. 6.35 James Clerk Maxwell (1831–1879). [From an engraving of James Clerk Maxwell by George J. Stodart (1784–1884) from a photograph by Fergus of Greenock, public domain]

where iron filings are sprinkled near a bar magnet marked with north (N) and south (S) poles, the iron filings align in a distinctive pattern between the two poles. Those iron filings are acting like small compass needles pointing along the bar magnet's magnetic field lines. A similar situation occurs with positive and negative charges, as shown in Fig. 6.36. Electric field lines trace out the direction that a positive charge would move. If a tiny positive charge is placed near a fixed positive charge then that tiny charge will be pushed directly away from the fixed charge in the direction that the electric field vector points at that location. Similarly, if a positive charge is placed near a fixed negative charge it will move toward the negative charge, again following the electric field vectors. For a small charge placed at a significant distance from either the positive or negative charge in Fig. 6.36, the charge will feel the combined effects of both fixed charges and follow the electric field vector lines that curve from one charge to the other. (It is also possible to talk about a gravitational field in the same way except that masses always attract one another.)

Recall that all scientific theories must be able to make testable predictions that advance our understanding of nature, otherwise a theory cannot be considered as scientific. Maxwell's marvelous equations predicted something previously unexpected. Not only did his equations summarize everything already discovered about electricity and magnetism, but they also revealed that a changing electric field can produce a changing magnetic field perpendicular to the electric field, *and* a changing magnetic field can produce a changing electric field perpendicular to the magnetic field, as shown in Fig. 6.37. In other words, once a changing electric field is generated, a changing magnetic field is generated, which in turn generates a changing electric field, which in turn generates a changing magnetic field, … . Moreover, the mutually generated fields propagate through space in a direction that is perpendicular to both the electric and magnetic fields, with a speed that involves the fundamental constants for electricity (k_E in Coulomb's law) and magnetism. The speed calculated for Maxwell's mathematical combination of the fundamental constants for electricity and magnetism exactly equals the measured value of the speed of light, c! Maxwell had not only succeeded in combining the theories of electricity and magnetism, but optics was integrated into the mix as well. Light is an electromagnetic wave.

The electromagnetic wave prediction of Maxwell went even further, however. The equation for an electromagnetic wave puts no restrictions on the wavelength that the self-propagating wave can have. The wavelengths of visible light range

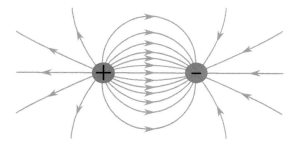

Fig. 6.36 Arrows on electric field lines (depicted here in green) point away from positive (+) charges and toward negative (−) charges.

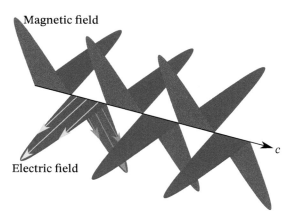

Magnetic field

Electric field

c

Fig. 6.37 An electromagnetic wave travels in a direction that is perpendicular to both the electric and magnetic fields. The electric (blue) and magnetic (red) fields are perpendicular to each other, and the speed of the self-propagating wave is the speed of light, *c*. The lengths of the field vectors at every point are the heights of the waves at those points.

from blue/violet light at 400 nm to red light at 700 nm, but Maxwell's result has no wavelength limitation. In fact, *any* wavelength is possible, from near zero to kilometers long, or longer. There is also nothing in the equations that says anything about what an electromagnetic wave must "wave" through. Unlike water or sound waves, electromagnetic waves don't require any medium for propagation, they can travel through empty space, with the electric and magnetic fields mutually generating each other at the speed of light *forever!*

In 1886, Heinrich Hertz (Fig. 6.38), for whom the unit of frequency is named, successfully constructed an apparatus that transmitted an electromagnetic wave that was received 12 meters away by a second apparatus. The wave that was predicted by Maxwell was invisible to the naked eye but was detected when it caused charges to move in the receiving apparatus, driven by the changing electric field of the electromagnetic wave. Unfortunately, Maxwell did not live to see the successful verification of the prediction of his electromagnetic wave theory.

Fig. 6.38 Heinrich Hertz (1857–1894). (Robert Krewaldt, public domain)

Light Polarization

When light is produced it is often generated with the electric field vectors pointing in arbitrary directions perpendicular to the direction that the waves are traveling, as depicted in Fig. 6.39 (*Top*). The magnetic field vectors also point in arbitrary directions perpendicular to both the direction of travel and their associated electric field vectors. In this case, the light is said to be unpolarized. However, in some cases the light is generated with all of the electric field vectors aligned (or at least predominately aligned). In this situation, the light is said to be polarized [Fig. 6.39 (*Bottom*)]. Knowing that light is polarized tells us something about how the light may have been produced.

Even if light starts out unpolarized, it is possible for light to become polarized due to reflections or some other process. You are probably familiar with this situation in everyday circumstances. If you have ever observed glare off the surface of

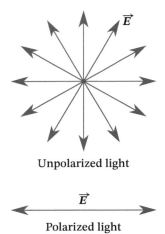

Unpolarized light

Polarized light

Fig. 6.39 *Top:* Unpolarized light coming out of the page with oscillating electric field vectors ($\vec{E}$) in arbitrary directions perpendicular to the direction that the light is propagating. *Bottom:* Polarized light with all of the electric field vectors pointed in the same direction.

a lake [Fig. 6.40 (*Left*)] or from the reflection off the hood or windshield of a car, you have observed polarized light. When the light bounces off the surface a particular direction is established for the oscillation of the electric field vectors. To cut down on glare, polarized sunglasses are often worn to block out electric field vectors that are oscillating in a particular direction [Fig. 6.40 (*Right*)]. If you wear polarized sunglasses you may have also noticed another curious effect; you may not be able to see the screen of your smartphone or a computer. This is because those screens may emit polarized light depending on the particular kind of screen you are looking at. If you turn your head (or your smartphone) 90°, you will be able to see the screen because now your sunglasses are aligned to allow the polarized electromagnetic waves to pass through rather than be blocked. Just for fun, you might try an experiment with two sets of polarized sunglasses. Look through both pairs at the same time and then slowly rotate one pair 90° to the other. What happens and why?

Fig. 6.40 Effects of polarization on reflected sunlight. *Left:* View of Wizard Island in Crater Lake, Oregon without a filter. Right: Same view taken five seconds later with a polarization filter. (Copyright 2018, Photon Acquisition)

The Wide Spectrum of Light

The wavelength of a particular wave (or alternatively the wave's frequency) is the result of how the wave is initially generated, something we will talk about extensively in the remainder of this text. After all, electromagnetic waves are the carriers of most of the information we know about the cosmos. Various ranges of electromagnetic waves are familiar to all of us today; gamma rays, x-rays, ultraviolet light, visible light, infrared light, microwaves, and radio waves. They are all the same, except for their wavelengths and corresponding frequencies. Visible light is just one tiny part of the electromagnetic spectrum. Electromagnetic waves are one of several types of *radiation* that we will be discussing. As a result, it is common to talk about electromagnetic waves as electromagnetic radiation.

The human eye has evolved to be sensitive to visible light, but thanks to our developing technologies, we are now able to "see" in wavelengths across the electromagnetic spectrum. This ability is critically important to understanding the tremendous variety of phenomena in the universe. Characteristic wavelength and frequency ranges are given in Table 6.1. You will certainly want to be familiar with the general wavelength ranges for the various kinds of electromagnetic radiation.

Table 6.1 Wavelength and frequency ranges of the electromagnetic spectrum

Type	Wavelength	Frequency (Hz)	Characteristic length
Gamma rays	less than 0.01 nm	greater than 3×10^{19}	smaller than 10% of an atom's diameter
X-rays	0.01 nm to 10 nm	3×10^{19} to 3×10^{16}	large molecules
Ultraviolet light	10 nm to 400 nm	3×10^{16} to 7.5×10^{14}	interstellar dust grains
Visible light	400 nm to 700 nm	7.5×10^{14} to 4.3×10^{14}	coal smoke or small bacteria
Infrared light	700 nm to 1 mm	4.3×10^{14} to 3×10^{11}	large bacteria to water mist
Microwaves	1 mm to 10 cm	3×10^{11} to 3×10^{9}	rain drops to length of two AA batteries
Radio waves	greater than 10 cm	less than 3×10^{9}	length of two AA batteries to city size or more

You can always calculate the frequency ranges from the frequency–wavelength–speed equation of light [Equation (6.9)] if needed.

6.6 Blackbody Radiation

The famous English pottery maker, Josiah Wedgwood (Fig. 6.41), in an effort to improve the quality and consistency of his ceramics and glazes, was concerned about the true temperature (T) of his kilns: "red, bright red, or white heat are indeterminate expressions ... [they] can neither be expressed in words nor discriminated by eye." Although he was aware that the temperature of his kilns affected the color at which they glowed, he did not have an effective way to measure their temperatures. Rather than simply accept that reality, he set out to develop a thermometer to make the necessary measurements. After more than 5000 meticulously documented experiments, the results of his work were published in 1782[6] (the paper was read by his colleague, Joseph Banks). Within the paper, Wedgwood wrote:

> we are told, for instance, that such and such materials were changed by fire into a fine white, yellow, green, or other coloured glass: and find, that these effects do not happen, unless a particular degree of fire has fortunately been hit upon, which degree we cannot be sure of succeeding in again.

Fig. 6.41 Josiah Wedgwood (1730–1795). [Painting by Joshua Reynolds (1723–1792), public domain]

You have undoubtedly noticed yourself the red glow of coals in a fireplace or a campfire. The colors of the coals or the kilns do not depend on the materials that are heated, nor on their sizes or shapes, but only on their temperature.

The Kelvin Absolute Temperature Scale

In the United States, temperature is commonly measured in Fahrenheit degrees, where 32°F is the freezing point of water and 212°F is water's boiling point near sea level. Most of the rest of the world uses the Celsius temperature scale, where 0°C is the freezing point of water and 100°C is its boiling point. Technically, these values are true only under specific conditions. For example, if you live or play at altitudes

---

[6]Wedgwood, Josiah (1782). "An attempt to make a thermometer for measuring the higher degrees of heat, from a red heat up to the strongest that vessels made of clay can support," *Philosophical Transactions of the Royal Society of London*, 72:305–326.

Fig. 6.42 Sir William Thomson (Lord Kelvin; 1824–1907). (Photo by Messrs. Dickinson, London, New Bond Street, public domain)

well above sea level, you may know that water boils at temperatures below 100°C or 212°F.

We will find it necessary to use yet another temperature scale, named for Sir William Thomson (Fig. 6.42). The reason this becomes necessary is that there is a lowest possible temperature at which the energies of all particles reach an absolute minimum; this temperature is appropriately known as absolute zero. It is impossible for any substance to have a temperature that is less than absolute zero, because it is impossible for particles to have energies less than they do at that temperature. On the Celsius scale absolute zero is −273.15°C and on the Fahrenheit scale it is −459.67°F. The Kelvin temperature scale uses the temperature intervals of the Celsius scale but starts with absolute zero defined as 0 kelvins (0 K). This means that the freezing point of water is 273.15 K and its boiling point is 373.15 K (the degree symbol, °, is not used with the Kelvin scale). To convert kelvins to °C or °F:

$$°C = K − 273.15°, \qquad °F = \frac{9}{5}K − 459.67°. \qquad (6.11)$$

The conversions from kelvins to Celsius or Fahrenheit degrees

Many equations in science demand an implicit reference to absolute zero such as calculating a ratio of temperatures. For example, using the Fahrenheit scale, the ratio of the boiling point of water to its freezing point is 212/32 = 6.63, but that does not mean that boiling water is 6.63 times hotter than water at the freezing point. Using the Celsius scale, that ratio would be 100/0, which is completely meaningless (dividing by zero gets you arrested by the math police). However, the same ratio in the Kelvin scale is 373/273 = 1.37; in this case it is appropriate to say that something with a temperature of 373 kelvins is 1.37 times hotter than something with a temperature of 273 kelvins. On a hot summer day in Minneapolis, Minnesota, the temperature might reach 100°F, but in the winter the temperature can plummet to −30°F or lower. Does it make sense to say that on the hot day Minneapolis was 100/(-30) = −3.33 times hotter than on the cold day? Obviously not. Such a negative ratio is impossible using the Kelvin temperature scale.

The Intensity of Blackbody Radiation With Wavelength

The light given off by one of Wedgwood's heated kilns, the coals of a campfire, or any hot solid object has a characteristic distribution for the brightness of the light at different wavelengths, as illustrated in Fig. 6.43. No matter what the wavelength, from gamma rays to radio waves, there is always some amount of light given off, with the characteristics that an object gets brighter at every wavelength with increasing temperature and the peak wavelength of the continuous spectrum also moves to shorter wavelengths. Along with solid objects, these patterns hold true for hot, dense gases as well, such as stars. The temperatures of thousands of kelvins used in Fig. 6.43 are typical of the surfaces of stars, including our Sun.

It was in 1860, the first year of the American Civil War and five years before Maxwell's marvelous equations, that Gustav Kirchhoff (Fig. 6.44) described a theoretically perfect source of this continuous spectrum as a blackbody:[7]

---

[7]Kirchhoff, Gustav Robert (1860). "On the relation between the radiating and absorbing powers of different bodies for light and heat," *Philosophical Magazine Series 1*, 20:1–21.

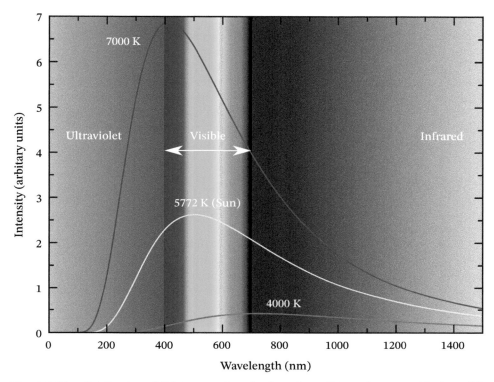

Fig. 6.43 The distribution of light at wavelengths from 0 to 1500 nm for stars with three different surface temperatures. The middle curve corresponds to the temperature of the Sun's surface (5772 K). Since the human eye cannot see in the ultraviolet or the infrared, the colors shown in those regions are artificial. Temperatures are expressed in kelvins.

> bodies can be imagined which ... completely absorb all incident rays, and neither reflect nor transmit any. I shall call such bodies *perfectly black*, or, more briefly, *black* bodies.

The important take-away from Kirchhoff's statement is the idealized concept that a true blackbody is a perfect absorber of electromagnetic radiation, a situation never achieved in reality. However, despite this idealized assumption of a perfectly absorbing object, the concept of a blackbody can be extremely useful, especially in describing planets and stars. This is because a blackbody is not only a perfect absorber of electromagnetic radiation that shines on it, it is also a perfect emitter of electromagnetic radiation.

The Color of Blackbody Radiation

In 1893 Wilhelm Wien (Fig. 6.45) was the first person to quantify the locations of the peaks of the blackbody radiation curves as a function of temperature through the relation

Fig. 6.44 Gustav Robert Kirchhoff (1824–1887). (Public domain)

$$\lambda_{\text{peak}} = \frac{2,900,000 \text{ nm K}}{T_{\text{surface}}}.$$ (6.12)

Wien's blackbody color law

Fig. 6.45 Wilhelm Wien (1864–1928). [Photo Gen. Stab. Lit. Anst. (Generalstabens Litografiska Anstalt), courtesy of AIP Emilio Segrè Visual Archives, Weber Collection, E. Scott Barr Collection]

Wien's blackbody color law states that the location of the peak for a specific temperature is inversely related to the temperature of the object emitting the electromagnetic radiation. In other words, *the hotter the star, the larger the value of $T_{surface}$ and the shorter the peak wavelength, λ_{peak}*. If the temperature is doubled, the peak wavelength is cut in half, or if the temperature is decreased by a factor of three, the wavelength of the peak light intensity increases by a factor of three.

As shown in Fig. 6.43, for stars with surface temperatures of 7000 K the curve peaks in deep violet, and for stars with surface temperatures of 4000 K the peak is deep red. Notice that the peak wavelength for the Sun is

$$\lambda_{\text{Sun's peak}} = \frac{2,900,000 \text{ nm } K}{5772 \text{ } K} = 502 \text{ nm},$$

which falls in the green portion of the visible spectrum. The middle of the visible spectrum is at 550 nm.

So why does the Sun look yellow? It is because the Sun emits all of the other colors of the rainbow as well, but at lower intensities. When averaged out, and combined with the sensitivity of the human eye, the sum of the colors appears yellow. It is hardly a coincidence that the human eye evolved to be most sensitive in the range where the Sun emits most of its light.

A tutorial for Wien's blackbody color law is available through the Chapter 6 resources page.

Example 6.10

Betelgeuse (recall Fig. 2.5), in the east shoulder of Orion has a surface temperature of 3600 K while Rigel, in Orion's west foot, has a surface temperature of 12,100 K (the directions correspond to the right shoulder and left foot as depicted in the drawing). (a) What are their peak wavelengths? (b) In which parts of the electromagnetic spectrum are their peaks located? (c) What colors do they appear to be to the human eye?

(a) From Wien's blackbody color law, Betelgeuse's peak wavelength is

$$\lambda_{\text{Betelgeuse's peak}} = \frac{2,900,000 \text{ nm } K}{3600 \text{ } K} = 806 \text{ nm}$$

and Rigel's peak wavelength is

$$\lambda_{\text{Rigel's peak}} = \frac{2,900,000 \text{ nm } K}{12,100 \text{ } K} = 240 \text{ nm}.$$

(b) According to Fig. 6.43, the peak of Betelgeuse's blackbody spectrum lies in the infrared portion of the spectrum and Rigel's is in the ultraviolet portion.

(c) Since the human eye cannot see either of these wavelengths, it will identify the brightest wavelengths that it can see, so Betelgeuse will appear red and Rigel will appear blue/violet.

If it's winter in the northern hemisphere you can always check your answer by going outside and looking!

Using Color to Determine Surface Temperature

Rather than using Wien's blackbody color law [Equation (6.12)] to find the peak wavelength from a known surface temperature, it is usually more useful in astronomy to use the equation to determine the temperature from the peak wavelength rather than the other way around. After all, the peak wavelength is, at least in principle, directly measurable. This is accomplished by rewriting Wien's law as

$$T_{surface} = \frac{2{,}900{,}000 \text{ nm K}}{\lambda_{peak, nm}}, \tag{6.13}$$

Temperature from the blackbody peak wavelength

A tutorial for determining temperature from the blackbody peak wavelength is available through the Chapter 6 resources page.

where the peak wavelength must be expressed in nanometers (nm).

Example 6.11

Deneb, the brightest star in the constellation Cygnus (the Swan), has a peak blackbody wavelength of 340 nm. In what part of the electromagnetic spectrum is this located, and what is Deneb's surface temperature?

The peak of Deneb's blackbody radiation curve is at a wavelength shorter than the human eye can detect (400–700 nm), and so its peak lies in the ultraviolet portion of the spectrum. Deneb's surface temperature is

$$T_{Deneb} = \frac{2{,}900{,}000 \text{ nm K}}{340 \text{ nm}} = 8500 \text{ K}.$$

It is convenient to remember that the peaks of the blackbody radiation curves bound the visible spectrum at temperatures of about 7000 K for 400 nm (blue/violet) and 4000 K for 700 nm (red).

The Luminosity Produced by Blackbody Radiation

If we are going to develop our understanding of stars and planets, one of the key pieces of information we need is how much energy these objects produce and emit over time. As we will see, answering that question is directly tied to the object's surface temperature and blackbody radiation.

Power is a quantity that you are very familiar with. Power is defined as the *rate* at which energy is produced, or the amount of energy generated over a specific period of time, usually one second:

$$\text{Power} = \frac{\text{energy}}{\text{time}}. \tag{6.14}$$

Definition of power

When you talk about the maximum amount of power that a car can produce, you usually use the odd unit of horsepower (hp). When you talk about a furnace or an air conditioner, you typically use another odd unit known as a BTU (British

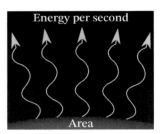

Fig. 6.46 Flux is the amount of energy emitted by the surface area in one second.

Thermal Unit). But the unit used more generally is the watt (W), named for James Watt (1736–1819), a Scottish engineer and chemist. Formally, one watt is one joule (J) of energy produced per second or 1 W = 1 J/s. You may also hear power discussed in thousands of watts (kilowatts, kW), millions of watts (megawatts, MW), or even billions of watts (gigawatts, GW). When you pay for the amount of electrical energy that you used in your home during the course of one month, you pay for the number of kilowatt-hours you used, which is the amount of energy you used each hour (energy = power × time, or kilowatts times hours), added up over the course of one billing cycle.

Unfortunately, the term "power" is often used in very unscientific ways as well. For example, in politics you may hear a phrase such as "The senator has a great deal of power in Congress." It is very important for you understand that the term as it is used in science has nothing at all to do with influence.

When you look at a portion of the surface of a star (Fig. 6.46) or some other glowing object, how bright that piece of the surface appears to you depends on how much light-energy per second (power) is coming from the total amount of area that the piece of surface covers, a quantity known as flux (F). The unit of flux reflects its definition, namely power per unit area, or in SI units, watts per square meter (W/m^2). (Brightness (b) is related to flux, except that brightness also takes into consideration how much surface area the star has and how far away the star is.)

In 1879 Jožef Stefan (Fig. 6.47) was able to show that the total flux from a blackbody added up over all wavelengths only depends on the surface temperature of the blackbody, and is given by

$$F = \sigma T_{surface}^4, \qquad (6.15)$$

Flux from blackbody radiation (Stefan's law)

where $\sigma = 5.67 \times 10^{-8}$ $W/m^2/K^4$ (sigma) is a constant that is defined in terms of nature's fundamental constants. The flux of a blackbody depends entirely on its temperature raised to the fourth power. This means that the flux increases very rapidly with an increase in surface temperature. You can get a sense of this behavior by reviewing Fig. 6.43 and seeing how rapidly the curves increase vertically as the temperature increases.

Fig. 6.47 Jožef Stefan (1835–1893). (Engraving by K. Schönbauer, public domain)

Example 6.12

A star can be treated approximately as one of Kirchhoff's perfect blackbodies. For this example, suppose that two stars are identical in size and both are the same distance from Earth. (a) If Star A has a surface temperature of 4000 K and Star B has a surface temperature of 8000 K, by what factor is the flux of Star B greater than the flux of Star A? (b) If the temperatures of Star A and Star C are 4000 K and 12,000 K, respectively, by what factor will the flux of Star C be greater than Star A?

(a) Since the temperature of Star B is twice the temperature of Star A, the amount of energy given off every second by one square meter of surface by Star B would be *greater* than Star A by a factor of $2^4 = 2 \times 2 \times 2 \times 2 = 16$.

(b) In this case Star C is three times hotter than Star A, meaning that the flux of Star C would be $3^4 = 81$ times greater than Star A's flux.

If we know how big a star or other object is, it is possible to calculate the total amount of energy that the object produces every second by using Stefan's law. This quantity is known as the luminosity, L, which is just another name for the total power output (in watts) of electromagnetic radiation over all wavelengths from gamma rays to radio waves emitted by the object's entire surface. Since flux (F) is the power output from just one square meter of surface (W/m^2), the luminosity of a blackbody is the flux multiplied by the object's total surface area (A), or from Equation (6.15), $L = A \times F = A \times \sigma T^4_{\text{surface}}$. Because we will be applying this law frequently to stars that are nearly spherical, it is common to replace A with the expression for the surface area of a sphere, which is $A_{\text{surface}} = 4\pi R^2$, where $\pi = 3.14159\ldots$ is the famous mathematical constant, pi (also a Greek letter), and R is the radius (one-half of the diameter) of the sphere. (The dots in the value of π indicate that there are an infinite number of digits that never form a repeating pattern.) Putting it all together:

$$L = 4\pi R^2 \sigma T^4_{\text{surface}}. \qquad (6.16)$$

The Stefan–Boltzmann luminosity law (for a spherical blackbody)

A Stefan-Boltzmann blackbody luminosity law tutorial is available through the Chapter 6 resources page.

It is rather remarkable that the luminosity of an object can be described by only two quantities: its total surface area and its surface temperature, even if we don't know ahead of time how the energy is produced. Although the Stefan–Boltzmann luminosity law may look a bit intimidating when you first see it, don't panic, we will only worry about how the luminosity depends on radius, R, and surface temperature, T_{surface} (in other words, we won't bother with using the constants 4, π, and σ). Again using the "trick" of writing equations in terms of values for the Sun, the Stefan–Boltzmann luminosity law can also be written as

$$L_{\text{Sun}} = R^2_{\text{Sun}} T^4_{\text{surface, Sun}}, \qquad \text{where} \qquad T_{\text{surface, Sun}} = T_{\text{surface}} / 5772 \text{ K}. \qquad (6.17)$$

The Stefan–Boltzmann luminosity law in units of the Sun

In this form, the Stefan–Boltzmann luminosity law gives luminosity in terms of the luminosity of the Sun, where R_{Sun} is written in terms of the radius of the Sun and $T_{\text{surface, Sun}}$ is written in terms of the surface temperature of the Sun, which is 5772 K. (Recall that $R^2 = R \times R$ and $T^4 = T \times T \times T \times T$.)

The blackbody luminosity law is something that will come up frequently when we talk about planets and stars, because it tells us a great deal about the amount of energy that is needed to heat up the planet or that must be generated inside the star every second. In the case of our parent star, the Sun, its luminosity is almost 4×10^{26} W, which is equivalent to nearly four trillion trillion 100-watt light bulbs. One of the great puzzles in astronomy and astrophysics up until the early 1900s was how to explain that amazingly large rate of energy production inside the Sun.

Fig. **6.48** Ludwig Boltzmann (1844–1906). (1902, public domain)

[Historical aside: Ludwig Boltzmann (Fig. 6.48) has his name associated with the luminosity law because he was successful in deriving it from the basic physics of heat and light five years after Stefan developed it by appealing to the experimental data obtained by other scientists.]

Example 6.13

The two stars Betelgeuse and Lalande 21185 have essentially the same surface temperature, but Betelgeuse is approximately 3.8 million times more luminous than Lalande 21185. Referring to the Stefan–Boltzmann luminosity law, explain how this could be the case. (Betelgeuse is in Orion and Lalande 21185 is in the southern part of Ursa Major.)

If the two stars have the same surface temperature, then their vastly different luminosities must be due to Betelgeuse having a surface area that is approximately 3,800,000 times larger than Lalande 21185. Since surface area is proportional to R^2, Betelgeuse must have a radius (R) that is the square root of 3,800,000, or $R = \sqrt{3,800,000} = 1950$ times larger than Lalande 21185. In fact, Betelgeuse is an immense star. with a radius of 3.6 au. If Betelgeuse was located where the Sun is, it would take up all of the space out to 2/3 of the way to Jupiter's orbit. This means that Mercury, Venus, Earth, Mars, and most of the asteroid belt would be inside Betelgeuse! On the other hand, Lalande 21185 is only about 40% of the size of the Sun. Clearly not all stars are the same.

Example 6.14

Sirius, in Canis Major, has a radius of $R_{\text{Sun}} = 1.711\ R_{\text{Sun}}$ and a surface temperature of $T_{\text{surface}} = 9940$ K. Calculate the luminosity of Sirius in terms of the luminosity of the Sun.

We must first calculate the surface temperature of Sirius in terms of the surface temperature of the Sun, or

$$T_{\text{surface, Sun}} = \frac{9940\ \cancel{K}}{5772\ \cancel{K}} = 1.722.$$

Now, the luminosity of Sirius in terms of the Sun's luminosity is

$$L_{\text{Sun}} = R_{\text{Sun}}^2 T_{\text{surface, Sun}}^4 = \left(1.711^2 \times 1.722^4\right) L_{\text{Sun}} = (2.928 \times 8.793)\ L_{\text{Sun}} = 25.7\ L_{\text{Sun}},$$

Sirius is almost 26 times more luminous than our Sun.

6.7 The Inverse Square Law of Light

The electromagnetic spectrum plays a central role in our quest for understanding the cosmos. One of light's important characteristics is how its intensity changes with distance from its source. It is not only how we observe the light that is important, but also how the intensity of light affects objects that are exposed to the light,

Fig. 6.49 A laser experiment (US Air Force).

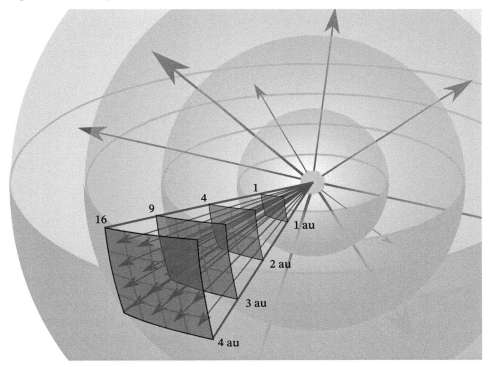

Fig. 6.50 As light spreads out in all directions from the Sun, the Sun looks less bright the farther away you get from it. The numbers above the red squares indicates the number of squares of equal area located at each distance within the diverging rectangular cone.

such as the surfaces of planets, moons, comets, and other Solar System objects, as well as the dust and gas around stars and galaxies.

Consider two different sources of light: one is a laser that emits a narrow beam of light that hardly spreads out at all (Fig. 6.49), and the other is a spherical source of light such as the Sun, emitting light that travels outward in every direction as depicted by the gold-colored rays in Fig. 6.50. In the case of a perfectly collimated laser, the intensity of the light doesn't diminish, no matter how far away you are from the laser. This is because all of the light is traveling in parallel, so all of the light arrives at your eye regardless of distance. The same amount of energy will strike your eye every second at any point along the beam, whether you are one centimeter away from the laser or one kilometer away. (Caution: Don't try this at home or anywhere else! Even a relatively weak laser can hurt your eye precisely because the light doesn't spread out. This is also why it is illegal and extremely dangerous to shine a laser light at an airplane, given the possibility that the pilot could accidentally look into the beam or become distracted.)

Now think about what happens when you move away from the spherical light source. Referring to Fig. 6.50, consider rays of light (represented by the 16 blue-colored rays) that are emitted by the Sun into the red rectangular cone. Because the light rays are spreading out as they travel, the farther away from the star you are, the less the amount of light that will actually strike your eye every second (unless your eye happens to double in diameter when you double your distance). Since the size of your eye certainly stays constant while the light rays get farther and farther apart with increasing distance, the Sun will appear to get dimmer.

A careful study of Fig. 6.50 reveals something else that is very important: not only do the light rays spread out over an increasingly greater area, they do so in a well-defined way. If your distance from the Sun is 1 au, all 16 rays pass through just one square. When you move out to 2 au from the Sun the amount of light is now spread out over four squares of the same size, with 4 rays per square. At a distance of 3 au, the same amount of light is spread out over nine squares. And at 4 au, sixteen squares are required for the 16 light rays. You should be seeing a pattern developing here: $1^2 = 1$, $2^2 = 4$, $3^2 = 9$, and $4^2 = 16$. Since the brightness of the light depends on how much light enters your eye every second, the brightness you observe decreases by an amount proportional to $1/(\text{distance})^2$, corresponding to $1/(\text{the number of squares})$.

In order to understand light's $1/(\text{distance})^2$ behavior it is important to realize that all of the light coming from the star is spreading out evenly over concentric hypothetical spheres that are centered on the Sun. This means that at a distance r from the Sun, all of the light that was emitted at a specific time will cross the surface of the same sphere of radius r at the same time. Now, rather than counting boxes as we did in Fig. 6.50, the area we are considering is the entire surface area of the sphere that the light crosses, where the surface area of a sphere of radius r is given by $A_{\text{sphere}} = 4\pi r^2$. So the amount of light *per* area crossing the sphere becomes proportional to $1/(\text{the area of the sphere}) = 1/4\pi r^2$.

This behavior for the brightness of the light with distance is the same inverse square law that you first encountered in connection with the universal law of gravitation in Section 5.3. The inverse square law is directly due to the fact that the surface area of a sphere is $A_{\text{sphere}} = 4\pi r^2$. In Newton's universal law of gravitation,

$$F = \frac{Gm_1 m_2}{r^2},$$

the inverse square law behaves in exactly the same way that the brightness of light does with respect to distance from the source of the gravitation (you can think of the constant 4π as being buried inside the universal gravitational constant, G).

The brightness we measure at a distance r from the Sun is the amount of light energy received *per square meter* of surface area per second. Recalling that luminosity (L) is the amount of light energy emitted per second, we now know that the brightness we measure is related to the luminosity by

$$b = \frac{L}{4\pi r^2}.$$ (6.18)

The inverse square law for light

Key Points
Luminosity is the total amount of energy given off by the light source every second (in watts) and brightness is the amount of light energy that crosses one square meter of area every second (flux; in W/m^2).

The equation can be read as "the brightness, b, of a light source is equal to its luminosity, L, divided by the surface area, $4\pi r^2$, of an enormous sphere centered on the source that reaches all the way out to the observing location, a distance r away from the object."

Example 6.15

Neptune is approximately 30 au from the Sun. How many times dimmer does the Sun appear to be as seen from Neptune than it does from Earth?

To answer the question you will want to note that you don't actually need to calculate the exact value of the brightness, b, as seen from both Earth and Neptune and then compare them. Instead, because L in the inverse square law equation [Equation (6.18)] is the Sun's intrinsic luminosity (its power output), which is the same value whether you observe the Sun from Earth or from Neptune, you just need to realize that Neptune is 30 times farther away from the Sun than Earth is. This means that the Sun's brightness as seen from Neptune will be $1/(\text{distance})^2 = 1/r^2$ as bright as when the Sun is observed from Earth. With $r = 30$ au for Neptune and $r = 1$ au for Earth, the Sun is *dimmer* when observed from Neptune by a factor of $30^2 = 900$.

An inverse square law for light tutorial is available through the Chapter 6 resources page.

Example 6.15 not only tells us that the Sun appears 900 times less bright from Neptune than from Earth, but it also implies that Neptune receives $1/900^{\text{th}}$ as much energy per square meter every second compared to the energy received on Earth. That decrease in energy has a big impact on the planet's temperature.

6.8 Heat as a Form of Energy

In our discussion of the conservation of energy (page 196) we considered the situation of pushing a filing cabinet across the floor, and how much work was required for the task. However, when you stop pushing, the filing cabinet isn't moving anymore. Assuming that the floor is level, the mere act of pushing the filing cabinet didn't change its potential energy, and when it stopped moving there wasn't any kinetic energy present either. In other words, the work you did didn't change its total mechanical energy (potential energy plus mechanical energy). But work requires an equivalent amount of energy be used; after all, you are certainly tired after pushing the filing cabinet all that way. So what happened to the energy? Conservation of energy demands that the energy you expended has to be hiding somewhere, so where did it go? The moving balls in Galileo's experiment ultimately stopped as well, so what happened to their kinetic energies?

Fig. 6.51 Sir Benjamin Thompson, Count Rumford (1753–1814). (Public domain)

Think about what happens when you rub your hands together on a cold day. What is it that you hope to accomplish by doing that? To warm up your hands of course. What you have done is convert the kinetic energy of moving your hands into heat, another form of energy, by means of work through the friction force.

The idea that kinetic energy could be converted into heat was put on a firm footing with the research of Benjamin Thompson (Fig. 6.51). Thompson was born in Woburn, Massachusetts where he obtained much of his education. He would also occasionally walk to Harvard College in Cambridge, about ten miles away, to attend lectures concerning mathematics and natural philosophy. When he was 21 he met and married a young widow, Sarah Roffe, who had been left with significant property in what is now Concord, New Hampshire, then known as Rumford. They moved to New Hampshire and thanks to his wife's influence he was appointed to the rank of major in the New Hampshire Militia.

When the American Revolution began, Thompson was a wealthy landowner who opposed the uprising. Confronted by the rebels, he escaped to the British army, leaving his wife behind forever. Thompson aided the British in the war by providing intelligence and conducting experiments in gunpowder. He was also appointed as a lieutenant-colonel for the Loyalist forces.

When the war ended, he moved to London where his success in the study of gunpowder established him as a successful scientist. His research also led him to explore the heating of cannon barrels when they were bored. By placing a barrel in a vat of water he was able to show that the friction from the boring process could cause the water to boil, and that there seemed to be no limit to the amount of heating as long as the boring process continued. This finding was in contradiction to the idea of the time that heat was some form of fluid that was ever-present in materials, and therefore limited in how much heat could be released. Associated with his research in heat, he improved the design of chimneys, developed ideas concerning thermal insulation, and is even credited with the invention of thermal underwear.

In 1784 he was knighted by King George III, becoming Sir Benjamin Thompson, and in 1791 he was made a Count of the Holy Roman Empire, taking the name of the town where he was first married, thus becoming Count Rumford. Later in life, he endowed the Rumford medals of the Royal Society, and he also endowed both the first award of the prestigious American Academy of Arts and Sciences and a professorship at Harvard University. His daughter in America inherited his estate.

Count Rumford's work in heat helped to establish the area of physics and chemistry known today as thermodynamics, the study of heat and its properties. Rumford's research also added another component to the idea that energy is conserved; energy may change between potential energy, kinetic energy, heat, light, and a variety of other forms, but the total amount of energy never changes. As we will see, energy can also exist as sound, chemical energy, atomic energy, nuclear energy, and even mass. The concept of conservation of energy plays a fundamental role in all aspects of science, including astronomy. Boltzmann later anchored thermodynamics to the solid foundation of statistical mechanics, that understands heat to be the result of organization and the average kinetic and potential energies of huge numbers of microscopic particles.

The Difference Between Heat and Temperature

Have you ever noticed how much more uncomfortable you feel on a hot day when it is very humid, compared to the same temperature on a day when the humidity is significantly lower? Even when the temperature is the same, the effect can be noticeable. Similarly, on a cold, damp day you can get chilled much more quickly than on a drier day. It is also possible to develop hypothermia in a matter of minutes if you fall into cold water, although you can easily survive in air at the same temperature even without insulating clothing.

It is common to confuse the concepts of temperature and heat. Heat is a form of energy associated with the total kinetic and potential energy of the individual particles contained within matter. When the amount of heat in a substance increases, the motion of the particles within the substance increases. For example, when highway concrete heats up on a hot summer day, it will sometimes buckle. This occurs because the motions of the particles comprising the concrete move about their fixed locations in the solid more rapidly, causing the concrete to expand. If the expansion is greater than the amount planned for by including expansion joints in the concrete, the increased length of the highway segment has nowhere to go but up, causing the buckling. Water similarly expands as the water molecules move about more rapidly in random directions, requiring more space. This, along with the added water from melting glaciers, is why ocean levels have risen due to global warming. Figure 6.52 illustrates these motions for a steaming cup of hot coffee and a cold cup of iced coffee.

Fig. 6.52 *Top:* Hot coffee molecules move rapidly as depicted by their velocity vectors. *Bottom:* Iced coffee molecules move slowly.

Temperature measures the average amount of kinetic energy of *individual particles* in a material. Increasing temperature means that the average amount of energy *per particle* increases. The actual amount of heat that an object contains depends on how many particles are in the material and the average amount of energy per particle.

Prior to Count Rumford's work, heat was thought to flow as a caloric[8] fluid from one substance to another. Today, we understand that heat transfer occurs when energy is exchanged between particles such as through collisions. When two objects of different temperatures come into contact, heat "flows" from the higher temperature object to the lower temperature object. This is because the particles with more energy in the hotter object collide with the particles in the colder object that have lower energies. These collisions transfer some of the excess energy of the hotter particles to the cooler particles until a common average amount of energy per particle is achieved for both objects, which is another way of saying that the two objects reach the same temperature. (Notice that we still use the imagery of heat "flowing" even though the old theory was debunked centuries ago.)

Moist air can contain more heat than dry air at the same temperature. This means that if the air is hot, moist air can transfer more heat to you, causing you to be more uncomfortable than during a drier day. Similarly, a cool, damp day can draw heat away from you more rapidly than on a drier day, causing you to chill more quickly.

---

[8]Caloric is a term that refers to heat. The calorie is actually a unit of heat energy that is commonly used in reference to the amount of energy contained in foods.

Fig. 6.53 A pot of boiling water. (Markus Schweiss, CC BY-SA 3.0)

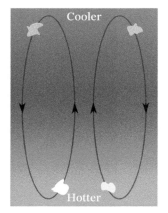

Fig. 6.54 Hot, rising blobs of material and cool, sinking material in a fluid.

Transporting Heat Energy

An important question in astronomy and in everyday life is how heat can be transferred between substances or transported from one place to another. Since heat is just one form of energy, this question is really about energy transfer from a region of higher temperature to a cooler region. There are three basic processes by which heat can move: convection, conduction, and radiation.

Convection

When you boil water (Fig. 6.53), the bubbling is due to hot steam rising to the surface and cooler water sinking back down to the bottom of the pan. When you heat your home on a cold winter day, the vents in the floor emit warm air that then rises toward the ceiling. The return vents allow the cooler air to return back to the furnace to be reheated. An important component in understanding both weather and climate change (they are not the same thing) are the motions of warm air rising and cool air sinking in the atmosphere, as well as warm water rising and cool water sinking in major bodies of water. All of these situations are examples of the transport of heat energy by the mass motion of warmer material to cooler regions and cooler material to warmer regions (see Fig. 6.54). Convection is a very efficient means of transporting energy within a fluid (either a gas or a liquid).

To minimize the effects of convection that contribute to you getting cold during a night of camping, a sleeping bag is typically used (Fig. 6.55). Within the bag is some form of fill (down or synthetic) that creates tiny air pockets which restrict the development of larger scale air currents, meaning that the fill inhibits convection. On a sufficiently small scale, air serves as an excellent insulator, but, if currents develop, that insulating property breaks down because the air itself carries the heat from one location to another. Contrary to common belief, it is not the fill itself that is the principal insulator, the fill simply keeps the air from moving in the sleeping bag. The fur on mammals and the down of birds also serve to minimize the flow of air, thereby providing protection from the cold. The double-pane glass in windows accomplishes the same purpose, by trapping a very thin layer of air or other gas between the panes to serve as insulation. Because of convective air currents set up around your body on a cold day, you need to consume fuel (food) so that you replace the energy lost, even if you aren't moving. The greatest loss of energy is actually

Fig. 6.55 A mummy-style sleeping bag. (Dale A. Ostlie)

from your head because it takes quite a bit of energy to keep that central processing unit (your brain) functioning correctly. If your feet are getting cold on a winter day, you should put on a hat! Your extremities get cold first as a defense mechanism, because your body is minimizing blood flow to those areas of your body in order to conserve precious heat in a bid to protect vital organs.

Conduction

You may have learned at some point the unpleasant consequence of using a metal spoon to stir a boiling pot of noodles in water. Once that lesson was learned, you have probably used a wooden spoon from that point forward.

We have already mentioned that the individual particles in a substance can exchange kinetic energy through collisions. As illustrated in Fig. 6.56, if the substance we are referring to is a solid, those particles are locked in place, unable to move beyond simply vibrating back and forth. In the case of a metal spoon, those vibrating particles can influence their neighbors very quickly, resulting in the entire length of the spoon being hot when you try to stir the pot. The large vibrations of the particles at the end of the spoon that is submerged in the boiling pot of noodles cause the adjacent particles to vibrate just as much, then the ones adjacent to them and so on, until all of the particles are vibrating quickly. When you touch the other end of the spoon, those rapidly vibrating particles try to increase the vibrations of the particles in your hand, resulting in damaged cells and burns if you hold on for too long.[9] But if the spoon is made of wood, the particles making up the wood do not have much of an effect on their neighbors, making wood a useful tool for the job (and a good insulator).

In the case of a liquid or a gas, particles are able to move around freely. When the particles bump into one another, they can exchange some of their energy through the collision, just like one billiard ball striking another one, causing the second ball to gain kinetic energy. With lots of billiard balls on the table, such as at the beginning of a game, all of the balls start colliding, distributing the energy of the cue ball with all of the others. Eventually, as balls strike the bumpers and move around, friction, sound, energy absorbed by the bumpers, and energy transferred to the particles that make up the billiard balls cause the balls to stop moving. If you were to measure the temperature of one of the balls after they stopped moving you would find that the balls got a little warmer, indicating that the kinetic energy of the particles in the interior of the billiard balls increased in the process.

Conduction is the process of transferring energy between particles through collisions or vibrations. Essentially, the bouncing of billiard balls off one another on the table is a macroscopic example of conduction, and the heating of the balls themselves is an example of conduction on the microscopic level.

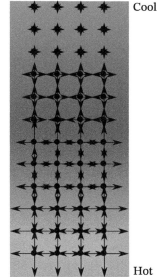

Fig. 6.56 Conduction transfers heat from the hot end of the bar to the cool end by means of vibrations. The lengths of the arrows represent the amount of vibration occurring in the particles that make up the bar.

Radiation

Even on a cold day, if you step outside and face the Sun (with eyes closed so you don't damage them!) you can feel the Sun's warmth on your face despite the fact that the Sun is 1 astronomical unit away. So how does heat from the Sun reach you?

---

[9]The conduction of heat energy through a metal spoon is also enhanced by the motions of electrons that can move freely through the metal, carrying energy with them.

Fig. 6.57 An infrared image of two buildings. The building on the right is far more energy efficient than the traditional building in the background. (Passivhaus Institut, CC BY-SA 3.0)

Fig. 6.58 An infrared image of a man. (NASA/IPAC)

Key Point
The meaning of never.

The answer is electromagnetic radiation (light). After all, light is just one of the many forms of energy that exist. Particles leaving a source and striking somewhere else are also considered to be forms of radiation.

If you want to see how energy leak-proof your house is, you can scan it with an infrared detector, as in Fig. 6.57. This is because, at the temperature of your house, blackbody radiation is emitted, primarily though windows and around doors. You radiate away energy as a blackbody as well, also in infrared light. Infrared goggles can see you in the dark because they are sensitive to the wavelengths longer than red light. You literally glow in the dark. Figure 6.58 is an infrared image of a man, with the varying colors depicting different temperature levels (the scale is in Fahrenheit degrees).

Entropy

Take a look back at Fig. 6.52. The molecules in the hot cup of coffee are moving around rapidly and the heat from the cup is escaping into the room. If the coffee was hot enough to boil, steam would be rising out of the cup as well. In the cup of iced coffee, the molecules are moving more slowly but the kinetic energy that they possess can still be absorbed by the ice in the cup, causing it to melt. In addition, the rapidly moving molecules in the hot cup each contain more kinetic energy on average than the molecules in the colder cup.

Closely related to heat is the concept of entropy, which, loosely speaking, is a measure of how disordered a system is. For example, ice is more ordered than cold coffee because the molecules in the ice are locked into a solid form, cold coffee is more ordered than hot coffee because the molecules in the hot coffee are moving around more rapidly, and hot coffee is more ordered than steam because the molecules in the steam can travel all over the room rather than being confined to the cup. In each case, disorder increases with the addition of heat. Entropy also increases with spontaneous events such as a building falling down; the rubble is obviously more disordered than the standing building before the collapse. Adding coffee to water increases entropy as the once-separated ingredients thoroughly mix. On the other hand, entropy "never" spontaneously decreases. You won't ever see the building suddenly rise up on its own out of the rubble, nor will you see the coffee separate from the water.

The word "never" is in quotes in the last paragraph because although it is not strictly impossible for the building to rise on its own, or the coffee to separate from the water, the odds of those things happening are almost incomprehensibly close to zero. The odds are much, much more unlikely than a monkey sitting at a keyboard, randomly hitting keys, and producing a string of characters that is Shakespeare's Hamlet. Even the monkey can't be expected to complete its task of writing Hamlet by accident in the age of the universe. It is fair to say that, in any meaningful sense, the quotes can be removed from "never."

Entropy is closely related to heat and to statistics. This means that entropy falls into the category of thermodynamics in general and statistical mechanics in particular. Technically, entropy, symbolized by S, is defined as the amount of heat Q a system has divided by its temperature T, expressed as

$$S \equiv \frac{Q}{T}. \qquad\qquad (6.19)$$

The definition of entropy

When the building fell, a large amount of heat was given off when potential energy gave way to kinetic energy and then to random heat and sound. When the building fell the degree of disorder also went up very dramatically. The erect building had a relatively few ways it could be put together, but there is an overwhelmingly larger number of ways its rubble can be arranged. When the heat produced went way up, so did the entropy.

Ludwig Boltzmann (Fig. 6.48) was able to make a connection between entropy and the number of ways a system can be ordered. His equation involves a constant known today as Boltzmann's constant (k_B), which is one of the fundamental constants of nature, and is the constant always associated with heat. In fact, the constant σ, that is found in both Stefan's law for flux [Equation (6.15)] and the Stefan–Boltzmann law for luminosity [Equation (6.16)], contains k_B within its definition. Boltzmann's equation for entropy became so important to our understanding of the physical world that it is engraved on the monument to him at his grave site in Vienna, Austria (Fig. 6.59):

Fig. 6.59 The grave of Ludwig Boltzmann (1844–1906), depicting his famous equation on top of the monument. (Daderot, CC BY-SA 3.0)

$$S = k_B \log W, \qquad\qquad (6.20)$$

Boltzmann's equation (the value of entropy from statistical mechanics)

entropy equals Boltzmann's constant times the logarithm[10] of the number of ways a system can be ordered, W. You won't need to apply either Equation (6.19) or Equation (6.20) directly, but they are useful for helping to develop an understanding of entropy.

There are four basic laws of thermodynamics that are fundamental to how nature behaves:

0. If two systems are in equilibrium with a third, then they are in equilibrium with each other. (This law is used to define temperature.)
1. The thermodynamic form of conservation of energy, relating work to heat.
2. The entropy of a *closed* system never decreases and usually increases.
3. As the temperature of a system approaches absolute zero, the entropy of the system approaches a constant minimum value.

Key Point
The laws of thermodynamics.

(If you are wondering about the weird numbering, the zeroth law was developed after the other three, but it is so fundamental to thermodynamics that it is now listed before the first law.)

Although all four laws are important, the second law of thermodynamics is perhaps the most well known but often the most misunderstood. What the second law essentially says is that no matter what happens in a system that is completely sealed from the rest of the universe, any physical process causes its entropy (its disorder) to either stay the same or increase. In other words, if you think things are mixed up

---

[10]A logarithm is a mathematical function that "pulls down" the exponent of a number; for example, $\log_{10}(10^2) = 2$.

now, just wait! Certainly the entire universe can be considered as a closed system, since nothing enters or leaves the universe (except in science fiction), so the entropy of the universe must be irreversibly increasing. The third law tells us that the entropy of the universe is also approaching a constant as its temperature appears to be inexorably approaching absolute zero (0 K); see Section 19.7.

Key Point
The proper understanding of the second law of thermodynamics requires an appropriate definition of the system.

Some have tried to argue that the second law of thermodynamics prohibits the evolution of more advanced species from those that are less advanced, or even that complex molecules necessary for life can't spontaneously develop. However, this is a false argument because a "system" defined to be molecules or life forms can't be considered to be closed; in other words, the system applicable for the second law is incorrectly defined. What is missing from the argument is the release of heat into its surroundings. For example, a water molecule is composed of three atoms, two of hydrogen and one of oxygen. Based on the incorrect application of the second law, a water molecule could never form because three individual atoms are less organized than a single, bound water molecule, which means that entropy must have decreased. Although a water molecule does have less entropy than the three separate atoms, a water molecule is still able to form because of the release of heat during its formation, which in turn increases the entropy of the new molecule's surroundings. When everything is properly accounted for, the total entropy of the correctly defined system does increase.

The laws of thermodynamics are also why it is impossible to build a perpetual motion machine. A perpetual motion machine is a device that, once started, can run forever by using the energy that it produces to power itself. Nothing about conservation of energy forbids such a machine, but increasing entropy does. A perpetual motion machine would radiate away heat, causing random motions of particles in the environment and a net increase in entropy. Every now and then an inventor will claim to have invented a perpetual motion machine; they didn't!

Key Point
Increasing entropy defines the arrow of time.

Yet another consequence of the second law of thermodynamics is that it defines the arrow of time. With the exception of the second law, all other physical laws allow for completely reversible behavior. For example, a collision between two American football players (page 194) must conserve both linear momentum and energy. Although it would certainly look weird, running the collision backward doesn't violate either conservation of linear momentum or conservation of energy. The same is true for the falling building. None of the conservation laws prohibits the building from rising back up from the rubble. The spontaneously rising building would need to have energy flow into it in order to lift concrete and steel to the top, straighten beams, and reconstruct the concrete, but energy is available from its surroundings, such as heat in the ground and in the air. What prohibits it from happening is the irreversibility of entropy change. Entropy specifies which way these events proceed; entropy defines which way time flows. It is entropy and the arrow of time that would make watching in reverse the collision of the football players or the building spontaneously rising from rubble seem very weird, because you know instinctively that the direction of time is wrong.

Although we won't spend any more time studying the details of entropy and the second law of thermodynamics in this text, know that it provides a very powerful theoretical tool for understanding our universe, from the motions of particles to the formation of stars, black holes, galaxies, and even the aftermath of the Big Bang and the ultimate fate of our universe discussed in Chapter 19.

? Exercises

All of the exercises are designed to further develop *your understanding* of the material by thinking carefully and critically about what you have read in this chapter. Answers to selected exercises are available in the back of the book.

True/False

1. The fundamental constants of nature have been shown to have different values, depending on what region of the universe is being considered.

2. (*Answer*) The fundamental constants of nature are believed to be time-independent, meaning that they have the same values today as they did immediately following the beginning of the universe 13.8 billion years ago.

3. The Moon is very roughly about 400 times closer to us than the Sun is.

4. (*Answer*) The gravitational constant, G ("Big G"), is assumed to have the same value everywhere in the universe.

5. The value for $g = 9.8$ m/s^2 ("little g," the acceleration due to gravity) is valid on the surface of Earth and on the surface of the Moon.

6. (*Answer*) The dialog in the 1995 film, *Apollo 13* (Universal Pictures) implied virtually no time delay in communications between the astronauts of the ill-fated mission to the Moon and ground control in Houston. Given that the Moon is only about 400,000 km from Earth, the movie was correct in that depiction.

7. If you lived on a planet with Earth's radius, but with twice Earth's mass, your weight would be twice what you weigh on Earth.

8. (*Answer*) If you lived on a planet with Earth's mass, but with twice the radius of Earth, your weight would be twice what you weigh on Earth.

9. If you are in the front seat of your car and you push hard on the dashboard, you can make the car move forward.

10. (*Answer*) Because a more massive object has more potential energy 10 m above the ground than a less massive object does at that height, in free-fall the more massive object is moving faster just before it hits the ground.

11. Both gravity and brightness from a star obey inverse square laws.

12. (*Answer*) A comet that is 5 au from the Sun will receive 10 times less light than when it is 0.5 au from the Sun.

13. Boiling water transports heat from the bottom of the pan to the surface by the process of conduction.

14. (*Answer*) Air can be an effective insulator if it isn't allowed to move.

15. Absolute zero is the coldest possible temperature.

16. (*Answer*) At absolute zero the energies of substances are as low as they can ever be.

17. Heat is a fluid, known as a caloric fluid, that can flow from a hotter material to a cooler one.

18. (*Answer*) Because the freezing point of water is 0°C and its boiling point is 100°C, the ratio of water's boiling point to its freezing point, 100°/0°, tells us that water is infinitely hotter when it is boiling than when it is freezing.

19. If the speed of a wave increases by a factor of ten while its wavelength is unchanged, then its amplitude must increase by a factor of ten as well.

20. (*Answer*) Constructive interference occurs when two waves interact in such a way that their peaks and troughs align.

21. Blue light has a wavelength that is about 4/7 of the length of red light, and so the frequency of blue light is about 7/4 times the frequency of red light.

22. (*Answer*) The kite experiment supposedly conducted by Benjamin Franklin was shown to be perfectly safe.

23. Electricity, magnetism, and optics are all related phenomena.

24. (*Answer*) Electromagnetic waves must have a medium through which they travel.

25. A star with a surface temperature of 40,000 K cannot be seen by the human eye because its peak wavelength is too short.

26. (*Answer*) Even though two stars have the same surface temperatures they don't necessarily have the same luminosities.

27. It is appropriate to say that 100°C is two times hotter than 50°C.

28. (*Answer*) Entropy is a measure of the number of ways a system can be ordered.

29. The second law of thermodynamics is violated during biological evolution because entropy decreases in the development of a more complex organism from a less complex form.

Multiple Choice

30. (*Answer*) 800°C is about _____ kelvins.
    (a) 1260   (b) 1073   (c) 527   (d) 340

31. About what time of day would it be for an observer at point O in Fig. 6.2?
    (a) Noon   (b) 6 p.m.   (c) Midnight   (d) 6 a.m.
    (e) Not enough information given to determine the time of the observer

32. (*Answer*) The distance from Earth to the Sun is approximately
    (a) 93 million miles.   (b) 150 million km.
    (c) one astronomical unit.   (d) 8.3 light-minutes.
    (e) 500 light-seconds.   (f) all of the above

33. Assume that the fastest that Galileo's assistant could react after seeing the lantern in Galileo's speed-of-light experiment would have been 2/3 of a second. How far does light travel from the moment he sees Galileo's lantern to the time he gets his own lantern uncovered?
    (a) 20 km   (b) 200 km
    (c) 2000 km   (d) 20,000 km
    (e) 200,000 km   (f) 2,000,000 km
    (g) none of the above

34. (*Answer*) The distance from Earth to the Moon is about 400,000 km. Roughly how long would it take to send a pulse of laser light to the Moon, bounce the light off of a mirror on the Moon's surface, and then see the returned pulse? (Timing the round-trip of a light pulse is actually used to measure the Moon's distance very accurately. Mirrors were left on the Moon by Apollo astronauts.)
    (a) 0.3 seconds   (b) 1.3 seconds
    (c) 2.7 seconds   (d) 5.4 seconds
    (e) none of the above

35. On July 14, 2015, the New Horizons spacecraft flew past Pluto after leaving Earth more than nine years earlier. At the time New Horizons reached Pluto the spacecraft was about 32.5 au from Earth. How long did it take a radio signal to reach us from the spacecraft?
    (a) instantaneously
    (b) 10 minutes
    (c) 30 minutes
    (d) 1.2 hours
    (e) 4.5 hours
    (f) 9 hours
    (g) none of the above

36. (*Answer*) Your author's mass is roughly 95 kg. What would his mass be if "Big *G*" suddenly increased by a factor of 5?
    (a) 19 kg
    (b) 48 kg
    (c) 95 kg
    (d) 190 kg
    (e) 475 kg
    (f) It is impossible to determine without knowing the radius of Earth.
    (g) none of the above

37. Your author's weight is roughly 910 N on Earth. What would his weight become if "Big *G*" suddenly increased by a factor of 5?
    (a) 182 N
    (b) 455 N
    (c) 910 N
    (d) 1820 N
    (e) 4550 N
    (f) It is impossible to determine without knowing the radius of Earth.
    (g) none of the above

38. (*Answer*) You are ice skating with your significant other who is standing still. In an attempt at a romantic gesture you skate up to give them a hug, but you don't stop moving completely. Immediately after the embrace the two of you are moving in the _____ direction with _____ speed relative to your speed before the embrace.
    (a) same, more
    (b) same, the same
    (c) same, less
    (d) opposite, more
    (e) opposite, the same
    (f) opposite, less
    (g) not enough information is given

39. Standing still, you are pushing a child on a merry-go-round when you decide to join her on the ride by sitting down on the edge. After sitting down, the merry-go-round will be spinning _____, an example of conservation of _____.
    (a) at the same rate, linear momentum
    (b) faster, linear momentum
    (c) slower, linear momentum
    (d) at the same rate, angular momentum
    (e) faster, angular momentum
    (f) slower, angular momentum

40. (*Answer*) Two satellites are in circular orbits around Earth with the same speed and at the same distance from the center of the planet. One satellite has three times more mass than the other one does. The more massive satellite has _____ the orbital angular momentum of the smaller one.
    (a) equal
    (b) three times
    (c) one-third
    (d) six times
    (e) one-sixth
    (f) nine times
    (g) one-ninth
    (h) not enough information is given

41. Two identical satellites are in circular orbits around Earth, but the orbital radius of one of the satellites (Satellite A) is twice the radius of Earth, while the orbital radius of the other satellite (Satellite B) is eight times the radius of Earth. Because Satellite B is in a higher orbit, its orbital speed is only one-half of the orbital speed of Satellite A in the lower orbit. Which satellite has the greater orbital angular momentum, and by what factor?

    (a) A, two times
    (b) A, four times
    (c) A, eight times
    (d) B, two times
    (e) B, four times
    (f) B, eight times

42. (*Answer*) Mars has a mass that is about 11% of the mass of Earth ($0.11 \, M_{Earth}$) and its radius is 53% of the radius of Earth ($0.53 \, R_{Earth}$). What is the escape speed from the surface of Mars? Is it faster, slower, or about the same as the escape speed from the Moon?
    (a) 1.2 km/s, slower
    (b) 2.3 km/s, almost equal
    (c) 5.1 km/s, almost equal
    (d) 5.1 km/s, faster
    (e) 24.6 km/s, slower
    (f) 24.6 km/s, almost equal
    (g) 54 km/s, faster

43. Is the escape speed greater, the same, or less for a 10,000 kg rocket compared to a 0.15 kg baseball?
    (a) greater
    (b) the same
    (c) less
    (d) The answer depends on what the rocket and the baseball are composed of.

44. (*Answer*) Carbon dioxide freezes at a temperature of $-78°C$ and becomes what is commonly called dry ice. What does that temperature correspond to in kelvins?
    (a) -78 K
    (b) -5 K
    (c) 0 K
    (d) 78 K
    (e) 195 K

45. Saturn is 9.5 au from the Sun. The brightness of the Sun as seen from Saturn would be about _____ of the brightness from Earth.
    (a) 1/9.5
    (b) 1/19
    (c) 1/90
    (d) 1/900

46. (*Answer*) One substance has a temperature of 200 K and another substance has a temperature of 400 K. The particles in the hotter substance have, on average, _____ energy as the particles in the cooler substance.
    (a) one-half as much
    (b) twice as much
    (c) 200 times less
    (d) 200 times more
    (e) I need to know the masses of the substances in order to answer this problem.

47. If a hotter material is brought into contact with a cooler material, heat is transferred from the hotter to the cooler material because collisions transfer _____ between the _____ particles in the hotter material and the _____ particles in the cooler material.
    (a) mass, faster, slower
    (b) fluid, faster, slower
    (c) kinetic energy, faster, slower
    (d) mass, slower, faster
    (e) fluid, slower, faster
    (f) kinetic energy, slower, faster

48. (*Answer*) A sound wave travels though air at a speed of about 343 m/s near the surface of Earth. When you hear a middle C played, you are listening to a frequency of 261.63 Hz. What is the wavelength of the sound wave moving through the air?
    (a) 0.13 cm
    (b) 0.76 cm
    (c) 1.3 m
    (d) 7.6 m
    (e) not enough information given

49. In a Young double-slit experiment (Fig. 6.26), a light wave from slit S_2 travels an extra two complete wavelengths relative to a light wave from slit S_1. The extra distance traveled by the wave from slit S_2 is 900 nm. The combination of the two waves produces a _____ spot, the wavelength of the light is _____, and the color of the wave is _____.
    (a) bright, 450 nm, blue
    (b) dark, 450 nm, violet
    (c) bright, 600 nm, red
    (d) dark, 600 nm, orange
    (e) bright, 900 nm, orange
    (f) dark, 900 nm, red
    (h) bright, 1800 nm, blue
    (i) dark, 1800 nm, violet

50. (*Answer*) In a Young double-slit experiment (Fig. 6.26), a light wave from slit S_2 travels an extra 1 1/2 wavelengths relative to a light wave from slit S_1. The extra distance traveled by the wave from slit S_2 is 900 nm. The combination of the two waves produces a _____ spot, the wavelength of the light is _____, and the wave's color is _____.
    (a) bright, 450 nm, blue   (b) dark, 450 nm, violet
    (c) bright, 600 nm, red   (d) dark, 600 nm, orange
    (e) bright, 900 nm, orange   (f) dark, 900 nm, red
    (h) bright, 1800 nm, blue   (i) dark, 1800 nm, violet

51. A certain radio wave has a wavelength of 21 cm. What is the wave's frequency? (If you need a refresher on scientific notation, see the textbook's Math_Review webpage.)
    (a) 7×10^{-10} Hz   (b) 7×10^{-8} Hz
    (c) 1.4×10^7 Hz   (d) 1.4×10^8 Hz
    (e) 1.4×10^9 Hz

52. (*Answer*) A specific infrared light wave has a wavelength of $1\,\mu m$ (1×10^{-6} m). What is the frequency of the wave? (If you need a refresher on scientific notation, see the textbook's Math_Review webpage.)
    (a) 3×10^{-14} Hz   (b) 3×10^{14} Hz
    (c) 9×10^{-14} Hz   (d) 9×10^{14} Hz
    (e) not enough information given

53. The frequency of your favorite radio station is 103.5 MHz (1.035×10^8 Hz). What wavelength does that correspond to?
    (a) 0.345 m   (b) 3.45 cm   (c) $2.9\,\mu m$
    (d) 2.9 m   (e) 29 m

54. (*Answer*) The order of the electromagnetic spectrum from the lowest frequencies to the highest frequencies is
    (a) gamma rays, x-rays, ultraviolet, visible, infrared, microwaves, radio waves
    (b) gamma rays, ultraviolet, infrared, visible, microwaves, x-rays, radio waves
    (c) radio waves, gamma rays, microwaves, x-rays, ultraviolet, infrared, visible
    (d) radio waves, microwaves, infrared, visible, ultraviolet, x-rays, gamma rays
    (e) visible, ultraviolet, infrared, x-rays, microwaves, gamma rays, radio waves

55. The brightest star in the summertime constellation of Lyra is Vega, with a surface temperature of 10,000 K. The peak of its blackbody spectrum is at a wavelength of _____ and it appears _____ to the human eye.
    (a) 2900 nm, red   (b) 2900 nm, blue/violet
    (c) 290 nm, blue/violet   (d) 290 nm, red

56. (*Answer*) If the peak blackbody wavelength of a star is 100 nm, the star will appear _____ to the human eye and the star's surface temperature is _____.
    (a) ultraviolet, 29,000 K   (b) blue/violet, 29,000 K
    (c) red, 2900 K   (d) infrared, 2900 K

57. The surface temperature of Proxima Centauri is approximately 3000 K. What is the peak wavelength of its blackbody spectrum, and in what part of the spectrum does that wavelength fall?
    (a) 2.9×10^9 nm, ultraviolet   (b) 2.9×10^9 nm, infrared
    (c) 967 nm, ultraviolet   (d) 967 nm, infrared

58. (*Answer*) You are comparing two stars. Star A has a surface temperature of 5000 K and Star B is four times hotter, with a surface temperature of 20,000 K. The peak wavelengths of

Star A and Star B are in the _____ and _____ parts of the electromagnetic spectrum, respectively.
    (a) infrared, ultraviolet   (b) infrared, visible
    (c) infrared, infrared   (d) visible, ultraviolet
    (e) visible, visible   (f) visible, infrared
    (g) ultraviolet, ultraviolet   (h) ultraviolet, visible
    (i) ultraviolet, infrared

59. A 100 W (0.1 kW) light bulb is left on in your house for thirty days without being turned off. Given that there are 24 hours in one day, for how many kilowatt-hours will you be billed for that one light bulb?
    (a) 24   (b) 30   (c) 72   (d) 100   (e) 3000
    (f) 72,000

60. (*Answer*) The Sun's luminosity is about 4×10^{26} W, or 4×10^{23} kW. Assume that the cost of one kilowatt-hour is $0.12 (12¢). The Sun produces about _____ kilowatt-hours of energy over the course of thirty days, so if you had to pay for all of that energy for one month, your bill would be _____. (If you need a refresher on scientific notation, see the textbook's Math_Review webpage.)
    (a) 3×10^{26}, 3.5×10^{25} ($35 trillion trillion)
    (b) 1×10^{25}, 1.4×10^{26} ($140 trillion trillion)
    (c) 3.1×10^{28}, 1.2×10^{27} ($1200 trillion trillion)
    (d) 3×10^{29}, 3.5×10^{30} ($3.5 million trillion trillion)

61. For the stars in Exercise 58, the surface flux produced by Star B is _____ times greater than the surface flux of Star A. (If you need a refresher on exponents, see the textbook's Math_Review webpage.)
    (a) 4   (b) 8   (c) 16   (d) 32   (e) 64
    (f) 128   (g) 256   (h) 512

62. (*Answer*) Two stars have the same surface temperatures but one star is three times larger than the other. The luminosity of the larger star is greater than that of the smaller star by a factor of _____. (If you need a refresher on exponents, see the textbook's Math_Review webpage.)
    (a) 3   (b) 6   (c) 9   (d) 12   (e) 15
    (f) 18   (g) 21   (h) 24

63. For the stars in Exercise 58, Star B has a radius that is twice the radius of Star A. Based on this information and the result of Exercise 61, how many times greater is the luminosity of Star B compared to the luminosity of Star A? (If you need a refresher on exponents, see the textbook's Math_Review webpage.)
    (a) 16   (b) 32   (c) 64   (d) 128
    (e) 256   (f) 512   (g) 1024 (h) 2048

Short Answer

64. Why is it impossible to make the measurement of the angle α shown in Fig. 6.2 for an observer at location O?

65. (a) Using a protractor, draw the angle in Fig. 6.3 at the location of Earth for a value of $\beta = 87°$ and again for an angle of $\beta = 89.9°$ at the same point. Make your best estimate of where 89.9° is on the protractor.
    (b) Explain why it was so difficult for Aristarchus to get a reasonable value for the distance to the Sun in terms of the distance to the Moon. Why was his answer off by such a large amount?
    (c) What distance would you get for the distance to the Sun if β was just 0.1° larger than its actual value of 89.9°?

66. At the end of Example 6.1 there was a comment about how Kepler's third law (the harmonic law), Equation (5.10), was written so that constants, including G, canceled out and the remaining variables had to be written in terms of the mass of the Sun, the orbital period of Earth, and the distance of Earth from the Sun.
    (a) Using the same approach, write a form of Kepler's third law that can be used for the Earth–Moon system, where the remaining variables must be in Earth masses, the orbital period of the Moon, and the distance of the Moon from Earth.
    (b) Assuming that the orbital distance of the Moon from the center of Earth was 25 R_{Earth} in the distant past, calculate its orbital period in terms of the orbital period of the Moon today.
    (c) Given that the present-day orbital period of the Moon is 27.32 days, what was the orbital period of the Moon (in days) when its orbital radius was 25 R_{Earth}? Compare your answer with the value quoted in Example 6.1.
67. (*Answer*) How much time elapsed between a pulse of light passing through Fizeau's spinning toothed wheel and returning to the wheel?
68. If light could be made to circle Earth, how many times could it go around Earth in one second? The circumference of Earth is approximately 40,000 km.
69. Why would it have been impossible to make a correction to the trajectory of the New Horizons spacecraft (see Exercise 35) if an error was detected by the spacecraft one hour before its closest approach to Pluto?
70. (*Answer*) If G suddenly decreased by 20%, what do you think would happen to the Moon's orbit? Explain your reasoning.
71. If G miraculously increased by a factor of two, what would happen to your weight? What would happen to your mass? Explain your answers.
72. Describe why it would be impossible to determine the masses of stars if G wasn't constant throughout the universe. Hint: Consider the effect that changing G would have on Newton's universal law of gravitation and the orbits of stars.
73. Considering the law of conservation of linear momentum, explain why there is a "kickback" when someone fires a gun. Hint: The total linear momentum before the gun is fired is zero.
74. A child was playing with a toy car in the aisle of an airplane in flight. While preparing to land, the parents forgot that the toy was in the aisle. What happened to the little car when the plane slowed down rapidly upon landing? Was the result due to conservation of linear momentum, conservation of angular momentum, or conservation of energy? Explain your answer. Hint: Think about the speed of the toy car relative to the ground before the plane started to slow down. Chalk it up to adventures in parenting!
75. Find a video of a diver jumping off a high-dive platform doing either a twisting move or a tuck move (pulling his legs to his chest and spinning rapidly). Describe how the diver starts the rotation. What does the diver do just before entering the water? Explain why.
76. Our Sun rotates on its axis about once per month. What would happen to the rotation rate of the Sun if it suddenly shrank in size to a radius of 10 km, about one-half of the size of New York City? Is your conclusion based on conservation of linear momentum, conservation of angular momentum, or

conservation of energy? Stars with masses greater than the mass of the Sun, but with such small radii do exist, and are known as neutron stars.
77. Why are race cars designed to crumple and fly apart during a crash while protecting the driver's compartment? Hint: Think about energy conservation.
78. From a purely conservation of energy perspective, would you expect the skier's mass to matter during the downhill, the fastest alpine ski race with the fewest turns? Why aren't all downhill racers very massive individuals? Explain your reasoning.
79. If du Châtelet had the necessary apparatus to carefully determine how much energy was required to deform the clay in her experiments, she would have discovered that it required less energy than the originally available potential energy of the balls that she dropped. Since energy cannot be either created or destroyed, speculate on where the rest of the energy went? Think about what you would observe and feel if you performed such an experiment.
80. The average surface temperature of the Sun is 5772 K and the average surface temperature of Earth is about 288 K. Explain why it doesn't make sense to say that the average temperature of the Sun–Earth system is $(5772\text{ K} + 288\text{ K})/2 = 3033\text{ K}$. (Unfortunately, at least one elementary school math textbook proposed such a problem.)
81. Explain why placing a 1 kg block of metal having an initial temperature of 300 K in 2 kg of water at 330 K will reach a different equilibrium temperature than if that same block at 300 K had been placed in 4 kg of water at 330 K. In which case will the final temperature be higher?
82. (a) Download an image of a wave form like the ones shown in Fig. 6.24 from the Chapter 6 resources page of the textbook's website. Refer to that wave as Wave A.
    (b) Sketch an identical copy of the wave over the original, but shifted to the right by one quarter of a wavelength. Refer to that wave as Wave B.
    (c) Sketch the wave that results by adding together the heights of waves A and B. Use at least 10 evenly spaced points along the waves.
    (d) Is the amplitude of the resulting wave greater or smaller than those of waves A and B?
    (e) Where does your sketch fit in the sequence shown in Fig. 6.24?
    (f) If waves A and B represent electromagnetic waves, would their combination be brighter or darker than either wave separately?
83. Explain why radio antennas on cars and trucks have characteristic lengths of roughly 1/2 to 1 meter or so. Hint: What is the broadcast wavelength of a radio station with a frequency of 100 MHz? Would it make sense to use antennas with wavelengths that are much, much larger or smaller than the broadcast wavelength?
84. A critical wavelength range in observational astronomy occurs in the vicinity of 21 cm. There are a number of other critical wavelength ranges as well.
    (a) In which part of the electromagnetic spectrum does that wavelength region lie?
    (b) What frequency does 21 cm correspond to?
    (c) Explain why broadcasts in that and other regions of the electromagnetic spectrum are prohibited around the world.

85. (*Answer*) Suppose that star A has a surface temperature of 7000 K and star B has a surface temperature of 3500 K.
    (a) How many times hotter is star A than star B?
    (b) The peak blackbody wavelength of star A is 410 nm. What would the peak blackbody wavelength of star B be? You shouldn't need to reach for your calculator for this one!

86. It is a challenging task to design the interior of a concert hall in such a way that it isn't too "live," meaning that there aren't distracting echos, but where listeners can still hear the orchestra easily. The design must also make sure that there aren't localized "dead spots" where it is difficult to hear. Based on what you know about constructive and destructive interference of light waves, explain how dead spots can occur in concert halls if acoustical engineers don't design them properly.

87. You and a friend are standing outside on a clear evening enjoying the stars, and you notice two stars that are both yellow in color, but one is much brighter than the other. With your vast knowledge of the distances to the stars you know that both stars are the same distance from Earth. How would you explain the difference in the brightnesses of the two stars to your friend?

88. The human body radiates as a blackbody with the peak of its electromagnetic energy distribution in the infrared portion of the spectrum. The solar wind, a very low-density stream of charged particles coming off the Sun and traveling through the Solar System, including past Earth, has a temperature of roughly 1 million kelvins. Neglecting all of the other nasty things that would happen to you if you were exposed to the vacuum of space without a spacesuit, you would eventually freeze to death in that 1 million Ksolar wind. How is that possible?

89. The peak of the blackbody radiation curve for Arcturus (in Boötes) is at 676 nm. What is the color of Arcturus and what is its surface temperature?

90. Astronomers often use color filters to observe stars in specific wavelength ranges, such as in blue light and in the middle of the visible spectrum; the filters effectively block all other wavelengths in the electromagnetic spectrum. If a star appears brighter in blue light than it does in the middle of the visible spectrum, what would that say about the general location of the peak of the blackbody spectrum? Would the star be hotter or cooler than the Sun?

91. Based on your answer to Exercise 90 how might you devise a process to determine the surface temperatures of stars based on the brightnesses of stars in a blue filter and a visible filter? How could you improve on the process if you also had information about the brightnesses of stars in a filter that was centered in the near ultraviolet (in this case "near" means not very far outside of the visible spectrum)?

92. (*Answer*) Assume that you are standing 10 m from a light bulb that emits light in every direction evenly. At your location the brightness of the light bulb is 0.08 W/m². What would the light bulb's brightness be if you were standing 50 m from it?

93. A common mistake that novice campers make when sleeping on the ground is to use an air mattress for comfort. Explain why the camper will get very cold even during only moderately cool nights.

94. The block-stacking game, Jenga™, starts with layers of crisscrossed blocks to make a rectangular tower. The length of each rectangular block is the width of the tower and the width of each block is one-third of the tower's width. The goal of the game is to remove blocks from the stack and place them on top to grow the tower's height without the structure falling down.
    (a) As the number of options for the arrangement of the blocks decreases while the tower gets taller, is the entropy of the tower increasing or decreasing? Explain how your answer can be correct in light of the laws of thermodynamics.
    (b) What happens to the entropy of the tower when it inevitably falls? Why?
    (c) The blocks make noise when they fall. What does this have to do with entropy?

95. (*Answer*) In 1783, John Michell (Fig. 6.10) suggested that perhaps stars could exist for which the speed of escape from their surfaces equals the speed of light.
    (a) By setting the speed of light, c, equal to v_{escape} in the escape speed equation [Equation (6.7)], solve for the radius of Michell's star if it has a mass, M.
    (b) Using a procedure similar to the discussion just prior to Example 6.5, show that the radius of Michell's star can be written in terms of solar units as
    $$R_{Sun} = M_{Sun} \times \left(\frac{v_{escape,\,Sun}^2}{c^2} \right),$$
    where $R_{Sun} = R_{Michell's\,star}/1\ R_{Sun}$ and $M_{Sun} = M_{Michell's\,star}/1\ M_{Sun}$, the radius of Michell's star divided by the radius of the Sun and the mass of Michell's star divided by the mass of the Sun, respectively.
    (c) The escape speed from the surface of the Sun is about $v_{escape,\,Sun} = 618$ km/s. What fraction of the radius of the Sun would Michell's star have if the mass of the star was equal to the Sun's mass, $M_{Michell's\,star} = 1\ M_{Sun}$?
    (d) The radius of the Sun is 696,000 km (109 times the radius of Earth). What is the radius of Michell's 1 M_{Sun} star?
    (e) Is that dimension closer to the size of Earth, the United States, Utah, Delaware, New York City, a small town, a city block, or a car? The radius of Earth is about 6400 km.
    Today we refer to Michell's stars as black holes.

Activities

96. For this activity you will want a ruler, a tape measure, a protractor, a compass (you can download a digital compass for your smartphone that will work well), a sheet of paper, and two sticks, stakes, or other markers that you can use to mark positions on the ground. At the end of the exercise you are going to determine the distance to an object outside by using data you will generate in the first part of the activity.

Part I: Generating necessary data

(a) On a sheet of paper draw a vertical line about 1 cm from the left edge of the paper that measures 25 cm long. This will be the longest leg of a series of right triangles that you will be drawing.
(b) Starting at the bottom of the line from part (a), draw a 1 cm long line that is perpendicular to the first leg (parallel to the the bottom of the sheet of paper), then complete the triangle by drawing the hypotenuse (the third leg, which is opposite to the 90° angle) from the end

of the 1 cm horizontal line to the top of the 25 cm vertical line.

(c) Calculate the ratio of the length of the longest, vertical, leg of the right triangle to its shortest, bottom, leg.

(d) Using the protractor, measure the angle between the shorter, bottom leg and the hypotenuse of the right triangle.

(e) Record your data in a table, with the following columns:

Bottom length (cm)	Length ratio	Angle (°)
1	25	87.7
2		

(f) Repeat parts (b), (c), and (d) for bottom legs of 2 cm, 3 cm, ..., 20 cm, and enter your results in the table.

A graphing tutorial is available at
Math_Review/Graphing

(g) Create a graph of your results by plotting your data with Length ratio on the vertical axis and Angle (degrees) on the horizontal axis. Draw a smooth curve through the twenty points that you put on your graph. You can download a prepared sheet of graph paper for this exercise from the Chapter 6 resources page, or you can generate a graph using a program such as Excel. (If you are familiar with trigonometry, the Length ratio is the tangent of the Angle.)

Part II: Determining the distance to an object outside

(h) Identify a tree or other object (perhaps a friend) that is somewhere between 10 m (about 33 ft) and 60 m (about 200 ft) away from you, and mark your location with one of your markers. Using your compass determine and record the direction to your target object in degrees from magnetic north. You should also take a picture of your target object from your location. The line between your marker and your target object will correspond to the long, vertical, leg of the right triangle that you constructed in Part I.

(i) Along a line that is perpendicular (90°) from the direction you just determined for your target, and using your tape measure, place your second marker 3 m (about 10 ft) away from the first one. This 3 m distance corresponds to the shorter, bottom leg of your triangles from Part I. From your new location, use your compass to determine the angle to your target object from magnetic north. Record your result.

(j) Sketch the right triangle comprised of your target object and the two markers you placed. Indicate the length of the shorter leg as being 3 m. The angle at the location of your target object equals the difference between your compass readings, while the angle at the location of your second marker equals 90° minus that angle. [Note: If one compass reading was on one side of 0° (north) and the other reading is on the other side, make sure that you are calculating the difference correctly!]

(k) Using your graph from Part I, determine the Length ratio of the two sides of your triangle.

(l) Given that the shorter leg is 3 m, how far is your target object from your first marker?

(m) Does your answer to the last question make sense? For an able-bodied individual you can get a rough estimate for comparison purposes by walking the distance from your first marker to your target object. The average stride for a man is about 3/4 m (30 in) and 2/3 m (26 in) for a woman.

Part III: Discussion

(n) Discuss possible sources of error in your determination of the distance to your target object. Note that even if you were extremely careful in all measurements, errors always exist. This means that a specific number for a result is never adequate in scientific publications of experimental and observational research; instead, the number must be accompanied by estimates of how small and large the number could actually be, with the reported value being a best estimate. (Recall the discussion of error bars on page 177.)

(o) Compare the process of determining the distance to your target object with the old geometric process of estimating the distance to the Sun based on knowledge of the distance to the Moon (Fig. 6.3).

(p) From the behavior of the curve you drew, as the angle approaches 90° explain why it becomes very difficult to accurately determine the distance to the Sun, based on the distance from Earth to the Moon.

Revealing Secrets Hidden in Light and Matter

7

Now my great plan, which was conceived of old, and quickens and kicks periodically, and is continually making itself more obtrusive ... is to let nothing be willfully left unexamined.

James Clerk Maxwell (1831–1879)

7.1	Prisms, Scattering, Diffraction, and Spectral Lines	238
7.2	The Periodic Table of the Elements	247
7.3	Detecting Helium and Other Elements in the Sun	251
7.4	Particles Smaller than Atoms	253
	Exercises	260

Fig. 7.1 The Sun's spectrum. (NSO/AURA/NSF; CC BY 4.0)

Introduction

Light is all around us. It permeates our existence and enriches our world. Not only does light allow us to see what is around us, but it also provides incredible beauty, from the blues reflected from ocean waters to the deep reds of sunsets, from the various shades of green seen in the leaves of trees during the summer months to their spectacular array of colors in the fall, and in the wonderful variety of colors in the skin pigments of humans. A spectacular rainbow lays out before us all of the colors that we are able to see with our eyes. To a visual artist, whether she is a painter, a potter, a photographer, or a film maker, it is the use of light that plays a fundamental role in her medium.

Visible light is just a small portion of a much broader spectrum of light that ranges from the very high-energy gamma rays to the very low-energy radio waves. It is light that is the principal conveyor of information about the objects we look at, whether they are on Earth or in space. Light is fundamental to the work of the artist and scientist alike.

7.1   Prisms, Scattering, Diffraction, and Spectral Lines

Prisms and Rainbows

Fig. 7.2 White light passing through a prism causing colors to separate, with the shortest wavelengths being bent the most. (Adapted from Suidroot, CC BY-SA 4.0)

Key Point
The speed of light in materials is always slower than in a vacuum, and the speed depends on wavelength.

When Newton sent sunlight through a prism, separating it into its component colors (similar to Fig. 7.2), little did he realize that he was opening a door to a vast storehouse of information. Today we know that he was separating the visible components of the electromagnetic spectrum according to wavelength, with the longest wavelengths of red light being bent the least and the shortest blue/violet wavelengths being bent the most. Although all wavelengths of the electromagnetic spectrum travel through the vacuum of empty space with exactly the same speed, the speed of light, c, when light passes through any medium such as air, water, glass, or quartz, the light's speed is always less than c. Moreover, the various wavelengths of light have different speeds, with red light traveling faster than blue/violet light. Light travels with a speed very close to c through air, but when it enters a glass or quartz prism its speed is reduced to about two-thirds of c, causing the light to bend at the interface. Since blue/violet light slows down the most, it bends the most. This is exactly the problem Newton was trying to solve when he designed his reflecting telescope, avoiding the spreading of the colors of white light when the light passes through a lens, even if he didn't understand the root cause (see Fig. 5.5).

Exactly the same thing happens when white light passes through water, although the wavelength speeds are slightly faster in water than in glass or quartz. Without worrying about the details of internal reflections, when sunlight passes into raindrops, it is again bent into its constituent colors, appearing as a rainbow, as was displayed in Fig. 6.27. Because the bending of light happens at specific angles depending on wavelength, rainbows always appear before you at specific angles. This is why you will never catch the leprechaun waiting by the pot of gold; as you move toward the rainbow, the angles cannot change and so the rainbow (and your get-rich-quick scheme) moves away at the same rate.

Blue Sky and Red Sunsets

On a clear day around midday, when we look up in any direction we see blue sky while the Sun appears yellowish in color [Fig. 7.3 (*Top*)]. However, the situation is very different when we observe a sunrise or sunset [Fig. 7.3 (*Bottom*)]. When the Sun is near our horizon, the Sun itself takes on a more reddish hue and the sky around it can turn a fiery red. You may have even had the chance to notice that on some days the sunsets are more spectacular than on other days. Although some of this perception has to do with clouds in the sky, sunrises and sunsets tend to be more impressive when there are more particulates in the air, such as dust from a dust storm, smoke from a forest fire, after a large volcanic eruption somewhere in the world, or if you live in an area with significant atmospheric pollution.

The correlation between an increased level of particulates in the air and the color of sunsets hints at the cause: because blue/violet light is the shortest wavelength in the visible portion of the electromagnetic spectrum, it is more prone to being scattered by small particulates than is red light, which has a wavelength that is almost twice as long. To understand this, think about what happens if a short-wavelength ripple in water perhaps 5 cm long runs into a rock the size of a baseball: the rock would tend to get in the ripple's way. But what happens if an ocean wave with a wavelength of 1 m runs into that same rock? The answer is, not much. The wave would hardly notice the rock as the wave moves past it.

Fig. 7.3 *Top:* Blue sky over Bryce Canyon National Park, Utah. (Dale A. Ostlie) *Bottom:* Red sunset over the Pacific Ocean. [Commander John Bortniak, NOAA Corps (ret.), public domain]

Key Point
Shorter wavelength blue/violet light scatters off dust and other particulates more than longer wavelength red light does.

The *top left* diagram in Fig. 7.4 depicts two people observing the Sun. The woman at the top of the diagram is looking at the setting Sun along her horizon, while the man on the right-hand side of the diagram sees the Sun over his head at local noon. Notice how much more atmosphere the woman must look through compared to the man's vantage point. This is because of the nearly spherical shape of our Earth and the thin atmospheric blanket that surrounds it. Obviously the two people in the diagram are not drawn in proportion to the size of Earth, but neither is the depiction of our atmosphere. In reality, our atmosphere is extremely thin relative to the size of the planet (see Fig. 7.5). As the Sun's light passes through the atmosphere, the shortest visible wavelengths are scattered out of the line of sight more effectively than are the longer wavelengths. That does not mean that the shorter wavelength light just disappears; instead it keeps bouncing around and some of it arrives at your eye from random directions in the sky. This is why, no matter which direction you look, during midday you see scattered blue light, and therefore a blue sky [Fig. 7.4 (*Lower Right*)]. Some of the blue/violet light will still be able to reach the observer directly, but the longer wavelength colors, including yellow and red, will be less affected, and so the Sun still looks yellowish. For the woman watching the sunset [Fig. 7.4 (*Lower Left*)], because of the much longer path of the light through the atmosphere, very little light remains in her line of sight except for the longest wavelength red light, with the result that the sky and the Sun become redder. Notice though that if she turns around and looks at the sky in the direction opposite to the Sun, she will still see those scattered blue wavelengths reaching her, and so the sky away from the Sun remains blue.

Something similar happens in space as well. Figure 7.6 shows a Hubble Space Telescope image of the Pleiades star cluster in the constellation of Taurus. This group of young stars formed together out of a cloud of gas and dust, termed a nebula, some of which is still in the region between and surrounding the stars. As

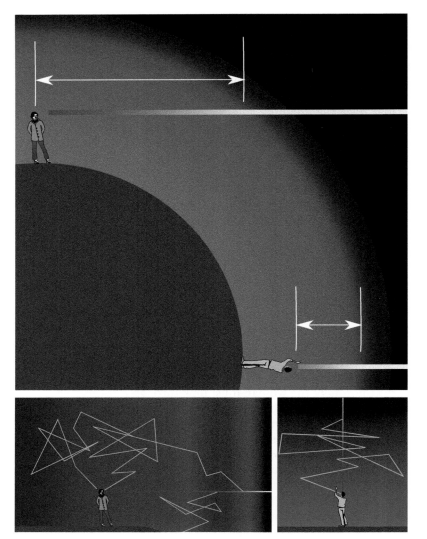

Fig. 7.4 *Top Left:* Two observers at different positions on Earth observing the Sun to the right. The woman at the top of the diagram is watching the Sun set while the man to the right sees the Sun overhead at local noon. The woman must look through more atmosphere to see the Sun compared to the man. *Lower Left:* The woman sees a red sunset because virtually all of the light, except the longer wavelength red light, is scattered out of her line of sight [compare to Fig. 7.3 (*Bottom*)]. *Lower Right:* The man sees a yellow Sun and blue sky all around him.

Fig. 7.5 The Himalayas seen from the International Space Station. Notice how thin our atmosphere is. (NASA)

the light from the stars passes through the gas and dust, the dust particles preferentially scatter blue light more than red light. Although the red light largely continues on its way, some of the blue light is scattered toward Earth and we see it as a blue reflection nebula.

Fig. 7.6 The Pleiades star cluster in Taurus exhibits a blue reflection nebula due to the scattering of short-wavelength visible light off dust surrounding the stars. (NASA, ESA, AURA/Caltech, Palomar Observatory)

Fraunhofer's Lines and Diffraction Gratings

Both parents of Joseph von Fraunhofer (Fig. 7.7) when he was eleven years old and as a result he had limited formal education. But in 1807, at age 20, he had the good fortune to be hired by an optics firm outside of Munich, Germany, that had been established three years earlier for the purpose of producing high-quality optical devices for military use and surveying. Seven years later he was co-owner of the company.

Fig. 7.7 Joseph von Fraunhofer (1787–1826). (Adapted from public domain)

In order to test the optical properties of the glass used in the company's devices, Fraunhofer began a careful study of the spectrum of sunlight. By passing the light through a prism and examining the colors using a small telescope, he discovered that superimposed on the bright colors were a large number of dark lines as illustrated in his original diagram in Fig. 7.8. He named the most prominent lines with capital letters from the red edge of the spectrum to the violet edge, while some of the less pronounced lines were given lower-case letters. In all he observed nearly 600 dark lines in the Sun's visible spectrum. The dark lines in the solar spectrum are still referred to today as Fraunhofer lines. A more modern representation of the lines on a colored spectrum is displayed in Fig. 7.9.

Based on the concept of Young's double-slit experiment, diffraction gratings were being developed that were composed of a large number of closely and evenly spaced slits. These diffraction gratings mimic the wavelength-spreading capabilities of prisms, but with much greater precision. Fraunhofer used them to determine the wavelengths of the most prominent dark lines in the solar spectrum.

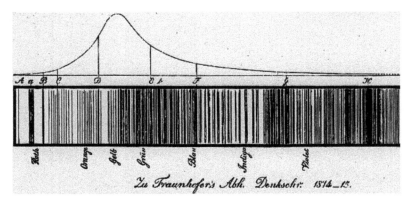

Fig. 7.8 Fraunhofer's 1814 sketch of the dark lines he observed in the Sun's visible spectrum. The curve above is his estimate of how bright the spectrum was as seen by the human eye. (Public domain)

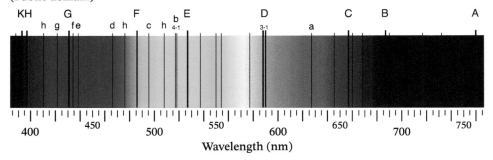

Fig. 7.9 The locations of some of the more prominent Fraunhofer lines in the Sun's visible spectrum. (Public domain)

You can think of a diffraction grating as a large number of Young double-slit experiments set side by side (recall Fig. 6.26). Diffraction gratings use densely and evenly spaced slits so that the interference between light rays of a specific wavelength passing through each of the slits produces constructive interference at a very localized position on the screen. This occurs because the distances traveled by every possible combination of two rays passing through any two of the slits must differ by an integer number of wavelengths in order for all of the peaks and troughs from all of the slits to arrive together on the screen. Remember that if you shift a wave forward or backward by one or two or three wavelengths, it will still look exactly the same, but if you shift it by any other amount the peaks and troughs won't align with their original positions.

Considering Fig. 7.10, think about what happens if the red light from the left-most slit travels a distance corresponding to one extra wavelength relative to the next slit to the right. With precise, even spacings between slits, the light from the left-most slit will travel a distance corresponding to two extra wavelengths relative to the third slit in line, three extra wavelengths for the fourth slit, and so on. This also means that the light from the second slit will travel an extra distance of one wavelength relative to the third slit, two extra wavelengths relative to the fourth slit, et cetera. Only a very specific wavelength can arrive at a location on the screen with all of the rays shifted by integer numbers of wavelengths.

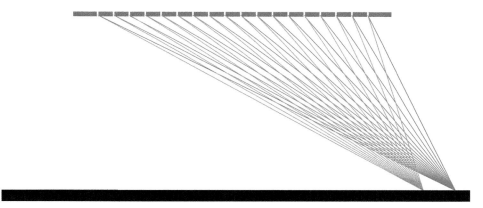

Fig. 7.10 A diffraction grating precisely separates the wavelengths of light.

A slightly different wavelength reaching that same spot will not have all of the peaks and troughs aligned properly. This is because the peaks and troughs arriving from two adjacent slits will be just slightly shifted, while the peaks and troughs arriving from two slits away will be shifted by a greater amount, from three slits away the shift will be even greater yet, and so on. If each of the waves is not shifted by exactly an integer numbers of wavelengths relative to every other wave, adding up all of the waves from tens or hundreds of thousands of slits or more will just end up as a large number of unaligned peaks and troughs, producing essentially complete cancellation (complete destructive interference), with no light of that wavelength being produced there. However, there will be another location on the screen where a different wavelength, such as the green one in Fig. 7.10, can have all of the rays from all of the slits travel an integer number of wavelengths relative to each other, so a bright spot will appear. The outcome is a spread of colors in a carefully engineered way. The specific location of a color, or lack thereof, can be measured on the screen, which then corresponds exactly to that color's wavelength.

Rather than transmitting light through slits, high-precision research-grade reflecting diffraction gratings are constructed by etching tens of thousands of thin grooves on a reflective material. If you have ever used a CD or DVD disc, you may have noticed that the grooves etched in the disc produce a visible spectrum of light (Fig. 7.11). Similarly, the light reflected from each groove of a research-grade reflecting diffraction grating converges to produce constructive interference for specific wavelengths at specific locations. In addition, rather than projecting the spectrum onto a screen, the spectrum is received by an electronic detector similar to the devices in digital cameras, including in your smartphone (a schematic diagram of a reflecting diffraction grating is shown in Fig. 7.12.

Fig. 7.11 The grooves on a CD or DVD act like a reflecting diffraction grating, separating the light into its constituent colors. (Luis Fernández Garcia, CC BY-SA 2.1 ES)

The study of spectra is known as spectroscopy and the optical instruments themselves are called spectrographs or spectrometers. Modern astronomical research relies heavily on spectrographs, and they are found in all professional astronomical observatories, both on the ground and in space. One of the reflecting diffraction gratings in the Cosmic Origins Spectrograph on board the Hubble Space Telescope contains 58,700 grooves per centimeter on its surface. It should be noted that astronomy is not the only discipline to employ spectroscopy; virtually all areas of science use this powerful investigative tool.

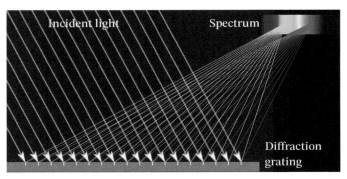

Fig. 7.12 A reflecting diffraction grating may be composed of tens of thousands of precision grooves per centimeter.

Fraunhofer didn't stop with his initial discovery of dark lines in the solar spectrum and the measurement of their wavelengths. Using a larger telescope, he observed the light coming from the Moon, Mars, and Venus, along with the bright stars Sirius, Castor, Pollux, Capella, Betelgeuse, and Procyon. Again he discovered dark lines in their spectra. For the spectrum of Venus he observed: "I have seen the lines D, E, b, F perfectly defined ... I have convinced myself that the light from Venus is in this respect the same nature as sunlight." However, when he turned his attention to the stars he noticed that there were some significant differences in their spectra when compared with the Sun and with each other.

Fraunhofer started us down the very productive road of spectroscopy, but he didn't have an explanation for what he was seeing in those dark lines. It is unfortunate for science that he died from tuberculosis at the age of 39.

Figure 7.13 shows a high-resolution visible spectrum of the Sun. Because the spectrum is so long, it is divided up and put into rows of increasing wavelength, with the shortest violet wavelengths in the upper left corner and the longest deep red wavelengths in the lower right corner.

Key Point
The Sun's spectrum generally follows that of a blackbody, but it is very jagged with dark and bright lines.

Since electronic detectors record the intensity of the light at every wavelength, it is more useful scientifically to plot the intensity of the light as a function of wavelength. Such a graph for the Sun is shown in Fig. 7.14. You should notice on the horizontal axis at the bottom of the graph that the wavelength range is from extremely short values (0 nm is on the left-most edge) to 2400 nm in the infrared portion of the electromagnetic spectrum. The small visible portion of the graph has been colored to help you compare the representation in the graph with the colorful high-resolution spectrum in Fig. 7.13 (the visible portion of the spectrum is enlarged in Fig. 7.15 so that you can see the complexity in more detail). A perfect, blackbody spectrum having the Sun's surface temperature is also shown on both graphs. Notice that the Sun's actual spectrum is very jagged but generally follows that of a perfect blackbody. The dips in the jagged spectrum over visible wavelengths correspond to the dark lines in Fig. 7.13. It is important to note that Figs. 7.14 and 7.15 clearly show that even under the deepest dips in the Sun's spectrum, there is still some light intensity, meaning that the dark lines you see in Fig. 7.13 aren't actually black, they're just not as bright as the surrounding wavelengths. If the dark lines were truly absent of light, the graph's dips would drop down to the horizontal axis where the intensity would be zero. Spikes upward in the spectrum are brighter than the surrounding spectrum.

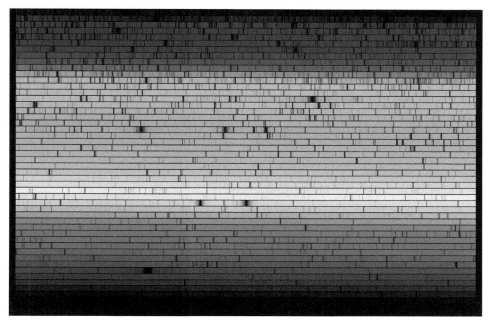

Fig. 7.13 A high-resolution spectrum of the Sun. The shortest wavelength is in the upper left-hand corner. Wavelength increases to the right by row with each successive row increasing in wavelength downward. The longest wavelength is in the lower right-hand corner. (N.A.Sharp, NOAO/NSO/Kitt Peak FTS/AURA/NSF, CC BY 4.0)

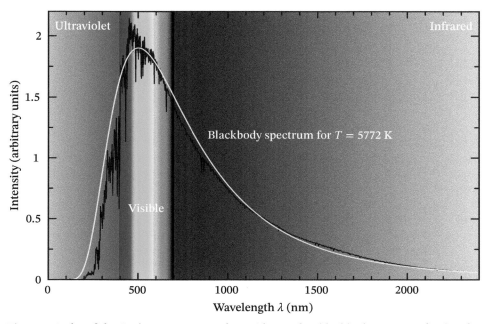

Fig. 7.14 A plot of the Sun's spectrum together with a perfect blackbody spectrum having the Sun's temperature of 5772 K. [Data courtesy of U.S. Department of Energy (DOE)/NREL/ALLIANCE]

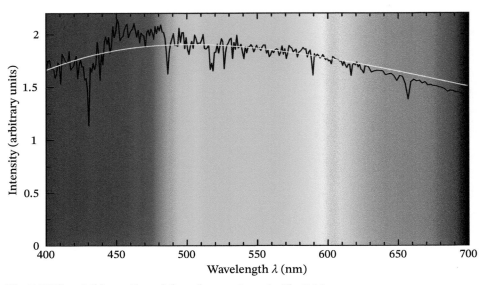

Fig. 7.15 The visible portion of the solar spectrum in Fig. 7.14.

Kirchhoff's Rules for Light Spectra

Fig. 7.16 A sketch of the first spectroscope developed by Bunsen and Kirchhoff. (Public domain)

With Fraunhofer's discovery of hundreds of dark lines across the bright, colorful spectrum of the Sun, scientists across Europe and United States began conducting research into how to produce those lines. Most notable among those were Gustav Kirchhoff (Fig. 6.44) and Robert Bunsen (1822–1899), after whom the old chemistry laboratory Bunsen burner is named. Kirchhoff and Bunsen are generally credited with developing the first research-worthy spectroscope, although the crude devices that Newton and Fraunhofer used could be classified as spectroscopes as well. As shown in Fig. 7.16, the spectroscope of Kirchhoff and Bunsen involved two small telescopes and multiple prisms to separate the colors of light.

By placing salts in a flame and observing the light that was produced, Kirchhoff and Bunsen discovered that unique bright lines could be seen, with the specific wavelengths depending on what type of salt was burned. These bright-line spectra are known as emission spectra. They also observed that if a bright, hot source of continuous light was placed behind the flame while the salts were burned, dark lines appeared on the continuous spectrum at the same wavelengths as the bright lines of the corresponding emission spectra. These dark-line spectra on a continuous background are known as absorption spectra. The corresponding emission and absorption spectra of hydrogen gas are shown in Fig. 7.17. The absorption spectra created in their laboratory were very reminiscent of Fraunhofer's solar spectra.

As a result of these studies, Kirchhoff articulated three rules for producing light spectra, illustrated in Fig. 7.18:

Key Point
Kirchhoff's rules for light spectra.

- A hot, high-density gas or a hot solid object emits a continuous spectrum (specifically a blackbody spectrum).
- A sufficiently warm, low-density gas produces bright lines at well-defined wavelengths (an emission spectrum).
- If a cooler, low-density gas is placed in front of a hotter continuous source of light, a dark line spectrum (an absorption spectrum) is produced.

Fig. 7.17 *Top:* The emission spectrum of hydrogen. *Bottom:* The absorption spectrum of hydrogen.

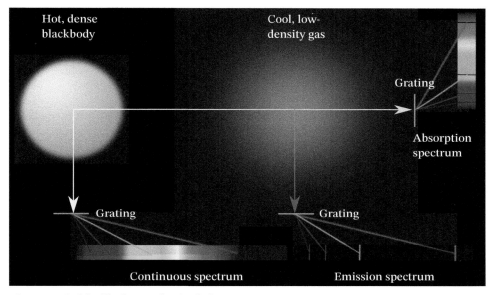

Fig. 7.18 Kirchhoff's three rules for light spectra.

7.2   The Periodic Table of the Elements

The Realization That Atoms Exist

Chemistry, the study of the properties of substances and how they react together to form other substances, has been around since prehistoric times. People have long known that the burning of wood and other materials changes the substance that was burned; the fermenting of wine and beer produces a sensation in the consumer; and plants have been (and often still are) used to produce medicines and perfumes (for examples, see Fig. 7.19). Techniques were developed to create pottery and glass, and combining copper and tin produced a harder metal that began the Bronze Age in the Ancient Near East during the 4th millennium BCE. But the underlying nature of chemical reactions largely remained a mystery for thousands of years.

Aristotelian science (page 63*ff*) argued that all materials were composed of just four terrestrial elements: earth, water, air, and fire, and the heavenly element, ether. It was further believed that the four terrestrial elements could be transformed into

Fig. 7.19 From top to bottom: Yellowstone fire of 1988 near Grant Village (National Park Service); a glass of stout (Jon Sullivan, public domain); Egyptian senna, a medicinal plant used as a laxative (Lalithamba from India, CC BY 2.0).

adjacent elements and that they could be divided infinitely many times. In other words, there was no smallest, indivisible quantity for any of the earthly elements. A competing view, apparently first proposed by Leucippus (5th century BCE) and later formulated by Democritus (c. 460–370 BCE), argued that every unique element is composed of corresponding small, indivisible, and imperishable units called atoms. As usual, the Aristotelian view held sway throughout most of history.

The lack of understanding of chemical processes led to success through trial and error, and in some cases had no hope whatsoever of reaching the goal that was set. Alchemy was one such case; a discipline that was even followed by Newton. The quest of the alchemist was to discover the so-called "philosopher's stone," believed to be capable of turning other metals, such as mercury, into gold. One of the tests employed by alchemists was to taste the product. Needless to say, given that mercury was one of the quantities involved, such an experimental approach was not particularly healthy; some of the effects of mercury poisoning include irritability and mood swings, insomnia, and headache. It has been suggested that Newton's aggressive attacks on some of his critics could have been attributable to mercury poisoning. In fact, in 1979 a sample of Newton's hair was analyzed and found to contain very high levels of mercury.

Throughout the 17th, 18th, and 19th centuries, considerable progress was made in chemistry regarding experimental techniques and reproducibility of results, along with the eventual abandonment of the search for the philosopher's stone. Scientists learned, for example, that our atmosphere is composed of several substances, including nitrogen, oxygen, carbon dioxide, argon, and other gases. Burning hydrogen in oxygen was discovered to produce water, and hydrogen and oxygen could be liberated from water in the presence of an electric current. Safe chemical analysis was slow to catch on, however, and one English experimenter, Sir Humphry Davy (1778–1829), who helped pioneer the process of separating substances through the use of electricity, nearly died on several occasions by inhaling the gases produced by his experiments. Davy did discover that one of his gases, nitrous oxide, had a curious effect on human behavior, leading him to nickname it "laughing gas."

Rather than the four terrestrial "elements" of the ancient Greeks, science was revealing that compounds could ultimately be separated into a small, but growing number of materials that had unique sets of chemical characteristics. These true elements can be extracted from more complex materials, but no new chemically unique materials can be extracted from pure samples of elements. Elements constitute the most fundamental building blocks of everything else in the chemist's laboratory.

In 1803 John Dalton (Fig. 7.20) reintroduced the theory that every element is composed of a collection of identical atoms that are unique to that element. Atoms are the smallest unit of matter that can have the chemical characteristics of the associated element. Furthermore, molecules such as carbon dioxide, water, and nitrous oxide are combinations of particular groupings of different types of atoms in *integer ratios*. Each carbon dioxide molecule is composed of one carbon atom and two oxygen atoms (CO_2), water is composed of two hydrogen atoms and one oxygen atom (H_2O), and nitrous oxide is composed of two nitrogen atoms and one oxygen atom (N_2O). Atmospheric nitrogen is actually a molecule composed of two nitrogen atoms (N_2) and atmospheric oxygen is also in a molecular form of two oxygen atoms (O_2). Ozone is a molecule with three oxygen atoms (O_3). See Fig. 7.21

Fig. 7.20 John Dalton (1766–1844), [Charles Turner (1773–1857) after James Lonsdale (1777–1839), public domain]

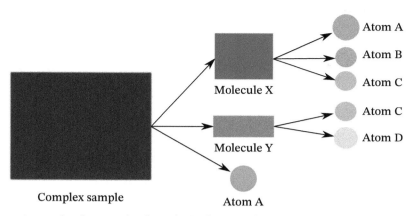

Fig. 7.21 A sample of a complex hypothetical material containing two types of molecules, call them X and Y, and atoms of element A that are not bound up in molecular form. The molecules may be further broken down into their constituent atoms. In this example, molecule X also contains atoms of element A, but it also contains atoms of elements B and C, while molecule Y contains atoms of elements C and D.

for a schematic diagram of the distinctions between complex materials, molecules, and elements.

But just how small are individual atoms? At about the same time that Maxwell (page 209) was formulating his four equations of electromagnetic theory, experimental evidence revealed that one gram of pure hydrogen gas contained approximately 6×10^{23} atoms, which is six hundred billion trillion! Today that number is known as Avogadro's number and, as with all such constants, it has a very precise modern definition. The mass of Avogadro's number of carbon atoms is 12 grams, for nitrogen atoms it is about 14 grams, and for oxygen atoms it is about 16 grams. It will be useful to remember that there are roughly 10^{24} atoms in a gram of material.

Mendeleev's Periodicity of Chemical Properties

Throughout the 18th and 19th centuries, researchers began to accumulate information on the chemical properties of the growing number of elements that were being isolated from more complex materials. By arranging the known elements according to the masses of their atoms, Dmitri Mendeleev (Fig. 7.22) noticed a developing pattern in their chemical behaviors. For example, the elements hydrogen (H), lithium (Li), sodium (Na), and potassium (K) all form molecules with chlorine (Cl), producing HCl (a gas), LiCl (a salt), NaCl (everyday table salt), and KCl (a salt), respectively. By grouping elements according to similar chemical properties he made a critical breakthrough in the field of chemistry. His first periodic table of the elements was developed in 1869. A modern version of Mendeleev's periodic table that includes the origins of the elements is displayed in Table 7.1. The numbers at the top of each element's box are that element's atomic number. As the atomic numbers increase, so do the masses of the atoms of each element. Elements with common chemical properties are arranged in columns with a new row starting every time an element has chemical properties similar to hydrogen.

Key Points
• An element is a fundamental chemical building block for more complex materials.
• An atom is the smallest unit of matter that possesses an element's unique chemical properties.
• A molecule is the smallest collection of atoms of a more complex substance that possesses its chemical characteristics.

Fig. 7.22 Dmitri Mendeleev (1834–1907). (Public domain)

Table 7.1 The periodic table of the elements. The color coding indicates how the elements were created, a topic to be discussed in later chapters. The very unstable, massive elements are not listed, but are all man-made. (Jennifer Johnson; CC BY-SA 4.0)

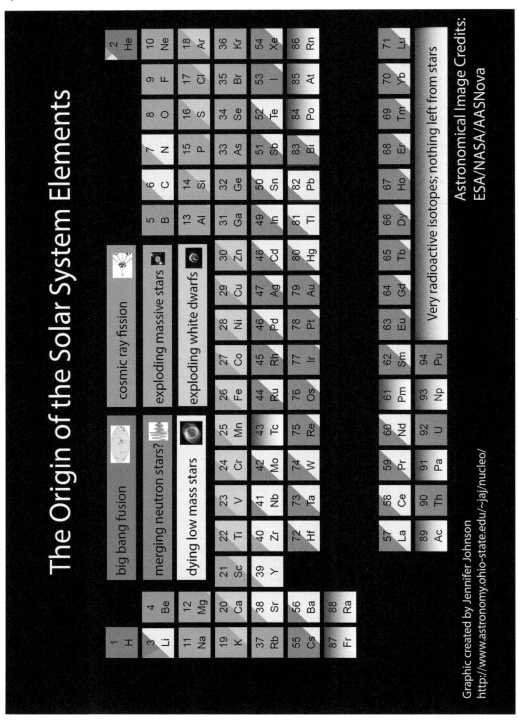

Mendeleev's periodic table not only serves as an organizational method for arranging the known chemical elements and their properties, it also predicted the existence and properties of three as-yet undiscovered elements: scandium (Sc), gallium (Ga), and germanium (Ge). The periodic table was, and remains, an important tool in guiding the development of our understanding of the atoms that all substances are composed of. Patterns in nature always call for explanations as to why the periodicity exists. That is certainly the case for the periodic table of the elements.

Spectroscopy As a Tool in Chemistry

The spectroscopic work of Kirchhoff, Bunsen, and others was leading to another realization about atoms and molecules: every pure element and every molecule produces a characteristic spectrum, essentially a unique fingerprint (see Fig. 7.23). By studying the spectrum of light produced when a salt is placed in flame, or when a gas is heated, the molecule or element can be identified. Spectroscopy became another tool in the chemist's toolbox that could be used to analyze the composition of an unknown sample. The presence of the specific set of wavelengths for sodium lines means that sodium is present; if chlorine lines are present then chlorine is present in the sample as well. Table 7.2 lists the visible wavelengths of some of the elements in the periodic table: hydrogen (1), helium (2), carbon (6), nitrogen (7), oxygen (8), and sodium (11); the numbers in parentheses are the elements' atomic numbers in the periodic table.

Fig. 7.23 The atomic spectra from top to bottom of hydrogen, helium, carbon, and iron, ranging from 400 nm (left) to 700 nm (right). (Public domain)

7.3   Detecting Helium and Other Elements in the Sun

Kirchhoff and Bunsen's Chemical Analysis of the Sun's Spectrum

The French philosopher and arguably the first modern philosopher of science, Auguste Comte (Fig. 7.24) wrote in 1835:[1]

> It is easy to describe clearly the character of astronomical sense ... only one [of our senses] perceives the stars. The blind could know nothing of them; and we who see, after all our preparation, know nothing of stars hidden by distance ... Of all objects, the planets are those which appear to us under the least varied aspect. We see how we may determine their forms, their distances, their bulk, and their motions, but we can never know anything of their chemical or mineralogical structure; and, much less, that of organized beings living on their surface. We may obtain positive knowledge of their geometrical and mechanical phenomena; but all physical, chemical, physiological, and social researches, for which our powers fit us on our own earth, are out of the question in regards to the planets. Whatever knowledge is obtainable by means of the sense of Sight, we may hope to attain with regard to the stars, whether we at present see the method or not; and whatever knowledge requires the aid of other senses, we must at once exclude from our expectations, in spite of any appearances to the contrary ... We can never learn their internal constitution, nor, in regard to some of them, how heat is absorbed by their atmosphere. We may therefore

Fig. 7.24 Auguste Comte (1798–1857). (Public domain)

[1]Comte, Auguste (1835). *Positive Philosophy.*

Table 7.2 Selected visible wavelengths of various elements (nm)

Hydrogen	Helium	Carbon	Nitrogen	Oxygen	Sodium
410	401	427	404	407	411
434	403	514	424	408	432
486	412	515	445	412	439
656	414	538	460	419	441
	439	589	461	432	446
	444	601	462	435	449
	447	658	463	441	450
	469	658	464	442	466
	471		480	459	467
	492		496	460	498
	502		499	464	515
	505		500	465	541
	541		501	466	568
	588		505	468	569
	656		567	471	589
	668		569	616	590
	687		571	646	615
			575		616
			593		651
			594		653
			648		654
			648		655
			661		

define Astronomy as the science by which we discover the laws of the geometrical and mechanical phenomena presented by the heavenly bodies.

If Comte had been correct about his less-than-positive description of astronomy, at the very least your textbook would be much shorter! You would still learn about astronomical mythology, the seasons of the year, the effect of latitude on what we see in the sky, and a bit about eclipses and orbits, but not much else. Observationally, we would be able to see the planets and their moons, and those dots and fuzzy patches in the sky, but we would know precious little about them.

Ironically, it was in 1802 when William Wollaston (1766–1828), an English doctor turned chemist and mineralogist, first noted dark lines in the spectrum of the Sun produced by a prism. Twelve years later Fraunhofer carefully sketched those dark lines, shown in Fig. 7.8. Little did Comte know when he wrote his words twenty years after Fraunhofer produced his sketch how wrong he would be proven to be.

In 1860, just twenty-five years after Comte's book was published, Kirchhoff and Bunsen wrote:[2]

Key Point
Spectra are critical to determining the compositions of planetary atmospheres, comets, stars, galaxies, and the gas and dust in interstellar space.

[2]Kirchoff, Gustav and Bunsen, Robert (1860). "Chemical Analysis by Observation of Spectra," *Annalen der Physik und der Chemie* (Poggendorff), 110:161–189, Heidelberg.

> It can be concluded that the spectrum of the sun with its dark lines is just areversal of the spectrum which the atmosphere of the sun would show by itself. Therefore, the chemical analysisof the sun's atmosphere requires only the search for those substances that produce the bright lines that coincidewith the dark lines of the solar spectrum.

Kirchhoff and Bunsen were pointing out that the absorption lines produced by the Sun's cooler atmosphere on top of a much hotter continuous source of blackbody radiation is simply the reverse of the emission spectrum that the Sun's atmosphere would produce if it were seen by itself.

It now became a matter of comparing the spectra of the elements obtained in terrestrial laboratories with the spectra of the Sun and the stars to determine their compositions. For example, by comparing the spectrum of hydrogen (Fig. 7.17) with Fig. 7.9, we can identify Fraunhofer's C, F, G, and h lines as belonging to hydrogen. Today those lines are known as hydrogen's Hα, Hβ, Hγ, and Hδ lines (α, β, γ, and δ being the first four letters of the Greek alphabet; alpha, beta, gamma, and delta, respectively).

A similar procedure could be carried out for the other elements as well. Some of the other stronger lines in the solar Fraunhofer spectrum are associated with the elements calcium (Ca), sodium (Na), magnesium (Mg), and iron (Fe). A host of other elements can also be identified in the myriad of lines in the solar spectrum, including, but certainly not limited to, carbon (C), nitrogen (N), and oxygen (O).

In 1868, a total solar eclipse occurred that was visible from India and the Malay peninsula; recall Fig. 4.23 and our previous discussion of eclipses. With the new-found power of spectroscopy, scientific expeditions were dispatched to observe the spectra produced by the ghostly glowing gas surrounding the Sun's bright disk.

The spectra revealed the now-expected lines, only this time they appeared as bright emission lines rather than dark absorption lines because the dense, hot blackbody source of the solar disk was blocked by the Moon. However, other lines that had never been seen in terrestrial laboratory experiments were observed as well. Norman Lockyer (Fig. 7.25) and his colleague, Edward Frankland (1825–1899), suspected that perhaps the new lines were produced by a previously unknown element. The element was named helium in honor of Helios, the Greek Titan god of the Sun. It wasn't until 1895 that helium was finally discovered on Earth by liberating the gas from the mineral lead uranate (a compound containing uranium). Comte was not only wrong about ever being able to know the composition of the Sun, but by studying the Sun's spectrum, a new element was discovered that hadn't even been detected on Earth yet.

Fig. 7.25 Sir Norman Lockyer (1836–1920). (Stereoscopic Co., Public domain)

7.4   Particles Smaller than Atoms

The Discovery of X-Rays

In 1895, Wilhelm Röntgen (Fig. 7.26) was studying the mysterious glow on a screen that was generated in an experiment when a beam of negatively charged particles called cathode rays struck a metal surface (Fig. 7.27).[3] While trying to block the

Fig. 7.26 Wilhelm Röntgen (1845–1923). (1900, Public domain)

[3]Cathode ray tubes were used in old-style television sets (before flat screens) and in old computer monitors. The cathode rays would strike a fluorescent screen, causing it to glow. The screen was "painted" by adjusting the beam's direction using electric fields. This glow was not due to x-rays however; the energies involved were much too low!

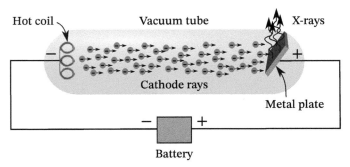

Fig. 7.27 X-rays produced by cathode rays striking a metal. The cathode rays are emitted by a hot, negatively charged coil and accelerated toward a positively charged metal plate with high voltage.

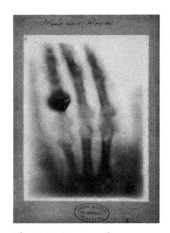

Fig. 7.28 An x-ray image of Anna Bertha Ludwig's hand, wife of Wilhelm Röntgen. Her wedding ring is clearly visible. December 22, 1895. (Public domain)

rays that were causing the glow on the screen, he put sheets of dark paper in their way to no effect. He then tried thin sheets of copper and aluminum but again they didn't block the rays. When he tried using a sheet of lead he was finally successful. But when he moved the sheet out of the way he was shocked to see an image of the bones in his hand showing up on the screen within the mysterious glow. Not knowing what the rays were, he called them x-rays, similar to using x as a symbol for an unknown quantity in an equation. Six weeks after his discovery he recorded the first medical x-rays image of his wife's hand in a photograph (Fig. 7.28). Röntgen received the very first Nobel prize in physics in 1901 for his discovery of x-rays.

The Discovery of Radioactivity

In 1896, one year after the discovery of x-rays, Henri Becquerel (Fig. 7.29), was experimenting with phosphorescent substances containing uranium. He had wrapped a photographic plate in thick black paper so that normal sunlight could not expose the plate and then set the uranium salt on top of the plate. When he unwrapped the plate he found that the shape of the substance formed an image on the plate. He also found that other phosphorescent materials did not expose photographic plates. From his experiments Becquerel concluded that some sort of emission must be spontaneously emanating from uranium, without any external need to activate it.

Fig. 7.29 Henri Becquerel (1852–1908). (ca. 1905, public domain)

Two years after Becquerel's discovery, Marie Curie (Fig. 7.30) and her husband, Pierre Curie, published a paper announcing the discovery of two more previously unknown substances that also emitted radiation spontaneously. The Curies named them polonium, in honor of Marie's native Poland, and radium, based on the Latin word for "ray." Both Curies, together with Becquerel, were awarded the Nobel prize in physics in 1903 for their discoveries. The Curies were the ones who coined the term radioactivity to describe the spontaneous emission of radiation from certain pure elements. In 1906, shortly after the Chair of the Department of Physics at the University of Paris was awarded to him, Pierre Curie was killed in a road accident, having been run over by a horse-drawn vehicle. The University then appointed Marie to the post and she became the first woman to hold a professorship at the University of Paris. In 1911, Marie Curie received a second Nobel prize, this time in chemistry, for chemically isolating pure radium. This meant that Marie Curie was not only the first woman to earn a Nobel prize, but she was also the first person to earn two Nobel prizes. The Curies' daughter, Irène Joliot-Curie (1897–1956), and her husband, Frédéric Joliot-Curie (1900–1958), also shared a Nobel prize in chemistry in 1935 for their discovery of artificial radioactivity.

Fig. 7.30 Marie Sklodowska Curie (1867–1934) and Pierre Curie (1859–1906). (1903, public domain)

During World War I, Marie Curie realized that radiological services were needed for the troops at the front lines. After learning a bit about automotive mechanics, anatomy, and the very young technology of radiology, she obtained x-ray equipment, generators, and trucks, and created mobile radiological centers. Additionally, she helped to set up, with the aide of her then 17-year-old daughter Iréne, about 200 radiological centers at field hospitals. Curie also captured the radioactive gas given off by her one-gram supply of radium, placing it into needles that were used to sterilize the infected wounds of soldiers. She didn't know it at the time, but the colorless gas was yet another new radioactive element, now called radon. It is believed that perhaps one million soldiers were aided through her efforts.

Marie Curie died of aplastic anemia, a disease associated with damaged bone marrow. It may be that she contracted the disease as a result of long-term exposure to radiation.

While working at Cambridge University and then in Canada at McGill University, Ernest Rutherford (Fig. 7.31) studied the nature of the radiations that were coming from uranium, polonium, and radium. In 1899, based on their penetrating power, he named the two types of particle being emitted by those elements alpha particles (α, least penetrating) and beta particles (β, more penetrating). By analyzing their movements in electric fields, he also discovered that alpha particles are positively charged and beta particles are negatively charged. In 1903 he found that a third type of radiation was also being produced in radium that had tremendous penetrating power, which he called gamma radiation (γ). He also discovered that his gamma radiation is unaffected by electric fields and therefore does not have an electric charge. In 1907, Rutherford and an assistant were able to trap the alpha particles in a glass container and then heat the resulting gas, producing an emission spectrum. What they observed was a spectrum identical to the spectrum of of the element helium, discovered in the Sun's spectrum almost forty years earlier. Atoms that possess a net amount of charge, either positive or negative, are called ions. Alpha particles turned out to be helium ions.

Fig. 7.31 Ernest Rutherford (1871–1937). (Public domain)

The Discovery of the Electron

In 1897, J. J. Thomson (Fig. 7.32), initially working with Rutherford, was trying to understand the nature of the cathode rays that were producing Röntgen's x-rays when they struck a metal. It had been known for some time that they were negatively charged, based on how they moved away from the negatively charged coil of the cathode ray tube and were attracted to the positively charged metal plate at the other end. In addition, by applying an electric field perpendicular to the direction that the cathode rays were traveling, it was possible to bend their path. How

Fig. 7.32 Sir Joseph John Thomson (1856–1940). (Public domain)

much the particles bend in an electric field depends on the strength of the field, the amount of charge on the particles, and their mass (more mass means that it is harder to deflect the particles according to Newton's second law, $\overrightarrow{F} = m\overrightarrow{a}$).

The paths of charged particles are also bent by magnetic fields. Not only does the strength of the magnetic field impact the amount of bending, but so does the amount of charge, the mass, and the velocity of the particles. By adjusting the electric and magnetic fields, it was possible for Thomson to determine not only that the cathode ray particles were negatively charged, but that they had very little mass relative to the amount of charge that they possessed. Based on his studies, Thomson concluded that these electrons must be very tiny.[4] He proposed that atoms must contain these tiny particles and that the rest of the atom was something like a positively charged pudding in which the electrons were distributed. Unfortunately, Thomson couldn't determine either the charge or the mass of the electron separately, but only the ratio of the two values. Not only were cathode rays determined to be electrons, but Rutherford's beta particles are also electrons.

It took another twelve years before Robert Millikan (Fig. 7.33) developed a clever way to measure the charge. It was done by spraying a very light mist of oil into a vertical electric field produced by a positively charged conducting plate on top and a negatively charged plate on the bottom, as illustrated in Fig. 7.34. Without the electric field turned on, the very tiny oil drops would just fall due to gravity, but by using an electric field with just the right strength, he could get some of the statically charged oil drops to hang in mid air, balanced by the electric field force pulling them up and gravity pulling them down. Based on the size of each oil drop and the density of oil, Millikan was able to estimate the mass of each oil drop and then calculate the force of gravity acting on them (their weights). The strength of the electric force pulling the oil drops up must equal the gravitational force pulling them down. Since the force due to the electric field depends only on the strength of the field and the charge on the oil drop, he could calculate the electric charge on each suspended oil drop.

What Millikan discovered was that every oil drop had an amount of negative charge that was always equal to an integer times the same basic value. Imagine that you were given the following list of numbers: -3, -9, -15, -12, -9, -15, -6, -3, and -18. You probably recognize that each number in the list is a negative integer multiple of 3: -1×3, -3×3, -5×3, -4×3, -3×3, -5×3, -2×3, -1×3, and -6×3. Millikan's data were similar, but rather than being negative integer multiples of 3, they were negative integer multiples of a constant, symbolized by e, that is referred to as the elementary charge. Every electron has a charge of $-e$. The elementary charge, e, is one of the fundamental constants of the universe, just as Newton's Big G, the speed of light, c, and Boltzmann's constant, k_B, are fundamental constants.

Because Thomson had already determined the ratio of the charge of an electron to its mass, once Millikan determined the electron's charge, its mass was also immediately known. As Thomson suspected, the electron's mass is extremely small. It would take about one million trillion trillion electrons to make one kilogram of material. The mass of the electron, m_e, is another of nature's fundamental constants.

Fig. 7.33 Robert Millikan (1868–1953). (Public domain)

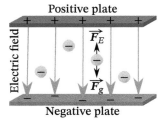

Positive plate

Electric field

$\overrightarrow{F}_E$

$\overrightarrow{F}_g$

Negative plate

Fig. 7.34 Millikan's oil drop experiment for measuring the elementary charge e. The negatively charged oil drops are attracted to the positively charged upper plate. The force of gravity is balanced by the electric force if the oil drop is suspended between the plates.

---

[4]The name "electron" was given to these as-yet undiscovered particles by the Irish physicist, George Johnstone Stoney (1829–1911) in 1891, six years before Thomson confirmed their existence. The term "electron" is a combination of "electric ion."

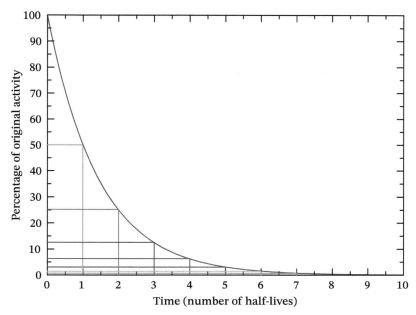

Fig. 7.35 The relationship between the amount of radioactivity of a sample and the number of half-lives since the start of the measurements. The colored straight lines are examples: After one half-life, 50% (1/2) of the original activity remains; after two half-lives, 25% (1/4) of the original activity remains; after three half-lives, 12.5% (1/8) of the original activity remains, and so on.

Radioactive Decay and Radiometric Dating

When Rutherford discovered that helium was a byproduct of the emissions of uranium, and Marie Curie used the colorless radioactive gas radon, a product of the radioactivity of radium, in medical treatments, they found that one type of atom can spontaneously decay into one or more other types of atom. Rutherford investigated the rate at which this process occurred and discovered that the rate of emission of a given sample diminished in a characteristic way. This happens because when uranium emits an alpha particle or when a radium atom changes into a radon atom, the original atom is no longer available to change, and so the rate of emission of the entire sample decreases as a result. Rutherford was also able to quantify the way in which the rate decreased, and found that the rate of emission in the sample decreased by a factor of two over a well-defined period of time. It didn't matter how big the original sample was, what the temperature of the sample was, or any other environmental factors; the rate of emission is always cut in half relative to its original rate over the same time period. The time required for the radioactive emission of a pure sample to decrease by a factor of two is known as its half-life.

The half-life is based on the *probability* that any random atom in a sample will decay into another type of atom during a given period of time. There is a 50% chance of the decay occurring in one half-life. There is then another 50% chance of decaying over another period of one half-life, and another 50% chance of decaying over the next period of one half-life, and so on. The graph in Fig. 7.35 shows that, starting with 100% activity, after one half-life the activity drops to 50%, after two half-lives the activity is one-half of 50% or 25%, after three half-lives the activity

is one-half of 25% or 12.5%, after four half-lives the activity is one-half of 12.5% or 6.25%, and so on. This behavior can be described by the equation

An exponents tutorial is
available at
Math_Review/Exponents

$$\text{Fractional amount remaining} = \left(\frac{1}{2}\right)^{t},\qquad(7.1)$$

Half-life equation

where t is the number of half-lives since measurement of the original sample began. Note that t can be any positive number; it need not be an integer. For example, if the half-life is 2 years and you wanted to know the fractional amount remaining after only $1/2$ year, which is one-quarter of one half-life, then $t = 1/4 = 0.25$.

Students are sometimes tempted to say, "But if there is a 50% chance of decaying during the first half-life and that didn't happen, then surely the atom must decay during the second half-life since two 50% chances give 100%." The problem with this argument is that an atom doesn't have a memory! It doesn't know that it didn't decay during the first half-life and therefore must decay during the next half-life. Think about what happens when you flip a coin. Excluding the very remote possibility that the coin could land on an edge and stay that way, there is a 50% chance of it landing heads or tails. If it lands heads on the first flip, the next time you flip the coin, it isn't guaranteed to land tails simply because it landed heads the first time; it doesn't remember what happened on the first flip. Even if the coin landed heads ten times in a row, or even one thousand times in a row, on the next flip there is still a 50% chance of landing heads and a 50% chance of landing tails.

Gamblers often fall into the trap of believing that the odds of a roll of the dice must be in their favor because the roll hasn't gone their way many times before, or conversely the roll will go their way because it *has* been going their way for a while; the gambler is "on a roll." In reality, the odds are exactly the same every time, regardless of past results. This flawed gambler's thought process is known as the "gambler's fallacy." If you gamble, don't let yourself get caught in this psychological trap.

Because the half-life of a particular type of atom is a basic characteristic of the atom, radioactive decay becomes a very powerful tool for determining the age of a sample. That sample could be an archaeological find such as a basket, a rock from a geological layer near the surface of Earth, or a Moon rock or a meteorite. Obtaining ages from radioactive decay has even been used as a way to determine how old some stars are. The method of determining age based on radioactive decay is known either as radioactive dating or radiometric dating.

Example 7.1

A rare type of carbon atom, called carbon-14, has a half-life of approximately 5730 years, making it a good candidate for determining the ages of samples of human origin. Suppose that a reed basket was unearthed at an archaeological site and then taken to a laboratory for analysis. It was discovered that the basket contained one-quarter of the amount of carbon-14 that it would have had when the basket was made. How old would the basket be today?

In order for the sample to have one quarter of the original amount of carbon-14, Fig. 7.35 and the half-life equation tell us that two half-lives must have occurred since the basket was woven:

$$\text{Fractional amount remaining} = \left(\frac{1}{2}\right)^2 = \frac{1}{2} \times \frac{1}{2} = \frac{1}{4}.$$

This means that the basket must be

$$\text{Age} = 2 \times \text{half-life} = 2 \times 5730 \text{ years} = 11{,}460 \text{ years.}$$

Atoms with longer half-lives can be used to date rock samples. For example, one form of uranium, uranium-235, has a half-life of 700 million years.

Example 7.2

Suppose that a hypothetical Moon rock was found to have 1.56% of its original uranium-235 present when the Apollo astronauts brought back the sample. How long ago did the rock form?

1.56% is the same as 0.0156. But $1/0.0156 = 64.1$, meaning that 0.0156 is nearly 1/64. By successive calculations of the half-life equation you can quickly show that

$$\left(\frac{1}{2}\right)^6 = \frac{1}{2} \times \frac{1}{2} \times \frac{1}{2} \times \frac{1}{2} \times \frac{1}{2} \times \frac{1}{2} = \frac{1}{64} = 0.015625.$$

This implies that the rock must be about 6 half-lives old, or $6 \times 700{,}000{,}000$ years = 4,200,000,000 or 4.2 billion years old.

Through essentially the same process, actual Moon rocks have been shown to be as old as 4.4 billion years.

All of this research into x-rays, radioactivity, and the nature of the electron demonstrated that there exist subatomic particles smaller than atoms, that are contained within the atoms themselves. Apparently atoms aren't the indestructible, smallest units of matter that scientists had previously believed.

Nature is presenting us with an intriguing puzzle, and you should always be asking the question, just like every two-year-old child: *Why?* Why is the shape of Kirchhoff's blackbody radiation curve the way it is? Why do dark lines appear in the spectrum of the Sun, and as Fraunhofer demonstrated, in other stars as well? Why is the spectrum of every element and molecule different? Why is the spectrum of hydrogen particularly simple? Why is there an elementary charge? Why are some types of atoms radioactive? You should also be asking yourself another question: What fundamental secrets of nature can we learn from these experimental and observational data?

Key Point
Why?

Exercises

All of the exercises are designed to further develop *your understanding* of the material by thinking carefully and critically about what you have read in this chapter. Answers to selected exercises are available in the back of the book.

True/False

1. The speed of light is always c, regardless of what medium light is passing through.
2. (*Answer*) Blue light bends more at the interface when moving from air into glass than does red light.
3. A rainbow is produced when light passes through raindrops because the speed of light in water is different for different wavelengths of light.
4. (*Answer*) It is impossible to know anything about the compositions of stars simply because they are too far away for us to ever sample them.
5. A research-grade reflecting diffraction grating can contain tens of thousands of grooves per centimeter.
6. (*Answer*) The spectrum of the Sun has the general shape of a perfect blackbody, but with superimposed emission and absorption lines.
7. It is because cesium (Cs) has a spectrum that looks almost identical to the spectrum of hydrogen (H) that cesium is located below hydrogen in the first column of the periodic table of the elements.
8. (*Answer*) Carbon (C) and silicon (Si) have similar chemical properties.
9. Helium was first discovered in the kingdom of Helios.
10. (*Answer*) Hα is an element first discovered in the solar spectrum.
11. If a radioactive material decays for three half-lives, the amount of activity will have decreased by a factor of three from its initial value.

Multiple Choice

12. (*Answer*) Which of the following cause rainbows?
    (a) Blue light scatters off raindrops less efficiently than red light does.
    (b) Blue light scatters off raindrops more efficiently than red light does.
    (c) Blue light travels faster through water than red light does.
    (d) Blue light travels slower through water than red light does.
    (e) The speed of light changes with elevation in the atmosphere.
    (f) Leprechauns
    (g) both (a) and (c)
    (h) none of the above
13. Which of the following is *not* ultimately due to the fact that blue light is scattered more readily by small particulates, such as dust particles?
    (a) blue sky
    (b) rainbows
    (c) red sunsets
    (d) the blue color of lakes
    (e) the blue hue of Earth's atmosphere as seen from space (e.g., Fig. 7.5)

(f) blue reflection nebulas near young star clusters like the Pleiades (Fig. 7.6)
(g) All of the above are the result of the preferential scattering of shorter wavelength blue light.
(h) None of the phenomena listed are due to the preferential scattering of shorter wavelength blue light.

14. (*Answer*) The doublet of Fraunhofer absorption lines labeled as D lines in Fig. 7.9 are due to the same element. Referring to Table 7.2, which element produces the Fraunhofer D lines in the spectrum of the Sun?
    (a) hydrogen (H)   (b) helium (He)   (c) carbon (C)
    (d) nitrogen (N)   (e) oxygen (O)   (f) sodium (Na)
15. Referring to Fig. 7.15, the deep absorption line in the violet portion of the solar spectrum, the narrow absorption line in the turquoise part of the spectrum shorter than 500 nm, and the deepest absorption line in the red portion of the spectrum are all due to the same element. Referring to Table 7.2, which element produces those lines in the spectrum of the Sun?
    (a) hydrogen (H)   (b) helium (He)   (c) carbon (C)
    (d) nitrogen (N)   (e) oxygen (O)   (f) sodium (Na)
16. (*Answer*) A hot, high-density gas produces
    (a) a continuous blackbody spectrum.
    (b) a bright line emission spectrum.
    (c) a dark line absorption spectrum.
17. A sufficiently warm, thin, low-density gas produces
    (a) a continuous blackbody spectrum.
    (b) a bright line emission spectrum.
    (c) a dark line absorption spectrum.
18. (*Answer*) Mendeleev's periodic table of the elements is organized by
    (a) common chemical characteristics of elements.
    (b) the masses of the atoms of each element.
    (c) the colors of the elements.
    (d) the progressively more complex spectra of elements in the visible portion of the electromagnetic spectrum.
    (e) all of the above
    (f) (a) and (b) only
    (g) (a), (b), and (c)
    (h) none of the above
19. Which of the following elements have chemical characteristics similar to that of fluorine (F)?
    (a) nitrogen (N)   (b) oxygen (O)   (c) neon (Ne)
    (d) chlorine (Cl)   (e) bromine (Br)   (f) iodine (I)
    (g) (a), (b), and (c)   (h) (d), (e), and (f)
    (i) none of the above
20. (*Answer*) Which of the following symbolize molecules rather than atoms?
    (a) CO_2 (b) H_2O (c) CO
    (d) Hg   (e) Fe   (f) Na
    (g) (a), (b), and (c)   (h) (d), (e), and (f)
    (i) none of the above

21. Cathode rays were shown to be _____ charged particles known as _____.
    (a) negatively, electrons   (b) positively, electrons
    (c) uncharged, electrons   (d) negatively, x-rays
    (e) positively, x-rays   (f) uncharged, x-rays

22. (*Answer*) Marie and Pierre Curie shared a Nobel prize with Henri Becquerel for their discoveries of
    (a) x-rays.   (b) electrons.   (c) α particles.
    (d) γ rays.   (e) radioactive materials.

23. Rutherford demonstrated that trapped _____ emit the spectral signature of _____.
    (a) x-rays, hydrogen   (b) α particles, helium
    (c) β particles, lithium   (d) γ rays, beryllium
    (e) electrons, boron

24. (*Answer*) JJ Thomson was able to determine the ratio of _____ to _____ of an electron, but not either quantity separately.
    (a) charge, color   (b) charge, mass
    (c) charge, size   (d) mass, color
    (e) mass, size   (f) size, color

25. Millikan balanced which two forces acting on a charged oil drop in order to determine the elementary charge, e?
    (a) economic, gravitational
    (b) magnetic, gravitational
    (c) electric, gravitational
    (d) electric, magnetic
    (e) dark, electric

26. (*Answer*) Soot discovered in an ancient cooking pit is found to have one-eighth of the radioactivity due to carbon-14 that it would have had immediately after the wood was burned. How old is the soot in the pit?
    (a) 5730 years   (b) 11,460 years
    (c) 17,190 years   (d) 22,920 years
    (e) 45,840 years

27. A particular radioactive material has been decaying for 10 half-lives. Approximately how much of the original radioactivity remains in the sample?
    (a) 100%   (b) 25%   (c) 10%   (d) 2.5%
    (e) 1%   (f) 0.25%   (g) 0.1%   (h) 0.025%

Short Answer

28. Describe why a beam of white light entering a prism ends up producing a continuum of colors ranging from red light to blue/violet light.

29. Explain why sunsets are reddish in color, but the sky around the Sun isn't reddish in color when the Sun is high in the sky.

30. Why does air pollution produce redder sunsets?

31. The James Web Space Telescope (JWST), launched on December 25, 2021, is designed to operate primarily in the infrared portion of the electromagnetic spectrum so that it can peer deep into interstellar dust and gas clouds. Based on the discussion of the scattering of light, why would the telescope need to operate in the infrared?

32. Download a copy of Fig. 7.15 from the `Chapter 7` resources webpage and label the Fraunhofer C, F, and G lines. Using Table 7.2, also indicate the wavelengths of the three absorption lines.

33. Explain in your own words how it is possible to identify the various elements contained in the atmosphere of a star that might be 1 million trillion kilometers away.

34. Referring to Kirchhoff's rules for light spectra, explain why absorption lines appear in the solar spectrum.

35. It is possible to rewrite the half-life equation [Equation (7.1)] in order to solve for the number of half-lives, t. This requires knowledge of logarithms, with the result being

$$t = \frac{\log_{10}(\text{fractional amount remaining})}{\log_{10}(1/2)}$$

which can be simplified (slightly) as

$$t = 3.322 \log_{10}\left(\frac{1}{\text{fractional amount remaining}}\right).$$

How many half-lives have occurred if the fractional amount remaining is
    (a) 0.125 (12.5%)?
    (b) 0.035 (3.5%)?
    (c) 0.00042 (0.042%)?

36. In a laboratory experiment, the activity of a radioactive substance was monitored over time using a Geiger counter, with the results given below.

Time (min)	Remaining (%)
0	100.0
2	87.8
4	77.2
6	67.8
8	59.6
10	52.3
12	46.0
14	40.4
16	35.5
18	31.2
20	27.4
22	24.0
24	21.1
26	18.6
28	16.3
30	14.3

(a) Plot the data in a manner similar to Fig. 7.35 except that the horizontal axis should be "Time (min)" ranging from 0 to 30. Always be sure that you label both the horizontal and vertical axes, including the units being used, so that the reader understands what is being presented. Should you need it, a standard (Cartesian) sheet of graph paper can be downloaded from `Graph Templates` on the textbook's website.

(b) Estimate the half-life of the material.

Activities

37. In this activity you will investigate the spectra produced by lights that you can find around you. You will want to obtain an educational transmission diffraction grating which your instructor can probably provide for you. Alternatively, you can purchase one through an online store (search for

"diffraction grating"), similar to the shown here (they cost a few dollars if purchased individually):

(a) On a sunny day, FACE AWAY from the Sun. Using a white sheet of paper as a screen, hold the paper up in your body's shadow. With your other hand, hold the diffraction grating off to the side and oriented at an angle away from the Sun. You will need to experiment with the orientation of the grating relative to your white paper screen, but you should be able to produce a nearly continuous spectrum (like a rainbow) on the sheet of paper. If it isn't obviously labeled on the grating, you may need to rotate it 90° so that the lines etched on the film are oriented vertically in order to produce a horizontal spectrum. Given the poor quality of the grating, you probably won't be able to see any of the Fraunhofer lines, but you may want to play around with the focus of the image, and perhaps you will get a glimpse of the most prominent ones.

If you have a friend assisting you, or if you can mount the paper on an easel or on a convenient wall, you should take a picture of the image. Otherwise, describe what you have seen. Include the time of day and the date of your observations.

(b) Locate an incandescent light bulb (the kind that glow by heating a thin filament). Start by looking through the diffraction grating toward the bulb, and then move the grating off to one side. As you do so you should see the spectrum produced by the bulb start to appear.

   i. What sort of spectrum do you see: continuous, emission, or absorption?

   ii. Explain what you observe in terms of Kirchhoff's rules for light spectra.

(c) The newer high-efficiency light bulbs used in many homes, and the fluorescent bulbs used in overhead lights and signs, use heated gases to produce light of various colors. Locate at least two such light sources (they should be of different types) and observe them using your diffraction grating. You should see multiple images of the sources, but in different colors, with the images spread out.

   i. Where did you make each set of observations? Describe the light sources.

   ii. Record the specific colors and their order for each source.

   iii. What sort of spectrum do you see: continuous, emission, or absorption?

   iv. Explain what you observe in terms of Kirchhoff's rules for light spectra.

   v. Are the gases used in the sources the same? How do you know?

38. In this activity you are going to simulate the statistical nature of radioactive decay by flipping coins. Heads will represent atoms that *do not* decay into another type of atom, and tails will represent atoms that do decay. Ideally, it would be helpful to have 100 coins, but that is not necessary, although you will start by simulating 100 coin flips. This can be done by flipping 50 coins twice, 25 coins four times, 10 coins ten times, or 1 coin one hundred times. You will also need a way to record the results of each coin flip, which could be on a sheet of paper or in a spreadsheet.

Your table should have two columns. The first column, labeled "Number of flips," will indicate which step you are on, and the second column, labeled "Number of heads," will indicate the number of heads for that trial. The first row of your table should read: 0, 100. This indicates the start of the coin flipping experiment, assuming that there are 100 "radioactive atoms" (coins that are all heads) present initially.

(a) Flip 100 coins and record the number of heads you obtained. This is recorded in the second row of your table as 1, # heads. Those coins that ended up tails are no longer part of the sample after 1 step (simulating decay within one half-life).

(b) Now re-flip your coins, but only the number that ended up heads in the last step. Those that end up heads after this step represent the number of atoms that survived unchanged after 2 half-lives. Again, record your results in the next row as 2, # heads.

(c) Continue to repeat the process by flipping only the number of coins that ended up heads in the previous step until none of the remaining coins land heads. Record the number of heads that survived each "half-life."

(d) Download Fig. 7.35 from the textbook's website, available from the Chapter 7 resources webpage. Plot your data from the coin flip experiment on the graph. The vertical axis of the graph, "Percentage of Original Activity," corresponds to your table's column, "Number of heads." The graph's horizontal axis, "Time (in half-lives)," corresponds to your column, "Number of flips."

(e) Explain why your data don't always, if ever, fall on the idealized curve.

(f) Would you expect your data to better fit the curve if, say, 10^{20} coins were flipped initially and you recorded the percentage of coins that landed heads? Why or why not? Hint: Study your data as progressively fewer coins remain in the sample. Do you always get close to 50% of the coins landing heads?

Modern Physics: New Science to Study the Universe

8

Look deep into nature, and then you will understand everything better.

Albert Einstein (1879–1955)

8.1	Blackbody Radiation and Planck's Energy Quanta	264
8.2	The Life of Albert Einstein	266
8.3	The Special Theory of Relativity	274
8.4	The Doppler Effect for Light	281
8.5	Einstein's Explanation of Planck's Energy Quanta	285
8.6	Bohr's Atom	288
8.7	Quantum Mechanics: What Are the Chances?	293
8.8	The Nucleus and Isotopes	298
	Exercises	301

Fig. 8.1 An artist's depiction of jets of material coming from a supermassive black hole at the center of a spiral galaxy. (ESO/M. Kornmesser, CC BY 4.0)

Introduction

Thus far you have studied the development of physics and chemistry up through the nineteenth century. Newton's laws of motion and his universal law of gravitation, Maxwell's electromagnetic theory that married electricity, magnetism, and light, and the study of heat that culminated in Boltzmann's statistical mechanics as the foundation of thermodynamics (the study of heat) are collectively known today as classical physics. Classical physics, combined with Mendeleev's periodic table of the elements and spectroscopy, were used to learn something of the motions and properties of objects in our Solar System, along with the surface characteristics of our Sun and a few other stars. However, despite the many successes in observational astronomy, experimental and theoretical physics, and chemistry through the 1800s, there remained many gaps in our understanding of the physical aspects of nature. In order to truly understand the planets, stars, galaxies, and the beginning and future of the universe itself, we must turn to the rapid pace of scientific advancement, and its decidedly unintuitive implications, that began with the dawn of the twentieth century.

8.1   Blackbody Radiation and Planck's Energy Quanta

As we saw in Section 6.6, experimentation revealed that hot solids or dense gases produce a characteristic rate of energy output that depends on temperature, surface area, and the wavelength of the emitted light. Although it is certainly interesting to learn that there exists a characteristic blackbody radiation curve, that is not the same thing as understanding what it is about nature that conspires to produce it. Blackbody radiation is obviously associated with the electromagnetic waves predicted by Maxwell's equations, but given its dependence on the temperature of the source, blackbody radiation must be related to the science of thermodynamics as well. Even though people like Wien, Kirchhoff, and Boltzmann studied blackbody radiation extensively, they were unable to fully explain its origin, although some aspects of its behavior could be described. A deeper understanding of blackbody radiation would be critical if we were ever going to fully understand the sources that produce the radiation.

As a child, Max Planck (Fig. 8.2) was a gifted musician who studied piano, organ, cello, and voice, and composed songs and operas. But Planck also had a talent for mathematics and a keen interest in understanding how nature worked. When he suggested to his physics teacher that he wanted to go into physics, he was actually discouraged from doing so, with his teacher saying that "in this field, almost everything is already discovered, and all that remains is to fill a few holes." The young Planck responded that he had no interest in discovering anything new, he simply wanted to understand what was known.

In graduate school, Planck was drawn to the discipline of thermodynamics and became a student of Kirchhoff at the Friedrich Wilhelm University in Berlin. After he completed his degree he had difficulty finding a teaching position for a time, but ultimately was awarded a position as Kirchhoff's successor after his death. Naturally, Planck was drawn to work on the solution to the problem that had eluded his mentor: namely to explain the fundamental nature of blackbody radiation, with

Fig. 8.2 Max Planck (1858–1947); Berlin, January 11, 1933. (Public domain)

the goal of being able to derive the equation that described its characteristic distribution of energy as a function of wavelength and temperature.

Those who came before him employed the physics of Newton, Maxwell, and Boltzmann in attempts to solve the blackbody radiation problem. This approach produced curves that seemed correct at the longest wavelengths in Fig. 6.43, but as the wavelength decreased the equation kept going higher and higher, never turning over to produce the peak at λ_{peak} as required by Wien's blackbody color law [Equation (6.12)]. In order to make progress on the problem, Planck realized that if he mathematically forced his electromagnetic waves to have only specific sets of energy, rather than any amount of energy whatsoever, then he could get to the right answer. Classical physics says no such thing about electromagnetic waves. What Planck proposed in 1900 was that the waves must have energies that are *integer multiples* of a constant times their frequencies, with the least amount of energy possible being

$$E = hf = \frac{hc}{\lambda}, \tag{8.1}$$

Planck's energy equation

where h was a constant that he simply invented and adjusted to make the equation fit the data. The second form of Planck's energy equation just replaces f by using the wavelength–frequency–velocity relationship for light waves, $f\lambda = c$ [Equation (6.9)]. Planck's energy equation can be read as Planck's energy bundle equals a constant multiplied by the frequency of the light, or, alternatively, Planck's bundle of energy equals that constant multiplied by the speed of light and then divided by the wavelength of the light. In other words, Planck's energy bundle increases in direct proportion to increasing frequency of the light and decreases in inverse proportion to the increasing wavelength of the light. If the frequency of the light doubles then the energy of the bundle doubles, but if the wavelength of the light doubles then the energy of the bundle is cut in half.

In some sense, Planck had no justification for what he did other than his mathematical intuition. He essentially admitted as much by saying that he was forced into "an act of desperation ... a theoretical explanation had to be found at any cost, whatever the price."

Planck referred to his bundles of energy as quanta, the plural of quantum. He postulated that for some frequency f, the energies of electromagnetic waves for blackbody radiation could only be hf, or $2hf$, or $3hf$, or $4hf$, and so on. The greater the frequency, or the shorter the wavelength of the wave, the more energy Planck's quanta could have.

When Planck inserted his mysterious h into his equations he did it with the hope that h would go away in the derivation somehow; it didn't. Although not zero, h is extremely small, having a value of only 6.63×10^{-34} J s (roughly 1.5 billion trillion trillion times h equals 1 J s). The units are such that when h is multiplied by f (which has units of hertz, or $1/s$), the result has the unit of energy, the joule. h is now universally called Planck's constant, and is another of nature's fundamental constants. The boy who stated that he wasn't interested in discovering anything new had opened the door at the dawn of the twentieth century to a new world of modern physics. His old physics teacher could not have been more wrong.

Tragically Planck experienced much loss in his life. With his first wife, Marie, Planck had four children: Karl, twin girls, Emma and Grete, and Erwin. His wife died, probably from tuberculosis, in 1909. During the First World War, Karl was killed in action at Verdun and Erwin was taken prisoner by the French, although he did survive the war. Grete died in childbirth in 1917 and Emma died the same way in 1919. During World War II, Erwin was involved in the failed 1944 plot to assassinate Adolf Hitler, and executed by the Gestapo in January 1945. Planck died two years later. He was survived by his second wife and a son by that marriage.

In honor of his great contributions to science, The Kaiser Wilhelm Society, for which he twice served as president, was renamed the Max Planck Society for the Advancement of Science in 1948. Today, the Max Planck Society supports fundamental research and scholarship in a wide range of fields, including the physical, life, and social sciences, and the arts and humanities. The Max Planck Institutes are among the premier research centers in the world.

8.2   The Life of Albert Einstein

His Formative Years

Fig. 8.3 Albert Einstein (1879–1955), with his younger sister, Maria 'Maja' Einstein (1881–1951), circa 1886. (Public domain)

In the year of James Clerk Maxwell's death, Albert Einstein (1879–1955) was born on March 14[1] in Ulm, Germany, into the Jewish family of Hermann and Pauline Einstein. Albert would be destined to play the preeminent role in helping to usher in modern physics, the radically new way of viewing the universe first hinted at by Planck's work in 1900. Fittingly, the motto of Einstein's birth town was "The People of Ulm Are Mathematicians."

At the time of Albert's birth, his father owned a featherbed store, but was a not particularly effective salesman. As a result, the next year Albert's Uncle Jakob convinced his father to move the family to Munich, where they started their own electrical engineering firm. Then, in 1881, Albert's sister Maria was born, whom he affectionately called Maja (Fig. 8.3). The two would remain very close until her death four years before her brother died.

When Albert was five years old, his father gave him a magnetic compass to cheer him up while he was suffering from a brief illness. Einstein recalled being fascinated by the mysterious force that caused the needle to always point north. The influence of the magnetic field, and the nature of fields in general, were to become central to his scientific pursuits throughout his career.

While in Munich Albert attended a Catholic school for three years, where he began to experience the insidious anti-Semitism that was prevalent in Germany, and which would eventually lead to the Holocaust during World War II. As Einstein would recall later in life, "Physical attacks and insults on the way home from school were frequent, but for the most part not too vicious. Nevertheless, they were sufficient to consolidate, even in a child, a lively sense of being an outsider." As anti-Semitism grew stronger in Germany, Einstein would come to embrace his Jewish cultural and ethnic heritage, although not the religion of Judaism.

---

[1] For self-proclaimed geeks, March 14 is celebrated for another reason as well: it is unofficially pi-day; 3.14 being the first three digits of the mathematical constant, π.

At the age of nine, Einstein entered the Luitpold Gymnasium (which was later renamed Albert Einstein Gymnasium), a public school that focused on math, science, Latin, and Greek, but Einstein rebelled against school's rigid method of teaching by rote memorization. After Einstein had become the world's most famous scientist, a journalist, assuming that the great man would immediately know the answer, asked him a question from Thomas Edison's (1847–1931) well-known list that the inventor would ask job applicants: "What is the speed of light?" Einstein's reply was that he did not "carry such information in my mind since it is readily available in books … The value of a college education is not the learning of many facts but the training of the mind to think." Einstein was also fond of emphasizing that "The true sign of intelligence is not knowledge but imagination."

The Prussian Luitpold Gymnasium also exhibited another trait that Einstein abhorred throughout his life; a strong sense of militarism. The other kids in the school would love to watch military parades and even march in lock step to the drum and fife music. Explaining to his parents why one such spectacle made him cry, he said that he didn't want to grow up "to be one of those poor people." Later, he would further elaborate by saying that "when a person can take pleasure in marching in step to a piece of music it is enough to make me despise him. He has been given his big brain only by mistake." Up until the start of World War II when Hitler made him reevaluate his position on war, Einstein was a strong and public pacifist. However, he would always remain a non-conformist.

There is a famous myth of Einstein's schooling that came out of his time at the Luitpold Gymnasium, namely that he was a poor student, even in mathematics and science. In fact, he was routinely a top student in all of his subjects, including those that he did not particularly enjoy, such as Latin and Greek. The myth was even included in a "Ripley's Believe It or Not!" newspaper column during Einstein's lifetime. When shown the headline, "Greatest Living Mathematician Failed in Mathematics," Einstein pointed out that he had mastered differential and integral calculus when he was fifteen years old. Unfortunately, the myth is still widely propagated today, including in a 2015 billboard campaign promoting confidence with a photo of the great scientist and a statement: "As a student, he was no Einstein." Actually, he was *the* Einstein.

Also instilled in the young Albert Einstein at an early age was his love of music. His mother was an accomplished pianist and she encouraged him to take violin lessons. The two of them would frequently play duets together, especially Mozart. His early duets with his mother turned into performing with others during social gatherings and occasionally at formal concerts and charity events (Fig. 8.4). Throughout his life, playing the violin would often help him settle his mind, allowing him to think about the problems that he was working on.

In the fall of 1894, when Albert was fifteen, the family business lost a major contract to provide lighting for Munich and other nearby municipalities, which resulted in the collapse of the company. Einstein's parents, his sister, and Uncle Jakob then decided to move to Milan, Italy, where they believed that prospects for a small electrical company would be more favorable. Albert was left behind in Munich with a relative to finish school. At Christmas break, the unhappy boy, who greatly disliked the rigidity of the the Luitpold Gymnasium, took a train to Milan and refused to return to Germany. Albert even had his German citizenship terminated, which foreshadowed another of his lifelong traits, a strong opposition to nationalism. Had

Fig. 8.4 Einstein playing a charity concert with teacher Lewandowszki in 1930. (CC BY-SA 4.0)

he remained in Germany, when he turned seventeen he would have been forced into required military service, something he dreaded. Einstein never actually graduated from the German gymnasium that would later take his name; in fact, Albert was officially a high school dropout. Perhaps that was the source of the persistent poor-student myth.

At sixteen years of age, Einstein was given permission to apply to the Zürich Polytechnic in Switzerland, but unfortunately he failed his entrance exam in the general subject areas (he passed the math and science portions). He then spent a year at a Swiss school near Zürich studying those subjects that he had struggled with in the entrance exam. Einstein was finally in an educational system that suited him. Rather than rote memorization and rigid authoritarian teaching, the Swiss system respected the student, placed more emphasis on independent thought, and encouraged personal responsibility. Einstein passed his Zürich Polytechnic entrance exams the next year.

It was during that year prior to his acceptance into the Zürich Polytechnic that Einstein conceived of a famous thought experiment that would affect his career eight years later. What would light look like if you could run at the speed of light alongside a light beam? That question gives a hint into his genius; the ability to pose conceptual questions that go to the fundamental core of physics. Later in his life, Einstein told a psychologist that he would typically think in terms of pictures before putting his ideas into words and equations.

Although he excelled in theoretical physics at the Zürich Polytechnic, Einstein completed his degree in 1900 with the lowest overall score in his graduating class. This was due largely to failing his elementary physics laboratory, combined with a less-than-stellar senior thesis, a topic that was largely forced on him by his professor. At one point in the lab he created an explosion that injured his right hand. Not only did it take a couple of weeks to heal, but it also meant that he couldn't play his violin for even longer.

Einstein's Marriages and Family

Fig. 8.5 Albert Einstein and his first wife, Mileva Marić (1875–1948) in 1912. (Public domain)

While attending the Zürich Polytechnic, Einstein fell passionately in love with the only woman in his physics class, Mileva Marić (Fig. 8.5), who would later become his first wife. Einstein didn't feel that he could marry Mileva until he found a permanent position. Unfortunately, following his graduation Einstein spent several years trying unsuccessfully to find a teaching appointment. During that time Albert and Mileva had a daughter, Lieserl, who was born while Albert was away. It seems that Einstein and his daughter never met; Lieserl may have been given to a friend of Mileva's to be raised or she may have died of scarlet fever in 1903, but details of the child's life have been lost to history.

Through the assistance of a friend and classmate from the Zürich Polytechnic, Einstein was finally able to land a permanent job at a patent office in Bern, Switzerland in 1902, and Albert and Mileva were married in 1903. From the marriage they had two sons, Hans Albert (1904–1973) and Eduard (1910–1965). Hans Albert would eventually become a professor of engineering at the University of California, Berkeley, while Eduard became a psychiatrist. Sadly, Eduard suffered from schizophrenia himself, and would end up living out much of his life under the care of his mother until she died, and then the remainder of it in a psychiatric clinic in Zürich.

Mileva apparently struggled with bouts of depression, and the marriage was not a happy one. This was due in no small part to Einstein's flirtatious behavior with other women. In 1912 he became reacquainted with his cousin, Elsa Löwenthal (Fig. 8.6) who had divorced her first husband four years earlier. Elsa had two daughters, Ilse and Margot. Ultimately, Albert and Mileva would go through a bitter divorce in 1918 and Elsa would become his second wife in 1919. As a part of the divorce settlement, Einstein agreed to give Mileva his Nobel Prize money should he receive the award one day. Elsa and her daughters remained close to Einstein throughout their lives, even though his flirtatious behavior and occasional affairs continued.

Fig. 8.6 Albert Einstein and his second wife, Elsa (1876–1936) in 1921. (Public domain)

1905: Einstein's *annus mirabilis*

In March, 1905, while working at the Swiss patent office, Einstein wrote a letter to an old friend that read in part:

> I promise you four papers ... The first deals with radiation and the energy properties of light and is very revolutionary ... The second paper is a determination of the true sizes of atoms ... The third proves that bodies on the order of magnitude 1/1000 mm, suspended in liquids, must already perform an observable random motion that is produced by thermal motion. Such movement of suspended bodies has actually been observed by physiologists who call it Brownian molecular motion. The fourth paper is only a rough draft at this point, and is an electrodynamics of moving bodies which employs a modification of the theory of space and time.

The contents of those four papers together with a fifth paper that was essentially an addendum to the fourth, were all published in the same year. That burst of scientific creativity hadn't been seen since Newton's time away from Cambridge University during the Plague, and as with Newton's *annus mirabilis* (miraculous year) 240 years earlier, physics would undergo another revolution.

Three of Einstein's *annus mirabilis* papers will be discussed in greater detail in Sections 8.3 and 8.5, together with their impacts on physics and astronomy. For the moment, it is enough to know that the first of those papers dealt with Planck's light quanta (Section 8.1) and was critical in giving birth to the new and strange theory of quantum mechanics that focuses on the smallest realm of nature. The second paper provided an estimate of Avogadro's number (see page 249). The third paper was a direct confirmation of the still-speculative concept that atoms exist. The fourth paper, on the "electrodynamics of moving bodies," addressed concerns Einstein had about aspects of Maxwell's electromagnetic theory (Section 6.5) by introducing the world to the concept that space and time are not fixed absolutes, but that spacetime itself is relative, a theory that became known as the special theory of relativity. The addendum to that fourth paper gave us science's most famous equation: $E = mc^2$.

"The Happiest Thought in My Life" — The General Theory of Relativity

In 1907, while still at the patent office in Bern, Einstein had what he would later call "the happiest thought in my life." When a person is falling he would not experience

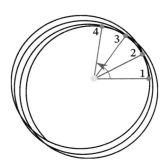

Fig. 8.7 The advance of perihelion of Mercury's orbit. The shift is greatly exaggerated to illustrate the effect. Points 1, 2, 3, and 4 indicate the locations of Mercury at the perihelion of its orbit on successive orbits.

his own weight. Eight years later, that simple realization would lead Einstein to an extension of his special theory of relatively. The special theory deals only with constant velocity motions (in a straight line at a constant speed), but his "happiest thought" provided insight into how to extend the special theory to one that also incorporated acceleration. In addition, because falling in a gravitational field is a form of acceleration (changing the magnitude and/or the direction of the velocity vector), this general theory of relativity, published in 1915, also resulted in a new theory of gravitation that superseded Newton's so-called universal law of gravitation [Equation (5.7)]: $F_{\text{gravity}} = GMm/r^2$.

One of the predictions of Einstein's general theory of relativity finally solved a long-standing problem with the orbit of the planet Mercury. Even after accounting for all of the other planets in the Solar System, the perihelion point of Mercury would shift very slowly so that the orbit would not retrace its path precisely, as illustrated in Fig. 8.7. The perihelion of Mercury's orbit would move forward at the very tiny rate of 43 arcseconds per century. You should recall that, when an unexplained deviation in the orbit of Uranus was discovered, a direct application of Newton's universal law of gravitation led to the prediction of the planet Neptune (Section 5.5). Astronomers had similarly tried to explain the advance of Mercury's perihelion as being due to an as-yet undiscovered planet orbiting closer to the Sun than Mercury, that they named Vulcan.[2]

Vulcan was never discovered of course, but the solution to Mercury's orbital anomaly came in an unanticipated way, through Einstein's general theory that spacetime around massive objects is curved by mass. The curvature is analogous to a bowling ball placed on a rubber sheet that is held up by its edges, causing the sheet to be stretched by the ball (Fig. 8.8). A much lighter tennis ball moving across the rubber sheet would follow the curved surface in much the same way that a planet would follow the curvature of spacetime. Rather than the curvature of an initially two-dimensional sheet into a third dimension (depth), mass "curves" the three dimensions of space and the fourth dimension of time. Even though we can't visualize the curvature of spacetime, mathematically it is completely analogous to the rubber sheet and the bowling ball. In essence, what Einstein proposed through his general theory of relativity is that gravity is nothing more than geometry in a

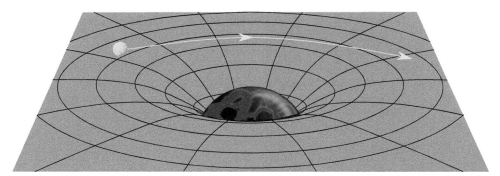

Fig. 8.8 A rubber sheet with a bowling ball at the center. The tennis ball must follow a curved path along the deformed sheet.

---

[2]The planet's name would ultimately become famous as the home planet of Spock, in the television series and movie franchise, *Star Trek*.™

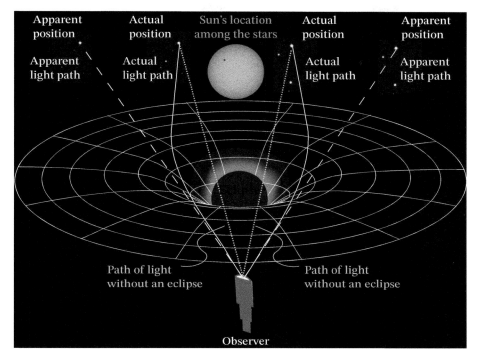

Fig. 8.9 Starlight following a curved path through curved spacetime around the Sun during a total solar eclipse.

four-dimensional spacetime! Not only is space curved, but time is affected as well, with clocks running more slowly the closer they get to a massive object.

The general theory of relatively also made another testable prediction, as all scientific theories must. If spacetime is bent near a massive object then not only do planets need to follow that spacetime curvature, but light must follow it as well. The theory implies that the path of starlight traveling past the Sun must be deflected, with the amount of deflection being exactly calculable (Fig. 8.9). That prediction was put to the test during a total solar eclipse in 1919, when the stars normally hidden by the Sun's glare during the daytime became visible for a few minutes. By taking pictures of the sky during the eclipse and comparing the stars' positions to their positions when the Sun is not in that region of the sky, the amount of deflection was measured. When the results were announced, Einstein's predictions fitted the observations. Afterward, when asked how he would have felt if the results of the observations had contradicted his predictions Einstein responded: "Then I would have been sorry for the dear Lord; the theory is correct."

Newspapers all over the world praised Einstein's theory and its eclipse confirmation. As the New York Times put it, "LIGHTS ALL ASKEW IN THE HEAVENS: Men of Science More or Less Agog Over Results of Eclipse Observations." Almost immediately, Einstein became world famous.

Einstein was finally awarded the Nobel Prize in 1921 "for his services to Theoretical Physics, and especially for his discovery of the law of the photoelectric effect," and Mileva Marić received the remainder of her divorce settlement.

The general theory of relativity will be discussed further in Section 17.4.

Key Points
- General relativity says that gravity is just a matter of geometry due to curved spacetime around a mass.
- The trajectory of light is bent in the vicinity of a massive object.
- Time runs more slowly near a massive object.

Einstein's Other Scientific Contributions

In addition to his work in 1905 and his general theory of relativity, Einstein also made important contributions to the theory of light emission, which led forty years later to the development of the laser (an acronym that stands for "light amplification by stimulated emission of radiation"), as well as to the developing theory of quantum mechanics. The latter contributions are ironic given his famous opposition to the theory, to be discussed in Section 8.7.

Much of his work over the final decades of his life was directed toward searching for a unified field theory that would tie together both his general theory of relativity and Maxwell's electromagnetic theory. Through the effort he also hoped to show that the strangeness of quantum mechanics was simply the result of an incomplete theory. It was an almost quixotic quest, that was largely ignored by the rest of the scientific community. That personal struggle to unveil nature's secret only ended with his death on April 18, 1955. At his bedside were equations that he was working on until the end, hoping that they might lead to the successful conclusion of his quest. Scientists are still working on unifying his gravitational theory with electromagnetism and the two other known forces of nature, the strong and the weak nuclear forces associated with nuclei and radioactive decay.

During the eight years that Einstein was working on and off on the ramifications of his "happiest thought" regarding general relativity, he finally landed his first university position, but not until after he was actually turned down for a high school teaching post; he wasn't even one of the three finalists! In 1909 he received an appointment at the University of Bern, but only after he wrote another paper specifically for that application. It was a bit of "yes you wrote those five papers in 1905 that revolutionized physics, but what have you done lately?" In fact, while working at the patent office six days a week and raising his three-year-old son, Einstein had written six papers and reviews in 1906 and ten more in 1907. By 1911 he was back at the Zürich Polytechnic, this time as an associate professor. Then in 1913 Einstein was elected to the prestigious Prussian Academy of Sciences and joined the faculty of the University of Berlin, the appointment requiring that he again become a German citizen.

A Citizen of the World

Following the unprecedented reception of Einstein's work the world over, he was a highly sought-after public figure. On trips to the United States, he was received much like a rock star or a championship local sports team would be today. He was now also a scientist to be courted for academic positions at major universities around the world. Both the California Institute of Technology (Caltech) and Princeton University competed for his attention. During his third trip to America, Adolf Hitler became the new chancellor of Germany on January 30, 1933 and Jewish scientists were no longer welcome to work at German institutions. Einstein never saw Europe again, and instead he made Princeton and the Institute for Advanced Study his new home. On March 10 Elsa's daughter, Margot, was back in Berlin, staying at Einstein's apartment when the Nazis raided it twice in one day. Fortunately, following the raids she was able to get his papers to the French Embassy and she relocated to Paris where she rendezvoused with her husband.

Being famous allowed Einstein the opportunity to voice his political and social views very publicly. He would routinely speak out against war and militarism, and lobby for a world government that would, he hoped, eliminate militaristic and nationalistic pride that he believed led to the atrocities of two world wars. Prior to the rise of Nazism he was an ardent supporter of pacifism, encouraging young men to refuse military service, but Hitler changed all of that. At least in some cases, Einstein came to admit that military opposition to nationalistic hatred and repression was sometimes necessary.

While Einstein may have struggled with empathy in intimate relationships, he was a gentle and loving soul toward friends and the breadth of humanity. He was passionately opposed to racism, antisemitism, and other forms of exclusion. In 1937, the celebrated African-American opera singer, Marian Anderson (Fig. 8.10), came to Princeton to perform, but was not allowed to stay at the Nassau Inn there. Einstein opened his own home on Mercer Street to her during the visit, and they remained friends thereafter. It was two years later that she was forbidden to perform in Washington's Constitution Hall, so she chose instead to give her famous concert for free on the steps of the Lincoln Memorial.

Fig. 8.10 *Top:* Marian Anderson (1897–1993). [Photographer Carl Van Vechten (1880–1964); public domain] *Bottom:* Marian Anderson's April 9, 1939, Easter Sunday concert in front of the Lincoln Memorial, Washington, D.C. (U.S. Information Agency, public domain)

Fig. 8.11 Leó Szilárd (1898–1964). (Department of Energy)

In 1939 Leó Szilárd (Fig. 8.11), a Hungarian physicist and old friend of Einstein's from Berlin, came to visit him at Einstein's summer rental cottage on Long Island, New York. Szilárd had been working on ways to create nuclear chain reactions at Columbia University in the United States following his escape from Europe. Upon learning of a successful experiment to split a uranium atom through the process of nuclear fission, Szilárd became deeply concerned that a tremendous amount of energy might be released if a new type of bomb could be developed that generated a chain reaction using fission. He was particularly concerned that, even though most of the top scientists had fled Germany when Adolf Hitler came to power, there were some who remained that might be capable to creating such a device. Szilárd was able to convince Einstein that they should write a letter to President Eisenhower in order to convince him that the United States must develop such a weapon first. With the aid of another Hungarian refugee, Edward Teller (Fig. 8.12), Szilárd and Einstein completed a letter dated August 3, 1939, that read in part: "A single bomb of this type, carried by boat and exploded in a port, might very well destroy the whole port together with some of the surrounding territory." They were correct about the destructive potential, but seriously overestimated the size required for the bomb. From the letter, and after some frustrating delays on the part of the administration, the Manhattan Project was born in Los Alamos, New Mexico.

In March 1945, when it was clear that Germany would be defeated, not having successfully developed an atomic bomb, Szilárd and Einstein wrote another letter to Eisenhower, asking him to halt development and not deploy the weapon. Eisenhower never got the letter, having died in office on April 12. The letter was passed on to President Truman but it remained unopened. In a Newsweek article after the war was over, Einstein stated that "had I known that the Germans would not succeed in producing an atomic bomb, I never would have lifted a finger." In the

Fig. 8.12 Edward Teller (1908–2003). (Lawrence Livermore National Laboratory)

late 1940s Einstein was asked what type of weapon might be used to fight World War III. He replied that he didn't know, but World War IV would be fought with rocks.

Ironically, Einstein had very little to do with the development of the atomic bomb, in part because nuclear physics wasn't his specialty, and in part because the FBI believed him to be a communist sympathizer. It turned out that when they were finally released to the public, J. Edgar Hoover's FBI (Federal Bureau of Investigation) had amassed a dossier containing 1427 pages of documents, none of which indicated that he had any communist-leaning tendencies. Amazingly, they did try to find an Albert Einstein, Jr. (Einstein never had a son named Albert Jr.) whom they believed had defected to the Soviet Union; meanwhile, Hans Albert Einstein was teaching engineering at Berkeley. The FBI also missed the fact that the then-widowed Einstein was romantically involved with Margarita Konenkova (1895–1980) from 1941 to 1945. The relationship ended when Konenkova returned home to the Soviet Union and he declined to go with her. Einstein never knew it, but Konenkova was a Russian spy although that didn't really matter because he had no national secrets to share. The affair only became public when a series of love letters were made public in 1998, forty-three years after Einstein's death.

Fig. 8.13 Albert A. Michelson (1852–1931). (Public domain)

8.3   The Special Theory of Relativity

The Ether and the Michelson–Morley Experiment

Back in Section 6.5 it was mentioned that Maxwell's electromagnetic waves don't require anything to "wave" through (page 211). Although this is true, at the time scientists were still convinced that there had to be some kind of medium in which electromagnetic waves propagated. That medium was dubbed the ether, resurrecting the idea that, if we can see the stars, then "empty" space must be filled with something like the fifth element of the ancient Greeks.

In an effort to detect and study the ether, numerous researchers first attempted to measure Earth's motion through the mysterious medium. The most famous of those experiments were performed by Albert A. Michelson (Fig. 8.13) and Edward W. Morley (Fig. 8.14) in 1887. Since the ether was supposed to provide the means by which electromagnetic waves traveled, it was thought that as Earth orbits the Sun, it should be moving through the ether, effectively creating a "wind" much like the wind you feel on your face as you ride a bicycle on a calm day.

In order to test their hypothesis, Michelson and Morley split a beam of light, sending the two beams in perpendicular directions, as shown in the diagram in Fig. 8.15. After traveling the same distance in each direction, mirrors reflected the beams back to where they started. This allowed the beams to then be recombined. By so doing, the two beams would interfere, producing constructive or destructive patterns (page 203*ff*), based on how much time it took one beam to make the round trip relative to the other one. If the round-trip along one leg of the apparatus takes longer than along the other leg, then the waves will arrive later from the slower leg and be shifted relative to the faster leg.

Fig. 8.14 Edward W. Morley (1838–1923). (Courtesy of Case Western Reserve University Archives; public domain)

To understand this, imagine yourself swimming in a river. If you swim upstream against the current, it will take you longer to travel 100 m than it would if you swim downstream with the current. On the other hand it will take you exactly

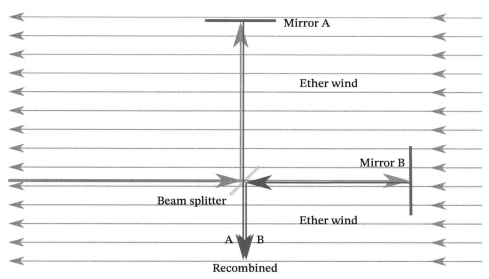

Fig. 8.15 The Michelson–Morley experiment. A light beam (represented by red arrows) strikes a beam splitter and the two beams travel the same distance in perpendicular directions, striking mirrors, and returning to be recombined (blue arrows). Traveling in an ether wind, the light striking Mirror A travels perpendicular to the wind in both directions while the light striking Mirror B travels "upwind" and then "downwind" before the two beams recombine.

the same amount of time to swim straight across the river in one direction as it would to swim across the river in the opposite direction. A similar situation should apply to light traveling through the ether. If the light goes against the ether wind it will take longer to travel one leg of the Michelson–Morley apparatus than it will to travel in the reverse direction, with the ether wind. If the light travels the other leg perpendicular to the ether wind, it will take the same amount of time on the return trip. Overall, traveling against and then with the ether wind takes more time to make the round trip than traveling perpendicular to the wind in both directions.[3]

When Michelson and Morley conducted their experiment they didn't detect any difference between the round-trip travel times of the two light beams. Did that mean that the ether doesn't exist? Rather than believing that light can actually move through a vacuum, researchers redid the experiment many times over the next few decades in attempts to detect some ether wind velocity. Perhaps the experiment was done at just the point in Earth's orbit when we were moving at a 45° angle relative to an ether wind blowing through the Solar System, making the travel times identical. The experiment was done at various times of the year, just to check that possibility. Still no difference. Perhaps ether gets trapped inside buildings, so the experiment was attempted outside. Perhaps the ether is attached to Earth due to gravity, causing the ether to rotate with the planet, so a version of the experiment was tried by sending light around circular paths in opposite directions. The result was always the same; no detection of an ether wind.

---

[3]Actually, in order for a swimmer to travel directly across the river perpendicular to the current direction, she would need to swim somewhat upstream with the current pushing her back downstream. The same would be true if an ether wind really did exist that light had to travel through. Try the swim sometime; it's an application of velocity vector addition!

At one point it was proposed that the ether somehow caused the apparatus to shrink in the direction of the ether wind by just the amount needed for the two beams of light to arrive at the same time. Even the measuring devices to check the length of the shortened arm shrank as well, meaning that a shorter arm could not actually be detected. This ad hoc solution could have caused the null result of the experiment but there was no way to know. Such a proposal is not only ad hoc, it is also unscientific because it cannot be tested.

Key Point
Michelson interferometers are used to measure lengths extremely precisely.

Although the apparatus was unsuccessful in measuring an ether wind, it turns out that it proved very useful for a different reason. By making tiny changes in the length of one of the arms of the apparatus, the recombined waves pass through constructive (bright) and destructive (dark) interference. By measuring the changing intensity of the light as the length of the arm changes, very tiny lengths can be measured, down to a small fraction of a wavelength of light. Used in this way, the Michelson interferometer provides an extremely accurate way to measure lengths. One of many applications of this process allows telescope mirror surfaces to be shaped very precisely. We will encounter this method again in future chapters.

Absolute Space and Absolute Time

An inherent assumption associated with the idea that the universe is filled with a mysterious ether is that space and time measurements must be made relative to absolute references. Newton built his ideas of physics on this assumption. In essence, Newton assumed, as did everyone who followed him, that it is always possible to tell if you are moving relative to some fixed universal point (an origin where $x = 0$, $y = 0$, $z = 0$); essentially one great coordinate system for the universe where every object's position could theoretically be specified relative to that unmoving point. Similarly, it was assumed that time always marches on; one second is the same for everyone, including any aliens, anywhere in the universe. These concepts are referred to as absolute space and absolute time. Since the ether was supposed to fill the entire universe, the ether became an obvious reference for absolute space. Absolute time seemed so obvious that there was no need for a universal grandfather clock for the universe.

Einstein's Postulates of Special Relativity

Recall Einstein's thought experiment of what it would be like to run alongside a beam of light. Would you see the beam frozen in space from your perspective? That certainly seems like an odd possibility. Recall also that his fourth paper in 1905 was about the "electrodynamics of moving bodies." In it, he questioned how a magnetic field could be generated by an electric current due to moving charges such as electrons if you are moving along with the electrons. In that case you wouldn't measure any speed for the electrons relative to you and so you wouldn't actually measure a current, which means, in turn, that you shouldn't be able to measure a magnetic field generated by a current. That inconsistency seemed to be major problem for Maxwell's beautiful electromagnetic theory. Einstein's paper was designed to solve that dilemma.

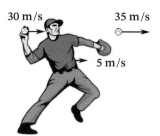 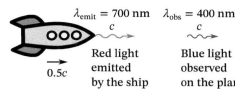

30 m/s   35 m/s

5 m/s

$\lambda_{\text{emit}} = 700$ nm $\lambda_{\text{obs}} = 400$ nm

c c

Red light   Blue light
emitted   observed
by the ship   on the planet

0.5c

Fig. 8.16 *Left:* A player running at 5 m/s throws a baseball at 30 m/s relative to himself. The result is the baseball traveling at 35 m/s relative to the ground; (© patrimonio) *Right:* A spaceship traveling at 50% of the speed of light (0.5c) fires a red laser beam toward a planet that it is approaching. The crew measures the speed of the laser beam to be the speed of light, c. Observers on the planet also see the laser beam approaching them at the speed of light, c. The only difference is that the planetary observers see the laser beam as being blue, instead of red. (Rocket: Maurice MacGyver, CC BY 4.0; Earth image from NASA)

To do so, Einstein started with two postulates (assumptions) from which he developed his theory:

1. All observers must agree on the results of an experiment, even if one observer is moving at a constant velocity relative to other observers.
2. All observers agree that the speed of light is always c in a vacuum, even if one observer is moving at a constant velocity relative to other observers.

Key Points
Einstein's postulates of special relativity.

The first postulate is basically saying that if someone sees something like a home run occur at a baseball game, then everyone else must also agree that the home run took place, even if one observer sees it happen from an airplane window while flying over the field. Nobody would see the ball fall short of the fence. The same would hold true for any measurable event: everyone has to agree that it happened.

The second postulate seems a bit stranger. As shown in Fig. 8.16, in a baseball game, if an outfielder is running in the direction of home plate at 5 m/s and throws the ball at 30 m/s, the catcher would see the ball coming at 5 m/s + 30 m/s = 35 m/s. Replacing the outfielder with a spaceship going 50% of the speed of light that fires a red laser beam at a planet that it is approaching, it would see the light beam leave the ship at the speed of light, c, but someone on the planet would see the laser light arriving at the planet at the speed of light as well, not at 150% of the speed of light, as you might expect. In fact any observer, no matter how fast they are moving relative to the ship, will see the laser beam traveling at the speed of light, c. In this way you can never actually run at the speed of light and see a beam of light frozen in space, because the beam of light will still be traveling at the speed of light relative to you. (In actuality, one implication of Einstein's special theory of relativity is that you can't ever run at the speed of light, no matter how fast you are.)

The End of Absolute Space and Absolute Time

In talking about Einstein's first postulate above, it was pointed out that everyone watching a baseball game, even from a plane flying past the stadium, has to agree that a home run was hit, but they don't have to agree on *the time* it took for the ball to reach the fence or *the distance* from home plate to the fence. From the vantage point of fans sitting in the stands, they will all agree on the time of flight and the size of

Key Points
The measurements of time and distance depend on who is making the measurements.

the stadium, but the people on the plane won't agree with the fans in the stands. If the plane is flying along the same line that the baseball travels, the observers on the plane will measure the field to be a bit shorter in that direction compared to what the fans in the stands measure it to be. In addition, the observers on the plane will measure the time of flight of the baseball to be less than the time measured by the people in the stadium.

All of this weirdness derives directly from the two postulates of special relativity. The second postulate states that everyone in the stadium and on the plane must agree that the speed of light is c in a vacuum, approximately 300,000 km/s. But nothing can travel faster than c, including information. This means that as the plane flies overhead, observers in the plane will have moved some distance themselves, relative to the stadium, between the time that the ball is hit and when it passes over the fence. So which observers, the ones in the stadium or the ones on the plane, are seeing the actual events at the right times and over the correct distance? The answer is they both are! Observations of time and distance are made *relative to the observer*, but all observers must still agree that a home run has occurred.

If this seems strange, it is not because of some type of optical illusion, it is because your intuition is failing you. Your intuition is based on what you have experienced throughout your life, and you probably haven't traveled anywhere near the speed of light. As a result, you don't notice this weirdness. Even on a plane traveling overhead at 0.22 km/s (500 mi/h), that plane is only moving at about $0.000\,000\,73c$ which is nowhere near fast enough for a casual observer to notice any of the effects just described. The observers on the plane will disagree with the fans in the stands about the distance to the fence by only about $0.000\,000\,000\,03\%$ (3×10^{-11} percent). They will disagree with the observers in the stadium about the time of flight by the same amount.

But what happens if the plane is replaced by a spaceship traveling at a substantial fraction of the speed of light, as in Fig. 8.17? If the spaceship has a speed of 50% of the speed of light relative to the stadium,[4] observers in the spaceship will see the distance from home plate to the fence as being 13% shorter than the distance that the people in the stadium measure it to be. If people in the stands measure the distance as 120 m (394 ft), then the spaceship fans will claim the distance is 104 m. If the spaceship is traveling at $0.9c$ its occupants will measure the distance to the fence as 52 m, at $0.99c$ the distance is 17 m, and at $0.999c$ that distance is a mere 5.4 m.

The same changes happen with the time of flight. Assuming that the ball leaves the bat at 44 m/s (100 mi/h), people in the stands will see that it takes 120 m̸ / 44 m̸/s = 2.7 s to reach the fence if it is a line drive. Fans on the spaceship traveling at $0.5c$ will see it take 2.3 s, but at $0.999c$ the time of flight will only be 0.12 s. After all, the fence is only 5.4 m away from home plate.

Einstein's special theory of relativity eliminates the absolutes of time and space. It is only possible to determine the elapsed times between events (e.g., the ball being hit and then reaching the fence) and lengths (the distance from home plate to the fence) in relation to who is making the observations and how fast that observer is traveling relative to the events or objects being measured. After all, the stadium isn't fixed in a location on some nonexistent universal coordinate system: Earth

Key Point
The measurements of time intervals and lengths depend on the velocity of the observer relative to the events and objects being measured.

[4]Of course, humans have not yet been able to build a spaceship that can travel even remotely close to the speed of the one described in this "thought experiment" (yet).

Fig. 8.17 *Left:* Fans watching the game from the stands; *Right:* Fans watching the game from a spaceship traveling close to the speed of light. (Image of Angels Stadium: Staff Sgt. Chad McMeen, public domain. Rocket: Maurice MacGyver, CC BY 4.0)

spins on its axis, Earth orbits the Sun, the Sun orbits the center of the Milky Way Galaxy, the Milky Way Galaxy moves among the Local Group of galaxies, the Local Group moves relative to other groups, clusters, and superclusters of galaxies. Only relative motions and times are measurable.

What About the Michelson–Morley Experiment and the Ether?

Not only does the special theory of relativity make measurements of time intervals and distances dependent on the relative motion of the observer, but it also made the Michelson–Morley experiment irrelevant. Since the second postulate assumes that the speed of light is always c, no matter how fast the observer is moving relative to the source of the light, the experiment is destined to always find no difference in the round-trip times of the light beams along the two arms of the apparatus, assuming that the two arms have the same length. The ether, if it exists, cannot constitute a universal reference system of absolute space. Since Maxwell's electromagnetic theory does not require any medium for light waves to propagate through, and since there is no absolute frame of reference in the universe, the ether simply serves no purpose. The idea of the ether eventually faded away, although the term was adopted for fiber optic cables and protocols used in many computer networks today (Ethernet).

When yet another version of the Michelson–Morley experiment was conducted after his 1905 paper was published, yielding one more null result, Einstein was asked what he thought about the result, to which he replied "Subtle is the Lord, but malicious He is not."

Testing Relativity

Recall from Chapter 1 that a scientific theory is built on a set of assumptions; in special relativity those assumptions are Einstein's two postulates. But a truly scientific theory must also make testable, falsifiable predictions based on those assumptions. In addition, even though the predictions can be tested and, hopefully, verified, a theory is still not proven to be true because it is impossible to know that some day one of the predictions will be shown to be wrong. In other words, a theory can never be proven to be true, it can only be disproved. However, if tests of predictions repeatedly verify the theory without ever being disproved, one can build confidence that the theory is useful in further advancing our understanding of nature. This was the case with Newton's universal law of gravitation until it was superseded by Einstein's general theory of relativity. The general theory of relativity was able to explain phenomena that Newton's theory could not, such as the advance of perihelion of Mercury's orbit and the bending of starlight around the Sun during a total solar eclipse. Today, the general theory is routinely used in the ubiquitous global positioning system (GPS), to make subtle corrections for more accurate location determinations on the surface of Earth by accounting for the curvature of space-time due to Earth's mass.

Key Point

GPS satellites use the general theory of relativity to make more accurate determinations of locations on the surface of Earth.

Einstein's special theory of relativity, which preceded the general theory by ten years and is part of the general theory, is tested routinely and has passed every experimental test with very high precision. The consequences of watching a baseball game while flying past a field at a significant fraction of the speed of light might seem fanciful and unrealistic (which they are, of course, since the ship would travel the length of the field long before the ball even left the bat), but the effects are real. Even if Einstein's special and general theories of relativity are superseded one day, the new theory that replaces them must incorporate the realities that relativity theory has revealed.

As depicted in Fig. 8.17, imagine yourself sitting in the stands and watching the ship fly by. If you could see a clock in the ship you would see the seconds tick by on that clock more slowly than you see them on your own watch. That doesn't mean that there is something wrong with the ship's clock, because people on the ship will see the clock tick along just as it should. The reality is that time is relative to the observer. That is why observers on the ship see the time of flight of the ball as being so short: according to their clock, the ball took less time to cross the fence. The effect not only applies to clocks or other mechanical devices, it applies to everything, including the passengers. According to people in the stands, the passengers are moving more slowly, thinking more slowly, and aging more slowly, right along with their clocks. Time itself is slower in the ship, as observed by the fans in the stands. The passengers in the ship will make the same claim of the people in the stands because, relative to those in the ship, it is the stadium that is moving.

Key Point

Time dilation causes clocks moving with respect to the observer to appear to run slow.

That time effect is known as time dilation, which is often described as "moving clocks run slow." Time dilation has been tested in a very straightforward way by actually buying a round-the-world airline ticket for an atomic clock. Before boarding the planes the clock was carefully compared to an identical clock that was left behind in a laboratory. After flying around the world the two clocks were brought back together and compared again. Although they were operating at exactly the same rate in the laboratory before and after the trip, the one that made the excursion was behind the one that remained in the lab by exactly the amount predicted. The moving clock ran slowly while flying around the world.

A more common test of time dilation is conducted using radioactive particles (radioactivity was discussed in some detail in Section 7.4). Radioactive particles transform into other types of particle with well-known characteristic half-lives. When radioactive particles travel at speeds close to the speed of light, either in laboratories or by cascading down through Earth's atmosphere as a result of charged particles from space striking the atmosphere, their half-lives are measured to be much longer than when they are not moving. The increase in their half-lives is directly predictable by how fast they are moving according to the special theory, and those predictions match the experimental results precisely.

Remember that the first postulate says that the fans in the stadium and the fans in the ship must both agree that a home run occurred, but according to the time of flight of the ball as recorded by those in the spaceship, the ball couldn't have gone very far. Their measurement of the distance to the fence was much shorter than the people in the stadium determine that distance to be. This effect is known as length contraction; for a moving observer the length of an object (in this case the distance to the fence) is shorter in the direction of motion than it is when measured by an observer who isn't moving relative to that object. Referring again to Fig. 8.17, the fans in the ship see the home run occur because the field in that direction is very short and so the time of flight to reach the fence was very short as well. According to the fans watching in the stadium, the distance traveled by the ball was much longer and took much longer, but it was a home run no matter who was watching. The same occurs for the traveling atomic clock or the radioactive particles. From their perspectives, time is behaving "normally," but they didn't have very far to go, so they made the trips, in agreement with the at-rest observers.

In reality, there is no correct view of length or time, but we do identify measurements made at rest with respect to a length or a clock as being "proper." For example, people in the stadium would measure the "proper length" of the distance between home plate and the outfield fence. On the other hand, the "proper time" of the home run would be measured by a hypothetical clock riding along with the baseball during its flight. The proper length of a wood plank is measured by a tape measure resting on the plank, and the proper half-life of a radioactive particle is measured when the particle isn't moving in a laboratory.

8.4   The Doppler Effect for Light

There is an important aspect of the relativistic spaceship in Fig. 8.16 that needs to be revisited. Note that the color of the laser light fired by the rocket was red but the color of that same light seen by observers on the planet that the rocket was approaching was blue. The light didn't undergo some transformation during its passage through space, the different colors have to do again with the postulates of special relativity and how the postulates impact measurements of time intervals and lengths.

You are likely familiar with the Doppler effect for sound. If you hear a siren from an emergency vehicle or a car horn coming toward you, the pitch is higher than it is when the vehicle is moving away from you. The higher pitch is due to a higher frequency of sound wave peaks arriving at your ear and the associated shorter wavelengths of those waves (recall the relationship between frequency, wavelength, and speed, $f\lambda = v$). This bunching up of the waves happens because between the times when each wavelength peak is emitted, the vehicle has moved. If the vehicle is coming toward you there is less distance between peaks, because the centers of the emitted waves keep changing.

The sound of a passing car horn on the Chapter 8 resources webpage. (Magickallwiz, CC BY-SA 3.0)

Fig. 8.18 The wake from a boat traveling faster than the speed of the waves. (Edmont, CC BY-SA 3.0)

A similar situation occurs if you move rapidly toward a fixed sound source. Instead of the waves bunching up because the source is moving, you simply encounter the waves more frequently than you would if you were standing still, and so you hear a higher pitch. Similarly, if you move away from the source it takes longer for each sound wave to pass you and so you hear a lower pitch. The difference between the two types of motion is that sound waves travel through air with a speed determined by the characteristics of the air (its temperature, density, and composition). It is also possible for the observer or the source to actually outrun sound waves, referred to as supersonic speed. If the source outruns the waves, the waves become very compressed, resulting in a high air pressure that creates a sonic boom (think of the waves at the bow of a boat as it travels quickly through water; Fig. 8.18).

Light waves behave similarly to sound waves when the source and the observer are moving, with one important exception. Since there is no absolute space through which light travels (there is no ether), it doesn't make sense to say whether it is the source or the observer that is moving; we can only specify how they are moving relative to each other. Recall that the second postulate of special relativity states that the speed of light is always c, meaning that light travels away from the source at the same speed that it approaches the observer, even if the observer and the source are moving toward or away from one another at speeds close to the speed of light.

Consider the situation in Fig. 8.19. A ship is traveling with a speed v through space relative to two observers. As the ship moves it emits a steady green light. According to the pilot of the ship, the light travels outward from the ship in every direction, with the ship at the center of the emitted light. If the pilot could see

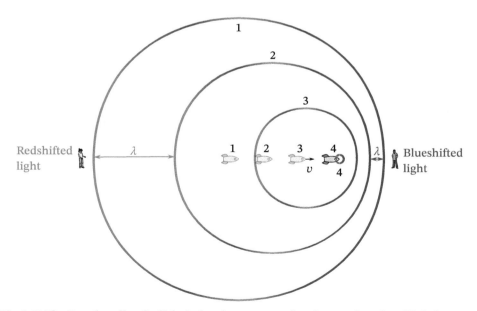

Fig. 8.19 The Doppler effect for light is the phenomenon that the wavelengths of light become shorter and the wave frequencies increase if the source and the observer are approaching each other. The wavelengths of light also become longer, with lower frequencies, when the source and observer are moving away from one another. Perpendicular to the direction of travel, the wavelengths are unchanged. The ship emitted green light in every direction.

the peaks of the propagating electromagnetic waves she would see them as being concentric with the source of the light.

However, consider what an observer sees if the ship is moving toward him. Because the ship moves between each emitted wave peak, the wave centers move as well. Notice that when the ship is at position 1 it emits a wave that expands outward from the position where the ship was at that time. Similarly for positions 2, 3, and 4. Just after the ship emits wave 4, the other waves have expanded outward to their indicated positions. Because wavelength is the distance between wave peaks, the observer that the ship is moving toward sees light of shorter wavelength and higher frequency. In other words, the color of the light is shifted toward the blue part of the visible spectrum and the light is said to be blueshifted. Similarly, if the ship is moving away from an observer, the wave peaks are spread out and the light's wavelength is longer. In this case the color of the light is shifted toward the red end of the visible spectrum and the light is said to be redshifted.

The terms blueshift and redshift simply indicate the direction that the color of the light is shifted (to shorter wavelengths or longer wavelengths, respectively), they are not meant to indicate the actual final colors. In fact, the wavelengths of the light can be shifted by any amount. For example, it is possible for ultraviolet light to be so redshifted that it ends up in the infrared part of the electromagnetic spectrum. As long as the source isn't moving too fast relative to the observer, the speed between the source and the observer is related to the amount of wavelength shift by

Key Point
Light from an approaching source is blueshifted, while light from a receding source is redshifted.

$$\frac{v}{c} = \frac{\Delta\lambda}{\lambda_{emit}} = \frac{\lambda_{obs} - \lambda_{emit}}{\lambda_{emit}}, \qquad (8.2)$$

The Doppler effect equation

A Doppler effect equation tutorial is available through the Chapter 8 resources page.

where λ_{emit} is the wavelength of the light when it is emitted from the source and λ_{obs} is the wavelength that is observed. The symbol $\Delta\lambda$, which equals $\lambda_{obs} - \lambda_{emit}$, is read as "delta lambda," where Δ is the Greek capital letter, delta. Δ (or the lower case δ) is almost universally used to represent the change in something; in this case, the change in wavelength caused by the source moving at a velocity v relative to the observer. The equation can be read as the ratio of the speed of the source to the speed of light equals the difference in the observed wavelength and the wavelength that would be measured in a laboratory (the emitted wavelength) divided by the laboratory wavelength.

If the source is moving away from the observer (redshift), then the velocity is positive, but if the source is moving toward the observer (blueshift), the velocity is negative. Absorption spectra that are blueshifted, not shifted, and redshifted are shown in Fig. 8.20 for sources that are approaching, at rest, and receding from the observer, respectively. Notice that the lines have shifted across the colors of the continuous spectrum.

Example 8.1

An astronomer is measuring the wavelength of one of the absorption lines of hydrogen in the spectrum of a galaxy. The absorption line normally has a wavelength

Blueshift

No motion

Redshift

Fig. 8.20 Absorption lines that are (*top*) blueshifted because the source is approaching the observer; (*middle*) not shifted because the source isn't moving relative to the observer; and (*bottom*) redshifted because the source is moving away from the observer.

of 656.470 nm in a laboratory, but the scientist finds the observed wavelength to be 662.844 nm. How fast is the galaxy moving toward or away from the telescope?

Since the observed wavelength is longer than the laboratory wavelength, the light is redshifted and the galaxy is moving away from the telescope. Taking the difference between the two wavelengths, the wavelength shift is

$$\Delta\lambda = \lambda_{\text{obs}} - \lambda_{\text{emit}} = 662.844 \text{ nm} - 656.470 \text{ nm} = 6.374 \text{ nm}.$$

This implies that the fractional change in wavelength is given by

$$\frac{\Delta\lambda}{\lambda_{\text{emit}}} = \frac{6.374 \text{ nm}}{656.470 \text{ nm}} = 0.0097.$$

Now, from the Doppler effect equation, the ratio of the speed of the galaxy to the speed of light is

$$\frac{v}{c} = \frac{\Delta\lambda}{\lambda_{\text{emit}}} = 0.0097,$$

meaning that the speed of the galaxy as it moves away from the telescope is

$$v = 0.0097c = 0.0097 \times 300{,}000 \text{ km/s} = 2900 \text{ km/s}.$$

The galaxy is moving away from the telescope at 0.97% of the speed of light.

The caveat, "as long as the source isn't moving too fast relative to the observer," mentioned just before the Doppler effect equation has to do with how close to the speed of light the source is moving relative to the observer. If the speed is more than about 10% of the speed of light then a more complex equation must be used to properly calculate the wavelength shift.

Today, it is possible to measure wavelengths so precisely that motions of stars can sometimes be determined to a few *meters per second*, which is close to walking speed. Such precise measurements led to the first detections of planets around other stars.

8.5 Einstein's Explanation of Planck's Energy Quanta

You may have noticed in the citation for Einstein's 1921 Nobel Prize that it was awarded to him "for his services to Theoretical Physics, and especially for his discovery of the law of the photoelectric effect." The citation says nothing directly about his relativity theories that most people associate with Einstein's career. The reason is somewhat complex and multifaceted, but involves in part the preference to experimental physics at the time and the absence of experimental verification of aspects of his special theory of relativity from 1905 (see Section 8.3), even though the much more complex general theory, published in 1915, had been verified through the explanation of the advance of Mercury's perihelion and the prediction of the eclipse results in 1919.

The Nobel Prize citation does explicitly reference Einstein's first paper in his *annus mirabilis* about the photoelectric effect. While the paper explained the results of that experiment, it also simultaneously provided a foundation for the mysterious ad hoc solution that Planck developed for the blackbody radiation curves (recall Section 8.1). More precisely, Einstein explained Planck's need for energy quanta, and then extended the idea to develop the law of the photoelectric effect.

When Planck inserted his energy quanta equation [Equation (8.1)], $E = hf = hc/\lambda$, to derive his solution for the blackbody radiation curves, he introduced a previously unknown fundamental constant, Planck's constant, h, without knowing why it was needed. He had assumed that there was some as-yet unexplained aspect of producing electromagnetic waves that required that the energy of the waves had to be integer multiples of $E = hf$. Einstein's paper instead argued that it isn't the production of electromagnetic waves that is quantized, but that light is composed of *particles of energy* each having an amount of energy equal to Planck's constant times the frequency of the light that would have been associated with Maxwell's electromagnetic waves. In other words, Einstein showed that, in some sense, Newton's claim that light is corpuscular was in fact correct, despite how well Maxwell's electromagnetic waves describe all other aspects of light. Einstein's light quanta would come to be called photons.

The idea that light can be both a wave and a particle *simultaneously* is very counter-intuitive. Our intuition is always based on our everyday experiences. The behavior of individual photons is not something we routinely experience at the macroscopic level. The fact that Planck's constant is so tiny, $h = 6.63 \times 10^{-34}$ J s, means that we only witness the behavior of photons on extremely small size and energy scales. On a macroscopic level, we see the collective behavior of enormous numbers of photons, the ensemble of which behaves as though light is a wave. Einstein not only introduced us to the concept of photons but he also introduced us to what is termed wave–particle duality, together with the birth pangs of quantum mechanics.

Key Points
Light is composed of particles called photons. Each photon has energy proportional to its wave frequency and inversely proportional to its wavelength: $E = hf = hc/\lambda$.

Example 8.2

An old-style, incandescent 100 W light bulb emits 100 J of energy every second. If we assume (incorrectly) that all of the photons are at the red edge of the visible spectrum, with wavelengths of 700 nm, how many photons are emitted

A scientific notation
tutorial is available at
Math_Review/Scientific_
Notation

every second? (Those old light bulbs actually emit about 90% of their energy in infrared light, where the human eye doesn't detect them. That means that they are much more efficient at heating up a room than they are at lighting up a dark space.)

The frequency of a photon with a wavelength of λ = 700 nm is given by the frequency–wavelength–velocity equation for light

$$f = \frac{c}{\lambda} = \frac{3 \times 10^8 \text{ m/s}}{700 \times 10^{-9} \text{ m}} = 4.3 \times 10^{14} \text{ Hz}.$$

[Remember that 1 nm = 1×10^{-9} m (one billionth of a meter) and 1 Hz = 1/s.] This means that the energy of a single 700 nm photon is only

$$E = hf = (6.63 \times 10^{-34} \text{ J s}) \times (4.3 \times 10^{14} \text{ Hz}) = 2.8 \times 10^{-19} \text{ J}.$$

Recalling that one watt is one joule per second, in order to determine how many photons are emitted per second by the 100 W light bulb we need to divide by how much energy each photon possesses:

$$\frac{100 \text{ J/s}}{2.8 \times 10^{-19} \text{ J / photon}} = 3.5 \times 10^{20} \text{ photons/s}.$$

That is 350 million trillion photons per second! No wonder we aren't used to seeing individual photons every day.

Since quantum energies are so small, it is common to use an energy unit that reflects those energies. An *electron volt* (eV) is defined in terms of the kinetic energy of an electron after it has "fallen" 1 volt (1 V). The value of an electron volt is 1 eV = 1.602×10^{-19} J: extremely small by human scales. It is also common to express wavelengths of light in the infrared, visible, and ultraviolet portions of the electromagnetic spectrum in units of nanometers, as was done in Table 7.2. Using these units, we can write the equation for a photons more conveniently as:

$$E_{\text{eV}} = \frac{1240}{\lambda_{\text{nm}}} \text{ eV}, \tag{8.3}$$

Energy of a photon in electron volts

assuming that the wavelength is given in nanometers (the subscript for λ_{nm} is to remind you that nanometers must be used). We will normally use this form of the equation of a photon whenever we need to calculate a photon's energy. The equation can also be rewritten to calculate the wavelength of a photon if its energy is known in electron volts (hence the eV subscript): $\lambda_{\text{nm}} = 1240/E_{\text{eV}}$ nm.

Example 8.3

The longest visible wavelength of a photon emitted by oxygen is 646 nm. What is the energy of the photon in electron volts?

Using the equation for the energy of a photon in electron volts,

$$E_{eV} = \frac{1240}{\lambda_{nm}} \text{ eV} = \frac{1240}{646} \text{ eV} = 1.92 \text{ eV}.$$

Einstein proposed a way to verify the existence of photons by shining light on a metal, as illustrated in Fig. 8.21. If the energy of the photons is great enough (meaning that the frequencies are high enough and the wavelengths are short enough), electrons in the metal can escape the surface, travel across a gap, and create a current in a wire. Think of the electrons as being "stuck" on the surface; if the photons don't have enough energy to knock the electrons off the surface, there won't be any current. Visible light photons don't have enough energy to get the job done, but ultraviolet photons do. It doesn't matter how bright the visible light is, if the frequency of the light isn't high enough, nothing happens. This contradicts what Maxwell's electromagnetic theory suggests, because according to the electromagnetic theory, brighter light means more energy. While that is true, only individual photons can eject individual electrons, so regardless of how many photons are arriving (the brighter the light, the more photons), if individual photons aren't energetic enough, there still won't be any current. However, if the photons do have enough energy, the brighter the light is, the more photons there are, and the greater the number of electrons that escape the metal. This means that the amount of current should be directly related to how bright the source is, *but only if the frequency is high enough*. This is the law of the photoelectric effect. The term "photoelectric" comes from the combination of light (photo) and the production of an electric current. The prediction of the law was successfully confirmed in 1914 by Millikan. It was

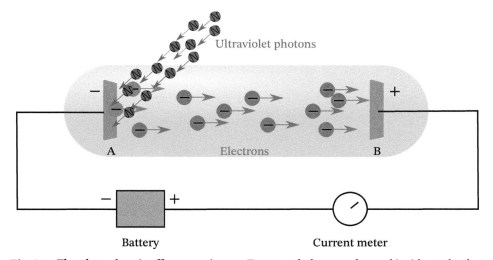

Fig. 8.21 The photoelectric effect experiment. Two metal plates are located inside a tube that has had the air removed. The two plates are connected to a battery. When ultraviolet photons with sufficiently high energy strike the negative plate (A), each photon ejects a single electron. The negatively charged electrons are then attracted to the positive plate (B), completing the circuit and producing a current that is measured by the meter.

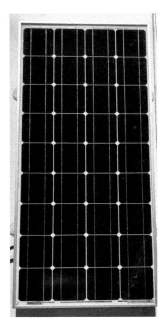

Fig. 8.22 A solar panel mounted on a truck camper. (Dale A. Ostlie)

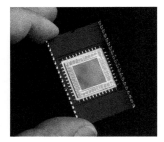

Fig. 8.23 A CCD designed for ultraviolet and visible wavelength photons. (NASA)

Millikan who would later try to recruit Einstein to Caltech, before Einstein decided to make Princeton his home.

It is the particle nature of light that is the basis for a number of modern advancements in technology. For example, the photovoltaic cell that is at the heart of solar panels generates electricity by absorbing photons that liberate electrons from their normal attachment to atoms (see Fig. 8.22). A charge-coupled device (CCD), like the one shown in Fig. 8.23, is used in numerous scientific instruments, as well as in digital cameras, such as the one in your smartphone. For digital cameras, photons again eject electrons, except in these devices the electrons are temporarily held at each pixel location in a two-dimensional grid of cells. The electrons are then counted, with the number of electrons being nearly equal to the number of photons that struck that pixel. The brightness of the image at that location is directly proportional to the number of trapped electrons.

8.6   Bohr's Atom

Planck and Einstein weren't the only scientists around the turn of the 20th century who were changing the way we understand nature and the universe. Recall that in the 19th century Dalton, Mendeleev, and others had already demonstrated that there appeared to be indivisible smallest possible quantities of matter, known as atoms, that retained the characteristics of specific elements. In addition, Kirchhoff, Bunsen, and others had succeeded in demonstrating that elements also had characteristic spectra that acted like fingerprints, allowing for the identification of some of the elements in the Fraunhofer spectrum of the Sun and other stars. Helium was even identified in the Sun, revealed by its spectrum during a total solar eclipse, before it was ever isolated on Earth. We have also seen that in the late 19th century and early 20th century, the determination of the properties of the electron, and the discovery of radioactivity with its emissions of alpha particles (producing the spectrum of helium), beta particles (electrons), and gamma rays (photons) were strongly hinting that atoms aren't actually the smallest units of matter after all. All of these discoveries and others described in this section were prelude to another major revolution in physics that developed more slowly than the relativity theories of Einstein, and involved more researchers, but would be no less profound. The rapid pace of discoveries would culminate in the development of quantum mechanics in the 1920s.

The Discovery of the Nucleus of the Atom

When Rutherford discovered that uranium emitted alpha particles, he carefully studied their properties. He learned that each alpha particle contains two positive elementary charges ($+2e$) and has a mass that is more than 7000 times greater than that of an electron. He also concluded (correctly) that the alpha particle is a helium atom, but with two electrons missing, hence its positive charge.

Recall from page 256 that J. J. Thomson had theorized that an atom is essentially like a smoothly distributed, positively charged pudding with negatively charged electrons embedded in it. Rutherford proposed to test that idea by using alpha particles as probes. As illustrated in Fig. 8.24, the experiment was to "shoot" alpha

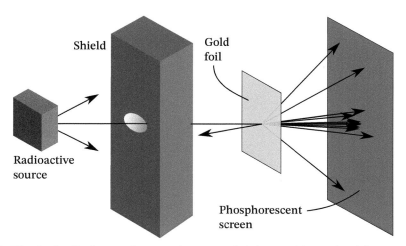

Shield

Gold
foil

Radioactive
source

Phosphorescent
screen

Fig. 8.24 The Rutherford scattering experiment used alpha particles emitted from a radioactive source as charged projectiles to probe the structure of the atoms in the gold foil. On rare occasions, an alpha particle would deflect at a high angle, or even backward, from the foil. In the actual experiment phosphorescent screens surrounded the gold foil to detect scattering in all directions.

particles at a very thin sheet of metal, and then detect how the alpha particles were scattered as they moved through Thomson's positively charged "pudding." The concept is not unlike throwing rubber balls into a thick fog and then determining which way they ricochet off unseen objects in the fog in order to determine the shapes of the objects. When Rutherford's assistants[5] conducted the experiment in 1909, they reported that most of the alpha particles passed straight through the foil with very little deflection, as expected for a uniformly distributed positive charge, but some of the alpha particles were deflected by large angles, including back toward the source of the alpha particles. Rutherford later recounted in a lecture at Cambridge University: "It was quite the most incredible event that has ever happened to me in my life. It was almost as incredible as if you fired a 15-inch shell at a piece of tissue paper and it came back and hit you."

Rutherford realized that the only way to explain the results of the scattering experiment was that atoms were actually made up of a very tiny, positively charged nucleus, with electrons orbiting around the nucleus, much like planets orbit around the Sun. Only instead of gravity holding the planets in their orbits, it was the electric force that was holding the electrons in their orbits around the nucleus. When the alpha particles, which must be helium nuclei, passed through the foil, most of them were only deflected slightly because they passed through empty space and were hardly pushed at all by the electric fields of the nuclei. However, every once in a while an alpha particle would pass close to a nucleus, or approach one virtually head on, and be deflected at large angles or even bounce straight back. In 1911, based on a careful analysis of his model for the nuclear atom, Rutherford was able to predict statistically how many alpha particles would be deflected in every direction. That prediction was confirmed by further experiments in 1913.

---

[5]Rutherford's two assistants in the scattering experiment were Hans Geiger (1882–1945) a physicist who Rutherford had asked to work with him and for whom the Geiger counter is named, and Ernest Marsden (1889–1970), an undergraduate student working with Geiger.

The dimensions of the atom are remarkable. The hydrogen atom, for example, has a diameter that is about 0.1 nm in diameter (the size of the electron's orbit), while the nucleus has a diameter that is less than 2 fm (one femtometer is 10^{-15} m, one thousandth of a trillionth of a meter). This means that the electron's orbit is 50,000 times larger than the nucleus. Thought of in another way, virtually all of the mass of an atom is contained in a nucleus that makes up only about 10^{-14} of the volume of the atom, or 0.000 000 000 001%. Using the Solar System as an analogy, if the nucleus were the size of the Sun, the orbiting planet would be more than 230 times farther from the Sun than Earth is. It has been pointed out that, even though our bodies contain something like 10^{27} (or one thousand trillion trillion) atoms, the vast majority of our bodies are simply empty space!

Balmer's Discovery of a Pattern in Hydrogen Line Wavelengths

Fig. 8.25 Johann Jakob Balmer (1825–1898). (Public domain)

Johann Balmer (Fig. 8.25) was a Swiss mathematician teaching at a girl's school in Basel, although he also taught a bit at the University of Basel as well. In 1885, at sixty years of age, Balmer set out to develop a formula for the wavelengths of hydrogen in the visible portion of the spectrum. His effort turned out to be successful. His equation wasn't based on any known fundamental aspect of physics, but he was able to develop an equation that successively involved using integers, starting with 3, then 4, 5, and 6. Each successive integer from 3 upward corresponded to the already determined wavelengths of hydrogen beginning with the red line to the bluest line. He was even able to predict a previously unknown wavelength of hydrogen by extending the sequence using 7 as the next integer, which turned out to be at the edge of the blue/violet and ultraviolet portions of the electromagnetic spectrum. Today, that sequence is known as the Balmer series and corresponds to the notation mentioned on page 253: 3 gives the wavelength of the Hα line, 4 for Hβ, 5 for Hγ, and so on through the Greek alphabet.

The Balmer series begs the questions, why does the sequence start with 3 and not 1 or 2, and why do the numbers used need to be discreet sequential integers?

Niels Bohr's Atomic Model

Fig. 8.26 Niels Bohr (1885–1962). (Public domain)

There was a problem with Rutherford's nuclear model that was recognized almost immediately. It was in conflict with Maxwell's electromagnetic theory equations. Maxwell's equations say that any accelerating charge must radiate electromagnetic waves and energy. Although Rutherford's electrons may be moving at constant speeds, they must also be accelerating because their direction of motion is constantly changing. The electric force causes them to move in an orbit, and so Newton's second law of motion, $\vec{F} = m\vec{a}$, requires acceleration. Because the electrons are accelerating they must radiate away energy, causing them to spiral into the very tiny nucleus within a tiny fraction of a second. This would mean that the dimensions of matter would immediately become 50,000 times smaller, which is obviously not the case!

In 1913 Niels Bohr (Fig. 8.26), a quiet Danish physicist working at the University of Copenhagen, combined Rutherford's nuclear model with the quantized light (photon) idea of Einstein. Essentially he argued that, since electrons in Rutherford's model would spiral into the nucleus, nature must somehow suspend that

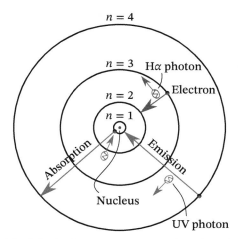

Fig. 8.27 The Bohr model of the hydrogen atom. A nucleus (much smaller than depicted) with one positive elementary charge is orbited by one electron (also much smaller than depicted), with a single negative elementary charge. The only orbits that the electron can occupy are those with specific integer multiples of a quantity of angular momentum and an associated energy. A photon is emitted when the electron "jumps" from a higher energy orbit to a lower one, and a photon is absorbed when the electron "jumps" from a lower energy orbit to a higher energy orbit. The orbital transitions must have the same energy change as the photon's energy.

"rule" for specific orbits at well-defined distances from the nucleus. In these special orbits, electrons can orbit forever. To specify which orbits have this unique characteristic, Bohr proposed that the orbital angular momentum must be quantized [Equation (6.3)] with the angular momentum of each orbit being an integer multiple of Planck's constant h (h has units of angular momentum) divided by 2π, the number of radians in a circle (2π radians $= 360°$).

An important implication of Bohr's model was that each orbit had a specific energy associated with it that also involved Planck's constant and integers. Bohr then proposed that when an electron jumps down from a higher orbit to a lower one the atom must emit energy in the form of a photon that is equal to the difference in orbital energy, as required by the conservation of energy. Bohr was able to use a combination of Coulomb's law of electrostatic force [Equation (6.10)] and the equation for quantized angular momentum to derive Balmer's formula for the visible wavelengths of light.

Figure 8.27 shows the orbits of an atom can be numbered starting with 1 as the orbit closest to the nucleus, 2 as the next orbit out, followed by 3, and so on. These numbers are symbolized by n and referred to as quantum numbers. To produce visible photons for Bohr's hydrogen atom, the jumps started at orbit $n = 3$ or above, with the electrons ending up in orbit $n = 2$. The jump from 3 to 2 produces a photon with an energy that corresponds to the wavelength of the Hα line ($E = hf = hc / \lambda$, Planck's energy equation). The jump from 4 to 2 produces the Hβ line, 5 to 2 corresponds to Hγ, ... Of course, electrons can also jump down to the first orbit, with jumps from 2 to 1, 3 to 1, 4 to 1, et cetera, producing ultraviolet (UV) photons. The bigger the jump, the more energy is given off by the atom in the form of a photon, and therefore the shorter the photon's wavelength. Jumps that land in orbit 3 or

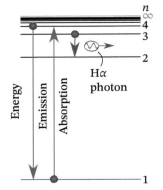

Fig. 8.28 Energy levels for the hydrogen atom. The quantum numbers on the right correspond to Bohr's orbit quantum numbers. Note that even though the orbits in Fig. 8.27 get farther and farther apart as n increases, the associated energy levels get closer and closer together, reaching a maximum value when n approaches infinity (infinity is universally symbolized by ∞).

higher (e.g., 4 to 3) produce infrared photons. The prediction of ultraviolet and infrared lines of the hydrogen spectrum was later confirmed. Higher orbits correspond to more energy, as illustrated in Fig. 8.28. The lowest electron energy level ($n = 1$) is commonly referred to as the ground state. For hydrogen, that energy is -13.6 eV, and the energies of the other energy levels are given by

$$E = \frac{-13.6 \text{ eV}}{n^2}. \tag{8.4}$$

Hydrogen atom energy levels

Example 8.4

An electron "jumps" from the $n = 3$ energy level to the $n = 2$ energy level, causing the atom to emit a photon. What is the photon's energy and wavelength?

The $n = 3$ orbit has an energy of $E_3 = -13.6 \text{ eV}/3^2 = -13.6 \text{ eV}/9 = -1.51$ eV. For the $n = 2$ orbit, $E_2 = -13.6 \text{ eV}/2^2 = -13.6 \text{ eV}/4 = -3.40$ eV. Since the $n = 3$ orbit is higher and has a less negative energy value, it has more energy than the $n = 2$ orbit. The difference in their energies is $E_3 - E_2 = -1.51 \text{ eV} - (-3.40 \text{ eV}) = 1.89$ eV.

The wavelength of the photon can be calculated by using the equation for the energy of a photon in electron volts [Equation (8.3)] after rewriting it to solve for λ_{nm}:

$$\lambda_{nm} = \frac{1240 \text{ eV}}{E_{eV}} = \frac{1240 \text{ eV}}{1.89 \text{ eV}} = 656 \text{ nm}.$$

You can find that wavelength listed in Table 7.2.

Kirchhoff's emission and absorption lines are explained by Bohr's model of the atom. Emission lines are produced when the electrons jump down from higher orbits to lower ones. On the other hand, the reverse is also possible. If a photon has just the right amount of energy to lift an electron from a lower orbit to a higher one, the atom absorbs the photon. When the photon is absorbed by the atom it is no longer available to be detected. The more atoms with electrons making the same downward transition, the brighter the emission line will be. Similarly, the more photons of the same energy that are absorbed by atoms causing their electrons to transition to higher orbits, the darker the absorption line is.

Although successful for hydrogen, Bohr's model didn't work for more complex atoms. It was also an uncomfortable combination of the classical physics of Maxwell combined with an ad hoc restriction on orbits that involved Planck's constant. Ad hoc theories are always found wanting in science, because they suggest that there is something deeper going on. The application of an ad hoc solution is simply a quick fix that doesn't rest on a fundamental understanding of nature. At this point, it would be well worth reviewing our discussion of the nature of science in Chapter 1.

8.7 Quantum Mechanics: What Are the Chances?

Synthesis Beyond Intuition: Wave–Particle Duality

In his 1924 Ph.D. dissertation, Louis de Broglie (Fig. 8.29) made a rather bold suggestion. If light can behave as both a wave and a particle (photon), then perhaps other particles, those having mass, can behave like waves. By thinking about the linear momentum of a photon, which can be written in terms of its wavelength, de Broglie solved for the wavelength in terms of the linear momentum instead. His resulting equation is given by

$$\lambda = \frac{h}{p} = \frac{h}{mv},$$

(8.5)

de Broglie's wavelength equation for matter

Fig. 8.29 Louis de Broglie (1892–1987). (AIP Emilio Segrè Visual Archives, Physics Today Collection)

the de Broglie wavelength for particles with mass equals Planck's constant divided by the particle's linear momentum. Since the wavelength of a particle is inversely related to its linear momentum, the greater the linear momentum, the shorter the wavelength. If the linear momentum doubles, then the wavelength is cut in half. If the linear momentum increases by a factor of 3, then the wavelength becomes shorter by a factor of 3. The wave–particle duality Einstein introduced us to with photons, de Broglie extended to all particles.

One implication of the particle wavelength concept is that it should be possible to conduct the equivalent of a Young double-slit experiment (page 206) with electrons instead of photons (Fig. 6.26). In the light version of the experiment, the interference pattern of alternating bright and dark spots corresponds to areas on the screen where many photons arrive (bright) or where no photons strike the screen (dark). The intensity of the light, as Einstein postulated in the photoelectric effect experiment, is associated with the number of photons. A version of the Young double-slit experiment for electrons should behave the same way. If a stream of electrons passes through the slits an interference pattern will appear, with alternating regions of many electrons or few electrons. Figure 8.30 shows that when the experiment was conducted, de Broglie's prediction was verified; electrons can behave as though they are waves, just like massless photons do. All particles, massive or massless, exhibit both wave and particle characteristics.

You, as a human being, certainly have mass, so do you behave like a wave as well? The answer, amazingly, is yes. However, before you get too worried about producing your own interference pattern when you pass through a slit, or a door, you should realize that your mass is a tad bit greater than that of an electron, and Planck's

Fig. 8.30 Electrons in a double-slit experiment building up over time. The number of electrons in each frame: (a) 11, (b) 200, (c) 6000, (d) 40,000, and (e) 140,000. (Provided with kind permission of Dr. Akira Tonomura (1942–2012); Belsazar, CC BY-SA 3.0)

constant is an extremely small number. Since wavelength is inversely proportional to mass times velocity (linear momentum), your much larger mass implies a very short wavelength. Your author's mass is approximately 95 kg, so when he is walking at a speed of 1 m/s, his wavelength is about 7×10^{-36} m, which is about 14 trillion trillion times smaller than the diameter of a hydrogen atom. On the other hand, the mass of an electron is extremely small, and so if an electron is moving at a speed of 10^6 m/s, its wavelength would be roughly 7×10^{-10} m, which is about 7 times greater than the diameter of a hydrogen atom. In the atomic realm, particle wavelengths become very important.

Schrödinger's Equation

One encouraging aspect of de Broglie's particle waves is that the circumference of electron orbits in Bohr's atom turn out to correspond not only to quantized angular momentum, but they are integer multiples of electron wavelengths. For the $n = 1$ orbit, the circumference of the orbit is one wavelength, for the $n = 2$ orbit, the circumference is two wavelengths, and so on. However, there were now other baffling questions: What physically does a particle wavelength actually mean, and how do they fit into a self-consistent theory for atoms and molecules?

In 1926, Erwin Schrödinger (Fig. 8.31) developed an equation that incorporated de Broglie's waves. In much the same way that Maxwell's wave equation described the wave nature of light, Schrödinger's equation described the wave nature of particles having mass. A different mathematical formulation that turned out be equivalent to Schrödinger's was developed at about the same time by Werner Heisenberg (Fig. 8.32).

Fig. 8.31 Erwin Schrödinger (1887–1961). (Public domain)

Schrödinger's equation marked the culmination of the bits and pieces of new ideas that had been developing over the 26 years since Planck introduced his mysterious constant, h. Schrödinger's equation not only describes the wave-like behavior of electrons, but it also provides for a self-consistent description of atoms. The orbital angular momenta of the Bohr hydrogen atom turned out not to be quantized in the way that Bohr proposed, although they are quantized, but the energies of the "orbits" were the same as Bohr's in special cases, as are the diameters of the "orbits" of the electrons. Furthermore, in addition to n, the principal quantum number, two other quantum numbers appeared out the equation when describing atoms: ℓ, the orbital angular momentum quantum number, and m_ℓ, the magnetic quantum number which plays a role when the atom is in the presence of a magnetic field. The three quantum numbers arise because of the three dimensions of space. When special relativity, with its four spacetime dimensions, was incorporated into Schrödinger's equation by Paul Dirac (Fig. 8.33), a fourth quantum number, m_s, also appeared, which is known as the spin quantum number.

The reason for putting "orbits" in quotes above is that they cannot be described as true orbits like those of planets around the Sun, but are instead more like clouds around the nucleus with the shapes of the clouds associated with particular combinations of the three quantum numbers as described by Schrödinger's equation. For this reason, the term "orbits" gets replaced by orbitals.

Fig. 8.32 Werner Heisenberg (1901–1976). (Public domain)

Revisiting the Periodic Table of the Elements

The four quantum numbers (n, ℓ, m_ℓ, m_s) explain Mendeleev's periodic table of the elements (Table 7.1). Each row of the table is associated with the principal quantum number, n, and each column is tied to the angular momentum quantum number, ℓ. Each element's atomic number (1 for hydrogen, 2 for helium, 3 for lithium, ...) represents the number of positive elementary charges in the nucleus of the atom as well as the number of electrons in each electrically neutral atom's orbitals. The atomic number also correlates with increasing mass for the atoms of each element. Furthermore, no two electrons in an atom can have the same combination of the four quantum numbers. This last aspect of the theory is known as the Pauli exclusion principle, named for Wolfgang Pauli (Fig. 8.34).

The Probabilistic Nature of the Universe

Calculating atomic orbitals and particle waves isn't the same thing as actually understanding what they mean physically, however. Trying to make sense of this new quantum mechanics brought together most of the leading physicists in the world for conferences. Figure 8.35 is the class picture of one such conference, held in Brussels in 1927. In that single photograph you will find Einstein, Marie Curie, Schrödinger, Planck, Lorentz, de Broglie, Bohr, Heisenberg, and Pauli. Other major names in the development of quantum mechanics are also in the photograph.

The interpretation of the particle waves in Schrödinger's equation that was developed through the various conferences is that the *waves represent probabilities* of where the particles can be detected. You already saw this in Fig. 8.30, as the electrons in the double-slit experiment hit the detector based on the probability of where they should end up. In the case of atomic orbitals, like those shown for hydrogen

Fig. 8.33 Paul Adrien Maurice Dirac (1902–1984) in his Nobel Prize portrait. (Public domain)

Fig. 8.34 Wolfgang Pauli (1900–1958) in 1933. (© CERN)

Fig. 8.35 Attendees at a quantum mechanics conference, held in 1927 at the Slovay Institute in Brussels. (Public domain)

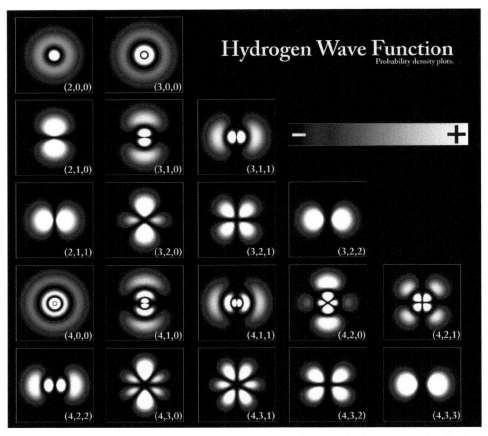

Fig. 8.36 Examples for hydrogen of the electron probability clouds, known as orbitals, that come out of Schrödinger's equation. The quantum numbers in parentheses are (n, ℓ, m_ℓ). The hydrogen ground state, $(1, 0, 0)$, is not shown, but is spherically symmetric with the greatest probability located in the center of the orbital. The horizontal bar indicates the color coding ranging from the lowest probability of finding the electron around the nucleus ($-$) to the highest probability ($+$). (PoorLeno, public domain)

in Fig. 8.36, the greatest probability of detecting an electron depends on its energy level, n, in combination with the other quantum numbers, ℓ and m_ℓ. The brightest regions in the various frames of Fig. 8.36 indicate the most likely places to find hydrogen's single electron, and regions where the cloud is less concentrated represent less likely locations for the electron. The shapes of the orbitals and the number of electrons in each atom govern how atoms combine with one another during chemical reactions. Although the orbits in Bohr's model depicted in Fig. 8.27 and the hydrogen atom energy levels equation [Equation (8.4)] were ultimately shown to be incorrect, the basic energy levels in Fig. 8.28 for the hydrogen atom produced by his theory did actually turn out to be right.

For this text, it is not important that you know any of the details of the shapes of the orbitals, or specific quantum numbers and their roles in the behavior of atoms; Fig. 8.36 is only provided to help you understand that electrons do not orbit the nucleus like planets orbiting around a star. Rather, simply know that there is a set

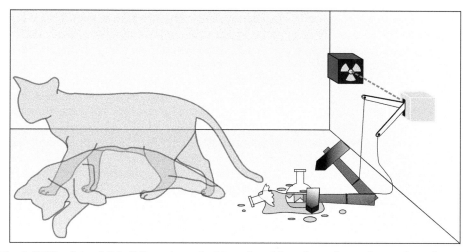

Fig. 8.37 The fate of Schrödinger's cat inside a closed box depends on the random decay of a radioactive atom. However, according to quantum mechanics, the actual outcome of the 50–50 chance of being alive or dead isn't actually established until an observer looks inside. (Dhatfield, CC BY-SA 3.0)

of quantum numbers that summarize probability clouds (orbitals) where electrons might be detected. Each of the quantum numbers can also affect the energies that the electrons have in the orbitals, and therefore the wavelengths that are produced when an electron makes a downward transition from a higher energy orbital to a lower one. Planck's introduction of his mysterious constant, h, led us down an amazingly rich new path. It can be said that general relativity and quantum mechanics are the two pillars upon which rests our understanding of nature.

The Strangeness of Quantum Mechanics

One of the many unintuitive aspects of the probabilistic nature of quantum mechanics is the idea that you can never know exactly where a particle is without directly detecting it. Before you detect it, the best you can assume is that the particle is somewhere in the probability cloud. In addition, once you detect your particle you can no longer know where it was going. Heisenberg argued that there is a limit to knowing where a particle is and what that particle's linear momentum is; the better you narrow down one of those quantities, the less accurately you can know the other one. This weird feature of quantum mechanics is known as Heisenberg's uncertainty principle.

Since quantum mechanics is based entirely on probabilities, other strange implications also exist. Figure 8.37 illustrates perhaps the most famous of those as a thought experiment known as Schrödinger's cat. The idea is that a live cat is placed in a box along with a device that can randomly poison the cat. Perhaps the device is a radioactive atom that has a 50% chance of decaying into another type of atom over one half-life. If the decay occurs, the device releases the poison, but if the decay doesn't occur, no poison is released. The question is, when you open the box is the feline alive or dead? The answer is that you simply cannot know until you open the box. By observing the state of the cat, you have essentially selected one

Key Point
Heisenberg's uncertainty principle.

of the two possible outcomes of the experiment. The thought experiment seems to suggest that the outcome of any experiment for which there is a set of possible results based on probabilities is only determined when it is observed. Before being observed the experiment is a mix of all of the possible outcomes, with the likelihood of any ultimate result being based on the various probabilities. In the case of Schrödinger's cat, the cat is simultaneously alive and dead until the box is opened and an observation is made! The experiment does raise another puzzling question: Who or what constitutes the observer? Does the observer have to be a person, or can a flee determine the ultimate fate of Schrödinger's cat? By the way, don't worry, no actual cat was ever harmed in this thought experiment.

Key Point
Young double-slit for electrons revisited.

So, can this bizarre aspect of quantum mechanics possibly be real? Think back to the electron version of the Young double-slit experiment. Even if the stream of electrons is reduced to just one electron at a time going through the slits, an interference pattern is still built up if the experiment is run long enough. But how could that happen? What did the electron interfere with? The answer is that it interfered with itself by going through both slits simultaneously. As strange as that sounds, if you try to determine which slit the electron actually went through, the interference pattern disappears. Nature is conspiring to never let you know which slit the electron went through. The same is true with using photons instead of electrons.

"God Does Not Play Dice"

Not every prominent physicist accepted the probabilistic nature of quantum mechanics. Ironically, Einstein was never convinced that quantum mechanics was a complete theory, despite the fact that he was in part responsible for its birth. He died believing that there must be a missing ingredient, an undetected variable, that was underlying all of the apparent randomness of the new theory.

One of the great, and friendly, academic debates in history was between Einstein and Bohr. Einstein would constantly raise arguments against the probabilistic interpretation of quantum mechanics. In fact, the first version Schrödinger's cat is due to Einstein. Einstein would pose one of his thought experiments to Bohr, and then Bohr would go off and try to figure out where the flaw was in Einstein's thinking. Of course, those flaws were never obvious. In effect, Einstein's constant questioning of quantum mechanics helped to focus and strengthen it. In one of Einstein's many quotable statements, he said to Bohr, "Quantum mechanics is certainly imposing. But an inner voice tells me that it is not yet the real thing. The theory says a lot, but it does not really bring us any closer to the secrets of the Old One. I, at any rate, am convinced that He does not play dice." Bohr's exasperated response: "Einstein, stop telling God what to do!"

8.8   The Nucleus and Isotopes

From Subatomic to Subnuclear

After Rutherford discovered that the alpha particle is emitted from radioactive material (page 288*ff*), he later discovered that the particle is an ionized helium atom with two positive elementary charges, $+2e$. He then used the alpha particle to probe the structure of the atom, discovering that the atom's nucleus is extremely small relative to the atom itself, and that the target atom's nucleus (gold) is also positively charged. In addition, since the alpha particle becomes a helium atom when it captures its two missing electrons, it was apparent that the alpha particle must be the nucleus of a helium atom. This led Rutherford to the conclusion that the alpha particle must have been a component of the radioactive atom's nucleus.

In 1919, Rutherford also observed something else when his alpha particles traveled into air: his detectors picked up another particle. He suspected that the alpha particles were colliding with atoms in the air, causing another type of particle to be liberated by the collisions. He rightly surmised that the alpha particles were interacting with the atoms in N_2 molecules. When he passed alpha particles through a pure gas of nitrogen, the new particle again appeared, as did oxygen. Further analysis revealed that the new particle has a single positive elementary charge and has a mass that was very close to the mass of a hydrogen atom.

The new particle, which he named a proton, is the nucleus of the hydrogen atom. Apparently the proton is a component of the nucleus of either the helium nucleus or the nitrogen nucleus, or perhaps both. Referring back to the periodic table (Table 7.1) and counting all of the positive elementary charges before the collision, the helium nucleus has two and the nitrogen nucleus has seven, giving a total of nine. Similarly, counting the number of positive elementary charges after the collision, hydrogen has one and oxygen has eight, again giving a total of nine. The total number of positive charges didn't change during the collision but the charges of the individual nuclei did change. Essentially,

$$\alpha + N \rightarrow O + p^+,$$

where $\rightarrow$ can be read as "produces," and p^+ represents the positively charged proton. In other words, the equation can be read as an alpha particle collides with a nitrogen atom and produces an oxygen atom and a proton.

Not only had Rutherford succeeded in discovering the proton, but he inadvertently produced the first man-made nuclear reaction, a transformation of atomic nuclei. Chemical reactions only involve the electrons in the atomic orbitals; they don't fundamentally change the atoms themselves.

Atomic Isotopes

The proton is a component of all nuclei and accounts for the positive elementary charges in them. The number of protons in the nucleus also corresponds to an atom's atomic number in the periodic table, with hydrogen having one proton, helium has two, lithium has three, and so on. But that doesn't give us the whole picture because there is still a discrepancy in mass. Although the most common type of hydrogen nucleus does have the mass of a proton simply because that is its only component, the most abundant types of nuclei for helium, nitrogen, and oxygen have masses that are about twice the mass of their combined protons.

It took until 1932 before the missing ingredient of the nuclei was found and again the alpha particle played a role. Alpha particles were "fired" at a boron target, and this time a nitrogen atom was produced. In addition, a particle not possessing a charge was produced that had a mass of slightly more than that of the proton. The particle was named the neutron, which comes from an Italian word that means the big, neutral one. [One year earlier, a "little neutral one" (the neutrino) was predicted to exist from the requirements that energy and linear momentum must always be conserved.]

The nucleus of every atom contains protons, and all nuclei except the most common form of hydrogen also contain neutrons. The number of protons in a nucleus, and therefore the number of positive elementary charges (e) it contains, is the atomic number (Z) and it designates the type of element that the atom represents. The atomic mass (A) is the total number of protons plus neutrons contained in the nucleus. This means that the total number of neutrons in a nucleus is given by $A - Z$. To fully describe a particular nucleus, both the atomic number and the atomic mass must be used. Although it is redundant, the chemical symbol (X) for the element is also included (in some books, the atomic number is left out because it gives redundant information). The general notation for a nucleus is expressed as $^A_Z X$.

Using this notation, we can rewrite the nuclear reaction involved in Rutherford's discovery of the proton. Since an alpha particle is a $^4_2$He nucleus (two protons and two neutrons) and the proton is a $^1_1$H nucleus, the reaction can be represented by

$$^4_2\text{He} + ^{14}_7\text{N} \rightarrow ^{17}_8\text{O} + ^1_1\text{H}.$$

Notice that if you add up all of the positive charges on the left-hand side of the reaction (the Zs, $2 + 7 = 9$) you get the same number as on the right-hand side ($8 + 1 = 9$). The same is true for the atomic mass (the As), with $4 + 14 = 18$ on the left-hand side and $17 + 1 = 18$ on the right-hand side. Requirements of nuclear reactions include conservation of electric charge and conservation of nucleon number (protons and neutrons are referred to collectively as nucleons, meaning that they are both components of atomic nuclei).

Remember that Z, the number of protons, determines the number of electrons in an electrically neutral atom, and therefore the atom's chemical characteristics. But it is possible for the atoms of an element to have different numbers of neutrons. Atoms with the same Z but different numbers of neutrons are referred to as isotopes of the element.

As examples, the lightest isotope of hydrogen, and the most common, has just one proton in its nucleus, and so is described by $^1_1$H: 1 proton for Z, and 1 proton plus 0 neutrons giving 1 for A. However, there are actually two other isotopes of hydrogen, one with one neutron in the nucleus and one with two neutrons in the nucleus, so they are represented by $^2_1$H and $^3_1$H, respectfully. [For the next time you are playing a trivia contest, these heavier isotopes have their own names based on their atomic mass, deuterium and tritium ("deu" for 2 and "tri" for 3). Tritium also turns out to be radioactive.] The first row of Fig. 8.38 depicts the three isotopes of hydrogen. The bottom row of Fig. 8.38 illustrates isotopes of three other elements, including $^{12}_6$C. Carbon comes in $^{12}_6$C, the most common isotope, as well as $^{13}_6$C, and $^{14}_6$C. Like tritium, $^{14}_6$C is radioactive and is the isotope that was used in the carbon-dating example, Example 7.1.

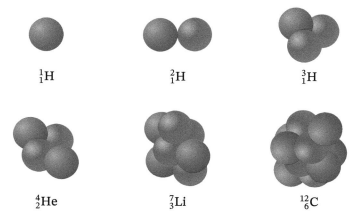

$^1_1$H $^2_1$H $^3_1$H

$^4_2$He $^7_3$Li $^{12}_6$C

Fig. 8.38 Examples of the nuclei of various isotopes. Protons are represented by the blue spheres and neutrons are represented by the red spheres. *Top row:* Three isotopes of hydrogen; $^1_1$H (normal hydrogen), $^2_1$H (deuterium), $^3_1$H (tritium); *Bottom row:* $^4_2$He (helium-4, the alpha particle), $^7_3$Li (lithium-7), $^{12}_6$C (carbon-12). Note that one of the neutrons in the $^{12}_6$C nucleus is hidden behind the other nucleons.

Anything Else?

Since atoms, once thought to be the smallest form of matter, turned out to be composed of electrons and a nucleus, and a nucleus proved to be composed of protons and neutrons, those discoveries beg the question, are electrons, protons, and neutrons the simplest forms of matter? The answer is yes for the electron, but no for both the proton and the neutron. Protons and neutrons are composed of three quarks each, with the quarks bound together by gluons. When you dive this deep, the researchers become very creative with naming. If you want to read more about this topic you will encounter terms like up and down quarks, charm and strange quarks, top and bottom quarks, flavors, and colors. It is in this realm of elementary particle physics where the Higgs boson lives, with the colorful if unfortunate nickname, "the God particle." We won't explore this particular cave any further, except to say that some of the myriad of truly elementary particles will be briefly mentioned in Chapter 9 and again in the text's final chapter, Chapter 19, when we discuss the Big Bang that occurred 13.8 billion years ago.

 Exercises

All of the exercises are designed to further develop *your understanding* of the material by thinking carefully and critically about what you have read in this chapter. Answers to selected exercises are available in the back of the book.

True/False

1. Planck had a solid theoretical basis for introducing quanta into his derivation of the blackbody radiation curve.
2. (*Answer*) A radio-wavelength photon has more energy than does an ultraviolet photon.
3. An x-ray photon has more energy than does a photon in the visible portion of the electromagnetic spectrum.
4. (*Answer*) The value of Planck's constant depends on the temperature of the material being investigated.
5. Planck's constant is one of the universal constants of nature.
6. (*Answer*) Light has both wave-like properties and particle-like properties.
7. Electrons have both wave-like properties and particle-like properties.
8. (*Answer*) Einstein was born on pi-day 1879, in the year of Maxwell's death.

9. Before the development of the general theory of relativity, it was proposed that an as-yet undiscovered planet, which was given the name Vulcan, must exist closer to the Sun than Mercury. This proposal was meant to explain Mercury's peculiar orbit.

10. (*Answer*) Einstein was fully involved in the technical aspects of developing a nuclear weapon during World War II.

11. The FBI (the United States' Federal Bureau of Investigation) had a massive dossier on Einstein, believing that he was a Communist sympathizer.

12. (*Answer*) The special theory of relativity finally proved the existence of the ether.

13. The effects of the general theory of relativity are only measurable around objects having extremely strong gravity, like black holes.

14. (*Answer*) A spaceship is moving toward you at 10% of the speed of light and shines a laser beam toward you that has a wavelength of 600 nm. You would measure the speed of the laser beam as being 110% of the speed of light.

15. A spaceship is moving toward you at 10% of the speed of light and shines a laser beam toward you that has a wavelength of 600 nm. You would measure the wavelength of the laser beam as being shorter than 600 nm.

16. (*Answer*) Einstein predicted the results of the photoelectric effect experiment by saying that if its wavelength is short enough, one photon will eject one electron from the surface of the negative plate of a cathode ray tube.

17. Solar panels produce electricity by using photons from the Sun in a fashion similar to the photoelectric effect.

18. (*Answer*) When an electron in an atom makes a transition from a higher energy level to a lower energy level a photon must first be absorbed.

19. We don't experience the wave-like behavior of macroscopic particles on an everyday basis because the wavelengths are much smaller than the diameter of an atom.

20. (*Answer*) According to quantum mechanics, the behavior of atomic and subatomic systems are based on probabilities.

Multiple Choice

21. Two photons, A and B, have wavelengths of 300 nm and 900 nm, respectively. Photon A's wavelength is found in the _____ portion of the electromagnetic spectrum, and it has _____ as much energy as photon B, which has a wavelength in the _____ portion of the electromagnetic spectrum.
    (a) ultraviolet, three times, visible
    (b) ultraviolet, one-third, visible
    (c) ultraviolet, three times, infrared
    (d) visible, one-third, infrared
    (e) visible, three times, visible
    (f) visible, one-third, ultraviolet
    (g) infrared, three times, ultraviolet
    (h) infrared, one-third, visible

22. (*Answer*) Two photons, A and B, have frequencies of 5×10^{15} Hz and 5×10^{10} Hz, respectively. Photon A's frequency is found in the _____ portion of the electromagnetic spectrum, and it has _____ times as much energy as photon B, which has a frequency in the _____ portion of the electromagnetic spectrum. Hint: You may find Table 6.1 helpful.
    (a) gamma ray, 0.00001, radio
    (b) gamma ray, 10, x-ray
    (c) x-ray, 100, ultraviolet
    (d) ultraviolet, 1000, visible
    (e) ultraviolet, 10,000, infrared
    (f) ultraviolet, 100,000, microwave
    (g) ultraviolet, 1,000,000, radio

23. Select the correct order of energies, from lowest to highest, for photons in each of the following regions of the electromagnetic spectrum.
    (a) gamma ray, x-ray, ultraviolet, visible, infrared, microwave, radio
    (b) x-ray, gamma ray, visible, ultraviolet, microwave, infrared, radio
    (c) radio, infrared, microwave, visible, ultraviolet, gamma ray, x-ray
    (d) radio, microwave, infrared, visible, ultraviolet, x-ray, gamma ray
    (e) x-ray, radio, infrared, visible, ultraviolet, microwave, gamma ray
    (f) none of the above

24. (*Answer*) Planck's constant
    (a) is a fundamental constant of nature.
    (b) is an extremely small number.
    (c) plays a critical role in the description and behavior of nature at atomic and subatomic length scales.
    (d) has a different value for every element in the periodic table.
    (e) produces readily measurable effects for all types of motion from subatomic particles to planetary orbits.
    (f) all of the above
    (g) (a), (b), and (c) only
    (h) (d) and (e) only
    (i) none of the above

25. An important wavelength of the hydrogen spectrum that is used to study the locations of clouds of hydrogen gas in galaxies is 21 cm long. The photons of the 21 cm line are found in the _____ region of the electromagnetic spectrum and they have _____ energies when compared with visible wavelength photons.
    (a) radio, very low   (b) radio, very high
    (c) infrared, low   (d) infrared, high
    (e) ultraviolet, low   (f) ultraviolet, high
    (g) x-ray, very low   (h) x-ray, very high

26. (*Answer*) Newton believed that space and time are _____ the observer while Einstein argued that space and time are _____ the observer.
    (a) absolute and independent of, absolute and independent of
    (b) absolute and independent of, relative to
    (c) relative to, absolute and independent of
    (d) relative to, relative to
    (e) Newton and Einstein made no assumptions about the nature of space and time.

27. Einstein's "happiest thought in [his] life" was
    (a) his method of measuring G for the first time.
    (b) realizing that moving clocks run slow.
    (c) that while someone is falling she doesn't feel her own weight.
    (d) that Vulcan actually exists.

28. (*Answer*) Einstein's general theory of relativity explains gravity as
    (a) the curving of space around a massive object while leaving absolute time unaffected.
    (b) the slowing of time near a massive object while leaving absolute space unaffected.
    (c) the curvature of spacetime in the vicinity of a massive object.
    (d) being closely related to Maxwell's electromagnetic theory.
29. Einstein's special theory of relativity describes
    (a) why an observer moving relative to a clock measures the clock running slowly.
    (b) why an observer moving relative to a long beam measures the beam as being shortened in the direction of the observer's motion.
    (c) why an observer moving relative to another person observes that person to be aging more slowly than normal.
    (d) a strategy for getting along with your relatives.
    (e) all of the above (except d of course!)
    (f) only one of the above
30. (*Answer*) At the core of the special theory of relativity is the assumption that
    (a) it doesn't matter if you are moving relative to someone else while you both perform the same experiment, you will both get the same physical result.
    (b) every person, no matter how fast they are moving relative to someone else, will always measure identical values for the speed of light.
    (c) both (a) and (b)
    (d) none of the above
31. In the baseball game and spaceship situation discussed in Section 8.3, assume that the fans in the stadium can see a clock in the spaceship as it passes by. The fans would notice that the clock is running _____ than their own clocks and therefore the people on the spaceship would see a home run taking _____ than the fans in the stadium would see the home run take. This consequence of special relativity is known as _____.
    (a) faster, more time, time dilation
    (b) faster, more time, length contraction
    (c) faster, less time, time dilation
    (d) faster, less time, length contraction
    (e) slower, more time, time dilation
    (f) slower, more time, length contraction
    (g) slower, less time, time dilation
    (h) slower, less time, length contraction
32. (*Answer*) If a light source is moving away from you, the light from the source will appear _____ than it would if the source wasn't moving relative to you.
    (a) dimmer   (b) brighter   (c) redder   (d) bluer
    (e) (a) and (c) only   (f) (b) and (d) only
    (g) none of the above
33. Astronomers are studying the Balmer Hδ absorption line for a particular star. In the laboratory, the wavelength is measured as 410 nm, but the line is measured to have a wavelength of 415 nm in the star's spectrum. The star is moving _____ Earth, causing the star's spectrum to be _____. Based on the measured wavelength, the star's speed relative to Earth is _____.

(a) toward, blueshifted, 0.012c
(b) toward, redshifted, 0.12c
(c) toward, blueshifted, 1.2c
(d) away from, redshifted, 0.012c
(e) away from, blueshifted, 0.12c
(f) away from, redshifted, 1.2c

34. (*Answer*) Scientists realized almost immediately that the atomic model of Rutherford's, with electrons orbiting the nucleus like tiny planets, could not be correct in detail because
    (a) electrons orbiting a nucleus must be constantly accelerating.
    (b) accelerating charges must emit electromagnetic radiation.
    (c) as orbiting electrons lose energy they must necessarily spiral into the nucleus, dramatically decreasing the sizes of atoms in a fraction of a second.
    (d) all of the above
    (e) none of the above
35. When an electron in a hydrogen atom makes a downward transition from an $n = 3$ orbital to an $n = 2$ orbital, a photon is _____ that has a wavelength in the _____ part of the visible electromagnetic spectrum.
    (a) emitted, blue   (b) emitted, red
    (c) absorbed, blue   (d) absorbed, red
36. (*Answer*) If a photon having the right amount of energy strikes an atom, the photon may _____, causing the electron to _____.
    (a) be absorbed, jump to a higher energy orbital
    (b) be absorbed, jump to a lower energy orbital
    (c) be reflected, speed up in its orbital
    (d) generate a second photon, split in two
    (e) generate a second photon, jump to a higher energy orbital
    (f) none of the above
37. If an electron in a hydrogen atom makes a transition from an $n = 3$ orbital to an $n = 2$ orbital, it produces a photon with _____ energy and a _____ wavelength than when an electron in a hydrogen atom jumps from an $n = 3$ orbital to an $n = 1$ orbital.
    (a) more, longer   (b) more, shorter
    (c) less, longer   (d) less, shorter
38. (*Answer*) According to the data in Table 7.2, the wavelength of the highest energy photon of sodium listed in the table is _____, corresponding to an energy closest to _____.
    (a) 411 nm, 510,000 eV   (b) 411 nm, 3.02 eV
    (c) 411 nm, 0.331 eV   (d) 655 nm, 1.89 eV
    (e) 655 nm, 0.528 eV
39. A photon emitted by a carbon atom has an energy of 2.1 eV. What is the photon's wavelength?
    (a) 427 nm   (b) 514 nm   (c) 538 nm
    (d) 589 nm   (e) 658 nm   (f) none of the above
40. (*Answer*) An electron in a hydrogen atom makes a downward transition from the $n = 5$ energy level to the $n = 2$ energy level. The energy of the emitted photon is _____, its wavelength is _____, and it is in the _____ portion of the electromagnetic spectrum.
    (a) 2.86 eV, 433 nm, visible
    (b) −2.86 eV, 433 nm, visible
    (c) 285 eV, 4.3 nm, x-rays
    (d) 4.08 eV, 304 nm, ultraviolet
    (e) 40.8 eV, 30.4 nm, gamma rays

41. A photon is absorbed by a hydrogen atom in which the electron starts out in the $n = 3$ energy level. The wavelength of the photon is 1094 nm. The absorbed photon was in the _____ portion of the electromagnetic spectrum, it had an energy of _____, and the electron ended up in the $n =$_____ energy level.
    (a) x-ray, 1240 eV, 12
    (b) ultraviolet, 1.13 eV, 4
    (c) visible, 0.0009 eV, 5
    (d) infrared, 1.13 eV, 6
    (e) radio, 0.0113 eV, 6

42. (*Answer*) An electron has a mass that is about 2000 times smaller than the mass of a proton. Both particles have the same amount of charge (the elementary charge, e), except that the electron is negatively charged and the proton is positively charged. If both particles are moving with the same speed the _____ would have a shorter wavelength by a factor of _____.
    (a) electron, 13.6   (b) electron, 2000
    (c) proton, 13.6   (d) proton, 2000

43. Ultimately, the organization of the periodic table of the elements is based on
    (a) the number of isotopes each element has.
    (b) the mass of each isotope of an element.
    (c) the electric charge of the nucleus of atoms of each element.
    (d) the number of protons in the nucleus of each atom of an element.
    (e) the distribution of electrons in the orbitals described by the four quantum numbers.
    (f) all of the above
    (g) (a) and (b) only
    (h) (c), (d), and (e) only.

44. (*Answer*) An important nuclear reaction in the core of the Sun involves the collision of two helium-3 nuclei. Two of the three products of the collision are hydrogen-1 nuclei. Symbolically, the reaction is represented as

    $$^3_2\text{He} + ^3_2\text{He} \rightarrow \text{??} + ^1_1\text{H} + ^1_1\text{H}$$

    Using the ideas of conservation of charge and conservation of nucleon number, together with referencing the periodic table, the missing product is
    (a) ^4_2He (b) ^4_2He (c) ^1_5B (d) ^3_3Li (e) ^6_4C

45. In the notation ^A_ZX, A represents the _____ and Z represents the _____.
    (a) element, isotope
    (b) atomic mass, isotope
    (c) atomic number, element
    (d) isotope, element
    (e) isotope, atomic mass
    (f) atomic number, atomic mass

46. (*Answer*) In the notation ^A_ZX, A corresponds to the number of _____ and Z corresponds to the number of _____.
    (a) neutrons, protons   (b) protons, neutrons
    (c) nucleons, protons   (d) protons, nucleons
    (e) the number of protons plus neutrons, protons
    (f) (a) and (c)   (g) (b) and (d)
    (h) (c) and (e)

Short Answer

47. In what sense can Planck's solution to the blackbody radiation curve be considered ad hoc and therefore not fully satisfactory?

48. Your author is fond of saying that "memorization does not equal understanding." Discuss that statement in the context of Einstein's quote that "the value of a college education is not the learning of many facts but the training of the mind to think."

49. A very common learning outcome espoused by colleges and universities is that students should develop critical thinking skills. What is meant by "critical thinking?" Discuss that concept within the context of Einstein's quote that "the value of a college education is not the learning of many facts but the training of the mind to think."

50. Why did Einstein's special theory of relativity come out of a paper about "electrodynamics of moving bodies?"

51. Explain the significance of the term spacetime.

52. The mathematical factor that describes how measurements of time intervals and lengths are affected by motion is given by

    $$\gamma = \frac{1}{\sqrt{1 - v^2/c^2}},$$

    which is always greater than or equal to one. The farther the departure from $\gamma = 1$ the greater the effect of relativity. As an example, for $v = 0.2c$ (a speed of 20% of the speed of light),

    $$\gamma = \frac{1}{\sqrt{1 - (0.2\cancel{c})^2/\cancel{c}^2}} = \frac{1}{\sqrt{1 - 0.04}} = 1.02.$$

    Calculate the γ factor
    (a) for a speed typical of a commercial aircraft ($v = 800$ km/h $= 500$ mi/h $= 0.22$ km/s $= 0.000\,000\,74c$),
    (b) for a particle traveling at 10% of the speed of light ($v = 0.1c$).
    (c) for a particle traveling at 50% of the speed of light ($v = 0.5c$).
    (d) for a particle traveling at 90% of the speed of light ($v = 0.9c$).
    (e) for a particle traveling at 99% of the speed of light ($v = 0.99c$).
    (f) Make a graph of your answers for parts (a) through (e). The horizontal axis should range from 0 to 1 (the fraction of the speed of light) and the vertical axis should range from 0 to 10 for γ. Don't forget to label your axes, indicating the quantities that are being plotted. You should also draw a smooth curve through the data points. If you need a standard (Cartesian) sheet of graph paper, it can be downloaded from the textbook's website at Math_Review/Graphing/Graph_Templates. A graphics tutorial is available at Math_Review/Graphing.
    (g) Based on these factors, explain why we don't experience the strange behavior of the special theory of relativity on a daily basis in our lives.
    (h) What happens to the impact of relativity on observations as the speeds of particles approach the speed of light?
    (i) Is it possible to calculate γ when $v = c$? Why or why not? This is an indication that nothing except light itself can travel at the speed of light.

53. Cold, very-low density hydrogen gas can produce an electromagnetic wave with a wavelength of 21 cm. That wavelength is used by astronomers to map out the location of hydrogen gas clouds found throughout our Milky Way Galaxy. 21 cm radiation can also be used to measure the speed that the gas clouds are moving toward or away from Earth. Briefly describe how that is accomplished.

54. A star is moving toward Earth with a speed of 0.01c (1% of the speed of light). You are measuring the wavelength of a spectral line that is normally at a wavelength of 500 nm.
    (a) By what factor has the wavelength been changed by the relative motion of the star?
    (b) Is the wavelength of the light shorter or longer than it would normally be?
    (c) What would the new wavelength be when you make the measurement?

55. On page 283 it was mentioned that the Doppler effect equation as presented was only valid "as long as the source isn't moving too fast relative to the observer." The fully relativistic equation for the observed wavelength when the source is moving at a more substantial fraction of the speed of light is given by

$$\lambda_{obs} = \lambda_{emit} \sqrt{\frac{1 + v_r/c}{1 - v_r/c}},$$

where v_r is positive if the source is moving *away* from the observer and negative if the source is moving toward the observer.
    (a) From Fig. 8.16, verify that if the ship emitted a wavelength of 700 nm while traveling at 0.5c, observers on the planet will see a wavelength of close to 400 nm.
    (b) Suppose that the ship was traveling at 0.9c toward the planet. What wavelength would the observers on the planet observe?
    (c) What about if the ship is traveling at 0.99c?
    (d) What about if the ship is traveling at 0.999c?
    (e) If the ship could travel at $v_r = -c$ (it can't), what would the observed wavelength be?

56. (a) A galaxy is moving away from Earth with a speed of 2900 km/s. Astronomers detect a wavelength emitted by the galaxy that normally has a wavelength of 656.470 nm in a laboratory. Using the fully relativistic Doppler effect equation given in Exercise 55, calculate the wavelength that the astronomers measure coming from the galaxy.
    (b) Compare your answer to the observed wavelength from Example 8.1. Does the fully relativistic equation agree with the approximate equation on page 283 for this relatively slow speed? What is the difference in wavelength between that quoted in the example and what you calculated?

57. The first exoplanets found orbiting other stars similar to the Sun were discovered by measuring the periodic changes in velocities of the stars that the planets were orbiting. Based on what you have learned in this chapter, together with Kepler's laws and the concept of center of mass, explain how this was accomplished.

58. Figure 8.28 shows an "energy level diagram" for hydrogen. If the top-most energy level ($n = \infty$) is set to 0 eV (electron volts), the bottom energy level is then −13.6 eV.

Note: These energies only apply for hydrogen, each element has its own set of energy levels, which is why they emit or absorb photons with a unique set of energies, wavelengths, and frequencies (the element's "fingerprint").
    (a) Draw an energy level diagram for hydrogen like the one shown in Fig. 8.28, but you should also specify the energy for each level in parentheses next to the principal quantum number, n. The spacings between the energy levels should accurately reflect the actual energies relative to the spacing between $n = 1$ and $n = \infty$. You should make the diagram large enough to clearly show the spacings up to $n = 5$, before all of the other lines merge into a densely packed set (there is an infinite number of lines between $n = 5$ and $n = \infty$).
    (b) Using arrows similar to the one in Fig. 8.28, show the transition jumps that produce the Balmer emission lines Hα, Hβ, and Hγ.
    (c) How much energy does an Hγ photon have? (It is equal to the energy difference between the electron's starting and ending energy levels.)
    (d) The Lyman hydrogen emission lines are similar to the Balmer lines, except that the electrons end up in the lowest energy level, $n = 1$. For example, the Lyα line (Lyman alpha) is produced when an electron makes a transition between $n = 2$ and $n = 1$. Indicate the transitions for Lyα, Lyβ, Lyγ, and Lyδ.
    (e) How much energy does a Lyγ photon have?
    (f) Given that higher photon energies have shorter wavelengths, which photon has the shorter wavelength, Lyγ or Hγ? [Recall Planck's energy equation, Equation (8.1), describing the energy of a photon.]
    (g) Hγ photons are in the blue/violet portion of the electromagnetic spectrum. Where would you expect Lyγ photons to be located in the spectrum?

59. (a) Create an energy level diagram like the one in Exercise 58, but this time indicate with arrows the transitions of electrons when the hydrogen atom has absorbed photons with energies equivalent to Lyα, Lyβ, Lyγ, and Lyδ.
    (b) What would happen to the electron if the atom absorbed a photon with an energy that was greater than 13.6 eV?

60. Describe in your own words how Rutherford was able to demonstrate that atoms are composed of a very tiny, positively charged nucleus with electrons surrounding the nucleus.

61. What does wave–particle duality refer to, and for what types of particles does it apply: photons, particles with mass, or both?

62. When we discuss the nuclear reactions that power the Sun in Chapter 9, the concept of quantum mechanical tunneling will become very important. Basically, tunneling means that a particle that would normally not have enough energy to pass "over" a barrier can still pass "through" the barrier if its wavelength is comparable to the width of the barrier, and the barrier isn't too high. Discuss how this phenomenon relates to Heisenberg's uncertainty principle.

63. Electron microscopes can be used to study very small objects in much the same way that ordinary laboratory microscopes can. Considering the wave-like behavior of electrons, why are they able to resolve much finer detail? Hint: Think about the wavelengths of visible light and of an electron traveling at 10^5 m/s.

64. We don't know all of the intricacies of how the human brain operates, but we do know that it uses electrochemical signals that work at the atomic level. Comment on whether you think the decision processes in the brain are deterministic or based on probabilities, and why. How does your answer impact the age-old question of whether or not humans have free will?

65. Going back the discussion in Chapter 1, The Nature of Science, discuss the failings of Bohr's model within the context of the characteristics of a scientific theory, and why the new paradigm of quantum mechanics developed in its place.

The Sun, Our Solar System, Exoplanets, and Life

PART II

Chapter 9   The Sun, Our Parent Star   309

Chapter 10   An Overview of the Solar System   358

Chapter 11   The Rocky Planets and Our Moon   379

Chapter 12   The Giant Planet Systems   460

Chapter 13   Dwarf Planets and Small Bodies   501

Chapter 14   Planets Everywhere and the Search for Extraterrestrial Life   541

Solar System. (NASA/JPL); Milky Way. (ESO/S. Brunier, CC BY 4.0)

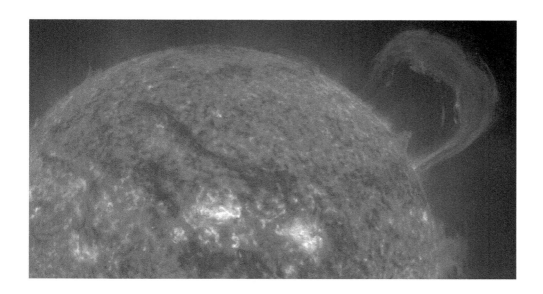

The Sun, Our Parent Star

9

*The Earth has received the embrace of the Sun
and we shall see the results of that love.*

Sitting Bull (1831–1890)

9.1	What Is Our Sun Like?	310
9.2	The Sun's Dynamic Atmosphere	317
9.3	The Solar Cycle	328
9.4	Using Computers for Exploration	334
9.5	How to Build a Star	336
9.6	The Sun's Nuclear Furnace	342
9.7	Test, Test, and Test Some More	349
	Exercises	354

Fig. 9.1 The Sun's active atmosphere with a large prominence. (ESA/NASA/SoHO)

Introduction

In Part I we began our study of astronomy by exploring the efforts of early cultures to uncover patterns and invoke meaning in what they saw in the night sky. The ancient Greek philosopher-scientists extended that work by applying careful and self-consistent logic in developing explanations for what they witnessed. When Copernicus challenged us to consider a far different view of the cosmos from what had been imagined for thousands of years, humanity began an intellectual journey that came to rely on empirical information, overarching and unifying theories with predictive power, and observational evidence and laboratory testing to support (and in some cases refute) existing scientific paradigms. Thanks to the scientific process, that is inherently self-correcting, ideas that do not prove to be consistent with all of the available evidence must be rejected in favor of theories that are broadly based and as simple as possible (recall Occam's Razor). Tycho, Kepler, Galileo, and Newton successively refined our fundamental understanding of celestial motions through careful observations, and the ability and willingness to set aside previously held beliefs about nature, combined with an insatiable desire to explain what they were seeing.

What started as a primarily astronomical endeavor to explain the motions of celestial objects transitioned into a marriage of seemingly disparate intellectual studies (mathematics, astronomy, physics, chemistry, geology, planetary science, and today microbiology as well) to become a quest for understanding that continues unabated to this day. In order to make additional progress we had to develop an understanding of the nature of light, gravitation, atoms and nuclei, and even the very fabric of spacetime. Today's study of the cosmos has long since moved past simple observations of lights on the celestial sphere to a deep understanding of all aspects of light, matter, space, and time. *Astronomy is now astrophysics.*

In Part II we will explore the realm of our own Solar System, looking for commonalities and differences among the bodies of the Solar System. We will also be searching for patterns across the Solar System. In what at first appears to be vague associations, and in some cases perplexing irregularities, a model will begin to emerge around how the Solar System began somewhat chaotically, and ultimately led to what we observe today. Help in developing that emerging view has come from the growing body of data associated with exoplanets found in other planetary systems. Throughout our exploration of the Solar System, our understanding of what we observe must remain rooted in the fundamental aspects of science and the universality of physical law.

We begin with the central member of our own Solar System, the Sun.

9.1 What Is Our Sun Like?

When you see the Sun up in the sky, such as in Fig. 9.2, it appears like a feature-less, bright yellowish-white ball that provides us with light and warmth. In reality, the Sun is a complex, constantly seething ball of gas that produces an astonishing amount of energy every second. The wide range of features seen on and around the

Fig. 9.2 The Sun in a blue sky. (Kreuzschnabel, CC BY-SA 3.0)

Sun change all the time; some change in just a few seconds, others may take days or months, and still others may require decades to observe characteristic patterns.

Figure 9.3 (*Left*) shows a full disk image of the Sun seen in yellow-red light that at first blush looks largely featureless with the exception of a visible group of sunspots. However, when the Sun is observed in extreme ultraviolet light, tremendous activity can be seen, indicating the truly dynamic nature of our nearest star. It is important to note that the images in Figs. 9.3 (*Center* and *Right*) appear yellow and red simply because they were colored that way during image processing; after all, the human eye cannot see extreme ultraviolet light. All three images in Fig. 9.3 were obtained by spacecraft dedicated to studying the Sun.

Summary of Basic Solar Characteristics

Table 9.1 lists the radius, mass, luminosity, surface temperature, and surface composition of the Sun. The following subsections will take you though the methods used to determine those values, relying only on what you have already studied in Part I. Those subsections contain slightly more mathematics, though no more complex, than you have seen so far. As always, you should be an active reader, striving to work through and understand the material presented. Hopefully, you have already developed that insatiable curiosity of a scientist or anyone else who strives to think critically and demands evidence.

The Sun's Radius

As you can see from Fig. 9.3 (*Right*), our Sun is enormous in size relative to Earth (and every other object in the Solar System). In a sense, the Solar System is just a byproduct of the Sun's formation. The Sun dominates the Solar System.

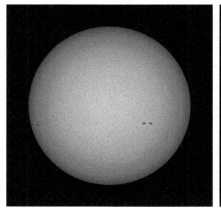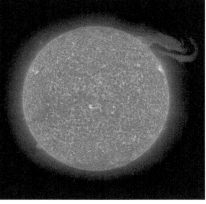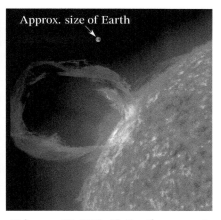

Fig. 9.3 *Left:* The solar disk as seen in visible light (at a wavelength of 617.3 nm) on February 28, 2013. Notice the sunspots below and to the right of the middle of the image. Darkening of the Sun toward the edges of the image is also apparent; an effect related to how photons escape its atmosphere. (NASA) *Center:* The Sun in extreme ultraviolet light (30.4 nm) on September 29, 2008. The surface is covered with dynamic, detailed structure. A loop of gas (a solar prominence) can be seen in the upper right. (NASA/SDO) *Right:* A solar prominence seen at a wavelength of 30.4 nm with an image of Earth shown to scale. (Adapted from NASA)

Table 9.1 Characteristics of the Sun

Quantity	Value	Comparisons and comments
Mass (M_{Sun})	1.99×10^{30} kg	330,000 M_{Earth}
Radius (R_{Sun})	6.96×10^{8} m	109 R_{Earth}, 696,000 km, 432,000 mi; volume of 1.3 million Earths
Luminosity (L_{Sun})	3.8×10^{26} W	30 trillion times more than consumed on Earth
Surface temperature ($T_{surface}$)	5772 K	5499°C, 9930°F; diamond melts at 3820 K
Surface composition	—	Percentages are by mass
Hydrogen	73.8%	
Helium	24.9%	
Heavy elements	1.3%	All elements other than hydrogen and helium
Rotation period	—	Counterclockwise as seen from above Earth's North Pole
Equator	25 d	
Poles	36 d	

So how big is the Sun? Recall the long history of effort that went into first determining the relative sizes of planetary orbits (Nicolaus Copernicus and Section 4.7), and finally measuring the length of the astronomical unit (au) accurately before it was formally defined by the International Astronomical Union (Section 6.1). With that knowledge and the angular size of the Sun in the sky (approximately 1/2°), the actual diameter of the Sun can be calculated directly.

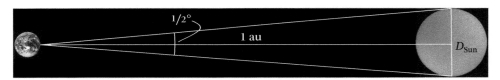

Fig. 9.4 The angular diameter of the Sun in the sky (1/2°) and its distance from Earth (1 au) can be used to determine the Sun's diameter (D_{Sun}) using geometry. The diagram is not to scale. (Images courtesy of NASA)

Example 9.1

Using the information in Fig. 9.4, determine the diameter of the Sun.

You can think of the Sun as sitting on a giant circle, centered on Earth, that has a radius of 1 au. The portion of the circle covered by the Sun has a length D_{Sun}, the diameter of the Sun. Since 1/2° is 1/720 of a circle (remember that a circle has 360°), then D_{Sun} must also be 1/720 of the circumference of the circle (technically this is an approximation because D_{Sun} is a straight line rather than an arc of a circle). However, the circumference of a circle is given by the familiar formula, $C = 2\pi r$, where r is the radius of the circle; in this case $r = 1$ au. This means that the circumference of the circle that is centered on Earth and passes through the Sun is

$$C = 2\pi \times 1 \text{ au} = 2\pi \text{ au}.$$

But 1 au $= 1.5 \times 10^{11}$ m, and $\pi = 3.14$ (approximately) giving

$$C = 2 \times 3.14 \times 1.5 \times 10^{11} \text{ m} = 9.4 \times 10^{11} \text{ m}.$$

Finally, since the diameter of the Sun is 1/720 of the circumference of the circle, the Sun's diameter is approximately

$$D_{\text{Sun}} = C/720 = 9.4 \times 10^{11} \text{ m}/720 = 1.3 \times 10^{9} \text{ m}.$$

It is more common to describe the size of the Sun in terms of its radius, which is just one-half of the diameter, giving

$$R_{\text{Sun}} = D_{\text{Sun}}/2 = 1.3 \times 10^{9} \text{ m}/2 = 6.5 \times 10^{8} \text{ m}.$$

A more precise determination gives $R_{\text{Sun}} = 7.0 \times 10^{8}$ m, or 700,000 km (430,000 mi).

To put this number in perspective, the radius of Earth, R_{Earth}, is approximately 6.4×10^{6} m. This means that the radius of the Sun is 7.0×10^{8} m / 6.4×10^{6} m = 110 times the radius of Earth. The Sun is so large that nearly 110 Earths would be required side by side to reach from one side of the Sun to the other [again recall Fig. 9.3 (Right)]. Considered another way, given that the volume of a sphere is

$$V = \frac{4}{3}\pi r^{3},$$

the ratio of the volume of the Sun to the volume of Earth is

$$\frac{V_{\text{Sun}}}{V_{\text{Earth}}} = \frac{\frac{4}{3}\pi R_{\text{Sun}}^{3}}{\frac{4}{3}\pi R_{\text{Earth}}^{3}} = \frac{R_{\text{Sun}}^{3}}{R_{\text{Earth}}^{3}} = \left(\frac{R_{\text{Sun}}}{R_{\text{Earth}}}\right)^{3} = 110^{3} = 1.3 \times 10^{6}.$$

If the Sun was a hollow ball, 1.3 million Earths would fit inside it!

Determining the Sun's Mass

In order to determine the Sun's mass we can use what we know about Earth's orbit around the Sun. You learned in Section 5.2 that in order to keep something traveling in circular motion, a centripetal force must be exerted on that object, which in this case is provided by the gravitational interaction between the Sun and Earth. From the centripetal acceleration equation, $a_{c} = v^{2}/r$, on page 161 and Newton's second law, $\vec{F}_{\text{net}} = m\vec{a}$, the centripetal force required to keep Earth moving in a (nearly) circular orbit around the Sun is given by

$$F_{c} = M_{\text{Earth}}\, a_{c} = M_{\text{Earth}} \times \frac{v^{2}}{r},$$

where v is the orbital speed of Earth around the Sun and r is the radius of Earth's orbit. But the centripetal force is provided by Newton's universal law of gravitation,

$$F_{c} = F_{\text{gravity}} = G\frac{M_{\text{Sun}}M_{\text{Earth}}}{r^{2}}.$$

Key Point

The radius of the Sun is about 7.0×10^{8} m or 700,000 km, which is about 110 times the radius of Earth.

A variables tutorial is available at
Math_Review/Variables

Since both forces are describing the same thing, namely the centripetal force acting on Earth, they can be set equal to one another, giving

$$M_{\text{Earth}} \times \frac{v^2}{r} = G \frac{M_{\text{Sun}} M_{\text{Earth}}}{r^2}.$$

Canceling the mass of Earth on both sides of the equation and doing a little bit of algebra allows us to solve for the mass of the Sun, giving

$$M_{\text{Sun}} = \frac{v^2 r}{G}.$$

A scientific notation tutorial is available at Math_Review/Scientific_Notation

---

Example 9.2

Calculate the mass of the Sun based on the equation immediately above.

In order to do so we must first determine how fast Earth is going around the Sun. This is just how far Earth travels in its orbit in one year divided by how much time it takes to complete one lap, or

$$v = \frac{\text{distance}}{\text{time}} = \frac{2\pi r}{P},$$

where the numerator is the circumference of Earth's orbit and P in the denominator is Earth's orbital period. Expressing $r = 1$ au in meters and $P = 1$ year in seconds gives

$$v = \frac{2\pi \times 1.5 \times 10^{11} \text{ m}}{3.16 \times 10^7 \text{ s}} = \left(\frac{2\pi \times 1.5}{3.16}\right) \times 10^{(11-7)}$$

$$= 3.0 \times 10^4 \text{ m/s} = 30 \text{ km/s}.$$

Earth orbits the Sun at an average speed of 30 km/s. Now we must find v^2:

$$v^2 = \left(3.0 \times 10^4 \text{ m/s}\right)^2 = 3.0^2 \times \left(10^4\right)^2 \text{ (m/s)}^2$$

$$= 9.0 \times 10^{4 \times 2} \text{ m}^2/\text{s}^2 = 9.0 \times 10^8 \text{ m}^2/\text{s}^2.$$

In order to calculate the mass of the Sun in kilograms, we must use the numerical value of Big G from page 192, which is $G = 6.67 \times 10^{-11}$ N m^2/kg^2; and, from page 164, 1 N = 1 kg m/s^2. Substituting gives

$$M_{\text{Sun}} = \frac{v^2 r}{G} = \frac{9.0 \times 10^8 \text{ m}^2/\text{s}^2 \times 1.5 \times 10^{11} \text{ m}}{6.67 \times 10^{-11} \text{ m}^3/\text{kg s}^2} = 2.0 \times 10^{30} \text{ kg}.$$

In comparison, the mass of Earth is 6.0×10^{24} kg, which means that

$$\frac{M_{\text{Sun}}}{M_{\text{Earth}}} = \frac{2.0 \times 10^{30} \text{ kg}}{6.0 \times 10^{24} \text{ kg}} = 330,000;$$

the Sun is about 330,000 times more massive than Earth.

Key Point
The Sun's mass is 2.0×10^{30} kg, or 330,000 times more massive than Earth.

What is the Sun's Luminosity?

In order to determine what the Sun's luminosity is (its power output as discussed on page 217), we must first know how much energy from the Sun reaches Earth every second. This value turns out to be about 1361 watts per square meter (1361 W/m^2) of surface that is oriented perpendicular to the direction to the Sun.

Since the Sun radiates its energy in every direction through space equally, we can calculate the Sun's power output by multiplying 1361 W/m^2 by the total surface area of a gigantic sphere that is centered on the Sun and has the radius of Earth's orbit. This is because we have measured the amount of energy per second that crosses just one square meter of that enormous spherical surface.

The area of a sphere is given by the expression

$$A = 4\pi r^2.$$

For a sphere the size of Earth's orbit, r is again 1 au. This means that the area of the sphere is

$$A = 4\pi \times \left(1.5 \times 10^{11} \text{ m}\right)^2 = 4\pi \times 1.5^2 \times 10^{2 \times 11} \text{ m}^2 = 2.8 \times 10^{23} \text{ m}^2.$$

As a result, the Sun's luminosity is

$$L_{\text{Sun}} = 1361 \text{ W/m}^2 \times 2.8 \times 10^{23} \text{ m}^2 = 3.8 \times 10^{26} \text{ W}.$$

According to the International Energy Agency, in 2013 the world averaged an energy consumption rate of 12.3 terawatts, or 1.23×10^{13} W. The Sun produces

$$\frac{3.8 \times 10^{26} \text{ W}}{1.23 \times 10^{13} \text{ W}} = 3 \times 10^{13}$$

Key Point
The Sun produces abundant renewable energy.

times that much, or 30 trillion times more energy per second than the entire population of our planet consumed in 2013, but remember that the Sun's energy is going out into space in every possible direction. The more important number for energy consumption on Earth is the total amount of energy produced by the Sun striking Earth's surface per second, which is about 1.7×10^{17} W, or 14,000 times more energy per second than we consumed in 2013; there is plenty of clean, renewable solar energy to support the human population.

The Sun's Surface Temperature

Now that we know both the radius of the Sun and the amount of energy it produces per second (its luminosity), we can determine its "surface" temperature using the Stefan–Boltzmann blackbody luminosity law [Equation (6.16)] discussed on page 219. However, to describe the Sun as having a surface is a bit misleading since it is entirely gaseous; there is no solid surface to identify as *the* surface. Instead, it is actually the Stefan–Boltzmann blackbody luminosity law that is used to define an effective surface for a star that is associated with an effective surface temperature.

The Stefan–Boltzmann blackbody luminosity law can be solved for the surface temperature algebraically, which produces the equation

$$T_{\text{surface, Sun}} = \left(\frac{L_{\text{surface, Sun}}}{4\pi\sigma R^2_{\text{surface, Sun}}} \right)^{1/4} \tag{9.1}$$

Surface temperature from the Stefan–Boltzmann law

[the power of ( )$^{1/4}$ means that you take the square root of the square root of the quantity in parentheses; the fourth root]. Note that for the purposes of this text it is not particularly important that you can manipulate the Stefan–Boltzmann black-body luminosity law in this way, it is only important that you realize that the law provides an important relationship between luminosity, surface radius, and surface temperature.

Plugging in all of the numbers, including the constant σ, and carrying out the somewhat tedious calculation produces the *approximate result*,

Key Point
The surface temperature of the Sun is 5772 K.

$$T_{\text{surface, Sun}} = \left(\frac{3.8 \times 10^{26}\ \text{W}}{4\pi \times 5.67 \times 10^{-8}\ \text{W}/\text{m}^2/\text{K}^4 \times (7 \times 10^8\ \text{m})^2} \right)^{1/4} = 5700\ \text{K}.$$

A detailed determination of the surface temperature of the Sun gives 5772 K.

Recalling that the boiling point of water is 100°C, or 373 K, the surface of the Sun is about 15.5 times hotter than boiling water. Diamond melts at 3820 K, about 66% of the temperature of the Sun's surface.

What is the Sun Made of?

Key Point
The Sun's surface composition is (by mass) 73.8% hydrogen, 24.9% helium, and 1.3% heavy elements.

Thanks to the contributions made to physics and astronomy by so many, including the founders of spectroscopy (Section 7.1) and quantum mechanics (Sections 8.6 and 8.7), and countless others, it has become possible to study the Sun's spectrum in great detail. When combined with the probabilities of energy level transitions and highly detailed computer modeling of the atmosphere of the Sun, astrophysicists can now determine the composition of the Sun's surface with some precision.

The Sun's surface is dominated by hydrogen and helium, the two lightest and least complex atoms in the universe. Today, we know that hydrogen comprises 73.8% of the mass of the Sun's atmosphere near its surface, with helium making up an additional 24.9% of the mass. The remaining 1.3% of the mass is a mixture of many of the elements in the periodic table, most notably carbon, nitrogen, oxygen, neon, magnesium, silicon, sulfur, and iron. Because hydrogen and helium play such dominant roles in astronomy, all of the other elements are referred to collectively as heavy elements. That is not to say that they aren't important to astronomy as well: to the contrary, we will learn that their presence provides a tremendous amount of information about how our Sun, our Solar System, other stars and their planets, and galaxies were formed and have developed over time. The dominance of hydrogen and helium also tells us important details about the nature of the early universe.

How Fast Does the Sun Rotate?

Because the Sun is not a solid object, it is possible to have different rotation rates at different locations within and on the "surface" of the Sun. This situation is somewhat analogous to Earth's atmosphere, where the jet streams move faster than the surface of the planet does. The rotation period at the Sun's equator is about 25 days, but increases to 36 days near its poles, with the rotation direction being counterclockwise when viewed above the Sun's north pole. This is also the same direction that all of the planets orbit the Sun and most of the planets spin, including Earth. This implies that the Sun's north pole and Earth's North Pole are on the same side of the ecliptic plane. The variation of rotation period with latitude is known as differential rotation and is a fundamental component of the solar cycle, to be discussed in Section 9.3.

Key Point
The Sun's rotation period ranges from 25 days at the equator to 36 days at its poles.

9.2 The Sun's Dynamic Atmosphere

Layers and Temperatures

When you studied Fig. 9.3, you may have been asking yourself why the Sun looks so different in yellow-red light (the left-hand image) from the other two images, which were taken in extreme ultraviolet light? The answer lies in which part of the Sun's atmosphere is being imaged.

The narrow photosphere (Fig. 9.5) at the base of the atmosphere is where the vast majority of the Sun's electromagnetic radiation is emitted, and it is emitted in visible wavelengths. The photosphere is also the coolest part of the atmosphere, with the temperature dropping with increasing height until the top of the photosphere reached, at which point the temperature begins to rise with increasing altitude.

The level in the atmosphere where the "surface" is located, and where the temperature we determined for the Sun is 5772 K, is located near the base of the photosphere. It is this "surface" where the radius of the Sun is also measured. The entire photosphere is only a little more than 500 km thick, and, as we discussed in the last section, the radius of the Sun is about 700,000 km, which means that the photosphere comprises only $500/700,000 = 0.0007$, or 0.07% of the radius. This is why the "surface" seems so well-defined and abrupt.

The chromosphere [Fig. 9.6 (*Left*)] sits above the photosphere, and its thickness is about four times greater than that of the photosphere. The temperature rises gently through the chromosphere, approximately doubling from its base to the top of the zone, which is defined to be where the temperature suddenly increases dramatically. Much of the radiation emitted in the chromosphere is ultraviolet light, with the wavelengths becoming shorter as the temperature increases.

Atop the chromosphere is the transition region [Fig. 9.6 (*Center*)]. The transition region is very narrow, about 250 km, but the temperature climbs from 10,000 K to near 200,000 K in that layer, an increase of a factor of 20. The amount of light emitted from the transition region is much less than from either the photosphere or the chromosphere. Given the high temperatures, much of the radiation is emitted in extreme ultraviolet light and low energy x-rays.

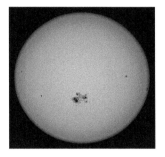

Fig. 9.5 The Sun's photosphere at 617 nm (yellow-red light) on October 18, 2014. A very large sunspot group is visible below the middle of the image. (NASA/SDO)

The outermost region of the atmosphere is the corona, where the temperature climbs more slowly with altitude, reaching about 1 million K. Because of its high temperature, what light is emitted from the corona tends to be in extreme ultraviolet light and x-rays. There is no well-defined outer edge to the corona.

Fig. 9.6 *Left:* The Sun at 30.4 nm (extreme ultraviolet) showing the chromosphere on August 24, 2015. The bright spot below center is a solar flare (an explosion on the Sun). (NASA/SDO) *Center:* An image in and above the transition region at a wavelength of 17.1 nm. The bright features in the distance are hot gasses moving around a sunspot at a temperature of 1,000,000 K and the dark, absorbing features are at 10,000 K. The reason for the very high temperatures may be due to rapidly changing magnetic fields funneling hot solar gasses. (NASA/TRACE) *Right:* A composite image of the Sun during the total solar eclipse of December 3, 2002, seen from Australia. The glow around the Sun is in visible light and the green image (artificially colored) is in extreme ultraviolet light. (NASA/ESA; CC BY-SA 3.0 IGO)

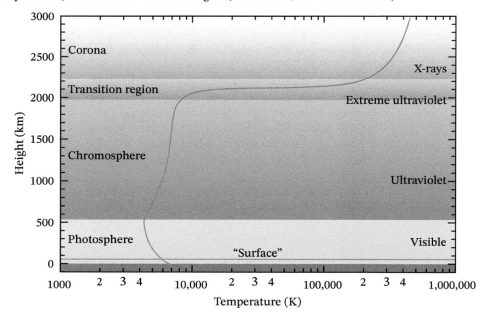

Fig. 9.7 The structure of the Sun's atmosphere, starting at the top of the Sun's interior (gray). The temperature at every point in the atmosphere is given by the red curve. The horizontal axis is logarithmic, meaning that the value increases by a factor of ten at each major tick mark. The small numbers indicate that you should multiple the last major tick mark value by the small number. For example, between 1000 K and 10,000 K, 2 corresponds to 2000 K, 3 corresponds to 3000 K, and so on. (Data from Avrett)

Figure 9.6 (*Right*) shows a composite of two images, one in visible light and the other of the inner corona obtained by the Solar and Heliospheric Observatory (SoHO) spacecraft at wavelengths between 17 and 30 nm. Those extreme ultraviolet wavelengths were produced by highly ionized iron and singly ionized helium.

The altitudes of the four layers that comprise the Sun's atmosphere are shown in Fig. 9.7, along with how the temperature of the atmosphere changes with altitude. The wavelength ranges of electromagnetic radiation emitted at the various altitudes in the atmosphere are indicated on the right-hand side of the figure.

Figure 9.8 shows images of the Sun in a variety of wavelengths. As the previous discussion suggests, it is possible to probe different levels in the Sun's atmosphere

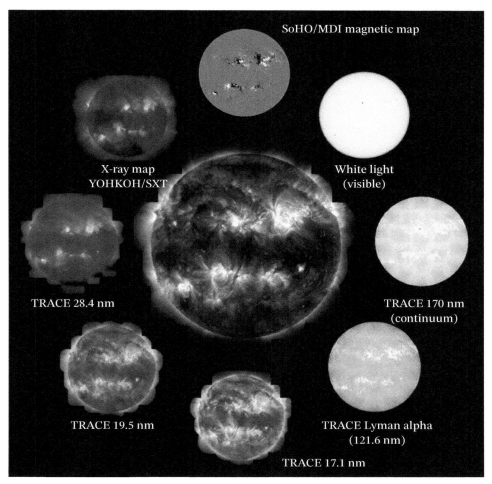

Fig. 9.8 A collection of solar images taken at different wavelengths. The top image is a map of the Sun's magnetic field, to be discussed in Section 9.3. The central composite image of the Sun's corona is a combination of 17.1 nm, 19.5 nm, and 28.4 nm. [Joe Covington (Lockheed-Martin Missiles and Space, Palo Alto) and Japan/NASA/UK; the Transition Region and Coronal Explorer, TRACE, is a mission of the Stanford-Lockheed Institute for Space Research, and part of the NASA Small Explorer program]

by studying the light produced by atoms and ions at different wavelengths. This is because the wavelengths that are produced depend on the temperature and density of the gas: higher temperatures and lower densities result in shorter wavelengths of light being emitted.

The density of the gas in the atmosphere drops rapidly through the photosphere and chromosphere, then plummets through the transition region. The density of the gas in the corona is very low, meaning that there aren't many particles in that diffuse outermost layer of the Sun. Because of its very low density, the corona takes on an almost ethereal glow and its shape is very amorphous and changes constantly. It is the corona that becomes visible to the unaided eye only when the rest of the Sun is obscured by the Moon during a total solar eclipse (e.g., Fig. 4.23). Astronomers can also observe the corona by creating an artificial eclipse using a disk to block out the light from the rest of the Sun. Even though the corona emits most of its light in x-rays, which the human eye cannot detect, we can see the corona during an eclipse because particles in the tenuous gas reflect a small fraction of the visible light produced in the photosphere.

Granulation and the Role of Convection

When observing the Sun's photosphere at high resolution, as in Fig. 9.9, bright regions and darker areas surrounding them become readily apparent. These ever-present features at the base of the photosphere are called granulation. The bright

Fig. 9.9 A closeup view of solar granulation due to convection cells emerging at the base of the photosphere. This is a portion of a first-light image of the 4-m Daniel K. Inouye Solar Telescope, obtained in December 2019 on the summit of Haleakala, Maui, in Hawai'i. The bright regions are hot gas rising up from the interior, and the dark lanes are cooler gas sinking back down into the star. Each rising cell is about the size of the state of Texas, and features as small as 30 km (18 mi) are visible. Tiny bright spots in the dark, sinking gas are markers of magnetic fields that may help channel energy up into the hot corona. The width of the region included in the image is 7800 km (4800 mi). A time-elapsed video of granulation is available on the textbook's website. (Adapted from NSO/AURA/NSF; CC BY 4.0)

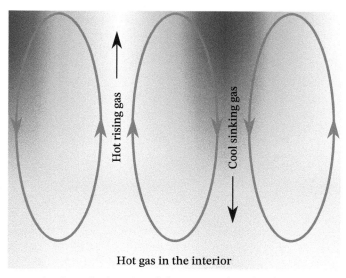

Hot rising gas

Cool sinking gas

Hot gas in the interior

Fig. 9.10 Hot gases rise from the interior of the Sun, cool, and sink back into the interior.

regions are typically about 1000 km across and have individual lifetimes of about ten minutes. Even though granulation is always present in the Sun's dynamic atmosphere, individual features are transient. A movie showing this ever-changing and fascinating environment on the Sun's surface is available with Fig. 9.9.

The reason for the varying coloration has to do with temperature differences in the gas. The darker regions have temperatures that are slightly cooler than the bright regions. Recall that the amount of light per square meter (the flux) emitted by a blackbody depends on temperature raised to the fourth power, $F = \sigma T^4$, which is Stefan's flux equation for blackbody radiation [Equation (6.15)]. The strong dependence of flux on temperature means that even if the temperature is just a little less than the surrounding material, it will produce significantly less light and therefore appear to be dark when compared with the hotter and much brighter material around it. If the cooler gases were isolated by themselves, they would actually shine brightly because the temperature is still typical of the base of the photosphere.

In Section 8.4 we discussed how the Doppler effect can be used to measure velocities toward or away from the observer. As illustrated in Fig. 9.10, by focusing on the bright regions of the granulation, the spectral lines produced by the gas are blueshifted, meaning that the gas is moving outward from the interior and toward the observer at about 0.4 km/s. Conversely, the spectral lines emitted by the gas in the darker regions are redshifted, indicating that the flow of the gas is away from the observer and directed inward into the star at approximately the same speed. Hot gases are rising up into the photosphere where light energy can be emitted into space. Then, as the gases lose energy, they cool off and sink back down into the Sun. In other words, granulation is just the visible top of a convection zone that extends deep into the interior of the Sun (recall our original discussion of convection on page 226).

Although the details aren't yet completely understood, it is clear that convection plays an important role in the very rapid rise in temperature with height through the Sun's transition region, and the nearly one million kelvins temperature in the

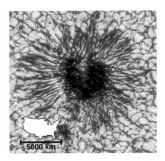

Fig. 9.11 A closeup view of a sunspot at a wavelength of 530 nm, obtained by the Daniel K. Inouye Solar Telescope on January 28, 2020. The umbra is the dark central region and the penumbra is the filamentary structure surrounding the umbra. The relative size of the United States (5000 km) is shown in the lower left corner. A movie of gas motions around a sunspot is available on the Chapter 9 resources webpage. (NSO/AURA/NSF, CC BY 4.0)

corona (recall Fig. 9.6). As hot convective cells move up through the interior and emit electromagnetic radiation into space from the base of the photosphere, those moving cells also send shock waves through the lower atmosphere into the chromosphere and the transition region. These shock waves are akin to sonic booms when planes fly faster than the speed of sound: the convection cells push the gases above them faster than sound waves (pressure pulses) can normally travel through the gas. The result is that when the shock waves slow down, the kinetic energy of the shocks transforms into heat when the waves collide with the atoms and ions in the gas, partially contributing to the rapid temperature increase seen in the transition zone.

Sunspots

You have seen two full-disk images of the Sun's photosphere that include sunspots in Figs. 9.3 (*Left*) and 9.5. Figure 9.11 shows a sunspot in greater detail, along with the usual presence of granulation. The darkest part of the sunspot is the umbra and the surrounding filamentary structure is the penumbra. These terms should sound familiar because the term "umbra" is also used to describe the darkest part of a solar or lunar eclipse, while the term "penumbra" describes the portion of a partial eclipse where the shadow is less dark (see Fig. 4.18). The dynamic environment around sunspots can be seen in the sunspot movie associated with Fig. 9.11 available from the Chapter 9 resources webpage.

Sunspots have typical lifetimes of several days to a month or more and can become enormous in size. Figure 9.12 shows a sunspot group with Earth's size for comparison.

The penumbra's filamentary pattern may look familiar because it bears a striking resemblance to the appearance of iron filings around a bar magnet, like the one shown in Fig. 9.13. However, there is an important difference between how iron filings line up with a bar magnet's magnetic field and how the filamentary structure of the penumbra develops. Iron filings have natural magnetic properties themselves; they can act as tiny bar magnets with their south poles attracted to the bar magnet's north pole and their north poles attracted to the bar magnet's south pole. But the same is not true of the charged particles in the Sun's atmosphere. In Section 6.5

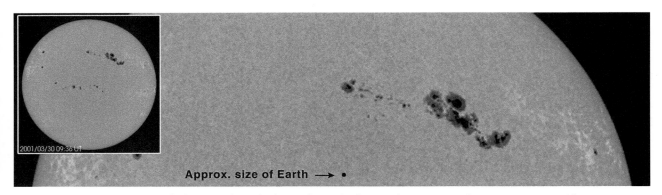

Fig. 9.12 A sunspot group, with the size of Earth shown for scale. [SoHO (ESA & NASA); CC BY-SA 3.0 IGO]

you learned that an electric current can produce a magnetic field and that a varying magnetic field can produce an electric current, so it is clear that moving charged particles (which is what an electric current is, after all) interact with magnetic fields. As illustrated in Fig. 9.14, moving charged particles in a gas are forced into circular motions around magnetic field lines if their motion is perpendicular to the field, or more generally they follow spiral patterns if there is some motion of the charged particles in the direction of the magnetic field. The result is, in essence, that charged particles are forced to follow magnetic field lines and so get trapped in magnetic fields. The filamentary structure of a penumbra is due to those trapped charged particles spiraling around magnetic field lines.

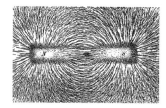

Fig. 9.13 Iron filings around a bar magnet. (Newton Henry Black, public domain)

Figure 9.15 shows two images of the Sun from July 15, 2002. The image on the left is a traditional visible wavelength view of the Sun obtained by the SoHO spacecraft, while the image on the right is a map of what the Sun's magnetic field looked like at the time. The light coloration of the magnetic field map represents magnetic north-pole-like fields while the darker coloration corresponds to magnetic south-pole-like fields. The image at the top of the composite of images in Fig. 9.8 is another magnetic field map, taken at a different time.

Fig. 9.14 An electron (in red) spiraling around a magnetic field line (in blue).

There are several aspects of the magnetic field map that are important to make note of. First, for every bright region there is a paired dark region. This pattern is just like the bar magnet, or any magnetic field for that matter. Nature never creates just a north pole or just a south pole; they always come in pairs. Magnetic fields coming out of magnetic north pole regions always enter into magnetic south pole regions. The iron filings in the bar magnet photograph (Fig. 9.13) show that same universal pattern. More than that, magnetic field lines never just stop; instead, magnetic fields always form closed loops. In the case of bar magnets, the magnetic field lines complete their loops inside the bar magnet. For the Sun, the loops are completed with magnetic field lines existing in the interior of the Sun.

Second, the large sunspot group above the middle of the visual image can easily be seen in the magnetic map as well. The magnetic field associated with that

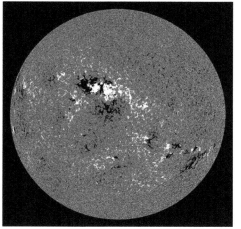

Fig. 9.15 *Left:* The Sun in visible light on July 15, 2002. *Right:* A magnetic field map of the Sun from the same date. Geographic north is at the top of both images. (NASA/SoHO/MDI)

huge group is very strong. There are also numerous smaller sunspots that are evident in the magnetic field map that are not as readily apparent in the visible image. Sunspots are associated with magnetic fields in some way, as first suggested by the filamentary penumbra.

Third, if you study the magnetic field map closely you will notice that on the top half of the image the bright regions are always to the right of their paired dark regions. However, on the bottom half of the map, just the opposite occurs. What is not apparent in the images is that the Sun rotates from left to right (counterclockwise when viewed from above the Sun's north geographic pole at the top of the image) with a rotation period of approximately one month, although that period increases with increasing latitude away from the Sun's equator. If a magnetic north pole region leads in the northern hemisphere, then a magnetic south pole region leads in the southern hemisphere, and vice versa.

Solar Flares

When large sunspot groups form they are often associated with unpredictable, and very violent, explosive events known as solar flares; two examples can be seen in Fig. 9.16. Solar flares can range in energy from the equivalent of a ten megatonne bomb (10^{17} joules) to over one billion megatonnes (10^{25} J), with the energy being released over a period ranging from a fraction of a second to an hour or more. The electromagnetic energy produced is typically in the extreme ultraviolet and x-ray regions.

But it is not just electromagnetic energy that is released when flares occur, a significant amount of matter can be blasted away from the Sun's surface as well.

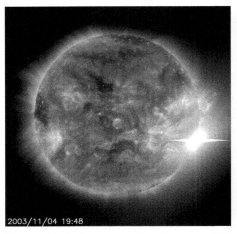 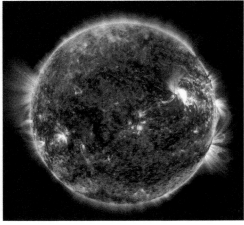

Fig. 9.16 *Left:* An extraordinarily powerful flare seen on the right-hand edge of the Sun. The flare occurred on November 4, 2003, as observed by the SoHO spacecraft. A movie available on the `Chapter 9` resources webpage that covers October 17 to November 5, 2003, shows the very active sunspot group rotating across the solar disk and emitting multiple flares. The most powerful flare occurs at the end of the movie. [SoHO/EIT (ESA & NASA; CC BY-SA 3.0 IGO)] *Right:* A June 25, 2015, flare can be seen on the solar disk (the bright area) above the Sun's equator and toward the right edge of the disk. (NASA/SDO)

Fig. 9.17 An eruptive solar prominence seen by the Solar Dynamics Observatory on April 21, 2015. The gold colors are associated with light emitted at an extreme ultraviolet wavelength of 17.1 nm and the red coloration is associated with a wavelength of 30.4 nm. A movie spanning six hours around the eruption is available from the `Chapter 9` resources webpage. (NASA's Goddard Space Flight Center/SDO)

For the most energetic solar flares that are directed toward Earth, the particles can reach us in as little as thirty minutes. Because those particles are electrically charged in the form of electrons and hydrogen and helium ions, and because they also have a tremendous amount of kinetic energy, solar flares have the potential to significantly impact radio communications.

Solar Prominences

Like sunspots and solar flares, solar prominences are also manifestations of the Sun's magnetic field. In Figs. 9.3 (*Center* and *Right*), prominences are evident toward the edge of the Sun, with Earth shown to scale in the right-hand image. A close-up view of a prominence region can be seen in Fig. 9.17.

The characteristic loop-like shapes that are evident for many prominences exist because the charged particles in the Sun's atmosphere are trapped along magnetic field lines, just like the penumbra of sunspots. When the field lines undulate over time the particles are pulled along with them, as can be seen in the movie associated with Fig. 9.17 that is available from the `Chapter 9` resources webpage. In addition, although the gas associated with solar prominences is hot (typically 8000 K), it is actually much cooler than the very thin, nearly 1 million kelvins gas in the Sun's corona into which the prominences protrude. As a result, charged particles can "rain" back down into the lower atmosphere along magnetic field lines following an eruptive solar prominence, which is also captured in the movie.

Figure 9.18 also shows the July 30, 2002, eruptive prominence as a whip-like structure (lower right corner). Other prominences can be seen across the Sun's disk as long, dark filaments. Because their gases are cooler than the underlying gas, they appear darker when superimposed on the brighter background. The whip-like prominence had disappeared when the next image was taken, six hours later, probably due to an eruption that ejected the gas out into space.

So why are the gases in prominences cooler than the gases that surround them? The answer lies in the support they receive by being elevated high into the corona via magnetic fields. Trapped along magnetic field lines, the gases associated with solar prominences are not heated rapidly enough by the very thin coronal gas to

Fig. 9.18 A whip-like prominence seen near the edge of the Sun, and dark, filamentary prominences seen across the Sun's bright disk in this 30.4 nm wavelength image. (ESA&NASA/SoHO)

cause their temperature to increase significantly. Think about what happens if you take a cool drink outside on a hot day. If you drink it over the course of a few minutes, it will still be cold, simply because the air couldn't heat it up very quickly.

Coronal Mass Ejections

Coronal mass ejections (CMEs) are events that can also send solar gases out into space. Figure 9.19 shows a CME that occurred on February 27, 2000, as recorded by the Solar and Heliospheric Observatory (SoHO). Time-elapsed movies of CMEs are available from the `Chapter 9` resources webpage.

Although CMEs are often associated with solar flares or eruptive prominences, they are separate phenomena. Blasting massive amounts of charged particles into space, the material from CMEs can take two to three days to travel one astronomical unit. Unlike solar flares, however, CMEs can damage communications satellites, affect the accuracy of GPS coordinates, and even overload ground-based power grids. Coronal mass ejections can occur with frequencies ranging from once a week to several times per day, depending on how active the Sun is.

More visually, solar storms can significantly energize the aurora borealis and the aurora australis (the northern and southern lights, respectively), producing beautiful displays like the one shown in Fig. 9.20. Auroras appear when charged

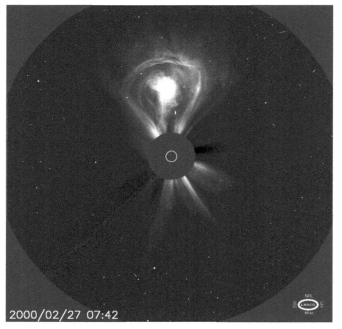

2000/02/27 07:42

Fig. 9.19 A coronal mass ejection on February 27, 2000, captured by the joint ESA/NASA SoHO mission in blue light (420–520 nm). Note that stars are visible in the background. A disk is placed in front of the Sun to block out its very bright surface, allowing the corona and CMEs to be visible (the location of the Sun is the white circle). A movie showing the so-called Halloween storm of CMEs between October 18 and November 7, 2003, is available from the `Chapter 9` resources webpage. (ESA&NASA/SoHO)

Fig. 9.20 Aurora seen in Denali National Park, Alaska. (NPS Photo/Kent Miller)

particles from the Sun spiral around Earth's magnetic field lines and collide with atoms and molecules in our atmosphere where the field lines get closer together near Earth's magnetic north and south poles. The trapped solar particles cause the atmospheric atoms and molecules to become excited; photons are emitted when electrons fall back down to lower energy levels, similar to the atomic transitions in Fig. 8.28.

Key Point
Auroras are produced by charged particles from the Sun getting trapped in Earth's magnetic field and colliding with atmospheric atoms and molecules near the magnetic north and south poles.

The Solar Wind and Coronal Holes

Earth is constantly being bombarded by charged particles originating from the Sun, whether or not a solar flare, an eruptive prominence, or a coronal mass ejection is violently sending particles out into space. There is also a much less energetic solar wind produced by the Sun which travels through the Solar System with a speed of hundreds of kilometers per second.

Although this wind from the Sun is ever-present, and the Sun is losing mass as a result, our parent star is certainly in no danger of disappearing any time soon. The rate of mass loss due to the solar wind is only about 3×10^{-14} M_{Sun} per year, meaning that at that rate it would take nearly thirty trillion years for the Sun to "evaporate" via the current solar wind; this length of time is roughly 2000 times greater than the present age of the universe of 13.8 billion years.

Some of the solar wind escapes from streamers in the corona and tends to be a bit gusty and slow, but a somewhat faster wind originates in regions of the corona known as coronal holes. Coronal holes, like the very large one near the Sun's north pole in Fig. 9.21, change position, and appear and disappear, depending on the Sun's magnetic field configuration at the time. They appear dark in extreme ultraviolet and x-ray images because the density of the already low-density coronal gas is less

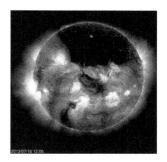

Fig. 9.21 A large coronal hole is visible near the Sun's north pole in this 28.4 nm image of its corona from July 19, 2013. A smaller coronal hole can be seen near the southern pole. (ESA&NASA/SoHO)

than average and the gas is cooler than its surroundings. More importantly in the context of the solar wind, magnetic field lines leaving or entering coronal holes extend vast distances through the Solar System. This means that, rather than being forced to stay close to the Sun, charged particles spiraling around magnetic field lines that are associated with coronal holes can travel out through the Solar System and beyond. The winds coming from coronal holes have speeds that are about twice that of their slower, gustier cousins that escape from streamers.

Space Weather

Because of the impact that solar storms can have on Earth, including disruptions to our technology, NASA and other organizations continually monitor solar activity. Based on these observations, if a coronal mass ejection is headed toward us, utility companies can adjust power loads in electrical grids to protect them, and astronauts may be warned to stay within the confines of their spacecraft. This continual monitoring is also how space weather forecasters can provide information to your local terrestrial weather forecaster so that she can alert you to the possibility that an especially impressive aurora may be visible in the nights ahead (assuming that you live at a sufficiently high latitude, of course).

9.3   The Solar Cycle

Observations

By now it should be apparent that much of what happens in the Sun's atmosphere is tied to our star's magnetic field: sunspots, solar flares, solar prominences, coronal mass ejections, coronal holes, and the component of the solar wind associated with coronal holes. But there is more to it than the existence of these phenomena. Over time, the number of sunspots increases and decreases, the latitudes at which sunspots form change in a systematic way, and the frequency and severity of explosive solar events also change periodically.

Ever since David Fabricius and Galileo Galilei first pointed their telescopes at the Sun in 1611 and 1612, respectively (a very bad idea without proper equipment by the way!), astronomers have been recording sunspots. Those early observations weren't particularly detailed, but over time the data became more reliable and precise. As evident in Fig. 9.22, the number of sunspots visible on the Sun's surface changes from a minimum number to a maximum value and then back to a minimum again over a period of approximately 11 years, known as the solar cycle. The times in the cycle when the number of sunspots is at a minimum or a maximum are referred to as solar minimums and solar maximums, respectively.

Somewhat ironically, not long after those first telescopic observations of sunspots, the number of sunspots decreased significantly and all but disappeared for almost 70 years. This extended period is known as the Maunder minimum, named for Annie Maunder, seen in Fig. 9.23 *Top*, and her husband E. Walter Maunder (Fig. 9.23 *Bottom*), who studied how sunspots change with latitude over time. The Maunder minimum has been associated with an unusually cool period on Earth called the "Little Ice Age." During that time, rivers froze over that normally

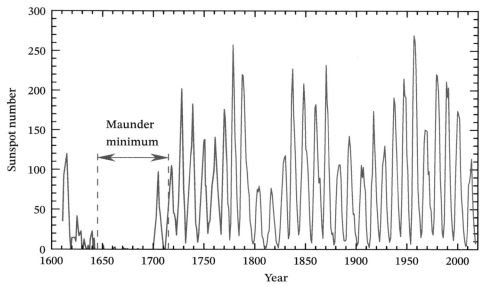

Fig. 9.22 Sunspot numbers vary with a periodicity of 11 years. The Maunder minimum was an unusually quiet period of time from about 1645 to 1715. (Data courtesy of WDC-SILSO, Royal Observatory of Belgium, Brussels, and Dr. David H. Hathaway, Marshall Space Flight Center, NASA)

don't, and snow remained at lower elevations in Europe than typically happens. There is evidence that further back in time similar stretches of solar inactivity have also occurred. The relationship between activity on the Sun and Earth's climate is an area of active research.

Even excluding the Maunder minimum, Fig. 9.22 demonstrates that although the periodicity in the number of sunspots is fairly constant, the minimum and maximum number of sunspots is not particularly consistent from one cycle to the next. The other phenomena already discussed (solar flares, solar prominences, and CMEs) also peak in number and intensity around solar maximums and reach a more quiet phase during solar minimums. The last solar maximum occurred in April, 2014, and the most recent solar minimum was in December, 2019.

Not only is there an 11-year periodicity in the number of sunspots, the same solar cycle is also associated with where they form and disappear on the Sun's surface. When a sunspot forms at a certain latitude north or south of the solar equator, it will rotate along with the Sun, staying at that latitude until it disappears. During solar minimums, sunspots tend to form at about 30° north and south of the solar equator. Over the course of the cycle, sunspots begin to form and disappear closer and closer to the equator until the cycle starts all over again during the next solar minimum. At solar maximums, sunspots tend to be found near the middle of the latitude range. As the number of sunspots increases from minimum to maximum, the amount of surface area covered by sunspots also increases. Both of these behaviors are recorded in Fig. 9.24. The top panel is called the "butterfly diagram" for the characteristic wing-like shapes that seem to repeat endlessly.

Fig. 9.23 *Top:* Annie Russell Maunder (1868–1947). (Courtesy of Dorrie Giles) *Bottom:* Edward Walter Maunder (1851–1928). (Hector Macpherson, public domain)

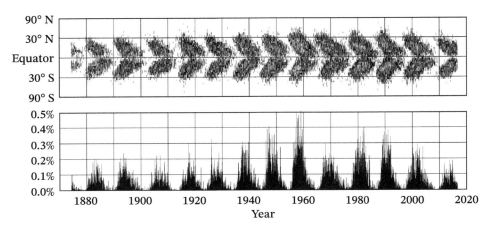

Fig. 9.24 *Top:* The butterfly diagram displays the latitudes at which sunspots form and disappear over time. *Bottom:* The amount of surface area facing Earth that is covered by sunspots. The 11-year solar cycle periodicity is evident in both diagrams. (Adapted from a graphic by Dr. David H. Hathaway, Marshall Space Flight Center, NASA)

There is another aspect of the solar cycle that is also important to note. Recall from Fig. 9.15 (*Right*) that if sunspot groups in the Sun's northern hemisphere lead with a north magnetic polarity, then sunspot groups in its southern hemisphere lead with a south magnetic polarity. This pattern reverses every 11 years, meaning that to get back to the original north magnetic polarity leading in the northern hemisphere, the 11-year cycle must occur twice. It is also the case, as illustrated in Fig. 9.25, that during solar minimums the Sun's magnetic field looks most like that of a bar magnet. At the start of one solar cycle the magnetic north pole may

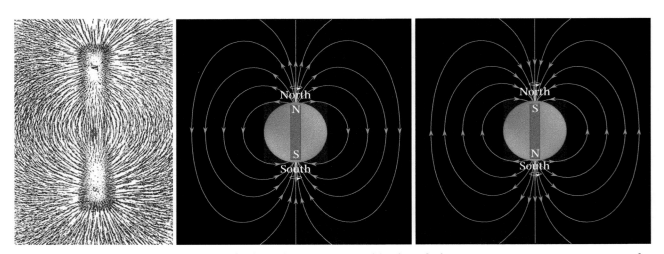

Fig. 9.25 The shape of the Sun's magnetic field at solar minimums is like that of a bar magnet. However, every 11 years the magnetic polarity reverses. (Of course, there isn't actually a bar magnet inside the Sun!) *Left:* A bar magnet with its north pole (N) on top. (Newton Henry Black, public domain) *Center:* The Sun with its magnetic north pole (N) at the geographic north pole. (adapted from a solar image of NASA/SDO) *Right:* The Sun with its magnetic south pole (S) at the geographic north pole. Magnetic field lines are considered to point away from magnetic north poles and toward magnetic south poles. (Adapted from a solar image of NASA/SDO)

be located at the Sun's geographic (rotational) north pole, but eleven years later the magnetic south pole is located at the Sun's geographic north pole. Eleven years after that the magnetic north pole has returned once again to the geographic north pole. The Sun's global magnetic field flips once every 11 years as one feature of the complete 22-year solar cycle.

The Magnetic Dynamo

Electric currents, which are simply moving charges, generate magnetic fields. If fact, all magnetic fields are the result of moving charges. In the case of seemingly static situations like bar magnets, the moving charges are at the atomic level with electron orbitals.

So where is the electric current that produces the Sun's magnetic field? That answer lies in (a) the Sun's rotation, (b) the ionized gas that makes up the Sun's very hot interior, and (c) the continual motion of gas through convection. As the Sun rotates on its axis, and as convection carries hot gases toward the surface and cooler gases back down, the motions of the ions and electrons generate an electric current which, in turn, produces a magnetic field. This process of generating a magnetic field is referred to as a magnetic dynamo.

As you will learn, the Sun is far from alone in generating magnetic fields in this way. Earth's magnetic field is also generated by a magnetic dynamo that involves its electrically conducting fluid outer core. Magnetic fields are common throughout the universe.

Key Point
The Sun's magnetic field is due to its rotation, convection, and the resulting movement of charged particles in its interior.

Differential Rotation

Because the Sun is so hot, there isn't any location inside the star that is solid; it is entirely gaseous, and most of it, especially in the interior, is also entirely ionized. This means that the Sun doesn't have to rotate at the same rate everywhere, and in fact it doesn't. The surface of the Sun at its equator rotates with a period of about 25 days, but near the north and south geographic poles the rotation period is much longer, at 36 days. This changing rotation period with latitude is called differential rotation.

Earth's atmosphere shows a somewhat similar phenomenon in its jet stream: a ribbon of atmosphere that moves much faster than the air around it. Ocean currents also exist on Earth, where portions of the ocean circulate faster than the surrounding water. In both cases, the behaviors are due in part to Earth's overall rotation.

Differential rotation in the Sun plays an important role in the solar cycle, because not only do moving charged particles generate magnetic fields, but magnetic fields also cause charged particles to spiral around them, something we saw numerous times in Section 9.2. The coupling of charged particles to the magnetic field means that as the ionized gas moves, it drags the magnetic field along with it. As illustrated in Fig. 9.26, the more rapidly moving gas near the equator pulls magnetic field lines along with it at a faster pace than near the poles, causing the magnetic field lines to stretch. Because magnetic fields exist below the Sun's surface as well, those magnetic field lines are also dragged along with the differential rotation. Over time, the magnetic field lines near the equator make many more

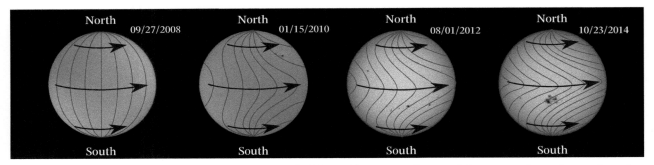

Fig. 9.26 The Sun's differential rotation drags magnetic field lines (shown in red) around the equator faster than at the poles. The drawings of the magnetic field lines are meant to be illustrative rather than correct in detail. A solar minimum occurred in December, 2008 and again in December, 2019, and a solar maximum occurred in April, 2014. (Left two images adapted from ESA&NASA/SoHO, and right two images adapted from NASA/SDO)

revolutions than they do at higher latitudes. This means that the simple-looking bar-magnet-like magnetic field that exists near solar minimum is gone.

Magnetic Fields Doing the Twist; The Source of Solar Activity

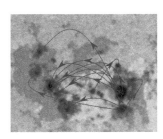

Fig. 9.27 A sunspot group superimposed on a magnetic field map. Hypothetical magnetic field lines are also shown. (NASA/SoHO/MDI)

Differential rotation alone doesn't cause sunspots to appear. Even though magnetic field lines get tightly wound by differential rotation, another step is required for the magnetic field lines to get twisted together, which is what results in sunspots, solar flares, prominences, and the rest of the activity in the Sun's atmosphere around solar maximums.

If you watched any of the movies discussed in Section 9.2 you would have seen just how chaotic the motions in the atmosphere are, including at the base of the photosphere where the convective cells from the interior can be seen rising to the surface and then falling back down into the Sun. These random motions of ionized gas also drag magnetic field lines along with them. Differential rotation winds up the magnetic field lines and chaotic convection twists them in knots. It is the tightly twisted magnetic field lines that buoy up the cool, darker gas of sunspot umbras.

Figure 9.27 shows a closeup of one of the active regions in Fig. 9.15 overlaid on the associated magnetic field map. The lighter background regions are north-pole-like areas of magnetic field while the darker background regions are south-pole-like areas. Examples of magnetic field lines leaving and entering sunspots are also shown. The umbras of the sunspots, where magnetic field lines converge, have the most intense fields.

Twisted magnetic field lines don't just hold up the cool gas of sunspots, they also contain a lot of pent up magnetic energy, much like twisting a rubber band. When too much energy is confined in that way, something has to give. It is the explosive release of that stored energy and the associated untwisting of magnetic field lines that cause solar flares, eruptive prominences, and the other spectacular events that occur during the middle of a solar cycle (Fig. 9.28).

With the releasing of energy when magnetic fields restructure themselves, the magnetic field begins to return to the more simplistic configuration it had at the beginning of the 11-year cycle in preparation for the next one. The only difference between the start of the new cycle and the previous one is that the overall bar-like magnetic field has reversed its polarity.

Key Point
Overall solar activity peaks and ebbs with the same 11-year periodicity exhibited by sunspots.

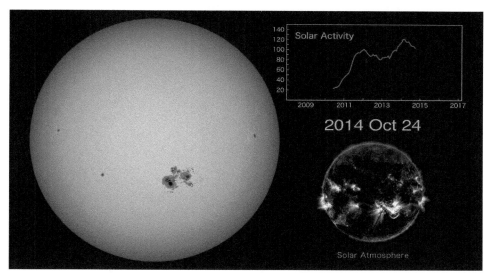

Fig. 9.28 A still from NASA's Solar Dynamics Observatory images compiled into a video showing the majority of one solar cycle from June 2, 2010, to January 15, 2017. In the movie, available from the Chapter 9 resources webpage, the sequence of images on the left are of the photosphere in visible light, and the sequence of images in the lower right are of the chromosphere in extreme ultraviolet light. One image was obtained every twelve hours. The graph on top-right is the number of sunspots. (NASA's Goddard Space Flight Center/SDO)

What's Next?

Although we currently understand a great deal about the solar atmosphere, there remains much that we do not know. For example, exactly how does the Sun's atmosphere get so hot? While it is true that convection cells that top out at the base of the photosphere do send shock waves up through the atmosphere, that by itself doesn't fully explain why the temperature climbs so dramatically through the transition region and in the corona, increasing from about 10,000 K to nearly 1 million K.

Another important question has to do with what relationship, if any, that the number of sunspots and total sunspot surface area have on Earth's climate. We know, for instance, that Europe experienced a "Little Ice Age" during the Maunder minimum. Was that a coincidence or is there a coupling between our climate and the solar cycle that remains to be explained? We do know that there is a small effect on the Sun's overall energy output associated with the surface area covered by sunspots, but is that enough to alter climate in a significant way? Are there longer term cycles associated with the Sun's magnetic field that may impact Earth's climate as well? Do cycles exist that could somehow influence the starting of major ice ages? There is some evidence to suggest that other cycles are superimposed on the 11-year cycle, with lengths of decades or perhaps centuries. But given that humanity has only been observing sunspots for 400 years, teasing out cycles that are a few centuries long is a very difficult, and error-prone prospect. If cycles exist that are even longer, the data are simply not available to uncover them by statistical means. There is much research that is still necessary to understand our Sun's atmosphere and its relationship to climate and life on our little, fragile planet. More will be said about this topic in Section 11.8.

Key Point
Much research remains to be done to truly understand the solar cycle and its relationship with space weather and Earth's climate.

9.4 Using Computers for Exploration

Astronomy is primarily an observational science, as opposed to an experimental one. In physics, chemistry, biology, and other laboratory sciences, it is possible to set up controlled experiments to test physical processes, study the structures of organisms, or investigate the properties of atoms, molecules, or semiconductors.

In general, astronomy doesn't have that same opportunity, except in very special circumstances such as sending robotic ambassadors to planets or other objects in our Solar System. Instead, observational data provide what amounts to a one-directional flow of information: we get what we get from whatever we are observing. That doesn't mean that we are restricted to only a small amount of information, however. As you have already learned, a tremendous amount of information is available through the radiation that a celestial object sends to us, assuming that we understand how that information is produced and how to decipher it.

Key to deciphering that information is the development of a theoretical model of the system that is being researched. To construct such a model requires a detailed understanding of the physical processes that are at play; for example, how masses interact gravitationally, how light is produced, absorbed, or scattered by ions, electrons, atoms, and molecules, or how energy is produced and transported from one place to another.

Although some work in theoretical physics, chemistry, and astrophysics can still be performed with pencil and paper, the very complex interactions between physical processes require the speed and memory of computers to create reasonably accurate models. It is through comparing those models, and the predictions they make about what should be observed, with actual observations that astronomers are able to decode our universe. Today, many of the most complex and sophisticated computer models in the world are associated with astronomical simulations.

Certainly, large-scale computational simulations are not only reserved for theoretical astrophysics. Computers, like the one shown in Fig. 9.29, have become very powerful tools in the scientist's toolbox. Another class of very large and complex simulations are those having to do with global climate change research, a subject that we will be discussing in Section 11.8. Data mining and computer modeling in the biological sciences, quantum chemistry, energy research, and other disciplines also demand immense computational resources.

At this point it is very important to emphasize once again that the phrase so often heard, "it is just a theory," has very little relevance to scientific research. Returning to the discussion in Chapter 1, The Nature of Science, recall that scientific theories, and the models (computer or otherwise) derived from those theories, must be self-consistent and devoid of ad hoc assumptions. This means that all of the components of a theoretical model must rely on the detailed understanding and data that are aspects of every piece of physics that go into the simulations. As a deeper understanding of the energy levels of iron atoms are developed, for example, those become incorporated into the overall computer model. With this highly focused approach to understanding nature, scientists are able to make progressively more detailed models of stars or galaxies, and their life histories. Such an approach is even able to help us develop a deeper understanding of the birth and life of the universe as a whole.

Fig. 9.29 Built by IBM™, the Summit supercomputer at Oak Ridge National Laboratory was the world's fastest supercomputer as of June, 2018. Summit has a peak performance speed of 200,000 trillion calculations per second (200 petaflops).[1] The entire machine weighs more than a large commercial aircraft and 4000 gallons of water must be pumped through its cooling system every minute to carry away the heat that is generated. (ORNL, CC BY 2.0)

It is also a critical point of theoretical work that if a model does not agree with the observations then either some important feature of the system has not been adequately included in the model, or perhaps there is an incomplete or even incorrect understanding of the physics involved. In the latter case, simply "fudging" the model is unacceptable; instead, the source of the problem must be further studied and ultimately revised. In the most extreme case, perhaps the very underlying theory upon which the source of the problem rests, such as quantum mechanics or general relativity, is flawed and needs to be replaced with a more complete and successful theory. Fortunately, at least so far, that most severe of problems has not occurred. Of course, that is no guarantee that this sort of scientific crisis won't occur in the future; in fact, it may well happen at some point. Remember that it is never possible to prove a theory, it is only possible to disprove one. Should that crisis occur, what ultimately replaces the disproven theory must incorporate all of the successes already seen, and extend our understanding of nature as well.

In the remainder of this chapter we will explore how such a modeling process, based on our understanding of fundamental physics, reveals the Sun's hidden secrets.

---

[1]In May, 2022, the Frontier supercomputer, also at Oak Ridge National Laboratory, broke the exoscale computing barrier with a peak performance of 1.1 exoflops (1100 petaflops), or 1.1×10^{18} floating point operations per second. All the people on Earth would have to work for four years to perform as many calculations as the Frontier supercomputer can perform in one second. Some of the work carried out by Frontier will be astrophysical. Frontier was built by Hewlett Packard Enterprise.

9.5   How to Build a Star

The Key Ingredients

Key Point
A wide range of physical processes must be included when constructing a realistic computer model of the Sun.

Virtually each aspect of physics that has been discussed throughout this text so far enters into "building" a computer model of the Sun. In a general sense, each of Newton's laws of motion enter into the calculations either directly or indirectly, and of course so does his universal law of gravitation. In addition, the universality of the conservation laws requires that energy, linear momentum, and angular momentum all be conserved throughout the Sun (and everywhere else in the universe, for that matter). Given that heat is also an important component of the nature of the Sun, the area of physics known as thermodynamics must be included as well.

By now it should also be abundantly clear that describing how light is produced and how it interacts with matter is crucial if we are to be able to explain how the Sun works. This means not only including details about electromagnetic radiation in general, and blackbody radiation in particular, but also atomic and molecular physics (quantum mechanics). As we will explore in more detail in Section 9.6, the physics of the atomic nucleus, (nuclear physics) is also fundamentally important.

Balancing Forces

Beyond the physics that you have already studied, understanding how a gas behaves is essential to all studies of the Sun, both in its atmosphere and in its interior. This includes understanding the balance between the forces due to pressure and gravity, referred to as hydrostatic equilibrium. You are likely already familiar with the implications of this universal balancing act on Earth. The atmospheric pressure at Earth's surface is greater than it is higher up, as is the water pressure at the bottom of a swimming pool compared to near the surface.

Figure 9.30 shows a diver in the ocean swimming horizontally. According to Newton's second law ($\vec{F}_{net} = m\vec{a}$), in order to continue swimming horizontally the forces acting on him must be balanced so that the net force is zero, implying that $\vec{a} = 0$. Obviously, one of the forces being exerted on the diver is gravity pulling him down toward the center of Earth. The other force is due to the pressure in the water.

Formally, pressure, P, is defined as the magnitude of the *perpendicular* force, $F_\perp$, acting on a surface divided by the area, A, of that surface, or

$$P \equiv \frac{F_\perp}{A}, \tag{9.2}$$

Definition of pressure

with units of newtons/meter^2 (N/m^2), also known as a pascal (Pa). In water or air, the pressure at a particular location is produced by the weight (a force) of the water and/or air above that location. To better understand this, imagine lying on the floor with a fish tank sitting on your chest. As someone fills the tank, you will certainly experience an increasing amount of pressure. That is precisely the situation for the diver; the weight of the water above the diver is producing the added pressure. Even at the surface of the water, air pressure exists for the same reason, specifically the weight of the air overhead that reaches out to the edge of space.

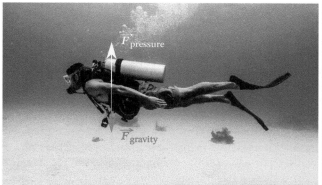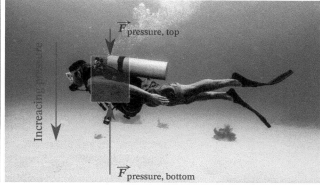

Fig. 9.30 *Left:* A diver with the correct amount of weight in his vest and tank can swim horizontally, being balanced by the force of gravity pulling him down and the force due to water pressure pushing him upward. *Right:* The overall upward force due to pressure is caused by increasing pressure with depth. There is less pressure pushing down on the diver from above than there is pushing the diver upward from below. (Adapted from a work by Summitandbeach; CC BY-SA 4.0)

On the right-hand side of Fig. 9.30, the weight of the water above the diver, combined with the weight of the air above the water's surface, is pushing down on the top of the diver, producing a downward pressure. But the diver has a finite thickness (indicated by the blue box) which means that on the bottom side of the diver the pressure is even greater, because there is a greater weight due to water sitting above that depth in the ocean. This extra weight, which would be equal to the weight of water in the blue box if the diver wasn't there, means an additional amount of pressure is exerted on the diver. Since pressure always exerts a force that is perpendicular to a surface, that bottom pressure results in a force that pushes up on the diver (the dark blue vector) to a greater extent than the force due to pressure on the top pushes down on him (the purple vector). The overall result is that there is a net upward force (the yellow vector in the left-hand figure) that is due to that pressure difference acting on the diver. It is that net upward force due to pressure that balances the force of gravity pulling the diver toward the center of Earth. (By the way, that net upward force due to pressure is what floats your boat.)

So why does pressure always apply a force that is perpendicular to a surface? Referring to Fig. 9.31, think about what happens when a tennis ball is thrown at a wall. When the tennis ball hits the wall, the direction of the tennis ball changes because the wall exerts a force on it, perpendicular to the surface of the wall. Since Newton's third law dictates that there must exist an equal and opposite force, the tennis ball pushes back on the wall in the opposite direction, which is also perpendicular to the wall. Lots of tennis balls, hitting the wall in different places, produce a total force on the wall. The pressure exerted on the wall is the sum of all of those individual forces divided by the wall's surface area. In the case of a gas or liquid, it is electrons, ions, atoms, or molecules bouncing off a surface, or one another, that are the source of the forces that produce pressure.

All of this discussion about pressure implies that pressure increases with depth toward the center of the Sun. Given the Sun's enormous mass, the gas pressure at its center is huge. The same situation applies for planets as well.

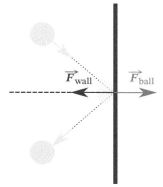

Fig. 9.31 When a tennis ball hits a wall, the wall exerts a force ($\vec{F}_{wall}$) on the tennis ball that is equal and opposite to the force exerted by the tennis ball ($\vec{F}_{ball}$) on the wall.

Gas Pressure and the Ideal Gas Law

Assuming that you have had the experience of blowing up a balloon, you have witnessed one component of the relationship between gas pressure, the density of a gas, and gas temperature that is called the ideal gas law. Initially, when the balloon is empty, as you add more air to the balloon, it begins to inflate, expanding outward. In other words, the volume of the gas is increasing as more air is blown into the balloon. However, at some point the balloon isn't able to continue expanding without stretching the balloon. When that happens you find that it gets harder and harder to blow into the balloon because the pressure inside is increasing, exerting more force on its walls while the walls push back on the air in the balloon. Eventually the force due to the high pressure causes the balloon to burst when the rubber is stretched too far.

Why did the pressure increase as more air was added to the balloon? Referring back to Fig. 9.31, when a particle bounces off a wall it pushes on the wall. If more particles bounce off the wall at the same time the force increases proportionally; twice as many particles produces twice the force, assuming that they are identical particles and they are all moving at the same speed.

But what would happen if the particles in the gas are still moving at the same speed but somehow suddenly become twice as massive? According to Newton's $\overrightarrow{F} = m\overrightarrow{a}$, if there is no difference in acceleration (the change in velocity over time) then the particles with twice the mass will produce twice the force, and therefore exert twice the pressure ($P = F_{\perp}/A$).

Both of these characteristics of the particles in the balloon (the number of particles and their individual masses) can be considered together by describing the mass density, ρ, of the gas, where ρ is the Greek letter rho (recall Table 4.1). The mass density is just the total mass of the particles in the gas, m, divided by the volume of the gas, V, or

$$\rho \equiv m/V, \tag{9.3}$$

Definition of mass density

with units of kg/m^3. (To get a sense of what the unit represents, water has a density of 1000 kg/m^3, which means that if you built a box that measured one meter (39 in) on each side and filled it with water, the mass of the water would equal 1000 kg (one tonne); this corresponds to about 2200 lb$_f$ on the surface of Earth.)

Even if the number of air molecules in a balloon doesn't change, you can do a quick experiment of putting the balloon in a refrigerator to discover that the size of the balloon shrinks, or conversely if you heat up a balloon that is already full, it will burst. Recall that when temperature increases, the speeds of particles increase, implying that the particles collide with each other and with the walls of the balloon with more momentum and energy. Since the average distance between particles doesn't change if the volume is constant, the frequency of collisions also increases simply because the particles are moving around faster. Therefore, as temperature increases so does the pressure of the gas.

Finally, recall that the kinetic energy of a particle, $mv^2/2$, is directly related to its mass and the square of its speed. But the temperature of a gas is also a measure of the average energy of particles in a gas. More specifically, for particles in the gas having a mass m, the average kinetic energy of the particles is given by

$$\frac{1}{2}mv^2_{\text{average}} = \frac{3}{2}k_BT,$$ (9.4)

Average kinetic energy of gas particles of mass m

A variables tutorial is available at
`Math_Review/Variables`

where k_B is Boltzmann's constant. (In case you are curious, and the author certainly hopes that you are, the factor of three is because there are three dimensions in space through which the particle can move.) The equation can now be solved for the average speed of the particles to give

$$v_{\text{average}} = \sqrt{\frac{3k_BT}{m}}.$$ (9.5)

Average speed of gas particles

A tutorial on the speeds of gas particles is available on the Chapter 9 ressources page.

The average speed of particles of mass m in a gas is equal to the square root of a constant ($3k_B$) times temperature, divided by the mass of the particles.

Assuming that the masses of the particles in the gas don't change, if the temperature *increases* by a factor of two, then the average speed increases by a factor of $\sqrt{2} = 1.41$. If the temperature increases by a factor of three, the average speed increases by a factor of $\sqrt{3} = 1.73$. On the other hand, if the temperature *decreases* by a factor of ten, then the average speed of particles in the gas is $\sqrt{10} = 3.16$ times slower.

Key Point
Be sure that you understand how mass, speed, and temperature affect the behavior of a gas.

The equation for the average speed of gas particles also says that if temperature doesn't change but the mass increases, then the average speed of the particles must decrease. Increasing the mass by a factor of two means that the average speed decreases by a fact of $\sqrt{2} = 1.41$, and so on. This should make sense to you, given that temperature describes the kinetic energy of the gas, and more massive particles can still have the same kinetic energy if they are moving more slowly (kinetic energy = $mv^2/2$).

So what happens to the gas if the average speed decreases? Since pressure is due to collisions, if the particles are moving more slowly they won't hit one another or the wall of the container as often or as hard, meaning that the pressure of the gas necessarily decreases. This is why the balloon shrinks when you put it in the refrigerator.

When the effects of mass density, temperature, and *average* particle mass are combined, gas pressure behaves according to

$$P = \frac{\rho k_BT}{m_{\text{average}}}.$$ (9.6)

Ideal gas law

An ideal gas law tutorial is available at Chapter 9 ressources page.

This equation can be read as saying that the pressure of a gas is directly proportional to the mass density, ρ, of the gas and its temperature, T, and inversely proportional to the mass of the particles in the gas (or average mass, if the gas contains a mixture of different kinds of particles). As either more gas particles are added to the gas or the gas is squeezed, the density increases and the pressure goes up. As the temperature of the gas goes up, the particles move faster and the pressure goes up. But, as the average mass of the particles increases they move more sluggishly and the pressure goes down.[2]

Example 9.3

(a) Imagine that two containers are filled with helium gas at the same mass density. If container A has a temperature of 300 K and container B has a temperature of 900 K, which container encloses the gas with the greater pressure, and by what factor is the pressure greater?

(b) Suppose that container C is filled with molecular hydrogen (H_2) with the same mass density and the same temperature as container A. Since a hydrogen molecule has about one-half of the mass of a helium atom, which container encloses a gas with the lower pressure, and by what factor is the density less?

(c) Container D encloses the same number of helium atoms as in container A and both containers have the same temperature of 300 K. However, container D is three times the size of container A. Which container has the greater pressure, and by how many times?

(a) Since both containers have helium gas in them at the same mass density, the gas in the hotter container has the greater pressure (as T increases, P increases proportionally). Therefore, the pressure inside container B is three times greater (900 K/300 K = 3). Physically, this is because the helium atoms in container B are moving around with three times as much kinetic energy as the atoms in container A.

(b) Gas pressure is inversely proportional to the average mass of the particles in the gas, meaning that if the average mass of particles is twice as great, then the pressure is one-half as much, and vice versa. In this case the average mass of the molecular hydrogen particles in container C is one-half the average mass of the helium atoms in container A. This means that the pressure in container A, that has the heavier helium atoms, must have the lower pressure by a factor of two.

(c) Since container A and container D both have helium in them, the average masses of the particles, $m_{average}$, are identical. Furthermore, both containers are at 300 K. This means that the only other quantity that can affect the pressure is the mass density. Given that container D is three times larger than container A, but both containers have the same number of helium atoms, the atoms in container D must be spread out over a volume that is three times greater. As a result, the mass density of the gas, $\rho = m/V$ (mass / volume), is three times less in container D. Consequently, the pressure of the gas in container A must be three times greater than the pressure of the gas in container D.

Although the ideal gas "law" is actually just an approximation to what really happens in a gas, it is an excellent approximation for most situations. There are important exceptions however, that will be discussed in Chapters 16 and 17.

---

[2] If you have taken other physics or chemistry courses you may be familiar with other forms of the ideal gas law, $PV = Nk_BT$ or $PV = nRT$. Physicists usually prefer the first form and chemists the second, although all of the forms are equivalent. R is another universal constant, the ideal gas constant, that is directly related to Boltzmann's constant, k_B. Equation (9.6) is more convenient when discussing a gas that isn't enclosed in a container.

Radiation Pressure

Even though photons don't have any mass, they still have linear momentum. Einstein taught us that the energy of a photon is given by $E = hf = hc/\lambda$, but he was also able to prove that since the photon has energy then it must also have linear momentum, given by

$$p = E/c = h/\lambda; \qquad\qquad (9.7)$$

The linear momentum of a photon

the linear momentum of a photon is given by its energy divided by the speed of light, or equivalently, by Planck's constant divided by the photon's wavelength.

Example 9.4

The energy of a photon with a wavelength of $\lambda = 410$ nm is 4.8×10^{-19} J. Calculate the photon's linear momentum.

From the equation for the linear momentum of a photon, and since the energy unit, the joule (J), is the same as $\mathrm{kg\,m^2/s^2}$, the momentum of the photon is

$$p = \frac{4.8 \times 10^{-19}\ \mathrm{J}}{3 \times 10^8\ \mathrm{m/s}} = \frac{4.8}{3} \times \frac{10^{-19}}{10^8} \times \frac{\mathrm{kg\,m^2/s^2}}{\mathrm{m/s}} = 1.6 \times 10^{-27}\ \mathrm{kg\,m/s}.$$

As with all things in quantum mechanics, the linear momentum of a photon is incredibly tiny. For a baseball to have such a small linear momentum, it would only be traveling at 1.1×10^{-26} m/s. At that speed the baseball would move one meter (39 in) in 3×10^{18} y, which is more than 200 million times the age of the universe!

Despite a single photon's extremely tiny linear momentum, a very large number of them can still have an influence on the interior of a star. Look back at the tennis ball bouncing off the wall in Fig. 9.31. When the tennis ball's direction changes, so does its linear momentum (page 193*ff*). Originally, the ball was moving down and to the right, but after the encounter with the wall it was moving down and to the left. As a result, the horizontal component of tennis ball's linear momentum vector flipped from pointing right to pointing left. Apparently, the force of the ball hitting the wall is related to the change in linear momentum of the tennis ball. In fact, in its most general form, rather than writing Newton's second law as $\vec{F}_{\text{net}} = m\vec{a}$, it can be written as

$$\vec{F}_{\text{net}} = \frac{\vec{p}_f - \vec{p}_i}{t};$$

the force exerted on an object equals the change in the linear momentum of the object (the final linear momentum minus the initial linear momentum) divided by the amount of time required for the momentum to change. This works because acceleration is just the change in velocity over time, $\vec{a} = (\vec{v}_f - \vec{v}_i)/t$, and linear momentum is $\vec{p} = m\vec{v}$, mass times velocity, giving $m\vec{a} = (m\vec{v}_f - m\vec{v}_i)/t$.

So what does all of this mean? If a photon bounces off an object or is absorbed, the change in linear momentum that occurs must result in the object that it hits getting a push (a force is applied), just like what happens with the tennis ball hitting the wall. Photons hitting the wall must exert a pressure on the wall, since $P = F_\perp / A$. This kind of pressure produced by photons is called radiation pressure. To account for all of the pressure acting inside a star, both gas pressure and radiation pressure must be included. Most of the time radiation pressure doesn't count for much, but sometimes it can be the most important pressure involved.

Moving Energy Around

Another critical component to understanding the structure of the Sun is energy transport, the movement of energy from one location to another. In Section 6.8 (page 226*ff*) we discussed convection, conduction, and radiation within the context of transporting heat energy. However, when we first discussed radiation we only considered electromagnetic radiation (photons). But radiation is not just limited to massless photons, it can apply to massive particles as well, like alpha particles ($^4_2$He nuclei) and beta particles (electrons).

In its most general form, radiation is the physical transport of particles from one place to another. The actual motion of material does happen in convection, but in that case the motion is cyclic, meaning that the particles essentially return to where they started, giving up heat along the journey. However, it is also possible of course for particles to move somewhere, like away from the Sun, and not return. Since kinetic energy and mass (see Section 9.6) are both forms of energy, particles moving from one place to another also transport energy. One type of particle, a neutrino (ν), first mentioned in Section 8.8, has the ability to travel directly from the center of the Sun out into interstellar space, which makes it an important way to carry energy away from the Sun's core and from the cores of other stars. Neutrinos will be discussed in more detail in Sections 9.6 and 9.7.

9.6   The Sun's Nuclear Furnace

How Can the Sun Be That Old?

Fig. 9.32 A Moon rock
returned by the Apollo 15
astronauts. (NASA/Johnson
Space Center)

When the Apollo astronauts returned with rocks from the Moon in the years between 1969 and 1972, the rocks were carefully studied with regard to their chemical composition and age (e.g., Fig. 9.32). Using the process of radiometric dating (page 257*ff*), the oldest rocks were determined to be about 4.4 billion years old, with the Moon itself estimated to be 4.51 billion years old. Using the same method, slightly older ages have been found by studying meteorites that have fallen to Earth, giving a Solar System age of 4.57 billion years. In 2014, scientists were also able to determine the age of an Earth-based crystal to be 4.4 billion years old, placing a lower limit on the age of Earth itself, now believed to be 4.54 billion years. Assuming that the Sun is at least as old as Earth, the Moon, and the oldest meteorites, this immediately begs the question of how the Sun has been able to shine for so long. What is the source of the Sun's impressive power output of 3.8×10^{26} W?

There are several possible sources of energy that could be responsible for the Sun's luminosity, including (a) chemical reactions, (b) the potential energy that was released when gases fell together due to their mutual gravitational attraction during the Sun's formation, or (c) the heat that is generated by radioactive decay. However, in each of these cases there simply isn't enough energy available to power the Sun for a sufficient length of time.

Key Point
Based on radiometric dating of the oldest meteorites, the Sun is approximately 4.57 billion years old.

Example 9.5

Consider chemical reactions as a possible energy source for the Sun. A single chemical reaction can release something like 10^{-18} J of energy per atom, based on electron energy levels [recall that 1 electron volt (eV) equals 1.6×10^{-19} J (page 286)]. Assuming, for simplicity, that the Sun is composed entirely of hydrogen, for how long can the Sun produce its measured luminosity?

To answer that question, we must first estimate how many atoms there are in the Sun. The mass of a single hydrogen atom is 1.67×10^{-27} kg while the mass of the Sun is nearly 2×10^{30} kg. This means that there would be roughly

$$N = \frac{2 \times 10^{30} \text{ kg}}{1.67 \times 10^{-27} \text{ kg/atom}} = \left(\frac{2}{1.67}\right) \times 10^{[30-(-27)]} \text{ atoms}$$

$$= 1.2 \times 10^{57} \text{ atoms},$$

or 1.2 billion trillion trillion trillion trillion atoms.

If one atom can provide 1×10^{-18} J of energy, then all of the hydrogen atoms in the Sun combined would be able to generate

$$E = \left(1 \times 10^{-18} \text{ J/atom}\right) \times \left(1.2 \times 10^{57} \text{ atoms}\right)$$

$$= (1 \times 1.2) \times 10^{(-18+57)} \text{ J} = 1.2 \times 10^{39} \text{ J}.$$

Although 1.2×10^{39} J of energy is a huge amount by human standards (equivalent to a one billion trillion megaton bomb), it doesn't come close to powering the Sun for its current 4.57 billion year age.

Recall from page 218 that one watt (W) is the same as one joule per second (J/s). To calculate how long the Sun could shine based on chemical energy alone, we only need to divide the available amount of energy by the rate at which the energy is being produced, or

$$\text{time} = \frac{\text{amount of energy}}{\text{rate of energy production}} = \frac{1.2 \times 10^{39} \text{ J}}{3.8 \times 10^{26} \text{ J/s}} = 3.2 \times 10^{12} \text{ s}.$$

Since there are just under 3.2×10^{7} s in one year, the length of time that the Sun can produce energy by chemical reactions is approximately 100,000 years, nowhere near as long as the Sun or the Earth has been around.

A scientific notation tutorial is available at `Math_Review/Scientific_Notation`

The other sources of energy already mentioned can't get the job done either, although gravitational potential energy comes closest. Gravitational energy could last about 10 million years, which is only 0.2% of the current age of Earth.

Einstein's $E = mc^2$

So what is the Sun's energy source? It was in Einstein's *annus mirabilis* (1905) that the makings of a solution to the energy source of the Sun first appeared. It was published in his addendum paper to the special theory of relativity in which

$$E = mc^2, \tag{9.8}$$

Einstein's mass–energy relationship

the most well-known equation in all of physics, was derived. What the equation tells us is that mass is just another form of energy, similar to light, kinetic energy, heat, and all the rest. The energy equivalent, E, of mass is calculated by multiplying mass, m, by the speed of light squared, c^2. Given that the speed of light is nearly 3×10^8 m/s,

$$c^2 = \left(3 \times 10^8 \text{ m/s}\right)^2 = (3 \times 3) \times \left(10^8 \times 10^8\right) \times (\text{m/s} \times \text{m/s})$$
$$= 3^2 \times 10^{(2 \times 8)} \times (\text{m/s})^2 = 9 \times 10^{16} \text{ m}^2/\text{s}^2.$$

Multiplying mass by this very large quantity means that a small amount of mass is equivalent to an enormous amount of energy.

Example 9.6

(a) What is the energy equivalent of a 1 kg mass (2.2 lb_m)?
(b) What is the energy equivalent of the Sun's mass?
(c) How long could the Sun shine at its present rate if all of its mass somehow disappeared over time?

(a) Applying $E = mc^2$,

$$E = 1 \text{ kg} \times \left(9 \times 10^{16} \text{ m}^2/\text{s}^2\right) = 9 \times 10^{16} \text{ J}.$$

In other words, if 1 kg of mass suddenly disappeared, 9×10^{16} J would be released, an amount of energy that would be capable of powering New York City for five months.

(b) The mass of the Sun is 2×10^{30} kg, so the total energy equivalent of the Sun's mass is

$$E = \left(2 \times 10^{30} \text{ k\cancel{g}}\right) \times \left(9 \times 10^{16} \text{ J}/\text{k\cancel{g}}\right) = 18 \times 10^{(30+16)} \text{ J} = 1.8 \times 10^{47} \text{ J}.$$

(c) With that amount of energy, the Sun could shine at its present luminosity for

$$\text{time} = \frac{1.8 \times 10^{47} \text{ \cancel{J}}}{3.8 \times 10^{26} \text{ \cancel{J}}/\text{s}} = 0.47 \times 10^{(47-26)} \text{ s} = 4.7 \times 10^{20} \text{ s} = 1.5 \times 10^{13} \text{ y}.$$

Fifteen trillion years is more than long enough to account for the 4.57 billion year age of the oldest meteorites. The Sun won't actually convert all of its mass into energy, but the estimated lifetime of the Sun is about 10 billion years.

The Sun certainly isn't going to just disappear in the future, but its mass is decreasing due to the conversion of mass into energy. Today, the Sun is losing mass at the rate of 4 billion kilograms per second or 130 million billion kilograms per year. At that rate, the Sun will convert the equivalent of the mass of Earth to energy in 46 million years. Over its lifetime so far, the Sun's mass has diminished by the equivalent of roughly 100 Earths. However, as was demonstrated in Example 9.2, the Sun's mass is about the same as 330 000 Earths; therefore, the Sun's mass has only been reduced by approximately 0.03% due to the conversion of mass into energy.

Transforming Hydrogen Into Helium

So what is the process by which the Sun converts mass into energy? In short, nuclear fusion: the combining, or fusing, of atomic nuclei to form other nuclei. Nuclear fusion is one of two general forms of nuclear reactions (see Section 8.8), the other being nuclear fission, where nuclei are split apart.[3] Because elements are defined by the number of protons in their nuclei, and isotopes of a given element have different numbers of neutrons, both fusion and fission are capable of transforming an isotope of one type of element into a different isotope, perhaps associated with a different element. In essence, the core of the Sun is a massive nuclear fusion reactor.

Our Sun is currently converting its abundant hydrogen, in the form of individual protons ($^1_1$H nuclei), into one of the isotopes of helium ($^4_2$He). When the sum of the masses of the four protons required to create helium is compared with the total mass of the products, a small amount of mass has vanished. Overall, the mass decrease amounts to only about 0.7% of the original mass of the protons; not much, but enough to do the job, given how much energy is associated with mass. In fact, energy associated with nuclear reactions is typically millions of electron volts (MeV), while chemical reactions involve a few electron volts (eV).

The process of converting hydrogen into helium can't all happen in one step, because the odds of four protons all crashing into one another in the chaotic environment of the Sun's interior is simply too low. Instead, the process happens in steps, with the overall result being

$$4\,^1_1\text{H} \rightarrow\, ^4_2\text{He} + 2e^+ + 2\nu_e + 2\gamma, \tag{9.9}$$

Converting four hydrogen nuclei into a helium nucleus

where e^+ represents a positively charged positron, ν_e symbolizes an electrically neutral electron neutrino, and γ designates an electrically neutral photon. As was discussed in Section 8.8, the notation ^A_ZX includes the atomic number (Z, the number of protons), the atomic mass (A, the number of nucleons), and the chemical symbol (X) of the element.

In Section 8.8 you also learned that nuclear reactions must obey certain conservation laws: in particular, conservation of electric charge and conservation of nucleon number (the total number of protons and neutrons). Since there are four positively charged protons on the left-hand side of the summary reaction (the atomic

---

[3]Nuclear fission is the source of the somewhat misleading phrase, "splitting the atom."

Key Point

The Sun is powered by nuclear fusion. Four hydrogen nuclei ($^1_1$H) are being converted into a helium nucleus ($^4_2$He) and additional particles, resulting in the conversion of 0.7% of the total mass of the hydrogen nuclei into energy.

number), there must also be four positive charges on the right-hand side of the re-action as well. Two of the four positive charges are the protons contained in the helium nucleus and the other two are the positrons. For every particle of normal matter, there exists an antimatter counterpart with identical properties except for some opposite quantum properties such as electric charge. The antimatter partner of the electron (e^-) is the positron (e^+).

Two of the other particles on the right-hand side of the summary reaction are electron neutrinos. These ethereal particles have no electric charge, an extremely small mass, and are very weakly interacting, capable of traveling through a light-year-long length of lead without stopping. They show up on the right-hand side of the reaction is because of another conservation law of nuclear physics known as lepton number. Leptons[4] are a class of elementary particles that include electrons (e^-), positrons (e^+), electron neutrinos (ν_e), and electron antineutrinos ($\bar{\nu}_e$). [For completeness, Table 9.2 lists all of the leptons, including the muon (μ^-), the tau (τ^-), and their associated neutrinos and antiparticles; we will briefly mention these last two groups in Section 9.7.] Normal matter leptons each count positively for the total lepton number and antimatter leptons count negatively.

Returning to the summary equation for converting four hydrogen nuclei into a helium nucleus, the total number of leptons on the left-hand side is zero (there aren't any leptons on the left-hand side), which means that the total lepton number on the right-hand side must be zero as well. Given that there are two positrons on the right-hand side with their negative lepton numbers, there must be two nor-mal matter leptons on the right-hand side with positive lepton numbers. The two positive lepton number particles can't be electrons because then electric charge wouldn't be conserved in the reaction. This means that those positive lepton num-ber particles must be the electron neutrinos. (As the names suggest, electron neu-trinos are associated with electrons, muon neutrinos are associated with muons, and tau neutrinos are associated with taus.)

The two photons in the summary equation carry away some of the energy from the mass conversion.

Although this may all sound a bit confusing, it really amounts to a counting game:

- Four nucleons on the left-hand side (4 protons) require four nucleons on the right-hand side (2 protons and 2 neutrons).
- Four positive charges on the left-hand side (protons) requires that four posi-tive charges must be on the right-hand side (2 protons and 2 positrons).
- Zero total lepton number on the left-hand side requires zero total lepton num-ber on the right-hand side (two positrons count negatively and two electron neutrinos count positively).

Key Point

Conservation laws of nuclear reactions:
- electric charge
- nucleon number
- lepton number.

Table 9.2 Leptons

Particle	Anti-particle
Electron group	
e^-	e^+
ν_e	$\bar{\nu}_e$
Muon group	
μ^-	μ^+
ν_μ	$\bar{\nu}_\mu$
Tau group	
τ^-	τ^+
ν_τ	$\bar{\nu}_\tau$

Example 9.7

One of the individual nuclear reactions that occurs in the Sun involves the cap-ture of an electron by a beryllium isotope. A portion of the reaction is shown:
$$^7_4\text{Be} + e^- \rightarrow {}^7_Z\text{X} + ?.$$
What are the values of Z and X, and what is the missing particle?

---

[4]The word *lepton* comes from a Greek word meaning "fine, small, thin", but not all leptons fit that description; taus are twice as heavy as protons and neutrons.

Since the element X has an atomic mass of 7, equaling the total number of protons plus neutrons, and since the beryllium isotope also had an atomic mass of 7, the missing particle cannot be either a proton or a neutron. Furthermore, since there is one lepton of normal matter on the left-hand side (a positive lepton number), there must also be a lepton of normal matter on the right-hand side. The conservation of lepton number suggests that the missing particle must either be an electron, in which case nothing changed, or it must be an electron neutrino, ν_e. Finally, since the beryllium isotope had four protons (the atomic number) and there is also a negative charge on the left-hand side (the electron), the total charge on the right-hand side must total 3 positive charges, and since we have already argued that the missing particle is a neutrino with zero electric charge, the atomic number of element X must be 3. Referring to the periodic table of the elements (Table 7.1), $Z = 3$ corresponds the the element, lithium (Li). Therefore, the complete nuclear reaction is

$$^{7}_{4}\text{Be} + e^{-} \rightarrow {}^{7}_{3}\text{Li} + \nu_e.$$

Although there are several ways in which the summary reaction of converting four hydrogen-1 nuclei into a helium-4 nucleus occurs simultaneously in the core of the Sun, the most common is a form of what is known as the proton–proton chain. The reaction chain involves three steps, with each step only requiring the collision of two nuclei, rather than the much less likely case of three or even four nuclei colliding with one another simultaneously. (Figure 9.33 is a graphical representation of the proton–proton chain.)

Step 1 of the proton–proton chain involves two protons colliding to form deuterium ($^{2}_{1}\text{H}$):

$$\text{Step 1:} \quad {}^{1}_{1}\text{H} + {}^{1}_{1}\text{H} \rightarrow {}^{2}_{1}\text{H} + e^{+} + \nu_e.$$

Step 2 uses the collision between a deuterium nucleus and a proton to form an isotope of helium that contains only one neutron:

$$\text{Step 2:} \quad {}^{1}_{1}\text{H} + {}^{2}_{1}\text{H} \rightarrow {}^{3}_{2}\text{He} + \gamma.$$

Finally, since all of the reactions are occurring within the Sun's deep interior simultaneously, a mixture of protons, deuterium, and helium-3 exists, along with a variety of other nuclei. When two helium-3 nuclei collide, Step 3 of the proton–proton chain is achieved:

$$\text{Step 3:} \quad {}^{3}_{2}\text{He} + {}^{3}_{2}\text{He} \rightarrow {}^{4}_{2}\text{He} + 2\,{}^{1}_{1}\text{H}.$$

Note that there is a total of six nucleons on the left-hand side of the final reaction, requiring six nucleons on the right-hand side. There are also four positive charges on the left-hand side, requiring four positive charges on the right-hand side. The result is the production of a helium-4 nucleus and the liberation of two protons. Since steps 1 and 2 must both happen twice in order to produce two helium-3 nuclei for the third step, six protons were involved in the overall process, with two being returned to the environment, meaning that a net of four protons were "consumed" in the production of the helium-4 nucleus. Two positrons, two electron neutrinos, and two photons were also created, giving the summary reaction [Equation (9.9)].

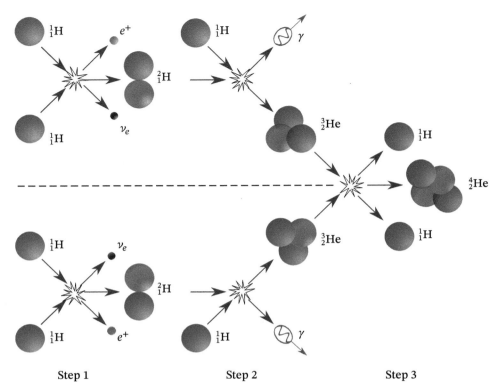

Step 1   Step 2   Step 3

Fig. 9.33 The proton–proton chain. Protons are represented by the blue spheres and neutrons are represented by the red spheres.

What Is The Sun's Nuclear Furnace Like?

An important aspect of the proton–proton chain has to do with the problem of two positively charged particles being able to collide with one another; a proton–proton collision involves two particles both with charge $+e$, as does a proton–deuterium nucleus interaction, while a helium-3–helium-3 collision occurs between two nuclei, each of charge $+2e$, where e represents the elementary charge. According to Coulomb's law [Equation (6.10)], two charges that have the same sign repel one another in direct proportion to the product of their charges. This means that they have to be moving rapidly in order to get close enough to actually touch.

If kinetic energy, and therefore speed, was all that was available to get two positively charged nuclei in the proton–proton chain to collide, then the kinetic energy needed would be far too high. In the first step of the proton–proton chain, think of one of the protons as having to climb a potential energy "hill" caused by a force of repulsion because both particles are positively charged (Coulomb's law) as depicted by Fig. 9.34 (*Top*). When the two protons get close enough together to effectively touch, the "hill" becomes a "well," because of the strong nuclear force, a very localized attractive force that holds nuclei together. Once inside the potential energy "well" occupied by the other proton, the two protons can interact. If the kinetic energy isn't great enough, the incoming proton will only be able to climb part-way up the hill before being pushed away by the force of repulsion. Even in the hot, dense interior of the Sun, the protons aren't moving around fast enough to climb the hill.

Fortunately, the bizarre nature of quantum mechanics comes to the rescue, due to the inherent wave–particle duality of all particles as described by the de Broglie wavelength equation [Equation (8.5)]. Heisenberg's uncertainty principle points out that there is an inherent uncertainty associated with knowing both the position and the linear momentum of a particle simultaneously. Essentially, there is some fuzziness as to where the protons are located, as depicted in Fig. 9.34 (*Bottom*). This allows the two protons to interact even though there isn't enough kinetic energy for the incoming proton to overcome the potential energy barrier. This apparent penetration through the Coulomb barrier is called quantum mechanical tunneling. Without quantum mechanical tunneling, nuclear reactions wouldn't occur in the Sun's interior, the Sun wouldn't shine, and you wouldn't exist to read this text.

In order for the protons, deuterium nuclei, and helium-3 nuclei to be moving fast enough and to be close enough for the nuclear reactions to occur, aided by quantum mechanical tunneling, the temperature of the gas must exceed about 10 million kelvins (10^7 K) and the density must be more than about 100,000 kg/m^3 (10^5 kg/m^3). Today the center of the Sun has a temperature of close to $T_c = 1.54 \times 10^7$ K and a density of almost $\rho_c = 1.49 \times 10^5$ kg/m^3, which is close to 150 times the density of water.

Because the Sun has been converting hydrogen into helium in its core for the past 4.57 billion years, the central composition of the Sun is no longer the same as it is at its surface. Today the Sun's nuclear furnace has a composition that is about 36% hydrogen by mass and 62% helium, with the remaining 2% being primarily carbon, nitrogen, and oxygen, with smaller amounts of other elements. At the rate that the Sun is converting hydrogen into helium, it will run out of hydrogen fuel in its core in a little over 5 billion years. The Sun has lived out nearly one-half of its life; the Sun is middle-aged.

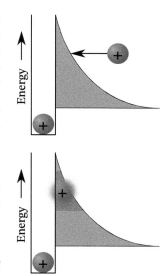

Fig. 9.34 *Top:* The potential energy hill due to the Coulomb repulsion of the two positively charged protons. *Bottom:* Due to Heisenberg's uncertainty principle, there is an inherent uncertainty (fuzziness) associated with where the protons are located. This allows the protons to collide even though there isn't enough energy to surmount the potential energy barrier.

The Sun in Cross-Section

Figure 9.35 illustrates our understanding of the Sun's interior when a self-consistent computer model is generated that incorporates all of the physical processes discussed. In the outer 29% of the Sun's radius, the most efficient way to transport energy is via convection. We already knew that a convection zone must exist in the outer portion of the Sun, given that granulation is obvious from observations of the photosphere. In the remainder of the Sun's interior, energy is transported outward by photons (light radiation). The inner 71% of the Sun is referred to as the radiation zone.

The source of the energy being transported to the surface, either by radiation or convection, is the nuclear core that encompasses the inner 30% of our star. It is in the nuclear core that the relentless squeezing of the gas by the Sun's own gravity causes the temperature and density to become great enough for fusion to occur, resulting in hydrogen being converted into helium through the proton–proton chain.

9.7 Test, Test, and Test Some More

It is one thing to develop a complex computer code that produces a model of the interior of the Sun based on our knowledge of the Sun's surface composition, mass, radius, surface temperature, luminosity, and age, but it is another thing to actually

Key Point
The process of science demands the continual testing of theories and models that derive from them.

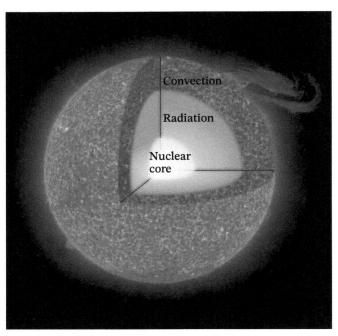

Fig. 9.35 The innermost 30% of the Sun is the nuclear core, where fusion reactions convert hydrogen into helium. In the inner 71% of the interior, which includes the nuclear core, energy is transported outward by photons; this region is called the radiation zone. Within the outermost 29% of the Sun's interior, convection transports the energy to the surface. (Adapted from an image by NASA)

believe the results of those elaborate calculations. Returning once again to the opening chapter of this text, a fundamental component of science requires that its core theories and resulting models must be continually tested in search of fallacies in the theories. In order to do so, every scientific theory must be able to make testable predictions; otherwise the theory cannot be considered as truly scientific.

In addition to experimentation, theories can be tested through computer modeling. If there are errors in our understanding of any of the fundamental physics that go into creating the models, such as the energy levels of atoms, how light interacts with matter, or the amount of energy produced by each of the nuclear reactions in the Sun's core, then a realistic solar model cannot be produced. But even when a model is produced, there are other ways it can be tested as well.

The Solar Neutrino Problem and New Physics

The first computer models of the Sun were constructed in the 1960s, proving to be remarkably successful despite less refined knowledge of some of the crucial physical processes involved and dramatically less computing power than exists today. Shortly thereafter, Raymond Davis (Fig. 9.36) began a series of observations in 1970 that led to a crisis in reconciling one of the predictions of the models.

Davis's observations were being made one mile below ground in the Homestake Mine in Lead, South Dakota, and his "observatory" (Fig. 9.37) was a tank containing

Fig. 9.36 Raymond Davis, Jr. (1914–2006). (Courtesy Brookhaven National Laboratory)

100,000 gallons of dry-cleaning fluid (tetrachloroethylene, C_2Cl_4), rich in the chlorine isotope $^{37}_{17}Cl$. His experiment was designed to detect the electron neutrinos produced by nuclear reactions in the center of the Sun. In particular, Davis's experiment was most sensitive to the electron neutrinos in the relatively rare decay reaction

$$^{8}_{5}B \rightarrow {}^{8}_{4}Be + e^{+} + \nu_e.$$

Those electron neutrinos would occasionally trigger a reaction with a chlorine atom in his tank, resulting in the production of an argon atom by

$$^{37}_{17}Cl + \nu_e \rightarrow {}^{37}_{18}Ar + e^{-}.$$

The downside of the reaction he designed his experiment around was that the far more abundant electron neutrinos from the first step of the proton–proton chain, $^{1}_{1}H + {}^{1}_{1}H \rightarrow {}^{2}_{1}H + e^{+} + \nu_e$, didn't have enough energy to produce the chlorine-to-argon reaction.

The electron neutrinos coming from the Sun's nuclear reactions are amazingly abundant. Remember that these particles are weakly interacting in the extreme with matter under ordinary conditions. This means that once they are created they simply escape the Sun. It also means that they can travel directly through Earth as well, which is why the neutrino observatory was one mile below ground; no other particles are going to pass through Earth to reach the tank of cleaning fluid.

The same also applies to you. If you are lying down there are approximately 500 trillion neutrinos passing through your body every second that originated in the nuclear core of the Sun 8.3 minutes earlier. It doesn't matter if the Sun is overhead or if it is midnight and you are sleeping in your bed, because at night they simply pass through Earth, come up through the floor, through your bed, through you, through the ceiling, and back out into space again, never interacting with you at all.

Although the electron neutrinos are extremely weakly interacting, one of them would strike one of Davis's chlorine atoms every couple of days, producing an argon atom. Once every few months the tank would be emptied and the argon atoms would be trapped and counted. The result was that there were one-third as many argon atoms present than expected. Given the very challenging nature of the experiment, the results were certainly suspect, but when repeated over and over again between 1970 and 1994, the outcome was always the same: one-third of the predicted number of argon atoms.

Since that initial experiment, other neutrino observatories have come online, using different methods of detecting the ethereal particles. For example, Japan's Super-Kamiokande detector is a giant tank of water with photoelectric detectors covering the walls that are designed to record flashes of blue light as neutrinos pass through the water. The neutrinos (rarely) interact with electrons, causing the electrons to move faster than the speed of light in water (Fig. 9.38).[5] The blue flashes occur when the electrons lose kinetic energy in order to slow to sub-light speed in the water. In every case, the findings of Davis's chlorine experiment were confirmed.

Fig. 9.37 The neutrino observatory in the Homestake Mine in Lead, South Dakota. (Courtesy Brookhaven National Laboratory)

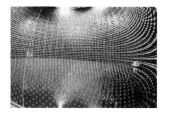

Fig. 9.38 Japan's vast Super-Kamiokande neutrino observatory being filled with pure water. The workers in the raft are inspecting the detectors along the wall. (Kamioka Observatory, The Institute for Cosmic Ray Research, The University of Tokyo)

[5]Electrons moving faster than the speed of light in water doesn't violate Einstein's special theory of relativity because the fastest anything can move is the speed of light *in a vacuum*, which is always faster than the speed of light in any physical substance.

This discrepancy between the predicted number of solar neutrinos and the number being detected became known as the solar neutrino problem.

There was clearly something happening that wasn't as predicted. Could the temperature and/or density at the center of the Sun be less than the models predicted, thus affecting the nuclear reaction rates? But of course we know what the luminosity of the Sun is by direct observation, so if the reaction rates are wrong, how is the luminosity explained? Perhaps there is another source of energy that wasn't being accounted for? If so, what is it? Perhaps our understanding of nuclear reactions is flawed and not all of the reactions are producing neutrinos as predicted? What about the rates at which neutrinos interact with matter? As low as the probability is, perhaps there were errors in estimating those probabilities?

Eventually the culprit was discovered and it led to a new understanding of neutrinos. As listed in Table 9.2, there are actually three kinds of neutrino; electron neutrinos, muon neutrinos, and tau neutrinos (and their antimatter relatives). The tau and muon neutrinos are associated with two other types of leptons, taus and muons, respectively. It is not particularly important for our purposes to worry about these other two types of neutrinos, it is only important for this story to know that there are three types of neutrino. Although the three types of neutrino had already been known to exist, what wasn't known is that it is possible for the three types to spontaneously swap identities. This could only happen in the presence of magnetic fields, such as those inside the Sun, and if the neutrinos have mass.

Up until the time of the solar neutrino experiments, it was thought that neutrinos, like photons, were massless. However, the "oscillations" between the three types of neutrinos demonstrated that their masses are not zero. This also meant that the original electron neutrinos created in the Sun's nuclear furnace became an equal mix of three types before they reached Earth. Since the detectors were only measuring electron neutrinos, the results were exactly one-third of the originally anticipated number.

The outcome of this story is that the computer models of the Sun were confirmed after accounting for the neutrino "oscillations," and we learned something new about the world of elementary particle physics at the same time. It is amazing to think that the detection of the solar neutrinos represents confirmation of our knowledge of nuclear reaction rates as well as verification that computer models accurately describe the physical conditions that exist at the center of the Sun.

Helioseismology

If you have ever played a musical instrument you are likely familiar with the idea of harmonics. Consider a guitar string for example (Fig. 9.39). Because the guitar string is attached at both ends, it cannot move at those locations. If you pluck the string it must vibrate in such a way that the string is stationary at those fixed endpoints (any position where the string doesn't move while it vibrates is referred to as a node). When the string vibrates, it can oscillate up and down in one or more harmonics (modes), with each harmonic having a unique wavelength, and therefore a unique frequency (the same relation between wavelength, λ, f, and speed, v, that was discussed back in Section 6.5 regarding electromagnetic waves ($f\lambda = v$). In this case, v is the speed that the wave can travel along the string. Since each harmonic

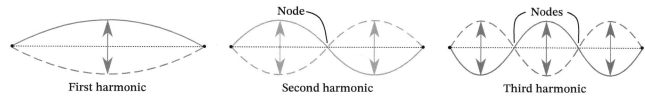

First harmonic   Second harmonic   Third harmonic

Fig. 9.39 A guitar string vibrating in three different harmonics. In each harmonic, the string vibrates back and forth between the solid and the dashed shapes. The first harmonic is the longest possible wavelength, the second harmonic is the second-longest wavelength, and the third harmonic has the shortest wavelength of the three. Each of the harmonics has nodes at the ends where the sting is fixed. The second harmonic also has one node in the middle while the third harmonic has two nodes, equally spaced between the ends.

has a different frequency, a different musical note is produced. Each harmonic also has a unique set of nodes located between the endpoints.

In general, multiple harmonics can all be present at the same time with different amplitudes. The only constraints on which waves can exist on the guitar string are those endpoint nodes. It is the presence of various combinations of harmonics and their associated amplitudes that gives different musical instruments their unique sound. (As an aside, voice recognition works by determining which harmonics, and the amplitudes of those harmonics, are present in a sound sample.) A wide variety of things other than traditional musical instruments also support various harmonics. For example, if you tap on a saw, it will "ring," as will a bell, of course. Even a suspension bridge can oscillate up and down, sometimes with disastrous consequences (the bridge could collapse if the oscillations become too large).

Stars can also be included in the list of objects that can oscillate with specific harmonics. In the case of our Sun, the oscillations are driven by the chaotic motions in the convection zone, which cause the Sun to "ring," much like a bell. Using the Sun's internal vibrations to study its interior is called helioseismology, similar to the method of seismology that geologists use to study the interior of Earth. (If you want to know what's inside a present you might shake it; the Sun shakes itself.) Just which frequencies and types of oscillation exist are very sensitive to the Sun's internal structure, the composition of the gas, rotation, and so on. Figure 9.40 shows just one of the sets of harmonics that is present in the Sun; there are literally thousands of different harmonics. If the depth of the convection zone were to change, or the composition or density of the gas in the nuclear core was different, the frequencies and their associated amplitudes observed at the surface would change as well. In other words, by recording the "ringing" of the Sun and comparing those observations with the theoretically expected frequencies and amplitudes from the computer models, astronomers have a very sensitive test of their models. If the observed frequencies don't match the models, the models must be revised by improving one or more of the many physical inputs upon which they are based.

In 2017, thanks to the collection of over sixteen years of oscillation data by the Solar and Heliospheric Observatory, scientists discovered a previously unknown aspect of the Sun's deep interior that had not been revealed by the highly successful computer models. Superimposed on the other oscillations was a very subtle signal produced by waves reflecting off a core that is rotating four times faster than the Sun's surface. Recall that the surface of the Sun rotates differentially with a period of about 25 days at the equator but much more slowly near the poles, with a rotation

Key Point
The core of the Sun rotates with a period of one week.

Fig. 9.40 One of the thousands of sets of harmonics present in the Sun. The dark blue regions represent nodes, and the light blue areas are moving outward while the red areas move inward and vice versa. Those oscillations have a well-defined frequency based on the detailed nature of the Sun's interior. The cutaway also shows the convection zone, the radiation zone, and the nuclear core. (Courtesy of the Global Oscillation Network Group, National Solar Observatory/AURA/NSF)

period of 36 days. The Sun's core, on the other hand, rotates with a period of just one week. This discovery provides information about the Sun's formation and its subsequent aging. Thanks to drag produced by the solar magnetic field, the outer portions of the interior have slowed dramatically over the past 4.57 billion years.

 Exercises

All of the exercises are designed to further develop *your understanding* of the material by thinking carefully and critically about what you have read in this chapter. Answers to selected exercises are available in the back of the book.

True/False

1. Astrophysics is a self-contained discipline that is independent of other sciences.
2. (*Answer*) Astrophysics, like all sciences, is inherently self-correcting.
3. The Sun displays different features when viewed at different wavelengths.
4. (*Answer*) Observing the Sun at different wavelengths allows astronomers to probe different parts of the solar atmosphere.
5. The Sun produces approximately 100 times more energy every second than Earth's population consumes in the same period of time.
6. (*Answer*) The temperature of the surface of the Sun is close to 5772 K (5500°C, or 9900°F).
7. The surface composition of the Sun is identical to the composition of its core.
8. (*Answer*) Granulation that is visible at the base of the photosphere is due to varying composition on a characteristic length scale of about 1000 km.
9. An individual sunspot typically lasts for 11 years, the length of the solar cycle.
10. (*Answer*) The umbra of a sunspot is cooler than the surrounding gas.
11. The filamentary appearance of a sunspot's penumbra is due to charged particles trapped along magnetic field lines.
12. (*Answer*) The umbra of a sunspot is composed of carbon, which is why it is so dark.
13. When viewed from above its geographic north pole, the Sun rotates counterclockwise.
14. (*Answer*) Magnetic field lines always form closed loops.
15. The rate at which the Sun is losing mass due to the solar wind means that the Sun can't be more than about 6000 years old.
16. (*Answer*) The solar cycle is roughly one month long.

17. Solar activity increases and decreases on average approximately once every eleven years.
18. (*Answer*) Pressure increases with depth in a swimming pool because there is more weight due to water above that depth.
19. Pressure and force represent the same quantity.
20. (*Answer*) Since photons don't have any mass, they cannot have linear momentum.
21. Photons apply a force to objects they reflect off or are absorbed by.
22. (*Answer*) $E = mc^2$ means that mass is another form of energy, and that the energy equals the mass times the speed of light squared.
23. The proton–proton chain only works in the core of the Sun because the temperature is more than one billion kelvins.
24. (*Answer*) Quantum mechanical tunneling is a direct consequence of Heisenberg's uncertainty principle.
25. You are constantly being bombarded by hundreds of trillions of solar neutrinos every second, that pass completely through you, day and night.
26. (*Answer*) The solar neutrino problem had to do with not being able to detect any neutrinos coming from the Sun's core.
27. Helioseismology probes the interior structure of the Sun by observing surface oscillations.
28. (*Answer*) Oscillations in the Sun's interior are caused by the gravitational pull of the planets in the Solar System.
29. The Sun and our Solar System formed 4.57 billion years ago.

Multiple Choice

30. (*Answer*) In order to make advances in our understanding, astrophysics relies on
    (a) data analysis and statistics.
    (b) computer modeling.
    (c) advances in theoretical physics.
    (d) advances in experimental physics.
    (e) advances in chemistry.
    (f) increasingly advanced observational techniques.
    (g) observations at all wavelengths of the electromagnetic spectrum.
    (h) peer review.
    (i) collaborative research and international conferences.
    (j) all of the above
31. A wavelength of 30.4 nm is found in which part of the electromagnetic spectrum?
    (a) radio   (b) microwave
    (c) infrared   (d) visible
    (e) near ultraviolet   (f) extreme ultraviolet
    (g) extreme x-ray   (h) gamma ray
32. (*Answer*) The radius and mass of the Sun are approximately _____ and _____ times greater than those of Earth, respectively.
    (a) 10, 300   (b) 110, 3300
    (c) 1100, 33,000   (d) 10, 33,000
    (e) 110, 330,000   (f) 1100, 3,300,000
33. The surface composition of the Sun is approximately (by mass), _____ hydrogen, _____ helium, and _____ heavy elements.
    (a) 80.2%, 20.8%, 10.2%
    (b) 73.84%, 24.85%, 1.31%
    (c) 50.8%, 35.6%, 15.9%
    (d) 24.9%, 1.3%, 73.8%
    (e) 1.3%, 73.8%, 24.9%
    (f) unknown; we have never returned a sample

34. (*Answer*) Suppose a solar panel on a roof has a surface area of 0.5 m², and has an efficiency of 10%, meaning that 10% of the visible wavelength photons that strike it get converted into usable electrical power. If the solar panel is pointed straight toward the Sun and nothing obstructs sunlight from reaching the panel, approximately how much power is produced?
    (a) 1361 W/m²   (b) 136.1 W   (c) 13.61 W
    (d) 681 W   (e) 68.1 W   (f) 6.81 W
35. Characteristic temperatures of the photosphere, the chromosphere, the transition region, and the corona are, respectively, _____.
    (a) 5000 K, 8000 K, 50,000 K, and 500,000 K
    (b) 10,000 K, 20,000 K, 30,000 K, 40,000 K
    (c) 500,000 K, 50,000 K, 8000 K, 5000 K
    (d) 40,000 K, 30,000 K, 20,000 K, 10,000 K
36. (*Answer*) Visible wavelength photons come from which of the following layers of the Sun's atmosphere?
    (a) chromosphere   (b) transition region
    (c) photosphere   (d) corona
37. Referring to Fig. 9.7, at an altitude of 1500 km above the Sun's "surface," the _____ has a temperature of approximately _____.
    (a) photosphere, 6 K
    (b) chromosphere, 200,000 K
    (c) transition region, 8000 K
    (d) corona, 5 K
    (e) photosphere, 3000 K
    (f) chromosphere, 7000 K
    (g) transition region, 10,000 K
    (h) corona, 2500 K
38. (*Answer*) The variation in coloration of granulation is due to
    (a) the different colors of hydrogen and helium.
    (b) temperature differences, with bright regions being hotter and dark regions being cooler.
    (c) temperature differences, with dark regions being hotter and bright regions being cooler.
    (d) magnetic fields.
    (e) the natural unevenness of the surface, with the darker regions being in shadow.
    (f) bright regions being deeper in the star where the temperature is greater.
39. The rapid temperature rise through the _____ is believed to be due to _____ produced by _____ at the base of the photosphere.
    (a) photosphere, electromagnetic radiation, photons
    (b) chromosphere, shock waves, electromagnetic radiation
    (c) transition region, shock waves, convection
    (d) corona, shock waves, conduction
40. (*Answer*) If the north magnetic pole of a sunspot group leads in the northern hemisphere, then the south magnetic pole _____ in the northern hemisphere and _____ in the southern hemisphere.
    (a) leads, leads   (b) leads, follows
    (c) follows, leads   (d) follows, follows
    (e) there is no predictable relationship
41. Which of the following are associated with magnetic fields?
    (a) sunspots   (b) solar flares
    (c) solar prominences   (d) coronal mass ejections
    (e) the solar cycle   (f) all of the above
    (g) none of the above

42. (*Answer*) Solar flares can release an equivalent of a _____ bomb and they emit most of their electromagnetic energy at _____ wavelengths.
    (a) one kilotonne, radio
    (b) one thousand megatonne, infrared
    (c) one million megatonne, visible
    (d) one billion megatonne, x-rays

43. Charged particles ejected from the Sun during a solar flare can reach Earth in _____ and affect _____.
    (a) 5 minutes, growing cycles
    (b) 5 minutes, local weather forecasting
    (c) 30 minutes, radio communications
    (d) 30 minutes, exposure before getting a sunburn
    (e) 2 to 3 days, solar eclipses
    (f) 2 to 3 days, power grids

44. (*Answer*) Which of the following *does not* exhibit an 11-year periodicity?
    (a) the number of visible sunspots at any time
    (b) the number and intensity of solar flares
    (c) the latitudes at which sunspots appear and disappear
    (d) the magnetic polarity of leading sunspots in sunspot groups
    (e) fluctuations in the weather conditions on Earth
    (f) all of the above exhibit an 11-year periodicity

45. Which observation(s) demonstrate a 22-year cycle?
    (a) the polarity of leading sunspots in a sunspot group in the Sun's northern hemisphere
    (b) the location of the Sun's magnetic north pole
    (c) the number of visible sunspots
    (d) the frequency of granulation seen at the base of the photosphere
    (e) the frequency and intensity of solar coronal mass ejections
    (f) all of the above
    (g) none of the above
    (h) (a) and (b) only
    (i) (a), (b), and (c)

46. (*Answer*) The temperature in a rigid container full of helium gas increases in temperature from 250 K to 1000 K. Does the pressure in the container increase or decrease, and by what factor?
    (a) increase, 4   (b) increase, 750
    (c) decrease, 4   (d) decrease, 750

47. Argon atoms are 10 times more massive than helium atoms. Container A contains argon at the same mass density and temperature as helium in container B. Which container has the greatest gas pressure, and by what factor?
    (a) A, $\sqrt{10} = 3.16$ (b) A, 10
    (c) A, $10^2 = 100$ (d) B, $\sqrt{10} = 3.16$
    (e) B, 10   (f) B, $10^2 = 100$

48. (*Answer*) A blue photon has a wavelength of 410 nm and an infrared photon has a wavelength of 820 nm. When the infrared photon is absorbed by a blackbody object it can apply _____ as much force on the object as the blue photon when it is absorbed.
    (a) twice   (b) one-half
    (c) four times   (d) one-fourth

49. Radiation pressure is caused by
    (a) the radioactive decay of uranium.
    (b) x-ray machines causing stress.
    (c) the absorption of photons.
    (d) the emission of photons.
    (e) photons reflecting off surfaces.
    (f) more than one of the above

50. (*Answer*) Which of the following is *not* a way in which energy can be transported from one location to another?
    (a) convection   (b) conduction
    (c) radiation   (d) particles
    (e) all of the above are ways of transporting energy

51. Each proton in the proton–proton chain has a mass of 1.67×10^{-27} kg. Given that 0.7% of the mass vanishes after the sequence of reactions is completed, how much energy was produced by one completed reaction sequence?
    (a) 4.7×10^{-29} J   (b) 4.2×10^{-12} J
    (c) 4.7×10^{-27} J   (d) 4.2×10^{-10} J
    (e) none of the above

52. (*Answer*) Which of the following *is not* conserved during a nuclear reaction?
    (a) lepton number   (b) electric charge
    (c) nucleon number   (d) mass
    (e) all of the above   (f) none of the above

53. Complete the following reaction: $^8_5X \rightarrow\ ^8_ZBe + a + b$.
    (a) $X = B, Z = 4, a = e^+, b = \nu_e$
    (b) $X = B, Z = 4, a = e^+, b = \bar{\nu}_e$
    (c) $X = C, Z = 4, a = e^+, b = e^+$
    (d) $X = N, Z = 8, a = \ ^1_1H, b = \gamma$
    (e) $X = O, Z = 5, a = e^-, b = \gamma$

54. (*Answer*) Complete the following reaction: $^{12}_6X + \ ^{12}_6X \rightarrow\ ^{23}_ZNa + a$.
    (a) $X = B, Z = 10, a = e^-$
    (b) $X = B, Z = 10, a = \bar{\nu}_e$
    (c) $X = C, Z = 11, a = \ ^1_1H$
    (d) $X = C, Z = 11, a = e+$
    (e) $X = F, Z = 12, a = 2e^-$

55. Complete the following reaction: $^4_8O + \ ^4_2He \rightarrow\ ^{20}_ZX + a$.
    (a) $A = 14, Z = 12, X = Mg, a = 2e^-$
    (b) $A = 16, Z = 10, X = Ne, a = \gamma$
    (c) $A = 16, Z = 10, X = B, a = e^+$
    (d) $A = 16, Z = 16, X = S, a = \nu_e$
    (e) $A = 18, Z = 17, X = O, a = \bar{\nu}_e$

56. (*Answer*) The composition in the Sun's core is changing over time due to
    (a) the conversion of iron into carbon.
    (b) settling of heavy elements to the center of the Sun.
    (c) spontaneous fission of carbon.
    (d) the fusion of four hydrogen nuclei into a helium nucleus.
    (e) chemical reactions.

57. _____ is required in the core of the Sun to overcome the _____ of positive charges in nuclear reactions.
    (a) Coulomb's law, electrical attraction
    (b) electrical attraction, kinetic energy
    (c) Heisenberg's uncertainty principle, kinetic energy
    (d) the Pauli exclusion principle, potential energy
    (e) quantum mechanical tunneling, electrical repulsion

58. (*Answer*) The solar neutrino problem was the result of
   (a) electron neutrinos "oscillating" between three types, electron neutrinos, tau neutrinos, and muon neutrinos.
   (b) the transformation of electron neutrinos while passing through the magnetic field in the Sun's interior.
   (c) previously undetected physics having to do with neutrinos.
   (d) neutrinos having mass, unlike photons.
   (e) all of the above
   (f) none of the above
   (g) (a) and (b) only

59. Helioseismology provides a way of
   (a) probing the detailed structure of the Sun's interior.
   (b) testing computer models of the Sun.
   (c) studying the Sun's orbit.
   (d) measuring the neutrinos being produced by chemical reactions in the Sun's core.
   (e) all of the above
   (f) none of the above
   (g) (a) and (b) only

Short Answer

60. Explain how astronomers are able to selectively observe different layers of the Sun's atmosphere.

61. Complete the following sentence, based on the last sentence of the subsection titled "Summary of Basic Solar Characteristics" on page 311. "Ideally, a scientist, or anyone else for that matter, should possess an _____, be able to think _____, and demand _____."

62. Describe what is meant by critical thinking, and how that concept applies to science.

63. Repeat the process in Example 9.2 for determining the mass of the Sun, except this time rather than substituting a numerical value for the speed of Earth in its orbit around the Sun, substitute the algebraic form of the speed and carry out the squaring of that quantity.

   [You have just arrived at the expression for Kepler's Third Law as derived by Sir Isaac Newton for the special case of a small mass (Earth) orbiting a much larger mass (the Sun) in a circular orbit as shown just above Example 5.13. The only difference is that your equation works for the units of kilograms, meters, and seconds, rather than solar masses, astronomical units, and years.]

64. A home roof-mounted solar panel converts about 14% of the sunlight that strikes it into electrical energy. Assume that the solar panel has a surface area of 0.5 m^2.
   (a) On a bright, sunny day, approximately how much electrical power can the solar panel produce (measured in watts)?
   (b) A 32-inch (81-cm) flat screen television uses approximately 50 W during operation. Could the solar panel be used to power the TV?

65. Along with the other nasty things that can happen to you if you are exposed to the near-vacuum of space somewhere outside of Earth's protective magnetic field is that you would freeze to death, despite being constantly bombarded by the nearly 1 million kelvins solar wind. Explain this paradox. (Hints: Remember what temperature is actually measuring, and don't forget about your own blackbody radiation.)

66. Explain the mottled appearance of granulation at the base of the photosphere. Why are some areas bright and others dark, and what is the source of those features?

67. How long is the solar cycle and how is the cycle length determined?

68. Explain why solar prominences often appear as loop-like structures.

69. Describe how activity on the Sun leads to producing the aurora borealis and the aurora australis.

70. (a) What was the Maunder minimum, when did it occur, how long did it last, and what does it seem to correlate with in Earth's history?
   (b) Is correlation the same as causation? Why or why not?

71. Briefly describe the magnetic dynamo.

72. In general terms, describe what happens to the Sun's global magnetic field over the course of one solar cycle and how that behavior gives rise to sunspots and solar flares. You may want to include sketches.

73. (a) What role does computer modeling play in developing our understanding of stars?
   (b) Should the results of computer models simply be accepted as fact? Why or why not?
   (c) What is required to give validity to computer models?

74. A common lecture demonstration in introductory physics courses is to pump the air out of a paint can, causing the can to compress like being squeezed by a giant hand. Based on the discussion of gas pressure, explain what happens.

75. Scuba divers carry a tank of compressed air and a pressure regulator. The pressure regulator controls the pressure of the air that enters the diver's mouthpiece.
   (a) Explain why the pressure regulator "feeds" air at increasing pressure as the diver descends. (Hint: Think about what would happen if you were lying underneath a 10-m (33-ft) tall aquarium tank full of water. What is the regulator effectively doing to help the diver breathe?)
   (b) Why is it extremely dangerous, and possibly deadly, to hold your breath while ascending when scuba diving? In an emergency ascent, say when the tank runs out of air, an experienced diver will exhale all the way up.

76. An argon atom has a mass that is 10 times greater than the mass of a helium atom. Container A holds argon at the same mass density and temperature as the helium in container B. Which container has a higher gas pressure? Explain physically why that is the case. ("Because of the ideal gas law" is not an acceptable answer!) Hint: Do both types of atoms have the same kinetic energy in this situation? The same speed? The same linear momentum?

77. Is it possible for a dust particle to move away from the Sun? Explain your reasoning.

78. Describe quantum mechanical tunneling and its relationship to Heisenberg's uncertainty principle.

79. Explain how helioseismology is able to provide detailed information about the Sun's structure and provide precise tests of sophisticated computer models of the Sun at the same time.

Activities

80. *Only perform this activity if you have appropriate solar-safe eye wear or equipment.* On a clear day, record the number of sunspots visible on the surface of the Sun and sketch their location. Be sure to record the date and time that you made your observations. Ideally, you should use a solar-safe telescope.

10 An Overview of the Solar System

The Solar System should be viewed as our backyard, not as some sequence of destinations that we do one at a time.

Neil deGrasse Tyson (b. 1958)

10.1   Taking Inventory: A Taxonomy of the Solar System   359
10.2   Putting It in Perspective: A Scale Model   367
10.3   The Importance of Tides   370
       Exercises   376

Fig. 10.1 Montage. [NASA/JPL-Caltech/Johns Hopkins University Applied Physics Laboratory/Southwest Research Institute/UMD/ESA/Rosetta/MPS for OSIRIS Team/MPS/UPD/LAM/IAA/SSO/INTA/UPM/DASP/IDA (CC BY-SA 3.0 IGO)]

Introduction

The Sun, the Moon, Earth, and the five "wandering stars" have been known since antiquity. Today we are aware of a much richer Solar System than the ancient Babylonians, Greeks, Native Americans, Mayans, Chinese, and the other pre-modern societies of the world could have ever dreamed of. Although much progress has been, and continues to be made by observing and studying the individual objects in our Solar System from the ground and low-Earth orbit (including Earth itself), much of the knowledge we have gained since the 1960s has come from sending astronauts to the Moon, and robotic missions to the Moon, to each of the eight known planets and some of their moons, to dwarf planets, and to several asteroids and comets.

In Chapter 9 we explored the central body of our Solar System, the Sun. In Chapter 10 we will get a brief overview of the other members of the Solar System, develop some understanding of the sizes and vast distances involved, and learn about tides, a gravitationally driven process that affects essentially everything in the Solar System. A more detailed exploration of Solar System bodies and the characteristics of planetary systems beyond our own are presented in Chapters 11–14.

10.1 Taking Inventory: A Taxonomy of the Solar System

As you will learn in more detail in Section 14.2, our Solar System formed 4.57 billion years ago when an interstellar cloud of dust and gas, a nebula, collapsed. The Sun was born at the center of the collapsing solar nebula, and, in a fashion similar to the spinning of pizza dough, conservation of angular momentum (page 194*ff*) resulted in a flattened protoplanetary disk of dust and gas around the Sun. It was within the disk that the Sun's progeny were born.

Key Point
The Sun and the rest of the Solar System formed from a spinning, collapsing cloud of dust and gas that flattened into a disk.

The Rocky Planets

If you examine the columns for the rocky planets in Table 10.1 you should notice that Mercury's mass is only about 6% of the mass of Earth, the mass of Venus is just over 80% of Earth's mass, and the mass of Mars is slightly more than 10% of the mass of Earth. At the same time, the radii of Mercury, Venus, and Mars are about one-third of the radius of Earth, just slightly smaller than Earth, and about one-half of Earth's radius, respectively (the relative sizes correspond to those in Fig. 10.2). Given that mass density is just mass divided by volume, we find that the average densities of Mercury, Venus, and Earth are all very similar (between 5200 and 5500 kg/m^3, where 1000 kg/m^3 is the density of water), but the density of Mars is substantially less, at about 3900 kg/m^3. For reference, the density of a typical rock on the surface of Earth is about 2700 kg/m^3. In addition, Earth and Mars rotate with periods of about one day, while Mercury and Venus rotate very slowly with Venus's rotation being retrograde.

As mentioned on multiple occasions in this text, it is not important that you memorize the data presented throughout this chapter because you can always look it up when needed, but you should strive to recognize and understand patterns

Key Points
The rocky planets:
• orbit closer to the Sun,
• are relatively small,
• have lower masses,
• have high average densities,
• rotate slowly,
• have few or no moons, and
• don't have planetary ring systems.

Table 10.1 Comparing the rocky and giant planets

Characteristics	Rocky planets				Gas giants		Ice giants	
	Mercury	Venus	Earth	Mars	Jupiter	Saturn	Uranus	Neptune
Mass[a] (M_{Earth})	0.055	0.81	1.00	0.11	317.91	95.21	14.56	17.18
Radius[a] (R_{Earth})	0.38	0.95	1.00	0.53	11.21	9.45	4.01	3.88
Average density[b] (kg/m^3)	5427	5243	5514	3933	1326	687	1271	1638
Rotation period[c] (d)	58.65	243.7	0.997	1.03	0.41	0.44	0.72	0.67
Rotation axis tilt[d] (°)	0.03	177.36	23.44	25.19	3.13	26.73	97.77	28.32
Number of known moons	0	0	1	2	79	82	27	14
Ring system	no	no	no	no	yes	yes	yes	yes
Semimajor axis (au)	0.39	0.72	1.00	1.52	5.20	9.54	19.19	30.07
Orbital period (y)	0.24	0.62	1.00	1.88	11.86	29.46	84.02	164.77
Orbit eccentricity	0.206	0.007	0.017	0.093	0.048	0.054	0.047	0.009
Orbit inclination (°)	7.00	3.39	0.00	1.85	1.31	2.48	0.77	1.77

[a] $M_{Earth} = 5.972\,365 \times 10^{24}$ kg, $R_{Earth} = 6378.1366$ km

[b] For comparison, the density of water is 1000 kg/m^3 and the density of a rock on Earth's surface is about 2700 kg/m^3.

[c] Earth's solar day is 24 hours long but its rotation period is about 4 minutes shorter (Fig. 4.10).

[d] Relative to orbit. Axis tilts greater than 90° imply retrograde rotation.

that the data present. When out of the ordinary exceptions to patterns appear, you will want to make note of them as well; and as always, ask why. What could be going on or what may have happened in the past, that can explain why the odd characteristic exists?

In the case of the rocky planets and the information in Table 10.1, you will certainly want to appreciate details (without worrying about the exact numbers) such as Earth being the largest and most massive, Venus is similar in size, and Mercury and Mars are the smallest. The order of the planets from the Sun is also worth knowing. However, as always, keep in mind that *memorization is not the same as understanding.*

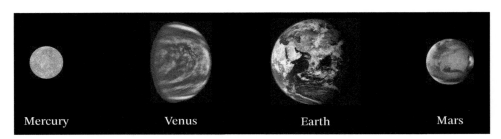

Fig. 10.2 The relative sizes of Mercury, Venus, Earth, and Mars. [Mercury: NASA/Johns Hopkins University Applied Physics Laboratory/Carnegie Institution of Washington; Venus: NASA/JPL; Earth: NOAA/NASA GOES Project; Mars: NASA, ESA, and Z. Levay (STScI), J. Bell (ASU), and M. Wolff (Space Science Institute)]

Mythology

Mercury was the Roman god of commerce, poetry, and travel, and a son of Jupiter (Zeus in Greek mythology). In Greek mythology, Mercury's counterpart is Hermes. Mercury was depicted with wings on his feet and on his hat, making the naming of the fastest planet in the Solar System very appropriate. The Roman goddess Venus (Aphrodite in Greek mythology) was the goddess of love, beauty, desire, and fertility, and the mother of the Roman people. Mars was the god of war in Roman mythology, corresponding to Ares in Greek mythology. In Greek mythology, Ares was a son of Zeus.

The Giant Planets of Gas and Ice

Again referring to Table 10.1 you should notice an abrupt change in the parameters between the rocky planets and the giant planets. As is apparent in Fig. 10.3, Jupiter is clearly the king of the Sun's planetary system, having a mass that is nearly 318 times greater than the mass of the largest rocky planet, Earth. Jupiter's radius is also more than 11 times greater that Earth's, meaning that if Jupiter were a hollow ball, more than 1300 Earths would be able to be placed inside it (recall that volume is proportional to the radius cubed, $V = 4\pi r^3/3$, and $11^3 = 11 \times 11 \times 11 = 1331$). In addition, unlike the rocky planets, Jupiter's density is only about 30% greater than the density of water.

Moving out to the second largest planet in the Solar System, Saturn may be 95 times more massive than Earth, but its mass is less than one-third of the mass of Jupiter. This leads to a peculiar feature of the planet, namely that Saturn's density is significantly lower than that of water. In order to float in water, the overall density of an object must be less than the density of water; *if* a bathtub existed on a sufficiently gigantic world somewhere in the universe, Saturn would float in it! (Of course, you would have to wrap it up first because it is gaseous.) Clearly Jupiter and Saturn, often referred to as gas giants, are very different from the rocky planets.

Uranus and Neptune are also enormous in size when compared with Earth, but they are much smaller than the two inner giant planets. With masses of just under and over 5% of the mass of Jupiter, respectively, Uranus and Neptune are tiny when compared with that behemoth. Indeed, their radii are about 36% of Jupiter's value and, as a result, their volumes are about 5% of Jupiter's volume ($0.36^3 = 0.047$).

Key Points

The giant planets:
- orbit far from the Sun,
- are very large,
- are massive,
- have low densities,
- rotate rapidly,
- have large numbers of moons, and
- possess ring systems.

Fig. 10.3 The relative equatorial sizes of Jupiter, Saturn, Uranus, and Neptune. [Jupiter: NASA, ESA, and A. Simon (NASA Goddard); Saturn: NASA/JPL/Space Science Institute; Uranus: NASA/JPL; Neptune: NASA/JPL]

In examining the densities of Uranus and Neptune, note that their densities are comparable to, and somewhat greater than Jupiter's density. As we will learn in Chapter 12, rather than being dominated by gases, these worlds are ice giants.

Mythology

Jupiter got its name from the Romans, who named it after the primary god of Roman mythology. As already mentioned, in Greek mythology, Jupiter corresponds to Zeus. Saturn, in Roman mythology, was the god of agriculture, and is Cronus in Greek mythology, with Saturn being the father of Jupiter. After its discovery by Sir William Herschel (Section 5.5), Uranus was named for the Greek god of the sky, who was both the son and husband of Gaia, Mother Earth. Uranus and Gaia became the parents of the first generation of the Titans, which included Cronus (Saturn). Helios, the Sun god (corresponding to the Roman god, Sol), was a member of the second generation of Titans. Neptune, the Roman god of the sea, was the brother of Jupiter. In Greek mythology, Neptune is Poseidon. In the Greek pantheon of gods, Zeus (Jupiter), Poseidon (Neptune), Ares (Mars), Aphrodite (Venus), and Hermes (Mercury) were members of the Olympians who went to war with the Titans.

Moons and Rings

There is a dramatic difference between the rocky planets and the giant planets when it comes to their moons and the absence or presence of orbiting ring systems. Earth's moon, the Moon, is by far the largest of the three moons belonging to rocky planets, and although it isn't the largest moon in the Solar System, it is easily the largest moon in relative size to its parent planet. On the other hand, the giant planets have myriads of moons with Saturn currently known to have the largest number (82). Among Jupiter's moons are the four that Galileo saw with his early telescope and recorded in his *Sidereus Nuncius* (Fig. 4.51). Two of those moons, Ganymede and Callisto, are larger than the Moon. One of Saturn's many moons, Titan, is also larger than the Moon, and even boasts a thick atmosphere. The giant moons will be explored in more detail in Section 12.2.

None of the rocky planets currently exhibit a ring system, while all of the giant planets do (Section 12.3). Of course, most spectacular and well known are the majestic rings of Saturn.

The Asteroid Belt

Giuseppe Piazzi (Fig. 10.4) was an Italian Catholic priest who also served as the Chair of Mathematics at the University of Malta for a time, after which he became Professor of Astronomy at the University of Palermo, starting in 1787. While at the University of Palermo, Piazzi was placed in charge of the Palermo Observatory, supervising the creation of the Palermo catalog of 7646 stars.

On the night of January 1, 1801 Piazzi discovered a "stellar object" that moved against the background stars. Given its motion, Piazzi correctly concluded that he had not in fact discovered a new star, but he had found what he believed might be a planet. Being a bit conservative in his public announcement, Piazzi speculated that perhaps the object was a comet (see page 24 and page 365*ff*). What Piazzi had

Fig. 10.4 Giuseppe Piazzi (1746–1826). (Smithsonian Institute Library; public domain)

 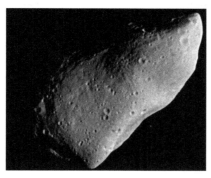 

Fig. 10.5 *Left:* 1 Ceres (NASA/JPL-Caltech/UCLA/MPS/DLR/IDA). *Center:* 951 Gaspra (NASA/JPL/USGS). *Right:* 243 Ida and its tiny moon, Dactyl, seen to the right. (NASA/JPL)

actually discovered was 1 Ceres [Fig. 10.5 (*Left*)], the first object found to be orbiting between the orbits of Mars and Jupiter in a region referred to as the asteroid belt (hence the designation 1 in front of the name). Ceres was named for the Roman goddess of agriculture, fertility, and motherly relationships, and corresponds to Demeter, another of the Greek Olympians. Today there are believed to be millions of asteroids in our Solar System, ranging in size from about 1 meter across to Ceres, which is 946 km in diameter, making it about the same size as Texas.

It is common in science fiction movies and television shows to see chase scenes with spaceships that depict an asteroid field as being perilous to navigate, with the heroes constantly dodging deadly rocks that could spell doom. In reality, at least in our Solar System, a spacecraft passing through the asteroid belt needs to go out of its way to come close to an asteroid. Despite the huge number of asteroids that exist in the asteroid belt, the distance between them is typically immense.

A number of asteroids have now been visited by spacecraft from the National Aerodynamics and Space Administration (NASA), the European Space Agency (ESA), and Japan's Institute of Space and Astronomical Science (ISAS). The first such visits occurred when the Galileo spacecraft flew past 951 Gaspra [Fig. 10.5 (*Center*)] in 1991 and then past 243 Ida and its tiny moon, Dactyl, [Fig. 10.5 (*Right*)] in 1993, on its way to an extended mission studying Jupiter.

An Abundance of Dwarf Planets

Following its discovery in 1930 (Section 5.5), Pluto was considered to be the ninth planet in the Solar System. The newly discovered object was named by Venetia Burney (1918–2009), an eleven-year-old girl in Oxford, England who was interested in Greek mythology. Pluto was the Greek god of the underworld and the brother of Zeus and Poseidon.

In 2006 the International Astronomical Union (IAU) officially developed a definition for a planet that removed Pluto from that exclusive club. It may seem odd that there wasn't an official definition before that time, but up until then astronomers simply assumed that we would know a planet when we saw it. What changed was the deployment of more powerful telescopes that led to the rapid discoveries of numerous significantly sized objects beyond the orbit of Neptune, referred to as trans-Neptunian objects.

Table 10.2 Dwarf planets and dwarf planet candidates

	Ceres[a]	Pluto[b]	Haumea[c]	Quaoar[†]	Make-make	Gong-gong[†]	Eris[d]	Sedna[e†]
Discovered	1801	1930	2004	2002	2005	2007	2005	2003
Mass (M_{Pluto})	0.072	1.000	0.31	0.11	0.34	0.13	1.27	—
Average radius (R_{Pluto})	0.40	1.000	0.69	0.47	0.60	0.52	0.98	0.45
Average density (kg/m^3)	2160	1850	1800	2000	2300	1700	2500	—
Rotation period (d)	0.38	6.39	0.16	0.74	0.32	0.93	1.08	0.43
Rotation axis tilt to orbit (°)	4	123	—	—	—	—	—	—
Number of known moons	0	5	2	1	1	1	1	0
Ring system	no	no	yes	no	no	no	no	no
Semimajor axis (au)	2.77	39.48	43.35	43.62	45.68	67.48	67.66	483.28
Orbital period (y)	4.6	247.9	285.4	288.1	308.7	554.3	556.6	10,624.6
Orbit eccentricity	0.076	0.249	0.190	0.038	0.156	0.501	0.441	0.843
Orbit inclination (°)	10.59	17.14	28.21	7.99	28.98	33.70	44.20	11.93

[a]Ceres is about the same diameter as the state of Texas.

[b]Pluto is only 0.22% (0.0022) the mass of Earth and has a radius that is 18.7% (0.187) the radius of Earth.

[c]Haumea is very elongated (2322 × 1704 × 1138 km), and spins rapidly end over end every 3.9 hours.

[d] Eris is more massive than Pluto but slightly smaller.

[e]Sedna has a very eccentric orbit, giving a perihelion of 76.1 au and an aphelion of 890.5 au.

[†]Quaoar, Gonggong, and Sedna are three of the objects proposed as dwarf planets but not yet formally adopted by the IAU.

Key Points

Definition of a planet:
• orbits a parent star(s),
• has sufficient mass to become spherical or ellipsoidal, and
• has cleaned out the region around its orbit.

Dwarf planets do not satisfy the third requirement.

With the potential of discovering hundreds, or perhaps even thousands of these small "planets," the IAU determined that a new classification of objects was required. The technical definition of a planet became an object that (a) orbits its parent star (or stars) directly, (b) has enough mass, and therefore sufficient gravity, to take on a spherical shape if not rotating or an ellipsoidal shape (meaning flattened at the poles) if the object is rotating, and (c) was able to sweep clean the region of the planetary system around its orbit. Dwarf planets satisfy the first two criteria but do not fully satisfy the third one. In other words, other objects may be orbiting their parent star at the same general orbital distance as a dwarf planet. The definitions of a planet and dwarf planet explicitly allow for bodies discovered to be orbiting other stars. Within our Solar System all dwarf planets except one are trans-Neptunian objects. Ceres, the asteroid belt's largest body, is, according to the definition, a dwarf planet. Table 10.2 lists the years of discovery, along with some physical and orbital characteristics, of the officially designated dwarf planets as well as three of the trans-Neptunian objects that have been proposed for dwarf planet status. Figure 10.6 shows the relative sizes of four of the trans-Neptunian objects.

In Table 10.2 you should note the semimajor axes of the dwarf planets' orbits and their orbital periods. As already mentioned, Ceres is fairly close to the Sun, while Pluto, Haumea, Quaoar, and Makemake have semimajor axes between about

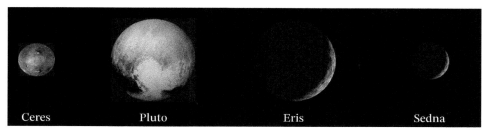

Fig. 10.6 The relative equatorial sizes of Ceres, Pluto, Eris, and Sedna. The representations of Eris and Sedna are artistic conceptions included to represent relative sizes only. (Ceres: NASA/JPL-Caltech/UCLA/MPS/DLR/IDA; Pluto: NASA/Johns Hopkins University Applied Physics Laboratory/Southwest Research Institute; Eris and Sedna: NASA/JPL-Caltech)

39 and 46 au from the Sun. Gonggong and Eris have average distances from the center of the Solar System of 67 and 68 au, respectively, and have large orbital eccentricities, while Sedna has a staggering 483 au semimajor axis with an orbital period of 10,625 years. Sedna's orbit is also extremely eccentric ($e = 0.84$), which means that it can get as close as 76 au from the Sun and as distant as 890 au. Recall that the speed of light implies that sunlight takes 8.3 minutes to travel 1 au; when Sedna is at its most distant, sunlight takes nearly 5.8 days to reach it. (Gonggong was named for a Chinese water god and Sedna was named for the Inuit goddess of the sea. All dwarf planets are named for characters in creation mythologies from various cultures.)

There are several breaks in the distances of the dwarf planets and dwarf planet candidates listed in Table 10.2. Pluto, Haumea, Quaoar, and Makemake are all members of a doughnut-shaped distribution of objects known as the Kuiper belt, named in honor of Gerard Kuiper (Fig. 10.7). The inner edge of the Kuiper belt is 30 au from the Sun (the location of Neptune's orbit) and the outer edge is at about 48 au. Eris, with its very eccentric orbit, lies farther out than the Kuiper belt and was apparently scattered outward gravitationally when it came too close to Neptune. Gonggong and Sedna appear to be objects that were scattered by a gravitational interaction with Neptune as well, along with many others. These scattered trans-Neptunian objects are distributed in another doughnut-shaped collection of bodies called the scattered disk. The inner edge of the scattered disk overlaps the Kuiper belt, making the scattered disk essentially an extended component of the Kuiper belt. Note that all of the objects listed in Table 10.2 have orbits that are significantly tilted (inclined) relative to the ecliptic, meaning that they can travel well above and below the plane of Earth's orbit.

Fig. 10.7 Gerard Kuiper (1905–1973). (Gelderen, Hugo van/Anefo, CC BY-SA 3.0 nl)

Comets

As was discussed in Section 5.1, Sir Isaac Newton's *Principia* was ultimately published thanks to the funding and encouragement of Sir Edmond Halley after Newton successfully described the orbit of a comet that reappears roughly every 75 years (it last made an appearance in 1986 and is predicted to return again in 2061). Understanding the orbit of Halley's comet (Fig. 10.8) was one of the great successes of early Newtonian physics.

Fig. 10.8 Halley's comet in 1986. (NASA)

Fig. 10.9 Jan Oort (1900–1992). [Joop van Bilsen, (Nationaal Archief NL Fotocollectie Anefo), CC BY-SA 3.0]

Fig. 10.10 Comet 67/Churyumov-Gerasimenko photographed in 2015 at an altitude of 86 km by ESA's Rosetta spacecraft. (ESA/Rosetta/NAVCAM; CC BY-SA 3.0 IGO)

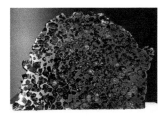

Fig. 10.11 A stony–iron meteorite on display at the Royal Ontario Museum. (Captmondo, CC-BY-SA-3.0)

Halley's comet is just one example of a very large collection of similar objects that orbit the Sun, some with relatively short periods, like Halley, and others with orbital periods of millions of years. In 1950 Jan Oort (Fig. 10.9) carried out a statistical study of the orbits of comets to determine where they come from. He determined that there must be a vast reservoir, known today at the Oort cloud, that may contain as many as two hundred billion (2×10^{11}) comets. Some researchers estimate that the inner edge of the Oort cloud is roughly 2000 au (0.03 ly) from the Sun, and its outer edge may be greater than 50,000 au (0.8 ly) from the Sun although other estimates place the inner edge farther out, at 5000 au, and the outer edge as far as 200,000 au from the Sun. The Oort cloud is the source of the very long-period comets.

The sources of the short-period comets, comets with orbital periods of less than 200 years, is less well understood, but most likely originate in either the Kuiper belt, its related scattered disk, or a group of icy bodies trapped in Neptune's orbit called trojans. Some short-period comets also appear to have originated in the Oort cloud, but had their orbits altered through gravitational interactions with the giant planets. Comet Halley probably came from the Oort cloud.

Comets are famous for their feathery, very long gas and dust tails and their giant, puffed up comas around a central nucleus. These famous features of comets only appear when they enter the inner Solar System, however. The nucleus, which is generally referred to as the actual comet, is composed of ices and dust. The tails and the coma are the result of gases and dust escaping from the nucleus of the comet as it heats up, with the gas and dust tails being pushed away from the Sun by the solar wind and radiation pressure, respectively. In essence, a comet contains original material left over from the Solar System's formation 4.57 billion years ago, that has been kept in deep freeze before the comet becomes active when it enters the inner Solar System. A close-up of Comet 67/Churyumov–Gerasimenko and gases escaping from it is shown in Fig. 10.10.

Meteoroids

Objects orbiting the Sun that are smaller than one meter in diameter and are composed of either rock or metal, or some combination of the two, are considered to be meteoroids. There are numerous sources of meteoroids including (a) pieces broken off asteroids that collide with one another, (b) remnants of disintegrating comets, (c) chunks of rock blasted into space from the Moon or Mars when impacting asteroids eject debris from their surfaces at speeds greater than the local escape speed (page 198ff), and (d) remnants of the original formation of the Solar System, in which case they represent ancient histories of the environment of the solar nebula when the Sun and the rest of the Solar System formed.

Technically, it is only when these objects are falling through Earth's atmosphere and produce a light show that they are referred to as meteors (recall Fig. 2.14), and any pieces that survive the decent and hit the ground, such as the stony–iron in Fig. 10.11, are known as meteorites. (Occasionally, individuals come across an unusual-looking rock, pick it up, and claim that it is a meteorite. Indeed, in some cases they do turn out to be bona fide meteorites, but frequently they are "meteorwrongs.") By the way, meteors are often confusingly called "falling stars" or "shooting stars." Let's hope not!

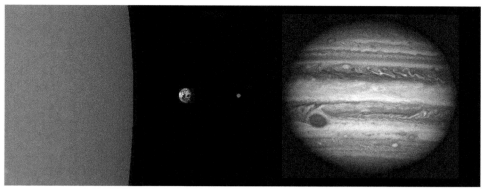

Fig. 10.12 The relative sizes of the Sun, Earth, the Moon, and Jupiter. Separations are obviously not to scale. [Sun: NASA/SDO, Earth: NOAA/NASA GOES Project, Moon: NASA/GSFC/Arizona State University, Jupiter: NASA, ESA, and A. Simon (NASA Goddard)]

10.2 Putting It in Perspective: A Scale Model

Figure 10.12 shows the relative sizes of our Sun, Earth (the largest of the rocky planets), our Moon (one of the larger moons in the Solar System), and Jupiter (the largest member of the Solar System other than the Sun). As is readily apparent, they are vastly different in size. However, Fig. 10.12, along with the other figures depicting relative sizes in this chapter, is very misleading. Absent from all of the figures are the relative *distances* between objects in the Solar System. In fact, when compared with the distances between them, the Sun, the planets, and the dwarf planets are tiny.

In order to better appreciate how *enormous and empty* our Solar System really is, it helps to construct a scale model based on objects and distances we are familiar with. Begin by imagining the Sun as a mid-sized, 65-cm diameter, exercise ball (Fig. 10.13). At that scale we need to reduce everything else in the Solar System by a factor of 2.1 billion (2.1×10^9). Table 10.3 lists relative sizes and distances if we were to shrink the Solar System that much. It also gives every day objects that would have the sizes of the planets and dwarf planets at that scale.

Let's first consider the inner Solar System, out to the orbit of the dwarf planet, Ceres, in the asteroid belt. Imagine placing two American football fields end to end as shown in Fig. 10.14. Each field is 120 yards long (110 m) including the two, 10-yd-deep end zones, making the total length of the two fields 240 yd (220 m).

We begin by placing our model Sun (the exercise ball) on the goal line of the field on the left. The image labeled "Actual scale" represents the approximate scaled size of the Sun, with an enlarged version, five times the size scaled to the exercise ball, also depicted. In this scale model, Mercury is on the 30-yard line (27.4 m), and would be the combined thickness of two dimes, 500 times smaller than the image in the figure. Venus and Earth are nearly the same size and would be about 60% of the diameter of a AAA battery, although their images in Fig. 10.14 are 500 times larger than they should be. The planets are located 55 yd (51 m) and 77 yd (70 m) from the Sun, respectively. Mars is 117 yd (107 m) from the Sun with a diameter corresponding to the combined thickness of three dimes, again expanded in size by a factor of 500. Finally, Ceres is 212 yd (194 m) away from the Sun and is the size of

Fig. 10.13 An exercise ball with a diameter of 65 cm serves as our model "Sun." (Public domain, CC0 1.0)

Table 10.3 Scale model of the Solar System

Object	Scaled diameter (cm)	Approximate representation	Scaled distance (km)	Scaled (comments)
Sun	65	Exercise ball	0.000	0 yd
Mercury	0.23	Thickness of two dimes	0.027	30 yd
Venus	0.57	57% diameter of AAA battery	0.051	55 yd
Earth	0.60	60% diameter of AAA battery	0.070	77 yd
Mars	0.32	Thickness of three dimes	0.107	117 yd
Ceres	0.04	Grain of salt (period on page)	0.194	212 yd
Jupiter	6.68	Baseball	0.365	399 yd
Saturn	5.63	Billiard ball	0.671	734 yd
Uranus	2.39	Diameter of one nickel	1.34	0.84 mi
Neptune	2.38	Diameter of one nickel	2.11	1.31 mi
Pluto	0.11	Thickness of one dime	2.77	1.72 mi
Eris	0.11	Thickness of one dime	4.74	2.95 mi
Sedna	0.08	Two grains of salt (two periods)	34.0	21.2 mi
Oort cloud[a] (inner edge)			140	87 mi
Oort cloud[a] (outer edge)			3500	2200 mi
				(New York to Las Vegas)
Proxima Centauri	19	Three baseballs	18,640	11,600 mi
				(47% of Earth's circumference)

[a]Assuming an inner edge of 2000 au and an outer edge of 50,000 au.

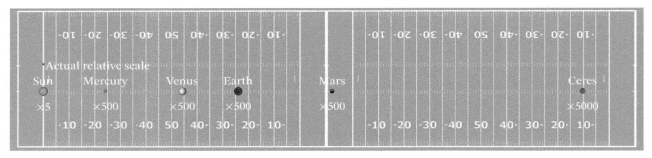

Fig. 10.14 The relative sizes and distances of the Sun, the rocky planets, and Ceres, assuming that the Sun is the size of an exercise ball. (Xyzzy n; CC BY-SA 3.0)

a single grain of salt or one of the periods on this page (the image of Ceres is 5000 times larger than it should be). Even for the relatively closely spaced inner Solar System, when compared to the vastness of the distances involved, the planets and Ceres are very tiny, leaving space virtually empty.

To see where the giant planets would be located, along with Pluto and Eris, it is necessary to zoom out. We now place the left-most football field of Fig. 10.14 inside the stadium in East Rutherford, New Jersey, where the professional teams, the New York Giants and the New York Jets, play football. As depicted in Fig. 10.15, Jupiter

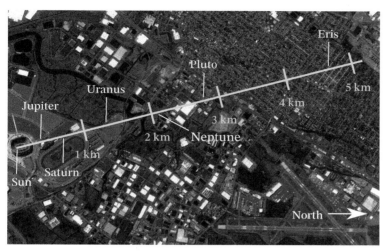

Fig. 10.15 The relative distances of the Sun, the giant planets, Pluto, and Eris, assuming that the Sun is the size of an exercise ball. The professional football stadium in East Rutherford, New Jersey is on the left side of the image and Teterboro airport is visible in the lower right-hand corner. North is to the right. (Adapted from NASA/USGS Landsat image gallery)

would be found outside of the stadium at a distance of 365 m (nearly 400 yards) from the Sun and would be the size of one baseball. Saturn would be the size of a billiard ball located 671 m (734 yd) from the Sun. Uranus and Neptune would each have diameters equal to about one nickel, and would be 1.34 km and 2.11 km (0.84 mi and 1.31 mi) from the Sun, respectively. Finally, Pluto and Eris would both be about the thickness of the edge of one dime, and would be 2.77 km and 4.74 km (1.72 mi and almost 3 mi) from the Sun, respectively.

This scale model gives some idea of why it took so long to discover Pluto, and particularly Eris, in the first place. Imagine trying to locate Eris by reflected sunlight when it is only equivalent to the thickness of one dime seen from 4.7 km away. Making things even harder, unlike the shiny dime, Eris doesn't reflect light nearly as efficiently. The discoveries of these extremely faint objects are often made by carefully comparing images taken of the same region of the sky at different times. By switching back and forth quickly, the human eye can see the changes in position, or the absence and then presence, of an object. This process was used by Clyde Tombaugh when he discovered Pluto in 1930 (Section 5.5). Examples of blink comparisons can be seen on the text's Chapter 10 resources webpage. With today's electronic imaging, one image can be subtracted from the other, revealing the motion of the object.

So how far away is Sedna, the most distant dwarf planet yet discovered at the time this text was written, and how far out is the inner edge of the Oort cloud (assuming a distance of 2000 au)? Figure 10.16 shows that in our scale model Sedna would be equivalent in diameter to two grains of salt located 34 km (21 mi) from the original goal line where our exercise-ball Sun was placed. The inner edge of the Oort cloud would form a sphere of radius 140 km (87 mi) from our model Sun. From Fig. 10.16 you can see that the inner edge of the Oort cloud would encompass most of Long Island.

Fig. 10.16 The relative distances of the Sun, Sedna, and the inner edge of the Oort cloud assuming that the Sun is the size of an exercise ball located on the goal line of the professional football stadium in East Rutherford, New Jersey. (Adapted from NASA/USGS Landsat image gallery)

What about the outer edge of the Oort cloud where the outermost extent of our Solar System can be considered to be? If we assume the outer edge is 50,000 au distant, in our scale model the outer edge of the Solar System would be a sphere of radius 3500 km (2200 mi), equivalent to the distance between New York City and Las Vegas, Nevada; most of the way across the United States of America.

Key Point
The Solar System is a vast and nearly empty place.

Finally, where would the next closest star be? Proxima Centauri is 266,000 au (4.2 ly) from the Sun. In our scale model that would put the star 18,600 km (11,600 mi) away, a distance corresponding to the distance between the South Pole and Longyearbyen, Norway in the Svalbard archipelago (a collection of islands halfway between Norway and the North Pole). In our model, the exercise ball-sized Sun's closest stellar companion measures three baseballs across and is nearly halfway around the world!

By now it should be evident just how empty space really is, and just how amazingly successful our human quest for understanding our cosmic neighborhood has been so far. To paraphrase Sir Isaac Newton, we have only been able to see this far by "standing on the shoulders of giants": the men and women of science whose revelations about mathematics, physics, chemistry, astronomy, and planetary science have paved the way for future exploration, discovery, and self-consistent, theory-based understanding.

10.3   The Importance of Tides

There are a great many aspects of physics, chemistry, geology, planetary science, and biology that become important when talking about the wide array of objects in our Solar System. But one in particular is so universal that it deserves further investigation now before proceeding to explore the details of the Sun's family.

We are all familiar with the so-called "Man in the Moon" (Fig. 10.17), the apparent face that is always directed toward Earth. The fact that one side of the Moon is constantly oriented toward Earth while the Moon orbits our planet approximately once each month is a natural consequence of an important aspect of gravity; the production of tides.

Fig. 10.17 From the silent film, *Le Voyage dans la lune* (A Trip to the Moon; 1902), by French director Georges Méliès. The film was very loosely based on Verne, 1865.[1] (Georges Méliès; public domain)

Revisiting Angular Momentum and Torque

Back in Section 6.4 you learned about three conservation laws: linear momentum, angular momentum, and energy. Regardless of what happens during any physical process, all three of these laws must be fully satisfied at all times. In order to understand the production of tides and their consequences, it is helpful to review briefly the conservation of angular momentum in particular.

Recall that the angular momentum equation, $L = mv_\perp r$, states that the angular momentum of an object is given by its mass, m, multiplied by the speed (the "length") of the *perpendicular* component of the velocity vector, $\vec{v}_\perp$, and then multiplied by the distance, r, to the rotation axis. You have already learned that linear momentum is a vector ($\vec{p} = m\vec{v}$), but angular momentum is also a vector ($\vec{L}$). This means that just like linear momentum, when angular momentum is conserved, the direction of its vector must be kept oriented in a constant direction at the same time that the magnitude of the vector is preserved.

Have you ever wondered why it is much easier to stay upright on a bicycle when it is moving? The answer lies in the angular momentum of the spinning wheels. The angular momentum vector for each wheel points along the axis of the wheel to your left when you are moving forward on your bike (see Fig. 10.18). As long as the wheels are spinning it is relatively difficult to tip over because that would cause the direction of the angular momentum vectors to change. When you stop, the length of the angular momentum vectors go to zero and there is no preferred direction. The result is that you must balance without the aid of $\vec{L}$.

Fig. 10.18 "Life is like riding a bicycle. To keep your balance, you must keep moving" (Einstein). The wheel's angular momentum vector, $\vec{L}$, is directed along the wheel's axle. (Public domain)

Of course, a bicycle isn't much help if you can't turn it, speed it up, or slow it down, all of which require that the angular momentum vector must be changed. In order to change a linear momentum vector, $\vec{p}$, a force, $\vec{F}$, is required. Similarly, changing an angular momentum vector, $\vec{L}$, requires a torque, $\vec{\tau}$. Without concerning ourselves with the direction of the torque vector, recall that the torque equation is $\tau = F_\perp r$, where, like $\vec{v}_\perp$ being perpendicular to $\vec{r}$ in the angular momentum equation (Fig. 6.14), $\vec{F}_\perp$ is perpendicular to $\vec{r}$ in the torque equation. To slow the bicycle, a force is applied by the brakes on the rim (or on the brake disk if so equipped), which results in a torque since the force is applied a distance r from the rotation axis. To speed up the bicycle a force is applied to the pedals away from the rotation axis of the pedals, thus producing a torque, which in turn drives the chain that produces a torque on the rear wheel. Turning requires leaning, causing gravity to exert the force and resulting torque, as demonstrated by Einstein in Fig. 10.18 (note that the angular momentum vector is tilted downward). In all cases, a change in the angular momentum vector requires a torque. Ultimately, total conservation of angular momentum is preserved even with a change in the angular momentum of the bike, because the angular momentum of the planet changes as well, and by the same amount but in the opposite direction (which of course has very little effect given the vastly different amounts of angular momentum in the bicycle and the rotating planet).

---

[1]Verne, Jules (1865). *From the Earth to the Moon*, Directed by Georges Méliès, [Short film], Star Film Company, France.

So what does the physics of riding a bicycle have to do with astronomy? A great deal. Conservation of angular momentum and torque will show up many times in the remainder of this chapter, and in future chapters.

The Constantly Changing Earth–Moon System

In Example 6.1, we considered how much the angular momentum of the Moon's orbit has changed over time. It is likely that at one time the Moon was orbiting just 25 R_{Earth} from the center of Earth and had an orbital period of 7 1/3 days, but it now orbits 60.3 R_{Earth} from Earth's center with an orbital period of 27 1/3 days. As a result, the Moon's orbital angular momentum is 50% greater today than it was long ago. (Don't forget that the Moon's orbital speed, $v_\perp$, decreased while it moved farther from Earth.)

The Moon's slow movement away from Earth continues today. When the Apollo astronauts explored the Moon between 1969 and 1972, a mirror was left on the lunar surface. Using powerful lasers, scientists have been able to send light pulses to the Moon, reflect them off the mirror and time how long it takes for the light to make the round trip back to Earth. Given the definition of the speed of light, a very precise determination of the Moon's distance over time can be made. From those measurements, we know that the Moon is moving away at a rate of 3.8 cm per year (1.5 in/y). This is roughly the speed at which fingernails grow.

As discussed earlier, the Moon also keeps the same side facing Earth throughout its orbit. This means that the Moon actually rotates exactly once on its axis for every orbit, as illustrated in Fig. 10.19. Astronomers refer to this behavior as synchronous rotation.

Perhaps you are aware that Earth's spin is slowing. The rate is extremely small by human standards, just 0.0016 seconds every century, or 16 microseconds per year. Despite this slow change, it is measurable. In fact, every once in a while you may hear on the news that atomic clocks must be adjusted. This is not because the clocks are inaccurate, it is because Earth itself is! Using Earth's rotation to measure time is significantly less precise than other means of measurement. However, it is easier to adjust our clocks than Earth's rotation rate.

We know today that (a) the Moon's rotation period is synchronous with its orbital period, (b) the Moon is slowly moving away from Earth, and (c) Earth's rotation rate is slowing. All of these facts are tied together, and in each case angular momentum is involved. Given that the lunar orbit is increasing in radius and Earth's rotation rate is slowing, the angular momentum of each body is changing, implying that there must be torques being applied.

By treating Earth and the Moon together, the changing angular momentum can be thought of as simply being exchanged between the rotational and orbital angular momentum for the combined system. So where is the torque that makes this happen? The answer is tides. Figure 10.20 shows the same dock in the Bay of Fundy, at Alma, New Brunswick, Canada during high tide and low tide. The ocean's waters have shifted toward the Bay of Fundy during high tide and away from the bay during low tide.

As Sir Isaac Newton first taught us, according to his universal law of gravitation, $F_{gravity} = GMm/r^2$, tides are produced because the force of gravity decreases with the square of increasing distance (the inverse square law), and one side of Earth is

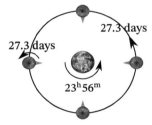

Fig. 10.19 The Moon's orbit around Earth, as seen from above the North Pole. The Moon orbits counterclockwise and both Earth and the Moon rotate counterclockwise as well. The Moon's orbital period equals its rotation period. The red "tent" marks the side of the Moon that is always facing Earth. (Moon: NASA/GSFC/Arizona State University; Earth: NASA/GSFC)

Fig. 10.20 *Left:* High tide and *Right:* low tide in the Bay of Fundy, at Alma, New Brunswick, Canada on June 29, 2019 when the phase of the Moon was waning crescent. The water in the right-hand photograph is the Salmon River flowing into the bay, which has retreated more than 0.5 km out of the frame to the right. (Dale A. Ostlie)

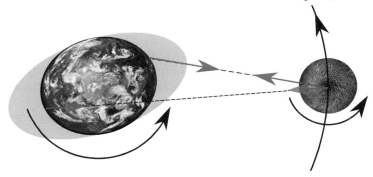

Fig. 10.21 The force of gravity due to the Moon pulling on Earth is equal and opposite to the force of gravity of Earth pulling on the Moon. The blue arrows are larger than the red arrows, signifying that the forces on the near-side bulge are greater than on the far-side bulge. The elongation of Earth and its ocean bulges are very highly exaggerated, and distances and sizes are not to scale. (Moon: NASA/GSFC/Arizona State University; Earth: NASA/GSFC)

closer to the Moon than the other side is. As a result, the side of Earth that is closest to the Moon is pulled toward the Moon with a greater force than the side away from the Moon. Just because the side of Earth farthest from the Moon is hidden from the Moon doesn't keep it from being affected by the Moon's gravitational pull. This difference in forces on the two sides of Earth causes Earth to become a bit elongated, with the oceans being most affected; hence the tides. But even the ground itself closest to the Moon is pulled up a bit more relative to the ground on the opposite side of Earth farthest from the Moon.

As depicted in Fig. 10.21, another factor also comes into play. Because Earth completes one rotation on its axis in just over $23^h 56^m$ while the Moon takes more than 27 days to complete an orbit, the tidal bulges that the Moon's gravity creates are slightly ahead of the line between the centers of the two bodies. This is caused by friction in and on the surface of Earth; the frictional force drags the bulges forward,

Key Point
Earth's tidal bulges are slightly ahead of the line between the center of Earth and the center of the Moon.

which is then balanced by the Moon trying to pull them back in line. If you have spent any time near an ocean you might have realized that high tide tends to arrive shortly before the Moon is directly overhead, although this can be affected by winds and the geography of the coastline. If you study Fig. 10.21, you will see that there are actually two high tides every 24 hours 53 minutes, one a little before the Moon is overhead, and one about 12 hours 26 minutes later. (Remember that the extra 53 minutes is due to the motion of the Moon in its orbit, implying that Earth has to spin more than 360° to bring the Moon overhead twice in a row; Fig. 4.17.)

Because the tides lead the line between the center of Earth and the center of the Moon, the equal and opposite forces between the Moon and the closest bulge are off-axis, and the same applies for the farthest bulge as well. Forces applied off-axis result in torques that affect angular momentum. The bulge closest to the Moon is being pulled back toward the Moon with a greater force than is the bulge on the far side, thus producing a greater torque than the other bulge. This means that the Moon is slowing Earth's rotation.

But something else is happening at the same time. Since for every force there is an equal and opposite force (Newton's third law, page 164*ff*), Earth's tidal bulge closest to the Moon is pulling the Moon forward in its orbit to a greater extent than the bulge farthest from the Moon is trying to pull it back. While the Moon affects Earth's rotation, Earth simultaneously affects the Moon's orbit. The result is that the rotational angular momentum of Earth is being transferred to the orbital angular momentum of the Moon, causing the Moon to drift away from Earth while Earth's rotation slows. Although the angular momentum of each body is changing, their total angular momentum remains constant as required by the conservation of angular momentum.

So how does all of this relate to the Moon being in synchronous rotation with its orbital period? Tides don't just happen to water, they occur within the planet and the Moon, meaning that rock gets pulled and deformed. The tidal bulges on the surface of Earth amount to about 10 cm, so you rise 10 cm every time one of the two bulges passes under your feet. However, since the bulges are very broad, you don't notice anything. The tidal bulges on the Moon are about 20 m since Earth is much more massive than the Moon is. Just as Earth's rotation is slowing today due to torques produced by forces acting on the tides, the Moon was once spinning faster than it is now and its rotation slowed for the same reason. Eventually the Moon's rotation slowed enough that it now keeps the same face toward Earth all of the time. One consequence of this natural process is that the heavier side of the Moon is facing Earth; essentially the Moon is "hanging down" toward Earth. Even today, the Moon's spin is gradually slowing precisely because its orbital period is decreasing as it moves farther from Earth. This slowing maintains the "Man in the Moon" facing Earth.

The constant deforming of the interior of Earth and the Moon has an additional implication, namely the generation of heat. Earth leaks heat through its surface and some of that heat comes from these tidal effects.

Finally, it is important to note that the Moon is not the only object in the Solar System that produces tides on Earth; all of the other masses in the Solar System do too, but the only other major producer of tides is the Sun. When the Sun and the Moon are aligned, either at full moon or at new moon (recall Fig. 4.16), their combined influence generates larger-than-average tides, known as spring tides;

Key Points
• Tides cause Earth's rotation rate to slow while simultaneously moving the Moon farther away.
• Tides occur in rock as well as in water.

Key Point
Some of Earth's internal heat is generated by tidal friction.

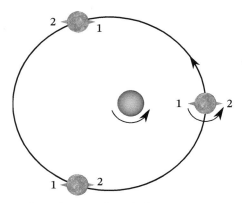

Fig. 10.22 Mercury rotates three times on its axis for every two times it orbits the Sun. The red "tents" show the orientation of Mercury beginning at perihelion (on the right side of the orbit) during the first orbit and the green "tents" show the orientation beginning with the second orbit. At the end of the second orbit, Mercury is back at perihelion again and oriented according to the red "tent." Mercury represents another example of tidal forces producing simple integer ratios of rotation and orbital period. (Sun: NASA/SDO; Mercury: NASA/Johns Hopkins University Applied Physics Laboratory/Carnegie Institution of Washington)

the ocean "springs forth." But when they are perpendicular to each other relative to the center of Earth, either during first quarter or third quarter, they partially cancel each other's influence; these smaller-than-average tides are called neap tides. When the photographs in Fig. 10.20 were taken at the Bay of Fundy on June 29, 2019, the phase of the Moon was waning crescent, three days before new moon; as a result, the tidal variations were about half-way between neap tide and spring tide.

In the distant future, Earth will finally slow to the point where it will keep one side always facing the Moon, just as the Moon keeps its "face" toward Earth. Calculations indicate that this will happen when the Earth day becomes about 47 current days long. When that happens, if the planet and humanity still exist, should a couple want to take a romantic moonlit walk they may need to travel half way around the world to do so.

A bit more complicated is the situation for Mercury in its orbit around the Sun. In this case it isn't a moon that is the culprit, but it is the Sun itself. From Table 10.1, Mercury's orbital period is 0.24 years, or more accurately, 87.97 days. Its rotation period is 58.65 days. By dividing Mercury's rotation period by its orbital period we find that $58.65/87.97 = 0.6667 = 2/3$. In other words, Mercury's rotation period is exactly two-thirds of its orbital period, implying that Mercury rotates on its axis three times for every two orbits around the Sun. This odd coupling of rotation and orbital period (spin–orbit coupling) means that at any given spot on its surface, the Mercurian day–night cycle is 176 Earth days long.

Mercury also has the largest orbital eccentricity (e) of any of the eight planets in the Solar System: $e = 0.21$. It was the advance of Mercury's perihelion that Einstein successfully explained with his general theory of relativity (page 269ff). As Fig. 10.22 shows, starting at perihelion, after 1/3 of an orbit, Mercury has completed one-half of a full rotation, after 2/3 of an orbit it has finished one rotation, and when it returns to perihelion to complete its first orbit, Mercury has completed

Key Points
• Mercury rotates on its axis three times for every two orbits around the Sun.
• Mercury's day–night cycle is 176 days long.

1 1/2 rotations. This means that the side facing the Sun at the start of the first orbit is facing away from the Sun at the end of the first orbit. It takes two orbits and three rotations to bring that face back toward the Sun at perihelion. This three-to-two relationship was forced because of Mercury's close proximity to the Sun and its very eccentric orbit. After all, it is when Mercury is closest to the Sun at perihelion that the Sun is able to exert the greatest gravitational force and the most significant tides.

? Exercises

All of the exercises are designed to further develop *your understanding* of the material by thinking carefully and critically about what you have read in this chapter. Answers to selected exercises are available in the back of the book.

True/False

1. The most massive planet in the inner Solar System is Earth.
2. (*Answer*) The only planets in the Solar System without any moons are Mercury and Venus.
3. The planet with the largest diameter in the Solar System is Venus.
4. (*Answer*) Mercury's rotation period exactly equals its orbital period.
5. Because Jupiter is almost entirely gaseous, if you could wrap it up and place it in a gigantic bathtub, it would float.
6. (*Answer*) Jupiter has more mass than all of the remaining planets in the Solar System combined.
7. All of the planets in the Solar System orbit in a prograde direction.
8. (*Answer*) Four of the planets in the Solar System rotate retrograde.
9. Mercury was named for the Roman god of travel.
10. (*Answer*) Jupiter and Saturn are dominated by gases, but Uranus and Neptune are dominated by ices.
11. Our Moon is the largest moon in the Solar System.
12. (*Answer*) The asteroid belt is so densely populated that it is challenging to navigate through it.
13. The largest object in the asteroid belt is the dwarf planet, Ceres.
14. (*Answer*) Ceres is roughly the diameter of the state of Texas.
15. Pluto is the ninth planet in the Solar System.
16. (*Answer*) Objects orbiting beyond the orbit of Pluto are known as plutoniums.
17. The most massive dwarf planet in the Solar System is Pluto.
18. (*Answer*) All of the dwarf planets rotate prograde.
19. The scattered disk is believed to be the source of all of the long-period comets.
20. (*Answer*) If the Sun were the size of a basketball (23 cm in diameter), Jupiter's diameter would be about 2.4 cm, the diameter of one nickel.
21. The rotation of Earth is slowing.
22. (*Answer*) The rotation of the Moon is speeding up.
23. Tides are the result of the force of gravity being different on the side of Earth closest to the Moon, compared to the side farthest from the Moon.

Multiple Choice

24. (*Answer*) Which of the following are *not* characteristics of the rocky planets in the inner Solar System?
    (a) They are the four smallest planets in the Solar System.
    (b) Their densities all exceed 3000 kg/m^3.
    (c) They all have moons.
    (d) They all have rotation periods longer than 23 hours.
    (e) All of the above are characteristics of the inner Solar System planets
    (f) (a) and (c) only
    (g) (b) and (d) only
    (h) none of the above
25. The order of the masses of the planets in the inner Solar System from lowest to highest is
    (a) Mercury, Venus, Earth, Mars.
    (b) Mercury, Mars, Earth, Venus.
    (c) Mercury, Mars, Venus, Earth.
    (d) Mars, Earth, Venus, Mercury.
    (e) Mars, Mercury, Venus, Earth.
    (f) Earth, Venus, Mars, Mercury.
26. (*Answer*) You pick up a shiny metal object that has a mass of 3.50 kg and you want to determine its density. Recalling a story about the ancient Greek scientist Archimedes and the king's crown, you have a eureka moment: you place the object in a container of water and find out that the object displaces 1.81×10^{-4} m^3 of water, meaning that the volume of the object must be 1.81×10^{-4} m^3. (This volume is equal to 0.181 liters or 0.383 pints.) The density of the object is about
    (a) 19,300 kg/m^3. (b) 67,550 kg/m^3.
    (c) 1930 m^3/kg. (d) 6750 m^3/kg.
    (e) 5.17×10^{-5} kg/m^3. (f) 5.17×10^{-5} m^3/kg.
27. Which planet in the Solar System has the most eccentric orbit?
    (a) Mercury (b) Venus (c) Earth (d) Mars
    (e) Jupiter (f) Saturn (g) Uranus (h) Neptune
28. (*Answer*) Which planet in the Solar System has the lowest density?
    (a) Mercury (b) Venus (c) Earth (d) Mars
    (e) Jupiter (f) Saturn (g) Uranus (h) Neptune

29. Which of the following are properties of the giant planets?
    (a) They are all at least ten times the mass of Earth.
    (b) They all have diameters that are at least ten times the diameter of Earth.
    (c) They all have densities that are less than the density of rocks found on Earth's surface.
    (d) They all rotate more slowly than Earth.
    (e) They all have ring systems.
    (f) They all have at least 10 moons.
    (g) They are all located beyond the main asteroid belt.
    (h) all of the above
    (i) (a), (b), (c), and (d) only
    (j) all except (b) and (d)
    (k) none of the above

30. (*Answer*) Which planet(s) in the Solar System rotate retrograde?
    (a) Mercury   (b) Venus   (c) Earth
    (d) Mars   (e) Jupiter   (f) Saturn
    (g) Uranus   (h) Neptune
    (i) none of the above   (j) (c) and (f)
    (k) (b) and (g)

31. In Greek mythology, Jupiter corresponds to the god
    (a) Hermes.   (b) Aphrodite.   (c) Ares.   (d) Zeus.
    (e) Cronus.   (f) Poseidon.   (g) Helios.

32. (*Answer*) Approximately how many Earths could fit inside the volume of Uranus?
    (a) 2   (b) 4   (c) 8   (d) 16   (e) 32
    (f) 64   (g) 128   (h) 256   (i) 512   (j) 1024

33. Which moon(s) is(are) larger than Earth's Moon?
    (a) Io   (b) Europa   (c) Ganymede
    (d) Callisto   (e) Titan   (f) all of the above
    (g) none of the above   (h) (a) and (b)   (i) (c), (d), and (e)

34. (*Answer*) According to the International Astronomical Union, dwarf planets are not considered to be planets because they
    (a) aren't round in shape.
    (b) don't have moons.
    (c) don't always orbit the Sun.
    (d) haven't cleared out their orbits.
    (e) all of the above
    (f) none of the above

35. The Kuiper belt is located between _____ and _____ from the Sun.
    (a) 2 au, 5 au   (b) 5 au, 39 au
    (c) 30 au, 50 au   (d) 200 au, 1000 au

36. (*Answer*) The Oort cloud may extend from _____ to more than _____ from the Sun.
    (a) 30 au, 50 au
    (b) 200 au, 1000 au
    (c) 2000 au, 50,000 au
    (d) It is impossible to know because no one has ever seen it.

37. Estimates place the comet population of the Oort cloud at
    (a) two hundred thousand
    (b) two hundred million
    (c) two hundred billion
    (d) two hundred trillion
    (e) It is impossible to know because no one has ever seen it.

38. (*Answer*) The age of the Solar System is
    (a) 10,000 years
    (b) 4.57 million years
    (c) 1.32 billion years

(d) 4.57 billion years
(e) 13.8 billion years
(f) It impossible to know because no humans were around to witness its birth.

39. Meteorites can originate from
    (a) the Moon or Mars.
    (b) broken pieces of asteroids.
    (c) remnants of disintegrating comets.
    (d) matter that directly condensed out of the solar protoplanetary disk.
    (e) all of the above

40. (*Answer*) If the Sun was scaled down to the size of a 65-cm diameter exercise ball, Earth would be located _____ meters away and would be roughly the size of _____.
    (a) 27, a baseball
    (b) 70, 60% of the diameter of a AAA battery
    (c) 365, a grain of sand
    (d) 1340, the thickness of two dimes

41. Vesta is a large asteroid with a semimajor axis of 2.36 au. In our scale model of the Solar System, how far from the Sun would Vesta be located? Recall that Earth is 1 au from the Sun.
    (a) 165 m   (b) 236 m   (c) 1.65 km   (d) 2.36 km

42. (*Answer*) If the Moon were twice as far from Earth as it is now, tides would be
    (a) larger than they currently are.
    (b) less than they are today.
    (c) the same as they are today.

43. In the distant future, when the Moon's orbit is 47 days long, what will its semimajor axis be in units of the radius of Earth?
    Hint: By dividing Kepler's third law (page 174) for the Earth–Moon system in the future by Kepler's third law today, the masses cancel and you are left with

$$\left(\frac{P_{\text{future}}}{P_{\text{today}}}\right)^2 = \left(\frac{a_{\text{future}}}{a_{\text{today}}}\right)^3,$$

which can be solved for a_{future}, giving

$$a_{\text{future}} = a_{\text{today}} \times \left(\frac{P_{\text{future}}}{P_{\text{today}}}\right)^{2/3}.$$

    (a) 25 R_{Earth} (b) 36 R_{Earth} (c) 47 R_{Earth} (d) 66 R_{Earth}
    (e) 87 R_{Earth}

Short Answer

44. (a) The distance that the Moon travels during each orbit is approximately equal to $2\pi r$, the circumference of a circle, with r equal to the Moon's semimajor axis, $a_{\text{Moon}} = 60.3\ R_{\text{Earth}} = 384,000$ km. The Moon's orbital period is 27.3 d, or 2.36×10^6 s. What is the Moon's speed in its orbit around Earth in km/s? Remember that speed is distance traveled divided by the time required; in this case $v_{\text{Moon}} = 2\pi a_{\text{Moon}} / P_{\text{Moon}}$.
    (b) At one time in the distant past, the Moon was orbiting 25 R_{Earth} from the center of Earth. Using Kepler's third law, what was the Moon's orbital period?
    Hint: An approach similar to the one in Exercise 43 can be used here, replacing future values with past values and solving for P_{past}.

Exercise 44 continues on the next page

44. *Continued*

(c) What was the Moon's orbital speed in km/s around Earth when it was only 25 R_{Earth} from Earth's center?

(d) The orbital angular momentum of the Moon is $L_{Moon} = m_{Moon} v_{Moon} a_{Moon}$. If the Moon's mass was the same in the past as it is today, we can compare the orbital angular momentum in the past to the orbital angular momentum today by calculating

$$\frac{L_{past}}{L_{today}} = \frac{v_{past} \, a_{past}}{v_{today} \, a_{today}}$$

because the Moon's mass simply cancels out. If the orbital angular momentum stayed constant, the result of the calculation would be 1. What does the result actually turn out to be when the Moon was 25 R_{Earth} away from Earth's center compared to today when it is 60.3 R_{Earth} away? Has orbital angular momentum been conserved? if not, why not?

45. Jupiter has a profound influence on the orbits of asteroids in the asteroid belt. For example, gaps exist in the asteroid belt at specific ratios of orbital periods with Jupiter. One of these gaps exists where the orbital period is 1/3 the orbital period of Jupiter. Using Kepler's third law, determine the distance of the gap from the Sun (in au).

46. Phobos has an orbital period of $7^h 39^m$ around Mars. With the aid of a diagram explain why tidal forces are causing Phobos to spiral in toward Mars.

Activities

47. In Exercise 26 you calculated the density of a mysterious shiny object.

(a) Using a web search engine or some other source, look up and record the densities of the following metals: aluminum, nickel, copper, zinc, silver, and gold.

(b) What was your mystery metal?

(c) Look up the current value of your mystery metal. How much is it worth?

The Rocky Planets and Our Moon

11

*There is but one Earth, tiny and fragile, and one must get 100,000 miles away
to appreciate fully one's good fortune in living on it.*

Michael Collins (1930–2021)

11.1	Mercury	380
11.2	Venus	386
11.3	Earth	394
11.4	The Moon	405
11.5	Mars	408
11.6	Their Interiors	421
11.7	Population Growth and the Exponential Function	431
11.8	Global Warming: Humans Impacting Environment	437
	Exercises	454

Fig. 11.1 Montage. (NASA/Johns Hopkins University Applied Physics Laboratory/Carnegie Institution of Washington, NASA/JPL, NOAA/NASA GOES Project, NASA/GSFC/Arizona State University, NASA/JPL)

Introduction

Although similar in many respects, at least when compared to the giant planets (recall Table 10.1), the rocky planets differ from each other in important ways, making each of the worlds truly unique in the Solar System. As noted in Table 11.1, Mercury, without an atmosphere to protect it, is heavily cratered, having endured countless impacts by micrometeorites, meteorites, asteroids, and comets over its roughly 4.54 billion year history. Venus has an oppressively dense and hot atmosphere at its surface, composed almost entirely of the greenhouse gas carbon dioxide. There is also abundant evidence of volcanic activity in its recent past and the possibility that it once possessed oceans of water. Earth, of course, has a very different atmosphere, that is composed primarily of nitrogen and oxygen molecules with a steadily rising contribution of carbon dioxide. Our home world is also rich in liquid water and ice, experiences significant weathering, and has an active geology. Finally, the atmosphere of Mars is also dominated by carbon dioxide, but its atmosphere is very thin. It has certainly had some volcanism in its history, has a significant amount of water ice below its surface, and may occasionally have liquid water on its surface.

11.1 Mercury

The Mercurian Atmosphere

Consider Mercury, seen in Fig. 11.2: it is only about 0.4 au from the Sun, it has an average global surface temperature of 440 K, and it has virtually no atmosphere to speak of. The gases that are present are temporary and the result of solar wind ions trapped in Mercury's very weak magnetic field along with gases being blasted off the surface by the solar wind and micrometeorites.

Fig. 11.2 Mercury as seen by the MESSENGER spacecraft. (NASA/Johns Hopkins University Applied Physics Laboratory/Carnegie Institution of Washington)

Table 11.1 Characteristics of the rocky planets

	Mercury	Venus	Earth	Mars
Mass (relative to Earth)	0.06	0.81	1.00	0.11
Radius (relative to Earth)	0.38	0.95	1.00	0.53
Semimajor axis (au)	0.39	0.72	1.00	1.52
Rotation period (d)	58.65	243.7	0.997	1.03
Rotation axis tilt to orbit (deg[d])	0.03	177.36	23.44	25.19
Blackbody temperature (K)	440[a]	227[b]	254	210
Average surface temperature (K)	440[a]	737	288	210
Surface pressure (relative to Earth[c])	$< 5 \times 10^{-15}$	92	1.00	0.006
Average density (kg/m^3)	5427	5243	5514	3933
Acceleration due to gravity (relative to Earth[c])	0.38	0.91	1.00	0.38
Escape speed (relative to Earth[c])	0.38	0.93	1.00	0.45
Atmospheric composition (% by volume)				
CO_2	trace	96.5	0.042[e]	95.1
N_2	trace	3.5	78.1	2.59
O_2	16	—	21.0	0.16
Ar	trace	0.007	0.93	1.94
H_2O	trace[f]	0.002	1[g]	0.021
Magnetic field (relative to Earth[c])	0.007	0	1.00	0
Magnetic field tilt to rotation axis (deg)	0.0	—	11.2	—

Note: Some data from Table 10.1 are reproduced here for convenience.

[a]Mercury's surface temperature can reach 725 K on the sunward side.

[b]Venus's blackbody temperature is low because so much sunlight is reflected off its clouds.

[c]For Earth: $P_{Earth} = 1.013 \times 10^5$ N/m^2, $g = 9.80$ m/s^2, $v_{esc} = 11.19$ km/s, magnetic field = 0.0306 mT.

[d]A tilt greater than 90° means that the planet rotates retrograde.

[e]Record high as of April, 2022; equivalent to 420 ppm (parts per million).

[f]Other constituents in Mercury's tenuous atmosphere: Na: 42%, Mg: 39%.

[g]Earth's atmospheric water content is highly variable and climate dependent.

Mercury lacks a substantial atmosphere because (a) its surface is very hot, meaning that any atoms or ions are moving very rapidly, (b) its gravity is fairly weak, leading to a low escape speed, and (c) it has a very weak magnetic field that doesn't effectively protect atmospheric particles from being hit by solar wind particles (recall that charged particles get trapped around magnetic field lines; Fig. 9.14). As described by the equation for the average speed of gas particles on page 339,

$$v_{\text{average}} = \sqrt{\frac{3k_B T}{m}} = \text{constant} \times \frac{\sqrt{T}}{\sqrt{m}},$$

the speed of gas particles increases with the square root of the temperature (in

Key Point
Mercury lacks an appreciable atmosphere because of its
- high surface temperature,
- low escape speed, and
- very weak magnetic field.

Fig. 11.3 A view of Mercury's cratered surface; July 17, 2012. (NASA/Johns Hopkins University Applied Physics Laboratory/Carnegie Institution of Washington)

kelvins) and decreases with the square root of the mass of the particle. On the sunward side of Mercury, the temperature reaches 725 K. For hydrogen atoms at 725 K, the average speed is approximately 4.2 km/s, very close to the escape speed of Mercury which is 4.3 km/s. Since some particles move faster than the average speed, and some move more slowly, even with an average speed that doesn't exceed the escape speed, the gas can eventually leak away (this is true for heavier, slower moving atoms or ions as well).

Because there is not a significant atmosphere to help retain heat or circulate it around the planet, and because the length of the day–night cycle on Mercury is so long (176 d), not only does the side of Mercury facing the Sun along its equator reach 725 K (over 452°C; 845°F), but during the long night on the opposite side of the planet the temperature drops to a very frigid 93 K (−180°C; −292°F).

Cratering on the Surface of Mercury

With almost no atmosphere to protect it, the surface of Mercury is subjected to constant bombardment of micrometeors, meteors, asteroids, and comets (Fig. 11.3). Over time this violent history has left a surface that is heavily cratered and covered with a fine, dusty layer of debris called regolith. As we will see throughout the remainder of Part II, impact cratering is a common theme across the Solar System, including on Earth.

Figure 11.4 (*Left*) shows Donne Crater, with a peak at its center. For craters of significant size, material is ejected from the impact site, but some of the surface is also liquefied (the rock becomes molten). When the molten rock rebounds after the compression, a peak forms that then solidifies. The process is similar to what happens when a drop of water strikes the surface of a pool of water, as shown in Fig. 11.4 (*Center*).

Figure 11.4 (*Center*) also shows waves moving outward from the center of the water drop's impact. When large impacts occur on the rocky surface of a planet, waves are generated in that situation as well. The result is that larger craters are often found to be double-rimmed, like the one in Fig. 11.4 (*Right*). Central peaks

Fig. 11.4 *Left:* Donne crater on Mercury with a well-defined peak at its center. The crater was named for the English poet, John Donne (1572–1631). *Center:* The splash immediately after a water drop struck the surface of the pool of water. (Luis nunes alberto; CC BY-SA 4.0) *Right:* A double-rim crater. (*Left* and *Right:* NASA/Johns Hopkins University Applied Physics Laboratory/Carnegie Institution of Washington)

tend to be absent in these larger craters though, probably because lava from below flowed into them and covered any central peak in the process.

The largest impact crater on Mercury is Caloris Basin (Fig. 11.5). Caloris Basin also dominates the upper left-hand quadrant of the image in Fig. 11.2. It gets its name from the fact that it is one of the two locations on Mercury that face the Sun at perihelion (you can think of Caloris Basin as being located on the side of Mercury represented by the red "tent" in Fig. 10.22). The name *Caloris* comes from the Greek word for heat (you may recognize a similarity to the word calorie, which is a unit of heat energy often associated with food). The impact that created Caloris Basin was so enormous that the waves that were generated traveled all the way to the other side of the planet and converged in a chaotic collection of hills. By studying the large number of craters that have been created on top of the original crater, it appears that Caloris Basin must be relatively old. After all, the craters on top must have formed after the basin.

Without an appreciable atmosphere on Mercury, weathering is absent on the surface. But the surface is still being altered by other processes, the most obvious being further cratering. Some clearly younger craters are evident on the planet's surface, such as Kuiper crater in Fig. 11.6 (*Left*). This image shows streaks, called rays, that resulted from ejected material raining back down on the surface after the impact. Older craters don't typically reveal ejecta rays, simply because the rays have been obliterated by many later impacts, which are typically smaller in size.

Fig. 11.5 Caloris Basin as seen by Mariner 10 during its initial flyby in 1974. The basin is 1300 km (800 miles) in diameter. Concentric waves are still evident. (NASA/JPL)

The Surface Has Been Influenced by Internal Cooling

In addition to cratering affecting the surface, internal influences are apparent as well. When Mercury formed it was very hot, and it has been cooling ever since. As the planet cools it shrinks slightly, which in turn puts stresses on the surface. There is evidence across the surface of the planet of features that were created as a result, such as the towering cliff in Fig. 11.6 (*Right*). Careful examination of Fig. 11.4 (*Left*) also shows that the floor and walls of Donne Crater have been similarly modified by the global shrinkage of Mercury. Crumpled ridges can also be seen on the planet's surface, likely produced by the same process.

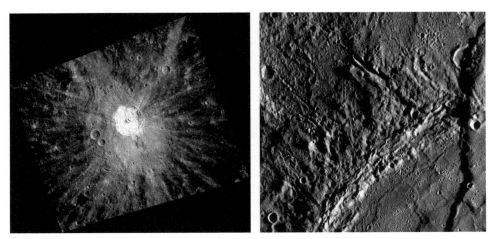

Fig. 11.6 *Left:* The ejecta rays from the relatively young Kuiper crater. *Right:* As Mercury cools, it shrinks slightly, producing towering cliffs. (NASA/Johns Hopkins University Applied Physics Laboratory/Carnegie Institution of Washington)

Lava Flowed in the Past

Another important geologic process that affected the surface of Mercury in the past was large-scale lava flows. These flows refreshed much of the surface, resulting in extensive, relatively smooth regions with few craters. This is because the lava simply covered up older craters. One such region is shown in Fig. 11.7. Notice that the lower left-hand corner of the image shows a crater with only a portion of its rim still visible. Another, much more localized lava flow smoothed out the central crater floor of the double-rim crater in Fig. 11.4 (*Right*).

The Age of Mercury's Surface

Without having landed on the surface of Mercury to directly analyze surface rocks, determining how old the planet's surface is requires an indirect approach. This is accomplished by counting the number of craters in various regions of the planet. The greater the surface density of craters, the older the surface must be, simply because it has been around long enough to experience so many impacts. By that same argument, if an area has fewer impacts over the same surface area relative to a more heavily cratered area, then the area with the fewest craters must be the youngest.

In Section 11.4 we will look at how radiometric dating is used to determine an absolute age for the various rock samples that were returned from the Moon. Based on that information and the density of cratering on the Moon, it is possible to estimate the rate of cratering throughout the rest of the inner Solar System. One of the conclusions that can be drawn from that information is that the inner Solar System went through a period of late heavy bombardment between about 3.5 and 4 billion years ago, or roughly from 500 million years to 1 billion years after the planets formed. It is called *late* heavy bombardment because prior to that time there was a period when impacts had been less frequent.

Fig. 11.7 Lava flows that buried older craters cover large areas of Mercury's surface. (NASA/ Johns Hopkins University Applied Physics Laboratory/Carnegie Institution of Washington)

From studying crater densities on Mercury, it seems that the entire surface was refreshed by an enormous amount of volcanic activity at about the same time that the late heavy bombardment began. It was during this period that areas like the one shown in Fig. 11.7 were formed.

Ice on the Surface

A particularly surprising discovery on the surface of Mercury came from radar studies using the very large Arecibo Observatory,[1] in Puerto Rico. Because of the tidal forces produced by the Sun, that also caused Mercury's three-to-two spin–orbit relationship, the planet's rotation axis is almost exactly perpendicular to its orbit. This means that near the poles sunlight is unable to shine into some parts of deeper craters, leaving them in perpetual darkness. Without an atmosphere to scatter sunlight, or winds to distribute heat across the planet, the temperature can be as low as 50 K (−223°C, −370°F) in those shadows. Earth-based radar observations made of the north polar region seemed to detect the presence of reflective water ice on crater walls. The highly successful MESSENGER orbiter, that was responsible for most of the images of Mercury shown in this text, obtained strong evidence to support the conclusion of water ice (Fig. 11.8). It is likely that the water was delivered to the surface by impacting comets rich in water molecules, and that water ice can survive for billions of years at such low temperatures, even when exposed directly to the vacuum of space.

A Weak Magnetic Field

When NASA's Mariner 10 mission made fly-by observations of the closest planet to the Sun in 1974, it was initially somewhat surprising to discover that Mercury has

---

[1]Unfortunately, the telescope experienced a catastrophic collapse in December, 2020, after having operated for nearly six decades. The category 5 Hurricane Maria in 2017, followed by earthquakes in 2019 and 2020 severely affected the telescope's structural integrity.

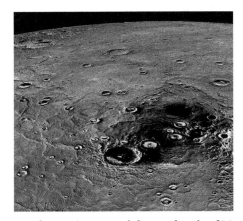

Fig. 11.8 Probable locations of water ice around the north pole of Mercury are highlighted in yellow. (NASA/Johns Hopkins University Applied Physics Laboratory/Carnegie Institution of Washington)

a magnetic field, albeit a very weak one (Table 11.1), a result that was later confirmed by the MESSENGER orbiter. As was discussed in Section 9.3 in relation to the Sun's magnetic field, moving charges are required to produce magnetic fields. It is the circulation of electrons and ions in the Sun's interior, caused by the Sun's rotation and convection, that are the culprits. So where are Mercury's moving charges? Remember that Mercury rotates very slowly. The existence of the planet's magnetic field provides a hint about its interior, to be discussed in Section 11.6.

11.2   Venus

A Thick Atmosphere Hides the Surface

Fig. 11.9 A true color image of Venus. (NASA/Johns Hopkins University Applied Physics Laboratory/ Carnegie Institution of Washington)

The surface of Venus is dramatically different from that of Mercury (see Table 11.1). As is readily apparent from Fig. 11.9, the planet has a very dense atmosphere. The ever-present thick clouds prohibit us from seeing the surface in visible light, even from orbit. In addition, the atmospheric pressure at the surface of Venus is 92 times greater than at the surface of Earth, equivalent to the pressure at a depth of over 900 m in Earth's oceans. It is also very hot at the surface with a temperature of about 740 K (460°C, 870°F), far greater than the average global surface temperature of Mercury, and even exceeds Mercury's sunward-side temperature (725 K). At first blush this may seem odd, given that Venus is 0.72 au from the Sun, almost twice Mercury's distance.

The answer to the surface-temperature paradox rests with the makeup of the atmosphere (Fig. 11.10), which is 96.5% CO_2 (carbon dioxide), with most of the remaining 3.5% being molecular nitrogen (N_2). There are also traces of sulfur dioxide (SO_2), the gas that gives rotten eggs their smell, argon (Ar), neon (Ne), hydrochloric acid (HCl), and carbon monoxide (CO). The clouds that make it impossible to see the surface in visible light are composed primarily of sulfuric acid droplets (H_2SO_4) and some water.

We have learned about this hellish environment from the Venera spacecraft of the former Soviet Union that was able to land on the surface of Venus before quickly succumbing to the intense heat and pressure. By using Earth-based telescopes along with spacecraft orbiting Venus, scientists have also investigated the atmosphere through the analysis of spectral lines from reflected sunlight, and from starlight passing through the thin upper atmosphere when the planet occults a star (Chapter 7).

The Evolution of the Venusian Atmosphere

The next question, of course, is *why* did the atmosphere of Venus end up the way it is today, which is dramatically different from Earth's atmosphere? After all, Venus is only 30% closer to the Sun than Earth is, the two planets are nearly the same size and mass, and they presumably formed out of the solar nebula with essentially the same material at nearly the same time. Despite those similarities, Earth's atmosphere is dominated by nitrogen and oxygen molecules (N_2 and O_2, respectively), with very little carbon dioxide (CO_2). The answer lies in how the atmospheres of the two worlds evolved over time. We'll begin by studying the evolution of the atmosphere of Venus.

When the Sun first formed, its luminosity was only about 2/3 of what it is today. This means that the amount of energy received by each of the planets every second was also about 2/3 of what it is today, and so the planets' surface temperatures were lower as well. As we will see repeatedly, water (H_2O) is a common molecule throughout the Solar System, and because of its lower surface temperature early in its history, Venus also had an abundance of water. In fact, there is growing evidence that Venus likely had liquid water oceans just as Earth does now (see Fig. 11.11), although the planet was quite a bit warmer than Earth was at the time. Venus may have even been habitable, although not necessarily supporting life, during its first two billion years.

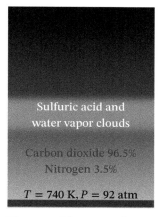

Sulfuric acid and water vapor clouds

Carbon dioxide 96.5%
Nitrogen 3.5%

$T = 740$ K, $P = 92$ atm

Fig. 11.10 The atmosphere of Venus is 96.5% carbon dioxide, with clouds of sulfuric acid and some water.

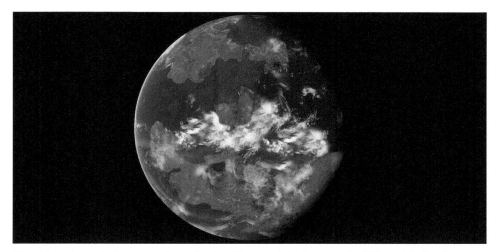

Fig. 11.11 An artist's conception of what Venus might have looked like with liquid water oceans. (NASA)

As the luminosity of the Sun increased over time, the temperature at the surface of Venus also increased. This caused the oceans to evaporate and much more water vapor entered the atmosphere. Water is an important greenhouse gas molecule, along with carbon dioxide and methane (CH_4).

The Greenhouse Effect Strongly Influences the Atmosphere

Recall from our discussion of spectral lines and quantum mechanics that every type of atom and molecule has a unique set of energy levels, corresponding to a unique set of emission and absorption lines that have wavelengths associated with the *differences* in energy between energy levels ($E = hf = hc / \lambda$) (see e.g., Fig. 8.28). It is the unique set of spectral-line "fingerprints" that allows scientists to determine what atoms and molecules are present in a gas.

Greenhouse gas molecules have the characteristic that they are largely transparent to visible light, meaning that they have few combinations of energy levels that allow visible-wavelength photons to be absorbed. However, as illustrated in Fig. 11.12, greenhouse gas molecules like water have an abundance of closely spaced molecular energy levels. In the water molecule example, the atomic energy levels of hydrogen and oxygen are too few and far between to absorb infrared light or visible light effectively, but the molecular energy levels above about −2 eV are a different story. Given their number and spacings, the molecular energy levels effectively make water vapor opaque to infrared light, because the closely spaced molecular energy levels easily absorb lower energy infrared photons. The same holds true for the other greenhouse gas molecules, including carbon dioxide (CO_2) and methane (CH_4).

This behavior of greenhouse gas molecules means that when the Sun brightened and put more water vapor into the atmosphere of Venus, visible-wavelength sunlight could penetrate the atmosphere to reach the surface and warm it. However, the surface also emitted blackbody radiation (photons) at primarily infrared wavelengths in accordance with its temperature as specified by Wien's blackbody color law (page 215),

$$\lambda_{peak} = \frac{2,900,000 \text{ nm K}}{T_{surface}}.$$

Those infrared photons were then immediately absorbed by the atmospheric water molecules [see Fig. 11.13 (*Left*)]. As a result, instead of a balance between sunlight energy striking the surface and blackbody radiation escaping from it out into space, there was more energy entering than leaving. This partial one-way "door" of energy flow caused the surface temperature to rise, leading to even more evaporation of liquid water and a steadily increasing temperature that eventually reached 1800 K, hot enough to vaporize the rest of the liquid water and even melt some rock on the surface. At that time the atmospheric pressure at the surface reached 300 times the present-day value at Earth's surface. Energy balance was restored only when the upper atmosphere radiated away just as much energy back into space as was entering the atmosphere and heating the surface.

It is this process that is famously known as the greenhouse effect. It gets its name because greenhouses work the same way [Fig. 11.13 (*Right*)]. Glass also allows visible light to easily penetrate (otherwise glass wouldn't be of much use

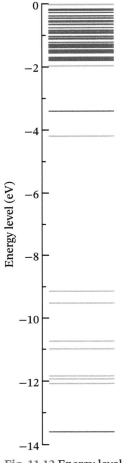

Fig. 11.12 Energy levels [in electron volts (eV)] of a water molecule (H_2O) can absorb and emit infrared photons. The red-colored energy levels are from the bonds holding the water molecule together, the blue-colored energy levels are those of atomic hydrogen, and the green-colored energy levels are those of atomic oxygen.

Fig. 11.13 *Left:* On Venus today, about 10% of the sunlight (shown in yellow) is able to reach the surface of the planet, with the remainder reflected back into space (white). When the heated surface emits infrared light as blackbody photons (indicated in red), those lower energy, longer wavelength photons cannot penetrate through the dense greenhouse gas (carbon dioxide) atmosphere. The trapped energy heats the planet's surface and its lower atmosphere. *Right:* A greenhouse allows visible light in through the glass ceiling and walls, but the heat inside cannot escape back out in the form of blackbody radiation, dominated as it is by infrared photons. (The greenhouse is adapted from Joi Ito; CC BY 2.0)

for windows) and the sunlight heats the interior of the greenhouse. However, the reradiated blackbody photons emitted by the interior are at infrared wavelengths and are blocked by the glass. This same effect is why cars get very hot inside when the windows are closed on sunny summer days. The greenhouse effect is why you must never leave small children or pets inside cars unattended; the temperature can reach deadly levels very quickly.

As we will learn when we study its surface, Venus has an abundance of volcanoes that can emit carbon dioxide into the atmosphere, adding to the carbon dioxide that is already present. CO_2 and N_2 are much heavier molecules than H_2O, which caused them to sink toward the planet's surface while the water molecules "floated" up to the upper atmosphere. This exposed the water molecules to the Sun's ultraviolet light, along with it's visible and infrared wavelengths. The higher energy ultraviolet photons are able to break water molecules apart, producing individual hydrogen atoms and OH molecules: $H_2O + \gamma \rightarrow OH + H$, where γ symbolizes the photon. Ultraviolet photons can also break apart the OH molecule in the same way: $OH + \gamma \rightarrow O + H$. The liberated hydrogen atoms, with their low mass, slowly escape into space, ridding the atmosphere of water vapor and leaving behind the abundant CO_2 we see today. A strong electric field in Venus's atmosphere also effectively lifts heavy ions like charged oxygen and water molecules directly into space. The dramatic evolution of Venus's atmosphere was the direct result of a runaway greenhouse effect.

Today, 90% of the incoming sunlight is reflected off the sulfuric acid droplets and water vapor clouds, a major reason why the planet shines so bright in the sky (Venus is the brightest object in the sky after the Sun and the Moon). The sunlight that does reach the surface is absorbed and reradiated as infrared light which is unable to directly escape through the very dense carbon dioxide atmosphere.

Key Point
Venus is so bright in the sky because 90% of sunlight is reflected off its clouds.

Testing the Hypothesis of Atmospheric Evolution

So how do we know that this particular scenario for the evolution of Venus's atmosphere is correct? Remember that any scientific theory or hypothesis must make testable predictions; otherwise it is just a good story without evidence. In this case, one powerful but subtle prediction has to do with the heavy isotope of hydrogen known as deuterium, ^2_1H; recall Fig. 8.38. Since deuterium is virtually identical chemically to regular hydrogen (^1_1H), both types of atoms can bind with oxygen to form water molecules. Because Venus formed out of the solar nebula (page 359) with essentially the same material that Earth formed from, the relative abundances of deuterium and normal hydrogen should have initially been the same. However, deuterium, with its neutron attached to the proton, is twice as heavy as normal hydrogen, and therefore would move more slowly and have a harder time escaping from Venus's atmosphere. If the normal hydrogen preferentially escapes, a higher abundance of deuterium relative to hydrogen should be left behind. In addition, because of their greater mass, the spectral lines of deuterium atoms differ slightly from those of the more abundant, lighter form of hydrogen, even though the two isotopes are virtually identical, chemically. The different spectral signatures allow scientists to tell the difference between the two types of hydrogen. As it turns out, the ratio of deuterium to hydrogen in the atmosphere of Venus today is indeed one hundred times greater than it is in Earth's atmosphere, just as would be expected from the process described above. A scientific theory must make testable predictions and always agree with *all* relevant data, no matter how subtle.

Retrograde Rotation and the Effect of Tides

Fig. 11.14 An ultraviolet image of the clouds of Venus, enhanced to show their rapid equatorial motions. Obtained by the Galileo spacecraft on February 14, 1990. (NASA/JPL)

Another curious feature of Venus's atmosphere was first detected in the 1960s. As shown in the ultraviolet light view of Venus in Fig. 11.14, the sulfuric acid clouds, which are located between 40 km and 75 km above the surface, travel nearly 100 m/s near the equator and more slowly farther north or south. This causes them to develop a characteristic ➤ shape. Those high velocity wind speeds slow to a gentle breeze at the planet's surface.

More interesting is the fact that the clouds are moving backwards (retrograde) relative to the planet's orbital direction. In other words, when viewed from above the north pole of the Sun, the planet's clouds are moving in a clockwise direction rather than a counterclockwise direction. This was originally detected by measuring the Doppler shifts of spectral lines in its atmosphere [recall that if wavelengths are shorter than they would be in a laboratory then the source is approaching you (blueshifted) and if the wavelengths are longer, then the source is moving away from you (redshifted)]. This hints that the planet itself is rotating retrograde, which was later confirmed by bouncing radar signals off the surface and determining their reflected Doppler shifts (radio wavelengths are able to penetrate the atmosphere even though visible wavelengths cannot).

Venus appears to have developed its retrograde rotation due to gravitational perturbations with the Sun and the other planets. Computer simulations suggest that even though Venus was originally spinning prograde (counterclockwise), the gravitational interactions caused its spin to slow to near zero and then go through a chaotic period where its axis of rotation could tilt between 0° (upright) and 90°

Fig. 11.15 The surface of Venus as seen by the Soviet Union's Venera 13 lander in 1982. A portion of the spacecraft is visible at the bottom. A lens cap is lying on the ground. (Venera 13 Lander, {VG00261,262})

(with its axis lying in the plane of its orbit). The very thick atmosphere also seems to have played an important role because it can be strongly affected by tides from the Sun's gravity and interact with the planet's surface through drag and bumping into the planet's volcanoes. The net outcome is an extremely slow retrograde rotation that takes 244 days to complete. (Measurements indicate that the atmosphere is still affecting Venus's rotation.)

Revealing a Hidden Surface

In the 1970s and 1980s, the former Soviet Union made attempts to land on the surface of Venus. Figure 11.15 shows what the Venera 13 lander saw on March 1, 1982 before it succumbed to the harsh environment after just 2 hours and 7 minutes of operation. It is difficult to know what the true colors of the surface should be, because the atmosphere filters out the blue light. However, various types of rock are evident, including flat, lava-based rocks called basalts, and what are likely ejecta from meteor impacts.

Then, from 1990 to 1994, the United States Magellan spacecraft orbited Venus, gathering most of what we know today about the planet's surface. Rather than trying to land on the surface, Magellan made observations through the dense sulfuric acid and water vapor clouds by bouncing radio-wavelength (radar) signals off the surface. Using this technique, Magellan's radar images are able to resolve features as small as 100 m (approximately the length of an American football field). Although the clouds are opaque to visible light, and the atmosphere effectively absorbs infrared light, Magellan was able to "see" the surface using the long wavelengths of radio waves.

Figure 11.16 shows a mosaic radar map of Venus in false color. The lower regions are darker while higher elevations are light brown and white. Venus has two major continents, Aphrodite Terra and Ishtar Terra. Aphrodite Terra is visible along the equator and is about the size of Africa. Ishtar Terra is about the size of Australia. Although all of the other features on Venus are named for women, the Maxwell mountains on Ishtar Terra are named for James Clerk Maxwell, whose theoretical work on electricity and magnetism made the radar mapping of the planet possible.

A History of Cratering

Despite its very thick atmosphere, Venus has not been immune to impacts capable of producing large-diameter craters, as evident by the three craters visible in

Key Point
The wide range of wavelengths in the electromagnetic spectrum provides "windows" into different environments.

Fig. 11.16 One hemisphere of Venus. The artificial colors denote elevation, with dark blues representing low areas and light brown to white indicating higher elevation. (NASA/JPL/USGS)

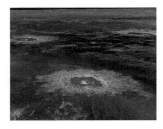

Fig. 11.17 Three large impact craters visible on Lavinia Planitia (Planitia means plain). (NASA/JPL)

Fig. 11.17. Howe crater, in the foreground, measures 37 km (23 mi) in diameter, while Danilova (upper left) and Aglaonice are even larger, with diameters of 47 km (30 mi) and 63 km (39 mi), respectively. The bright colors are indicative of rougher terrain. Characteristic peaks at the centers of the craters can also be seen, just like craters on Mercury. As with all of the other Magellan images that follow, the coloration is artificial and based on the surface colors seen by the Venera landers.

The smaller craters we saw on Mercury are absent on Venus because smaller impactors cannot survive the descent through the atmosphere. In fact, few craters smaller than 10 km in diameter have been detected. Given that the energy of an in-falling meteor is what determines the size of a resulting crater, the absence of craters smaller than 10 km across implies that rocks smaller that 1 km across are probably destroyed when they enter the atmosphere. Even the large impacts are sometimes found in groups, suggesting that very large in-falling meteors can break apart due to the strain of passing through the dense atmosphere before reaching the surface.

By counting the number of craters and comparing that value with the number of impacts expected over time that could produce those craters, it is possible to estimate the age of the planet's surface. This somewhat imprecise method places the surface age at between 300 and 600 million years. Although long by human standards, that age is very short on planetary timescales. The 300–600 million year estimate for the age of the surface does not imply that that is the age of Venus itself, which is closer to 4.54 billion years; instead, the younger age of the surface means that it was refreshed that long ago, erasing previous surface features.

How then was the surface refreshed so recently? We will return to that question again in Section 11.6, when we discuss the interiors of the rocky planets and our Moon.

Key Point
The surface of Venus was refreshed 300–600 million years ago.

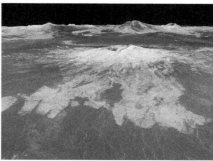

Fig. 11.18 *Left:* Maat Mons towers 5 km (3 mi) above the surrounding terrain. The vertical scale is exaggerated by a factor of 22.5. *Center:* Sapas Mons in the center of the image with Maat Mons seen along the horizon. The vertical scale is exaggerated by a factor of 10. *Right:* A portion of western Eistla Regio. The volcanoes Sif Mons (left) and Gula Mons (right) are visible in the background. (NASA/JPL)

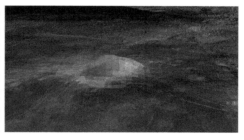

Fig. 11.19 *Left:* Idunn Mons is 200 km (120 mi) in diameter. The vertical scale is exaggerated by a factor 30. *Right:* A thermal scan of Idunn Mons with red being warmest. (NASA/JPL-Caltech/ESA)

A Surface Dramatically Affected by Volcanism

Arguably the most significant discovery made by Magellan is that there is abundant evidence of wide spread volcanism across the surface in the recent past, and it is likely still going on today. Figure 11.18 shows several volcanoes rising up above the surrounding terrain. Although these volcanoes can be very tall (Maat Mons rises 5 km), the images are somewhat misleading because the vertical relief has been exaggerated. For example, Fig. 11.18 (*Left*) is stretched vertically by a factor 22.5 in order to highlight the surface features. This implies that for the height shown, the horizontal axis should be stretched by the same amount in order to get a true picture of the relative dimensions. Said another way, although the volcanoes are tall, they are *very* wide. Sif Mons in the right-hand image actually has a base with a diameter of 300 km (190 mi).

The light colors in the images of Fig. 11.18 are more reflective of radar than are the darker regions, and clearly show extensive lava flows that can stretch for hundreds of kilometers. Venus's volcanoes are of a type known as shield volcanoes. Rather than having explosive eruptions, the lava flows gently down the sides, slowly building up the height and the base of the volcano, resulting in the very wide bases seen on the planet. Kilauea Volcano on Hawai'i's Big Island is an example of a shield volcano.

It is likely that at least a few of the volcanoes are still active today. Figure 11.19 shows Idunn Mons (*Left*) and an overlaid temperature map (*Right*). The map illustrates that the top of the volcano is measurably warmer than regions around its broad base. Other indirect evidence of recent eruptions includes the concentration of sulfur dioxide in the upper atmosphere, that has been measured to change by as much as a factor of 100 over a few decades. Sulfur dioxide is a common product in the eruption of volcanoes on Earth.

Volcanoes aren't the only example of geologic activity on Venus due to the presence of lava. There are locations where the surface has cracked or been pushed upward by forces within the planet. Eistla Regio in Fig. 11.18 (*Right*) is one large area that has been uplifted.

The Lack of a Magnetic Field

Perhaps its very sluggish rotation is the reason that Venus doesn't have a magnetic field. However, since Mercury also rotates very slowly but with a period four times shorter than Venus's, perhaps there is more to the story regarding Venus's interior.

11.3   Earth

Earth's Evolving Atmosphere

Although Earth may seem impressively large to us on a human scale, in reality it is extraordinarily tiny, isolated, and fragile within the cosmos, as dramatically illustrated in Figs. 11.20 and 11.21. Even its atmosphere, which we often take for granted, is easily susceptible to human modification.

Fig. 11.20 "The Day the Earth Smiled;" July 19, 2013. Earth as seen from the Cassini spacecraft orbiting Saturn, with knowing Earthlings waving at the camera! (NASA/JPL-Caltech/Space Science Institute)

Has Earth's atmosphere evolved since the planet was born even without human tampering? Yes, but not in the same way that Venus's atmosphere did. Because Earth is farther from the Sun, its atmosphere did not experience a runaway greenhouse effect caused by evaporating oceans. Instead, much of the carbon dioxide was absorbed into the oceans, producing carbonate rocks such as limestone (Fig. 11.22). If all of the carbon dioxide currently locked up in our oceans and rock was released, Earth would have about the same amount of atmospheric CO_2 as Venus does.

Even so, Earth still experiences a small greenhouse effect due to water molecules and about 420 parts per million (ppm) of carbon dioxide molecules in its atmosphere as of April, 2022 (420 CO_2 molecules for every 1 million atoms or molecules in the atmosphere). Without this greenhouse effect, the average surface temperature of the planet would be below the freezing point of water, which would have prohibited the development of life as we know it. However, the amount of carbon dioxide is rising rapidly today, enhancing the greenhouse effect significantly. Much more will be said about this crucial topic in Section 11.8.

That is not the whole story, however. A process unique to our world within the Solar System (at least as far as we know today) was that the development and evolution of life also contributed to the evolution of our atmosphere. Initially, most of the oxygen was tied up in carbon dioxide and water molecules, but early photosynthetic organisms, such as blue–green algae known as cyanobacteria, and small sea creatures helped to lock up the carbon dioxide in marine sediments through the formation of organic material such as shells (again refer to Fig. 11.22). Cyanobacteria also generated small amounts of oxygen through photosynthesis that added to the then nitrogen-rich atmosphere. Because other forms of life at the time found oxygen to be poisonous, this stage of the evolution of Earth's atmosphere resulted in a mass extinction. Non-life-based processes also contributed oxygen to the atmosphere through the interaction of light, carbon dioxide, and water molecules. Fortunately some of that initial oxygen resulted in the production of ozone (O_3) in the upper atmosphere, protecting the surface from the destructive ultraviolet light from the Sun. It was only then that more complex life was able to develop, adding more and more oxygen to the atmosphere, ultimately resulting in the 78% N_2 and 21% O_2 atmosphere that we enjoy today.

Fig. 11.21 The iconic view of Earthrise over the Moon as seen by the Apollo 8 astronauts on December 24, 1968. (NASA)

Fig. 11.22 Limestone with fossils embedded in the rock. From the Okagi Castle Wall. (Taken by Miya.M in March 2005, CC BY-SA 3.0)

Weather

The weather on Earth is far more complex than it is on any of the other rocky planets, due in large part to its abundance of liquid water, snow, and ice on its surface, and water vapor in its atmosphere. The variations in temperature across its surface and Earth's rotation also contribute in important ways to its weather.

Figure 11.23 shows two of the more dramatic weather events that can occur. The image on the left is of Hurricane Harvey making landfall along the Texas Gulf coast on August 25, 2017. Due to unusually warm waters in the Gulf of Mexico that summer, Hurricane Harvey produced significant damage, not just from its 210 km/h (130 mi/h) wind, but also by producing the most rainfall from a single storm ever to strike the United States. The right-hand image is of a tornado, a much more localized, but often extremely destructive weather event. Hurricanes are able to spawn tornadoes and thunderstorms, although those types of storms can exist independent of hurricanes as well. (Cyclones and typhoons are the same weather phenomena as hurricanes; it is only the location where the storm forms

Fig. 11.23 *Left:* Hurricane Harvey in the Gulf of Mexico on August 25, 2017. Harvey dumped 1.539 m (60.58 in) of rain near Nederland, Texas, 130 km (80 mi) east of Houston. City lights can be seen in this nighttime image. (CIRA) *Right:* A tornado's funnel cloud. (Public domain, CC0 1.0)

that determines what it is called. Hurricanes are tropical storms that form over the North Atlantic Ocean and Northeast Pacific Ocean, typhoons form over the Northwest Pacific Ocean, and cyclones form over the South Pacific and Indian Ocean.)

Earth's 23^h56^m rotation period sets up movements in the atmosphere. This is because of varying surface speed with latitude. Since the circumference of Earth at the equator is $C = 2\pi R_{Earth} = 40,075$ km (24,901 mi), the rotation speed of a point on the equator is 40,075 km/23.93 h = 1675 km/h (1041 mi/h). On the other hand, at either the North or South Pole, the rotation speed is 0, because all you would do at one of those locations is spin around slowly in place. The atmosphere responds to those differing speeds through air movement. The left-hand illustration of Fig. 11.24 shows four "rivers" of air high in the atmosphere, known as jet streams.

Air movement is also set up by warmer, rising air and cooler, sinking air. This can happen fairly locally such as breezes near large bodies of water when the water is warmer than land in the evening. It can also happen on a global scale with warm air rising near the equator and cooler air sinking near the poles.

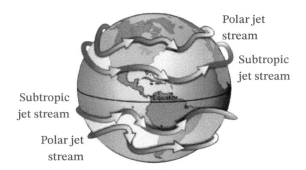
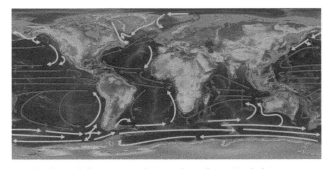

Fig. 11.24 *Left:* Northern and southern hemisphere jet streams. (Adapted from NWS/NOAA) *Right:* Wind-driven ocean currents. The red paths represent warm water being carried away from the equator and blue paths represent cool water being carried toward the equator. (NWS/NOAA)

Fig. 11.25 *Left:* Sunset over the Indian Ocean as seen from the International Space Station. (NASA) *Right:* A view of weather and Earth's thin atmosphere as seen from the Space Shuttle. (NASA/JPL/UCSD/JSC)

The combination of Earth's rotation and rising and sinking air combine to form regions of high and low pressure. When storms develop, it is because air is "falling" into low pressure areas from areas of higher pressure. Rotation, together with temperature and pressure differences, as well as the amount of heat energy and moisture that is present in the atmosphere, drive weather patterns. Land forms such as mountains and bodies of water also contribute to the complex process of weather.

To complicate matters further, currents also transport warm and cold water across the world's oceans, as shown in Fig. 11.24 (*Right*). Since ocean temperatures are not uniform, the amount of water evaporating into the atmosphere and the amount of heat energy that an ocean can feed into a storm vary significantly by location and time of year. Hurricane Harvey [Fig. 11.23 (*Left*)] grew rapidly into a major category 4 hurricane, and Hurricane Marie devastated Puerto Rico as a category 5 hurricane because they crossed over unusually warm waters in the Gulf of Mexico in August and September 2017, respectively.

It is important to note just how thin Earth's atmosphere actually is. The yellows and oranges in Fig. 11.25 (*Left*) are located in the lower 6 to 20 km of the atmosphere. This layer contains about 80% of the atmosphere's mass and almost all of its water vapor, clouds, precipitation and complex weather. The white and pinkish portion extends up to 50 km above the surface, and the blue-to-black layer marks the top of the atmosphere as it fades into space. Figure 11.25 (*Right*) also shows just how thin our fragile atmosphere is. Within that thin blanket of air and the world's waters exist the planet's enormous diversity of plant and animal life, including creatures that developed a certain level of self-awareness and, in many individuals, the curiosity to learn as much as possible about ourselves, the other creatures that live among us, our environment, our planet, and the vast universe beyond.

A Complex and Varied Surface

Earth is a planet with a rich array of surface features, including vast plains, deserts, an abundance of water, canyons, mountains, volcanoes, and craters. Its surface is also constantly changing. It experiences earthquakes, mountain building (such as the still-rising Tetons range in northwest Wyoming, Fig. 11.26), frequent volcanic

Fig. 11.26 The Tetons mountain range in northwest Wyoming is rising, while the Jackson Hole valley sinks. (Little Mountain 5, CC BY-SA 3.0)

Fig. 11.27 Examples of volcanic activity. *Left:* Old Faithful Geyser, Yellowstone National Park, with the Milky Way as a backdrop. (NPS/Jacob W. Frank) *Center:* Augustine Volcano, Cook Inlet, Alaska. (Cyrus Read, USGS, Alaska Volcano Observatory) *Right:* The interaction of lava and the ocean at Hawai'i Volcanoes National Park. (NPS Photo/Janice Wei)

activity (examples are seen in Fig. 11.27), and ongoing impacts by meteorites and, more rarely, asteroids and comets (recall Fig. 2.14). Unlike Mercury with its nearly complete lack of an atmosphere, and dry Venus with low wind speeds at its surface, Earth also experiences continual weathering. Although volumes could (and have been) written about each of these aspects of our home planet, we are focused here on comparing Earth with the other rocky planets.

The feature of Earth that clearly sets it apart from Mercury, Venus, and Mars is the existence of large amounts of liquid water on its surface. Due to its distance from the Sun, and fortunately for our existence, Earth is in what is sometimes called the Goldilocks zone, also known as the habitable zone; not too hot, not too cold, but just right for liquid water to exist.

About 71% of our planet's surface is covered with water, with about 91% of the water being contained in saltwater oceans. The rest of the water is contained in the atmosphere as water vapor, and in lakes, rivers, glaciers, ice caps, snow, underground aquifers, and in living things. Most of the water not contained in the oceans is fresh water, with the exception of a few large inland bodies such as the Dead Sea in Israel, Jordan, and Palestine (Fig. 11.28 *Left*), and the Great Salt Lake in Utah.

As liquid water evaporates into the atmosphere and rains back out again, it is able to erode the landscape; sometimes subtly and sometimes very dramatically. The famous Grand Canyon (Fig. 11.28 *Right*) is an example of deep erosion by the Colorado River over millions of years as the Colorado Plateau was uplifted by forces in Earth's interior.

Areas that see repeated freezing and melting cycles can also affect the landscape. When water seeps into cracks in rock and freezes, the expanding ice can open the cracks a bit more each time. The result can be pieces of rock, small and large, ultimately breaking away from rock slabs, as happened in Zion National Park, Utah (Fig. 11.29). Exactly the same process creates potholes in roadways, sometimes

Fig. 11.28 *Left:* The Dead Sea and evaporation ponds as seen from the International Space Station. (NASA/Chris Hadfield) *Right:* The Grand Canyon and the Colorado River. (NPS Photo, CC BY 2.0)

causing people in northern climes to complain that there are really only two seasons: winter and road repair.

Water, ice, wind, and impacts, along with plant and animal life (most notably humans), affect the surface of Earth. Erosion of various kinds works to tear down mountains and volcanoes over millions of years, but other forces are also continually at play that create new land forms.

Plate Tectonics: The Paradigm of the Geosciences

As early as the late sixteenth century, people had noticed intriguing similarities between the continents, such as the form of the eastern coast of South America and the western coast of Africa. It was almost like they were once connected together, but then separated to form the southern Atlantic Ocean. Other major land masses also look like they may have once fit together like pieces from a jigsaw puzzle.

In 1912, (Fig. 11.30), a polar explorer and meteorologist, put the idea on a more solid footing by studying fossilized plants along matching sides of the continents, and noted remarkable similarities. Then, in 1915, he suggested that perhaps at some point in the past all of the continents were connected in a giant land mass only to begin drifting individually across Earth's surface. Unfortunately, his continental drift theory didn't garner much support at the time.

In what was a decades-long effort, Marie Tharp (Fig. 11.31) and her colleague Bruce C. Heezen, mapped the entire ocean floor. Heezen collected the necessary data aboard a ship owned by Columbia University for the first 18 years of their research, while Tharp used the data to generate maps. Because women were not allowed on board the ship at the time, Tharp was forced to stay ashore. It wasn't until 1965 that she was finally allowed to go on a data-gathering expedition.

Fig. 11.29 A major rockfall in Zion National Park, Utah, in 2016. The massive boulders broke off the cliff face high above the road. (NPS Photo)

Fig. 11.30 Alfred Wegener (1880–1930) during an expedition to Greenland in 1912–1913. (Loewe, Fritz; Georgi, Johannes; Sorge, Ernst; Wegener, Alfred Lothar; Archive of Alfred Wegener Institute; public domain)

Fig. 11.31 Marie Tharp (1920–2006) and Bruce Charles Heezen (1924–1977) studying their map of the ocean floor. (Marie Tharp Maps, CC BY 2.0)

A rise in the mid-Atlantic has been known for about 150 years, following its discovery in 1872 when researchers were investigating possible locations for a trans-Atlantic cable. By 1925, it was determined that the rise stretched the entire north–south length of the Atlantic. But our knowledge of the ocean floor became much more complete in 1957, when Tharp and Heezen published their first map of the North Atlantic. It took another twenty years before the last of their maps was published in 1977, the year of Heezen's death.

Careful research of the ocean floor revealed not only long ridges, underwater volcanoes, and deep trenches, but measurements also determined that the seafloor is spreading away from the Mid-Atlantic Ridge at the rate of about 2.5 cm (1 in) per year. Not only are South America and Africa actually moving apart, but other portions of Earth's surface are moving as well.

Today, the old theory of continental drift has been replaced by the broadly encompassing theory of plate tectonics, in which the surface of Earth is broken up into a collection of moving plates depicted in Fig. 11.32. As these plates move, plate boundaries collide or slide past one another, creating enormous amounts of stress and heat that can lead to earthquakes, volcanoes, deep ocean trenches, and mountain building. A vivid example of this geologic activity is the Ring of Fire, highlighted in red in Fig. 11.32, that stretches from the southern tip of South America, up the west coast of the Americas, along the Aleutian Islands chain of Alaska, down through Japan to the islands of the south Pacific, and ending in New Zealand. The theory of plate tectonics is the overarching theory of terrestrial geology today, much like Newton's laws were the culmination of the Copernican revolution, and marked the beginning of modern physics and astronomy.

Another dynamic example of plate tectonics in action is the ongoing growth of the highest mountain range on Earth, the Himalayas (Fig. 11.33). The Himalayas are the result of the northward migration of the Indian plate (containing the Indian subcontinent) and the southward motion of the enormous Eurasian plate. At the intersection of those two plates the ground buckles, forcing land skyward.

As our understanding of plate tectonics has developed, ongoing research clearly indicates that Wegener's original concept of a supercontinent was correct. From roughly 335 million years ago until 185 million years ago, the supercontinent of Pangaea existed, as depicted in Fig. 11.34. When the supercontinent formed, plants and animals existed, and some species, especially insects, were even able to thrive on the supercontinent. Then, 252 million years ago a mass extinction resulted in the loss of some 96% of marine species and 70% of land-based vertebrates. The Great Dying was the most severe extinction event known today. Following the extinction event, reptiles were able to evolve. When Pangaea began to break apart, fossils of these plants and animals remained which, along with other geologic data, provides evidence across present-day continents that Pangaea existed. Pangaea is just the most recent of many supercontinents that have existed over the eons. Plate tectonics began more than 3.5 billion years ago and has been going on ever since.

The question of what mechanism(s) cause these massive plates to move across Earth's surface demands an explanation. A complete description of the theory of plate tectonics requires knowledge of the forces that exist in the interior of our planet, a subject we will return to in Section 11.6.

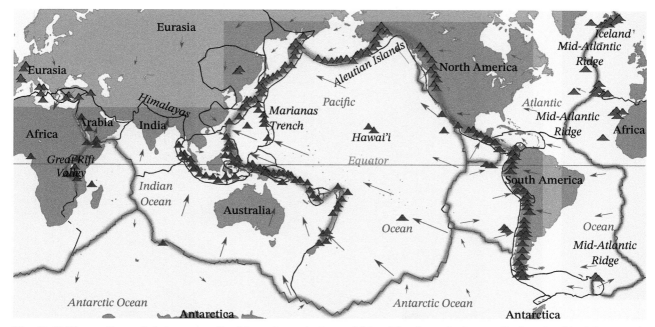

Fig. 11.32 The motions of plates across Earth's surface relative to Africa. Plate boundaries are displayed as black lines with their relative velocity vectors shown as blue arrows. The world's volcanoes are also depicted. Ocean ridges are highlighted in blue, the Pacific Ring of Fire is traced out in red, and the Great Rift Valley of Africa is displayed in green. [Adapted from a work by Eric Gaba (Sting - fr:Sting); CC BY-SA 2.5]

Fig. 11.33 The Himalayas, including Mt. Everest. (CC0; public domain)

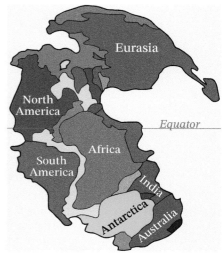

Fig. 11.34 The Pangaea supercontinent. (Adapted from Kieff; CC BY-SA 3.0)

Earth's Magnetic Field

Earth's magnetic field is more than 140 times stronger than the magnetic field associated with Mercury. Figure 11.35 is an illustration of the shape of Earth's magnetic field *near the planet*; its field is similar in shape to the magnetic field of a bar magnet, or the Sun's magnetic field at solar minimum (recall Fig. 9.25).

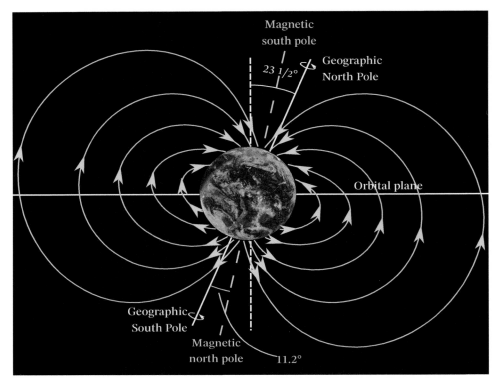

Fig. 11.35 Near Earth its magnetic field looks a lot like that of a bar magnet. (Earth image courtesy of NASA)

Close inspection of Fig. 11.35 also shows that the magnetic poles where the field vectors come together do not exactly align with Earth's geographic North and South Poles. In fact, Earth's magnetic field poles are shifted in latitude by about 11.2° relative to Earth's geographic poles (its rotation axis). If you are a hiker, you may already be aware of this alignment issue because at the bottom of every United States Geological Survey topographic map, such as the one in Fig. 11.36, an arrow points toward the magnetic pole relative to true geographic north. This is so that readings from a compass pointing toward the magnetic pole can be adjusted to orient the hiker to geographic north.

Farther from Earth the magnetic field gets distorted. As charged particles from the Sun's solar wind blow past Earth, they are deflected around the planet and they drag the magnetic field along with them, causing the field to stretch out away from the Sun as illustrated in Fig. 11.37. On Earth's sunward side, the field becomes compressed. The process is similar to charged particles in the Sun's interior dragging and twisting magnetic field lines (page 331). The deflection of the solar wind around Earth because of its magnetic field helps to protect the planet's atmosphere.

Not all of the solar wind particles move past Earth, some of them get trapped spiraling around magnetic field lines closer to Earth where the field lines look more like those of a bar magnet. These energetic particles form radiation belts that were discovered in 1958 by James Van Allen (Fig. 11.38) when he sent sounding rockets beyond Earth's atmosphere with radiation detectors on board. There are two major Van Allen radiation belts: one is located between 1000 km (600 mi) and 6000 km

Key Point

Earth's magnetic field deflects solar wind particles, which helps to protect its fragile atmosphere from escaping into space.

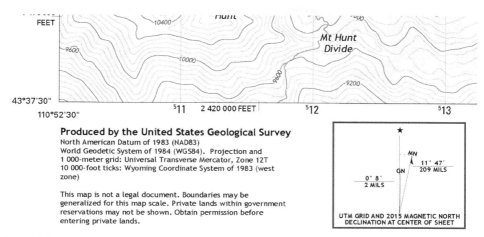

FEET

Hunt

10400

Mt Hunt
Divide

9600

10000

9600

9200

43°37'30"
110°52'30"

⁵11   2 420 000 FEET   ⁵12   ⁵13

Produced by the United States Geological Survey
North American Datum of 1983 (NAD83)
World Geodetic System of 1984 (WGS84). Projection and
1 000-meter grid: Universal Transverse Mercator, Zone 12T
10 000-foot ticks: Wyoming Coordinate System of 1983 (west
zone)

This map is not a legal document. Boundaries may be
generalized for this map scale. Private lands within government
reservations may not be shown. Obtain permission before
entering private lands.

★
MN

11° 47'
209 MILS

GN

0° 8'
2 MILS

UTM GRID AND 2015 MAGNETIC NORTH
DECLINATION AT CENTER OF SHEET

Fig. 11.36 The lower left portion of an United States Geological Survey (USGS) topographic map. The region within the red box shows the directions toward magnetic north (MN) and geographic north (GN). The directions differed by 11° 47′ in 2015. (Adapted from USGS)

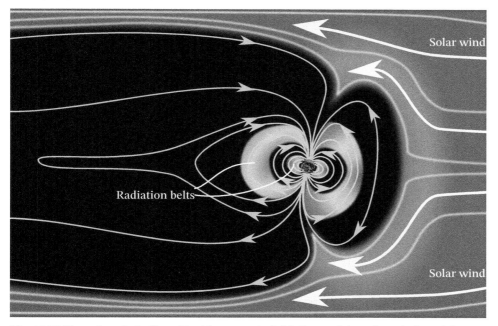

Solar wind

Radiation belts

Solar wind

Fig. 11.37 The solar wind affects Earth's magnetic field. Some of the solar wind particles can become trapped, producing the Van Allen radiation belts.

Fig. 11.38 James Van Allen (1914–2006) and one of his "Rockoon" sounding rockets. With radiation detectors on board, the rockets were lifted into Earth's upper atmosphere by balloons and then launched into space, minimizing the amount of rocket fuel required. (NASA/JPL and the Frederick W. Kent Collection, University of Iowa Archives)

(3700 mi) above Earth's surface, and the other one is between 13,000 km (8000 mi) and 60,000 km (37,000 mi) above the surface. For astronauts in low Earth orbit, they travel below the inner belt, but for astronauts who might travel beyond low Earth orbit, they can be exposed to potentially unhealthy levels of radiation. Electronics can also be affected by these very energetic belts.

Particularly energetic solar wind particles, like those ejected by solar flares, can follow magnetic field lines into Earth's upper atmosphere near the north and south magnetic poles (note the infringement of the solar wind near the poles in Fig. 11.37).

Fig. 11.39 *Left:* The aurora borealis seen over Alaska. (NASA/Terry Zaperach) *Right:* The aurora australis seen from above Australia by the International Space Station. (NASA)

When these charged solar wind particles strike Earth's atmosphere, they excite the electrons in atmospheric atoms to higher energy levels. When the electrons jump back down to lower energy levels they emit photons to create what we identify as the aurora borealis in the northern hemisphere and the aurora australis in the southern hemisphere (Fig. 11.39).

For the most energetic solar flares, the ions and electrons that are fired our way can even reach sensitive electronics and power grids, causing damage and power outages. On March 10, 1989 an extremely powerful solar flare launched a billion-tonne cloud of ionized gas toward Earth. By the evening of March 12, the geomagnetic storm caused by the gas striking Earth's magnetic field produced an aurora borealis that could be seen as far south as Florida and Texas. Electrical currents were even produced in the ground beneath much of North America. Then, in the early morning of March 13, the currents knocked out the entire power grid of the province of Quebec, Canada, putting millions of people in the dark. The blackout lasted for 12 hours.

If you studied Fig. 11.35 carefully, you may have noticed what appeared to be a typo in the labeling of Earth's magnetic poles. The magnetic south pole is shown as being close to the geographic North Pole and the magnetic north pole is shown near the geographic South Pole. In actuality, the labeling is correct (at the moment). It is the magnetic north pole of a compass needle that points toward north. But a compass needle is nothing more than a thin bar magnet that is allowed to pivot. Since opposite magnetic poles attract, Earth's magnetic south pole must located near the geographic North Pole. This fact is also reflected in the direction of the magnetic vector arrows, which by convention always point away from north magnetic poles and toward south magnetic poles.

The reason that this detail is important to mention is that the Earth's magnetic poles flip roughly every 800,000 years. You should recall that the Sun's magnetic field also flips, but on a much shorter 11-year cycle (the solar cycle). An interesting consequence of seafloor spreading away from the Mid-Atlantic Ridge is that new ocean floor must be created at the ridge. The iron contained in new seafloor material is affected by the existing magnetic field of the planet, as though tiny bar magnets are located in the iron. Before the rock that contains the iron hardens completely, the iron magnetic material aligns with the direction of Earth's magnetic field. By mapping the fossilized magnetic fields of the seafloor, researchers have discovered that the fields alternately point north, then south, then north, et cetera, flipping direction roughly every 800,000 years. At some point in the future, the labels in Fig. 11.35 will be wrong, but not at the time this text was written and almost certainly not at the time you are reading this paragraph.

11.4   The Moon

It may seem surprising that we are including our Moon in a chapter about the characteristics of the rocky planets, but the Moon does share many features that are similar to the rocky planets, while also differing significantly from the other large moons of the Solar System. The Moon is also only about 30% smaller than Mercury (Fig. 11.40).

Our Moon may not be the largest moon in the Solar System, but it is among the largest (for a direct comparison, see Fig. 12.16). It is also the largest moon *relative to* the planet that it orbits, with a diameter that is 27% (more than one-quarter) of Earth's diameter. Taken together, Earth and the Moon are sometimes thought of as a double-planet system.

The Surface Records a History of Bombardment

Given the heavy cratering evident in Fig. 11.41, it is clear that the Moon has no atmosphere to protect it from continuing impacts by meteorites, asteroids, or comets. The Moon also has no magnetic field to protect it from the solar wind. In other words, weathering is absent on the Moon, except for its surface being completely exposed to the effects of large and small impacts, solar radiation, and other forms of radiation from space that have left a layer of regolith on its surface.

Key Points
- The Moon doesn't have an appreciable atmosphere.
- The Moon doesn't possess a global magnetic field.

Fig. 11.40 The relative sizes of Mercury, Earth, and the Moon. (Mercury: NASA/Johns Hopkins University Applied Physics Laboratory/Carnegie Institution of Washington; Earth: NASA's Earth Observatory; Moon: NASA/GSFC/Arizona State University)

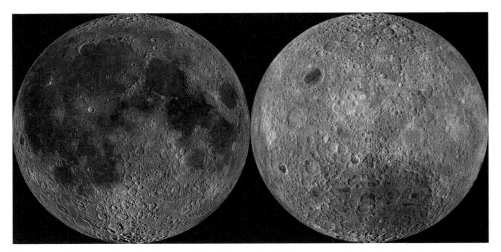

Fig. 11.41 The two hemispheres of the Moon. *Left:* The near side facing Earth. *Right:* The far side, most of which is never visible from Earth. A video showing a rotating view of the Moon is available on the textbook's `Chpater 11` resources webpage. (NASA/GSFC/Arizona State University)

The Ages of the Lunar Surface

Recall that radiometric dating (page 257*ff*) has been used to determine the ages of Moon rocks returned by the Apollo astronauts between 1969 and 1972 (see Example 7.2). During the Apollo missions, all of the landings took place on the Moon's near side so that the astronauts could remain in constant contact with Earth. Rocks returned from the darker, relatively smooth maria ("seas") were found to be between 3.2 and 3.8 billion years old, while rocks returned from the much lighter colored and more heavily cratered mountainous lunar highlands are typically 4.4 billion years old. Rocks were not returned from all of the maria, however, and based on estimating relative ages using crater counting (see page 384*ff*), it appears that other impact basins are 4 billion years old or older.

These data seem to imply that the lunar surfaces within the maria were refreshed by lava flows between 500 million and 1.3 billion years after the Moon formed, 4.51 billion years ago. This conclusion is consistent with the type of rock being basalt, which forms from hardened lava. The age estimate is also consistent with there being far fewer craters in the maria. Apparently the Moon experienced the same late heavy bombardment after its formation that Mercury experienced between 3.5 and 4 billion years ago. These impacts allowed lava to flow out from the interior, filling the enormous impact craters with fresh material. Not all of the basin-forming impacts occurred during the episode of late heavy bombardment, however; a few are more recent and some remnant basins are older, but most of the maria did form during the late heavy bombardment uptick in the frequency of large impacts.

One of the consequences of volcanism on the Moon appears to be the creation of enormous lava tubes. Lava tubes are common on Earth, such as the Thurston lava tube in Hawai'i Volcanoes National Park [Fig. 11.42 (*Top*)], which are produced by flowing rivers of lava. The surface of the flow solidifies quickly, much like ice forming on the surface of a river in the winter months. As lava below the solidified crust continues to flow, it can melt the floor and walls, slowly increasing the size of the tube. Over time, terrestrial lava tubes can become tall enough for people to comfortably walk through. The lunar lava tubes likely formed in a similar fashion but, at least in one case, became large enough to enclose an entire major city [Fig. 11.42 (*Bottom*)]. Because the Moon doesn't have an atmosphere or a magnetic field to protect astronauts from constant exposure to the high-speed charged particles in the solar wind and the even more energetic and intense barrage from solar flares and coronal mass ejections, lunar lava tubes may prove to be an important discovery for future bases as a means of shielding inhabitants from long-term exposure to that deadly radiation.

Density of the Moon and the Apollo Lunar Samples

Based on its mass and radius, the Moon is much less dense than Earth, having an overall density of 3344 kg/m^3 (about 3.3 times the density of water), compared to Earth's average density of 5514 kg/m^3. When the Apollo astronauts returned rocks from the lunar surface (e.g., Fig. 11.43) we learned that their densities ranged from about 2800 kg/m^3 for samples from the highlands, to between 3000 and 3300 kg/m^3 for some of the samples from the maria. In the case of the maria samples, their compositions have higher concentrations of iron oxide (FeO) than the samples from the highlands, or for typical Earth rocks. Although similar in many ways to rocks on Earth, there are subtle details that provide important hints about how our Moon may have formed.

Fig. 11.42 *Top:* Nāhuku (Thurston lava tube) in Hawai'i Volcanoes National Park. (NPS/D. Boyle) *Bottom:* The Marius Hills Skylight is the entrance into a city-sized lava tube below the Moon's surface. (NASA/Goddard/Arizona State University)

The Presence of Water

The Apollo moon rocks are also less abundant in compounds that are easily vaporized relative to rocks on Earth's surface. There is an important exception, however; analysis of the basalts returned by the Apollo missions shows small quantities of water (H_2O) molecules trapped inside rock samples. Using orbiting spacecraft, observations of reflected light and the analysis of its spectra suggest that significant deposits of water molecules may also be contained in the ancient lava flows of the maria. Additionally, as with Mercury, water ice has been observed within craters near the Moon's north and south poles. If it turns out that an abundance of water does exist in lunar basalts that can be effectively extracted, water could play an important role in the future, should permanent colonies be established on our nearest celestial neighbor. If present, how did water find its way into basaltic rock, and where did the water ice come from in the polar lunar craters?

Fig. 11.43 A rusty rock sample returned from the highlands by the Apollo 16 mission. Rust is due to iron in the presence of water. (NASA)

Did the Moon Have an Atmosphere in the Past?

Although there is no atmosphere on the Moon today, it is possible that an atmosphere may have existed briefly in the ancient past. Again, based on the composition of the rocks returned from the Moon, including the abundances of the various isotopes of easily vaporized elements, an atmosphere could have been created by gases escaping from the interior through volcanic activity in the Moon's early history. However, because of its weaker gravity and the lack of a magnetic field, the solar wind would have blown the atmosphere away fairly quickly after its formation.

The Two-Faced Moon

Key Points
• The Moon's crust is thicker on the far side.
• There are subtle, smoothly varying changes in surface composition from the near to the far side of the Moon.

Another factor to consider when trying to understand the nature of our Moon is the striking difference between the number of maria on the Moon's near side and its far side, as evident in Fig. 11.41. There are a few maria on the far side, but they are much smaller, and far less abundant. It is certainly not the case that the far side is devoid of impacts, and it seems highly unlikely that the far side avoided all of the largest collisions during the period of late heavy bombardment that produced the near-side maria. So what caused the stark contrast between the two hemispheres? There is strong evidence to suggest that the far side of the Moon has a "thicker skin," its crust, allowing it to absorb impacts without lava coming out of the interior to refresh the surface. Tracking the orbits of spacecraft around the Moon reveals that some parts of the Moon produce a stronger gravitational pull than other areas, demonstrating that the Moon is not the same everywhere. Observations made by orbiters also show that there are smooth changes in the composition of the Moon's surface from the near side to the far side. Of course, these observations immediately beg the questions: Why is the crust on the near side thinner than it is on the far side? What produced the changing composition across the surface? And, are these two features of the lunar surface related somehow?

The Demands of Getting to the Answer

Key Point
Explanations must satisfy all physical laws and account for *all known data*.

Remember that all physical laws must be satisfied and *all data* must be included when explaining what we observe. Science doesn't allow us to pick and choose what evidence we want to consider or which physical processes to employ, while ignoring everything else when seeking answers to our questions. It is also the case that, as more information is gathered, the set of possible explanations becomes more restrictive.

11.5 Mars

Where are the Martians?

Historically Mars, the "Red Planet," has received more attention, and inspired more stories of aliens, than any other body in the Solar System. The public's fascination with Mars was born in 1888, when an Italian astronomer, Giovanni Schiaparelli (Fig. 11.44), reported what he termed to be *canali*, meaning naturally occurring channels, on the planet's surface, as depicted in his sketch in Fig. 11.45. Even from the Hubble Space Telescope images in Fig. 11.46 you may be able to imagine how Schiaparelli thought he saw canali in his much less powerful telescope in the second half of the 1880s.

Percival Lowell, the American astronomer and wealthy businessman whom we first encountered on page 177, became very excited about Schiaparelli's canali. However, he misinterpreted canali to mean canals, believing that they could have been artificially manufactured by intelligent beings to carry water from the polar ice caps for irrigation purposes. To support his conjecture and the further study of Martian canals, he built the observatory that bears his name. Lowell also published a book on the subject in 1895.

Fig. 11.44 Giovanni Virginio Schiaparelli (1835–1910). (Mondadori Publishers; public domain)

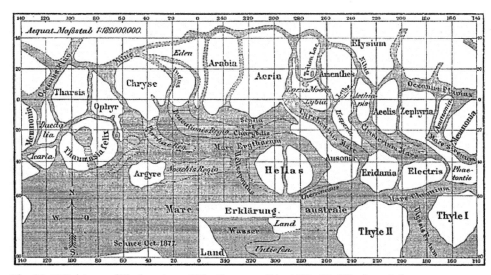

Fig. 11.45 Schiaparelli's drawing of Martian canali in 1888. (Public domain)

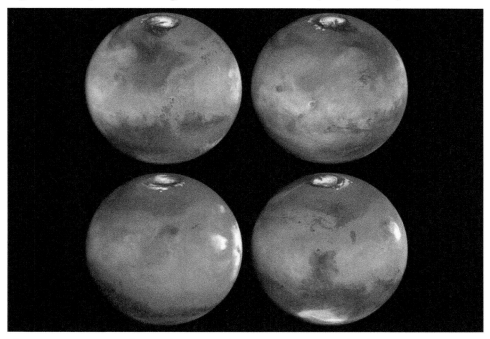

Fig. 11.46 Four views of Mars during northern summer, as seen by the Hubble Space Telescope on March 30, 1997. Note the small northern polar ice cap relative to the southern ice cap. [Phil James (Univ. Toledo), Todd Clancy (Space Science Inst., Boulder, CO), Steve Lee (Univ. Colorado), and NASA]

In 1897, the English author Herbert George (H. G.) Wells (1866–1946) published a book entitled, *The War of the Worlds*, in which astronomers saw explosions occur on Mars, followed later by the arrival of a meteorite that turned out to be an alien

Fig. 11.47 George Orson Welles (1915–1985) in 1938, trying to explain to reporters that no one associated with the radio broadcast of *The War of the Worlds* had any idea that the show would cause listeners to panic. (Public domain)

spaceship with others arriving soon afterward. The aliens were determined to take over Earth, but when all seemed lost, the aliens ultimately succumbed to Earth-based microbes for which they had no immunity.

There have been many adaptations of *The War of the Worlds* since its publication, but arguably the most famous version was the live 1938 Halloween night radio presentation by Orson Welles (Fig. 11.47) on the Columbia Broadcasting System (CBS). The broadcast was set as a series of news-bulletin interruptions to a regular show, first announcing explosions on Mars, then meteors in the skies, only later to go on the air with what appeared to be live coverage of a Martian invasion. There were some panicked listeners who didn't hear the clear message at the beginning of the program stating that it was only a radio drama. After the broadcast there were demands that the Federal Communications Commission (FCC) regulate such broadcasts.

Of course, nothing like that reaction could possibly happen in our enlightened modern scientific age, right? In a 1976 mission, the Viking 1 spacecraft sent back an image of what appeared to be a face two kilometers in length on the surface of the Red Planet [Fig. 11.48 (*Left*)]. Scientists understood that the image was the result of shadows on a mesa, but that didn't stop conspiracy theorists from claiming that NASA was hiding the fact that there really are (or at least were) Martians. Some believed that they could see the remnants of a city and a pyramid near the face. A higher resolution image of the "face" was returned by the Mars Global Surveyor in 2001 [Fig. 11.48 (*Middle*)], and then an ultra-high resolution image was obtained by the High Resolution Imaging Science Experiment (HiRISE) aboard the Mars Reconnaissance Orbiter in 2007 [Fig. 11.48 (*Right*)]. A close-up and colorized view of the "face" is seen in Fig. 11.49. Even the more recent data don't deter some conspiracy theorists from believing that there is a cover-up at NASA and within the government. Unfortunately, there are those who believe that the absence of confirmable evidence only strengthens their argument regarding the

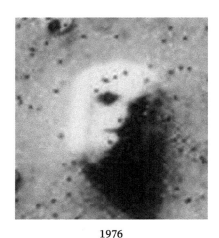
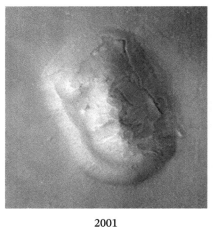
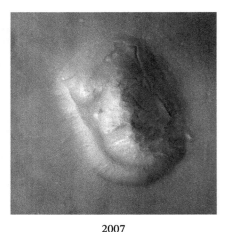

1976　　　　　　　　　　　　2001　　　　　　　　　　　　2007

Fig. 11.48 The "Face on Mars" as seen with increasing resolution and differing lighting. *Left:* Viking 1, 1976. The black dots are data errors. (NASA/JPL) *Center:* Mars Global Surveyor, 2001. (NASA/JPL/MSSS) *Right:* HiRISE on the Mars Reconnaissance Orbiter, 2007. (NASA/JPL/University of Arizona)

Fig. 11.49 A close-up view of the "Face on Mars," taken by HiRISE on the Mars Reconnaissance Orbiter, 2007. Comparing this image to Fig. 11.48 (*Right*), also from HiRISE onboard MRO, can you imagine the eyes, nose, and mouth within the natural terrain? (NASA/JPL/University of Arizona)

depth and extent of the conspiracy. Seeing the Face on Mars, the Man in the Moon, animals in cloud patterns, or hidden messages when recordings are played backward, are examples of a well-known psychological phenomenon referred to as pareidolia, where our minds try to construct recognizable patterns when none exist. Always think critically and demand evidence, especially when the claims seem extraordinary.

Key Point
A lack of evidence is not itself evidence of a conspiracy to cover up information, whether it is scientific, political, or some other subject.

What is Mars Really Like?

Mars has been studied with ever greater resolution with Earth-based and orbital telescopes, and with an onslaught of spacecraft (see, e.g., Fig. 11.50). Humanity's first successful mission to Mars took place in 1964, with NASA's Mariner 4 flyby of the planet (six previous missions, five by the Soviet Union and one by NASA had either failed at launch or failed en route). Since then, the Red Planet has been visited by orbiters, landers, rovers, and even a tiny helicopter, trying to uncover its secrets. Today, a remarkable amount of information has been gathered about Mars' atmosphere, its surface, and even beneath its surface, with data gathering continuing around the clock. Much of this effort is simply to satisfy the human need to continually explore and learn, and some is in preparation for potential manned missions.

Fig. 11.50 A mosaic of one side of Mars. Valles Marineris, a 2000-km-long canyon system, is in the center and three volcanoes are near the left edge. (NASA/JPL-Caltech)

Mars shares some characteristics with both Venus and Earth, although it is much smaller (0.53 M_{Earth}) and less massive (0.11 M_{Earth}). Given its mass and radius, the acceleration due to gravity on Mars is approximately 38% of what it is on Earth, meaning that your weight would be just 38% of what it is now (unless you happen to be reading this text somewhere other than on planet Earth!). Mars' axis tilt is very close to Earth's tilt (25.2° vs. 23.4°) and the length of its day is nearly the same as well (24^h37^m vs. 23^h56^m).

Mars' Thin Atmosphere

Like Venus, Mars has an atmosphere that is 95.1% carbon dioxide with most of the rest being molecular nitrogen (N_2, 2.6%) and argon (Ar, 1.9%). Unlike Venus, where the average surface temperature is 740 K, the average temperature of Mars is only 210 K, well below the freezing point of water (273 K, 0°C, 32°F). Clearly, the strong greenhouse effect on Venus does not operate on Mars. The reason is that, although the atmosphere is dominated by CO_2, the atmosphere is very thin, implying that there are relatively few carbon dioxide molecules available to absorb the infrared photons that are emitted from the surface as blackbody radiation. The atmospheric pressure at the surface of Mars is only 0.63% of the atmospheric pressure on Earth and 0.007% of the atmospheric pressure on the surface of Venus. As a result, the infrared photons are largely free to escape directly into space, despite the

Fig. 11.51 *Left:* A swirling dust devil towering above the surface of Mars. The plum is about 30 m in diameter, with its shadow visible on the ground. *Right:* Sand dunes with large boulders lying in flat areas. Frost is visible on the tops of the dunes. (NASA/JPL-Caltech/University of Arizona)

high *proportion* of carbon dioxide molecules relative to other constituents in Mars' atmosphere.

Dust Storms

Despite having a very thin atmosphere, Mars is still capable of generating significant weather in the form of wind and dust storms. But, unlike what is sometimes depicted in science fiction movies or novels, the winds are not particularly ferocious, peaking at about 100 km/h (60 mi/h), significantly slower than the winds in Earth's hurricanes or typhoons, and those speeds are even slower than some straight-line winds in more normal weather systems. Furthermore, the Martian atmosphere is very thin so there isn't much atmosphere to push around at those speeds in any case. Even so, the fine-grained dust on the surface of Mars can be picked up to form dust storms, towering dust devils (narrow, spinning columns of dust), and sand dunes on the surface (Fig. 11.51). Roughly once every five years or so, those dust storms can grow large enough to engulf the entire planet, as dramatically illustrated in Fig. 11.52. The blowing dust has the effect of getting into moving parts in Martian rovers and settling onto solar panels. During larger storms, dust can even block out the Sun, making the use of solar panels during human exploration of the planet problematic.

All of this atmospheric turbulence tends to peak when Mars is at perihelion. Recall that Mars' orbit is more eccentric than any of the planets in the Solar System except Mercury, and that it was the study of Mars' orbit that led Kepler to realize that planetary orbits are elliptical (Fig. 4.10). The activity in Mars' atmosphere also tends to occur during the summer months in its southern hemisphere, which is when Mars is at perihelion and when there is more heating from sunlight.

The Search for Martian Water

Even though its atmosphere is thin and dominated by CO_2, some water vapor is still present. As shown in the ultraviolet image in Fig. 11.53, the atmosphere is capable of producing clouds. The three clouds forming a diagonal across the image are all located above the three tall volcanoes that are also visible in Fig. 11.50.

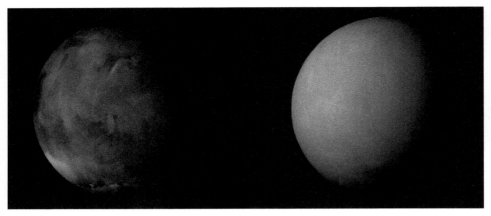

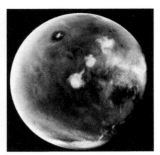

Fig. 11.52 Two images from the Mars Color Imager onboard NASA's Mars Reconnaissance Orbiter in 2018. The image on the left was taken on May 28 and the image on the right was obtained on July 1 showing the same hemisphere during a global dust storm. The dust storm was one of the largest weather events ever seen on Mars. (NASA/JPL-Caltech/MSSS)

Fig. 11.53 An ultraviolet image of Mars taken by the MAVEN orbiter on July 9–10, 2016. The planet's southern polar ice cap is visible at the bottom of the image. (NASA/MAVEN/University of Colorado)

A fourth, smaller cloud appears above Olympus Mons in the upper left portion of the image; Olympus Mons is the largest volcano in the Solar System. Over the course of the day when the image was taken, the three diagonally oriented clouds grew to form a single, very large cloud whose length stretched across an appreciable fraction of the diameter of the planet. Martian clouds are even capable of producing snow at night, with evidence of a fresh layer of snow seen on mountain tops when light returns to the peaks in the morning.

It has long been known that the prominent polar ice caps grow and shrink during the seasons, being smallest in the hemisphere's summer and largest during the winter. The permanent component of the ice caps is composed of frozen carbon dioxide (dry ice) and the variable component is water ice that sublimates in the summer and refreezes out of the atmosphere in the winter. Figure 11.54 shows the northern ice cap, which is at least 2 km thick.

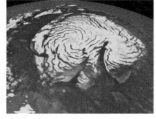

Fig. 11.54 The northern polar ice cap. The darker regions are troughs in the ice sheet that are in shadow. To the right of center is Chasma Boreale, a chasm in the ice that is as long as the Grand Canyon on Earth and up to 2 km deep. (NASA/JPL-Caltech/MSSS)

The lack of any visible water on the surface beyond the ice caps, along with some snow or frost, doesn't necessarily imply that water has otherwise vanished from the planet or never existed in the first place. High resolution imagery provided tantalizing evidence that perhaps liquid water has been seen indirectly, through apparent erosion on the surface and the seasonal appearance of dark streaks on steep, equator-facing slopes (Fig. 11.55). The dark streaks were initially thought to be the result of briny (salty) water running downslope when the temperature warms to near the freezing point of water. Salts in water lower the freezing point, causing ice to melt at less than 273 K (this is why salt is sometimes put on roadways in anticipation of a winter storm). While the streaks appear and lengthen between late spring and early fall, they fade from late fall until the following spring. While intriguing, no flowing water has yet been observed, so perhaps there are other explanations for what appears to be evidence of some sort of fluid. More recent analysis suggests that these features may simply be fine-grained sand sliding down steep slopes.

Images like the one in Fig. 11.56 hint that ice may be hidden below the surface; similar terrain can be found in the Canadian Arctic. To investigate that possibility

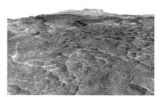

Fig. 11.55 *Left:* Dark streaks appear on this southern hemisphere equator-facing slope between late spring and early fall, then fade during the remainder of the Martian year. *Right:* Dark seasonal streaks in Hale Crater, located in the southern hemisphere. The vertical dimension is exaggerated by a factor of 1.5. (NASA/JPL-Caltech/University of Arizona)

Fig. 11.56 Scalloped terrain that suggests the possibility of buried ice. The foreground is about 1.8 km (1.1 mi) across. The vertical dimension is exaggerated by a factor of five to bring out the terrain features. (NASA/JPL-Caltech/University of Arizona)

NASA's Mars Reconnaissance Orbiter (MRO) carries a suite of instruments, one of which is the ground-penetrating Shallow Radar (SHARAD), provided by the Italian Space Agency. SHARAD is able to detect geologic boundaries and electrical properties of materials. In this way, it can distinguish between rock, sand, water ice, and carbon dioxide ice, as well as determine depth. Today it is known that a great deal of water ice exists below the surface. In one area the size of New Mexico, located halfway between the equator and the north pole, radar measurements suggest that a deposit of 50% to 85% water ice may exist, ranging in depth from about 80 m (260 ft) to 170 m (560 ft). If true, there is enough water in that one deposit to fill Lake Superior, the largest of the Great Lakes along the United States–Canadian border. As large as the deposit is, it only represents a small percentage of the subsurface ice known to exist on Mars.

In late 2018 the European Space Agency released Fig. 11.57, a composite image obtained by the Mars Express orbiter, showing Korolev crater filled with water ice. The crater is 81.2 km (50.6 mi) across and 2 km (1.2 mi) deep. There is no longer any doubt that water ice exists on and below the surface of Mars.

Imagery has also revealed strong evidence that, although an abundance of liquid water is absent on the surface today, water did flow freely in the past. Figure 11.58 shows an apparent ancient river bed in the southern hemisphere. Similar features, along with what may be ancient lakes or oceans, are located across the planet. Given the amount of water currently trapped below the Martian surface, it seems that the planet must have once had an abundance of water on its surface, perhaps as imagined in Fig. 11.59.

What Happened to Mars' Atmosphere?

The thicker atmosphere that was required to support a warmer planet with liquid surface water has since evolved into the thin, and much colder, carbon-dioxide-dominated atmosphere of today. This happened in part through the escape of some carbon into space in much the same way that hydrogen escaped into space by the ionization of H_2O in the ancient Venusian atmosphere: $CO_2 + \gamma \rightarrow CO + O$ and $CO + \gamma \rightarrow C + O$. Because carbon atoms are lighter than oxygen atoms, it is easier

Fig. 11.57 Korolev crater is approximately 81 km (51 mi) in diameter and 2 km (1.2 mi) deep. The crater is almost completely filled with water ice. The crater is located near the northern polar ice cap. (©ESA/DLR/FU Berlin; CC BY-SA 3.0 IGO)

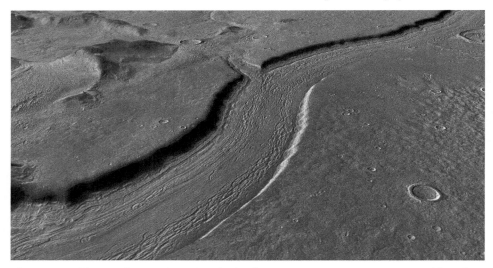

Fig. 11.58 Evidence of an ancient river bed. The image was computer-generated based on data from the European Space Agency's Mars Express. [©ESA/DLR/FU Berlin (G. Neukum); CC BY-SA 3.0 IGO]

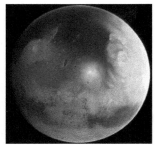

Fig. 11.59 An artist's impression of what Mars might have looked like four billion years ago. (ESO/M. Kornmesser; CC BY 4.0)

for carbon atoms to escape from the atmosphere (recall the periodic table of the elements; Table 7.1). Carbon dioxide also became trapped in rock on Mars, just as it did on Earth. As the amount of carbon dioxide was reduced, the greenhouse effect began to weaken and the planet cooled. This led to temperatures dropping below the freezing point, causing surface water to become ice. Over time the ice was buried by blowing dust.

The hypothesis that carbon escaped into space can be tested by comparing isotope ratios in the same way that the amount of $^2_1$H (deuterium) was compared to $^1_1$H (normal hydrogen) for the evolution scenario of Venus's atmosphere on page 390, only this time, since it was carbon that escaped, the comparison can be carried out between $^{13}_6C$ and $^{12}_6C$ (carbon-12 being the most abundant form of carbon). Again, because carbon-13 is slightly heavier than carbon-12, it is a bit harder for the heavier isotope to escape relative to the lighter one, implying that $^{13}_6C$ should increase in abundance relative to $^{12}_6C$ over time, which is just what is observed.

When the temperature of the atmosphere dropped below the freezing point of water, very little water vapor remained in the atmosphere, but the hydrogen in H_2O molecules that are still in the atmosphere is very slowing escaping after the water molecules are broken apart by ultraviolet photons from the Sun. NASA's Mars Atmosphere and Volatile EvolutioN (MAVEN) orbiter actually detected this process directly in 2016. Thanks to the planet's particularly eccentric orbit, ten times more hydrogen escapes at perihelion than at aphelion.

The lack of a global magnetic field around Mars also made the slow escape of much of the planet's atmosphere easier. The presence of a magnetic field helps to retain planetary atmospheres because magnetic fields trap charged particles like ionized hydrogen, carbon, oxygen, and electrons.

Key Point
Global magnetic fields help to retain planetary atmospheres.

An Arid Landscape of Canyons, Plains, and Volcanoes

Today, Mars is an arid world that in some places resembles the desert southwest of the United States (see Fig. 11.60). The planet also has vast plains, highland regions, low mountains, deep canyon networks, craters, and towering volcanoes. There is also a distinct difference between the northern and southern hemispheres of the planet: the north is primarily composed of lowland plains while the southern hemisphere is heavily cratered and covered with countless volcanoes.

Near the equator of Mars is Valles Marineris, a canyon system 4000 km (2500 mi) long, that in places reaches a depth of 8 km (5 mi), and can be as wide as 200 km (120 mi). The equatorial portion of Fig. 11.50, including Valles Marineris, is shown in Fig. 11.61. Valles Marineris dwarfs Earth's Grand Canyon by comparison; it is nearly equal to the length of the continental United States, five times deeper than the Grand Canyon, and about twelve times as wide. To the west of Valles Marineris and in the north-central region are what may be channels produced by ancient water flows.

In stark contrast to Valles Marineris are the giant volcanoes of Mars. The biggest of them all is Olympus Mons, measuring 22 km (13.6 mi; almost 72,000 ft) in height and 600 km (370 mi) wide. That makes Olympus Mons about 2 1/2 times taller than Mount Everest and about as wide as the State of Utah is long. Depictions of Olympus Mons can be seen in Fig. 11.62 (*Left*).

Fig. 11.60 *Left:* A mosaic panorama of Vera Rubin Ridge with Mount Sharp in the background. The image was obtained by the Curiosity rover. *Right:* Butte M9a in the Murray Buttes region, also captured by Curiosity. (NASA/JPL-Caltech/MSSS)

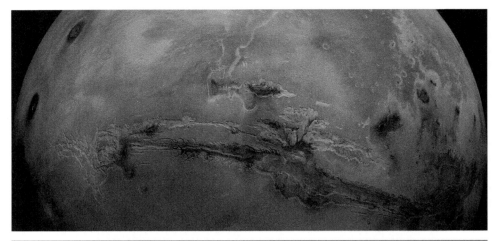

Fig. 11.61 *Top:* The equatorial region of Fig. 11.50 showing Valles Marineris and shield volcanoes to its west. (NASA/JPL-Caltech) *Bottom:* A closeup of a central portion of Valles Marineris. (NASA/JPL/Arizona State University)

To better appreciate the topography of this region of Mars, Fig. 11.62 (*Right*) displays a color-coded vertical relief map of most of the Martian surface. The map includes large portions of Fig. 11.50 and Fig. 11.53; Olympus Mons is on the left side of the map and Valles Marineris is center-right. The large red-coded part of the map is the Tharsis region, an uplifted portion of Mars that was volcanically active in the ancient past. To the north are vast lowland plains, and in the lower right-hand corner of the map is Argrye Planitia, a nearly circular plain within an immense impact crater.

We still don't know for sure how Valles Marineris formed, although it was almost certainly not formed by flowing water. The two most likely mechanisms proposed at the time of the textbook's writing are either the result of lava flows from the Tharsis region or a crack in Mars' surface crust, possibly associated with past

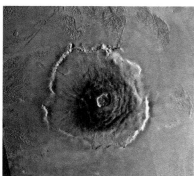

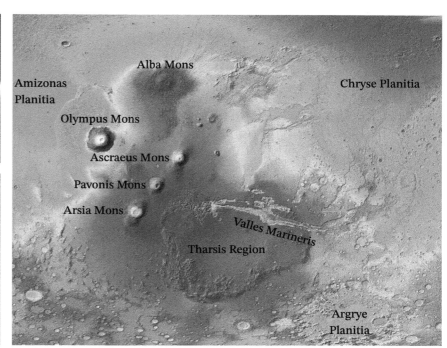

Fig. 11.62 *Top Left:* A computer-generated view of Olympus Mons. (NASA/MOLA Science Team/O. de Goursac, Adrian Lark); *Bottom Left*: Olympus Mons as seen from the Viking 1 Orbiter in 1978. (NASA/JPL) *Right:* A color-coded relief map of much of Mars. White represents the highest elevations (atop the four volcanoes left of center). Progressively lower elevations are displayed in browns, reds, golds, yellows, greens, and blues. (NASA/JPL-Caltech/Arizona State University)

tectonic plate motions. In either scenario, Valles Marineris is an ancient feature of the planet, likely billions of years old. On the other hand, Olympus Mons and the other giant volcanoes are shield volcanoes, much like the Hawai'ian Islands and the volcanoes of Venus.

No Magnetic Field

It may appear strange that Mars has no global magnetic field. After all Mars does rotate rapidly (just slightly slower than Earth), but it is much smaller and farther from the Sun. Again, magnetic fields, or lack thereof, provide clues to internal structure.

Two Moons

Mars does possess two small, irregularly shaped moons, Phobos and Deimos, pictured in Fig. 11.63. These two natural satellites of the Red Planet have an interesting pre-discovery history. Johannes Kepler had actually "predicted" their existence based only on numerology. Kepler argued that since Venus didn't have any moons, Earth had one, and Galileo had just discovered that Jupiter had four, it seemed natural that Mars ought to have two! (At last count, Jupiter actually has 79 moons.) Numerology may have been important to Kepler at the time, but today we know

Fig. 11.63 The two tiny moons of Mars. *Left:* Phobos. (NASA/JPL/Arizona State University) *Right:* Deimos. Phobos orbits closer to Mars than Deimos does. (NASA/JPL-Caltech/University of Arizona)

that the coincidental grouping of numbers is meaningless unless there is an underlying rational explanation for the apparent coincidence. Unfortunately, there are many who still believe in numerology's supposedly mystic meanings.

The other interesting pre-discovery reference was also simply a lucky imagination, although the story does reveal that even those who were not necessarily scientists could be aware of scientific discoveries. In 1726, 150 years before the discovery of the moons by Asaph Hall (1829–1907), Jonathan Swift (1667–1745) wrote *Gulliver's Travels.* In his novel, Swift had astronomers who discovered two satellites orbiting Mars. According to his astronomers, one of the moons had an orbital period of 10 hours and the other, 21 1/2 hours, "so that the squares of their periodical times are very near in the same proportion with the cubes of their distance from the centre of Mars, which evidently shows them to be governed by the same law of gravitation that influences the other heavenly bodies." In actuality, the orbital period of Phobos is 7^h39^m and the orbital period of Deimos is 30^h17^m; remarkably close given the fact that Swift simply made them up! It is also interesting that Swift clearly had some understanding of Kepler's third law.

It is not known how Mars' two moons came to be orbiting the planet, although they could have been captured from the nearby asteroid belt. Another possibility is that they somehow formed in place around Mars. Due to the tidal forces exerted by Mars, both Phobos and Deimos are in synchronous rotation around the planet. Furthermore, since Phobos orbits Mars faster than Mars rotates, the same interaction that is causing our Moon to move away from Earth is causing Phobos to slowly spiral into the planet. Deimos is moving slowly away from Mars.

Does Mars Support Life Today or in the Past?

Although we are fairly confident that there aren't any "little green men" wandering around the surface of Mars today, that doesn't mean that there aren't tiny life forms such as bacteria or worms in the soil now or perhaps in the ancient past when the planet was wetter and warmer.

In 1976, NASA sent two identical lander–orbiter pairs to Mars; Vikings 1 and 2 (a model is shown in Fig. 11.64). One of the two landers touched down near the equator and the other was farther north, making them the first spacecraft to successfully land on the surface. Along with taking pictures and collecting data, the landers conducted three biological experiments designed to test for the presence of active microbial life on the Red Planet. While two of the experiments found no evidence of life, the third experiment on both landers added nutrients to soil samples and detected gases characteristic of what microbial life on Earth would produce under similar conditions. Most scientists today believe that the apparently positive results were due to unanticipated chemical reactions not associated with life, but there is a minority, including the principal investigators of the experiment, who argue that a positive detection of microbial life in the soil of Mars was the correct conclusion of the results.

Fig. 11.64 A model of the Viking landers that landed on Mars in 1976. The long arm in the front is a scoop that brought soil samples into the lander to one of three onboard laboratories. (NASA)

Another tantalizing discovery occurred, not on Mars, but on Earth in Antarctica. When energetic meteorites strike Earth, the Moon, or Mars, they can eject material out into space and occasionally land on another body in the inner Solar System. In 1984, when meteorite hunters were exploring Antarctica, they came across a Martian meteorite, ALH84001 (ALH stands for Allan Hills, the region of the continent where the meteorite was discovered). It turns out that Antarctica is a great place to look for meteorites because they are dark and will be sitting on the snow waiting to be picked up! As seen in Fig. 11.65, when ALH84001 was cut open it revealed a surprising set of microscopic structures that NASA scientists thought could be fossil evidence of ancient life on Mars. The discovery immediately became world-famous; even President Bill Clinton talked about it in a speech. However, later studies suggested that those same features could have been produced by non-biological means. Once again, careful, peer-reviewed science led to a different conclusion regarding life (past or current) on Mars.

Fig. 11.65 Microscopic chain structures seen in the Allan Hills 84001 meteorite. (NASA)

NASA landed the rover, Perseverance, on Mars in February 2021. Perseverance also brought with it a small helicopter named Ingenuity, which is the first aerial explorer of the Red Planet. The primary missions of the two machines are to look for signatures of past life and study the planet's geology. Perseverance will also take core samples of interesting rocks, place them in containers, and deposit them on the planet's surface for a future mission to pick up and return to Earth for intensive study. No results were available from the mission at the time of writing.

11.6 Their Interiors

Earth

The planet we know the most about its interior is Earth. First, we know that it must be more dense in the interior than it is at the surface because Earth's average density is 5514 kg/m^3 (Table 10.1) even though the density of rocks on the surface is only about 2700 kg/m^3 (recall that the density of water is 1000 kg/m^3). If the surface rock is below the planet's average density, then somewhere in the interior there must be material with a density that is greater than the average of 5514 kg/m^3.

There are other hints about what the interior might be like, as well. Measurements tell us that heat is coming from the interior, indicating that a heat source must exist. Of course, we already know about volcanic activity, earthquakes, and the fact that surface tectonic plates are in constant motion.

Earth also possesses a significant magnetic field, that must be generated in the interior. In Chapter 9 we studied the Sun's magnetic field and discussed the magnetic dynamo (page 331). Three ingredients are necessary to generate a magnetic field in the Sun: rotation, convection, and an electrically conducting fluid. After all, all magnetic fields are caused by moving electric charges (current). For the Sun, rotation and convection provide the motion for electric charges, while electrons and ions in the hot interior provide the charge. Earth is obviously rotating fairly rapidly, with heat rising from the interior that is capable of producing convection in fluids, so the other ingredient, some sort of electrically conducting fluid, must also be present in the interior.

In the Sun the conducting fluid is ionized hydrogen and helium, but a gas interior can't exist for Earth. Instead, possible candidates would be molten metals, such as iron and nickel. They have the added characteristics of being abundant components of the rocky planets and they have high densities; the density of iron is about 7800 kg/m^3, significantly higher than the average density of the planet. Assuming that Earth was completely molten very early in its history, the lighter elements would have floated to the surface while the heaviest elements sank into the interior.

How can we test to see if these ideas are correct? We certainly can't drill or travel to the center of the planet like the adventurers in Jules Verne's (1828–1905) fanciful 1864 novel, *Journey to the Center of the Earth*, or the numerous other science fiction stories and movies, but fortunately we don't have to. This is because earthquakes (and below-ground nuclear weapons tests) cause waves to move through the planet. As the waves encounter different densities within solids and fluids and the boundaries that separate them, the speeds and directions of the waves are altered. Detecting the arriving waves at seismic stations around the world provides the necessary data to determine what the interior is like.

There are two types of wave that pass through Earth's interior. The first to arrive at seismic stations are the primary waves (P waves). Primary waves propagate through material by compressing and expanding the material as they move, just

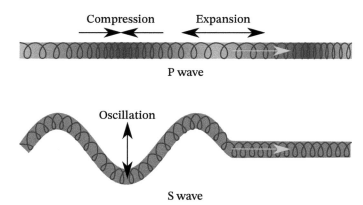

Fig. 11.66 *Top:* A primary (P) wave is a compression of material as the wave travels along, temporarily causing the density to increase slightly. When the P wave pulse moves past, the material relaxes back to its normal density. *Bottom:* A secondary (S) wave is an oscillation of material perpendicular to the direction that the wave travels. Both waves are traveling to the right.

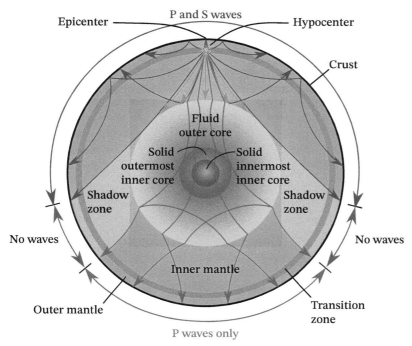

Fig. 11.67 Some of the many P (primary, pressure) and S (secondary, shear) seismic waves traveling through Earth from an earthquake site (the hypocenter). The red arrows represent both P and S waves; the green arrows represent S waves only; the blue arrows represent P waves only. Radii: Earth, 6378 km; transition zone, 5710 km–5960 km; outer core, 3485 km; outer inner core, 1216 km; innermost inner core, 650 km.

like a pulse moves through a toy slinky as illustrated in Fig. 11.66 (*Top*). This type of wave is actually a sound wave of traveling pressure pulses; when sound travels through air, water, a rock, or any other substance, the waves travel just like the pulses through the slinky. When pulses moving through air (or water, if you are underwater) reach your eardrum, they push and pull on the eardrum, causing it to vibrate back and forth, which your brain registers as sound. The louder the sound, the harder the pulses push on your eardrum. This is why very loud sounds, including music, can permanently damage your eardrums, and consequently your hearing.

The secondary waves [S waves; Fig. 11.66 (*Bottom*)], which are shear waves, arrive at seismic stations after the P waves get there, unless they don't arrive at all. *S* waves are oscillations that are perpendicular to the direction of travel of the waves, much like shaking a slinky or a rope up and down; the oscillation starts at your hand and then moves the length of the slinky or rope. Although *P* waves can travel through any kind of material, *S* waves can only travel through solids (after all, you can't shake a lake up and down from the surface to the bottom). This means that when an S wave encounters a fluid in the interior of Earth, it simply dies and doesn't make it through the fluid layer.

Figure 11.67 illustrates what happens when an earthquake occurs. Notice that both P and S waves propagate away from the hypocenter, which is the location of the earthquake (the epicenter is located on the surface directly above the hypocenter). As the waves pass through constantly changing densities their speeds change and their paths bend. If they hit a discontinuity, say between the inner mantle and the fluid outer core, some of the energy reflects off the discontinuity. The reflection is like yelling into a canyon and hearing your echo after the sound bounces off the rock walls. Because of bending and reflection, both P and S waves can be measured at seismic stations on Earth's surface labeled as "P and S waves" in Fig. 11.67.

P waves not only reflect off discontinuities, some of the P wave energy continues into the fluid outer core, similar to being underwater in a pool but still being able to hear someone above the surface. Since S waves are unable to pass through the fluid outer core, they are not detected in the areas labeled "shadow zone" and "P waves only." The shadow regions occur because when P waves enter into the fluid outer core, they bend abruptly at the inner mantle–fluid outer core boundary, but they also bend more gently as the density of the fluid outer core changes with depth.

As some of the P waves continue through the fluid outer core, they encounter another discontinuity at the solid outermost component of the inner core, again causing abrupt bending. There is also evidence for a solid innermost inner core. The paths of P waves bend abruptly as they exit the solid two-component inner core and the fluid outer core on their way to the surface on the opposite side of the planet from where the earthquake occurred ("P waves only").

By piecing together all of the arrival time data for P and S waves gathered from earthquakes around the world, the paths of the waves through Earth can be calculated, together with the densities that the waves passed through. This allows scientists to construct a detailed model of Earth's interior. It is even possible to measure rotation in the interior based on how rotation affects the waves. The core is actually rotating just slightly faster than the rest of the planet.

As expected, the outer core is composed primarily of iron and nickel. Given the electrically conducting characteristics of iron and nickel and its fluid nature, the outer core is the source of Earth's magnetic field; Earth's rotation and convection in the metallic outer core power a magnetic dynamo.

Because the pressure produced by overlying material increases so much toward the center (page 336ff), the extra squeezing turns the mix of iron and nickel in the inner core into a solid. The temperature of the solid inner core is about 5700 K (5400°C, 9800°F), very close to the Sun's surface temperature of 5772 K (5499°C, 9930°F). The existence of the solid innermost inner core may be the result of different forms of iron and nickel in the increasingly severe conditions at the center of Earth, or it may be telling us something about how Earth cooled in the early Solar System.

The mantle is primarily composed of minerals called silicates that make up the most common type of rock-building material in the Solar System. These minerals are primarily built on various combinations of silicon and oxygen, such as SiO_4, SiO_5, or Si_2O_7. The basic silicon and oxygen groups can then bond with one or more atoms of various other elements, such as beryllium (Be), magnesium (Mg), iron (Fe), or zirconium (Zr). SiO_2, known simply as silica, is a special type of silicate that doesn't combine with anything else and is commonly found in quartz, sand, and those moisture-absorbing gel packets that come with electronics, pills, leather goods, and other products.

Key Points
- The outer core is fluid.
- The outermost inner core is solid.
- The innermost inner core is also solid.
- All parts of the core are dominated by iron and nickel.
- The magnetic dynamo operates in the outer core.

A principal component of the mantle is a mineral known as olivine (Fig. 11.68) that is composed of the silicates Mg_2SiO_4 and Fe_2SiO_4. Olivine is also frequently found in certain types of meteorites and asteroids. When an olivine sample is subjected to high pressure and temperature, its internal structure morphs first into a form called wadsleyite, and then at even greater pressure and temperature the wadsleyite morphs into ringwoodite. At yet higher pressure and temperature, the ringwoodite decomposes into other minerals.

A detailed analysis of the speeds of S and P waves passing through the mantle indicates that there is a transition region in the mantle between 418 km and 668 km below the surface. Above the transition zone, olivine dominates, but in the transition zone wadsleyite and ringwoodite exist. Below 668 km the mineral decomposes.

Fig. 11.68 A gem-quality sample of olivine. Peridot is the name used for a gemstone composed of olivine. (Rob Lavinsky, iRocks.com, CC BY-SA 3.0)

In 2014, researchers conducting work in the Juína district of Mato Grosso, Brazil, were trying to find ways of determining diamond ages. Quite by accident, one of the worthless-looking diamonds they came across contained an inclusion of ringwoodite, the first sample ever discovered outside of a laboratory or in meteorites (Fig. 11.69). A volcano had ejected the ringwoodite during an eruption. Before its discovery, theory and experiment had shown that although olivine isn't able to store significant amounts of water, wadsleyite and ringwoodite can. The ringwoodite sample that was pulled up from the transition zone was shown to contain 1.4% H_2O (water). Assuming that the sample is characteristic of the transition zone, the transition zone within the mantle contains about as much water as all of Earth's oceans combined.

Fig. 11.69 A diamond containing an inclusion of water-rich ringwoodite. (Richard Siemens © University of Alberta)

As depicted in Fig. 11.70, although the mantle is essentially solid, it is really an extremely stiff fluid. This is because the mantle is hot enough to flow, very slowly, carrying heat from the core to the surface through convection. (Everyday window glass is also a stiff fluid; very old windows tend to be thicker at the bottom because gravity has caused material to slowly flow downward over time.)

The convection in Earth's mantle takes between 50 and 200 million years to complete one cycle, and is only moving about 2 cm (1 in) per year near the surface. Heat is also carried from the interior to the surface via conduction. The abundant water in the transition zone may play an important role in keeping the mantle fluid.

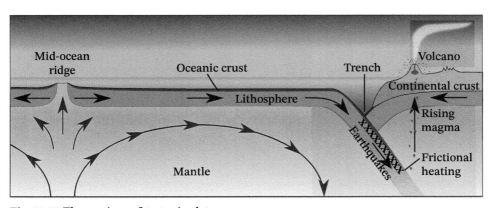

Fig. 11.70 The motions of tectonic plates.

Above the mantle is a layer called the lithosphere, on which the tectonic plates and the approximately 20 km thick surface crust ride. Warm mantle material is able to push through breaks between tectonic plates forming mid-ocean ridges, which leads to seafloor spreading away from the ridges. It is solidifying mantle material that forms new tectonic plate material at these locations, recording the direction of Earth's magnetic field at the time (page 404).

The seafloor spreading away from mid-ocean ridges drives lithospheric material and oceanic crust toward plates containing continental crust and its underlying lithosphere. As the oceanic crust and lithosphere cool and shrink, that material becomes more dense and sinks below the continental crust and lithosphere because continental crust is lighter (less dense) than oceanic crust. It is where these plates collide that deep ocean trenches can form. Although the convective mantle may play a small role in the motion of tectonic plates, it is the push coming from rising magma at mid-ocean ridges and the weight of the oceanic lithosphere moving back down into the mantle, pulling on the rest of the lithosphere that are the primary drivers of plate tectonics.

As the oceanic material continues to dive into the mantle, friction between the lithospheric and mantle material creates earthquakes and also leads to melting of the rock. The magma is lighter than the mantle and rises toward the surface, creating volcanoes. It is this process that is the cause of the Ring of Fire's earthquakes, volcanoes, and some of the deepest trenches in the world (Fig. 11.32). The recycling of Earth's crust and lithosphere is also why oceanic crust is younger than Earth itself. Because continental crust doesn't recycle into the mantle, some portions of continental crust are still very old, although still not as old as Earth overall.

Not all volcanoes occur at plate tectonic boundaries. A classic example is the Hawai'ian Island chain, located in the middle of the Pacific plate (Fig. 11.71). The Hawai'ian islands are examples of shield volcanoes, so named because they have very broad bases and are broadly rounded like the shields of Hawai'ian warriors. Shield volcanoes are produced when magma makes its way up from the upper mantle through a weak point in a tectonic plate, called a hotspot. Rather than explosive eruptions, shield volcanoes build up slowly over time as lava reaches the surface and oozes down the sides, as depicted in Fig. 11.72. If the volcano grows in an ocean, once it reaches the surface the lava may flow down to the water's edge, producing spectacular visual displays, including steam [Fig. 11.73 (*Left*)].

Key Point
Plate tectonics is driven by mantle material rising at mid-ocean ridges and lithospheric material sinking at ocean trenches.

Key Point
Oceanic crust gets recycled back into the mantle, making it younger than Earth itself.

Fig. 11.71 The Hawai'ian Islands seen from the International Space Station. (NASA)

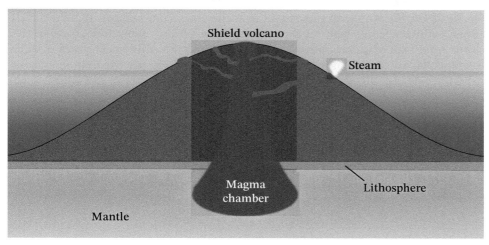

Fig. 11.72 A shield volcano arises from magma coming up through a hotspot in the lithosphere and crust.

Fig. 11.73 *Left:* Lava from the Kilauea Volcano reaching the ocean in Hawai'i Volcanoes National Park on the Big Island of Hawai'i. (NPS Photo/Janice Wei) *Right:* A thermal pool in the Yellowstone National Park caldera. (Dale A. Ostlie)

Mauna Kea volcano on the Big Island of Hawai'i is the tallest mountain on Earth when measured from its base. The volcano stands 4205 m tall (13,800 ft) above the surface and another 5000 m is below sea level. Because of its height and favorable atmospheric conditions, Mauna Kea is the site of one of the world's largest collections of major astronomical observatories. Mauna Kea is also considered sacred to some native Hawai'ians, which can lead to political, religious, and legal challenges when new construction is considered for the top of the mountain.

Because the Pacific plate is in motion, the Hawai'ian shield volcanoes eventually go dormant when they move off the hotspot. Over time, erosion wears the older volcanoes down and they disappear below the surface again, while new volcanoes grow from the hotspot. This is why the Hawai'ian islands are a chain, with the Big Island being the newest of the chain's members.

Another famous volcanic hotspot is in northwest Wyoming. Much of Yellowstone National Park sits atop a magma chamber and inside a volcanic caldera (the crater on the top of a volcano). It is for this reason that there is an abundance of thermal features in the park, which draw visitors from all over the world [e.g.,

Fig. 11.73 (*Right*)]. As with movement over the Hawai'ian hotspot, there is a trail of old, extinct volcanic features that run southwest from Yellowstone, because the North American plate has moved across the hotspot for millions of years. (A different explanation for Yellowstone's geothermal features suggests that an ancient tectonic plate is diving down into the deep interior, similar to Fig. 11.70.)

The Moon

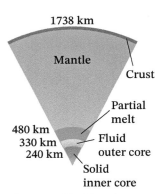

Fig. 11.74 The Moon's interior contains a mantle, a partially melted shell, a fluid outer core, and a solid inner core.

During the Apollo missions, the astronauts from five of the missions set up seismic stations at their various landing sites that operated flawlessly until they were shut off in 1977. The stations collected data from moonquakes. The analysis of the data revealed that moonquakes occur for a variety of reasons:

- impacts by meteorites produce weak quakes on the surface,
- heating and cooling of the surface between the two weeks of night followed by two weeks of daylight cause stresses that also produce surface quakes,
- some shallow moonquakes are produced 20 to 30 km below the surface, for which the cause is still unknown, and
- deep quakes originating about 700 km below the surface occur periodically due to the tidal stresses (Fig. 10.3) produced by Earth as the Moon travels in its eccentric orbit.

Unlike Earth, there are no tectonic plates capable of producing moonquakes.

Of the four types of moonquakes, the shallow quakes are the strongest, measuring up to 5.5 on the Richter scale and continuing for more than 10 minutes in each instance as though the Moon rings like a bell. Twenty-eight of these quakes were measured between 1972 and 1977. Given that quakes of this magnitude on Earth are capable of moving furniture and cracking plaster, it seems that future lunar settlements will need to be quake proof.

More than thirty years after the original data were collected from the lunar seismic stations, the data set was processed again by scientists using more sophisticated analytical methods. The influence of mass distribution on the orbit of the Gravity Recovery and Interior Laboratory spacecraft between 2011 and 2012 provided additional information. The results give a fairly detailed picture of what the interior of the Moon is like. As with Earth, the Moon has a thin crust made of silicates, ranging in thickness from a few kilometers in the maria to more than 50 km on the Moon's far side. The Moon also possesses a thick mantle composed of silicates, a fluid outer core primarily made of iron and nickel, and a solid inner iron–nickel core. What is different is that there is also a partially melted shell around the fluid outer core.

Unfortunately, although there is sufficient information to determine the structure and general composition of the Moon's interior, there is not enough detail to tell if a significant amount of water exists in the mantle. The important question for future lunar bases, "is the Moon 'wet' or 'dry'?", remains to be answered, although the general consensus is that it is dry.

From Fig. 11.74 it is apparent that the Moon's core is fairly small, which is consistent with its average density being only 3344 kg/m^3. After all, the density of material on its surface ranges from 2800 to 3300 kg/m^3.

Mercury

The evidence of ancient lava flows on the surface, the relative paucity of cratering compared to the Moon, Mercury's overall density, and the presence of a weak magnetic field are all clues to the planet's interior. There is no evidence of any plate tectonic activity on Mercury, but there is evidence that the planet contracted when it cooled. Since there are no seismic data available to analyze the propagation of P and S waves through its interior, the only other information we have is how Mercury affected the MESSENGER spacecraft's orbit around the planet during its mission between 2011 and 2015, together with how the planet rotates. From all this information, scientists have been able to construct a model of how material is distributed inside the planet as illustrated in Fig. 11.75.

One oddity of Mercury's interior was known even before the MESSENGER mission: its overall density is very high (5427 kg/m^3), second only to Earth's density in the Solar System. This is peculiar, because Mercury is also the smallest planet in the Solar System, with a radius that is just over 1/3 of Earth's radius and only slightly larger than the Moon. Earth owes part of its high average density to the fact that it is the most massive of the rocky planets, and therefore is being squeezed more by gravity, which compresses its interior. That is not the case for Mercury.

The answer to that paradox was found by analyzing MESSENGER's orbital data. Mercury has a substantial silicate crust that may average 50 km thick, making it several times thicker than Earth's crust. Below the crust is a surprisingly thin silicate mantle compared to the other rocky planets and the Moon. Immediately below the mantle there appears to be a solid shell of iron sulfide (FeS). Dominating its interior, Mercury has a very large fluid metallic core encompassing 85% of the planet's radius. There is also the possibility of a small solid metallic inner core. It is in the fluid iron–nickel metallic core that the magnetic dynamo operates to generate Mercury's very weak magnetic field, despite the planet's slow rotation. The size of the core should immediately raise the question, "how did Mercury end up with such a large core compared to its overall size?"

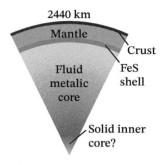

Fig. 11.75 Mercury's interior contains a fluid metallic core that comprises 85% of its radius. It also contains an FeS shell and perhaps a solid inner core.

Venus

The only seismic data available for Venus came from the Soviet Union's short-lived Venera landers in 1981 (page 391). Unfortunately, wind and the vibrations of the landers themselves made the data useless for determining the interior structure of Venus. The only other way to get measurements is to measure the effect of mass distribution on orbiting spacecraft, but to do so also requires that the planet is rotating. Since Venus rotates extremely slowly, that approach doesn't provide any usable data either. As a result, the best scientists have been able to do so far is to compare Venus with Earth, since they are very nearly the same size and have similar densities. One possible structure of the planet is depicted in Fig. 11.76.

In Section 11.2, it was mentioned that the entire surface of Venus appears to have been refreshed sometime between 300 and 600 million years ago; so how could that have happened so recently? It seems that a cataclysmic global event occurred sometime during that period. The lack of water in the atmosphere and on the surface of Venus likely extends to the planet's interior as well. Without water to help keep the mantle fluid, convection can't take place and there isn't any tectonic plate

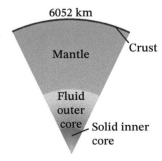

Fig. 11.76 Venus's interior likely has a mantle, a fluid outer core, and a solid inner core.

activity. This also means that the interior is unable to move heat to the surface and out into space very efficiently. The result is that interior heat builds up, causing the crust to melt. When the crust melts, some heat is released and a new crust forms.

Without any tectonic plate activity, all of the volcanism on Venus today and in the past must be due to hotspots. This implies that all of the volcanoes must be shield volcanoes, some of which have grown over millions of years to become very tall.

Mars

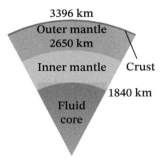

Fig. 11.77 The interior of Mars includes an outer and inner mantle along with a large fluid core.

In quest of seismic data for Mars, NASA's InSight mission landed on the surface in November 2018. After recording its first marsquake in April 2019, hundreds of quakes have been detected, but when this textbook was written none of the quakes had measured greater than 4 on the Richter scale. Despite these being relatively weak seismic events, early results indicate that some of the quakes only travel through the crust while stronger quakes do travel through the mantle and in some cases reflect off the planet's core. A handful of events have the signature of quakes generated by fault-line slippage, similar to earthquakes that are generated along fault lines on our planet, but on a much smaller scale.

To construct an interior model of Mars, other data are also available to augment the seismic data, including spacecraft orbital data, contributing to our knowledge of its density distribution, its surface composition, and whether or not a global magnetic field is present. In addition, unlike Mercury and Venus, an armada of orbiters, landers, rovers, and one tiny helicopter have explored Mars starting in the 1960s. All of the available data led to the interior model in Fig. 11.77.

The Red Planet gets its title from an unusually large amount of oxidized iron (rust) in the soil and rock on its surface. This would also imply that there is probably more iron in its mantle as well. Given that Mars is only about 1/2 the size of Venus and Earth, and it is 50% farther from the Sun than Earth is, it seems likely that it cooled down more quickly after formation, giving less time for iron and nickel to sink to the center. This idea seems to be supported by the fact that the Red Planet's average density is only about 3933 kg/m^3, which is much less than the other three rocky planets (Table 10.1), but still larger than the Moon's average density. Did Mars cool down too quickly to allow more of the heavier iron and nickel to sink to the center of the planet?

Another question to be answered has to do with the nature of Mars' core. Is the core completely solid, or is there a component that is still fluid? The absence of a measurable global magnetic field would suggest that if there is a fluid part to the core, then it probably isn't undergoing convection. This is because there isn't a magnetic dynamo in operation. Remember that dynamos require three parts: an electrically conducting fluid, rotation of the planet, and convection in the fluid. Mars does rotate just slightly slower than Earth does, which means that either none of the iron–nickel core is fluid or, if there is a fluid component, then convection must be absent. Once again, the universal nature of physical law allows us to analyze a situation based on fundamental processes and comparison to other, similar situations.

Finally, as with the Moon and all of the other rocky planets except Earth, Mars does not appear to have any current tectonic plate activity: either the entire crust is one tectonic plate or there are plates that no longer move around, suggesting that convection in the mantle stopped sometime in the past. As was the case with Venus, the enormous volcanoes of Mars, including Olympus Mons, must be shield volcanoes that have never moved off the hotspots that produced them, so they just kept growing. But could they still be growing? Based on the existence of cratering across all volcanic regions, volcanism on Mars seems have been occurring as recently as a few tens of millions of years ago, which is very recent relative to the Solar System's age of 4.57 billion years.

In Summary

Although the rocky planets and the Moon appear to have many characteristics in common (and they do), throughout Chapter 11 we have seen a range of important differences as well. Venus, Earth, and Mars currently have atmospheres, but Mercury and the Moon do not. The ones that do have atmospheres have very different atmospheres from each other. Although all of the planets are rocky, their interior structures can be quite different. Did something unique happen to Mercury during, or subsequent to, its formation that resulted in a core that makes up 85% of its radius? What about Mars, with its core comprising more than 50% of its radius, or the Moon with a very small core? Why does the Moon have rocks on its surface that seem similar to those on Earth and yet come up short when looking for compounds that can easily be vaporized, such as water? Is the Moon's interior "dry"?

One planet, the only one known to be harboring life, has liquid water and plate tectonic activity. That planet is also the only one with a large moon orbiting it. Certainly water is important for the development and sustainment of life on Earth, but does the existence of our large Moon and plate tectonics play any role?

There are still many questions about our own world and the other major bodies of the inner Solar System that remain unanswered. Answering those and many other questions not only satisfies our natural human curiosity, but also help us to understand Earth and our impact on it more fully. Perhaps finding answers will also help us understand how our Solar System formed, and how other planetary systems came into existence. Perhaps finding answers will even lead to discovering whether or not we are alone in the universe.

11.7 Population Growth and the Exponential Function

> *The greatest shortcoming of the human race is our inability*
> *to understand the exponential function.*
>
> Albert Allen Bartlett (1923–2013)

As the only body in our Solar System (as far as we know today) with a rich diversity of life, and the only one dominated with homo sapiens capable of intentionally modifying its regional and global environments, it important to understand the explosive rate of growth in the human population in order to fully understand Earth.

Figure 11.78 shows Earth's population, measured in billions, since 1000 CE. Prior to 1000 CE, hundreds of thousands of years passed before the human population reached 250 million. Another 600 years passed before the population doubled to 500

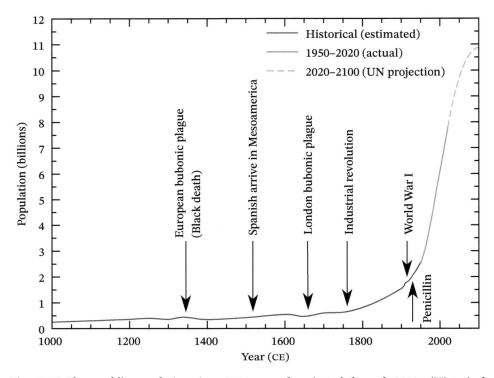

Fig. 11.78 The world's population since 1000 CE and projected through 2100. (Historical data from the United States Census Bureau; projections from the United Nations Population Diivision)

million worldwide during the lives of Galileo and Newton. Along the way, and for some two hundred years following, there were devastating setbacks in population growth due to disease, such as the Black Death (bubonic plague) in Europe in the late 1300s that reduced the European population by nearly one-third. As already discussed in this text on page 44, Mesoamerica was also devastated by disease when the Spanish arrived from Europe in 1519, and the plague that hit across England during Newton's time at Cambridge University, causing the university to close for eighteen months. Although not shown in Fig. 11.78, the 1918 influenza pandemic began during the last year of World War I and lasted until 1920, killing an estimated 17 to 100 million people world wide (the pandemic is also referred to as the Spanish flu, although it probably started elsewhere).

The world human population began to grow more rapidly from the time of the industrial revolution in the latter half of the eighteenth century. At the start of the industrial revolution, the world's population was still just 600 million. With the discovery of penicillin shortly after World War I, when the population had grown to nearly 1.8 billion (triple what it was at the start of the industrial revolution), population growth began to increase even more rapidly. When your author was born, in 1955, the human population of the planet was roughly 2.8 billion. As of March 2021, the population was almost 7.8 billion, nearly tripling in sixty-five years, and increasing by 200 million in the year preceding March 2021. The "medium variant" projection from the United Nations (UN) suggests that the world's population will reach 9.7 billion by 2050 and 10.9 billion by 2100.

Making predictions like these, especially over 80 years, requires numerous assumptions. Making predictions is a bit like driving a car down the road by only looking in your rear-view mirror; you base where you're going on where you've been, the behavior of the road so far, and some educated guesses as to what may be coming up in the future. But what if there is an unexpected cliff in the way and the road must change to adjust for the drop-off? You may have noticed that the UN's "medium variant" starts to rise more slowly going forward, based in part on effects such as diminishing resources, lower birthrates, war, and disease. The UN's "high variant" puts fewer restrictions on population growth, with the number of human beings reaching 15.6 billion by 2100. On the other hand, their "low variant" reaches a maximum population of 8.9 billion in 2054, then drops back down to 7.3 billion by 2100; a decrease of 1.6 billion in less than 50 years, and a smaller population than exists on the planet today.

The Exponential Function

Human population growth since the industrial revolution is a real-world, approximate example of the mathematical exponential function. Formally the exponential function always has the form

$$y = a^t, \tag{11.1}$$

Exponential function

where a is a constant and t is a variable, such as time. As an example, consider the function $y = 2^t$. When $t = 0$, $y = 2^0 = 1$ (any non-zero number raised to the 0 power is always one); when $t = 1$, then $y = 2^1 = 2$; for $t = 2$, $y = 2^2 = 2 \times 2 = 4$; when $t = 3$, $y = 2^3 = 2 \times 2 \times 2 = 8$, ... Figure 11.79 shows what the example function looks like graphically. You should compare the shape of that curve to the population curve in Fig. 11.78.

Professor Albert Allen Bartlett (1923–2013) used to present the exponential function applied to population in the form of an allegorical example. Imagine a bottle that is initially populated with self-aware bacteria that reproduce once every ten minutes, as illustrated in Fig. 11.80. After the first ten minutes, one bacterium divides into two bacteria, which then divide ten minutes later into four bacteria, and so on. Next, suppose that a population of bacteria occupy just 1/1024 of the volume of the bottle (lots of elbow room, plenty of space) and they also know that they have an abundance of resources on which they can feed and support themselves. After ten minutes they only occupy 1/512 of the volume, still a vast amount of space to live in.

Fifty minutes after the experiment began, the bacteria occupy just 1/32 of their bottle, so very few are worried at all about their living arrangements and the resources they have available to sustain themselves. However, the scientist bacteria are starting to sound an alarm that their population can't keep going like this or they are doomed! Given that the political cycle of the bacterial culture is only one minute between elections, why would the politician bacteria bother to worry about such a distant and highly speculative prediction; after all, there are many more urgent and immediate issues to deal with.

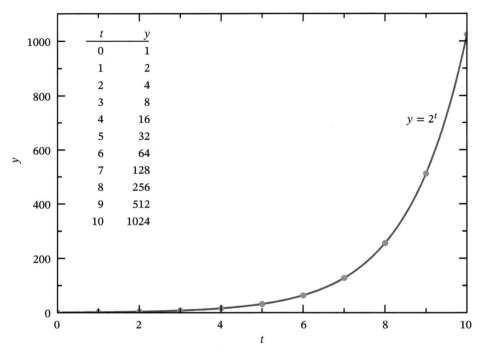

Fig. 11.79 The exponential function, $y = 2^t$.

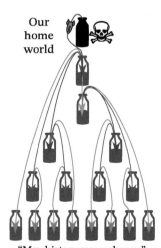

Our home world

"May history remember us"

Fig. 11.80 The bacteria laboratory.

One hour after the experiment began, still only 1/16 of the bottle is occupied, but the scientists' pleas are getting stronger: "we must do something or we won't survive." After another twenty minutes, one-quarter of the volume is occupied by the bacteria, and more of the population is starting to realize that resources are dwindling and that it is starting to feel claustrophobic in their little world. Meanwhile, the astronomer bacteria have been exploring beyond their bottle world and discover 16 other identical and habitable bottles elsewhere in the laboratory which they could colonize. Some politician bacteria are also starting to take notice and they fund the construction of spaceships that can take adventurous bacteria colonists out to other bottle worlds.

Ninety minutes into the experiment, the home bottle is 50% filled. At that moment, one-half of the population (one-quarter of the bottle's volume) heads out to a new bottle. When the ship arrives at its destination, one-quarter of the new bottle world's volume is immediately filled. Meanwhile, the home world bottle has already sent their intrepid explorers out into the laboratory universe, and so they simply await their fate. Twenty minutes after the colonists leave, the home world bottle is completely filled, all resources are gone, and the entire population dies off.

Ten minutes after arriving at their destination the population of the new world must send out another ship, and every ten minutes thereafter if they don't want to meet the same fate as the home bottle world; always with 1/2 of the current population, or 25% of the bottle volume. Every new bottle world must do the same thing, otherwise they will face catastrophe. This means that ten minutes after the first new bottle world is colonized, there are now two populated bottle worlds, then four more bottle worlds ten minutes after that, and then eight populated bottle worlds.

Unfortunately, 50 minutes after the first brave lab-faring bacteria left their home bottle world, all 16 bottles have been populated, each with 25% of their worlds filled. In another 20 minutes (2 1/2 hours after the experiment began), all 16 bottles must face the fate as their beloved, ancient, home bottle world.

At the end of this sad tale, the once proud and adventurous population of bacteria that believed their species would live on forever dies off, only to be one experiment in the notebooks of the laboratory. The laboratory scientists do try to keep their memories alive by writing a remarkable journal article, only to have it rejected for publication because the reviewers couldn't believe this amazing story and no other researcher could replicate the results.

The point of this story should be obvious; our planet Earth is a small place with very limited resources. It is impossible to support an exponentially growing population forever. If the human population were to double every 50 years (a slower pace than from 1950 to 2000), 50 years from now there may be 15 billion people on Earth. Even the most optimistic estimates suggest that 15 billion is very close to the maximum number of people that our planet can support, while many estimates suggest that we are currently very near, or have already exceeded, the carrying capacity of our home world. Mass starvation is already occurring on the African continent in the second decade of the 21st century.

Wildlife managers are very familiar with this story. When an elk herd grows too large, for example, there becomes too much competition for food, resulting in a die-off of much of the population. This is why an appropriate predator–prey balance is required for a healthy ecosystem. Humans are not immune to the simple reality that a finite planet has finite resources and cannot sustain boundless population growth.

Human population growth also threatens other species beyond our own. There have been five major mass extinction events in Earth's history; the Ordovician–Silurian extinction 439 million years ago, the Late Devonian extinction 364 million years ago, the Permian–Triassic extinction 251 million years ago, the Triassic–Jurassic extinction between 199 and 214 million years ago, and the Cretaceous–Paleogene extinction 66 million years ago when an asteroid impact created the Chicxulub crater (Fig. 6.19) near the present-day Yucatán peninsula that resulted in the extinction of the dinosaurs. Scientists believe that we are living in the sixth mass extinction today, brought about by human overpopulation and over-consumption. It is estimated that billions of species have vanished locally or regionally and that species are becoming extinct at a faster rate than has happened in millions of years. Many mammal species have already lost 80% of their range since 1900. The new geologic age in which human activity has a dominant role in shaping our climate and environment is termed the Anthropocene.

In order to maintain the current human population of the planet, the birth rate must equal the death rate. Given different mortality rates in different parts of the world, the number of children per woman would need to be about 2.1 in developed countries, and 3.0 in some developing counties. These numbers reflect the fact that not all girls will grow up to have children of their own due to death, infertility, or by personal choice. As of March 2021, the United States (the world's third most populous nation after China and India) was adding one net new individual to the population every 49 seconds through births.

Key Point
We are living during the sixth mass extinction in Earth's history; the Anthropocene extinction.

A Global Pandemic As a Real-World Exponential Function Application

In December 2019 a novel coronavirus, SARS-CoV-2, appeared in Wuhan Province, China. Within six months, more than 127,000 people in the United States had died of Covid-19, the disease caused by SARS-CoV-2. By the beginning of June 2022, more than one million Americans had died, with almost 6.3 million dead worldwide although those numbers are thought to be too low by perhaps 30% due to unreported and indirect deaths. Large numbers required hospitalization, seriously taxing health care systems. At times in the United States, refrigerated trucks were parked outside of hospitals to serve as temporary morgues. In 2 1/2 years more than 525 million people are believed to have been infected worldwide. The rapid increases in infections and resultant deaths were the direct consequence of individuals infecting more than one additional person, leading to exponential growth in infection rates, just as predicted in the earliest days of the pandemic.

In an effort to control the spread of the virus, in early 2020 governments ordered all but essential workers to stay home, causing world economies to enter into severe recessions. As countries started to reopen after several months of lockdowns, individuals were asked to wear masks when unable to stay more than 2 meters (6 feet) away from others (a similar strategy was used in the 1918 influenza pandemic). Within the United States, many people ignored the social distancing guidelines, the requests or orders to wear masks, and the restrictions on large gatherings, citing personal freedoms, or simply not believing the warnings of scientific experts. Still others believed that the pandemic was a hoax, even as they were dying of the disease. The skeptics included President Donald Trump and many in his administration, as well as governors of conservative states and their administrations. The unfortunate result was that infections continued to rise rapidly. In October 2020 President Trump became very seriously ill from the coronavirus and was rushed to the hospital, being released several days later after receiving some of the best medical care in the world (President Trump had previously held multiple maskless rallies and other, smaller gatherings).

Thanks to dramatic advances in medical science, by the end of 2020 and the first quarter of 2021, multiple vaccines started to become available and an intense effort was underway to vaccinate as many people as possible in most developed countries. Unfortunately, a significant percentage of the population refused to get shots because of a distrust of science and the spread of misinformation through social media platforms. In addition, after declining infection rates and deaths, rates began to rise again in parts of Europe and the United States during vaccine distribution when restrictions were lifted prematurely. In fact, several peaks in death rates occurred during the first 2 1/2 years of the pandemic within the United States and in other countries.

Although a firm estimate hadn't yet been determined by June 2022, 85 to 90% of the world's population may need to be vaccinated in order to halt virus spread, an outcome referred to as "herd immunity" (that number for measles is 95%). On the other hand, it may be that continuing vacine booster shots may be required similar to those that are given annually for flu.

Because pandemics can spread rapidly around the world, the World Health Organization (WHO) established an international Covid-19 Vaccines Global Access (COVAX) Facility, supported by most of the world's richest countries, to provide vaccines to developing nations.

11.8   Global Warming: Humans Impacting Environment

The Difference Between Weather and Climate

Human impact on planet Earth goes beyond overpopulation; homo sapiens are also directly affecting its global climate. There is a common confusion between the terms weather and climate. On February 26, 2015, United States Senator James Inhofe of Oklahoma (Fig. 11.81) famously brought a snowball into the Senate chamber to declare that the planet is not warming due to human CO_2 emissions. His argument was that it was unusually cold in Washington, D.C. that day, and therefore the climate can't be changing. Weather is a term that applies locally and at a specific time. We all know that the weather can change rapidly; for example, in northern Utah on September 4, 2017, the high temperature in Ogden was 35°C (95°F) but just twenty days later, the high temperature was only 9°C (48°F), with snow on the local mountain tops. Mark Twain (1835–1910) once said, "If you don't like the weather in New England now, just wait a few minutes." The same can be said for other parts of the United States as well. When Utah weather was unseasonably cool in late September, 2017, it was also unseasonably warm in Chicago (34°C, 93°F) on the same day.

Climate is an average behavior of the weather over much longer time periods, such as over decades. Climate is also averaged over larger areas that should be specified, such as the climate in northern Utah, southern Idaho, and Wyoming, or in the desert southwest. Climate is very insensitive to short-term fluctuations in the local weather, such as the temperature changes in Utah during September, 2017 or a snowstorm in Washington, D.C. in February 2015. The climate in northern Utah during the summer months is hot and dry, but cold and snowy during the winter. Washington, D.C. is hot, with high humidity during the summer months, and cooler with occasional snow during the winter.

Global climate *change* references a trend in the planet-wide climate, including average global and regional temperature variations and precipitation distributions, behavior of ocean currents, changes in the frequency and severity of major weather events, changes in sea level, impact on growing seasons and animal habitat, and so on. Global warming focuses specifically on the average temperature of the planet's atmosphere and oceans over time.

There is often confusion within the general population, and frequently among politicians, concerning the differences between global climate change, global warming, the greenhouse effect (page 388*ff*,) and weather. For informed conversations regarding global and regional climate change, it is important that definitions be properly understood.

Fig. 11.81 United States Senator James Inhofe of Oklahoma, February 26, 2015, on the floor of the Senate "proving" that climate change isn't real because it was cold and snowy in Washington, D.C. on that day. (C-SPAN)

Key Points
• Weather occurs at a given time in a specific location.
• Climate is an average of the weather across broad regions and over long periods of time.

Global Warming: The Evidence and the Cause(s)

Scientists have been raising concern for decades that the planet's oceans and atmosphere are warming at an alarming rate and yet there are many politicians and members of the general public who either remain climate-change deniers or don't consider climate change to be a serious or imminent issue. Being skeptical of claims is an important attribute of critical thinking; one of the principal learning outcomes that all higher education institutions strive to instill in their students. More

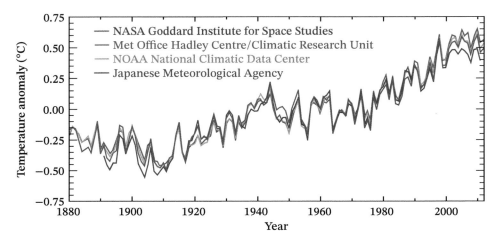

Fig. 11.82 Global warming data from four independent research groups. (Adapted from NASA image by Robert Simmon, based on data from the NASA Goddard Institute for Space Studies, NOAA National Climatic Data Center, Met Office Hadley Centre/Climatic Research Unit, and the Japanese Meteorological Agency.)

is required than simply being skeptical, however: a critical thinker must always demand and evaluate the evidence, regardless of whether or not the evidence fits their preconceived notions about the outcomes.

So, are the claims of global warming real? One way to answer that question is to look for independent verification. Figure 11.82 shows what is known as the global temperature anomaly from 1880 to 2012 as determined by four separate research groups. The temperature anomaly is simply the deviation, hotter or cooler, relative to an average temperature over a specified period of time, in this case from 1951 to 1980. If the temperature anomaly reads $-0.1°C$ that means that the global average temperature at that time was cooler than the average value between 1951 to 1980 by $0.1°C$. Apparent from the graph is that all four independent research groups are in very good agreement as to what is happening, with the global temperature over a period of more than 130 years. There is nearly unanimous agreement among scientists that the planet is warming. The more interesting and urgent question is *why is the planet warming?*

Throughout Chapter 11 you have seen how carbon dioxide and water play important roles in the development of rocky planet atmospheres, their surfaces, and subsurfaces. On Venus, a run-away greenhouse effect caused the once wet environment to lose most of its water, leaving behind a very hot, dense atmosphere composed mostly of CO_2. On Earth, most of the carbon dioxide ended up trapped in carbonaceous rock such as limestone, or dissolved in the oceans, although a small amount of CO_2 remains in the atmosphere, contributing to a small greenhouse effect. On Mars, a once-wet environment lost much of its carbon dioxide, with the result that almost all of the water ended up in large reservoirs of ice below the planet's surface or escaped into space.

The evolutionary story of Earth's atmosphere is far from complete today. Carbon dioxide (CO_2) and water (H_2O), along with methane (CH_4) and nitrous oxide (N_2O), are greenhouse gases that continue to affect our planet's atmosphere.

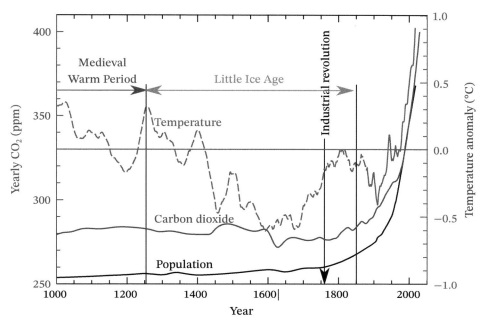

Fig. 11.83 The amount of carbon dioxide in the atmosphere (blue) and the global average temperature anomaly (red) since 1000 CE. The amount of CO_2 is read from the left-hand axis and the global temperature anomaly (relative to the average temperature between 1951 to 1980) is read from the right-hand axis. The dashed line for temperature is based on estimates from indirect sources such as tree rings, and the solid temperature line is from direct instrument measurements. The shape of the population curve from Fig. 11.78 is also included in black for comparison. (Scripps CO_2 Program; Loehle and McCulloch,[2] 2008

Figure 11.83 shows a graph of the amount of carbon dioxide in our atmosphere since 1000 CE (the blue curve). The amount of carbon dioxide is measured in terms of the number of CO_2 molecules that exist in the atmosphere in a sample of one million atoms and molecules (parts per million; ppm). In order to study the contributions of various gases in the past, scientists use a variety of techniques, including extracting ice cores (Fig. 11.84) from the Greenland and Antarctic ice sheets and testing air bubbles trapped in the ice. Since the ice is built up from snow falling over millions of years, the deeper researchers drill to obtain the ice, the farther back in time they are exploring.

You should note the exponential rise in atmospheric CO_2 beginning with the industrial revolution. It was this point in time when coal, oil, and later on, natural gas (primarily methane) began to be burned to produce energy at ever-increasing rates in order to power the machines being built (e.g., Figures 11.85 and 11.86). These energy sources are referred to as fossil fuels because they were formed from the decay of ancient plants and organisms. If you observe rising bubbles from a pond with decaying plants at the bottom, that is methane gas being released. When ancient plants and tiny organisms were covered under many layers of rock, they

Fig. 11.84 An ice core containing air bubbles. (Photo by Lonnie Thompson, Byrd Polar Research Center, Ohio State University/NOAO)

[2]Loehle, Craig and McCulloch, J. Huston (2008). Correction to: "A 2000-year global temperature reconstruction based on non-tree ring proxies," *Energy and Environment*, 19(1):93–100.

Fig. 11.85 Smoke stacks. (Photo by Tony Webster from San Francisco, CA; CC BY-SA 2.0)

Fig. 11.86 Traffic jam in Miami, Florida. (Photo by B137; CC BY-SA 4.0)

Fig. 11.87 An eruption on Krakatoa Island, Indonesia in 2008. (Photo by flydime; CC BY 2.0)

were transformed under pressure and heat. When these fossil fuels are burned they produce carbon dioxide, carbon being a fundamental ingredient for life on Earth.

Also shown, as the red curve in Fig. 11.83, is the global temperature anomaly. To read the values of the curve you need to refer to the vertical axis on the right-hand side of the graph (the red horizontal line at 0.0 marks the average global temperature between 1951 and 1980). The solid portion of the curve beginning in 1880 is based on actual instrument measurements of temperature at various sampling locations around the planet. The dotted portion of the curve prior to 1880 is derived from proxies for temperature, such as tree rings, isotope variations in the air bubbles of ice cores, fossil pollen, corals, mineral deposits in natural caves, and lake and ocean sediments. A proxy is some quantity that strongly correlates with another quantity that cannot be measured directly, such as global average temperature. If a proxy is used it is very important that its correlation is understood and carefully determined. Remember that peer review of research findings, and various approaches from multiple research groups help to validate scientific studies.

Global temperature variations can occur due to a variety of factors, such as volcanic eruptions. When volcanic ash enters the atmosphere and is circulated around Earth by winds, the ash cuts down on the amount of sunlight reaching the surface. On August 27, 1883, the volcano on Krakatoa Island, Indonesia (Fig. 11.87) experienced four huge explosions that were so violent they could be heard as far as 4800 km (3000 mi) away, in the Indian Ocean. Then, just 29 years later, Novarupta exploded on June 6, 1912 in a remote part of the Alaskan Peninsula (part of the Ring of Fire). Novarupta was the largest explosion of the twentieth century. The precipitous drop in the average global temperature due to the ash from those eruptions is apparent in Fig. 11.83.

We still don't fully understand all of the contributors to global temperature variations, such as the Little Ice Age between roughly 1250 CE and 1850 CE[3] or the Medieval Warm Period between 950 CE and 1250 CE, but it is worth recalling that the Maunder minimum in the sunspot cycle did occur in the deepest part of the Little Ice Age (page 328 and Fig. 9.22). Despite this fact, variability of the solar cycle doesn't account for the change in global temperature rise since 1880, but the average global temperature does coincide very closely with the rise in CO_2 emissions by the human population. For reference, the exponential population growth curve from Fig. 11.78 has also been included in Fig. 11.83.

Historically, the correlation between carbon dioxide concentration in the atmosphere and global temperature is very apparent, going back more than 400,000 years (see Fig. 11.88). The drops in temperature correspond to past ice ages. Although the temperature variations are larger than those shown in Fig. 11.83, it is important to remember that the time span is 400 times longer. In other words, the time span of Fig. 11.83 is the width of one-tenth of one minor tick mark interval in Fig. 11.88. You should also notice that, although the temperature anomaly graph ranges up to 4°C, most of the data are below the average global temperature of present day, and they go as low as −9°C; much colder than today's temperatures. In addition, the CO_2 concentration never went above 300 parts per million over at least the past 800,000 years, but the concentration exceeded 420 ppm in April, 2022 and is rising very rapidly.

---

[3] Estimates of the start and end of the Little Ice Age vary. Some studies suggest a start date in the 1500s and end dates as late as the early 1900s.

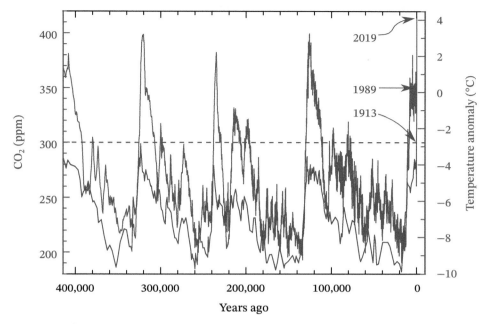

Fig. 11.88 The correlation between carbon dioxide concentration (blue) and temperature anomaly (red) since 414,000 BCE. The horizontal, dashed blue line is the maximum level of CO_2 over the past 800,000 years, until 1913. (Data courtesy of Petit, et al., 2001[4])

When searching for explanations of patterns in nature, it is important to understand that correlation does not necessarily imply causation, meaning that just because two phenomena seem to behave similarly doesn't mean that they are necessarily directly (or even indirectly) related. But in the case of CO_2 and global average temperature we do understand how their trends are coupled through the greenhouse effect.

There still remains the question of what other influences may be forcing the rise in global temperature that we have seen since humans began to increase the proportion of carbon dioxide content in the atmosphere. Before CO_2 can be established as the primary forcer, other potential contributions must be studied as well.

We have already seen that volcanoes can decrease global temperatures through the spewing of ash, but they also emit a host of gases, including carbon dioxide and sulfur-based compounds. CO_2 can certainly increase global temperatures, but sulfur compounds can lower it. In any case, human-caused emissions of CO_2 have been approximately 100 times greater than from volcanoes since 1880, as shown in Fig. 11.89.

Perhaps variability in the Sun's luminosity or changes in Earth's orbit might alter how much sunlight strikes the planet. Those possibilities have also been investigated carefully, and found to be insignificant compared to how CO_2 is affecting today's climate. For example, a 405,000 year cycle exists that is driven by very long-term patterns in the orbits of Earth, Venus, and Jupiter. Their gravitational interactions increase the eccentricity of Earth's orbit from nearly circular to about

[4]Petit, J.R., et al. (2001). "Vostok Ice Core Data for 420,000 Years," IGBP PAGES/World Data Center for Paleoclimatology Data Contribution Series #2001-076. NOAA/NGDC Paleoclimatology Program, Boulder CO, USA.

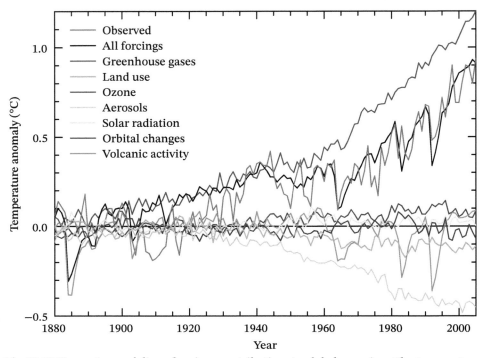

Fig. 11.89 Computer modeling of various contributions to global warming. The temperature anomaly is based on the average global temperature between 1880 and 1910. An animation of the graph illustrating the effects of each of the contributions separately and their combined contributions is available on the `Chapter 11` resources webpage. (Based on data from Bloomberg and ModelE2 climate simulations of NASA/GISS)

$e = 0.05$ and back again (recall from Table 10.1 that Earth's current orbital eccentricity is only $e = 0.017$, meaning that it is nearly circular).

Research that involved extracting very long rock cores from ancient lake beds in New York (6900 m long, 4.3 mi) and New Jersey (2500 m, 1.6 mi), similar to the ice core in Fig. 11.84, revealed cyclic patterns of lake beds drying up and refilling over hundreds of thousands of years, due to periods of severe drought followed by much wetter periods. Another 500 m (1500 ft)-long core extracted from a butte in Petrified Forest National Park in Arizona allowed scientists to study Earth's shifting magnetic field and correlate that with radiometric dating (page 257) of the three cores in order to tie together the cyclic patterns in climate with Earth's magnetic field changes, and then couple those changes with the 405,000-year eccentricity cycle. The eccentricity cycle correlates with climatic cycles back at least 215 million years, to the time of the dinosaurs in the Late Triassic age. This is 149 million years before dinosaurs became extinct due to the Chicxulub meteor impact (page 200).

So is this 405,000-year cycle impacting our climate today? The answer is no, because we are very near the middle of the cycle when Earth's orbit is nearly circular; we are only slightly closer to the Sun during the northern hemisphere winter than we are during northern hemisphere summer. However, it is important to include all orbital patterns in assessing impacts to our climate, in order to determine what the major culprit(s) is (are) in forcing climate change.

There is also a host of human causes beyond carbon dioxide, such as deforestation, ozone pollution, and aerosol pollution. Again, none of those potential sources explain the rapid rise in global temperature since the industrial revolution. However, by combining all of the possible contributors, both natural and human-driven, the results agree very well with the temperature rise that is observed, which is dominated by human carbon dioxide emissions.

Key Point
Science goes to great lengths to solve problems as definitively as possible, and must incorporate all relevant information and data.

Computer Modeling of Global Warming, Climate Change, and Weather

Back in Section 9.4 we briefly discussed how computer modeling has become critical in helping scientists understand complex physical systems. Section 9.5 and the sections that followed provided a short introduction to how computers are used to explore the interior of our Sun. Computers are also fundamental in helping scientists understand our planet's climatological past to the present day, and forecasting both the weather and what we can expect from climate change going forward.

Some of the world's most powerful supercomputers are brought to bear in studying global warming, climate change, and weather forecasting. This is because these subjects are critically important, both for our near- and long-term future and because the calculations are extremely complex.

Consider the issues raised in the last subsection on potential forcers of global warming. In order to fully understand what is happening to our environment, all of these components must be included in any successful model of climate change. In addition, a great many other aspects of physics, chemistry, geology, meteorology, and plant biology must be included as well. A partial list of what should be incorporated in the modeling of climate change and its effects includes:

Key Point
Drivers and effects of climate change.

- all of the various influences on Earth's orbit and shifts in Earth's axis tilt,
- atomic and molecular details of energy levels, the efficiency of photon absorption and emission per particle for all wavelengths of light, how light can break molecules apart, and the chemistry of how molecules form and interact,
- how long various greenhouse gases survive in the atmosphere,
- the relationships between temperature, air and water density, pressure, and composition (the most simplified form of this for a gas is the ideal gas law, discussed in Chapter 9),
- the temperatures, densities, pressures, and compositions with height in the atmosphere and depth in the oceans,
- the surface temperatures at many locations for land masses, oceans, and other large bodies of water,
- the abundance of various isotopes of the elements and how their concentrations are affected by physical processes that depend on temperature (remember that different isotopes of the same element have different masses and therefore move at different speeds with the same kinetic energy),
- how much heat can be absorbed at various levels in the atmosphere and in bodies of water,
- how heat is transferred both around Earth and throughout the atmosphere and oceans,
- the various influences that affect atmospheric and ocean currents, including Earth's rotation,

- how much carbon dioxide, methane, oxygen, nitrogen, and other gases can be absorbed in the oceans, and how quickly,
- the salinity and acidification of oceans resulting from gas absorption,
- the release of H_2O through evaporation,
- the rise in sea levels due to land-based glacial and ice cap melting, along with expansion of water as it warms,
- the influence of wildfires, including the release of carbon dioxide and soot,
- the emissions, both ash and gases, from volcanoes,
- the amount of sunlight striking Earth at locations all around the planet (recall that this is strongly influenced by latitude; Section 4.4),
- the amount of sunlight reflected back into space, rather than being absorbed, by ice, snow, water, clouds, and dust in the atmosphere,
- the shapes of land masses, mountain ranges, bodies of water, deserts, forests, and large cities (cities are sometimes referred to as "heat islands" because they can get hotter than surrounding areas),
- the absorption and release of carbon dioxide, methane, and other greenhouse gases by living and decaying plants, as well as deforestation and how rapidly plants grow in different conditions,
- the release of trapped methane as permafrost melts (Fig. 11.90),
- the production of methane by various human activities, including agricultural (livestock produce methane in their digestive tracts, and rice production also produces significant quantities of methane), coal mining, natural gas extraction, and waste disposal,
- and of course, the burning of fossil fuels.

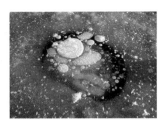

Fig. 11.90 Methane bubbles trapped in a frozen pond after the melting of permafrost below the surface. (U.S. Geological Survey)

Not only do these physical processes need to be included, but they must be included by describing those processes in a language that computers can understand; mathematics converted into computer languages.

Beyond the physical processes, the success of computer models depends heavily on how many locations are sampled in the three-dimensional climate system. Computers are unable to calculate what happens in every cubic centimeter of ocean, bodies of water, and the atmosphere, nor can computers sample everything that happens on every square centimeter of the planet's surface every second of every day for tens or hundreds of years; there simply aren't computers big enough, with enough data storage, and fast enough computationally to accomplish that overwhelming task. Instead, discrete sampling is required that (hopefully) accurately approximates the real world while still being computationally possible to accomplish in a reasonable amount of time. What constitutes a *reasonable amount of time* largely depends on patience, the need to get reliable results, and funding.

Still another aspect of computer modeling of physical systems, including climate, is the use of the most efficient and accurate computer programming techniques possible. With computer-based computations there are typically many different ways to write parts of the program; some may be fast and produce excellent results, but others may be very slow and/or give inaccurate results. The reason for this has to do with the fact that computers use approximate numbers instead of values that are infinitely accurate. For example, the fraction 2/3, expressed digitally is 0.666 666 66 ..., going on forever. How many sixes the computer uses can have a big influence on what the answer turns out to be after trillions of subsequent calculations.

Fig. 11.91 Hurricane Irma as seen by the NOAA satellite GOES-16. The outlines of Cuba, southern Florida, and other islands can be seen (NOAA/CIRA)

This sensitivity to how many digits are used in computer calculations was first realized in the early days of weather forecasting, when a calculation was stopped while the scientist went to lunch. When he returned he entered numbers into the computer from his printout to restart the calculation from where it was stopped. In reality the printout didn't display as many digits for the numbers as the computer itself was using in the calculations. When the calculation finished, the results were very different from the same calculation performed without stopping the computer in the middle. This result became known as the butterfly effect. The name derives from the idea that a butterfly flapping its wings somewhere in the world will subtly change the local weather conditions by changing the wind speed near the butterfly by a very tiny amount. The change will then be amplified tremendously by the very sensitive nature of these physical systems, possibly resulting in a storm somewhere else.

On September 6, 2107, Hurricane Irma (Figures 11.91 and 11.92) reached wind speeds of 285 km/h (180 mi/h) as a category 5 storm while passing through the Caribbean on its way to Florida, with forecasts of tremendous damage, especially along Florida's southeast coast. As a result, warnings went out to evacuate Miami and other coastal areas in its path. By the time the hurricane made landfall on September 10, its direction had shifted slightly to the west and its strength had decreased by a small amount, making it technically a category 4 storm. Despite the downgrade, Irma still did very significant damage to the Florida Keys and the west coast of the state.

Although the weather forecasts for this huge and complex storm were remarkably accurate, given that its track was initially forecast twelve days in advance of making landfall on the Florida coast and refined in subsequent days, there were those, including some talk show celebrities, who tried to claim that the scientific

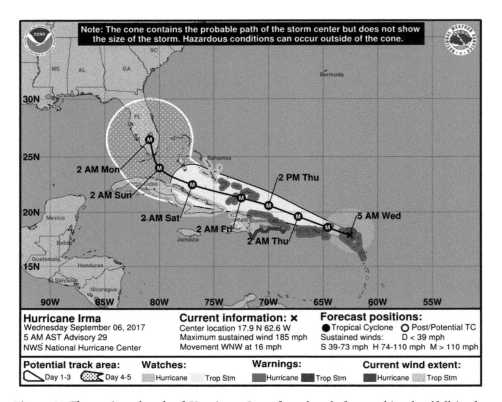

Fig. 11.92 The projected path of Hurricane Irma four days before making landfall in the Florida Keys on September 10, 2017. The white cone is the range of uncertainty one to three days into the future from the hurricane's current position. The hashed region is the uncertainty in forecast models four to five days ahead. (NOAA)

experts didn't know what they were talking about. Some even claimed prior to the storm's arrival that forecasters always make predictions that are far more severe than what ultimately occurs. Sadly, those pre-storm statements had the potential of getting people killed when this devastating storm struck. One of the most well-known media celebrities also claimed that there is a conspiracy to promote the *climate change hoax* through forecasting powerful hurricanes. He further suggested there is a goal to increase panic buying of supplies such as food, water, fuel, and generators during such emergencies. Ironically, the individual making those claims had a residence in the projected path and at the last minute inferred that he had to evacuate because he wasn't sure that he could legally remain in place.

Just eight days after Hurricane Irma made landfall in Florida, Hurricane Maria, another category 5 storm, passed close to the Dominican Republic and Haiti, before it struck Puerto Rico on September 20, 2017. In all, Hurricane Maria killed 3057 people, 2975 of them in Puerto Rico, during and in the wake of the storm. Both storms followed Hurricane Harvey [Fig. 11.23 (*Left*)], a category 4 storm, that struck Texas on August 25, dropping as much as 1.539 m (60.58 in) of rain on locations east of Houston and killing at least 68 people.

One year after three devastating storms struck the Caribbean and the United States, Hurricane Florence, at one point a category 4 storm, made landfall in North Carolina as a category 1 hurricane on September 14, 2018, shattering state rainfall records with more than 77.67 cm (30.58 in) of rain near Swansboro, 80 km (50 mi) northeast of Wilmington, and leaving at least 13 dead. At the same time Florence was wreaking havoc on the Mid-Atlantic coast of the United States, Typhoon Mangkhut was tearing through the Philippines and Hong Kong as a storm 890 km (550 mi) wide, with sustained winds of 190 km/h (120 mi/h), and killing at least 64 people. The same media celebrity who tried to claim a climate change hoax when forecasters warned of the dangers of Hurricane Irma in 2017, made the same claim about Hurricane Florence in 2018.

The increasing strength and rainfall potential of tropical storms, hurricanes, cyclones, and typhoons arises because of more heat energy in warming oceans and greater water-carrying capacity of warmer air. Additional potential for damage due to storm surge results from increased wind speeds and rising sea levels. The storm surge from Hurricane Florence was reported to be 3 m (10 ft) in some places.

Testing Computer Models

Short-term weather forecasting (several days to one week or more) is very difficult given the tremendous complexities involved, but those forecasts are constantly improving with further understanding of the physical processes involved and the continual increase in computational power and sophistication. Climate change forecasting is no less difficult, but it is not particularly sensitive to the butterfly effect, given the long-term nature of the problem.

So how are the results of these very complex calculations to be believed? The answer lies in (a) comparing the results of separate, independent studies (a form of peer review), and (b) testing the models' predictions, just as scientists are always required to do with any other claim coming from a theory or a hypothesis derived from theory.

As of 2018 there were roughly thirty independent research groups in eleven countries around the world with more than 60 climate models that are studying global climate change. These groups take slightly different approaches based on how to handle various components of their models, such as cloud formation or ice coverage. Considering all of the best data and computer modeling available, there is overwhelming evidence to support the conclusions that global warming is occurring, that CO_2 is driving global warming, and that human-caused emissions are to blame. These conclusions are supported by more than 97% of global warming experts and nearly 200 scientific organizations worldwide.

Key Point
Computer models must always be tested against experimental or observational data. Computer models should also lead to predictions of new results.

What Is Happening Today and Forecasts For the Future

In addition to rising temperatures due to the greenhouse effect, there are many other expected consequences of global climate change that include

- melting and retreating glaciers worldwide,
- sea level rise,
- stronger storms,

Fig. 11.93 Alaska's Muir glacier as seen in 1941 and 2004. (National Snow and Ice Data Center, W. O. Field, B. F. Molnia)

- extended and more extreme heat waves and cold weather,
- longer and more severe droughts in historically dry parts of the world, along with more precipitation in other, historically wetter regions,
- changes in the length of growing seasons,
- shifts in crop-growing regions northward (or southward in the southern hemisphere), and
- migration of animals and insects to higher latitudes and/or higher elevations.

Each of these predictions appears to be occurring.

Although glaciers go through periods of growth and retreat, the ice ages being major examples, today's glaciers are retreating, and in some cases disappearing, very quickly. In some sense, glaciers are the "canary in the coalmine"[5] for climate change. One dramatic example is the Muir glacier in Alaska, shown in photographs taken from the same location in 1941 and again in 2004 (Fig. 11.93).

Sea ice is also a sensitive measure of climate change. In the Arctic ocean, the amount of sea ice waxes and wanes with the seasons, but for decades the ice thickness has been diminishing and the annual minimum extent of coverage has generally been decreasing. As of 2020, the fourteen lowest extents have occurred in the last fourteen summer seasons, with 2012 holding the record and 2020 finishing in second place. In Antarctica, an enormous rift developed in 2016 along an ice shelf floating on the ocean [Fig. 11.94 (*Left*)], which then broke away completely in 2017, producing an iceberg the size of the State of Delaware in the northeastern United

Key Point
Only water from land-based melting glaciers contributes to sea level rise.

Fig. 11.94 *Left:* A rift in the Larsen C ice shelf in Antarctica, 2016 that broke off in 2017, creating an iceberg the size of the state of Delaware. (John Sonntag/NASA) *Center:* A dry lake bed due to drought. (NOAA) *Right:* A wildfire approaching power lines. (NPS)

[5]Early in the coal-mining industry, canaries were used underground to test for poisonous gases such as carbon monoxide; if the canary died the coal miners needed to get out immediately.

States. Although the iceberg was huge, it didn't result in a sea level rise because it was already in the ocean; only the water from land-based glaciers, such as those in Greenland, or on the continent of Antarctica, can actually result in sea level rise due to melting.

In addition to water being added to the oceans by the melting of land-based ice, water also expands when it is heated. Since 2000, sea levels have been rising at a rate of 0.3 cm (0.1 in) per year. As sea levels continue to rise, low-lying areas are at risk of permanent flooding unless expensive construction projects are undertaken to mitigate the problems. In the United States, and throughout much of the world, a significant fraction of the human population lives in coastal areas. Unfortunately, some regions of the world such as Bangladesh and some islands in the South Pacific simply do not have the financial resources for major mitigation efforts.

Disappearing ice worldwide has a direct influence on global warming, through a feedback loop. Snow and ice help to reflect sunlight back out into space without being absorbed by the ground. This means that there is less heat trapped by greenhouse gases. Unfortunately, as glaciers and ice in the Arctic and Greenland disappear, less light is reflected and more light is absorbed, further raising global temperatures and melting even more ice.

Key Point
Melting ice accelerates global warming through a feedback loop.

As already mentioned, hurricanes can become stronger due to global warming by tapping into the additional energy contained in the warmer waters that they pass over, and by being able to hold more water in the storms due to warmer air. But other storms, such as thunderstorms and tornadoes, can also strengthen because of warmer air conditions. Even snowstorms can become more powerful because of increased energy, warmer oceans, and more water in the atmosphere.

From late 2011 to early 2017, California experienced the most severe drought in recorded history. The drought was so severe that aquifers (naturally occurring below-ground supplies of water) were getting depleted because water was being pumped out of them for irrigation purposes. As a result, in some areas the ground sank by more than 33 cm (13 in) in just eight months. The drought paused when record snowfall hit the mountains, only to resume the following summer with continuing record-breaking drought throughout California and the American West [Fig. 11.94 (*Center*)]. As of 2022, Lake Mead and Lake Powell, America's two largest reservoirs, both of which are on the Colorado River that passes through Grand Canyon National Park, are critically low. The reservoirs are important sources of both water and electricity for Las Vegas and southern California. Emergency policies are being put in place to keep the reservoirs viable water and energy sources for as long as possible.

With drought comes an increased volume of dead trees and grasses that can burn faster and hotter than normal [Fig. 11.94 (*Right*)]. Not allowing natural fires to occur over time causes a build-up of fuels on the forest floor. The result is more devastating wildfires, such as those experienced in the American West and other areas of the world during the twenty-first century. Through May 2022, six of the seven largest wildfires in California's history occurred since August, 2020, and the seventh was in July, 2018. The irony of fire suppression in wild areas is that it interrupts the natural cycle of clearing forest floors and the destruction of aging, unhealthy trees, which in turn interferes with the creation of open areas and new growth needed for a diversity of animal species. Some trees, such as the lodgepole pine and the jack pine, even require fire to destroy the resins that keep their cones sealed. Only when very hot fire occurs can the cones open to release their seeds. All of this can dramatically alter ecosystems.

On the other side of the ledger, global warming may have a beneficial effect on certain regions, as long as the temperature rise isn't too excessive. In some agricultural areas, warmer weather means extended growing seasons and potentially more food production. However, if the average temperature in those regions increases too much then heat waves can become more deadly to humans and agricultural products.

Dangerously high sustained temperatures have been seen thoughout the world, including India, Pakistan, Africa, Australia, and western North America. Short-term temperature spikes have been recorded in Lytten, British Columbia, Canada, where the temperature reached 49.6°C (121.3°F) on June 29, 2021, and in Antarctica, where the temperature was 38 celsius degrees (70 fahrenheit degrees) above normal on March 18, 2022. On June 30, 2021, the day after Lytten's record-setting temperature, a wildfire destroyed most of the small village. It is important to understand, however, that individual weather events cannot be attributed to climate change, but continually increasing global average temperatures raise the baseline from which higher record temperatures can be expected.

Another feedback loop of great concern with increasing global temperatures is the thawing of permafrost, like the area shown in Fig. 11.95. Recall that Earth's tectonic plates have been moving around the planet continuously, including into regions that are now far enough north that the ground became permanently frozen. When those areas were in warmer climates, plant and animal life was prevalent. The decay of those plants and animals left large amounts of methane trapped in what is now permafrost. As the permafrost thaws, the methane gets released into the atmosphere. As a greenhouse gas, methane is 82 times as potent as carbon dioxide, but is shorter lived in the atmosphere. While carbon dioxide can survive for centuries, methane only persists for a decade or two. Unfortunately, the release of methane leads to more rapid thawing of the permafrost and the release of more methane into the atmosphere. The ground temperature of permafrost regions is rising more rapidly than the air temperature, having increased by 1.5 to 2.5 celsius degrees (2.7 to 4.5 fahrenheit degrees) over a period of 30 years.

Fig. 11.95 Lakes formed by thawing permafrost in Alaska's Arctic National Wildlife Refuge. (U.S. Fish and Wildlife Service)

To further illustrate the issue of melting permafrost, the northeastern Siberian town of Verkhoyansk recorded a temperature of 38°C (100°F) on June 20, 2020. Verkhoyansk is located north of the arctic circle and has an average June temperature of 20°C (68°F). Three weeks earlier, a diesel holding tank in Siberia ruptured after sinking into once-frozen permafrost. The spill released 150,000 barrels of fuel that drained toward the Arctic Ocean.

International Climate Agreements

The Intergovernmental Panel on Climate Change (IPCC) was established jointly in 1988 by the United Nations Environmental Programme and the World Meteorological Organization. The organization's mission is to assess the science of climate change, including potential impacts, future risks, adaptation, and potential mitigation. Their best estimates for the future of global population, global temperature, and sea level rise are given in Table 11.2. The projections become more uncertain farther into the future, as one would expect, with projections depending

Table 11.2 Historical data and future projections. (United Nations, 2017;[6] Temperature and sea level projections for 2100 from IPCC, 2014, Table 2.1[7])

Date	Global population (billions)	Global temperature change (°C)	Global sea-level rise (cm)
1990	5.5	0	0
2000	6.1	0.2	2
2020	7.7	0.6	8
2046–2065	[a]8.7–8.6 / 10.4–12.3	[b]0.4–1.6 / [c]1.4–2.6	[b]17–32 / [c]22–38
2081–2100	[a]8.2–7.3 / 14.2–16.5	[b]0.3–1.7 / [c]2.4–4.8	[b]26–55 / [c]45–82

[a]Population estimate: first–last year of range; lowest projection / highest projection.

[b]Uncertainty range for the lowest population estimate.

[c]Uncertainty range for the highest population estimate.

on various assumptions made in the forecasts. Note that in 2100 the lowest population estimate is below the population in 2021, reflecting the possibility that resource depletion, disease, and other factors could act to reduce the population; without those potential limiting factors, the upper end of the population projection exceeds 16 billion.

In light of these and past predictions, the world has worked to establish agreements between nations to try to limit greenhouse gas emissions. In 1992 the United Nations Conference on the Environment and Development met to begin the first of these efforts. Then in 1997, the participating nations met in Kyoto, Japan, to agree on broad targets for emissions. Known as the Kyoto Protocol, the agreement set legally binding targets relative to 1990 (or earlier) emissions levels for 38 industrialized countries and the European Union, while developing countries could participate through an alternative process. The goal was to keep temperature rise to less than 2°C above pre-industrial levels, in the hope of avoiding feedback-loop processes like ice melt and melting permafrost. Less industrialized countries were given less stringent emissions targets, given their less significant contributions to greenhouse gas emissions in the past. The United States of America's target was 93% of 1990 emissions during the 2008–2012 commitment period, for Canada the target was 94%, and for the United Kingdom the target was 87.5%. Most of the participating countries ratified the agreement by 2008, indicating their intention to participate. However, in the United States, President George W. Bush was elected in 2000, and his administration chose not to ratify, citing negative economic impact and the fact that China and India, the two most populous nations, were not participating. This made the United States the only original signatory not to ratify the agreement. Given that the United States produced 26% of all greenhouse gas emissions in 1990, that country's decision not to ratify was a major setback for the

[6]United Nations (2017). *World population prospects: The 2017 revision*, available at https://www.un.org/development/desa/publications/world-population-prospects-the-2017-revision.html

[7]IPCC (2014). *IPCC Climate Change 2014: Synthesis Report* available at https://www.globalchange.gov/browse/reports/ipcc-climate-change-2014-synthesis-report

Kyoto Protocol. In 2011 Canada withdrew from the agreement during the Great Recession that began in 2008; the nation was concerned that they would be unable to meet their targets, which would trigger significant financial penalties.

The Paris Climate Agreement was the next major effort to limit greenhouse gas emissions. This agreement was written within the United Nations Framework Convention on Climate Change, representing a negotiated agreement between 196 nations, and adopted in 2015. In the Paris accord, countries that did not participate in the Kyoto Protocol were brought into the agreement, including China and India. Again, the target is to limit the global average temperature to less than 2 celsius degrees above pre-industrial levels, but strive to limit the temperature rise to just 1.5 celsius degrees. As with the Kyoto Protocol, several months after the United States President Donald Trump took office in January 2017, he announced his intent to withdraw from the agreement, although the withdrawal couldn't take place until November 4, 2020, due to language of the agreement (the withdrawal took effect one day after Trump was defeated in the presidential election). President Trump stated that "the Paris accord will undermine [the U.S.] economy ... [and] put [the U.S.] at a permanent disadvantage." President Trump also made it part of his energy policy to roll back many climate change and environmental policies established by his predecessor, President Obama. Following President Trump's defeat in the elections, President Joe Biden signed an executive order for the United States to rejoin the Paris Climate Agreement on February 19, 2021. President Biden also set an aggressive agenda to lower greenhouse gas emissions in the United States by 50% by 2030 although much of the agenda requires the approval of Congress.

The percentage of CO_2 emissions by country in 2019 is shown in Fig. 11.96. Between the drafting of the Kyoto Protocol and the Paris Climate Agreement, China rapidly grew its industrial sector, especially steel production. It also expanded its generation of electricity by building coal-burning power plants. Because of the resulting poor air quality, it is estimated that 1.1 million Chinese die annually from air-pollution-related ailments. However, in May 2017 China made the decision to dramatically improve its air quality. To that end, they announced the closure or cancellation of 103-coal burning power plants, to be replaced by solar and wind energy generation through the world's largest financial investment in these renewable technologies. As of 2019, China is the world's leader in solar power production. China is also improving auto emission standards, and its citizens have purchased more electric vehicles than all of the rest of the world combined. Despite these efforts, because China currently generates 30% of the world's greenhouse gas emissions, its targets for the Paris Climate Agreement are inadequate to meet the goal of keeping the global temperature rise below 2°C of pre-industrial levels. At the same time, China is providing loans to other countries so that they can construct new coal-burning power plants.

In a March 2021 survey of more than 700 economists, conducted by the Institute for Policy Integrity at the New York University School of Law, nearly three-quarters of those surveyed believed that the cost of not ending our dependence on fossil fuels will likely exceed the cost of reaching net-zero greenhouse gas emissions by 2050, something scientists have been arguing for decades. Projected costs of dealing with the economic damage of climate change could be as much as $1.7 trillion by 2025 and $30 trillion annually by 2075.

Iraq joined the Paris Climate Agreement in January 2021, making a total of 192 of 197 countries, the most notable holdouts being Iran and Turkey.

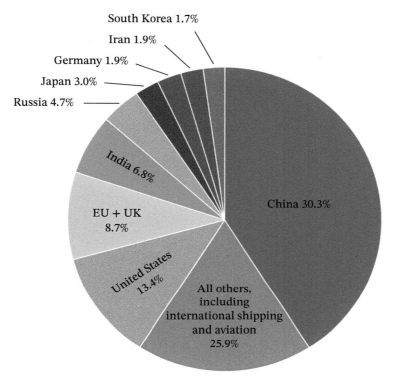

Fig. 11.96 The top ten CO_2 emitting countries in 2019. In 2010, China's share was 27.0% and the United States' was 16.4%. (Data from Crippa, et al, 2020.[8])

On Valentine's Day, February 14, 1990, at the urging of Carl Sagan, a planetary scientist who worked on the mission, NASA pointed the camera of Voyager 1 back toward Earth from a distance of 40 au and obtained the iconic picture shown in Fig. 11.97. In his book *A Pale Blue Dot* (Random House, 1994), Sagan reminded us that everyone we know or have ever known — indeed all of human history — is tied to that "pale blue dot." Wars are often fought over small portions of that tiny orb. We live on a small and fragile planet; there is no "planet B" for humanity and the other endangered species that reside on that "pale blue dot."

Someone cannot claim to know more than the experts until that person first knows everything the experts know.

[8]Crippa, M., Guizzardi, D., Muntean, M., et al. (2020). "Fossil CO_2 emissions of all world countries - 2020 Report," JRC science for policy report EUR 30358 EN (Publications Office of the European Union).

Fig. 11.97 The "pale blue dot" image of Earth from a distance of 6 billion kilometers (40 au), obtained by Voyager 1 on Valentine's Day, February 14, 1990 (republished with modern image-processing software for the 30th anniversary in 2020). The streaks are scattered sunlight in the camera. (NASA/JPL-Caltech; circle added)

? Exercises

All of the exercises are designed to further develop *your understanding* of the material by thinking carefully and critically about what you have read in this chapter. Answers to selected exercises are available in the back of the book.

True/False

1. The reason that Mercury doesn't possess an appreciable atmosphere is because it moves too quickly around the Sun and any atmosphere is swept away.
2. (*Answer*) A carbon atom is 12 times more massive than a hydrogen atom. This means that a carbon atom in a hot gas moves 12 times faster than a hydrogen atom.
3. Cliffs formed on the surface of Mercury because the planet shrank.
4. (*Answer*) Peaks are commonly found in moderately sized, younger craters.
5. The relative ages of craters can be estimated by studying which craters overlay other craters.
6. (*Answer*) Given that Mercury is only about 0.4 au from the Sun, it is impossible for water ice to exist on its surface.
7. The atmosphere of Venus is currently dominated by carbon dioxide because the planet formed that way out of the original solar nebula.
8. (*Answer*) A prediction of the runaway greenhouse effect on Venus is that the heavy version of hydrogen, deuterium, should be many times more abundant than normal hydrogen; a prediction that was confirmed.
9. It is possible that Venus once had oceans on its surface.
10. (*Answer*) Tides, including atmospheric tides, appear to have caused Venus to rotate very slowly retrograde.

11. We know a great deal about the surface of Venus thanks to robotic landers from the 1990s, that continue to move across its surface.
12. (*Answer*) Venus's volcanoes have undergone explosive eruptions in the past, but are completely inactive today.
13. The surface of Venus appears to have been completely refreshed between 300 and 600 million years ago.
14. (*Answer*) Without some greenhouse effect, Earth's average temperature would be below the freezing point of water.
15. Most of the nitrogen (N_2) in Earth's atmosphere is due to the development of life on the planet.
16. (*Answer*) An important cause of global circulation of Earth's atmosphere is the rising of warm air near the equator and the sinking of cold air near the poles.
17. Only about 20% of the mass in Earth's atmosphere lies below an altitude of 20 km.
18. (*Answer*) South America and Africa are moving apart at the rate of 2.5 cm per year.
19. The Himalayas are still growing in height due to the continuing collision of the India subcontinent with the Eurasian tectonic plate, causing the crust to buckle.
20. (*Answer*) Pangaea was proposed to be a supercontinent that existed hundreds of millions of years ago to explain the similar shapes of the east coast of South America and the west coast of Africa. However, the concept was abandoned when no other evidence could be found of its existence.

21. Earth's magnetic north pole is about 11° south of true north (geographic north).
22. (*Answer*) It is possible that satellite electronics can be damaged and planetary power grids can be affected by material blasted into space from especially powerful solar flares.
23. Geomagnetic storms are high-altitude thunderstorms.
24. (*Answer*) Earth is 4.54 billion years old, but the Moon is younger, at 4.51 billion years old.
25. Our Moon is roughly 30% larger than Mercury, making it the fourth largest body in the inner Solar System.
26. (*Answer*) The Moon's maria range in age from 3.2 to 4.4 billion years old.
27. There is evidence that liquid water exists on the lunar surface.
28. (*Answer*) When Schiaparelli reported that he believed canali existed on Mars, he was convinced that they were artificially created by an intelligent species.
29. The radio drama, *War of the Worlds* was aired less than one year before the start of World War II, when Nazi Germany invaded Poland.
30. (*Answer*) Wind speeds on Mars typically reach 1000 km/h.
31. Clouds have been seen above the giant volcanoes on Mars.
32. (*Answer*) Valles Marineris is a canyon system that dwarfs the Grand Canyon by comparison.
33. Valles Marineris probably formed in the same way that the Grand Canyon formed on Earth, through water erosion over millions of years.
34. (*Answer*) Similar to canyons in the desert southwest of the United States, the side canyons of Valles Marineris are subject to intense flash flooding caused by water runoff from its giant volcanoes.
35. The giant volcanoes on Mars can be seen erupting explosively today.
36. (*Answer*) The presence or absence of a global magnetic field gives us some information about the interior of a planet or moon.
37. Both P and S waves can travel through fluids.
38. (*Answer*) P waves travel faster through solids than S waves do.
39. A P wave is a sound wave.
40. (*Answer*) Earth's innermost inner core is composed of molten iron and nickel.
41. Silicates are molecules containing silicon and oxygen along with other atoms such as magnesium or iron.
42. (*Answer*) The temperature at the center of Earth is close to the temperature at the surface of the Sun.
43. There doesn't appear to be any water in the interior of Earth.
44. (*Answer*) The Mid-Atlantic Ridge is a place where mantle material pushes through the oceanic crust, contributing to seafloor spreading.
45. The Hawai'ian Islands are the result of the Pacific plate moving over a hotspot.
46. (*Answer*) The Moon's strong magnetic field means that the body contains a large, molten iron–nickel core.
47. If the density of the crust on a rocky body is 3000 kg/m³ but the overall density of the body is 5000 kg/m³, there must be material somewhere inside the body with a density that is greater than 5000 kg/m³.
48. (*Answer*) Like Earth, Venus and Mars have tectonic plates that move slowly across their surfaces today.
49. The CO_2 concentration in Earth's atmosphere has reached 400 ppm on several occasions over the past 400,000 years.

50. (*Answer*) One way in which scientists study Earth's atmosphere over hundreds of thousands of years into the past is to analyze air bubbles trapped deep in glacial ice.
51. Computer models of global climate change do not make testable predictions and therefore cannot be considered to be scientific.
52. (*Answer*) Sea level on Earth has risen by more than 8 cm since 1990.

Multiple Choice

53. Which of the following reasons contribute to a lack of any significant atmosphere on Mercury.
    (a) The planet moves too rapidly around the Sun.
    (b) The planet's rotation rate is too fast.
    (c) Its surface temperature is very high, especially on the sunward side.
    (d) Mercury has a low escape speed.
    (e) Its magnetic field is too weak to protect it from the solar wind.
    (f) all of the above
    (g) (a) and (b) only
    (h) (c), (d), and (e) only
54. (*Answer*) An oxygen atom has a mass that is 16 times greater than a hydrogen atom. The average speed of an oxygen atom on the sunward side of Mercury is closest to _____ km/s.
    (a) 0.06   (b) 0.25   (c) 0.5   (d) 1
    (e) 2   (f) 4   (g) 16
55. Caloris Basin on Mercury
    (a) is very old.
    (b) faces the Sun once every 176 days.
    (c) is associated with ripples that traveled across Mercury to the other side of the planet.
    (d) is covered with craters.
    (e) all of the above
56. (*Answer*) The reason that water ice exists in deep craters near Mercury's north and south poles is because
    (a) Mercury's rotation axis is perpendicular to its orbit.
    (b) Mercury doesn't have any meaningful atmosphere.
    (c) water ice exists below the surface and when the craters were formed, the ice was exposed.
    (d) all of the above
    (e) none of the above
    (f) (a) and (b) only
57. According to Wien's blackbody color law, most of the photons emitted from the surface of Venus have wavelengths near _____ nm, which is in the _____ portion of the electromagnetic spectrum.
    (a) 200, ultraviolet   (b) 400, infrared
    (c) 600, visible   (d) 800, radio
    (e) 1000, infrared   (f) 2000, ultraviolet
    (g) 4000, infrared   (h) none of the above
58. (*Answer*) Which of the following *is not* a consequence of the runaway greenhouse effect on Venus.
    (a) Atmospheric deuterium is 100 times more abundant than normal hydrogen.
    (b) With the exception of small amounts in its clouds, water has virtually vanished from the planet's atmosphere.
    (c) Sulfuric acid (H_2SO_4) makes up 3.5% of Venus's atmosphere.

Exercise 58 continues on the next page

58. *Continued*
    (d) The planet once had liquid water oceans, but it is completely dry today.
    (e) The surface temperature is hotter than the Sunward side of Mercury.
    (f) All of the above are consequences of Venus's runaway greenhouse effect.
59. We are able to view the surface of Venus in which part(s) of the electromagnetic spectrum?
    (a) radio   (b) infrared
    (c) visible   (d) ultraviolet
    (e) x-rays   (f) gamma rays
    (g) both (a) and (c) (h) none of the above
60. (*Answer*) Craters on the surface of Venus
    (a) range in size from less than 1 km in diameter up to 30 km across.
    (b) often contain central peaks.
    (c) indicate that only very large impactors are able to make it through the dense atmosphere.
    (d) provide a crude means of estimating the age of the planet's surface.
    (e) all of the above
    (f) (a) and (b) only
    (g) (b), (c), and (d)
61. The two major constituents in Earth's atmosphere, and their relative percentages, are
    (a) N_2 (78%), O_3 (21%)
    (b) N_2 (21%), CO_2 (78%)
    (c) N_2 (78%), O_2 (21%)
    (d) O_3 (78%), N_2 (21%)
    (e) CO_2 (96.5%), N_2 (3.5%)
    (f) none of the above
62. (*Answer*) What happened to most of the carbon dioxide on Earth?
    (a) It escaped into space because the molecule is unusually light.
    (b) It dissolved into the oceans where it is still located today.
    (c) It was absorbed into the oceans, ultimately producing carbonaceous rock, such as limestone.
    (d) It was all absorbed into Earth's mantle.
    (e) It was destroyed during the gigantic collision that created the Moon.
63. Earth's jet streams and other weather features are driven in part because
    (a) the planet's rotation means that the equator moves at more than 1500 km/h but the poles only spin in place.
    (b) of drag on the atmosphere while Earth orbits the Sun.
    (c) of tides produced by the Moon and the Sun.
    (d) of differential rotation of Earth's mantle.
64. (*Answer*) Approximately _____ of the surface of Earth is covered by water.
    (a) 3.5%   (b) 10%   (c) 21%   (d) 50%   (e) 71%
    (f) 96.5%
65. Which of the following are consequences of plate tectonics on Earth?
    (a) earthquakes   (b) volcanoes
    (c) the Ring of Fire   (d) mid-ocean ridges
    (e) deep ocean trenches   (f) motions of continents
    (g) all of the above   (h) (a) and (b) only
    (i) (d) and (e) only

66. (*Answer*) Earth's magnetic field
    (a) is more than 140 times stronger than Mercury's field.
    (b) is compressed on the sunward side and stretched out on the side opposite to the Sun.
    (c) traps charged solar wind particles.
    (d) helps to protect life on the surface from dangerous charged particles from space.
    (e) has its north pole 11° away from Earth's geographic South Pole.
    (f) helps to retain Earth's atmosphere.
    (g) all of the above
67. Earth's magnetic field flips (reverses polarity) approximately once every _____ years.
    (a) 11   (b) 22   (c) 10,000
    (d) 100,000   (e) 800,000   (f) 800 million
    (g) 800 billion
68. (*Answer*) Radiometric dating of Moon rocks revealed that most maria are between _____ and _____ billion years old, although some older maria exist as well.
    (a) 3.2, 3.5   (b) 3.2, 3.8   (c) 3.2, 4
    (d) 3.5, 4.4   (e) 3.5, 4.54
    (g) None of the above; all of the maria are 4.51 billion years old.
69. The maria on the lunar surface are dark and fairly smooth because
    (a) they are significantly younger than the lighter colored regions of the Moon.
    (b) when they formed, lava flowed to the surface, filling in impact craters.
    (c) the maria managed to avoid the impacts that occurred at the same time in the lunar highlands.
    (d) even though they are the same age, the rock of the maria contain a large amount of carbon, which is darker than the iron-based material in the highlands.
    (e) (a) and (b) only
    (f) (c) and (d) only
70. (*Answer*) Which of the following are evidence of ancient lava flows on the Moon?
    (a) Long, narrow paths from the extinct volcanoes in the highlands to the lower elevation maria.
    (b) The maria are mostly smooth and largely devoid of cratering.
    (c) The rocks collected from the maria by the Apollo astronauts are made of basalts.
    (d) Lava tubes have been discovered on the Moon.
    (e) all of the above
    (f) none of the above; there is no evidence of past lava on the Moon's surface
    (g) (b), (c), and (d)
71. The claim that the "Face on Mars" was an example of a monument built by an intelligent civilization that had lived on, or perhaps visited, Mars is a consequence of a well-known psychological phenomenon of _____, and is known as _____.
    (a) identifying patterns where none exist, pareidolia
    (b) creating conspiracy theories, X-Files syndrome
    (c) imagination triggering, Pangaea
    (d) facial recognition, Wellesianism
    (e) none of the above

72. (*Answer*) Mars has a mass that is roughly _____ and a radius that is _____ times that of Earth.
    (a) 0.11, 0.8   (b) 0.11, 0.53   (c) 0.2, 0.53   (d) 0.4, 1.2
73. Like Venus, Mars has an atmosphere that is composed of _____, but unlike Venus, Mars' atmosphere _____.
    (a) 0.0420% N_2, also contains 3.5% CO_2
    (b) more than 95% CO_2, is very thin
    (c) more than 95% CO_2, is very hot
    (d) 78% N_2, is more than 21% O_2
74. (*Answer*) The winds on Mars can
    (a) cause dust devils.
    (b) create sand dunes.
    (c) fill in smaller, older craters.
    (d) sometimes lead to planet-wide dust storms.
    (e) cause solar panels on rovers to get covered with dust.
    (f) all of the above
75. It seems that Mars once had a large amount of liquid water on its surface. What happened to the water?
    (a) Almost all of it escaped directly into space.
    (b) All of the molecules broke apart due to the Sun's infrared radiation.
    (c) It is all contained in the polar ice caps.
    (d) Much of it turned to ice and the rest broke down in the upper atmosphere from ultraviolet radiation.
    (e) (a) and (b)
76. (*Answer*) The permanent component of the polar ice caps on Mars is composed of _____ and the seasonal component, the part that is there during the winter, is _____.
    (a) carbon dioxide ice, nitrogen ice
    (b) carbon dioxide ice, water ice
    (c) nitrogen ice, carbon dioxide ice
    (d) nitrogen ice, water ice
    (e) water ice, carbon dioxide ice
    (f) water ice, nitrogen ice
77. Which body has the densest atmosphere?
    (a) Mercury   (b) Venus   (c) Earth   (d) the Moon
    (e) Mars
78. (*Answer*) Rank the densities of the five bodies discussed in Chapter 11 from high to low.
    (a) Mercury, Venus, Earth, Moon, Mars
    (b) Mars, Moon, Earth, Venus, Mercury
    (c) Earth, Venus, Mercury, Mars, Moon
    (d) Earth, Mercury, Venus, Moon, Mars
    (e) Earth, Mercury, Venus, Mars, Moon
79. Which of the following can provide information about the interior structure of a rocky body?
    (a) Subtle changes in the orbits of satellites.
    (b) Heat flow from the interior.
    (c) The propagation of seismic waves.
    (d) The presence or absence of magnetic fields.
    (e) The average density of the body.
    (f) The composition of rocks and minerals found on the surface.
    (g) all of the above
80. (*Answer*) The interior of Earth has been studied by analyzing P and S waves. P waves travel _____ than S wave, and both types of waves can travel through solids but _____ waves are unable to travel through liquids.
    (a) more slowly, P   (b) more slowly, S   (c) faster, P
    (d) faster, S

81. If a rocky planet rotates slowly but still has a global magnetic field, what can you conclude about the interior?
    (a) It must have a large molten metallic inner or outer core.
    (b) There is a solid core composed of iron and nickel that possesses a permanent magnetic field.
    (c) A small molten metallic outer core exists that is rotating at least ten times faster than the rest of the planet.
    (d) (a) and (b)
    (e) (b) and (c)
82. (*Answer*) Earth's mantle differs from those of Venus and Mars because
    (a) Earth's mantle is convective.
    (b) Earth's mantle is associated with the movement of tectonic plates, whereas tectonic plates on Venus and Mars (if they exist) aren't moving today.
    (c) Earth's mantle is primarily composed of silicates while Venus and Mars have mantles composed of iron and nickel.
    (d) The mantle of Venus has very little to no water in it, and Mars' mantle is composed almost entirely of saltwater.
    (e) The mantles of Venus and Mars make up a much smaller percentage of their volumes than Earth's mantle does.
    (f) (a) and (b)
    (g) (c), (d), and (e)
    (h) all of the above
83. Shield volcanoes have been found on
    (a) Mercury   (b) Venus   (c) Earth
    (d) the Moon   (e) Mars
    (f) all of the above
    (g) (b) and (c)
    (h) (b), (c), and (e)
84. (*Answer*) Which of the following appear to be required for tectonic plates to move across the surface of a planet?
    (a) a fluid mantle   (b) convection in the mantle
    (c) water in the mantle
    (d) all of the above   (e) none of the above
85. For the exponential function, $y = 2^t$, if $t = 20$, what is the value of y?
    (a) 2056   (b) 20,480   (c) 204,800
    (d) 1,048,576   (e) 1.7244×10^{30}
86. (*Answer*) In the bacteria parable, how much longer could the self-aware bacteria society have survived if there were 32 bottles that they could have occupied instead of 16?
    (a) 10 minutes   (b) 20 minutes   (c) 30 minutes
    (d) 40 minutes   (e) 50 minutes   (f) 1 hour
87. The Medieval Warm Period ended in about 1250. The CO_2 concentration in the atmosphere at that time was about _____ and the global temperature anomaly was close to _____.
    (a) 355 ppm, $-0.6°C$ (b) 282 ppm, 0.3°C
    (c) 282 ppm, 355°C   (d) 2820 ppm, 0.03 °C
    (e) none of the above
88. (*Answer*) The highest the global temperature anomaly has been in the past 400,000 years is about
    (a) $-2.8°C$ (b) 300 ppm   (c) 400 ppm
    (d) 3.2°C   (e) 4°C   (f) 8°C
89. What has made the greatest contribution to cooling our atmosphere since the beginning of the twentieth century?
    (a) greenhouse gases   (b) land use   (c) ozone
    (d) aerosols   (e) solar radiation
    (f) orbital changes   (g) volcanic activity

90. (*Answer*) The primary concern raised by melting permafrost on Earth is
    (a) the release of large quantities of methane gas.
    (b) the creation of large mud flats.
    (c) mosquitoes.
    (d) loss of the cooling effect of frozen ground.
    (e) There is no significant concern.

91. As of 2021, the only major country (countries) not to ratify the Paris Climate Accord is (are)
    (a) China (b) European Union (c) India
    (d) Iran (e) Iraq (f) Russia
    (g) Turkey (h) United Kingdom (i) United States
    (j) (a) and (c) (k) (a), (f), and (i) (l) (d) and (g)

Short Answer

92. Explain why the lack of a magnetic field contributes to the absence of any significant atmosphere around Mercury and the Moon.

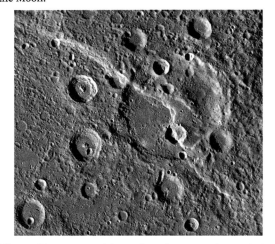

Fig. 11.98 A portion of the surface of Mercury. (NASA/Johns Hopkins University Applied Physics Laboratory/Carnegie Institution of Washington)

93. Figure 11.98 shows a small portion of the surface of Mercury. Is the 2-km-tall cliff older or younger than the large crater? Explain your reasoning.

94. With the aid of a diagram that includes light coming from the Sun, explain how it is possible for craters at the north and south poles of Mercury to be in constant darkness.

95. In your own words, explain how carbon dioxide, water, and methane produce the greenhouse effect.

96. (*Answer*) The relationship between the energy of a photon and its wavelength can be rewritten as $\lambda = hc/E_{\text{photon}}$. If the photon is emitted by an atom then $E_{\text{photon}} = \Delta E$, where $\Delta E = E_i - E_f$ is the initial energy of an atom, ion, or molecule, and E_f is its final energy after the photon is emitted. By writing hc in a convenient form, the equation for the wavelength of the photon becomes

$$\lambda = \frac{1240 \text{ nm eV}}{\Delta E}.$$

Two of the energy levels in the water molecule of Fig. 11.12 are -0.453 eV and -0.893 eV (electron volts). If a transition

occurs in a water molecule from the higher energy level to the lower energy level, determine the emitted photon's energy (ΔE), its wavelength, and the region of the electromagnetic spectrum that it resides in.

97. Explain why craters smaller than 10 km in diameter are absent on the surface of Venus.

98. What evidence exists to suggest that Venus's volcanoes have been active in the recent past and are probably still active today?

99. Despite lacking a magnetic field, explain why Venus has been able to hang onto its atmosphere when Mercury couldn't.

100. What happened to much of the life on Earth in the ancient past when photosynthetic organisms first appeared?

101. List at least four ways in which the surface of Earth can be altered.

102. Describe how the aurora borealis and aurora australis are produced, and how they are related to the Sun.

103. (*Answer*) Explain how it is possible to know that the darker maria are younger than the lighter colored highlands, simply by looking at Fig. 11.41 (*Left*).

104. What is the most likely reason that the Moon has so few maria on the side facing away from Earth?

105. Explain why the surface temperature of Mars is so much colder than the surface temperature of Venus, even though both of their atmospheres are more than 95% carbon dioxide. Include two reasons in your answer.

106. In your own words, briefly describe how the Martian atmosphere evolved to its present-day conditions.

107. Explain how the ideas that H_2O and CO_2 broke apart and escaped from the atmospheres of Venus and Mars, respectively, can be tested by studying isotopes.

108. (a) Describe at least three ways in which scientists have been able to learn something about the interior of Earth.
     (b) Have any of those methods been used to study the Moon's interior? If so, which one(s)?

109. Explain how Earth's outer core could be fluid while its inner core is solid, and give two possible reasons for why the inner core appears to have two components.

110. Do you think that the apparent correlation in Fig. 11.83 between carbon dioxide concentration in the atmosphere and human population growth is coincidental or causal? Explain your reasoning.

111. What are the current United States and World populations? See, for example, the United States Census Bureau's population clock at https://www.census.gov/popclock/.

112. Use a web search engine to identify the three major events between 1960 and 2010 that caused the drops in the volcanic activity contribution to the temperature anomaly curve in Fig. 11.89.

113. On page 428 the statement was made that the Moon's core must be fairly small, given that the Moon's average density is only 3344 kg/m^3 and the density of rock on its surface ranges from 2800 to 3300 kg/m^3. Explain how this conclusion must be true.

114. In the parable of the bacteria society, suppose that there were evil bacteria who were determined to survive, along with their offspring. Could they permanently stop the inevitable by killing other bacteria in their bottle world? Explain your answer.

115. Explain the difference(s) between weather and climate.

116. What is the butterfly effect?
117. Describe a simple experiment you can do in your kitchen to demonstrate that melting sea ice does not contribute directly to sea level rise.
118. Sea levels are rising on Earth for two reasons resulting from rising global temperatures. What are they?

Activities

119. (a) Create planetary "slice" diagrams, like those in Figs. 11.74–11.77, for all four planets and our Moon, placing them side by side. You should also scale each of them relative to the size of Earth.
(b) Comment on similarities and differences between the interiors of the five bodies.

12 The Giant Planet Systems

We must believe then, that as from hence we see Saturn and Jupiter; if we were in either of the Two, we should discover a great many Worlds which we perceive not; and that the Universe extends so in infinitum.

Savinien de Cyrano de Bergerac (1619–1655)

12.1	Comparing the Giant Planets	462
12.2	The Moons of the Giant Planets	476
12.3	Rings around the Planets	491
	Exercises	497

Fig. 12.1 Montage. (NASA/JPL/SwRI/MSSS)

Introduction

As we continue our exploration of the Solar System, we now move out to the giant planets: Jupiter, Saturn, Uranus, and Neptune. Recall from Table 10.1 that there are very profound differences between the rocky planets and the giant planets. Not only are the giant planets much larger, more massive, significantly farther from the Sun, with very different compositions from their rocky cousins, but they are also more than just individual planets with perhaps a moon or two; they are major systems of bodies, consisting of many moons and possessing ring systems.

Exploration

The first visitors from Earth to the giant planets were the early flyby missions to Jupiter and Saturn by Pioneers 10 and 11 in the 1970s and the spectacularly successful Voyagers 1 and 2 between 1979 and 1981, seen as an artist's rendition in the center of Fig. 12.1. Voyager 2 also flew past the ice giants, Uranus and Neptune, in 1986 and 1989, respectively, while Voyager 1 traveled up out of the plane of the ecliptic after its encounter with Saturn. Remarkably, Voyagers 1 and 2 were still operating as late as 2021, becoming the first two messengers from Earth to leave our Solar System and become ambassadors to the stars. Onboard the Voyagers are golden records like the one shown in Fig. 12.2 that contain images, natural sounds, songs, and messages from President Jimmy Carter and U.N. Secretary-General Kurt Waldheim representing the inhabitants of planet Earth. The records were conceived by Carl Sagan with the contents selected by a committee that he chaired.

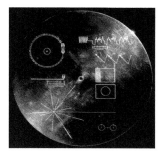

Fig. 12.2 The golden record carried on Voyagers 1 and 2. (NASA/JPL)

The gas giants, Jupiter and Saturn, have also been the targets of extended orbital missions. Galileo (1995–2003) and Juno (2016–continuing) both studied Jupiter, and Cassini (2004–2017) investigated Saturn. Cassini was also accompanied to the Saturn system by the Huygens probe, a lander that explored Titan, Saturn's large, atmospheric moon. Both Galileo and Cassini were sent plummeting into the two gas giants at the ends of their missions to ensure that they didn't someday inadvertently crash into moons. Of course, not a moment of data gathering was wasted during their suicide plunges because they continued to sample composition, gas pressure, temperature, wind velocities and other quantities, and send information back to Earth for as long as possible. Juno will eventually meet the same fate when it is intentionally sent into Jupiter to burn up. The ice giants, Uranus and Neptune, have yet to be studied by orbital missions.

Jupiter was also briefly visited by the New Horizons spacecraft in 2007. New Horizons used Jupiter to get a gravity assist that increased the spacecraft's speed for its long journey to the dwarf planet Pluto, the Kuiper belt, and then out to the stars.

The various missions have returned spectacular images of the planets, their moons, and ring systems. They have also returned light spectra that allow scientists to identify various constituents in their atmospheres and in the atmosphere of Titan. Spectra also provide information about the surfaces of some of the moons and the compositions of ring particles. In addition, the missions have gathered data about magnetic fields of the four giant planets, and internal mass-distribution data are obtained by measuring subtle changes in spacecraft orbits produced by the planets' gravitational fields.

In this chapter we will be exploring the giant planets based on the information gathered by these emissaries to the outer Solar System. Additional important information about the planets has been obtained by Earth-based and space-based telescopes.

12.1   Comparing the Giant Planets

General Characteristics

The giant planets are displayed in relative size in Fig. 12.3 and some of their characteristics are given in Table 12.1. The largest planet in the Solar System, both by mass and size, is Jupiter. At nearly 318 M_{Earth}, Jupiter contains more than 71.1% of the total mass of the planets, and its enormous size implies that more than 11 Earths can be placed side by side across its equator. Saturn, the second largest planet, contains an additional 21.3% of the mass, while Uranus and Neptune, combined, make up another 7.1%. This leaves the rocky planets with just over 0.4% for the Solar System's planetary mass [our home planet, Earth, contains a meager 0.2% (0.002) of the total mass]. Little wonder why the outer planets are referred to as giants.

Fig. 12.3 The giant planets and their relative sizes from left to right: *Gas giants:* Jupiter and Saturn; *Ice giants:* Uranus and Neptune. (NASA/JPL)

Table 12.1 Characteristics of the giant planets

Characteristic	Gas giants		Ice giants	
	Jupiter	Saturn	Uranus	Neptune
Mass (relative to Earth)	317.91	95.21	14.56	17.18
Radius (relative to Earth)	11.21	9.45	4.01	3.88
Semimajor axis (au)	5.20	9.54	19.19	30.07
Rotation period (h)	9.93	10.55	17.24	16.11
Rotation axis tilt to orbit (deg[a])	3.13	26.73	97.77	28.32
Blackbody temperature (K)	109.9	81.0	58.1	46.6
Heat radiated (relative to solar energy received)	1.7	1.8	1.1	2.6
Average density (kg/m^3)	1326	687	1271	1638
Acceleration due to gravity (relative to Earth[b])	2.53	1.07	0.91	1.12
Escape speed (relative to Earth[b])	5.32	3.17	1.90	2.10
Atmospheric composition (by volume[c])				
H_2 (hydrogen molecules; %)	89.8	96.3	82.5	80.0
He (helium; %)	10.2	3.25	15.2	19.0
CH_4 (methane; %)	0.3	0.45	2.3	1.5
NH_3 (ammonia; ppm)[d]	260	125	—	—
Aerosols (ices)	NH_3	NH_3	NH_3	NH_3
	H_2O	H_2O	H_2O	H_2O
			CH_4 (?)	CH_4(?)
Magnetic field (relative to Earth[b])	14.1	0.70	0.75	0.46
Magnetic field tilt to rotation axis (deg)	9.4	0.0	58.6	46.9

Note: Some data from Table 10.1 are reproduced here for convenience.

[a] A tilt greater than 90° means that the planet rotates retrograde.

[b] For Earth: $g = 9.80$ m/s^2, $v_{esc} = 11.2$ km/s, magnetic field is 0.0306 mT.

[c] Percentage totals may not equal 100% because of measurement uncertainties.

[d] Parts per million (ppm) numerically equals percentage times 10^4; for example, 260 ppm = 0.0260% by volume.

As with other tables in this textbook, it is not important that you memorize the data (after all, these data can always be looked up when needed), but you should understand the relationships and patterns reflected in the data, and develop a sense of scale. Throughout this section we will be exploring how these data translate into a comprehensive view of the giant planets individually and as groups.

Planet Temperatures and Heat

An immediate implication of their large distances from the Sun relative to the rocky planets is that the giant planets receive significantly less sunlight because of the inverse square law of light, discussed in Section 6.7. This means that the blackbody temperatures of the giant planets are much lower than for the rocky planets. Table 12.1 lists the blackbody temperatures of the giant planets, from Jupiter's 110 K to Neptune's 47 K temperature. For comparison, the blackbody temperature of Earth is 254 K.

From the temperature conversion equations found on page 214, Earth's blackbody temperature corresponds to $-19°C$ or $-2°F$. Recall from Section 6.6 that the blackbody temperature indicates how much energy per second is radiated into space by an object, as described by the Stefan–Boltzmann blackbody luminosity law on page 219. Fortunately for us, Earth's average surface temperature is not at the blackbody temperature, which is well below the freezing point of water; the slight greenhouse effect that has always been present keeps the base of our atmosphere warmer than the blackbody temperature. Applying the same temperature conversion equations to Jupiter and Neptune tells us that they have blackbody temperatures of $-163°C$ ($-262°F$) and $-227°C$ ($-376°F$), respectively.

Key Point
The giant planets radiate significantly more energy out into space than they receive from the Sun.

The blackbody temperatures given in Table 12.1 are the temperatures the planets would have if they simply absorbed all of the energy coming from the Sun that isn't reflected off their clouds and then reradiate that absorbed energy back out into space. However, there is a curious feature of the giant planets: they radiate more heat than they absorb. For example, Jupiter radiates 1.7 times the heat it receives from the Sun (70% more than it receives), and Neptune radiates 2.6 times as much (160% more is radiated than received). Although the rocky planets do this as well, the extra amounts are extremely small compared to the energy that they receive from sunlight.

Rotation and Planet Shapes

Key Point
The giant planets rotate more quickly than any of the rocky planets despite being much larger. This causes them to be somewhat flattened.

Jupiter and Saturn rotate significantly faster than Earth, which is the fastest rotator of the rocky planets. Given the enormous sizes of these gas giants, their rotation periods of 9.9 h and 10.5 h, respectively, seem remarkably short compared to Earth's 23.9 h rotation period. As can be seen from Fig. 12.4, these fast rotation rates cause Jupiter and Saturn to be noticeably wider at their equators than their distances from pole to pole (by about 6.5% and almost 10%, respectively). Although Earth is also very slightly wider at its equator than from pole to pole, the effects for Jupiter and Saturn are much greater. The ice giants, Uranus (17.2 h) and Neptune (16.1 h), also exhibit non-spherical shapes, but the flattening is not nearly as significant as it is for the gas giants; the pole-to-pole distances of Uranus and Neptune are slightly greater than, and slightly less than 2% of their equatorial diameters, respectively.

Atmospheres

Unlike the rocky planets, the giant planets don't have solid "surfaces" except for their very deep cores. As a result, defining where the atmosphere stops and the interior starts is completely arbitrary. The bottom of the atmosphere is typically assumed to be somewhere between where the gas pressure equals the pressure at sea level on Earth and perhaps one to two hundred times that value.

Fig. 12.4 *Left:* Jupiter with well-defined cloud bands and its famous Great Red Spot observed by the Hubble Space Telescope in June, 2019. [NASA, ESA, A. Simon (Goddard Space Flight Center), and M.H. Wong (University of California, Berkeley)] *Right:* Saturn and its spectacular ring system, as seen by the Cassini orbiter in 2008. Cloud bands are present, but more muted than those on Jupiter. The planets are shown in relative size. (NASA/JPL/Space Science Institute)

The Gas Giants

Because of the low atmospheric temperatures and high escape speeds of Jupiter and Saturn, hydrogen and helium can't drift off into space as easily as they can from the rocky planets. As a result, instead of the heavier molecules of carbon dioxide (CO_2), oxygen (O_2), nitrogen (N_2), and water (H_2O) being major components of gas-giant atmospheres, the atmospheres of both Jupiter and Saturn are mostly composed of hydrogen molecules (H_2) and helium (He) atoms.

If a bottle full of Jupiter's atmosphere was collected, almost 90% of *the volume* of the gas would be occupied by H_2 and the remaining 10% would be helium. Only tiny amounts of methane (CH_4), ammonia (NH_3), water (H_2O) and other compounds would be present in the bottle. A bottle full of Saturn's atmosphere would contain an even higher percentage of H_2 (96%) and less helium (3%) than the Jupiter sample.

In Chapter 9 we learned that the *mass* of the Sun's atmosphere is composed of 73.8% hydrogen atoms and ions, and 24.9% helium atoms and ions. All other elements make up the remaining 1.3% of the solar atmosphere's mass. So how does this compare with Jupiter? Assume (incorrectly) for the moment that a sample of Jupiter's atmosphere contains nine H_2 molecules for every helium atom. This means that there would actually be $9 \times 2 = 18$ hydrogen atoms for every helium atom. Since helium atoms are about four times as massive as hydrogen atoms, for every nine hydrogen molecules and one helium atom, the mass would be about the same as $18 + 4 = 22$ hydrogen atoms. The percentage that is actually hydrogen by mass would then be 18/22, or about 82%, with helium being 4/22, or 18% of the mass.

Although closer to the percentages by mass for the composition of the Sun's atmosphere, there is still a difference. Most of the remaining difference comes from the fact that for the same temperature helium atoms move more slowly than H_2 molecules. From the average speed of gas particles equation on page 339, helium, with twice the mass of H_2, moves more slowly by a factor of $1/\sqrt{2}$ and therefore takes up less of the volume of the gas. Said another way, $\sqrt{2}$ more helium atoms relative to hydrogen molecules are required to make up the volume of the gas. This is why the word "incorrectly" was used in the last paragraph; 90% of the volume being H_2 doesn't mean 90% of the number of particles are H_2 molecules. Adding the additional helium atoms, the fraction of the mass composed of hydrogen becomes $18/(18 + \sqrt{2}) = 0.76$, or 76%, and the remainder, 24%, is due to the mass of helium atoms. When the other, even heavier, atoms and molecules that make up a small fraction of the number of particles are included, the atmosphere of Jupiter turns out to be about 75% hydrogen, 24% helium, and 1% metals by mass; very close to the composition of the Sun's atmosphere.

The differences in composition between Jupiter, Saturn, and the Sun immediately beg the question, why do these differences exist? Consider the fact that Saturn's atmosphere is significantly cooler than Jupiter's. Again, the average speed of gas particles equation tells us that the helium particles move more slowly than the hydrogen molecules in Saturn's atmosphere, and that they both move more slowly than their counterparts in Jupiter's atmosphere. Assuming the two planets formed with similar compositions out of the solar nebula, much of Saturn's helium sank into the planet's interior. The heavier atoms and molecules also sank toward the interior in both planets, leaving only traces behind or transported upward by convection. The same effect also explains why there is slightly less helium relative to hydrogen in Jupiter's atmosphere compared to the Sun's atmosphere.

The sinking of heavier atoms and molecules has another effect beyond just affecting atmospheric composition; as they sink their gravitational potential energy turns into kinetic energy, which in turn produces heat through friction. In Saturn, the heat generated from sinking helium and other heavier atoms and molecules helps to explain why it radiates 1.8 times more energy into space than it receives from the Sun.

The fact that Jupiter radiates 1.7 times more energy than it receives is primarily due to the fact that it is still shrinking under its own gravity, although some sinking of helium and other heavier atoms also contributes to the energy released. In the early days of the Solar System, Jupiter was likely almost twice as large as it is today. By shrinking so much, Jupiter is still radiating away the heat generated by that compression. As it has compressed it has also become more dense, until today Jupiter is about 1.3 times denser *on average* than water. Saturn, on the other hand, hasn't compressed as much because its gravity is weaker, due to the fact that it is less than one-third of Jupiter's mass. As a result, its density is slightly less than 70% of the density of water. Anything having a density that is less than water can float in water. If Saturn could be placed in a gigantic ocean, it would float. Saturn's average density is even less than water ice.

Jupiter's ever-changing and complex cloud patterns, evident in Fig. 12.5, are driven in part by the planet's rapid rotation. Although most of the small-scale features can change quickly by human standards, the cloud bands are permanent features in Jupiter's atmosphere [Fig. 12.5 (*Left*)]. In contrast to small-scale features, Jupiter's famous Great Red Spot [Fig. 12.5 (*Right*)] is a long-lived storm that has been present continuously since at least the seventeenth century. Multiple Earths could fit along this cyclonic storm's long axis, but observations over centuries have documented that the Great Red Spot is slowly getting smaller.

Figure 12.6 shows that Saturn displays many of the same kinds of atmospheric features that are present on Jupiter, although more muted because they exist deeper in the atmosphere. Colored bands are present that circle the planet, and complex, swirling vortices are visible when examined close-up. A particularly odd pattern in the form of a hexagon is also visible around Saturn's north pole.

Key Points
• Sinking heavy elements explains why Saturn radiates more energy than it receives from the Sun.
• Jupiter is still shrinking, causing it to radiate more energy than it receives from the Sun.
• Saturn is less dense than water.

Key Point
Jupiter's Great Red Spot is an enormous storm that has been present since at least the seventeenth century.

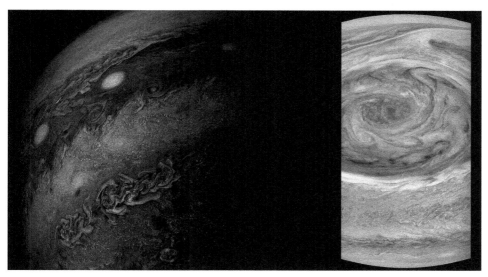

Fig. 12.5 *Left:* A view of Jupiter's cloud tops. Several of the eight white oval-shaped "string-of-pearls" storms are visible in the dark red band near the top. The image was taken by the Juno spacecraft in 2017. (NASA/JPL-Caltech/SwRI/MSSS/Gerald Eichstadt/Sean Doran © CC NC SA) *Right:* A close-up of Jupiter's Great Red Spot from Juno in 2017. [Enhanced image by Kevin M. Gill (CC-BY) based on images provided courtesy of NASA/JPL-Caltech/SwRI/MSSS.]

Fig. 12.6 *Upper left:* A portion of Saturn's atmosphere, showing muted bands of clouds obtained by the Cassini orbiter in 2017. *Upper right:* A view above Saturn's northern hemisphere, with the unusual hexagon-shaped pattern at the pole as seen by Cassini in 2016. A portion of Saturn's ring system is visible at upper-right. *Bottom:* A close-up view of turbulence in Saturn's atmosphere with a small bluish storm visible near bottom-center, from Cassini, 2011. (NASA/JPL-Caltech/Space Science Institute)

Venus's long rotation period of 243 days results in very broad cloud patterns (Fig. 11.14), while Earth's much more rapid 23^h56^m rotation period produces complex weather patterns, including cyclones (Fig. 11.23) and jet streams (Fig. 11.24). Storms like the Great Red Spot and the "string-of-pearls" white ovals on Jupiter, and similar storms on Saturn, are cyclones much like our hurricanes and typhoons, and the cloud bands on Jupiter and Saturn are similar to terrestrial jet streams. Wind speeds of more than 640 km/h (400 mi/h) have been recorded in Jupiter's atmosphere. In addition, because Jupiter and Saturn have no solid surfaces except very deep at their cores, their atmospheres rotate differentially, with rotations being faster at their equators than near their poles. We saw this same pattern of differential rotation in the Sun (page 331*ff*).

As with Earth's atmospheric and ocean currents, rapid rotation isn't the only driver of weather patterns on Jupiter and Saturn. Just as heat from the oceans and atmosphere drive hurricanes and typhoons on Earth, the heat rising from their interiors via convection helps to power the storms of the gas giants. Variations in temperature within the clouds due to rising and falling convective cells, depth in the atmosphere, and slight composition variations combine to produce the spectacular colors in Jupiter's atmosphere, and the more muted tones in Saturn.

A video of Jupiter's dynamic atmosphere is available through the Chapter 12 resources page.

The Ice Giants

The ice giants, Uranus and Neptune, are similar in some ways to the gas giants, Jupiter and Saturn, but they are also very different in other ways. As can be seen by referring back to Table 12.1, although the atmospheres of the ice giants are dominated by hydrogen molecules and helium atoms just as the gas giants are, the ice giants have significantly more methane (CH_4) in their atmospheres. Because methane efficiently absorbs red light, the reflected sunlight from the ice giants becomes bluer in color, as seen in Fig. 12.7. However, the cyan color of Uranus does differ from Neptune's deeper blue color. Scientists aren't yet sure what accounts for the difference, but it is likely that Neptune's atmosphere contains still undetected molecules that contribute to the different shades of blue.

The atmospheres of the ice giants also differ from their much larger gas giant cousins by the presence of ice crystals of ammonia (NH_3) and water (H_2O), as well as ices of more complex molecules and compounds, including ammonium hydrosulfide (NH_4SH) and methane hydrate, which is composed of multiple methane molecules trapped in a frozen cage of water molecules. (It is methane hydrate that is trapped beneath Earth's arctic permafrost and under ocean floors, and is the primary contributor to natural gas. Methane hydrate may contain more energy than all of the rest of Earth's fossil fuels combined.)

Clouds made up of various molecules also exist at different levels in the atmospheres of the ice giants. Methane clouds are present at the highest altitudes, with ammonia, ammonium hydrosulfide, and water (H_2O) clouds forming at increasing depths. By knowing the physical conditions under which these clouds can form, together with their unique spectral signatures (Fig. 7.1), scientists can study different layers in their atmospheres.

Although the two ice giants are very similar in many ways, there is a profound difference in the amount of heat coming from their interiors: Uranus radiates just slightly more energy back into space than it receives from the Sun while Neptune

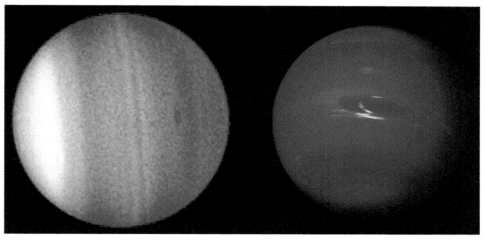

Fig. 12.7 Uranus and Neptune are shown in relative sizes. *Left:* Uranus, as observed by the Hubble Space Telescope in 2006. A dark spot is visible toward the right-center of the image. Its geographic north pole is near the 3 o'clock position. [NASA, ESA, L. Sromovsky and P. Fry (University of Wisconsin), H. Hammel (Space Science Institute), and K. Rages (SETI Institute)] *Right:* Neptune with its Great Dark Spot, smaller storms, and cloud bands seen by Voyager 2 in 1989. (NASA/JPL)

radiates 2.6 times more energy than it receives. Because of this fact, the atmosphere of Neptune is only slightly cooler than Uranus's, despite being much farther from the Sun. Why there is such a dramatic difference between the two remains a mystery, but whatever the cause, that difference likely helps to explain in part why Neptune appears to have a much more active atmosphere than Uranus. It is also the case that Neptune is slightly smaller than Uranus even though it is more massive, implying that it is more dense than Uranus. In fact, Neptune is also denser than Jupiter and Saturn.

Another important contributor to atmospheric activity differences has to do with another puzzling characteristic of Uranus. You may have noticed from Fig. 12.7 that Uranus appears to be almost lying on its side, given that its cloud bands are close to vertical while Neptune's cloud bands are nearly horizontal. This is not an error in relative orientation between the two worlds displayed in the figure; instead, Uranus rotates about an axis that is tilted 97.8°, placing the axis almost in the plane of the ecliptic, while Neptune's rotation axis is tilted just 28.3°. Because the tilt of Uranus is greater than 90°, it technically rotates retrograde (Venus is the only other planet in the Solar System that rotates retrograde).

Because Uranus's axis alignment is so close to the ecliptic, it has seasons that are unique in the Solar System. Uranus's orbital period is 84 years long. At one point in its orbit, Uranus's south pole (on the left in Fig. 12.7) is facing directly toward the Sun. Forty-two years later its north pole is facing directly toward the Sun. The first orientation amounts to Uranus's winter solstice and the second orientation is its summer solstice. This means that Uranus's north pole is in darkness for 42 years while its south pole is in sunshine for 42 years, and vice versa. As a result, the poles can get very cold while they are in shadow for so long, causing weather activity to be suppressed. When Voyager 2 flew past Uranus in 1986 (Fig. 12.8) it happened

Key Point
Uranus's rotation axis is almost in the plane of the ecliptic, and it rotates retrograde.

Fig. 12.8 Uranus photographed by Voyager 2 in 1986. (NASA/JPL-Caltech)

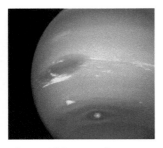

Fig. 12.9 Neptune from Voyager 2 in 1989 showing the Great Dark Spot and a small, fast-moving storm called Scooter near its south pole. The color contrast has been exaggerated to highlight features. (NASA/JPL)

to be near winter solstice so Uranus's south pole was facing almost directly toward the Sun. The image in Fig. 12.7 was taken 20 years later when the equator of the planet was oriented toward the Sun during its vernal equinox. Thanks to improved image processing and more uniform exposure to the Sun because of its 17-hour rotation period, cloud bands became readily visible. Close examination of Uranus in Fig. 12.7 shows the presence of a dark spot in the northern hemisphere, indicating cyclonic storm activity.

A Great Dark Spot was visible on Neptune when Voyager 2 flew past that planet in 1989 (Fig. 12.9), completing the spacecraft's incredible grand tour of all four giant planets. Voyager 2 also caught a smaller white oval moving more rapidly around Neptune that was nicknamed Scooter. Although present when Voyager 2 made its flyby, the spots proved to be temporary, unlike the very long-lived Great Red Spot of Jupiter. Different cyclonic storms have been observed by the Hubble Space Telescope since Voyager 2's visit, demonstrating the planet's ongoing dynamic atmospheric behavior. The highest wind speeds anywhere in the Solar System have been recorded on Neptune: 2100 km/h (1300 mi/h).

Axis Tilts, Magnetic Fields, and Aurora

Figure 12.10 illustrates the orientations of the giant planets and Earth's rotation axes (marked in yellow) relative to their orbital planes. Of course the giant planets don't actually orbit exactly on the ecliptic, but they have been aligned to make comparisons easier. Earth, Saturn, and Neptune all have very similar axis tilts, while Jupiter's axis is nearly perpendicular to its orbital plane. All four of these planets also rotate counterclockwise when viewed from above their north poles. Uranus's axis is tilted a little more than 90°, meaning that its rotation axis is nearly parallel to its orbital plane and it rotates retrograde, or clockwise when viewed from above its north pole.

Figure 12.10 also shows the orientations of the magnetic fields of the five planets. Jupiter, Saturn, Earth, and Uranus all have magnetic fields that are tilted by close to the same amount, while Neptune's field axis is tilted closer to its orbital plane. In addition, all five of the planets have their south magnetic poles on the same side of the planet as their north rotational poles. Another curiosity of Jupiter, and especially of the ice giants, is that their magnetic field axes don't actually go through the middle of planets, suggesting that the magnetic dynamos operating in those planets are not perfectly centered. In the cases of Uranus and Neptune, the magnetic field centers are off by 1/3 and 1/2 of the planet's radius, respectively.

By far the strongest magnetic field in the Solar System is Jupiter's, with a field strength that is more than fourteen times greater than Earth's, compared to the Sun's *average* magnetic field, which is roughly twice the strength of Earth's field (however, the magnetic field of sunspots can be thousands of times greater than Earth's field). Among the planets, Earth has the second strongest field, while Saturn, Uranus, and Neptune have field strengths ranging between about one-half and three-quarters of Earth's field. (Recall that the three remaining planets, Mercury, Venus, and Mars, have nearly non-existent magnetic fields.)

As is the case for Earth's magnetic field (Fig. 11.37), the charged particles in the solar wind interact with the giant planets' magnetic fields, causing the fields to be compressed on their Sunward sides and elongated on the sides opposite from the

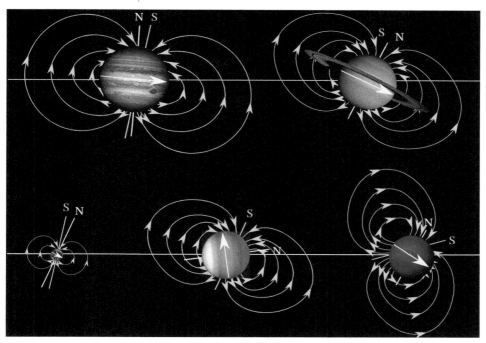

Fig. 12.10 The rotations and magnetic field orientations of the five planets with strong magnetic fields. The yellow axes and yellow arrows indicate the axis tilts and directions of rotation. The aqua axes indicate the orientations of the magnetic fields. The white lines are the planes of the planetary orbits. The south poles of all of the magnetic fields are north of their orbital planes. *Top:* Jupiter and Saturn in relative size. *Bottom:* Earth, Uranus, and Neptune. Their sizes have been doubled relative to Jupiter and Saturn. The magnetic field lines are idealized close to the planets. The solar wind causes the fields to be compressed in the direction toward the Sun and stretched away from the Sun as in Fig. 12.11. (All images courtesy of NASA/JPL)

Sun. Recall that this happens because charged particles get trapped by magnetic fields, causing them to spiral around field lines. The charged particles are also traveling very fast when they encounter the magnetic fields, meaning that they have linear momenta directed away from the Sun. In essence, although the magnetic fields trap the charged particles, the charged particles also push against and stretch the magnetic fields. An example for Jupiter is shown in Fig. 12.11. The locations of the four Galilean moons are also shown, although their sizes have been increased by a factor of ten relative to Jupiter.

As is also the case for Earth (Fig. 11.39), the charged particles from the solar wind can follow the magnetic field lines of the giant planets toward their magnetic poles. When the particles strike atoms in their atmospheres, auroras can be produced. In fact, auroras have been observed for all four of the giant planets. Figure 12.12 shows an ultraviolet observation of aurora on Jupiter, combined with an image of the planet obtained in visible light.

The Juno spacecraft entered orbit around Jupiter in July 2016 for a five-year mission, with a sophisticated magnetic field detector that led to an unexpected discovery: Jupiter's magnetic field is very different from the idealized bar-magnet-like magnetic field that is depicted in Figs. 12.10 and 12.11. Although the southern

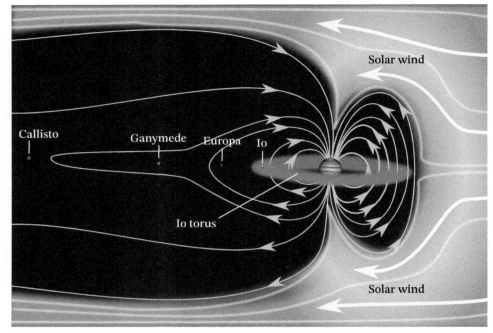

Fig. 12.12 An ultraviolet image of aurora on Jupiter, combined with a visible light image of the planet itself. [NASA, ESA, and J. Nichols (University of Leicester)]

Fig. 12.11 The solar wind interacts with Jupiter's magnetic field to elongate it in the direction away from the Sun. The locations of the four Galilean moons' orbits are also indicated. Note that each moon has been enlarged by a factor of ten relative to Jupiter. The Io torus is a ring of charged particles located in the orbit of Io. (All images courtesy of NASA/JPL)

Fig. 12.13 A simulation of Jupiter's magnetic field. (NASA/JPL-Caltech/Harvard/Moore et al.)

hemisphere field looks fairly bar-like, the northern hemisphere is anything but (Fig. 12.13). Along with a southern magnetic pole near its rotational north pole, there is an intense magnetic north pole located north of the planet's equator. The Juno mission also detected variations in Jupiter's magnetic field over time. It has been proposed that the strong winds in Jupiter's atmosphere may extend down into the region where the magnetic dynamo operates, creating currents that significantly complicate the external magnetic field structure.

Just as our discussion of the Sun's magnetic field and the presence or absence of magnetic fields around the rocky planets and our Moon gave us hints about their interiors, the magnetic fields of the giant planets provide hints about their interiors as well.

Interiors

Understanding what the interiors of the giant planets are like is in many ways more difficult than it is for the rocky planets. There are two major reasons that give rise to these challenges. The first has to do with determining how mass is distributed inside the planets. Obviously, measuring quakes is impossible because there are no solid surfaces except perhaps very near their centers, so it becomes necessary to use other means to determine the distribution of mass.

As was done with Mercury (page 429*ff*), measuring the subtle changes in spacecraft trajectories helps scientists infer how mass must be distributed in order to cause the changes. Scientists can measure small wavelength shifts of the spacecrafts' radio signals caused by the Doppler effect, to make very precise determinations of tiny variations in their velocities. Complicating those efforts are the gravitational influences caused by all of the masses in the planets' systems and by the other bodies in the Solar System. Fortunately, knowing the motions of these other masses does allow scientists to subtract their influences on the spacecraft trajectories. For the gas giants, extensive orbital missions gathered a great deal of valuable and detailed information about how the planets affected the orbits of Juno and Galileo (Jupiter), and Cassini (Saturn). The situation is different for the ice giants Uranus and Neptune, however; both planets have only been visited by one spacecraft, Voyager 2, during fast flybys rather than through long-term orbital missions. This means that very little information was collected about how internal mass distributions affected Voyager 2's trajectory on its way past.

It is also possible to see how the planets' moons (Section 12.2) and ring systems (Section 12.3) respond to the mass distributions of planets. However, using moons and rings is not generally as accurate as the information returned from spacecraft. Nevertheless, tiny oscillations in the inner edge of Saturn's C ring (Fig. 12.35) have yielded remarkably detailed data, used to deduce information about the structure of that planet's interior.

The other major reason that it is so challenging to determine the interior structures of the giant planets has to do with the difficulties in using quantum mechanics to calculate theoretically how the various gases behave under the extreme conditions of pressure, density, and temperature found deep in their interiors. Under those exotic conditions, molecules can form and break apart, and gases can change to liquids and solids. Unfortunately, there is very little experimental laboratory data to test those theoretical calculations.

Despite the challenges, both observational and theoretical, models like the ones shown in Fig. 12.14 have been developed as possible representations of what the interiors of the giant planets might be like. Given the vastly larger amount of data about the gas giants and the higher quality of those data, the state-of-the-art for the gas giant computer models are more sophisticated than for the ice giants. The gas giant models include composition gradients (changes with increasing depth) and even convection zones, while the ice giant models are treated more simply as three

Key Points
Understanding the interior structures of the giant planets is challenging because of
• uncertainties in the distribution of mass, and
• the difficulty in calculating how gases respond under conditions found in giant-planet interiors.

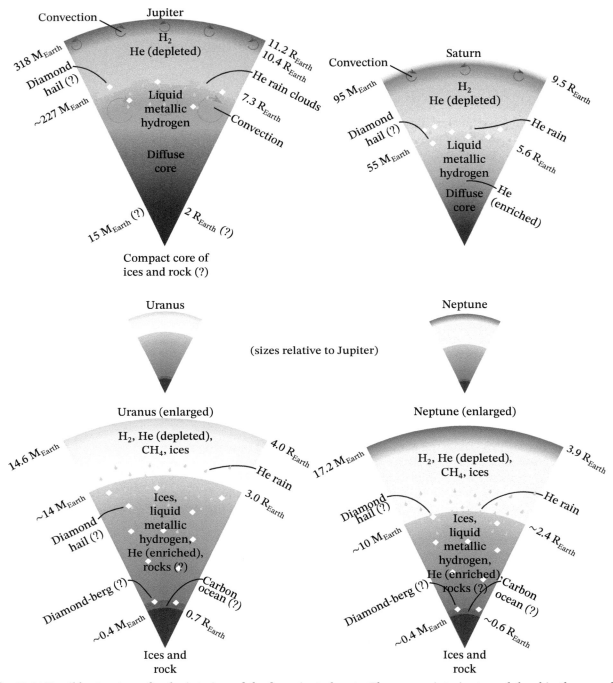

Fig. 12.14 Possible structures for the interiors of the four giant planets. The masses interior to each level in these models are indicated to the left of each "slice" and are expressed in terms of the mass of Earth. The radii at each level are indicated to the right of each "slice," expressed in terms of the radius of Earth. General descriptions of composition are also shown. *Top:* The gas giants. Jupiter and Saturn are drawn to scale. *Center:* The ice giants. Uranus and Neptune are drawn to scale. *Bottom:* Uranus and Neptune enlarged to show detail. (The symbol ∼ can be read as "is approximately equal to.")

distinct parts: atmospheres that extend well down into their interiors, mantles as their middle regions, and cores of rock and ices. Until 2019, models of the gas giants were also simplified as three separate parts, the innermost of those being a distinct rock-and-ice core, but improvements in data collection and analysis, and advances in our theoretical understanding continue to move science forward. Today, it appears that the cores of Jupiter and Saturn are a diffuse (or "fuzzy") mixture of rocky material, ices, and gases (primarily hydrogen and helium). It may still be possible that Jupiter has a relatively small distinct rock and ice core that is much more massive than Earth, but it seems unlikely that Saturn has a similar distinct core.

Recall from page 465*ff* that Jupiter's upper atmosphere has slightly more hydrogen relative to helium than the Sun does, and Saturn has even more hydrogen relative to helium in its upper atmosphere. In order for this to be true, helium, which is about twice as heavy as molecular hydrogen (H_2), has been sinking down into the interiors of the gas giants, enriching their deep interiors while leaving the outer regions of the planets slightly depleted of helium. When the pressure inside the planets gets sufficiently great, the helium gets squeezed so much that it turns into a liquid, causing helium rain. This process may also be occurring in the ice giants.

Heavier molecules, like methane (CH_4), also sink into the interior. It has been proposed that, under great pressure, methane and other carbon-based molecules can be broken apart into carbon and hydrogen atoms, and the carbon can then be compressed enough to form diamonds (diamond is just a crystallized form of carbon created under great pressure). As a result, the interiors of all four planets may have diamond hailstones falling through them. It may even be the case, as suggested by very high-pressure experiments in laboratories, that liquid carbon oceans could exist on top of the cores of the ice giants, and that "diamond bergs" are floating in those oceans! Good luck mining them, though.

Theory predicts that pressures and temperatures within the giant planets provide the conditions for hydrogen to become a liquid metal. One liquid metal that you are likely familiar with under normal terrestrial conditions is mercury. Old-style thermometers like the one shown in Fig. 12.15, and some barometers, used mercury in tubes. In thermometers, when the mercury is heated, it expands and rises in the tubes in a predictable way so that the level of the mercury corresponds to the temperature. When measuring air pressure in the United States, weather forecasters often report how high mercury would climb in a tube when a reservoir of mercury is pushed down on by the air, typically in inches (air pressure at sea level is about 29.9 in of mercury). In other parts of the world, air pressure is usually reported in more conventional units of millibars (mb), where 1010 mb is roughly the air pressure at sea level; 1000 mb = 1 b ≡ 100,000 Pa (pascals). Given its toxic properties, mercury thermometers and barometers have largely been replaced by electronic devices today.

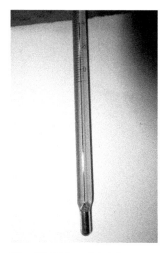

Fig. 12.15 An old-style laboratory glass thermometer using mercury. (Palagiri; CC BY-SA 3.0)

As a liquid metal, hydrogen can take on metallic properties, such as being electrically conducting. The presence of magnetic fields around all four of the planets suggests that magnetic dynamos must be in action. Since magnetic dynamos require rotation, convection, and electrically conducting fluids (page 331), their existence lends support to the idea that hydrogen can become a liquid metal in the mantles of the giant planets.

In the ice giants, with their lower masses and with cooler temperatures due to their greater distances from the Sun, many of the molecules, such as CH_4, H_2O, and NH_3, can form ices in their mantles and cores, and to a lesser extent in their lower atmospheres. It is largely this aspect of their structures that distinguish the ice giants from the much larger gas giants.

One of the many unanswered questions concerning the interiors of the ice giants has to do with heat being radiated out into space. Recall from Table 12.1 that Uranus only radiates 1.1 times as much energy as it receives from the Sun, while Neptune radiates 2.6 times as much energy as it receives. This is particularly puzzling because Uranus and Neptune otherwise seem to be almost like twins in the Solar System. Since convection is generally more efficient at transporting energy than conduction is (page 226*ff*), perhaps there is something in Uranus that partially blocks the transport of heat by convection from the deep interior to the surface? Perhaps this puzzle relates to the weird tilt of Uranus's rotation axis?

In creating models of the giant planets, it is important that those models are the natural outcome of 4.57 billion years of planetary evolution since they first formed out of the original solar nebula. Although the other three giant planet models do a reasonable job of satisfying that requirement, models of Uranus take too long to cool down compared to the current age of the Solar System by about a factor of two! Obviously, something is missing with our construction of models of Uranus.

It is often the case in science that discovering answers to some questions leads to new questions that need answering. Such is the case for the models of the giant planets. More will be said about how the giant planets, and the rest of the Solar System, formed and evolved in Section 14.2.

12.2   The Moons of the Giant Planets

Key Point
Ganymede and Titan are larger than Mercury, but no moon is more than 45% of Mercury's mass.

If you go back to Table 10.1, you will note that one of the many distinct differences between the rocky and giant planets is the vastly larger number of moons that the planets of the outer Solar System have. In total, there are at least 202 moons distributed among the giant planets, with the gas giants having 79% of them. Although many of the moons are small, some of them are comparable in size to Mercury and our Moon, as shown in Fig. 12.16. In fact, Jupiter's Ganymede and Saturn's Titan are both larger than Mercury, and Jupiter's Callisto is just a tiny bit smaller than Mercury. However, because of its unusually high density, none of the moons is more massive than about 45% of the mass of that planet (see Table 12.2). Even among the lesser moons in the giant planet systems there are many fascinating and bewildering worlds. Given the sheer number of moons in the outer Solar System, it is impossible to discuss them all. We will, however, briefly investigate some of the more significant and intriguing natural satellites of the giant planets.

Jupiter's Galilean Moons

When Galileo first turned his small telescope to the heavens he discovered many wondrous features of the universe never before seen by inhabitants of Earth, which he reported in his *Sidereus Nuncius* in 1610. Among them were four moons clearly orbiting Jupiter, one of the mysterious naked-eye planets that had been observed

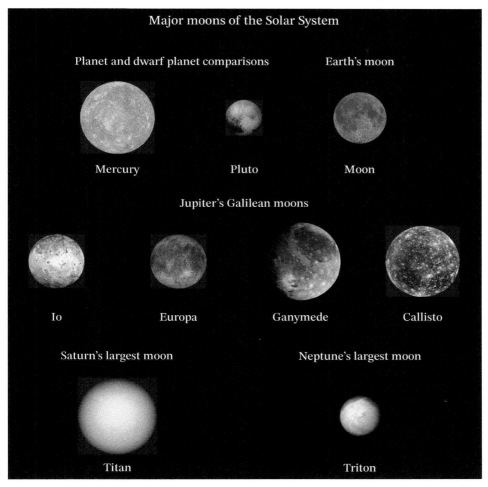

Fig. 12.16 The relative sizes of major moons in the Solar System, with Mercury and Pluto shown for comparison. (NASA, Johns Hopkins University Applied Physics Laboratory, Carnegie Institution of Washington, Southwest Research Institute, GSFC, Arizona State University, University of Arizona, JPL, DLR, Space Science Institute)

throughout human history. Although studied by ground-based telescopes of ever-increasing size and sophistication since that time, it wasn't until the two Voyager missions flew past Jupiter in 1979 that we became aware of how spectacular and complex Galileo's four moons really are.

In order from the closest to Jupiter outward are Io, Europa, Ganymede, and Callisto. The moons were named by Simon Marius, who discovered them independently, and nearly at the same time as Galileo. The names come from mythology associated with Jupiter (Zeus); Io, Europa, and Callisto were lovers of Zeus, and Ganymede was a beautiful mortal boy captured by Zeus to be his wine-bearer.

Table 12.2 Characteristics of the Solar System's largest moons

Characteristic	Moon	Io	Europa	Ganymede	Callisto	Titan	Triton
Parent planet	Earth	Jupiter	Jupiter	Jupiter	Jupiter	Saturn	Neptune
Radius (R_{Mercury})	0.71	0.75	0.64	1.08	0.99	1.06	0.55
Mass (M_{Mercury})	0.22	0.27	0.15	0.45	0.33	0.41	0.07
Density^a (kg/m³)	3344	3530	3010	1940	1830	1881	2050
Orbital/Rotation period^b (d)	27.32	1.77	3.55	7.15	16.69	15.95	−5.88
Semimajor axis (R_{planet})	60.27	6.10	9.60	15.47	27.22	21.32	14.49
Mean temperature (K)	253	118	103	113	118	93.2	37.8
Atmosphere	no	no	no	no	no	thick	thin
Global magnetic field	no	possible	no	yes	no	unknown	unknown

^{a}From Table 10.1, Mercury's density is 5427 kg/m³.

^{b}All moons are in synchronous rotation. A negative period indicates retrograde rotation and a retrograde orbit.

Io: The Volcanic Moon

Io (pronounced "eye-oh") was a shock when it was first imaged by Voyager 1. The moon is covered with over 400 active volcanoes that deposit sulfur compounds on its surface, giving it the mottled appearance that is visible in Fig. 12.17. The plume of the erupting volcano, Loki, can be seen on the limb in the Voyager 1 image on the left. Some of the plumes can reach heights of 500 km (300 mi) above the surface. The local dark spots are the calderas of Io's many volcanoes, and lava flows are also visible that extend for up to 500 km. Io also seems to have less water than any other object in the Solar System, undoubtedly the result of the constant volcanic activity ejecting the relatively light water molecules into space. The Io torus depicted in Fig. 12.11 is a ring of charged particles in Io's orbit that were ejected from its volcanoes.

Based on gravity measurements similar to those made of the giant planets themselves, the inner one-third of the moon likely contains a metallic core composed of iron, nickel, and possibly sulfur compounds. The core is surrounded by a mantle and a crust of rock (silicates). Since there doesn't appear to be a magnetic field, or at best a very weak one, it is likely that there is at most only a small amount of convection occurring in the conducting core. Io's internal structure results in a density that is greater than any moon in the Solar System, including our Moon.

Unlike most other bodies in the Solar System, there are very few impact craters on Io, implying that the surface is very young. This is because the surface is constantly being refreshed by molten material from the interior, burying any recent craters.

The logical question to ask is "what is causing so much volcanic activity on Io?" The answer is a combination of its close proximity to Jupiter and its relationship with two of its siblings, Europa and Ganymede.

First, consider the fact that Io is the closest Galilean moon to Jupiter. In Section 10.3 we learned that tides occur on Earth and our Moon because both bodies have finite sizes; they are not simply point masses. The sides of Earth and the Moon that

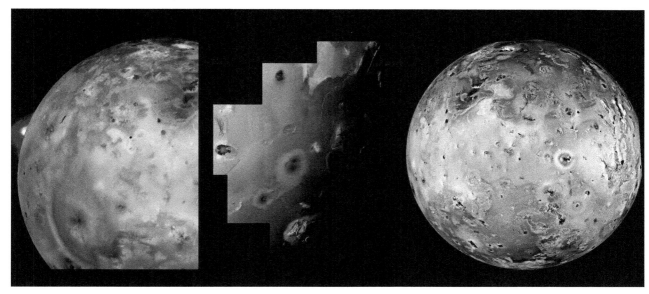

Fig. 12.17 *Left:* Io as seen by Voyager 1 in 1979. The volcanic plume from Loki is visible on the limb. (NASA/JPL/USGS) *Center:* A mosaic of Voyager 1 images near Io's south pole, showing Haemus Mons, a mountain 10 km (32,000 ft) tall. Volcanic plains, eroded plateaus, and calderas are also easily visible. (NASA/JPL/USGS) *Right:* The Galileo mission obtained this image in 1996. (NASA/JPL/University of Arizona)

are closest to each other are pulled harder by gravity than the sides farther away, causing them to be stretched slightly. The same situation holds for Jupiter and Io, except in a more extreme way. Io is only 6.1 Jupiter radii away from the center of the planet, while our Moon is 60.3 Earth radii away from Earth. Add to that the fact that Io is nearly the same size as our Moon and Jupiter is 318 times more massive than Earth, you can see that tidal forces due to Jupiter alone are enormous.

Just as happened to our Moon, the tidal forces caused Io's rotational period to become equal to its orbital period, meaning that one side of Io always faces Jupiter. Io's orbit is also slightly elliptical, bringing it closer than average to Jupiter at one point in its orbit and farther away than average one-half orbit later. The tidal bulges produced on Io when it is closest to Jupiter may be much as 100 m (330 ft) higher during its closest approach than when it is farthest from Jupiter. As is characteristic of most moons in the Solar System, including all of the major moons, the four Galilean moons are all in synchronous rotation.

Also as is the case with Earth spinning on its axis faster than the Moon's orbital period, Jupiter spins on its axis about once every 9.9 hours, whereas Io orbits roughly once every 1.77 days. For the Earth–Moon system, this causes the Moon to move farther from Earth while Earth also slows its rotation rate. But this isn't happening for Jupiter and Io because they don't exist in isolation from everything else in the much more complex Jupiter system. Europa and Ganymede are prohibiting Io's outward migration from happening (Fig. 12.18).

In what at first blush may seem like a big cosmic coincidence, Europa's orbital period is exactly twice as long as Io's (2:1), and Ganymede's is exactly four times as long (4:1). As a result, once every Ganymede orbit the three moons and Jupiter all line up along Io's major orbital axis twice. This phenomenon is known as an orbital

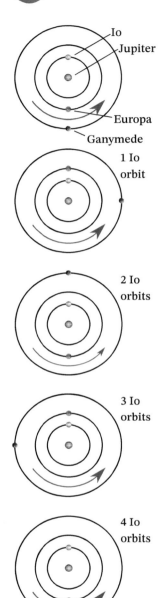

Fig. 12.18 The orbital resonances between Io, Europa, and Ganymede.

resonance. Orbital resonances are common throughout the Solar System and have been detected within other planetary systems as well. It is this synchronization of the orbits of these three Galilean moons that keeps Io locked in its orbit and not spiraling away from Jupiter.

Resonances of various kinds are common throughout nature. Many musical intervals are comprised of two notes that have simple integer ratios of frequencies. A washing machine with an unbalanced load can bang around with the rotation speed of the cylinder. Sometimes disastrous and potentially deadly results can occur if a suspension bridge or platform is bounced at certain frequencies, such as by soldiers marching in unison, couples dancing to music with a specific beat, or even with the wind blowing just right as in Fig. 12.19. The rhythmic motions amplify the oscillations until the structure collapses.

With Io caught in a gravitational tug-of-war between Jupiter, Europa, and Ganymede, it is constantly getting twisted slightly back and forth rather than simply facing Jupiter directly all the time. Callisto, the second most massive Galilean moon and the farthest of the four from Jupiter, does not share the same orbital resonance as the other three, and so it doesn't line up with them except on extremely rare occasions. This means that it tries (unsuccessfully) to break the orbital resonance. Callisto's unaligned and constantly changing gravitation pull on Io, together with the continuous tidal flexing by the other two moons and Jupiter, produces a great deal of heating in Io's interior. The result is a partially molten interior and possibly a molten-rock ocean just below the moon's surface.

Fig. 12.19 Extreme resonance oscillations caused by wind resulted in the collapse of the Tacoma Narrows Bridge in 1940 near Tacoma, Washington. A video is available from the Chapter 12 resources webpage. (Public domain. Barney Elliot, Prelinger Archives)

Europa: Ice World

Europa is a very different world from the volcanic Io, as is immediately apparent if you compare Fig. 12.17 with Fig. 12.20. Europa's crust is composed of water ice that is heavily fractured due to the tidal flexing caused by Jupiter and the other Galilean moons. The cracks appear to be filled with a water slush contaminated with rocks. Because of the flexing, it is likely that a liquid water ocean exists below the surface that icy plates float on and move past one another. Estimates indicate that the thin ice crust and ocean are about 100 km (60 mi) thick.

Given the subsurface ocean, it is possible that the rocky mantle and metallic iron core beneath the ocean rotate at a slightly different rate from the surface. This difference between rotation rates drives ocean currents that cause the icy plates to separate, allowing warm ocean water to come to the surface and freeze between the plates, similar to mid-ocean ridges on Earth (page 399*ff*). In other places the plates may be forced back into the interior and recycled, not unlike terrestrial trenches. Europa experiences plate tectonic activity that is similar in many ways to tectonic activity on Earth, except that water and water ice plates on Europa play the same role as molten rock and rocky plates on Earth.

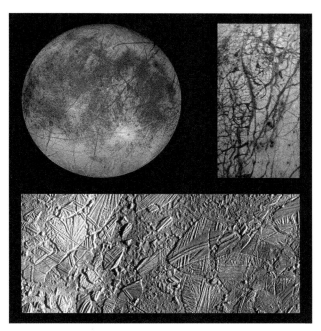

Fig. 12.20 *Upper Left:* Europa as seen by the Galileo Orbiter in 1996. (NASA/JPL/DLR) *Upper Right:* A portion of Europa's icy crust from near its equator. (NASA/JPL) *Bottom:* A view of a small region of the surface crisscrossed by fractures in the thin ice crust. Small impact craters are also visible. (NASA/JPL/University of Arizona)

Fig. 12.21 A water vapor plume from an icy geyser is visible on the lower left limb of Europa. The image was obtained by the Hubble Space Telescope in 2014. [NASA, ESA, W. Sparks (STScI), and the USGS Astrogeology Science Center]

The subsurface ocean also seems to drive the cryogeysers (icy-cold geysers) of water vapor and ice observed on Europa (Fig. 12.21). In addition, the Hubble Space Telescope detected an extremely thin atmosphere composed of atomic oxygen, possibly associated with the breaking up of water molecules (H_2O) exposed to the Sun's ultraviolet light. The much lighter hydrogen atoms simply escape into space.

Because Europa's surface is constantly being refreshed, only a small number of impact craters are visible. The impact-crater count implies that the surface is geologically very young, ranging from 20 to 180 million years. This constant refreshing of the surface by liquid water also makes Europa the smoothest body in the Solar System.

There is some evidence to suggest that Europa has a weak magnetic field, that is the result of its interaction with the extremely strong magnetic field of Jupiter. If that is the case, there would need to be conducting fluid in the interior, a condition that could be satisfied if the water ocean is a bit salty.

The strong possibility that a liquid water (or saltwater) ocean exists just below the surface of Europa has led to speculation that some form of life may exist in the ocean. Of course, movie makers and science fiction writers can't pass up a good scenario, so several movies have come out with a life-on-Europa theme, including *2010: The Year We Make Contact* (1984)[1] [the sequel to the classic film, *2001: A Space Odyssey* (1968)[2]] and *Europa Report* (2013)[3].

The possibility that Europa may host life is also the reason that the Galileo Orbiter was sent diving into Jupiter to burn up rather than possibly contaminating Europa with any terrestrial hitchhikers. The European Space Agency and NASA are planning future missions to continue research into this fascinating moon.

Ganymede: The Solar System's Largest Moon

Ganymede, shown in Fig. 12.22, is the third Galilean moon from Jupiter, and the largest and most massive moon in the Solar System. It is also the eighth-largest object in the Solar System, excluding the Sun.

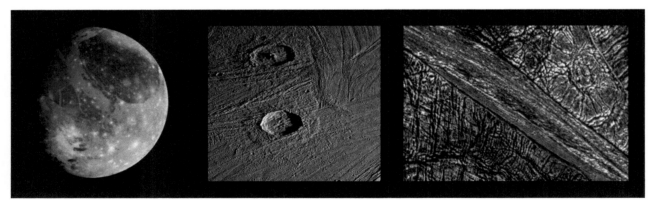

Fig. 12.22 *Left:* A view of the side of Ganymede, facing away from Jupiter. (NASA/JPL) *Center:* Two impact craters. Gula (top) has a diameter of 40 km (25 mi). (NASA/JPL/Brown University) *Right:* Fractured terrain caused by plate tectonics in the ancient past. The band running from upper left to lower right is approximately 15 km (9 mi) wide. (NASA/JPL/Brown University)

---

[1] *2010: The Year We Make Contact* (1984). Directed by Peter Hyams. [Feature film]. Beverly Hills, CA: MGM/UA Entertainment Company.
[2] *2001: A Space Odyssey* (1968), Directed by Stanley Kubrick. [Feature film]. Beverly Hills, CA: Metro-Goldwyn-Mayer.
[3] *Europa Report* (2013). Directed by Sebastián Cordero. [Feature film]. New York, NY: Magnolia Pictures.

Ganymede's surface has two distinct regions: the darker region is the oldest and the lighter region is somewhat younger, but still ancient. The estimation of their ages comes from the number density of impact craters, with more craters implying an older surface. Tectonic activity that ended long ago is probably responsible for refreshing the younger surface. In support of this conjecture is the presence of fracture lines, including a set of parallel fractures seen in Fig. 12.22 (*Right*). Ganymede's surface also shows water frost at both poles, although they cover a much larger fraction of the surface than those on Earth or Mars.

Like Europa, Ganymede's surface crust is composed of water ice, but the crust is much thicker than Europa's. Below the crust there is a saltwater ocean, and perhaps several alternating layers of liquid water oceans and different types of water ice, with the bottom-most layer being an ocean sitting atop a rocky mantle. At the center is a molten outer iron core surrounding a solid iron core. Ganymede seems to have cooled slowly after its formation, providing time for the heaviest elements to sink to the center, much like the interiors of the rocky planets. The slow cooling of the interior was probably caused by the tidal forces exerted on it by Jupiter and the other large moons.

While the Galileo Orbiter was exploring Ganymede it detected a magnetic field, lending support to the conclusion that a molten iron outer core exists in the moon, and that convection is present there.

Callisto: The Ancient, Cratered World

The outermost of the four Galilean moons is Callisto (Fig. 12.23), the second largest moon in Jupiter's vast system of moons, and the third largest moon in the Solar System.

Callisto's surface crust is composed of an extensive layer of water ice 80 to 150 km (50 to 90 mi) thick that clearly shows the battle scars of a large world exposed to impacting meteors and comets without any significant atmosphere to protect it and without any plate tectonic activity to refresh its surface. As a result, cratering is extensive, covering virtually every part of the moon. Where more recent impacts have occurred, fresh ice has been exposed that leaves white spots. Very large impacts have also occurred. One such impact, shown in Fig. 12.23 (*Bottom*), resulted in a basin with multiple rings of mountains centered on the impact site.

The darker surface of Callisto is composed of non-ice materials, including rock and organic compounds that contain rings or long chains of carbon atoms commonly found in living organisms on Earth. Considering that the surface is geologically inactive, it is likely that the rock and organic material was delivered by impacting comets. Studies of the reflected spectra of Callisto have also revealed the presence of frozen carbon dioxide (CO_2) on the surface, along with sulfur dioxide (SO_2). There is also a very tenuous atmosphere made up of carbon dioxide.

Callisto seems to respond to Jupiter's strong magnetic field in the same way as Europa, suggesting that there must be an electrically conducting liquid in the interior, most likely a saltwater ocean below the surface that may also contain some ammonia that acts like antifreeze. The ocean is located below the icy crust and is some 200 km (120 mi) deep. Below the ocean sits a mantle that is mostly a mixture of ices and rock, with the possibility of a very small iron metallic core.

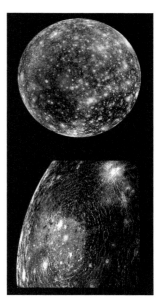

Fig. 12.23 *Top:* Callisto, as observed by the Galileo Orbiter in 2001. (NASA/JPL/DLR) *Bottom:* Concentric mountainous rings produced by an enormous impact was imaged by Voyager 1 in 1979. (NASA/JPL)

Unlike Ganymede, it seems that Callisto did cool off fairly quickly after formation, which didn't provide much time for the heaviest elements to sink to the center. Given that Callisto is the farthest giant moon from Jupiter, and outside the orbits of the other three, it doesn't experience the same level of tidal flexing, and therefore internal heat generation, that the others endure.

Jupiter's Other Moons

The Galilean moons are the only ones in the Jupiter system that are large enough for gravity to force them into spherical shapes. All of the others have irregular shapes, some orbit closer to Jupiter than Io, and some have regular prograde orbits while others orbit retrograde with very unusual orbits. Some of the small moons may have been captured by Jupiter and/or resulted from collisions that caused the breakup of other small moons in the past.

Fig. 12.24 Amalthea (inset) shown in comparison to Io. This image of Amalthea was captured by the Galileo Orbiter in 1998. (NASA/JPL/Cornell University)

As one example, Amalthea is shown in the inset in Fig. 12.24. Approximately 150 km (90 mi) in length, or about the size of Long Island, New York, Amalthea is the sixth-largest moon orbiting Jupiter, and is located inside Io's orbit. Amalthea has a distinctive reddish tint, perhaps caused by sulfur compounds. Just like all of the other moons orbiting Jupiter, Amalthea is tidally locked, meaning that it is in synchronous rotation, so it keeps the same side facing Jupiter at all times. It has been speculated that Amalthea was captured by Jupiter or migrated to its current position from a different orbit around the planet.

Moons of Saturn

Titan: A Giant Atmosphere-Shrouded Moon

Titan, seen in Fig. 12.25, is Saturn's largest moon, and the second largest moon in the Solar System. Titan is unique in the Solar System because it has a thick atmosphere, with a pressure at the surface that is nearly 50% greater than the atmospheric pressure at the surface of Earth. Titan is also unique in that it is the only other body in the Solar System besides Earth known to have a large amount of liquid on its surface.

Key Point
Titan has a thick atmosphere primarily composed of methane.

Long before it was ever visited by spacecraft, studies of the spectra of Titan in the 1940s revealed that the giant moon had an atmosphere composed of methane (CH_4). Then, in successive years, Pioneer 11 and Voyagers 1 and 2 provided images of the haze-shrouded moon and some information about the density, composition, and temperature of its atmosphere. Observations by Voyager 1, followed by infrared studies by the Hubble Space Telescope, further hinted at large land forms.

Intrigued by the possibilities of what lays below the orange–yellow haze, NASA and the European Space Agency (ESA) partnered in a combined mission to create the Cassini Orbiter and the Huygens probe. The former would remain in orbit around Saturn, closely studying the gas giant and some of its many moons including Titan, while the Huygens probe would parachute into Titan's atmosphere and land on the surface, obtaining images and sampling environmental conditions during its decent and while on the surface. The two spacecraft were launched together in 1997, with Huygens piggybacking on Cassini until one month prior to reaching Saturn. At that point Huygens separated, and parachuted to the surface of Titan

Fig. 12.25 *Top Left:* Titan in natural color by combining images taken in red, green, and blue (RGB) filters by the Cassini Orbiter in 2017. *Top Right:* Same as *Top Left* except that the red filter was replaced by an infrared filter that can detect light reflected from the surface through the thick hydrocarbon haze. (NASA/JPL-Caltech/Space Science Institute) *Bottom:* A false-color radar mosaic of a large liquid sea of hydrocarbons near Titan's north pole, comparable in volume to three times that of the Great Lakes' Lake Michigan. The mosaic was created by combining data obtained from February 2006 to April 2007. (NASA/JPL-Caltech/ASI/Cornell)

on January 14, 2005. Cassini was finally deorbited on September 15, 2017, sending it plunging into Saturn one month short of 20 years after launch. Cassini continued to send back information about Saturn's atmosphere until it finally burned up. Thanks to Cassini–Huygens we have come to learn a great deal of the mysterious Titan.

Figure 12.25 (*Top*) shows two images of Titan captured by the Cassini Orbiter. The image on the left is in natural color. The image on the right used an infrared filter because infrared light is capable of penetrating through Titan's methane atmosphere. Visible near the north pole are white, wispy, summer clouds of methane.

The Cassini Orbiter bounced radar signals off Titan's surface just like the Magellan spacecraft did, in order to peer through Venus's thick atmosphere (page 391). One of the radar images produced by Cassini is shown in Fig. 12.25 (*Bottom*). When

Table 12.3 Freezing and boiling points of some substances

Substance	Freezing [K (°C)]		Boiling [K (°C)]	
Water (H$_2$O)	273	(0)	373	(100)
Carbon dioxide^a (CO$_2$)	195	(−178)	—	—
Ammonia (NH$_3$)	195	(−178)	240	(−33)
Methane (CH$_4$)	91	(−182)	112	(−161)
Ethane (C$_2$H$_6$)	90	(−183)	185	(−88)
Carbon monoxide (CO)	68	(−205)	82	(−191)
Nitrogen (N$_2$)	63	(−210)	77	(−196)

^{a}Carbon dioxide sublimates directly from a solid to a gas.

a strong radar signal was returned after reflection from the surface, the region was assigned colors ranging from yellow to white. When very little or none of the signal was returned, the coloration is dark to black. Liquid hydrocarbons [primarily methane (CH$_4$) and ethane (C$_2$H$_6$)] effectively absorbed the radar signal. What is being depicted in the figure is a vast sea of liquid methane located near the north pole of Titan. Further analysis of the data indicated that the sea is more than 200 m (660 ft) deep in places. The entire volume of this second-largest body of liquid on the moon is roughly equal to three times the volume of Lake Michigan (one of the five Great Lakes along the east-central Canada–United States border).

Figure 12.26 shows an image obtained by Huygens during its decent to the surface and an image taken after the probe successfully landed. It is apparent that Titan is a complex world that in some ways is similar to Earth, except that water is replaced by methane and ethane as the active liquids. Of course, it is also much colder than Earth, with a surface temperature of 93 K (−180°C, −292°F).

In order to understand the composition of the surfaces of bodies in the outer Solar System, it is helpful to refer to Table 12.3 for the freezing and boiling points of common substances that we have encountered, and will continue to encounter, in the remainder of Chapter 12 and in Chapters 13 and 14. Water ice, liquid water, and water vapor are common on Earth, while water ice and frozen carbon dioxide exist on Mars. Three of the Galilean moons have water ice surfaces and subsurface oceans. But by the time we move out to the moons of Saturn and beyond, we start to find contributions from other substances that would otherwise be in gaseous form closer to the Sun. For methane (and to a lesser extent ethane) on Titan to behave like water does on Earth, their characteristic freezing and boiling points must be similar to Titan's surface temperature.

Remember that methane is a greenhouse gas. This causes the surface temperature to be quite a bit warmer than it would be without the thick atmosphere. Titan's thick atmosphere also cuts down on the amount of sunlight reaching the surface. Because of the inverse square law of light, Titan only receives about 1% of the amount of light at the top of its clouds compared to the amount of light that Earth receives. The clouds reduce that by another factor of 10, so that at the surface the amount of light present is only 0.1% of the amount of light on Earth.

Fig. 12.26 *Top:* A view of Titan's surface during the decent of the Huygens probe on January 14, 2005. A movie of the decent is available on the Chapter 12 resources webpage. (ESA/NASA/JPL/University of Arizona) *Bottom:* A view from the Huygens probe after it landed on Titan. (NASA/JPL/ESA/University of Arizona)

Because the surface temperature is so low, the "bedrock" beneath Titan's surface is made of rigid water ice. Titan's surface also appears to be fairly young (between 100 million to 1 billion years old), implying forces may be at work, or were at work, to refresh the surface. Possibilities include cryovolcanic activity (icy-cold volcanoes) that uses water and ammonia (NH_3) rather than molten rock as is the case on Earth. Methane rain and liquid methane on the surface may also erode features.

While a few impact craters have been observed, they are relatively rare compared to the other moons of the outer Solar System. This is certainly due to meteors burning up as they fall through Titan's atmosphere, and others being erased by weathering.

In general, the surface of Titan is quite smooth. Along with the few craters and liquid lakes and seas, Titan does have "mountains," although the tallest don't exceed more than about 1 km (0.6 mi). According the International Astronomical Union, all of the mountains named on Titan are taken from J. R. R. Tolkien's (1892–1973) books about the realms and characters of Middle-Earth.

Titan's interior likely contains a crust of water ice with a water (or perhaps a water–ammonia mix) ocean below the crust. There may also be layers of different forms of ice below the ocean. But the bulk of the moon is probably composed of a water ice–rock mix, similar to that of Callisto. It is unlikely that even a small metallic iron core exists, because Titan would have cooled off too fast for the heaviest elements to sink to the center of the moon.

Enceladus: The Moon of Water Vapor Geysers

Even before Cassini's careful study of this small moon, Enceladus (pronounced "en-sel-uh-dus") seemed to be unique in part because it is the most reflective body in the Solar System (see Fig. 12.27). Enceladus has a diameter of about 500 km (310 mi), making it roughly the size of the state of Wyoming. Enceladus orbits close to Saturn at a distance of just 4.1 Saturn radii, with an orbital period of 1.4 days (Titan's orbit has a semimajor axis of 21.3 Saturn radii). Enceladus is also caught in a 2:1

Fig. 12.27 *Left:* The Saturn-facing side of Enceladus seen from Cassini in 2015. The north is heavily cratered, but transitions to a younger surface toward its equator and southern hemisphere. (NASA/JPL-Caltech/Space Science Institute) *Center:* A view from above the south pole of Enceladus. (NASA/JPL-Caltech) *Right:* The geyser basin near the southern pole of Enceladus. (NASA/JPL-Caltech/Space Science Institute)

Fig. 12.28 Dione is one of Saturn's mid-sized moons. (NASA/JPL-Caltech/Space Science Institute)

orbital resonance with Dione (Fig. 12.28), a moon that is farther from Saturn and a little more than twice Enceladus's diameter. And, as is true with virtually all of the moons in the Solar System, Enceladus is in synchronous rotation, keeping the same face toward Saturn at all times.

When Cassini began imaging Enceladus, cryogeysers were observed near its south pole, ejecting ices and dust into space from cracks in the surface. Some of the ejected material snows back down on the surface, refreshing the south-polar region of the moon and covering up impact craters. Salts are also observed in the cracks, suggesting that the subsurface ocean itself may be "salty."

The oldest parts of the surface are near Enceladus's north pole, and the youngest are near the south pole. Estimates of the smooth regions in the south place their ages at perhaps half a million years, while material in some of the surface cracks could be as young as 1000 years. Enceladus is a geologically active moon, with on-going tectonic activity implying circulation in the subsurface, salty ocean.

By now it shouldn't be surprising to find another moon of the outer Solar System that is covered with ice, or that a subsurface ocean may exist below the ice crust, but what is special about Enceladus is that there is a hot-spot around the south pole that is as-yet unexplained. None of the proposed heating sources, including heating due to tidal forces, are enough to explain why the interior is so warm. It is likely that a rocky (and possibly small metallic) core exists below the 10 km (6 mi)-deep ocean.

Especially intriguing is that those water vapor cryogeysers and the material in the surface cracks on Enceladus contain a variety of organic hydrocarbon molecules, molecular hydrogen, salts, and an energy source that could be suitable for the kind of chemistry believed to be required for the development of life on Earth and potentially for the formation of microbial extraterrestrial life. As a result, along with Mars and Jupiter's Europa, Enceladus has become a focus of the search for life beyond Earth. Just as was the case for the Galileo Orbiter after it completed its mission to Jupiter, the Cassini Orbiter was intentionally sent crashing into Saturn to avoid accidental contamination with Earth microbes.

Mimas: That's No Space Station, That's a Moon

Fig. 12.29 Mimas with its enormous impact crater. (NASA/JPL-Caltech/Space Science Institute)

No tour of Saturn's moons would be complete without a call out to Mimas, shown in Fig. 12.29. Mimas is slightly smaller than Enceladus, and orbits a little closer to Saturn than Enceladus does. Mimas is also substantially less dense than Enceladus suggesting that it is primarily composed of water ice. A major difference between the two is that Mimas is clearly not geologically active. Its surface is very old and saturated with craters. But what really makes it stand out is a very prominent crater, that is enormous in size relative to the moon. Based on the size of the crater, it appears that the impact that produced it was almost large enough to completely shatter the little moon.

Pioneer 11 flew past Mimas in 1979, with Voyagers 1 and 2 passing by in 1980 and 1981, respectively. Those visits were shortly after the original *Star Wars* movie came out in 1977 (George Lucas). Fans couldn't help but notice the similarity of Mimas to the fictional Death Star featured in the film. (The film was later renamed *Star Wars: Episode IV — A New Hope*[4].)

[4]*Star Wars: Episode IV — A New Hope* (1977). Directed by George Lucas. [Feature film]. Los Angeles, CA: 20th Century Fox.

Uranus's Miranda: A Moon by Committee

Miranda, shown in Fig. 12.30, is the smallest of Uranus's five round moons, averaging just 470 km (290 mi) in diameter. As with the other named moons of Uranus, including Ariel, Umbriel, Titania, and Oberon, Miranda is named for a character in one of William Shakespeare's (1564–1616) or Alexander Pope's (1688–1744) works; in this case, Shakespeare's *The Tempest*. Along with the other major moons of Uranus, Miranda orbits in a plane closely aligned with the planet's equator. Since Uranus is tilted 98° (essentially laying on its side compared to its orbital plane), its major moons are orbiting nearly perpendicular to the ecliptic.

Fig. 12.30 Miranda is Uranus's smallest spherical moon. (NASA/JPL-Caltech)

Miranda is the smallest object in the Solar System that is still large enough for its own gravity to force it into a spherical shape. With a density that is only 1200 kg/m³, 1.2 times greater than liquid water, Miranda is most likely composed of 60% water ice, with the remainder being rocky material in its center.

When Voyager 2 passed Miranda in 1986, it witnessed a moon that is a collection of varying terrains, including concentric oval shapes that are canyons 20 km (12 mi) deep, seen on the left side of Fig. 12.30 and nicknamed the "racetrack," a check mark-like chevron (lower right-center), and a cliff (bottom) more than 5 km (3 mi) high. Judging from the amount of cratering seen, the surface maybe as young as one hundred million years. Having only seen one side of Miranda during the Voyager 2 flyby, we don't know what the other side looks like. (Miranda has sometimes been referred to as a moon that looks like it was put together by a committee.)

Several hypotheses have been proposed for the jumble of features on Miranda's surface, including that the moon was hit so hard that it broke into pieces and gravity pulled it back together again but without everything quite fitting right afterward. However, the more likely explanation is that blocks of ice were uplifted by tectonic forces associated with warmer ice beneath in the interior. Given what appears to be a relatively young age for the surface, the moon may still be geologically active today.

Neptune's Triton: Where Did This Moon Come From?

The last object that Voyager 2 visited on its "Grand Tour" of the Solar System was Triton, Neptune's largest moon, shown in Fig. 12.31. Triton is the seventh largest moon in the Solar System and is slightly larger than the dwarf planet Pluto.

Even before Voyager 2 arrived at Triton, the moon was known to be bit unusual for such a large body. Rather than orbiting more or less above the planet's equator and in the same direction that the planet spins, Triton is in a retrograde orbit that is tilted 157° to Neptune's equator and its ring system (Fig. 12.32). Triton's orbit is also nearly perfectly circular, and again as with virtually all of the moons in the Solar System, it is in synchronous rotation, keeping the same face toward Neptune at all times. Since tidal forces are required to both circularize the orbit and produce synchronous rotation, heating must have occurred in the interior for an extended period of time, which probably caused the interior to separate into an icy mantle and a rock–metal core.

This odd behavior strongly suggests that Triton did not form in the same region as Neptune, but was instead captured somehow. Given that Triton does share many characteristics with other dwarf planets in the Kuiper belt, it is almost certainly a captured dwarf planet itself.

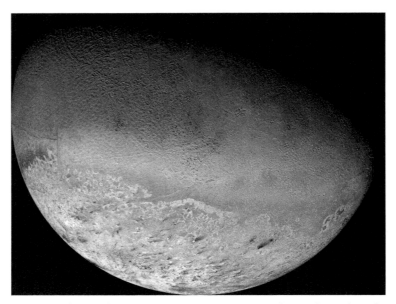

Fig. 12.31 Triton, as observed by Voyager 2 in 1989. A portion of Triton's southern hemisphere is visible at the bottom of the image. (NASA/JPL/USGS)

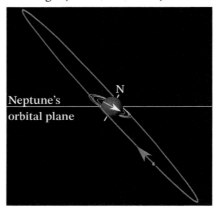

Fig. 12.32 Triton orbits Neptune retrograde in an orbit tilted with respect to Neptune's equator and its ring system. Triton's size is exaggerated by a factor of four. (Images: NASA/JPL/USGS)

In order for a planet to capture a moon, some of the moon's kinetic energy must be lost in the process so that its speed is below the planet's escape speed (page 198). If an object simply came close to a planet, it would speed up as it approached the planet rather than slow down, swing around it, and then slow down while escaping the planet's gravitational pull. If the object is to be captured it must either collide with something or give up some of its kinetic energy to another object that gets thrown clear of the planet's attraction. It seems that Triton used the second strategy. It is not uncommon for Kuiper belt objects to be in orbits around other Kuiper belt objects. If Triton had a binary companion when they came close to Neptune, the interaction could have caused Triton to be caught while its partner escaped with some of Triton's kinetic energy.

When Voyager 2 passed Triton we learned that the moon is odd for reasons other than it being a captured dwarf planet; it has a surface unlike anything that had been seen up to that time. Its surface is also the coldest yet measured in the Solar System, with a temperature of about 38 K ($-235°C$), or $-392°F$. From the location of the shadow in Fig. 12.31 the Sun was primarily illuminating the moon's southern hemisphere at the bottom of the image, which appears pinkish in color. The coloration is caused by nitrogen frost on a surface of water ice that contains a mixture of other ices as well (again recall the freezing and boiling points of various substances in Table 12.3). The southern hemisphere also shows dark streaks that appears to be the result of dust contained in ices ejected by cryovolcanoes.

Figure 12.33 (*Top*) shows some of the volcanic flows toward the top of the image, with the volcanic material being ices. Toward the bottom of the image is some of the so-called "cantaloupe" terrain, named for its resemblance to the surface of a cantaloupe melon. The cantaloupe terrain seems to have formed when the surface overturned due to rising blocks of ice from underneath. Figure 12.33 (*Bottom*) reveals fault lines running across the surface through the cantaloupe terrain.

From the almost complete lack of impact cratering, Triton's surface must be quite young, with regions having been refreshed as recently as 8 million years ago. The cryovolcanic activity, the icy lava flows, the cantaloupe terrain, and fault lines all suggest a geologically active world at the outer reaches of the Solar System. Given that the volcanic streaks occur on the sunlit hemisphere of the moon, solar heating may be partially responsible. However, additional heat also seems to be required, leading to the possibility that radioactive decay of elements deep in the rock–metal core may still be producing enough energy to power some convection in the mantle.

Finally, Triton does have a very thin, but measurable atmosphere dominated by nitrogen, with small amounts of carbon monoxide and methane. The atmosphere is probably the result of sublimation of nitrogen ice from the moon's surface. A haze in the atmosphere comes from hydrocarbons and nitrogen-based molecules. Clouds of nitrogen were also detected by Voyager 2.

With only getting a glimpse of about 40% of the surface during the Voyager 2 flyby, we are left to wonder what the rest of the moon looks like.

Fig. 12.33 *Top:* Volcanic plains on Triton with "cantaloupe" terrain visible toward the bottom of the image. (NASA/JPL/Universities Space Research Association/Lunar & Planetary Institute) *Bottom:* Faults are seen crossing the surface. (NASA/JPL)

12.3 Rings around the Planets

Discoveries

One feature that all four giant planets share is the presence of a ring system. Certainly the most famous are the spectacular rings of the Saturnian system, first observed by Galileo in 1610 (Section 4.12). But Galileo wasn't sure what he was seeing with his small telescope, believing that he might be observing two "planets" orbiting Saturn. He became especially confused when Saturn's "planets" apparently disappeared, only to reappear later. We now know that Galileo was seeing Saturn's rings edge-on when Earth was aligned with the plane of the ring system; a similar situation occurred in 1995 and was recorded in the Hubble Space Telescope images displayed in Fig. 12.34. It wasn't until 1655 that Christiaan Huygens first proposed that the strange appendages to Saturn were rings. Then, twenty years later, as telescopes became larger and their magnifications increased, Giovanni Cassini was able

Fig. 12.34 Saturn's rings are very thin and virtually disappear when observed edge on. The dark line across Saturn in the bottom image is the shadow cast by the rings. Two mid-sized moons, Tethys (left) and Dione (Fig. 12.28), are also visible on the left side of the bottom image. (NASA/JPL/STScI)

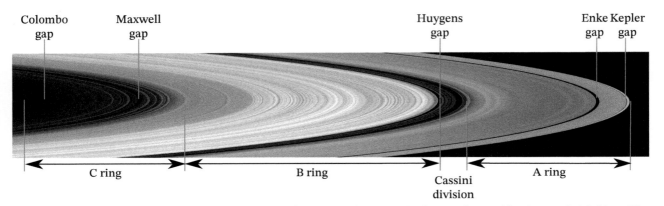

Fig. 12.35 A portion of Saturn's ring system showing thousands of narrow ringlets as observed by the Cassini Orbiter. The ring particles are composed almost entirely of water ice, making the rings highly reflective. The D ring is inside the C ring. The F ring is just beyond the edge of this image and the E ring is still farther from Saturn. (Adapted from NASA/JPL/Space Science Institute)

to resolve smaller rings rather than one wide, smooth, ring. Today, thanks to constantly improving observational techniques, much larger telescopes, and the beautiful observations of the Cassini Orbiter (e.g., Fig. 12.35), we know that the Saturnian ring system is vast and very complex. Cassini would be amazed at the intricacy of the rings seen in the images that his namesake spacecraft provided.

In his notes on Uranus, Sir William Herschel wrote: "February 22, 1789: A ring is suspected." It is very doubtful that Herschel was actually able to resolve a ring system around Uranus given the telescopic capabilities of the day, but they were definitely observed in 1977, nearly two centuries later, becoming the second system to be seen around a giant planet. Jupiter's system was discovered by Voyager 1 two years after the observations of the Uranian rings, and the Neptunian system was indirectly deduced in 1984 when its rings blocked the light from a star as the rings passed in front of the star during the planet's orbit around the Sun; the Neptunian ring system was confirmed five years later during the Voyager 2 flyby of the planet. The rings of Jupiter, Uranus, and Neptune are seen in Fig. 12.36, and Uranus, its rings, and eight of its moons are seen in infrared light in Fig. 12.37. Recall from Fig. 12.10 that Uranus's equator is inclined more than 90° (97.8°) to its orbital plane, meaning that the system rotates retrograde; the same holds true for its rings and its moons, which are located above the planet's equator.

Today we know that ring systems are not just reserved for the giant planets, but they can also exist around some dwarf planets, asteroids, perhaps moons, and potentially other planets beyond our Solar System.

General Characteristics and Sources

Rings are not solid, but are composed of particles ranging in size from a few kilometers across (10^3 m) to as small as micrometers (10^{-6} m) or less. Each of these particles orbits around their parent body like tiny moons and are primarily influenced by the forces of gravity associated with the parent object, other ring particles, and the presence of moons within and external to the ring system. It is also possible that some ring particles can be influenced by a planet's magnetic field and by

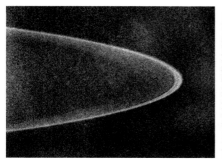

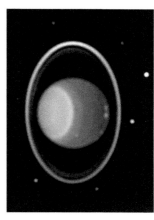

Fig. 12.36 *Left:* Jupiter's rings. There is a narrow ring on the outside and a very thin sheet of dust inside it that reaches to Jupiter's cloud tops. (NASA/Johns Hopkins University Applied Physics Laboratory/Southwest Research Institute) *Center:* Uranus's rings are primarily composed of water ice with some dark, carbon-based compounds that were processed by the Sun's ultraviolet radiation. The streaks are caused by stars as the spacecraft stayed focused on the ring system. (NASA/JPL) *Right:* Neptune's very faint, dusty rings are made of very dark material, probably carbon-based compounds. The bright glow is the over-exposed Neptune. (NASA/JPL)

Fig. 12.37 The Uranian system seen in infrared light by the Hubble Space Telescope in 1997. (NASA/JPL/STScI)

electrostatic charges if the particles are electrically charged or can conduct electricity themselves.

Saturn's main rings are made up of particles of almost pure water ice (99.9%), with some amount of contamination by silicate-based dust and hydrocarbons, and with a combined mass that is slightly less than the mass of Mimas. Saturn's three main rings (A, B, and C in Fig. 12.35) extend from the inner edge of the C ring at 1.2 R_{Saturn} from the planet's center (7000 km above its cloud tops) to 2.3 R_{Saturn} at the outer edge of the A ring. However, although the rings are more than 1/2 of Saturn's diameter wide, they are incredibly thin vertically, ranging from just 10 m thick to 1 km thick. It is for this reason that the rings seemed to disappear when Galileo viewed them edge on. Seen from 9.5 au away, 1 km is only 0.000 000 04° wide, or about one ten-millionth of the angular width of a full moon! They are so prominent when not seen edge on only because water ice reflects sunlight very effectively.

Not all ring systems are created equal, because they can develop in a variety of different ways. How rings form, and the influences on them over time, govern what they are made of and how they appear today.

One mechanism for creating rings is for a moon, asteroid, or comet to be torn apart by extreme tidal forces (Section 10.3). The difference in the force of gravity between the near and far sides of the object can exceed its internal strength and the object can be ripped apart. Édouard Roche (1820–1883) was able to calculate how close to a planet an object would need to be for its destruction to occur in this way. Mars may acquire a ring system in the distant future by this process when one of its moons, Phobos, finally spirals in too close to the planet due to tidal drag, and gets torn apart.

When ring systems are within Roche's distance limit the individual particles can't combine to form a small moon, because the tidal forces simply prevent the moon from coalescing. The reason spacecraft (or people) don't get ripped apart is because their internal structural strength exceeds the tidal force, but objects held together by gravity are subject to this shredding force. For Saturn, Roche's distance limit is close to 2.2 R_{Saturn}, near the outer edge of the A ring.

A second way for rings to form is from collisions of meteors with moons, effectively chipping off parts of the moons' surfaces. The material ejected from the surface of a moon ends up in the same orbit as the moon itself. More catastrophically, a moon may be completely destroyed by a collision and its remnants end up being ring material.

In some cases, moons like Saturn's Enceladus can eject material from their interiors by processes like cryovolcanism. Enceladus orbits in the middle of a tenuous outer ring of Saturn, known as the E ring (Fig. 12.38). Unlike other rings in the Saturnian system, the E ring is primarily composed of very tiny micrometer-sized (0.001 mm) particles of water ice, with some contributions of dust, carbon dioxide ice, ammonia ice, molecular nitrogen, and hydrocarbons.

Enceladus also happens to be located in the narrowest part of the E ring. Since the E ring would not last very long without being refreshed constantly, Enceladus appears be the source of the ring material; as the particles drift away from the moon, they spread out and the ring gets wider. In fact, the widest part of the ring is on the opposite side of Enceladus's orbit. Assuming that Enceladus is the source of the ring material, the ring's composition tells us something about the composition of the moon's interior.

Finally, ring systems may form as remnants of planet formation, to be discussed in Section 14.2. When a planet forms out of the rotating nebula of gas and dust, some of the material might escape from being pulled in and instead end up orbiting the planet. Tidal forces from the rotating, slightly flattened planet cause the rings to be drawn into a thin plane above the planet's equator.

Fig. 12.38 Cassini's view of Saturn's E ring, backlit by the Sun. Enceladus is the bright dot in the middle of the ring. (NASA/JPL/Space Science Institute)

The Effects of Moons on Ring Systems

Orbital Resonance and the Cassini Division

Figure 12.4 and Fig. 12.34 show what appear to be significant gaps in Saturn's rings. The widest "gap," seen about three-quarters of the way out in the ring system, is called the Cassini division. However, when you carefully study the Cassini division in Fig. 12.35, it doesn't seem quite as apparent; although the division is still visible, some rings exist within the gap. In reality, there are no true gaps in Saturn's main rings, just regions where the material isn't as dense.

But what causes the enhancement or thinning of the rings that give rise to the thousands of ringlets? Scientists can't account for all of them, but many are due to orbital resonances with Saturn's many moons. The inner edge of the Cassini division, in particular, has a 2:1 orbital resonance with Mimas (Saturn's "death star" moon), visible in Fig. 12.39. Any particle in orbit at the inner edge of the Cassini division makes exactly two orbits around Saturn for every one orbit of Mimas around the gas giant. The orbital plane of Mimas is also nearly the same as the orbital plane of the ring system. The process is completely analogous to the 2:1 orbital resonance of Io and Europa around Jupiter, illustrated in Fig. 12.18; a ring particle corresponds to Io, and Mimas corresponds to Europa. Referring to Fig. 12.40, when the particle is between Mimas and Saturn, Mimas tugs the particle away from Saturn. But when Mimas has completed one-half of an orbit and the particle has completed

Fig. 12.39 Mimas is visible in the top-right of this image of Saturn's rings. Mimas interacts gravitationally with the rings, resulting in the sharp inner edge of the Cassini division. (NASA/JPL-Caltech/Space Science Institute)

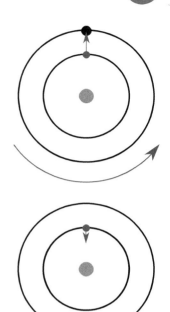

Fig. 12.40 The 2:1 orbital resonance of Mimas with the inner edge of the Cassini division. *Top:* Mimas (black) pulls a ring particle (red) at the inner edge of the Cassini division away from Saturn. *Bottom:* One-half Mimas orbit later, Mimas pulls the ring particle toward Saturn, but much more weakly.

a full orbit, Mimas is now on the opposite side of Saturn from the particle and the moon is helping Saturn pull it back toward the planet. However, being farther from the particle when Mimas is on the opposite side of Saturn, Mimas can't pull back as hard as when it was on the same side of Saturn as the particle. The result is that the particle gets forced into an increasingly elliptical orbit and it eventually collides with particles either closer to Saturn or farther away. The particle has effectively been removed from the inner edge of the Cassini division. Mimas is the reason why the inner edge of the Cassini division is so sharp.

Gap Sweeps

In some cases, very small moons can produce gaps by essentially sweeping out material within their orbits. One such example is the tiny moon Pan, orbiting within the Encke gap (Fig. 12.41). Pan measures just 28 km (17 mi) across. Similarly, Daphnis, orbits within the Keeler gap, governing its structure. The reason that these moons aren't torn apart by the tidal forces inside the Roche distance limit is simply because they are too small to experience enough of a difference in gravitational pull from Saturn to do the job (recall from Newton's universal law of gravitation that the force of gravity is an inverse square law that decreases in strength as $1/r^2$). The pull of Saturn on the near side of a tiny moon is just not different enough from the pull on the moon's far side.

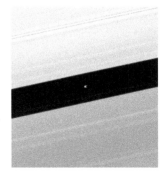

Fig. 12.41 Pan in the Encke gap in Saturn's rings. (NASA/JPL-Caltech/Space Science)

Shepherd Moons

The F ring in Saturn's ring system, shown in Fig. 12.42, is particularly delicate. Orbiting beyond the A ring, the F ring is comprised of ringlets that are essentially intertwined. If it weren't for two tiny moons, Prometheus and Pandora, the F ring would soon spread out because of collisions between ring particles. Acting like shepherds guiding their flock, Prometheus and Pandora orbit just inside and outside of the F ring, respectively. Since Prometheus has a slightly shorter orbital period than material in the F ring, if particles in the ring drift outward, Prometheus passes the wayward particles, giving them a gentle tug back toward Saturn. Conversely, if particles drift inward resulting in more rapid orbits, when those particles pass Pandora the moon pulls them back out again.

Uranus also has shepherd moons, Cordelia and Ophelia, that escort its ϵ ring.[5]

Fig. 12.42 The outermost F ring of Saturn is a series of twisting strands that are shepherded by two tiny moons, Prometheus (closer to Saturn) and Pandora. A movie of the interplay is available from the Chapter 12 resources webpage. (NASA/JPL/Space Science Institute)

Doing the Waves

As one last example of the intricacies present in ring systems, especially in the complex Saturnian system, consider Fig. 12.43. The moon, Janus, is in a 2:1 orbital resonance with ringlets in the innermost portion of the B ring. The periodic gravitational pull of Janus on that part of Saturn's B ring causes waves to propagate through that area. Although it may look like the upper left part of the image is farther from the camera than the lower right portion is, that is only an optical illusion; in fact, all parts of the image are essentially the same distance away. The deception is because the waves are more closely spaced in the upper left part of the image.

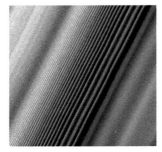

Fig. 12.43 Waves moving through Saturn's B ring. (NASA/JPL-Caltech/Space Science)

Ring System Lifetimes

Scientists have struggled with determining how old the ring systems of the giant planets are. With the exception of rings like Saturn's E ring, that is constantly replenished by eruptions from the moon Enceladus, the formation of particular rings is still unclear. It is possible that some rings may have formed with planets, or perhaps from small moons, comets, or asteroids that got torn apart when they ventured inside Roche's distance limit at a time when they were more prevalent, during the early days of the Solar System. However, this raises the question of how they could survive for nearly 4.6 billion years, given that collisions between ring particles would bounce them into different orbits, shepherd moons aside. There are forces that can cause at least the innermost ring particles to spiral into the planet as well. Over time, these processes would lead to the dissipation of the ring system.

Alternatively, it may be that planetary rings are simply transient and relatively young. In Saturn's case, perhaps the main rings (A, B, and C) were created when a small, icy moon drifted inside Roche's distance limit; certainly icy moons are quite common today. If rings are young, the problem of them dispersing goes away simply because there hasn't been enough time for them to disappear.

This leads to the debate over very old ring systems, because they are easier to create, and young ring systems, because they are still around. Scientists certainly don't have all of the answers yet, but that is what science is all about: searching for answers to questions not yet solved, and even asking new questions that had not been thought of before.

---

[5]You can find the Greek alphabet in Table 4.1.

? Exercises

All of the exercises are designed to further develop *your understanding* of the material by thinking carefully and critically about what you have read in this chapter. Answers to selected exercises are available in the back of the book.

True/False

1. All four giant planets have been visited by orbiting spacecraft that conducted observing campaigns lasting at least one year.
2. (*Answer*) All four giant planets rotate in a prograde direction.
3. Uranus's rotation axis is almost in the plane of the ecliptic.
4. (*Answer*) An object with Saturn's average density could float in water.
5. The only planet in the Solar System with a magnetic field that is stronger than Earth's is Jupiter.
6. (*Answer*) The Great Red Spot is a storm on Jupiter than has been growing in size over the past few hundred years.
7. At the time this textbook was written, the reason that Uranus only radiates just slightly more energy than it receives from the Sun, while Neptune radiates more than 2 1/2 times more energy than it receives is well understood by scientists.
8. (*Answer*) The north pole of Uranus is in sunlight for about 1/2 of its orbit and in darkness for the other half.
9. Neptune's Great Dark Spot, first seen by Voyager 2 in 1989, is a storm system that has existed for hundreds of years, and still exists today.
10. (*Answer*) The axes of the magnetic fields for Earth and the four giant planets all pass directly through their centers.
11. We can estimate the distribution of mass in the interiors of the giant planets by seeing how the orbits of spacecraft, moons, and ring systems are affected by the planets.
12. (*Answer*) When hydrogen is subjected to sufficiently high pressure, it can become a liquid that has metallic properties, including becoming electrically conducting.
13. Water in either liquid or ice form is common throughout the Solar System.
14. (*Answer*) Ganymede, the largest moon is the Solar System, is larger but less massive than Mercury.
15. The seven largest moons in the Solar System are all in synchronous rotation with their orbital periods.
16. (*Answer*) Although ocean currents likely exist below the ice surface of Europa, the ice is unaffected.
17. Tectonic activity on Ganymede ended in the ancient past after it cooled sufficiently and its ice surface became thick and hard.
18. (*Answer*) Organic compounds are not known to exist on the surfaces of bodies in the Solar System other than Earth.
19. Titan has oceans and lakes of liquid water on its surface.
20. (*Answer*) Enceladus has been observed ejecting water vapor through cryogeysers.
21. Triton's surface is the coldest ever recorded in the Solar System.
22. (*Answer*) Triton is a frozen, dead world with no evidence of current geologic activity.
23. When Galileo Galilei observed Saturn's rings, they appeared to vanish for a period of time because the plane of the ring system was directly in line with Earth.
24. (*Answer*) Shepherd moons guide ring particles and help to keep the ring from spreading out and disappearing.

Multiple Choice

25. A gravity assist
    (a) is used to increase the speed of a spacecraft when it passes a planet.
    (b) aids a large rocket during liftoff.
    (c) helps someone get out of a chair.
    (d) moves an orbiting spacecraft to a higher orbit without using any fuel.
26. (*Answer*) The Galileo and Cassini spacecraft were sent plunging into Jupiter's and Saturn's atmospheres, respectively, at the end of their missions to
    (a) avoid discovery by aliens.
    (b) avoid contaminating the environments of their moons with microbes from Earth.
    (c) study the planets' atmospheres at increasing depth before burning up.
    (d) all of the above
    (e) (b) and (c) only
27. The most massive planet in the Solar System is
    (a) Jupiter   (b) Saturn   (c) Uranus   (d) Neptune
28. (*Answer*) The planet in the Solar System with the fastest rotation period is
    (a) Jupiter   (b) Saturn   (c) Uranus   (d) Neptune
29. The giant planet with the highest percentage of H_2 and the lowest percentage of He in its atmosphere is
    (a) Jupiter   (b) Saturn   (c) Uranus   (d) Neptune
30. (*Answer*) Referring to Table 12.2, if you weigh 700 N (157 lb) on Earth, approximately what would you weigh if you were standing on a floating platform on Jupiter? (700 N is the weight of a 71.5 kg mass on Earth.)
    (a) 700 N   (b) 1250 N   (c) 1770 N   (d) 2830 N
    (e) 3720 N   (f) 6860 N
31. Which giant planets are known to have diffuse cores composed of rock, ices, hydrogen, and helium?
    (a) Jupiter   (b) Saturn   (c) Uranus   (d) Neptune
    (e) all of the above   (f) none of the above
    (g) (a) and (b) only   (h) (c) and (d) only
32. (*Answer*) Two ways that significantly more heat can be released by giant planets than they receive from the Sun are through _____ and _____.
    (a) nuclear reactions, chemical reactions
    (b) nuclear reactions, sinking heavy elements
    (c) chemical reactions, planet shrinkage
    (d) nuclear reactions, planet shrinkage
    (e) planet shrinkage, sinking heavy elements
33. The cyclonic storms of the gas giants are caused by
    (a) the rapid rotation of the planets.
    (b) heat rising from their interiors.
    (c) swirling of gases around submerged mountains.
    (d) all of the above
    (e) (a) and (b) only
    (f) (a) and (c) only

34. (*Answer*) Ice giants differ from gas giants in that they
    (a) are significantly smaller.
    (b) are significantly less massive.
    (c) contain a much higher percentage of methane in their atmospheres.
    (d) rotate more slowly.
    (e) all of the above
    (f) (a) and (b) only
35. Which planet has its magnetic field axis tilted closest to the ecliptic?
    (a) Earth   (b) Jupiter   (c) Saturn   (d) Uranus
    (e) Neptune
36. (*Answer*) The aurora of Jupiter are produced
    (a) from lightning.
    (b) in the same way that Earth's aurora are produced.
    (c) by charged particles rising up from Jupiter's interior and interacting with its strong magnetic field.
    (d) when large meteorites impact the atmosphere.
    (e) more than one of the above
37. The magnetic fields of the giant planets likely originate in their
    (a) cores of rock and ices.
    (b) carbon oceans that sit atop their cores of rock and ices.
    (c) liquid metallic hydrogen mantles.
    (d) atmospheres.
38. (*Answer*) Which moon is the largest in the Solar System?
    (a) Moon   (b) Io   (c) Europa   (d) Ganymede
    (e) Callisto   (f) Titan   (g) Triton
39. The only moon with a thick atmosphere is
    (a) Moon   (b) Io   (c) Europa   (d) Ganymede
    (e) Callisto   (f) Titan   (g) Triton
40. (*Answer*) What keeps Io almost entirely molten in its interior?
    (a) tidal interactions with Jupiter and the other Galilean moons
    (b) the heat that Io absorbs from Jupiter
    (c) its many volcanoes
    (d) the Io torus
    (e) Jupiter's intense magnetic field
41. Which moon has an ocean of water below its icy surface?
    (a) Moon   (b) Io   (c) Europa   (d) Ganymede
    (e) Callisto   (f) Titan   (g) Triton   (h) Enceladus
    (i) more than one of the above
42. (*Answer*) Why does Io's surface appear reddish-yellow?
    (a) rust stains from iron
    (b) sulfur ejected from its volcanoes that falls back down onto its surface
    (c) Io is completely covered with lava
    (d) the surface of Io is very hot so it has yellow fever
43. Only a small number of impact craters are visible on Europa because
    (a) Jupiter shields the moon from meteoroids and comets.
    (b) even small meteorites punch holes in the moon's icy crust, allowing water to fill in the holes and turn to ice.
    (c) the surface is constantly being refreshed by tectonic activity and cryogeysers.
    (d) the entire surface periodically melts and then refreezes again because of the gravitational tug-of-war with Jupiter, Io, Ganymede, and Callisto.

44. (*Answer*) Of all the Galilean moons, Callisto exhibits the greatest number of craters on its surface. What does that fact say about the moon?
    (a) It has the oldest surface among the Galilean moons.
    (b) It has experienced the largest number of impacts since formation.
    (c) It is farthest from Jupiter and therefore the least protected from impacts.
    (d) all of the above
45. The ages of the surfaces of moons in the outer Solar System are estimated by
    (a) radiometric dating.
    (b) how bright or dark the surface is.
    (c) the surface density of impact craters.
    (d) the size and depth of impact craters.
    (e) none of the above; all surfaces are nearly the same age given that all of the moons formed when the Solar System formed.
46. (*Answer*) Titan has a thick atmosphere largely composed of
    (a) carbon dioxide   (b) oxygen   (c) nitrogen
    (d) methane   (e) ammonia   (f) sulfur dioxide
47. Titan's surface is composed of _____ with oceans of liquid _____.
    (a) rock; water
    (b) rock; carbon dioxide
    (c) rock; methane and ethane
    (d) water ice; water
    (e) water ice; carbon dioxide
    (f) water ice; methane and ethane
48. (*Answer*) The composition of Titan's oceans was determined by
    (a) reflecting radar signals off the surface. Hydrocarbons effectively absorb the radar signals rather than reflecting them.
    (b) knowing the evaporation and freezing temperatures of methane and ethane, combined with knowing the temperature and atmospheric pressure at the surface of Titan.
    (c) the Huygens lander sampling the oceans directly when it landed in them.
    (d) all of the above
    (e) (a) and (b) only
49. Which of the following are observations associated with Enceladus?
    (a) cryogeysers of water vapor, ices, and dust
    (b) organic compounds and salts in cracks on its surface
    (c) a subsurface ocean
    (d) the brightest surface currently known in the Solar System
    (e) a hot-spot near its southern pole
    (f) tectonic activity
    (g) all of the above
    (h) (a), (b), and (c) only
50. (*Answer*) Scientists generally believe that the most likely explanation for the unusual surface features of Miranda is
    (a) that it was hit by a massive asteroid that broke it apart, but gravity pulled it back together again as a jumbled mess.
    (b) shifting blocks of ice due to tectonic activity.
    (c) that it was torn apart by tidal forces.
    (d) that it was built by an alien committee.

51. Among the largest moons in the Solar System, Triton is unique because
    (a) it orbits in a retrograde direction.
    (b) it rotates retrograde.
    (c) its orbit is tilted significantly relative to its planet's equator.
    (d) it is in synchronous rotation.
    (e) all of the above
    (f) (a), (b), and (c) only
52. (*Answer*) The pinkish-colored surface of Triton is attributed to frozen
    (a) carbon dioxide (b) oxygen (c) nitrogen
    (d) methane (e) ammonia (f) sulfur dioxide
53. Saturn's rings are composed of particles made up of almost pure _____ ice.
    (a) water (b) carbon dioxide (c) methane
    (d) ammonia (e) ethane (f) dry
54. (*Answer*) The rings of Saturn are only about _____ to _____ thick.
    (a) 1 cm, 1 m (b) 1 m, 10 m (c) 10 m, 1000 m
    (d) 1000 m, 1000 km
55. The Roche distance limit
    (a) is how close a small body can get to a much larger one before tidal forces tear the smaller one apart.
    (b) for the Saturnian system is near the outer edge of the A ring.
    (c) explains why particles orbiting too close to a planet cannot coalesce into a small moon.
    (d) all of the above
56. (*Answer*) Which of the following could be a source of a planetary ring system?
    (a) a moon that was torn apart by tidal forces when the moon moved inside the Roche distance limit
    (b) material left over from planet formation that never came together to form a moon because the debris orbits too close to the planet
    (c) collisions between small moons that caused them to fragment
    (d) dust and ices emitted from a moon through cryovolcanoes
    (e) all of the above
57. Moons orbiting outside of ring systems can affect the structure of the rings gravitationally
    (a) by causing waves to propagate through the rings.
    (b) through orbital resonances with particles in the rings.
    (c) by making ring particles crash onto the surfaces of the moons.
    (d) all of the above
    (e) (a) and (b) only
58. (*Answer*) Tiny moons can survive inside the Roche distance limit because they are
    (a) all made of pure iron.
    (b) composed of crystalline carbon.
    (c) too small to experience significant tidal stress.
    (d) protected gravitationally from the planet by the planet's ring system.
    (e) none of the above

Short Answer

59. (a) Which famous scientist mentioned in this text died during Cyrano de Bergerac's (1619–1655) lifetime?
    (b) Referring to the quote in the opening of Chapter 12, what was the religious environment like with regard to celestial discoveries during Cyrano's lifetime?
    (c) Explain how Cyrano could have been inspired to write those words.
60. Explain how it is possible that Saturn radiates 70% more energy into space than it receives from the Sun.
61. Why are Jupiter and Saturn both noticeably wider at their equators than they are from pole to pole?
62. Explain why measuring the composition of a gas based on mass gives different percentages than measuring composition based on volume.
63. Why do Uranus and Neptune appear bluish in color when Jupiter and Saturn don't have the same coloration?
64. Explain why Jupiter's magnetic field is compressed in the direction of the Sun and stretched out on the side opposite from the Sun.
65. Based on your knowledge of Io, explain how the Io torus shown in Fig. 12.11 could have been created. Is the Io torus likely to interact with Jupiter's magnetic field? Why or why not?
66. How does diamond form in the lower atmospheres and mantles of the giant planets?
67. Based on the trend in densities for the Galilean moons, what can you conclude about the ratio of water and water ice to rock in these bodies with increasing distance from Jupiter? In order to develop a complete understanding about how the moons formed and evolved, is this pattern something that needs to be included?
68. From Table 12.2, calculate the ratios of periods, P_{Europa}/P_{Io}, $P_{Ganymede}/P_{Io}$, and $P_{Callisto}/P_{Io}$. What do those period ratios tell you about the relationships in their orbits? You may want to refer to Fig. 12.18.
69. Why did Callisto cool off more quickly than the other Galilean moons after formation?
70. Why are the smallest moons in the Solar System not spherical in shape?
71. The International Astronomical Union is responsible for the official naming of astronomical objects and features on Solar System bodies. Typically, naming surface features on Solar System bodies follows certain themes. What is the basis for selecting the names of mountains on Titan?
72. Why are scientists so intrigued by Saturn's moon Enceladus?
73. Explain why Mimas is assumed to no longer be geologically active?
74. How might the dark streaks of dust from cryovolcanoes have formed in Triton's very thin atmosphere rather than the dust simply landing in random distributions?
75. Why couldn't Triton have simply been captured by Neptune without the involvement of a third body?
76. Explain how Mimas is responsible for the sharp inner edge of Saturn's Cassini division.
77. Shepherd moons have been discovered orbiting Saturn and Uranus. Describe the effect they have on specific rings in their planets' ring systems, and how that process works.

Activities

78. For this activity you will need to determine if Jupiter is visible at night, and where in the sky to look for it. You may want to consult an online resource such as https://skyandtelescope.org/observing/sky-at-a-glance/ or planetarium software such as ©*Stellarium* at https://stellarium.org/.

    (a) If Jupiter is visible, use a telescope or a good pair of binoculars to observe the Galilean moons. Weather permitting, record the locations of the moons relative to Jupiter and each other *every night for ten consecutive nights* by drawing diagrams. Try to be as accurate as possible, including relative apparent distances from Jupiter. You should also record the times and dates when you made the observations. It may be that sometimes one or more of the moons are in front of or behind Jupiter, making it impossible to record their positions accurately. That's okay, simply record the moons you can see.

    (b) In each diagram, label each moon by its name. It may take a few nights to sort out which moon is which. One moon may seem closest to Jupiter only because it is almost in line with Jupiter at that time in its orbit.

    (c) Estimate how many orbits, including fractions of an orbit, were made by each moon during the ten-day period.

    (d) Based on your data, estimate each moon's orbital period. Be sure to explain how you arrived at your results, and show your work.

79. For this activity you will need to determine if Saturn is visible at night, and where in the sky to look for it. You may want to consult an online resource such as https://skyandtelescope.org/observing/sky-at-a-glance/ or planetarium software such as ©*Stellarium* at https://stellarium.org/.

    If Saturn is visible, use a telescope or a good pair of binoculars to observe the planet. Sketch the orientation of the planet's ring system.

Dwarf Planets and Small Bodies

13

I came in with Halley's Comet in 1835. It is coming again next year, and I expect to go out with it. It will be the greatest disappointment of my life if I don't go out with Halley's Comet. The Almighty has said, no doubt: "Now here are these two unaccountable freaks; they came in together, they must go out together."

Mark Twain (1835–1910)

13.1	Dwarf Planets	502
13.2	Asteroids and Meteorites	510
13.3	Comets	520
13.4	Impacts throughout the Solar System	530
	Exercises	536

Fig. 13.1 A portion of the Bayeux tapestry with Comet Halley and King Harold. (Myrabella; CC0 1.0 Universal Public Domain Dedication)

Introduction

As we learned in Chapter 10, the Solar System contains far more than just the Sun, the major planets, their moons, and ring systems. There are also numerous dwarf planets, asteroids, meteoroids, and comets. On very rare occasions, impacting asteroids and comets can do major damage on Earth, but the vast majority of cases are smaller meteors that burn up in the atmosphere, strike harmlessly on land or in the ocean, or simply travel unobstructed through the Solar System or impact other bodies.

Comets that enter the inner Solar System can develop majestic tails that may be visible to the naked eye. To pre-modern societies these appearances were sometimes taken as omens of future events. Figure 13.1 shows a portion of the Bayeux tapestry that was created in the 1070s. The tapestry is almost 70 m (230 ft) long and 50 cm (20 in) high, depicting the Norman conquest of England and the victory of William the Conqueror over King Harold at the Battle of Hastings, October 14, 1066. The portion of the tapestry in the opening shows Halley's comet, which appeared in England in 1066, months before Harold died at the Battle of Hastings. The artist seems to suggest that the appearance of Halley's comet was a bad omen for King Harold (although apparently a good omen for William the Conqueror). Notice the ghost-like ships at the bottom of the tapestry: William arrived in ships ahead of the invasion. The story told in the tapestry is from the point of view of the Normans, and was created some years after the events; the "omen" of Halley's comet was based on one hundred percent hindsight.

Modern science has provided us with a much deeper, and less mysterious, understanding of comets and the other objects in the universe. However, a lack of mysticism doesn't make them any less fascinating. The stories told today by celestial bodies continue to challenge us to unravel their deepest secrets and integrate them into an overall understanding of nature.

13.1   Dwarf Planets

When the International Astronomical Union established the dwarf planet designation in 2006 and classified Pluto as one member of the group, it was done in recognition of the growing number of similar bodies (trans-Neptunian object) being discovered in the Solar System beyond Neptune. Located in the asteroid belt between Mars and Jupiter, Ceres (Fig. 13.2) was also inducted into the new club based on the dwarf planet definition (Ceres is the only object closer to the Sun than Neptune that is so classified). Table 10.2 lists a few of these objects, many of which are comparable to, and in one case more massive than, Pluto. Thus far, the only dwarf planets or dwarf planet candidates that have been visited by spacecraft are Ceres and Pluto.

Ceres

Ceres was discovered in 1801, nearly 130 years before Pluto was first detected. However, despite being observed from Earth for over 200 years, it wasn't until NASA's

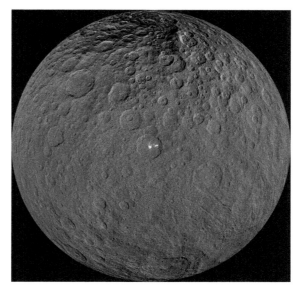

Fig. 13.2 Ceres. Occator crater is visible at center and Ahuna Mons is at the top of the image. (NASA/JPL-Caltech/UCLA/MPS/DLR/IDA)

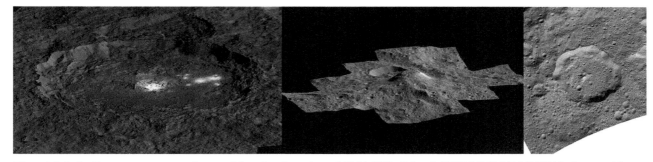

Fig. 13.3 *Left:* Occator crater on Ceres with salty deposits. (NASA/JPL-Caltech/UCLA/MPS/DLR/IDA) *Center:* Ahuna Mons is believed to be a cryovolcano. (NASA/JPL-Caltech/UCLA/MPS/DLR/IDA/PSI) *Right:* Organic material near Ceres's Ernutet crater. (NASA/JPL-Caltech/UCLA/MPS/DLR/IDA)

Dawn spacecraft entered orbit around Ceres in 2015 that we started to uncover many of the secrets of this enigmatic world.

Figure 13.2 shows an image of Ceres obtained by the Dawn spacecraft on approach to the dwarf planet. As with other bodies in the Solar System, Ceres is heavily cratered, although not as heavily cratered as would be expected for an ancient world. This fact alone indicates that Ceres has experienced some geologic activity during its existence. The surface appears to be composed of a mixture of water ice and a clay-like material.

The highly reflective bright spots at the center of Fig. 13.2 are located in the Occator crater and also hint at some form of geologic activity. A perspective view of the Occator crater is shown in Fig. 13.3 (*Left*). The material that makes up the bright regions is likely ice and salts that rose to the surface from the dwarf planet's interior.

Another clue as to what lies beneath is Ceres's only mountain, Ahuna Mons, seen near the top of Fig. 13.2 and in Fig. 13.3 (*Middle*). It appears that Ahuna Mons is a cryovolcano, similar to those seen on some of the moons of the outer Solar System. Dawn has even observed an increase in water ice on a crater wall during its time orbiting the dwarf planet, indicating that Ceres remains geologically active today.

Given Ceres's density of 2160 kg/m^3, slightly more than twice the density of water, combined with evidence of salts and ice on the surface, there may be a large layer of ice below the surface that comprises perhaps one-quarter of its mass and one-half of its volume. This hypothesis is supported by the subtle changes in the gravitational force exerted on Dawn as it orbits Ceres that are caused by the distribution of mass in the interior of the dwarf planet. Overall, the interior likely contains a rocky core, an icy mantle, and possibly a liquid water ocean sandwiched between the two. (By the way, if you found yourself walking around on the surface of Ceres, your weight would be less than one-half of your weight on Earth.)

Key Point

Ceres has a rocky core, a water ice mantle, and possibly a liquid water ocean between the core and the mantle.

Another intriguing discovery is the existence of organic compounds on the surface, as depicted in Fig. 13.3 (*Right*). It does not appear likely that the material was delivered to the surface by the collision of a comet or another asteroid because these violent collisions would be too energetic to allow the fragile molecules to survive. Instead, it seems that the material probably originated in the body's interior. If so, perhaps the interior of Ceres could be a place where some form of primitive life may have developed.

Pluto

In 1992 Robert Staehle, a scientist with the Jet Propulsion Laboratory (JPL), called Clyde Tombaugh (Fig. 5.24), the discoverer of Pluto, to ask him if it would be alright to visit his planet, to which Tombaugh reportedly replied, "[you are] welcome to it, though [it will be] one long, cold trip." In 2001 the New Horizons mission was selected by NASA to conduct a flyby of the-then ninth planet from the Sun. New Horizons was launched on January 19, 2006 about seven months before Pluto received its "demotion" to dwarf planet status. On July 14, 2015, nine and one-half years after launch, New Horizons flew past Pluto, returning the image shown in Fig. 13.4 along with many other high-resolution images and scientific data. Thanks to that brief flyby we now have far more information about this dwarf planet, trans-Neptunian object, and member of the Kuiper belt than we had obtained in the more than 90 years, since its discovery in 1930.

The most obvious feature that stands out in images of Pluto is the "heart" seen in Fig. 13.4. The west (left) lobe of the "heart," named Sputnik Planitia after the first artificial satellite put into Earth orbit, is a vast plain that is nearly devoid of cratering, indicating that it formed less than 10 million years ago, and perhaps much more recently. The region is larger than Texas and was likely produced by a large impact with another body that formed the 4 km (2 1/2 mi) deep depression. As should be expected from the freezing and melting points in Table 12.3, over time the crater was filled in with frozen nitrogen (N_2), carbon monoxide (CO), and methane (CH_4).

Overall, Pluto's surface appears to be composed of 98% nitrogen ice, with the remainder being carbon monoxide and methane ices. Beneath the surface lies a "bedrock" of rigid water ice (H_2O). This can occur on Pluto because at the

Fig. 13.4 Pluto with its "heart" on display. The left, western lobe of the "heart" is Sputnik Planitia and the entire light-colored region, including below the "heart," is Tombaugh Regio. (NASA/Johns Hopkins University Applied Physics Laboratory/Southwest Research Institute)

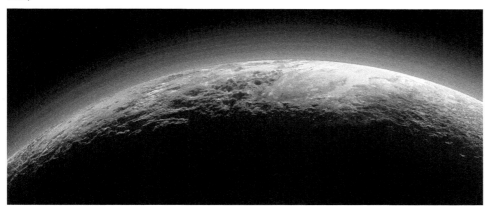

Fig. 13.5 Pluto's thin atmosphere is dominated by N_2, with minor amounts of various molecules composed of carbon and hydrogen. Some of Pluto's mountains and flat icy plains are also visible. (NASA/Johns Hopkins University Applied Physics Laboratory/Southwest Research Institute)

distance of the Kuiper belt's inner edge the inverse square law for light (page 223) implies that the brightness of sunlight is only about 1/1600 of what it is on Earth, resulting in a very cold surface at about 38 K (−235°C, −391°F). At that temperature, N_2, CO, and CH_4 behave much like water does on Earth; those molecules can exist in solid and gaseous forms in equilibrium. As a result, Pluto's thin atmosphere (seen in Fig. 13.5) is composed of about 99% N_2 with the remainder being CH_4, and smaller amounts of other molecules containing carbon and hydrogen (CO hasn't been detected in the atmosphere but is probably present as well). In equilibrium, the ices on the surface can sublimate into the atmosphere and the gases in the atmosphere can condense back onto the surface. [On Earth, even when the temperature stays below the freezing point of water, water ice can sublimate, causing it to slowly disappear without melting first (essentially evaporating directly from solid to gas).

Fig. 13.6 *Left:* Heavily cratered rugged terrain on Pluto. (NASA/Johns Hopkins University Applied Physics Laboratory/Southwest Research Institute) *Right:* A glacial flow from higher elevations down into Sputnik Planitia is indicated by the red arrow. (Adapted from NASA/Johns Hopkins University Applied Physics Laboratory/Southwest Research Institute)

Conversely, water vapor on Earth can condense on cold surfaces to form ice (so-called "black ice" on roadways can form this way under cold, foggy conditions).]

Unlike Sputnik Planitia, other regions of Pluto's surface, such as the portion shown in Fig. 13.6 (*Left*), may be older than 4 billion years, based on crater counts. Along with being heavily cratered, mountains with heights of several kilometers also exist, that are composed of the very rigid water ice that exists below the surface. The reddish material on the surface in Fig. 13.6 (*Left*) is likely composed of organic molecules that form from carbon, hydrogen, oxygen, and nitrogen when exposed to the Sun's ultraviolet radiation over long periods of time, similar to the reddish material found near Ceres's Ernutet crater seen in Fig. 13.3 (*Right*).

In Fig. 13.6 (*Right*), glaciers of frozen nitrogen, methane, and carbon monoxide can be seen flowing down into Sputnik Planitia from the highlands to the east. The ices built up in the highlands as they condensed out of the atmosphere. Although no active eruptions have been observed, there is also evidence of cryovolcanoes on Pluto with the appearance of streaks downwind of dark spots on the surface.

All of this surface evidence, combined with measurements of the gravitational pull of Pluto on New Horizons as it flew past, suggests that the interior of the dwarf planet is probably not dissimilar to Ceres. Pluto likely has a core composed of rock, with a thick water ice mantle (the "bedrock") and a surface veneer of nitrogen, methane, and carbon monoxide ices. As with Ceres, it is possible that a liquid water boundary could exist between the rock core and the water ice mantle. A question remains regarding whether or not Pluto is still geologically active; in other words, are cryovolcanoes still erupting and is water ice still being pushed up from the interior? If so, what is the source of the energy that could be driving this activity?

Charon (shown in Fig. 13.7) is Pluto's largest moon. Unlike Pluto, Charon's surface does not appear to have been refreshed in more than 4 billion years, implying that it is no longer geologically active. However, given that there is significant evidence of tectonic fractures and perhaps even cryovolcanism, Charon was probably active in its very early history.

Overall, Charon's surface, unlike Pluto's, is dominated by water ice instead of nitrogen, methane, and carbon monoxide ices. Reddish organic material is also visible near Charon's north pole that seems to be inconsistent with its surface composition. One possible explanation is that the necessary nitrogen, carbon, hydrogen, and oxygen required to produce the red-colored organic materials may have escaped from Pluto's atmosphere and captured by Charon.

Fig. 13.7 Charon is Pluto's largest moon. The color has been enhanced to bring out the moon's surface features. (NASA/Johns Hopkins University Applied Physics Laboratory/Southwest Research Institute)

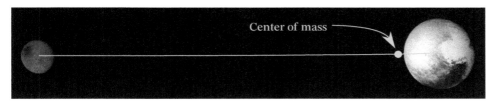

Fig. 13.8 Pluto and Charon showing their relative sizes and distance. (Adapted from NASA/Johns Hopkins University Applied Physics Laboratory/Southwest Research Institute)

As can be seen in Fig. 13.8, Pluto and Charon are fairly similar in size and relatively close together. Charon's mass is also large for a moon relative to the body it is orbiting, being about one-eighth of the mass of Pluto. This means that the center of mass of the Pluto–Charon system (recall Fig. 5.19) is located at a point in space slightly above the surface of Pluto. The system's center of mass can be thought of as orbiting the Sun according to Kepler's laws, while Pluto and Charon orbit their mutual center of mass. Because of their comparable sizes and masses, and because their mutual center of mass isn't located inside either body, Pluto and Charon are sometimes thought of as a double dwarf planet system, although Charon has not been officially designated as a dwarf planet. This situation illustrates again the challenges of trying to force nature into human classification schemes.

The orbital period of the Pluto–Charon system about their mutual center of mass is 6.39 days, which exactly equals the rotation periods of both bodies. Of course, this isn't a random coincidence: as we learned in Section 10.3, this mutual synchronous rotation is the result of tidal forces acting between Pluto and Charon over their history. Just as our Moon keeps the same face toward Earth at all times, Pluto and Charon are forever locked in position so that only one side of Charon ever faces Pluto and one side of Pluto forever faces Charon. Pluto's "heart" is on

Key Point
Pluto and its moon, Charon, are completely tidally locked.

the side of Pluto facing away from Charon. Eventually the Earth–Moon system will also reach this configuration of ultimate tidal lock, but not for billions of years.

Pluto also has four other moons, but they are tiny by comparison and irregular in shape. Styx and Kerberos are about 16 km and 19 km long, respectively, while Nix and Hydra are several times larger with lengths of about 50 km and 65 km, respectively. All four of the moons reflect light very effectively, suggesting that they may be dominated by ices, most likely water ice.

All five of Pluto's moons have circular orbits that are above Pluto's equator, with Charon's orbit being the closest to Pluto. Pluto's rotation axis and the orbits of its moons are tilted by about 123° with respect to its orbit, which means that Pluto is rotating in a retrograde direction. It has been proposed that the Pluto–Charon system formed from a collision between two Kuiper belt objects, and the four tiny moons are debris from the collision that did not end up in one of the two larger bodies.

Finally, consider the orbit of Pluto and its relationship to Neptune in Fig. 13.9. Pluto's orbit is very unusual compared to the Solar System's eight planets. In three dimensions, the upper part of Pluto's elliptical orbit in Fig. 13.9 (*Right*) comes out of the page toward you and the lower portion is behind the page.

Given Pluto's eccentric orbit ($e = 0.25$), it is actually closer to the Sun than Neptune for twenty years during its 247.9 year orbit. The last time this happened was

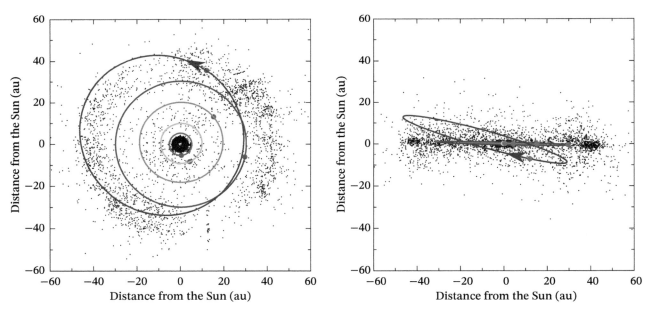

Fig. 13.9 The locations of all asteroids, trans-Neptunian objects, and dwarf planets for which orbits are known beyond 6 au from the Sun and out to 60 au on September 26, 2019. The orbits and positions of Jupiter, Saturn, Uranus, Neptune, and Pluto are also shown. The main asteroid belt within Jupiter's orbit is visible as well. *Left:* From a perspective above the Sun's north pole with the vernal equinox to the right. The Kuiper belt is apparent beyond the orbit of Neptune. All objects are orbiting counterclockwise. *Right:* The same data set seen along the plane of the ecliptic with north being up and the vernal equinox is again to the right. The top of Pluto's orbit is in front of the page and the bottom of the orbit is behind the page with Pluto currently traveling northward and away from the vernal equinox. (Asteroid data provided by the International Astronomical Union's Minor Planet Center and planetary orbital parameters provided by NSSDC/GSFC/NASA)

between February 7, 1979 and February 11, 1999. Pluto's orbit is also inclined significantly (17.1°) and tilted relative to the ecliptic so that it passes through the plane of Neptune's orbit during its long journey around the Sun. Despite this orbital geometry, Pluto and Neptune are not in any danger of colliding any time in the foreseeable future. This is because Pluto is in an orbital resonance with Neptune. From Table 10.1, Neptune's orbital period is 164.8 years. Calculating the ratio of the two orbital periods we find

$$\frac{P_{\text{Pluto}}}{P_{\text{Neptune}}} = \frac{247.9 \text{ y}}{164.8 \text{ y}} = 1.504,$$

which is very close to a ratio of 3:2 ($3/2 = 1.5$). For every three orbits that Neptune makes, Pluto completes two orbits. Because of this orbital resonance, Pluto always keeps a safe distance from Neptune when the dwarf planet crosses the outermost planet's orbit.

Key Point
Pluto is in a 3:2 orbital resonance with Neptune.

Pluto is only one of many objects in the Kuiper belt that have this same 3:2 orbital resonance with Neptune. As a group, they are referred to as plutinos.

The Others

Referring back to Table 10.2, you see that all of the dwarf planets and dwarf planet candidates listed have densities that range between 1700 and 2500 kg/m³. Excluding Ceres, which is a member of the asteroid belt between Mars and Jupiter, all of the others are very reflective objects, suggesting that they are covered with ices and probably contain an appreciable amount of rocky material in their interiors (the density of water ice is about 930 kg/m³ while rock is typically about 2700 kg/m³). The implication is that these trans-Neptunian objects may all have formed similarly in the early Solar System and they could have similar internal structures. Pluto, Haumea, Quaoar, and Makemake are all members of the Kuiper belt, while Gonggong, Eris, and Sedna are probably members of the scattered disk.

Table 10.2 indicates that, of the eight dwarf planets and dwarf planet candidates listed, six are known to have at least one moon (Pluto has 5 and Haumea has 2). The table also indicates that the rapidly spinning and elongated Haumea has a ring. The ring was discovered when the dwarf planet occulted a star; the same method that led to the discovery of Neptune's rings (page 492). The star's light dimmed very briefly when the ring blocked the light, and then for a longer period of time when Haumea itself passed in front of the star. After Haumea finished blocking the star's light, the ring caused the light to dim again very briefly.

The Kuiper Belt

Figure 13.9 displays the locations of asteroids, planets, dwarf planets, and Kuiper belt objects within 60 au of the Sun. Note that, along with a sparse distribution of objects outside of Jupiter's orbit, there is also a vast collection of objects orbiting beyond Neptune (the trans-Neptunian objects), which is the Kuiper belt. When looking along the plane of the ecliptic, the doughnut-shaped distribution becomes evident. Pluto just happened to be the first member of the Kuiper belt and second dwarf planet to be discovered (after 1 Ceres). The Kuiper belt ranges from about 30 au (Neptune's orbit) to approximately 50 au. The so-called "classical Kuiper belt"

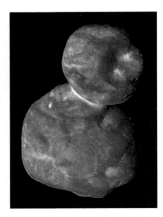

Fig. 13.10 Arrokoth, a small Kuiper belt object measuring 33 km (21 mi) long, as seen in true color during the January 1, 2019 flyby of the New Horizons spacecraft. The semimajor axis of its orbit is 44.6 au, 5 au greater than the semimajor axis of Pluto. (NASA/Johns Hopkins University Applied Physics Laboratory/Southwest Research Institute/ Roman Tkachenko)

Key Point

Small Solar System bodies are objects in the Solar System that are smaller than dwarf planets.

lies between the 3:2 orbital resonance with Neptune associated with plutinos that have semimajor axes of 39.5 au, and the 2:1 orbital resonance (semimajor axes of 47.7 au), although some Kuiper belt objects are found outside of those limits.

In its ongoing mission, on January 1, 2019, New Horizons flew past the small Kuiper belt object, Arrokoth, shown in Fig. 13.10. The object was nicknamed Ultima Thule by the New Horizons team, which means "beyond the borders of the known world," but it was formally named Arrokoth ("sky" in the Native American Powhatan/Algonquian language) by the International Astronomical Union. The contact binary is the result of a slow (walking speed) collision between two primitive Solar System bodies. The larger lobe appears to be composed of many much smaller objects that have clumped together. It seems likely that the reddish appearance of Arrokoth is due to the same process involving organic materials and ultraviolet radiation that accounts for the reddish coloration of Pluto, Charon, and other Solar System bodies. New Horizons may fly past other bodies in the years ahead, on its way out of the Solar System.

13.2 Asteroids and Meteorites

Small Bodies of the Solar System

If there is anything that the "demotion" of Pluto to dwarf planet status taught us, it is that nature doesn't care about our artificial classification schemes. Pluto was the same object before and after the change in its status. As we have learned more about the objects in our Solar System we have come to realize that there is more of a continuum of objects rather than distinct groups. Is Ceres a dwarf planet or an asteroid, or both? Are the two Martian moons, Phobos and Deimos, asteroids? Was Neptune's moon, Triton, once a dwarf planet? Should the ice-covered objects beyond the orbit of Jupiter be considered asteroids, comets, or something else? Is Sedna a dwarf planet, a member of the scattered disk, or is it perhaps an interloper from another star system?

Our traditional designations remain today, but the question is raised as to how much these designations help us to understand what it is we are studying. The International Astronomical Union has formally designated all objects that are not either planets or dwarf planets as being small Solar System bodies, although the terms asteroid (also known as minor planet) and comet continue to be used to subdivide the small bodies designation.

The Asteroid Belt

The asteroid belt is the vast collection of asteroids located between the orbits of Mars and Jupiter. Figure 13.11 shows the positions of one in every 30 asteroids in the inner Solar System with known orbits on September 26, 2019. The orbits and positions of the four rocky planets and Jupiter are also included, along with the position of 1 Ceres in the asteroid belt. The orientations of the two plots are like those of Fig. 13.9.

You probably noticed that there are also numerous asteroids found inside Mars' orbit, some of which cross Earth's orbit. These near-Earth objects are carefully

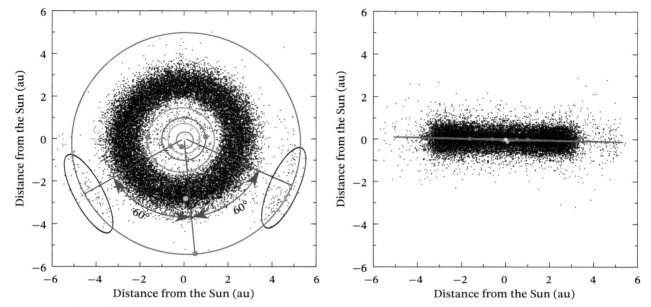

Fig. 13.11 Plots of the positions of objects in the inner Solar System on September 26, 2019, with the vernal equinox to the right. Only three percent of all asteroids with known orbits are indicated by black dots, Ceres is represented by an orange dot, and the four rocky planets and Jupiter are marked by red dots. *Left:* From a perspective far above the Sun's north pole. The trojans are circled. *Right:* The same data set seen along the plane of the ecliptic with north being above the ecliptic. (Asteroid data provided by the International Astronomical Union's Minor Planet Center and planetary orbital parameters provided by NSSDC/GSFC/NASA)

tracked by scientists in the event that they may be on a collision course with our little blue home world.

The Trojans

It is obvious that most of the asteroids are bunched closer to Mars than they are to Jupiter. However, there are also two groupings of asteroids in Jupiter's orbit with orbital periods that exactly match Jupiter's orbital period: one group leads Jupiter by 60° and the other group trails by 60°. Those two groups of asteroids are referred to as **trojans** because the first asteroids discovered in the groups were named after heroes of the Trojan War. The presence of the trojans is a consequence of the gravitational pulls of Jupiter and the Sun as they travel in their orbits. The combination of those effects produce regions that are essentially "wells" of gravity. Objects can fall into the wells and remain trapped there as the wells move around the Sun 60° in front of and behind Jupiter. Other (probably very large) groups of trojans also exist in the orbit of Neptune, although they are more cometary in nature (see page 524).

Key Point
Some small Solar System bodies, referred to as trojans, are trapped in 1:1 orbital resonances with their host planets, leading and trailing the planets by 60°.

Kirkwood Gaps and Asteroid Groups

Whenever two objects with different semimajor axes orbit around a very massive body, an orbital resonance could exist between the smaller bodies. Small integer ratios of orbital periods (3:2, 2:1, et cetera) result in periodic alignments of the three

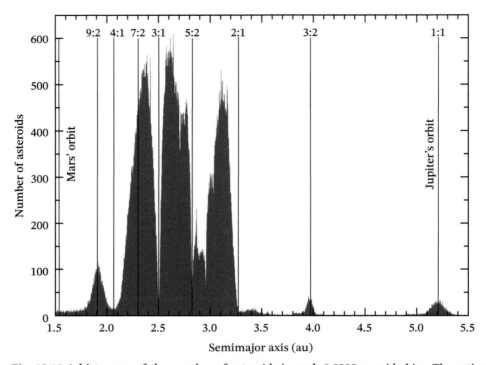

Fig. 13.12 A histogram of the number of asteroids in each 0.0005-au-wide bin. The ratios indicate the number of orbits an asteroid would make with the indicated semimajor axis relative to the number of orbits Jupiter would make; in other words, the ratios of Jupiter's orbital period to the orbital period of an asteroid. (Asteroid data provided by the International Astronomical Union's Minor Planet Center and planetary orbital parameters provided by NSSDC/GSFC/NASA)

objects that occur at exactly the same locations in their orbits each time (recall the 2:1 orbital resonance of Mimas with a particle on the inner edge of the Cassini division in Saturn's rings; Fig. 12.40). If one of the orbiting objects is tiny compared to the other one, the sum of the gravitational pulls on the tiny object due to the other two bodies always acts in the same direction at the same position in its orbit with the result that repeated tugs on that tiny object add up over time, altering its orbit. Just as Mimas affects Saturn's rings, an identical process produces the Kirkwood gaps in the asteroid belt due to orbital resonances with Jupiter. Figure 13.12 is a histogram of the number of asteroids between the orbits of Mars and Jupiter. Notice that the distribution of asteroids is certainly not smooth and continuous across the asteroid belt. Instead, there are gaps and, in some cases, groupings of asteroids at integer ratios of orbital periods.

The outer edge of the main belt is about 3.28 au from the Sun. This location is where the orbital period of an asteroid would be exactly one-half of the orbital period of Jupiter, meaning that the asteroid makes two orbits for every orbit that Jupiter makes (2:1). Other major gaps occur at ratios of 5:2, 3:1 and 4:1. If 5:2 seems out of place, note that 1:1 is the same as 2:2, 2:1 is the same as 4:2, 3:1 is 6:2, and so on. Other gaps also exist that do not correspond to those ratios, but of course there are many other ratios of small integers, such as 7:3, 8:3, … .

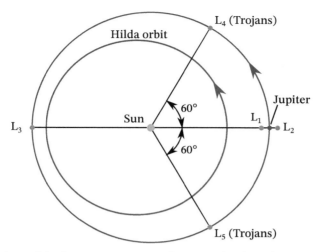

Fig. 13.13 Locations of the five Lagrange points for Jupiter's orbit. The points all move counterclockwise in lockstep with Jupiter. The trojans are located at L_4 and L_5. The orbit of 153 Hilda is also shown, the prototype of the Hilda asteroids.

There are also a few groupings of asteroids (rather than gaps) in the histogram at specific orbital resonances, one of which is the 1:1 resonance corresponding to Jupiter's trojan asteroids. The bump in the number of asteroids at the 3:2 orbital resonance is associated with the Hilda asteroids, named for 153 Hilda, the first asteroid found to have a 3:2 orbital resonance with Jupiter. Both sets of asteroids owe their existence to the combined gravitational forces of the Sun and Jupiter while Jupiter orbits the Sun.

There are five equilibrium points, referred to as Lagrange points, where tiny masses can orbit a massive object in sync with a less massive object, like an asteroid orbiting the Sun in combination with Jupiter as depicted in Fig. 13.13. Those equilibrium points exist because of the combined gravitational pulls of the Sun and Jupiter, together with accelerations associated with Jupiter's orbital motion. The first three of these Lagrange points, L_1, L_2, and L_3, were computed by the brilliant Swiss mathematician, Leonhard Euler, (Fig. 13.14 *Top*) using Newton's three laws of motion and his universal law of gravitation. A few years later Joseph-Louis Lagrange (Fig. 13.14 *Bottom*), a friend and colleague of Euler's who was also a very talented mathematician, calculated the locations of L_4 and L_5. Both men conducted their work during the Enlightenment (the Age of Reason; Section 5.6).

The 1:1 orbital resonance for Jupiter's trojan asteroids seen in Fig. 13.12, and their locations in relation to Jupiter in Fig. 13.11, correspond to the L_4 and L_5 Lagrange points in Fig. 13.13. The 3:2 orbital resonance for the Hilda asteroids results from a more complex interaction. First, notice that L_4 and L_5 are 120° apart. This is exactly 1/3 of a full circle of 360°. There are also 120° between L_4 and L_3, and between L_3 and L_5. All of the Lagrange points move counterclockwise with Jupiter, which means that they are not fixed in space but they are fixed relative to Jupiter. Now, consider what happens to an asteroid that is at aphelion in its elliptical orbit at the same time that it is near L_3. After one complete orbit of the asteroid, Jupiter and its Lagrange points will have moved through 2/3 of an orbit,

Fig. 13.14 *Top:* Leonhard Euler (1707–1783), portrait by Jakob Emanuel Handmann, 1753. (public domain) *Bottom:* Joseph-Louis Lagrange (1736–1813). (public domain)

meaning that L_5 will now be near the asteroid's aphelion point. After a second asteroid orbit, Jupiter will have completed 1 1/3 orbits and L_4 will now be near the asteroid's aphelion point. Finally, after the asteroid has completed three orbits and Jupiter has completed two orbits, L_3 will be near the asteroid's aphelion point again. This all means that the asteroid always finds itself close to one of those three Lagrange points when it is at aphelion during every orbit, causing the asteroid's orbit to remain stable, and in a sense protected by Jupiter, which explains the grouping of asteroids at the 3:2 resonance in the histogram in Fig. 13.12. There is no reason that the cycle couldn't have started at L_4 or L_5 instead of L_3. This implies that there are essentially three groups of Hilda asteroids, with three different aphelion positions near Jupiter's orbit, forming an equilateral triangle (a triangle with three equal sides).

Lagrange points are sometimes used to "park" spacecraft in stable orbits for long-term observations. In the Sun–Earth system, one of those points, L_1 between Earth and the Sun, has been the location of the Solar and Heliospheric Observatory so that it can study the Sun uninterrupted. SoHO produced many of the beautiful images seen in Section 9.2. L_2 is the target location of the James Web Space Telescope (JWST), depicted in Fig. 13.15, which was launched on December 25, 2021. L_2 will give JWST an unobstructed view of deep space while being close to Earth for easy communication. Both L_1 and L_2 are beyond the Moon's orbit.

The Hungaria asteroids are the collection of asteroids around the 9:2 resonance in Fig. 13.12. They are actually more influenced by Mars than by Jupiter. The inner edge of that group corresponds to a 3:4 orbital resonance with Mars, and the outer edge is near both the 4:1 orbital resonance with Jupiter and the 2:3 orbital resonance with Mars. Mars has been very slowly ejecting them from that region since the formation of the Solar System.

As should be obvious by now, orbital resonances play important roles throughout the Solar System. We now know that orbital resonances were critical in producing the overall structure of the Solar System we see today. Orbital resonances are also important in exoplanetary systems (systems of planets orbiting stars other than the Sun) and in galaxies and systems of galaxies. Physical processes always apply universally.

Fig. 13.15 An artist's impression of the James Webb Space Telescope deployed in space. (Northrop Grumman)

Vesta

Key Point

Vesta is the second largest asteroid after Ceres.

Before the Dawn spacecraft went into orbit around Ceres, it first rendezvoused with 4 Vesta, the second largest body in the asteroid belt (Fig. 13.16), and the fourth asteroid discovered. Dawn spent nearly 14 months in orbit around the giant asteroid. Vesta is slightly larger than 2 Pallas, the third largest asteroid in the belt, but Vesta is significantly more massive, comprising about 9% of the mass of the asteroid belt, although still only about one-fourth of the mass of Ceres.

4 Vesta was named for the Roman's Trojan virgin goddess of hearth, home, and family. Initially, 4 Vesta, like 1 Ceres, 2 Pallas, and 3 Juno, was classified as a planet, along with the then-known planets, Mercury, Venus, Earth, Mars, Jupiter, Saturn, and Uranus, giving 12 planets in all. Almost four decades went by before the next asteroid was discovered, and then they started accumulating quickly (Neptune was discovered after the fifth asteroid was found, 5 Astraea). Eventually, given the large

number of bodies being discovered between Mars and Jupiter, all of which were much smaller than any of the planets in the original group, the bodies became classified as asteroids rather than planets. Sound familiar? The term "asteroid" derives from an ancient Greek term for lights in the sky such as stars, planets, meteors, and comets. As of April 2022, more than 1,200,000 have been discovered.

Based on Vesta's overall density and its effect on Dawn's orbit, Vesta has a different internal structure from Ceres. While Ceres has a rocky core, Vesta has a core composed of iron and nickel. Whereas Ceres has a water ice mantle, Vesta has a mantle composed of olivine (page 425), a crystallized silicate that formed from magma.

Like the planets, major moons, and dwarf planets in the Solar System, Vesta is large enough, and stayed warm long enough, for the heaviest elements to sink to the center of the asteroid. The source of the heat that kept Vesta from cooling off too quickly for chemical differentiation to occur was the radioactive decay of $^{26}_{13}$Al, which has a half-life of 717,000 years. While the heat source was active, the molten mantle was convective, much like Earth's mantle today, but when the heat source was depleted, the magma crystallized, forming olivine, and the interior of Vesta was frozen in place.

From the images in Fig. 13.16, it is obvious that the surface has taken a beating over time. Figure 13.16 (*Left*) shows a surface that has been impacted throughout the asteroid's roughly 4.57-billion-year lifetime. Figure 13.16 (*Center*) shows the product of an immense collision at another location that sent shock waves through the asteroid, resulting in significant troughs on the surface (ripples frozen in time). An image of the south pole of Vesta [Fig. 13.16 (*Right*)] shows Rheasilvia, an enormous impact crater that is some 19 km (12 mi) deep and 500 km (300 mi) wide. It appears that Rheasilvia was created less than one billion years ago. The crater may be deep enough to have exposed some of the mantle material. The impact also produced a central peak that towers 20 to 25 km (12 to 16 mi) above the crater floor (Fig. 13.17). Rheasilvia was named for the vestal virgin mother of Romulus and Remus.

Key Point
The radioactive decay of aluminum-26 slowed the cooling of Vesta, allowing heavy elements to sink to the center.

  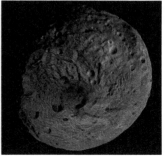

Fig. 13.16 Dawn gray-scale images of 4 Vesta obtained between July 2011 and September 2012. *Left:* A mountain twice the size of Mount Everest is seen at the bottom and the triplet of craters called the "snowman" is visible near the upper-left portion of the image. *Center:* A computer-generated view from thousands of individual images. The enormous groves can be hundreds of kilometers long, up to 15 kilometers wide, and 1 kilometer deep. *Right:* The south pole of Vesta shows the 500 km wide, 19 km deep Rheasilvia crater. (NASA/JPL-Caltech/UCLA/MPS/DLR/IDA)

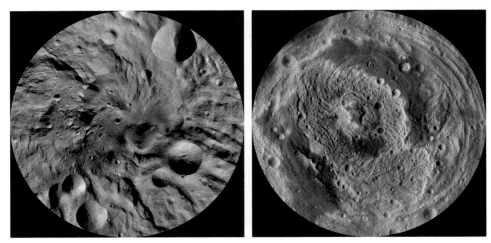

Fig. 13.17 *Left:* A composite image closeup of Rheasilvia crater near Vesta's south pole. (NASA/JPL-Caltech/UCLA/MPS/ DLR/IDA) *Right:* Rheasilvia crater with elevations color-coded; red is highest and blue is deepest. Veneneia crater is visible to the lower-right of Rheasilvia. Rheasilvia partially overlays Veneneia. (NASA/JPL-Caltech/UCLA/MPS/DLR/IDA/ PSI)

Energetic asteroid collisions can eject a large amount of material that then becomes smaller asteroids and meteoroids. Because they come from the same body, these objects can end up in similar orbits around the Sun, and they will have the composition of their parent. These groups of asteroids are known as families. It has been suggested that one asteroid family could be the product of a collision that shattered a planet in the early Solar System.

Vesta's orbit has a semimajor axis of about 2.4 au with a moderate eccentricity of 0.09, which means that it can travel as far from the Sun as almost 2.6 au. This causes some of its family of asteroids and meteoroids to pass through the 3:1 orbital resonance with Jupiter (Fig. 13.12). Eventually, Jupiter will alter their orbits, sending some out of the asteroid belt and in some cases on collision courses with Earth. Based on their compositions, scientists have determined that about 5% of all the meteorites that strike our planet come from Vesta, probably having been ejected when the Rheasilvia crater was formed or when the Veneneia crater formed two billion years ago; you can tell that Rheasilvia is younger than the Veneneia crater because the impact that created Rheasilvia destroyed a portion of Veneneia. Veneneia was named for one of the early Roman vestal virgins.

Of the many millions of asteroids in the asteroid belt, Vesta is the last largely intact body that is a fully differentiated rocky protoplanet. Technically, Ceres is considered to be planetary embryo, because it never developed a fully differentiated structure that resulted in a metallic iron–nickel core, and Pallas is probably the remnant of the exposed core of a differentiated protoplanet. Other protoplanets that existed in the asteroid belt have been destroyed by violent collisions during the history of the Solar System. Metallic meteorites that sometimes fall to Earth attest to that conclusion. This class of meteorites is composed of iron–nickel alloys that were forged in the cores of giant asteroids. Although relatively rare among meteorites striking Earth, they are very over-represented in meteorite collections, simply because they have a very distinctive appearance and are easier to identify on the ground: for example, the Tamentit meteorite (Fig. 13.18). They also have

Fig. 13.18 The Tamentit meteorite is a metallic meteorite that has a mass of 450 kg and weighs more than 4400 N (1000 lb). It was discovered in the Algerian Sahara Desert in 1864. (Ji-Elle, public domain)

Fig. 13.19 *Left:* A composite image of one hemisphere of 433 Eros. A 5.3 km (3.3 mi)-wide crater is visible on the right-hand side of the image and the "saddle" is seen on the left. *Right:* A section of the surface of Eros seen from an altitude of 250 m (820 ft) during NEAR–Shoemaker's descent. (NASA/JPL/JHUAPL)

a better chance of surviving the descent through Earth's atmosphere than stony meteorites.

Small Asteroids and Asteroid Types

433 Eros and Stony Asteroids

In 1998, NASA's Near Earth Asteroid Rendezvous–Shoemaker mission flew past the asteroid 433 Eros, returning again in 2000 to enter orbit around the peanut-shaped world. Figure 13.19 (*Left*) shows one hemisphere of Eros. NEAR–Shoemaker was renamed after its launch in 1996, to honor Eugene Shoemaker (Fig. 13.20), one of the founders of planetary geology. Shoemaker studied terrestrial craters formed by meteorite impacts. He also trained Apollo astronauts in geology for their lunar missions. He had been very disappointed to not have had the opportunity to go to the Moon and "[bang] on it with my own hammer." In 1999, Shoemaker became the first person to have his ashes deposited on another Solar System body when the Lunar Prospector mission carried his ashes to the Moon. Gene Shoemaker was also a co-discoverer of the Shoemaker–Levy 9 comet (page 524), together with his wife Carolyn S. Shoemaker (1929–2021) and their colleague David H. Levy (b. 1948).

Fig. 13.20 Eugene M. Shoemaker (1928–1997). (USGS)

NEAR–Shoemaker became the first spacecraft to actually land on an asteroid, when it set down on Eros in 2001. Figure 13.19 (*Right*) shows one image of its surface during NEAR–Shoemaker's descent. Not only is Eros heavily cratered, but its surface shows signs of being hit repeatedly, resulting in a surface covered with a fine layer of regolith. Eros measures 34.4 km × 11.2 km × 11.2 km (21.4 mi × 7.0 mi × 7.0 mi).

On January 31, 2012, Eros came within 27 million kilometers (17 million miles) of Earth; roughly 70 times the distance to the Moon. A possible Earth impactor sometime in the distant future, Eros is about five times larger than the asteroid that formed the Chicxulub crater (page 199), causing the extinction of the dinosaurs 66 million years ago. Eros was named for the Greek god of physical desire.

Eros has a density of 2700 kg/m^3, or 2.7 times greater than the density of water, making its density similar to Earth's crust. Eros is also similar to many other small stony asteroids like Gaspra, Ida, and Ida's moon, Dactyl, all seen in Fig. 10.5.

Key Point
Eros is a stony asteroid.

101955 Bennu, Carbonaceous Asteroids, and Rubble Piles

101955 Bennu, seen in Fig. 13.21 from an altitude of 24 km (15 mi), is another near-Earth object that is being carefully tracked as a potential Earth impactor because it approaches Earth about once every six years. Bennu is nowhere near as large as Eros, measuring an average of about 500 m or 0.5 km (1600 ft) wide and having a mass of 8×10^{10} kg (about 2×10^{11} lb$_m$), but it has a much higher probability of striking Earth in the near future. Being one of the most potentially hazardous asteroids, current estimates place the chance of impact with Earth at 1-in-2700 (0.037%) between 2175 and 2196. Although very small, the risk is not zero. Over the next 300 million years, its chances of hitting Earth are about 10%, while the odds of it simply falling into the Sun are about 48%.

Bennu has been the target of a unique mission conducted by the OSIRIS-REx spacecraft (Fig. 13.22). Launched in September 2016, OSIRIS-REx arrived at Bennu just over two years later (December 2018). The spacecraft then spent nearly 1 1/2 years orbiting and mapping the surface. On October 20, 2020, OSIRIS-REx also descended close to the asteroid, but did not land on the surface to avoid contaminating it, and then scooped up regolith for return to Earth. If all goes according to plan, it will parachute its precious cargo down to the west desert of the Utah Test and Training Range in September 2023.

There are several reasons why Bennu is the focus of such an ambitious project. First, it is a potential threat to Earth. In order to better predict its orbit into the

Fig. 13.21 The asteroid Bennu, from a distance of 24 km (15 mi). (NASA/Goddard/University of Arizona)

Fig. 13.22 Artist's conception of the rendezvous of OSIRIS-REx with Bennu. (NASA/GSFC/University of Arizona/Lockheed Martin)

future, more information is needed about subtle forces not related to gravity that can affect its orbit, such as uneven heating of the rotating asteroid by the Sun's radiation. Second, its orbit makes it relatively easy to reach and return from in a fairly short period of time. And third, studies of its spectrum indicate that Bennu is a carbonaceous asteroid, meaning that it contains carbon-based (organic) molecules on its surface. Even more important, asteroids are much like time capsules in that they contain material that dates back to the first 10 million years of the Solar System's formation. It is for this reason that scientists want to bring some of the material home for careful study in a laboratory. The data will certainly provide valuable information about how the Solar System, and asteroids, formed.

Bennu appears to be a member of a family from a larger carbonaceous asteroid in the inner asteroid belt. Bennu's orbit was probably altered by orbital resonances with Jupiter and perhaps Saturn. It is also telling that Bennu's density is only about 1300 kg/m^3, less than one-half of the density of Eros. This indicates that Bennu is a rubble-pile asteroid, meaning that it is a weak collection of reassembled debris held together by its own gravity. The debris from which Bennu formed was the result of its parent asteroid experiencing a major collision. Bennu isn't unique in this regard: many, perhaps most, of the smaller asteroids are rubble-pile asteroids. These objects are very porous, somewhat akin to sand piles.

Bennu got its name after an international "Name That Asteroid!" contest. A third-grade student from North Carolina, Michael Puzio, submitted the winning name out of eight thousand entries. Bennu is a mythical Egyptian bird that he thought resembled the OSIRIS-REx spacecraft. The name of the mission was chosen because the asteroid is a possible threat to Earth and Osiris is the Egyptian god of the dead, the underworld, and the afterlife. Rex is Latin for "king." (OSIRIS-REx was also turned into a cleverly devised acronym designed to fit both the nature of the mission and the Egyptian god: Origins, Spectral Interpretation, Resource Identification, Security, Regolith Explorer.)

16 Psyche and Metallic Asteroids

Along with stony and carbonaceous asteroids, there is a third asteroid type known as metallic asteroids. 16 Psyche is thought to be the exposed iron–nickel core of what was a protoplanet after the protoplanet experienced a huge collision that tore off most of its stony mantle (alternatively, a series of less catastrophic collisions could have also done the job). Psyche may be the source of the stony–iron meteorites that strike Earth (Fig. 13.23). NASA approved a mission to Psyche in 2017 with a projected launch date in 2022, arriving at the asteroid in 2026; the objective being to study this unique remnant of a protoplanet.

An important trend across the asteroid belt is the variation in the concentration of stony and carbonaceous asteroids. The asteroids that are closest to Mars are mostly stony asteroids, while carbonaceous asteroids tend to dominate in the outer asteroid belt. In any general model of how the Solar System formed, it is necessary to incorporate this variation in the asteroid belt's population.

Fig. 13.23 A portion of a stony–iron meteorite found in 1951 near Esquel in Chubut Province, Argentina. Liquid metal infused the olivine crystalline structure before solidifying. (James St. John, CC BY-SA 2.0)

Primitive Meteorites

At 1:05 a.m. on February 8, 1969, a bolide was seen over the Mexican state of Chihuahua. On its descent through Earth's atmosphere, the automobile-sized stone exploded and broke into thousands of individual pieces that fell over a huge area in the vicinity of the municipality of Allende.

Fig. 13.24 A sample of the Allende meteorite showing calcium–aluminum inclusions (CAIs). (James St. John, CC BY 2.0)

The Allende meteorite, a fragment of which is shown in Fig. 13.24, became one of the best studied meteorites ever to fall to Earth. Coincidentally, the fall in 1969 happened only a few months before the Apollo astronauts would be bringing samples back to Earth from the surface of the Moon, and the meteorite samples provided an excellent opportunity to test laboratory equipment and techniques before the lunar samples arrived. But the samples were also very important in their own right, because the Allende meteorite is of a very rare class called carbonaceous chondrites, that comprise only about 4% of all meteorites. Chondrites in general are meteorites that contain small, round grains, called chondrules, that formed from molten droplets. Carbonaceous chondrites contain inclusions that are calcium and aluminum-rich, referred to as CAIs. Dated radiometrically, as explained on page 257*ff*, the CAIs have been determined to be nearly 4.57 billion years old. It is carbonaceous chondrites with CAIs, like the Allende meteorite, that helped us to establish the age of the Solar System and Sun itself.

As the name of the meteorite class suggests, these primitive meteorites also contain carbon-based materials, including graphite, diamond, and organic compounds that include amino acids. [Amino acids are building blocks of deoxyribonucleic acid (DNA) and ribonucleic acid (RNA).] By studying these ancient remnants of the Solar System, it is possible to glean valuable information about the composition of the solar nebula and the formation of planetesimals from which planets, dwarf planets, moons, and asteroids ultimately formed.

13.3 Comets

When a comet becomes visible to the naked eye it is referred to as an apparition. That term is also used to describe an unexpected sight or a ghostly figure. Each of these definitions are appropriate for these visitors from the outer Solar System. Figure 13.25 shows Comet Hale–Bopp as seen from Pazin, Croatia in March, 1997. Because it was visible to the naked eye for 18 months, it became known as the Great Comet of 1997. The previous record holder was the Great Comet of 1811, which was only visible for less than 9 months.

Fig. 13.25 Comet Hale–Bopp in March 1997. (Philipp Salzgeber, CC BY-SA 2.0 AT)

Alan Hale (b. 1958) and Thomas Bopp (1949–2018) discovered Hale–Bopp independently on the night of July 22–23, 1995, using amateur telescopes of 41 cm (16 in) and 44 cm (17.5 in), respectively. The discovery came two years before its apparition. Comets are traditionally named for their discoverers.

Recalling the portion of the Bayeux tapestry shown in Fig. 13.1, comets have long been associated with mystical powers or foretelling of important events. Sadly, Hale–Bopp is a reminder that even at the end of the twentieth century (and continuing into the twenty-first century), our scientific understanding has not brought an end to these mystic views among some in modern culture. Heaven's Gate, a cult that believed in the religious significance of UFOs (unidentified flying objects), was formed in 1974 in San Diego, California by Marshall Applewhite (1931–1997) and Bonnie Nettles (1927–1985). They believed that a spacecraft would arrive upon "graduation" from their current Human Evolutionary Level to take them to the Evolutionary Level Above Human (the "Kingdom of Heaven"). On March 26, 1997, police found the bodies of 39 members of Heaven's Gate who had committed mass suicide so that they could reach the spacecraft they believed was following Comet Hale–Bopp.

The Structure of a Comet

When an active comet enters the inner Solar System, it develops a large coma and two tails, although both of the tails aren't always well defined. Figure 13.26 clearly shows the dust tail of Comet Lovejoy but the gas tail is barely visible. Figure 13.27 (*Left*) displays the spectacular coma and both tails of Comet Hale–Bopp during its 1997 visit.

The coma can range in size from thousands to millions of kilometers across (the Sun is 1.4 million km in diameter), and the tails can exceed 1 au in length (the radius of Earth's orbit). Although their dimensions are immense, the coma and the tails contain very little material. The coma can be thought of as an extremely thin atmosphere surrounding the relatively tiny nucleus of the comet, while the tails are material that has been pushed away from the coma due to forces associated with the Sun. The nucleus ranges in size from tens of meters to tens of kilometers across (e.g., Comet Churyumov–Gerasimenko in Fig. 13.28).

The forces pushing on the tails always cause both of them to point away from the Sun regardless of which direction the comet is moving in its orbit. Referring back to Fig. 13.27, the gas tail (sometimes called the ion tail) is always

Fig. 13.26 Comet Lovejoy as seen from the International Space Station on December 22, 2011. The comet is behind the airglow of Earth's upper atmosphere. (NASA/Dan Burbank)

Fig. 13.27 *Left:* A close-up of Comet Hale–Bopp. (E. Kolmhofer, H. Raab; Johannes-Kepler-Observatory, Linz, Austria; CC BY-SA 3.0) *Right:* The anatomy of a comet. (Adapted from NASA Ames Research Center/K. Jobse, P. Jenniskens)

Fig. 13.28 The nucleus of Comet Churyumov–Gerasimenko while inert. The nucleus can be seen outgassing in Fig. 10.10. A time-elapsed video is available on the Chapter 13 resources webpage. (ESA/Rosetta/NAVCAM, CC BY-SA IGO 3.0)

straight and tends to have a bluish glow while the dust tail curves and tends to be yellowish–white in color.

The ions in the gas tail are driven away from the Sun by the electrostatic forces from the ions in the solar wind. Because charged particles spiral around magnetic field lines (page 322*ff*), the gas tail follows the lines away from the Sun. The gas tail's blue hue comes from light emitted by the ions after becoming excited via collisions with solar wind ions (recall the discussions in Sections 8.6 and 8.7).

The dust tail is pushed away from the Sun by the radiation pressure of sunlight (page 341*ff*). The curvature of the dust tail develops because as the dust particles move away from the Sun they also become tiny masses in orbit around the Sun, with dust particles moving more slowly with increasingly larger orbits. Essentially, this means that as the distance along the dust tail increases, the dust lags farther and farther behind, producing the curve. The dust tail's color comes from reflected sunlight.

The nucleus of a comet is the source of the coma and the tails. In 1950 Fred Whipple (1906–2004) coined the phrase "dirty snowball" to describe a cometary nucleus. Figure 13.29 is an image of Comet Hartley 2's nucleus with gases and dust escaping from it. An image of the nucleus of Comet Churyumov-Gerasimenko while outgassing (Fig. 10.10) was displayed in Section 10.1.

Ever since the first international armada of spacecraft shared resources to study Comet Halley during its return in 1986, numerous other, increasingly sophisticated, spacecraft have undertaken flyby, orbital, lander, impactor, and dust-return missions to various comets. Spectroscopic studies from Earth-based and Earth-orbiting telescopes have also provided valuable data about a comet's composition and behavior. Based on those explorations of cometary nuclei we have come to realize that rather than simply being "dirty snow-

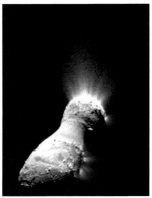

Fig. 13.29 The nucleus of Comet Hartley 2. (NASA/JPL-Caltech/UMD)

balls," it may be more appropriate to think of the nuclei of periodic comets that have made multiple trips through the inner Solar System as more like "deep fried ice cream" with a crust containing dust and loose rocky material covering an "icy dirt ball."

In the outer reaches of the Solar System the nucleus travels as an inert body through the deep freeze of space. But, as the comet ventures close to the Sun, the nucleus heats up causing some of the ices in the interior to sublimate. The build-up of gases inside the nucleus can create sufficient pressure that they burst through the surface carrying dust particles with them. The crusty surface develops because not all of the dust and rocky material is blown off into space. As an analogy, think of a large pile of melting snow on Earth. Initially, the snow appears white but there is some dirt contained in the pile as well. As the snow melts the pile becomes much darker on the surface, because the dirt throughout the pile is left behind. Eventually, a dirt pile is all that remains. When comets make repeated passes through the inner Solar System they will eventually run out of most of their ices, leaving inactive, loosely compacted dust and rock piles in orbit. It is likely that some of the asteroids moving through the inner Solar System are ancient, ice-depleted cometary nuclei.

It is the outgassing of cometary nuclei that makes the behavior of comets notoriously hard to predict. Determining a comet's orbit is complicated by the fact that when gases shoot away from the nucleus they act like little rocket engines, slightly pushing the comet in the opposite direction, as required by Newton's laws.

It is also very difficult to predict whether or not the apparition of a comet will be visually spectacular or a disappointment. How bright a comet appears as it approaches perihelion depends in part on its orientation with respect to the Sun and Earth; is Earth located in its orbit at a vantage point that provides excellent viewing? That piece of the visibility prediction can be calculated very accurately once the comet's orbit in the inner Solar System is determined. What can't be predicted with any certainty is how much dust and gas will be ejected from the nucleus near perihelion. If there is relatively little dust content, there won't be much sunlight reflected from the coma or the dust tail. Similarly, if there is relatively little gas escaping from the nucleus the gas tail won't stand out.

Such was the case with Comet Kohoutek in 1973 (Fig. 13.30). Based on its orbit, it seemed likely that its 1973 apparition would have been its first entrance into the inner Solar System. As such, it was thought that the comet probably contained large quantities of frozen ices that would release significant amounts of dust and gas. The comet was even billed, before its apparition, as the "comet of the century." Although Kohoutek did brighten to become an easily visible naked-eye object, the comet didn't come close to living up to its hype.

Long-Period Comets

When Jan Oort carried out his statistical study of comets in 1950 (see page 366*ff*) and concluded that there must exist a spherical distribution of cometary nuclei at great distances from the Sun, he based his analysis on the facts that a great many comets have orbital eccentricities that are extremely close to one and they seem to come from virtually any direction in space. (For a review of the geometry of ellipses, refer to Fig. 4.43.)

These long-period comets from the Oort cloud can have orbital periods of millions of years and may have spent their entire existence in the deep freeze of the outermost regions of the Solar System. When a passing star or perhaps a rogue planet makes a close approach to our Solar System, these comets can have their orbits perturbed enough for them to come raining down on the inner Solar System. They can also be sent inward when the Sun periodically passes through the midplane of the Milky Way Galaxy, where the changing strength of the gravitational field perturbs the comets' orbits (the oscillation period of the Sun above and below the galactic midplane is 68 million years; see Section 18.2). In any case, long-period comets probably fall inward in bunches rather than at a constant rate throughout the age of the Solar System simply because of the unique and very rare requirements that cause them to start their long journeys toward the Sun in the first place.

Long-period comets have eccentricities so close to one that their orbits are almost straight lines to the principal foci of their ellipses, which is the location of the Sun. As they fall into the inner Solar System, they can come perilously close to our central star. These sungrazers may survive the encounter and head back out to the deep freeze or they may come apart due to the intense heat. In some cases they even dive bomb straight into the Sun. Figure 13.31 shows a sungrazer that evaporated when it approached the Sun.

Key Point
The orbits and visual nature of comets are difficult to predict accurately.

Fig. 13.30 Comet Kohoutek on January 11, 1974. (NASA)

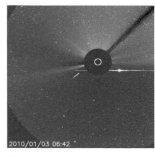

Fig. 13.31 A 2010 sungrazing comet seen to the lower left of SoHO's occulting disk that blocks the Sun's light. The bright object to the lower right of the occulting disk is Venus. A time-elapse video of the comet approaching the Sun is available from the Chapter 13 resources webpage. [SoHO (ESA & NASA)]

Comet Hale–Bopp is a long-period comet with an orbital period of 2456 y, an orbital eccentricity of 0.9950 (an orbit that is very long and narrow), and an inclination to the ecliptic of 89.2°, meaning that it's orbit is almost perpendicular to the ecliptic. Although Hale–Bopp's period is very long, its orbit must have been tweaked by gravitational interactions with the Solar System's planets. This is because its aphelion distance of 363 au indicates that Hale–Bopp will never get back out to the Oort cloud.

Short-Period Comets

Short-period comets have orbital periods of less than 200 years and orbital eccentricities from near zero to close to one (nearly circular to long and narrow). In order to produce comas and tails, these comets must have entered their current orbits within the past few thousands of years or so, otherwise repeated passages into the inner Solar System would have resulted in the depletion of their ices and they would have become inert dust and rock piles. Short-period comets are believed to originate in one of three regions: (a) the Oort cloud (page 366), (b) the collection of Neptune's icy trojans (page 511), or (c) the Kuiper belt or its related scattered disk (page 365).

Short-period comets from the Oort cloud, as with the long-period comets, can have any inclination, including perpendicular to the ecliptic. Comet Halley is one of the rare comets that fits into this category and is the prototype of a class of short-period comets referred to as Halley-type comets. In order to have relatively short orbital periods, Halley-type comets must have had their orbits altered by one of the giant planets, forcing them into elliptical orbits with eccentricities measurably less than one, and reducing their orbital periods from millions of years to less than 200 years. Comet Halley's orbital period is 75.3 years, its orbital eccentricity is 0.9671, and its orbital inclination is 162.3°(Halley's inclination being greater than 90° indicates that it is in a retrograde orbit). The first documented apparition of Comet Halley was in 240 BCE, more than 2200 years ago.

Given the orbits of the Neptune trojans and objects in the Kuiper belt or the scattered disk, short-period comets originating from these sources naturally have orbits that are close to the ecliptic.

Sometimes Jupiter affects comets enough to reduce their orbital periods to 20 years or less. These comets are referred to as Jupiter-family comets. In some cases, the rocky planets can also affect a comet's orbit so that it never makes it out as far as Jupiter's orbit. These comets are known as Encke-type comets, named for Comet Encke, which has an orbital period of only 3.3 years, an orbital inclination of 11.8°, and it travels from 0.34 au (inside Mercury's orbit) to 4.1 au (Jupiter orbits at 5.2 au).

Shoemaker–Levy 9

In 1993 Carolyn Shoemaker, her husband Eugene Shoemaker, and their colleague, David Levy, discovered a comet that had been so affected by Jupiter that it was actually in orbit around the planet rather than the Sun, the first comet ever observed to be orbiting a planet. Based on the calculated orbit of the comet, Shoemaker–Levy 9 had probably been captured by Jupiter 20 to 30 years before its discovery. Besides orbiting Jupiter, two other aspects of the discovery were also unique. The first was

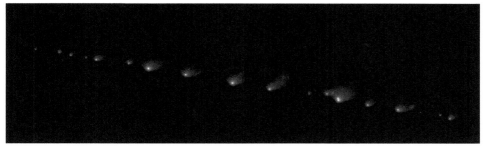

Fig. 13.32 Shoemaker–Levy 9. The "string of pearls" as observed by the Hubble Space Telescope on May 17, 1994, two months before colliding with Jupiter. [NASA, ESA, and H. Weaver and E. Smith (STScI)]

that Shoemaker–Levy 9 had broken apart into 21 fragments during its last close pass of Jupiter in July 1992, seen as the "string of pearls" in Fig. 13.32. Apparently the comet had been severely weakened by its many previous passes through the inner Solar System and the stress of a close approach with Jupiter tore it apart. The second surprise came when astronomers realized that the fragments would crash into Jupiter during its next visit to the giant planet. Between July 16 and July 22, 1994, a fascinated public and professional astronomers alike watched the spectacle of one fragment after another crash into the gas giant. Telescopes around the world, both ground- and space-based, were used to witness the unique event and gather as much information as possible. Even the Galileo Orbiter, then on its way to a rendezvous with Jupiter the following year, was able to focus its instruments on the impacts.

Figure 13.33 shows some of the impact sites as seen by the Hubble Space Telescope. The dark coloration was caused by dust in the cometary fragments that dissipated over the course of several weeks. The fragments themselves had densities of only about 500 kg/m^3, one-half the density of water. The world-wide focus on the collisions not only provided information about the comet, but also helped astronomers learn more about the atmosphere of Jupiter by analyzing spectra when the fragments plunged through the outer cloud decks to reveal the atmosphere farther down. The collisions also produced waves, similar to seismic waves, running through Jupiter's atmosphere, that yielded additional information about the atmosphere's structure.

Fig. 13.33 Dark impact sites from Shoemaker–Levy 9 fragments; July 1994. (Hubble Space Telescope Comet Team and NASA)

Composition

The great scientific interest in comets lies primarily in the fact they are fossils of the Solar System's formation 4.57 billion years ago. Having been in astronomical deep freeze since the Solar System formed means that they contain the raw materials of formation at the locations where they were born. Astronomers have gone to great lengths to sample that material directly and indirectly. The Stardust mission successfully collected dust from the coma of Comet Wild 2 and returned it to Earth in a capsule that was parachuted down to the Utah Test and Training Range. Rosetta landed on Comet 67P/Churyumov–Gerasimenko (recall Fig. 13.28). Deep Impact (Fig. 13.34) even smashed a high-energy impactor into Comet Temple 1 in order to get through the outer crust so that pristine material in the comet's interior could be studied (the main spacecraft recorded the collision and captured spectra).

Fig. 13.34 Deep Impact recorded the collision of its impactor. This image was captured 67 seconds after impact. A time-elapse video of the impact is available on the Chapter 13 resources webpage. (NASA/JPL-Caltech/UMD)

Fig. 13.35 The Geminids meteor shower during the night of December 12–13, 2017. Thirty-seven separate images were combined, taken over a period of 8.5 hours. The dish is a solar-dedicated radio telescope at Mingantu Station in Inner Mongolia, China. (Yin Hao)

Comets contain a variety of materials, including rock, dust, water ice, carbon dioxide ice, frozen methane, and frozen ammonia, many of the materials found on moons and dwarf planets of the outer Solar System. Comets also contain various organic compounds like methanol, hydrogen cyanide, and formaldehyde. They may even contain amino acids. The Stardust mission did confirm the presence of the amino acid glycine in the dust of Temple 1. The Deep Space 1 mission to Comet Borrelly found a very dark surface, possibly composed of complex organic compounds not unlike tar or crude oil. It seems that the building blocks of life may have been present at the Solar System's beginning.

Meteor Showers

The early discovery of the poisonous gas cyanogen in the spectrum of Halley's comet before its 1910 apparition caused people to buy "anti-comet pills," "comet umbrellas," and gas masks when the public learned that Earth would be passing through the comet's tail. Just in case you are worried about the effects of passing through the tail of a comet, it is completely benign given the extremely low density of cometary tails and the protection of Earth's atmosphere.

When Earth passes through the orbit of a comet we can be rewarded with a fascinating nighttime display. Because short-period comets release dust from their nuclei, they leave dust along their orbits. When Earth passes through these orbits, the dust particles enter our atmosphere and quickly burn up, but in so doing produce what are called meteor showers. The Geminids meteor shower of 2017 is shown in Fig. 13.35.

Meteor showers get their names from the constellation in which the meteors seem to originate. The radiant is the point in the sky that Earth happens to be moving toward at the time we pass through the comet's tail. In Fig. 13.35 that constellation is Gemini. Its two brightest stars, Castor and Pollux, are easily visible near the radiant. The constellation of Orion can be seen in the center-right of the image, and the stars Sirius and Procyon can be seen immediately above the telescope's dish and center-left, respectively. The reddish-orange star Betelgeuse (in Orion), and the two stars Sirius, and Procyon, comprise the winter triangle. The constellation Taurus, along with the bright yellowish star, Aldebaran, and the two star clusters, the Hyades (V-shaped) and the Pleiades are located in the upper-right portion of the image.

The impression that meteors in a shower seem to come from a common point in space is simply an illusion. The effect is exactly like looking down the length of a straight stretch of railroad tracks as in Fig. 13.36. The farther the tracks are away from the observer, the closer together they appear, until at great distances the tracks seem to converge to a point. The same is true of meteor showers; the meteors are essentially moving parallel to one another (similar to railroad tracks) as Earth moves through the dust in a comet's orbit, but they seem to originate from a single point in the vast distances of space.

Fig. 13.36 A long, straight stretch of railroad tracks seems to converge to a single point in the distance. (Public domain, CC0 1.0 Universal)

Table 13.1 Meteor showers throughout the year

Shower	Approximate date	Parent body
Quadrantids	January 3	Asteroid 2003 EH_1
Lyrids	April 22	Comet Thatcher
Eta Aquariids	May 5	Comet Halley
Alpha Capricornids	July 26	Comet NEAT
Delta Aquariids	July 29	Comet Machholz (?)
Perseids	August 12	Comet Swift–Tuttle
Draconids	October 8	Comet Giacobini–Zinner
Orionids	October 21	Comet Halley
Southern Taurids	November 4	Comet Encke
Northern Taurids	November 11	Comet Encke
Leonids	November 16	Comet Tempel–Tuttle
Andromedids	December 5	Comet Biela
Geminids	December 13	Asteroid 3200 Phaeton
Ursids	December 20	Comet Tuttle

It is difficult to predict how active a particular meteor shower will be in a specific year, because the number of meteors that fall per hour depends on the local density of the debris in the comet's orbit. Don't forget that the cometary debris is orbiting the Sun just like the parent body is (or was, if it broke apart), so just because the Earth passed through a dense clump of debris one year is no guarantee that another dense clump of debris will be present when Earth crosses the orbit the next year. About once in a generation a particular shower can produce so many meteors that the event is sometimes called a meteor storm. The Andromedids famously produced spectacular outbursts in 1872 and 1885, with thousands of meteors visible per hour; in Chinese records it was described as "stars fell like rain." Since that time, the Andromedids had been very weak or absent, until they returned again in 2011.

Table 13.1 lists the dates of major meteor showers during the year, as well as the parent body responsible for leaving the dusty debris behind. You should notice that two of the showers listed, the Quadrantids and Geminids, indicate that the parent body is designated as an asteroid rather than a comet. It is not always a simple matter to classify a body as one or the other, because in some cases what today appears to be an asteroid is simply the largely inert remains of a once-active comet. In the case of 2003 EH_1, it has been suggested that it could be the same body that was observed in China, Korea, and Japan as a comet (C/1490 Y1). On the other hand, 3200 Phaeton may indeed be an actual asteroid that gets very hot when it moves through perihelion only 0.14 au from the Sun, which is much closer than Mercury's orbit. Phaeton was observed to be ejecting dust in 2010, perhaps because the extreme heat was causing the surface to fracture, much like "mud cracks in a dry lake bed." The dust is then pushed away from the asteroid by sunlight, producing a dust tail. Phaeton has been called a "rock comet."

If you wish to view a meteor shower in the future, the best time of night to do so is about 2 a.m. local time. This is because, at that time of night, not only is Earth passing through the cometary debris, but the rotation of Earth is carrying you toward the debris in the direction of the radiant. In planning a viewing session you will want to consult up-to-date predictions of exact dates and projected meteor counts per hour at an internet site such as the International Meteor Organization. You will also want to note the phase of the Moon on your observing date; a full Moon or a waning gibbous Moon can cause the sky to be too bright to view faint meteors. If you have the opportunity in a dark sky to set up a chair and look in the direction of the radiant, it can be well worth the effort.

Centaurs and Main-Belt Comets

In 1920, a body named 944 Hidalgo, with a diameter of about 50 km, was discovered to be orbiting between Jupiter and Saturn. Then, in 1977, a second body was found orbiting in the outer Solar System, this time between Saturn and Uranus, and, with a diameter of 206 km, was designated as 2060 Chiron. Yet another object found orbiting between Saturn and Uranus, 10199 Chariklo, is even larger than Chiron, and has a diameter of 232 km. Chariklo is also known to have rings.

These objects are a few of what could be a large collection of bodies with diameters greater than 1 km orbiting the Sun between Jupiter and Neptune. Named collectively as centaurs, for the half-human, half-horse creatures from mythology, their classification as either comets or asteroids is not always clear-cut. Several centaurs have exhibited comet-like comas, including Chiron. As such, Chiron and others have been classified as both asteroids and comets, although they tend to be much larger than typical comets. It has been suggested that the centaurs were once members of the scattered disk or perhaps they had been members of Neptune's population of trojans in the past.

Another type of object that defies simple classification are the main-belt comets that orbit in the outer part of the asteroid belt between about 2 au and 3.2 au from the Sun (recall that the 2:1 resonance with Jupiter is at 3.2 au; Fig. 13.12). At times they develop dusty tails and may be ice-covered asteroids. They are also sometimes referred to as active asteroids rather than comets. The existence of these objects suggests that perhaps a large number of asteroids in the outer part of the asteroid belt may be ice covered.

Once again, nature doesn't always want to conform to our classification schemes.

Rogue Objects

Key Point
1I/'Oumuamua was the first known visitor to pass through our Solar System from somewhere else in the Milky Way Galaxy.

In 2017, a completely unexpected body was observed. 1I/'Oumuamua[1] (pronounced oh MOO-uh MOO-uh) became the first object known to pass through our Solar System that originated from somewhere else (Fig. 13.37). Astronomers believe that such interstellar visitors may pass through the Solar System on average once per year but, because they are moving so rapidly (faster than the escape speed

---

[1]Its formal designation by the International Astronomical Union is 1I/2017 U1; the first object discovered in our Solar System from interstellar space (1I) in 2017, and the first object discovered in the 21st half-month of that year. Given the enormously large number of objects in the universe, catalog designations need to be systematic.

Fig. 13.37 An artist's rendering of 1I/ʻOumuamua. First thought to be an asteroid, ʻOumua-mua appears to have undergone outgassing, making it either the first comet or a shard from a Pluto-like dwarf planet from another planetary system known to pass through our own Solar System. [ESA/Hubble (CC BY-SA 3.0 IGO), NASA, ESO, M. Kornmesser; CC BY 4.0]

from the Solar System), and they are very small and dark, they have been hard to de-tect. ʻOumuamua was only detected because Pan-STARRS, an advanced survey tele-scope in Hawaiʻi, was able to spot it. The name chosen for this object, ʻOumuamua, has a Hawaiʻian origin, meaning to reach out in advance of, or as a scout or mes-senger from, the distant past. Based on its orbit, ʻOumuamua has likely been wan-dering through the Milky Way Galaxy for hundreds of millions or even billions of years before it got temporarily deflected by our Sun's gravitational pull. Because of its unique nature, the International Astronomical Union created a new class of objects for ʻOumuamua, giving it the prefix of 1I, which indicates that it is the first interstellar comet or asteroid to be discovered.

1I/ʻOumuamua is at least 400 m (0.25 mi) long, and ten times longer than it is wide. It also rotates every 7.3 hours and is very dark and red, caused by space weath-ering over millions of years. When first discovered it was thought that ʻOumuamua may have been a stony or metallic asteroid, or perhaps a hybrid stony–iron asteroid, but as it was on its way out of the Solar System it seemed to have gotten a boost from venting gases, suggesting that the object is actually a comet rather than an asteroid. Additional analysis indicates that it could even be a portion of the sur-face of a dwarf planet, much like the surface of Pluto. It is apparent that the same physical processes that created our Solar System have been active elsewhere.

Just in case ʻOumuamua was an interstellar spaceship rather than an asteroid or comet, the SETI Institute spent some 70 hours using its radio telescopes to "listen" for artificial signals coming from the object. There have been no indications that ʻOumuamua is anything other than a wandering natural body.

It didn't take long for the second interstellar interloper to be discovered. On August 30, 2019, an amateur astronomer and Crimean telescope builder, Gennadiy Borisov (b. 1962), discovered a comet with an orbital eccentricity of 3.4 (any eccen-tricity greater than or equal to 1 is not bound by the Sun). While 1I/ʻOumuamua

Fig. 13.38 Comet 2I/Borisov. [NASA, ESA, and D. Jewitt (UCLA)]

seems to possess a confusing mix of characteristics that challenge our classification schemes, there is little doubt that 2I/Borisov is a traditional comet, having developed a clear coma as it approached the Sun, as shown in Fig. 13.38. At its closest approach, on December 8, 2019, Comet Borisov came within 2 au of the Sun. Analysis of its spectrum indicates that the comet's coma contained far more carbon monoxide (CO) than comets in our Solar System. Since CO sublimates at much lower temperatures than water (H_2O) or carbon dioxide (CO_2), it doesn't survive in a typical comet's nucleus by the time the comet reaches the inner Solar System. 2I/Borisov, on the other hand, still had a large amount of carbon monoxide in its coma as it rounded the Sun, suggesting that it probably formed in a colder region of space than the Oort cloud comets; possibly in a protoplanetary disk around a new-born star that is much cooler than the Sun. This would have allowed 2I/Borisov to collect much more icy carbon monoxide.

Additional interstellar visitors are sure to be discovered in the future.

13.4　Impacts throughout the Solar System

June 30 every year is designated as International Asteroid Day by a United Nations resolution approved in 2016. The designation was the result of a combined effort by filmmaker Grigorij Richters (b. 1987), Danica Remy, the-then chief operating officer and later president of the nonprofit group, the B612 Foundation,[2] which is dedicated to planetary defense against near-Earth-object impacts, Apollo 9 astronaut Rusty Schweickart (b. 1935), and Brian May (Fig. 13.39), astrophysicist and lead guitarist for the British rock band, Queen. International Asteroid Day is meant to bring awareness of the potential for a catastrophic event sometime in the future. June 30 is the anniversary of the largest fireball and explosion ever recorded in modern times.

Fig. 13.39 Brian May (b. 1947) with his "red special" guitar in the Queen Live concert in 2008, San Carlos de Apoquindo Stadium, Chile. (By Compadre Edua'h, public domain)

Tunguska Event

What is termed the Tunguska event took place on June 30, 1908 over a very sparsely populated region of northern Russia near Lake Cheko and the Podkamennaya Tunguska River. Incredibly, there were no known confirmations of human deaths as a result of the above-ground explosion that destroyed the asteroid, but there were numerous reports of people being knocked off their feet and windows being shattered hundreds of kilometers from the explosion. An eyewitness report of the explosion from 65 km (40 mi) away claimed that the heat was so great that he "couldn't bear it, as if [his] shirt was on fire." The shock wave from the explosion was equivalent to an earthquake measuring 5.0 on the Richter magnitude scale and atmospheric pressure changes were strong enough to be felt in Great Britain. It was reported that there was a glow in the night sky across Asia and Europe for several days after the explosion. Although there were no reported human casualties, the burned carcasses of many reindeer were discovered.

Winter lasts a long time in that region of the world and summer comes with boggy terrain that made it very difficult to get near the location directly below the detonation. On top of that, 1908 was a time of great political challenge for the Tsarist

---

[2]B612 gets its name from the book, *The Little Prince* by Antoine de Saint-Exupéry (1943), Gallimard, France. The prince visits Earth from his tiny, house-sized asteroid, B612.

Fig. 13.40 Photograph from the Kulik expedition. (Public domain)

autocracy, with the start of World War I just six years away (1914–1918) and the Russian Revolution occurring three years later (1917). It wasn't until 1921 that Leonid Kulik (1883–1942) made the first expedition to the region as part of a survey project for the new Soviet Academy of Sciences, after he learned of the giant meteorite impact. He then convinced the Soviet government to let him return in 1927 for a more careful study of the area, based on the argument that he may find meteoritic iron that could help the new nation's industry.

To his surprise, Kulik never found the large impact crater he was expecting, but he did find incredible devastation even two decades after the event. As shown in Fig. 13.40, the Kulik expedition found trees flattened and pointing away from the epicenter of the blast over an area of more than 2100 km² (830 mi²). Close to the epicenter, trees were left standing but their tops were burnt and all of the branches had been stripped off. It is estimated that 80 million trees were felled by the blast.

Due in part to the fact that no impact crater was found, various fanciful explanations for the explosion were proposed: the captain of an alien spacecraft knew that their ship was about to explode while on an Earth reconnaissance mission, so the captain flew the craft to a remote location in an effort to minimize human deaths. In variants of the story, an alien craft fired a laser at an asteroid so that it didn't strike the planet; there was a matter–antimatter explosion (think *Star Trek*®); or perhaps a black hole hit Earth. Of course, none of these ideas could stand up to actual facts and careful analysis.

During descent, the fireball exploded as an airburst estimated to have occurred 6 to 10 km (4 to 6 mi) above ground. Supercomputer simulations conducted by scientists at Sandia National Laboratories in Albuquerque, New Mexico (Fig. 13.42) suggest that the energy released by the explosion was equivalent to between 3 and 5 megatonnes of TNT, roughly 140 to 240 times the energy of the Fat Man nuclear bomb that was dropped on Nagasaki, Japan, by the United States on August 9, 1945, near the end of World War II. The supercomputer estimate of the energy was significantly smaller than prior estimates because the linear momentum (page 193) of the asteroid and numerous other complicating factors were incorporated into the simulations that had not previously been considered. The simulations assumed an entry speed of 20 km/s (45,000 mi/h) and a stony rubble-pile asteroid with a mass of about 100,000,000 kg (2×10^8 lb$_m$) and a diameter of roughly 40 m (130 ft).

Fig. 13.41 Leonid Alekseyevich Kulik (1883–1942). (Public domain)

Fig. 13.42 A supercomputer simulation of the Tunguska event airburst. A movie of the simulation is available on the Chapter 13 resources webpage. (Sandia National Laboratories/Mark Boslough; photo by Randy Montoya)

Fig. 13.43 The Chelyabinsk fireball on February 15, 2013. (Aleksandr Ivanov, CC BY 3.0)

Chelyabinsk Fireball

While the world anticipated the close approach of asteroid 367943 Duende on February 15, 2013, which had long been predicted to pass only 27,700 km (17,000 mi) from Earth (the Moon is on average 14 times farther away), a previously undetected asteroid entered Earth's atmosphere and exploded over Chelyabinsk in west-central Russia near the Kazakhstan border (Fig. 13.43). The asteroid that exploded over the city of more than 3 million people was only about one-half the diameter of the Tunguska asteroid and yielded an energy equivalent of 440 kilotonnes of TNT. Although the explosion happened almost 30 km (19 mi) above ground, the blast still managed to damage 7200 buildings across the region and injured more than 1600 people seriously enough, mostly from flying glass, that they sought medical attention. Extensive observations of the fireball occurred thanks to security cameras and car dash-cams.

Fig. 13.44 Two meteorites from the Chelyabinsk airburst on February 15, 2013, with visible fusion crusts. (Muséum de Toulouse, Didier Descouens, CC BY-SA 4.0)

Unlike the Tunguska event, small fragments of the Chelyabinsk asteroid did survive and were recovered on the ground, including the fragments shown in Fig. 13.44. Visible on the meteorites are the dark fusion crusts produced by the extreme heat from friction as they fell through the atmosphere. The largest piece to be recovered was found at the bottom of Lake Chebarkul, where it had punched a hole through the ice. The fragment has a mass of 654 kg (1400 lb$_m$), although it broke into three pieces when it was being weighed.

Thousands of fragments of the Chelyabinsk meteorite were collected by school children and others because they were easy to spot lying on top of the snow. Fragments of the meteorite even made it into the gold medals that were awarded on February 15 during the 2014 Winter Olympic Games held in Sochi, Russia, marking the one-year anniversary of the Chelyabinsk event.

Following the fall, the "Church of the Chelyabinsk Meteorite" was formed, with the founder, Andrey Breyvichko, claiming that the meteorite recovered from the lake contains, in coded form, a "set of moral and legal norms that will help people live at a new stage of spiritual knowledge development" that can only be deciphered by "psychic priests" of the new church. The group proposed building a temple in Chelyabinsk where they say the meteorite should be housed.

Craters on Earth and Throughout the Solar System

Throughout our discussion of the airless worlds of the Solar System we have seen evidence of impact cratering. Cratering has also been visible on Earth and Mars, although processes such as weathering and plate tectonics eventually hide the evidence of past impacts (Figures 13.45 and 13.46). The oldest known impact crater on Earth dates back 3 billion years and is located near the town of Maniitsoq, on the southwest coast of Greenland. The impactor is estimated to have been 30 km (19 mi) wide and left behind a crater more than 100 km (62 mi) across. Although the crater itself has long since vanished, the shock compression left telltale signs deep in Earth's interior. Had that impact occurred today, it would have likely wiped out all higher forms of life. Then, of course, there is the impact that produced the Chicxulub crater (page 200) 66 million years ago, marking the extinction of the dinosaurs.

Fig. 13.45 Meteor crater, also known as Barringer crater, near Winslow, Arizona, is approximately 50,000 years old; see also Fig. 2.14 (*Right*). [Image courtesy of the National Map Seamless Server (USGS)]

Impact Frequency

It is possible to estimate how often impacts occur on Earth and throughout the inner Solar System based on factors such as the number of near-Earth objects, the rate at which meteoroids are observed entering Earth's atmosphere using satellites, sound waves produced by meteors moving through our atmosphere, and the cratering on Mercury, the Moon, and Ceres. Figure 13.47 shows impacts of varying sizes and ages on the three bodies. Relative ages of craters can be estimated by seeing which ones lie on top of others.

The locations and energies of asteroid impacts on Earth that resulted from fireballs between April 15, 1988, and August 21, 2018, are shown in Fig. 13.48. The Chelyabinsk event can be seen in Russia as the largest reddish-orange dot on the map.

What is not depicted in the asteroid fall map in Fig. 13.48 are all of the meteoroids and micrometeoroids that enter Earth's atmosphere that are smaller than 1 m across. The meteor showers that we see and the random meteors seen in the night sky are produced by rocks often no larger than a grain of sand. When they enter the atmosphere with speeds sometimes much greater than Earth's escape speed of 11.2 km/s (page 199) they carry a great deal of kinetic energy, despite their small size. This causes them to heat up to temperatures around 1600°C (nearly 3000°F), producing a streak across the sky before they disintegrate. Still smaller particles also fall to Earth, but are too small to produce enough light to be visible to the naked

Fig. 13.46 Intrepid crater on Mars. (NASA/JPL-Caltech/Cornell University)

Fig. 13.47 Impact craters on *Left:* Mercury. (NASA/Johns Hopkins University Applied Physics Laboratory/Carnegie Institution of Washington) *Center:* Lunar highlands. (NASA/GSFC/Arizona State University) *Right:* Ceres. (NASA/JPL-Caltech/UCLA/MPS/DLR/IDA)

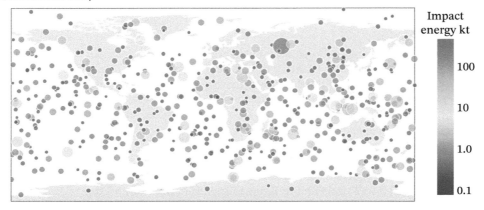

Fig. 13.48 The distribution and energies of fireballs between April 15, 1988, and August 21, 2018. Energies, in kilotonnes (kt) of TNT, are represented by size and color on a logarithmic scale. The Chelyabinsk fireball is the large reddish-orange dot over east-central Russia. (Adapted from Alan B. Chamberlain, JPL/Caltech)

eye. Dust-sized particles are falling into our atmosphere almost continuously and eventually make it to the surface. Based on core samples taken from the Antarctica ice sheet, it is estimated that the amount of cosmic dust falling to Earth is about 14 tonnes *per day* (about 31,000 lb_m).

Given that Earth is constantly being bombarded by micrometeorites, meteorites, asteroids, and comets, it is natural to want to know what the chances are of being hit by a very large asteroid or comet, and how catastrophic the outcome would be. Examples like Chelyabinsk, Tunguska, Chicxulub, and the impact under Maniitsoq, Greenland, certainly get our attention.

Figure 13.49 is a graph of the interval between impacts associated with the size of the impactor. Because the range of numbers is so large, the graph is a log–log plot, meaning that the powers of ten in scientific notation are evenly spaced instead of the numbers themselves. Meteorites of only 1 cm across are plotted at the bottom of the graph, but at the top of the graph asteroids of 100 km across are indicated.

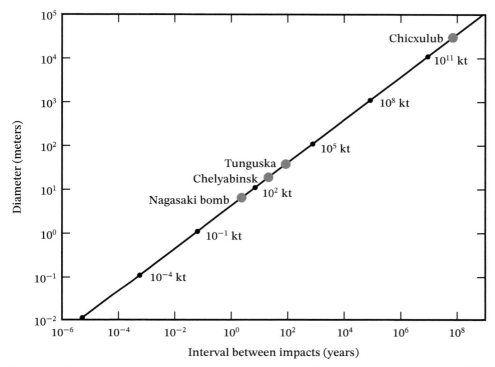

Fig. 13.49 The approximate frequency, size, and energies of meteor or asteroid impacts. The labels of the black dots are energy released in kilotonnes (kt) of TNT.

Similarly, on the left end of the graph, the time interval between impacts is only one one-millionth of one year, or about 30 seconds, but on the right end, the interval between impacts is 10^9 (one billion) years. The black dots indicate how much energy is released when the asteroid is either destroyed in the atmosphere as an airburst, or it strikes the ground. The dot at the very bottom represents meteorites only 1 cm across that hit Earth about once every 2 1/2 minutes with an energy of roughly 10^{-7} kilotons of TNT (roughly the kinetic energy of a 1000 kg car going 100 km/h; 2200 lb_m at 65 mi/h). At the other extreme end of the graph (upper right-hand corner) is the not good, very, very bad catastrophe that could end all but the simplest forms of life on the planet (just ask the dinosaurs)! Fortunately, those events only come around once every one hundred million to one billion years or so. It is also important to keep in mind that the numbers are so small in that part of the graph that it is difficult to make anything like a reasonable prediction based on statistics alone.

The middle part of the graph is where the most concern rests: events like Chelyabinsk and Tunguska or somewhat larger come around once every 100 years to 100,000 years or so (20 m to 1 km in diameter, respectively). They may not end life on Earth, but they can certainly do significant damage. Whether they end as an airburst in the atmosphere or hit the ground depends on what kind of asteroid (or comet) strikes. If the impactor is a loose, rubble-pile stony asteroid or a comet it won't penetrate as far through the atmosphere as a solid stony asteroid would. A metallic asteroid of the same mass would be able to penetrate even farther before ending in an airburst, or if big enough, hitting the ground. The more massive the impactor, the more likely it is to survive the trip through our atmospheric shield.

It is because of the potential for impacts of these mid-sized asteroids that intensive near-Earth object surveys are conducted to identify potentially hazardous objects, and why organizations like the B612 Foundation and NASA's Planetary Defense Coordination Office exist. If potential impactors could be identified early enough, it may be possible to deflect orbital paths to avoid a collision in the future.

Despite the efforts to identify potentially hazardous objects, every once in a while one does surprise us. On April 15, 2018, an asteroid several times larger than the Tunguska asteroid passed Earth at a distance of 192,000 km (119,000 mi) which is approximately 1/2 of the distance to the Moon. The asteroid was only discovered 22 hours prior to its closest approach Earth.

As a first test of a possible planetary defense system, NASA launched the Double Asteroid Redirect Test mission (DART) in November 2021. The mission is designed to ram the 500 kg spacecraft into a moonlet, Dimorphos, at a speed of almost 24,000 km/h (15,000 mi/h), sometime between September 26 and October 1, 2022. Dimorphos is a perfect target for the test because it is orbiting the larger asteroid, Didymos. While neither asteroid is a threat to Earth, the binary nature of the system will allow scientists to determine how much the collision affected the orbit of Dimorphos around Didymos. The idea is that if a comet or asteroid on a potentially catastrophic impact trajectory with Earth can be detected early enough, perhaps such a nudge from a massive spacecraft can deflect the incoming object into an orbit that will miss us. A science fiction scenario come to life.

 # Exercises

All of the exercises are designed to further develop *your understanding* of the material by thinking carefully and critically about what you have read in this chapter. Answers to selected exercises are available in the back of the book.

True/False

1. Comet Halley foretold the downfall of King Harold when it appeared in 1066.
2. (*Answer*) The surface of Ceres has never been geologically active and is as old as the Solar System itself.
3. Organic material found on the surface of Ceres probably originated from its interior.
4. (*Answer*) Like some of the Galilean moons, Ceres likely has a water ice mantle and rocky core.
5. Pluto's heart has been on its surface since the dwarf planet first formed.
6. (*Answer*) Glaciers of frozen, N_2, CH_4, and CO have been detected on Pluto.
7. If you stood on the surface of Charon, you would see Pluto's "heart" once every orbit around the dwarf planet.
8. (*Answer*) It is inevitable that Pluto and Charon will one day collide with Neptune, since the Pluto–Charon system moves closer to the Sun than Neptune for 20 years during each orbit around the Sun.
9. Plutinos are a subgroup of objects within the classical Kuiper belt that happen to be locked in a 3:2 orbital resonance with Neptune, just as Pluto is.

10. (*Answer*) Most Kuiper belt objects are located in orbits between the 3:2 and 2:1 orbital resonances with Neptune.
11. The distinction between asteroids and comets is always easily apparent.
12. (*Answer*) Trojan asteroids are found in a planet's orbit and either lead or follow the planet by 60°.
13. Because there are so many asteroids in the asteroid belt, it is extremely difficult to navigate through the belt safely.
14. (*Answer*) The Kirkwood gaps in the asteroid belt, and many of the gaps in the rings of Saturn, are caused by the same physical processes, but on much different size scales.
15. The Lagrange points of Jupiter are equilibrium locations that move around the Sun in sync with the planet.
16. (*Answer*) Earth's Lagrange points are sometimes utilized to "park" orbiting observatories for long-term observing programs.
17. Orbital resonances are only associated with Jupiter's orbit and the rings of Saturn.
18. (*Answer*) Eugene Shoemaker's ashes were deposited on the surface of the Moon.
19. The Allende meteorite is a very rare and very old carbonaceous chondrite.

20. (*Answer*) Stony-iron meteorites form when a stony asteroid collides with a purely metallic asteroid and pieces fuse together.
21. The Heaven's Gate cult members believed that a spaceship was hiding behind Comet Hale–Bopp, waiting to transport them to a higher evolutionary level.
22. (*Answer*) The tails of comets always trail behind the nucleus during its orbit.
23. The tails are permanent features of a comet, even beyond the orbit of Neptune.
24. (*Answer*) Comet Shoemaker–Levy 9 provided a way for scientists to study Jupiter's atmosphere below its top layer of clouds.
25. The explosion of an asteroid over Chelyabinsk, Russia, in 2013 was expected, because NASA had been monitoring its orbital trajectory well ahead of its entry into Earth's atmosphere.
26. (*Answer*) Dust from space is constantly falling into Earth's atmosphere at a rate of up to a few hundred metric tons per day.

Multiple Choice

27. The surface of Ceres is composed of
    (a) a clay-like material.　(b) water ice.
    (c) frozen methane.　(d) frozen nitrogen.
    (e) all of the above　(f) (a) and (b) only
    (g) (c) and (d) only
28. (*Answer*) Observations suggesting that Ceres has been geologically active in the fairly recent past and could still be geologically active today include
    (a) the relatively low density of cratering compared to the Moon.
    (b) a mountain that appears to be a cryovolcano.
    (c) ice and salts in Occator crater.
    (d) an observed increase of ice on a crater wall.
    (e) all of the above
29. Both Triton and Pluto have surfaces covered with _____ and a "bedrock" of _____.
    (a) water ice, silicates　(b) carbon dioxide ice, silicates
    (c) water ice, nitrogen ice　(d) nitrogen ice, water ice
30. (*Answer*) Pluto's heart is _____ that _____ within the past 10 million years.
    (a) a methane ocean, froze over
    (b) an enormous impact crater, filled in with nitrogen and other ices
    (c) a deep valley, formed from plate tectonic activity
    (d) an as-yet unexplained feature, seems to have appeared
31. The atmosphere of Pluto is almost entirely composed of _____ that is in equilibrium with the same substance in ice form on its surface.
    (a) water　(b) carbon dioxide　(c) ammonia
    (d) methane　(e) ethane　(f) carbon monoxide
    (g) nitrogen
32. (*Answer*) Which of the following substances has the lowest freezing point?
    (a) water　(b) carbon dioxide　(c) ammonia
    (d) methane　(e) ethane　(f) carbon monoxide
    (g) nitrogen
33. Pluto and Charon are an example of a system that
    (a) is tidally locked.
    (b) has a center of mass that is not located within either body.
    (c) is sometimes thought of as a double dwarf planet.

(d) is covered with nitrogen ice.
(e) has organic material on both bodies.
(f) all of the above
34. (*Answer*) The orbit of the Pluto–Charon system around the Sun differs from the orbits of the planets because the
    (a) orbit is retrograde.
    (b) orbit is tilted significantly to the ecliptic.
    (c) system is in a 3:2 orbital resonance with a planet.
    (d) system travels inside the orbit of Neptune.
    (e) all of the above
    (f) (a), (b), and (c)
    (g) (b), (c), and (d)
35. Where have tidal forces *not* led to spin–orbit couplings, synchronous rotations, or orbital resonances?
    (a) in Mercury's orbit and its rotation rate
    (b) between Earth and the Moon
    (c) between the orbits of Io, Europa, and Ganymede
    (d) in the spins and orbits of all major moons and most smaller moons
    (e) in the ring system of Saturn
    (f) across the asteroid belt
    (g) within the Kuiper belt
    (h) Tides have affected all of the systems mentioned.
36. (*Answer*) Some of the gaps in Saturn's rings are produced by the moon _____ in a process that is identical to the Kirkwood gaps in the asteroid belt being produced by _____.
    (a) Titan, Saturn　(b) Miranda, Uranus
    (c) Mimas, Jupiter　(d) Triton, Neptune
37. There are three groupings of Hilda asteroids because
    (a) they formed from collisions with three different, much larger, asteroids.
    (b) the Lagrange points L_3, L_5, and L_4 provide stable aphelion locations for asteroids in a 3:2 orbital resonance with Jupiter.
    (c) they get trapped in three Lagrange points that have a 1:1 orbital resonance with Jupiter
    (d) Jupiter's Lagrange points force the asteroids into a 2:1 orbital resonance.
38. (*Answer*) Vesta
    (a) is the second largest body in the asteroid belt.
    (b) has a core made up of iron and nickel.
    (c) has a mountain on it that is twice the size of Mount Everest.
    (d) has an impact crater that is almost 20 km deep and 500 km wide.
    (e) is the source of some of the meteorites found on Earth.
    (f) is the last largely intact protoplanet in the asteroid belt.
    (g) all of the above
39. Vesta became a fully differentiated body because
    (a) of its tidal interaction with Mars.
    (b) it formed very close to the Sun.
    (c) it contained enough radioactive aluminum to keep the interior hot long enough for heavy elements to sink to the center.
    (d) it got very hot when it collided with Ceres.
40. (*Answer*) Metallic meteorites come from
    (a) rock bands on Titan.
    (b) one or more shattered protoplanets.
    (c) the surface of Vesta.
    (d) an as-yet-to-be-discovered source deep in the Oort cloud.

41. An asteroid family is
    (a) a group of asteroids that resulted from a larger body getting hit with enough energy to shatter part of its surface.
    (b) asteroids that have the same orbital period, but may have very different orbital eccentricities or inclinations from the ecliptic.
    (c) a collection of asteroids that formed in the same general location at the beginning of the Solar System.
    (d) all of the above

42. (*Answer*) Regolith is a very fine, powdery material that is found on the surfaces of the Moon, asteroids, and smaller moons. Regolith is the result of
    (a) tidal lock.
    (b) rocket engine exhaust.
    (c) cryovolcanoes.
    (d) ices of methane and nitrogen.
    (e) constant impacts of meteorites ranging in size from micrometeorites to large rocks.

43. Asteroids composed of silicates with densities similar to Earth's crust are collectively referred to as _____ asteroids.
    (a) stony   (b) metallic   (c) rubble-pile   (d) trojan

44. (*Answer*) Carbonaceous asteroids have _____ on their surfaces.
    (a) soot   (b) organic compounds
    (c) frozen diamond   (d) metallic ices

45. 16 Psyche is a _____ asteroid believed to be the _____ of a protoplanet.
    (a) stony, mantle
    (b) carbonaceous, core
    (c) metallic, chemically differentiated core
    (d) stony–iron, mantle
    (e) rubble-pile, remnants

46. (*Answer*) Stony asteroids are generally found _____ and carbonaceous asteroids are found _____.
    (a) in the outer regions of the asteroid belt, near Mars
    (b) throughout the asteroid belt, near Ceres
    (c) in the inner portions of the asteroid belt, in the outer asteroid belt
    (d) in the inner asteroid belt, only among the Hilda asteroids
    (e) among the trojan asteroids, throughout the asteroid belt

47. Asteroids with densities not much greater than water, such as 101955 Bennu, are
    (a) icy asteroids.
    (b) composed of loosely packed debris from asteroids that were broken apart by collisions.
    (c) rubble-pile asteroids.
    (d) all of the above
    (e) (a) and (b) only
    (f) (b) and (c) only

48. (*Answer*) Carbonaceous chondrites
    (a) are the oldest remnants from the formation of the Solar System.
    (b) are nearly 4.57 billion years old.
    (c) contain inclusions that are rich in calcium and aluminum.
    (d) include amino acids.
    (e) provide information about the original solar nebula from which the Solar System formed.
    (f) all of the above

49. The nucleus of a periodic comet that has made multiple trips through the inner Solar System contains
    (a) ices.
    (b) dust.
    (c) a crust of dust and loose rocky material left behind as ices sublimate.
    (d) all of the above
    (e) (a) and (b) only

50. (*Answer*) Comets typically display _____ tails when they pass through the inner Solar System.
    (a) two
    (b) three
    (c) four
    (d) The number depends on the strength of solar radiation.

51. Comet tails can sometimes be as long as
    (a) 1 km   (b) 100 km   (c) 1,000,000 km
    (d) 1 au   (e) 100 au

52. (*Answer*) A comet's coma can sometimes reach the size of
    (a) the Moon   (b) Earth   (c) Neptune
    (d) Jupiter   (e) the Sun   (f) Earth's orbit
    (g) none of the above

53. The mass of a comet's coma typically contains the mass of
    (a) the Moon   (b) Earth   (c) Neptune   (d) Jupiter
    (e) the Sun   (f) none of the above

54. (*Answer*) The orbits of comets are a bit unpredictable because
    (a) they don't naturally follow Kepler's laws.
    (b) the solar wind pushes them toward Mercury.
    (c) jets of gases can escape from the interior, pushing the comets off course.
    (d) their dust tails produce a drag on the nucleus.

55. The dust tail of a comet is curved because
    (a) the radiation pressure from sunlight pushes the tail away from the Sun.
    (b) the dust particles follow different elliptical orbits, with those farthest from the Sun moving slowest relative to the nucleus.
    (c) the solar wind drives the dust away from the Sun and the Sun's highly curved magnetic field lines trap the dust.
    (d) (a) and (b) only
    (e) (b) and (c) only

56. (*Answer*) The bluish color of the gas tail is caused by _____ and the yellowish-white color of the dust tail is due to _____.
    (a) negatively charged nuclei of atoms, ions
    (b) excited ions, reflected sunlight
    (c) reflected sunlight, negatively charged nuclei of atoms
    (d) reflected blue light, reflected yellow light

57. Comets that have orbital eccentricities that are extremely close to, but still less than one, originate from
    (a) Jupiter's trojans.   (b) Neptune's trojans.
    (c) plutinos.   (d) the Kuiper belt.
    (e) the scattered disk.   (f) the Oort cloud.
    (g) more than one of the above

58. (*Answer*) Sungrazer comets started out
    (a) from among Jupiter's trojans.
    (b) from among Neptune's trojans.
    (c) as broken fragments of plutinos.
    (d) in the Kuiper belt.
    (e) in the scattered disk.
    (f) in the Oort cloud.
    (g) more than one of the above

59. Short-period comets can come from
    (a) among Jupiter's trojans.
    (b) among Neptune's trojans.
    (c) broken fragments of plutinos.
    (d) the Kuiper belt.
    (e) the scattered disk.
    (f) the Oort cloud.
    (g) more than one of the above
60. (*Answer*) According to Kepler's third law, for a short-period comet to have an orbital period of less than 200 years, the semimajor axis of its orbit must be
    (a) less than 34.2 au.
    (b) greater than 10.5 au.
    (c) greater than the semimajor axis of Pluto.
    (d) less than the semimajor axis of Jupiter.
    (e) none of the above, because there isn't enough information given
61. Shoemaker–Levy 9
    (a) was a Jupiter-family comet.
    (b) was the first comet discovered to be orbiting a planet instead of the Sun.
    (c) broke into pieces when it passed close to Jupiter during one of its orbits.
    (d) ultimately crashed into Jupiter.
    (e) all of the above
62. (*Answer*) The compositions of comets provide helpful information about
    (a) the composition of the solar nebula from which the Solar System formed.
    (b) the presence or absence of organic compounds in the early Solar System.
    (c) whether or not our Solar System was once part of a larger planetary system.
    (d) all of the above
    (e) (a) and (b) only
    (f) (b) and (c) only
63. Meteor showers occur near specific dates during the year because
    (a) Earth is passing through debris left in the orbits of specific comets, or asteroids that are the inert remains of once active comets.
    (b) when the Solar System formed, swarms of micrometeoroids were left behind in well-defined orbits.
    (c) asteroids in some orbits broke apart when they were hit by meteorites, thereby leaving pieces of the asteroids in orbits that Earth intersects annually.
    (d) all of the above
    (e) (a) and (b) only
    (f) (b) and (c) only
64. (*Answer*) 1I/Oumuamua was first thought to be an asteroid from another planetary system, but it may actually be an interplanetary comet or a piece from the surface of a Pluto-like dwarf planet from another system because
    (a) astronomers detected a large coma developing as it exited the Solar System.
    (b) it sped up unexpectedly, presumably because of ejected gases.
    (c) Earth experienced a meteor shower when we passed through its orbit.
    (d) gas and dust tails were briefly observed.

65. The Tunguska event in 1908 was probably
    (a) an asteroid that reached the ground and blasted out a large crater.
    (b) a metallic asteroid that exploded high up in the atmosphere.
    (c) a stony rubble-pile asteroid that exploded in the atmosphere with 140 to 240 times more energy than the World War II bomb that was dropped on Nagasaki, Japan.
    (d) a comet that exploded with an energy 10 times greater than any nuclear weapon ever used in a war.
66. (*Answer*) Fusion crusts exist around meteorites because
    (a) of the intense heat caused by traveling very rapidly through Earth's atmosphere.
    (b) the original meteoroids were forged in the center of the Sun through nuclear fusion reactions.
    (c) they formed that way out of the original solar nebula due to the very high temperatures involved.
    (d) all of the above
    (e) none of the above
67. Fireballs with the power of Tunguska or Chelyabinsk happen on average roughly once every
    (a) 10 days.   (b) year.   (c) 100 years.
    (d) 10,000 years.   (e) 100,000 years.
68. (*Answer*) An asteroid with a diameter of 1000 m impacts Earth roughly once every _____ with an energy of _____ kilotonnes of TNT
    (a) 10 days, 0.001   (b) year, 1
    (c) 100 years, 1000   (d) 10,000 years, 1,000,000
    (e) 100,000 years, 100,000,000

Short Answer

69. What is the general composition of the dark reddish material found on Ceres, Pluto, Charon, and numerous smaller bodies in the outer Solar System?
70. Explain why Pluto has a thin atmosphere composed almost entirely of N_2.
71. Describe how Pluto and Charon orbit one another, along with their orientations with respect to one another. Is this situation a random coincidence? If not, why not?
72. Doing a bit of internet research, describe the mythological gods and creatures behind the naming of Pluto and its five moons.
73. What is unique about the group of Kuiper belt objects referred to as plutinos?
74. (a) How are the Kirkwood gaps analogous to some of the gaps in the rings of Saturn, most notably the Cassini division?
    (b) Are those two situations related in any way to the orbital relationships between Io, Europa, and Ganymede?
    (c) Explain what is meant by the universality of physical law.
75. The five Lagrange points associated with the Sun–Jupiter system are unique locations where what special circumstance occurs?
76. Are the Lagrange points of the Sun–Jupiter system fixed in space or are they fixed relative to the position of Jupiter in its orbit around the Sun?
77. Would you expect Lagrange points to exist for any body orbiting another body? Why or why not?
78. The term "trojan" is not specific to Jupiter or Neptune. Could trojans exist in the orbits of other planets? Explain.

79. There are three locations in space where the number of Hilda asteroids is greater than average. Those locations form an equilateral triangle just inside the Lagrange points, L_3, L_5, and L_4, of Jupiter's orbit. Although individual members move in and out of the groupings of asteroids, the increased numbers in those locations remain and travel in sync with Jupiter. Thinking about what happens to orbital speeds throughout an elliptical orbit, explain why the three groupings exist.

80. (*Answer*) Assume that Earth and the Moon don't actually orbit their mutual center of mass, but are simply stationary relative to one another. (This would be impossible, of course, because they would just fall toward one another due to their mutual force of gravity; recall Newton's universal law of gravitation on page 168.)

    (a) Explain why there is a point between the two bodies where a small mass, like a spaceship, could remain at rest and not fall toward either Earth or the Moon.

    (b) Is that point closer to Earth or the Moon? Why?

    (c) What would happen if the spaceship moved just a little closer to the Moon?

    (d) Although not a perfect analog to one of the Lagrange points because Earth and the Moon do orbit their center of mass, which Lagrange point is most like this situation?

81. (a) Why is Vesta considered to be a protoplanet, Pallas the exposed core of a protoplanet, and Ceres a planetary embryo? What distinguishes the first two from the third body?

    (b) How did Vesta's and Pallas's cores evolve?

82. Why is an active comet's orbit difficult to predict accurately?

83. Sketch a comet in three locations in its orbit: after it has entered the inner Solar System (inside the orbit of Mars); at perihelion; and while it is outbound but still in the inner Solar System. Make sure that the gas and ion tails are shown with their orientations relative to the comet's orbit and the location of the Sun. The curvature of the dust tail should be properly drawn as well. Explain your reasoning for the shapes and directions of the tails in the three locations.

84. How is it that Comet Halley has an orbital period of 75.3 years when it originated in the Oort cloud?

85. Explain why the streaks in a meteor shower all appear to originate from one place in the sky, called the radiant.

86. What was it about 1I/ʻOumuamua that led astronomers to propose that the body should be reclassified as an interstellar comet, or perhaps a piece from the surface of a Pluto-like dwarf planet from another planetary system, instead of an interstellar asteroid?

87. Why do comets and some asteroids explode high up in Earth's atmosphere?

88. By comparing the data in Table 13.2 with those in Table 12.3, and based on everything you have learned about the Solar System, briefly discuss the general trends in liquids and ices on the rocky planets, major moons, dwarf planets, asteroids, and comets. What types of liquids and ices exist on body surfaces and/or in interiors? Explain why, including any significant deviations from the trends.

Table 13.2 Blackbody temperatures, T_{bb}, across the Solar System as a function of semimajor axes, a

Object	a (au)	T_{bb} (K)
Mercury	0.4	440
Venus	0.7	332
Earth	1.0	278
Mars	1.5	227
Jupiter	5.2	122
Saturn	9.5	90
Uranus	19.2	63
Neptune	30.1	51
Pluto	39.5	44
Eris	67.7	34
Sedna	483.3	13
Oort cloud (inner)	2000	6

Planets Everywhere and the Search for Extraterrestrial Life

14

One in [5][†] stars has habitable Earth-like planets surrounding it — in the galaxy, [40] billion stars have Earth-like planets going around them — that's huge, [40] billion. So when we look at the night sky, it makes sense that someone is looking back at us.

Michio Kaku (b. 1947)

†Numbers were increased significantly to reflect current estimates.

14.1	Exoplanets in Abundance	542
14.2	The Development of Planetary Systems	551
14.3	In Search of Extraterrestrial Life	566
	Exercises	582

Fig. 14.1 Artist's impression of Proxima Centauri B, one member of the triple star system, and its orbiting planet. Proxima Centauri A is in the lower left-hand corner. [ESO/L. Calçada/Nick Risinger (skysurvey.org); CC BY 4.0]

Introduction

In 1989, a possible detection of the first exoplanet (also called an extrasolar planet) was announced, believed to be orbiting the binary star system, γ Cephei[1] (in the constellation of Cepheus). The claim was originally suspect but was finally confirmed in 2002. Three years after the original announcement regarding γ Cephei, three planets were confirmed to be orbiting an exotic dead star in Virgo known as a pulsar. Finally, in 1995, a planet was detected and confirmed around the first Sun-like star, 51 Pegasi, in Pegasus. Since that time, thousands of planets have been discovered and we now know that planetary systems are ubiquitous throughout our Milky Way Galaxy and presumably throughout the billions of other galaxies scattered across the observable universe.

With so many planetary systems what do you think about Michio Kaku's quote in this chapter's opening?

14.1 Exoplanets in Abundance

Fig. 14.2 An artist's rendition of the system KOI-961. (NASA/JPL-Caltech)

Since the first confirmed detection of a planet orbiting a Sun-like star in 1995, the number of confirmed exoplanets has increased to 5036 as of May 13, 2022 (specifying the date on which the statistics are recorded is necessary, simply because the numbers change almost on a daily basis). The number of planetary systems having at least one planet in them has reached 3712, with 823 of those being multiple-planet systems. An additional 2640 planets, considered as candidates or unconfirmed, are found in 2446 planetary systems, 162 of which are multiple-planet systems.

The majority of confirmed planets and candidate or unconfirmed planets have been detected by NASA's Kepler Space Telescope mission that was launched in 2009. Over its four-year primary mission the telescope focused on more than 150,000 stars in a very small region of the sky contained in Cygnus and Lyra. A depiction of one of the Kepler target stars and its suite of three Earth-like planets is shown in Fig. 14.2. All three planets in KOI-961 are very close to their parent star and orbit it in less than two days. After failures in its pointing system, Kepler began a second phase of observations that scanned areas of the sky in the ecliptic plane.

Based on the success of Kepler, NASA launched the Transiting Exoplanet Survey Satellite (TESS) in April 2018, a more advanced mission that is scanning nearly 85% of the sky and studying more than 200,000 stars for at least four years. TESS's unique and highly elliptical orbit around Earth places the observatory in a 2:1 orbital resonance with the Moon, which makes the orbit very stable. The results of that mission are expected to significantly increase the number of known exoplanets.

When the James Web Space Telescope (Fig. 13.15) was launched in 2021 as a replacement for the Hubble Space Telescope, it became the most powerful space telescope ever built. Among its many missions, JWST continues to add to the number of known exoplanets. The telescope also investigates exoplanet atmospheres, giving us information about their compositions and chemistries, and perhaps detect telltale signs of life.

[1]Remember that you can find the Greek alphabet in Table 4.1.

How Do We Find Them?

Direct Imaging Method

Because planets emit very little visible light of their own, observing planets directly with telescopes is extremely difficult. In fact, only 194 confirmed planets have been detected by direct imaging. Figure 14.3 shows the first multiple-planet system to be imaged directly. The infrared image of four planets orbiting HR 8799 was obtained using the Large Binocular Telescope Observatory in Arizona by first removing the glare of the host star.

JWST is designed to "see" the infrared light that planets emit due to their cool temperatures (Table 6.1 and Section 6.6). This will allow for many more direct detections in the future.

Fig. 14.3 Four planets orbiting the star HR 8799. The star's light has been subtracted from the image in order to feature the planets. (Courtesy of Andrew J. Skemer)

Gravitational Microlensing Method

According to Albert Einstein's general theory of relativity, when light from a distant source passes a massive object, the light bends when moving through curved spacetime; recall Fig. 8.9. This effect can focus the light of the distant source much like a glass lens can focus light. As a result, the distant light source will seem to get brighter as the foreground object passes in front of it, dimming again as the foreground object continues on its way, as depicted in Fig. 14.4. Because the amount of bending increases with increasing mass, this effect is more significant when a

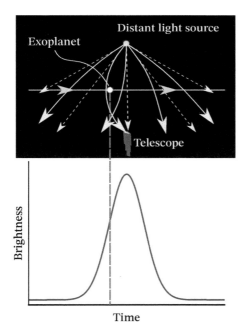

Fig. 14.4 Light from a distance source bends around a rogue exoplanet, focusing the light. The dashed arrows are the directions light would travel without the presence of a rogue planet. The brightness reaches a maximum when the planet is in line with the light source and the telescope. Of course, the bending is extremely exaggerated.

star moves in front of the light source, but with modern technology, a tiny amount of brightness increase is detectable when Jupiter-sized masses are involved. The process that causes these tiny increases in brightness is called microlensing.

As of May 13, 2022, 192 exoplanets within our Milky Way Galaxy have been confirmed this way. The detections not only include exoplanets orbiting other stars, but many of the detections involve so-called rogue exoplanets; exoplanets that are wandering through space unattached to any star. From the data gathered so far, the number of rogue exoplanets may outnumber stars by a factor of two. However, as we will learn in Chapter 16 it is not always a clear-cut determination if an object is a massive Jupiter-like planet, a very small and very cool star, or an intermediate type of object called a brown dwarf.

The Radial Velocity Method

Most of the early exoplanet discoveries were accomplished using the radial velocity method that relies on the Doppler effect for light. Radial velocity is the component of the exoplanet's velocity vector that is directed straight toward or away from the telescope. As depicted in Fig. 14.5 (*Left*), as a planet orbits around a star, or more accurately orbits around the center of mass (page 173) of the star–planet system, the star wobbles as a result. This causes the spectrum of the star to periodically shift to longer wavelengths (redshift) and then shorter wavelengths (blueshift) and back again. The amount of redshift and blueshift immediately tells the researcher how fast the star is moving toward and away from Earth and its orbital period. Kepler's third law then provides some information about the semimajor axis of star's orbit and the mass of the planet relative to the star.

So how fast does a star orbit around the center of mass of a star–planet system? This is what the radial velocity method is trying to measure, after all. Consider the Sun–Jupiter system as an example: the Sun moves around the Sun–Jupiter center

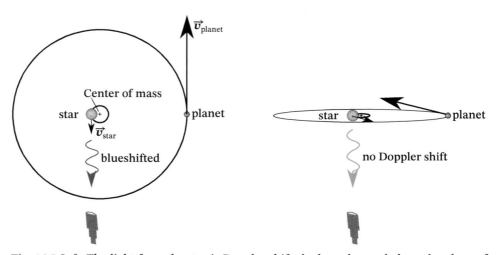

Fig. 14.5 *Left:* The light from the star is Doppler shifted when observed along the plane of the star's orbit. *Right:* No Doppler shift is observed when the plane of the star's orbit is perpendicular to the line of sight.

of mass with a speed that is a factor of $M_{\text{Jupiter}}/M_{\text{Sun}} = 1/1000 = 0.001$ slower than Jupiter's orbital speed around their center of mass. Since Jupiter's orbital speed is about 2.1 km/s (2100 m/s), the Sun's orbital speed about their mutual center of mass is just 2.1 m/s. For comparison, a fast walk or slow jog is roughly 1 1/2 m/s and the world's fastest humans can run at speeds of about 10 m/s. Measuring the radial velocity speeds of stars that are only a few meters per second is very challenging.

But there is another complication. In Section 9.2 we learned that the Sun's atmosphere is very turbulent, with convection zones rising and falling at 0.4 km/s (400 m/s); there are also sunspots, solar flares, solar prominences, and coronal mass ejections. All of this activity means that gases in the Sun's atmosphere produce chaotically redshifted and blueshifted spectral lines. With convection zones rising and falling at speeds 200 times faster than the Sun's orbital speed, teasing out the Sun's Doppler shift due to its orbital motion is extremely difficult. For other stars the same thing can happen, which means that the tiny periodic tugs of planets can get buried in the "noise" of all the other stuff that is going on. It wasn't until 1995 that telescopes became powerful enough, the instruments studying the light became sensitive enough, and the computer programs used to analyze the data became sophisticated enough to determine the subtle changes in the spectrum of 51 Pegasi, resulting in the first confirmed discovery of an exoplanet, named 51 Peg b, orbiting a Sun-like star (the star itself is referred to as 51 Peg a).

Another complication that makes determining the masses of the planets difficult has to do with the tilt of the planet's orbit. This is because radial velocities are only the portion of the star's velocity vector toward or away from the telescope, and not the entire motion of the star. As depicted in Fig. 14.5 (*Right*), it is only when the plane of the planet's orbit is in line with the view of our telescopes that the maximum radial velocities over an orbit reflect the true orbital speed of the planet; all other tilts produce lower radial velocities and therefore lower apparent masses for the planets. If the plane of the orbit is perpendicular to our line of sight, no radial velocity variations will be measured at all.

Because of the complications of the radial velocity method, the first planets that were discovered tended to have masses much like Jupiter's and they were found very close to their parent stars. These results were favored because, according to Newton's universal law of gravitation, massive planets closer to their host stars can pull harder on their stars, and being close to the stars implies faster orbits (larger radial velocities). Planets being close to their parent stars also means that many orbital periods can be measured easily. If an orbital period is only a few days, one hundred orbits could be recorded in one year, but if an orbital period is like Earth's, only one orbit can be observed in the same period of time. These first planets to be discovered were referred to as "hot Jupiters" because of their masses and proximity to their host stars.

The Transit Method

Although the radial velocity method started the avalanche of exoplanet discoveries, the transit method has produced the largest number of finds by far (3537 by May 13, 2022). The transit method involves monitoring the light of a star over time with the goal of detecting periodic dimming of the starlight as a planet passes in front of it. The Kepler and TESS missions employ the transit method.

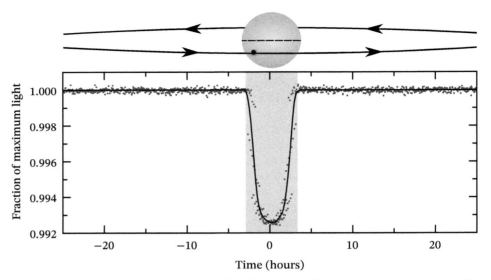

Fig. 14.6 The dimming of starlight as Kepler 7b passes in front of the star, Kepler 7a. The orbital period of Kepler 7b is 4.890976 days or 117.3834 hours. The dots are the actual data and the solid curve is a best fit curve through the data. The blue portion of the graph is the only time when the planet is in front of the star. Note that rather than starting from zero, the bottom of the vertical axis is at 0.992 (99.2%) of maximum light. (Data courtesy of *NASA Exoplanet Archive*, an online database)

Figure 14.6 shows the data collected by Kepler for one of its first planet detections, Kepler 7b. The diagram above the graph shows most of the orbit of Kepler 7a around its parent star. When the planet passes in front of the star the star's apparent brightness decreases very slightly. Notice that the greatest decrease in the light is only 0.75% [(1.000 − 0.9925) × 100%] of the star's brightness when the planet is not moving in front of the star.

The horizontal axis of Fig. 14.6 is shown as "phase" rather than time. This simply means that the data of many orbits were plotted on top of each other, with 0 h being chosen as the time when the light is at its minimum. This process of combining data leads to a more accurate determination of quantities like the orbital period, how tilted the orbit is relative to our line of sight to the star, and the eccentricity of the orbit.

Notice that the lowest points of the light curve aren't flat for very long. This is a telltale sign that we are not seeing the planet cut across the middle of the star from our vantage point. If we are looking directly along the plane of the planet's orbit, the bottom of the curve would be flat for a longer period of time. If our observing angle is too far above or below the plane of the planet's orbit, we wouldn't see any dip in the light curve at all, because the planet would appear to pass above or below the star. For this reason, statistics must be used that include the tilt of the orbit relative to our line of sight in order to make determinations about how many stars have planetary systems around them.

Videos of exoplanet-searching methods are available from the Chapter 14 resources page.

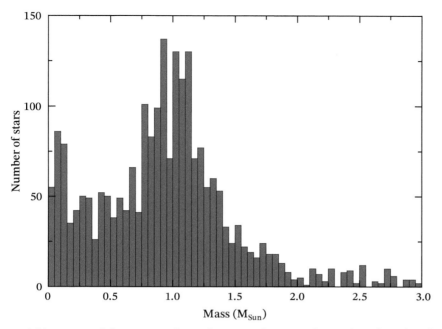

Fig. 14.7 A histogram of the masses of stars known to host exoplanets, based on data from May, 2022. Each bin is 0.05 M_{Sun} wide. (Data courtesy of *The Extrasolar Planets Encyclopaedia*, an online database)

What Do We Know About Them?

The increasing number of known exoplanets has allowed astronomers to conduct meaningful statistical studies of the types of planets that are out there beyond our Solar System. We can also learn something of the stars that host planetary systems.

The histogram in Fig. 14.7 tells us that the vast majority of stars harboring planets have masses very similar to the mass of the Sun, although there are low-mass stars with families of planets as well. Beyond about 1.2 M_{Sun} the number of stars known to host planets drops off quickly until there are very few planets known to be orbiting stars with masses greater than 2 M_{Sun}. As we will be learn in Chapters 15 and 16, low-mass stars have surface temperatures that are approximately half the Sun's temperature and they can live for much longer than the current age of the universe; in other words every low-mass star that ever formed is still around today. Stars with twice the Sun's mass have surface temperatures that are roughly 50% greater than the Sun, and their lifetimes are about 1/9 the Sun's lifetime (the Sun is currently middle aged with a life expectancy from birth to death of about 10 billion years). When you also consider the fact that many more low-mass stars form from nebulas, relative to higher mass stars, it is not surprising that stars less than 2 M_{Sun} are home to most exoplanets discovered so far. What is curious is that there is a peak in the number of stars at very close to the mass of the Sun.

Figure 14.8 shows two histograms of the masses of exoplanets. The histogram on the left clearly shows that most of the exoplanets discovered have masses that are much smaller than the mass of our Solar System's behemoth, Jupiter. However, there certainly are exoplanets that are similar to, and even exceed Jupiter's

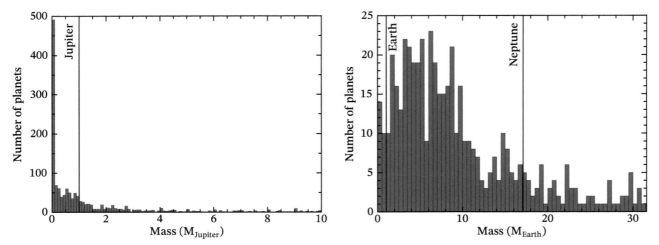

Fig. 14.8 *Left:* A histogram showing the distribution of known exoplanet masses from May, 2022 in terms of the mass of Jupiter. There are also exoplanets more massive than those shown. Each bin is 0.2 $M_{Jupiter}$ wide and the mass of Jupiter is indicted by the vertical line. *Right:* Details of the first half of the left-most bin of the left panel. The locations of Earth's and Neptune's masses in the histogram are shown by vertical lines. Each bin is 0.5 M_{Earth} wide. (Data courtesy of *The Extrasolar Planets Encyclopaedia*, an online database)

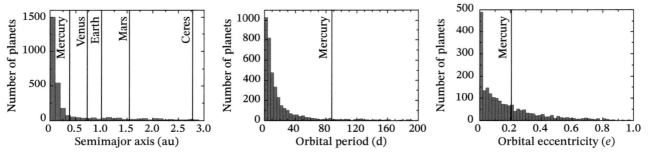

Fig. 14.9 *Left:* A histogram showing the distribution of known exoplanet semimajor axes from data in May, 2022. Each bin is 0.1 au wide. *Center:* The distribution of orbital periods for most of the known exoplanets. Each bin is 4 days wide. *Right:* The orbital eccentricities of the known exoplanets. Each bin is 0.02 wide. (Data courtesy of *The Extrasolar Planets Encyclopaedia*, an online database)

mass. Many of these are the "hot Jupiters" that were first discovered using the radial velocity method.

The right-hand histogram in Fig. 14.8 zooms in on the left half of the left-most bin of Fig. 14.8 (*Left*). Examining the exoplanets with masses less than 1/10 of Jupiter's mass reveals a peak in the number of planets between about 3 times Earth's mass and 9 Earth masses. Rocky planets in that range are referred to as "super Earths." There is also a second peak around 14 to 17 M_{Earth}, which is a bit less than Neptune's mass. Ice giants with masses less than Neptune's are called "mini Neptunes."

Particularly curious is the fact that most of the exoplanets discovered thus far have orbits that are very close to their parent stars, as reflected in the left panel of Fig. 14.9. Recalling that Mercury's orbit around the Sun has a semimajor axis of 0.39 au, most of the exoplanets discovered have semimajor axes that are less than half that value. For those exoplanets that are tucked in close to their parent star and are orbiting stars similar in mass to the Sun, this implies that these exoplanets must be very toasty indeed. That is not as true if the host star is one of the very cool, low-mass ones, however.

The central panel of Fig. 14.9 shows that most exoplanets have very short orbital periods (consistent with their small orbits). Very few exoplanets have been discovered with orbits longer than about one-quarter of one Earth year (Mercury's orbit is 88 d long, just under one-quarter of one Earth year). The highest peak in the histogram contains exoplanets with orbital periods of between 2 and 4 days. If you lived on one of those worlds you would constantly be having birthdays.

Finally, consider the orbital eccentricities of exoplanets (the right-most panel in Fig. 14.9). Again, using Mercury's $e = 0.21$ as an example, most exoplanets do have eccentricities that are smaller than this, but there are a few with surprisingly high values: more like the orbits of short-period comets in our Solar System rather than our eight planets (even the dwarf planet Pluto has an orbital eccentricity of just under 0.25). Those exoplanets that are tucked in very close to their host star with short orbital periods would naturally end up in synchronous rotation with their orbital periods because the star's tidal influence would be significant. Tides would also tend to circularize the exoplanets' orbits, leading to very low eccentricities.

Earth-Like Planets and the Habitable Zone

These data beg the question: Is our Solar System somehow unique in the universe? When the data started coming in, it sure seemed to be the case. At the very least, most of the exoplanetary systems are different from our own, but that doesn't necessarily imply that ours is unique. In some sense this would violate the idea that Copernicus first implied, namely that Earth is not the center of the Solar System, and therefore does not possess a special place among the planets. Extending that philosophy beyond our Solar System suggests that Earth and our Solar System should not be unique in the universe (an argument known as the Copernican principle). The data presented do indicate that most stars hosting planets are similar to our own; do at least a few of them host planets similar to Earth?

Key Point
The Copernican principle.

Although most of the systems studied so far don't appear to be similar to our own, there have been some that appear to show similarities to our Solar System. More to the point, are there rocky, Earth-like planets orbiting their hosts in significant numbers? From the data presented in Fig. 14.8, the answer to that question is yes. Studying the densities of those objects also supports the idea that many planets with masses similar to Earth's contain a significant amount of rocky material. Based on data obtained by the Kepler mission and extrapolating to the rest of the Milky Way Galaxy, it is estimated that almost 6% of Sun-like stars have Earth-sized planets in orbit at distances that result in orbital periods of between 200 and 400 days. Asking a different question: How many Sun-like stars harbor Earth-like planets at distances where liquid water can exist on the planet's surface? That answer appears to be about 22%; more than one in five Sun-like stars host rocky, Earth-like planets where liquid water can exist. Given that water is a major requirement for life (at least on Earth), these planets are orbiting within the habitable zone of their host planets. With some 200–400 billion stars in our own Milky Way Galaxy, there

Key Point
1 in 5 Sun-like stars host Earth-like planets within the stars' habitable zones.

Fig. 14.10 An artist's impression of what it might look like to stand on TRAPPIST-1f. The planets of the system are all close together; three of the planets are visible in the middle of the picture. [NASA/JPL-Caltech/T. Pyle (IPAC)]

may be as many as 40 billion Earth-like planets orbiting at distances from their parent stars where life could have developed.

With such large numbers, it seems likely that some of those systems could be in the celestial neighborhood of our Solar System. In fact, in 2016 an Earth-like planet was detected orbiting in the habitable zone of Proxima Centauri, the closest star to Earth (4.2 light-years) and a member of a triple star system known as Alpha Centauri. The star, Proxima Centauri a, is only about 0.12 M_{Sun} and Proxima Centauri b is in an 11.2 d orbit just 0.05 au from the star.

TRAPPIST-1

A particularly interesting system was discovered in 2017. At least seven planets are now known to be orbiting TRAPPIST-1, a star 40 light-years from Earth in the constellation of Aquarius (Fig. 14.10). The system was first detected by the transit method using the TRAnsiting Planets and PlanetesImals Small Telescope (TRAPPIST) at the La Silla Observatory in Chile. Since the initial discovery of this system from the ground, it has been observed by a host of ground-based and space-based observatories, including continuously for 22 days by the Spitzer Space Telescope that operated at infrared wavelengths.

All seven of the TRAPPIST-1 planets orbit within the star's habitable zone, with periods ranging from 1.5 days to 20 days. Most of the planets have densities that are slightly less than Earth and have radii that are comparable to Earth.

The six innermost planets are also in near orbital resonances with each other, much like Io, Europa, and Ganymede around Jupiter (page 479). It is also likely that all, or at least most, of the TRAPPIST-1 planets are in synchronous rotation or in some form of rotational spin–orbit coupling (page 375).

Exoplanet Atmospheres

An important effort to study the atmospheres of exoplanets is under way. Because these planets are so far away, the process involves observing changes in the spectra of their host stars when the planets pass in front of them. As the planets transit the stars, starlight passes through their atmospheres, introducing new absorption lines in the spectra (recall the discussions in Sections 7.1, 8.6, and 8.7). By subtracting the star's spectrum when the planet isn't in front of it, the atmospheric spectrum of the planet remains. This technique has been used in infrared light to detect the spectra of sodium (Na), potassium (K), water (H_2O), carbon dioxide (CO_2), and methane (CH_4) in exoplanetary atmospheres. One of the goals of the James Web Space Telescope is to expand that capability, with the possibility of detecting evidence of life on distant planets.

14.2 The Development of Planetary Systems

The Need to Satisfy All Evidence and All Physical Laws

Throughout Part II we have been focusing on the current characteristics of the various types of bodies in our Solar System and the exoplanets that are present around other stars. We have also been gathering clues about how our Solar System came into existence and how old it is. Although it is not a complete list, some of those clues include:

Key Point
Some characteristics of our Solar System and exoplanetary systems that must be incorporated into a complete model of planetary system formation.

- The Sun is a gaseous body that is chiefly composed of hydrogen (73.8%) and helium (24.9%), but it also contains a wide range of other elements totaling 1.3% of its surface composition by mass.
- The four rocky planets are relatively close to the Sun while the locations of the gas and ice giants are much farther from the Sun.
- All eight planets orbit the Sun very close to the plane of the ecliptic and have small orbital eccentricities.
- All eight planets orbit in the same prograde direction (counterclockwise when viewed from above the Sun's north pole).
- The Sun and six of the eight planets rotate prograde; Venus and Uranus rotate retrograde.
- All major moons, except Neptune's Triton, are in prograde orbits.
- Many of the smaller moons are in irregular orbits that may be either prograde or retrograde, their orbits may be highly inclined to the equator of the host planet, and they can be orbiting far from the planet.
- The Sun is slightly older than Earth, and Earth is slightly older than the Moon, with ages of 4.57, 4.54, and 4.51 billion years, respectively.
- Earth and the Moon have very similar surface compositions, although there are important subtle differences.
- The late heavy bombardment began 4.0 billion years ago with a significant uptick in the number of major impacts in the inner Solar System, then the rate slowed over the next 500 million years.
- Mars has a relatively low mass and a density that is significantly less than the other rocky planets.

- Variations are found in the ratio of the abundance of deuterium ($^2_1$H) to ordinary atomic hydrogen ($^1_1$H) in Solar System bodies, including the Sun, as well as in water (H_2O) that is present in planets, meteorites, asteroids, comets, and moons (recall that deuterium has twice the mass of ordinary hydrogen).
- Easily vaporized water is present in the inner Solar System.
- Liquid water and water ice exist on Earth, abundant water ice has been discovered on Mars, and there is evidence of past liquid water on Venus.
- Water ice and other ices exist on the surfaces of most of the moons of the giant planets.
- The gas giants contain diffuse cores of rocky material, ices, hydrogen, and helium.
- The asteroid belt exists between Mars and Jupiter, the Kuiper belt starts just beyond Neptune, the scattered disk begins near the outside edge of the Kuiper belt, and the Oort cloud has a roughly spherical distribution far beyond the scattered disk.
- Stony asteroids are found in the inner part of the asteroid belt and carbonaceous asteroids containing water are common in the outer portion of the belt.
- Some bodies in the outer asteroid belt demonstrate comet-like activity (the main-belt comets).
- Some centaurs demonstrate comet-like activity.
- Hilda asteroids and trojans are associated with Jupiter and Neptune.
- There is evidence that heavy elements, particularly iron and nickel, settled to the centers of some asteroids.
- A variety of different types of meteorites have been discovered on Earth: stony, metallic, stony–iron, and carbonaceous.
- Primitive carbonaceous chondrite meteorites exist with calcium and aluminum-rich inclusions (CAIs).
- Vesta appears to be a remnant protoplanet, Pallas is the exposed core of a former protoplanet, and Ceres is a planetary embryo; all are examples of planetary building blocks.
- Dwarf planets and smaller trans-Neptunian objects have been discovered orbiting the Sun farther out than Neptune.
- "Hot Jupiters" orbit other stars.
- Most exoplanets orbit very close to their parent stars.
- The vast majority of exoplanets discovered to date orbit stars similar in mass to the Sun.
- Orbital resonances, synchronous rotation, and more generally, spin–orbit coupling, are common throughout our Solar System and other planetary systems.
- Planetary systems are common throughout the Milky Way Galaxy, and there is some evidence through gravitational microlensing that they might be common in other galaxies as well.
- As of early 2022, two rogue objects, one a comet and the other perhaps a piece of a Pluto-like dwarf planet, have been discovered passing through our Solar System. By inference, a great many others are also moving through the Galaxy unattached to any planetary system.
- Rogue planets have been detected wandering through the Galaxy without a host star.

It is one thing to develop an imaginative all-encompassing model that describes the many features of our Solar System and other planetary systems, but it is something else entirely to require that the model must be both self-consistent *and* obey all of the physical processes upon which the universe operates. Adding ad hoc, physics-violating assumptions, such as appealing to supernatural causes, does not constitute a scientific solution.

As has been discussed in this text on numerous occasions going back to Chapter 1, a truly viable scientific theory or model must be able to (a) explain all of the data that are relevant, (b) make predictions capable of explaining future discoveries, (c) have the ability to be disproven should contradictory evidence become available, and (d) satisfy all physical laws. In the context of our present discussion about the Solar System, a model that explains the features we observe locally must also be able to explain exoplanetary systems. With the wealth of available data, these requirements for a complete model set a very high bar for success.

Key Point
Scientific models must explain all of the observable phenomena *and* be consistent with all physical laws without ad hoc assumptions that cannot be disproven.

The Creation of Protoplanetary Disks

Within the constellation of Orion [Fig. 14.11 (*Left*)] is a massive cloud of dust and gas known as the Orion Nebula (a nebula is a cloud composed of atoms, molecules, and microscopic dust particles in interstellar space). Hubble Space Telescope views of the Orion Nebula are shown in Fig. 14.11 (*Center* and *Right*).

Like many nebulas, portions of the Orion Nebula are collapsing under their own gravitational forces. As gravity squeezes the locally collapsing regions of the nebula, the gas pressure and temperature begin to rise. When the central temperature gets high enough, the microscopic grains of dust in the cloud are vaporized and reduced to molecules and atoms. As the temperature continues to increase, the electrostatic forces holding the molecules together become overwhelmed by the rising heat energy (Section 6.8), causing the molecules to dissociate into their individual atoms. In the innermost regions of the compressing cloud, the pressure,

Key Point
The Solar System formed from a collapsing nebula 4.57 billion years ago.

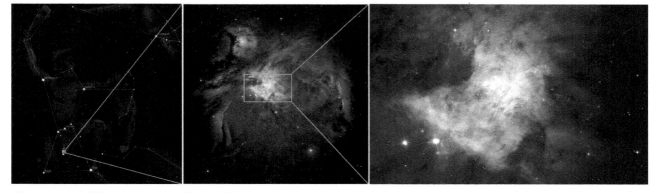

Fig. 14.11 *Left:* The winter constellation of Orion contains the Orion Nebula, located in his sword. (Adapted from a *Stellarium* screenshot, GFDL) *Center:* A Hubble Space Telescope image of the Orion Nebula. The nebula contains star formation regions where the dust and gas are collapsing due to the pull of gravity. The total mass of the nebula is estimated to be about 2000 times the mass of the Sun. *Right:* A Hubble Space Telescope closeup of a portion of the Orion Nebula. Young stars embedded in the nebula are visible. [*Center* and *Right*: Adapted from NASA, ESA, M. Robberto (Space Telescope Science Institute/ESA) and the Hubble Space Telescope Orion Treasury Project Team]

density, and temperature increase enough that the atoms are ionized, leaving a gas composed of nuclei and electrons. Finally, when the temperature rises to a few million kelvins and the density and pressure become sufficiently great, the kinetic energies of the nuclei become high enough and the spacing between the nuclei becomes so small that nuclear reactions can occur, converting abundant hydrogen into helium (Section 9.6). At that point, the central region of the once-collapsing cloud has become a new-born star. Young stars can be seen in the closeup image of a portion of the nebula in Fig. 14.11 (*Right*). Our own Sun began in much the same way about 4.57 billion years ago.

The Role of Angular Momentum

When the nebula collapsed that gave birth to our Sun, it undoubtedly had some overall spin caused by the collective motions of all of the dust and gas particles in the cloud; in other words, it possessed some angular momentum. As the nebula continued to collapse, its rotation rate increased as a direct consequence of the universal law of conservation of angular momentum (page 194*ff*), just like the figure skater in Fig. 6.15 when he draws his arms and legs in.

Figure 14.12 shows one particle in the nebula moving in a direction symbolized by the velocity vector $\vec{v}$. Even though the particle isn't necessarily traveling in a circular path around the center of the cloud, one component of the velocity vector is, namely $\vec{v}_\perp$, which means that there is angular momentum associated with that particle (remember that a vector can be broken up into components, one for each dimension, as discussed Section 5.2; the $\perp$ subscript indicates that $\vec{v}_\perp$ is perpendicular to the direction to the center of the circular motion). The other two components of this three-dimensional diagram of a velocity vector are shown in brown, but they have nothing to do with angular momentum. (If this all seems a bit confusing, go back to the two-dimensional diagram, Fig. 6.14, which is very similar to Fig. 14.12 when looking down the rotation axis.)

The angular momentum of the particle in Fig. 14.12 is $L = mv_\perp r$, where m is the particle's mass and r is the distance from the center of rotation to the particle. This means that as gravity pulls the particle to the center of the collapsing cloud, $\vec{r}$ decreases in length, which causes $\vec{v}_\perp$ to increase because the angular momentum, $L = mv_\perp r$, can't change (conservation of angular momentum in action). This is just a way of saying that the particle's speed around the center increases (faster rotation) as it gets closer to the center (think of the particle as one of the hands of the figure skater).

There was more to the nebula's collapse, however, than simply becoming more compressed and spinning faster; some of the mass of the nebula also flattened into a disk. As shown in Fig. 14.13 (*Left*), any clumps of material that were gravitationally attracted to the center of the nebula along the nebula's rotation axis, like the white clumps near the top and bottom of the sketch, simply fell straight in toward the center because their distances from the axis were zero ($\vec{r} = 0$ in Fig. 14.12), implying zero angular momentum.

Of course, most of the material in the collapsing nebula didn't fall in along the axis; instead, material was attracted toward the center from every possible direction. Given the random motions of the particles before the collapse started, the particles had velocity vector components that were like those illustrated in Fig. 14.12,

Key Point
The formation of a protoplanetary disk around a newly forming star is the direct result of conservation of angular momentum and energy lost due to collisions between particles.

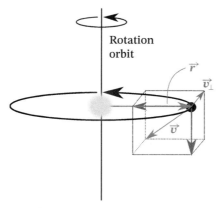

Fig. 14.12 The velocity vector, $\vec{v}$ (red), represents the velocity of a particle in the collapsing solar protoplanetary disk. The particle is located in the front top-right corner of the imaginary box and the velocity vector points from the particle down to the back bottom-left corner of the box. $\vec{v}_\perp$ (blue) is perpendicular to the position vector $\vec{r}$ (green), and is the only part of $\vec{v}$ that contributes to the particle's angular momentum. The brown vectors are the other two component vectors of the original three-dimensional velocity vector, $\vec{v}$.

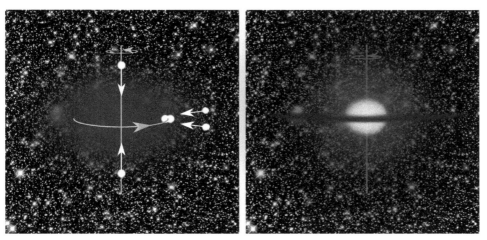

Fig. 14.13 *Left:* As the nebula that formed the Sun collapsed, conservation of angular momentum required that a protoplanetary disk be formed along the middle of the cloud due to the collisions among particles that were not falling along the rotation axis. *Right:* An illustration of the newly formed Sun with a disk around its equator. [Star fields: The Hubble Heritage Team (AURA/STScI/NASA)]

where only the components of their velocities that were directed around the rotation axis resulted in angular momentum that needed to be conserved. As the particles were being pulled closer and closer together, collisions between the particles occurred, with the result that motions parallel to the rotation axis canceled out while the total angular momentum of the particles around the rotation axis was preserved. With the continuing collapse, as depicted in Fig. 14.13 (*Right*), a rotating hot ball of gas formed the Sun, accompanied by a rotating disk of dust and gas around it. As the mass of the disk grew, its own gravitational field helped to attract even more mass to it. In essence, within the disk each particle was orbiting the newly formed Sun in the same direction, somewhat analogous to how a pizza chef gets dough to flatten into a disk when he throws it in the air while spinning it.

The Births of the Sun's Progeny

Key Points

Dust (microscopic up to mm/cm)

 melting

CAIs (mm/cm)

 *pebble accretion*

Clumping silicates, carbonaceous materials, and ices (up to 1 km)

continued pebble accretion

Planetesimals (1 km to 1000 km)

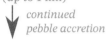 *mergers and pebble accretion*

Planetary embryos (1000 km to 5000 km)

mergers, gases, ice, and pebble accretion; internal heating by Al-26

Protoplanets (differentiated planetary embryos)

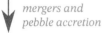 *mergers, gases, ice, and pebble accretion; swept out orbit*

Planets

With the formation of a thin protoplanetary disk around the newborn Sun, there was a lot of material swirling around the star in a flat plane and in the same direction as the star's rotation. The result is that today the planets orbit in the same prograde direction as the Sun's rotation, which is counterclockwise when viewed from above the Sun's north pole. The planets also orbit very close to the plane of the ecliptic which was very near the central plane of the protoplanetary disk. The fact that the net angular momentum vector (page 371) points north in association with counterclockwise rotation means that planet rotations ended up being prograde as well, unless subsequent unusual circumstances locally affected the rotation direction of individual planets. One curious feature, which is still unexplained, is that the Sun's equator is tilted 7.3° with respect to the ecliptic. Why isn't the Sun's equator much more closely aligned with the ecliptic plane?

Although the protoplanetary disk was dominated by molecular hydrogen (H_2) and helium (He), it also contained every other element currently found in the Solar System, with some of the most abundant being carbon (C), nitrogen (N), oxygen (O), magnesium (Mg), aluminum (Al), silicon (Si), sulfur (S), calcium (Ca), iron (Fe), and nickel (Ni). (Why these elements are particularly abundant will be discussed within the context of the lives of stars in Chapters 16 and 17.) It is carbon and silicon, in combination with other elements, that form dust particles. Dust only made up about 1% of the total mass of the disk and yet played the key role in getting planet formation started.

It was also certainly the case that the disk was not perfectly smooth, but contained local regions that were a bit more dense than others. Initially, when dust particles bumped into one another they would stick together, because of static electricity: essentially the "static cling" that happens when stuff sticks to your clothes or the growth of clumps of dust ("dust buddies") under your bed or in the corners of a room. When the clumps became massive enough they started to attract other clumps toward them gravitationally, causing more dust particles to stick together and grow over time. When the clumps grew large enough, collisions became more energetic, leading to some melting. The radioactive decay of $^{26}_{13}\text{Al} \rightarrow {}^{26}_{12}\text{Mg} + e^+ + \nu_e$, with its half-life of 717,000 years, also contributed to melting, due to the energy released in radioactive decay. (Because its half-life is much shorter than the age of the Solar System, no naturally occurring $^{26}_{13}\text{Al}$ remains today.) The radioactive decay of aluminum-26 created the calcium and aluminum-rich inclusions (CAIs) that are now found in carbonaceous chondrite meteorites. Being the first solids to form in the protoplanetary disk that are larger than dust grains, CAIs are used to establish the age of the Solar System.

Not all of the dust grains in carbonaceous chondrite meteorites melted, however; some of those that survived could themselves be used to determine ages. Researchers have been able to extract and study dust grains made up of silicon and carbon (SiC) from the Murchison meteorite that fell to Earth in Australia in 1969. Some grains were determined to be as much as 2 1/2 billion years older than the Solar System itself, predating the collapse of the solar nebula. The 7-billion-year-old grains likely came from aging stars that ejected dust before they died, thus contributing material to the formation of the solar nebula.

In the inner portion of the Solar System, the CAIs, silicate melt, and dust of millimeter-to-centimeter size continued to grow through gravitational attraction, a process termed pebble accretion, eventually forming planetesimals ranging in size from about 1 km (0.6 mi) to 1000 km (600 mi). Planetesimals were essentially the building blocks of planets. As these planetesimals traveled around the Sun, many of them collided with other planetesimals in their neighborhood. If the collisions weren't too energetic, meaning that the relative speed of the colliding bodies wasn't too great, they could stick together, ultimately creating planetary embryos that measured from 1000 km to 5000 km across, or roughly the size range from Ceres to Mars. On the other hand, if planetary embryos underwent high-energy collisions with other planetary embryos, they could be destroyed in the process, leaving smaller solid bodies of meteoroid and asteroid sizes.

Some of these planetary embryos grew large enough, and contained enough radioactive aluminum, that their interiors remained hot for some time, allowing the heaviest elements of iron and nickel sufficient time to sink to the center, producing chemically differentiated bodies. These objects are referred to as protoplanets, with Vesta being perhaps the last remaining example from the earliest days of the Solar System. Although Ceres is larger than present-day Vesta, Ceres never developed a fully differentiated core. It is likely that Vesta was once significantly larger than it is today but it suffered one or more energetic collisions after differentiation. The remnants of shattered protoplanets are the origin of stony, stony-iron, and metallic asteroids and meteoroids.

Close to the Sun, the temperature was so high that the gases had too much energy to exist as molecules, which is why the planetesimals forming in the inner Solar System were largely devoid of easily destroyed molecules. Farther out, it was possible for water (H_2O), carbon dioxide (CO_2), methane (CH_4), and ammonia (NH_3) molecules to form, along with other more complex molecules, such as the organic molecules, including amino acids, that are critical building blocks for life on Earth.

Sufficiently far from the Sun, some of the molecules were able to form ices (recall Table 12.3). The distance from the Sun where it became cool enough for water ice to form is known as the snow line, depicted in Fig. 14.14. The snow line is believed to have been at about 3.5 au in the solar protoplanetary disk. Beyond that line, when the local temperature in the disk was low enough, carbon dioxide ice was able to form, and at even cooler temperatures methane, ethane, ammonia, and nitrogen ices showed up.

With the formation of ices beyond the snow line came additional planet-building material. The pebbles of CAIs, melted silicates, and dust could now contain water ice as well. This new ingredient led to the rapid formation of icy planetary embryos and protoplanets. Carbonaceous meteoroids and asteroids resulted from shattered planetesimals and planetary embryos that formed beyond the snow line, including the Murchison carbonaceous chondrite meteorite, that struck Earth 4.57 billion years later.

As the ice-enhanced planetary embryos and protoplanets beyond the snow line merged, they grew much faster and became much more massive than their rocky counterparts nearer the Sun. Because of the greater masses of the ice-enhanced

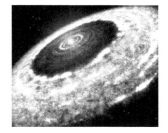

Fig. 14.14 An artist's depiction of a proto-planetary disk around the young star, V883 Orionis, illustrating the lack of ice interior to the snow line and an abundance of ices farther from the new-born star. [A. Angelich (NRAO/AUI/NSF)/ALMA (ESO/NAOJ/NRAO); CC BY 4.0]

planetary embryos and protoplanets, together with the cooler temperatures of the protoplanetary disk (implying that the gas molecules and atoms weren't moving very quickly), these icy bodies began to attract gases as well. Not far beyond the snow line, where the density of the protoplanetary disk's gas was greater than farther out, the most massive rock-and-ice core collected an enormous quantity of gas around it, principally hydrogen and helium, resulting in the formation of Jupiter. Moving farther from the Sun, where the protoplanetary disk became thinner, there was progressively less building material available. As a result, Saturn formed beyond Jupiter but with less than one-third of Jupiter's mass.

The diffuse cores of Jupiter and Saturn depicted in Fig. 12.14 developed because, as their envelopes of gases grew larger, infalling pebbles, meteors, and asteroids started to vaporize just like meteors falling through Earth's atmosphere. The vaporized rocky material then became absorbed into the lower envelopes of the gas giants in much the same way that water vapor is absorbed into Earth's atmosphere. The dew point is the temperature of complete saturation in the atmosphere at a specific location on Earth where the atmosphere can't hold any more water, resulting in fog, mist, and rain. A similar situation occurs for vaporized rocky material in the gas giants, but at much higher temperatures and pressures, producing a dense, diffuse mix of silicates, ices, metallic hydrogen, and helium.

Beyond Saturn, Uranus and Neptune formed with even less material, leaving them as giants dominated by ices with small rock and ice cores. And farther out yet, there simply wasn't enough material spread across vast distances to form anything of significant size except icy asteroids, comets, and dwarf planets of planetary embryo size.

As the giant planets grew by pulling in more dust, gas, pebbles, planetesimals, and occasionally some remaining planetary embryos and protoplanets in their region, portions of the protoplanetary disk started to be swept clean, leaving gaps in the disk. Figure 14.15 shows a protoplanetary disk around the young, active star TW Hydrae, with gaps in the disk where planet formation appears to be occurring. Figure 14.16 shows a planet forming around the newborn star PDS 70.

Within a few million years after its formation, the Sun's protoplanetary disk dissipated. This happened because most of the dust and gas that didn't end up in planetesimals by that time either ended up spiraling into the Sun, evaporating due to the Sun's high-energy ultraviolet radiation, or was blown away by a very strong solar wind. In any case, planetary growth from the collection of dust and gas slowed dramatically. It is worth pointing out, however, that planet-building had never completely stopped; Earth, Jupiter, the other planets, and even the Sun itself, are still gaining mass from dust and meteoroids, and the occasional asteroid or comet.

Key Point
The formation of moons.

In a process that was similar to the formation of the planetary system itself, many of the moons of the giant planets formed from localized disks swirling around the massive planets as those planets grew to control their immediate vicinities. As a result, the most massive moons and many of the smaller moons of the giant planets now orbit in a prograde direction very near to the planes of the planets' equators. Those moons also grew from the material in the regions of their hosts. The singular exception for a large moon is the unique retrograde orbit of Triton around Neptune.

Because Jupiter grew at a location not far beyond the snow line, its huge mass controlled the planetesimals, planetary embryos, and protoplanets between it and the much smaller Mars, not allowing them to combine to form larger bodies, in

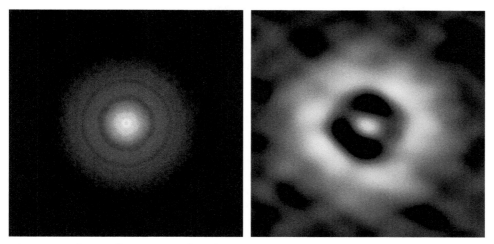

Fig. 14.15 *Left:* A millimeter-wavelength image of the very young star TW Hydrae, with its protoplanetary disk, obtained by the Atacama Large Millimeter Array (ALMA) in the high Atacama desert of northern Chile. Apparent in the disk are rings largely devoid of material, due to being swept up in the process of planet formation. *Right:* A closeup of the central portion of the disk, showing what appears to be a gap in the disk with a radius of about 1 au, the size of Earth's orbit around our Sun. [*Left* and *Right:* S. Andrews (Harvard-Smithsonian CfA); B. Saxton (NRAO/AUI/NSF); ALMA (ESO/NAOJ/NRAO); CC BY 4.0]

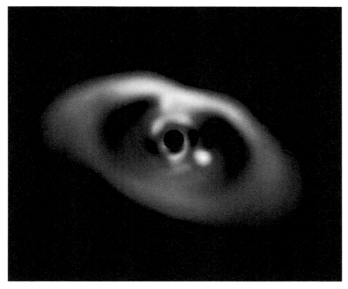

Fig. 14.16 A planet caught in the act of forming around the newborn star PDS 70. [ESO/A. Müller (Max Planck Institute for Astronomy, Heidelberg, Germany) et al.; CC BY 4.0]

what today is the asteroid belt. As collisions between those ancient bodies continued over time, most of the planetary embryos and protoplanets were eventually ground down to smaller asteroids and sub-meter-sized meteoroids.

In the innermost part of the protoplanetary disk, without the aid of abundant ices, the rocky planets grew much more slowly than the giants did. When the gases

Key Point
The rocky planets were slow in forming.

in the protoplanetary disk dissipated, the only food on which to feed were any remaining planetesimals and the occasional planetary embryo or protoplanet. While the giant planets formed within a few million years after the Sun's formation 4.57 billion years ago, Earth, for example, didn't complete its formation until 30 million years after the Sun's birth.

More Questions That Demand Answers

Although the Solar System formation model just described seems to explain major aspects of the Sun's planetary system, there are still many important pieces still missing from our story. Where did the water come from that currently exists in the inner Solar System, given that the protoplanetary disk was too hot in that region for water molecules to form? Why is the Moon 30 million years younger than Earth, and why is its composition not identical to Earth? How come Mars is only about the size of the largest planetary embryos that formed in the early Solar System and why is it so much younger in age than Earth as determined from Martian meteorites that were blasted off the planet's surface by impacts? Where did the Kuiper belt and scattered disk come from? Why is there a vast reservoir of cometary nuclei at the distant reaches of the Oort cloud? And how does the late heavy bombardment mentioned on pages 384 and 406 fit into the picture?

With the discovery of many thousands of exoplanets come other questions. Why are most of the exoplanets discovered to date so close to their hosts? How did so many Jupiter-sized gas giants ("hot Jupiters") end up very close to their central stars? And finally, what is it about our Solar System that caused it to form differently from most of the systems discovered so far?

Once again, science demands complete, self-consistent solutions to nature's puzzles, and the solutions must include all of the available data, satisfy all physical laws without exception, and cannot invoke ad hoc additions that are unable to be disproven.

Planet Migration

Key Point
Planets migrate through protoplanetary disks due to interactions with the gas and planetesimals.

The answers to all of these questions are associated with the fact that the orbits of planets are not fixed within protoplanetary disks during the early formation of planetary systems. When a giant planet forms in a protoplanetary disk and creates a gap in it (e.g., Fig. 14.17), the planet interacts with the gases in the disk, transferring orbital angular momentum to the gas. This causes the giant planet to spiral in toward the central star rather rapidly. Once the giant planet gets close to the star it finds a region devoid of gas, because the star's magnetic field creates an inner boundary for the gas. As a result, when the giant planet passes the inner edge of the disk it stops migrating; hence "hot Jupiters." If a giant planet hasn't fully swept out a gap in the disk, it will migrate inward even more rapidly because it also needs to plow through the disk material, causing it to lose angular momentum more quickly. In some circumstances, it is also possible for planets to migrate outward through a gas disk by drawing angular momentum from the gas.

Giant planets can also migrate either inward or outward in a protoplanetary disk, even after all of the protoplanetary disk's gas has disappeared. This is because once they get big enough they can start throwing their weight around, so to

Fig. 14.17 A possible initial configuration of the Solar System shortly after the formation of the solar protoplanetary disk. The orbits of Jupiter, Saturn, Uranus, and Neptune are shown, although Saturn, Uranus, and Neptune aren't fully grown yet. Jupiter formed first about 3.5 au from the Sun and carved out a gap in the nebular disk. The frame measures 30 au by 20 au.

speak, by flinging planetesimals and planetary embryos through space. If objects are thrown inward through gravitational interactions with a planet, the planet must move out, and if objects are flung outward, the planet migrates inward. Again, angular momentum is exchanged, resulting in altered orbits. In some cases, the planetesimals are flung so hard by their interactions with giant planets that they are ejected from the protoplanetary system altogether.

So why didn't our Jupiter wander in close to the Sun like hot Jupiters in other systems? Although not all of the details had been fully accounted for at the time this textbook was written, the answer appears to lie with Saturn. Computer simulations suggest that in our home planetary system the two gas giants formed fairly close to one another. Jupiter formed with a final mass of 318 M_{Earth} just beyond the snow line at about 3.5 au (Fig. 14.17). It then swept out the disk in its orbit and started to migrate inward. When Jupiter started its migration Saturn was still forming at perhaps 4.5 au, where the disk gas density was lower, and it had only accumulated about 30 M_{Earth}, or roughly one-third of its final mass. After Saturn reached about two-thirds of its final mass it hadn't yet completely cleared a gap in the disk but it had reached sufficient mass where it too started to migrate inward, but at a faster rate than Jupiter because it was plowing through the gas nebula and still growing as a result. When Jupiter reached an orbit 1.5 au from the Sun, Saturn had gotten close enough to get trapped in a 3:2 orbital resonance with its bigger sibling in the large gap that Jupiter had created during its migration [(Fig. 14.18). Saturn was now full grown and about 2 au from the Sun.

Key Point
Jupiter formed just beyond the snow line at about 3.5 au from the Sun and Saturn started forming 1 au farther out.

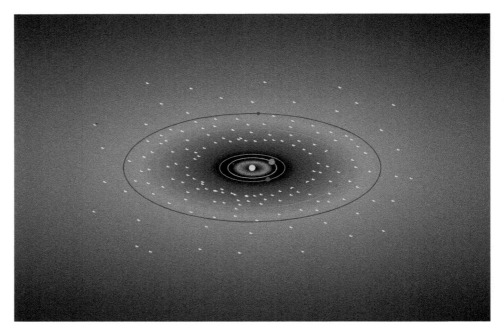

Fig. 14.18 One hundred thousand (100,000) years after Fig. 14.17, Jupiter has migrated inward to 1.5 au from the Sun with Saturn following. Saturn became locked in a 3:2 orbital resonance with Jupiter as the nebula was slowly dissipating. The gas giants then started their outward migration. The frame measures 30 au by 20 au, identical in size to Fig. 14.17.

Key Point
The formation of the Oort cloud and the small Mars.

Key Points
• The generation of the asteroid belt.
• The delivery of water to the terrestrial planets.

Because of the inward migration, Jupiter had scattered most of the planetesimals and planetary embryos out of the Solar System, sent some of them into the inner Solar System, and others much farther from the Sun. By doing so, Jupiter effectively starved Mars from being able to grow any larger. Scattering most of the bodies to great distances from the Sun or out of the Solar System entirely also contributed to Jupiter's inward migration to 1.5 au. The very distant objects that still remained attached to the Sun became what is today the Oort cloud; the reservoir of long-period comets.

With Jupiter and Saturn now locked in the 3:2 orbital resonance, that unique interaction caused the two planets to migrate outward together, scattering more bodies as they went. Some of the water-rich carbonaceous objects that had formed beyond the snow line ended up in the outer part of the asteroid belt and others struck the forming terrestrial planets. In the case of Earth, some of the water that was delivered ended up in its wet mantle and some of it formed the oceans. This is also why the inner part of the belt is now occupied by rocky (silicate) asteroids and the outer portion contains water and carbon-rich carbonaceous asteroids and main-belt comets. In addition, being thrown around like that explains why the asteroids and main-belt comets are in orbits that can range well above and below the plane of the ecliptic, producing an asteroid belt that resembles a thick torus (recall Fig. 13.11). When the protoplanetary disk's gases finally dissipated Jupiter ended up at, or close to, 5.2 au, where it continues to orbit today.

Some computer simulations support the idea that the Solar System may have initially formed with a third ice giant just beyond the orbit of Saturn and interior to the orbits of Uranus and Neptune. The inclusion of a third ice giant in the simulations also helps to explain some subtleties in the capture of Jupiter's trojans around its L_4 and L_5 Lagrange points. According to the third-ice-giant scenario, as Jupiter and Saturn migrated outward they interacted with that additional planet. Saturn caused the ice giant to move in a chaotic orbit while the gas giant moved farther out. The 3:2 orbital resonance of Saturn with Jupiter was also broken through gravitational interactions with the third ice giant. Eventually, a close encounter of the third ice giant with Jupiter resulted in the third ice giant being completely ejected from the Solar System and becoming a rogue planet wandering among the stars. At the same time, Saturn moved rapidly through a 2:1 Jupiter resonance to finally reach its present-day orbit with an orbital period that is very close to a 5:2 orbital resonance with the Solar System's most massive planet. The ice giant's ejection also caused Jupiter to move slightly closer to the Sun than it was before, while also gathering up its collection of trojan asteroids.

Key Point
The Solar System may have originally had three ice giants.

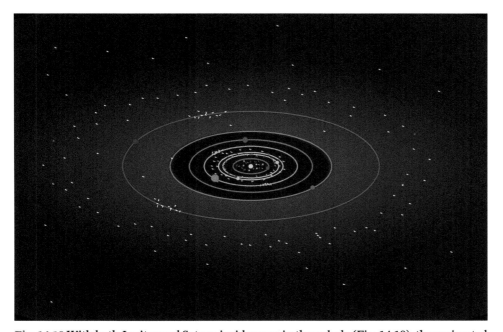

Fig. 14.19 With both Jupiter and Saturn inside a gap in the nebula (Fig. 14.18), they migrated outward, still locked in a 3:2 orbital resonance. Approximately 300,000 years after Jupiter fully formed and carved a gap in the nebula, it had migrated out to 4 au from the Sun as it scattered planetesimals. Mars is now a starved planetary embryo with its orbit depicted in red. The asteroid belt has been formed, with the orbits of Vesta (a protoplanet) and Ceres (a planetary embryo) shown in gray. Uranus has become locked in a 3:2 resonance with Saturn, and Neptune became locked in a 2:1 resonance with Uranus. At this point, Neptune was now driving planetesimals outward that are locked between a 3:2 resonance and a 2:1 resonance with the outermost ice giant. Neptune also collected trojans in its 1:1 resonance. Some planetesimals were scattered farther out into the scattered disk. This frame is larger than Figs. 14.17 and 14.18, measuring 45 au by 30 au.

Uranus and Neptune may have formed about 6 au and 8 au from the Sun, respectively, with the outer edge of the protoplanetary disk having a radius of perhaps 30 au. But, as depicted in Fig. 14.19 on the previous page, as Jupiter and Saturn moved outward, Uranus and Neptune became trapped in orbital resonances that forced them to move outward as well.

Even though the protoplanetary disk's gas had disappeared, icy planetesimals and planetary embryos still remained in the outer Solar System. As the ice giants encountered them, most of those remnant bodies were sent inward toward Saturn, Jupiter, and the inner Solar System, resulting in the late heavy bombardment that scarred the rocky planets and moons of the Solar System. The inward scattering of the smaller bodies also caused the ice giants to migrate farther from the Sun.

Continuing its outward migration, Neptune trapped planetesimals and planetary embryos in 3:2 orbital resonances, including Pluto and the other plutinos. The outer edge of the collection of bodies that Neptune shepherded during its migration is the 2:1 orbital resonance with the ice giant seen in Fig. 13.9 at 48 au from the Sun. Kuiper belt objects are located between Neptune's orbit at 30 au and the 2:1 resonance at 48 au, although most Kuiper belt objects are located between the 3:2 and 2:1 resonances (39 au–48 au). Objects beyond 48 au, and some objects within the Kuiper belt, are there because of orbital encounters that flung them outward rather than being more gently shepherded during Neptune's migration; these objects constitute the scattered disk. Neptune also trapped many icy bodies as its trojans in the planet's L_4 and L_5 Lagrange points. It has been suggested that the source of most of the short-period comets and centaurs is Neptune's population of trojans. After the planetary migrations ended, the Solar System's present-day configuration is depicted in Fig. 14.20.

Key Point

The formation of the Kuiper belt, the scattered disk, and the capture of Neptune's trojans.

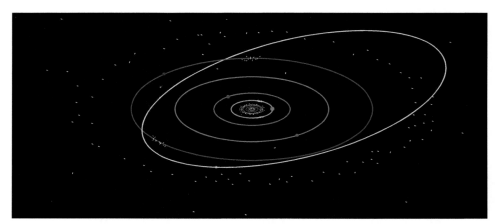

Fig. 14.20 As the nebula dissipated and most of the planetesimals had been scattered into the inner Solar System or out into the Oort Cloud or beyond, the current structure of the Solar System was established. A few planetesimals are depicted occupying the scattered disk, the Kuiper belt is evident outside of Neptune's orbit, Neptune's trojans are shown, and a few centaurs are inside Neptune's orbit. The asteroid belt is seen in the inner Solar System. The elongated and inclined orbit of Pluto is also shown, with Pluto being a planetary embryo and Kuiper belt object. The frame is 135 au by 60 au.

Another consequence of the chaos created by Jupiter and Neptune scattering bodies throughout the Solar System is that many of these objects became trapped by the gas and ice giants, creating their collections of irregular moons. Neptune also captured a planetary embryo in a retrograde orbit that is now Triton.

As should be abundantly clear, the formation and the evolution of our Solar System to its present-day configuration was a messy process. Complex computer simulations have been remarkably successful in explaining both broad and detailed features of the planetary system that we call home. But do the simulations represent past reality? In 2004 astronomers discovered an object in the Kuiper belt that appears to be a remnant of the Solar System's chaotic youth. Based on an analysis of its spectrum, 2004 EW_{95} (envisioned in Fig. 14.21) seems to be a carbonaceous asteroid that was probably flung into the Kuiper belt when the marauding Jupiter raised havoc during its inward migration phase.

Fig. 14.21 An artist's depiction of the Kuiper belt object 2004 EW_{95}, believed to be a carbonaceous asteroid. (ESO/M. Kornmesser; CC BY 4.0)

The Formation of The Moon

The formation of Earth's Moon doesn't fit well with the general description of the formation of the Solar System and the largest moons of the giant planets as just outlined. Although the composition of the Moon is similar to Earth, there are subtle differences, like the near absence of many of the compounds that are easily vaporized, such as water. Based on the radiometric dating of Moon rocks, the Moon is also 30 million years younger than Earth (4.51 and 4.54 billion years old, respectively). Adding to that the fact that the Moon's orbit is inclined 5° to the ecliptic suggests that the Moon's formation was not simultaneous with the formation of Earth, but that the Moon or its progenitors probably formed somewhere in the general vicinity of Earth in the solar protoplanetary disk.

Our Solar System was a violent place during the early days of planet formation, with planetary embryos and protoplanets colliding frequently, sometimes producing mergers and at other times resulting in destruction. The most widely accepted model of the Moon's origin is that 30 million years after Earth formed, a planetary embryo the size of Mars, named Theia, crashed into the still hot, largely molten Earth. (Theia was named for the Greek Titaness who gave birth to Selene, the goddess of the Moon.) An artist's concept of a cosmic collision between a planet and a planetary embryo is illustrated in Fig. 14.22. Numerous computer simulations conclude that the tremendous collision stripped away some of Earth's surface while destroying the impactor. Some of the debris from the two bodies ended up in a low Earth orbit and quickly merged to produce the Moon. Given its initially tight orbit of a few days, the Moon would have ended up in synchronous rotation almost immediately after it solidified. Over time, tidal effects caused the Moon to migrate out to its current location, 4.51 billion years after the catastrophic collision between Theia and Earth.

Assuming that the impactor struck the young Earth at an angle above the ecliptic, the model explains the Moon's present orbit. Thorough mixing of the debris from both bodies helps to explain current similarities and yet subtle compositional differences between Earth and the Moon. The very violent collision also explains why easily vaporized molecules like water are in short supply on the Moon.

Fig. 14.22 An artist's depiction of a collision between a planet and a planetary embryo in the early stages of the formation of a planetary system. (NASA/JPL-Caltech)

A Model That Is Still Incomplete

Despite all of the successes of the planetary system formation model just described, it is important to remember once again that explaining *some* of the observations and data is not sufficient in science. There still remain unanswered questions, some subtle, like details of composition, and some more dramatic. For example, on page 556 it was noted that the Sun's equator does not align with the ecliptic; in fact, it is misaligned by more than 7°. Were there external forces and torques acting on the Solar System as it was forming? Perhaps a passing star or massive planet managed to tilt the plane of the protoplanetary disk? It has been proposed that there may be a ninth, as yet undiscovered, planet in the Solar System that is very far from the Sun. Was it once a rogue planet or a planet captured from another star? Science always demands a complete, self-consistent explanation that satisfies all physical laws. Work remains to be done in understanding our Solar System.

14.3   In Search of Extraterrestrial Life

Shortly after Nicolaus Copernicus put forward his proposal in 1543 that the Sun, rather than Earth, is the center of the Solar System, Giordano Bruno (page 118) suggested that perhaps all of the stars in the sky were suns like our own, and perhaps each of those suns might have planets around them like Earth. He went on to further speculate that if there are Earth-like planets around other stars, perhaps those planets might have people living on them. It has only been in the last few decades that we have been able to confirm an abundance of exoplanets, with some of them having Earth-like characteristics, but we have yet to be able to draw any conclusions about the second part of his radical idea that life, let alone intelligent life, exists on other worlds.

There has certainly been no shortage of fictional ideas about life elsewhere in our Milky Way Galaxy or beyond. The origin of the general genre of science fiction is debated, with aspects of it argued to date back as far as 2000 BCE, with the ancient Mesopotamian *Epic of Gilgamesh*, in which there is a quest for immorality. Carl Sagan (Fig. 14.23) and Isaac Asimov (Fig. 14.24) considered the first work of science fiction to be Johannes Kepler's novel, *Somnium*, in which humans travel to the Moon. Kepler's work even contains mention of what happens when passing from the influence of Earth to the influence of the Moon, known today as the L_1 Lagrange point (Fig. 13.13), despite the fact that the concept of gravitation applied to orbits had not yet been developed by Newton. Others assign the honor of writing the first science fiction novel to Mary Wollstonecraft Shelley (Fig. 14.25) for her Gothic work, *Frankenstein*, and the depiction of the "mad scientist." Other early important contributions came from Jules Gabriel Verne (1828–1905) with his novels *Journey to the Center of the Earth* and *Twenty Thousand Leagues Under the Sea*, and H. G. Wells' *The Time Machine*, *The Island of Doctor Moreau*, *The War of the Worlds* (page 410), and *The Invisible Man*.

Fig. 14.23 Carl Sagan (1934–1996). (NASA/JPL)

Regardless of when the genre of science fiction began, it is safe to say that it is overwhelmingly popular today, often with an imaginative universe teeming with extraterrestrial life both benevolent and hostile. Beyond science fiction, actual reports of alien abductions, sightings of flying saucers and other kinds of unidentified flying objects (UFOs, also referred to as unidentified aerial phenomena), and claims that alien spacecraft have crashed on Earth. There are also those who believe that ancient cave paintings depict extraterrestrial visitors, or that crop circles found in farmers' fields, the enormous and ancient line art of Nazca, Peru (e.g., Fig. 14.26), the pyramids of Giza, Egypt (page 49), and all manner of other large constructions or artifacts could only have been made by extraterrestrial visitors. In many cases, claims of massive governmental cover-ups have been invoked to explain why information about these alien activities has not been released by authorities. Of course, there are still groups who are convinced that Earth is flat (The Flat Earth Society, https://www.tfes.org/), the Moon landings were faked, biological evolution doesn't occur, the universe is very young rather than being 13.8 billion years old (Chapter 19), and that global warming and climate change (Section 11.8), along with the Covid-19 pandemic that began in 2019, are worldwide scientific hoaxes. As Carl Sagan stated in his book, *Broca's Brain: Reflections on the Romance of Science* (Random House, 1979), "Extraordinary claims require extraordinary evidence." This is clearly in contradiction to conspiracy theorists' views that a lack of evidence only strengthens the claims of conspiracy or that it is up to others to disprove conspiratorial claims.

Fig. 14.24 Isaac Asimov (1920–1992). (Phillip Leonian from New York World-Telegram & Sun; public domain)

The Copernican Principle Revisited

Science always demands evidence that is both empirical and verifiable. There isn't any confirmation that Earth has been visited by extraterrestrial beings, but that is a different issue from whether or not extraterrestrial life exists. So what is the truth about the existence of extraterrestrial life? The simple answer so far is that we just don't know (at least at the time of this writing). The discipline of astrobiology, the merger of astronomy, physics, planetary sciences, chemistry, and biology, has the goals of exploring the process of how life originated on our planet, whether

Fig. 14.25 Mary Wollstonecraft Shelley (1797–1851) by Reginald Easton, sometime between 1851 and 1893, purportedly painted from her death mask. (Bodleian Library; public domain)

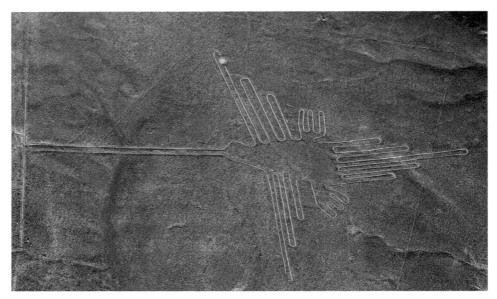

Fig. 14.26 A depiction of a hummingbird in the Nazca desert of Peru. Some of the figures are 370 m (1200 ft) long and were created sometime between 500 BCE and 500 CE. The location is a UNESCO World Heritage site. (Diego Delso, delso.photo, License CC-BY-SA)

or not life could have developed on other worlds, and how life might be detected elsewhere. The somewhat more restrictive subdiscipline of exobiology focuses explicitly on life beyond Earth.

In essence, the human quest for detecting the existence of extraterrestrial life is a straightforward extension of the Copernican principle that Earth and humanity don't occupy some unique or favored place in the universe. The very fact that life arose on our "little blue dot" suggests that the same universal laws of physics could have resulted in the development of life on other planets or moons somewhere among the hundreds of billions to trillions of planets that probably exist in our Galaxy, as well as within the confines of the other billions of galaxies scattered across the cosmos.

When and How Did Life Begin on Earth?

Fig. 14.27 High-temperature hydrothermal vents with microbial mats surrounding them; $T > 100°C$. (Image courtesy of Submarine Ring of Fire 2004 Exploration, NOAA Vents Program)

The time when life first appeared on Earth isn't precisely known, but in 2015 a sample of zirconium silicate ($ZrSiO_4$) was found in the Jack Hills region of western Australia that dates to 4.41 billion years ago and contains chemical evidence that life may have been present a mere 130 million years after Earth formed. Another discovery was made in 2017 along the eastern shore of Hudson Bay in eastern Canada, where fossilized microorganisms were identified in the precipitates of underwater hydrothermal vents (Fig. 14.27) that are between 3.77 and 4.29 billion years old. For reference, the oceans formed about 4.41 billion years ago. Regardless of when life first appeared on Earth, it is apparent that life began surprisingly quickly.

Necessary Ingredients

Life on Earth is fundamentally based on both water and carbon. Water acts as a solvent in which a wide variety of compounds can be dissolved. Carbon is an element unique in the universe for its ability to form more kinds of molecules than all other elements combined. Carbon atoms can also bond with each other, an important part of its chemical flexibility. In addition, the bonds of carbon are strong but can be broken and new bonds formed at temperatures typically found on Earth. The vast subdiscipline of chemistry known as organic chemistry features carbon as the central player. There is also a subdiscipline of organic chemistry that focuses on the intersection of biology and chemistry, referred to as biochemistry.

It has been proposed, especially in science fiction, that silicon (Si) could be used as a replacement for carbon elsewhere in the universe. After all, it is located immediately below carbon in the periodic table (Table 7.1), implying similar chemical properties. While that is true, silicon cannot come close to producing the wide diversity of chemical links that carbon is capable of making. For example, carbon is able to interact with hydrogen, oxygen, nitrogen, phosphorus, sulfur, iron, magnesium, and zinc. Furthermore, carbon is much more abundant in the cosmos than silicon is.

It may be more plausible that water could be replaced by another solvent, such as ammonia (NH_3), methane (CH_4), or ethane (C_2H_6), which are common in ice form throughout the outer Solar System, and exist as liquids and ices on Titan. But in order for water and carbon to work together effectively to form biological molecules, a planet must be within its star's habitable zone, where water can exist as a liquid.

However, it is one thing to have the available ingredients, it is another for nature to be able to produce the very complex, replicating molecules necessary for life to exist.

The Miller–Urey Experiment

A famous experiment was performed in 1952 by Stanley Miller (Fig. 14.28) and his doctoral advisor, Harold Urey (1893–1981), in which they placed water (H_2O) in a small glass flask, and methane (CH_4), ammonia (NH_3), and hydrogen (H_2) in a large flask that was connected to the smaller one. The water was supposed to simulate an ocean and the other gases were meant to be Earth's primitive atmosphere. The water was boiled, the vapor was mixed with the other gases, and then condensed and cycled back into the "ocean." While in the large flask, the mixture was subjected to simulated lightning by creating continuous sparks that passed through the gaseous mixture for one week. At the end of the experiment, the Miller–Urey experiment had produced a tar-like compound that Sagan later termed a tholin, meaning "muddy," although Sagan and his colleague were tempted to use the phrase "star-tar." Twenty-two amino acids have been identified in the gooey material.

Fig. 14.28 Stanley Miller (1930–2007) in 1999, with a recreation of the Miller–Urey experimental apparatus. (NASA)

Amino Acids and Proteins

Amino acids are important organic molecules that are present in all life, but by themselves are not sufficient to produce life. All amino acids contain two groups of atoms, NH_2 and COOH, combined with an additional grouping of atoms called a "side chain" that makes each type of amino acid unique. In other words, all amino acids must contain hydrogen, carbon, nitrogen, and oxygen; four of the six most abundant elements in the Solar System [the other two are helium (2nd) and neon (5th), neither of which are able to bond with other atoms]. Amino acids are, in effect, simple building blocks for forming more complex proteins, that are composed of long chains of amino acids.

We now know that amino acids are prevalent throughout the Solar System. The Murchison carbonaceous chondrite meteorite that fell to Earth in Australia in 1969 was found to contain at least 70 different kinds of amino acids. Tholins similar to the results of the Miller–Urey experiment and other experiments that have been performed since 1952, have been identified on moons, dwarf planets, Kuiper belt objects, centaurs, comets, and asteroids across the outer Solar System, giving those bodies a characteristic dark, reddish appearance. It is the Sun's ultraviolet light acting over a long period of time that turns simpler organic compounds into the "star-tar" that contains amino acids.

Revealed by their spectra, amino acids have also been discovered out in interstellar space within the very thin clouds of dust and gas similar to the solar nebula from which our Sun and the Solar System were born. Naturally forming dust grains appear to play an important role in serving as sites where the necessary elements can come together in order to aid the formation of these complex building blocks of life.

The Nucleic Acids RNA and DNA

The big step required for the development of life is the formation of the nucleic acids that are found within the cells of all living things on Earth: ribonucleic acid (RNA) and deoxyribonucleic acid (DNA). The single-stranded helix structure of RNA and the double-stranded double helix structure of DNA are illustrated in Fig. 14.29, along with their associated nucleobases. For simplicity, the nucleobases are represented by their first letters: uracil (U), adenine (A), cytosine (C), guanine (G), and thymine (T).

Nucleobases (or simply bases for short) are nitrogen-containing groups of atoms that also include various combinations of hydrogen, carbon, and, with the exception of adenine, oxygen. The backbone of RNA and the two backbones of DNA are formed from alternating groups of sugars and phosphates (sugars are formed from hydrogen, carbon, and oxygen, while phosphate groups contain the element phosphorus, P). The sugar in RNA is ribose and the sugar in DNA is deoxyribose, giving the two nucleic acids their names. In DNA, thymine and adenine combine to form one base pair and cytosine and guanine combine to form the second base pair, with the pairs binding the double helix strands together.

It has been proposed that the earliest life on Earth was likely based on RNA only, but today DNA structures contain coding for all of the genetic information about a living organism. One role of RNA and an associated protein is to "unzip" portions of the double helix of DNA, replicate the sequence of bases, and ultimately produce new functional proteins. Specific proteins are formed from sequences of amino acids that are in turn coded as triplets (groupings of three sequential bases) in the original DNA. In RNA, uracil is able to bond with adenine in the same way

RNA

Uracil
Adenine
Cytosine
Guanine

DNA

Thymine
Adenine
Cytosine
Guanine

Fig. 14.29 The single helix of RNA and the double helix of DNA.

that thymine does in DNA so that an exact copy of the information in the unzipped portion of DNA can be replicated in the RNA molecule. For example, if a triplet of bases in a portion of DNA are AGT (adenine-guanine-thymine), the base pair copies in RNA will be UCA (uracil-cytosine-adenine). UCA is a code for the amino acid serine. For redundancy, and to help avoid errors, there are multiple codings for the same amino acid.

Genes are groupings in DNA of between one thousand and one million base pairs, that carry coding for producing proteins through RNAs or to make RNAs that have functions other than producing proteins. With four bases able to be ordered in so many ways along the RNA and DNA backbones, a tremendous amount of information can be coded in the molecules. This is not unlike using zeros and ones in computers to store information, except that in RNA and DNA there is the equivalent of zeros, ones, twos, and threes, and a lot of them!

Chromosomes contain a DNA molecule with a portion or a complete set of an organism's genetic information, that can replicate information during cell division. Chromosomes also pass on the genetic information of a parent to an offspring. In the human genome there are 23 pairs of chromosomes (each parent contributes one chromosome in a pair), approximately 21,000 protein-coding genes, possibly more than 25,000 non-protein-coding genes, and more than 3 billion base pairs that can be ordered in any possible pattern. Human chromosome 1 alone contains 2000 protein-coding genes, comprised of 247 million base pairs, and it would be 85 mm long if straightened out. Through the DNA in chromosomes, genes can be passed down through generations, resulting in inherited traits. However, mutations of genes do occur, sometimes producing new, positive traits inherited by future generations (evolution), and at other times mutations can produce results that may be harmful to the organism, such as birth defects.

An immediate jump from naturally forming amino acids to molecules as complex as RNA, DNA, and proteins is extremely unlikely, so intermediate steps were certainly required. In 2015, NASA scientists were able to reproduce three of the nucleobases, uracil, cytosine, and thymine, in a laboratory when they replicated deep-space conditions (see Fig. 14.30). The scientists infused pyrimidine, a ring-shaped molecule made up of hydrogen, carbon, and nitrogen, into a mixture of water ice with some ammonia (NH_3), and either methane (CH_4) or methanol (CH_3OH). The mixture was kept at a temperature of 11 K ($-262°C$; $-440°F$) and irradiated with ultraviolet light.

It seems that many of the building blocks of life were available in space when Earth formed. The chemical analysis of some primitive meteorites, including the Murchison carbonaceous chondrite meteorite, has revealed the presence of all five of the nucleobases required for RNA and DNA: uracil (U), adenine (A), cytosine (C), guanine (G), and thymine (T). This implies that the same ingredients would have also been available during the formation of the other planets, moons, and dwarf planets. Those same building blocks would almost certainly have been present during exoplanet formation as well.

Key Point
All five nucleobases contained in DNA and/or RNA have been detected in primitive meteorites.

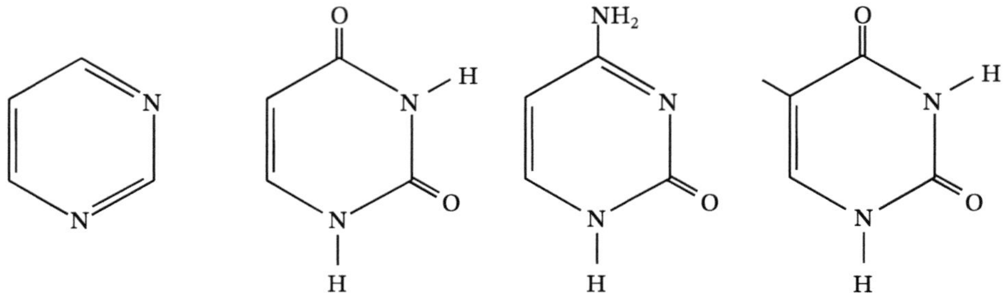

Pyrimidine   Uracil   Cytosine   Thymine

Fig. 14.30 The chemical structures of pyrimidine, uracil, cytosine, and thymine. The corners of the hexagons ("rings") that are not explicitly labeled are the locations of carbon atoms. A carbon atom is also located at the end of the bond (represented by a line) in thymine. Carbon atoms form four bonds. Anywhere there is a carbon atom and fewer than four bonds, the remaining bonds connect to hydrogen atoms that are not explicitly labeled. You may note that nitrogen forms three bonds and oxygen forms two bonds. Referring to the periodic table (Table 7.1), the three elements are located next to each other. The pattern continues with fluorine (F) only forming one bond, while neon (Ne) doesn't form any bonds.

Panspermia

Fig. 14.31 A tardigrade. [Schokraie E, Warnken U, Hotz-Wagenblatt A, Grohme MA, Hengherr S, et al. (2012); CC BY 2.5]

If the building blocks of life can be transported to the surface of a planet by meteorites and comets, it is natural to ask the question of whether or not living organisms can be transported too. Based on their compositions we know, for example, that meteorites originating from the Moon and Mars have landed on Earth. For a time the Martian meteorite, Allan Hills 84001 (Fig. 11.65), found in Antarctica in 1984, was thought to contain microscopic fossils. However, it was later demonstrated that the features seen in the meteorite could be explained without the necessity of invoking fossils of Martian organisms. To date, there remains no definitive evidence that life ever existed on Mars.

That negative result doesn't prohibit the possibility of transfer of life from one astronomical body to another; the hypothesis was only rejected in that particular case. Even on Earth, we know of organisms that can survive under extreme conditions for long periods of time, including tiny creatures named tardigrades, that are only 0.5 mm long when fully grown (Fig. 14.31). These animals can survive extreme temperature variations, radiation, dehydration, and even the vacuum of space, only to fully revive later.

The idea that life could have developed elsewhere and then deposited on a previously lifeless body is called panspermia. Even if this can occur, it is not a solution to the development of life *somewhere*; panspermia only describes how microbial life may spread from one world to another after it has developed at least once in the universe.

As has been illustrated time and time again throughout this textbook, the human quest for understanding continues to shed light on previously mysterious phenomena. Whereas ancient cultures invoked deities to explain the unknown, science strives to understand nature through fundamental physical laws, and does so with remarkable success.

Just because something isn't yet known does not mean that it is ultimately unknowable.

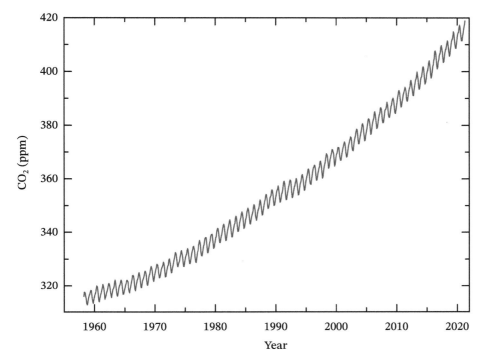

Fig. 14.32 The annual oscillations in carbon dioxide content of Earth's atmosphere due to growing seasons in the northern hemisphere, where most of Earth's land masses are located. The rapid rise is due to human causes. (Data courtesy of NOAA)

Markers of Life Beyond the Solar System

With the exception of being able to visit bodies within our own Solar System and sample their atmospheres and surfaces directly, the only way we can search for life on exoplanets (at least currently) is through the spectral signatures that life might leave in their atmospheres (unless of course they contact us directly; see below). One possible atmospheric biosignature might be the large amount of oxygen that could be produced by life, similar to the dramatic increase in oxygen content in our atmosphere when photosynthetic organisms developed (page 394*ff*). Decaying biomass on Earth also produces methane (CH_4), a molecule that could also be detected in the spectrum of an exoplanet's atmosphere. In addition, as is evident in Fig. 14.32, because most of the planet's land mass is located north of the equator, the carbon dioxide content of our atmosphere oscillates in response to the northern hemisphere's growing season. If a technological civilization exists on an exoplanet that burns biomass for energy, rising carbon dioxide levels may also be detectable over time. Many other atmospheric biosignatures have also been proposed, although it is important to remember that non-biological causes must be ruled out first before biological causes can be accepted.

The Search for Extraterrestrial Intelligence

The Drake Equation

In 1961 Frank Drake (Fig. 14.33), was organizing an inaugural scientific meeting on the search for extraterrestrial intelligence (SETI) and wanted a way to start a conversation about how likely it might be to detect a radio signal from intelligent life elsewhere in the Milky Way Galaxy. Rather than having the goal of coming up with a precise number, the expression he developed was the product of estimated values associated with the potential for intelligent civilizations to evolve to the point where they could transmit a signal into space. Formally, the Drake equation is expressed as

$$N = R_\star \, f_p \, f_s \, f_\ell \, f_i \, f_c \, T, \tag{14.1}$$

The Drake equation

Fig. 14.33 Frank Drake (b. 1930). (By Raphael Perrino; CC BY 2.0)

where the various factors are:
- N, the total number of civilizations capable of sending radio signals.
- $R_\star$, the average *rate of star formation* in the Milky Way Galaxy (units of number/time).
- f_p, the average number of *planets*[2] orbiting each star in the Galaxy.
- f_s, the fraction of planets capable of *supporting* life.
- f_ℓ, the fraction of planets capable of supporting life that actually do or did have *life* develop on them.
- f_i, the fraction of life-bearing planets where *intelligent life* has developed.
- f_c, the fraction of planets with intelligent life that have developed the ability to *communicate* by sending radio signals out into space.
- T, the period of *time* that such a civilization can exist (units of time).

If you are checking units, and this author hopes that you are, every term in the equation is simply a number without units except $R_\star$ and T, with units of number/time and time, respectively, so that multiplying one by the other cancels the time unit, leaving only a final number without units.

When Drake first proposed his equation, virtually all of the terms were just educated guesswork. Today the first few terms have been, or are becoming, more accurately determined, but the remaining terms are still guesstimates.
- The rate of star formation is reasonably well known for the Milky Way Galaxy and is roughly $R_\star = 2$ stars/year, although that number has varied over galactic timescales.
- Based on the latest results of exoplanet searches, it seems that the number of stars with planets around them could be close to 100%. Furthermore, the results of planet detection efforts suggest that there are, on average, multiple planets per star, and each planet could have multiple moons; and of course there could be many dwarf planets as well. As a result, a conservative estimate for the term might be $f_p = 3$.
- The fraction of worlds that could exist in the habitable zone around the host star is a bit more difficult to assess, particularly given the tendency of planets

---

[2]"Planets" in this context could also include dwarf planets and any moons orbiting the planets or dwarf planets.

to migrate during the early days of a planetary system, but perhaps 10% might be a reasonably conservative guess, or $f_s = 0.1$.

The remaining terms become even more speculative at this point in our understanding of extraterrestrial intelligence. After all, we only have one example from which to extrapolate our guesstimates.

- At least on Earth, we know that primitive microbial life began shortly after the planet's formation (Fig. 14.34), leading to the possible conclusion that primitive life may develop relatively easily on planets capable of supporting life. If that is the case then perhaps $f_\ell = 0.5$ might not be unrealistic. However, more pessimistic guesses suggest only one in one hundred thousand, or $f_\ell = 10^{-5}$.

- The possibility of intelligent life evolving from primitive microbial life may be significantly more difficult to achieve. This term is certainly the most speculative, and guesstimates range widely because so many different situations could influence the development of intelligent life. Do special circumstances need to exist such as a large, tidally locked moon that can help stabilize the rotation axis tilt of the host planet so that seasons don't fluctuate wildly? How fragile is life once it forms? How much time is necessary to move from single-celled organisms to multi-celled organisms? There have been billions of microbial, plant, and animal species on Earth, but only one has developed sufficient intelligence to create technologies. Does this mean that getting to a level of sufficient intelligence is extremely unlikely or, given all of the chances, is it inevitable? Estimates of f_i have ranged everywhere from extremely low at one in one billion ($f_i = 10^{-9}$) to inevitable ($f_i = 1$).

- Once a sufficiently advanced intelligence has developed, what is the likelihood that it will develop technology capable of sending a radio signal into space? Given that the capability happened quickly on Earth, this number may be something like $f_c = 0.2$. It assumes that the intelligent beings have sufficient curiosity and motivation to explore and understand the universe.

Fig. 14.34 A present-day cyanobacterial-algal mat along the White Sea on the northwest coast of Russia near Finland. (Aleksey Nagovitsyn; CC BY-SA 3.0)

- The last item, T, is also extremely controversial. There are those who advocate for a very low number because of the evidence on our own planet of unchecked resource use, pollution, and human-caused climate change. Political forces and our history of violence could also imply that we might destroy ourselves through nuclear war. After the start of the industrial revolution we reached our current technological state in only about 250 years, which is an extremely tiny fraction of the time since life first appeared on the planet. If one subscribes to that very pessimistic view, $T = 300$ years might be appropriate for the lifetime of a technologically advanced society. On the other hand, if a society is successful in managing resources, not destroying itself, and setting off into the cosmos, perhaps that value could be one billion years, or $T = 10^9$ years. In addition, a catastrophic impact from a large asteroid or comet could terminate advanced life, including intelligent life, at any time. But perhaps this scenario isn't especially significant, given that an advanced technological society would likely develop a version of B612 (page 530) or NASA's Planetary Defense Coordination Office (page 536), together with the ability to deflect or destroy a threatening object.

If the most pessimistic values of the Drake equation are used, the result could be

$$N_{low} = 2 \text{ stars/year} \times 3 \times 0.1 \times 10^{-5} \times 10^{-9} \times 0.2 \times 300 \text{ years} = 4 \times 10^{-13}.$$

Since we know empirically that the answer for our own galaxy is actually at least one (us!), that very small value (0.4 trillionths) could be interpreted as humans being alone in the Milky Way Galaxy and perhaps in the entire universe. That result is a very anti-Copernican principle conclusion. On the other hand, adopting the most optimistic values gives

$$N_{high} = 2 \text{ stars/year} \times 3 \times 0.1 \times 0.5 \times 1 \times 0.2 \times 10^9 \text{ years} = 6 \times 10^7 = 60,000,000,$$

leading to the conclusion that there ought to be many tens of millions of civilizations that are currently sending, or have sent, radio signals throughout the Galaxy that we could potentially detect. That is a staggering range of possible outcomes of the Drake equation. Even with our improved estimates of the first few terms of the equation, Drake certainly succeeded in generating discussion with his now almost 60-year-old equation.

One final aspect of the question regarding how many technologically advanced civilizations might exist in the Milky Way Galaxy today has to do with the issue of timing. The Drake equation only suggests how many technologically advanced civilizations might develop over a period of T years, but not *which* T years. The disk of our Galaxy is roughly 100,000 light-years across, meaning that it would take a radio signal traveling at the speed of light 100,000 years to traverse the disk. Relative to the age of the Galaxy (about 13.6 billion years) that period of time is minuscule. If a signal is or was powerful enough, and our radio telescopes are sensitive enough, any signal sent by a civilization in our direction within the last 100,000 years would be detectable. But what if such a civilization has already come and gone; even $T = 10^9$ years is only 8% of the age of the Galaxy. It is also possible that civilizations haven't yet reached a sufficiently technologically advanced stage, given that stars are still forming in the Galaxy and it took life on our planet 4.54 billion years to reach that level of technological sophistication.

The Wow! Signal

Astronomers have been listening for any radio signal evidence of intelligent life for more than 50 years without reproducible success. However, on August 15, 1977, a stunningly strong and brief signal was received from the constellation of Sagittarius by the Ohio State University Big Ear radio telescope. Figure 14.35 shows a printout of the signal, along with the excited comment by astronomer, Jerry R. Ehman, who discovered it on the printout. The symbols 6EQUJ5 do not represent a code buried in the signal, they are simply indications of the strength of the signal over time, starting at level 6, increasing to U, and then decreasing to 5 before vanishing. The signal lasted for 72 seconds, the entire period of time that the telescope was tuned to that frequency, but the signal was never repeated, despite many attempts to reacquire it.

The frequency at which the Wow! signal was detected happened to be very close to a key frequency (1.42 GHz[3]) used in astronomy to study cold hydrogen gas in the universe. That particular frequency is so important to astronomy that a worldwide broadcasting ban exists, prohibiting any transmissions in that portion of the electromagnetic spectrum; only reception is allowed. It has been proposed that since any technologically advanced civilization would certainly be aware of this naturally occurring hydrogen frequency, the civilization might want to broadcast a signal out into space at a frequency related to it under the assumption that other technological civilizations would probably be listening. In Carl Sagan's book, *Contact*[4], and the 1997 film by the same name[5], a message is detected from an alien civilization at a frequency of $\pi = 3.14159\ldots$ times the hydrogen frequency; apparently the aliens were avoiding "polluting" that fundamental hydrogen frequency.[6]

Was the Wow! signal from some technologically advanced civilization, or was it from some undetermined natural source? Numerous tests were performed to verify that it was not of terrestrial origin which strengthens the likelihood that it was indeed extraterrestrial, although not necessarily from an extraterrestrial intelligence.

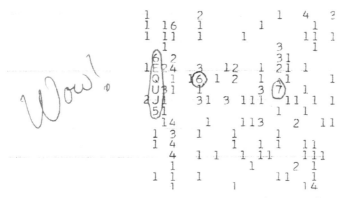

Fig. 14.35 The Wow! signal, received on August 15, 1977 from the constellation of Sagittarius. [Big Ear Radio Observatory and North American AstroPhysical Observatory (NAAPO), public domain]

---

[3]Recall prefixes in Table 5.1.
[4]Sagan, Carl (1985), *Contact* (Simon & Schuster).
[5]*Contact* (1997). Directed by Robert Zemeckis. [Feature film]. Burbank, CA: Warner Bros.
[6]Sagan was the science advisor for the film but died during its production.

In 2017 a paper was published suggesting that the Wow! signal could have come from comet 266/P Christensen, which happened to be passing through Sagittarius at the time the signal was detected. Perhaps the signal came from the cold hydrogen gas in the comet's coma. Observations of the comet were made in late 2016 and early 2017, confirming a radio signal of the correct frequency originated from the comet.

Again, "extraordinary claims require extraordinary evidence" (Sagan).

Where Is Everybody?

In 1950, while walking to lunch at Los Alamos National Laboratory in New Mexico, Enrico Fermi (Fig. 14.36), and his colleagues were discussing the recent claims about UFOs and how easily they were discredited when he suddenly asked "Where is everybody?" The question was raised regarding some of the very optimistic estimates of the number of technologically advanced civilizations that could exist in our Galaxy. If millions of civilizations exist, then where is the indisputable evidence that they have visited Earth recently or in the past? Despite the vast distances between the stars, if such a civilization did exist, wouldn't it have moved out from its home world to explore the Milky Way Galaxy? The question has come to be known as the Fermi paradox.

But why would a civilization set off into the cosmos? Perhaps curiosity and the adventure of exploration, or are those traits unique to our human civilization? After all, even without knowing where they were going ahead of time, humans populated islands across the oceans. Although the distances are large, for a species with a sufficiently long life span, traveling from one star to another is certainly possible within one or a few generations. The closest rocky planet to Earth known to orbit another star is Proxima Centauri b, only 4.2 light-years away. Once that first step is taken, moving throughout the Galaxy would theoretically be possible within millions of years. At least such a civilization would know where they are headed before leaving home, an advantage early human explorers didn't have. That possibility would require a large value for *T* in the Drake equation.

Fig. 14.36 Enrico Fermi (1901–1954). (By Department of Energy. Office of Public Affairs; public domain)

One obvious solution to the Drake equation, and an answer to Fermi's question, is that technically advanced civilizations simply don't exist elsewhere in the Galaxy. Another possibility may be that traveling between the stars is simply too difficult. Yet another reason that we have not yet seen evidence of extraterrestrial visitors is that Earth is designated a galactic wilderness area, "Take only pictures, *don't* leave footprints."

Escaping Large Rocky Planets

Another possible reason why we haven't been visited, based on the data of exoplanets obtained so far, is that rocky planets several times more massive than Earth appear to be far more common than Earth-sized planets. If the densities of rocky planets are similar, then larger planets have significantly greater gravities. This is because the amount of mass increases with volume, and the volume of a sphere is given by $V = 4\pi R^3/3$, where R is the sphere's radius. Writing density as ρ, the mass of a sphere is density times volume, or

$$M = \rho V = \rho \times \frac{4}{3}\pi R^3.$$

But the acceleration due to gravity (page 169) is given by

$$g = G\frac{M}{R^2}.$$

Substituting the equation for mass into the acceleration due to gravity equation gives

$$g = G\frac{4\pi\rho R^3/3}{R^2} = \text{constant} \times \rho \times \frac{R^{\cancel{3}}}{\cancel{R^2}} = \text{constant} \times \rho R,$$

where the constant contains all of the other terms in the equation that don't change: $4\pi G/3$. As a result, if the density, ρ, of Earth and the exoplanet are the same, the weight, mg, of someone, or something, on the exoplanet increases in proportion to the radius, R, of the planet. The conclusion of this math is that bigger, more powerful rockets are needed in order to escape a larger rocky planet's surface. But that in turn would then mean that more fuel is needed, which makes the rocket even heavier, making it even more difficult to escape. It has been suggested that perhaps many advanced civilizations are simply trapped on their super-sized rocky planets (perhaps super-Earths; Fig. 14.8) because they can't build sufficiently powerful rockets that are light enough to get away!

Reaching Out

Unintentional Contact: Radio and Television

The first radio transmission of a human voice took place on June 30, 1900, and some Christmas Eve programming was broadcast in 1906, but widespread radio broadcasts didn't begin until 1920. The first live television broadcast of a sporting event took place from Berlin, Nazi Germany (Fig. 14.37) during the 1936 Summer Olympic Games led by Adolf Hitler, in which a black American, Jesse Owens, won four gold medals. The first American television news broadcast took place in 1940 near the beginning of the Second World War. Beginning with those early broadcasts, Earth has been sending coded signals into space ever since. By now, the very first signals have traveled nearly 125 light-years, although those first transmissions were very weak.

Fig. 14.37 A television camera being used in the 1936 Summer Olympic Games in Berlin, Nazi Germany. [Rübelt, Wien (gem. Quelle); CC BY-SA 3.0]

Arecibo Message

On November 16, 1974, the first intentional broadcast of a powerful, coded radio signal for extraterrestrial "ears" was directed toward the vast and very dense cluster of stars known as the Hercules Cluster (Fig. 14.38). The target was chosen because there are about 300,000 stars in that one region of space, increasing the likelihood that an advanced civilization might receive the message.

The message, graphically depicted in Fig. 14.39, was created by Frank Drake, with assistance by Carl Sagan and others, and beamed toward the cluster using the Arecibo radio telescope (Fig. 14.40). The message consisted of 1679 bits of information because 1679 is the product of two prime numbers, 73 and 23. It was assumed

Fig. 14.38 Messier 13, the Great Cluster in Hercules. (KuriousGeorge; CC BY-SA 4.0)

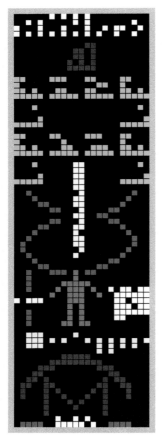

that an intelligent civilization advanced enough to decipher the message would certainly understand the universality of the language of mathematics and the significance of prime numbers; numbers that are only divisible by themselves and 1.

The Arecibo message was ordered in 73 rows with 23 columns per row, with one bit of information in each cell, either on (bright) or off (dark). The color coding is only for easy reference and was not included in the original message. The information contained in the message is as follows:

- The first four rows contain the numbers 1 through 10 written in columns.
- Once the counting scheme is deciphered, the purple symbols represent, in columns from left to right, the atomic numbers of hydrogen (1), carbon (6), nitrogen (7), oxygen (8), and phosphorus (15), the elements that comprise DNA.
- The green symbols represent formulas for the sugars and bases that make up the chemical structures upon which DNA is built.
- The famous double helix of DNA is next depicted in blue and white.
- The red figure of a man is apparent immediately below the double helix. On either side of the man are typical dimensions of a human on the left and the world's population (in 1974) on the right.
- A schematic of the Solar System is depicted in yellow, with Earth elevated one row immediately below the human figure.
- The Arecibo telescope is shown in purple, centered on Earth, and the dimensions of the huge dish are contained in the final two rows.

Fig. 14.40 The Arecibo Observatory in Puerto Rico. The observatory's receiver/transmitter platform collapsed in December, 2020 after extensive damage from recent hurricanes and earthquakes. The main spherical dish was 305 m (1000 ft) in diameter and was constructed within a natural sinkhole. The receiver/transmitter platform weighed 900 tonnes. [Mariordo (Mario Roberto Durán Ortiz); CC BY-SA 4.0]

The Pioneers and Voyagers

Radio transmissions haven't been the only messages sent out into space for extraterrestrial civilizations to intercept and interpret. Messages have also been attached to our robotic ambassadors that are venturing out to the stars.

Figure 14.41 (*Left*) is an illustration of gold-covered aluminum plates that were attached to the Pioneer 10 and 11 spacecraft when they were launched in 1972 and 1973, respectively. The two circles at the top with marks pointing in opposite directions depict the process that produces an important emission line of hydrogen at a frequency of 1.42 GHz. The drawings of a man and a woman are placed in front of a line drawing of the spacecraft for scale.[7] The man's raised hand symbolizes a greeting, the presence of the opposable thumb (a key feature of human evolution), and an illustration of how limbs move. The lines emanating from a central point

[7]The female's genitalia were not made explicit so that the plaque would be approved by NASA.

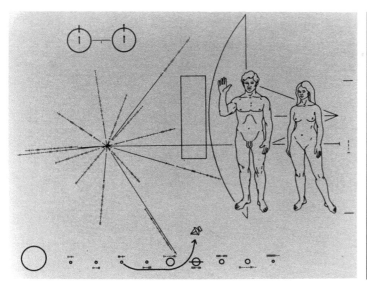 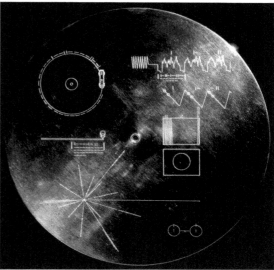

Fig. 14.41 *Left:* An illustration of the plaques that were mounted on the Pioneer 10 and 11 spacecraft. (NASA/JPL; designed by Carl Sagan and Frank Drake; artwork by Linda Salzman Sagan; photograph by NASA Ames Research Center; public domain) *Right:* The cover of *The Sounds of Earth* golden records, that are attached to Voyagers 1 and 2. (NASA/JPL)

depict the center of the Milky Way Galaxy, the locations and periods of blinking stars called pulsars, and the location of Earth. Finally, at the bottom is the Solar System, along with a depiction of Pioneer's flyby of Jupiter and then out into deep space.

The golden records attached to Voyagers 1 and 2 [Fig. 14.41 (*Right*)] are significantly more ambitious and information dense. The symbols on the cover are meant to explain how to play the record and decode the pictures that are embedded in the messages (a record needle was even provided). The bottom of the cover contains the corresponding information that was included on the Pioneer plaques. The record itself is a sort of time capsule of life on Earth. It contains 115 images, a collection of natural sounds, musical selections from different eras and cultures, greetings spoken in 55 languages, and messages from American President Jimmy Carter and United Nations Secretary-General Kurt Waldheim. The record also contains the Latin phrase "Per aspera ad astra," meaning "through difficulties to the stars."

Carl Sagan, who chaired the committee that designed the golden records, said of them

> The spacecraft will be encountered and the record played only if there are advanced space-faring civilizations in interstellar space, but the launching of this 'bottle' into the cosmic 'ocean' says something very hopeful about life on this planet.

Exercises

All of the exercises are designed to further develop *your understanding* of the material by thinking carefully and critically about what you have read in this chapter. Answers to selected exercises are available in the back of the book.

True/False

1. The Kepler mission discovered planets orbiting every one of the 150,000 stars it studied.
2. (*Answer*) The radial velocity method was used to make the first successful detection of a planet orbiting a Sun-like star.
3. Direct imaging is the easiest and most effective way of detecting exoplanets.
4. (*Answer*) Rogue planets, unattached to any star, have been discovered wandering through space by using the method of gravitational microlensing.
5. The Sun's mass is about 1000 times greater than Jupiter's mass. As a result, the Sun orbits their mutual center of mass 1000^2, or one million, times slower than Jupiter does.
6. (*Answer*) Most of the stars with planets orbiting them have masses that are between about 3 and 8 times the mass of the Sun.
7. Proxima Centauri b lies outside of its parent star's habitable zone.
8. (*Answer*) All known relevant data have now been explained in a comprehensive model of the formation of the Solar System.
9. The Sun's equator lies within 2° of the ecliptic plane.
10. (*Answer*) Triton is the only large moon in the Solar System that is in a retrograde orbit.
11. Orbital resonances played important roles in the present structure of the Solar System.
12. (*Answer*) The late heavy bombardment occurred when planetesimals and planetary embryos were scattered inward during planetary migrations.
13. Amino acids have been found in meteorites and in interstellar space.
14. (*Answer*) Tholins ("star-tar") have only been produced in complex laboratory experiments, and have not yet been discovered to be naturally occurring.
15. A triplet of nucleobases in a strand of DNA codes for a specific amino acid.
16. (*Answer*) Pyrimidine formed naturally in a 2015 experiment simulating deep-space conditions that also created uracil, cytosine, and thymine.
17. The Allan Hills 84001 meteorite proved that microbial life once lived on Mars.
18. (*Answer*) It may be possible for life to be transported from one planet to another through meteorites.
19. The Drake equation was derived from fundamental ideas of astrobiology.
20. (*Answer*) The identical records attached to the two Voyager spacecraft are designed to give intelligent beings a sense of what Earth and its human inhabitants are like, and warn them to stay away.

Multiple Choice

21. Even if every star that the Kepler mission observed had planets orbiting them, Kepler wouldn't be able to detect them because
    (a) not all planets pass in front of their parent stars' disks as seen from the vantage point of Kepler.
    (b) only gas giants are detectable.
    (c) only rocky planets are detectable.
    (d) planets may only take a few seconds to pass across the equators of their star' disks, making the odds of detecting them very remote.
22. (*Answer*) In order to use a minimum of fuel to maintain a stable orbit, TESS is
    (a) orbiting in Earth's L_2 Lagrange point.
    (b) orbiting in Earth's L_5 Lagrange point.
    (c) locked in a 2:1 orbital resonance with the Moon.
    (d) locked in a 3:2 orbital resonance with Mars.
23. The radial velocity method tends to preferentially detect
    (a) planets with long orbital periods.
    (b) planets with orbits that are nearly perpendicular to the line-of-sight of the telescope.
    (c) planets orbiting close to their parent stars.
    (d) massive planets.
    (e) all of the above
    (f) (a) and (b) only
    (g) (c) and (d) only
24. (*Answer*) The Kepler mission and the Transiting Exoplanet Survey Satellite (TESS) use which method to find exoplanets?
    (a) direct imaging
    (b) gravitational microlensing
    (c) radial velocity measurements
    (d) transits
25. A star with a mass of 0.1 M_{Sun} has a Jupiter-mass planet (0.001 M_{Sun}) traveling in a circular orbit around it with a speed of 70 km/s about their mutual center of mass. What is the star's orbital speed around their mutual center of mass? This is an example of a "hot Jupiter." (*Hint*: How many times more massive is the star compared to the planet?)
    (a) 70 m/s   (b) 700 m/s   (c) 7 km/s
    (d) 70 km/s   (e) 700 km/s   (f) 7000 km/s
26. (*Answer*) Suppose that you are living on Triton, Neptune's largest moon, and you observe Jupiter transit the Sun. Estimate how much the light from the Sun dims when Jupiter passes in front of it. Jupiter's radius is about 10% of the radius of the Sun. *Hint*: Treat the disks of the Sun and Jupiter as perfect circles, where the area of a circle is given by πr^2 (*r* is the radius of the circle). Although the effect would be slightly greater because Jupiter is closer to Triton than the Sun is, don't worry about that complication.
    (a) 0.01%   (b) 0.1%   (c) 1%   (d) 10%
27. Along with some exoplanets having masses similar to, or greater than Jupiter, there is a mass distribution peak between roughly 3 to 8 times the mass of _____.
    (a) Earth   (b) Saturn   (c) Neptune   (d) Pluto

28. (*Answer*) Most exoplanets appear to have orbital periods of _____ and semimajor axes of _____.
    (a) a few days, a few astronomical units
    (b) one year or more, less than the semimajor axis of Mercury
    (c) one year or more, a few astronomical units
    (d) a few days, less than the semimajor axis of Mercury

29. The habitable zone is the
    (a) range of distances from a star for which atmospheric oxygen is abundant.
    (b) location on a planet that is warm enough to support liquid water.
    (c) range of distances from a star where liquid water can exist on an orbiting planet.
    (d) altitude above the surface where the atmospheric temperature is comfortable for human explorers.
    (e) outer ring of a spinning space station where the simulated gravity equals surface gravity on Earth.

30. (*Answer*) The fact that the Sun rotates prograde, and all of the planets as well as all but one of the major moons are in prograde orbits, is a direct consequence of
    (a) coincidence.
    (b) conservation of energy.
    (c) conservation of linear momentum.
    (d) conservation of angular momentum.
    (e) blackbody radiation.

31. What variation in the solar protoplanetary disk resulted in small, rocky inner planets and much more massive gas and ice giants farther out in the Solar System?
    (a) The composition of the disk varied with distance from the new-born Sun.
    (b) The density of the disk was greatest at the locations of the giant planets.
    (c) Rocks only existed close to the Sun.
    (d) The temperature of the disk was greatest close to the Sun and decreased with increasing distance from the Sun.

32. (*Answer*) Planets exist in the Solar System because
    (a) of small density variations in the protoplanetary disk and gravitational attraction.
    (b) they already existed in the giant nebula that formed the Sun.
    (c) they were captured by the Sun's gravity as they traveled randomly through space.
    (d) science is unable to answer that question

33. Earth and the Moon
    (a) formed together out of the solar nebula.
    (b) differ in age by 30 million years.
    (c) have identical compositions.
    (d) formed before the giant planets.
    (e) all of the above
    (f) (a) and (b) only
    (g) (c) and (d) only

34. (*Answer*) The philosophical Copernican principle implies that
    (a) the Sun is the center of the Milky Way Galaxy.
    (b) Earth and its inhabitants came into existence through physical processes that are unique among all of the planets in the universe.
    (c) the Solar System does not occupy a unique position in the universe.
    (d) all physical laws in operation in the Solar System must also apply throughout the universe.

    (e) all of the above
    (f) (a) and (b) only
    (g) (c) and (d) only

35. In order for a planet to migrate inward toward its parent star, it must
    (a) gain angular momentum.
    (b) collide with a protoplanet or planetary embryo.
    (c) lose angular momentum.
    (d) transfer angular momentum to other bodies or to the surrounding protoplanetary disk.
    (e) (a) and (b) only
    (f) (c) and (d) only

36. (*Answer*) What stopped Jupiter's inward migration?
    (a) the asteroid belt
    (b) the depletion of gas in the solar nebula
    (c) an orbital resonance with Saturn
    (d) magnetic fields

37. Mars
    (a) is either a partially differentiated planetary embryo or a protoplanet.
    (b) is younger than the other rocky planets.
    (c) is significantly less dense than the other rocky planets.
    (d) was starved by Jupiter.
    (e) all of the above
    (f) (a) and (b) only
    (g) (c) and (d) only

38. (*Answer*) Stony-iron and metallic meteorites originated
    (a) directly from the original solar nebula.
    (b) from the collision of Theia with Earth.
    (c) from shattered planetesimals.
    (d) from shattered protoplanets.
    (e) from Jupiter's trojans.

39. Jupiter grew so large primarily because it
    (a) formed just beyond the snow line where ices were abundant.
    (b) absorbed other planets during its inward and outbound migrations.
    (c) pulled in so many comets from the Oort cloud, just as it did with Shoemaker–Levy 9.
    (d) broke off from the Sun when our star was spinning much faster than it is today.

40. (*Answer*) The trojans of Neptune
    (a) were captured during the migration of the planet.
    (b) are a possible source of the short-period comets.
    (c) are planetesimals that formed in the same orbit as the ice giant.
    (d) are trapped in the planet's L_2 Lagrange point.
    (e) all of the above
    (f) (a) and (b) only
    (g) (c) and (d) only

41. The Kuiper belt developed
    (a) in place when the solar nebula collapsed into a disk.
    (b) when Jupiter threw asteroids out into deep space.
    (c) when Neptune migrated outward with smaller bodies trapped in orbital resonances with the ice giant.
    (d) as a result of a collision between two dwarf planets.
    (e) when plutinos became trapped in a 1:1 orbital resonance with Pluto.

42. (*Answer*) Neptune's moon, Triton,
    (a) formed with the planet.
    (b) is probably a captured planetary embryo.
    (c) was once a moon of Pluto.
    (d) was thrown out of the asteroid belt by Jupiter.

43. The late heavy bombardment occurred
    (a) when the rocky planets migrated into the asteroid belt.
    (b) when Jupiter migrated inward through the asteroid belt, throwing most of the bodies out of the inner Solar System.
    (c) as Uranus and Neptune migrated outward, tossing bodies into the inner Solar System and taking some of their angular momentum.
    (d) because Mars escaped as one of Jupiter's former moons.

44. (*Answer*) There is some evidence to suggest that primitive microbial life may have begun on Earth as early as 130 million years after the planet formed. What percentage of the present age of Earth does that date most closely correspond to?
    (a) 0.03%   (b) 0.13%   (c) 3%   (d) 13%

45. It is more likely that carbon-based life would develop elsewhere rather than silicon-based life because
    (a) carbon is more abundant.
    (b) carbon is capable of creating a far richer family of compounds.
    (c) carbon compounds are often soluble in water which is abundant in our Solar System and likely in other exoplanetary systems as well.
    (d) all of the above

46. (*Answer*) All amino acids contain
    (a) hydrogen, helium, carbon, silicon
    (b) hydrogen, helium, oxygen, iron
    (c) hydrogen, carbon, nitrogen, oxygen
    (d) carbon, nitrogen, oxygen, silicon
    (e) carbon, nitrogen, oxygen, iron

47. The Miller–Urey experiment
    (a) produced at least 22 amino acids.
    (b) was designed to mimic the primitive Earth environment as understood at the time of the research.
    (c) required a complex series of steps performed by the scientists.
    (d) all of the above
    (e) (a) and (b) only
    (f) (b) and (c) only

48. (*Answer*) Proteins are composed of long chains of
    (a) protoplasm   (b) bars   (c) protons
    (d) amino acids   (e) RNA and DNA   (f) genes

49. Life on Earth may have been _____ based rather than _____ based, as is the case today.
    (a) RNA, DNA   (b) DNA, RNA   (c) UCA, RNA
    (d) QNA, DNA   (e) QNA, FAQ

50. (*Answer*) Humans possess _____ containing _____.
    (a) tens of thousands of chromosomes, several billion base pairs
    (b) 23 pairs of chromosomes, tens of thousands of genes
    (c) 46 pairs of genes, several billion base pairs
    (d) 46 chromosomes, 1 unique triplet of nucleobases

51. The ring molecule pyrimidine, is the core structure of
    (a) three nucleobases.   (b) uracil.
    (c) thymine.   (d) cytosine.
    (e) all of the above

52. (*Answer*) Scientists may be able to detect signatures of life on exoplanets by
    (a) oscillating or rising carbon dioxide content in the atmosphere.
    (b) unusually high levels of methane in the atmosphere.
    (c) higher than expected levels of oxygen relative to other gases.
    (d) all of the above
    (e) none of the above, because it is impossible to detect these gases in exoplanet atmospheres

53. The last term, T, in the Drake equation represents
    (a) how long a technological society can survive.
    (b) the average age of an exoplanet.
    (c) how much time is required to form a star from an interstellar nebula.
    (d) the time required to send a signal across the Milky Way Galaxy.

54. (*Answer*) The Wow! signal was
    (a) detected very close to an important hydrogen frequency.
    (b) only detected for 72 seconds.
    (c) probably hydrogen gas emission from the cold coma of a passing comet.
    (d) all of the above

55. According to Fig. 14.8 (*Right*), the mass distribution of "super Earths" peaks between about 3 M_{Earth} and 9 M_{Sun}. Assume that the density of an 8 M_{Earth} "super Earth" is identical to the density of Earth. The radius of such an exoplanet would be _____ and the acceleration due to gravity on its surface would be _____.
    (a) 2 R_{Earth}, 4.9 m/s^2 (b) 2 R_{Earth}, 9.8 m/s^2
    (c) 2 R_{Earth}, 19.6 m/s^2 (d) 4 R_{Earth}, 4.9 m/s^2
    (e) 4 R_{Earth}, 9.8 m/s^2 (f) 4 R_{Earth}, 19.6 m/s^2
    (g) 8 R_{Earth}, 4.9 m/s^2 (h) 8 R_{Earth}, 9.8 m/s^2
    (i) 8 R_{Earth}, 19.6 m/s^2

56. (*Answer*) The Arecibo message that was beamed toward the Great Cluster in Hercules in 1974 contained information about
    (a) the elements that comprise DNA.
    (b) nucleobases.
    (c) sugars.
    (d) the structure of DNA.
    (e) typical dimensions of a man.
    (f) the world population at the time the message was sent.
    (g) all of the above

Short Answer

57. Go to *The Extrasolar Planets Encyclopaedia* at http://exoplanet.eu/catalog/ and record the date of your visit, the total number of currently known exoplanets, exoplanet planetary systems, and planetary systems possessing multiple exoplanets. How many more exoplanets have been confirmed since May 13, 2022, the last date for the data used in this textbook. Be sure that the drop-down menu under "Status" only has "Confirmed" selected.

58. Figure 14.4 depicts the temporary increase in the brightness of a distant light source caused by gravitational microlensing when an exoplanet passes between the source of the light and the telescope. In the figure the exoplanet has not yet reached the line connecting the light source to the telescope. Make a sketch of the situation at the moment that the exoplanet is

located on that line. *Hint:* You can use the symmetry of the situation to estimate what happens to the light rays on their way to the telescope. Explain in words why the the brightness of the light source is a maximum at that time.

59. (*Answer*) Assume that the plane of the planet's orbit in Exercise 25 is in line with the direction to the telescope. Also assume that the *average* velocity of the star toward or away from the telescope is 0 km/s over a complete orbit of the planet.
   (a) At what fraction of the speed of light is the star moving when the planet is moving directly away from the telescope?
   (b) Astronomers are measuring one of the spectral lines of hydrogen in the star's atmosphere. When there are no motions involved in a laboratory, the wavelength of the line is 656.469609 nm. Using the Doppler effect equation, calculate how much the star's wavelength was shifted when the observation was made.
   (c) Was the measured wavelength blueshifted or redshifted?
   (d) What was the measured wavelength?

60. Explain why the radial velocity method for detecting planets requires very precise measurements for wavelengths of light produced by atoms or molecules.

61. Referring to Fig. 14.10, describe the locations of the three planets in their orbits [TRAPPIST-1e (top left crescent), TRAPPIST-1d (middle crescent), and TRAPPIST-1c (bright disk)] relative to the the host star and TRAPPIST-1f. Are they all on the far side of the star, on the same side of the star as TRAPPIST-1f, or some combination? Explain how the depictions of the planets support your conclusions.

62. (a) Using the data in Table 14.1, calculate the following orbital period ratios for the TRAPPIST-1 planets: P_c/P_b, P_d/P_c, P_e/P_d, P_f/P_e, and P_g/P_f.
   (b) These ratios are also approximate ratios of small integers (less than 10). Which integer ratios do they most closely correspond to?
   (c) What do your results imply about relationships between orbits for the six innermost TRAPPIST-1 planets?
   (d) What physical process led to those relationships?

63. Again using the data from Table 14.1, calculate the mass of TRAPPIST-1a (the parent star) by using Kepler's third law. Don't forget to convert the orbital periods from days to years (1 y = 365.25 d).

64. Explain in your own words how it is possible to measure the spectra of transiting exoplanet atmospheres even when the exoplanets themselves can't be seen.

Table 14.1 Orbits of TRAPPIST-1 planets

Planet	Period (d)	Semimajor axis (au)
b	1.51	0.0115
c	2.42	0.0158
d	4.05	0.0223
e	6.10	0.0293
f	9.21	0.0385
g	12.35	0.0469
h	18.55	0.0619

Note: TRAPPIST-1a is the parent star.

65. How does the formation of the Sun and the protoplanetary disk that formed around it explain the fact that all of the major planets orbit in the same direction and are all very close to the ecliptic (recall Fig. 4.29)?

66. Why did the rocky planets only form close to the Sun while the gas and ice giants formed farther out?

67. Describe at least two pieces of evidence as to why the Moon could not have formed in orbit around Earth at the same time that Earth formed from the solar nebula.

68. (a) Based on 4.57 billion years for the Solar System, estimate the number of half-lives of $^{26}_{13}\text{Al}$ that have occurred between the birth of the Solar System and today.
   (b) $^{26}_{13}\text{Al}$ is believed to have played an important role in the formation of calcium and aluminum-rich inclusions found in carbonaceous chondrite meteorites. From your result for part (a), explain why there is no evidence of naturally occurring $^{26}_{13}\text{Al}$ in the Solar System today.
   (c) What other role is aluminum-26 thought to have played in the formation of the Solar System, specifically in the hierarchy of building planets?
   (d) What indirect elemental evidence might scientists search for that would suggest that aluminum-26 once existed in the early solar nebula? Explain.
   (e) Comment on the use of a radioactive isotope to explain important aspects of the model of the Solar System for which there is no direct evidence of its existence in the Solar System today. Does its use constitute an ad hoc explanation in contradiction to the process of science discussed throughout this book? Explain. Note: $^{26}_{13}\text{Al}$ is a direct byproduct of a type of supernova explosion that is responsible for the demise of some types of stars.

69. (a) Using Kepler' third law, when Jupiter migrated in to 1.5 au from the Sun during the early formation of the Solar System, calculate its orbital period.
   (b) What was Saturn's orbital period when Jupiter captured it in a 3:2 resonance at that time?
   (c) After Jupiter and Saturn reached their present-day orbits, use the data in Table 10.1 to calculate the ratio of orbital periods, $P_{\text{Saturn}}/P_{\text{Jupiter}}$.
   (d) What is the closest small-integer ratio corresponding to the ratio of orbital periods in part (c)?

70. How was it possible to identify amino acids in very low-density interstellar clouds?

71. What is meant by the Fermi paradox?

72. The Netflix® science fiction movie *The Titan* (2018) was based on the premise that humanity had overpopulated Earth and consumed its resources, forcing its inhabitants to look for another home. Given Titan's Earth-like environment, albeit with methane replacing water, the movie contends that it would be necessary to evolve humans in a laboratory on Earth to be able to live on that moon, including flying in its thick atmosphere and swimming in its methane lakes without requiring spacesuits for protection. The evolved human would also be able to travel back and forth between Titan and Earth, and exist easily on both worlds. Given the molecular composition of the human body, why would that be impossible?

73. Supporters of the flat earth theory use various ad hoc assumptions to support their claims, such as the Sun only shines like a flashlight beam to explain day and night, or that Antarctica is a giant wall of ice that surrounds the Earth to

keep water from falling off the edge (e.g., Flat Earth Society Frequently Asked Questions).

(a) How does such an approach differ from a truly scientific one?

(b) Even without rocket launches it is possible to visually see Earth's curvature from a commercial airliner, or with high-altitude weather balloons. College physics students doing undergraduate research projects can photograph obvious curvature using balloons that travel to altitudes near 30 km (100,000 ft). How do supporters of a flat earth argue against such observations?

(c) In your view, does the flat earth theory satisfy Occam's Razor? Explain.

74. The flat earth theory proposes to explain the fact that objects dropped near the surface of Earth accelerate downward at a rate, g, equal to 9.8 m/s^2 by claiming that gravity does not exist on Earth but, rather, that the flat earth is accelerating upward at a rate g (see https://wiki.tfes.org/General_Physics. This solution to flat earth acceleration is termed "universal acceleration" as opposed to the "celestial gravitation" that attracts masses on Earth toward heavenly bodies. The latter is required to explain "tides and other gravimetric anomalies across the Earth's plane."

(a) If Earth was constantly accelerating toward the stars, increasing its speed by 9.8 m/s every second, how might that affect stellar spectra according to the Doppler effect?

(b) If acceleration of Earth is invoked to explain falling objects, what else must the flat earth theory explain according to Newton's laws?

75. Individuals espousing fantastical claims or conspiracies sometimes state that it is up to others to prove their claims false rather than for them to provide evidence for their statements. How does such a viewpoint compare to the role of evidence in science?

76. Comment on Carl Sagan's statement that "extraordinary claims require extraordinary evidence" in terms of both scientific discovery and conspiracy theories targeting science.

Stars and the Universe Beyond

PART III

Chapter 15   Measuring the Stars   589

Chapter 16   The Lives of Stars   638

Chapter 17   The End of a Stellar Life   684

Chapter 18   Galaxies Galore   726

Chapter 19   The Once and Future Universe   776

Spiral Galaxy Pair NGC 4302 and NGC 4298. [NASA, ESA, and M. Mutchler (STScI)]

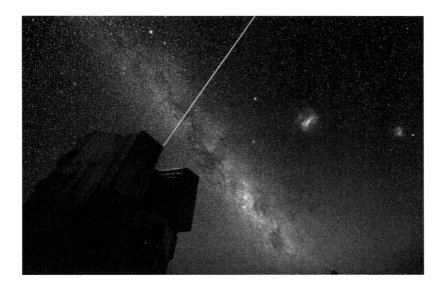

Measuring the Stars

15

Teaching man his relatively small sphere in the creation, it also encourages him by its lessons of the unity of Nature and shows him that his power of comprehension allies him with the great intelligence over-reaching all.

Annie Jump Cannon (1863–1941)

15.1	Observing the Cosmos across Wavelengths	590
15.2	The Brightness of Stars	599
15.3	How Far Away are they?	604
15.4	Determining Stellar Masses in Binary Star Systems	611
15.6	Calculating the Radii of Stars	623
15.5	The Harvard Spectral Classification Scheme	616
15.7	The Universality of Physical Law Revisited	624
15.8	The Birth of the Hertzsprung–Russell Diagram	627
15.9	Finding Patterns and Asking Why	633
	Exercises	634

Fig. 15.1 Observing the southern sky using the laser guide star facility at Yepun, one of the four unit telescopes of the Very Large Telescope. [G. Hüdepohl (atacamaphoto.com), CC BY 4.0]

Introduction

In Part I you learned about much of the science needed to develop an understanding of the stars that extends beyond simply seeing them as dots of light in the sky. One of those powerful and fundamental theories, quantum mechanics, has provided us with a means for analyzing the spectra of stars, leading to knowledge of their atmospheric and "surface" temperatures, pressures, compositions, motions, and magnetic fields. By measuring the distances to the stars we are also able to determine their radii and luminosities.

In Part II, the physical tools we had developed thus far were applied to studying our Sun, including deducing what its interior is like. Those same universal tools were also used to study the other objects in our Solar System and the exoplanets orbiting other stars.

In this final part, Part III, we continue our exploration of the cosmos by investigating interstellar nebulas, the nature and lives of the stars, galaxies, and the universe itself. What we see and know about the universe in which we live today flows naturally from its beginning 13.8 billion years ago, all due to the fundamental laws of nature.

We begin in Chapter 15 by introducing additional tools along with some terminology that is unique to astronomy. As we continue our quest for human understanding, you will find that the information we gather leads to patterns among the characteristics of the stars which invariably require us to ask the central question in astronomy, and indeed all of science: *Why?*

15.1   Observing the Cosmos across Wavelengths

Optical Astronomy

Ever since humans first looked up at the sky at night we have wanted to understand more about what we are seeing. The first major advance in our ability to see what's up there in greater detail came with the invention of the telescope, sometime around 1600 (page 131), and its use by Galileo and others. Those first telescopes were constructed using lenses, a design called a refracting telescope. Nearly 70 years later, Newton developed a different design that was based on mirrors rather than lenses. His mirror-based telescope displayed in Fig. 5.5 is called a reflecting telescope, which again revolutionized how we peer out into space.

With the realization that there was far more to see in the sky than is visible to us with the unaided eye, our quest to understand led to the need for ever-larger telescopes, in order to collect more light from those distant objects. Fig. 15.2 shows the world's largest telescope until 1910, the 1.8 m (6 ft) diameter Leviathan of Parsonstown.

In one sense, telescopes can be thought of as "photon buckets," collecting as much light as possible. As the diameter of a telescope lens or mirror increases, the number of photons collected increases with the cross-sectional area of the lens or mirror. Since the area of a circle is given by $A = \pi r^2$, where r is the radius of the circle, doubling the diameter (or radius) means that $2^2 = 4$ times as many photons

Fig. 15.2 The Leviathan of Parsonstown was the largest telescope in the world until 1910. Built by William Parsons, third Earl of Rosse (1800–1867), the telescope measures 6 ft (1.8 m) in diameter. Lawrence Parsons, fourth Earl of Rosse (1840–1908), can be seen looking through the eye piece. Photograph by Robert French (1841–1917), circa 1885. (Courtesy of the Library of Northern Ireland; public domain)

are collected in the same amount of time. An increase in diameter from 1 m to 10 m means that $10^2 = 100$ times as many photons can be collected. This gain in collecting area becomes critical when trying to see faint sources of light far away.

As with the Leviathan of Parsonstown, today all major telescopes are reflecting telescopes, although generally not of Newton's original design. There are many reasons for this:

- Reflecting telescopes don't suffer from the spreading of light into its component colors (technically known as chromatic aberration), as first pointed out by Newton.
- Reflecting telescopes only require one surface to be smooth, rather than two surfaces as lenses do.
- Unlike mirrors of reflecting telescopes, since light passes through lenses, the entire body of the lens of a refracting telescope must be free of imperfections. This becomes progressively harder to achieve as lenses get increasingly larger.
- As lenses get larger they also get much heavier. Although mirrors get heavier too, it doesn't matter what happens behind the surface, so it becomes possible to support mirrors from behind instead of only by their edges as is the case for lenses.

Key Point
Reasons for building telescopes with mirrors rather than with lenses.

Fig. 15.3 Heat from a roadway affecting the air above. Light passing through the hot air is bent as a result. (Dan Brown, CC BY 2.0)

- Only holding lenses by their edges means that massive lenses can sag under their own weight, affecting the shape of the lens.
- Massive pieces of glass also take much longer to adjust to the temperature of the air, which means the shape of the lenses can again be affected as they cool down during the night. The heat from the glass also affects the air above it, creating heat currents that distort the light's path, similar to seeing distortions in the air above a hot road in the summer (Fig. 15.3).

The use of mirrors in building telescopes doesn't solve every problem, however. The surfaces of large research-grade telescope mirrors require highly precise grinding so that imperfections in the surface are significantly smaller than the wavelength of light. With the wavelength of blue/violet light being only 400 nm long (0.0004 mm), no imperfection on the surface of a mirror can be any larger than about one ten-thousandth of a millimeter in size. This kind of precision is accomplished using an interferometer, that detects tiny shifts in length using interference patterns in the reflected light, much like the apparatus that Michelson and Morley used in their attempt to detect an ether wind (Fig. 8.15). Today the process is accomplished using laser light. Needless to say, this is a very challenging task.

A second problem mirrors face is making sure that the overall shape of the mirror is as precise as possible, so that all of the light is focused onto the right location. A very famous mistake was made in the grinding of the Hubble Space Telescope main mirror, when a reference mirror used during the grinding process was placed in the wrong location, essentially giving the Hubble Space Telescope a case of spherical aberration, not unlike a problem some people require glasses for. The Hubble's main mirror was ground very precisely to the wrong shape. Ultimately the mirror deformation was compensated for by placing corrective lenses in the light path of the telescope during a space shuttle servicing mission. Future instruments on the telescope were designed to account for the problem internally, resulting in the spectacularly successful results produced by the orbiting telescope.

As telescopes get larger, even mirrors suffer from distortion due to their weight as the telescope moves around during the night. This can be alleviated by placing pressure plates behind the mirror that push in just the right way to maintain an optimal shape. Today, many of the world's largest telescopes actually use segmented mirrors, which are individual smaller mirrors that are combined to produce a much larger mirror. Each segmented mirror is pushed from behind to keep the entire collective mirror aligned correctly (e.g., Fig. 15.4). Lasers and interferometry are again used for high-precision, real-time positioning.

Fig. 15.4 The segmented mirror of the Southern African Large Telescope (SALT). The overall mirror is hexagonally shaped, measures 11 m (36 ft) across, and is composed of 91 individual 1-m hexagonal mirrors. (Mark J. Roe / Janusz Kałużny, Copyrighted free use)

Besides being able to gather more photons, the larger the telescope, the better it can resolve fine detail. Unfortunately, there is a limit to how much detail can be seen, regardless of the size of the telescope. This is because we live at the bottom of a sea of turbulent atmosphere that bounces light around even under the most ideal conditions. Thanks to ongoing innovation this limitation has become much less significant with the development of a new technology that involves creating artificial "guide stars" high in the atmosphere. Lasers are used to excite sodium atoms, causing them to emit light at an altitude of 90 km (56 mi), as seen in Fig. 15.1 and Fig. 15.5. By monitoring the effect turbulent air has on the light coming from these artificial stars, computers can remove much of the impact that atmospheric turbulence has on the light arriving from the real objects that the telescopes are trained on. This technology wasn't available at the time the Hubble Space Telescope was launched, in 1990, and so getting above our atmosphere was critical for getting the best observations possible. Today, much of that atmospheric limitation has been removed from ground-based observing with the most advanced telescopes. As a result, enormous new telescopes are being envisioned and built, including the 24.5 m (80 ft) diameter Giant Magellan Telescope being constructed at the Las Campanas Observatory in Chile at an elevation of 2550 m (8500 ft) and the Thirty Meter Telescope International Observatory, planned for Mauna Kea, an extinct volcano in Hawai'i, at an elevation of 4050 m (13,300 ft). By the way, it is atmospheric turbulence that causes stars to apparently "twinkle"; the twinkling is not inherent in the stars themselves. Planets don't twinkle, because their angular sizes are larger than the scale of atmospheric turbulence, but virtually all stars look like points of light even with the largest telescopes.

Yet another way to increase the photon-gathering capabilities of an observatory while also improving resolution is to use multiple telescopes simultaneously to make the observations, such as the four unit telescopes of the Very Large Telescope (VLT) in Fig. 15.6. In this way, the effective collecting area of the combined telescope is the area of each mirror added together. In addition, since the resolution depends on the diameter of the effective mirror, the distance between the separated mirrors becomes that effective diameter. However, it is not a simple matter to add the data from the telescopes. This is because although the speed of light is very high, it is not infinite. This means that the light doesn't arrive at every telescope at exactly the same time; the time difference has to be accounted for, as do the slightly different angles that each telescope has to point in order to see the same object because of the curvature of Earth. In addition, any systematic differences between detectors of the multiple telescopes must also be taken into consideration. All of that plus the effects of curved spacetime due to Earth's gravity (general relativity, page 269) means that some very complex computations must be performed in order to combine the data from multiple telescopes for a single observation.

Fig. 15.5 The Four Laser Guide Star Facility operating at the Yepun unit telescope of European Southern Observatory's Very Large Telescope. (G. Brammer/ESO, CC BY 4.0)

Fig. 15.6 Sunset at Paranal Observatory where its Very Large Telescope array prepares for the night's work. Each of the four large unit telescopes has an 8.2 m (27 ft) diameter mirror. They are named after celestial objects in Mapuche, an ancient native language of the indigenous people of Chile and Argentina. From left to right, we have Antu (the Sun), Kueyen (the Moon), Melipal (the Southern Cross) and Yepun (Venus). There are also four movable auxiliary telescopes, each with a 1.8 m (5.9 ft) diameter mirror, that are visible to the left. The telescopes can operate individually or collectively. Paranal Observatory is operated by the European Southern Observatory and is situated at an elevation of 2635 m (8645 ft) in the northern Atacama Desert of Chile. [ESO/B. Tafreshi (twanight.org), CC BY 4.0]

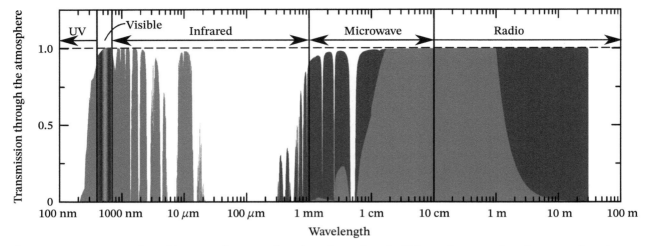

Fig. 15.7 Transparency of Earth's atmosphere as a function of wavelength. Values of 1 mean that the atmosphere is completely transparent to those wavelengths of electromagnetic radiation. Values of 0 mean that those wavelengths cannot penetrate the atmosphere down to the location of the telescope. Intermediate values mean some of the radiation can get through. The light blue color is for observations near sea level. The visible wavelength band (400 nm to 700 nm) is colored from blue to red. The dark blue at longer wavelengths is for an elevation of 5060 m (16,600 ft), where the Atacama Large Millimeter/submillimeter Array (ALMA) in the northern Chilean Andes is located. (Data courtesy of NASA, Steve Lord, the Harvard–Smithsonian Center for Astrophysics, and ALMA.)

Windows Through the Atmosphere

Key Point
Our atmosphere only provides a few "windows" through which we can look to see the universe beyond.

You have probably noticed that a great deal of attention has been given to the altitude of ground-based observatories. The reasons for this are two-fold: getting above as much atmosphere as possible means less atmosphere to distort observations, and getting as high as possible also minimizes the effects of water vapor, carbon dioxide, and other molecules in the atmosphere. If you want to study water molecules in space, you don't want the ones in Earth's atmosphere confusing your observations. As you learned during the discussions of climate change in Section 11.8, water and carbon dioxide effectively absorb infrared light, making it harder to observe in those wavelengths. In fact, at certain wavelengths it is impossible to make any observations at all; if your eyes only worked at those wavelengths you wouldn't see anything in the sky, it would be completely black.

As illustrated in Fig. 15.7, our atmosphere provides only a few wavelength "windows" that we can look through to see the stars from sea level. Of course, one of those wavelength ranges is in visible light (400 nm to 700 nm). A second major window exists at microwave and radio wavelengths (roughly 1 cm to 1 m). There are a few narrow infrared bands that are transparent as well, along with some transparency at long ultraviolet and short microwave wavelengths; otherwise the atmosphere is opaque.

Radio and Microwave Astronomy

For obvious reasons, astronomers only observed in the visible wavelength window from the time of Galileo to the early 1930s. That there were other regions of the electromagetic spectrum to explore wasn't even known until 1865, when Maxwell developed his electromagnetic theory predicting that visible light is just a special case of an infinitely broader electromagnetic spectrum (Section 6.5).

The next region to be explored was at radio wavelengths. The first detection of radio waves from space came in 1933, when Karl Jansky (Fig. 15.8) built a radio antenna to study static radio signals related to communications, while working at Bell Telephone Laboratory. His radio antenna (Fig. 15.9) was built on Ford Model-T wheels so that it could rotate in a circle to help locate the source of radio signals. One of the radio signals he detected peaked in intensity every 23^h56^m (one sidereal day; recall Fig. 4.10), suggesting that the source must be celestial in origin. If his "merry-go-round" antenna doesn't look much like a telescope, remember that it was designed to pick up radio waves, just like radio antennas on cars are designed to do. Different wavelengths require different technologies for detection. Because Bell Laboratory was only concerned with applied research at that time during the Great Depression, the curious discovery of Jansky was not pursued further. Considered the father of radio astronomy, Jansky may have won the Nobel Prize for his work had he not died at a relatively young age, due to a massive stroke (Nobel Prizes are only awarded to living recipients).

Fig. 15.8 Karl Guthe Jansky (1905–1950). (NRAO/AUI/NSF; CC BY 3.0)

Major advances in radio astronomy came after World War II, thanks in part to research in radar technology during the war. Today, radio astronomy plays a critical role in astronomical research, from studies of the Sun, the intense radiation from

Fig. 15.9 Karl Jansky and the first radio telescope, built by in 1933. The telescope was nick-named "Jansky's merry-go-round." (NRAO/AUI/NSF, CC BY 3.0)

Fig. 15.10 The Karl G. Jansky Very Large Array near Socorro, New Mexico. Most of the array is visible in this view of its most compact configuration. (NRAO/AUI/NSF, CC BY 3.0)

Jupiter's magnetic field, emissions in the regions around black holes at the centers of galaxies, the rotations of galaxies, and much more.

On the desert plains of San Agustin, near Socorro, New Mexico, sits the Karl G. Jansky Very Large Array (VLA), a Y-shaped configuration of 27 radio dishes with one spare, shown in Fig. 15.10.[1] Referring back to the wavelength "windows" in Fig. 15.7, the VLA operates at wavelengths between 0.7 cm and 4 m. Each dish measures 25 m (82 ft) across and contains 10 receivers designed to detect different frequencies within that wavelength band (recall that $f = c/\lambda$). The detectors are supercooled so that their own blackbody radiation and electronics don't swamp the signals coming from space. The dishes can also be moved along the Y-configuration tracks to various locations for different observing goals. In its most compact configuration (shown), the array measures less than 1.6 km (1 mi) across, but in its most extended configuration it is more than 35 km (22 mi) across. Just as with combining multiple optical telescopes to work in unison, the same is true of the 27 dishes of the VLA. In fact, that technique was pioneered with radio telescopes long before it was applied to optical telescopes. The distance across the array produces a resolution of fine detail similar to a single telescope dish of the same diameter.

Complementing the VLA is ALMA, the Atacama Large Millimeter/submillimeter Array, located 5060 m (16,600 ft) above sea level in the Atacama Desert of the northern Chilean Andes, as seen in panoramic view with the Milky Way in Fig. 15.11. ALMA consists of 66 dish antennas in two arrays, operating at wavelengths between 0.3 mm and 1 cm. As can be seen from the difference between the light blue and dark blue transmission graphs in Fig. 15.7, ALMA's site was selected because it is very dry and above most of the microwave-absorbing molecules in Earth's atmosphere, especially water vapor.

The VLA and ALMA are certainly not the only microwave/radio-wavelength telescopes in operation today. For example, China began observations with the Five-hundred-meter Aperture Spherical Telescope (FAST) in 2016, operating between 10 cm and 4.3 m (Fig. 15.12). FAST is similar to the smaller, and now decommissioned, Arecibo observatory in Puerto Rico[2] (Fig. 14.40), in that it was constructed in a natural depression in the terrain. FAST is currently the world's largest single-dish radio telescope.

---

[1]You may recognize the VLA from the movie *Contact* (1997) [Directed by Robert Zemeckis. (Feature film). Burbank, CA: Warner Bros.], based on the novel by Carl Sagan, in which humanity receives a message from an extraterrestrial intelligence.

[2]Arecibo was also featured in the film *Contact*, along with the VLA.

Fig. 15.11 A panoramic view of the Atacama Large Millimeter/submillimeter Array (ALMA) with the Milky Way in the background. ALMA is located in the Atacama Desert of the northern Chilean Andes at an elevation of 5060 m. [NESO/B. Tafreshi (twanight.org); CC BY 4.0]

Because radio astronomy investigates the cosmos at much longer wavelengths than visible light, to reach the same resolution that optical telescopes can achieve they must necessarily be much larger in diameter. The VLT, discussed on page 593, can achieve an optical resolution of about 0.001 arcsecond (one arcsecond is 1/3600 of one degree), which is the equivalent of being able to see both headlights of a car from the distance of the Moon (384,000 km, 240,000 mi). If a radio telescope observing at a wavelength of 4 m were able to achieve the same resolution, it would need to be 1 million kilometers (620,000 mi) in diameter, which is about 1.3 times larger than the Moon's orbit! Although this is impractical with a single radio telescope, it is possible to create a radio telescope with nearly the diameter of Earth, by combining the data from widely separated individual telescopes. One example of this approach is the Very Long Baseline Array (VLBA) in the United States, comprised of ten 25-m dishes spread across the country from Mauna Kea in Hawai'i to St. Croix in the U.S. Virgin Islands. The VLBA and other very long baseline interferometers (VLBIs) achieve their impressive resolutions in the same way that the VLT does; with precise timing at each telescope as the data are recorded locally, knowledge of the precise position of each telescope using global positioning systems, accounting for general relativistic effects, utilizing the wavelength interference patterns that occur because of different light-travel times to each telescope, just like the Michelson interferometer, and combining the data of each telescope by using powerful computers.

If you refer back to Fig. 15.7, you can see a sharp cutoff in the transparency of Earth's atmosphere for wavelengths longer than 30 m (100 ft). This is due to the ionosphere, a layer high in the atmosphere where there is a high concentration of ions and free electrons. Radio waves longer than 30 m reflect off the ionosphere, both coming in from space and coming up from the ground. This is why AM radio signals can sometimes be picked up as far as 1600 km (1000 mi) away. AM radio wavelengths are from 187 m to 561 m (614 ft to 1840 ft). FM signals don't reflect off the ionosphere because their wavelengths are much shorter (2.8 m to 3.4 m; 9 ft to 11 ft). AM wavelengths lie well to the right of the abrupt cutoff in Fig. 15.7, but FM wavelengths are in the transparent radio band. Television broadcasts range from 0.4 m to 5.5 m (FM wavelengths are situated between channels 6 and 7 of VHF TV broadcasts). As mentioned on page 579, Earth has been sending radio-wavelength signals out into space since not long before the first television broadcast

Fig. 15.12 The Five-hundred-meter Aperture Spherical Telescope (FAST) is the world's largest single-dish radio telescope. FAST is located in southwestern China's Guizhou Province. (Absolute Cosmos, CC BY 3.0)

Key Point
The ionosphere blocks all radio wavelengths longer than 30 m.

of the 1936 Olympics in Nazi Germany. Because of the importance of the radio portion of the electromagnetic spectrum to astronomy, some key wavelengths are restricted internationally from being used for terrestrial broadcasting.

Infrared Astronomy

Fig. 15.13 The Stratospheric Observatory for Infrared Astronomy (SOFIA) uses a 2.7-m (8.8 ft) diameter telescope and operated at altitudes of 11.5 km to 13.8 km (38,000 ft to 45,000 ft). The observatory ceased operations in 2022. (NASA/Jim Ross)

The other wavelength region where observations can be made within Earth's atmosphere are in the infrared, but only in a very limited way. Because of water vapor, carbon dioxide, and other molecules, "windows" are narrow and certainly not ideal. For the best observing, infrared observatories are necessarily in the highest and driest locations on Earth. NASA, in partnership with the German Aerospace Center, even operated an infrared observatory (SOFIA) out of the side of a converted Boeing 747SP aircraft, in order to reach altitudes not achievable from mountain tops (Fig. 15.13).

With such efforts to get as high in the atmosphere as possible, the next logical step is to observe from space.

Taking Observing Into Space

In order to fully explore the cosmos, astronomy must be able to make observations across the entire electromagnetic spectrum. The reason for this is simple: different wavelengths correspond to different energies. From Planck's energy equation, $E = hf = hc/\lambda$, the energies of photons depend inversely on wavelength, so that

Fig. 15.14 The Hubble Space Telescope. (NASA)

the shortest wavelengths correspond to the most energetic photons and the longest wavelengths describe the lowest-energy photons. Different physical phenomena produce photons of different energies, not just for discrete spectral lines, but for continuum radiation as well. The extreme environments around black holes or the fusion of atomic nuclei produce the very short-wavelength gamma rays and x-rays, while cold interstellar space or electrons spiraling around magnetic field lines are some of the sources of radio-wavelength radiation.

Beginning in the late 1960s, the United States, the former Soviet Union (and subsequently Russia), the European Space Agency, the Japan Aerospace Exploration Agency, and other countries began to launch spacecraft into orbit and throughout the Solar System, designed to study phenomena in new wavelength regions. Certainly the two most famous orbiting observatories are the Hubble Space Telescope (Fig. 15.14) and the James Web Space Telescope. For more than three decades, the Hubble Space Telescope, with its 2.4-m-diameter mirror, has been an orbiting workhorse in the visible and near ultraviolet wavelength portions of the electromagnetic spectrum, and the James Web Space Telescope, with its 6.5-m-diameter mirror, was launched in late 2021, and operates at infrared wavelengths. But HST and JWST are far from the only orbiting observatories that have had tremendous success. There are far too many important orbiting observatories to list, but the data collected by many of them will be discussed throughout the remainder of this textbook.

One of the driving forces in the explosion of new knowledge in astronomy comes from revolutionary new techniques, observatories, and advances in electronic digital detectors attached to telescopes, both on the ground and in orbit. Another driver is the rapid advancement of computational power for data analysis and theoretical modeling. As a result, the oldest of the natural sciences is today one of the most rapidly advancing, producing on an almost daily basis new discoveries that not only thrill astronomers, but fascinate the general population as well.

Observational astronomy is indeed a very precise and complex science.

15.2   The Brightness of Stars

The Magnitude Scale

In Section 3.2 we discussed the work of Hipparchus of Nicaea, arguably one of the greatest astronomers of antiquity. One of his enduring contributions is the magnitude scale. More than 2100 years later we are still using his basic system: the brighter the star, the smaller the value of the star's magnitude. As Hipparchus first defined his system, in a dark sky (away from the light pollution produced by humans), those stars that were just barely visible to the naked eye were classified as magnitude 6 and the brightest stars in his catalog were classified as magnitude 1. The magnitudes of some stars in the region of Orion are shown in Fig. 15.15.

Today's modern version of the Hipparchus magnitude scale has been defined to be more precise by relating magnitudes to stellar brightnesses. Although he didn't realize it, when Hipparchus established the magnitude system his scale was based on how the human eye and brain interpret the amount of light that the eye receives from the stars. In reality, we *perceive* brightness in a way that is logarithmic (a term

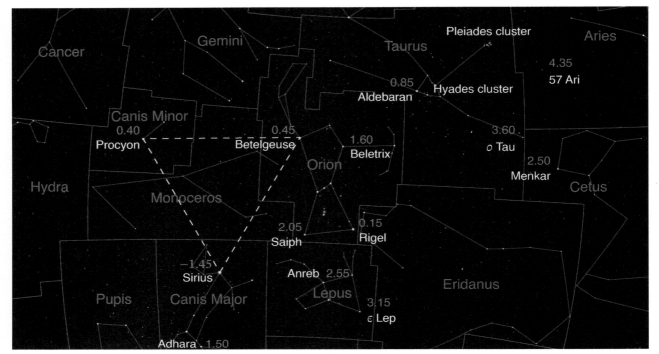

Fig. 15.15 The region of the sky in and around Orion. The magnitudes of some of the stars are indicated. Constellation names are in blue, and star and star cluster names are in white. The brightest star shown is Sirius at −1.45 magnitudes (near lower left in Canis Major) and the dimmest star labeled is 57 Aries at 4.35 magnitudes near upper right. Procyon, Betelgeuse, and Sirius comprise the "winter triangle" (dashed yellow). (Adapted from a *Stellarium* screenshot; GFDL)

derived from the mathematical function known as a logarithm). This means that if one source of light is twice as bright as another, we don't perceive it as being twice as bright, but rather only slightly brighter.

The evolutionary explanation for this behavior of the human eye is simple; if we really did perceive something that is twice as bright as being twice as bright, the brightest objects in our view would so overwhelm the dimmer ones that we wouldn't be able to discern them. In effect, the logarithmic behavior of the eye gives us the ability to see a very wide range of brightnesses. Our hearing behaves in much the same way, allowing us to perceive very faint sounds and still tolerate very loud sounds as well. The human ear is sensitive to a range of sound intensities of more than 1 trillion without too much discomfort (the decibel scale behaves much like the magnitude scale).

Although an amazing instrument, the human eye is not ideally suited for scientific measurements. Modern detectors, like the CCD in your digital camera, first discussed on page 288, actually count the number of photons arriving at the detector (see for example, Fig. 15.16). As a result, if one star is twice as bright as another one, twice as many photons will be counted in the same period of time.

To translate between what the human eye sees, and what a modern detector actually measures, astronomers have modernized the magnitude scale to reflect those different types of measurement. It turns out that, in Hipparchus's original

Fig. 15.16 The Hubble Space Telescope's **Wide Field Camera 3** onboard the Hubble Space Telescope uses CCD technology similar to digital cameras. The instrument is located in one of the bays of the telescope. The remaining bays are occupied by other detectors. (NASA/Amanda Diller)

magnitude scale, if one star is classified as magnitude 1 and another star is magnitude 6 (a difference of 5 magnitudes), then the magnitude-1 star is about 100 times brighter than the magnitude-6 star. To quantify that rough correspondence, the relationship has been made exact. Expressed mathematically,

$$\text{if} \quad \frac{b_1}{b_2} = 100 \quad \text{then} \quad m_2 - m_1 = 5. \tag{15.1}$$

Magnitude differences as brightness ratios

Key Point
Astronomy defines the magnitude system to correspond to the natural response of the human eye.

where b_1 and b_2 are the brightnesses of stars 1 and 2 associated with the number of photons per second received at the CCD detector, and m_1 and m_2 are their magnitudes. In words, the equation can be read as: "If dividing the brightness of star 1 by the brightness of star 2 equals 100 (star 1 is 100 times brighter than star 2) then the magnitude of star 2 minus the magnitude of star 1 equals 5 magnitudes."

Another peculiar aspect of Hipparchus's magnitude scale was also retained in our precise, modern relationship between brightness ratios and magnitude differences, namely that the *brighter* a star is, the *smaller* its magnitude. For example, Sirius, located in the constellation Canis Major and sometimes referred to as the Dog Star, is the brightest star in the night sky, with a magnitude of −1.45 (a negative value). It is also still the case that those stars which are just barely visible to the naked eye in a very dark night sky are classified as magnitude 6, just as Hipparchus originally classified them. However, much dimmer magnitudes are also possible. The dimmest object yet detected is an extremely distant galaxy that is fainter than magnitude +31.

Key Point
The brighter a star is, the smaller the value of its magnitude. The brightest stars have negative magnitudes.

Table 15.1 The relationship between magnitude differences and brightness ratios

$m_2 - m_1$	b_1/b_2		Standard notation	Scientific notation
0	$2.512^0 =$	$1 =$	$1 =$	1
1	$2.512^1 =$	$2.512 =$	$2.512 =$	2.512
2	$2.512^2 =$	$2.512 \times 2.512 =$	$6.310 =$	6.310
3	$2.512^3 =$	$2.512 \times 2.512 \times 2.512 =$	$15.85 =$	1.585×10^1
4	$2.512^4 =$	$2.512 \times 2.512 \times 2.512 \times 2.512 =$	$39.81 =$	3.981×10^1
5	$2.512^5 = 2.512 \times 2.512 \times 2.512 \times 2.512 \times 2.512 =$		$100 =$	1×10^2
6	$2.512^6 =$	$2.512^1 \times 2.512^5 =$	$2.512 \times 100 =$	2.512×10^2
7	$2.512^7 =$	$2.512^2 \times 2.512^5 =$	$6.310 \times 100 =$	6.310×10^2
8	$2.512^8 =$	$2.512^3 \times 2.512^5 =$	$15.85 \times 100 =$	1.585×10^3
9	$2.512^9 =$	$2.512^4 \times 2.512^5 =$	$39.81 \times 100 =$	3.981×10^3
10	$2.512^{10} =$	$2.512^5 \times 2.512^5 =$	$100 \times 100 =$	1×10^4
15	$2.512^{15} =$	$2.512^5 \times 2.512^5 \times 2.512^5 =$	$100 \times 100 \times 100 =$	1×10^6
20	$2.512^{20} =$	$2.512^5 \times 2.512^5 \times 2.512^5 \times 2.512^5 =$	$100 \times 100 \times 100 \times 100 =$	1×10^8

An exponents tutorial is available at
Math_Review/Exponents

So how do we apply this relationship to the difference in magnitude for any two stars? The answer lies in first determining the *brightness ratio*, b_1/b_2, when the *magnitude difference*, $m_2 - m_1$, is 1. That ratio is approximately 2.512, which is the fifth root of 100, or $100^{1/5} = 2.512$: for every change in magnitude of 1 the brightness changes by a factor of 2.512. This can be verified by realizing that

$$2.512 \times 2.512 \times 2.512 \times 2.512 \times 2.512 = 2.512^5 = 100.$$

In general, the equation that produces the brightness ratio for any difference in magnitudes is

A brightness ratio–magnitude difference equation and Table 15.1 tutorial is available from the Chapter 15 resources page

$$\frac{b_1}{b_2} = 2.512^{(m_2 - m_1)}. \tag{15.2}$$

Brightness ratio–magnitude difference equation

If the difference in magnitude between star 2 and star 1 is $m_2 - m_1$, then star 2 appears brighter than star 1 by a factor of 2.512 raised to the difference in magnitude.

Table 15.1 shows the magnitude differences and brightness ratios for a number of cases. However, it is important to note that, in Table 15.1, for each case it is assumed that star 1 appears brighter than star 2, and so the difference in magnitude is always positive (don't forget that the dimmer star has the larger value for its magnitude). This certainly doesn't need to be the case, of course; if star 1 is fainter than star 2, then the exponent in the brightness ratio–magnitude difference equation becomes negative and b_1/b_2 is less than 1.

Key Point
Apparent magnitudes are observed from Earth.

Since the magnitudes we have discussed so far correspond to how bright objects *appear to be* from Earth, these magnitudes are termed apparent magnitudes. The magnitudes that are indicated in Fig. 15.15 for stars in the region of Orion, Canis Major, and Canis Minor, are apparent magnitudes.

Example 15.1

Betelgeuse in Orion (Fig. 15.15) has an apparent magnitude of 0.45, while the star that is closest to Earth, Proxima Centauri, has an apparent magnitude of about 11. How many times brighter does Betelgeuse appear to be when compared with Proxima Centauri as seen from Earth?

The question asks you to determine how many times brighter Betelgeuse is relative to Proxima Centauri. This means that Betelgeuse should be represented by b_1 and m_1 in the brightness ratio–magnitude difference equation and Table 15.1, and Proxima Centauri should be represented by b_2 and m_2. As a result,

$$m_2 - m_1 = m_{\text{Proxima Centauri}} - m_{\text{Betelgeuse}} = 11 - 0.45 = 10.55.$$

Now that you know the magnitude difference between the two stars you can use the brightness ratio–magnitude difference equation to find your answer. In this case

$$\frac{b_1}{b_2} = \frac{b_{\text{Betelgeuse}}}{b_{\text{Proxima Centauri}}} = 2.512^{10.55}.$$

You can do the calculation by using the exponential function on your calculator, typically labeled x^y, with the result

$$\frac{b_{\text{Betelgeuse}}}{b_{\text{Proxima Centauri}}} = 16,600.$$

Betelgeuse is about 16,600 times brighter than the very faint Proxima Centauri as seen from Earth.

You can at least verify that your answer is reasonable by checking it against Table 15.1. Notice that, for a magnitude difference of $m_2 - m_1 = 10$, $b_2/b_1 = 2.512^{10} = 10,000$ or 1×10^4. For $m_2 - m_1 = 11$, $b_2/b_1 = 2.512^{11} = 2.512^1 \times 2.512^{10} = 25,120$. This means that a magnitude difference of $m_2 - m_1 = 10.55$ giving a brightness ratio of $b_1/b_2 = 16,600$ is between the brightness ratios for $m_2 - m_1 = 10$ and $m_2 - m_1 = 11$.

A scientific notation tutorial is available at
`Math_Review/Scientific_Notation`

The Inverse Square Law of Light and Magnitude Differences

Although the brightness ratio–magnitude difference equation and Table 15.1 describe how we move back and forth between ways of describing the *apparent brightness* of an object in the sky, that is only one part of the story. It would be much more useful to our *understanding* of what we are seeing if we can translate that information into what is actually happening at the source of the light. For example, why does Vega appear to be so much brighter than Proxima Centauri if Proxima Centauri is really the star closest to Earth? You can probably guess that Vega produces much more light than Proxima Centauri does, which is indeed the case. However, when comparing any two stars, in order to determine which star is intrinsically brighter than the another, we must first develop a way to truly compare apples to apples. In other words, we must be able to compensate for the familiar dimming of a light source with increasing distance from the observer as described by the inverse square law from page 223.

Example 15.2

Neptune (Fig. 15.17) is approximately 30 au from the Sun. The Sun's apparent magnitude as seen from Earth is about −27 (the more negative the magnitude of a light source is, the brighter that source is; the Sun appears as *very bright* from Earth). What is the Sun's *approximate* apparent magnitude as seen from Neptune?

From Example 6.15 we learned that the Sun appears to be 900 times less bright than it does from Earth ($30^2 = 900$). In order to determine the apparent magnitude of the Sun as seen from Neptune, go back to Table 15.1. In the b_1/b_2 column you can see that if $b_1/b_2 = 631.0$ then the difference in magnitudes is 7, and if the ratio, b_1/b_2 is 1585 then the difference in magnitudes is 8. This means that we have found magnitude differences that bracket the brightness ratio of 900. Therefore, the difference in the apparent magnitudes of the Sun as seen from Neptune and from Earth must be between 7 and 8 magnitudes. Since an increasing value for the magnitude corresponds to a dimmer object, the apparent magnitude of the Sun as seen from Neptune must be somewhere between $-27 + 8 = -19$ and $-27 + 7 = -20$. If you play around with the brightness ratio–magnitude difference equation directly by trying different values for $m_2 - m_1$ you can find out quickly that a magnitude difference of 7.4 gives a brightness ratio that is close to 900.

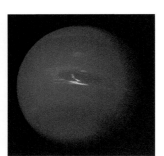

Fig. 15.17 Neptune as seen from Voyager 2 in 1989. (NASA/JPL)

Thanks to the brightness ratio–magnitude difference equation we now know how to go back and forth between apparent brightness and apparent magnitude. Furthermore, thanks to the inverse square law equation, we also know how to go back and forth between apparent brightness and luminosity *assuming that we know the distance to the star.*

15.3 How Far Away are they?

In Chapter 3, along with the discussion of Hipparchus of Nicaea and his magnitude scale, there was also a discussion of Aristarchus of Samos, the first person known to ever develop a heliocentric model of the universe. The reasons for the lack of acceptance of his model were simple: it seemed preposterous that Earth was actually moving, and there was simply no evidence of parallax; the apparent shifts in positions of stars in the sky as Earth moves around the Sun.

As we know today, the essential idea of Aristarchus regarding heliocentrism was correct, preceding the work of Nicolaus Copernicus (1473–1543) by nearly sixteen hundred years. The suggestion that the stars are simply too far away to measure parallax with the human eye was correct as well. In fact, the first time that stellar parallax was measured didn't occur until 1838, when Friedrich Bessel (1784–1846) was able to observe the apparent annual wobble of the star 61 Cygni, almost three hundred years *after* Copernicus proposed his heliocentric model.

Stellar Parallax

Even today, determining distances to celestial objects remains one of the most important and difficult problems in all of astronomy. Given the enormous distances involved, astronomers have had to become very creative. In Section 6.1 you learned that distances within the Solar System were first calculated by using geometric methods to ascertain the distances to Mars and Venus. Once the distance to one of the planets was determined, geometry led to the distances of each of the other planets from the Sun, including Earth. This Earth–Sun distance gave us the value of the astronomical unit (au). Today, various very precise measurements yield distances within the Solar System. For example,

- measurements of the distance to the Moon are made by timing the round-trip time of laser light bounced off of retroreflectors (Fig. 15.18) that were left on the lunar surface by the Apollo astronauts,
- the distance to Venus has been obtained by the reflection of radar signals off the surface of that planet,
- timing signals sent back to Earth by spacecraft studying the planets, their moons, and other objects in the Solar System,
- and geometry based on very precise angle measurements still provides important distance determinations.

The accurate determination of the astronomical unit represents the first rung on what is referred to as the distance ladder. Because light-travel time measurements became so accurate, and because the Earth–Sun distance is never constant, even when taking into consideration Earth's elliptical orbit, the au is now a defined quantity (recall the discussion on page 189).

The second rung of the distance ladder comes from stellar parallax measurements, as depicted in Fig. 15.19 (also recall Fig. 3.20). Using parallax to directly determine distances to nearby stars depends only on knowledge of the astronomical

Fig. 15.18 A retroreflector placed on the lunar surface in 1971 (Dave Scott, NASA)

Key Point
The distance ladder is the process of determining successively greater distances using methods that rely on shorter distances determined by other techniques.

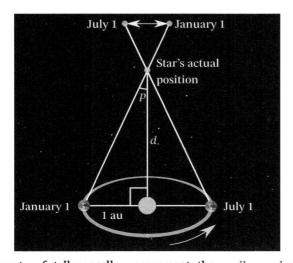

Fig. 15.19 The geometry of stellar parallax. p represents the parallax angle and the box inside the triangle at the Sun's location indicates the right angle (90°). The dimensions of the right triangle are not to scale.

unit and the measurement of the parallax angle, p. The parallax angle is *one-half* of the maximum apparent shift in the position of the nearby star, relative to a background of much more distant objects when observed from Earth during our orbit around the Sun. The parallax angle leads directly to the distance, d, of the star. As you can see from the figure, a right triangle is formed by the nearby star being observed, the Sun, and Earth, with the 90° angle located at the Sun. Obviously, Fig. 15.19 is not drawn to scale; instead, the distance, d, from the Sun to the star is very much greater than 1 au and the parallax angle, p, is extremely small. If the star in the figure represents Proxima Centauri, the Sun's closest stellar neighbor, and the distance between the Sun and Earth in the figure represents 1 au, then the figure would need to be stretched vertically until d measures 4.5 km (2.8 mi)!

Given the very small angles that are measured when determining distances to nearby stars, it is customary to represent the parallax angle in arcseconds, symbolized by p_{arcsec}. To give you a sense of how small one arcsecond is, the angular diameter of a full moon is approximately one-half of one degree (30 arcminutes, or 30′). You would need to subdivide the apparent angular width of the full moon into 1800 equally thin slivers before just one of those slivers would have a width of one arcsecond. Not even Proxima Centauri is close enough to have a parallax angle that is that large; its parallax angle is just 0.77″ (0.77 arcseconds), while every other star in the night sky has a parallax angle that is even smaller (most are much smaller). It's no wonder that it took until 1838 before anyone was able to measure such a small angle and definitively prove that Earth does in fact orbit the Sun.

Using the properties of a right triangle and the fact that all parallax angles are extremely small, the distance d from the Sun to the star can be written as

A parallax equation tutorial is available from the `part_3/chapter_15` resources page

$$d_{pc} = \frac{1}{p_{arcsec}} \text{ pc,} \qquad (15.3)$$

Parallax equation

where p_{arcsec} **must be given in units of arcseconds and the distance to the star,** d_{pc}**, is in units of parsecs (pc).** The term "parsec" is actually a concatenated word that comes from **par**allax **sec**ond, indicating how the distance is determined. The parallax equation can be read as "the distance to a star in units of parsecs is equal to 1 divided by the parallax angle in units of arcseconds."

Given that there are 2π radians[3] in one complete circle of 360° and that there are 3600 arcseconds in one degree, there are $(360°/2\pi \text{ radian}) \times (3600″/1°) = 206{,}265$ arcseconds per radian. This leads to the fact that one parsec equals 206,265 astronomical units; 1 pc = 206,265 au.

A units conversion tutorial is available at `Math_Review/Unit_Conversions`

Example 15.3

Proxima Centauri has a parallax angle of $p_{arcsec} = 0.77″$ and Betelgeuse has a parallax angle of $p_{arcsec} = 0.00595″$. (a) How far away is Proxima Centauri in parsecs and in astronomical units? (b) Do the same calculations for Betelgeuse.

(a) To determine the distance to Proxima Centauri, you can use the parallax

---

[3]The *radian* is the natural mathematical unit of angular measure.

equation to find that

$$d_{pc} = \frac{1}{p_{arcsec}} \text{ pc} = \frac{1}{0.77''} \text{ pc} = 1.3 \text{ pc}.$$

Since there are 206,265 au in one parsec, the distance to Proxima Centauri in astronomical units is approximately

$$1.3 \text{ pc} \times \frac{206{,}265 \text{ au}}{1 \text{ pc}} = 268{,}000 \text{ au}.$$

It is very important that you truly understand what the numbers are saying in any calculation. In this case, the answer tells us that Proxima Centauri is about 268,000 times farther from the Sun than Earth is. For comparison, Pluto is, on average, only about 39 au from the Sun, or about one-seven-thousandth of the distance between the Sun and Proxima Centauri.

(b) You may have noticed that the parallax angle for Betelgeuse is less than 1/100 of Proxima Centauri's parallax angle, which means that Betelgeuse is more than 100 times farther from the Sun than Proxima Centauri is. Carrying out the same calculations for Betelgeuse, you will find that $d_{pc} = 168$ pc $= 34{,}670{,}000$ au. Although Betelgeuse is more than one hundred times farther away, it is still much, much brighter.

It is also possible to relate parsecs to the more familiar unit of **light-years (ly)**. However, before moving on it is important to point out that there is a common misunderstanding about what one light-year represents; it is *not* a unit of time, it is a unit of distance, specifically the distance that light can travel in one year. Similarly, one light-minute is the distance that light can travel in one minute, one light-second is the distance that light travels in one second, and so on.

To find out how many light-years are in one parsec, start by determining how long it takes light to travel from the Sun to Earth (1 au). Since the speed of light is the universal constant $c = 3 \times 10^8$ m/s and 1 au $= 1.5 \times 10^{11}$ m, the time required for light to travel one astronomical unit is

$$\text{time} = \frac{\text{distance}}{\text{speed}} = \frac{1.5 \times 10^{11} \text{ m}}{3 \times 10^8 \text{ m/s}} = 500 \text{ s};$$

light takes 500 seconds to travel from the Sun to Earth, or 1 au $= 500$ light-seconds. Since there are 60 seconds in one minute, 1 au also equals 8.3 light-minutes. Finally, with one year having 526,000 minutes, one parsec equals

$$1 \text{ pc} \times \frac{206{,}265 \text{ au}}{1 \text{ pc}} \times \frac{8.3 \text{ light-minutes}}{1 \text{ au}} \times \frac{1 \text{ ly}}{526{,}000 \text{ light-minutes}} = 3.26 \text{ ly};$$

light requires 3.26 years to travel a distance of one parsec.

Listing all three of these important units of distance together:

$$1 \text{ pc} = 3.26 \text{ ly} = 206{,}265 \text{ au.} \tag{15.4}$$

Relationships between astronomical lengths

Key Point
The light-year *does not* refer to an interval of time; it is a unit of length.

A scientific notation tutorial is available at `Math_Review/Scientific_Notation`

Key Point
The relationship between parsecs, light-years, and astronomical units.

Although we will primarily use light-years as the unit of choice for large distances, the parsec is typically used by astronomers because of its direct tie to the parallax method for determining distances. Occasionally, we will also refer to the parsec because of its natural definition.

Example 15.4

What is the distance to Proxima Centauri in units of light-years?

By converting from parsecs in the previous example to light-years

$$1.3 \, \text{pc} \times \frac{3.26 \, \text{ly}}{1 \, \text{pc}} = 1.3 \times 3.26 \, \text{ly} = 4.2 \, \text{ly}.$$

Light takes 4.2 years to travel from Proxima Centauri to Earth. Therefore, when you look at Proxima Centauri through a telescope you are not seeing the star as it is today, but instead, you are seeing it as it was when it emitted the light that took 4.2 years to reach us. You are literally looking into the past, seeing Proxima Centauri 4.2 years ago!

Because of the finite speed of light, the universe provides us with a natural time portal to the past (Fig. 15.20). We may not be able to go into the past ourselves, but the farther from Earth we look, the farther back in time we are looking. Said another way, the farther away something is, the younger it appears to be and the younger the universe appears to be. This simple observational reality will play a huge role when we explore how stars, galaxies, and the universe itself age over time.

Fig. 15.20 An astronomical observatory is a time portal to the past. [Background image by NASA, ESA, and J. Lotz (STScI) and the HFF team]

The method of stellar parallax is limited by our ability to measure very tiny parallax angles. However, advances in technology have allowed for the measurement of ever-smaller angles. Launched by the European Space Agency in December 2013, the Gaia mission (Fig. 15.21) is determining the distances, positions, and motions for nearly 2 billion stars down to 20.7th magnitude (about 1% of all stars in the Milky Way Galaxy). To accomplish this amazing feat, Gaia is theoretically capable of measuring parallax angles as small as 10 *micro*-arcseconds (0.000 01″) for the brightest sources, corresponding to distances as great as $d_{\text{pc}} = 1/0.000\,01″ = 100{,}000$ pc or $100{,}000 \, \text{pc} \times 3.26 \, \text{ly}/1 \, \text{pc} = 326{,}000$ ly! By way of comparison, the diameter of the Milky Way Galaxy's disk is about 100,000 ly. Combining the distance to the stars with their directions in the sky (right ascension and declination), a three-dimensional map of our Galaxy can be constructed. During its mission, Gaia is also obtaining other valuable scientific information about the objects it measures and their environments in space.

Fig. 15.21 An artist's impression of Gaia. (© ESA–D. Ducros, 2013)

To give you a better sense of the remarkable capabilities of the Gaia spacecraft, refer back to Fig. 15.19. For Proxima Centauri we estimated that the figure would need to be stretched 4.5 km to represent the parallax angle correctly. Given Gaia's parallax angle limit of 10 microarcseconds compared to Proxima Centauri's parallax of 0.7 arcseconds or 700,000 microarcseconds, Fig. 15.19 would need to be stretched vertically by more than 315,000 km (almost 200,000 mi), which is a distance more than 4/5 (80%) of the way from Earth to the Moon.

The Relationship Between Absolute Magnitude and Luminosity

Although luminosity (L) is the most basic way of describing how much energy is given off by a star every second, in order to have a parallel terminology with apparent magnitude, astronomers have traditionally defined the absolute magnitude. The absolute magnitude is what the apparent magnitude *would be* if the star were located 10 pc or 32.6 ly from Earth. You can think of the absolute magnitude as a hypothetical, mathematical way of lining all stars up at the same distance and comparing them side by side. Knowing the absolute magnitude of a star is essentially the same thing as knowing its luminosity, because both the absolute magnitude and the luminosity provide the same intrinsic information about the star, just in different units of measurement. (By the way, there is nothing special about using a distance of 10 pc as the standard distance, it is just a convenient round number.)

To calculate the absolute magnitude of a star, we need to

1. determine the distance to the star,
2. use the inverse square law for light (page 223) to calculate the ratio in brightnesses for the star based on its actual brightness and the theoretical brightness that it would have if it was 10 pc from Earth,
3. use the brightness ratio–magnitude difference equation to determine how much the magnitude would change if the star were moved to 10 pc, and
4. use the difference in magnitude to change from the apparent magnitude to the absolute magnitude.

Example 15.5

As we learned in Examples 15.1 and 15.3, Proxima Centauri is located just 1.3 pc from Earth. Proxima Centauri is also very faint, with an apparent magnitude of only 11, making it much dimmer than the human eye can detect. On the other hand, Betelgeuse is one of the brightest stars in the sky and is 168 pc from Earth, almost 130 times farther from Earth than Proxima Centauri, and it has an apparent magnitude of 0.45.

(a) Do you expect the absolute magnitude of Proxima Centauri to be greater than 11 (meaning fainter) or less than 11, and why?

(b) Using the inverse square law on page 223, how many times brighter or dimmer would Proxima Centauri be if it was moved (mathematically of course) from 1.3 pc to 10 pc?

(c) By using the brightness ratio–magnitude difference equation on page 602, or Table 15.1, estimate the difference in magnitudes of Proxima Centauri when it is at 1.3 pc compared to what it would be at 10 pc.

(d) Estimate the absolute magnitude of Proxima Centauri.

Key Point
Reminder: Be an active reader; don't just read the example, you should actively work through it.

(e) Carry out the same calculations for Betelgeuse.

(f) Which star is intrinsically brighter, Betelgeuse or Proxima Centauri?

(g) Finally, from the magnitude difference–brightness ratio equation or Table 15.1, how many times more luminous is the intrinsically brighter star than the less bright star?

(a) Proxima Centauri is closer to Earth than 10 pc, so if it could be moved out to 10 pc it would appear *dimmer*. As a result, the value of the absolute magnitude should be *greater* than 11.

(b) Since Proxima Centauri is 1.3 pc from Earth, based on the inverse square law of light, moving Proxima Centauri out to 10 pc would make it

$$\frac{b_1}{b_2} = \frac{\cancel{L}/4\pi r_1^2}{\cancel{L}/4\pi r_2^2} = \frac{1/r_1^2}{1/r_2^2} = \frac{r_2^2}{r_1^2} = \frac{1.3^2}{10^2} = \frac{1.69}{100} = 0.0169 = 1/59$$

times less bright, where b_1 refers to the brightness at $r_1 = 10$ pc and L is the star's intrinsic luminosity (its power output), which is the same no matter how far away the star is.

(c) From Table 15.1, a brightness ratio of 59 lies between magnitude differences of 4 and 5. By experimenting with the the brightness ratio–magnitude difference equation, a brightness ratio of 59 corresponds to a magnitude difference of about 4.4; $2.512^{4.4} = 59$.

(d) Since Proxima Centauri would be dimmer at 10 pc, this means that its magnitude would be larger (remember larger is dimmer). Therefore the absolute magnitude of Proxima Centauri is $11 + 4.4 = 15.4$.

(e) Going through the same calculations for Betelgeuse at 168 pc gives a brightness ratio of $b_1/b_2 = 0.01/0.000\,0354 = 182$ and a difference in magnitudes of 5.6. Since Betelgeuse is farther from Earth than 10 pc its absolute magnitude must be smaller than its apparent magnitude, giving $0.45 - 5.6 = -5.1$.

(f) With its smaller absolute magnitude Betelgeuse is intrinsically brighter than Proxima Centauri.

(g) The difference in absolute magnitudes of Betelgeuse and Proxima Centauri is 20.5. From Table 15.1, a difference of 20 magnitudes gives a brightness ratio of 100 million, while a brightness ratio of 21 is 2.152×10^8, or 251 million. More accurately $2.512^{20.5} = 158$ million; Betelgeuse gives off more than 158 million times as much energy every second than Proxima Centauri does. In other words, $L_{\text{Betelgeuse}} / L_{\text{Proxima Centauri}} = 158{,}000{,}000 = 1.58 \times 10^8$. Clearly these must be very different stars.

Note that the brightness ratio–magnitude difference equation and Table 15.1 work exactly the same way for differences in absolute magnitudes as they do for differences in apparent magnitudes. In addition, when absolute magnitudes are used, the ratio of brightnesses becomes exactly equal to the ratio of luminosities, because the common distance of 10 pc cancels out in the brightness ratio. This means that the brightness ratio–magnitude difference equation can be written in terms of luminosities and absolute magnitudes as

$$\frac{L_1}{L_2} = 2.512^{(M_2 - M_1)}. \tag{15.5}$$

Luminosity ratio–absolute magnitude difference equation

Although there has been some mathematics in Section 15.3, the important take-aways are:

(a) It is possible to determine distances to nearby stars using the geometry of right triangles (triangles that contain one 90° angle); a process known as stellar parallax.

(b) Once the distance to a star is known, its absolute magnitude can be calculated, which corresponds to critical knowledge regarding the star's luminosity (power output).

In the next three sections you will learn how astronomers determine three other important parameters of stars: their masses, their surface temperatures, and their radii. Armed with these data, astronomers have uncovered patterns that have illuminated the road to a deep understanding of how stars are formed, live out their lives, and die.

A luminosity ratio–absolute magnitude difference equation tutorial is available from the Chapter 15 resources page

15.4 Determining Stellar Masses in Binary Star Systems

Back in Section 5.4, you learned that Newton was able to derive Kepler's laws by using his three laws of motion and his universal law of gravitation. Although Kepler originally developed his "laws" for planets orbiting our Sun, Newton's more general derivations of Kepler's laws apply to any two objects in orbit about one another, whether the objects are planets orbiting the Sun, moons orbiting planets, the orbits of comets around the Sun, stars orbiting each other, or even two galaxies orbiting each other. The complete set of Kepler's laws as developed by Newton are listed on page 175.

Key Point
The sum of the masses of two stars orbiting one another can be determined by knowing the average distance between the stars and their orbital period.

The Sum of the Stars' Masses

Newton's derivation of Kepler's third law plays a critical role in astronomy. If the orbital period (P) of the stars in a two-star (binary) system is known, combined with the average distance between the two stars (a), it is straightforward to calculate the sum of their masses. In fact, this is the *only direct way* to measure the masses of stars. By dividing both sides of Kepler's third law by the orbital period squared (P^2) we find

A tutorial for using Kepler's total mass equation is available from the Chapter 15 resources page

$$m_{\text{total, Sun}} = m_{1,\text{Sun}} + m_{2,\text{Sun}} = \frac{a_{\text{au}}^3}{P_{\text{y}}^2}, \tag{15.6}$$

Kepler's total mass equation

where the subscript "Sun" for m_1 and m_2 reminds us that the masses must be in units of the mass of the Sun [solar masses (M_{Sun})], and the subscripts "au" and "y" for a and P, respectively, remind us that the average distance between the stars must be expressed in astronomical units (au) and the orbital period must be in years (y) for this form of Kepler's third law. Kepler's total mass equation can be read as "The sum of the masses (in solar masses) of two objects orbiting one another is equal to the average distance (in astronomical units) between the objects cubed divided by their orbital period (in years) squared."

Fig. 15.22 Sirius B is the very faint star seen to the lower left of Sirius A. [© NASA/ESA/H. Bond (STScI) and M. Barstow (Univ. Leicester)]

Example 15.6

Sirius, in Canis Major (the Big Dog), is the brightest star in the night sky (recall Fig. 15.15). In 1844, after ten years of observation, Friedrich Bessel discovered that Sirius is actually a binary star system. The brightest star in the system, Sirius A, is known as the "Dog Star" and its very dim companion is Sirius B, the "Pup" (see Fig. 15.22). Although Sirius B could not be seen in Bessel's day, he was able to deduce its existence through the wobble of Sirius A that was caused by the gravitational force exerted on it by the "Pup." Today, the orbital period of the Sirius system is known to be 50.13 y and the average distance between them (their combined semimajor axis) is 19.78 au. Determine the sum of the masses of the two stars.

From Kepler's total mass equation

$$m_{1,\,\text{Sun}} + m_{2,\,\text{Sun}} = \frac{a_{\text{au}}^3}{P_{\text{y}}^2} = \frac{(19.78\ \text{au})^3}{(50.13\ \text{y})^2} = 3.08\ \text{M}_{\text{Sun}}.$$

The sum of the masses of Sirius A and Sirius B is 3.08 M_{Sun} (3.08 times the mass of the Sun).

Note that since Kepler's total mass equation uses a different system of units (au, y, and M_{Sun}, rather than m, s, and kg), the typical cancellation of units doesn't work. This is because the equation directly compares everything to the orbit of Earth around the Sun. As a result, you simply need to understand how the units are used together.

Determining the Individual Masses of Stars

Now that it is possible to determine the sum of the masses of the stars in a binary star system, the question naturally arises: "How can we find the mass of each star separately?" The answer relates to the concept of the center of mass that we first considered on page 174.

Fig. 15.23 A father and his daughter playing on a teeter-totter

Center of mass

Consider a father and daughter on a teeter-totter (also known as a seesaw) as depicted in Fig. 15.23. In order for the game to be fair, the fulcrum (the triangular balance point) must be positioned closer to the father, giving the daughter more leverage. If the fulcrum is at just the right location, the father and daughter can sit balanced on the beam because their weights are exerting equal and opposite torques (page 195). If the fulcrum had been placed exactly in the middle between the father and his daughter, the girl would get frustrated very quickly because she would be stuck in the air and her father would be resting with his end of the beam sitting on the ground.

Now study Fig. 15.24. In this case the father has been replaced by a block of mass $m_{\text{father}} = 80$ kg and the daughter is represented by a block of mass $m_{\text{daughter}} = 20$ kg. Note that the total mass of the father and the daughter is 100 kg, so the father has 80 kg/100 kg = 4/5 of the total mass and his daughter has 20 kg/100 kg = 1/5 of the total mass. If the beam is 5 m long, then in order for the 80 kg father and the 20 kg daughter to be balanced, the fulcrum must be placed $a_{\text{father}} = 1$ m from the father and $a_{\text{daughter}} = 4$ m from the daughter. This means that the father, who has

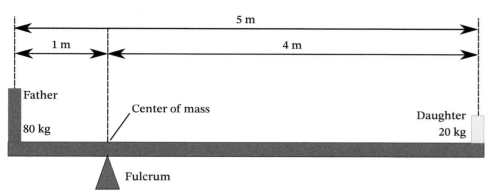

Fig. 15.24 The father and daughter from Fig. 15.23 represented as blocks of 80 kg and 20 kg, respectively, on a 5 m long teeter-totter beam. The two are exactly in balance.

$4/5$ of the mass, gets $1/5$ of the beam and his daughter, with $1/5$ of the mass, gets $4/5$ of the beam. This can also be illustrated mathematically by realizing that when you multiply the father's mass by the length of his portion of the beam you get

$$m_{father}\, a_{father} = (80 \text{ kg}) \times (1 \text{ m}) = 80 \text{ kg m}.$$

Similarly, for the girl you find that

$$m_{daughter}\, a_{daughter} = (20 \text{ kg}) \times (4 \text{ m}) = 80 \text{ kg m}.$$

In other words, the father's mass times his distance from the fulcrum equals the daughter's mass times her distance from the fulcrum.

More generally, for any two masses m_1 and m_2, it is the case that m_1 times its distance from the balance point, a_1, equals m_2 times its distance from the balance point, a_2, or

$$m_1\, a_1 = m_2\, a_2. \tag{15.7}$$

Center of mass equation

The center of mass equation gets its name from the location of the fulcrum on the teeter-totter beam when it is balanced, which is the center of mass between m_1 and m_2. (An aside: It is common in many physics textbooks to refer to this point as the "center of gravity" because technically the weights of the masses are being balanced due to the forces from Earth's gravitational pull on those masses. However, the phrase "center of mass" is both more general and more appropriate in astronomical situations.)

Finally, imagine that the two masses, m_1 and m_2, are not people on a teeter-totter, but instead are stars orbiting one another. Just as in the father–daughter discussion above, the position of the center of mass of the two stars is located along a line between the two stars such that $m_1\, a_1 = m_2\, a_2$ (center of mass equation). In addition, each star makes its own elliptical orbit around their mutual center of mass, and the distances a_1 and a_2 turn out to be the semimajor axes (average distances) of those

A tutorial for Fig. 15.24 and the center-of-mass equation is available from the Chapter 15 resources page

Key Points
• Stars orbit around their common center of mass.
• The value of a that goes into Kepler's third law for a binary system is the sum of the semimajor axes of the sytem's two components as they orbit around their common center of mass.

elliptical orbits from the center of mass (the center of mass is the principal focus of the elliptical orbits; recall the geometry of an elliptical orbit in Fig. 4.43. The value of a that is in Kepler's third law and Kepler's total mass equation is the sum of a_1 and a_2, or

$$a = a_1 + a_2. \tag{15.8}$$

Binary star separation equation

By combining Kepler's total mass equation with the center of mass equation and the binary star separation equation, it is now possible to determine the mass of each star in a binary star system rather than just their total mass. Doing a little algebra results in the pair of equations,

$$m_{1,\,\text{Sun}} = \left(\frac{a_2}{a}\right) \times m_{\text{total, Sun}}. \tag{15.9}$$

$$m_{2,\,\text{Sun}} = \left(\frac{a_1}{a}\right) \times m_{\text{total, Sun}}. \tag{15.10}$$

Equations for the individual masses in a binary star system

The mass of one star in a binary star system equals the distance of the other star from the center of mass divided by the total separation between the two stars, then multiplied by the total mass of the system (which is obtained from Kepler's third law in the form of the total mass equation).

These equations are just like the process used with the father and daughter on the teeter-totter. The total mass of the father and the daughter was 100 kg, the daughter was 4 m from the fulcrum (the center of mass), the father was 1 m from the center of mass, and the total length of the teeter-totter was 4 m + 1 m = 5 m. Applying the equations above for the teeter-totter example,

$$m_{\text{father}} = \left(\frac{a_{\text{daughter}}}{a_{\text{total}}}\right) m_{\text{total}} = \left(\frac{4\ \cancel{m}}{5\ \cancel{m}}\right) \times 100\ \text{kg} = 80\ \text{kg}.$$

The calculation for the daughter is done the same way and gives her mass of 20 kg.

Figure 15.25 shows the orbits of Sirius A and Sirius B over their 50.13 y period. The last time the two stars were as close as they ever get to one another was in 1994. In 2019 (25 years later) they were as far apart from one another as they ever get. Study the figure carefully and draw lines between the stars for successive five-year intervals, as is shown for 1994. In each case you will find that the lines pass directly through the center of mass, effectively the "fulcrum" of their orbits.

An illustration of how this is accomplished is presented in Example 15.7 for Sirius A and Sirius B (seriously!). You may also want to study the center-of-mass tutorial available from the Chapter 15 resources page that is associated with Fig. 15.24 and the center of mass equation.

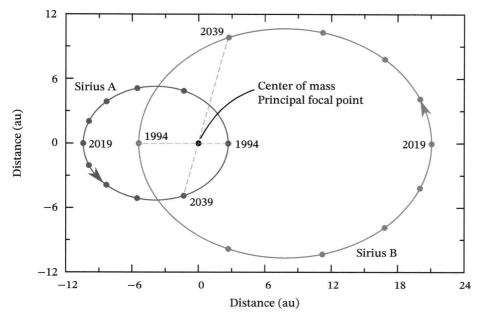

Fig. 15.25 The orbits of Sirius A (blue orbit) and Sirius B (red orbit) about their mutual center of mass, which is also the principal focal point for both orbits. The positions of each star are indicated every five years, beginning in 1994, with both stars orbiting counterclockwise. Note that at any time during their 50.13 year orbits, a line connecting Sirius A to Sirius B always passes through the center of mass; the center of mass *must* always be between the two stars. The eccentricity of the orbits is $e = 0.59$.

Example 15.7

(a) By simply inspecting Fig. 15.25, which star, Sirius A or Sirius B, is the *less massive*?

(b) From Example 15.6, the combined semimajor axis of the orbits of Sirius A and Sirius B is $a_{au} = 19.78$ au. Based on observations of Sirius A's orbit, its semimajor axis has been determined to be $a_{A,\,au} = 6.53$ au. What is the semimajor axis of Sirius B's orbit about their mutual center of mass?

(c) Determine their individual masses.

(a) Sirius B clearly has the larger orbit about the center of mass. As with the girl and her father on the teeter-totter, the least massive star is farthest from the center of mass. Therefore Sirius B must have less mass than Sirius A.

(b) According to the binary star separation equation,

$$a_{au} = a_{A,\,au} + a_{B,\,au}.$$

Solving for the semimajor axis of Sirius B:

$$a_{B,\,au} = a_{au} - a_{A,\,au} = 19.78 \text{ au} - 6.53 \text{ au} = 13.25 \text{ au}.$$

(c) Using the equations for the individual masses of a binary star system,

$$m_{B, \text{Sun}} = \left(\frac{a_{A, \text{au}}}{a_{\text{au}}} \right) \times m_{\text{total, Sun}}$$

$$= \left(\frac{6.53 \, \cancel{\text{au}}}{19.78 \, \cancel{\text{au}}} \right) \times 3.08 \, M_{\text{Sun}}$$

$$= 0.33 \times 3.08 \, M_{\text{Sun}} = 1.02 \, M_{\text{Sun}}.$$

Sirius B has very close to 1/3 of the total mass of the system, implying that Sirius A must have about 2/3 of the total mass. Carrying out the same calculation for Sirius A gives 2.06 M_{Sun}. Alternatively, once the mass for Sirius B is known, the mass for Sirius A can be found more directly by $m_{A, \text{Sun}} = m_{\text{total, Sun}} - m_{B, \text{Sun}}$.

In general, the process of determining the mass of each star individually can become quite complex and a complete solution may not even be possible. Although the orbits of the two stars in a binary must always be in the same plane (a consequence of the behavior of gravity), that common plane can be oriented in any arbitrary way relative to Earth. This means that the overall geometry of the orbits must also be determined. To see this, imagine that you are looking at a circular coin from directly above it. In this orientation you clearly see a circle. However, if you look at the coin from the side, but still above its edge, the coin now looks elliptical. If you look at the coin edge on, all you see is a straight line. The same is true when observing the orbits of binary star systems; if you look directly from above you will see a different shape for the orbits than if you looked from a viewpoint closer to the plane of the orbits. Since binary orbits are often elliptical rather than perfectly circular, astronomers must also be concerned with the direction of the major axes of the orbits. It is typically very difficult to sort out the orientation of the orbits relative to our point of view.

Ideally, it is also very helpful to be able to see each star separately, but that is often not the case; the two stars may be too close together and too far from Earth to see anything other than a single point of light, or, as was the case with Sirius A and Sirius B for many years, one star may be so bright that the view of the other star is overwhelmed by the brilliance of the brighter star. However, all is not lost. Careful analysis of the light coming from the two stars can sometimes reveal their combined spectral lines (Section 7.1), providing astronomers with additional information about the relative speeds of the two stars from the Doppler effect (Section 8.4). If one star passes in front of the other star while they orbit, then even more information can be obtained, ultimately allowing astronomers to determine a_1 and a_2 indirectly.

15.5 The Harvard Spectral Classification Scheme

Edward Pickering (Fig. 15.26) served as director of the Harvard Astronomical Observatory from 1877 through the remainder of his life. Pickering was an expert in studying the spectra of stars. He accomplished this work by being a pioneer in the

Fig. 15.26 Edward Pickering (1846–1919). (Public domain)

use of photography to record the spectra of multiple stars at the same time. The light from numerous stars was passed through a large prism that produced the spectra.

His method of capturing spectra on photographic images meant that Pickering had a great deal of data to process. To do so, he hired a number of male assistants to help him. However, as legend has it, he became frustrated with their performance and stated that his maid could do a better job. Pickering then went on to hire his maid, Williamina Fleming (Fig. 15.27), to do clerical work for him, and in fact she did do a better job. While performing the "clerical work," Fleming devised a scheme to classify stars according to the prominence of the hydrogen lines in their spectra. She classified stars with the strongest hydrogen lines as A, the group of stars with the next strongest set of hydrogen lines as B, and so on.

Fleming's superb work demanded patience and great attention to detail. Given her success, Pickering decided to hire more than a dozen other women to assist Fleming in performing the "clerical work" of classifying stellar spectra for him. Fleming was then asked to serve as supervisor of the group, that came to be known as Pickering's "computers" because of the mathematical nature of the classification process (Fig. 15.28). Unfortunately, other astronomers at the time sometimes referred to them less respectfully as "Pickering's Women" or "Pickering's Harem."

Fig. 15.27 Willimaina Fleming (1857–1922), circa 1890s. (Public domain)

Fig. 15.28 A group of Pickering's Harvard "computers," circa 1890s. Antonia Maury is center, foreground holding a magnifying glass. Fleming is standing in the back. Annie Jump Cannon is seated at the far right. Visible in the back left, seated below the chart, is Henrietta Leavitt, who established another important rung of the distance ladder. (Harvard College Observatory; public domain)

Fig. 15.29 Antonia Caetana de Paiva Pereira Maury (1866–1952). (Harvard College Observatory, courtesy of AIP Emilio Segrè Visual Archives)

Pickering's "computers" were paid 50 cents an hour (roughly $15 an hour today) for seven hours of work per day, six days each week.

Another of Pickering's "computers," Antonia Maury (Fig. 15.29), devised an alternative, but more complex, classification scheme to that of Fleming, based on the widths of spectral lines (not just hydrogen lines) on the photographic images. Although there were some important benefits to Maury's scheme, Pickering deemed it too complex for the work that was being carried out. At one point, Maury became so frustrated with Pickering and the debate over classification schemes that she left Harvard Observatory, only to return again about a decade later.

Despite Pickering's concerns, Maury did publish an important catalog of stars[4] in 1897 (with Pickering as co-author). The catalog was part of a memorial project to an American doctor and amateur astronomer, Henry Draper (1837–1882), that was funded by the Draper estate (Draper was Maury's uncle). The catalog and its successor volumes became known universally as the Henry Draper catalog. Stars within the catalog are recorded by a numerical designation and their "HD" numbers are still in wide use today (for example, Sirius A is also known as HD 48915A). Although Pickering was not particularly supportive of Maury's classification scheme, years later Ejnar Hertzsprung (page 627*ff*) found merit in her work and used it to identify groups of stars that shared a common classification according to Fleming's scheme, but with important differing characteristics.

In addition to her stellar classification work, and the Henry Draper catalog, Maury calculated the first orbit of a spectroscopic binary star system discovered by Pickering, in which it was possible to identify the shifting spectra of the two stars because of the Doppler effect. She also discovered a second spectroscopic binary system and determined its orbit. Sir John Herschel, a prominent British astronomer of the day, learned of Maury's work on spectroscopic binary stars and called it "one of the most notable advances in physical astronomy ever made." Unfortunately, Pickering received credit for both discoveries with only a brief mention of Maury's contributions.

Annie Jump Cannon (Fig. 15.30) was hired by Pickering in 1896. She graduated in physics from Wellesley College in 1884 and became a "special student" of astronomy at Radcliffe College (now a part of Harvard) in 1894. Cannon received a Masters of Arts degree from Wellesley College in 1907 and was the first woman to receive a doctorate in astronomy, which she earned from Groningen University in the Netherlands in 1921. Having contracted scarlet fever while on a trip in Europe to photograph a solar eclipse in 1892, Cannon became nearly deaf and remained so for the rest of her life.

Fig. 15.30 Annie Jump Cannon (1863–1941) sitting at her desk. (Smithsonian Institution; public domain)

While working for Pickering, Cannon took Fleming's alphabetical organization based on spectral line strengths of hydrogen and Maury's line widths for several elements and reorganized the letters in order of decreasing temperature (remember that the colors of stars provide information about temperature). Cannon also combined some redundant classes and tossed out others. When she was done, the sequence from the hottest stars to the coolest became OBAFGKM. Because catalogs had already been published using Fleming's classifications, the letter designations remained and Cannon's system became known as the Harvard spectral classification scheme. In order to remember the order, Cannon offered up a mnemonic to

---

[4]Maury, A. C. and Pickering, E. C. (1897). "Spectra of Bright Stars," *Annals of the Astronomical Observatory of Harvard College*, Vol XXVIII. — Part I.

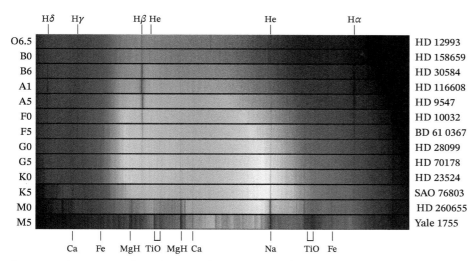

Fig. 15.31 The spectra of a selection of stars ranging from the hottest star on the top row to the coolest star at the bottom. The shortest wavelengths in the spectrum of each star are on the left (blue) and the longest wavelengths are on the right (red). The spectral class of each star per the Harvard spectral classification scheme is indicated on the left, and the name of the star that produced the spectrum is on the right. HD indicates that the star is contained in the Henry Draper catalog, BD stands for the Bonner Durchmusterung catalog of the Bonn Observatory, SAO designates the Smithsonian Astronomical Observatory Star catalog, and Yale is the Yale University Star catalog. A few of the atomic absorption lines and molecular absorption bands are identified: Hα, Hβ, Hγ, and Hδ are four of the Balmer lines of hydrogen, He designates helium, Ca is calcium, Na is sodium, Fe is iron, and MgH and TiO are molecules. Notice that the dark hydrogen absorption lines are most prominent for A-type stars and get fainter for the hottest and coolest stars. The M type stars display wide bands of absorption due to molecules in their atmospheres. (Adapted from a figure by NOIRLab/NSF/AURA; CC BY 4.0)

help others learn the new sequence. Ever since, generations of students have memorized "**O**h, **B**e **A** **F**ine **G**irl/**G**uy, **K**iss **M**e!" (the second designation for G was added more recently, but not by Cannon).

Cannon became known as the "Census Taker of the Sky," having classified more than 425,000 stars, at a rate of up to three stars per minute, just by looking at the photographic images of their spectra. With a magnifying glass she was even able to distinguish subclassifications, labeled from 0 through 9. The subclass designated as 0 is hotter than the subclass 9. For example, within the G spectral class, running from hotter to cooler the subdivided sequence becomes G0, G1, ..., G9. Our Sun is a G2 star. Cannon's work was published in the Henry Draper catalog. Samples of stellar spectra with their Harvard classifications are shown in Fig. 15.31.

In recognition of her tremendous contributions to astronomy, Cannon received numerous awards during her lifetime. In 1923 she was voted one of the 12 greatest living women in America, in 1925 she became the first woman to be a recipient of the honorary doctorate from Oxford University for her contributions to astrophysics, and she received honorary doctorate degrees from several other institutions as well. Annie Jump Cannon was finally awarded a professorship at Harvard in 1938, just three years before her death.

Key Point
Harvard spectral classification scheme for stars from high to low surface temperatures: OBAFGKM.

Using the $1000 she received for winning the Helen Richards Research Prize from the Association to Aide Scientific Research by Women, Cannon endowed the Annie Jump Cannon Award within the American Astronomical Society. Antonia Maury received the Annie Jump Cannon Award in 1943. Originally awarded only once every two to three years, due to a lack of women in astronomy, today the award is given annually to an outstanding woman astronomer in North America who received her PhD within the previous five years.

Cecilia Payne-Gaposchkin (Fig. 15.32) was a British-born astronomer and astrophysicist who emigrated to the United States in 1923 (she was married in 1934, at which time she changed her last name from Payne to Payne-Gaposchkin). Payne earned her PhD in 1925 from Radcliffe College, writing a dissertation that would become a historic turning point in our understanding of what the universe is made of. Before 1925 it was generally believed that the Sun had a composition that was similar in many ways to that of Earth. In particular, astronomers believed that elements such as carbon, oxygen, and silicon must be very abundant in the Sun while hydrogen and helium were much less abundant. It was also generally assumed that the variations in the absorption lines seen in the spectral classification sequences, such as those in Fig. 15.31, are likely due to changes in abundances among the stars, with A stars having the most hydrogen and therefore the darkest lines. In that interpretation of the data, Fleming's original spectral classification scheme would amount to a composition sequence. However, this idea was in conflict with the understanding that the hottest stars appear blue in color while the coolest stars are red, suggesting some sort of relationship between spectral lines and temperature rather than composition.

In Section 8.7 our developing understanding of the nature of atoms was discussed and the notion of atomic energy levels was introduced. You should recall that when an electron undergoes an atomic transition from a higher energy level to a lower energy level, a photon is emitted with an energy that precisely matches the difference in energy between the two levels. Similarly, when a photon strikes an atom, an electron can make a transition from a lower energy level to a higher energy level, but only if the difference in energy between the two levels corresponds to the energy of the absorbed photon. If there aren't two energy levels with an energy difference corresponding to the energy of the incoming photon, the photon cannot be absorbed and must continue on its way. These processes involving the emission and absorption of photons are what produce the emission-line and absorption-line spectra discussed in Kirchhoff's rules (page 246).

It is the underlying blackbody radiation of a star that is the source of the photons that can be absorbed to produce dark absorption lines in its spectrum (recall the three perfect, idealized blackbody curves in Fig. 6.43). The underlying blackbody spectrum of the star in Fig. 15.33 is clearly evident, but it also appears choppy, with some spikes above the general curve (emission lines) and other spikes dropping below the curve (absorption lines). All of that choppiness is due to the presence of atoms and ions in the star's atmosphere. Those dark absorption lines seen in Fig. 15.31 appear as dips in intensity below the continuous blackbody radiation curve in Fig. 15.33.[5]

Payne approached the problem of stellar abundances by incorporating science's evolving understanding of atomic physics, including some recent work on atomic

Fig. 15.32 Cecilia Payne-Gaposchkin (1900–1979). (Smithsonian Institution, public domain)

A graphing tutorial is available at
Math_Review/Graphing

---

[5]If these figures are still confusing to you, now would be a good time to review Section 7.1 carefully.

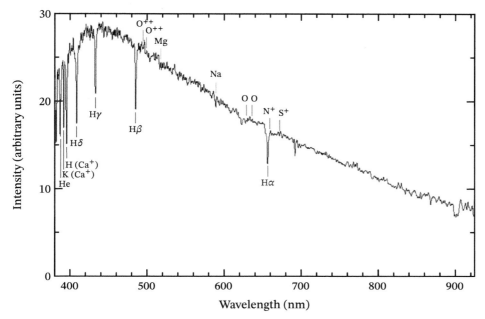

Fig. 15.33 A portion of the spectrum of a star from the visible to near infrared wavelengths. A similar spectrum of the Sun was depicted in Fig. 7.14. A few of the many thousands of lines are labeled. You should compare this graphical representation with the spectra displayed in Fig. 15.31. The dark absorption lines show up as deep dips in the graph. (Data courtesy of Sloan Digital Sky Survey SkyServer)

electron transitions and ionization. Since electrons require a specific amount of energy to make a transition from a lower energy level to a higher one, if the gas is too cool there is not enough energy available for the electrons to jump up to the higher energy level. As a result, certain absorption lines that appear in hotter stars won't be present in cooler ones. This effect explains why M stars don't exhibit hydrogen absorption lines, for example. But what about the absence of hydrogen lines in the hot O stars?

If an incoming photon has too much energy, or if the gas is too hot, meaning that the atoms are running around very rapidly and colliding too hard with one other, an electron can be kicked out of the atom altogether and end up flying around entirely on its own until captured by another ion. The minimum amount of energy required of a photon or from the collision of atoms that will result in an electron's escape is known as the ionization energy.

In 1921, Meghnad Saha (Fig. 15.34) developed an equation that was to become a cornerstone of astrophysics, just four years before Payne completed her dissertation. Saha's equation relates the temperature and density of a gas to the amount of ionization that occurs for different atoms based on their ionization energies. His equation specifies how hot a gas must be in order for electrons to escape from the atoms so they are no longer available to absorb incoming photons. Although other electrons may remain in multi-electron atoms, the removal of electrons alters the energy levels of those that remain. You can see this effect in Fig. 15.31, with two lines for neutral oxygen atoms (O) and two lines due to doubly ionized oxygen ions

Fig. 15.34 Meghnad Saha (1893–1956). (Pubic domain)

(O^{++}) where two electrons have escaped. Lines from other ions are also labeled, including the so-called H and K lines for singly ionized calcium ions (Ca^+), named because of their labeling in Fraunhofer's spectrum of the Sun; e.g., Fig. 7.9.

Payne's dissertation incorporated these complex ideas, allowing her to not only show that temperature is responsible for the presence or absence of particular spectral lines, but also to determine, for the first time, what the relative abundances of the elements must be in order to produce the strengths of the lines. After all, if the temperature of the gas is appropriate to produce hydrogen lines, those lines still won't occur if there isn't any hydrogen present in the star! Payne concluded, as expected, that many of the elements present in the Sun were in similar relative proportions to those on Earth, but she also found that hydrogen and helium are *far more abundant* than anyone had previously believed. No longer are hydrogen and helium relegated to insignificant roles in astrophysics, they are the most prominent players.

Unfortunately, when Payne's dissertation was finished, the initial reaction was one of great skepticism. A very well-respected astronomer at that time, Henry Norris Russell (Fig. 15.35), dissuaded her from publishing her results. However, four years later he came to the same conclusions about the abundance of hydrogen and helium. To Russell's credit, he fully acknowledged that Payne had arrived at this critically important result prior to his work. Otto Struve (1897–1963), another highly respected astronomer who would become Editor-in-Chief of the *Astrophysical Journal*, the preeminent journal for astrophysics in the United States, referred to Payne's work as "undoubtedly the most brilliant PhD thesis ever written in astronomy."

Fig. 15.35 A portrait of Henry Norris Russell (1877–1957). (Public domain)

Payne-Gaposchkin received much well-deserved recognition for her work, including her later studies on the structure of the Milky Way Galaxy, and on variable stars within the Milky Way and the Magellanic Clouds (two small galaxies orbiting the Milky Way). She spent her entire career at Harvard, being named "Astronomer" in 1938 (along with Annie Jump Cannon) and in 1956 she became the first female full professor promoted from within the faculty. She would later become the first woman to head a department at Harvard. Cecilia Payne-Gaposchkin was also the first recipient of the Annie Jump Cannon award, in 1934.

With the theoretical understanding of stellar spectra now in place, measurements identifying which spectral lines are present and with what strength provide astronomers with a precise method for determining the surface temperatures of stars. In addition, spectral lines serve as the means for determining the compositions of stars. As we will learn later, spectral lines also allow astronomers to probe different layers in the extended atmospheres of stars [it is this technique that is used to study the various layers of the Sun's atmosphere: the photosphere, the chromosphere, the transition region, and the corona (recall Section 9.2)].

Astronomy, and all of science, is deeply indebted to the women of the Harvard Astronomical Observatory who worked as "computers" in the late 1800s and the early years of the 1900s, along with Payne-Gaposchkin who arrived at Harvard after Pickering's death. Their work was spectacularly important in its own right, but their work also set the stage for tremendous advances in the years that followed. It is worth noting that both Annie Jump Cannon and Cecilia Payne-Gaposchkin became important role models and strong advocates for women in science throughout their careers.

15.6 Calculating the Radii of Stars

When we used a geometric method to determine the radius of the Sun in Section 9.1, that only worked because we can see the disk of the Sun. Unfortunately, with very few exceptions, no other stars are close enough or large enough for that method to work in general. However, having learned how to determine both a star's luminosity using its brightness combined with the inverse square law for light, and its surface temperature using either the peak wavelength of its blackbody radiation curve (page 217) or through the analysis of its spectrum, it becomes a relatively straightforward process to calculate a star's radius.

In Section 6.6, the Stefan–Boltzmann blackbody luminosity law was introduced as one of the fundamental characteristics of blackbody radiation. For a spherical object, such as a planet or a star, the law is given by the equation $L = 4\pi R^2 \sigma T^4$. Although the equation often looks a bit intimidating to students, we saw that it can be simplified to give the Stefan–Boltzmann luminosity law in units of the Sun (page 219) as

$$L_{Sun} = R_{Sun}^2 T_{surface, Sun}^4, \quad \text{where} \quad T_{surface, Sun} = T_{surface}/5772 \text{ K}.$$

An exponents tutorial is available at
`Math_Review/Exponents`

Example 15.8

Procyon A (shown in Fig. 15.15) is the brightest star in the constellation of Canis Minor, the small dog, which, along with Sirius A and Betelgeuse, make up the prominent winter triangle that is easily seen in the northern hemisphere sky in mid winter. Procyon A is an F5 star (based on the Harvard spectral classification scheme) with a surface temperature of 6629 K and a radius that is 2 times the Sun's radius, $R_{star} = 2\,R_{Sun}$. Determine the luminosity of Procyon A in terms of the luminosity of the Sun.

Writing Procyon A's surface temperature in terms of the surface temperature of the Sun, T_{Sun} in the Stefan–Boltzmann law in solar units is approximately

$$T_{Sun} = \frac{6629 \ \cancel{K}}{5772 \ \cancel{K}} = 1.15.$$

Since the star's radius is 2 times the radius of the Sun, this means that the radius of the star should be written as

$$R_{Sun} = \frac{2 \ \cancel{R_{Sun}}}{1 \ \cancel{R_{Sun}}} = 2.$$

Using these values, the luminosity (in solar units) is approximately

$$L_{Sun} = R_{Sun}^2 T_{Sun}^4 = 2^2 \times 1.15^4$$

$$= (2 \times 2) \times (1.15 \times 1.15 \times 1.15 \times 1.15) = 7.$$

Therefore, Procyon A has a luminosity that is 7 times greater than the luminosity of the Sun, or $L_{star} = 7\,L_{Sun}$.

Like Sirius A, Procyon A is one member of a binary star system, with the other member referred to as Procyon B.

Of course, it is possible to rewrite the solar units version of the Stefan–Boltzmann blackbody luminosity law to solve for the radius, which is what we are actually looking for at this point. Doing so gives

Key Point
The radius of a star depends on the star's luminosity and its surface temperature.

$$R_{\text{Sun}} = \frac{\sqrt{L_{\text{Sun}}}}{T_{\text{Sun}}^2},$$

(15.11)

The stellar radius equation in solar units

which means that the star's radius can be calculated by taking the square root of its luminosity and then dividing that result by the star's surface temperature squared, as long as R_{Sun}, L_{Sun}, and T_{Sun} are all written in terms of the Sun's values.

Example 15.9

A certain B1 star has a surface temperature of 25,400 K and a luminosity of 9950 L_{Sun}. What is the radius of the star in terms of the Sun's radius?

You first need to write the star's surface temperature in terms of the Sun's surface temperature:

$$T_{\text{Sun}} = \frac{25,400 \ \text{K}}{5772 \ \text{K}} = 4.40.$$

Since the star's luminosity was already given in terms of our Sun's luminosity,

$$L_{\text{Sun}} = \frac{9950 \ L_{\text{Sun}}}{1 \ L_{\text{Sun}}} = 9950.$$

Now, from the stellar radius equation in solar units, the star's radius is

$$R_{\text{Sun}} = \frac{\sqrt{L_{\text{Sun}}}}{T_{\text{Sun}}^2} = \frac{\sqrt{9950}}{4.40^2} = \frac{99.7}{19.4} = 5.15.$$

The star's radius is 5.15 times the radius of the Sun, or $R_{\text{star}} = 5.15 \ R_{\text{Sun}}$.

15.7   The Universality of Physical Law Revisited

At this point you may be feeling a bit overwhelmed by the rather dense amount of information and mathematics contained thus far in Chapter 15, but the reward for all of that hard work will soon begin to unfold, beginning in Section 15.8. However, before moving on, it is worthwhile reviewing what you have learned so far regarding how astronomers are able to determine so much about the characteristics of stars. The fact that this is even possible, given the vast distances to the stars, may seem rather remarkable; after all, with the exception of our parent star, the Sun, it would take about 74,000 years to reach the closest stars, traveling at 17 km/s (38,000 mi/h), which is the current speed of Voyager 1, our most distant robotic ambassador (unless, of course, warp drive or worm-hole travel are invented, which are the popular solutions in science fiction). In actuality, Voyager 1 isn't even headed in the direction of Proxima Centauri, it is headed in the direction of the constellation Ophiuchus.

As it turns out, we have just begun to scratch the surface regarding what can be learned about stars. Moving forward we will continue to make a rather profound assumption about the universe, namely that the same physical laws operating on and near Earth apply everywhere and have so since the beginning of time. Newton first made that assumption when he concluded that the same force of gravity that holds us to the surface of Earth, and that controls how fast objects fall, also applies to the Moon orbiting Earth, and planets and comets orbiting the Sun. It is for that reason that his equation concerning how the force of gravity behaves is known as the *universal* law of gravitation. The same holds true for the physical constants that you have seen so far in this textbook, including G (the universal gravitational constant), c (the speed of light), σ (the Stefan–Boltzmann constant), h (Planck's constant), e (the elementary charge), and so on. In fact it would be impossible for astrophysics to make any progress without this basic assumption of the universality of physical law.

Of course, as with everything in science, it is critical that assumptions are tested. For example, is G really the same everywhere in the universe and for all time? To conduct such a test it is assumed that the same behavior of gravity must occur wherever we look, including in deep space and deep time. If different results are observed than were expected, such as orbits behaving differently than anticipated, that would suggest that the universal gravitational "constant" isn't really constant. However, it isn't quite as simple as that since other more mundane solutions must also be considered, such as unobserved masses being present. In other cases, it may be that the theory itself is flawed or incomplete. This last possibility led to a profound change in our understanding of gravity in 1915, when Einstein finished the development of his general theory of relativity (page 269). When all possible explanations are taken into consideration, to date there has not yet been any indication that even one of the fundamental physical constants differs by any measurable amount from the values determined in laboratories on Earth. Although it is impossible to guarantee that discrepancies won't be found in the future, science continues to operate with great success under the assumption of the universality of physical law.

The most incomprehensible thing about the world is that it is comprehensible.

Albert Einstein (1879–1955)

Reviewing Our Knowledge of Stars

Throughout Chapter 15, and indeed throughout the entire text, we have assumed implicitly the universality of physical law. We have applied ideas from terrestrial experiments and geometry to deduce the luminosity and absolute magnitude, distance, mass, color, peak blackbody wavelength, surface temperature, chemical composition, and radius of a star. Before continuing, it is important that you understand how each of these quantities is determined:

Key Point
Fundamental physical laws and constants are assumed to be the same everywhere in the universe and unchanged since the beginning of time.

- The direct observation of the brightness of a star in the night sky, without any adjustment for distance, gives its apparent magnitude.
- The *difference* in the apparent magnitudes of two stars allows us to determine the *ratio* of their brightnesses by use of the brightness ratio–magnitude difference equation.
- The inverse square law for light is needed to determine a star's absolute magnitude and its luminosity based on its apparent magnitude and brightness, respectively. The inverse square law takes into consideration that the farther a star is away from Earth, the more spread out the star's light becomes, resulting in less light being available to be captured by the human eye or a telescope.
- In order to use the inverse square law for light, we need to determine how far a star is away from us. Although this is very challenging to do in general, for stars that are "close enough" to Earth the geometric method of parallax is used. As Earth orbits the Sun our view of a star changes as we change our position in space (Fig. 15.19). This change in perspective causes the star to appear to shift back and forth over the course of one year. Measuring the shift, combined with knowledge of the average distance from Earth to the Sun (approximately equal to the astronomical unit) gives the distance to the star via the parallax equation. When the parallax angle is measured in arcseconds, the resulting distance is in units of parsecs.
- Once the star's distance is known, its apparent magnitude can be used to calculate its absolute magnitude (the apparent magnitude that the star *would have if* it was located 10 pc from Earth). Since absolute magnitude and luminosity are directly related to one another, the luminosity can be determined as well. The luminosity is simply the amount of energy the star gives off every second (its power output).
- Newton's universal law of gravitation depends on the product of the masses of the two attracting objects and the inverse square of the distance between them. From his universal law of gravitation, Newton derived Kepler's laws, including the orbital mass–period–distance equation (Kepler's third law). If the average distance between two stars in a binary star system is known, along with the orbital period, then the sum of the masses of the two stars can be calculated by using Kepler's third law in the form of Kepler's total mass equation. (Don't forget that in order to use Kepler's third law in the form it is written, the orbital period must be given in years and the average distance of separation must be given in astronomical units. The sum of the masses will then be given in solar masses.)
- If the location of the center of mass of the system is known, then the center of mass equation can be used, combined with Kepler's third law, to determine the individual masses of each star in a binary system.
- The surface temperature (T_{surface}) of a star can be discerned by one of two methods: (a) determining the wavelength of the peak intensity of the underlying continuous blackbody spectrum and then calculating the temperature from the blackbody peak wavelength equation, or, alternatively, by (b) conducting a careful analysis of the absorption and emission lines that are superimposed on the underlying blackbody spectrum (Fig. 15.31 and Fig. 15.33).
- A star's composition can also be determined by the careful analysis of its absorption and emission spectra. Line strengths (depth and width) are used

to determine relative abundances of the atoms, ions, and molecules that are present at the surface of the star.

- Finally, the radius of the star is determined by using the star's luminosity and surface temperature via the stellar radius equation. (Again, it is important to remember that this form of the equation requires that the luminosity and surface temperature be given in terms of the Sun's values. The radius will then be returned in solar radii.)

15.8   The Birth of the Hertzsprung–Russell Diagram

The late 1800s through the early 1900s was a remarkable time in the development of both astronomy and physics. In Section 15.5 you learned about the ground-breaking work in astronomical spectroscopy carried out by Edward Pickering and the women of the Harvard Astronomical Observatory. During that same period, Max Planck discovered the quantized nature of energy when he derived the blackbody radiation curve, Albert Einstein developed his special theory of relativity, proposed quantized light particles that were later referred to as photons, and created his general theory of relativity, and Meghnad Saha developed his ionization equilibrium theory that Cecilia Payne-Gaposchkin built upon to explain stellar spectra. In addition, the theory of quantum mechanics was taking shape thanks to the contributions of numerous scientists, most notably Niels Bohr, Louis de Broglie, Erwin Schrödinger, Werner Heisenberg, Wolfgang Pauli, and Paul Dirac. Our knowledge about the structure of the atom, the existence of the atomic nucleus, and radioactivity was also developing over that same period thanks to the work of scientists like Ernest Rutherford and Marie and Pierre Curie.

As these new doors of insight were being opened into the realm of fundamental physical processes and their applications to the study of stars, scientists began to consider possible relationships between various stellar characteristics, in the hope of shedding light on how the universe works. One of the many major questions to be answered was how stars form and live out their lives.

In 1905, Ejnar Hertzsprung (Fig. 15.36) began to investigate whether or not there was a relationship between the spectral classification of stars and their absolute magnitudes. Although he didn't have access to stellar parallax data (measuring distances by that method was still in its infancy and not yet readily available) he was able to convert apparent magnitudes to approximate absolute magnitudes through a statistical approach based on how fast stars appear to move across the sky. He assumed that if stars appear to be moving relatively quickly across the sky then they must be comparatively close to Earth, and if they appeared to move more slowly then, on average, they must be farther away.

To understand Hertzsprung's reasoning, think about watching someone running down the street in front of your house and then compare that to watching a passenger jet 10 km away. Even though the plane is certainly moving much faster, the runner will traverse your field of view significantly more quickly than the plane will. If, instead of a plane, you are watching another runner 10 km away (with great binoculars!) it will take a very long time for that person to exit your field of view. The effect is simply due to the relative distances to the runner on the street and the runner 10 km away. The same holds true, at least in an average sense, for stars

Fig. 15.36 A photograph of Ejnar Hertzsprung (1873–1967). (Dr. Dorrit Hoffleit, Yale University, courtesy AIP Emilio Segrè Visual Archives, Tenn Collection)

nearby versus those much farther away. If, like the runners, all stars are moving at the same speed across the sky, the more distant ones simply won't appear to be moving as rapidly. In reality, stars can have very different speeds, whether they are close by or very far away (like the plane or the runners), so by using this approach Hertzsprung was certain to have very crude estimates of relative distances, at best.

As Hertzsprung carried out his study he discovered something unexpected: for yellow and red stars there seemed to be two types of star, those that were fairly similar to the Sun, while others were much more luminous but still of the same Harvard spectral class. He also realized that Maury's classification scheme, which included consideration of the widths of spectral lines, seemed to correlate with the two types of star. Groups that she referred to as c or ac stars had very narrow spectral lines and seemed to correspond to the very luminous stars, and those that did not have the narrow-line characteristic were more Sun-like. This effect held true, even though both types of star had very prominent hydrogen lines in their spectra. Upon realizing the association between stellar luminosity and line width for stars of the same spectral class, Hertzsprung wrote to Pickering in 1908 in support of the Maury classification scheme:

> In my opinion the separation by Maury of the c and ac-stars is the most impor-tant advancement in stellar classification since the trials by Vogel and Secchi ...To neglect the c-properties in classifying stellar spectra, I think, is nearly the same thing as if the zoologist, who has detected the deciding differences be-tween a whale and a fish, would continue in classifying them together.

Unfortunately for Hertzsprung, who began his research as an amateur astron-omer after having been trained as a chemical engineer, he published his results in rather obscure journals not typically read by astronomers. However, he did publish graphical results in 1911 in the Potsdam Astrophysical Observatory's Publications in which he showed the apparent magnitudes and peak blackbody wavelengths of stars in the stellar clusters of the Pleiades and the Hyades (Fig. 15.37), both in the constellation of Taurus; his original graphs are displayed in Fig. 15.38 (*Left*). One

Fig. 15.37 *Left:* The Pleiades star cluster (NASA, ESA, AURA/Caltech, Palomar Observatory) *Right:* The Hyades star cluster. The brightest star in the field, Aldebaran, is not a member of the cluster. (NASA, ESA, and STScI; CC BY-SA 4.0 IGO)

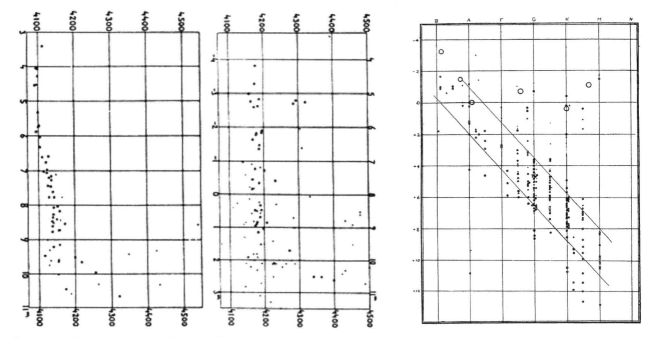

Fig. 15.38 *Left:* Hertzsprung's diagrams illustrating the apparent magnitudes of stars in the Pleiades (left) and the Hyades (right) on the vertical axes versus the peak blackbody wavelengths of the stars on the horizontal axes. The brightest stars are at the tops of the graphs and the hottest stars (shortest wavelengths) are on the left. The wavelengths are given in units of angstroms, where 1 angstrom (Å) equals 0.1 nm, meaning that 4100 Å is the same as 410 nm. The graphs are rotated 90° clockwise relative to Hertzsprung's original presentation so that they can be compared directly with more modern representations that always place the temperature parameter along the horizontal axis and magnitude or luminosity along the vertical axis. (Hertzsprung, 1911[6]) *Right:* Russell's original diagram. Absolute magnitudes are plotted on the vertical left axis and spectral class is on the horizontal top axis. Note that the O spectral class is not included while the now-defunct N spectral class is shown. (Russell, 1914[7])

common feature of both graphs is the near vertical line of stars toward the hot (left) sides of the graphs. In both graphs, dim stars (near the bottom) tend to have cooler surface temperatures (longer peak wavelengths). But one feature that is present in the Hyades cluster (right graph) and absent in the Pleiades is a small number of bright stars in the upper middle of the graph. These are stars that fit into Maury's c and ac classes. Later, the highly respected astronomer, Sir Arthur Stanley Eddington (1882–1944) teased Hertzsprung by saying "One of the sins of your youth was to publish important papers in inaccessible places."[8]

Initially unaware of Hertzsprung's work, Henry Norris Russell began his own investigation in 1910 but with the benefit of stellar parallaxes, which allowed him to more precisely plot the absolute magnitudes of stars against their spectral types. His first published diagram from 1914 is shown in Fig. 15.38 (*Right*). Note that the vast

---

[6]Hertzsprung, E. (1911). "Über die Verwendung Photographischer Effektiver Wellenlaengen zur Bestimmung von Farbenaequivalenten," *Publikationen des Astrophysikalischen Observatoriums zu Potsdam*, 63.

[7]Russell, Henry Norris (1914), "Relations Between the Spectra and Other Characteristics of the Stars," *Nature*, 93(2323):252–258.

[8]It was Eddington who led the solar eclipse expedition in 1919 that confirmed Einstein's general theory of relativity.

majority of stars are found in the diagonal band running from upper left (bright and hot) to lower right (dim and cool). Just as in Hertzsprung's diagram for the Hyades, Russell found some cooler stars that were unusually bright, suggesting that those stars are different from most of the rest. Russell's diagram also shows two hot stars, apparently identified as spectral class A, that are unusually dim, having absolute magnitudes of only about +9 and +11.

Apparent in Hertzsprung's diagrams of the Pleiades and the Hyades (the nearly vertical line of stars) and in Russell's diagram (the diagonal grouping of stars) is what is known as the main sequence. The fact that the main sequence is much more vertical in Hertzsprung's diagrams is simply due to the way the horizontal axes are presented. Stars that make up the main sequence are logically referred to as main-sequence stars.

Also present is a much smaller number of cool, but bright, stars. As you will recall from the discussion of the stellar radius equation, if two stars have the same temperature (or spectral type, or peak blackbody wavelength), the brighter star necessarily has the larger radius. These cool, bright stars have come to be known as giant stars. Stars like Betelgeuse, with their enormous radii that may exceed the orbit of Mars or even Jupiter, are referred to as supergiant stars (we made a comparison between the radii of Betelgeuse and Lalande 21185, a main-sequence star, in Example 6.13).

At the other extreme, in the lower left corner of the Russell diagram, are the two hot, but dim, stars. Again recalling the stellar radius equation, in order for a star to be both hot (large T) and dim (small L), the star must be very small. In fact, these stars have radii that are approximately the same as Earth's radius. These stars are known as white dwarf stars even though they are not necessarily white in color (both Sirius B and Procyon B are white dwarfs).

The labels of main-sequence stars, giant stars, supergiant stars, and white dwarf stars constitute what are known as luminosity classes of the stars. Luminosity classes provide additional information about the stars not contained in the Harvard spectral classes alone. It was luminosity classes that were hinted at in Maury's classification scheme.

In addition to patterns involving surface temperature and luminosity, there are also trends in radius and mass *along the main sequence*. The hot, bright B stars shown in Russell's diagram have radii that are roughly ten times the radius of the Sun, while the cool, dim M stars have radii that are as small as 20% of the radius of the Sun. The most luminous stars along the main sequence, not indicated on any of these diagrams, are the O stars, which are also the hottest stars. O stars have masses that range from more than $17.5\,M_{Sun}$ to perhaps hundreds of solar masses, although the most massive stars are very unstable. The M-type main-sequence stars have masses down to about $0.075\,M_{Sun}$. Just as stars farther down the main sequence are cooler, dimmer, and smaller than those higher on the main sequence, they are also less massive.

It is important to note that such a continuous change in masses does not, in general, exist among giant stars and supergiant stars (important exceptions to this general statement will be discussed in Section 16.5). However, it is the case that supergiant stars as a group tend to be more massive than giant stars. White dwarf stars have a very limited range of masses; they never exceed $1.4\,M_{Sun}$ and may have masses as low as about $0.5\,M_{Sun}$.

Table 15.2 Properties of main-sequence stars in the Harvard spectral classification scheme.

Class	Temperature (K)	Radius (R_{Sun})	Luminosity (L_{Sun})	Mass (M_{Sun})
O	> 30,000	> 7.4	> 4×10^4	> 17.5
B	30,000 – 10,000	7.4 – 2.4	4×10^4 – 52	17.5 – 2.9
A	10,000 – 7000	2.4 – 1.5	52 – 4.9	2.9 – 1.6
F	7000 – 6000	1.5 – 1.1	4.9 – 1.4	1.6 – 1.05
G	6000 – 5000	1.1 – 0.85	1.4 – 0.4	1.05 – 0.79
K	5000 – 3800	0.85 – 0.60	0.4 – 0.03	0.79 – 0.51
M	3800 – 2400	0.60 – 0.13	0.03 – 0.0005	0.51 – 0.075

At the time Hertzsprung and Russell were conducting their studies, a popular model of the birth and aging of stars involved them collapsing due to gravity, starting out very large and cool, heating as they contracted and continued to gather mass, and then cooling off and slowly losing mass until they faded from view. Initially these diagrams appeared to support such a model, with the stars starting out as supergiants or giants, heating up and changing spectral type from M through F and perhaps becoming O or B stars on the main sequence and then simply cooling off and losing mass as they progressed back down toward the dim, low-mass, M stars. Today we know that this model is entirely incorrect, but was certainly suggestive based on the data available at the time. However, our understanding of the physical processes occurring in the interiors of stars was seriously limited in the first half of the twentieth century. A more complete picture would have to wait for further theoretical, laboratory, and computational advances.

Russell did become aware of Hertzsprung's work in 1910, when Hertzsprung's good friend and mentor, Karl Schwarzschild (1873–1916), visited with Russell at a conference at Harvard. Hertzsprung and Russell also met at subsequent professional meetings. For several decades the diagram was simply referred to as the Russell diagram, but a fellow Dane of Hertzsprung's, Bengt Strömgren (1908–1987), introduced the label of the Hertzsprung–Russell diagram in a paper he wrote in 1933. Subrahmanyan 'Chandra' Chandrasekhar (1910–1995) also referred to the diagram as such in his classic stellar structure book written in 1939, and the name was universally accepted in the 1940s when the *Astrophysical Journal*, which Chandrasekhar was co-editing, adopted it as the standard designation. Today the diagram is often simply referred to as the H–R diagram for short. As you will see in Chapter 16, the H–R diagram plays the central role in developing our modern understanding of how stars are born, live out their lives, and die.

Figure 15.39 shows a modern version of the H–R diagram, based on more than four million stars from the Gaia catalog. The main sequence is readily apparent, running from the blue stars in the upper left corner to the red stars in the lower right corner, giant and supergiant stars are located above the main sequence toward the upper right corner, and the white dwarf stars can be seen in a diagonal in the lower left portion of the diagram. Properties of main-sequence stars are listed in Table 15.2.

Key Point

The Hertzsprung–Russell diagram is a fundamental starting point in developing our understanding about how stars are born, age, and die. As such, it is critical that you are able to *fully describe what it is representing* rather than simply being able to sketch the diagram.

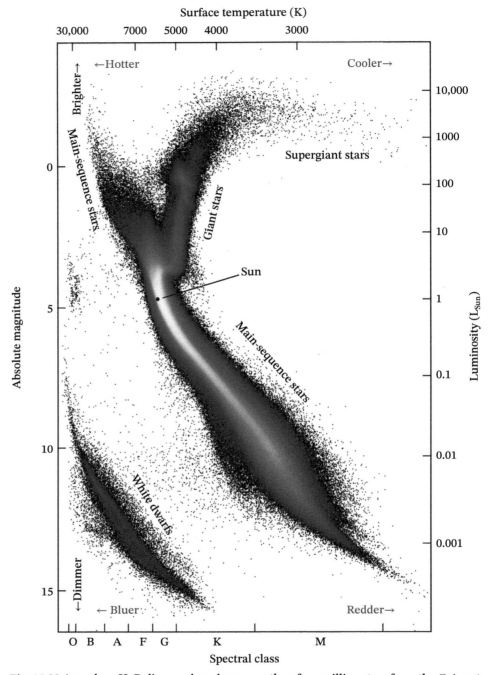

Fig. 15.39 A modern H–R diagram based on more than four million stars from the Gaia catalog that are within five thousand light-years of Earth. The brightest colors indicate the largest density of stars in the diagram, trailing off to individual stars as black dots. The large black dot on the main sequence marks the Sun's location on the diagram. (Adapted from ESA/Gaia/DPAC, CC BY-SA 3.0 IGO)

15.9   Finding Patterns and Asking Why

Patterns are ubiquitous throughout nature and in our human experience. We see patterns in the leaves of a flower or in the web of a spider. We hear the repetition of waves crashing on the shore or the meter of a piece of music. We experience the daily rising and setting of the Sun, and on our birthdays we celebrate another successful trip around the Sun. Patterns even exist in the double helix of our DNA. In so many ways we are rooted in the patterns of our lives.

When patterns previously unknown are discovered, we naturally assume that there must be some underlying cause. After all, patterns don't arise from pure chance; in fact, random chance is the antithesis of pattern. We are so ingrained in searching out patterns that sometimes we want to create them where no pattern actually exists. Our ancestors played connect-the-dots with the stars in the sky to construct gods that oversee the seasons governing when we plant our crops, hunt for game, expect the annual flooding of rivers, or when we must prepare for winter.

Throughout this chapter and in the rest of the textbook to this point we have seen that relationships exist at nature's most fundamental levels. For example, the energy levels of electrons in atoms provide signatures in the form of spectra, revealing the presence of specific atoms, ions, and molecules in the atmospheres of stars, planets, and comets. We have seen how surface temperature directly influences which spectral lines appear. The apparent brightness of a star is directly affected by its distance from Earth, and is precisely described by the inverse square law of light. The luminosity, radius, and temperature of a star are tightly related through the Stefan–Boltzmann blackbody luminosity law.

Now, thanks to the development of the Hertzsprung–Russell diagram, we have uncovered previously unanticipated patterns tied to certain characteristics of stars. As with all patterns, there must be one or more underlying causes that create the patterns we observe. One of the great driving forces throughout the history of science is to seek out those underlying causes. We continue to ask the same question that every two-year-old child asks incessantly — "Why?" Why is there a clear relationship between luminosity and surface temperature for the vast majority of stars in the sky that results in the main sequence? Why do some stars appear to be extreme outliers in that pattern, such as giants, supergiants, and white dwarfs? Is there a connection between all of these types of star yet to be uncovered, or are they indeed entirely different and unrelated objects?

Whenever patterns are detected, whether obvious or previously hidden, science strives to find answers to the question "Why?" The process of science builds upon previous work, sometimes finding itself headed down wrong paths such as a failed description of the life cycle of stars, and sometimes uncovering previously unimagined explanations that may initially defy our preconceived notions about how nature ought to work.

It is incumbent upon us as a species with the ability to think critically, that we ask questions and carefully evaluate and constantly challenge the answers that are presented to us. After all, science is a truly human enterprise.

? Exercises

All of the exercises are designed to further develop *your understanding* of the material by thinking carefully and critically about what you have read in this chapter. Answers to selected exercises are available in the back of the book.

True/False

1. Only the visible wavelength region can be used to observe the cosmos from the ground.
2. (*Answer*) Most of the infrared portion of the electromagnetic spectrum cannot be used for observations from the ground because of molecules in the atmosphere.
3. All of the world's largest optical telescopes are reflecting telescopes.
4. (*Answer*) When the Hubble Space Telescope's main mirror was ground, it achieved the most perfect shape possible in order to avoid spherical aberration.
5. Lasers are beamed into the sky during observations at the world's most advanced optical telescopes to aid in precision pointing.
6. (*Answer*) It is possible to combine radio telescopes from around the world to effectively create a telescope with the diameter of Earth.
7. ALMA is located high up in the Andes mountains so that it can be as close as possible to the stars it is observing.
8. (*Answer*) It is theoretically possible to make observations from the ground using radio wavelengths as long as 100 m.
9. The semimajor axis of Mars' orbit is about $a = 1.5$ au. All else being equal, if you were an astronomer on Mars the smallest stellar parallax angle, p, that you would be able to measure from the ground would be approximately 1.5 times smaller than the smallest angle measurable from Earth.
10. (*Answer*) When comparing the brightnesses of two stars, the star with the larger apparent magnitude will appear brighter in the sky.
11. The absolute magnitude of a star is what its apparent magnitude would be if it was exactly 10 pc from the Sun.
12. (*Answer*) A star that is closer to Earth than 10 pc will have a larger numerical value for its absolute magnitude than it will for its apparent magnitude.
13. The center of mass of a binary star system must always be along a line that connects the centers of the two stars.
14. (*Answer*) In a binary star system, the star with the greater mass has the smaller orbit.
15. The two stars in a binary star system orbit about the center of the more massive star.
16. (*Answer*) The Harvard spectral classification scheme is a sequence of stellar spectra ranging from the most luminous stars to the least luminous ones.
17. Annie Jump Cannon classified the spectra of more than 425,000 stars.
18. (*Answer*) For the most massive main-sequence stars (O stars), the peak of the blackbody radiation curve is in the blue portion of the electromagnetic spectrum.
19. Stars in the upper right-hand corner of the Hertzsprung–Russell diagram are cool and have very large radii.
20. (*Answer*) If two stars have identical radii, the star with the greater surface temperature will have the lower value for the absolute magnitude.

21. Our Sun is a G2 main-sequence star.
22. (*Answer*) Stars in the lower left-hand corner of the Hertzsprung–Russell diagram are hot and have radii that are comparable to that of our Sun.

Multiple Choice

23. One telescope has a 1 m diameter mirror and another telescope has a 4 m diameter mirror. How many times more photons can be "collected" by the larger telescope over the same period of time?
    (a) 2   (b) 4   (c) 8   (d) 16
24. (*Answer*) Which of the following reasons explains why all major optical research telescopes use mirrors instead of lenses?
    (a) Mirrors can be made lighter than lenses with the same diameter.
    (b) Only one surface of a mirror needs to be ground to the correct shape.
    (c) A lens must be free of defects throughout the entire body of the lens, not just on one surface.
    (d) Lenses must be supported from their edges while mirrors can be supported from behind.
    (e) Large lenses take much longer to reach an equilibrium temperature when temperature changes occur during observing.
    (f) all of the above
    (g) (a) and (b) only
    (h) (c) and (d) only
25. An astronomer is interested in observing an object at a wavelength of 21 cm. Can that observation be made from the ground? In which portion of the electromagnetic spectrum is that wavelength located?
    (a) no, ultraviolet   (b) yes, ultraviolet
    (c) no, visible   (d) yes, visible
    (e) no, infrared   (f) yes, infrared
    (g) no, radio   (h) yes, radio
26. (*Answer*) Which of the following wavelength regions has the most energetic photons?
    (a) gamma rays   (b) x-rays   (c) ultraviolet
    (d) visible   (e) infrared   (f) microwave
    (g) radio
27. Most of the infrared spectrum is unavailable for ground-based observing because
    (a) of the ionosphere.
    (b) water vapor and other molecules effectively absorb infrared light.
    (c) blue sky makes it hard to see infrared light.
    (d) of interference with ground-based communications.
    (e) more than one of the above
28. (*Answer*) Strategies to minimize the blurring caused by atmospheric turbulence at visible wavelengths include
    (a) constructing observatories on tall mountains.
    (b) using lasers to create artificial "stars" high in the atmosphere that can be monitored to see how they flicker.

Continued on the next page

28. *Continued*
    (c) making observations from space.
    (d) building ever-larger diameter telescopes.
    (e) using holographic imagery.
    (f) all of the above
    (g) (a), (b), and (c) only
29. Multiple radio telescopes can be combined to effectively make one large telescope through a technique known as _____ .

    (a) polarization   (b) duplication
    (c) interferometry   (d) collective replication
    (e) photosynthesis
30. (*Answer*) The star visible from Earth with the smallest numerical value for apparent magnitude is
    (a) Sirius   (b) Betelgeuse   (c) Procyon
    (d) Vega   (e) the Sun
31. Canopus, in the constellation Carina, is the second brightest star in the night sky. Canopus has a parallax angle of $0.010\,43''$ and an apparent magnitude of -0.62. Its distance from Earth is approximately _____ pc or _____ ly.
    (a) 95.9, 29.4   (b) 95.9, 312.6   (c) 0.01043, 0.0340
    (d) 0.01043, 0.0043   (e) 0.62, 2.0
32. (*Answer*) Based on the data presented in Exercise 31 and your answer, approximately how many times dimmer or brighter would Canopus be if it was magically moved to 10 pc?
    (a) 920,000 times brighter   (b) 920,000 times dimmer
    (c) 960 times brighter   (d) 960 times dimmer
    (e) 92 times brighter   (f) 92 times dimmer
    (g) 9.6 times brighter   (h) 9.6 times dimmer
33. A star that is located 4 pc from Earth has an apparent magnitude of 3. If it was seen from 10 pc away by alien astronomers who happened to use our magnitude system, its apparent brightness would be _____ by a factor of _____ relative to what Earth-based astronomers measure. Given that each change in magnitude of 1 corresponds to a brightness ratio of 2.512, its absolute magnitude is closest to _____ .
    (a) less, 2.5, 2   (b) greater, 2.5, 2   (c) less, 2.5, 4
    (d) greater, 2.5, 4   (e) less, 6.25, 1   (f) greater, 6.25, 1
    (g) less, 6.25, 5   (h) greater, 6.25, 5
34. (*Answer*) If two stars have the same absolute magnitude, and star A is located 10 pc from Earth while star B is located 1000 pc from Earth, star A will appear _____ by a factor of _____ .
    (a) brighter, 10,000   (b) dimmer, 10,000
    (c) brighter, 100   (d) dimmer, 100
35. For the two stars in Exercise 34, if star A has an apparent magnitude of 5, then star B will have an apparent magnitude of _____ .
    (a) −5   (b) 0   (c) 5   (d) 10   (e) 15
36. (*Answer*) ε Eridani has an absolute magnitude of 6.18 and Procyon A has an absolute magnitude of 2.68. _____ is the more luminous star by a factor of _____ . (Use either the brightness ratio–magnitude difference equation or Table 15.1.)
    (a) ε Eridani, 3.50   (b) ε Eridani, 25.1
    (c) Procyon A, 3.50   (d) Procyon A, 25.1
37. Vega, the bright star in the constellation of Lyra, has an absolute magnitude of 0.6, an apparent magnitude of 0.0, and a parallax angle of $0.129''$. Vega is also an A0 star. Estimate the surface temperature of Vega and its distance from Earth.

Referring to Fig. 15.39, is Vega a supergiant, a main-sequence star, or a white dwarf?
    (a) 12,900 K, 7.75 pc, supergiant
    (b) 10,000 K, 7.75 pc, main-sequence
    (c) 10,000 K, 0.129 pc, white dwarf
    (d) 5000 K, 0.129 pc, supergiant
    (e) 5000 K, 12.9 pc, main-sequence
    (f) 3000 K, 12.9 pc, white dwarf
38. (*Answer*) Two stars have identical radii, but one star is four times hotter than the other. The hotter of the two stars is _____ times more luminous than the cooler one.
    (a) 4   (b) 16   (c) 32   (d) 64   (e) 256
39. Two stars have identical surface temperatures, but one star has a radius that is four times greater than that of the other star. The larger star is _____ times more luminous than the cooler star.
    (a) 4   (b) 16   (c) 32   (d) 64   (e) 256
40. (*Answer*) Assume that star A has a radius that is 500 times greater than the radius of star B. Assume also that star A has a surface temperature that is only one-half the surface temperature of star B. Which star is more luminous, and by how many times?
    (a) star A, 1.5625×10^4 (b) star B, 1.5625×10^4
    (c) star A, 1.5625×10^{10} (d) star B, 1.5625×10^{10}
    (e) We can't tell because no information was given about the distances of the two stars from Earth.
41. A star has a luminosity of $10\,L_{Sun}$ and a temperature of 8000 K. What is the radius of the star, expressed in terms of the radius of the Sun (R_{Sun})?
    (a) 4.9×10^{-8} (b) 1.1   (c) 1.65   (d) 11.6
42. (*Answer*) Our Sun is classified as a _____ star.
    (a) A3   (b) G2   (c) M8   (d) K6   (e) O5
43. The correct order of the Harvard spectral classification scheme from cool stars to hot stars is
    (a) ABCDEFGKM   (b) MKGFEDCBA   (c) OBAFGKM
    (d) MKGFABO   (e) AHOTSFR
44. (*Answer*) The masses of main-sequence stars range from _____ .
    (a) $1.5 \times 10^4\,M_{Sun}$ to $20\,M_{Sun}$
    (b) $2020\,M_{Sun}$ to $0.001\,M_{Sun}$
    (c) at least $150\,M_{Sun}$ to $0.075\,M_{Sun}$
    (d) about $15\,M_{Sun}$ to $0.075\,M_{Sun}$
45. According to the data in Fig. 15.39, supergiant stars in the upper right-hand portion of the H–R diagram have absolute magnitudes between about _____ and _____ , while white dwarfs have absolute magnitudes between _____ and _____ .
    (a) −3, 0, 8, 16
    (b) 3000, 4000, 10,000, 20,000
    (c) 100, 1000, 0.001, 0.00001
    (d) 3, −6, 14, −6

Short Answer

46. Explain why refracting telescopes are susceptible to chromatic aberration while reflecting telescopes do not have this problem.
47. What behavior of light that was discussed in Section 6.5 can be used to measure imperfections in the surface of a mirror down to a fraction of one wavelength?
48. Describe what is meant by spherical aberration.

49. (a) What advantage(s) exist for constructing very large mirrors with segmented mirrors?
    (b) Are there any special challenges to that strategy? If so, explain.
50. What advantage(s) are provided by building very large telescopes?
51. Explain why radio telescopes must be much larger than optical telescopes in order to have the ability to see the same level of detail.
52. What challenges exist in combining the data from many individual radio telescopes in order to create the equivalent of a telescope with a diameter equal to the separation of the most widely separated individual telescopes?
53. In Section 11.8 the problem of global warming was discussed. What does that issue have in common with the challenge of making astronomical observations at infrared wavelengths?
54. Explain why it is important for astronomers to be able to make observations in all parts of the electromagnetic spectrum.
55. Aldebaran, in the constellation of Taurus, the Bull, has an apparent magnitude of 0.85 while 134 Tau, also in Taurus, has an apparent magnitude of 4.85. How many times brighter is Aldebaran than 134 Tau as seen from Earth?
56. (*Answer*) Castor and Pollux are the brightest stars in Gemini (the Twins). Castor has an apparent magnitude of 1.90 and Pollux has an apparent magnitude of 1.15. Which star is brighter, as seen from Earth?
57. Based on the information for Castor and Pollux given in Exercise 56, what is the ratio in brightness of the two stars?
58. (*Answer*) Mira, in Cetus, is a long-period variable star that changes its apparent brightness on a regular basis with a period of 332 days. Over one period its apparent magnitude in visible light changes from about 2 to 10 and back again. Is Mira visible to the naked eye every time it is up in the night sky? Why or why not? (Mira is Latin for wonderful.)
59. Assuming that Mira (see Exercise 58) varies in magnitude in such a way that it spends one-half of its time brighter than its average apparent magnitude, approximately how long is Mira visible to the naked eye during one period?
60. Again using the information on Mira given in Exercise 58, how many times brighter is Mira in visible light at the peak of its brightness than when it is at its dimmest?
61. (*Answer*) Cassiopeia, the seated queen, is a constellation in the northern sky with an easily recognizable W shape that represents the Queen's throne. The two stars that compose the top of the throne's back are Shedir and Caph, having nearly identical apparent magnitudes of 2.2. Shedir is located 228 ly from Earth while Caph is 55 ly from Earth. Which star has the greater luminosity, and by how many times? (Requires a calculation using the inverse square law for light, although a "guesstimate" can be made by noticing that 228 ly is only slightly larger than 220 ly, which is four times greater than 55 ly.)
62. Crux, the Southern Cross, is a prominent constellation in the southern hemisphere. Mimosa and Acrux, both in Crux, have apparent magnitudes of 1.25. Mimosa is 288 ly from Earth and Acrux is 321 ly from Earth. Which star has the greater luminosity, and by how many times?
63. (*Answer*) Betelgeuse (in the shoulder of Orion) is very near the end of its life and, in astronomical terms, will soon destroy itself as an impressive supernova explosion that will be easily visible from Earth. The parallax angle of Betelgeuse is $0.005\,95''$.

    (a) How far away is Betelgeuse, measured in parsecs and in light-years?
    (b) If Betelgeuse were to explode today, in what year would Earthlings learn about it?
    (c) Is it possible that Betelgeuse has already blown up in a supernova explosion? Explain your answer.
64. (a) Why do astronomers go to such great lengths to determine the distances to stars?
    (b) How is Gaia contributing to our understanding of stars on the other side of the Milky Way Galaxy?
    (c) How is Gaia contributing to our understanding of the shape and size of our Galaxy?
65. You are out camping on a moonless summer night in southern Utah. You and a friend naturally sit out under the stars to enjoy the spectacular view of very dark skies. Knowing that you had taken this astronomy course, your friend asks you how we really do know that Earth orbits the Sun. How would you explain it to her? (Unfortunately, about one-quarter of all adults in the United States still believe that the Sun orbits Earth.)
66. The apparent magnitude of Fomalhaut in Piscis Austrinus is 1.17 and its absolute magnitude is 1.74. Is Fomalhaut closer to Earth than 10 pc or farther away? Explain your reasoning.
67. The Sun has an absolute magnitude of 4.74 and Rigel, in Orion's west foot, has an absolute magnitude of −7.84. Use the luminosity ratio–absolute magnitude difference equation to determine how many times more luminous Rigel is than the Sun.
68. Explain in your own words how astronomers are able to determine the masses of the individual stars in a binary star system.
69. (*Answer*)
    (a) Referring to Fig. 15.25, calculate the semimajor axes of the orbits of Sirius A and Sirius B.
    (b) What is the sum of the two semimajor axes? Compare your answer with the value for a given in Example 15.6.
70. (a) Referring to Fig. 15.25 and Kepler's laws, which law describes the changing distances traveled by Sirius B (or Sirius A) over equal five-year intervals? Explain.
    (b) Explain how the changing distance traveled over equal time intervals during different parts of the orbit can be explained, based on what Newton's universal law of gravitation equation is telling you about this situation.
71. In order to help you understand the complexities of observing the orbits of stars in a binary system, draw on a sheet of paper the circular orbits of two stars about their mutual center of mass. Assume that one star is exactly twice the mass of the other and label the orbit of the more massive star A and the less massive star B. Your diagram should accurately reflect the relative sizes of the orbits of the two stars. You should also pick and mark a position for star A at some arbitrary time in its orbit and then locate and mark the position of star B at that same time. Note that the paper represents the plane of the orbits of the two stars.

    Tilt the paper in some direction. Clearly describe how the paper is oriented and carefully sketch how the orbits now appear from your vantage point. What can you conclude about the appearance of the orbits compared to their actual shapes as drawn on the paper? You should do this for three different orientations of the paper.

72. Explain why, as a consequence of the force of gravity, the orbits of both stars in a binary star system must lie in the same plane. *Hints*: Think about the direction of the gravitational pull on one star due to the other one. You should also consider the center of mass of the system.

73. (*Answer*) If star A has six times the mass of star B in a binary system and the average distance between the two stars is 140 au,
    (a) on average, how far is star A from the center of mass of the system?
    (b) on average, how far is star B from the center of mass of the system?

74. If star A has four times the mass of star B in a binary system and the average distance between the two stars is 150 au,
    (a) on average, how far is star A from the center of mass of the system?
    (b) on average, how far is star B from the center of mass of the system?

75. In your own words, describe how absorption lines can be used to determine the temperature of a star.

76. (*Answer*) Based on its absorption spectrum, a star is classified as K5.
    (a) Make a rough estimate of its surface temperature using Table 15.2.
    (b) From Wien's blackbody color law, what would the peak wavelength of the star be?
    (c) From Fig. 6.43, what color would you expect the star to be?

77. Write a paragraph or two summarizing the contribution(s) that Edward Pickering's "computers" made to furthering our understanding of stars.

78. (a) Make an estimate of the peak wavelength of the underlying blackbody spectrum in Fig. 15.33.
    (b) What surface temperature corresponds to that peak wavelength?
    (c) Estimate the Harvard classification for that star. You should estimate the subclass as well.
    (d) Compare your estimate with the absorption lines of the closest spectral class shown in Fig. 15.31. Which lines are present in both figures?
    (e) Now compare your estimate of the temperature with the curves in Fig. 6.43. Does your estimate seem consistent with the curves in that figure? Why or why not?

79. α Centauri A is a G2 star with an absolute magnitude that is very nearly the same as that of the Sun. What can you conclude about the star's radius?

80. A certain star has a surface temperature of 5750 K and a luminosity of 29,700 L_{Sun}. Without doing any calculations, what can you conclude about the radius of the star relative to the radius of the Sun?

81. (*Answer*) Use the data in Exercise 80 to calculate the star's radius (express your answer in units of the radius of the Sun, R_{Sun}).

82. In your own words, explain what is meant by "the universality of physical law." Why is that assumption so important to how we understand our universe?

83. For this exercise you will need to download a blank copy of a Hertzsprung–Russell diagram, BlankHR.pdf, from the Chapter 15 resources webpage.
    (a) Carefully sketch the main sequence as a thick line, the location of the supergiants as an elongated loop, another elongated and tilted loop where the giants are, and one more long, skinny loop indicating the location of the white dwarfs.
    (b) Mark and label the location where the Sun would be on the H–R diagram. The Harvard classification for the Sun is G2, its surface temperature is 5772 K, its absolute magnitude is 4.74, and its luminosity is 1 L_{Sun}.
    (c) Vega is an A0 main-sequence star with an absolute magnitude of about +0.6. Mark and label its position on the diagram.
    (d) Gl 887 is an M2 main-sequence star with an absolute magnitude of +8.6. Mark and label its position on the diagram.
    (e) Arcturus is a K2 star with an absolute magnitude of −0.9. Mark and label its position on the diagram.
    (f) Which luminosity class is Arcturus a member of?

84. On the same diagram you created for Exercise 83, specify the *approximate* masses and radii of a B0 main-sequence star and an M5 main-sequence star, in solar units. Do the same for the radius and typical mass of a white dwarf star (you only need to specify characteristic values rather than specific values). What might the radius of a supergiant star be?

85. For this exercise you will need to download a blank copy of a Hertzsprung–Russell diagram, BlankHR.pdf, from the Chapter 15 resources webpage. Plot the absolute magnitude (or luminosity) and the Harvard classification (or temperature) for each of the stars listed in Table 15.3. In each case indicate the name of the star and specify which luminosity class it belongs to.

Table 15.3 Spectral classes or surface temperatures and absolute magnitudes or luminosities for selected stars.

Star	Spectral class (Temperature)	Abs. mag. (Luminosity)
Sun	G2	4.74
Proxima Centauri	M5	+11.1
Gleise 411	M2	+8.5
Sirius A	A1	+1.2
Sirius B	(25,970 K)	(0.03 L_{Sun})
ε Eri	K2	+5.7
61 Cyg A	K5	+6.8
Procyon A	F5	+2.5
EGGR 555	(4747 K)	+15.4
τ Cet	G8	+5.3
Canopus	F0	−5.5
Capella	K0	+0.3
Spica	B1	−3.6
Betelgeuse	M2	−5.9
Altair	A7	+2.2
GJ 3016	(6018 K)	+14.3

Activities

86. Identify and photograph or describe at least three naturally occurring patterns found in nature. The patterns need not be astronomical in origin. Suggest possible reasons for the patterns that you see.

16

The Lives of Stars

Only in the darkness can you see the stars.

Martin Luther King Jr (1929–1968)

16.1	The Interstellar Medium: The Realm of Gas and Dust	640
16.2	A Star is Born	646
16.3	The Life of an Aging Star	653
16.4	The Pulsating Stars	665
16.5	Clusters as Tests of Stellar Evolution Theory	673
	Exercises	678

Fig. 16.1 The ancient globular cluster, ω Centauri, with about 10 million stars. (ESO; CC BY 4.0)

Introduction

Back in Chapter 9 we studied the closest star to Earth: our own Sun. Because it is only one astronomical unit away, astronomers and physicists have been able to study the Sun in great detail through telescopic observations on the ground, neutrino observatories underground, and continuous observations from spacecraft. Observations have even been made from within the outermost regions of the Sun's atmosphere.

Thanks to the myriad of high-precision observations and ever-increasing computational power, scientists have been able to develop a deep understanding of the Sun's interior and the physical processes at work there. The Sun serves as a laboratory for studying how nuclear reactions deep in the interior produce the necessary power to maintain it as a main-sequence star for the last 4.57 billion years. Oscillations ringing through the interior also allow us to use helioseismology to map the Sun's interior structure very accurately. For example, we know that a radiation zone exists in the inner 72% of its radius that transports the energy produced by those nuclear reactions via photons up to the convection zone, and convection then carries the energy the remaining distance to the surface. It is the convection zone that drives the solar oscillations.

As the convection zone deposits heat at the base of the Sun's atmosphere, the hot, rising convective blobs provide the visible evidence of its presence, creating the granulation that can be seen at the base of the photosphere. It is the base of the photosphere that is the lowest and coolest part of the solar atmosphere. The very thin photosphere is where the Sun's surface temperature and radius are defined, and from which most of the sunlight originates. The upward motion of the convective cells pushes the atmosphere in front of them, causing shock waves to move through the atmosphere, which, together with the Sun's magnetic field, heats the chromosphere, the transition region, and the corona to increasingly higher temperatures with height, reaching nearly one million kelvins in the ghostly corona. The ever-changing structure of the corona, which is tied to the solar cycle and the Sun's magnetic field, in turn produces coronal holes that are a major source of the solar wind, and where coronal mass ejections can send energetic particles out into space, including toward Earth, resulting in the spectacular aurora seen at higher latitudes on our planet.

It is our knowledge of blackbody radiation, atomic physics, the production of emission and absorption spectral lines, and the nature of magnetic fields that makes it possible to study the Sun's atmospheric composition, temperature structure, and motions in great detail. Atomic physics and the solar spectrum also provide us with the tools necessary to measure the strength of the Sun's magnetic field across the star, along with characteristics of sunspots, solar flares, prominences, and the 11-year solar cycle. The Sun's magnetic field, generated by the magnetic dynamo operating in its interior, together with the Sun's differential rotation and the turbulent motions of its convection zone, combine to produce the solar cycle.

All of this knowledge of the Sun, and our understanding of the physical processes upon which it is based, serve as one check on our overall understanding of other stars, how they are born, what happens to them as they age, and how they die.

Armed with the additional tools developed in Chapter 15, we now focus our attention on the study of stellar evolution. (It is important to point out here that, unlike its use in biology, the term "evolution" in astronomy refers to the life cycle of a single object, such as a star, and not of the change in collective characteristics of a species over time, or the development of a new species from a predecessor.)

16.1   The Interstellar Medium: The Realm of Gas and Dust

Not only do particles like photons and neutrinos traverse the vast distances between the stars, but gases and dust can be found in the so-called "vacuum" of space as well. Although the space between the stars isn't truly as empty as Fig. 16.2 might suggest, the typical density of particles is incredibly low. It is common in the interstellar medium to have only one atom or molecule in one cubic meter of space ($1/m^3$) or one in every 35 cubit feet ($0.03/ft^3$). The interstellar medium also hosts a very weak but ever-present complex magnetic field.

Gas and Dust Exist Everywhere

Key Point
Hydrogen and helium make up most of the mass of the interstellar medium.

As with the Sun and the overwhelming majority of stars, by far the most dominant element in the interstellar medium is hydrogen, which comprises about 70% of the mass, while helium makes up most of the rest. Hydrogen by itself can exist in a variety of forms: neutral hydrogen (H), ionized hydrogen with its electron missing

Fig. 16.2 A region of the sky in the constellation of Sagittarius. Note the wide variation in the colors of the stars captured in the image as a consequence of their surface temperatures. Many of the objects in the deep background are galaxies. (ESA/Hubble & NASA; CC BY 4.0)

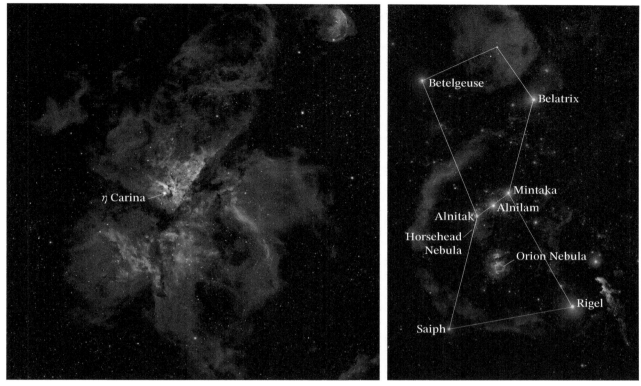

Fig. 16.3 *Left:* The Carina Nebula and the massive star η Carina (eta Carina) in the constellation of Carina. (Adapted from a work by Harel Boren; CC BY 4.0) *Right:* The Orion Molecular Complex. The Orion Molecular Complex fills much of the constellation of Orion because it is relatively close to Earth. Most of the complex isn't visible to the naked eye, however. (Adapted from a work by Rogelio Bernal Andreo; CC BY-SA 3.0)

(H$^+$), molecular hydrogen (H$_2$), and (rarely) as negatively ionized hydrogen (H$^-$) if the conditions are just right (as in part of the Sun's atmosphere). Helium (He) is one of the noble gases on the far right column of the periodic table, meaning that its lowest-energy orbital is completely filled and its electrons are paired up (recall the discussion of atomic energy levels and the Pauli exclusion principle; Section 8.7). As a result, it does not easily form molecules and so is almost always found as a neutral atom or in one of its ionized forms, He$^+$ and He^{++}. [The molecular ion helium hydride (HeH$^+$) was detected for the first time in 2019 after having been theoretically predicted to be the first molecule to form following the Big Bang at the beginning of the universe.]

Not all of interstellar space is as vacant as the region shown in Fig. 16.2. Some regions contain vast nebulas of dust and gas, like the Carina Nebula in the southern constellation of Carina, or the Orion Molecular Complex in the northern sky (Fig. 16.3). Both are nebulas of immense size: the Carina Nebula is 8500 light-years from Earth and nearly 500 ly across (it takes 500 years for light to traverse the cloud complex) while the Orion Molecular Complex is between 1000 ly and 1400 ly away, and measures about 400 ly from north to south.

Within interstellar nebulas, in addition to hydrogen and helium, a large number of other atoms, ions, and molecules have also been identified through their spectral

Fig. 16.4 *Top:* The central portion of the Gum nebula, located about 3000 ly away in the constellation of Vela. Dust lanes are very apparent, cutting across the glowing gases behind. The bright star in the horizontal dust lane left of center is a foreground star. (ESO, CC BY 4.0) *Bottom:* The Horsehead Nebula just south of Alnitak in Orion; see Fig. 16.3. Dust obscures the glowing gases behind the nebula. The Horsehead Nebula is 1400 ly away. (ESO, CC BY 4.0)

signatures. Carbon, nitrogen, and oxygen are common, along with the molecules they can form, including many molecules familiar to us on Earth and throughout the Solar System, such as water (H_2O), carbon monoxide (CO), carbon dioxide (CO_2), methane (CH_4), ammonia (NH_3), molecular nitrogen (N_2), and molecular oxygen (O_2); the last two being the major constituents in our atmosphere. Molecules involving chlorine, silicon, sulfur, iron, potassium, phosphorus, and other elements have also been found. There are also some surprisingly complex organic molecules, including the amino acid glycine (CH_2NH_2COOH), and a soccer-ball shaped molecule of carbon nicknamed a buckyball (C_{60}); buckyballs are one type of a class of molecules called fullerenes, because they reminded scientists of the geodesic domes popularized by the American architect, Richard Buckminster Fuller (1895–1983).

Dust is present in interstellar nebulas too, as can be seen in the Gum and Horsehead nebulas (Fig. 16.4). Dust cores can be composed of silicates (rock-building material) or graphite (carbon), upon which ices and other molecules can attach themselves. Dust will also obscure the light from stars and gases behind them. Dust blocks light most effectively at shorter wavelengths like blue light, but becomes more transparent with longer wavelengths such as red and infrared light. You have studied this effect before in the discussion on page 239*ff* about why the sky is blue and sunsets are red. The wavelength-dependent transparency of dust is seen very clearly in Fig. 16.5 where the image on the left was taken in visible light and the image on the right was taken in infrared light. The towering features are about 4 ly tall and have been called the Pillars of Creation because stars are in the process of forming in and near the tops of the pillars. Very young and hot stars of spectral classes O and B are beyond the tops of the images and their ultraviolet light is evaporating the gas and dust in the nebula except where the nebula is shielded by the most dense regions, hence the pillars.

Rock towers known as hoodoos, like those in Bryce Canyon National Park, Utah (Fig. 16.6), are found in various places around the world. Their formation is somewhat analogous to the formation of the Pillars of Creation. Hoodoos form when a stronger layer of rock sits atop softer, more easily eroded rock. As slightly acidic water seeps through cracks in the harder layer it can eat away at the softer rock below the cracks. The process leaves a hard rock cap that helps to protect rock below from weathering, much like the denser regions of dust and gas are shielding the gases in their shadows from the intense ultraviolet radiation of the massive young stars beyond the top of Fig. 16.5. (The formation of Bryce Canyon hoodoos is also aided by repeated cycles of freezing and melting of water in cracks: when the water freezes it expands as ice, breaking up the rock in the cracks in the same way that potholes form in roads.)

Because of the penetrating power of infrared light, the successor to the Hubble Space Telescope (HST), the James Web Space Telescope (JWST) illustrated in Fig. 13.15, is designed to operate at infrared wavelengths in order to see star-forming regions, exoplanets, new details in galaxies, and features of the early universe.

Fig. 16.5 The Pillars of Creation, one of the Hubble Space Telescope's most iconic images. The Pillars are part of the Eagle Nebula in the constellation Serpens. The image on left was taken in visible light and the image on the right is the same region of space as seen in infrared light. Infrared light is able to penetrate much of the dust and gas, except in the most dense regions, revealing previously hidden stars. [NASA, ESA, and the Hubble Heritage Team (STScI/AURA); CC BY-SA 3.0 IGO]

Fig. 16.6 Hoodoos at Bryce Canyon National Park, Utah. The tallest hoodoo at the center of the picture is known as Thor's hammer. (NPS)

Observations of Stellar Nurseries

Figure 16.7 shows an infrared view of the Orion Nebula, a vast cloud of neutral and ionized hydrogen, helium, dust, and other atoms and molecules. The hydrogen is ionized because of the intense ultraviolet radiation from massive, new-born stars. The mass of the nebula is estimated to be about 2000 times the mass of the Sun, and is a very active region of star formation. It is estimated that some 700 stars are in the process of forming within the nebula.

Although the density of particles in interstellar space far from nebulas is typically just one to ten particles per cubic meter, cloud densities can be between 100 million and 10 billion times greater (10^8–10^{10} particles/m^3), and even as high as 10^{15} particles/m^3 in the most dense cores of nebulas. Although far greater than interstellar space, the densities are still very low compared to our atmosphere, where the particle density at sea level is 2.5×10^{25} particles per cubic meter. In other words, the particle density of our atmosphere at sea level is some 25 billion times greater than the most dense cores of interstellar nebulas, and up to roughly 100 thousand trillion times greater than typical cloud densities. The favorite science fiction trick of hiding from a nearby enemy spaceship inside a nebula really wouldn't work very well! The temperatures of interstellar nebulas are also very low by Earth standards, usually ranging from 10 K to 50 K, while room temperature is roughly 300 K (remember that absolute zero, 0 K, is the limit for the lowest possible temperature in the universe).

Fig. 16.7 A deep infrared view of the Orion Nebula, one of the three objects comprising Orion's sword; see Fig. 16.3 (*Right*). (ESO/H. Drass et al.; CC BY 4.0)

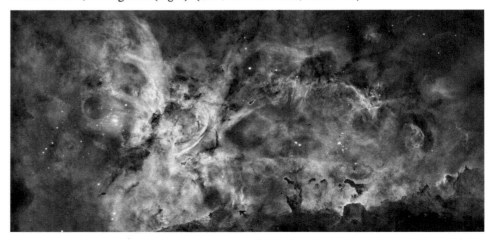

Fig. 16.8 A small portion of the Carina Nebula. [Hubble Image: NASA, ESA, N. Smith (University of California, Berkeley), and The Hubble Heritage Team (STScI/AURA); CTIO Image: N. Smith (University of California, Berkeley) and NOAO/AURA/NSF]

The regions within interstellar nebulas where stars are forming are chaotic and violent places. Figure 16.8 shows a portion of the Carina Nebula (Fig. 16.3 *Left*) where intense stellar winds and ultraviolet radiation are reshaping the nebula. Dust is still visible, along with bubbles and walls of gas. Figure 16.9 shows a region of the

Orion Nebula where an explosion appears to have occurred. Gases are streaming out of a central location at different speeds, symbolized by colors ranging from blue for those gases moving fastest toward us and red for those moving more slowly. Some young stars are ejecting jets of gas, like Herbig–Haro 212 in Fig. 16.10, while others eject whole envelopes, like the very massive η Carina (Fig. 16.11). While η Carina was ejecting its envelope in the 1800s, this binary system was the second brightest star in the sky for about 20 years. One member of the η Carina system is estimated to be 90 M_{Sun} and the other is 30 M_{Sun}.

Fig. 16.9 A portion of the Orion Nebula (Fig. 16.7) where the Atacama Large Millimeter/sub-millimeter Array (ALMA) measured Doppler-shifted motions of the gas from an explosion. The blue streaks are moving toward us the fastest while the red streaks are moving more slowly. [ALMA (ESO/NAOJ/NRAO), J. Bally/H. Drass et al.; CC BY 4.0]

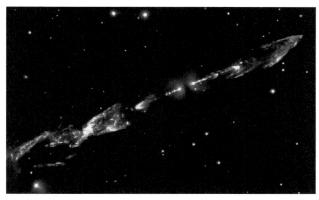

Fig. 16.10 Herbig–Haro 212 is a very young star that is ejecting jets of material perpendicular to the disk that formed around the star. The disk can be seen just to the right and slightly above the center of the image obscuring the star itself. (ESO/M. McCaughrean; CC BY 4.0)

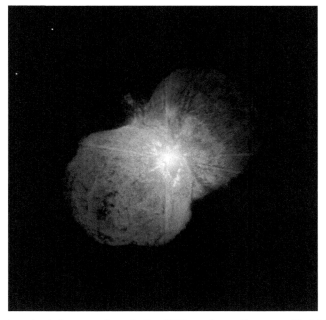

Fig. 16.11 η Carina ejected an envelope in the 1800s and was the second brightest star in the sky for a time. The location of η Carina in the Carina Nebula is shown in Fig. 16.3. (ESO; CC BY 4.0)

16.2   A Star is Born

Fig. 16.12 A dark, dusty Bok globule in the nebula NGC 281, located about 9500 light-years away in the constellation of Cassiopeia. The Bok globule is very close to the center of a young cluster of stars. [NASA, ESA, and the Hubble Heritage Team (STScI/AURA)]

An entire nebula, such as the Carina Nebula or the Orion Nebula, may be hundreds of light-years across with a typical temperature of perhaps 50 K and a mass that is thousands of times greater than the mass of the Sun. But within the nebula small, dark, dusty cocoons called Bok globules can be found, like the one shown in Fig. 16.12. Named after Bart Bok (Fig. 16.13), who first discovered them in the 1940s, Bok globules have temperatures of about 10 K, masses typically between 2 and 50 M_{Sun}, and they measure 1 to 2 light-years (60,000 to 120,000 au) across. Unfortunately, what is happening inside these wombs of star formation is hidden in visible light because of the opaqueness of dust. Although modern infrared telescopes can unveil some of what's hidden inside, astronomers still cannot probe the most opaque, dusty globules. Because of this, our knowledge of much of the earliest stages of star formation comes largely from computer models of the process.

Free-Fall of a Collapsing Nebula

You would certainly expect to find some clumpiness within a Bok globule; these are the seeds of star formation. A typical clump might have a temperature of 10 K, be perhaps 5000 au across, and despite being relatively dense compared to its surroundings, it may still have a mass of only 0.01 M_{Sun} or so. The upper left portion of Fig. 16.14 illustrates such a clump of gas and dust. Given their temperatures and sizes, when placed on a Hertzsprung–Russell diagram like the one in Fig. 16.15, these clumps would be located far below the lower right-hand corner of the diagram.

Fig. 16.13 Bartholomeus Jan 'Bart' Bok (1906–1983). (AIP Emilio Segrè Visual Archives, Physics Today Collection)

In a region that is slightly more dense, gravity can attract more dust and gas, causing the seed to grow in mass. The gas and dust would initially be in free-fall because, even with the slightly greater density in the region, collisions between particles are extremely rare. But as the collapse continues, the density increases, more and more collisions occur, and the internal pressure and temperature begin to rise. During the free-fall, gravitational potential energy is released, that goes in part into heating the gas and dust, while the rest of the released energy goes into producing photons (light). The changing temperatures and luminosities of collapsing clouds over time, plotted on a Hertzsprung–Russell diagram, are known as evolutionary tracks, and are depicted in Fig. 16.15.[1]

Free-fall in this case is nothing like free-fall on Earth, where the acceleration due to gravity on the surface is about $g = 9.8 \text{ m/s}^2$. Instead, in parts of the collapsing cloud, the acceleration due to gravity may be as small as one-trillionth of Earth's acceleration due to gravity. Despite this slow pace, the release of energy causes the cloud's temperature to rise slowly while the luminosity (the power output due to the production of heat and light) increases fairly rapidly over time. Because the density is so low at this point, the rapidly increasing luminosity occurs simply because light

---

[1] Some students become a bit confused by the "motion" of data representing a forming or aging star through the H–R diagram. There is no physical motion of the star through space involved here; the language used simply refers to a star's changing temperature and luminosity that cause the *displayed data point to move* through the diagram. An evolutionary track is simply a tracer of the movement of the $(T_{surface}, L)$ data point as the star's physical condition changes over time.

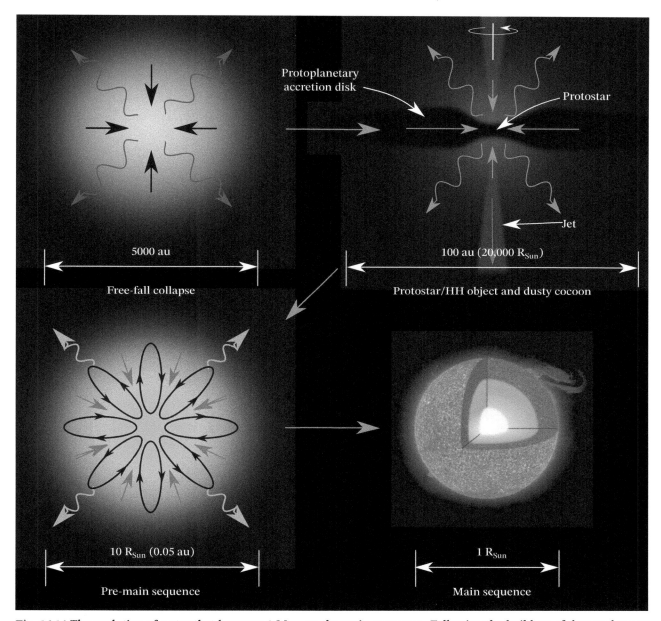

Protoplanetary accretion disk

Protostar

Jet

5000 au

Free-fall collapse

100 au (20,000 R_{Sun})

Protostar/HH object and dusty cocoon

10 R_{Sun} (0.05 au)

Pre-main sequence

1 R_{Sun}

Main sequence

Fig. 16.14 The evolution of a star that becomes 1 M_{Sun} on the main sequence. Following the build-up of dust and gas at the center of the collapse, rising temperatures vaporize the dust, break molecules apart, and ionize atoms. The changing temperature and luminosity over time are shown on the H–R diagram in Fig. 16.15. The rough sizes in each stage are given.

can easily escape from most of the locally collapsing cloud but it is not yet successful in escaping the dark, dusty cocoon that surrounds the local cloud. As the gas and dust fall inward, the mass accumulates in the center, a process we call accretion.

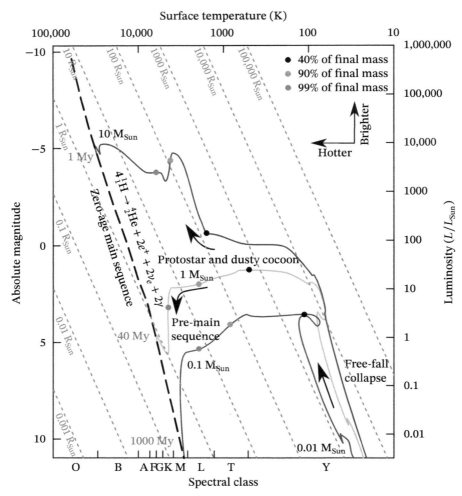

Fig. 16.15 The changing temperatures and luminosities (evolutionary tracks) depicted on the H–R diagram during free-fall collapse, the protostar/HH object and dusty cocoon phase, and pre-main-sequence evolution. The phases of stellar evolution indicated on the diagram for a growing Sun-like star of 1 M_{Sun} correspond to the ones depicted in Fig. 16.14. The evolution starts with 0.01 M_{Sun} cores having temperatures of 10 K surrounded by nebulas of gas and dust. Cases of star formation for stars that ultimately reach masses of 0.1 M_{Sun} (dark red), 1 M_{Sun} (orange), and 10 M_{Sun} (navy) are shown (the colors are indicative of the stars' colors when they reach the main sequence). The dots indicate the approximate locations on the tracks where the growing stars have accumulated 40%, 90%, and 99% of their mass. The dashed red lines are lines of constant radius, labeled in terms of the radius of the Sun with the smallest radii in the lower left and the largest radii in the upper right (the lines tell you the sizes of the growing stars as they cross the lines). The times required to reach the main sequence are one billion years for a 0.1 M_{Sun} star, 40 million years for a Sun-like star, and one million years for a 10 M_{Sun} star. The L, T, and Y spectral classes are for brown dwarfs and free-floating planets. (Data courtesy of Wuchterl and Tscharnuter, 2003,[2] and Bernasconi and Maeder, 1996[3])

[2]Wuchterl, G., and Tscharnuter, W. M. (2003). "From clouds to stars. Protostellar collapse and the evolution to the pre-main sequence I. Equations and evolution in the Hertzsprung-Russell diagram," *Astronomy & Astrophysics*, 398, 1081–1090.

[3]Bernasconi, P. A., and Maeder, A. (1996). "About the absence of a proper zero age main sequence for massive stars," *Astronomy & Astrophysics*, 307:829–839.

Protostars and Dusty Cocoons

As depicted in the upper right illustration of Fig. 16.14, when the collapsing cloud around the growing central mass becomes sufficiently dense, light can no longer escape easily. Essentially, the growing protostar at the center is shrouded in a dusty shell. Because the light is trapped by the dust, the luminosity stops increasing.

At this point the dust in the cloud starts to vaporize, as the temperature of the accreting central mass continues to rise. The dust vaporizes because the weak molecular bonds holding the dust particles together are easily broken. As the obscuring dust begins to disappear it is possible to see progressively deeper into the protostar, revealing the much smaller and hotter mass growing inside. This phase of protostar evolution is depicted in the H–R diagram in Fig. 16.15 as a roughly horizontal path of increasing temperature, without much change in luminosity. An object that was originally 1000 to 10,000 times larger than the Sun when the luminosity stopped rising is now much more compact, with a radius closer to 1–10 R_{Sun}. A surface temperature that was once near 10–100 K is now thousands of kelvins. In the meantime, the central regions continue to accrete infalling material.

Because the original cloud certainly had some rotation associated with it, and therefore angular momentum, as the cloud collapsed it formed a protoplanetary accretion disk of material perpendicular to its rotation axis. This is just the planetary disk formation process we saw in Section 14.2. Not only can planets likely form in the disk, but the disk can also continue to feed material to the growing protostar, allowing it to become more massive than it otherwise might. The disk also restricts the direction that some material can be ejected from the cloud that results from the chaotic nature of the collapse and the interaction with magnetic fields. The result is jets of material leaving the cloud along the rotation axis. An example of this was shown in Fig. 16.10. Figure 16.16 shows another example, where the disk is seen edge-on across the protostar, and a jet is seen shooting out from the center, perpendicular to the disk. Other examples are seen in Fig. 14.15 and Fig. 14.16.

During this nearly horizontal phase of protostar formation in the H–R diagram the easiest nuclear reaction to activate, known as deuterium burning ($^2_1H + {}^1_1H \rightarrow$ $^3_2He + \gamma$), begins in the cores of protostars [recall that deuterium (^2_1H) is an isotope of hydrogen and γ symbolizes a photon]. During deuterium burning, the rate of collapse of the cloud is slowed as nuclear energy is released. Eventually, deuterium burning stops because deuterium is not a very abundant fuel in collapsing clouds; nuclear reactions don't resume again until the full hydrogen-to-helium reaction sequence starts. It is deuterium burning that is the second step in the conversion of hydrogen into helium occurring in the Sun today; see Fig. 9.33.[4]

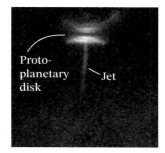

Fig. 16.16 HH 30 with its protoplanetary disk, seen edge on. A jet can be seen shooting out from the center of the protostar. Jets and a protoplanetary disk can also be seen in Fig. 16.10. [Adapted from C. Burrows (STScI & ESA), the WFPC 2 Investigation Definition Team, and NASA; CC BY 4.0]

Pre-Main-Sequence Evolution

Eventually, the gases will become dense enough that photons can no longer readily escape from the cloud's interior. Because the energy cannot easily get to the surface and escape into space, the newly forming pre-main-sequence star gets dimmer, but still gets hotter as well, with convection playing the key role in transporting

---

[4]As a reminder, the term "burning" for nuclear reactions is not the same as the everyday burning that you usually think of with fires, or when your author cooks dinner. Nuclear burning is millions of times more energetic and results in different elements or isotopes rather than different molecules, as chemical burning does.

energy to the surface as illustrated in the sketch in the lower left corner of Fig. 16.14. This behavior is displayed on the H–R diagram in Fig. 16.15, with the temperature increasing slightly and the luminosity decreasing significantly, creating tracks that still move slowly to the left but are now generally getting lower over time.

Main Sequence: Nuclear Reactions and a Star Is Born

Although the gases are still collapsing in on themselves, the rate of collapse is starting to slow as the density continues to increase. At the same time, the interior is getting hotter and hotter, especially at the very center of the collapse. Eventually, the central temperature and density get so high that all of the atoms are completely ionized and the nuclei are moving around very quickly, causing them to randomly run into one another. As discussed in detail in Section 9.6, when the temperature in the core reaches a few million kelvins, the collisions between ions, combined with quantum mechanical tunneling, produce nuclear fusion reactions. The steady release of nuclear energy finally helps the new-born star fight against gravity's relentless pull. As a result, the star starts to stabilize and its surface temperature and brightness change only very slowly in response to nuclear reactions changing the composition of the star's interior.

Summary of the Star Formation Process

Without worrying about the various "wiggles" in the evolutionary tracks through the H–R diagram in Fig. 16.15 for stars of different masses, the behaviors are more or less similar. To summarize:

1. The temperature slowly increases and the luminosity rises fairly rapidly during free-fall collapse.
2. The evolutionary tracks across the H–R diagram become roughly horizontal as
   (a) the protostar becomes opaque,
   (b) a planetary accretion disk forms,
   (c) jets develop,
   (d) deuterium burning starts, and
   (e) dust vaporizes.
3. The luminosity decreases just before reaching the main sequence, when convection carries energy to the surface of the pre-main-sequence star. (Technically, more massive pre-main-sequence stars change from convection to radiation near the surface, shortly before reaching the main sequence.)

What is not similar between stars of different masses is the time it takes from the start of the collapse of the cloud to the start of main-sequence nuclear reactions. A $10 \, M_{Sun}$ star requires less than one million years for those nuclear reactions to start, a $1 \, M_{Sun}$ star like the Sun takes 40 times longer, and a $0.1 \, M_{Sun}$ star takes 1000 times longer than a $10 \, M_{Sun}$ star, or roughly one billion years. The greater the mass, the faster a star lives out its life.

Why Does the Main Sequence Exist?

What is common with stars of all masses is that the gravitational collapse is essentially halted when nuclear reactions start: specifically, the conversion of four hydrogen atoms into a helium atom through a series of steps summarized by $4{}_1^1\text{H} \rightarrow {}_2^4\text{He} + 2e^+ + 2\nu_e + 2\gamma$. This is the process that is going on in the interior of the Sun today, as we discussed in Section 9.6. For stars a bit more massive than the Sun, up to the most massive main-sequence stars, the chain of reactions to get from hydrogen to helium is different from lower mass stars, but the result is the same. The set of points on the H–R diagram where each star reaches that stage of its life forms the left edge of the main sequence, referred to as the zero-age main sequence.

Key Point
All stars on the main sequence are converting hydrogen into helium in their cores.

If you go back and look at the H–R diagram in Fig. 15.39, which displays 4 million stars, you will see that the main sequence is not a thin line; instead, it has a thickness to it. There are two reasons for this. First, not all main-sequence stars have exactly the same composition. Although they are all primarily composed of about 70% hydrogen by mass give or take, with helium making up most of the rest, stars can have a small range of heavier elements of between nearly 0% –3% or so when they form. These differences in heavy element abundances affect the exact locations of new-born stars in the H–R diagram.

The second reason for the thickness of the main sequence in Fig. 15.39 is that all of the stars are not the same age on the main sequence, even if they have the same mass and initial composition. As stars age on the main sequence, the nuclear reactions occurring in their cores are changing their core compositions from more hydrogen and less helium to less hydrogen and more helium, while other abundances change a bit as well, including the proportions of carbon, nitrogen, and oxygen in stars more massive than the Sun. The observable impact of those internal changes is that stars slowly get a bit cooler and their luminosities increase slightly over time. The changes are very slow, especially for the lowest mass stars, but they do give the main sequence its width.

Key Point
Changing composition in the interior of a star changes its surface temperature, luminosity, and radius.

Imagine that you are at a sporting event or a large concert and you are able to record where all of the people in the stadium are, at a specific moment in time. Of course you will find the players or bands on the field, but you will also find most of the spectators in the stands, simply because that is where they spend most of their time. However, you will also find some spectators elsewhere: getting refreshments, wandering around the corridors, using the restrooms, or in the parking lot. Since most spectators don't spend most of their time in those other locations, the odds of finding them there is small, but not zero. As the game goes on, some spectators will return to their seats while others get up and move around. A snapshot of the stadium reflects the lives of the spectators at a specific moment in time.

When Gaia made the observations of those 4 million stars in Fig. 15.39, it was essentially taking a snapshot of the current lives of all of those stars. About 90% of the stars that the spacecraft observed were found on the main sequence, but at different points in time during the "hydrogen-burning" phase of their evolution. This is because stars spend about 90% of their lifetime converting hydrogen into helium. This is just like finding most of the spectators at the game in the stands, simply because that is where they spend most of their time.

Key Point
About 90% of the stars in the sky are main-sequence stars.

If you look at the upper end of the main sequence, there are very few stars of spectral classes O and B. This is because not many of those massive stars form in

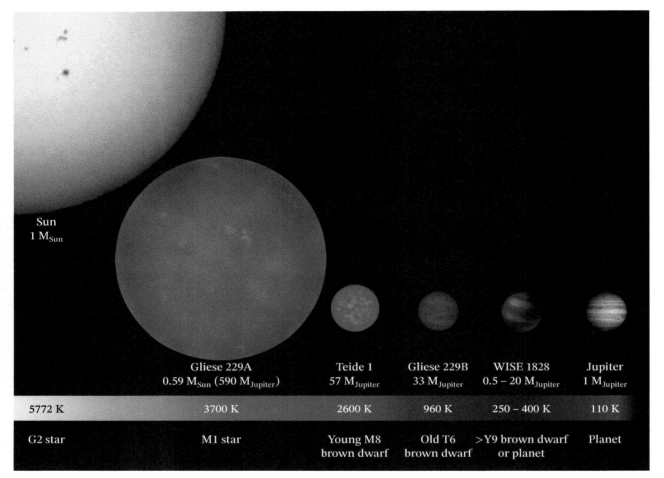

Sun 1 M$_{Sun}$	Gliese 229A 0.59 M$_{Sun}$ (590 M$_{Jupiter}$)	Teide 1 57 M$_{Jupiter}$	Gliese 229B 33 M$_{Jupiter}$	WISE 1828 0.5 – 20 M$_{Jupiter}$	Jupiter 1 M$_{Jupiter}$
5772 K	3700 K	2600 K	960 K	250 – 400 K	110 K
G2 star	M1 star	Young M8 brown dwarf	Old T6 brown dwarf	>Y9 brown dwarf or planet	Planet

Fig. 16.17 A comparison of the Sun, a low-mass main-sequence star, brown dwarfs, and Jupiter. Gliese 229B orbits Gliese 229A. The objects other than the Sun and Jupiter are artists' renditions. (Adapted from MPIA/V. Joergens; CC BY 3.0)

Key Point
The most massive and largest main-sequence stars are at the upper end of the main sequence and the least massive and smallest are at the lower end.

the first place, and because once those that do form arrive on the main sequence they consume their core hydrogen very quickly, a consequence of their cores being very hot and dense. The result is that a snapshot of even 4 million stars isn't going to find very many massive O and B main-sequence stars. This is reflected in the sparse numbers in Fig. 15.39. At the other end of the main sequence, very low-mass stars take an extremely long time to reach the main sequence, and so not all of these stars have reached that point in their evolution. The smallest stars that can get hot enough to convert hydrogen into helium have masses of about 0.075 M$_{Sun}$, or 75 times the mass of Jupiter (75 M$_{Jupiter}$).

Brown Dwarfs

Although objects smaller than 75 M$_{Jupiter}$ cannot convert hydrogen into helium, as long as they are more massive than 13 M$_{Jupiter}$ they can still undergo deuterium burning. Deuterium burning doesn't occur with sufficient vigor to stabilize them from further collapse, however. These objects finally stop collapsing simply because gravity can't compress them any farther. These very low-mass "failed stars" are known as brown dwarfs. (For historical reasons, main-sequence stars are referred to as dwarf stars, so the classification of brown dwarf is simply an extension of that classification.) Because they are very dim and cool, and they emit most of their radiation in infrared light, it took until 1988 before the first one was discovered. Since that time thousands have been found, necessitating the development of the spectral classes L, T, and Y seen in Fig. 16.15. The more massive brown dwarfs can arrive below the bottom end of the main sequence as M-class objects but then cool off to become progressively L-, T-, and Y-class objects. Their atmospheres are cool enough to contain dust, and they can have spectral lines of water (H_2O), methane (CH_4), ammonia (NH_3), and other molecules found among the planets in our Solar System. Objects with less mass than 13 $M_{Jupiter}$ are classified as planets, meaning that free-floating planets without a parent star can exist. Figure 16.17 shows a comparison of various stellar and sub-stellar objects.

To extend the mnemonic for remembering the order of spectral classes that was first suggested by Annie Jump Cannon, perhaps consider "Oh, Be A Fine Girl/Guy, Kiss Me. Less Talk, Yes?"

Key Point
Brown dwarfs have masses between 13 and 75 times the mass of Jupiter. Objects less massive than brown dwarfs are considered to be planets.

16.3 The Life of an Aging Star

If you refer back once again to Gaia's Hertzsprung–Russell diagram, there are obviously many stars that are not currently located on the main sequence. For example, there are giant stars, supergiant stars, and white dwarf stars labeled on the diagram. So how do these stars differ from main-sequence stars, and how did they get to be that way? The remainder of this chapter and Chapter 17 are devoted to answering those questions and what happens when stars reach the ends of their lives.

Stellar Evolution Is a Constant Fight Against Gravity

The main sequence exists because hydrogen is being converted into helium in the cores of all main-sequence stars. Before the nuclear reactions started, the birth and life of the star was governed by the relentless pull of gravity. Even on the main sequence, gravity still plays an important role, in part because it is gravity that caused the core to get hot and dense enough for nuclear reactions to start in the first place, and it continues to squeeze the core while core hydrogen burning is occurring.

What stops pre-main-sequence stellar collapse is the fresh source of abundant energy that the conversion of hydrogen into helium provides. In essence, the rate of hydrogen consumption is a self-regulating process. If gravity compresses the core too much, the temperature and density increase, causing the nuclear reactions to go faster, which in turn produces more heat. The added heat means that the nuclei and electrons in the core are moving faster and collide more frequently; in other words, the pressure in the core has increased. The increased pressure causes the core to expand, leading to decreasing temperature and density, and so the reactions throttle back.

Nuclear reactions are a means of fighting the relentless pull of gravity inside the star. But nuclear reactions aren't a permanent solution, because they require

Key Point
Core hydrogen burning is a self-regulating process.

Key Point
When nuclear reactions are occurring, changes in the structures of stars are relatively slow throughout the vast majority of their lifetime. When gravity dominates, changes happen much more quickly.

fuel to "burn." For example, as hydrogen is depleted in the core of a main-sequence star and helium becomes dominant, hydrogen burning must come to an end, leaving the star without that avenue to stop further compression. **Stellar evolution is essentially a battle between the energy released by nuclear reactions and the energy released or absorbed by gravitational potential energy.** As long as nuclear reactions are occurring, changes on the surface are generally slow, but when gravity dominates, changes occur much more quickly. There are exceptions to these general rules to be discussed in Sections 16.4 and 17.2, but those periods in a star's life are short-lived.

Main-Sequence Lifetimes

Once stars reach the main sequence, they spend about 90% of their nuclear-burning lifetimes there. The reason is simple, hydrogen is by far the most abundant element in zero-age main-sequence stars, and hydrogen burns relatively slowly. However, for stars above about 0.3 M_{Sun}, only the hydrogen contained in the nuclear core gets consumed, leaving the rest of the star's hydrogen unburned. The exceptions are the lowest mass stars, where convection extends throughout the entire star: like a conveyor belt, fresh hydrogen is fed into the central part of the star from as far away as the star's surface.

How long a star remains on the main sequence depends very much on its mass. The most massive stars burn up all of their core hydrogen in less than one million years; a 5 M_{Sun} star requires about 93 million years to consume its hydrogen; the Sun will be about 10 billion years old by the time its core hydrogen is depleted; a 0.8 M_{Sun} star will need 25 billion years to burn through all of its hydrogen; and a 0.075 M_{Sun} star will last for trillions of years on the main sequence. However, with the age of the universe being about 13.8 billion years, stars with masses up to just over 0.8 M_{Sun} haven't had a chance to use up their allotment of hydrogen fuel in the core because they would need to be older than the age of the universe for that to happen. All stars that have ever reached the main sequence with these low masses are still there, and will be for a very, very long time!

The Evolution of an Intermediate-Mass Star of 5 M_{Sun}: A Case Study

So what happens when the core's hydrogen is used up? Let's first look at what happens to a 5 M_{Sun} star, as an example. A 5 M_{Sun} star is considered to be an intermediate-mass star, where intermediate masses range from 2 M_{Sun} up to about 9 M_{Sun} or so. (Note: Because the material in the remainder of this section relies on what you learned in Section 9.6 about the Sun's interior, you may find it useful to review that section before proceeding. In any case, it will be helpful in preparing for a comprehensive final exam!)

Key Point
Intermediate-mass stars on the main sequence have convective cores and radiative envelopes.

The left-hand diagram in Fig. 16.18 shows the interior structure of a 5 M_{Sun} star on the main sequence and Fig. 16.19 shows its location on the H–R diagram between points (a) and (b). Unlike our Sun, the core is convective and the envelope transports energy to the surface by radiation (photons). The reason the core is convective is that hydrogen burning happens by a more temperature-dependent set of reactions than the proton–proton chain used by the Sun. These reactions use $^{12}_{6}C$,

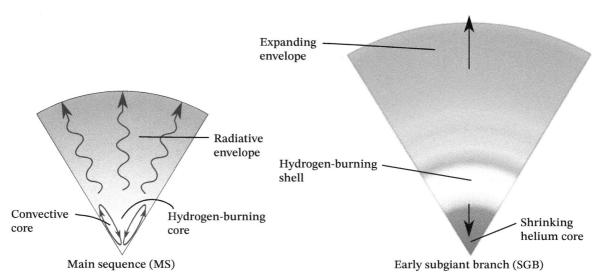

Fig. 16.18 *Left:* The structure of a blue 5 M_{Sun} star of spectral class B while on the main sequence between points (a) and (b) in Fig. 16.19. *Right:* The same star after its core hydrogen has been consumed, leaving a helium core and a hydrogen-burning shell. The star evolves to the right in the H–R diagram, as shown in Fig. 16.19 between points (c) and (d).

$^{14}_{7}$N, and $^{16}_{8}$O in the sequence of steps from hydrogen to helium, even though the carbon, nitrogen, and oxygen nuclei are not ultimately consumed in the process; they are burned in one step and regenerated again in another step (borrowing a term from chemistry, they are *catalysts* for the reactions). This sequence of reactions is known as the CNO cycle, because carbon, nitrogen, and oxygen cycle through the reaction sequence. The strong dependence of the CNO cycle on temperature means that the temperature drops off rapidly from the center of the star outward, making convection a much more efficient way to transport energy out of the cores of intermediate mass stars. The CNO cycle is actually the primary process that converts hydrogen into helium in all main-sequence stars with masses greater than about 1.2 times the mass of the Sun. In any case, when the hydrogen is depleted, the result is the same for all stars; the ash of hydrogen burning is helium, and so a helium core now exists in our 5 M_{Sun} star.

Without hydrogen fusion taking place in the core, gravity can now get the upper hand, causing the entire star to briefly shrink [between points (b) and (c) in Fig. 16.19]. However, not all of the star's hydrogen has been used up, only the hydrogen in the core. As the star shrinks, the temperature and density in the interior increase, and soon a layer of hydrogen sitting above the helium core gets hot enough to burn at point (c) in Fig. 16.19, resulting in what is termed a hydrogen-burning shell. When the shell ignites, the energy produced causes the envelope above the shell to puff up and the surface also cools off. The 5 M_{Sun} star now leaves the main sequence behind and moves to the cooler, redder portion of the H–R diagram to the right. The star's structure at this point is illustrated in Fig. 16.18 (*Right*). Its evolutionary track on the H–R diagram is shown as the subgiant branch (SGB) in Fig. 16.19 between points (c) and (d). The luminosity of the star decreases because energy is needed to inflate the envelope that would otherwise go into light escaping from the star, just like energy that is required to lift a barbell. The energy gets stored

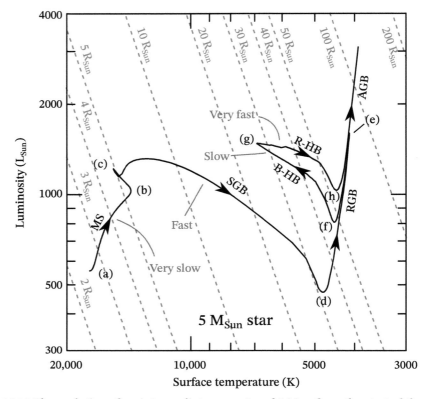

Fig. 16.19 The evolution of an intermediate-mass star of 5 M_{Sun} from the start of the main sequence to near the end of its life. Phases: MS (main sequence), SGB (subgiant branch), RGB (red giant branch), B-HB (blueward horizontal branch), R-HB (redward horizontal branch), AGB (asymptotic giant branch). Points: (a) core hydrogen burning on the zero-age main sequence, (b) end of core hydrogen burning, (b)–(c) star shrinks, (c) hydrogen-burning shell, (d) deep envelope convection develops, (e) start of core helium burning and weakening of the hydrogen-burning shell, (f) contraction of the star strengthens the hydrogen burning shell again, (g) core helium burning ends and a helium-burning shell develops and the hydrogen-burning shell weakens, (h) very deep envelope convection resumes as the hydrogen-burning shell strengthens and the helium-burning shell continues. (Data courtesy of Ekström et al., 2012[5])

as gravitational potential energy that can later be recovered (recall the discussion of conservation of energy starting on page 196).

As the star's envelope expands and cools, the cooler gases near the surface make it harder for energy to move through the star by radiation, which means that convection must carry the load. As a result, a deep convective envelope develops and energy is efficiently transported to the surface. This causes the luminosity to rise rapidly along a track known as the red giant branch [labeled RGB in Fig. 16.19 between points (d) and (e)]. The interior structure of the red giant branch star is illustrated in Fig. 16.20 (*Left*). The track that the red giant branch star follows in the H–R diagram is nearly the same track that the star followed during its pre-main-

[5]Ekström, S., et al. (2012). "Grids of stellar models with rotation. I. Models from 0.8 to 120 $M_{\odot}$ at solar metalicity ($Z = 0.014$)," *Astronomy & Astrophysics*, 537:A146–164.

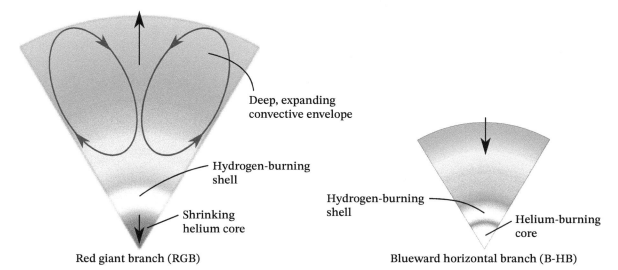

Deep, expanding
convective envelope

Hydrogen-burning
shell

Shrinking
helium core

Hydrogen-burning
shell

Helium-burning
core

Red giant branch (RGB)

Blueward horizontal branch (B-HB)

Fig. 16.20 The 5 M$_{Sun}$ star from Fig. 16.18. *Left:* The structure of the star as it rises up the red giant branch between points (d) and (e) in Fig. 16.19. *Right:* The star's structure on the blueward portion of the horizontal branch, moving from point (f) toward point (g) in Fig. 16.19.

sequence evolution phase, when the envelope was deeply convective then as well, except that the red giant branch star is moving upward instead of downward. That track is characteristic of all stars with large, deep convective envelopes.

As the 5 M$_{Sun}$ star continues to climb the red giant branch its helium core is still shrinking under the relentless force of gravity. Since the start of the subgiant branch, both the central temperature and the central density have been increasing. Helium nuclei are four times more massive than hydrogen nuclei, which means that they require four times as much kinetic energy to move at the same speed ($KE = mv^2/2$). Helium nuclei also have two positively charged protons instead of just one, making it four times harder (2×2) to overcome electrostatic repulsion between two helium nuclei before they can collide, as described by Coulomb's law of electrostatic force.

Stars actually accomplish helium burning in two steps, that result in the production of carbon-12:

Step 1: $^4_2\text{He} + ^4_2\text{He} \rightleftharpoons ^8_4\text{Be}$,

Step 2: $^4_2\text{He} + ^8_4\text{Be} \rightarrow ^{12}_6\text{C} + \gamma$.

The double arrows in step 1 indicate that the beryllium nucleus created from the two helium nuclei can simply decay back into two helium nuclei if the beryllium nucleus doesn't react with another helium nucleus quickly enough. If the central temperature and density are only great enough for Step 1 to occur, then for all practical purposes helium burning doesn't take place. This means that even higher temperatures and densities are needed for helium and beryllium to move fast enough to overcome electrostatic repulsion between the doubly charged helium nucleus and the four protons contained in the even more massive beryllium nucleus. At point (e) in Fig. 16.19, the temperature in the center is about 100 million kelvins (10^8 K) and the density is roughly one million times greater than the density of water (10^9 kg/m^3) for a 5 M$_{Sun}$ star. Even under these extreme conditions, if not for the aid of quantum mechanical tunneling that we first discussed with hydrogen burning in the center of the Sun in Fig. 9.6, helium burning still wouldn't happen in the core of the star at the tip of the red giant branch.

Because three helium nuclei ultimately lead to carbon, the two steps can be summarized by:

$$3\,{}^4_2\text{He} \rightarrow {}^{12}_6\text{C} + \gamma.$$

Since helium nuclei are Rutherford's stable alpha particles from his investigations of radioactivity (page 255), this two-step process is referred to as the **triple alpha process**. During helium burning, some of the carbon-12 that is produced also combines with helium to create oxygen-16, which in turn can form a small amount of neon-20:

$$\begin{aligned} {}^{12}_6\text{C} + {}^4_2\text{He} &\rightarrow {}^{16}_8\text{O} + \gamma, \\ {}^{16}_8\text{O} + {}^4_2\text{He} &\rightarrow {}^{20}_{10}\text{Ne} + \gamma. \end{aligned}$$

Key Point
The core-helium-burning phase (the blueward horizontal branch) is the helium-burning version of the hydrogen-burning main sequence.

With the start of core helium burning, the hydrogen-burning shell gets pushed out to lower temperatures and quiets down a bit, which causes the star's envelope to shrink. Observationally, this means that the star rapidly descends the H–R diagram of Fig. 16.19 between points (e) and (f). As the envelope gets smaller, convection weakens and the star's surface starts getting warmer again, sending its evolutionary track blueward [toward the left from point (f) to point (g)]. Meanwhile, the compression of the star causes the hydrogen-burning shell to strengthen again, leading to a rise in luminosity. This phase of the 5 M_{Sun} star's structure is shown in Fig. 16.20 (*Right*). In essence, this phase can be thought of as the core helium-burning version of the hydrogen-burning core on the main sequence, although it happens much more rapidly than main-sequence evolution. This relatively slow blueward track is known as the **blueward horizontal branch** (B-HB). With core helium burning taking place, the blueward horizontal branch phase is the second longest phase of evolution after the main sequence.

Key Point
The redward horizontal giant branch is the helium-burning shell analog to the hydrogen-burning shell subgiant branch.

Of course, as with core hydrogen burning, core helium burning must also come to an end when the helium fuel in the core is used up. At this point, the core is composed of carbon, oxygen, and perhaps a small amount of neon. Since carbon and oxygen are three and four times more massive than helium, respectively, and they have three and four times as many positive charges, the core is not dense or hot enough for those nuclei to undergo a significant amount of nuclear burning. As a result, just like when hydrogen ran out in the center at the end of the main-sequence phase and the core started to shrink, the same thing happens again to the carbon–oxygen core. However, instead of hydrogen burning outside of an inert core, there is still helium remaining where it wasn't dense and hot enough to burn before. With the shrinking core, that helium-rich shell can start to burn and when it does so, the hydrogen-burning shell shuts down temporarily, again because it was pushed outward. In a fashion similar to the subgiant branch following the main sequence, the **redward horizontal branch** (R-HB) carries the helium-shell-burning star back toward the deep convective envelope region on the right-hand side of the H–R diagram. The redward horizontal branch structure of the 5 M_{Sun} star is depicted in Fig. 16.21 (*Left*), and its observational evolutionary track is seen in Fig. 16.19 between points (g) and (h).

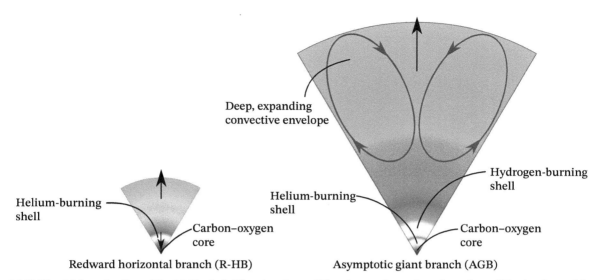

Fig. 16.21 The 5 M_{Sun} star from Fig. 16.20. *Left:* The structure of the star on the redward portion of the horizontal branch between points (g) and (h) in Fig. 16.19. *Right:* The asymptotic-giant-branch structure of the star following point (h) in Fig. 16.19.

When the star's evolutionary track reaches the deep convective envelope portion of the H–R diagram, the core has become hot enough again for a hydrogen-burning shell to add its energy output to that of the helium-burning shell following point (h). As a result, the star becomes very luminous again and follows an evolutionary track that parallels the red giant branch. This phase is referred to as the asymptotic giant branch (AGB), named for the mathematical term that describes one curve approaching but never quite getting to a line as the curve goes toward infinity. The structure of the AGB star with its deep, expanding convective envelope, its hydrogen- and helium-burning shells, and its carbon–oxygen core is shown in the diagram in Fig. 16.21 (*Right*).

Key Point
The asymptotic giant branch with its carbon–oxygen core and two burning shells is analogous to the helium core and hydrogen-burning red giant branch.

Confession of a Deception

After leaving the main sequence, as the 5 M_{Sun} star evolves along its evolutionary track in Fig. 16.19, the diagrams in Figs. 16.18, 16.20, and 16.21 become increasingly misleading. This is because the central regions of the star become increasingly out of proportion to the size of the star as depicted. In fact, the core becomes a smaller and smaller part of the star's overall volume because the core is continually shrinking while the envelope has been getting larger. Studying the red-dashed lines in Fig. 16.19, notice that the star starts out on the main sequence with a radius that is about 2 1/2 times greater than the radius of the Sun, but by the time it arrives at the upper part of the asymptotic giant branch the star is closer to 110 times larger than the Sun. To give you an idea of just how small the core is, compared to the overall size of the AGB star, take a look at Fig. 16.22: all of the nuclear reactions and their remnant ash had to be expanded by 250 times to see the tiny, innermost nuclear-burning region of the star. The region from the helium ash inward is roughly the size of Earth, while the overall size of the star is larger than the orbit of Mercury.

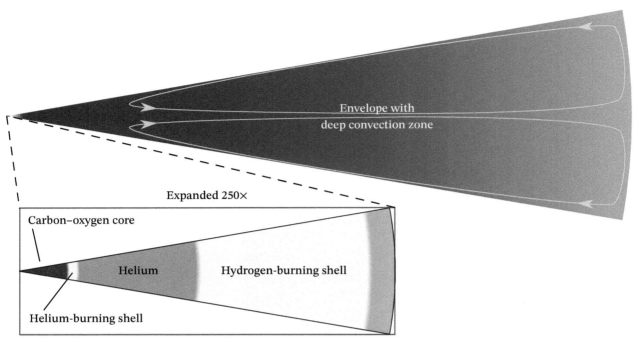

Fig. 16.22 The 5 M_{Sun} star near the top of the asymptotic giant branch in Fig. 16.19 with the core-nuclear-burning region enlarged by a factor of 250.

A Quick Summary

There is a lot of information in the discussion of the 5 M_{Sun} star above, but overall the star is constantly working to find nuclear reactions that can offset the relentless pull of gravity that is trying to compress the star. As a result:

1. Burning hydrogen into helium starts in the core when the temperature and density become great enough, together with the aid of quantum mechanical tunneling, to overcome Coulomb's electrostatic repulsion between positively charged nuclei.

2. When core hydrogen is depleted, the core resumes contracting, but unburned hydrogen just above the helium core now gets hot and dense enough to burn.

3. When hydrogen burning starts in a shell above the helium core, enough energy is released to inflate the star's envelope, which also causes the surface to get cooler.

4. When the star's surface gets cool enough (around 3000 K), a deep convective envelope develops that efficiently carries energy to the surface, dramatically increasing the star's luminosity.

5. When the contracting helium core gets hot and dense enough to cause the heavier and more positively charged helium nuclei to collide (again with the aid of quantum mechanical tunneling), the core is able to fight gravity by converting helium, and then beryllium, into carbon, oxygen, and a small amount of neon.

6. Core helium burning pushed the hydrogen-burning shell outward, causing it to weaken. This lowers the luminosity.

7. The weakened hydrogen-burning shell allows gravity to compress the envelope and the star gets hotter (bluer) at the surface. This blueward horizontal giant branch phase is analogous to the hydrogen-burning main sequence.

8. Helium is soon depleted, and just like when the helium core shrank after hydrogen burning stopped, the carbon–oxygen core shrinks as well. This allows helium to burn in a shell around the carbon–oxygen core, just like when hydrogen started burning around the helium core after leaving the main sequence. This redward horizontal branch phase is analogous to the subgiant branch.

9. The envelope again expands, gets redder, and reaches the region where deep convection occurs. With efficient convection, the star gets much brighter again. The contracting core also causes the hydrogen-burning shell to become more powerful. This climb up the asymptotic giant branch with two burning shells is analogous to the first climb up the red giant branch with one burning shell.

In general, for stars with no nuclear burning in the core, the core tends to shrink until it can get hot enough to start the next set of nuclear reactions. For the 5 M_{Sun} star, this stops when it can no longer compress enough to burn the higher mass, more highly charged nuclei of carbon, oxygen, and neon (if present). When core burning occurs after the main sequence, the star tends to shrink. Fuel around the core can also burn in shells when they get hot and dense enough. When shells ignite without a burning core, they tend to inflate the envelope, the star cools, and deep convection develops, driving the star to greater luminosity.

At this point we will stop following the 5 M_{Sun} star's evolution and wait for the ending in Chapter 17. Stay tuned!

The Helium Core Flash for Stars Less Than 2 M_{Sun}

Most of the time, gases behave themselves as we would expect them to from the ideal gas law of page 339 ($P = \rho k_B T / m_{average}$). However, in some situations that simple expectation falls short. One situation where this occurs is at the tip of the red giant branch for stars with masses less than about 2 M_{Sun}.

In a normal gas, squeezing the gas causes the pressure to increase. This is part of the self-regulating nature of most nuclear reactions discussed earlier in this section. But for these stars at the tip of the red giant branch, that response doesn't happen. In a regular gas, there is room for nuclei and electrons to run around and interact in a normal way. But if the gas gets too dense, the typical distribution of particle energies is dramatically changed. Recall our discussion of the Pauli exclusion principle on page 295*ff*, which says that no two electrons in an atom can have the same set of all four quantum numbers. It is this feature of quantum mechanics that leads to the structure of the periodic table of the elements. Electrons end up in different orbitals having different energy levels because they can't all occupy the same orbital.

The same Pauli exclusion principle applies in the cores of stars at the tip of the red giant branch with masses less than 2 M_{Sun}. With their electrons moving about in a sea of positively charged nuclei, when the gas is squeezed to an extreme by gravity, the electrons all occupy the lowest possible energy levels that they can reach without violating the Pauli exclusion principle. This degenerate electron gas has the strange property that heating up the gas doesn't cause it to expand by giving the electrons more kinetic energy. Instead, the energy first goes into "lifting" electrons into a more normal distribution of energies, rather than the lowest possible energies. This means that the normal throttling process of nuclear reactions — a hotter and denser gas causes the reactions to go faster, thereby heating the gas, which increases the pressure, causing the gas to expand and cool off — doesn't apply for a degenerate gas. At the tip of the red giant branch, when helium burning starts, stars with masses less than about 2 M_{Sun} release energy into their degenerate cores, which causes the temperature (but not the pressure) to go up, which causes the reactions to go faster, which causes the temperature (but not the pressure) to go up, which causes the reactions to go faster, ... This runaway helium core flash doesn't slow down until the electron degeneracy is lifted and a more normal distribution of of electron energies returns.

While most nuclear reaction processes happen very slowly, this particular situation is extremely fast, lasting only a few seconds. This phase of a star's evolution is so quick, in fact, that computers can't actually track the process efficiently. Notice that the computer tracks for these stars in Fig. 16.23 don't show the horizontal branches following the tip of the red giant branch. It isn't that the horizontal branches don't exist for these stars, it is just that the computer calculations were terminated for the tracks shown. Other calculations that restart after the helium core flash has subsided show that the star has moved down from the tip of the red giant branch onto the horizontal branches and up the asymptotic giant branch in a fashion similar to our 5 M_{Sun} example.

Even though the helium core flash is over in a matter of seconds, during that brief period of time so much energy is released that if the energy made it to the surface, the star would outshine all of the other stars in the Milky Way Galaxy combined! However, most, if not all, of the energy is simply absorbed by the overlying layers, with the possibility that a small amount of mass might get blown off the surface of the star.

Post-Main-Sequence Evolution of Massive Stars

Figure 16.23 shows a collection of post-main-sequence evolutionary tracks for stars of varying masses. What you might notice for the three most massive stars is that they look fundamentally different from the others. They tend to zigzag back and forth without the rapid rise along the asymptotic giant branch. The reason for the difference has to do with their immense gravities being able to squeeze their cores to much higher temperatures and densities than those achieved in low and intermediate mass stars. For stars up to around 9 M_{Sun} or so, nuclear reactions don't progress much beyond burning helium into carbon, oxygen, and neon, but high-mass stars don't have that limitation; with increasing mass, stars can burn heavier and more highly charged nuclei up to, and including, iron-56 and similar nuclei ($^{56}_{26}Fe$ has 26 protons and 30 neutrons).

As examples of what nuclei can be produced in high-mass stars, when one $^{12}_6C$ nucleus reacts with another $^{12}_6C$ nucleus, the result can be one of a host of possible nuclei:

$$^{16}_8O, \,^{20}_{10}Ne, \,^{23}_{10}Na, \,^{23}_{11}Mg, \text{ or } ^{24}_{12}Mg.$$

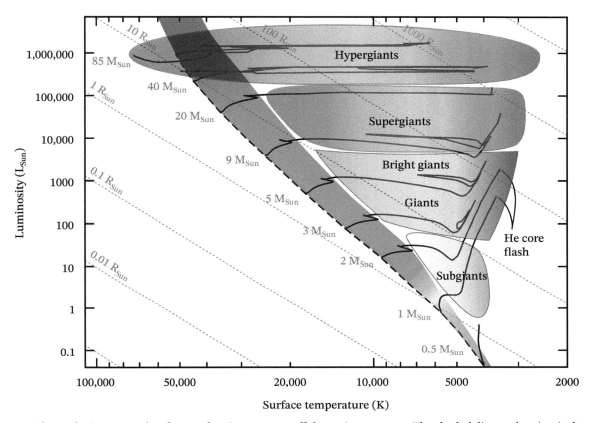

Fig. 16.23 The evolutionary tracks of stars of various masses off the main sequence. The shaded diagonal region is the main sequence where the dashed black line is the zero-age main sequence from Fig. 16.15, and the right edge is where hydrogen is exhausted in the cores of the stars. The approximate locations of the luminosity classes, subgiants, giants, bright giants, supergiants, and hypergiants are also displayed for reference. Once again, lines of constant radius are included as red dashed lines. (Data courtesy of Ekström et al., 2012)

When two $^{16}_{8}O$ nuclei interact, they can produce

$$^{24}_{11}Mg, \ ^{28}_{14}Si, \ ^{31}_{15}P, \ ^{31}_{16}S, \ or \ ^{32}_{16}S.$$

This list is far from complete, however.

When a new nuclear energy source is activated in the core of a massive star, the internal structure changes and nuclear-burning shells form as well. During this sequence of steps, the nuclear-burning region of the star starts to build up an onion-like layering of increasingly more massive isotopes going inward. But with each new energy source, the rate of nuclear burning increases and the time spent consuming that newfound energy source gets progressively shorter. While it takes the Sun nearly 10 billion years to use up its hydrogen, and about a tenth as long to burn its helium in the core, the core reaction $^{28}_{14}Si + {}^{4}_{2}He \rightleftharpoons {}^{32}_{16}S + \gamma$ only lasts about two days!

The Enormous Size of High-Mass Stars As They Age

Fig. 16.24 An image of Betelgeuse obtained by the Atacama Large Millimeter/submillimeter Array (ALMA) from a distance of 548 light-years. [ALMA (ESO/NAOJ/NRAO)/E. O'Gorman/P. Kervella; CC BY 4.0]

Going back to the H–R diagram in Fig. 16.23 that shows evolutionary tracks for stars of varying masses, notice the sizes indicated in the red-dashed diagonal lines. Betelgeuse, in Orion's shoulder, is a huge red supergiant shown at millimeter wavelengths in Fig. 16.24. But Betelgeuse certainly isn't the largest star out there: some of the tracks of the most massive stars reach regions where the star's radius can exceed 1000 R_{Sun}, or about 5 au. These so-called hypergiants can have radii that are greater than the orbit of Jupiter. To give a sense of these dimensions, the top portion of Fig. 16.25 shows the Sun, an O-type main-sequence star, Betelgeuse, and UY Scuti (a hypergiant, and the largest star yet discovered in the Milky Way Galaxy, as of June 2021). The bottom portion of the figure shows them side by side. Stars can be a really, really big deal.

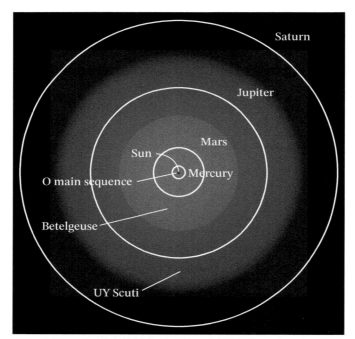

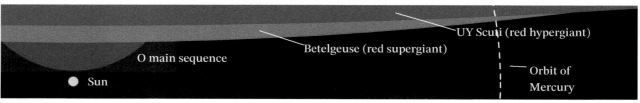

Fig. 16.25 *Top:* The sizes of the Sun, an O main-sequence star (10 R_{Sun}), Betelgeuse in Orion's shoulder (a red supergiant, 764 R_{Sun}), and UY Scuti (a red hypergiant, 1700 R_{Sun}), relative to the orbits of Mercury, Mars, Jupiter, and Saturn. *Bottom:* A side-by-side comparison of the stars.

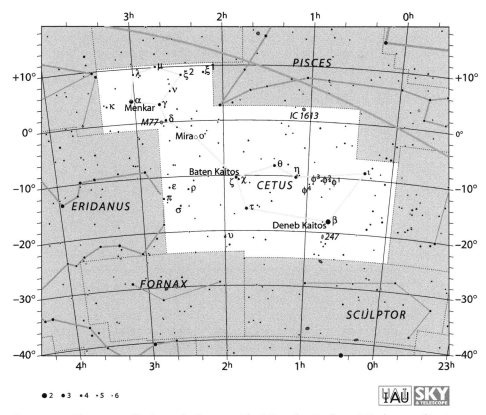

Fig. 16.26 The constellation of Cetus with Mira (o Ceti). Mira's coordinates are (02^{h}19^{m}20.792^s, −02°58′39.50″). Mira rises about 3 hours before Bellatrix in Orion. Right ascension is along the horizontal axis and declination is along the vertical axis. The dots in the lower left are a legend of stellar magnitudes on the chart. [IAU and *Sky & Telescope* magazine (Roger Sinnott & Rick Fienberg); CC BY 3.0]

16.4 The Pulsating Stars

An amateur astronomer and German Lutheran pastor, David Fabricius (1564–1617), began observing the star o Ceti in the constellation of Cetus, the sea monster, in August 1595 (o is omicron; don't forget the Greek alphabet in Table 4.1). The location of o Ceti is shown in the star chart (also referred to as a finding chart) in Fig. 16.26. Initially, the star was second magnitude, making it a fairly bright star in the southern sky, but over the next few months it completely faded from view, only to reappear again several months later. Over the course of 11 months this mysterious star had gone from second magnitude, to disappearing, then returning to its former brightness. We know today that the star decreased in brightness to an apparent visual magnitude of 10 before brightening again. The star was named Mira, which is Latin for "wonderful." Sadly, Fabricius's own life did not end wonderfully; after accusing a goose thief from the pulpit, the man hit Fabricius over the head with a shovel.

Today, even after eliminating stars whose light variability results from eclipsing binary star systems or non-repeating causes such as explosions or eruptions, tens of thousands of intrinsically variable stars of various types are known to exist, including Polaris, the North Star. These stars pulsate periodically or semi-periodically, expanding and contracting in radius almost as though they are breathing, while their intrinsic luminosities and surface temperatures also change. Because of their periodically changing radii, these variable stars are more specifically referred to as pulsating variable stars or simply pulsating stars.

Using Pulsating Stars to Measure Distances

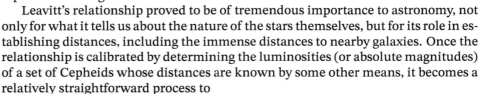

In Section 15.5 you were introduced to the remarkable work of the Harvard "computers," the group of women who analyzed data for Edward Pickering at the end of the nineteenth and early twentieth centuries. Among their many contributions was the development of the Harvard spectral classification scheme.

However, that wasn't the only important work they were doing. Henrietta Leavitt (Fig. 16.27) was busy studying variable stars found in two satellite galaxies of the Milky Way: the Large Magellanic Cloud (LMC) and the Small Magellanic Cloud (SMC). In 1908 she published a catalog of 1777 variable stars and noted what appeared to be a relationship between the luminosities of some of the stars and the periods of their luminosity variations. Four years later, after adding additional stars to her list, Leavitt and Pickering published a short three-page paper that contained the first documentation of the period–luminosity relationship for a class of variable stars known as Cepheids, named for δ Cephei (delta Cep), which was first discovered by John Goodricke (Fig. 16.28) be variable in 1784. (Polaris is also a Cepheid variable star.) The light curve (the time-changing apparent magnitude) of one Cepheid, V1154 Cygni, is shown in Fig. 16.29, and Leavitt's historic diagram of the period–luminosity relationship for Cepheids in the Small Magellanic Cloud is reproduced in Fig. 16.30.

Fig. 16.27 Henrietta Swan Leavitt (1868–1921). (Public domain)

Leavitt's relationship proved to be of tremendous importance to astronomy, not only for what it tells us about the nature of the stars themselves, but for its role in establishing distances, including the immense distances to nearby galaxies. Once the relationship is calibrated by determining the luminosities (or absolute magnitudes) of a set of Cepheids whose distances are known by some other means, it becomes a relatively straightforward process to

1. measure how much time is required (the pulsation period) for a star to go from maximum brightness to minimum and back to maximum again,
2. determine the luminosity of that star from the period–luminosity relationship,
3. determine the absolute magnitude from the luminosity using the luminosity ratio–absolute magnitude difference equation on page 610,
4. determine the apparent magnitude from direct measurements, and
5. from the apparent and the absolute magnitudes, use the inverse square law of light (Section 6.7) to tell us how far away the star is.

Fig. 16.28 John Goodricke (1764–1786). (ESA; CC BY-SA 3.0 IGO)

The period–luminosity relationship provides us with standard candles of known luminosity, which in turn provide the distances to those standard candles.

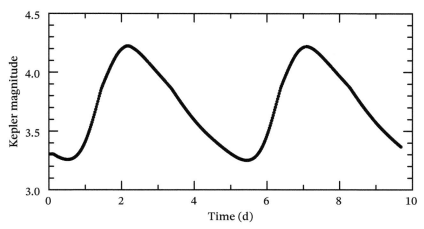

Fig. 16.29 The light curve of the Cepheid variable star V1154 Cygni over two pulsation periods. "Kepler magnitude" is a special magnitude scale used by the Kepler spacecraft. (Data courtesy of the Kepler team and the NASA Exoplanet Archive)

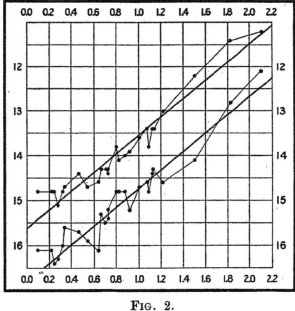

FIG. 2.

Fig. 16.30 The original figure from the 1912 article illustrating the period–luminosity relationship for 25 short-period Cepheid variable stars in the Small Magellanic Cloud. The vertical axis represents apparent magnitude and the horizontal axis is the logarithm of the pulsation period in days; 0.0 corresponds to $10^{0.0} = 1$ day, 1.0 corresponds to $10^{1.0} = 10$ days, 2.0 corresponds to $10^{2.0} = 100$ days, and 2.2 corresponds to $10^{2.2} = 158.5$ days. The top and bottom lines are best-fit straight lines drawn through the maxima and minima of the light curves, respectively. (Leavitt and Pickering, 1912;[6] public domain)

[6]Leavitt, Henrietta S. and Pickering, Edward C. (1912). "Periods of 25 Variable Stars in the Small Magellanic Cloud," *Harvard College Observatory Circular*, 173:1–3.

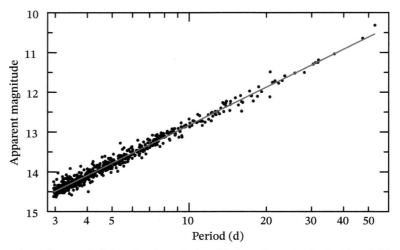

Fig. 16.31 A modern period–luminosity relationship based on 683 Cepheid variable stars in the Large Magellanic Cloud. The dots are individual Cepheid variable stars and the red line is a best fit through the data. The apparent magnitudes are in infrared light. Apparent magnitudes can be used instead of absolute magnitudes because all of the Cepheids in the LMC are essentially the same distance from us, so changing to absolute magnitudes would only shift the line upward, but it would not change the slope of the line. (Data from Macri et al., 2015[7])

The most direct means of determining the distances to stars is by the parallax method, first discussed in Section 15.3; parallax being the first rung of the distance ladder for distances beyond the Solar System. The period–luminosity relationship for Cepheids is another rung of the ladder, one that reaches much farther out into space. Cepheids are intermediate-mass stars on either the subgiant branch, the blueward horizontal branch, or the redward horizontal branch, with those on the horizontal branches being especially bright. Because they are so bright, they can be seen across hundreds of thousands and even millions of light-years. When Cepheids are observed in relatively nearby galaxies, determining the distances to the Cepheids implies that the distances to the galaxies in which they are located have been measured as well. An infrared period–luminosity relationship for 683 Cepheids in the Large Magellanic Cloud is shown in Fig. 16.31.

The Importance of Heavy Elements In Stars

There are actually two types of Cepheids, ones that are relatively rich in heavy elements (elements other than hydrogen and helium) and those that are heavy-element poor. The heavy-element poor Cepheids have a period–luminosity relationship that parallels the one in Fig. 16.31 but is slightly lower. Although heavy elements make up a small fraction of the composition of stars, in some cases they can play a significant role in how stars behave. In fact, the details of how heavy elements are treated in terms of their orbitals and electron energy levels play a critical role in the computer modeling of stellar pulsation.

[7]Macri, Lucas M., Ngeow, Chow-Choong, Kanbur, Shashi M., Mahzooni, Salma, and Smitka, Michael T. (2015). "Large magellanic cloud near-infrared synoptic survey. I. Cepheid variables and the calibration of the Leavitt law," *The Astronomical Journal*, 149(4), 117–133.

Why there are heavy-element rich and heavy-element poor stars has to do with their ages and the environments in which they formed. With the exception of hydrogen, most of the helium, and a tiny bit of the lithium in the universe, all of which formed very shortly after the Big Bang (Section 19.4), the remaining elements in the interstellar medium originated from previous generations of stars. In order for a star to have any heavy elements at all, it must have formed from the products of already-dead stars. When stars die, some or all of their forged elements are sent back out into the interstellar medium and new stars that form incorporate those heavy elements along with the still much more abundant hydrogen and helium. As a result, the more heavy-element rich a star is, the greater the number of generations of stars that came before it. Astronomers continue to look for the very first generation of stars that only contain hydrogen and helium, although still unsuccessfully as of June 2021. These stars would have formed very shortly after the Big Bang 13.8 billion years ago. Extremely heavy-element poor stars have been detected, that may represent the second generation of stars. One of the oldest stars currently known, which goes by the nickname SM0313,[8] is believed to be 13.6 billion years old and formed about 200 million years after the Big Bang. The iron content of SM0313 is just one-10-millionth of the iron content of our heavy-element rich Sun.

Key Point
The relative abundance of heavy elements in a star depends on how many generations of stars came before it.

Why Do Some Stars Pulsate?

The clear relationship between the luminosities of Cepheid variable stars and their pulsation periods, as illustrated in Fig. 16.31, strongly suggests that all of these stars must have a common reason for their pulsations. As mentioned earlier, when stars pulsate it is not just their luminosities and absolute magnitudes that change periodically over time, but so do their surface temperatures, their spectral classes, and their radii. In some cases, the radii of pulsating stars can change by as much as 25% over the course of one pulsation period. It is important to note that this type of pulsation is not like the oscillations that the Sun experiences (page 352*ff*). The Sun's oscillations are just the "ringing" of the star, like the ringing of a bell when it is struck, with the solar oscillations being driven by the chaotic motions of the Sun's convection zone.

Key Point
For some classes of pulsating stars, their radii can change by as much as 25% over one cycle.

The driving mechanism that causes Cepheids to expand has to do with the periodic trapping and releasing of heat inside the star. To understand how this works, think about a similar situation that you may be familiar with: a piston of an automobile engine, shown schematically in Fig. 16.32 (*Left*). Fuel is first injected into the region between the bottom of the cylinder and the movable piston. A spark plug then fires, causing the fuel to explode, which increases the pressure of the gas, causing the piston to be forced outward. The rod connected to the piston ultimately leads to power being delivered to the wheels. After the piston has been extended to the maximum height of the cylinder, the expended fuel is exhausted (with its CO_2 and ozone contaminants), fresh fuel is placed in the cylinder, the piston compresses, and the spark plug fires again, repeating the cycle.

The piston is analogous to how pulsating stars operate, but instead of a piston, it is a layer in the star's outer envelope that temporarily prevents light from reaching the surface [Fig. 16.32 (*Right*)]. The trapped light heats the gas, causing the pres-

[8]SM0313's formal name is SMSS J031300.36−670839.3, in case you are curious; all of the digits are its right ascension and declination coordinates in the constellation of Hydrus, and SMSS stands for the fully automated SkyMapper telescope at Siding Springs Observatory in Australia.

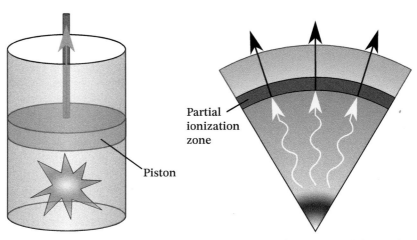

Fig. 16.32 *Left:* A piston of an automobile engine. An explosion of fuel drives the piston outward, ultimately delivering power to the wheels. *Right:* A partial ionization zone near the surface of a pulsating star traps energy from the interior, which pushes the partial ionization zone and the overlying layers outward.

sure of the gas to increase, which pushes the outer envelope and the atmosphere outward. After the star's outer layers pass through their equilibrium levels and cool, the trapped heat is released as light, the expansion slows down, and gravity pulls the layers back down again. Because the falling layers pick up speed as they move toward the equilibrium point, they go too far past it, compressing the gas, which then traps light again, heating the gas and increasing the pressure, causing the cycle to repeat.

You can think of the oscillations back and forth past equilibrium like someone being pushed on a swing as in Fig. 16.33. The equilibrium point would be the lowest point of the motion, where the rider could just sit if he wasn't being pushed. While swinging, the rider picks up speed while heading downward until he reaches the equilibrium point, then gravity slows him down until he stops going upward. At the highest point of his motion, his direction reverses, causing him to pass through equilibrium again. Well-timed pushes keep him from ultimately slowing down and stopping, just as well-timed pushes against the atmosphere do the same thing for the oscillations of pulsating stars.

It takes special circumstances for stellar pulsations to occur. As temperature decreases from the core of a star to the surface, levels within the star can exist where some particular element experiences ionization; if the gas is too deep and hot, all of the electrons have been stripped from their orbitals around the nucleus, and if the gas is too cool there isn't enough energy for the electrons to escape from the grasp of the nucleus. At a critical temperature, the particular element is in an in-between state, where some of the atoms are ionized and others aren't. The layer where this intermediate situation occurs is known as a partial ionization zone. In the case of Cepheids, the element involved is helium. Helium can exist as a neutral atom with two electrons (He), singly-ionized atom one electron missing (He^+), and doubly-ionized with both electrons missing (He^{++}, which is just the bare nucleus, an alpha particle). What powers Cepheid pulsations is the transition from singly ionized to doubly ionized helium, which happens at a temperature of about 40,000 K. In order

Fig. 16.33 A rider being pushed on a swing.

to go from singly to doubly ionized helium, the ion must absorb a photon (γ) with enough energy to free the electron, trapping the energy that the photon had in the layer of partial ionization as the layer becomes more ionized:

$$He^+ + \gamma \rightarrow He^{++} + e^- \quad \text{(energy absorbed and pressure increases)}.$$

The trapped energy raises the temperature and pressure of the gas, forcing the layer to expand outward. As the layer moves outward, it cools and the electron can be recaptured to create a new singly ionized helium atom, releasing a photon in the process:

$$He^{++} + e^- \rightarrow He^+ + \gamma \quad \text{(energy released and pressure decreases)}.$$

When the energy is released, the layer is no longer being pushed outward, and so its outward motion slows, reverses, and falls back in. (Note that because these reactions only involve atomic orbitals and are not nuclear reactions, electron neutrinos or electron antineutrinos are not involved.)

However, for this mechanism to actually cause a star to pulsate, the partial ionization zone must be at just the right depth in the star. If the layer is too deep, there is too much envelope and atmosphere to push outward (essentially the overlying layers are too heavy and can absorb too much energy). If the layer is too close to the surface, very little happens because there isn't much atmosphere to push around. In fact, if the star is too hot at the surface the partial ionization zone won't even exist, because the electrons have too much energy to be bound to the nucleus. It's a bit of a Goldilocks situation, not too hot and not too cold, not too deep and not too shallow. The other requirement is that there can't be any significant convection in the ionization region, otherwise convection will just carry the energy to the surface without being blocked by a partial ionization zone.

Figure 16.34 shows the locations of some of the classes of pulsating stars on the H–R diagram. The bright yellow giant and supergiant Cepheids can exist on the subgiant branch, and on the blueward and redward horizontal branches as they pass through the instability strip marked on Fig. 16.34. The instability strip exists because that is the region in the H–R diagram where the stars' singly to doubly ionized helium partial ionization zones are at the right depth to power their pulsations and where convection isn't strong enough to squelch the pulsations. While on the subgiant branch, stars enter the instability strip from the blue or hotter edge as the ionization zone moves deep enough to drive pulsations. When it exits the instability strip at the red or cooler edge, convection becomes too effective to allow pulsations to continue. After core helium burning starts at the tip of the giant branch, Cepheids enter the instability strip on the blueward horizontal branch, but in the reverse direction, and pass through the strip one more time on the redward horizontal branch. Because the blueward horizontal branch is the slowest of the three phases, it is much more likely to observe Cepheids during that part of their evolution. Cepheids are also fairly massive (4–20 M_{Sun}) so their post-main-sequence evolution is fast in general, which means that Cepheids are very rare objects. Of the 200–400 billion stars in our Milky Way Galaxy, only a few thousand Cepheids are known to exist.

Below the Cepheids are the subgiant RR Lyras, stars which can also serve as distance indicators. RR Lyras are typically very old, heavy-element poor, and of low mass, originally being about 0.8 M_{Sun} on the main sequence. Like Cepheids,

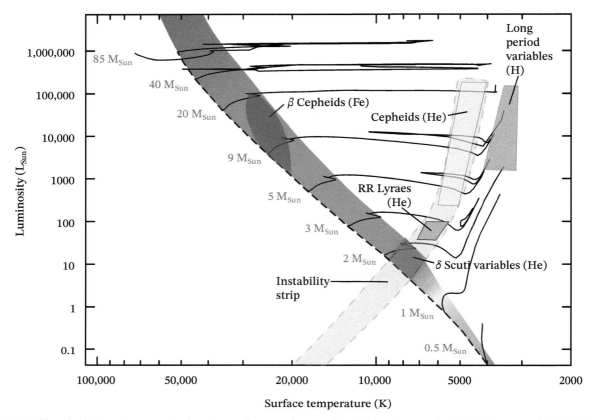

Fig. 16.34 The shaded regions are the locations of some classes of variable stars on the H–R diagram. β Cepheids and δ Scuti stars are main-sequence stars, Cepheids are yellow giants and supergiants, long period variables are red giants and supergiants, and RR Lyras stars are subgiants. Long period variables pulsate because of hydrogen (H); Cepheids, RR Lyras, and δ Scutis pulsate because of helium (He); and β Cepheids pulsate because of iron (Fe). The instability strip is shaded in light gray. (Evolutionary-track data courtesy of Ekström, et al., 2012)

RR Lyras stars are also powered by singly to doubly ionized helium, as are the main-sequence δ Scuti stars. To the right of the Cepheids are the long period variable (LPV) stars, that include the quasi-periodic semi-regular variables (Betelgeuse is one example), and the slightly lower luminosity, more periodic Mira variable stars. Because the LPVs' surfaces are quite a bit cooler than Cepheids, the helium partial ionization zone is too deep to drive their pulsations; instead, at between 10,000 and 15,000 K, hydrogen and neutral helium are partially ionized at just the right depth to cause them to pulsate. Finally, on the far left of the diagram, hot B spectral class main-sequence stars can undergo small pulsations, and are referred to as β Cephei stars (not to be confused with the star δ Cephei, for which the Cepheids are named). β Cephei stars are powered by the ionization of iron at a depth where the temperature is about 200,000 K. Although iron isn't particularly abundant in the atmospheres of stars, that element has 26 protons in its nucleus, and 26 electrons when it is electrically neutral. With so many electrons, iron has a lot of energy levels, making it very efficient at absorbing light when partially ionized. The intricate details of the energy levels of all elements are critical to understanding stars.

A Natural Probe of Structure and Underlying Physics

When we discussed helioseismology in Section 9.7, it was pointed out that studying solar oscillations is a very precise way of comparing the theoretical computer models of the star's interior with observations, just as earthquakes allow us to probe the interior of Earth. Helioseismology is just one of many tests of the computer models, with others being the rate at which neutrinos are detected coming from the core of the Sun, the age of the Sun based on computer models that are compared to the age of the Solar System based on the dating of Moon rocks and meteorites, the existence of granulation at the Sun's surface due to convection in the outer 28% of the Sun's radius, and the matching of computer models of the Sun's luminosity, surface temperature, radius, and composition with observations. In short, there are many ways to test theoretical models with observations to ensure that astronomers truly understand the Sun's interior and evolution.

The properties of pulsating stars, such as the period–luminosity relationship for Cepheids, the locations of pulsating stars on the H–R diagram, the stars' masses and compositions, and so on, allow us to test stellar evolution models. The tool of asteroseismology is a very powerful approach to testing our understanding computationally. Just as the wavelengths, periods of oscillation, and harmonics of guitar strings depend of the lengths of the strings and their masses, the same is true for pulsating stars; the pulsation periods of stars and their harmonics depend very sensitively on their sizes and internal characteristics.

What may seem strange at first glance is that studying the structure and evolution of stars, including through the investigations of stellar pulsation, also allows scientists to test our theoretical understanding of what is sometimes called microphysics, including nuclear reactions and the intricate details of electron energy levels in molecules, atoms, and ions. Studying the largest single objects in the universe provides a laboratory for investigating the smallest aspects of nature. Science always demands that we test our theoretical understanding with observations and experimental results in every way possible.

16.5 Clusters as Tests of Stellar Evolution Theory

Stars are rarely born in isolation. The giant clouds of dust and gas in which stars are born, such as the Orion Nebula or the Carina Nebula (Fig. 16.7 and Fig. 16.8, respectively), tend to form clusters of stars. Because all of the stars in a cluster form at about the same time and out of the same mix of elements, they play a special role in testing our understanding of stellar evolution. They can also play critical roles in understanding the formation and evolution of our Milky Way Galaxy and other nearby galaxies. One such stellar cluster, ω Centauri, shown in Fig. 16.1, is an ancient globular cluster that currently contains perhaps 10 million stars orbiting about their common center of mass.

Fig. 16.35 *Top:* Pleiades, an open (galactic) cluster. *Center:* M67, an open (galactic cluster). *Bottom:* NGC 6293, a globular cluster. (Digitized Sky Survey 2, STScI/NASA via Aladin)

Figure 16.35 shows images of three different clusters, the Pleiades, M67, and NGC 6293. The Pleiades, also known as the "seven sisters," is the famous naked-eye cluster in the constellation of Taurus. You can see immediately that the Pleiades must be very young astronomically, because it still contains much of the gas of the nebula from which it formed. The blue color of the gas in the reflection nebula arises for the same reason the the sky is blue; the shorter wavelengths of light are scattered more easily, including toward our eyes and telescopes. The Pleiades is an example of an open cluster, which are characterized by having relatively few stars, that tend to be spread out. Open clusters are preferentially found in or near the plane of the Milky Way Galaxy and so they are also known as galactic clusters.

M67, one of the members of the famous Messier catalog, is another example of an open cluster, although it is much older than the Pleiades and has long since lost its remaining nebular gas. The Messier catalog was compiled by Charles Messier (1730–1817). His motivation for recording the members of the catalog was that their fuzzy appearance in his telescopes of the day led to confusion with his true goal, which was to find comets. Messier compiled his catalog so he would know *not* to study those objects again! The Orion Nebula is included in his list as M42 (Orion Nebula) and the Pleiades is M44. The Messier catalog is very popular with amateur astronomers because its members are relative bright and easy to observe.

NGC 6293 is another globular cluster, like ω Centauri, and is composed of about 100,000 stars in a compact group. Globular clusters are very old, dating back to within 1–1 1/2 billion years after the Big Bang that was the beginning of the universe. NGC 6293 is one of the nearly 8000 entries in the *New General Catalog of Nebulae and Clusters of Stars* (1888);[9] the Orion Nebula is also NGC 1976. Globular clusters can be found well above or below the plane of the Milky Way, or they may be passing through the plane in orbits around the center of the Galaxy.

Given the cross-references among the growing number of catalogs that exist, care must be taken to make sure that astronomers are not referring to the same object as though it is multiple different objects. The online tool, SIMBAD, was developed for just that purpose. Other online tools exist to access images from databases (Aladin), or the many archives of data generated from astronomical observatories or computer simulations, such as VizieR. All of these tools and more can be accessed from the Strasbourg astronomical Data Center (CDS) portal at `http: //cdsportal.u-strasbg.fr/`.

Color–Magnitude Diagrams

Because of the large number of stars that can exist in a cluster, especially in globular clusters, it is essentially impossible to study every member star individually in great detail, such as through careful study of their spectra. It is possible to gather some data relatively quickly, however. Astronomers use detectors that are sensitive to light in specific wavelength ranges, or they can place colored filters in the telescope's light path that only allow certain colors to pass through. For example, a blue (B) filter allows light to reach the detector at wavelengths centered around 435 nm and a filter centered near the peak wavelength of the Sun, called a visible (V) filter, transmits light around 548 nm.

In Fig. 6.6 you learned that stars of different temperatures have blackbody radiation curves that peak at different wavelengths, with the hottest stars peaking in ultraviolet light, the Sun peaks near 550 nm, and the coolest stars peak in the infrared portion of the electromagnetic spectrum. This fact of blackbody spectra means that

---

[9]Dreyer, J. L. E. (1888). "A New General Catalogue of Nebulae and Clusters of Stars, being the Catalogue of the late Sir John F.W. Herschel, Bart., revised, corrected, and enlarged," *Memoirs of the Royal Astronomical Society*, 49(1):1–237.

subtracting the brightness of a star in one filter from the brightness of the same star in another filter gives information about what the peak wavelength of its blackbody curve is. If a star is relatively brighter in a blue filter than it is in a visible filter, then the peak of its blackbody spectrum must be toward the bluer or hotter part of the spectrum, but if the star is dimmer in the blue filter compared to the visible filter, then its peak must be toward the redder or cooler part of the spectrum; color differences provide information about a star's surface temperature. In other words, this color difference, called a color index, can replace the surface temperature or spectral type along the horizontal axis of the H–R diagram. Plotting the apparent magnitudes of the stars on the vertical axis completes the diagram. These special forms of H–R diagrams are known as color–magnitude diagrams.

Figure 16.36 shows color–magnitude diagrams of the clusters in Fig. 16.35. A quick look at the three diagrams shows that they are very different from one another. The Pleiades has a long main sequence with no cool giant stars in the upper right-hand corner. It does show stars near the top of the diagram deviating from the main sequence however, indicating that those stars have just exhausted their core hydrogen. The point where stars start breaking away from the main sequence is labeled as the main-sequence turn-off point (MSTO). The main sequence of M67 is missing the most massive and brightest stars down to stars just a little more massive than the Sun, resulting in a shorter main sequence than the Pleiades has, and so it's main-sequence turn-off point is closer to the Sun. M67 also has well-defined subgiant and red giant branches (recall the labels in Fig. 16.19). NGC 6293 has the shortest main sequence of all, with a main-sequence turn-off point that is below where the Sun would be on the diagram. It too has well-defined subgiant and red giant branches, but it also shows a horizontal branch (primarily the

Key Point
Color–magnitude diagrams provide a convenient way to test stellar evolution theory and determine cluster ages.

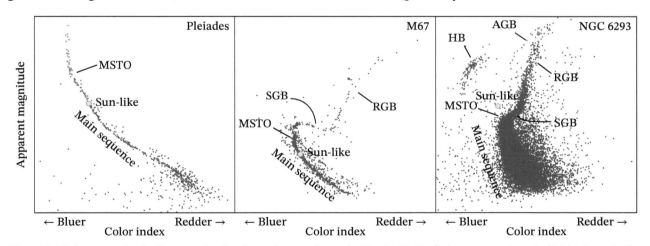

Fig. 16.36 Color–magnitude diagrams for the three clusters shown in Fig. 16.35. Each dot represents an individual star in the cluster. The abbreviations are: Main-sequence turn-off (MSTO), subgiant branch (SGB), red giant branch (RGB), horizontal branch (HB), asymptotic giant branch (AGB). The orange dots are where stars similar to our Sun would be placed on the diagrams. *Left:* The Pleiades, a young open cluster approximately 110–160 million years old; blue dots from Gaia, red dots from Hipparcos (ESA). *Center:* M67, an intermediate-age open cluster approximately 4 billion years old (Gaia/ESA). *Right:* NGC 6293, an ancient globular cluster approximately 12.5 billion years old (Piotto, et al., 2002;[10] HST/NASA)

[10]Piotto, G., King, I. R., Djorgovski, S. G., Sosin, C., et al. (2002). "HST color–magnitude diagrams of 74 galactic globular clusters in the HST," *Astronomy & Astrophysics*, 391(3):945–965.

blueward branch because it is the slowest) and an asymptotic giant branch. As always, a scientist is interested in solving nature's puzzles; in this case, what is causing these differences in color–magnitude diagrams?

Stars With Different Masses and the Same Ages

Key Point
Assumptions that can be made about stellar clusters that make them useful for testing stellar evolution theory.

There are three assumptions that can be made about clusters that allow astronomers to use them as tests of stellar evolution theory: all stars in a cluster

- are nearly the same distance from us,
- are essentially the same age, and
- formed out of the same material, meaning that their initial compositions are the same.

While none of these assumptions are strictly true, variations between the stars are relatively small. Consider the first assumption, for example. Obviously, stars in the front of a cluster are closer to us than stars in the back, but the distance to a cluster is much larger than that difference. If you live in New York City, the distance to John F. Kennedy International Airport is significantly different, depending on where you are, but if you live in London the difference of a few kilometers (or miles) is largely irrelevant; you are more concerned about the distance between the two cities. The same is true for clusters. The uniform nature of the distance means that using apparent magnitudes on the vertical axis instead of absolute magnitudes doesn't affect the comparison between stars. If the distance to the cluster is known and the apparent magnitudes can be converted to absolute magnitudes, then the entire diagram would just move vertically upward but the overall shape wouldn't change. It is only when comparing one cluster with another on the same diagram that distances are needed in order to make the conversion from apparent to absolute magnitudes.

Now imagine that a research team has generated a series of evolutionary tracks of stars of different masses, similar to the tracks in Fig. 16.23, with all of the models starting with the composition of a specific cluster of stars. Because the most massive stars evolve the fastest, they will leave the main sequence before less massive stars. By picking a particular age, say 4 billion years as is the case for M67, the various evolutionary tracks can be searched to determine the absolute magnitude and color index for theoretical stars of each mass at that age. The data for each mass can then be plotted on a color–magnitude diagram and compared with the actual cluster. If the evolutionary models are correct, the artificial color–magnitude diagram should agree with the real cluster. Some discrepancies do exist, but overall the agreement is excellent. As models have gotten more sophisticated with the inclusion of rotation, improved treatment of the complexities of convection, mass loss, and so on, the agreement has steadily improved. Stellar evolution modeling, with its reliance on our understanding of physics at the atomic and subatomic levels, has been one of the great successes of theoretical astrophysics since the 1960s.

Key Point
In general, open clusters are significantly younger than globular clusters.

Because stellar evolution calculations have been so successful, it is possible to use them to determine cluster ages. This is done by noting the location of the main-sequence turn-off point for a cluster's color–magnitude diagram and pairing it up with the theoretical star that is leaving the main sequence with the same color index. The age of the theoretical star can then be assumed to be the age of the actual cluster. Figure 16.37 shows several clusters all placed on the same color–magnitude

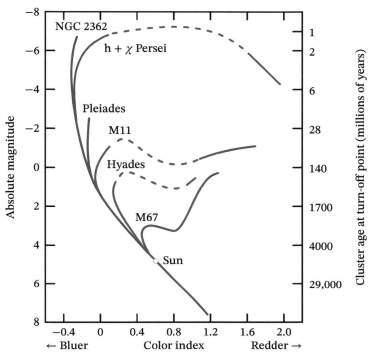

Fig. 16.37 Color–magnitude diagrams for a number of open (galactic) stellar clusters. The dashed lines represent gaps in the data where no stars are observed due to rapid evolution. Approximate ages of the clusters based on their main-sequence turn-off points are given on the right. Closer looks at the Pleiades are shown in Fig. 16.36. (Adapted from Fig. 1 of Allan Sandage, 1958[11]) and M67

diagram with approximate main-sequence turn-off ages given on the vertical axis to the right. The youngest clusters have their main-sequence turn-off points highest up on the main sequence, because their more massive stars haven't yet had time to consume all of their core hydrogen.

Determining Cluster Distances

One way to "measure" the distance to a cluster is by determining how far the cluster's main sequence must be moved vertically in a color–magnitude diagram to match the absolute magnitudes of theoretical color–magnitude diagrams. The difference between the actual apparent magnitudes and the theoretical absolute magnitudes leads directly to the distance to the cluster. Remember that the absolute magnitude of a star is what the apparent magnitude would be if the star was exactly 10 pc (32.6 ly) away from us. The same technique can be used to compare one cluster with another whose distance is already known, but care must be taken to make sure that the clusters have similar chemical compositions, since composition affects evolution. The vertical shifting technique to determine distance is called

Key Point
Shifting a cluster's color–magnitude diagram vertically until its apparent magnitudes align with the absolute magnitudes of a theoretical color–magnitude diagram can be used to determine the distance to the cluster.

[11]Sandage, A. (1958). "The Color–Magnitude Diagrams of Galactic and Globular Clusters and their Interpretation as Age Groups," *Ricerche Astronomiche*, 5:41–56.

spectroscopic parallax. The term "spectroscopic" is used in the name because the masses of stars on the main sequence, and therefore their temperatures and color indices on the horizontal axis of the H–R diagram, relate to their spectra. The use of the word "parallax" is only meant to indicate that the method gives distance, just as actual parallax measurements do. Spectroscopic parallax is more accurate than using individual stars because the method is using the results of large numbers of stars simultaneously, resulting in very good statistical averages.

Another method for finding the distances to some clusters, especially globular clusters, is by identifying RR Lyra variable stars in them. All RR Lyras have similar absolute magnitudes, which means that they can be used as standard candles in the same way that the much-brighter Cepheid variable stars can. By measuring their apparent magnitudes and knowing their absolute magnitudes, distances can be calculated. This also serves as a way to test the distance result obtained by spectroscopic parallax.

Key Point
RR Lyra stars are pulsating stars that serve as standard candles for determining distances to globular clusters.

Exercises

All of the exercises are designed to further develop *your understanding* of the material by thinking carefully and critically about what you have read in this chapter. Answers to selected exercises are available in the back of the book.

True/False

1. The typical number density of particles in the interstellar medium is millions of atoms or molecules per cubic meter of space.
2. (*Answer*) Hydrogen is by far the most dominant element in the universe.
3. Helium easily forms molecules with itself and other elements.
4. (*Answer*) Nebulas can be hundreds of light-years across.
5. Soccer-ball shaped molecules of carbon that contain 60 atoms have been detected in the interstellar medium.
6. (*Answer*) The Pillars of Creation are permanent, never-changing features found in the Eagle Nebula of Serpens.
7. The James Web Space Telescope is designed to operate primarily at infrared wavelengths.
8. (*Answer*) The massive binary star system, η Carina, is too far from Earth to have ever been visible to the naked eye.
9. Star formation regions are known to be turbulent and violent places.
10. (*Answer*) Bok globules have temperatures of thousands of kelvins because stars are forming inside them.
11. During the initial stage of free-fall collapse inside a Bok globule, collisions between particles are extremely rare.
12. (*Answer*) The protostar phase of star formation is characterized by a roughly horizontal track of increasing surface temperature in the H–R diagram.
13. A protoplanetary accretion disk develops during star formation because of conservation of energy alone.
14. (*Answer*) No nuclear reactions of any kind occur inside a forming star before it reaches the main sequence.
15. Deuterium burning inside protostars produces new molecules.
16. (*Answer*) The more massive a star is, the faster it lives out its life.

17. The lowest mass that a main-sequence star can have is about 7.5% of the Sun's mass.
18. (*Answer*) Free-floating planets cannot exist.
19. Stars only evolve because they interact gravitationally with other stars in their cosmic neighborhood.
20. (*Answer*) Stars spend about 90% of their lives as main-sequence stars.
21. The CNO cycle is the primary reaction sequence that converts helium into carbon, nitrogen, and oxygen.
22. (*Answer*) All stars on the red giant branch and the asymptotic giant branch of the H–R diagram have deep convective envelopes.
23. When hydrogen or helium burning is taking place in the center of a star, evolution tends to move slowly, but when core burning stops, evolution goes much more quickly.
24. (*Answer*) Core helium burning starts along the asymptotic giant branch.
25. Some stars can get to be larger than Jupiter's orbit.
26. (*Answer*) A subgiant branch star has a helium core and a hydrogen-burning shell.
27. A helium core flash blasts most of the star's envelope out into space.
28. (*Answer*) A 5 M_{Sun} blueward horizontal-branch star has a helium-burning core and a hydrogen-burning shell.
29. A 5 M_{Sun} star can produce iron in its core.
30. (*Answer*) The first pulsating variable star was discovered at the end of the sixteenth century.
31. Mira has a pulsation period of 11 days.
32. (*Answer*) Cepheid variable stars play a critical role in determining distances to nearby galaxies.
33. A heavy-element-rich star is likely to be older than a heavy-element poor star.
34. (*Answer*) A star can pulsate by periodically trapping and releasing energy that is moving outward through the star.

35. One way to test the interior structure of a star is to study how it pulsates (if it does).

36. (*Answer*) The most likely phase of evolution of an observed Cepheid variable star is the redward horizontal branch.

37. Cepheids are yellow giant or supergiant stars.

38. (*Answer*) Stellar clusters form when individual stars fall toward one another because of their mutual force of gravity.

39. Globular clusters can contain hundreds of thousands to millions of stars.

40. (*Answer*) Globular clusters tend to be younger than open clusters.

41. The main-sequence turn-off point corresponds to the age of a stellar cluster.

42. (*Answer*) Stellar clusters provide a means to test stellar evolution calculations.

43. The Pleiades is a globular cluster.

44. (*Answer*) By comparing the apparent magnitudes of stars in a color–magnitude diagram with the absolute magnitudes of a theoretical diagram, the distance to the cluster can be inferred.

Multiple Choice

45. Hydrogen makes up about _____ percent of the mass of the interstellar medium.
   (a) 20 (b) 40 (c) 60 (d) 70 (e) 80 (f) 90

46. (*Answer*) The Orion Molecular Complex is between _____ and _____ light-years from Earth.
   (a) 10, 100   (b) 100, 1000   (c) 1000, 1400
   (d) 1400, 10,000   (e) 8500, 10,000

47. Three relatively common elements in the interstellar medium after hydrogen and helium are _____, _____, and _____.
   (a) carbon, nitrogen, oxygen
   (b) uranium, neptunium, plutonium
   (c) lithium, beryllium, nepotisium
   (d) titanium, vandalisium, chromium
   (e) lithium, sanctimonium, einsteinium
   (f) iron, krypton, zirconium

48. (*Answer*) The best wavelength range for seeing deep inside dusty star-formation regions is in
   (a) gamma rays and x-rays.
   (b) ultraviolet light.
   (c) visible light.
   (d) infrared light.
   (e) radio waves.

49. The temperature of a typical interstellar cloud ranges from _____ to _____.
   (a) 0 K, 10 K   (b) 10 K, 50 K   (d) 50 K, 100 K
   (e) 100 K, 300 K   (f) 300 K, 1000 K   (g) −10 K, −300 K

50. (*Answer*) The young binary star system, η Carina, has a combined mass of about
   (a) 10 M_{Sun}.   (b) 20 M_{Sun}.   (c) 50 M_{Sun}.
   (d) 80 M_{Sun}.   (e) 100 M_{Sun}.   (f) 120 M_{Sun}.
   (g) 1200 M_{Sun}.   (h) 2000 M_{Sun}.   (i) 12,000 M_{Sun}.

51. Bok globules have masses that typically range from _____ to _____ and have temperatures of near _____.
   (a) 1 M_{Sun}, 2 M_{Sun}, 10 K
   (b) 2 M_{Sun}, 50 M_{Sun}, 10 K
   (c) 10 M_{Sun}, 20 M_{Sun}, 100 K
   (d) 20 M_{Sun}, 50 M_{Sun}, 100 K
   (e) 20 M_{Sun}, 500 M_{Sun}, 1000 K
   (f) 200 M_{Sun}, 5000 M_{Sun}, 10 K

52. (*Answer*) The early rate of gravitational collapse inside a Bok globule is typically _____ the acceleration due to gravity on Earth.
   (a) trillions of times less than
   (b) millions of times less than
   (c) thousands of times less than
   (d) ten times less than
   (e) about the same as
   (f) ten times greater than
   (g) thousands of times greater than
   (h) millions of times greater than
   (i) trillions of times greater than

53. Dust vaporizes during the protostar phase of star formation because
   (a) carbon and silicon atoms that form the basis of dust grains are destroyed.
   (b) very fast rotation of the collapsing cloud tears the dust particles apart through centrifugal forces.
   (c) the molecular bonds that hold dust particles together are very weak and easily broken as the collapsing cloud heats up.
   (d) long-wavelength radio photons are very powerful and blast the dust particles apart.

54. (*Answer*) The protoplanetary disk that forms during star formation
   (a) does so because of conservation of angular momentum.
   (b) helps to feed the growing star at the center of the disk.
   (c) contributes to the formation of jets that eject material along the rotation axis of the collapsing cloud.
   (d) all of the above
   (e) (a) and (b) only
   (f) (b) and (c) only

55. The products of nuclear deuterium burning are
   (a) heavy water and carbon dioxide.
   (b) methane and a proton.
   (c) tritium and a photon.
   (d) helium-3 and a photon.
   (e) lithium-7 and an antineutrino.

56. (*Answer*) What is the most important method of transporting energy from the interior of a pre-main-sequence star to the surface?
   (a) conduction   (b) convection   (c) photon radiation
   (d) neutrinos   (e) (a) and (b)   (f) (c) and (d)

57. Outside of stellar clusters, approximately _____ of all stars in the sky are main-sequence stars.
   (a) 10% (b) 20% (c) 30% (d) 40% (e) 50%
   (f) 60% (g) 70% (h) 80% (i) 90% (j) 100%

58. (*Answer*) The main sequence has some width to it because
   (a) not all main-sequence stars have exactly the same composition.
   (b) even for stars of the same mass and initial composition, the internal composition changes over time as a consequence of nuclear reactions.
   (c) otherwise identical stars at different distances from Earth have different luminosities.
   (d) all of the above
   (e) none of the above
   (f) (a) and (b) only

59. Brown dwarfs range in mass from about _____ to _____ times the mass of Jupiter.
    (a) 1, 13
    (b) 10, 100
    (c) 13, 75
    (d) 75, 130
    (e) The answer depends on the amount of lithium in the object.

60. (*Answer*) Under most circumstances, if the fuel is available but nuclear reactions in the core of a star were to somehow run too slowly
    (a) the core would shrink and heat up, causing the nuclear reactions to speed up.
    (b) the core would cool down, causing the reactions to run even more slowly.
    (c) the star would collapse.
    (d) the star would expand to enormous size.
    (e) the star would explode.
    (f) the core of the star would split in two.
    (g) (b) and (c) only
    (h) (d), (e), and (f) only

61. It doesn't make sense to calculate the entire evolution of a 0.1 M_{Sun} star from the main sequence to the end of its life because
    (a) no 0.1 M_{Sun} star has been in existence long enough to evolve off the main sequence.
    (b) there is no way to test the validity of the theoretical computer calculations through observations.
    (c) these calculations are computationally too difficult and require too much computer time.
    (d) it is theoretically impossible for 0.1 M_{Sun} stars to exist in nature.
    (e) (a) and (b) only
    (f) none of the above

62. (*Answer*) The CNO cycle
    (a) is a reaction sequence that converts hydrogen into helium.
    (b) is the dominant process of hydrogen burning in the cores of stars more massive than about 1.2 M_{Sun}.
    (c) depends strongly on temperature.
    (d) drives convection in the cores of intermediate mass stars.
    (e) all of the above
    (f) (a) and (b) only
    (g) (c) and (d) only

63. When hydrogen is depleted in the core of a main-sequence star, what happens next?
    (a) The star's core shrinks and cools down.
    (b) The star's core shrinks and gets hotter.
    (c) Hydrogen can start to burn around the core of helium ash.
    (d) After some initial adjustment, the star becomes less luminous and its surface gets cooler.
    (e) The star immediately becomes a red supergiant.
    (f) (a), (c), and (e) only
    (g) (b), (c), and (d) only

64. (*Answer*) During the main sequence and blueward horizontal branch phases of stellar evolution
    (a) hydrogen is burning in the core.
    (b) hydrogen is burning in the main-sequence core and helium is burning in the blueward horizontal-branch core.

    (c) shell burning is occurring in both phases.
    (d) a hydrogen-burning shell exists during the blueward horizontal branch phase, but not in the main-sequence phase.
    (e) (a) and (c) only
    (f) (b) and (d) only

65. Which phase of stellar evolution is the slowest?
    (a) protostar
    (b) pre-main sequence
    (c) main sequence
    (d) subgiant branch
    (e) red giant branch
    (f) blueward horizontal branch
    (g) redward horizontal branch
    (h) asymptotic giant branch

66. (*Answer*) It is much harder for helium burning to take place compared to hydrogen burning because
    (a) the helium nucleus is four times heavier than a proton.
    (b) a helium nucleus has two protons in it instead of just one.
    (c) a helium nucleus must collide with a beryllium nucleus.
    (d) much higher temperatures are required to get the nuclei going fast enough.
    (e) much higher densities are needed to get the nuclei close enough together.
    (f) all of the above
    (g) (a) and (b) only

67. What is the size of the carbon–oxygen core of an asymptotic-giant-branch star?
    (a) about the size of a city block
    (b) about the size of a city
    (c) about the size of Mimas
    (d) about the size of Earth
    (e) about the size of Jupiter
    (f) about the size of the Sun
    (g) none of the above

68. (*Answer*) The underlying reason that a helium core flash occurs at the tip of the red giant branch in stars of less than about 2 M_{Sun} is that
    (a) the density of the gas gets very high.
    (b) electrons are forced into their lowest allowed energy states.
    (c) increasing the temperature of the gas in the core doesn't cause the core to expand.
    (d) all of the above
    (e) (a) and (b) only

69. Assuming that Mira spends one-half of its time dimmer than its average magnitude, how long is the star unseen during its pulsation period?
    (a) 4 months   (b) 4 1/2 months   (c) 5 months
    (d) 5 1/2 months   (e) 6 months   (f) 6 1/2 months
    (g) 7 months   (h) 7 1/2 months   (h) 8 months

70. (*Answer*) From the star chart in Fig. 16.26, which star is located at (3^h02^m, $+04°05'$)?
    (a) Menkar (α Ceti)   (b) Deneb Kaitos (β Ceti)
    (c) γ Ceti   (d) δ Ceti
    (e) π Eridani   (f) none of the above

71. The period–luminosity relationship for Cepheids is useful because it allows astronomers to determine
    (a) the period of a star's pulsation once the star's luminosity is known.
    (b) how many Cepheids are needed to produce the luminosity of a nearby galaxy.

(c) the absolute magnitude of a Cepheid once its period is measured.
(d) the distance to the star.
(e) the distance to nearby galaxies.
(f) all of the above
(g) (a) and (b) only
(h) (c), (d), and (e) only

72. (*Answer*) According to Fig. 16.31, a Cepheid in the Large Magellanic Cloud with a pulsation period of 20 days has an infrared apparent magnitude closest to
(a) 9 (b) 10 (c) 11 (d) 12
(e) 13 (f) 14 (g) 15 (h) 16

73. Stars that are heavy-element rich
(a) had the elements in their cores pushed up to the surface when they became red-giant stars.
(b) have very little hydrogen and helium in them.
(c) formed from material that was processed by multiple generations of stars that came before them.
(d) are all very massive.
(e) all pulsate.

74. (*Answer*) The element and change in ionization states responsible for driving pulsations in Cepheid variable stars is
(a) hydrogen, neutral to singly ionized
(b) hydrogen, singly ionized to doubly ionized.
(c) helium, neutral to singly ionized.
(d) helium, singly ionized to doubly ionized.
(e) all of the above
(f) (a) and (c) only
(g) (b) and (d) only

75. The element and change in ionization states responsible for driving pulsations in Mira variable stars is
(a) hydrogen, neutral to singly ionized.
(b) hydrogen, singly ionized to doubly ionized.
(c) helium, neutral to singly ionized.
(d) helium, singly ionized to doubly ionized.
(e) all of the above
(f) (a) and (c) only
(g) (b) and (d) only

76. (*Answer*) Partial ionization zones in pulsating stars power their pulsations by
(a) periodically trapping and releasing photons.
(b) causing the gas pressure in that region of the star to temporarily increase before decreasing again.
(c) blocking photons that then push the partial ionization zone outward through radiation pressure.
(d) periodically speeding up and slowing nuclear reactions.
(e) all of the above
(f) (a) and (b) only
(g) (c) and (d) only

77. A stellar cluster with 200,000 stars is likely to be _____.
(a) a galactic cluster (b) an open cluster
(c) a globular cluster (d) (a) and (b)

78. (*Answer*) A _____ cluster is most likely to be _____ cluster.
(a) heavy-element poor, an open
(b) heavy-element poor, a globular
(c) heavy-element rich, an open
(d) heavy-element rich, a globular
(e) (a) and (d) only
(f) (b) and (c) only

79. The Messier catalog was created to
(a) record comets.
(b) avoid studying comets.
(c) avoid revisiting "fuzzy" objects in the sky that weren't comets.
(d) record "fuzzy" objects in the sky for further investigation.

80. (*Answer*) In a color–magnitude diagram, ω Centauri would have a main-sequence turn-off point that is closest to
(a) the turn-off point of the Pleiades.
(b) the turn-off point of M67.
(c) where the Sun would be located on the diagram.
(d) the turn-off point of NGC 6293.
(e) There is not enough information to answer the question.

81. The color–magnitude diagram of M67 shows a subgiant branch and a red giant branch because
(a) its stars that are somewhat more massive than the Sun have had time to exhaust the hydrogen in their cores.
(b) it has the right composition to create these stars.
(c) it is a globular cluster that is extremely old.
(d) none of the above
(e) (b) and (c) only

82. (*Answer*) The double cluster h + χ Persei is closest to being _____ million years old.
(a) 1 (b) 2 (c) 6 (d) 28
(e) 140 (f) 1700 (g) 4000 (h) 29,000

83. If a cluster with the Sun's composition had a main-sequence turn-off point corresponding to an absolute magnitude of 4.7, the age of the cluster would be closest to _____ billion years.
(a) 1 (b) 2 (c) 3 (d) 5 (e) 8
(f) 10 (g) 11 (h) 12 (i) 13 (j) 15

Short Answer

84. (*Answer*) The number density, n, of particles in a gas can be written in terms of the mass density, ρ, and the average mass of each particle, $m_{average}$, as $n = \rho / m_{average}$.
(a) Referring back to the ideal gas law on page 339, rewrite the law in terms of the number density, n.
(b) Assume that the number density of a particular interstellar cloud is 1×10^8 particles/m^3 and the temperature is 10 K. Calculate the gas pressure in the cloud. (Boltzmann's constant, k_B, is 1.38×10^{-23} N m/K.)
(c) The gas pressure at sea level for Earth's atmosphere is about 1×10^5 N/m^2. Approximately how many times greater is atmospheric pressure at sea level than the pressure you calculated in part (b)?

85. Describe in your own words how the Pillars of Creation developed their iconic shapes.

86. What is the main wavelength region that the James Web Space Telescope was designed to study, and why?

87. The captain of an interstellar spaceship decides to evade and then surprise a nearby enemy ship by hiding in a nebula. Is this a good strategy? Why or why not?

88. Why does the luminosity of a collapsing cloud of gas and dust initially increase rapidly?

89. What causes the surface temperature to rise while the luminosity stays roughly constant during the protostar phase of star formation?

90. (a) Explain why the pre-main-sequence evolution of a star causes it to get dimmer and slightly hotter over time.
    (b) What is the general direction that a 1 M_{Sun} pre-main-sequence star takes through the H–R diagram?
91. (a) What is common to all main-sequence stars?
    (b) Why do approximately 90% of all stars in the sky reside on the main sequence in the H–R diagram?
92. Why are brown dwarfs referred to as failed stars?
93. In a sense, the reason that all stars must evolve can be summed up in one sentence. What is it?
94. Explain how nuclear burning is a self-regulating process as long as the ideal gas law is a reasonably good approximation of the relationship between pressure, density, and temperature.
95. What is the common characteristic of all stars either moving up or down the nearly vertical track on the H–R diagram around 3000 K?
96. There is a pattern among the nuclei $^4_2$He (the alpha particle), $^8_4$Be, $^{12}_6$C, and $^{16}_8$O. What is it?
97. The blueward horizontal branch can be thought of as being analogous to the main sequence. Explain why and describe any differences.
98. The main sequence and the blueward horizontal branch in Fig. 16.19 are labeled as very slow and slow evolutionary phases, respectively, while the subgiant branch and the redward horizontal branch are labeled as fast and very fast. What is common about the main sequence and the blueward horizontal branch, and about the subgiant branch and the redward horizontal branch, and why do those aspects affect the speed of evolution across the H–R diagram?
99. Describe in words what the structure of a 5 M_{Sun} star is on the asymptotic giant branch.
100. What is meant by a degenerate electron gas? Why does it lead to a helium core flash at the tip of the red giant branch?
101. Explain why only stars more massive than about 9 M_{Sun} can create iron in their cores.
102. The Strasbourg astronomical Data Center (CDS) portal at `http://cdsportal.u-strasbg.fr/` provides access to an enormous amount of astronomical data. Go to the site and enter RR Lyra, the star for whom the RR Lyra variable stars are named.
    (a) What are its celestial coordinates in right ascension and declination?
    (b) What is its spectral type range?
    (c) What are its apparent magnitudes in the blue (B) and red (R) filters?
    (d) Is RR Lyra visible to the naked eye?
103. Go to the homepage of the American Association of Variable Star Observers (`https://www.aavso.org/`), scroll down on the homepage and click on the box "(VSX) Variable Star Index" to access the International Variable Star Index. Select search and enter "del Cep" in the search bar for delta Cephei, the namesake of the Cepheid variable stars. Scroll down on the results page to answer the following questions.
    (a) What are the star's J2000.0 right ascension and declination coordinates (right ascension is listed first in hours, minutes, seconds)?
    (b) What is the star's pulsation period?
    (c) What is the star's range in visual apparent magnitudes during its pulsation period?
    (d) Is δ Cephei among the brightest naked eye stars in a dark night sky, among the dimmest, or somewhat average?

(e) What is the range in spectral classes during its pulsation period? (Don't worry about the Ib appended to the spectral classes; that luminosity class indicates that the star is at the dimmer end of the supergiants.)
104. Describe how the pulsation mechanism works for Cepheids.
105. (a) A Cepheid variable star is most likely to be observed during which phase of stellar evolution?
    (b) Describe the star's internal structure during that phase of evolution.
106. Why is it generally more accurate to develop a period–luminosity diagram in infrared light, as in Fig. 16.31, instead of visible light?
107. The formula for determining the distance to a star, in parsecs, if its apparent magnitude (m) and absolute magnitude (M) are both known, is

$$d = 10^{(m-M+5)/5} \text{ pc.} \tag{16.1}$$

Distance determination from apparent and absolute magnitudes

The equation derives from the inverse square law and the fact that a difference of 5 magnitudes corresponds to a brightness ratio of 100.

In this exercise you will be creating a table with $m - M$ in one column and d (pc) in a second column. The column for $m - M$ should have rows for 0, 1, 2, 3, 4, 5, 10, 15, 20, 25, 30. The header and the first two rows should look like

$m - M$	d (pc)	d (ly)
0	10.0	32.6
1	15.8	51.5

Be sure you carry out the entire calculation for the exponent of 10, $(m - M + 5)/5$, before raising the result to the power of 10. For example, for $m - M = 1$, the exponent is $(1 + 5)/5 = 6/5 = 1.2$, so that the final result for $m - M = 1$ is $10^{1.2} = 15.8$. (Your calculator probably has a key that looks like 10^x or something similar, where x is the result of the exponent calculation.)

For the third column you will want to refer to the relationship between pc, ly, and au from page 607.

Astronomers call the quantity $m - M$ the distance modulus, because it directly relates the difference between apparent and absolute magnitudes to distance.
108. Suppose a Cepheid variable star is observed to have an average apparent visual magnitude $m = 22$. From its pulsation period of 15.3 d the period–luminosity relationship indicates that its absolute visual magnitude is $M = -4.5$. Using the distance from apparent and absolute magnitudes equation in Exercise 107, calculate the distance to the star in
    (a) parsecs (pc).
    (b) light-years (ly). (You will want to refer to the relationship between pc, ly, and au from page 607.)
109. Explain how the distance determination from apparent and absolute magnitude equation of Exercise 107 provides astronomers with the ability to determine the distance to a star *if* the star's absolute magnitude is known by some independent process. These standard-candle stars represent rungs on the distance ladder.)

110. Suppose that the color–magnitude diagram of a stellar cluster must be moved upward by 17 magnitudes to match a theoretical diagram that uses absolute magnitudes ($m - M = 17$).

 (a) Assuming that the cluster is identical to the Pleiades, sketch a color–magnitude diagram that shows the cluster with its absolute magnitudes from Fig. 16.37 and its apparent magnitudes shifted appropriately downward. The two curves should parallel one another. You will want the vertical axis to range from −4 magnitude at the top to 25 magnitude at the bottom.

 (b) Using the distance from apparent and absolute magnitudes equation of Exercise 107, determine the distance to the cluster in parsecs and light-years.

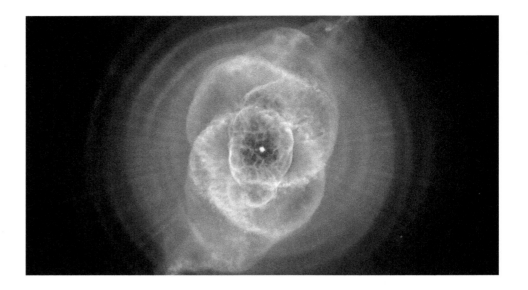

17

The End of a Stellar Life

We are a way for the universe to know itself. Some part of our being knows this is where we came from. We long to return. And we can, because the cosmos is also within us. We're made of star stuff.

Carl Sagan (1934–1996)

17.1	Winds and White Dwarfs	685
17.2	Supernovas: Going out with a Bang	691
17.3	Neutron Stars	700
17.4	Black Holes: Gravity's Ultimate Victory	709
	Exercises	721

Fig. 17.1 The Cat's Eye Nebula, a planetary nebula in the constellation of Draco. [ESA, NASA, HEIC and the Hubble Heritage Team (STScI/AURA); CC BY 4.0]

Introduction

The saying goes that "all good things must come to an end." One can quibble about the use of the word "all" in the saying, but it certainly applies to all stars. It was pointed out many times in the last chapter that stellar evolution is the consequence of a constant fight against gravity. In the collapse of a portion of a nebula to the dusty cocoon of a Bok globule, to the pre-main-sequence evolution leading to hydrogen burning in the core of a new star, gravity keeps driving changes in the star. When all of the available hydrogen fuel is consumed in its core, the star resorts to burning helium in the core and hydrogen in a shell around the core. But of course there is only a finite amount of helium available that was produced by the burning of hydrogen, so when the helium runs out, the star is forced to burn helium in a shell around the carbon–oxygen or oxygen–neon core and hydrogen in a shell above the helium-burning shell.

All of this internal activity deep in the heart of the star is reflected on its surface by sometimes huge changes in its radius, color changes due to changes in the surface temperature, and even compositional changes as the products of nuclear reactions in the central regions of the star are carried to the surface by convection. At times, if the conditions are right, the star can even experience episodes of periodic pulsation.

Underlying it all is the star's constant quest for new energy sources to combat gravitational collapse. What happens to a star at the end of its quest depends on its initial mass.

17.1 Winds and White Dwarfs

Planetary Nebulas

Planetary nebulas (or alternatively, planetary nebulae) are delicate and intricate gas clouds that are among the most beautiful objects seen in deep space. The term "planetary" is used to describe these expanding clouds of gas because of the fuzzy, planet-like appearance that they presented in the small telescopes that existed when they were first discovered. Today we know that they are the result of dying stars with masses of less than about 9 M_{Sun} or so. The Cat's Eye Nebula (Fig. 17.1) is a planetary nebula located in the constellation of Draco. Other examples of planetary nebulas are shown in Fig. 17.2.

When a star below about 7 M_{Sun} nears the end of its life it is an asymptotic-giant-branch star with a tiny carbon–oxygen core. If the original star was between about 7 and 9 M_{Sun}, an oxygen–neon core could exist while on the asymptotic giant branch because more carbon and oxygen are able to burn in the higher mass stars. In either case, the core is surrounded by helium– and hydrogen–burning shells along with an extremely extended and very low-density envelope composed primarily of hydrogen and helium. Because the envelope is so large and tenuous, it is only loosely bound to the core that is far from the surface. The star may also experience long-period pulsations that cause the envelope to rise and fall over periods of months to years. The relatively cool gases at the surface are also able to form dust grains around

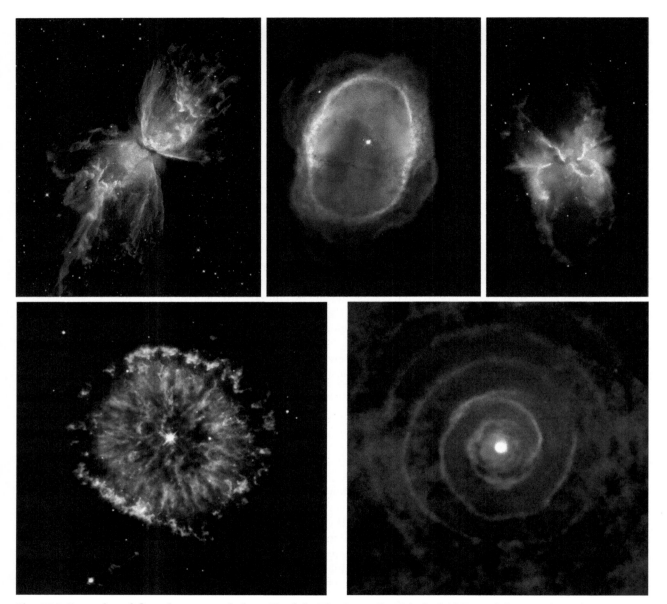

Fig. 17.2 Examples of five planetary nebulas. *Top left:* The Butterfly Nebula (NGC 6302) in Scorpius. (NASA, ESA, and the Hubble SM4 ERO Team) *Top center:* The Eight-Burst Nebula (NGC 3132) in Vela. [The Hubble Heritage Team (STScI/AURA/NASA)] *Top right:* NGC 2818 in Pyxis. [NASA, ESA, and the Hubble Heritage Team (STScI/AURA)] *Bottom left:* NGC 6751 in Aquila. [NASA, The Hubble Heritage Team (STScI/AURA)] *Bottom right:* LL Pegasi is an old star ejecting a developing planetary nebula in a binary star system. [ALMA (ESO/NAOJ/NRAO)/H. Kim et al., ESA/NASA & R. Sahai; CC BY 4.0]

water ice or carbon-based molecules that can be pushed outward by the radiation pressure of photons from the very bright asymptotic-giant-branch star.

The result of these conditions is that a superwind develops which causes the star to lose mass at a rate of about one solar mass every 10,000 years or so. This superwind is the mechanism that ultimately creates planetary nebulas. Planetary nebulas tend to be fairly short-lived phenomena because the expanding clouds eventually dissipate as they spread out into interstellar space, returning some of the gases from which they originally formed.

The complex structures seen in many planetary nebulas may be because the dying star is a member of a binary or multiple star system with the motions and gravitational interactions stirring up the expanding gases, or the dying star could be rotating rapidly. The dying star could have experienced previous mass-loss events or stellar pulsations, and there are likely to be magnetic fields present. The Cat's Eye is particularly complex and appears to show concentric shells of gas that were ejected prior to the formation of the planetary nebula inside the shells. Episodic shell ejection may in fact be common during the final stages of a star's life.

If you look at the Cat's Eye Nebula or several of those in Fig. 17.2, the central star is clearly visible. As the central star expels the remnants of its tenuous envelope, the burning shells shut off because there is no longer any gas squeezing down on them to create the high pressures and temperatures needed for nuclear fusion. What is left is the tiny sphere of carbon and oxygen or oxygen and neon that is the fossil remnant of nuclear burning.

White Dwarf Stars

The dead core that once was a living star is a called a white dwarf star. A white dwarf is the final stage of evolution of a star with an initial mass of less than 9 times the mass of the Sun. Figure 15.39 shows the locations of white dwarf stars on the H–R diagram from the Gaia catalog and the full evolutionary track of a 1 M_{Sun} star is shown in Fig. 17.3.

If you go back and take a look at Fig. 16.22, you can appreciate just how dim and small a white dwarf star is. Sirius B, the "Pup," is a white dwarf star and the binary companion to Sirius A, which happens to be the brightest star in the sky. At one time the Sirius system would have been enveloped in an expanding planetary nebula.

Key Point

White dwarf stars started out as stars with initial masses of less than 9 M_{Sun}.

Example 17.1

In Section 15.4 we learned, by analyzing the binary star orbits of the Sirius system, that Sirius B has a mass of 1.02 M_{Sun} (Example 15.7). Table 15.3 indicates that the surface temperature of Sirius B is 25,970 K and its luminosity is just 0.03 L_{Sun}. Determine the star's (a) radius, (b) volume, (c) density, and (d) acceleration due to gravity at its surface.

(a) To determine the star's radius we can make use of the stellar radius equation in solar units on page 624, but first we must calculate T_{Sun}, the temperature of Sirius B as a multiple of the temperature of the Sun:

$$T_{Sun} = 25{,}970 \text{ K} / 5772 \text{ K} = 4.499 \text{ T}_{Sun}.$$

Now the radius of Sirius B is given by

$$R = \frac{\sqrt{0.03 \text{ L}_{Sun}}}{(4.499 \text{ T}_{Sun})^2} = 0.0086 \text{ R}_{Sun}.$$

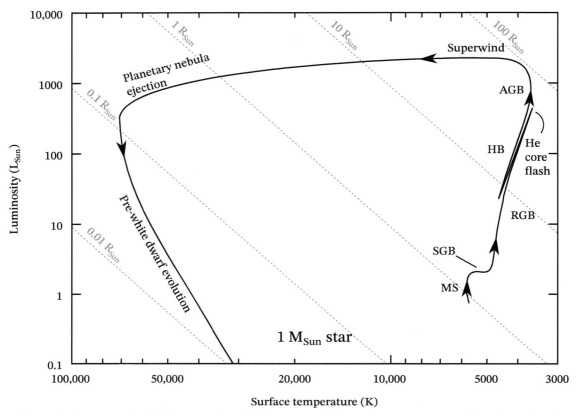

Fig. 17.3 The evolutionary track of 1 M_{Sun} star from the beginning of the main sequence to the formation of a white dwarf star. The diagonal red-dashed lines are lines of constant radius.

Since the radius of the Sun is 6.96×10^8 m, the radius of Sirius B is

$$R = 0.0086 \times 6.96 \times 10^8 \text{ m} = 6 \times 10^6 \text{ m}.$$

For comparison, the radius of Earth is 6.38×10^6 m. This 1.02 M_{Sun} white dwarf star is smaller than Earth!

(b) The volume of a sphere is given by the expression

$$V = \frac{4\pi}{3} R^3,$$

so the volume of Sirius B is

$$V = \frac{4\pi}{3} (6 \times 10^6 \text{ m})^3 = \frac{4\pi}{3} (6^3 \times 10^{3 \times 6} \text{ m}^3) = 9 \times 10^{20} \text{ m}^3.$$

(c) Density is given by mass / volume. The mass of Sirius B is

$$1.02 \text{ M}_{Sun} = 1.02 \times (1.99 \times 10^{30} \text{ kg}) = 2.03 \times 10^{30} \text{ kg}.$$

As a result, dividing the star's mass by its volume gives its density:

$$\rho = \frac{M}{V} = \frac{2 \times 10^{30} \text{ kg}}{9 \times 10^{20} \text{ m}^3} = \frac{2}{9} \times 10^{(30-20)} \text{ kg/m}^3 = 2.2 \times 10^9 \text{ kg/m}^3.$$

The density of water is 1000 kg/m^3, which means that the density of the remnant carbon–oxygen core from stellar evolution is about 2.2 million times greater than water, or about 700,000 times greater than a rock on the surface of Earth!

(d) To determine the surface gravity of Sirius B in terms of the surface gravity of Earth, we can use Equation (5.8) back in Section 5.3 and divide the equation using Sirius B values by the equation using Earth values, which has the added value of canceling Big G:

$$\frac{g_{\text{Sirius B}}}{g_{\text{Earth's surface}}} = \frac{\cancel{G}M_{\text{Sirius B}}/R_{\text{Sirius B}}^2}{\cancel{G}M_{\text{Earth}}/R_{\text{Earth}}^2} = \frac{M_{\text{Sirius B}}/M_{\text{Earth}}}{R_{\text{Sirius B}}^2/R_{\text{Earth}}^2} = \frac{M_{\text{Sirius B}}/M_{\text{Earth}}}{(R_{\text{Sirius B}}/R_{\text{Earth}})^2}.$$

The last two steps use algebra to rewrite the expression into a more convenient form, where each Sirius B value is divided by its equivalent Earth value. For example, $M_{\text{Sirius B}}/M_{\text{Earth}}$ tells us how many times greater the mass of Sirius B is compared to the mass of Earth. A similar "trick" has been used multiple times before in this textbook, such as writing Kepler's third law (page 174) in terms of Earth's orbit and the mass of the Sun.

We already have the mass of Sirius B in kilograms, and the mass of Earth is 5.97×10^{24} kg. We also have the radii of both objects. Putting everything into the equation,

$$\frac{g_{\text{Sirius B}}}{g_{\text{Earth's surface}}} = \frac{2.03 \times 10^{30} \cancel{\text{kg}}/5.97 \times 10^{24} \cancel{\text{kg}}}{(6 \times 10^6 \cancel{\text{m}}/6.38 \times 10^6 \cancel{\text{m}})^2} = \frac{3.4 \times 10^5}{0.94^2} = 3.85 \times 10^5.$$

All of this math means that the acceleration due to gravity on the surface of Sirius B is 385,000 times greater than it is on the surface of Earth; if you weigh 200 lb (890 N) on Earth, you would weigh $385,000 \times 200$ lb = 77 million pounds on the surface of Sirius B! That is a seriously weighty issue.

White dwarf stars are incredibly dense and are composed of carbon and oxygen in a crystallized form, meaning that they are solid objects. As you may know, one form of crystallized carbon is used extensively on Earth, including in expensive jewelry; it is known as diamond. Recall the nursery rhyme by Jane Taylor (1783–1824):

Twinkle, twinkle little star,
How I wonder what you are,
Up above the world so high,
Like a diamond in the sky.

In essence, these little white dwarf stars that are dominated by crystallized carbon are not unlike diamonds in the sky, but you wouldn't want something of that density and temperature around your neck or on your finger!

Although white dwarf stars are extremely dense, they don't continue to collapse under gravity's relentless pull. The reason for this is that they are supported by the pressure of the degenerate electron gas. This is the same electron degeneracy pressure that leads to the helium core flash in stars with masses of less than about 2 M_{Sun} first discussed on page 661. The intense gravity has caused the electrons of the carbon and oxygen atoms to occupy the lowest possible energy levels and the resulting electron degeneracy pressure is able to resist gravity's crushing force.

Fig. 17.4 The nova of 1901, GK Persei. The debris from the explosion is expanding outward at about 300 km/s (200 mi/s) and the temperature of the gas is about 1 million kelvins. (X-ray: NASA/CXC/RIKEN/D.Takei et al.; Optical: NASA/STScI; Radio: NRAO/VLA)

So what happens to white dwarf stars after they form? For isolated white dwarfs not bound in orbits with other stars, the answer is not much. These end-products of stars with initial masses of less than about 9 M_{Sun} simply cool off and get progressively dimmer, meaning that they will move slowly downward and to the right in the H–R diagram. Far, far into the distant future, white dwarfs will become black cinders with the temperature of the average background of the universe.

For white dwarf stars in binary or multiple star systems, the future can be much brighter.

Novas

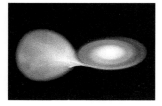

Fig. 17.5 An artist's conception of a binary star system where gases from a companion star are flowing through an accretion disk onto a white dwarf star. (NASA/CXC/M.Weiss)

On February 21, 1901, Thomas David Anderson (1853–1932), a Scottish amateur astronomer who had originally studied theology at the University of Edinburgh but chose to focus his attention on astronomy full-time, discovered a new star that suddenly appeared in the constellation of Perseus. GK Persei soon became brighter than almost any other star, reaching an apparent magnitude of +0.2. Today, GK Per is about magnitude 14. A modern image of GK Per is shown in Fig. 17.4 as a combination of x-ray, optical, and radio wavelength observations. GK Per is one example of a class of cataclysmic variable stars known as novas (from Latin meaning "new"; when they were first discovered novas, or alternatively, novae, were thought to be new stars).

As depicted in Fig. 17.5, a nova occurs in a close binary star system when the outer atmosphere of the companion star (a main-sequence, subgiant, or red-giant star) becomes more attracted to the white dwarf star than to the host star itself, and flows onto an accretion disk that eventually spirals onto the surface of the white dwarf. This process forms a thin atmosphere around the white dwarf that becomes very hot and dense under the strong gravitational pull of the white dwarf. Because the gas flowing from the companion star comes from its outer atmosphere, the gas will be primarily composed of hydrogen, with most of the rest being helium. Under the immense gravity, the gas becomes degenerate and a runaway nuclear explosion occurs on the surface. The violent explosion expels the products of the surface nuclear reactions out into space, which is what you see in Fig. 17.4. Since the explosion occurs on the surface, the white dwarf star itself survives the event. The explosion is also too weak to have any significant impact on the companion star, meaning that the process could, and likely will, start all over again. When the successive explosions happen often enough to be detected repeatedly by humans, they are referred to as recurrent novas. Although the recurrent explosions of GK Per are nowhere near as energetic as the 1901 event, GK Per does experience explosive events about once every three years, typically increasing by about 2 to 3 magnitudes during an event.

Novas are seen throughout the Milky Way Galaxy about ten times per year, but only become naked eye objects roughly once every 18 months to two years.

17.2 Supernovas: Going out with a Bang

In 185 CE, Chinese astronomers recorded a "guest star" in the southern hemisphere, in what is today the constellation of Circinus, the Compass. It is also possible that the object may have been recorded in Roman literature. As reported by the Chinese astronomers, "The size was half a bamboo mat. It displayed various colors, and gradually lessened. In the 6th month of the succeeding year it disappeared." Rather than being a nova, the guest star was what we refer to today as a supernova, designated SN 185. An expanding gaseous shell 9100 ly away is believed to be the remnant of the tremendous explosion that the Chinese saw.

A second, even brighter, supernova appeared in the sky 821 years later. Only 7500 ly from Earth, SN 1006 in Lupus is estimated to have reached an apparent magnitude of −7.5, making it sixteen times brighter than Venus and the brightest celestial event in recorded history. It would have also been easily visible during the daytime. The supernova was noted in China, Japan, the Middle East, Europe, and possibly in North American petroglyphs. An x-ray image of the remnant of SN 1006 is shown in Fig. 17.6.

Only 48 years after SN 1006, SN 1054 appeared in the constellation of Taurus the Bull (the remnant of SN 1054 is seen in Fig. 17.7). Its appearance in the sky was again noted by naked-eye observers in several cultures, including the Ancestral Puebloans of Chaco Canyon (Fig. 2.23), the Chinese, the Japanese, and in the Middle East. Since that time, only three other naked-eye supernovas (or alternatively, supernovae) are known to have been recorded; the supernova studied by Tycho Brahe (SN 1572), the supernova named for Johannes Kepler (SN 1604), and the supernova of 1987 (SN 1987A).

The remnant of the massive explosion that was SN 1054 is now listed as the first entry in Messier's catalog (M1) and is commonly called the Crab Supernova

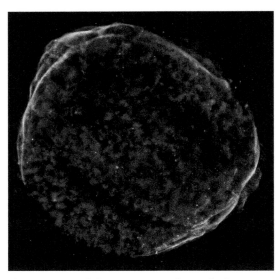

Fig. 17.6 SNR 1006, the remnant of the supernova of 1006 in Lupus seen in x-rays. (NASA/CXC/Middlebury College/F. Winkler)

Fig. 17.7 The Crab Supernova Remnant (M1, SNR 1054) in the constellation of Taurus at a distance of 6500 ly. The nebula is expanding at 1500 km/s (930 mi/s) and measures 11 ly across. [NASA, ESA, J. Hester and A. Loll (Arizona State University)]

Fig. 17.8 A sketch of M1 made by William Parsons, third Earl of Rosse. (1844;[1] public domain)

Remnant. The nebula gets its name from William Parsons, third Earl of Rosse, who also built the Leviathan (Fig. 15.2). In making a sketch of the supernova remnant from views through a 36-inch diameter telescope, Parsons thought that the object looked like a crab (Fig. 17.8).

[1] Earl of Rosse (1844). "Observations on some of the nebulae," *Philosophical Transactions of the Royal Society of London*, 134:321–324.

There are two basic types of supernovas: those that arise from the destruction of white dwarf stars and those from massive stars that occur as a consequence of producing iron and nickel in their cores. Although naked-eye supernovas are exceedingly rare, it is estimated that roughly one supernova occurs every 50 years or so among the 200 to 400 billion stars within the Milky Way Galaxy.

White Dwarf Supernovas

Fig. 17.9 Subrahmanyan 'Chandra' Chandrasekhar (1910–1995). (AIP Emilio Segrè Visual Archives, Gift of Kameshwar Wali)

In 1931, while only 21 years old, the Indian physicist, Subrahmanyan Chandrasekhar (Fig. 17.9), or Chandra for short, showed that an upper limit existed for the mass of a white dwarf star. Although the star is supported against its intense gravity by degenerate electron pressure, that resistance has a breaking point. White dwarf stars have the odd characteristic that the more massive they get, and the stronger their gravity becomes, the smaller they get. When the mass of a white dwarf star reaches just over 1.4 M_{Sun}, even degenerate electron pressure is unable to support the star any further.

For a white dwarf star of close to 1.4 M_{Sun} in a binary star system like the one shown in Fig. 17.10, it is possible that enough mass can be transferred to the white dwarf from its companion that the Chandrasekhar limit is exceeded. At that point, the highly compressed white dwarf star starts runaway nuclear burning in its core that completely destroys the star, blasting the

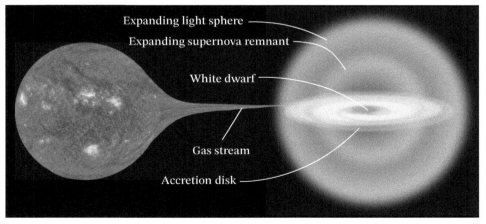

Fig. 17.10 A companion star being stripped of the gas in its outer atmosphere by a white dwarf star. The gas spirals onto the white dwarf through an accretion disk with the temperature of the disk increasing toward the center. A white dwarf supernova occurs when the mass of the white dwarf star reaches 1.4 M_{Sun}. Because massless photons travel faster than any other type of particle, the light from the explosion moves outward faster than the expanding shell of the supernova remnant. (Image of the star adapted from NASA/SDO)

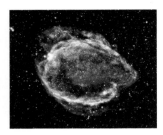

Fig. 17.11 The remnant of a white dwarf supernova (G299.2−2.9) that exploded 4500 years ago and about 1600 ly away in the constellation of Musca. The image is a composite of images taken in x-rays and infrared light. (X-ray: NASA/CXC/U. Texas/S. Post et al.; Infrared: 2MASS/UMass/IPAC-Caltech/NASA/NSF)

Fig. 17.12 The white dwarf supernova SN 1994D (lower left) in the outskirts of the galaxy NGC 4526, located 50 million light-years away in the constellation of Virgo. (NASA/ESA, the Hubble Key Project Team and the High-Z Supernova Search Team; CC BY 4.0)

remnants of the white dwarf out into the cosmos, thereby enriching interstellar space with elements created by the former star's nuclear reactions. It is likely that SN 1006 in Fig. 17.6 was a white dwarf supernova, as were G299 (Fig. 17.11), Tycho's supernova (SN 1572), and Kepler's supernova (SN 1604).

White Dwarf Supernovas and the Distance Ladder

When the destruction of the white dwarf star occurs it can briefly outshine the entire galaxy of which it is a member. Figure 17.12 shows a white dwarf supernova and the galaxy that hosts it. Because the vast majority of all white dwarf supernovas are the result of 1.4 M_{Sun} carbon–oxygen explosions, they are all very similar in their peak brightnesses and can be detected over immense distances. This means that they are useful as standard candles to determine distances to remote galaxies. White dwarf supernovas become a major rung in the distance ladder after they have been calibrated by observing Cepheids or the brightest stars at the tip of the red giant branch in relatively nearby galaxies that also contain this type of supernova.

Gravitational Waves and White Dwarf Mergers

A rare type of white dwarf supernova can occur when two white dwarf stars orbit one another with a small separation between them. Over a long period of time, their orbits can get closer and closer together until they finally collide. The combined mass of the two stars can cause them to exceed the Chandrasekhar limit of 1.4 M_{Sun}, resulting in a supernova. The process that causes the orbital decay is the generation of gravitational waves, a prediction of Einstein's general theory of relativity that we first encountered in Section 8.2 (page 269ff). Gravitational waves can be thought of as waves moving through spacetime, carrying energy and angular momentum away from the source that generated them. The generation of gravitational waves requires accelerating masses, just like the continual acceleration of changing directions for stars in orbits, as described by the definition of acceleration equation on page 158. More will be said about gravitational waves in Section 17.3.

The Supernovas of Massive Stars

Supernovas produced by massive stars exceeding about 9 M_{Sun} are caused by an entirely different process from white dwarf supernovas. For lower mass stars on the asymptotic giant branch of the H–R diagram, the temperatures and densities at their centers never get hot enough to burn the carbon–oxygen or oxygen–neon cores that exist there, but the same is not true for more massive stars. With the heavier, more positively charged nuclei of carbon, oxygen, and neon, much higher temperatures (meaning more kinetic energy) and more tightly packed nuclei (higher densities) are needed to overcome the electrostatic repulsion that works to keep them apart (recall Coulomb's law on page 209). This means that the potential energy "hill" of Fig. 9.34 is a much higher and more formidable barrier than it is for proton–proton collisions in hydrogen burning. For example, in a nuclear reaction between two carbon nuclei, $^{12}_{6}C + {}^{12}_{6}C$, the barrier between the two nuclei of six positively charged protons each is $6 \times 6 = 36$ times higher than for the proton–proton collision, $^{1}_{1}H + {}^{1}_{1}H$. At the same time, carbon nuclei are close to 12 times heavier than protons and therefore much more sluggish at the same temperature, with average speeds that are $1/\sqrt{12}$ the speeds of protons, or about 3 1/2 times slower (see the average speed of gas particles equation on page 339). So, instead of 15 million kelvins required for the proton–proton chain to operate in the Sun, temperatures of 600 million kelvins are needed for carbon burning. In order for oxygen burning to occur, $^{16}_{8}O + {}^{16}_{8}O$, the temperature at the center of the star must reach about 1 billion kelvins. Only very massive stars have gravities that are strong enough to be up to the task.

Energy Released Per Nucleon

Despite all of that effort, as nuclei get bigger and bigger, the amount of energy that comes out of the reaction divided by the total number of protons and neutrons that they contain (the atomic mass, A) starts to go down. When a nucleus is constructed out of its individual protons and neutrons (collectively called nucleons), the final mass of the nucleus is less than the total mass of the nucleons individually. This is because energy is given off in the process of binding the nucleons together, just like the situation when energy is given off as an electron combines with a proton to form a hydrogen atom, except that for nucleons the energies involved are millions of times greater. Remember that energy released is the same thing as mass reduced, according to Einstein's $E = mc^2$.

Figure 17.13 is a graph depicting this situation. The vertical axis is the amount that the mass–energy has been reduced in forming the nucleus divided by its number of nucleons. This is interpreted as meaning that nuclei lower on the graph have released more energy per nucleon than those higher up on the graph. The horizontal axis is just the total number of nucleons in the nucleus; the smallest nuclei are to the left and the more massive, larger nuclei are to the right. Starting with hydrogen on the far upper left corner, as the number of nucleons in the nucleus increases to the right, the mass–energy per nucleon of the nuclei decreases. But this only happens up to a point. When the atomic mass reaches about 28, corresponding to silicon-28 ($^{28}_{14}Si$), the curve starts to flatten noticeably. This means that building larger nuclei only gives off a small amount of extra energy. As a massive star fights the relentless pull of gravity, the star is not getting much extra help by burning elements heavier than silicon. This is why it only takes a couple of days for silicon to burn in the center of a massive star, as mentioned in Section 16.3; a lot of silicon must be burned rapidly for the star to prolong its inevitable fate.

The curve in Fig. 17.13 becomes very flat around iron-56 ($^{56}_{26}Fe$) and reaches a minimum at nickel-62 ($^{62}_{28}Ni$), meaning that $^{62}_{28}Ni$ is the most stable nucleus in the universe, with $^{56}_{26}Fe$ being a very close second. This is the reason why iron and nickel are so prevalent in nature, including in the cores of the rocky planets of our Solar System.

Key Point
Iron-56 and nickel-62 are the two most stable nuclei in the universe.

The Most Abundant Elements

You may have also noticed in Fig. 17.13 that $^{4}_{2}He$, the second most abundant isotope in the universe after $^{1}_{1}H$, and the product of hydrogen burning in stars, is far lower on the diagram than all of its immediate neighbors, indicating that helium-4,

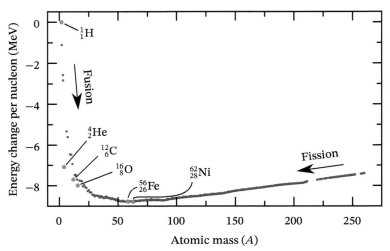

Fig. 17.13 The mass–energy decrease per nucleon in building a nucleus out of its constituent protons and neutrons. The red dots highlight labeled isotopes. The vertical axis is in units of millions of electron volts (MeV); see page 286. (Data courtesy of NIST)

Rutherford's alpha particle (page 255), is an unusually stable isotope. Two multiples of the alpha particle, $^{12}_{6}$C and $^{16}_{8}$O, are also significantly more stable than their neighbors. This is in part why carbon-12 and oxygen-16 are the primary end-products of nuclear burning for stars below 9 M_{Sun} and are the isotopes that make up most white dwarf stars. These isotopes are also among the most common in the universe, along with $^{14}_{7}$N, $^{20}_{10}$Ne, $^{24}_{12}$Mg, $^{28}_{14}$Si, $^{32}_{16}$S, $^{36}_{18}$Ar, $^{40}_{20}$Ca, $^{56}_{26}$Fe, and $^{62}_{28}$Ni, all of which are produced by various stages of nuclear burning in massive stars. Notice that neon-20, magnesium-24, silicon-28, sulfur-32, argon-36, and calcium-40 are again multiples of $^{4}_{2}$He. Nitrogen-14 owes its abundance to the fact that it plays an important role in hydrogen burning for stars slightly more massive than the Sun, through the CNO cycle of reactions. You may have also already realized that the most abundant molecules in the Solar System (and throughout the universe) are composed of these elements; for example, N_2 and O_2 are the dominant molecules in Earth's atmosphere, H_2O (water), CO_2 (carbon dioxide), NH_3 (ammonia), CH_4 (methane), silicates that form rock, and the calcium compounds in our bones. With the exception of the elements that formed as a consequence of the Big Bang at the beginning of the universe — hydrogen, some of the helium, and a very small amount of lithium — the rest of the helium, lithium, and all of the other elements in the periodic table and their isotopes are the result of nuclear reactions in the cores of stars, products of the cataclysmic destruction of stars, or through interactions with high-energy charged particles flying through the universe (cosmic rays).

Figure 17.14 shows the relative abundances of the elements found in the Sun and meteorites relative to hydrogen. The vertical axis is logarithmic, meaning that when helium is close to 11 and hydrogen is set to be exactly 12, hydrogen is not 12/11 or 1.09 times more abundant than helium, it is about $10^{12\text{-}11} = 10^1 = 10$ times more abundant. The vertical scale is marked this way to allow for the huge range in abundances relative to hydrogen and helium. Note that some of the abundances are so small that they do not show up on the graph. It would also be useful to go

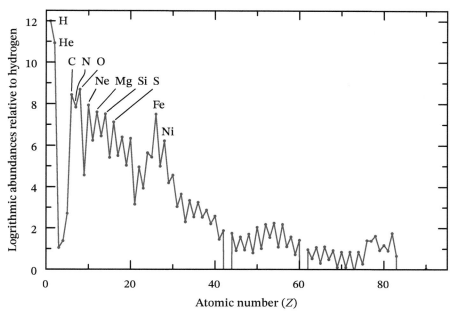

Fig. 17.14 The chemical abundances in the Sun and meteorites. The vertical scale is logarithmic relative to hydrogen, which is defined as 12. (Data courtesy of Asplund, et al., 2009[2])

back to the periodic table on page 250 that shows where the elements came from. One of the many successes of astrophysics has been the ability to explain the origins of the elements and their relative abundances.

Paraphrasing the quote of Carl Sagan in the chapter opening:

We are all made of star stuff.

Fusion and Fission

As first mentioned in Section 9.6, there are two general forms of nuclear reactions: fusion (the process of creating larger nuclei out of smaller ones) and fission (the breaking up of nuclei). As illustrated by the arrows in Fig. 17.13, fission reactions release energy when more massive elements like uranium split into smaller decay products that are closer to the bottom of the curve but still to the right of iron and nickel. Today's commercial nuclear reactors and the earliest atomic bombs produce energy through fission reactions. On the other hand, when fusion reactions build larger nuclei while moving toward the bottom of the curve from the left, energy is also released, which is just what stars spend their lives doing.

The End — The Collapsing Core

Disaster for massive stars strikes when the bottom of the curve in Fig. 17.13 is finally reached. When massive stars try to burn iron through fusion, instead of releasing energy, energy must be *absorbed* from its surroundings. The available energy is the

[2]Asplund, Martin, Grevesse, Nicolas, Sauval, A. Jacques, and Scott, Pat (2009). "The Chemical Composition of the Sun," *Annual Review of Astronomy and Astrophysics*, 47:481–522.

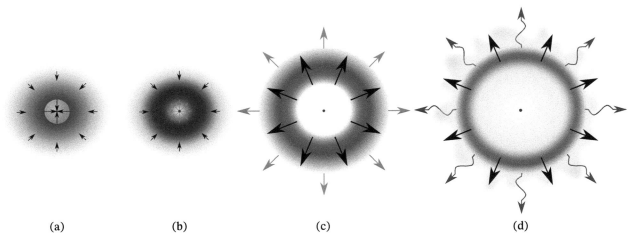

(a)   (b)   (c)   (d)

Fig. 17.15 (a) The iron core collapses as nuclei disintegrate down to protons and neutrinos. The rest of the star starts to collapse. (b) The neutrinos (red arrows) are trapped behind an extremely dense shell of gas falling inward faster than the local speed of sound (a shock front), causing the neutrinos to push the shock outward. The proto-neutron star is at the center. (c) When the shock wave thins, neutrinos escape into space (again, red arrows). (d) The shock reaches the surface (black arrows) ejecting a supernova remnant and allowing photons to escape. Note that the neutrinos escape *before* the photons do.

star's heat generated by gravitational potential energy. As that energy is taken from the star, the star's interior shrinks a bit, causing the temperature and density of the core to increase. The increased temperature and density accelerate iron burning, which leads to even faster nuclear reactions, even faster shrinking, and even higher temperatures and densities. This viscous cycle happens very, very fast. The collapse of the outer part of the iron core reaches a speed close to 70,000 km/s, almost 25% of the speed of light! In about one second a core of roughly the size of Earth has collapsed to a radius of 50 km (30 mi).

Key Point
The production of neutrinos.

In reality, the nuclear reactions are not creating more massive elements. Instead, the very high-energy gamma-ray photons in the core are destroying the iron. After the star spent a lifetime getting to iron, the nuclei disintegrate back down to individual protons and neutrons. The protons then combine with electrons to produce even more neutrons, together with electron neutrinos (ν_e), through the reaction

$$p^+ + e^- \rightarrow n + \nu_e.$$

Recall that neutrinos under normal conditions are able to pass through matter with ease, which is why detecting the ones coming from the center of the Sun requires the enormous underground observatories discussed in Section 9.7. However, in the extremely high-density environment at the center of the collapsing iron core of a massive star, the neutrinos slam into a wall. When combined with the bounce that occurs at the end of the core collapse when infalling overlying material collides with the extremely dense core, the neutrinos blast the overlying material out into space. The massive star's impressive neutrino-driven detonation is a core-collapse supernova. The stages of a core-collapse supernova are illustrated in Fig. 17.15.

Fig. 17.16 SN 1987A is the bright star (a core-collapse supernova) near the top of the image and at the fringe of the Tarantula Nebula in the Large Magellanic Cloud. North is toward the left. (Cropped image by ESO; CC BY 4.0)

How Do We Know?

It is one thing to have a sophisticated and elaborate computer model of how a core-collapse supernova works, but as with all scientific explanations, our ideas must be tested and tested again by comparing as many details as possible with observational or experimental data. It is this constant testing of the predictions that derive from our fundamental theories (including computational modeling), against empirical, verifiable data, that sets science apart as a way of knowing, and has made scientific exploration so successful.

On the night of February 23, 1987, Ian Shelton (b. 1957), a Canadian astronomer, was taking telescopic photographs of the Large Magellanic Cloud (LMC) while working at Las Campanas Observatory in Chile when a wind came up and slammed the roll-off roof closed. In a bit of luck, he decided to compare the photograph he had been taking with the one from the night before, and immediately noticed a new, bright light source in the LMC near the fringe of the Tarantula Nebula (Fig. 17.16). Running outside, he looked up to confirm the "new star," realizing that he must be looking at a naked-eye supernova, which became designated SN 1987A as the first supernova discovered in 1987. At only 160,000 ly from Earth, it was the first naked-eye supernova since Kepler's supernova of 1604. It was also the closest supernova since the advent of large ground- and space-based telescopes, covering nearly the entire electromagnetic spectrum, the development of modern electronic detectors, and the existence of neutrino observatories. An alert was sent out to astronomers around the world that began an intense observing program of the core-collapse supernova.

Although not known until after the data were analyzed, three neutrino observatories had recorded small bursts of neutrinos coming from the LMC that arrived several hours before the light reached Earth. This is consistent with a core-collapse supernova's neutrinos driving the outward explosion of material. Once the density of the material had dropped sufficiently, neutrinos were free to escape out into space. However, the light was still trapped behind the expanding shock wave of the explosion until the material became thin enough at the surface for photons to escape. In this way, the neutrinos had a several-hour head start on the photons as they traveled to Earth's observatories. The neutrino detections and timing relative to the light were just what was expected from the computer simulations. The energies of the neutrinos were also consistent with predictions, as were the number of neutrinos detected, given the size of the detectors relative to the area of a sphere centered on SN 1987A, and the very low probability of detecting an elusive neutrino as it passed through the observatory. In fact, neutrinos carry away 99% of the energy of a core-collapse supernova, not the light.

As the shock wave of SN 1987A expanded out into space at high speed, astronomers predicted that a second cosmic light show would occur when the shock wave caught up to, and plowed through, material previously ejected by the now-destroyed Sanduleak −69 202, a blue supergiant star that had been studied prior to the unexpected explosion. Figure 17.17 shows the ring that lit up when the shock wave hit clumps of material that had been emitted about 20,000 years before the supernova occurred.

Of course, there are other ways to test of our understanding of both white-dwarf and core-collapse supernovas, such as analyzing the composition and abundances of the ejected material. Our Solar System even holds evidence that a supernova may have triggered the collapse of the original solar nebula from which the Sun and the Solar System formed. Aluminum-26 ($^{26}_{13}Al$) is a radioactive isotope that decays with a half-life of 717,000 years and is produced by supernovas. Although long by human standards, the half-life of aluminum-26 is very short compared to the age of the Solar System, which is 4.57 billion years old. Contained within meteorites is a high overabundance of the stable isotope magnesium-26 ($^{26}_{12}Mg$), that is produced by the radioactive decay of aluminum-26 (recall the discussion on page 556). The ancient meteorites must have formed very soon after the supernova that created the aluminum radioisotope; otherwise the aluminum-26 would have decayed away before becoming incorporated into the meteorites. Once inside the meteorites, the aluminum-26 transformed into magnesium-26. The composition of the meteorites strongly suggests that the solar nebula was contaminated with remnant material from a supernova at about the time that the nebula was collapsing. There is also strong evidence that the asteroid Vesta, a protoplanet remnant, once contained a substantial abundance of aluminum-26 as well (pages 515 and 557).

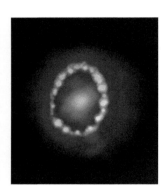

Fig. 17.17 The ring from SN 1987A seen in a combination of wavelengths. (ALMA (ESO/NAOJ/NRAO)/ A. Angelich. Visible light image: the NASA/ESA Hubble Space Telescope. X-Ray image: The NASA Chandra X-Ray Observatory; CC BY 4.0)

17.3   Neutron Stars

Key Point
When a star of up to 25 M_{Sun} undergoes a core-collapse supernova, a neutron star is formed at the center.

When a massive star of up to 25 M_{Sun} or so undergoes a core collapse that triggers a supernova, the core itself survives the explosion while the overlying material is blasted into space, along with an enormous number of neutrinos and photons. At this point, the core has been converted into a sphere of neutrons with a density greater than the density of an atomic nucleus (see page 698). As it cools off, the star shrinks down to a radius of about 10 km (6 mi), with a typical mass of 1.4 times the mass of the Sun (1.4 M_{Sun}). This corresponds to a density of about 7×10^{17} kg/m^3, or 700 trillion times the density of water! This ultra-dense sphere of neutrons is known simply as a neutron star.

As another way to think about what the extreme environment of a neutron star is like, imagine falling from a height of 1 m above the neutron star's solid surface. With its immense acceleration due to gravity you would hit the surface going about 0.7% of the speed of light (2 million meters per second or 4 1/2 million miles per hour). That 1 m fall could ruin your whole day. Of course, under that enormous gravity your body would be squeezed down to a microscopic ball of neutrons anyway.

Can such a bizarre theoretical object actually exist? Again, all theoretical predictions must be tested against empirical evidence, as science always demands.

Pulsars

Along with shrinking to such a tiny size for something with more mass than the Sun has, conservation of angular momentum (page 194) requires that decreasing the radius of a star must increase its spin rate, just like when the figure skater in Fig. 6.15 pulls his arms in. If the angular momentum remains constant, the rotation period is approximately proportional to the square of the radius, or

$$P_{rotation} = \text{constant} \times r^2,$$ (17.1)

Rotation period from angular momentum conservation

where the constant depends on the spinning object's geometry.

Example 17.2

The Sun currently spins about once every 25 days at its equator. For simplicity, let's assume that the entire star spins at that rate. What would the Sun's rotation period be if its radius suddenly decreased from its current value of 7.0×10^8 m down to 10 km or 10,000 m?

Using the rotation period from the angular momentum conservation equation, and dividing the rotation period of the Sun if it is the size of a neutron star to its rotation period today, the constant will cancel. This is the same "trick" we have used many times before to eliminate pesky constants, like in Example 17.1 of Section 17.1, Kepler's third law on page 174, or the Stefan–Boltzmann law in solar units on page 219. Doing so gives

$$\frac{P_{neutron\,star}}{P_{Sun}} = \frac{\cancel{constant} \times r^2_{neutron\,star}}{\cancel{constant} \times r^2_{Sun}} = \left(\frac{r_{neutron\,star}}{r_{Sun}}\right)^2$$

$$= \left(\frac{10,000\,\cancel{m}}{7 \times 10^8\,\cancel{m}}\right)^2 = \frac{1}{70,000^2} = \frac{1}{4.9 \times 10^9} = 2 \times 10^{-10}.$$

The star's rotation period would become shorter by a factor of 4.9 billion.

Converting the Sun's rotation period to seconds and multiplying by 2×10^{-10}, we find that after shrinking to the size of a neutron star the Sun's rotation period would become

$$25 \cancel{d} \times \left(24 \cancel{h}/1 \cancel{d}\right) \times \left(3600 \text{ s}/1 \cancel{h}\right) \times \left(2 \times 10^{-10}\right) = 0.00044 \text{ s} = 0.44 \text{ ms}.$$

This result is the same as saying that the Sun would rotate $1/0.00044 \text{ s} = 2270/\text{s}$ or 2270 times *every second*; a direct consequence of conservation of angular momentum.

It's not just the spin rate that would be affected, the Sun's magnetic field would also be tremendously compressed and strengthened as well. Just as with the spin rate, the magnetic field strength also increases by one divided by the radius squared. If the Sun was compressed to the size of a neutron star, its magnetic field would suddenly become $70{,}000^2 = 4.9$ billion times stronger.

Little Green Men

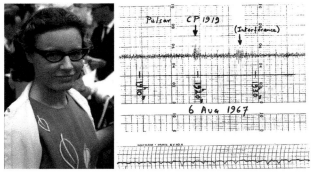

Fig. 17.18 Susan Jocelyn Bell Burnell (b. 1943) and portions of the paper charts showing the discovery of the first pulsar. The top chart labels the "scruff" of pulsar CP1919 and the bottom chart shows the regular pulses separated by very regular time intervals. (*Left:* Roger W. Haworth; CC BY-SA 2.0; *Right:* Two charts recording the first known pulsar, 1967–1974, GBR/0014/HWSH Acc 355. Churchill Archives Centre)

In July 1967, as a graduate student, Jocelyn Bell Burnell (Fig. 17.18) was using a radio telescope that she spent two years constructing, together with her thesis advisor Antony Hewish (1924–2021), designed the telescope array. In those days, the signals were recorded on paper tape that was printed steadily over time (Fig. 17.18). Each day, Bell (she was married one year later) would review many meters of paper charts looking for signals from distant objects, known as quasars. She soon realized that some "scruff," as she called it, appeared on the paper chart once every $23^{\text{h}}56^{\text{m}}$, or one sidereal day, meaning that the source was coming from a point in space that passed over the array once every rotation period of Earth. To understand what the scruff might be, Bell used a faster signal detector that would be able to do a better job of resolving the details of the signal. What she discovered was a signal that was very precise, with very narrow spiked peaks that repeat exactly once every 1.337 s; the original signal is shown at the bottom-right of Fig. 17.18.

At first, Bell and Hewish jokingly nicknamed the signal LGM-1, standing for "Little Green Men 1." Soon a second source with similar, very evenly spaced signals was detected in another part of the sky with a slightly different repeating interval. Since it seemed highly unlikely that two civilizations would be sending signals to Earth on the same radio frequency at the same time, it was obvious that a physical explanation for the pulses from space was needed.

Instead of jumping to improbable or fanciful possibilities (like little green men), science first tries to find more likely scenarios. Could pulsating stars be the answer? Since stellar pulsations occur with periods related to how fast sounds waves can move through stars, that solution wasn't viable. Although white dwarf stars can pulsate with short periods, even they can't pulsate with periods as short as the signal intervals detected by Bell and Hewish. What about rotating white dwarf stars? Could they rotate fast enough to account for the signals? Again the answer is no, because rotations that fast would cause the surfaces of white dwarfs to fly off; gravity just isn't strong enough to provide the necessary centripetal force to hold them together. Although theoretical at that point, could neutron stars spin rapidly enough? With their tremendously strong gravities, and the ability of the cores of core-collapse supernovas to spin very fast due to conservation of angular momentum, the answer is yes; spinning neutron stars can account for the signals that Bell and Hewish were receiving at their radio telescope antenna array. These spinning neutron stars are today referred to as pulsars, not to be confused with pulsating stars.

Nobel Prize Controversy

For their discovery Antony Hewish received the Nobel Prize in Physics in 1974, but his graduate student, Jocelyn Bell Burnell, did not receive the award. That episode in the awarding Nobel prizes has sometimes been referred to as the No-Bell prize. In 1977, during her after-dinner speech of the Eighth Texas Symposium on Relativistic Astrophysics, Bell Burnell said:[3]

> It has been suggested that I should have had a part in the Nobel Prize awarded to Tony Hewish for the discovery of pulsars. There are several comments that I would like to make on this: First, demarcation disputes between supervisor and student are always difficult, probably impossible to resolve. Secondly, it is the supervisor who has the final responsibility for the success or failure of the project. We hear of cases where a supervisor blames his student for a failure, but we know that it is largely the fault of the supervisor. It seems only fair to me that he should benefit from the successes, too. Thirdly, I believe it would demean Nobel Prizes if they were awarded to research students, except in very exceptional cases, and I do not believe this is one of them. Finally, I am not myself upset about it — after all, I am in good company, am I not!

Bell Burnell's career wasn't seriously hurt by being passed over; the astronomical community certainly understood her contribution. She was elected as a Fellow of the Royal Society (the same society that Sir Isaac Newton presided over), she served as President of the Royal Society of Edinburgh, and she was the recipient of numerous awards, including receiving the honorific title of Dame. In 2018, 51 years after the discovery of pulsars, Dame Jocelyn Bell Burnell received another honor, the $3 million Special Breakthrough Prize in Fundamental Physics "for fundamental contributions to the discovery of pulsars, and a lifetime of inspiring leadership in the scientific community." She donated the funds to assist women, underrepresented ethnic minorities, and refugee students become physics researchers. In 2018 Dame Jocelyn Bell Burnell was named Chancellor of the University of Dundee in Scotland.

---

[3]Bell Burnell, S. Jocelyn (1977). "Petit Four," *Eighth Texas Symposium on Relativistic Astrophysics (Annals of The New York Academy of Sciences)*, 302(1):685–689.

The Pulsar Mechanism

A pulsar is a bit like the lighthouse in Fig. 17.19, that produces a narrow beam of light sweeping through space as the lighthouse's mirrors rotate; at night, you only see the light when the beam is pointed at you. Figure 17.20 illustrates a model of how the signal from a pulsar is generated. The intense magnetic field of a rapidly spinning neutron star is tilted relative to its rotation axis, just as Earth's magnetic field axis is tilted relative to geographic north (Fig. 11.35). As the neutron star rotates, a beam of electromagnetic radiation from the magnetic poles points briefly toward Earth once every rotation period.

The pulsar beam is produced by charged particles spiraling around its magnetic field lines in the same way that charged particles get trapped in the Sun's magnetic field (Fig. 9.14) or in the magnetic fields of Earth and other planets in our Solar System. In the case of the neutron star, however, because the magnetic field is so intense and the gravitational field is so strong, the particles are moving around the field lines at close to the speed of light. The relativistic speeds beam the photon radiation very narrowly along the magnetic field axis.

Based on observational data obtained by the Neutron star Interior Composition ExploreR (NICER), an x-ray instrument attached to the International Space Station, researchers have obtained the most precise information ever obtained for a pulsar. PSR J0030+0451 has a mass of 1.4 M_{Sun} and a diameter of 26 km, just what was predicted theoretically. On the other hand, two independent groups of researchers have found a far more complex magnetic field than the simplified model

Fig. 17.19 Cape Hatteras lighthouse shining at sunrise. (Adapted from an NPS photo)

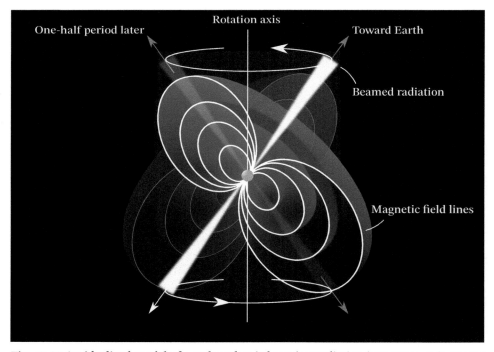

Fig. 17.20 An idealized model of a pulsar that is beaming radiation into space as it rotates very rapidly.

Fig. 17.21 *Left:* The complex magnetic field of pulsar PSR J0030+0451. *Right:* One model of the pulsar showing two hotspots; one is nearly circular and the other has an arc shape. (NASA's Goddard Space Flight Center)

in Fig. 17.20 suggests; see Fig. 17.21 (*Left*). The groups also found hotspots in the star's southern hemisphere where high energy particles strike its surface along magnetic field lines, and where the magnetic field lines beam radiation out into space (Fig. 17.21 *Right*). No hotspots were detected in the pulsar's northern hemisphere.

The Mystery of the Crab Nebula

A lingering question existed for a long time regarding the power source for the Crab Nebula, the remnant of the core-collapse supernova, SN 1054 (recall Fig. 17.7). Given that the supernova occurred almost 1000 years ago, how come the nebula still shines so brightly? The answer came when very precise measurements revealed that its central pulsar, which rotates nearly 30 times per second, is slowing down ever so slowly. As its spin rate slows, the rotational kinetic energy gets converted into energy that powers the nebula. The pulsar is the bright dot at the center of Fig. 17.22. The glowing gases around the pulsar are produced by a wind that is blowing outward along its equatorial plane at close to the speed of light, carrying the rotational kinetic energy with it. The pulsar is also the first object identified to be emitting ultra-high-energy gamma-ray photons, with energies that are about 100 million times more energetic than gamma-ray photons associated with the nuclei of atoms.

Star Quakes

Many other pulsars are also known to slow down as energy is radiated into space and particles in the surrounding gas are dragged along by the magnetic field lines. However, every so often the steady slowing is interrupted by an abrupt speed-up. Although not a large change, the speed-up indicates that the neutron star shrank by a very small amount. As the star continues to cool, its *solid surface* can suddenly crack; the star experienced a star quake. Again, just as when a figure skater pulls her arms in to spin faster through conservation of angular momentum, the tiny change in size following the quake causes the star to spin faster.

Fig. 17.22 The environment around the Crab pulsar (center), as seen in a combined image in x-rays and optical light. This image is from deep inside the remnant of SN 1054 seen in Fig. 17.7. (X-ray image: NASA/CXC/ASU/J. Hester et al.; Optical image: NASA/HST/ASU/ J. Hester et al.)

Runaway Neutron Stars

In some cases the neutron star that forms from a core-collapse supernova can get shot out of the supernova remnant like a cannonball. For example, the pulsar PSR J0002+6216, whose trail can be seen in Fig. 17.23, left the supernova remnant about 5000 years ago and is traveling fast enough to someday escape the Milky Way Galaxy. The pulsar spins 8.7 times per second and was discovered using the downloadable software Einstein@Home (https://einsteinathome.org/) that utilizes the other-wise available idle time on the personal computers of volunteer citizen scientists to help analyze very large data sets.

Fig. 17.23 A pulsar that formed 10,000 years ago was shot out of the core-collapse supernova remnant CTB1 at greater than 1100 km/s (2.5 million mi/h). CTB1 is the bubble in the image and the pulsar's path is the orange trail to the lower left of the remnant. (Cropped from composite by Jayanne English, University of Manitoba; F. Schinzel et al.; NRAO/AUI/NSF; DRAO/Canadian Galactic Plane Survey; and NASA/IRAS. CC BY 3.0)

Marking Our Place in the Universe

When our robotic messengers, Pioneers 10 and 11 and Voyagers 1 and 2, were sent out through the Solar System and then out into the cosmos, they carried greetings from Earth, depicted in Fig. 12.2. In each case the spacecraft contain a sort of "We are here" diagram. Each line originates from a central point (Earth) with different lengths and directions. The endpoint of each line represents a beacon in the cosmos and our distance from it. These universal beacons, that would surely be known to any advanced civilization capable of coming across one of the spacecraft, are pulsars.

What Holds a Neutron Star Up Against Gravity?

White dwarf stars can only have masses up to the Chandrasekhar limit of 1.4 M_{Sun}, because otherwise gravity overwhelms the force produced by electron degeneracy pressure and the white dwarf collapses to become a neutron star. This is why so many neutron stars have masses of 1.4 M_{Sun}. There is a similar limit for neutron stars. Although neutron degeneracy pressure helps to support a neutron star, it is the strong nuclear force, or also known as the strong force, that supports a neutron star, but only up to about 2.7 M_{Sun}. (The limit is somewhat uncertain because of our lack of a complete understanding of the exotic nature of neutron-star material.)

The strong nuclear force is the same force that holds the nucleus of an atom together. Since protons are all positively charged and neutrons don't have an electric charge, without the strong nuclear force, the repulsive force between charges of the same sign would push the protons away from one another while not affecting the neutrons. As depicted in Fig. 17.24, over distances characteristic of the radius of an atomic nucleus, the strong nuclear force is strongly attractive for both protons and neutrons, regardless of electric charge. But the strong nuclear force has another characteristic as well: if the nucleons get too close together, the strong nuclear force becomes strongly repulsive, keeping the particles apart so that they don't fall into one another. It is the second aspect of the strong force that supports neutron stars, but only up to a point. Above about 2.7 M_{Sun} even the strong nuclear force isn't strong enough to support the neutron star against the overwhelming force of gravity.

Key Point
The maximum mass of a neutron star is about 2.7 M_{Sun}.

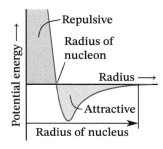

Fig. 17.24 The strong force of the nucleus is repulsive across the size of a nucleon, and becomes attractive outside of a nucleon and across the size of an atomic nucleus.

Neutron Star Mergers

A Test of the General Theory of Relativity

Just like when two white dwarf stars in orbit around one another can spiral in and collide due to energy and angular momentum being carried away by gravitational waves, two neutron stars in a binary system can do the same thing. If one of those neutron stars happens to be a pulsar with its beam striking Earth once every rotation period, a natural laboratory exists to test Einstein's general theory of relativity. This is because the pulsar provides a very precise clock for timing events like the orbital period of the pulsar. As the two neutron stars spiral in toward one another, as with white-dwarf binary systems, the orbits of neutron-star binary systems get smaller, their orbital speeds increase, and their orbital periods decrease in a very predictable way. These natural laboratories have been used to test the prediction of gravitational waves with incredible accuracy. The decay rate of binary neutron-star orbits is exactly as predicted, spectacularly verifying the general theory of relativity and indirectly demonstrating that gravitational waves exist.

Kilonovas

When two neutron stars spiral into one another, one of two things can happen: they can merge to form a more massive neutron star, or they can collapse to form a black hole (Section 17.4). Since most neutron stars have masses of about 1.4 M_{Sun} or slightly greater, when two neutron stars spiral into one another and merge, their combined mass will exceed the 2.7 M_{Sun} limit, meaning that the merged object will collapse into a black hole. When that happens, a different kind of stellar explosion happens, called a kilonova. The name comes from the fact that these explosions have energies that are typically about 1000 times greater (kilo) than a nova, but

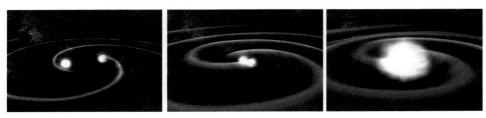

Fig. 17.25 An artist's impression of the merger of two neutron stars. The ripples represent gravitational waves moving away from the accelerating stars. (NASA)

they are still ten to one hundred times less energetic than a core-collapse supernova. Figure 17.25 depicts the merger of two neutron stars.

Again questions arise that demand answers: Do neutron stars really merge in this way? Where is the evidence?

Gravitational Waves

Fig. 17.26 The gravitational wave detector of LIGO–Hanford. Each arm of the L-shaped configuration is 4 km (2.5 mi) long. (LIGO Lab/Caltech/MIT)

Ever since Einstein proposed his general theory of relativity in 1915, scientists have worked to either confirm the theory or show that it is flawed. From Einstein's explanation of the advance of perihelion of Mercury, the prediction of curved spacetime around the Sun that was tested during the total solar eclipse of 1919 (page 271), the indirect evidence for gravitational waves from a pulsar in a two-neutron-star system, and numerous other examples, the general theory of relativity has been confirmed every time. The remaining direct observational test had to wait until September 14, 2015, when two groups of researchers detected gravitational waves passing through their detectors. The Laser Interferometry Gravitational-wave Observatory (LIGO) Scientific Collaboration operates two widely separated installations as a single observatory, one in Hanford, Washington (Fig. 17.26), and the other in Livingston, Louisiana. Virgo operates its observatory near Pisa, Italy. Each of the installations are L-shaped, and work in the same way that the Michelson–Morley experiment does, to measure shifts in the wavelength of light (Fig. 8.15). Laser light is sent down each arm and reflected off a mirror at the other end. When a gravitational wave passes through the apparatus, one arm changes length by about 1/1000 of the diameter of a proton due to spacetime distortion! That first detection in 2015 was the result of two black holes in orbit around one another, spiraling in to merge into a single black hole. Following the observation, the 2017 Nobel Prize in Physics was awarded to three scientists for their work in gravitational wave detection [Rainer Weiss (b. 1932), Barry C. Barish (b. 1936), and Kip S. Thorne (b. 1940)].

On August 17, 2017, the observatories detected the merger of two neutron stars in the constellation Hydra, similar to the blue signal in Fig. 17.27. Since gravitational waves travel at the speed of light, the different arrival times of the signals received at each observatory provided information about the direction of the merger. Optical observations and observations in other electromagnetic wavelength ranges were also made of the same event, revealing a kilonova explosion. This was the first time that humans detected both gravitational waves and electromagnetic waves carrying information to Earth. This event marked the dawn of multimessenger astronomy; a new, powerful way to study the universe.

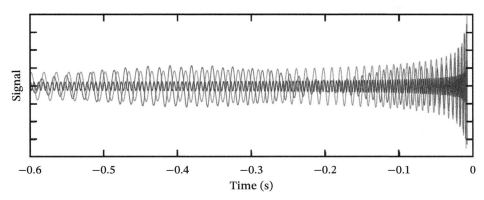

Fig. 17.27 Examples of signals for inspiraling objects that can be detected by gravitational wave observatories as the ripples in spacetime pass through their arms. Zero seconds on the right-hand side is the moment of merger. Red is for two non-rotating black holes of masses 10 M_{Sun} and 15 M_{Sun}, green is for spinning black holes with the same masses, and blue is for two non-rotating neutron stars, each with a mass of 1.4 M_{Sun}. (Adapted from a graphic by the LIGO Scientific Collaboration)

Completing the Periodic Table

If you go back and study the periodic table of the elements (Table 7.1), the table's color-coding indicates how each of the elements is formed in nature. Along with elements produced by low-mass stars, most of the heaviest elements are also produced by neutron-star mergers. In fact, it appears that the heaviest isotopes of those elements are all produced by neutron-star mergers. This is because, during a kilonova, the extremely neutron-rich environment can make isotopes that have more neutrons than protons. Even the gold in your jewelry (Au, $Z = 79$) was produced during the merger of two neutron stars in the distant past, before our Solar System formed. Through processes involving the Big Bang, low-mass stars, white dwarf supernovas, core-collapse supernovas, kilonovas, and cosmic rays, the origins of the naturally occurring elements and their isotopes are now largely understood.

17.4   Black Holes: Gravity's Ultimate Victory

In Section 17.1 we studied white dwarf stars, the final product of stellar evolution for stars with initial masses of less than about 9 M_{Sun}. Then in Section 17.3 we considered neutron stars that are produced when stars of masses between 9 and 25 M_{Sun} undergo core-collapse supernova explosions. But what happens for stars that are even more massive? The answer is that during the core-collapse supernova, or shortly afterward, if the neutron star was very close to its maximum mass and has had a chance to cool a bit, a black hole forms where the star used to be. A summary of the end-products of the three mass ranges is shown in Fig. 17.28.

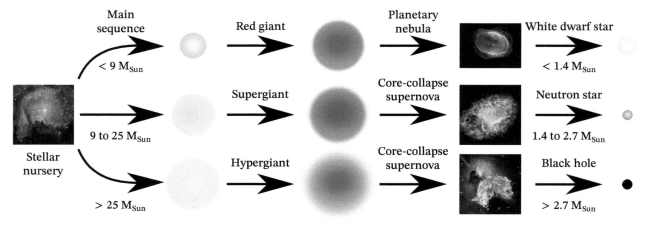

Fig. 17.28 A summary of the main phases and final outcomes of stellar evolution by mass range. Sizes are not to scale. (Images courtesy of NASA and ESA; CC BY 4.0)

The Formation of Stellar-Mass Black Holes

If the surviving core of a core-collapse supernova exceeds about 2.7 M_{Sun}, the strong nuclear force isn't strong enough to oppose the tremendous pull of gravity, and so all of the neutrons that were produced when the heavy nuclei disintegrated inevitably fall into one another in an ultimate collapse. With no other force left to support the star, it literally collapses down to a geometrical point in space. The result is an object with no size at all, containing all of the remaining mass of the star after the core-collapse supernova has blown the overlying layers of the star out into space. With density defined as mass divided by volume, and with the volume being zero, the density of that point is literally infinite if the effects of quantum mechanics are neglected. In mathematical terms, when a finite number is divided by a quantity that goes to zero, the answer is referred to as a singularity. The result of this ultimate core collapse is a massive **singularity**.

That is not to say that the influence of all of that mass has disappeared; the mass is still there and so is its gravitational force. Recalling Newton's universal law of gravitation equation, $F = GMm/r^2$, as the distance between two masses approaches zero ($r \rightarrow 0$), the force between the masses becomes infinite. The same holds true for the acceleration due to gravity, $g = GM/r^2$, and the escape speed, $v_{escape} = \sqrt{2GM/r}$.

Since nothing can travel faster than the speed of light there is a "sphere of no return" centered on the singularity, where the speed of light, c, equals the escape speed anywhere on that sphere. Once inside the sphere of no return, everything is destined to fall into the singularity. Even at the distance where the escape speed equals the speed of light, an infinite amount of energy would be needed to break away. This sphere of no return is known as the **event horizon**, with the event horizon being the outer boundary of the black hole.

So what is the radius of a black hole? To find out, set $v_{escape} = c$, or

$$c = \sqrt{\frac{2GM}{r}}.$$

Now solve for the distance r by first squaring both sides of the equation and then multiply by sides by r/c^2, which leads to

$$r_{\text{bh}} = \frac{2GM}{c^2}.$$

If a singularity has the mass of the Sun ($M = 1\ \text{M}_{\text{Sun}}$ in the equation), then a solar-mass black hole has a radius of $r_{\text{bf, Sun}} = 2.95$ km (or 1.83 mi). To simplify the equation, if the mass is expressed in terms of the mass of the Sun, the radius of the black hole in kilometers can be written as

$$r_{\text{bh}} = 2.95\, M_{\text{Sun}}\ \text{km} \approx 3\, M_{\text{Sun}}\ \text{km}, \tag{17.2}$$

Radius of a black hole using solar mass units

where $\approx$ can be read as "approximately equal to." (The last expression can be used for quick estimates without the need for a calculator.)

Since there is no limit to the mass of a black hole, there is no limit to its size. A black hole could theoretically be from as small as an individual neutron, or smaller, to as large as all of the mass and energy ($E = mc^2$) the universe would allow. However, there is no known way today to form black holes with masses smaller than 2.7 M_{Sun}, except perhaps at the time of the Big Bang when the universe began, or very shortly thereafter.

This derivation for the radius of a black hole was first described by John Michell in 1783.[4] As well as being a clergyman, Michell was a polymath with interests in a variety of scientific fields, including geology, gravitation, magnetism, optics, and astronomy. He accepted Newton's idea that light was composed of tiny particles and should therefore be subject to Newton's law of gravitation. Although the derivation above is technically wrong, because the correct approach requires the use of Einstein's general theory of relativity rather than Newton's "universal law" of gravitation under such extreme conditions, somewhat remarkably the final result does turn out to be right. Michell referred to his hypothetical objects as "dark stars." (Michell was also the person who first designed the apparatus that Cavendish later used to measure G and the mass of Earth; page 191.)

Example 17.3

Although there is no limit to the size of a black hole, the smallest mass that can be formed from a core-collapse supernova is the maximum mass of a neutron star, or 2.7 M_{Sun}.

The radius of the smallest stellar black hole is about

$$r_{\text{bh}} = 2.95 \times 2.7\ \text{km} = 8\ \text{km}\ (5\ \text{mi}).$$

---

[4]Michell, John (1784). "On the Means of Discovering the Distance, Magnitude, &c. of the Fixed Stars, in Consequence of the Diminution of the Velocity of Their Light, in Case Such a Diminution Should be Found to Take Place in any of Them, and Such Other Data Should be Procured from Observations, as Would be Farther Necessary for That Purpose. By the Rev. John Michell, B. D. F. R. S. In a Letter to Henry Cavendish, Esq. F. R. S. and A. S." *Philosophical Transactions of the Royal Society of London.* 74:35–57.

Fig. 17.29 An artist's impression of the binary star system Cygnus X-1, with its black hole pulling gas from a massive blue supergiant. [NASA, ESA, Martin Kornmesser (ESA/Hubble)]

Black Holes in Binary Star Systems: Cygnus X-1

Can something as strange as a black hole really exist? Scientists were initially very skeptical that neutron stars could exist until pulsars were discovered. That opened the door to the possibility that there may be black holes as well.

The initial evidence for black holes came in binary star systems. Discovered in 1964, the first strong candidate was Cygnus X-1, an x-ray source northeast of the center of the swan's neck and near the constellation of Lyra (an artist's impression of the system is shown in Fig. 17.29). In order for the gases spiraling toward the unseen companion of the blue supergiant to become hot enough to emit x-rays, the unseen companion must be a very compact object, either a neutron star or a black hole. The blue supergiant is an O9 star with an estimated surface temperature of 31,000 K and a mass of between 20 and 40 times the mass of the Sun. Since stellar evolution goes faster the more massive the star is, the dead star that is the compact object must have started out with even more mass than the blue supergiant. Today, it is estimated that the mass of the compact object is almost 15 M_{Sun}, far above the 2.7 M_{Sun} upper limit for the mass of a neutron star. The only known possibility for the compact object is a black hole.

Since the discovery of Cygnus X-1, other binary systems make even stronger cases that one of the members is a black hole.

The Curvature of Spacetime

Revisiting Einstein's Happiest Thought

Back in Section 8.2 we looked briefly at what Einstein considered to be the happiest thought of his life, namely that when a man is falling he doesn't feel his own weight. This simple thought led to Einstein's development of the general theory of relativity. But how does that basic observation get us to the strange notion of curved spacetime?

First, think about what happens to a man riding in an elevator with no view of the outside. Suddenly the cable of the elevator breaks, sending the man and the elevator car plunging downward as depicted in the top frame of Fig. 17.30. The falling man is accelerating downward at exactly the same rate as the elevator car, and so relative to the elevator car he is simply floating, not experiencing his own weight. Had he not known prior to the cable breaking that he was actually in an elevator, he would have no way of distinguishing between the freely falling elevator and being in an identical elevator car floating or moving at a constant velocity somewhere in space away from any measurable gravitational forces. Not only could he not tell the difference visually, but there would be no experiment that he could conduct that would allow him to determine which situation he found himself in. Falling freely in a gravitational field is no different from floating freely in deep space. What Einstein realized is that gravitation can be replaced by an *accelerating frame of reference*, where the acceleration equals the acceleration due to gravity, as is the case for the elevator car example where the elevator car serves as the accelerating frame of reference. The notion that the two situations are equivalent is known as the equivalence principle. (The special theory of relativity is just a special case of the general theory of relativity with zero acceleration, hence the names.)

Fig. 17.31 Long hair in near-weightlessness. Astronaut Marsha Ivins on the Space Shuttle Atlantis in 2001. (NASA)

The same situation applies while in orbit around Earth. In orbit, the spacecraft and everything in it is constantly falling toward the center of Earth, just as Sir Isaac Newton described with his thought experiment of a giant cannon on the top of a mountain that fires a cannon ball fast enough to cause it to orbit the planet (Fig. 5.18). It is incorrect to say that there is no gravity in orbit, it still exists because of Earth's mass, something the occupants would quickly discover if the forward motion of the spacecraft was stopped. (Technically, there is a very tiny measurable difference, termed *microgravity*, between the force of gravity on the side of the ship closest to Earth relative to the side farthest from Earth. This means that stuff on the ship closest to Earth accelerate toward Earth ever so slightly faster than stuff on the far side of the ship.) Figure 17.31 shows astronaut Marsha Ivins onboard the Space Shuttle Atlantis with her floating hair.

Fig. 17.30 *Top:* A man in a free-falling elevator car. His downward acceleration is identical to the downward acceleration of the elevator car. *Bottom:* The man in an identical elevator car floating freely in space away from any source of gravity. There is no way to distinguish between the two situations without looking beyond the cars. No cartoon characters were injured in this thought experiment.

The Accelerating Laboratory

You may be thinking: "But surely the equivalence principle can't be correct, Albert." Suppose we design an experiment in a laboratory onboard a spaceship that is accelerating at $g = 9.8 \text{ m/s}^2$ while a green laser shines a beam of light parallel to the floor from one side of the lab to the other. According to an astronaut on a spacewalk watching the ship go by, she would see the beam of light travel in a straight line. However, in the time it takes the beam to travel the length of the laboratory, the ship would have moved, causing the beam to strike the opposite wall farther down on the wall relative to where it was emitted. This thought experiment is depicted in Fig. 17.32 (*Left*). The same experiment would be seen differently by a scientist onboard the ship, however; it would appear that the beam of light took a curved path downward while traversing the laboratory, simply because it took some time for the light to travel across the accelerating laboratory (Fig. 17.32 *Middle*).

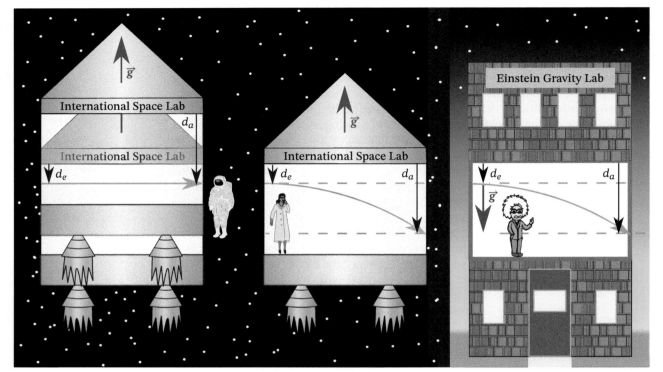

Fig. 17.32 *Left:* A rocket accelerating forward at $\vec{g}$ as seen by an astronaut outside the ship. A beam of green light is emitted a distance d_e below the ceiling and crosses the ship's laboratory during the time the ship moves some distance, arriving at the other side at a distance d_a below the ceiling. The black arrows indicate how far the beam is from the ceiling of the laboratory when it is emitted and when it arrives at the other side. *Center:* The same ship and light beam as viewed from inside the accelerating ship. *Right:* An identical laboratory on Earth in a gravitational field of acceleration $\vec{g}$. The light beam "falls" as it moves across the laboratory because it is following curved spacetime.

Einstein's response according to the general theory of relativity would be that in fact the same thing would happen in an identical laboratory on Earth where the acceleration due to gravity is g. The light would bend in a gravitational field in exactly the same way that it does in the ship's laboratory as viewed from within the lab. In the Earth-based laboratory, the beam of light simply follows the straightest possible line in spacetime, which is curved by the mass of Earth (Fig. 17.32 *Right*). There is no experiment that can distinguish between a constantly accelerating laboratory and a laboratory in a gravitational field with the same free-fall acceleration.

The Nature of Black Holes and Observational Tests

The gravitational environment around a black hole is in some sense no different from around a bowling ball, a planet, the Sun, a white dwarf star, or a neutron star, except that in each case the effects become increasingly more extreme. The same general theory of relativity applies in every instance but the effects reach their ultimate limit when approaching a black hole's event horizon.

Figure 17.33 shows a two-dimensional representation of curved spacetime in and around a black hole, similar to the general relativistic environment around the

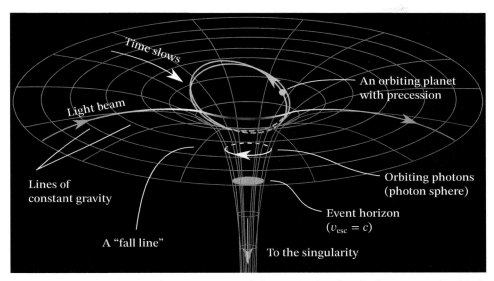

Fig. 17.33 A two-dimensional representation of the gravitational well of a non-rotating black hole.

Sun depicted in Fig. 8.9. In both figures, the concentric circles represent contours of constant gravity similar to contours of constant elevation that you find on a topographic map like the one in Fig. 17.34. Perpendicular to the gravity contours, are lines that converge at the center of the black hole. On a topographic map of a ski area, skiers and snowboarders would call lines that are perpendicular to elevation contours "fall lines," the fastest route down a ski run and the path that a rolling snowball would follow. Streams and rivers always follow "fall lines" on topographic maps between rising elevations on either side. Figure 17.35 shows a three-dimensional view from far above the black hole in Fig. 17.33.

We have already seen in Section 8.2 that the general theory of relativity was spectacularly verified in 1919 during a total solar eclipse, when starlight arriving at Earth had been deflected slightly after passing close to the Sun. This is the same deflection discussed in the thought experiment of Fig. 17.32. In the case of a black hole, the bending becomes more and more severe, the closer the light beam gets to the event horizon. Figures 17.33 and 17.35 both show a light beam passing close to

Fig. 17.34 A topographic contour map of a portion of Grand Teton National Park in Wyoming. The lines represent contours of constant elevation. (USGS)

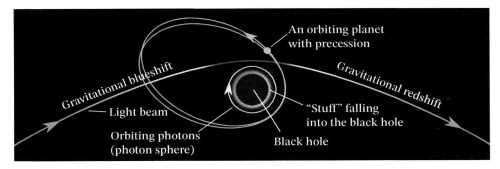

Fig. 17.35 Looking down on the black hole of Fig. 17.33 from far above it.

the black hole and being deflected in the process. As the beam approaches the black hole it finds itself deeper and deeper in the black hole's gravitational well until the point of closest approach, after which it climbs out of the well.

Bending Light and the Photon Sphere

Fig. 17.36 An astronaut near the event horizon of a black hole that is below her feet. She is pointing lasers in four directions from vertical to horizontal.

The closere one gets to the event horizon deep in the gravitational well of a black hole, the more severely light of every wavelength is bent. Figure 17.36 shows an astronaut shining four lasers in different directions near, but still above, the event horizon that is below her feet. The vertical beam is able to escape into space, as is the beam pointed slightly away from vertical, even though the path of the second beam bends significantly. The remaining two beams aren't able to escape the gravitational well and end up crossing the event horizon, adding energy, and therefore mass ($E = mc^2$), to the black hole.

If the astronaut were to move farther away from the black hole (assuming that she has an amazingly powerful rocket), she will find a distance from the event horizon where the horizontally directed laser beam would actually go in orbit around the dead star. This radius from the singularity, where light moving parallel to the event horizon orbits the black hole is, known as the photon sphere. If your eyes are at that radius and you look straight ahead you will see the back of your own head! A great place to open a do-it-yourself hair salon. The photon sphere is shown in Figs. 17.33 and 17.35.

Gravitational Time Dilation

It is not just space that is curved in the vicinity of any massive object, but time is affected as well (including, but not restricted to, black holes). Figure 17.33 points this out with an arrow pointed toward the singularity, indicating that time gets progressively slower in that direction. Suppose that a clock is sent falling into the black hole. As with the effect of velocity on time in the special theory of relativity, what is observed is entirely in the eye of the beholder. An observer at rest some distance from the black hole will see the passage of time on the clock slow down and finally stop at the event horizon, a phenomenon referred to as gravitational time dilation. But that wouldn't be the case for someone falling into the black hole along with the clock; to that person, it would behave completely as expected, even after crossing the event horizon.

As with all predictions of a scientific theory, such a claim must be tested in as many different ways as possible. One such test was conducted on November 25, 1976, when Mars moved behind the Sun. Utilizing Vikings 1 and 2, the first NASA missions to land on Mars, signals were sent from Earth to the landers with instructions to send the signals back to Earth. After accounting for effects of radio signals passing through the Sun's corona, the signals were delayed by exactly the amount of time predicted by the theory compared to the time it would have taken if gravity didn't affect the passage of time. Gravitational time dilation is an observed effect of spacetime curvature, reaching its limit at a black hole's event horizon.

Gravitational Redshift

In Section 8.4 you learned how the wavelength of light is affected by the source of the light moving toward or away from the observer, resulting in a blueshift or a redshift, respectively. This shift due to motion is known as the relativistic Doppler effect. But the wavelength of light is also affected by a gravitational field. Figure 17.33 and Fig. 17.35 show the light beam skirting the black hole and changing from red to deep blue as it approaches the black hole and then back to red again as it moves away.

Fundamental to Einstein's general theory of relativity is the fact that the speed of light in a vacuum is always exactly $c = 299{,}792{,}458$ m/s (or about 300,000 km/s), but there are no restrictions on the wavelength of light. To understand the shift in wavelength of light in a gravitational field recall that a photon of wavelength λ has an energy given by Planck's energy equation, $E = hc/\lambda$, where a shorter, or bluer, wavelength corresponds to a more energetic photon and a longer, or redder, wavelength corresponds to a less energetic photon. For an object of mass falling into a gravitational well, that object loses gravitational potential energy and gains gravitational kinetic energy (it speeds up). Conversely, a mass moving out of a gravitational well without the aid of an engine gains gravitational potential energy and loses kinetic energy (it slows down). Although a photon is a massless particle that always moves at the speed c, as it moves deeper into a gravitational well its energy increases, causing its wavelength to shift blueward, and when it moves farther away from the mass it loses energy and its wavelength shifts redward. Unlike the relativistic Doppler effect, this second type of wavelength shift is entirely due to the gravitational field and has nothing to do with the motion of a source relative to an observer.

Again, we must always test the predictions of science; theories cannot simply be accepted on faith, even the theories of Albert Einstein. In one such test in 1960, gamma rays were emitted upward along a tall tower. The wavelength received at the top of the tower was redshifted by the amount predicted by the general theory of relativity. The gravitational redshift was confirmed by experiment.

Black Hole Mergers and Gravitational Waves

We have already seen that gravitational waves have been detected from the merger of two neutron stars that produced a kilonova (page 707). The first direct detection of gravitational waves was actually the merger of two black holes in 2015. The shape of the signal was consistent with the collision of a 29 M_{Sun} black hole with another of 36 M_{Sun}, as predicted by the general theory of relativity. After they merged, the signal was also consistent with a "ring-down" of the new, combined black hole as it settled into its new size and shape.

Detections like these are confirmation of both the existence of black holes and the predictions of the general theory of relativity.

Other Observational Tests of the General Theory of Relativity

There are other tests of the general theory of relativity that give further support for the prediction that black holes do exist.

The very first confirmation of the general theory of relativity was Einstein's use of his theory to explain precisely the advance of the perihelion of Mercury's orbit that was otherwise unexplained by Newtonian gravity. The precessing orbit of a planet is depicted in both Fig. 17.33 and Fig. 17.35.

Rotation axes can also precess, much like that of a toy top (Fig. 4.14). In fact, the 25,772 year period of this effect for Earth was first detected by the ancient Greek astronomer, Hipparchus. In the environment of a very strong gravitational field, precession becomes a test of the general theory of relativity, as was verified in 2019 for a pulsar in a binary system with another neutron star.

But perhaps one of the strangest predictions of the general theory of relativity is that spacetime itself can be dragged along by a rotating mass. No object is actually involved except for the rotating mass; spacetime is simply pulled along with the rotation, a phenomenon known as frame dragging. As weird as that may sound, the effect was confirmed with high precision during a NASA mission called Gravity Probe B. The spacecraft was pointed carefully toward one star as the spacecraft orbited Earth from pole to pole.

It is important to remember that, as with quantum mechanics, just because something seems too weird to be true doesn't mean that it isn't the way nature behaves; it just means that our everyday intuition is failing us in the extremes of small size and/or intense gravity. This assumes, of course, that the required tests confirm the theory.

Black Holes Have No Hair!

Although strange by our everyday experience, black holes are actually extremely simple objects in the sense that there are only three characteristics that completely describe every black hole. The most obvious is mass, but we have already mentioned a second, namely rotation, or more accurately angular momentum. The third is electric charge. This last characteristic is almost guaranteed to be unimportant, however. This is becaue if a star somehow had a net positive or negative charge when it collapsed, which itself is highly unlikely, the black hole would almost immediately attract enough opposite charges to become electrically neutral anyway. The notion that only mass, angular momentum, and electric charge are needed to describe a black hole is tongue-in-cheek known as the black holes have no hair theorem.

The theorem gets its name from the idea that it doesn't matter what formed a black hole, all of those details vanish inside the event horizon. It doesn't matter if the collapsing star was made of pure hydrogen, plutonium, garbage bags, toasters, or hair, the only remaining information is mass, angular momentum, and electric charge.

The General Theory of Relativity and Quantum Mechanics

In a general sense, the two major theories of physics under which all other ideas lie are the general theory of relativity and quantum mechanics. But there is a problem in that they seem to be inconsistent with one another. An enormous amount of theoretical research has gone into trying to reconcile the two theories at their intersection of very small size and immense gravity, such as near a singularity or at the event horizon of a black hole.

In 1974, Stephen Hawking (Fig. 17.37) proposed that given a sufficiently long period of time, black holes could actually evaporate! The idea is based on quantum mechanics and the Heisenberg uncertainty principle, which argues that it is impossible to know both the position and velocity of a particle simultaneously to arbitrary accuracy, or the amount of energy something has and the amount of time that it has that energy, again to arbitrary accuracy (the better you know one quantity the worse you know the other). We first encountered this idea in Section 8.7. According to the uncertainty principle, it would be possible for a particle and its antimatter twin to simply pop in and out of existence as long as they do so within the confines of the time uncertainty (Fig. 17.38). Although this strikes our intuition as extremely bizarre, if not impossible, it is an important component in a complete understanding of the subatomic world. If these particles make their appearance just outside the event horizon, one of the particles could fall into the black hole before they have a chance to recombine again, leaving the one outside the event horizon without a partner. As that particle moves away from the event horizon, it appears to have come from inside of the event horizon, carrying mass and energy away from the black hole. Assuming that nothing else falls through the event horizon, the escaping unpaired particles effectively mean that the black hole is evaporating, very, very slowly. This idea is known as Hawking radiation. As evaporation proceeds, with decreasing mass the black hole's gravitational pull and inherent energy constantly decreases and the evaporation process accelerates. At the end, the evaporation becomes so fast that it is essentially an explosion, releasing its final energy in a blazing burst of light.

Hawking radiation is tied to a notion raised in the last paragraph of the Black Holes Have No Hair! subsection regarding disappearing information. Information is intimately intertwined with quantum mechanics and presents a theoretical quandary for physicists as they try to marry the general theory of relativity with quantum mechanics.

Fig. 17.37 Stephen Hawking (1942–2018). (Public domain)

Fig. 17.38 A particle and its antimatter twin form close to the event horizon of a black hole. One particle falls into the singularity and the other escapes into space as Hawking radiation.

That Poor Astronaut

Let's wrap up Chapter 17 with another thought experiment. What would it be like to fall into a black hole? We will assume that the black hole isn't rotating and is electrically neutral so only its mass matters.

The future Global Space Alliance is looking for a volunteer for this (literally) once-in-a-lifetime adventure and an eager young recruit steps forward. Bob has always been fascinated by black holes and is anxious to see what one is like, while advancing science as well. And besides, Bob believes himself to be indestructible.

The commander maneuvers the ship into a safe parking orbit around the black hole with the rest of the crew there to monitor Bob's progress. Bob climbs into a specially designed space suit with a very powerful beacon on the top of the helmet that blinks on and off once every second, and an equally powerful radio transmitter and receiver. Bob is ready to go.

When Bob leaves the air lock he first stops his orbital motion so that he can free-fall directly toward the singularity (see Bob fall; Fig. 17.39). Initially, not much seems to be happening as he carries on a pleasant conversation with his fellow crew

Fig. 17.39 Astronaut Bob getting stretched and squeezed as he falls into a black hole while his crew mates look on from a safe distance.

mates. From the ship they begin to notice that his originally blue, blinking beacon is starting to get a bit redder. This is due in part to the fact that Bob is moving away from them with increasing speed (relativistic Doppler redshift), but also because he is falling deeper into the black hole's gravitational well, and so the photons from the beacon are decreasing in energy as they climb out of the well (gravitational redshift). Bob's crew mates notice something else happening as well: the rate at which the beacon is blinking is slowing down because of the speed of his motion (moving clocks run slow), his increasing distance from the ship, and gravitational time dilation.

As Bob nears the event horizon, his crew mates onboard ship see the light from his beacon is getting deeply red, then it moves into infrared wavelengths, microwave wavelengths, and finally into progressively longer radio wavelengths. The time between signals stretches from seconds to minutes, to hours, days, and years, until no final signal is ever sent; in fact, Bob appears frozen just outside the event horizon. Centuries and even eons later, as the ongoing experiment continues, there is Bob, apparently suspended infinitesimally close to the event horizon where time stops, the escape speed equals the speed of light, and light is gravitationally redshifted to zero energy and infinite wavelength. Bob joins any other material that has fallen toward the event horizon since the black hole formed, including any remnants of the star that didn't get blasted out into space during the core-collapse supernova that formed the black hole in the first place.

On the other hand, this is certainly not what Bob is experiencing from his own vantage point. His beacon is working perfectly, blinking its blue light steadily every second. He is starting to feel a bit uncomfortable however; since his feet are closer to the black hole than his head, he is feeling a bit stretched. Unfortunately, as he gets closer and closer to the black hole, that difference in forces between his feet and his head continues to get more and more severe. He is also feeling squeezed, more so at his feet than his head, because the "fall lines" are converging toward the singularity at the center of the black hole.

Bob doesn't notice anything special when he crosses the event horizon, nor does he see an accumulation of stuff that never fell through, because from his perspective that stuff fell through long before he ever got there. His beacon also continues to blink blue once every second by his measurements. He is feeling increasingly stretched and squeezed, however.

Sadly, Bob is doomed. Once he crossed the event horizon there is nothing he can do to avoid falling directly to the singularity. Not to put too fine a point on it, but Bob, along with all of the rest of the black hole's mass, is destined to become a geometrical point in spacetime with zero volume and infinite density.

(There is some debate about Bob's ultimate fate. Including quantum mechanics with the general theory of relativity suggests that the event horizon may not be such a benign location after all. As stuff falls into the black hole, temperatures near the event horizon reach truly "astronomical" levels. If true, then when Bob reaches the event horizon he encounters a wall of intense heat and meets his fate as a mass of dissociated elementary particles, rather than as "gore at the core.")

The End
(of Bob)

? Exercises

All of the exercises are designed to further develop *your understanding* of the material by thinking carefully and critically about what you have read in this chapter. Answers to selected exercises are available in the back of the book.

True/False

1. A planetary nebula is the ejected atmosphere of an exoplanet.
2. (*Answer*) White dwarf or pre-white dwarf stars exist at the centers of planetary nebulas.
3. Planetary nebulas are collapsing clouds of dust and gas.
4. (*Answer*) Planetary nebulas are always nearly perfectly spherical in shape.
5. The material from a planetary nebula ultimately dissipates and becomes part of the interstellar medium from which future generations of stars can form.
6. (*Answer*) All stars with masses of less than about 19 M_{Sun} end their lives as white dwarf stars.
7. White dwarf stars tend to be about the size of Earth.
8. (*Answer*) Most white dwarf stars are composed of crystallized carbon and oxygen.
9. The Sirius binary star system was likely engulfed in a planetary nebula in the past.
10. (*Answer*) The mass of a white dwarf star can be up to 2.7 M_{Sun}.
11. A nova is an explosion on the surface of a white dwarf star, caused by accreting gases from a binary star companion.
12. (*Answer*) Novas are extremely rare events in our Milky Way Galaxy, only occurring about once every 50 years or so.
13. For a time SN 1006 was the brightest object in the sky after the Sun and the Moon.
14. (*Answer*) Over the past 2000 years, at least six supernovas have been visible with the naked eye and their appearance recorded.
15. SN 1054 was first seen by William Parsons.
16. (*Answer*) Given that perhaps as few as five naked-eye supernovas have been seen within the Milky Way Galaxy over the past 2000 years, it is estimated that one supernova occurs in our Galaxy roughly once every 300 to 400 years. (SN 1054 occurred in the LMC, a satellite galaxy of the Milky Way.)
17. As a white dwarf star gains mass in a binary star system, its volume decreases and its density increases.
18. (*Answer*) Virtually all white dwarf stars that produce supernovas have masses of 1.4 M_{Sun}.
19. White dwarf supernovas are important distance indicators.
20. (*Answer*) Only massive stars are capable of causing carbon nuclei to fuse with other carbon nuclei.
21. $^4_2$He nuclei have much lower masses per nucleon than their immediate neighbors in a list of increasing isotope masses.
22. (*Answer*) A silicon-28 nucleus moves about 28 times slower than a proton in a gas with a temperature of 1 billion kelvins.
23. The most abundant elements in the universe are, in order: H, He, C, N, O, Ne.
24. (*Answer*) In order to form elements more massive than $^{62}_{28}$Ni, energy must be absorbed in the process.
25. Core-collapse supernovas are triggered because of an oxygen core flash that occurs as a result of degenerate electron pressure.
26. (*Answer*) Iron can break down into protons and neutrons in the cores of massive stars.

27. Neutrinos immediately escape from the core of a core-collapse supernova.
28. (*Answer*) Neutrinos carry away about 1% of the energy of a core-collapse supernova.
29. A star with an initial mass of between 9 M_{Sun} and 25 M_{Sun} will experience a core-collapse supernova that results in a neutron star.
30. (*Answer*) If a star suddenly collapses down to the size of a neutron star, its rotation period will remain constant, because of the conservation of angular momentum.
31. Jocelyn Bell Burnell realized that the radio signals she detected must have come from a point in space because she could see the visible light blinking on and off.
32. (*Answer*) Jocelyn Bell Burnell was a recipient of the Special Breakthrough Prize in Fundamental Physics.
33. All neutron stars are observable as pulsars.
34. (*Answer*) Pulsars can potentially be used as reference points by an advanced space-faring alien civilization, to keep track of where they are, or to share positions with another advanced alien civilization.
35. It is possible for a pulsar to travel through space at high speed, far from the site of the supernova explosion that created it.
36. (*Answer*) The maximum mass of a neutron star is 1.4 M_{Sun}.
37. The strong nuclear force is repulsive when nucleons get close enough to touch one another.
38. (*Answer*) A kilonova is a nova that destroys a white dwarf star.
39. The generation of gravitational waves requires that a mass must be accelerating.
40. (*Answer*) Theoretically, the smallest mass a black hole can have is 2.7 M_{Sun}.
41. The escape speed from the event horizon of a black hole is c, the speed of light.
42. (*Answer*) Cygnus X-1 is an excellent candidate for a black hole in a binary star system, only because it emits x-rays.
43. The equivalence principle states that it is impossible to distinguish between an accelerating frame of reference such as a rocket and acceleration due to gravity.
44. (*Answer*) There is no gravity in orbit around Earth.
45. We have no way of knowing if black holes actually exist.
46. (*Answer*) The Doppler effect for light and gravitational redshift are two names for the same physical process.
47. Time runs more slowly as the event horizon is approached according to someone from a distant location observing a clock falling into a black hole.
48. (*Answer*) Time runs more slowly as the event horizon is approached, according to someone holding a clock while moving toward the black hole.
49. It is possible for a photon to orbit a black hole.
50. (*Answer*) In order to obtain the most accurate positions possible for the Global Positioning System (GPS), the general theory of relativity must be used.
51. A rotating mass drags the fabric of spacetime with it, even if the spacetime is empty of light and matter.

Multiple Choice

52. (*Answer*) When a star with an initial mass of less than about 9 M_{Sun} nears the top of the asymptotic giant branch,
    (a) it experiences a violent explosion that destroys the star.
    (b) a slow wind dissipates the envelope at a rate of about 1 M_{Sun} every one million years.
    (c) a superwind develops that carries mass away from the star at a rate of about 1 M_{Sun} every 10,000 years.
    (d) the hydrogen-burning shell blasts the entire envelope out into space as a hydrogen shell flash.
    (e) the entire star finally succumbs to gravity and collapses.

53. A planetary nebula is the ejected
    (a) atmosphere of a planet.
    (b) atmosphere of a main-sequence star.
    (c) envelope of a red-giant star.
    (d) envelope of an asymptotic-giant-branch star.
    (e) none of the above

54. (*Answer*) Because of how they form, planetary nebulas are primarily composed of
    (a) hydrogen and helium.
    (b) helium and carbon.
    (c) carbon and oxygen.
    (d) oxygen and neon.
    (e) silicon and iron.
    (f) the composition varies significantly from one planetary nebula to another.

55. A planetary nebula is ejected from a star
    (a) during pre-main-sequence evolution.
    (b) on the subgiant branch.
    (c) on the red giant branch.
    (d) during a hydrogen core flash.
    (e) on the blueward horizontal branch during Cepheid pulsations.
    (f) on the redward horizontal branch.
    (g) following the asymptotic giant branch.

56. (*Answer*) The complex shapes of planetary nebulas can be caused by
    (a) being a member of a binary or multiple star system.
    (b) the existence of magnetic fields.
    (c) one or more prior mass ejection events.
    (d) stellar pulsations.
    (e) all of the above
    (f) (a) and (b) only
    (g) (c) and (d) only

57. At the end of the process of ejecting a planetary nebula,
    (a) the helium- and hydrogen-burning shells shut down.
    (b) the remnant core of nuclear burning becomes exposed.
    (c) a white dwarf star is unveiled at the center of the expanding gas and dust cloud.
    (d) all of the above
    (e) none of the above

58. (*Answer*) A white dwarf star is about the size of
    (a) a small moon.   (b) Earth.
    (c) Jupiter.   (d) a brown dwarf.
    (e) a 0.075 M_{Sun} star.

59. Relative to a 1 M_{Sun} main-sequence star, a white dwarf is
    (a) cool and dim   (b) hot and dim
    (c) cool and bright   (d) hot and bright

60. (*Answer*) The density of a white dwarf star is
    (a) about the same as diamond.
    (b) 1000 times greater than the density of a rock on the surface of Earth.
    (c) 100,000 times the density of water.
    (d) more than 2 million times the density of water.
    (e) more than 1 trillion times the density of water.

61. Most white dwarfs are stars that are composed primarily of
    (a) hydrogen.
    (b) hydrogen and helium.
    (c) helium and carbon.
    (d) carbon and oxygen.
    (e) neon and silicon.
    (f) silicon and iron.
    (g) Disney™ characters that whistle while they work. (Not the correct answer!)

62. (*Answer*) The form of the material that a white dwarf star is composed of is
    (a) gas.   (b) liquid.
    (c) disorganized solid.   (d) a crystallized solid.
    (e) none of the above.

63. The fate of isolated white dwarf stars is that they
    (a) slowly cool off and get dimmer.
    (b) eventually explode.
    (c) experience surface nuclear reactions that periodically blast material out into space.
    (d) disintegrate.
    (e) ultimately turn into iron.

64. (*Answer*) Sirius is actually a binary star system containing the brightest star in the sky, Sirius A, and a dim companion, Sirius B. Sirius B is a white dwarf with a mass closest to
    (a) 0.5 M_{Sun}.   (b) 0.75 M_{Sun}.   (c) 1.0 M_{Sun}.
    (d) 1.25 M_{Sun}.   (e) 1.5 M_{Sun}.

65. A 0.6 M_{Sun} white dwarf star in a binary star system that slowly collects gas from its partner will
    (a) likely experience periodic hydrogen nuclear flash burning.
    (b) probably blast surface hydrogen and helium out into space.
    (c) appear as a nova or a recurrent nova.
    (d) explode as a white dwarf supernova.
    (e) all of the above
    (f) (a) and (b) only
    (g) (a), (b), and (c) only

66. (*Answer*) The Sirius star system would have been surrounded by a _____ in its past.
    (a) planetary nebula
    (b) hypergiant atmosphere
    (c) supernova remnant
    (d) none of the above

67. If SN 1006 reached an apparent magnitude of −7.5, approximately how many times brighter was that supernova than Venus at its brightest? The apparent magnitude of Venus at its brightest is −4.4.
    (a) −3.1   (b) 3.1   (c) 17.4   (d) −17.4   (e) 1260
    (f) 3100

68. (*Answer*) The maximum mass of a white dwarf star is about
    (a) the mass of Earth.   (b) the mass of Jupiter.
    (c) 100 times the mass of Jupiter.   (d) 1 M_{Sun}.
    (e) 1.4 M_{Sun}.   (f) 2.7 M_{Sun}.
    (g) There is no upper limit to the mass of a white dwarf star.

69. The Chandrasekhar limit for white dwarf stars exists because
    (a) there is a limit to the pressure that a degenerate electron gas can produce.
    (b) white dwarf stars decrease in size with increasing mass.
    (c) the carbon–oxygen star slowly converts into an iron star.
    (d) carbon and oxygen are the most stable elements found in nature.
    (e) all of the above
    (f) (a) and (b) only
    (g) (a), (b), and (c) only
70. (*Answer*) Which conservation law explains why the gas being stripped from a binary star member onto a compact object forms an accretion disk around the compact member?
    (a) energy   (b) linear momentum
    (c) angular momentum   (d) leptons
71. A white dwarf supernova serves as a useful standard candle because
    (a) most exploding white dwarfs have masses at the Chandrasekhar limit.
    (b) they are extremely bright.
    (c) they can be seen from enormous distances.
    (d) we know exactly when they will occur.
    (e) all of the above
    (f) (a) and (b) only
    (g) (a), (b), and (c) only
72. (*Answer*) A silicon-28 nucleus ($^{28}_{14}$Si) would move _____ than a proton ($^1_1$H) at the same temperature in a gas.
    (a) 3.8 times slower   (b) 3.8 times faster
    (c) 5.3 times slower   (d) 5.3 times faster
    (e) 14 times slower   (f) 14 times faster
    (g) 28 times slower   (h) 28 times faster
73. The Coulomb potential energy "hill" is _____ times higher for an oxygen-16–oxygen-16 reaction than it is for a proton–proton reaction.
    (a) 8   (b) 16   (c) 32   (d) 64   (e) 128
    (f) 256
74. (*Answer*) The nucleus that would release the most energy per nucleon if it was formed directly from individual protons and neutrons is
    (a) $^1_1$H   (b) $^4_2$He   (c) $^8_4$Be   (d) $^{12}_6$C   (e) $^{16}_8$O
    (f) $^{56}_{26}$Fe   (g) $^{62}_{28}$Ni
75. The amount of energy given off when helium-4 is created from protons and neutrons is 7.0739 MeV (millions of electron volts) per nucleon. How much energy is given off overall?
    (a) 7.0739 MeV   (b) 14.1478 MeV   (c) 28.2956 MeV
    (d) 42.4434 MeV
76. (*Answer*) The six most abundant elements in the universe, in order, are
    (a) H, He, Li, Be, B, C   (b) H, He, C, N, O, Ne
    (c) H, He, C, O, Ne, N   (d) H, He, O, N, C, Ne
    (e) H, He, O, C, Ne, N
77. The reaction $^4_2$He + $^{12}_6$C → $^{16}_8$O + γ is a _____ reaction that _____ energy.
    (a) fission, absorbs   (b) fission, releases
    (c) fusion, absorbs   (d) fusion, releases
78. (*Answer*) The energy released in the building up of an iron-56 nucleus from its constituent protons and neutrons is 8.7903 MeV per nucleon. How much energy must be absorbed by the nucleus to liberate the protons and neutrons?
    (a) 8.7903 MeV   (b) 228.548 MeV
    (c) 263.709 MeV   (d) 492.257 MeV

79. Suppose that a massive star has an iron core of 2 M_{Sun}, or about 4×10^{30} kg. One iron-56 nucleus has a mass of 9×10^{-26} kg. Roughly how many iron nuclei would there be in the core of the star?
    (a) $2 \times 10^{1.15}$ (b) $2 \times 10^{-1.15}$ (c) 4×10^5
    (d) 4×10^{-5} (e) 4×10^{55} (f) 4×10^{-55}
    (g) 4×10^{781} (h) 4×10^{-781}
80. (*Answer*) Although neutrinos were produced in huge numbers during the disintegration of iron nuclei in SN 1987A, why were only a small handful detected by neutrino observatories on Earth?
    (a) Neutrinos are very difficult to detect.
    (b) The neutrinos spread out in every direction once they escaped from the supernova's core.
    (c) Because of the very large distance of 160,000 ly from the exploded star to Earth.
    (d) all of the above
    (e) (a) and (b) only
81. Aluminum-26 decays into magnesium-26:

    $^{26}_{13}$Al → $^{26}_{12}$Mg + two particles.

    What are the two particles? (Recall the conservation laws of nuclear reactions.)
    (a) e^-, ν_e (b) $e^-, \bar{\nu}_e$ (c) e^+, ν_e (d) $e^+, \bar{\nu}_e$
82. (*Answer*) Suppose that the Sun suddenly shrank down to the size of Sirius B (Example 17.1), the rotation period would become roughly
    (a) 1 second.   (b) 3 minutes.   (c) 4 hours.
    (d) 1 day.   (e) 25 days.
83. White dwarf stars can't be the source of pulsar signals because
    (a) they can't pulsate fast enough.
    (b) they can't spin fast enough.
    (c) there aren't enough white dwarf stars to account for all of the pulsars observed.
    (d) white dwarf stars are too dim.
    (e) all of the above
    (f) (a) and (b) only
    (g) (c) and (d) only
84. (*Answer*) Pulsars
    (a) can spin thousands of times per second.
    (b) are neutron stars.
    (c) have very strong magnetic fields.
    (d) have magnetic fields with axes tilted relative to their rotation axes.
    (e) all of the above
    (f) (a) and (b) only
    (g) (c) and (d) only
85. The force that holds the nucleus of an atom together is the _____ force.
    (a) electromagnetic   (b) strong nuclear   (c) weak
    (d) gravitational   (e) elastic
86. (*Answer*) The force that keeps a neutron star from collapsing down to a singularity is the _____ force.
    (a) electromagnetic   (b) strong nuclear   (c) weak
    (d) gravitational   (e) elastic
87. The LIGO and Virgo observatories detect gravitational waves by
    (a) detecting extremely tiny changes in the lengths of perpendicular arms that are many kilometers long.

Exercise 87 continues on the next page

87. *Continued*
    (b) sensing the flicker in light as the waves pass through a beam of light coming from a distant star.
    (c) measuring gamma rays.
    (d) detecting atmospheric oscillations.
88. (*Answer*) Gravitational waves
    (a) are generated when accelerating masses distort spacetime.
    (b) have specific "signatures" depending on the masses involved.
    (c) can reveal whether or not neutron stars or black holes were rotating prior to collisions.
    (d) can tell us about the "ringing" of a newly formed black hole after a collision.
    (e) all of the above
    (f) (a) and (b) only
    (g) (c) and (c) only
89. Stellar-mass black holes are produced by stars with initial masses that are
    (a) greater than $2.7 M_{Sun}$.   (b) less than $9 M_{Sun}$.
    (c) greater than $9 M_{Sun}$.   (d) less than $15 M_{Sun}$.
    (e) greater than $15 M_{Sun}$.   (f) less than $25 M_{Sun}$.
    (g) greater than $25 M_{Sun}$.
90. (*Answer*) A black hole with a mass of $40 M_{Sun}$ has a radius of
    (a) 2.95 km.   (a) 11.8 km.   (a) 29.5 km.
    (a) 40 km.   (a) 118 km.
91. If two $30 M_{Sun}$ black holes orbiting one another inspiral and merge,
    (a) the resulting black hole will have a mass of $60 M_{Sun}$.
    (b) gravitational waves will reveal the inspiral and merger.
    (c) the radius of the resulting black hole will be 177 km.
    (d) all of the above
92. (*Answer*) Objects appear to float inside a spacecraft in orbit because
    (a) there is no gravity in space.
    (b) everything is in free-fall while in orbit.
    (c) neutrinos push on objects inside the spacecraft with enough force to cause them to float.
    (d) of the spacecraft's engines.
    (e) all of the above
    (f) (a) and (b) only
    (g) (c) and (d) only
93. A laser beam with a wavelength of exactly 500 nm is sent from the sidewalk next to the Empire State Building to the observing deck near the top of the skyscraper. A wavelength detector on the observing deck will measure the wavelength of the beam to be
    (a) shorter than 500 nm.   (b) exactly 500 nm.
    (c) longer than 500 nm.
94. (*Answer*) Light bends in a gravitational field because it
    (a) follows the shortest path in curved spacetime.
    (b) starts to run out of energy.
    (c) changes color.
    (d) none of the above
95. A black hole can be characterized by its
    (a) mass.   (b) angular momentum.
    (c) electric charge.   (d) composition.
    (e) all of the above   (f) (a) and (b) only
    (g) (b) and (c) only   (h) (a), (b), and (c) only

96. (*Answer*) A probe is sent free-falling into a black hole that is designed to send out a light signal of precisely 700 nm exactly once every second. What happens to the signal being received by a distant research ship?
    (a) The wavelength increases because of gravitational redshift.
    (b) The wavelength increases because of the relativistic Doppler effect.
    (c) The time interval between signals increases because, according to the special theory of relativity, moving clocks run more slowly.
    (d) The time interval between signals increases due to gravitational time dilation.
    (e) The time interval between signals increases due to increasing distance from the research ship.
    (f) all of the above
    (g) (a) and (b) only
    (h) (c), (d), and (e) only

Short Answer

97. Sirius A has a mass of $2.06 M_{Sun}$ and Sirius B, the white dwarf star, has a mass of $1.02 M_{Sun}$. It is certainly the case that the two stars must have formed together out of the same interstellar cloud. Since the more massive a star is, the faster it evolves, how is it that Sirius B reached the end of its lifetime before Sirius A, which is a main-sequence star of spectral class A1?
98. Why is the diagonal track on the left-hand-side of Fig. 17.3 labeled as "pre-white dwarf evolution"? What is happening during that period, based on the nature of the track?
99. In your own words, describe the evolution of an isolated $5 M_{Sun}$ star from the top of the asymptotic giant branch to its final stage of evolution.
100. White dwarf supernovas are an important rung on the distance ladder. Explain why that is the case.
101. Is a supernova that occurs because of the inspiral collision between two white dwarf stars in a binary star system likely to be a useful distance indicator? Why or why not?
102. Describe the difference between fission and fusion, and the implication for stellar evolution.
103. (a) Using the results of Exercises 78 and 79, how much energy is produced by the disintegration of a $2 M_{Sun}$ iron core, expressed in millions of electron volts (MeV)?
    (b) 1 MeV equals 1.6×10^{-13} J of energy. How many joules of energy are produced in the collapse of a $2 M_{Sun}$ iron core?
    (c) The Sun currently produces about 3.8×10^{26} J of energy every second. How many seconds would be required for the Sun to produce the amount of energy that is generated in the one second required for the $2 M_{Sun}$ iron core to collapse?
    (d) One year equals 3.16×10^7 s. How many years would it take for the Sun, at its current luminosity, to produce as much energy as the $2 M_{Sun}$ iron core collapse generates in one second?
    (e) The present age of the Sun is 4.57×10^9 y, how many times longer or shorter than the age of the Sun is the result of part (d)?
104. Explain why neutrinos arrived at Earth from SN 1987A before the light did.

105. Explain how finding an overabundance of magnesium-26 in ancient meteorites leads to the suggestion that a supernova may have triggered the collapse of the solar nebula 4.57 billion years ago.

106. In the second paragraph of Section 17.3, it was mentioned that if you fell from a height of one meter above the surface of a neutron star you would hit the surface going 2 million m/s. Following Example 17.1 for Sirius B and using typical values for the mass and radius of a neutron star, estimate what the acceleration due to gravity would be near its surface.

107. Why do pulsars that are otherwise slowing their spin rates, abruptly speed up from time to time?

108. Are all neutron stars observed as pulsars? Why or why not?

109. Where does the energy come from that keeps the Crab Supernova Remnant glowing brightly enough to be in the Messier catalog, even almost 1000 years after SN 1054 that created it?

110. Explain why kilonovas are the likely source of isotopes of heavy elements that have many more neutrons than protons.

111. Examining Fig. 17.27, why do the oscillations from the signals get progressively more closely spaced as the neutron stars or black holes approach collision?

112. Although his equation for the size of a black hole turned out to be correct, John Michell's derivation was invalid. What was wrong with his derivation?

113. Explain how x-rays can be emitted in the vicinity of a black hole if the escape speed from the event horizon equals the speed of light.

114. What is the equivalence principle?

115. You decide to use an extremely powerful ship to explore the environment near the event horizon of a black hole while your colleagues remain in a distant orbit. After spending a few hours near the black hole you return to the mother ship. When you get back, you discover that your colleagues are all 25 years older than when you left a few hours earlier (they are a patient group).
    (a) Could this actually happen, at least in theory? Why or why not?
    (b) A similar scenario was part of the movie, *Interstellar* (2014).[5] Using an internet search engine, who was the scientific consultant and what is that person's level of expertise?

116. In your own words, describe how it is possible for a black hole to evaporate, given enough time.

---

[5]*Interstellar* (2014). Directed by Christopher Nolan. [Feature film]. Hollywood, CA: Paramount Pictures.

18 Galaxies Galore

Our galaxy, the Milky Way, is one of [two trillion]† other galaxies in the universe. And with every step, every window that modern astrophysics has opened to our mind, the person who wants to feel like they're the center of everything ends up shrinking.

Neil deGrasse Tyson (b. 1958)

†Numbers were increased significantly to reflect current estimates.

18.1	Historical Studies of the "Universe"	727
18.2	The Milky Way Galaxy	733
18.3	A Taxonomy of Galaxies	749
18.4	Galaxies Everywhere	754
18.5	Supermassive Black Holes and Active Galaxies	760
18.6	How to Build a Galaxy	765
	Exercises	769

Fig. 18.1 NGC 1187. (ESO; CC BY 4.0)

Fig. 18.2 The Milky Way over the Tetons in Grand Teton National Park, Wyoming. [Photo by Robert C. Hoyle, National Park Ranger and former Professor of Astronomy (Copyright Robert C. Hoyle)]

Introduction

Throughout history, humans have turned their gaze upward and observed the band of light that stretches from horizon to horizon across the night sky (e.g. Fig. 18.2). Although the Milky Way is not uniformly smooth, it appeared to be different from anything else in the heavens. More than one culture believed it to be a pathway to the gods or the afterlife. As you learned in Section 4.12, it wasn't until Galileo pointed his small telescope to the sky that humanity learned that the Milky Way was actually composed of countless numbers of stars. Later, larger telescopes revealed that the dark lanes across the Milky Way are caused by intervening clouds of dust and gas. Armed with our knowledge of stars, dust, and gas from previous chapters, together with our ability to study spectra and measure distances and velocities, it is now time to investigate the Milky Way Galaxy and the myriad of other galaxies, large and small, that exist throughout our universe. As we do so we will take much larger strides out into the universe; instead of referring to distances between objects as being mere astronomical units or light-years, it will become common to talk in terms of thousands, millions, and even billions of light-years.

18.1   Historical Studies of the "Universe"

Early Attempts To Map the "Universe"

It may seem a bit ironic that trying to determine the shape, size, distribution of stars, rotation, and other characteristics of the Milky Way Galaxy has proven to be a tremendous challenge. After all, we live *in* the Galaxy.[1] In fact, until the early twentieth century astronomers believed that the Milky Way *was* the entire universe.

[1]"Galaxy" with a capital "G" is typically used to mean the Milky Way Galaxy, while "galaxy" with a small "g" refers to a galaxy other than the Milky Way.

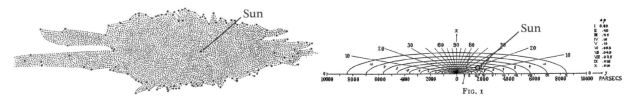

Fig. 18.3 *Left:* An annotated version of Herschel's 1785 map of one vertical slice of the Milky Way with the flattened disk of the Milky Way running horizontally across the diagram. The Sun is represented by the larger star near the center. (Public domain; label for the Sun added) *Right:* An annotated version of Kapteyn's 1922 map of the Milky Way showing lines of constant density of stars above the Galactic disk. The Sun is represented by an open circle just to the right of center and above the midplane of the Galaxy. Distances are shown in parsecs (pc), where 1 pc = 3.26 light-years (ly). (Public domain; label for the Sun added)

It is precisely because we live in the Milky Way that makes understanding it so difficult. The problem is not unlike trying to determine what the overall shape and size of a dense forest is by looking around you. "You can't see the forest for the trees." Cartographers faced the same problem in developing an accurate map of the world without the ability to look down on the planet or have precise surveys of Earth's entire surface (recall the imprecise map by Ptolemy and the challenges it created for Christopher Columbus; page 68). Given the size of the Milky Way, we certainly won't be getting an extragalactic view of our Galaxy anytime soon, so we are forced to sort things out from the inside with the aid of observations of other galaxies.

Herschel's Map of the Milky Way

Fig. 18.4 Herschel's 12 m-long reflecting telescope, completed in 1789. Etching by J. Pass, 1819. (Public domain)

It was in 1785 that Sir William Herschel (Fig. 5.20) made the first attempt to map the shape and extent of the Milky Way. His map, reproduced in Fig. 18.3 (*Left*), was based on counting all of the stars he could see in 685 regions of the sky through a 6.1 m (20 ft) long reflecting telescope of his own design. [Herschel completed a 12 m (40 ft) long reflecting telescope four years later that was the largest ever built at that time; Fig. 18.4.] From those data he selected a slice through the data that represented a plane cutting through the center of the celestial sphere and tilted 35° to the celestial equator. Assuming that (a) all stars are evenly distributed in space, (b) he could see out to the edge of the Galaxy, and (c) all stars had nearly the same absolute magnitude, Herschel estimated the edge of the Galaxy in each direction of the slice based on how many stars were in that region of his survey. He found:

- the Sun to be very near the center of the Milky Way,
- the Galaxy is much longer than it is thick,
- the thickest part of the Galaxy is around its center in the form of a bulge, and
- there is a region near the midplane of the Galaxy that appeared to be much shorter than regions immediately above and below it.

Kapteyn's "Universe"

Professor J. C. Kapteyn[2] (Fig. 18.5), a Dutch astronomer, published a new map of the Milky Way Galaxy in 1922, shown in Fig. 18.3 (*Right*). As with Herschel's map, Kapteyn's map shows a slice through the Galaxy that is perpendicular to the disk and passes through the Galaxy's center in Sagittarius. The map represented a significant improvement in Herschel's map from almost 150 years earlier, because it was based on estimated distances to stars while also accounting for their varying absolute magnitudes when estimating the densities of stars in space (M main-sequence stars far outnumber brighter stars in a typical volume of space, for example).

Kapteyn's "universe," as it came to be known, shows lines of constant overall number density of stars in space with distance from the center of the Galaxy and the density of stars decreases outward away from the Galactic center. At the time, the total number of stars in the Galaxy was believed to be almost 50 billion. Kapteyn estimated the Sun's location to be about 2100 ly (640 pc) from the center and 120 ly (37 pc) above the midplane. The length of the Galaxy's disk was determined to be just over five times greater than its maximum thickness. Kapteyn also argued that the disk of the Galaxy must be rotating in order to maintain the stellar distribution.

Fig. 18.5 Jacobus Cornelius Kapteyn (1851–1922) portrait by Jan Veth (1864–1925). (Public domain)

Shapley's Globular Clusters

Fig. 18.6 Harlow Shapley (1885–1972). (Smithsonian Institution Archives, Accession 90-105, Science Service Records, Image No. SIA2008-5931; public domain, cropped)

Harlow Shapley (Fig. 18.6) was a prolific astronomical researcher and an effective administrator. When Edward Pickering, the director of the Harvard Astronomical Observatory, died in 1919, an acting director was appointed for two years, after which Shapley was selected to be its permanent director, a position he held until 1952. During that time he encouraged Cecilia Payne (Fig. 15.32) to pursue a Ph.D. in the graduate program that he had just established. Shapley had also served as president of the prestigious American Association for the Advancement of Science in 1947. Shapley had even been referred to unofficially as the "Dean of American astronomers" by the New York Times.

Along with his other responsibilities, Shapley was also very politically active. Shapley helped to establish funding for governmental scientific organizations like the National Science Foundation, one of the primary sources of federal funding for the sciences in the United States. Apparently, he even singlehandedly put the S in UNESCO (the United Nations Education, Science, and Cultural Organization). Shapley was also a member of the Independent Voters Committee of the Arts and Sciences for [Franklin Delano] Roosevelt (Fig. 18.7). Shapley was a member of other committees as well, including the Independent Citizens Committee of the Arts, Sciences, and Professions, that advocated for FDR's New Deal, various socialist-leaning agendas, and world peace. For being affiliated

Fig. 18.7 Members of the Independent Voters Committee of the Arts and Sciences for Roosevelt with President Franklin D. Roosevelt. Shapley is on the right. (Photo by George Skadding)

[2]Kapteyn's daughter, Henrietta (1881–1956), married Ejnar Hertzsprung and they had a child whom they named Rigel after the bright blue star in the west foot of Orion (Fig. 2.5).

with the committee, Shapley was called before the House Committee on Un-American Activities in 1947 as fears of the Soviet Union were growing, following World War II. Of Shapley's appearance, the chairman, John E. Rankin stated "I have never seen a witness treat a committee with more contempt." A host of famous individuals were members of the Independent Citizens Committee of the Arts, Sciences and Professions, including then-actor and future United States President, Ronald Reagan.

Much of Shapley's research was focused on pulsating variable stars, bright stars, globular clusters (Section 16.5), open clusters, and the "spiral nebulas" (to be discussed in Section 18.3). His wife, Martha (1890–1981), also worked with him on several projects and published many papers on her own findings. Shapley realized that his study of the distribution of globular clusters could be used to help shed light on the size of the Milky Way Galaxy. He reasoned that knowledge of their locations in space, together with some information regarding their velocities, could tell us where the center of the system of globular clusters is. Since all of the globular clusters must orbit around the center of mass of the Galaxy, the center of the globular cluster system must correspond to the center of the Galaxy itself.

Shapley used several methods to determine the distances to the globular clusters, although he relied primarily on the period–luminosity relationships for Cepheid and RR Lyra variable stars (Section 16.4). Recall that, because of their brightnesses, these pulsating stars, especially Cepheids, can be seen over immense distances. What Shapley discovered was that the system of globular clusters, and then presumably the Milky Way Galaxy, is much larger than previously believed. This was in direct conflict with what Kapteyn would find just a few years later by counting stars.

While Kapteyn concluded in 1922 that the disk of the Galaxy is some 5500 ly (1700 pc) across, Shapley's data indicated that the nearly spherical distribution of globular clusters is 360,000 ly (110,000 pc) in diameter; their estimates differed by a factor of about 6 1/2 in overall size. Clearly, the two studies came to very different conclusions. Kapteyn also found that the Sun is located relatively near the center of the Galaxy at only 2100 ly (650 pc), while Shapley's data led him to conclude that the Sun is 65,000 ly (20,000 pc) from the center of the Cepheid distribution; the two studies differed in their estimates of the Sun's distance from the center of the Galaxy by a factor of 31! At least they roughly agreed that the Sun is north of, and close to, the midplane of the Galaxy. [The solid line in Fig. 18.8 (*Left*) is the line-of-sight from the Sun to the center of the globular cluster distribution and the dashed line is Shapley's estimate of the major (long) axis of the distribution.]

Figure 18.8 (*Right*) shows Shapley's edge-on view along the Galaxy's midplane, and the solid arc is his estimate of the distance limit from Earth of his most reliable determinations of distances. According to Shapley's observations, the hashed region above and below the midplane is apparently devoid of globular clusters, something referred to as the zone of avoidance. Shapley argued, incorrectly, that the lack of globular clusters in the disk was because they can be disrupted by the Galaxy's mass when passing through it, causing them to appear as open clusters, which are in fact found preferentially in the disk.

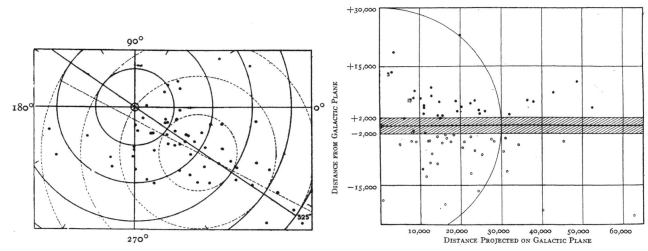

Fig. 18.8 *Left:* Shapley's 1918 determination of the distribution of globular clusters (dots) as viewed from above the disk of the Milky Way. The radii of the circles increase in intervals of 32,600 ly (10,000 pc) except for the smallest circle around the Sun, which has a radius of 3260 ly (1000 pc). The solid circles are centered on the Sun, and the dashed circles are centered on the middle of the globular cluster distribution. *Right:* The distances of the globular clusters from the Sun (located on the left edge and midway up the vertical axis) and their distances above and below the Galactic midplane. The hashed region is the zone of avoidance. Distances are indicated in parsecs. [Shapley (1918);[3] public domain]

Resolving Differences

As pointed out in Section 1.1, science is very much a human enterprise, subject to errors, false starts, retreating, and following new paths on the quest for human understanding. The history of determining the size and shape of the Milky Way Galaxy is just one example of that reality. Understanding conflicting results and resolving them is a major part of the scientific process.

We know today that both Kapteyn and Shapley were wrong in their analyses of their data, in part, surprisingly, for the same reason. Kapteyn's "universe" proved to be too small and Shapley's was too large. What both men failed to fully appreciate was just how effectively interstellar dust can obscure objects.

Kapteyn had chosen stars that were preferentially in the disk of the Galaxy where the obscuring effects of interstellar dust is most severe; dust lanes are clearly visible in Fig. 18.9. As a result he couldn't see the entire distribution of stars in the Galaxy. This problem can be seen on the left side of Herschel's map (Fig. 18.3 *Left*) as the shortened distribution near the midplane of the Galaxy in the direction of Sagittarius. Kapteyn was aware of the obscuring effects of dust, but couldn't find any evidence that it was significantly affecting his analysis.

Shapley's choice of studying globular-cluster distances meant that he was looking at objects that can be far above or below the disk of the Galaxy where the problems of dust are much less significant, but not completely absent. As was mentioned in Section 16.4, there are two types of Cepheid variable stars: heavy-element rich Cepheids have a greater abundance of elements heavier than helium compared to

Fig. 18.9 A man looking at the central bulge and a portion of the disk of the Milky Way Galaxy in the direction of Sagittarius. The location is the Chajnantor plateau in Chile, where the Atacama Large Millimeter/submillimeter Array (Fig. 15.11) is located. (ESO/P. Horálek; CC BY 4.0)

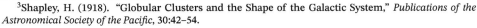

[3]Shapley, H. (1918). "Globular Clusters and the Shape of the Galactic System," *Publications of the Astronomical Society of the Pacific*, 30:42–54.

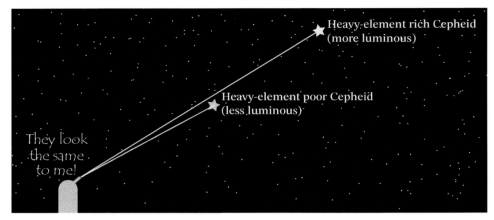

Fig. 18.10 A heavy-element rich Cepheid and a heavy-element poor Cepheid with the same pulsation period. Since the heavy-element rich Cepheid is more luminous (meaning it has a *smaller* absolute magnitude), it must be twice as far away in order to have the same apparent magnitude as the heavy-element poor Cepheid. Shapley mistook heavy-element poor Cepheids for heavy-element rich ones, causing him to overestimate the size of the Galaxy.

the heavy-element poor Cepheids found in globular clusters.[4] As a result, heavy-element rich and heavy-element poor Cepheids have different period–luminosity relationships; without any dust in the way, if a heavy-element rich and a heavy-element poor Cepheid have the same pulsation period and the same apparent magnitude, the heavy-element rich Cepheid would be twice as far away (Fig. 18.10). Heavy-element rich Cepheids are also found preferentially near the disk of the Milky Way where the effects of dust are much more significant. Henrietta Leavitt and those who followed, including Shapley, calibrated their Cepheid period–luminosity relationships using heavy-element rich Cepheids. Because astronomers hadn't yet discovered that there are two different populations of Cepheids, Shapley used his calibration of heavy-element rich Cepheids for the heavy-element poor Cepheids in globular clusters, causing him to overestimate the size of the Galaxy.

Just as dust along the disk of the Galaxy is revealed in Herschel's and Kapteyn's maps, it is also evident in Shapley's data. The zone of avoidance in Fig. 18.8 (*Right*) exists because objects, including globular clusters, simply cannot be seen in visible light near and on the other side of the Galactic center. Many of the globular clusters also tend to move relatively rapidly through the disk in their high orbits around the center of the Milky Way Galaxy, making the probability very small of finding many globular clusters in the disk at any given time.

Even before the obscuring effects of dust were understood, the "universe" maps of Herschel and Kapteyn each had a hidden red flag; in both cases the Sun is found to be very close to the center of the stellar system. Copernicus taught us that not only is Earth not at the center of the Solar System, but it is only one of many planets orbiting the Sun and therefore doesn't occupy any unique position in the universe. This argument suggests that humans likely don't occupy a special place in the universe either. Astronomers tend to be suspicious of studies that seem to imply that

[4]Astronomers also refer to heavy-element rich Cepheids as classical or Type I Cepheids, and heavy-element poor Cepheids as W Virginis stars or Type II Cepheids.

Earth, the Solar System, and humans occupy a position in space and time that would violate the Copernican principle.

18.2   The Milky Way Galaxy

We have come a long way in our understanding of the Galaxy. All of the observational tools and physics presented in this textbook have been and continue to be used to study the Milky Way and other galaxies throughout the universe. Parallax, the period–luminosity relationships of various classes of pulsating stars, and other components of the distance ladder have revealed distances within the Galaxy. Using spectroscopy, we have been able to determine the compositions of stars and the interstellar medium, radial velocities from the Doppler effect, and the presence and strength of magnetic fields. In addition, observations in infrared, microwave, and radio wavelengths have made it possible to look through the obscuring dust in Shapley's zone of avoidance to see what lies near the Galactic center and on the other side of the Milky Way's disk.

Our Modern Map

Figure 18.11 shows a full-sky map of 1.7 billion stars that was produced using data obtained by the Gaia spacecraft (there are an estimated 200 billion to 400 billion

Fig. 18.11 A modern, full-sky color view of the Milky Way Galaxy and neighboring galaxies as seen from Earth. The center of the Milky Way (located in Sagittarius) is in the middle of the image, the Milky Way's bulge is the bright region around the center, and the disk is seen as the horizontal band of light. Two of our satellite galaxies, the Large and Small Magellanic Clouds, are visible in lower right. (© ESA/Gaia/DPAC; CC BY-SA 3.0 IGO)

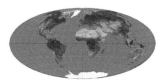

Fig. 18.12 A projection map of Earth similar to the projection map of the cosmos in Fig. 18.11. (Strebe; CC BY-SA 3.0)

stars in the entire galaxy). You can think of the map as the celestial sphere being "unwrapped" and flattened, with the midplane of the Milky Way placed horizontally across the middle, and the north and south galactic poles at the top and bottom, respectively. The center of the Galaxy in the constellation of Sagittarius is positioned at the center of the map, and the anticenter (the opposite direction from the center) in Auriga is on the right and left edges. For comparison, a similar projection of Earth is depicted in Fig. 18.12 with 0° latitude and 0° longitude at the center of the map. The midplane of the Milky Way is tilted 63° relative to the celestial equator and intersects the celestial equator in Aquila (north of Sagittarius) and in Monoceros (between Canis Minor and Orion). Because the celestial equator is simply an extension of Earth's equator projected out into space, you can think of Fig. 18.11 as being tilted 63° relative to Fig. 18.12 and rotated along Earth's equator.

Figures 18.9 and 18.11 give us an idea of what the brightest parts of the Milky Way Galaxy look like when viewed from our location near its midplane. However, to fully appreciate the size of the Galaxy we must also include distances and an outside perspective. Fig. 18.13 illustrates what the Milky Way would look like viewed edge on from far outside the Galaxy.

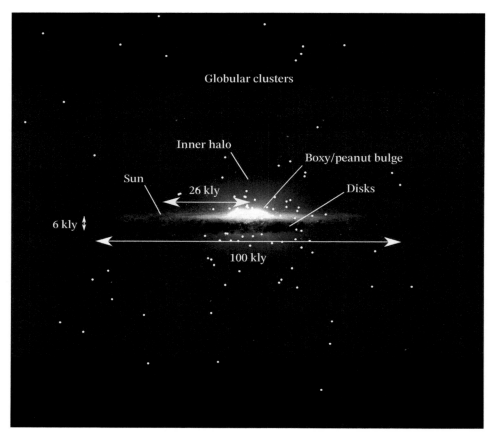

Fig. 18.13 An artist's depiction of the Milky Way viewed edge on from a distant vantage point. (Adapted from ESO/NASA/JPL-Caltech/M. Kornmesser/R. Hurt; CC BY 4.0, labels and globular clusters added)

Table 18.1 Properties of the Milky Way and its components

Property	Value	Property	Value
Sun's distance from the center	26 kly	Size (full diameter)	1.8 Mly
Sun's orbital period	210 My	supermassive black hole diameter	0.16 au
Midplane tilt to celestial equator	63°	boxy/peanut bulge bar length	4.6 kly
Number of stars	200–400 billion	boxy/peanut bulge bar thickness	1.2 kly
Mass (total)	1.3×10^{12} M$_{Sun}$	long bar length	34 kly
stars	4%	disk diameter	100 kly
cold gas and dust	1%	thin disk thickness	2 kly
hot gas ($T = 2 \times 10^6$ K)	2%	thick disk thickness	6 kly
dark matter	93%	inner stellar halo diameter	160 kly
supermassive black hole	4.1×10^6 M$_{Sun}$	outer stellar halo diameter	460 kly
Luminosity (50% in infrared)	5×10^{10} L$_{Sun}$	dark matter halo diameter	1800 kly
boxy/peanut bulge	16%	Ages of components	
long bar	14%	oldest stars	13.5 Gy
thin disk	67%	boxy/peanut bulge	12.5 Gy
thick disk	1%	inner halo	12.5 Gy
stellar halos (inner and outer)	2%	outer halo	0.1–11 Gy
hot corona halo (x-rays)	< 0.1%	thick disk	9–10 Gy
dark matter halo	0%	thin disk	7–8 Gy
Magnetic field strength	0.2–1 nT	long bar	3–6 Gy

The Milky Way Galaxy is composed of many parts that can be distinguished by their shapes and dimensions, their masses and luminosities, the motions of their stars and gas clouds, whether or not they contain globular clusters, and their heavy element content (which gives us some information about their ages). All of these features provide clues as to how the Milky Way formed, something we will look at in Section 18.6. In this section we will take a look at the properties of the Milky Way's various parts.

Some characteristics of the enormous collection of gas, dust, stars, light, magnetic fields, and unseen matter that is the Milky Way are summarized in Table 18.1. As with other detailed tables presented in this textbook, the table is provided for reference and for comparisons, rather than to be memorized.

The Milky Way's Total Mass and Luminosity

Adding up all of the mass contained in stars gives a result that is nearly 57 billion times the mass of the Sun (5.7×10^{10} M$_{Sun}$). Gas and dust in the Galaxy adds close to 34 billion solar masses. Combined, everything that we can see amounts to roughly 91 billion times the mass of the Sun. The stars, gas, and dust also produce 50 billion times the luminosity of the Sun across all electromagnetic wavelengths, with roughly 50% of the luminosity being in infrared light and most of the rest in visible and near-ultraviolet wavelengths. Since, in the case of our Sun, a mass of 1 M$_{Sun}$ currently produces 1 L$_{Sun}$ across all wavelengths of light, the fact that the mass of

Key Point
The visible mass of the Milky Way is about 90 billion M$_{Sun}$ and its luminosity is about 50 billion L$_{Sun}$.

all visible matter and the amount of light that it produces are similar in numerical value shouldn't be much of a surprise (it is important to remember, though, that since mass and luminosity are fundamentally different quantities with different units, we are really comparing apples to oranges; 1 M_{Sun} does not equal 1 L_{Sun}).

The Visible Galactic Halos

Key Point
The Milky Way has three visible halos that are roughly spherical but a bit flattened:
• the inner stellar halo,
• the outer stellar halo, and
• the hot coronal halo.

By far the largest components of the Milky Way Galaxy are its halos. More than 90% of the nearly 160 known globular clusters in the Galaxy are found in orbits around the Galaxy's center that keep them within a significantly flattened spheroidal shape that is the inner stellar halo depicted in Fig. 18.13. The shape of the inner halo is a bit like that of a rugby ball and it has a diameter of about 160 kly, approximately 60% larger than the diameter of the Galaxy's disks.

Studying the distribution of ancient stars and the most distant globular clusters indicates that there is also a much larger outer stellar halo that has a diameter of 460 kly. The shape of the outer stellar halo is more spherical than the inner halo, but it is still slightly flattened.

Curiously, there doesn't appear to be a preferred orbital direction for the stars and clusters in either stellar halo. This means that, overall, the motions are essentially random, although groups of stars and gas within the outer halo move collectively in different directions. This is very different from what happens in the rest of the Galaxy, raising questions about how the halos formed, and when.

Key Point
The hot corona halo is composed of ionized gas at about 2×10^6 K.

Although there is relatively little cold gas or dust in the halos, there is a significant amount of hot, ionized gas having a characteristic temperature of about two million kelvins. Of course, the density of this hot halo gas is extremely low, being only about 10 particles in every cubic meter of space: at sea level in Earth's atmosphere there are about 10^{25} molecules per cubic meter (1 trillion trillion times greater than the ionized gas in the halos). The distribution of this gas is known as the hot corona halo, named for the hot gases that make up the outermost atmosphere of the Sun and other stars. Where did this gas come from?

Despite their enormous dimensions, the stars in the two stellar halos are only responsible for about 1% of the mass of all stars in the Galaxy. On the other hand, despite its very low density, the hot, ionized gas in the corona halo has a total mass of about 50% of the mass of all of the Galaxy's stars.

The Galaxy's Disks

Fig. 18.14 The spiral galaxy NGC 4565 with its disk seen edge-on and its bulge clearly visible. (ESO; CC BY 4.0)

Most of the visible mass and luminosity of the Milky Way is located in a disk that is centered on the Galaxy's midplane, similar to NGC 4565 seen edge-on in Fig. 18.14. Although there isn't a distinct cutoff, most of the disk's material is confined to a total thickness of 6000 ly (or roughly 2000 pc). The horizontal extent of the disk is also not clearly defined, but most of the visible mass is found within about 50,000 ly (15,000 pc) of the Milky Way's center, meaning that the bulk of the disk measures roughly 100,000 ly (30,000 pc) across. However, some disk stars have been found much farther from the Galactic center, and gas has been found 130,000 ly (40,000 pc) from the center, implying a disk diameter of 260,000 ly (80,000 pc). Defining an "edge" of the disk is somewhat arbitrary.

Technically, there are two distinct and overlapping disks: a thin disk that contains mostly heavy-element rich stars, and a thick disk that contains stars that are more heavy-element poor. As their names imply, the thick disk has a larger vertical extent than the thin disk. The thin disk contains virtually all of the disk's dust and gas, the vast majority of disk stars, most of the disk mass, and it is far brighter than the thick disk. In general, thin-disk stars are also younger than thick-disk stars, as evidenced by their heavy element content. Remember that the more heavy-element rich a star is, the more generations of stars came before it to enrich the interstellar medium and the gas cloud that the star was born in. The existence of two disks provides more clues about how the Galaxy formed over time, the topic of Section 18.6.

The Sun is a member of the thin disk and is located 26,000 ly (8 kpc) from the center and only about 20 ly (5.5 pc) north of the midplane. The Sun's distance from the Galactic center means that we are not seeing the center as it looks today, but how it looked 26,000 years ago, when Earth was experiencing an ice age. The Sun's orbital speed around the center of the Milky Way is almost 240 km/s (860,000 km/h or 540,000 mi/h), but even at that speed it still takes the Solar System 210 million years to make one lap around the Galaxy. Thinking about that amount of time in another way, the Solar System has only gone 30% of the way around the Galaxy since the dinosaurs went extinct 66 million years ago, and it has completed about 22 orbits since the Solar System formed 4.6 billion years ago.

Key Point
The Sun is 26 kly from the center of the Galaxy and requires 210 My to complete one orbit.

Example 18.1

Knowing both the distance of the Sun from the center of the Milky Way Galaxy and its orbital period means that we are able to apply Kepler's third law (page 174) to calculate the Galaxy's mass *inside the Sun's orbit*.

Since the mass of the Sun is very tiny relative to the mass of the Galaxy, we can safely neglect it in Kepler's third law. This means that the law becomes

$$M_{\text{Sun, MW}} = \frac{a_{\text{au, Sun}}^3}{P_{\text{y, Sun}}^2},$$

where a_{Sun} = 26,000 ly [which must be converted to astronomical units (au)], $P_{\text{y, Sun}}$ = 210 million years = 2.1×10^8 y, and $M_{\text{Sun, MW}}$ means the mass contained inside the Sun's orbit in units of the mass of the Sun.

To convert from light-years to astronomical units, recall the relationships between astronomical lengths summarized on page 607. Carrying out the conversion, we find that $a_{\text{au, Sun}}$ = 1.7 billion au; in other words, the distance from the Sun to the center of the Galaxy is 1.7 billion times greater than the distance from Earth to the Sun. The Milky Way is big!

Substituting the values into Kepler's third law gives

$$M_{\text{Sun, MW}} = \frac{(1.7 \times 10^9)^3}{(2.1 \times 10^8)^2} \, M_{\text{Sun}} = \frac{1.7^3 \times 10^{9\times3}}{2.1^2 \times 10^{8\times2}} \, M_{\text{Sun}}$$

$$= \frac{4.9 \times 10^{27}}{4.4 \times 10^{16}} \, M_{\text{Sun}} = \frac{4.9}{4.4} \times 10^{27-16} \, M_{\text{Sun}}$$

$$= 1.1 \times 10^{11} \, M_{\text{Sun}}.$$

A scientific notation tutorial is available at
`Math_Review/Scientific_Notation`

The portion of the Milky Way's mass *within* 26,000 ly of the Galaxy's center is 110 billion times the mass of the Sun.

There is something of a mystery about the result we just found in Example 18.1: if we add up all of the mass due to the stars in the thin and thick disks, including outside of the Sun's orbit, that total only amounts to 4.1×10^{10} M_{Sun}, or 41 billion solar masses. Even adding the gas and dust in the disk (0.7×10^{10} M_{Sun}) only gives 4.8×10^{10} M_{Sun}. But Kepler's third law tells us that there must be more than twice that amount of mass. What are we missing?

Mapping Cold Hydrogen Gas Using the 21-cm Line

A particularly important wavelength for mapping clouds containing cold hydrogen gas is hydrogen's extremely low-energy spectral line, with a radio wavelength close to $\lambda = 21$ cm and a corresponding frequency of $f = 1.42$ GHz (recall that $f\lambda = c$, the speed of light). Unlike other emission and absorption lines we have discussed, that involve the transitions of electrons between the energy levels associated with electron orbitals, the 21-cm line is associated with the quantum mechanical "spins" of the electron and the proton in neutral hydrogen atoms. These spins are the reason for the spin quantum number, m_s, that plays such an important role in the Pauli exclusion principle and the structure of the periodic table of the elements discussed on page 295. Because both electrons and protons have electric charge, their spins imply that they have tiny magnetic fields as well (recall that moving charges produce magnetic fields).

Consider Fig. 18.15. Suppose that an everyday bar magnet is attached to a table so that it can't move, and another identical magnet is brought close to it so that their north poles and south poles are almost touching. Because the two north poles and the two south poles repel while opposite poles attract, when the second bar magnet is released it will immediately be pushed away from the fixed magnet, flip around, and be attracted back to the fixed magnet with opposite poles now touching. When this happens, the two magnets move from a higher potential energy state to a lower one. An analogous situation can happen in hydrogen, when the spin orientation of the electron "flips" relative to the proton, causing the electron's magnetic field to flip relative to the proton's magnetic field, which produces a photon with an energy equal to the energy difference between the two spin orientations. The associated wavelength of the photon is 21 cm. The reverse can also happen when a cold hydrogen atom absorbs a 21-cm photon.

The spontaneous hydrogen spin-flip is extremely rare, taking an average of several million years for an isolated hydrogen atom to make the transition. This means that the atom cannot undergo a collision with another particle during that time (including photons), otherwise the delicate arrangement will be disrupted before the spontaneous spin-flip can occur. Only in the phenomenally low densities found in the interstellar medium can the spin-flip take place. Despite being so rare, the spin-flips of the enormous number of hydrogen atoms in an interstellar cloud of gas and dust can produce a detectable signal.

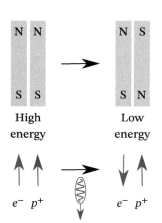

Fig. 18.15 *Top:* The transition from aligned to anti-aligned bar magnets. *Bottom:* The spin-flip of an electron relative to a proton in a hydrogen atom changes the relative orientations of their magnetic fields, which produces a photon with a wavelength of 21 cm.

Because of the way in which the 21-cm line is produced, other gases and dust are essentially invisible at that wavelength and the photons arrive at our radio telescopes without being absorbed or deflected. This means that cold hydrogen clouds can be detected on the other side of the Galaxy. Measuring either a blueshifted or a redshifted 21-cm line from a particular hydrogen cloud tells us if that cloud is approaching or moving away from Earth, respectively, and how fast (recall the Doppler effect in Section 8.4). By making velocity measurements in different directions along the disk of the Milky Way we can determine the orbital speeds of clouds around the center of the Galaxy. The strengths of the signals also tell us about the amount of hydrogen in the clouds.

The Long Bar and the Spiral Arms

Figure 18.16 on the following page is an artist's depiction of what the Milky Way would look like if we could magically transport ourselves to a location far above its north galactic pole. It is immediately apparent that the disk isn't featureless with a smooth distribution of stars, gas, and dust, but instead has a long bar in the middle and spiral arms that appear to start at the ends of the bar.

The brightest, bluest, youngest, and most massive stars are found in the spiral arms, which is what makes the arms stand out so prominently. Using the 21-cm hydrogen line reveals that these massive stars are also found very near relatively dense, cold clouds of gas and dust in the arms. Lower mass, redder stars are also found in the spiral arms, but they are found between the spiral arms as well, where the massive young stars are largely absent. Our Sun is located very near, but not actually in, one of the Milky Way's principal spiral arms.

Galaxies similar to our own Milky Way are found throughout the universe. A face-on image of M61, one of the famous Messier objects, is displayed in Fig. 18.17. M61 shows the same bright spiral arms and long bar that are present in our Galaxy. Its spiral arms also contain its brightest, most massive stars and associated cold gas clouds. Figure 18.18 illustrates what happens when a spiral galaxy is viewed in infrared light: the spiral structure that is dominated by blue light almost vanishes, bringing out the older, cooler, and redder stars that can be found between the spiral arms.

Density Waves and Star Formation

The existence of spiral arms poses two problems for our understanding of the structure of the Milky Way and other spiral galaxies. How did this spiral structure form in the first place, and how is it maintained? An answer to the first question seems to require that spiral structure spontaneously forms somehow, with a greater density of stars, gas, and dust along the arms. As for the second question, if stars, gas, and dust relatively close to the center of the Galaxy have shorter orbital periods than matter farther away (similar to Mercury having a shorter orbital period around the Sun than Neptune), then, as illustrated in Fig. 18.19, the matter closer to the center can complete multiple orbits in the time that more distant material completes just one orbit. As this mismatch in orbital periods continues, the spiral arms would get tighter and tighter until the pattern becomes so tight that it essentially disappears when the spiral arms, with their significant thickness, start to overlap. Does this

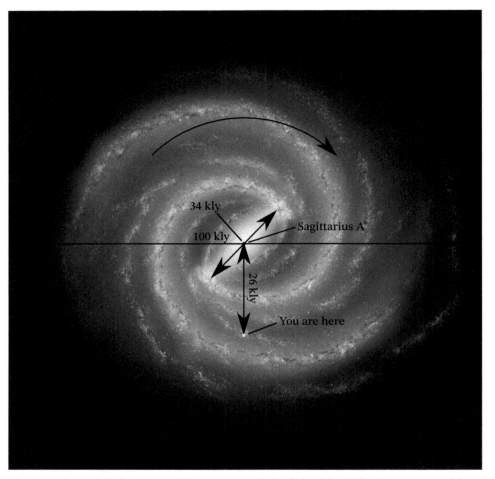

Fig. 18.16 An artist's depiction of the structure of the disk of the Milky Way as viewed from above the Galactic north pole. The bulk of the disk is 100 kly in diameter, the long bar is 34 kly in length, and we are 26 kly from the center of the Galaxy, which is Sagittarius A*. (Adapted from NASA/JPL-Caltech; labels added)

mean that the spiral structure is just temporary? If so, why are there so many spiral galaxies in the universe? Surely we don't see them all at a unique time in their histories. In essence, this would be a violation of the Copernican principle, in the sense that we just happen to live at a time when spiral structure exists in disk galaxies. This is known as the winding problem.

The answer to our first question given in the last paragraph is hardly satisfying. It is never enough in science to say the equivalent of "oh, just because." While it is true that spiral arms develop on their own, science demands that the mechanism by which that happens must also be understood. Sir Isaac Newton taught us that all masses attract all other masses gravitationally. If a region of locally greater mass density exists, then it can pull other material toward it, further enhancing the density. In a very large, rotating disk of material, gravitational instabilities naturally develop due to gravitational attraction, resulting in spiral patterns. However,

Fig. 18.17 Messier 61 is located 50 million light-years from Earth in the constellation Virgo. Other, more distant galaxies can be seen behind it, including another spiral galaxy, seen nearly edge on, along the upper left edge of the image. (ESO; CC BY 4.0)

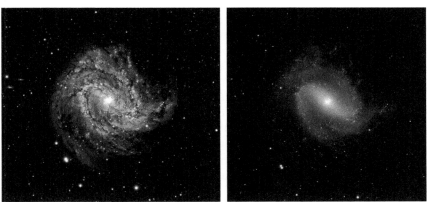

Fig. 18.18 *Left:* The barred spiral galaxy, Messier 83, referred to as the Southern Pinwheel, is seen in visible light. M83 is located 15 million light-years away in Hydra. [ESO/IDA/Danish 1.5 m/R. Gendler, S. Guisard (www.eso.org/ sguisard) and C. Thöne; CC BY 4.0] *Right:* M83 in infrared light. (ESO/M. Gieles; CC BY 4.0)

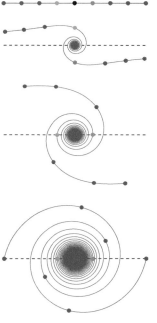

Fig. 18.19 *Top:* Stars are initially in a straight line orbiting clockwise around the center of the galaxy as viewed from above the north galactic pole. *Second:* The innermost marked stars have completed 1/4 orbit. *Third:* The innermost marked stars have completed one orbit. *Bottom:* The innermost marked stars have completed four orbits, but the outermost stars have still only completed 1/2 orbit.

material is not locked in the density enhancements of spiral arms, which is what would lead to the winding problem. Instead, stars and interstellar material pass through the regions of increased density so that the material in the spiral arms is constantly changing. These enhanced density patterns are referred to as density waves, and are similar to sound waves in the sense that sound waves move *through* water or air rather than carrying the water or air along with them.

This idea of a density wave is analogous to a traffic jam that can develop in the vicinity of an accident as depicted in Fig. 18.20; vehicles enter the traffic jam, move slowly through it, and then exist the traffic jam while the congestion remains in

Fig. 18.20 *Top:* A traffic jam with the blue car entering our field of view from the right. *Bottom:* The same traffic jam seen one minute later, with the blue car having moved across our field of view. The cars move through the traffic jam, while the traffic jam itself doesn't move relative to the accident.

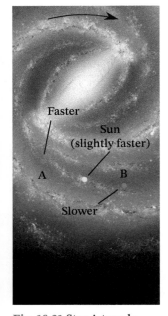

Fig. 18.21 Star A travels faster than the spiral pattern and so outruns the spiral arms. Star B travels slower than the spiral pattern, meaning that the spiral arms overtake it. (Adapted from NASA/JPL-Caltech; cropped and labels added)

place near the accident. Spiral arms do in fact rotate around the center of the galactic disk, maintaining their structure, but material in the disk is not trapped in them. Stars and clouds at different distances from the center of the disk galaxy have different orbital periods, but the entire shape of the spiral arm rotates with a single period. In the Milky Way's case, its spiral arms have a rotation period of 220 million years, compared to the Sun's 210 million years. As can be seen in Fig. 18.21, this means that the Sun is traveling slightly faster than the arms so the Sun approaches the spiral arms from behind, passes through them, and then moves ahead. Only a little farther from the center of the Galaxy, where orbital periods are longer, the spiral arms actually overtake the stars and clouds rather than the other way around.

So why are hot, massive stars preferentially found in spiral arms? In Chapter 16, we learned that star formation occurs when gas clouds collapse. When density waves pass through clouds or vice versa, the density of the clouds increases, which can trigger cloud collapse and star formation. This is because increasing density means that atoms, molecules, and dust grains are closer together and the force of gravity between them is greater. We also learned in Chapter 16 that the more massive a star is, the shorter its lifetime, with the most massive stars only living about one million years or less. In the vicinity of the Sun, this means that a very massive star would barely have any time at all to orbit the center of the Galaxy. Remember that it takes the Sun 210 million years to complete one lap. In other words, when a very massive star is born, it lives out its life and dies in a supernova explosion without having time to leave the spiral arm in which it formed. When it does explode, the shock wave from the expanding supernova remnant can trigger the collapse of other nearby gas clouds, generating more star formation. It is this secondary star-formation process that is believed to have caused our Sun and the Solar System to form. On the other hand, the lowest mass, reddest stars have lifetimes of hundreds of billions of years, meaning that once formed, these reddest stars are all still around and orbiting the center of the Milky Way. The Sun, with its nearly

10-billion-year life span, will complete more than 40 orbits before it ejects a planetary nebula and slowly cools off for all eternity as a white dwarf star. Stars of a few solar masses or less exist throughout the Galaxy, but the most massive stars have bright, glorious, and explosive lives, all while confined to spiral arms.

The long bar does not actually participate in a density wave. Instead, the stars in the long bar move with the bar structure. The rotation period for the long bar is also much shorter than for the spiral arms or the Sun's orbital period; its rotation period is only 160 million years.

The Bulge

If you study Figures 18.9 and 18.11 carefully, you can see a bulge in the distribution of stars in the direction of Sagittarius. The bulge is also labeled in Fig. 18.13. The distinctive boxy/peanut shape of the bulge is clearly visible in the infrared image in Fig. 18.22, protruding above and below the thin disk. The bulge has its own inner disk and a short bar that keeps its shape as the bulge rotates. There is also a cluster of stars roughly 10 pc in diameter that is located deep within the bulge and centered on the center of the bulge (and the center of the Galaxy).

As can be seen in Table 18.1, the entire bulge is the second-brightest feature of the Milky Way, barely beating out the long bar. However, even the combined luminosity of the bulge and the long bar is still less than half the brightness of the Galaxy's disk.

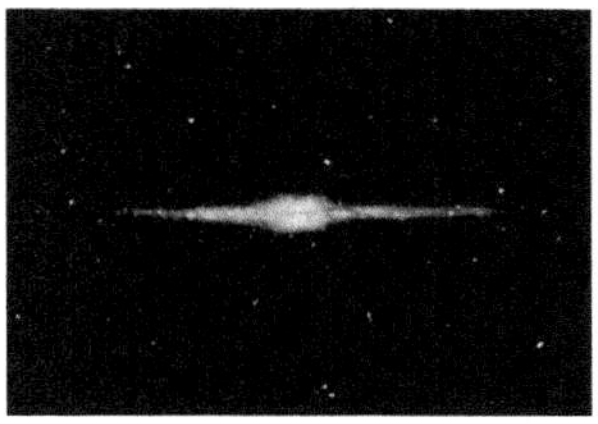

Fig. 18.22 The central region of the Milky Way and its boxy/peanut bulge, as seen in infrared light. (DIRBE, COBE, NASA)

Sagittarius A*: The Beast at the Center

As we look deep into the heart of the Milky Way Galaxy and its boxy/peanut bulge, we find a turbulent and violent environment. Because of the tremendous amount of dust between us and the Galaxy's center, observing that region in the visible light that the human eye can detect is impossible. Instead, we must rely on other wavelengths to see the core of the Milky Way.

In 1931 Karl Jansky (Fig. 15.8), a pioneer of radio astronomy, detected a radio signal in the direction of Sagittarius, which he named Sagittarius A because it was the first source of radio waves detected in that constellation. Forty-three years later, a very compact but bright radio source was discovered within Sagittarius A, that came to be called Sagittarius A* (pronounced "Sagittarius A star" and abbreviated Sgr A*). The "star" superscript was chosen to indicate excitement about the discovery, taking the idea for the designation from the notation that is used for nuclei of atoms that are in excited states). Based on studies of the motion of Sgr A* it appears that the radio source is extremely close to, if not at the exact center, of the Milky Way. Figure 18.23 shows the region surrounding Sgr A* that was obtained by the Atacama Large Millimeter/submillimeter Array.

Radio and millimeter wavelengths aren't the only wavelengths generated near the center of the Galaxy: Fig. 18.24 shows an x-ray view of the region surrounding Sgr A*. X-rays are produced by high-energy processes, including those associated with environments having temperatures of millions of kelvins. On January 5, 2015, a record-breaking x-ray flare was observed originating from Sgr A*.

As astronomers continued to investigate ever-smaller volumes of space around Sgr A*, individual stars and gas clouds were discovered, that exhibited very high

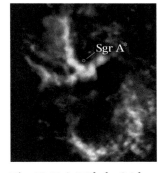

Fig. 18.23 A 7.5 ly by 8.2 ly view of molecular hydrogen clouds near the center of the Milky Way, obtained at millimeter wavelengths. The red circle marks the location of Sgr A*. [Adapted from ALMA (ESO/NAOJ/NRAO) /J. R. Goicoechea (Instituto de Física Fundamental, CSIC, Spain); CC BY 4.0, label added]

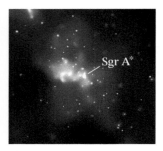

Fig. 18.24 An x-ray view of the region around the center of the Galaxy obtained by the Chandra X-Ray Observatory. The width of the image is about 45 ly. (Adapted from NASA/CXC/Caltech/M. Muno et al.; cropped and labels added)

velocities. One group of stars has been particularly well studied in near-infrared light for over 30 years and their orbits are shown in Fig. 18.25. One of those stars, S2, has an orbital period of 16.1 years and made its closest approach to Sgr A* in May, 2018. S2 has a very elliptical orbit with a semimajor axis of 1018 au that brought it within 117.5 au of the mass that it is orbiting. At the time of closest approach it had a speed of 7641 km/s (17,090,000 mi/h). Bear in mind that in our Solar System, the dwarf planet Sedna (Table 10.2) has a semimajor axis of 483 au. Whatever Sgr A* is, it must be smaller than Sedna's orbit.

Example 18.2

Given the data we have for the star S2, we can estimate the mass inside S2's orbit by carrying out the same type of calculation that we conducted in Example 18.1.

Again from Kepler's third law,

$$M_{\text{Sun, Sgr A}^*} = \frac{a_{\text{au}}^2}{P_{\text{y}}^2} = \frac{(1018)^3}{(16.1)^2}\,M_{\text{Sun}} = \frac{1.05 \times 10^9}{259}\,M_{\text{Sun}}$$

$$= 4.1 \times 10^6\,M_{\text{Sun}}.$$

There are 4.1 million solar masses inside S2's orbit.

Given how much mass is confined to such a relatively small region of space, today there is very little doubt that the source of the powerful signals coming from Sgr A* is a supermassive black hole (the * superscript has become the standard symbol to indicate that an object is a supermassive black hole).

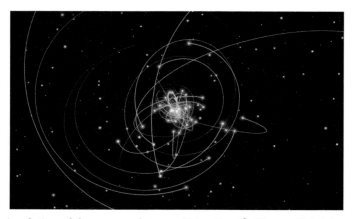

Fig. 18.25 A simulation of the swarm of stars orbiting Sgr A*. The small, bright green orbit at the center is that of S2. (Adapted from ESO/L. Calçada/spaceengine.org; CC BY 4.0, S2 orbit highlighted)

Example 18.3

What is the size of the Milky Way's supermassive black hole?

We can use the equation for the radius of a black hole given on page 711 to calculate the supermassive black hole's size just as we did in Example 17.3 for the smallest stellar-mass black hole:

$$r_{bh} = 2.95 \, M_{Sun} \text{ km} = 2.95 \times (4.1 \times 10^6) \text{ km} = 1.2 \times 10^7 \text{ km}.$$

Since there are 1.5×10^8 km in one astronomical unit, the radius of the supermassive black hole is also

$$r_{bh} = 1.2 \times 10^7 \text{ km} \times \frac{1 \text{ au}}{1.5 \times 10^8 \text{ km}} = 0.08 \text{ au}.$$

This means that the diameter of the supermassive black hole is 0.16 au, which is about 17 times larger than the Sun, but only about 20% of the size of Mercury's orbit.

The powerful emissions from Sgr A* don't actually come directly from the supermassive black hole; after all, nothing can escape from a black hole, not even light. Instead, much of the light comes from an accretion disk of material spiraling into the black hole. Figure 18.26 shows an image of Sgr A*, produced by the Event Horizon Telescope (EHT). The EHT image was created by combining very precisely timed data from radio telescopes all over the world, from Antarctica to Spain, using

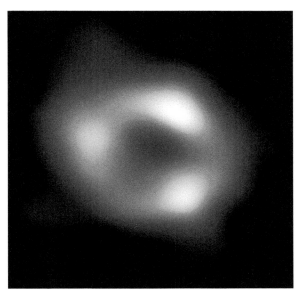

Fig. 18.26 A radio-wavelength image of the accretion disk around Sagittarius A* obtained using the Event Horizon Telescope. The supermassive black hole is the black region in the middle of the image, and the event horizon is the inner boundary of the accretion disk. Gas near the event horizon is orbiting at close to the speed of light, meaning that it completes an orbit in a matter of minutes. The image was published in May, 2022. (EHT Collaboration)

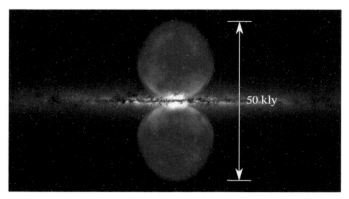

Fig. 18.27 An artist's sketch of the Fermi bubbles of x-ray and gamma-ray radiation stretching a combined distance 50,000 ly above and below the disk of the Milky Way. (Adapted from NASA's Goddard Space Flight Center; labels added)

the technique of very long baseline interferometry (see page 597). The combination of eight telescopes results in what is effectively a radio telescope the size of Earth.

As material gets closer and closer to the event horizon of Sgr A*, it heats up to extreme temperatures, causing it to emit x-rays (local hot spots can be seen as brighter regions in the accretion disk). Magnetic fields associated with the black hole and its accretion disk can generate radio signals like those first detected by Jansky almost a century earlier. The accretion disk and magnetic fields can also channel most of the radiation perpendicular to the disk; something we will come back to in Section 18.5.

In 2010, the orbiting Fermi observatory discovered previously unknown bubbles of extremely high-energy x-ray and gamma-ray emissions that are depicted in Fig. 18.27. The Fermi bubbles were also seen in visible light in 2020. The bubbles are centered on the Milky Way's supermassive black hole and are expanding outward perpendicular to the disk. The full extent of the bubbles is about half of the diameter of the Galaxy's disk. There is some evidence to indicate that the bubbles are the result of a very powerful explosive flare produced by Sgr A* perhaps 3 1/2 million years ago. What did our Galaxy eat at that time? We don't yet know. Science keeps exploring questions.

The Milky Way's Magnetic Field

An important global feature of the Milky Way is its very weak magnetic field, with a typical strength of 0.2–1 nT (roughly 5000 times weaker than Earth's magnetic field). Not only do magnetic fields exist around individual objects like ordinary stars, pulsars, stellar black holes, and Sagittarius A*, but fields also exist along the spiral arms and throughout the stellar halos. Even though it is weak, the Milky Way's magnetic field is still strong enough to play a role in star formation, the motions of cosmic rays, and perhaps the structure and formation of the Galaxy itself. Astronomers still have a long way to go before fully understanding how the complex magnetic field affects the Galaxy.

The Mysterious Dark Matter Halo

Fritz Zwicky (Fig. 18.28), a Bulgarian-born astrophysicist who spent his career at the California Institute of Technology (Caltech), was a flamboyant and brilliant scientist (Fig. 18.28). It was in 1933 that Zwicky first suggested that dark matter might exist. He was studying the speeds of distant galaxies in a large cluster of galaxies (the Coma cluster). After estimating how much mass should be in the cluster based on the matter that he could see, Zwicky realized that there was no way that the cluster should exist, simply because the galaxies were moving faster than the cluster's escape speed as implied by the visible mass; the galaxies should be flying away from one another, not swarming around their mutual center of mass. Zwicky coined the term "dark matter" to account for the matter that he couldn't see but knew had to be there to hold the cluster together.

Zwicky also worked with Walter Baade (1893–1960) in identifying supernovas as a distinct new class of astronomical objects, they proposed that supernovas may be transition types to neutron stars, and that supernovas may be the source of cosmic rays (they are one source). In addition to his astronomical successes, Zwicky was also a talented and daring mountaineer who completed challenging first ascents. He would also occasionally hurl insults, such as referring to a colleague as a "spherical bastard" because the person was a bastard no matter how you looked at him. Zwicky was awarded the Presidential Medal of Freedom in 1949 for his work on rocketry during World War II.

Fig. 18.28 Fritz Zwicky (1898–1974). (Courtesy of The Fritz Zwicky Stiftung)

In the 1970s, Vera Rubin, shown in Fig. 18.29, was measuring the rotation speeds of gas around spiral galaxies. Recalling Kepler's laws, objects orbiting a central mass should have slower orbital speeds, the farther they are from the center. This is exactly how the planets and other objects behave in the Solar System based on their distances from the Sun; of the planets, Mercury has the fastest orbital speed and Neptune has the slowest. Orbits that follow Kepler's laws as derived by Newton are referred to as Keplerian. What Rubin and her team discovered was something quite different.

Figure 18.30 shows a rotation curve of orbital speed versus distance from the center of the Milky Way. In the inner 22 kly of the Galaxy, orbital speeds *increase* with distance, which was expected. This increasing speed is because the amount of mass inside these orbits increases significantly with increasing distance, due to the large sizes of the boxy/peanut bulge, the long bar, the inner portions of the disk, and the stellar and hot corona halos. This type of rotation curve is called rigid-body rotation, because orbits increase in speed from the center much like points on a solid disk increase in speed of rotation the farther the points are from the rotation center.

Fig. 18.29 Vera Rubin (1928–2016) sitting with part of her collection of antique globes.(Photograph by Mark Godfrey, courtesy of AIP Emilio Segrè Visual Archives. Gift of Vera Rubin.)

Between about 22 kly and 30 kly, the orbital speeds start to behave like Kepler's laws predict: the Sun is in this region of the Galaxy's disk at 26 kly from the center. But beyond 30 kly, the orbital speeds stop decreasing and remain almost constant at approximately 225 km/s. The great surprise was that by this distance from the center, almost all of the visible mass is already inside the orbits and so the orbits should be Keplerian. In order to have the flat rotation curve beyond 30 kly there must be *a lot* of Zwicky's dark matter. Between the rotation curve of the Milky Way's disk and the effect that the Milky Way has on very distant stars and globular clusters in the Galaxy, as well as on the motions of neighboring satellite galaxies, astronomers can

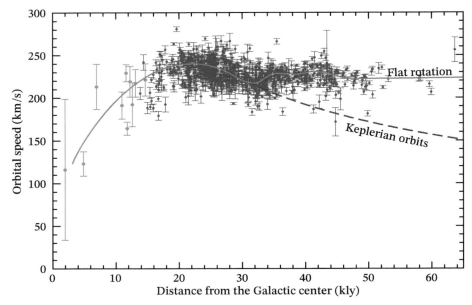

Fig. 18.30 Orbital speeds in the disk of the Milky Way from near the center out to 65 kly (20 kpc). The Sun is the large yellow dot and the dark red dashed curve is what would be expected if all of the mass calculated in Example 18.1 was the vast majority of the Galaxy's mass. The bright red line is an estimate of the actual rotation curve based on the data. The vertical bars indicate the uncertainty in the measurements. (Data courtesy of Reid et al., 2014,[5] and Mróz et al., 2019[6])

Fig. 18.31 The Milky Way, its globular clusters, and satellite galaxies sit inside a sphere of invisible dark matter. [Milky Way art from ESO/NASA/JPL-Caltech/ M. Kornmesser/R. Hurt, CC BY 4.0; background image from NASA, ESA, the GOODS Team and M. Giavalisco (STScI/University of Massachusetts), CC BY 4.0]

deduce the amount of mass and its distribution from the center of the Milky Way outward. The Milky Way has an enormous, unseen dark matter halo that extends across a diameter of 1800 kly (1.8 million light-years). It would take light 1.8 million years to travel from one side of the dark matter halo to the other. The distribution of dark matter is crudely illustrated in Fig. 18.31. The total gravitational mass of the Galaxy is approximately 1.3 *trillion* times the mass of the Sun, or 1.3×10^{12} M_{Sun}. This value is much larger than the combined mass of everything that we can see: 93% of the mass of the Milky Way doesn't give off or absorb any light whatsoever at any wavelength!

At the time this book was written, we don't know what the particles are that make up 93% of the mass of the Milky Way Galaxy! It has been suggested that perhaps dark matter is composed of neutrinos or some other more exotic and as yet undiscovered type of weakly interacting massive particle (WIMP), but we do know today that at least neutrinos aren't massive enough to account for a significant fraction of the dark matter in the Galaxy, or in the rest of the universe (remember that we can detect neutrinos, as discussed in Section 9.7). In order to form a dark matter halo, dark matter particles would have to be *cold*, meaning that they have to be moving slowly relative to the speed of light, as opposed to *hot* (fast moving).

[5]Reid, M. J., Mentien, K. M., Brunthaler, A., Zheng, X. W., et al. (2014). "Trigonometric Parallaxes of High Mass Star Forming Regions: The Structure and Kinematics of the Milky Way," *The Astrophysical Journal*, 783:130–143.

[6]Mróz, Przemek, Udalski, Andrzej, Skowron, Dorota M., Skowron, Jan, et al. (2019). "Rotation Curve of the Milky Way from Classical Cepheids," *The Astrophysical Journal Letters*, 870:L10–L14.

The cold dark matter (CDM) requirement does not support neutrinos as being the mysterious dark matter, because they have extremely low mass and are fast moving.

After WIMPs were suggested as possible dark matter candidates, and before theoretical constraints argued against objects formed from protons, neutrons, and similar particles, scientists suggested that if WIMPs can have their theory, then MACHOs should as well (massive compact halo objects). MACHO candidates would be hard-to-see dim objects like very low-mass main-sequence stars, brown dwarfs, rogue planets, white dwarfs, neutron stars, and black holes. Because they are compact and massive, MACHOs are able to bend light around them and intensify starlight, as predicted by Einstein's general theory of relativity (this is the same phenomenon as gravitational microlensing used in the search for exoplanets, as discussed on page 543). Although there have been MACHO detections, again not nearly enough have been discovered to account for the measured dark matter in the Milky Way Galaxy.

Vera Rubin received many honors for her work on dark matter, including having the Large Synoptic Survey Telescope renamed in her honor in 2020. When completed in 2022, the Vera C. Rubin Observatory on Cerro Pachón in Chile will obtain images of the entire sky visible from that location every few nights. Over the course of its initial 10-year survey, the telescope will image the same locations in the sky about 825 times. One of its major science goals is to study dark matter.

18.3   A Taxonomy of Galaxies

The Great Debate: What Are Those Spiral Nebulas?

Until the middle of the second decade of the twentieth century, astronomers puzzled over the true nature of the "spiral nebulas" that were seen scattered across the sky. One of them, the Andromeda nebula shown in Fig. 18.32, is bright enough to be seen with the naked eye. Charles Messier included the fuzzy object in his catalog as M31 (the brightest entry in his catalog). With the relatively low-resolution images of the day, there were those who suggested that perhaps these spiral nebulas were spinning gas clouds that may be planetary systems in the making. There was also a question as to how far away they are.

Fig. 18.32 An 1899 picture of the Andromeda nebula, M31, by Isaac Roberts (1829–1904). (Public domain)

On April 26, 1920, Harlow Shapley, then at Mount Wilson Observatory, and Heber Doust Curtis (1872–1942) of Lick Observatory, both in California, got together at the National Academy of Sciences in Washington D.C. to have what has become known as the Great Debate regarding the distance question. Shapley believed that the spiral nebulas are a part of the Milky Way. One of his arguments was based on the apparent brightnesses of novas in Andromeda and his significant overestimate of the size of the Milky Way's disk. If Andromeda was as large as the Milky Way it would have to be extremely far away to have the apparent size it has in the sky. But if that were the case then the novas would have to be much brighter than those in the Milky Way in order to have the apparent brightnesses observed. Curtis believed that the spiral nebulas must be outside of the Milky Way. His argument was also based on the apparent magnitudes of novas. For the Andromeda novas to be as faint as they are, they must be at least 500,000 light-years away (150 kpc), which would make Andromeda about the size of Kapteyn's Milky Way. In the end, neither man proved sufficiently convincing to settle the question.

Edwin Hubble and Cepheids in Andromeda

Fig. 18.33 Edwin Powell Hubble (1889–1953) by photographer Johan Hagemeyer (1884-1962). (Public domain)

Edwin Hubble (Fig. 18.33), for whom the Hubble Space Telescope was named, went to the University of Chicago to study law at the request of his father. Hubble was very athletic, and while studying for his baccalaureate degree, he led the University of Chicago's basketball team to its first conference championship in 1907. After graduation, Hubble pursued a master's degree in jurisprudence at the University of Oxford as a Rhodes scholar. But upon his return to the United States, not finding the practice of law particularly enticing, Hubble taught high school mathematics, physics, and Spanish for a time while also coaching basketball. With a life-long interest in astronomy, Hubble decided to enroll again at the University of Chicago, this time to earn a Ph.D. in astronomy while working at the famed Yerkes Observatory, that was operated by the University. Following the completion of his degree in 1917, Edwin Hubble joined the military during World War I, but never saw combat. After the war ended in 1919, Hubble went to work at Mount Wilson Observatory where Shapley was also working, and where Hubble would spend his career until his sudden death in 1953 from a blood clot in his brain. While at Mount Wilson, Hubble had the opportunity to use the world's then-largest telescope and, with its photon-gathering capability, he would make multiple profound contributions to extragalactic astronomy and modern observational cosmology (the study of the creation, evolution, and possible futures of the universe itself).

Thanks to Hubble, the clear answer to the Great Debate question came three years after the debate took place, when he discovered Cepheid variable stars in Andromeda. Although it still wasn't understood that there are two classes of Cepheids, those that are heavy-element rich and those that are heavy-element poor, the distance that was established by using the Cepheid period–luminosity relationship clearly indicated that the spiral nebula of Andromeda is in fact a system of stars, gas, and dust separate from and far beyond the Milky Way Galaxy. The universe was suddenly immensely larger than previously believed. A modern view of Andromeda from combined images of two all-sky surveys is shown in Fig. 18.34 (*Left*) and a small portion of an extremely high resolution Hubble Space Telescope image of Andromeda is displayed in Fig. 18.35.

Hubble's Tuning Fork Diagram

We know today that the Milky Way and Andromeda are just two of perhaps two trillion (2×10^{12}) galaxies in the universe. Galaxies come in all shapes and sizes; there are large spiral galaxies like the Milky Way (barred) and Andromeda (regular), there are enormous elliptical galaxies like M87 [Fig. 18.34 (*Center*)] and much smaller elliptical galaxies, and there are large lenticular galaxies like NGC 5866 [Fig. 18.34 (*Right*)] that have elliptical galaxy shapes along with disks, but no significant spiral structure.

Edwin Hubble made an early attempt at classifying galaxies based on their general characteristics (morphology); see Fig. 18.36. Because of its shape, the scheme came to be called the tuning fork diagram, also known as the Hubble sequence. The tuning fork diagram organizes large galaxies into ellipticals, lenticulars, spirals, and barred spirals. Galaxies that do not fit into one of those categories are grouped into separate classes.

Fig. 18.34 *Left:* The Andromeda Galaxy, M31 (Sb; see Fig. 18.36). Two of its satellite galaxies are also visible: M32 (E2) is below M31's central bulge and M110 (E5) is above and to the right of the bulge. [The Digitized Sky Surveys (NASA/Space Telescope Science Institute) and the Pan-STARRS survey (University of Hawaii). Image downloaded from the Aladin Sky Atlas (Bonnarel, et al., 2000[7])] *Center:* The supergiant elliptical galaxy, M87 (E0, also classified as cD), has ten times as many stars and 100 times as many globular clusters as the Milky Way does. [NASA, ESA and the Hubble Heritage Team (STScI/AURA)] *Right:* NGC 5866 is a lenticular galaxy with a visible dust lane. [NASA, ESA and the Hubble Heritage Team (STScI/AURA)]

Fig. 18.35 A cropped portion of a very high resolution Hubble Space Telescope image of M31, the Andromeda Galaxy. [NASA, ESA, and the Hubble Heritage (STScI/AURA)-ESA/Hubble Collaboration]

Starting from the left-hand side of Fig. 18.36, the elliptical galaxies in the "handle" of the tuning fork are rated according to their degree of flattening, from spherical (E0) to nearly flat (E7). After E7 the sequence "forks" into regular and barred classes. At the fork are the lenticular galaxies; if no bar is present, they are classified as S0, but if a central bar is identified they are SB0. In both the regular and barred spiral branches, galaxies with (a) tightly wound spiral arms, (b) relatively smooth spiral arms, and (c) large central bulges are Sa and SBa, respectively. As the central bulges become smaller and the spiral arms get clumpier and looser, the classes move

[7]Bonnarel, F., Fernique, P., Bienaymé, Egret, D., et al. (2000). "The ALADIN interactive sky atlas," *Astronomy & Astrophysics Supplement Series*, 143:33–40.

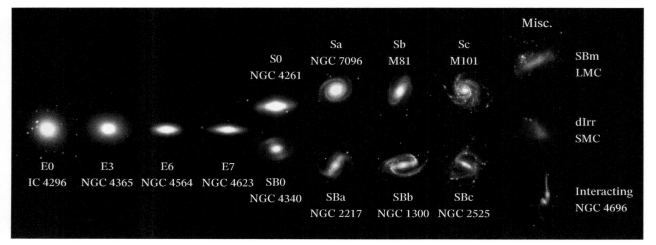

Fig. 18.36 The Hubble tuning fork diagram. [The Digitized Sky Surveys (NASA/Space Telescope Science Institute); NGC 2217 and NGC 4696 from the Pan-STARRS survey (University of Hawaii). All images downloaded from the Aladin Sky Atlas (Bonnarel, et al., 2000)]

Key Point

There is a zoo of galaxy types that don't fit Hubble's tuning fork diagram.

from b, to c, and d (not shown). The Milky Way is classified as SBbc (between b and c) and Andromeda is Sb.

As with any classification scheme that tries to group different types of objects, there always seem to be objects that refuse to conform to our simple categories. In galaxy classifications, many miscellaneous galaxy types do not really fit on the tuning fork diagram. For example, the largest of the Milky Way's satellite galaxies, the Large Magellanic Cloud (LMC), has a discernible bar and a hint of one spiral arm; as such, it is considered to be a barred galaxy of the magellanic type (SBm). The Milky Way's Small Magellanic Cloud (SMC) is a small, irregularly shaped galaxy referred to as a dwarf irregular (dIrr). Other similarly small galaxies exist with more spherical shapes, that are referred to as dwarf spheroidal galaxies (dSph). There are also small, ultra-faint dwarf galaxies (UFD), that are extremely heavy-element poor and are the most common type of satellite galaxy orbiting the Milky Way. At the other end of the size range, very large ultra-diffuse galaxies (UDG) exist, with very few stars and no detectable structure. Then there are galaxies where portions of the galaxies are undergoing episodes of very rapid star formation: such galaxies are known as starburst galaxies. Additionally, there are many gravitationally interacting galaxies that produce significantly distorted objects, because they tug on each other's stars, gas, and dust.

Dark Matter Halos

One thing that virtually all galaxies have in common is a dark matter halo. Our understanding of galaxies today essentially requires that dark matter halos exist and dominate the galaxies' mass while enveloping all of the galaxies' stars, gas, and dust. For all practical purposes, galaxies are essentially dark matter halos salted with stuff that gives off light (meaning anything from radio waves to gamma rays). Giant ellipticals and spirals have enormous dark matter halos. The reason that ultra-faint galaxies or ultra-diffuse galaxies hold together despite few stars and almost no gas and dust is because of the gravitational hold of dark matter. When galaxies interact, they do so under the influence of their dark matter halos. When galaxies merge during collisions, their dark matter halos merge as well. Understanding the nature of dark matter is one of the "Holy Grails" of astrophysics.

General Characteristics of Elliptical and Spiral Galaxies

Elliptical Galaxies

Elliptical galaxies have the largest range in size of any general category of galaxies (see, e.g., Fig. 18.37). The largest galaxies in the universe are the giant and supergiant ellipticals that can have masses as much as 100 times the mass of the Milky Way (10^{14} M_{Sun}). Many of the largest ellipticals have very diffuse stellar halos far from their centers, with the stellar density increasing significantly toward their centers; these galaxies are given the special designation cD, where c means supergiant and D stands for diffuse [M87, Fig. 18.34 (*Center*), is one example of a cD galaxy]. Other elliptical galaxies can be much smaller than the Milky Way.

Fig. 18.37 The elliptical galaxy M60 (E2), and the spiral galaxy NGC 4647 (Sc). [NASA, ESA, and The Hubble Heritage (STScI/AURA)-ESA/Hubble Collaboration]

In general, elliptical galaxies tend to have relatively little cold gas and therefore don't have much star formation occurring today. This is because only atoms, molecules, and dust in cold gas clouds move slowly enough for the clouds to remain confined in local regions and collapse under the force of gravity. The lack of significant star formation means that the vast majority of an elliptical galaxy's most massive, short-lived stars have long since died away, leaving older, redder, and less-massive stars. Because of this, elliptical galaxies are sometimes characterized as being "red and dead." That is not to say that elliptical galaxies don't contain any gas; in fact, elliptical galaxies contain significant amounts of hot gas with temperatures of more than 100,000 kelvins, as evidenced by the emission of x-rays that are always associated with high-energy phenomena such as fast-moving gas particles. The hot gas is usually spread throughout an elliptical galaxy and even beyond its distribution of stars, but still only accounts for a little less than 1% of the galaxy's mass. However, because the gas is so hot, all of that gas is moving too fast to clump gravitationally to form stars.

In addition to an elliptical galaxy's distribution of individual and multiple star systems, it typically also contains a large number of globular clusters. M87, for example, is estimated to possess 15,000 globular clusters, compared to about 150 in the Milky Way.

As we will see in Section 18.5, elliptical galaxies also normally contain one or more supermassive black holes at or near their centers. In some cases these supermassive black holes can have masses that are more than 10,000 times greater than Sagittarius A*, the supermassive black hole that resides at the center of the Milky Way (page 743*ff*).

Spiral Galaxies

The Milky Way is fairly typical of spiral galaxies, both regular and barred. Like the Milky Way, the stellar halos of spiral and barred spiral galaxies are similar to elliptical galaxies, including containing predominately old, red, low-mass stars and hot, x-ray-emitting gas. However, the stellar halos of spiral galaxies differ from

Fig. 18.38 M51, the Whirl-pool galaxy, is a grand-design spiral (Sbc). The red regions in the spiral arms are star-forming clouds of gas and dust. [NASA, ESA, S. Beckwith (STScI), and the Hubble Heritage Team (STScI/AURA)]

ellipticals by having significantly fewer globular clusters, especially compared to the giant and supergiant ellipticals. In addition, as with large ellipticals, spiral galaxies contain supermassive black holes in their centers.

Of course, what distinguishes spiral and barred spiral galaxies from the largely featureless ellipticals are their beautiful spiral arm structures (e.g., Fig. 18.38), and bars in the case of barred spirals. As was pointed out with the Milky Way, it is in the spiral arms where most star formation occurs, as density waves pass through cold gas clouds, triggering their collapse. Because of the star formation, and because the most massive stars burn hot, bright, and blue, the spiral arms stand out. But, as was also pointed out with the Milky Way, older, redder, lower mass, long-lived stars, as well as cold gas and dust, exist between the spiral arms, but are overwhelmed in blue light by the young, massive stars of the arms.

One way in which the Milky Way does differ from many spiral galaxies is its apparent lack of a spheroidal bulge. Instead, the Milky Way has a boxy/peanut bulge, sometimes referred to as a pseudo-bulge. The Milky Way also has a relatively small supermassive black hole at its center when compared to other, similar galaxies; for example, the Andromeda Galaxy has a central supermassive black hole that is estimated to be 35 times more massive than the Milky Way's supermassive black hole.

Spiral galaxies tend to be larger than most galaxies, except for supergiant and some giant ellipticals. The Milky Way and Andromeda are among the largest of the spiral galaxies, but are still considered as intermediate mass when compared to the much larger giant and supergiant ellipticals.

18.4   Galaxies Everywhere

The Local Group

Galaxies are not found in isolation from one another. Instead of moving through space on paths uninfluenced by others, galaxies are gravitationally attracted to one another just like any other massive body. We live in a cluster of galaxies known as the Local Group. By far the largest members of the Local Group are the great spirals of the Milky Way and Andromeda. The two galaxies have similar masses and are gravitationally bound to one another. However, despite similar masses, Andromeda may have in excess of 1600 globular clusters (more than ten times as many as the Milky Way), and perhaps as many as one trillion stars (compared to the Milky Way's 200 billion to 400 billion stars). Figure 18.39 shows the relative locations of the members of the Local Group that are known as of late 2019.

The only other spiral galaxy in the Local Group is the much smaller Triangulum Galaxy (M33; Fig. 18.40). Triangulum is a satellite galaxy of Andromeda, just as the Large and Small Magellanic Clouds are satellite galaxies of the Milky Way. In fact, Triangulum may only be about 20% more massive than the LMC and both galaxies are roughly 10%–15% of the mass of their parent galaxies.

Both large spirals in the Local Group have their own collections of satellite galaxies. It is not yet known how many there are, given that many are ultra-faint dwarfs (UFD) or perhaps ultra-diffuse galaxies (UDG). The Milky Way's dusty disk also makes finding satellite galaxies on the other side of the Galaxy difficult — faint or otherwise. Since 2015, many previously unknown satellite galaxies of Andromeda

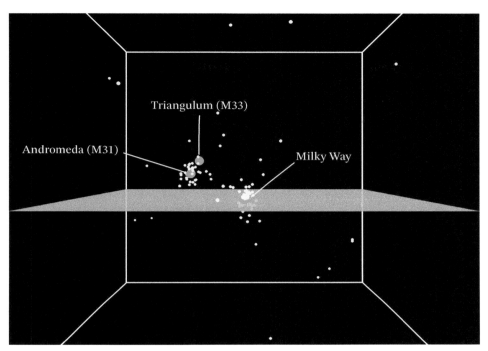

Fig. 18.39 The Local Group of galaxies. Each dot indicates the location of one galaxy; the blue dots represent the three spiral galaxies. The smaller the dot, the farther away the galaxy is from your perspective. The plane passes through the Milky Way's disk.

and the Milky Way have been detected, thanks to all-sky surveys using dedicated telescopes such as the Sloan Digital Sky Survey (SDSS), the Panoramic Survey Telescope and Rapid Response System Observatory (Pan-STARRS), and the orbiting Gaia telescope. In some cases, the newly discovered satellites couldn't have been identified without computer analysis. A sampling of dwarf galaxies orbiting the Milky Way is shown in Fig. 18.41.

As these massive data sets and others are analyzed, it is likely that many more satellite galaxies will be discovered. Even the Milky Way's satellite galaxy, the Large Magellanic Cloud, is known to possess at least five satellite galaxies of its own, the brightest of which is the Small Magellanic Cloud.

Fig. 18.40 The Triangulum Galaxy, M33, with glowing red gas clouds and active star formation. [ESO; CC BY 4.0]

Fig. 18.41 A sampling of dwarf satellite galaxies of the Milky Way. *Left to right:* Leo A (IrrBm), Leo T (UFD), Sculptor dwarf (dSph/E3), Ursa Minor dwarf (dSph/E4), and Phoenix dwarf (dIrr). (PanSTARRS/Aladdin; NASA/ESA Hubble Space Telescope, Hubble Legacy Archive; PanSTARRS/Aladdin; Digitized Sky Surveys/Aladdin; ESO, CC BY 4.0)

Fig. 18.42 *Left:* A portion of the Virgo cluster, whose center is located 54 Mly (16 Mpc) away, in the constellations of Virgo and Coma Berenices. The cD galaxy, M87, from Fig. 18.34 (*Center*), is at lower left. The black dots are where bright foreground stars have been removed. [Chris Mihos (Case Western Reserve University)/ESO; CC BY 4.0] *Right:* Abell 2744 (nicknamed Pandora's Cluster) is located 4000 Mly (1200 Mpc) in the constellation Sculptor. There is a huge number of extremely distant galaxies in the image that are not members of Abell 2744. (NASA/ESA/STScI)

Clusters Upon Clusters of Galaxies

Compared with the Local Group, the Virgo cluster [Fig. 18.42 (*Left*)], is a much larger and richer collection of galaxies. Estimates place the number of galaxies in Virgo at 2000. The mass of the entire cluster exceeds 10^{15} M_{Sun} (1 quadrillion solar masses).

Most of the mass of the Virgo cluster is composed of the mysterious dark matter, as revealed by the motions of the galaxies within the cluster. Some of the remaining mass is contained in the stars, dust, gas, and supermassive black holes of the galaxies, but far more matter exists *between* the galaxies in the form of extremely hot intracluster gas. The mass of the intracluster gas can amount to almost ten times the mass of all the other visible matter in the cluster combined, and can have temperatures of 100 million kelvins. This hot gas is common in clusters of galaxies.

The Virgo cluster's diameter is greater than 14 Mly (4.2 Mpc); light takes 14 million years to traverse the diameter of the cluster. The giant cD galaxy, M87, with its several trillion stars, rules the center of the Virgo cluster. M87 is estimated to have a diameter of 980 kly (300 kpc) and a mass of perhaps 200 times the mass of the Milky Way. As with Andromeda and the Milky Way in the Local Group, the Virgo cluster has several subgroups where some of its galaxies orbit around dominant members.

Abell 2744 [Fig. 18.42 (*Right*)] is a much richer cluster than Virgo and also much farther away. Light from the cluster took 3.4 *billion* years to reach Earth, meaning that the light started on its trip when our Solar System was only one-quarter of its current age. Given the distance and time scales involved, we are now beginning to explore what can be considered as truly deep space and deep time; we are seeing Abell 2744 not as it is today, but as it was 3.4 billion years ago.

Superclusters of Galaxy Clusters

Large galaxies can possess satellite galaxies, and satellite galaxies can have satellite galaxies of their own, but what about going the other direction in a hierarchy of galaxy groupings? Is the Local Group of galaxies a member of a larger cluster of clusters? The answer is yes, and our small Local Group, along with the larger Virgo cluster, are two members of the Local Supercluster of galaxies (also known as the Virgo Supercluster).

The Local Supercluster measures 110 million light-years (34 Mpc) across and contains some 100 groups and clusters of galaxies that include tens of thousands of individual galaxies. The Local Group is located near one edge of the supercluster and the Virgo cluster, one of its largest members, is located at the center of the supercluster. If an obituary had been sent out from Earth to the universe 66 million years ago announcing the loss of the dinosaurs, it would have only had time to traverse 60% of the diameter of the Local Supercluster by now. Observers on the other side of the supercluster won't have a chance to mourn their demise for another 44 million years! There are millions of superclusters in the observable universe.

But the hierarchy doesn't seem to stop there. The overall velocities of the Local Supercluster and other "nearby" superclusters indicate that they are all moving toward a region called the Great Attractor. The Great Attractor turns out to be the center of an even larger structure referred to as Laniakea. The Local Group of galaxies is at the edge of the Local Supercluster, which in turn is one filament toward the edge of Laniakea. Laniakea has an estimated diameter of 520 Mly (160 Mpc) and a mass of 10^{17} M_{Sun} (one hundred thousand trillion, or one hundred quadrillion, times the mass of the Sun)! The name "Laniakea" comes from the Hawai'ian words *lani* for heaven and *akea*, meaning spacious or immeasurable.

These incredibly large collections of galaxies, clusters of galaxies, superclusters of galaxy clusters, and beyond, don't themselves occupy random positions scattered throughout the universe. Instead, they exist in walls that surround enormous, virtually empty, bubble-like voids. These structures of filaments, walls, and voids make up what is collectively called the cosmic web. How all of this structure came to be is a topic we will return to in Chapter 19.

Galaxies in Collision

Andromeda Coming Our Way

From wavelength-shift measurements caused by the Doppler effect, astronomers discovered that the light from the Andromeda Galaxy is blueshifted, meaning that the Milky Way and Andromeda are approaching one another at 300 km/s (670,000 mi/h). Although not on a direct collision course, the two great spirals will side-swipe each other in several billion years and begin a celestial death dance. As depicted in Fig. 18.43, they will interact gravitationally to pull out streams of each others' stars, clusters, gas, and dust, trigger more active star formation, and eventually destroy their iconic spiral structures. As their centers get closer and closer together, they will merge into a single giant elliptical galaxy in about six billion years, with a merged dark matter halo and a combined collection of globular clusters. The collision will ultimately heat what gas didn't form into stars, resulting in a

Key Point
The Milky Way and Andromeda will begin to merge in 4 to 5 billion years.

Fig. 18.43 Two frames extracted from a computer-generated movie of the future merger of the Milky Way and Andromeda that is available on the `Chapter 18` resources webpage. *Left:* The two great spiral galaxies approaching one another as seen from a planet in the Milky Way 3.75 billion years into the future. *Right:* The Milky Way and Andromeda distort one another as they gravitationally interact 4 billion years from now. The merger will be completed in about 6 billion years. [NASA; ESA; Z. Levay and R. van der Marel (STScI); T. Hallas; and A. Mellinger]

hot corona halo that emits x-rays. Further star formation will largely stop, and the combined galaxy will become red and dead. The supermassive black holes of the two original galaxies will also merge into a single, larger, supermassive black hole.

It is extremely unlikely that the Sun and the Solar System will be destroyed by this colossal collision simply because stars are so far apart, making a head-on or near collision between two stars extremely remote. The Sun and its children will probably be thrown into a very large orbit near the outer portion of the new Milky Way/Andromeda mashup, with a very slight possibility that the Solar System could be flung out of the new galaxy entirely. Either way, if there are still Earth inhabitants that far into the future, they will have already had to deal with, or soon will be dealing with, the transformation of the Sun into a white dwarf star. Regardless, the sky will look amazing from anywhere in either galaxy.

However, before that great crash happens, another, smaller collision will occur about 2.4 billion years from now. The Large Magellanic Cloud, along with the Small Magellanic Cloud and the other, smaller satellites of the LMC, are falling into the Milky Way. As a prelude, the Large and Small Magellanic Clouds have already been interacting with one another, producing the Magellanic Stream (Fig. 18.44). When the SMC passed through the LMC during one of its many orbits around the larger galaxy, gas was ripped away through the tug of gravity, assisted by the force produced by increased gas pressure that occurred when their gases collided. Because the LMC is so much more massive than the SMC, the LMC largely won the tug-of-war and so most of the gas is from the SMC. Thanks to the impressive capabilities of Gaia, a small stellar cluster that is only about 117 million years old has been discovered at the head of the Magellanic Stream. Given its age and the motion of the Stream, it seems that the cluster formed the last time the Stream passed through the Milky Way's midplane. Perhaps gas in the Stream was compressed when it interacted with the gas in the thin disk, triggering the star formation. Over time, the LMC, the SMC, and other satellites will simply become a part of the Milky Way Galaxy.

Fig. 18.44 An visible-light all-sky survey combined with a radio map (pink) that shows the Magellanic Stream as an arc at the bottom of the image. Both the LMC and the SMC are visible in white near the head of the stream. (Radio/visible-light image: David L. Nidever et al., NRAO/AUI/NSF and A. Mellinger, LAB Survey, Parkes Observatory, Westerbork Observatory, and Arecibo Observatory; Radio image: LAB Survey)

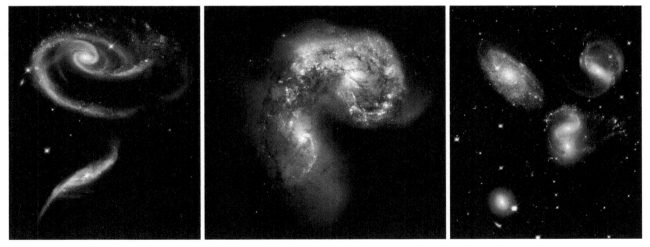

Fig. 18.45 *Left:* The Rose Galaxy. [NASA, ESA, and The Hubble Heritage Team (STScI/AURA)] *Center:* The Antennae Galaxies. [NASA, ESA, and the Hubble Heritage Team (STScI/AURA)-ESA/Hubble Collaboration] *Right:* Stephan's Quintet, (the spiral galaxy in the upper left is a foreground galaxy). (NASA, ESA, and the Hubble SM4 ERO Team)

Other Examples of Galaxy Interactions

With clusters upon clusters of galaxies spread throughout the universe, there are literally trillions of galaxies that are orbiting and gravitationally interacting with other galaxies. With so many chances for interactions, it is hardly surprising that the upcoming Milky Way/Andromeda collision and the currently active infall of the LMC and its satellites into the Milky Way are far from unique. We end this section with Fig. 18.45, a collection of galaxies in collision.

18.5 Supermassive Black Holes and Active Galaxies

It is likely that supermassive black holes live in most, if not all, large galaxies. They reveal their presence by the rapid motions of stars in small volumes at the centers of galaxies and by the radiation those regions emit, including x-rays and radio waves, presumably from infalling matter in accretion disks (recall Fig. 18.26).

Since black holes are small (even supermassive ones) and galaxies are very far away, actually resolving the region around a supermassive black hole is extremely challenging, which means that determining a black hole's size is typically done indirectly (one exception is using the image obtained by the Event Horizon Telescope to measure the angular diameter of Sagittarius A*). One way to make an indirect determination of a supermassive black hole's size is to measure the time scale of brightness changes associated with radiation coming from the center of the galaxy being studied. Since nothing can travel faster than the speed of light, the soonest that any part of the region around a black hole can learn about a change in energy output from another region is when the light reaches it, as illustrated in Fig. 18.46. If that region is to participate in the brightening it must wait for its signal.

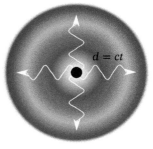

Fig. 18.46 The farthest a signal can travel in time t is $d = ct$, where c is the speed of light.

Example 18.4

Suppose that the x-ray output in a region around a black hole varies over a time span of 100 seconds. Estimate the minimum size of the region involved.

Since the fastest that any signal can travel is c (the speed of light), the farthest the signal can travel in 100 seconds is

$$\text{distance} = \text{speed} \times \text{time} = ct$$

$$= c \times 100 \text{ s} = (3 \times 10^8 \text{ m/\cancel{s}}) \times 10^2 \text{ \cancel{s}}$$

$$= 3 \times 10^{10} \text{ m} = 30 \text{ million km}.$$

For comparison, the radius of the Sun is 0.7 million km (700,000 km) and the radius of Mercury's orbit is 60 million km. The radius of the region is about 40 times larger than the Sun and one-half the size of Mercury's orbit.

Fig. 18.47 The 5000-ly-long jet of M87. [NASA and the Hubble Heritage Team (STScI/AURA)]

Supermassive Black Holes and Jets

In 1918, Heber Curtis was observing the supergiant elliptical galaxy, M87, when he noted a "curious straight ray ... apparently connected with the nucleus by a thin line of matter." If you studied the image of M87 carefully in Fig. 18.34 (Center) you may have noticed that "curious straight ray." The jet of material is not unlike the jets we encountered during the formation of stars (Herbig–Haro objects; Fig. 16.16) but much, much larger and far more energetic. A higher resolution image of the central region of M87 and its 5000-light-years-long jet can be seen in Fig. 18.47. A time-elapsed movie of the changing knots seen in the jet over 13 years of Hubble Space Telescope observations are available on the Chapter 18 resources webpage. M87 is also known as Virgo A because it was the first radio source detected in that constellation.

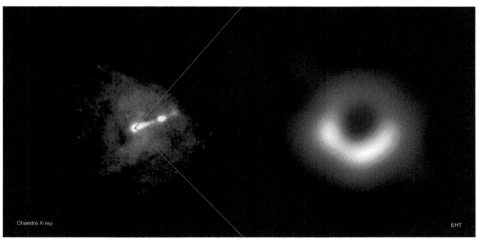

Fig. 18.48 *Left:* The core of M87 seen in x-rays by Chandra. *Right:* A closer view with the Event Horizon Telescope in radio wavelengths. (X-ray: NASA/CXC/Villanova University/J. Neilsen; Radio: Event Horizon Telescope Collaboration)

Diving deeper into M87 using the Chandra X-Ray Observatory, astronomers were given a hint as to the origin of the jet. Figure 18.48 (*Left*) is a view of the central region of the cD galaxy. The jet is visible leaving the vicinity of the tiny black dot in the middle of the image. Figure 18.48 (*Right*) is a radio view of the very center of the galaxy obtained by the Event Horizon Telescope (EHT). You are seeing the shadow of the supermassive black hole that resides at the heart of M87. According to the results of the EHT, M87*, the supermassive black hole, has a mass of 6.5 billion times the mass of the Sun (6.5×10^9 M_{Sun}), or 1500 times greater than the mass of Sagittarius A*.

M87 certainly doesn't have the only galaxy-sized jet out there. Jets can be seen in Fig. 18.49 shooting out in opposite directions from the unusual elliptical galaxy, Centaurus A (also a radio source). The elliptical galaxy is unusual because it has a very well-defined dust lane across its middle. More important for this discussion are the two jets leaving the region of its supermassive black hole at close to the speed of light. Although the x-ray jets are thousands of light-years long, the radio-wavelength jets stretch for millions of light-years! The mass of Cen A's supermassive black hole is about 55 million solar masses, or about 13 times greater than Sagittarius A*.

A video of the M87 jet is available from the Chapter 18 resources page

Active Galaxies

In the 1950s, radio sources were discovered that appeared to be at enormous distances from Earth. Optical views of the sources showed them to be point-like in size from our distant viewpoint. They were also discovered to vary in luminosity on very short time scales. Using calculations similar to Example 18.4, the short time scales imply that their sizes are remarkably small for such strong energy sources. These mysterious objects were dubbed quasi-stellar radio sources, or quasars. Later, more powerful optical observatories such as the Hubble Space Telescope revealed that they live in the centers of very young galaxies. Since the discovery of quasars, other types of galaxy were discovered that exhibited various degrees of short-time-scale

Fig. 18.49 Centaurus A with its characteristic dust lanes. The blue light is x-ray light (high-energy photons) and the orange color is at millimeter wavelengths (low-energy photons). [X-ray: NASA/CXC/CfA/R.Kraft et al.; Submillimeter: MPIfR/ESO/APEX/A.Weiss et al.; Optical: ESO/WFI]

activity. Today they are all understood to be examples of a general class, termed active galaxies.

Essentially, an active galaxy has a thick, clumpy, doughnut-shaped torus of material that feeds a supermassive black hole, possibly through a thin accretion disk, as depicted in the artist's interpretation in Fig. 18.50. The rapid changes in luminosity come from clumps of material falling into the black hole. This central region of an active galaxy is known as an active galactic nucleus (AGN). Various types of active galaxies exist, because of differing sizes and distances of the toruses from the supermassive black holes, the presence or absence of accretion disks, viewing angles of the toruses relative to the supermassive black holes, and the rates at which material is falling into the black holes. Quasars are situations where the jets produced by the black holes and their accretion disks are pointing directly toward Earth. An artist's depiction of a quasar is shown in Fig. 18.51.

Remember that looking deep into space is equivalent to looking far back in time (page 608). The photons from the most distant quasars that we see today were emitted 13 billion years ago, meaning that we see them as they were when the universe was only about 700 million years old. At the great distances of quasars, galaxies were in their earliest days, and so a lot of material was falling onto them and their central supermassive black holes, causing the quasars to blaze away. The number of quasars decreases in regions of space closer to us and not as far back in time.

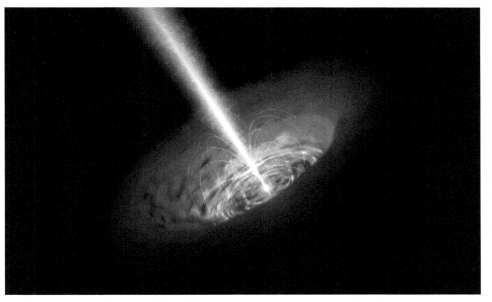

Fig. 18.50 Jets produced by material falling onto an accretion disk that spirals into the super-massive black hole at the center. A thick torus of material feeds the accretion disk that has a radius smaller than the inner radius of the torus, and the accretion disk in turn feeds the supermassive black hole. The jet is beamed because of the confinement caused by the torus and the strong magnetic fields associated with the spinning black hole. (ESO/L. Calçada; CC BY 4.0)

Fig. 18.51 An artist's depiction of a distant quasar. (ESO/M. Korn-messer; CC BY 4.0)

Because quasars are so far away, after removing the effects of the rotation of Earth, the motion of Earth around the Sun, the movement of the Sun around the center of the Galaxy, and the velocity of the Milky Way through the Local Group, it is impossible to detect any motions of quasars perpendicular to our line of sight (remember that it is always possible to measure radial velocities using the Doppler effect). This means that quasars can be very reliably used as fixed markers in space, relative to which very precise positions and motions of other celestial objects can be measured. The International Celestial Reference System (ICRAS) is a coordinate system based on the positions of hundreds of quasars. Quasars can even be used to measure the rate of slowing of Earth's spin or the shift of tectonic plates, through the technique of very long baseline interferometry (page 597).

Quasars can also be used as probes of the intergalactic medium, which is gas found between clusters and superclusters of galaxies that hasn't yet collapsed into galaxies. As a quasar's light passes through cold clouds of hydrogen gas, absorption lines appear as the light excites the hydrogen atoms (recall the discussion of absorption and emission lines in Section 8.6). It is common to see the Lyman alpha (Lyα) line of hydrogen when an electron absorbs a photon and jumps from the ground state ($n = 1$) to the first excited state ($n = 2$); see Exercise 8.58. With many clouds of cold hydrogen having different redshifts, a large number of closely spaced lines can be seen against the background spectrum; this collection of Lyman alpha lines across wavelength is referred to as the Lyman alpha forest.

The Long Arc of History — Gravitational Lensing

As we discuss the universe in deep space and deep time, Einstein's general theory of relativity plays a crucial role, and not just with the supermassive black holes found at the hearts of large galaxies. During the discussion on searching for exoplanets, one of the discovery methods involved looking for temporary increases in brightness due to gravitational microlensing resulting from curved spacetime around the exoplanets (page 543). Gravitational lensing also happens on much larger scales. Figure 18.52 (*Left*) is a remarkable image of a single quasar seen four times because it happened to fall directly behind a galaxy that is much closer to us. We know that the four objects forming the "Einstein cross" are the same quasar, because they are all moving away from us with exactly the same speed, their spectral lines are the same, and their light variations happen the same way, although not at exactly the same time due to different distances traveled by the light associated with each image of the quasar around the closer galaxy.

Figure 18.52 (*Right*) shows Abell 370, a cluster of galaxies containing hundreds of members. However, there are also an enormous number of galaxies lying behind Abell 370. The paths of the light from those background galaxies traversing the cluster and its dark matter halos are bent as the light follows the straightest possible path through curved spacetime. Unlike the Einstein cross where the alignment is almost perfect, the background galaxies are not aligned with the center of some nearly spherical mass distribution, and so the galaxy images are stretched into arcs. The misshapen background galaxy images are analogous to aberrations in optical lens systems.

Fritz Zwicky first suggested that galaxies could act as gravitational lenses.

Fig. 18.52 *Left:* The Einstein cross in Pegasus. A quasar as it appeared 8 billion years ago is seen four times thanks to a relatively nearby galaxy directly in front of the quasar that gravitationally focused its light. (NASA, ESA, and STScI) *Right:* The galaxy cluster Abell 370 in Cetus is at a distance of 6.4 billion light-years and contains hundreds of galaxies. There are numerous arcs in the image due to gravitational lensing of galaxies behind the cluster. [NASA, ESA, and J. Lotz and the HFF Team (STScI)]

18.6 How to Build a Galaxy

The Age(s) of the Milky Way Galaxy

Some estimates place the evolutionary age of the Galaxy's oldest, most heavy-element poor stars at close to 13.5 billion years. If true, this age is only about 300 million years younger than the entire universe (the evolution of the universe itself will be discussed in Chapter 19). This doesn't represent the age of the entire Galaxy, however; in fact, the Milky Way, with its halos, globular clusters, open clusters, boxy/peanut bulge, bars, and disks has been under construction ever since those oldest stars formed. Even today, our great spiral galaxy is growing more massive by adding stars, gas and dust, and dark matter.

Key Point
The oldest stars in the Galaxy may be 13.5 billion years old.

Excluding individual stars, it is generally believed today that the oldest components in the Galaxy are the globular clusters that are members of its boxy/peanut bulge and its inner stellar halo. As was discussed in Section 16.5, because it is possible to compare theoretical evolutionary tracks on the Hertzsprung–Russell (H–R) diagram with the observed turn-off points of color–magnitude diagrams (e.g., Fig. 16.36), astronomers can effectively use all of the stars to determine a single age for the cluster. In the case of globular clusters in the boxy/peanut bulge and in the inner stellar halo, they are typically 12–13 billion years old. The much larger, outer stellar halo is younger overall than the inner stellar halo, having an average age of 11 billion years, but there are also streams of gas and groups of stars that are sometimes much younger (recall the Magellanic Stream and its 117-million-year-old cluster in Fig. 18.44 as one example).

Key Point
The boxy/peanut bulge and the inner stellar halo are the Galaxy's oldest components at 12–13 Gy. The outer stellar halo is typically 11 Gy but also contains some far younger objects.

Although it is younger than the inner stellar halo, the age of the thick disk is between 8.7 and 9.9 billion years old. The thin disk, on the other hand, contains more than four times as many stars as the thick disk, and the thin disk's stars are also more heavy-element rich. The age of the thin disk is estimated to be between 6.8 and 8.2 billion years old and may have subcomponents that are even younger. It is also important to remember that the thin disk includes most of the Galaxy's cold gas, dust, and current star formation. Compared to the thin disk, the boxy/peanut bulge and the inner stellar halo are roughly twice as old.

Key Point
The Milky Way's thick disk is older than the thin disk.

Finally, the Milky Way's long bar seems to be much younger than any of the other components of the Milky Way; it may be younger than 3 billion years and is certainly younger than 6 billion years. However, the fact that the long-bar structure is relatively young doesn't necessarily mean that all of its stars are that young. The long bar developed over time from gravitational instabilities in the rotating disks that naturally create bars. The long bar can also draw gas and dust from the thin disk inward into the region of the boxy/peanut bulge, resulting in new star formation. This means that these new stars find themselves mixed in with some of the oldest stars in the Galaxy.

Key Point
The long bar is the most recent major structure that formed in the Milky Way.

Given that different components of the Galaxy have vastly different ages, it is clear that the Milky Way didn't develop all at once, but instead formed and continues to evolve over time.

Dark Matter, the First Stars, and Primeval Galaxies

Key Point
Small dark matter halos attracted gas in the early universe.

If we are to understand how galaxies of all types are assembled over time, we must first consider their dominant dark matter halos. As will be discussed in more detail in the last chapter, it is thought that small clumps of dark matter formed that then attracted gas, causing it to fall into the center of the dark matter clump. There was only hydrogen and helium (and a very tiny amount of lithium) in the earliest years of the universe and therefore there couldn't have been any dust (which requires carbon, silicon, or other heavier elements). As the clumps of dark matter merged together and more gas fell into them, the first stars were born, made entirely of hydrogen and helium.

Key Point
The first stars were massive, short lived, and made entirely of hydrogen and helium.

Our best understanding of these earliest stars suggests that they were probably massive, being 100 M_{Sun} or more. Massive stars evolve very quickly while converting their hydrogen and helium into heavier elements, and they end their lives in core-collapse supernovas. The high-speed supernova remnants then enrich their local dark matter halos with elements heavier than hydrogen and helium, from which the next generation of stars formed. If some dark matter halos didn't have enough mass, much or all of the ejected material from the supernovas could be traveling faster than the escape speed of the halo, and could even push some of the remaining gas out as well. Even with newly created heavy elements that did not escape their halos, the next few generations of stars would still be extremely heavy-element poor, but they could also be less massive. Stars that formed with masses less than about 0.8 M_{Sun} age so slowly that they are still around today. The oldest stars in the Milky Way may be these early generation stars that formed in small dark matter halos.

In the early universe, when small dark matter halos were close to one another, enough of them could merge with their slightly enriched gases to produce thousands of stars in these primeval galaxies. The ultra-faint dwarf galaxies with their very heavy-element poor stars may be fossils from that earliest phase of galaxy formation.

Key Point
Galaxies in the early universe grew rapidly through mergers and the accumulation of gas.

Many of the small, young galaxies in close proximity to one another would have grown quickly through mergers. By combining their dark matter halos and adding stars that had already formed, along with gas, and now some dust, the masses of galaxies would increase rapidly, allowing them to pull in other gas that had not yet settled into other galaxies, or that had been ejected by supernovas. Figure 18.53 shows the most distant galaxy ever discovered at the time this textbook was written. Given its immense distance, this means that we are also seeing GN-z11 when it was very young, only 400 million years after the universe began. This primeval galaxy is about 1/25 of the Milky Way's size and only 1% of our Galaxy's mass. Still in its extreme youth, GN-z11 is ablaze with hot, massive, bright blue stars, and is making stars 20 times faster than the Milky Way is producing stars today.

GN-z11 is the result of many prior mergers, with little time to relax into a typical shape that we see in present-day galaxies. As it continues to undergo mergers with smaller, young galaxies, it may evolve into a spiral galaxy if its dust and gas settle into a disk. Stars that have already formed may find themselves in random orbits, resulting in an inner stellar halo and central spherical bulge, or possibly a boxy/peanut pseudo-bulge. Gas heated by supernovas and the fast winds of massive stars would lead to an x-ray-producing hot corona.

Fig. 18.53 The primeval galaxy, GN-z11, located in Ursa Major. The light from the galaxy had been traveling for 13.4 billion years before it was captured by the Hubble Space Telescope; we are seeing GN-z11 as it appeared just 400 million years after the Big Bang. The overall image contains tens of thousands of galaxies. [NASA, ESA, and P. Oesch (Yale University); CC BY 4.0]

If, instead of absorbing much smaller, evolving galaxies, GN-z11 collides with another similarly sized young galaxy, the collision between their gas clouds could trigger more star formation, heat much of the low-density gas to millions of degrees, and destroy any forming disks, leaving an elliptical galaxy instead of a spiral galaxy. As the massive stars formed from the collision of gas clouds explode, their energy also heats up the new galaxy's gas, adding to a hot corona halo. With little to no cold gas remaining, star formation ends and galaxy becomes "red and dead."

Over time, very young galaxies like GN-z11 may continue to grow into large elliptical or spiral galaxies, or they may remain much smaller, like dwarf spheroidal galaxies or ultra-faint dwarfs. They would find themselves as parts of clusters of galaxies and superclusters of galaxies, with the smaller galaxies orbiting their more dominant members. If merging dark matter halos with few stars and gas merge together, they may produce the ultra-diffuse galaxies seen in the nearby universe today. Figure 18.54 shows an artist's impression of the mega-merger of young galaxies in the protocluster SPT2349-56, 1.4 billion years after the beginning of the universe.

As smaller galaxies continue to orbit their much larger hosts, they will repeatedly pass through the hot corona halos of the larger galaxies. Gas in these smaller galaxies will be stripped away and their dark matter halos will become part of the much more massive halos of their hosts. Stars from the smaller galaxies will also end up orbiting in the outer stellar halos of the larger galaxies. Most large, nearby galaxies show evidence of galactic cannibalism (when a large galaxy eats its own), including streams of gas and globular clusters. One example within the Milky Way Galaxy is the long and massive Magellanic Stream shown in Fig. 18.44.

Even today, if two relatively gas-rich spiral galaxies collide, the resulting chaos will destroy their beautiful spiral structures, drive star formation, heat their remaining cold gas, and produce a giant elliptical galaxy. Such is the future fate of our Milky Way Galaxy and the Andromeda Galaxy, beginning in four to five billion years.

As galaxies collide and merge, so too do their supermassive black holes. Three supermassive black holes were caught in the act of merging within the colliding

Key Point
Colliding spiral galaxies are one path to creating giant elliptical galaxies.

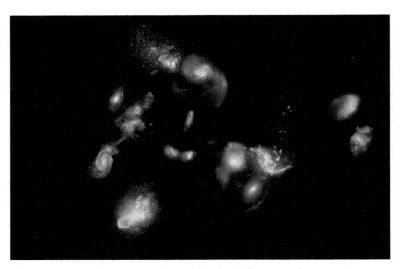

Fig. 18.54 An artist's depiction of merging galaxies in the protocluster SPT2349-56, 12.4 billion years ago. The drawing is based on observations made in the southern constellation, Phoenix, by the ALMA and APEX telescopes. (ESO/M. Kornmesser; CC BY 4.0)

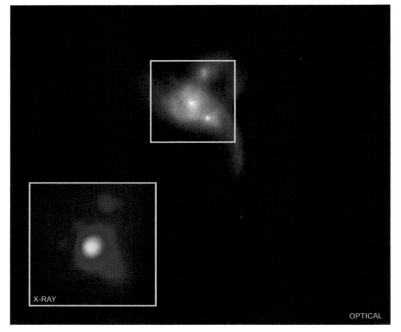

Fig. 18.55 Three merging supermassive black holes from colliding galaxies caught in the act in these optical and x-ray (inset) images of SDSS J0849+1114. (X-ray: NASA/CXC/George Mason Univ./R. Pfeifle et al.; Optical: SDSS & NASA/STScI)

galaxy system of SDSS J0849+1114; see Fig. 18.55 The merging system will ultimately result in a single, larger supermassive black hole embedded in a giant elliptical galaxy. SDSS J0849+1114 was first identified as a possible supermassive black hole merger candidate by amateur citizen scientists participating in the Galaxy Zoo project (`https://www.zooniverse.org/projects/zookeeper/galaxy-zoo/`).

Building galaxies is a messy and sometimes violent process that continues to the present day. We see evidence throughout the universe from the distant past to the present. How it all started is the topic of our final chapter.

A great many people have made important contributions to our understanding dark matter and the evolution of galaxies, including Sandra Faber (Fig. 18.56). Her work has included identifying an important relationship between galaxy brightness and the rotation speeds of spiral disks, studying how dark matter leads to the formation of galaxies, and that dark matter must be composed of "cold" particles, which is now considered to be the cold dark matter (CDM) paradigm. In 1972, Faber became the first woman on the staff of Lick Observatory of the University of California, which began operations in 1888. Faber and her team also helped to analyze the problem with the Hubble Space Telescope's main mirror after launch and determined that it had spherical aberration.

Fig. 18.56 Sandra Faber (b. 1944) receiving the National Medal of Science from President Barack Obama in 2013. (Public domain)

Exercises

All of the exercises are designed to further develop *your understanding* of the material by thinking carefully and critically about what you have read in this chapter. Answers to selected exercises are available in the back of the book.

True/False

1. There are an estimated 2 trillion galaxies in the universe.
2. (*Answer*) Kapteyn estimated the number of stars in the Milky Way Galaxy to be four to eight times greater than the current estimate.
3. Harlow Shapley used the distribution of globular clusters to estimate the size of the Galaxy.
4. (*Answer*) Dust in the disk of the Milky Way didn't play an important role in Shapley's determination of the size of the Milky Way Galaxy.
5. The midplane of the Milky Way is 37° from the north celestial pole.
6. (*Answer*) The corona halo is composed of ionized gas with a temperature of about 2×10^6 K.
7. If an astronomer from another galaxy trained its telescope on the Milky Way, it would see most of the Galaxy's light coming from the boxy/peanut bulge, the long bar, and the thin disk.
8. (*Answer*) There is a clearly defined outer edge to the Milky Way's thin and thick disks.
9. The Sun is a member of the thin disk.
10. (*Answer*) The structure of the Milky Way's long bar is due to a spiral density wave.
11. All stars in the thin disk are found in spiral arms.
12. (*Answer*) The vast majority of all the dust and cold gas in the Milky Way is found in the thin disk.
13. Spiral arms are most prominent in infrared light.
14. (*Answer*) Density waves eliminate the winding problem.
15. At its closest approach to Sgr A*, S2 was traveling at about 2.5% of the speed of light.

16. (*Answer*) Dark matter emits electromagnetic radiation at radio wavelengths that we are unable to see with the human eye.
17. The tuning fork diagram is used to classify galaxies according to their state of evolution, from round ellipticals to loose spirals.
18. (*Answer*) The Milky Way Galaxy is classified as SBbc.
19. Elliptical galaxies are generally "red and dead."
20. (*Answer*) Dark matter halos, like the one surrounding the Milky Way, are relatively rare.
21. The type of galaxy that is most common among the Milky Way's satellites is the barred galaxy of the magellanic type (SBm).
22. (*Answer*) Spiral galaxies tend to be larger and more massive than most galaxies, but are smaller than many giant ellipticals and all supergiant ellipticals.
23. Andromeda's spiral satellite, Triangulum (M33), and the Milky Way's LMC are similar in size.
24. (*Answer*) The Small Magellanic Cloud is a satellite galaxy of the Large Magellanic Cloud, which in turn is a satellite galaxy of the Milky Way Galaxy.
25. Many of the ultra-faint dwarf galaxies orbiting the Milky Way were discovered in automated all-sky surveys with the aid of computer analysis.
26. (*Answer*) The Local Supercluster contains five clusters of galaxies.
27. The Local Supercluster is the largest identifiable structure in the universe.
28. (*Answer*) Laniakea is the entire collection of superclusters of galaxies throughout the universe.

29. The Milky Way Galaxy and the Andromeda Galaxy will eventually merge to form a giant elliptical galaxy.
30. (*Answer*) The Milky Way is still growing by devouring smaller satellite galaxies.
31. The Magellanic Stream is a ribbon of gas that was stripped from the Small Magellanic Cloud when it passed through the Large Magellanic Cloud.
32. (*Answer*) The only way to determine the mass of a galaxy's supermassive black hole is to see it with a telescope.
33. The mass of M87* is more than one thousand times greater than Sagittarius A*.
34. (*Answer*) Quasars are produced by active galactic nuclei.
35. Because of their extreme distances, quasars appear to be very old.
36. (*Answer*) For all practical purposes, quasars appear as fixed points in space.
37. The Lyman alpha forest is a set of closely spaced absorption lines produced by light passing through many cold hydrogen gas clouds having differing redshifts.
38. (*Answer*) The entire Milky Way Galaxy formed from the collapse of an enormous cloud of hydrogen and helium 300 million years after the birth of the universe.
39. The merger of dark matter halos plays a critical role in galaxy building.
40. (*Answer*) Galaxies grow through galactic cannibalism.

Multiple Choice

41. It is estimated that there are _____ galaxies in the universe.
    (a) 2×10^6 (b) 2×10^8 (c) 2×10^9 (d) 2×10^{12}
    (e) 2×10^{15}
42. (*Answer*) Kapteyn _____ the size of the Milky Way and Shapley _____ the Galaxy's size.
    (a) underestimated, also underestimated
    (b) underestimated, overestimated
    (c) overestimated, underestimated
    (d) overestimated, also overestimated
    (e) none of the above; they were both remarkably accurate
43. Ignoring any effects of dust, if a heavy-element rich Cepheid and a heavy-element poor Cepheid have the same apparent magnitude and the same pulsation period, then the heavy-element poor Cepheid must be _____ and have an absolute magnitude that is numerically _____ the absolute magnitude of the heavy-element rich Cepheid.
    (a) closer, less than   (b) closer, greater than
    (c) farther away, less than (d) farther away, greater than
44. (*Answer*) The zone of avoidance is
    (a) along the plane of the Milky Way.
    (b) a region of the sky where observation of distant objects is difficult due to obscuring dust.
    (c) completely devoid of globular clusters.
    (d) perpendicular to the plane of the Galaxy, where there are far fewer stars.
    (e) a region on Earth where nighttime observing is very poor.
    (f) (a) and (b) only
    (g) (a), (b), and (c) only
    (h) none of the above
45. It is estimated that the Milky Way contains how many stars?
    (a) 200 thousand–400 thousand
    (b) 200 million–400 million

(c) 200 billion to 400 billion
(d) 200 trillion–400 trillion
(e) 200 quadrillion–400 quadrillion
46. (*Answer*) Astronomers have concluded that the Milky Way contains
    (a) an inner stellar halo.   (b) an outer stellar halo.
    (c) a hot corona halo.   (d) a dark matter halo.
    (e) a cold gas halo.   (f) a dust halo.
    (g) all of the above   (h) (a)–(d) only
    (i) (a)–(e) only
47. The thin disk is _____, contains _____ stars, and its stars are generally more _____ than stars in the thick disk.
    (a) brighter, more, heavy-element rich
    (b) dimmer, more, heavy-element rich
    (c) brighter, fewer, heavy-element rich
    (d) dimmer, fewer, heavy-element rich
    (e) brighter, more, heavy-element poor
    (f) dimmer, more, heavy-element poor
    (g) brighter, fewer, heavy-element poor
    (h) dimmer, fewer, heavy-element poor
48. (*Answer*) The Sun is _____ light-years from Sgr A*.
    (a) 6000   (b) 10,000   (c) 26,000   (d) 100,000
    (e) 260,000
49. The wavelength of the photon that is emitted when an electron in a hydrogen atom "spin-flips" relative to the atom's proton is about
    (a) 21 μm   (b) 21 mm   (c) 21 cm   (d) 21 m
    (e) 21 km
50. (*Answer*) Massive main-sequence stars are found in the spiral arms of galaxies, while low-mass main-sequence stars can be found either in spiral arms or between them in the disk. This is because
    (a) massive stars form predominantly in spiral arms.
    (b) massive stars die before they have a chance to move out of spiral arms.
    (c) low-mass stars have very long lifetimes, giving them plenty of time to move through and between spiral arms.
    (d) all of the above
51. The central bulge of the Milky Way has
    (a) a spiral pattern.   (b) a spherical shape.
    (c) a short bar.   (d) a boxy/peanut shape.
    (e) all of the above   (f) (b) and (c) only
    (g) (c) and (d) only
52. (*Answer*) The size of the Milky Way's supermassive black hole (Sgr A*) is
    (a) larger than a major city but smaller than Earth.
    (b) larger than Earth but smaller than Jupiter.
    (c) larger than Jupiter but smaller than the Sun.
    (d) larger than the Sun but smaller than Mercury's orbit.
    (e) larger than Mercury's orbit but smaller than Neptune's orbit.
    (f) about the size of Sedna's orbit.
53. The dark matter in the Milky Way makes up more than _____ of the Galaxy's mass. (Caution: Think carefully about your answer.)
    (a) 1%   (b) 2%   (c) 4%
    (d) 5%   (e) 50%   (f) 90%
    (g) all of the above   (h) none of the above

54. (*Answer*) Flat rotation curves for spiral galaxies, including the Milky Way, are evidence
    (a) of the existence of dark matter.
    (b) of how matter is distributed throughout the galaxies.
    (c) of a large amount of mass beyond the disks of spiral galaxies.
    (d) that most matter in spiral galaxies is concentrated toward the galaxies' centers.
    (e) all of the above
    (f) (a), (b), and (c) only
    (g) (b), (c), and (d) only

55. After the Milky Way's dark matter, which component contains the most mass?
    (a) stars   (b) cold gas and dust
    (c) hot gas   (d) Sagittarius A*

56. (*Answer*) Dark matter is probably composed of
    (a) dim, red dwarf main-sequence stars.
    (b) white dwarf stars.
    (c) neutron stars.
    (d) black holes.
    (e) neutrinos.
    (f) all of the above
    (g) none of the above

57. Who first determined unambiguously that "spiral nebulas" are located outside of the Milky Way?
    (a) Harlow Shapley   (b) Heber Curtis
    (c) Edwin Hubble   (d) Vera Rubin
    (e) Sandra Faber

58. (*Answer*) An SBa galaxy has
    (a) a large bulge or boxy/peanut bulge.
    (b) tight spiral arms.
    (c) smooth spiral arms.
    (d) all of the above
    (e) (a) and (b) only

59. A supergiant elliptical galaxy with a dense core and diffuse stellar halo is classified as
    (a) cD   (b) E5   (c) S0   (d) SBm
    (e) UDG   (f) UFD

60. (*Answer*) The most common type of galaxy orbiting the Milky Way is
    (a) cD   (b) E5   (c) S0   (d) SBm
    (e) UDG   (f) UFD

61. Giant and supergiant elliptical galaxies
    (a) are typically "red and dead."
    (b) can contain tens of thousands of globular clusters.
    (c) contain a lot of hot gas.
    (d) have one or more supermassive black holes at or near their centers.
    (e) contain a lot of cold gas and dust.
    (f) can show a lot of star formation.
    (g) all of the above
    (h) (a), (b), and (c) only
    (i) (a), (b), (c), and (d) only
    (j) (d), (e), and (f) only
    (k) (c), (d), (e), and (f) only

62. (*Answer*) Large spiral galaxies like the Milky Way and Andromeda
    (a) are typically "red and dead."
    (b) usually contain tens of thousands of globular clusters.

(c) contain a lot of hot gas.
(d) have one or more supermassive black holes at or near their centers.
(e) contain a lot of cold gas and dust.
(f) can show a lot of star formation.
(g) all of the above
(h) (a), (b), and (c) only
(i) (a), (b), (c), and (d) only
(j) (d), (e), and (f) only
(k) (c), (d), (e), and (f) only

63. Which of the following is *not* a member of the Local Group?
    (a) Large Magellanic Cloud   (b) Leo T
    (c) M31   (d) M32
    (e) M33   (f) M87
    (g) M110   (h) Milky Way Galaxy
    (i) Small Magellanic Cloud   (j) Triangulum Galaxy

64. (*Answer*) After dark matter, the second most massive component in the Virgo cluster is
    (a) cold gas and dust.   (b) supermassive black holes.
    (c) stars.   (d) intracluster gas.
    (e) rogue planets.

65. The mass of the Virgo cluster is about _____ times the mass of the Milky Way.
    (a) 10   (b) 100   (c) 1000   (d) 10,000
    (e) 100,000

66. (*Answer*) The Local Supercluster measures _____ across.
    (a) 14 Mly   (b) 66 Mly   (c) 110 Mly   (d) 220 Mly
    (e) 3400 Mly

67. The Great Attractor is
    (a) a mysterious mass that is pulling the Andromeda Galaxy and the Milky Way together on a collision course.
    (b) a region of space toward which the Local Group, the Virgo cluster, the Local Supercluster, and other superclusters are moving.
    (c) the center of Laniakea.
    (d) all of the above
    (e) (a) and (b) only
    (f) (b) and (c) only

68. (*Answer*) The order of structure sizes from smallest to largest is:
    (a) Solar System, Milky Way, Local Group, Local Supercluster, Laniakea, cosmic web.
    (b) Solar System, Local Group, Milky Way, Andromeda cluster, Laniakea, Local Supercluster, cosmic web.
    (c) cosmic web, Solar System, Local Group, Milky Way, Andromeda cluster, Local Supercluster, Laniakea.
    (d) cosmic web, Solar System, Milky Way, Local Group, Local Supercluster, Laniakea.
    (e) Solar System, Milky Way, Local Group, Great Attractor, Local Supercluster, Laniakea.

69. Quasars are
    (a) one type of active galactic nucleus.
    (b) at extreme distances.
    (c) apparently very young relative to the age of the universe.
    (d) essentially fixed points in space.
    (e) all of the above
    (f) (a) and (b) only
    (g) (c) and (d) only

70. (*Answer*) Suppose that one of the images of the quasar in the Einstein cross in Fig. 18.52 changes brightness exactly thirty days before a second image does. How much farther did the light travel for the second image than for the first? (You should be able to select the correct answer without any calculations. Hint: What is the definition of one light-year?)
    (a) 0.0082 ly   (b) 0.082 ly   (c) 0.82 ly
    (d) 8.2 ly   (e) 82 ly
71. The visible arcs in Fig. 18.52 (*Right*) are due to
    (a) gravitationally interacting galaxies that distort each other.
    (b) gravitational lensing of distant background galaxies.
    (c) irregularly shaped galaxies in the early universe.
    (d) all of the above
72. (*Answer*) The relative ages of the components of the Milky Way Galaxy from oldest to youngest are
    (a) dark matter halo, inner stellar halo and boxy/peanut bulge, outer stellar halo, thick disk, thin disk, long bar.
    (b) inner stellar halo and boxy/peanut bulge, thick disk, dark matter halo, outer stellar halo, thin disk, long bar.
    (c) dark matter halo, inner stellar halo and boxy/peanut bulge, thick disk, outer stellar halo, thin disk, long bar.
    (d) long bar, inner stellar halo and boxy/peanut bulge, thin disk, thick disk, outer stellar halo, dark matter halo.
    (e) inner stellar halo and boxy/peanut bulge, thick disk, thin disk, outer stellar halo, dark matter halo, long bar.
73. The merging of galaxies can
    (a) trigger star formation.
    (b) produce a more massive supermassive black hole by merging existing supermassive black holes.
    (c) destroy existing spiral structure.
    (d) heat cold gas to very high temperatures.
    (e) result in a merged dark matter halo.
    (f) all of the above
    (g) (a) and (b) only
74. (*Answer*) When the Milky Way Galaxy and M31 collide, the resulting mashup will be
    (a) a larger spiral galaxy.
    (b) a giant elliptical galaxy.
    (c) an irregular galaxy of the Magellanic type (Irrm).
    (d) a complete destruction of the galaxies, sending their stars, gas, and dust scattered throughout intergalactic space.
75. GN-z11 is
    (a) a nearby spiral galaxy.
    (b) a nearby giant elliptical galaxy.
    (c) an old dwarf spheroidal galaxy.
    (d) the youngest-appearing galaxy discovered so far.
    (e) the most distant galaxy discovered so far.
    (f) about 400 million years old as it appears to us today.
    (g) (a), (d), and (f) only
    (h) (c) and (e) only
    (i) (d), (e), and (f) only
76. (*Answer*) Supermassive black holes grow
    (a) by consuming stars that get too close.
    (b) by devouring gas and dust that fall in through their accretion disks.
    (c) though mergers.
    (d) all of the above

Short Answer

77. Herschel made three assumptions when he tried to map the Milky Way Galaxy. What were those assumptions, and how was each one flawed?
78. Explain how the Copernican principle suggested that the maps of the Milky Way by Herschel and Kapteyn were almost certainly wrong.
79. (*Answer*)
    (a) If there are 300 billion stars in the Milky Way Galaxy, what would the average mass of a single star be?
    (b) What is the main-sequence spectral class of this "average" star? You may want to refer back to Table 15.2.
80. The mass-to-luminosity ratio is defined as M_{Sun}/L_{Sun}, where M_{Sun} and L_{Sun} are mass and luminosity, both expressed in units of the Sun's values. For example, an O-type star may have $M = 17.5\ M_{Sun}$ and $L = 40{,}000\ L_{Sun}$, giving $M_{Sun}/L_{Sun} = 17.5/40{,}000 = 0.000\,44$. The mass-to-luminosity ratio plays an important role in understanding what kinds of stars make up certain components of a galaxy, or of a galaxy on average.
    (a) Make a table of the ranges of mass-to-light ratios for each broad spectral class of main-sequence stars by referring back to Table 15.2.
    (b) If astronomers find an average mass-to-light ratio of 0.2 in one region of a galaxy, what type of main-sequence star might be most common in that region?
    (c) What type of main-sequence star would be most common in a region where the mass-to-light ratio is 3?
    (d) Which region, $M_{Sun}/L_{Sun} = 0.2$ or $M_{Sun}/L_{Sun} = 3$, is likely to be older? Explain your answer.
81. You have plenty of oxygen, but the heater/air conditioning unit in your space suit stopped working while floating around in the hot corona halo. Would you ultimately freeze or cook? Explain your reasoning.
82. Explain why a star that contains a greater percentage of heavy elements (elements heavier than hydrogen and helium) is likely to be younger than a heavy-element poor star.
83. (*Answer*) The Ordovician–Silurian mass extinction on Earth occurred 439 million years ago. How many trips around the center of the Galaxy has the Sun completed in that time?
84. What is the answer to the question posed at the end of the paragraph following Example 18.1?
85. Explain how clouds of cold hydrogen gas can be detected on the other side of the Milky Way Galaxy.
86. (a) What is the speed of light in km/s?
    (b) What is the orbital speed of the Sun around the center of the Galaxy?
    (c) At what percentage of the speed of light is the Sun traveling around the center of the Galaxy?
87. The wavelength of the 21-cm line of hydrogen is, more precisely, 21.106 cm. Suppose that an astronomer measures the wavelength to be 21.120 cm.
    (a) Is the measured wavelength redshifted or blueshifted?
    (b) Is the hydrogen cloud that generated the spectral line approaching or moving away from Earth?
    (c) How much has the wavelength changed as a result of the Doppler effect?
    (d) At what fraction of the speed of light is the hydrogen cloud moving relative to Earth?
    (e) What is the cloud's speed relative to Earth?

88. How do density waves trigger star formation in spiral galaxies?
89. Explain why massive stars are found in spiral arms but not between them, while low-mass stars are found throughout the thin disk.
90. Do cosmic rays travel through the Milky Way in straight lines? Explain your answer.
91. (*Answer*) Refer back to Exercise 80 and the definition of the mass-to-light ratio. It is estimated that the mass of the Local Supercluster is $M = 1.4 \times 10^{15}$ M_{Sun} and its total luminosity is $L = 6 \times 10^{12}$ L_{Sun}.
   (a) Calculate the mass-to-light ratio $(M_{\text{Sun}} / L_{\text{Sun}})$ for the Local Supercluster.
   (b) What does that value suggest about the major component of mass in the supercluster? Explain your answer.
92. (a) If a spiral galaxy was observed to have a rotation curve that was Keplerian, like the dashed curve in Fig. 18.30, what would that say about the distribution of most of the galaxy's mass?

(b) How does your answer differ if the rotation curve is flat, similar to the solid curve in Fig. 18.30?
93. What component dominates virtually all galaxies? What does that imply for the formation and evolution of galaxies?
94. List at least three ways in which most giant elliptical galaxies differ from spiral galaxies.
95. If the fastest that any signal can travel is the speed of light, how is it possible that the Hubble Space Telescope was able to observe changes in knots along the 5000-ly-long jet of M87 during just 13 years of observations?
96. Explain how quasars can serve as the basis for a precise coordinate system?
97. Describe at least two ways in which the ionized gas in a galaxy's hot corona halo can reach hundreds of thousands to millions of kelvins.
98. If the first stars were made entirely of hydrogen and helium, how did heavier elements appear in the early universe?
99. Describe the formation process of the first primeval galaxies.
100. What is meant by the cold dark matter paradigm?

Activities

Investigating the Milky Way Galaxy's flat rotation curve.

Table 18.2 Data for an idealized Keplerian orbit and an idealized flat rotation curve

Case I: Idealized Keplerian orbits			Case II: Idealized flat rotation curve		
r (kly)	M_r (10^9 M_{Sun})	v_r (km/s)	r (kly)	M_r (10^9 M_{Sun})	v_r (km/s)
26	110	240	26	110	240
30	110		30		240
35	110		35		240
40	110		40		240
45	110		45		240
50	110		50		240
55	110		55		240
60	110		60		240
65	110		65		240

Exercises 101–103 make use of the artificial data in Table 18.2. The Sun's orbital data are in the first row.

In preparation for the following activities we need to develop a few mathematical expressions using some algebra and equations that you have already encountered elsewhere in this textbook. The results are in the highlighted boxes.

101. The acceleration of a star in a circular orbit around the center of the Galaxy is described by the equation for the magnitude of centripetal acceleration on page 161. From Newton's second law, $\vec{F} = m\vec{a}$, the general expression for the centripetal force on the star that keeps it in orbit is

$$F_c = m\frac{v_r^2}{r},$$

where m is the star's mass, r is the radius of its orbit from the center of the Galaxy, and v_r (pronounced "v sub r") is its orbital (circular) speed at radius r. Physically, what provides the centripetal force is the force of gravity due to all of the mass *interior* to the star's orbit, which we will label as M_r.

From Newton's universal law of gravitation on page 168, the centripetal force due to gravity is expressed as

$$F_c = G\frac{M_r\, m}{r^2}.$$

The two expressions can be set equal to one another so that

$$m\frac{v_r^2}{r} = G\frac{M_r\, m}{r^2}.$$

This equation can be solved for the orbital speed by canceling the star's mass, m, on both sides of the question, multiplying both sides by the radius, r, and taking the square root of both sides, giving

$$v_r = \sqrt{\frac{GM_r}{r}}.$$

To calculate the orbital speed for the Sun, we can substitute the mass of the Milky Way inside the Sun's orbit that we found in Example 18.1; $M_{Sun,MW} = 1.1 \times 10^{11} \, M_{Sun} = 110$ billion solar masses. Since the distance of the Sun from the center of the Galaxy is 26 kly, the Sun's orbital speed is

$$240 \text{ km/s} = \sqrt{\frac{G \times 110 \text{ billion solar masses}}{26 \text{ kly}}}.$$

Finally, using a "trick" that we have used several times before in this textbook, we can divide the general equation for orbital speed by the values for the Sun's orbit, which cancels "Big G," the universal gravitational constant. Solving for v_r gives

$$v_{r,\text{km/s}} = 240 \times \sqrt{\frac{M_{r,\text{billion}}/110}{r_{\text{kly}}/26}}, \tag{18.1}$$

Velocity of orbiting object (in km/s)

where $v_{r,\text{km/s}}$ is in units of km/s.

In Equation (18.1), $M_{r,\text{billion}}$ is simply entered as billions, so 110 billion is entered as 110, and r_{kly} is entered in thousands of light-years, so 26 kly is entered as 26. As an example, suppose that the mass of the Galaxy inside 52 kly is the same as it is inside 26 kly. Then the equation would give:

$$v_{r,\text{km/s}} = 240 \times \sqrt{\frac{110/110}{52/26}} = 240 \times \sqrt{1/2} = 170,$$

or $v_r = 170$ km/s.

If, instead of mass, the orbital speed of stars or hydrogen gas clouds is known from observational data, Equation (18.1) can be solved for the mass interior to the orbiting object. Rewriting,

$$M_{r,\text{billion}} = 110 \times \left(\frac{v_{r,\text{km/s}}}{240}\right)^2 \times \left(\frac{r_{\text{kly}}}{26}\right). \tag{18.2}$$

Mass interior to orbiting object (in billions of M_{Sun})

For example, if $v_{r,\text{km/s}} = 170$ and $r_{\text{kly}} = 52$, then Equation 18.2 gives

$$M_{r,\text{billion}} = 110 \times \left(\frac{170}{240}\right)^2 \times \left(\frac{52}{26}\right)$$

$$= 110 \times (0.708)^2 \times 2$$

$$= 110 \times 0.5 \times 2 = 110.$$

Therefore, $M_r = 110 \times 10^9 \, M_{Sun} = 1.1 \times 10^{11} \, M_{Sun}$.

(a) For Case I, complete Table 18.2 by calculating the missing values in column v_r (km/s) using Equation (18.1). Case I is the case where we assume (incorrectly) that all of the Milky Way's mass is inside the orbit of the Sun.

(b) Download a standard (Cartesian) sheet of graph paper from Graph Templates, orient the graph paper in landscape (long axis is horizontal), and make a graph of orbital speed as a function of distance from the center of the Galaxy for Case I. v_r (km/s) should be on the vertical axis and r (kly) is on the horizontal axis. You can start the horizontal axis at 20 on the left-hand end and label every *fourth* tick mark or vertical line as 30, 40, 50, 60, and end it at 70 on the right-hand end (every grid line along the horizontal axis marks an interval of 2.5 kly). On the vertical axis, start at the bottom with 140 and label *every other* tick mark as 150, 160, et cetera, up to 240 at the top (every grid line marks an interval of 5 km/s). For any graph you must *always* label your axes. Label the axes in the same way that the corresponding table columns are labeled.

(c) For Case II, complete the M_r ($10^9 \, M_{Sun}$) column in Table 18.2 using Equation (18.2). In Case II you are determining the mass interior to the radius r_{kly}, where the dark matter halo is contributing an increasing amount of mass inside orbits farther and farther from the Galaxy's center.

(d) On a separate sheet of graph paper, make a graph of M_r ($10^9 \, M_{Sun}$) as a function of r_{kly}. Label the tick marks on the horizontal axis as you did for part (b). On the vertical axis, label the bottom line as 100, then label every *fifth* line as 150, 200, 250, and 300 (each grid line along the vertical axis marks an interval of 10 billion M_{Sun}). Again, label the axes in the same way that the corresponding table columns are labeled.

(e) Case I is for a Keplerian orbit (e.g., the curve is like planets orbiting the Sun). Describe what the v_r vs. r curve looks like.

(f) Case II is a flat rotation curve. What does the graph of M_r vs. r look like in order to produce a constant value of v_r with increasing distance?

(g) What can you conclude about the distribution of mass in a dark matter halo that produces a flat rotation curve?

102. In Exercise 101 you investigated the difference between a galaxy rotation curve where all of the mass is at the galaxy's center (Case I, Keplerian) and one where the mass increases with distance (Case II, flat). However, examining orbital speed, v_r is not the same as examining orbital period, P_{orbit}. Calculating the orbital period requires dividing the distance traveled around the orbit by the orbital speed (think of driving 120 km at 60 km/h, which takes 120 km/60 km/h = 2 hours).

The circumference of a circle is given by $C = 2\pi r$ where r is the radius of the circle and π ("pi") is the famous mathematical constant, 3.14159 .... This means that orbital period is $P_{\text{orbit}} = 2\pi r / v_r$. In order to get an answer in *millions* of years (My), which is characteristic of orbital periods around the center of the Galaxy, a series of unit conversions must be performed, or we can invoke our "trick" again. The orbital period of the Sun, assuming a circular orbit, is 210 million years. This means that after the appropriate unit conversions,

$$210 \text{ My} = \frac{2\pi \times 26 \text{ kly}}{240 \text{ km/s}}.$$

Dividing the general equation for orbital period by the one

for our Sun leads to

$$P_{\text{orbit}} = 210 \times \left(\frac{r_{\text{kly}}}{26}\right) \times \left(\frac{240}{v_{r,\,\text{km/s}}}\right). \qquad (18.3)$$

Orbital period (in millions of years)

(a) From the data in your completed Table 18.2, create and complete a new table that shows the orbital periods at each distance for Case I and Case II.

(b) Create two graphs of P_{orbit} (My) vs. r (kly), one for each case.

(c) Comment on any differences between the two graphs.

103. Average mass density, ρ ("rho"), is defined as mass, M, divided by the volume, V, that the mass occupies. The volume of a sphere is given by $V = 4\pi r^3 /3$, where r is the radius of the sphere (in our case, the distance from the center of the Galaxy). As a result,

$$\rho_r = \left(\frac{3}{4\pi}\right) \times \left(\frac{M_r}{r^3}\right),$$

where ρ_r is the average mass density inside r. For the orbit of the Sun from Table 18.2,

$$\rho_r = \left(\frac{3}{4\pi}\right) \times \left(\frac{110}{26^3}\right) \text{ billion } M_{\text{Sun}}/\text{kly}^3$$

$$= 0.0015 \text{ billion } M_{\text{Sun}}/\text{kly}^3$$

$$= 1.5 \text{ million } M_{\text{Sun}}/\text{kly}^3$$

$$= 1.5 \times 10^6 \, M_{\text{Sun}}/\text{kly}^3.$$

[Notice that the answer is in *millions* of solar masses per kly^3 ("cubic kilolight-year") even though *billions* of solar masses was entered into the equation.] Once again, using our "trick" of dividing equations,

$$\rho_{r,\,\text{million}} = 1.5 \times \left(\frac{M_{r,\,\text{billion}}}{110}\right) \times \left(\frac{26}{r_{\text{kly}}}\right)^3. \qquad (18.4)$$

Interior average density (in millions of $M_{\text{Sun}}/\text{kly}^3$)

As an example, return to the Case-I example from Exercise 101. At 52 kly, where $M_{r,\,\text{billion}} = 110$, the average density inside the orbit is

$$\rho_{r,\,\text{million}} = 1.5 \times \left(\frac{110}{110}\right) \times \left(\frac{26}{52}\right)^3$$

$$= 1.5 \times 1 \times \left(\frac{1}{2}\right)^3$$

$$= 1.5 \times 1 \times \frac{1}{8} = 0.19,$$

so $\rho_r = 0.19 \times 10^6 \, M_{\text{Sun}}/\text{kly}^3$.

(a) Using the data from your completed Table 18.2, calculate the average mass density inside each radius for both Case I and Case II, and display your results in a table. The columns for each case should be labeled r (kly) and ρ_r ($10^6 \, M_{\text{Sun}}/\text{kly}^3$).

(b) Create graphs of ρ_r ($10^6 \, M_{\text{Sun}}/\text{kly}^3$) vs. r (kly) for both cases.

(c) Comment on the differences between the Case-I Keplerian graph and the Case-II flat rotation curve graph. Which one decreases more rapidly?

(d) For Case II, explain why the M_r ($10^9 \, M_{\text{Sun}}$) vs. r (kly) graph from Exercise 101 increases while the ρ_r ($10^6 \, M_{\text{Sun}}/\text{kly}^3$) vs. r (kly) graph behaves differently.

You have just made an approximate map of the dark matter halo density distribution of the Milky Way Galaxy without having any idea what dark matter is other than it interacts gravitationally with other masses!

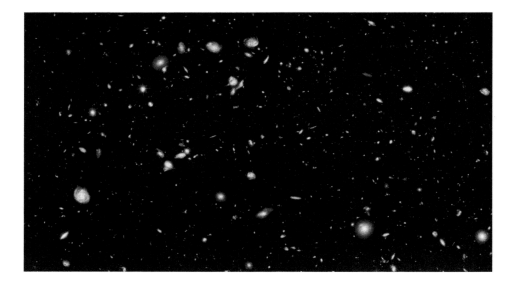

19 The Once and Future Universe

We shall not cease from exploration, and the end of all our exploring
will be to arrive where we started and know the place for the first time.

T. S. Eliot (1888–1965)

19.1	The Expanding Universe	777
19.2	The Big Bang and Spacetime	779
19.3	The Accelerating Universe	786
19.4	The Early Universe	790
19.5	The Λ-Cold Dark Matter Model	806
19.6	The Meaning of "The Universe"	810
19.7	The Scientist's Crystal Ball: Looking to the Future	815
	Exercises	821

Fig. 19.1 A portion of the Hubble eXtreme Deep Field. [NASA, ESA, G. Illingworth, D. Magee, and P. Oesch (University of California, Santa Cruz), R. Bouwens (Leiden University), and the HUDF09 Team]

Introduction

Throughout this text we have explored how we have come to understand the physical laws that undergird all of nature. Those laws have been applied to explain the many kinds of object in our Solar System, the formation and evolution of our Sun and other stars, and the nature of galaxies. In this final chapter we use what we have learned, combined with new and surprising discoveries, to investigate the nature and evolution of the universe as a whole, from its beginning to possible scenarios for the very distant future.

19.1 The Expanding Universe

The Hubble–Lemaître Law

Georges Lemaître (Fig. 19.2), seen in Fig. 19.2, was a Belgian priest who earned his doctorate in 1920 and was ordained in 1923 after first serving as an artillery officer in World War I. In 1923 he worked at the University of Cambridge with the highly respected astronomer, Sir Arthur Eddington, who introduced him to astronomical research, including the disciple of cosmology, which has to do with the beginning and evolution of the universe. In 1924, Lemaître traveled to the United States for one year to work at Harvard Astronomical Observatory during the time when Harlow Shapley was head of the famous facility.

Fig. 19.2 Georges Lemaître (1894–1966) seen with Robert Millikan (left) and Albert Einstein (right) at Caltech in 1933. (Public domain)

After returning to Belgium, Lemaître published a paper[1] in 1927 in a journal that was not widely read in the United States.[2] In his paper, Lemaître used Einstein's general theory of relativity to show that the universe could be expanding, and if it is doing so, the expansion would explain why most galaxies are observed to be redshifted (moving away from the observer). He also argued that the farther away a galaxy is, the faster the galaxy would be moving away from the observer. Later, Lemaître recalled Einstein telling him that "Your calculations are correct, but your physics is atrocious."

Alexander Friedmann (Fig. 19.3), a Russian and Soviet physicist and mathematician, had already proposed a cosmological model of an expanding universe in 1922 that was similar to Lemaître's, for which he received a similar response from Einstein. Friedmann's equations also allowed for the possibility of a universe that could collapse to a singular point. Like Lemaître, Friedmann served in World War I, but as a Russian bomber pilot. Sadly, Friedmann died of typhoid fever at 37 years of age, apparently having contracted the disease while on his way home from his honeymoon.

Fig. 19.3 Alexander Friedmann (1888–1925). (Public domain)

---

[1]Lemaître, M. l'Abbè G. (1927). "Un Univers homogène de masse constante et de rayon croissant rendant compte de la vitesse radiale des nébuleuses extra-galactiques" (translation: "Expansion of the universe, A homogeneous universe of constant mass and increasing radius accounting for the radial velocity of extra-galactic nebulae"), *Annals of the Scientific Society of Brussels*, 47:49–59.

[2]Ironically, it was Eddington, Lemaître's mentor at the University of Cambridge, who teased Ejnar Hertzsprung about publishing in obscure journals (page 629).

Don't forget that a scientific theory must, by definition, be able to make testable predictions. Although Lemaître did use a scattering of data available in 1927 to support his model of an expanding universe, the data were far from convincing.

That situation changed two years later when Edwin Hubble, while using the 100-inch telescope at Mount Wilson Observatory in California, combined previously available recessional velocity data of galaxies (redshifts) with their distances that he determined through the use of a variety of standard candles. Hubble's data included the period–luminosity relationship for Cepheid variable stars (page 667), novas, the brightest stars in galaxies, and the brightest galaxies in the Virgo cluster of galaxies. Figure 19.4 is the original figure from Hubble's 1929 paper. Hubble commented in his paper that "For such scanty material, so poorly distributed, the results are fairly definite." After Hubble's paper was published, Einstein reversed his criticisms of Friedmann's and Lemaître's original work.

The linear relationship between the recessional velocities of galaxies and their distances had been referred to as Hubble's Law for almost 90 years, but in 2018 the International Astronomical Union officially changed the name to the Hubble–Lemaître law, in recognition of the contributions both men made to the revolutionary relationship that permanently altered our view of the universe.

Mathematically, the linear relationship can be represented by the equation

$$v = H_0 d, \tag{19.1}$$

The Hubble–Lemaître law

where v is the recessional velocity, d is the distance from the observer, and H_0,

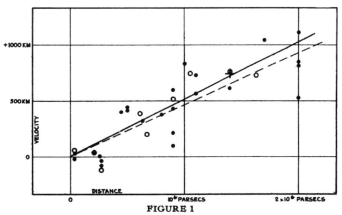

FIGURE 1

Velocity-Distance Relation among Extra-Galactic Nebulae.

Fig. 19.4 Edwin Hubble's original diagram demonstrating a straight-line relationship between the distance to a galaxy and the speed with which the galaxy is moving away from us. The solid line is based on individual measurements and the dashed line is from grouping the data. The recessional velocities on the vertical axis are in units of km/s even though Hubble indicated a distance unit of km. The distances along the horizontal axis are in units of millions (10^6) of parsecs. (Hubble, 1929[3])

---

[3] Hubble, Edwin (1929). "A Relation between Distance and Radial Velocity among Extra-Galactic Nebulae," *Proceedings of the National Academy of Sciences of the United States of America*, 15(3):168-173.

known as the Hubble constant, is the slope of the straight-line (linear) relationship revealed in Fig. 19.4. Simply stated, according to the Hubble–Lemaître law, the recessional velocity of a galaxy is directly proportional to its distance from us, where the constant of proportionality is the Hubble constant.

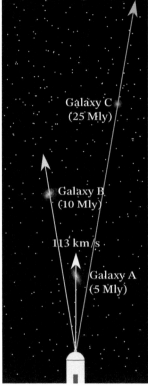

Fig. 19.5 Receding galaxies in Example 19.1.

Example 19.1

Figure 19.5 shows that Galaxy A is five million light-years (5 Mly, or about 1.5 Mpc) from Earth and receding from us at a speed of 113 km/s. Galaxy B is 10 Mly away, and Galaxy C is 25 Mly distant. How fast are Galaxy B and Galaxy C moving away from us?

Galaxy B is two times farther from Earth than Galaxy A is, and therefore moving away from us at twice the speed of Galaxy A. This means that Galaxy B has a recessional velocity of

$$v_B = 2v_A = 2 \times 113 \text{ km/s} = 226 \text{ km/s}.$$

Similarly, Galaxy C is 25 Mly/5 Mly = 5 times as far from us as Galaxy A, and so its recessional speed is

$$v_C = 5v_A = 5 \times 113 \text{ km/s} = 565 \text{ km/s}.$$

Einstein's Biggest Blunder

When Einstein developed his general theory of relativity in 1915 he believed that the universe must be static; neither expanding nor contracting. By 1917 he realized that his equations implied a universe that can't be static; it must either expand forever if the galaxies are moving fast enough, or it would ultimately collapse under its own gravitational pull, but it couldn't just sit there doing nothing. In order to stop either of those inevitable outcomes (at least on paper), Einstein introduced an arbitrary constant into his equations that forced a static universe. That constant became known as the cosmological constant, historically symbolized by Λ (Lambda). But when Hubble determined that galaxies beyond our immediate neighborhood (the Local Group) were all moving away from us, Einstein was forced to admit that introducing his cosmological constant was a mistake, perhaps even calling it his "greatest blunder."

It was because Einstein believed stronly at the time, in line with the belief of most of the astronomical community, that the universe is static that he was so critical of both Friedmann and Lemaître. Both of these men had predicted an expanding universe based on the general theory of relativity without a cosmological constant.

19.2   The Big Bang and Spacetime

An immediate implication of an expanding universe is that it must have started expanding at some time in the distant past. If the expanding universe could be viewed as a movie, then running the movie backward would show all of the galaxies coming together rather than spreading apart, and they would converge at a specific time; the start of the expansion. Lemaître had proposed that perhaps the universe began from a quantum "primeval atom" that contained all of the mass and energy in the universe. Lemaître argued that his "primeval atom" contained so much energy that it was unstable and blew apart, throwing the mass out into the cosmos. Lemaître's proposal was essentially the first scientific creation cosmology of the universe.

Determining when the expansion started depends on the value of H_0, which allows us to trace the line in Fig. 19.4 (the Hubble–Lemaître law) back to its origin.

The Hubble Constant and the Age of the Universe

When Hubble first published the results shown in Fig. 19.4, he estimated the slope of his solid line to be 500 km/s per Mpc, or about 150 km/s per Mly. We know today that his value was almost 7 times larger than the measured slope of

$$H_0 = 22.4 \text{ km/s per Mly} = 73.2 \text{ km/s per Mpc} \tag{19.2}$$

The Hubble constant (from distance-ladder measurements)

as determined from the Cepheid period–luminosity relationship, the brightnesses of white dwarf supernovas, and other distance-ladder indicators. (However, the universe still has a puzzle for us, waiting in Section 19.5.)

For a car moving on a highway at a constant speed, v, the distance the car will travel in a specified period of time, t, is given as $d = vt$. You are certainly familiar with this relationship, since if you are driving at a speed of 100 km/h (62 mi/h) for 2 hours you would travel a distance of 100 km/$\cancel{h}$ × 2 $\cancel{h}$ = 200 km. The relationship between distance, speed, and time can also be written to calculate the time it takes to travel a specified distance at a constant speed, or $t = d/v$; traveling a distance of 200 km at a speed of 100 km/h requires 2 h [$t = 200 \cancel{\text{km}}/(100 \cancel{\text{km}}/\text{h})$]. The Hubble–Lemaître law has both distance and speed in it and can be rewritten as $t = d/v = 1/H_0$.

It may not be immediately obvious, but $1/H_0$ is the time necessary for a galaxy traveling away from us at a *constant* speed v to be a distance d from us. The strange units of the Hubble constant are really just units of 1/time, because km/s is distance/time and Mly (or Mpc) is a distance, so (distance/time)/distance is simply 1/time once the distance units are canceled [the "trick" is converting one million light-years (Mly) to km; 1 Mly = 9.46×10^{18} km or 9.46 million trillion km]. Since all galaxies beyond the Local Group share the same Hubble–Lemaître law, if we assume that galaxies have neither increased or decreased in speed (assumptions that will turn out to be wrong), then Equation (19.1) and the value of H_0 from Equation (19.2) imply that those galaxies have been moving away from us for

$$t_H = \frac{1}{H_0} = 4.22 \times 10^{17} \text{ s} = 13.4 \text{ billion years.} \tag{19.3}$$

The Hubble time (from distance-ladder measurements)

(However, again stay tuned for a different result from other measurements, to be discussed in Section 19.5).

By this crude estimate, all of the galaxies in the universe were at the same place at the same time about 13.4 billion years ago! Is that when the universe began, and has it been expanding ever since?

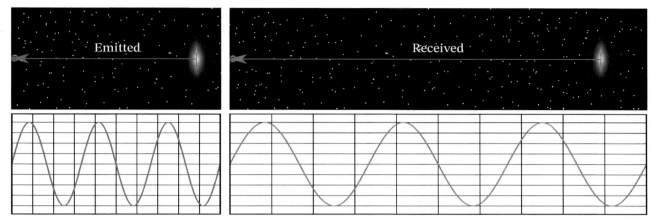

Fig. 19.6 Imagine a grid (*bottom*) attached to points in space (*top*). As the universe expands so does the grid attached to it, indicating that those same points in space are also spreading out. Imagine blue light with a wavelength of 400 nm is emitted by a galaxy when the universe was smaller. When the universe has doubled in size, so has the wavelength, meaning that we detect infrared light on Earth with a wavelength of 800 nm when the galaxy's distance from us has doubled.

The Beginning of Spacetime and the Big Stretch

The name, the Big Bang, conjures up imagery of a gigantic explosion *in* space, at a specific time and location; the point where Lemaître's primeval atom existed. The problem with that view is that the explosion wasn't at a specific location at a specific time, *it was an explosion throughout the entire universe at the beginning of time!* There was no time and there was no universe before the Big Bang. The expansion of the universe observed today is literally the expansion *of* the universe, including space itself. Furthermore, it is generally thought by cosmologists that the universe is infinitely large. As difficult as it is to imagine, the infinite universe is expanding into infinity.

There is also another misleading aspect to the imagery suggested by the term, the "Big Bang." As illustrated in Fig. 19.6, the observations of galaxy redshifts that led to the Hubble–Lemaître law are not due to galaxies and galaxy clusters moving through space, the redshifts are the result of galaxies and galaxy clusters being carried along *by* space as the space itself expands. In other words, the redshifts *are not* Doppler shifts due to motion from one location in space to another, but are cosmological redshifts that occur because the expanding space is stretching the wavelengths of light emitted by galaxies. This lengthening of wavelengths of light also means that the observed frequencies and energies of the photons are lower when they reach our telescopes [remember that $f = c/\lambda$ and $E = hf$, Equations (6.9) and (8.1), respectively]. It may be more appropriate to think of the Big Bang and the resulting universal expansion as the *Big Stretch;* the stretching out of coordinates attached to points in space.

For convenience, astronomers think of cosmological redshifts, distances, and time in terms of a redshift parameter, z, defined as

$$z \equiv \frac{\lambda_{\text{obs}} - \lambda_{\text{emit}}}{\lambda_{\text{emit}}} = \frac{\lambda_{\text{obs}}}{\lambda_{\text{emit}}} - 1 \qquad (19.4)$$

The redshift parameter

Key Point
The Big Bang is thought to be the beginning of spacetime, with the universe likely infinite in extent.

Key Point
Cosmological redshifts are caused by galaxy clusters being carried along with the expansion of space itself, *not* motion through space.

Key Point
The redshift parameter measures the fractional change in wavelength due to the Big Stretch of the expanding universe.

where λ_{obs} is the wavelength of a particular spectral line as observed at the telescope and λ_{emit} is the wavelength of the same spectral line as it would be measured if the source of the emitted photon isn't moving relative to the detecting instrument, say in an Earth-based laboratory. Said another way, the redshift parameter is the *fractional change* in wavelength of a spectral line due to expanding space (the Big Stretch of the length of the electromagnetic wave associated with the photon). For galaxies at immense distances from us, observed wavelengths can be many times longer than those measured in a laboratory. For example, the Lyman lines of hydrogen (page 763) that are at ultraviolet wavelengths in a laboratory can be seen in visible light, infrared light, or even at microwave wavelengths for large enough values of z.

You may have noticed that this definition for the redshift parameter looks like the Doppler effect equation presented on page 283, with z replaced by velocity divided by the speed of light, or $z = v/c$. It does turn out that the cosmological redshift can be approximated by v/c, but only if v is much less than c ($v \ll c$), meaning that z must be much less than 1 (the fully general relativistic relationship between z and v is much more complex mathematically). However, it is important to understand that the cosmological redshift is fundamentally different from the Doppler redshift and they should not be confused with one another. Again, *Doppler redshifts are measured for objects moving through space* from one location to another while *cosmological redshifts are due to locations in space (coordinates) moving away from one another*.

Example 19.2

The Lyman-α line of hydrogen has an ultraviolet wavelength of 121.6 nm when measured in a laboratory, but while observing a very remote galaxy the line is measured to have a wavelength of 832.8 nm, placing it in the infrared portion of the electromagnetic spectrum. What is the corresponding value of the redshift parameter in this case?

From the redshift parameter equation,

$$z = \frac{832.8 \text{ nm}}{121.6 \text{ nm}} - 1 = 6.85 - 1 = 5.85.$$

You may be wondering why we measure the spectrum of the Andromeda Galaxy as blueshifted, meaning that it is coming toward us rather than being carried away by the cosmological expansion of the universe. The reason is because, on relatively local scales, gravitational attraction wins out over cosmological expansion. More accurately, galaxy clusters get farther from each other over time while the clusters themselves stay bound together. The strength of gravitational attraction is also why galaxies and planetary systems aren't stretched along with the expansion (at least at this point in time; however, stay tuned for Section 19.7).

The Three Redshifts

It is important to note that you have now encountered three distinctly different types of redshift, caused by three different physical processes:

Key Point
Three different kinds of redshifts.

- **Doppler redshifts** result due to objects moving away from the observer from one location in space to another location in space,
- **Gravitational redshifts** occur because photon energies decrease as the photons climb out of gravitational wells, and
- **Cosmological redshifts** arise from the expansion of space itself.

The Steady-State Theory of Cosmology

With the theoretical framework of an expanding universe put forth by Friedmann and Lemaître, and with the confirming discovery of the Hubble–Lemaître law, the "hypothesis of the primeval atom" became the standard view of the origin of the universe. But not all astronomers agreed with that view. In 1948, three young astronomers at Cambridge University, Sir Hermann Bondi (1919–2005), Thomas Gold (1920–2004), and Sir Fred Hoyle (Fig. 19.7) developed an eternal steady-state theory of the universe that had no beginning, despite the indisputable observational evidence of an expanding universe. To reconcile the steady-state theory with observations, they proposed that matter could pop into existence in empty space. As this new matter accumulated it would lead to the formation of galaxies that would then fill in the growing gaps caused by the expansion. In this way, the universe would always look essentially the same over long time scales, assuming that we don't worry about which galaxies or galaxy clusters occupy a specific region of space. In order to avoid violating the law of conservation of energy, the group proposed a "creation field" of negative pressure and negative energy that would balance the addition of positive energy from new matter. Providing the necessary new mass required an extremely small rate of matter creation of just a few hydrogen atoms per cubic meter (35 cubic feet) per 10 billion years; far too slow to ever be measured experimentally.

Fig. 19.7 Sir Fred Hoyle (1915–2001). (By permission of the Master and Fellows of St John's College, Cambridge)

Along with being an astronomer, Hoyle was a radio and television personality who popularized science. During a BBC radio broadcast in 1949, Hoyle was discussing the "hypothesis of the primeval atom" at the beginning of the universe and referred to the destruction of the primeval atom as a "big bang"; unfortunately, the term stuck and has led to endless confusion among the general public and introductory astronomy students ever since. Sir Fred Hoyle was also a prolific science fiction and mystery writer, a proponent of the panspermia proposal for the beginning of life on Earth (page 572), a writer of operas, and a mountain climber.

A Scientific Theory Must Make Testable Predictions

As has been pointed out repeatedly in this textbook, scientific theories must be testable; otherwise they cannot be considered as truly scientific. So how might we distinguish between the Big Bang theory and the steady-state theory?

The theory that the likely infinite universe began as a Big Bang with all of its mass and energy initially packed into an infinitely dense space would also require that the average temperature of the universe must have been essentially infinite as well. With the Big Bang came a universe-wide fireball, but as the universe expanded, the theory suggests that the wavelengths of light associated with the fireball should stretch (lengthen) in proportion to the size of the universe, just like the illustration in Fig. 19.6. In addition, as the light bounced around off charged particles in the hot ionized gas, the distribution of wavelengths of the light would take on the shape of an extremely hot blackbody spectrum. The steady-state theory of an eternal, unchanging universe, on the other hand, does not provide for the possibility of a primeval fireball that produces a blackbody spectrum.

The Cosmic Microwave Background

In 1948, Ralph Asher Alpher (1921–2007) and Robert Herman (1914–1997) used the Big Bang theory and the Hubble constant to propose that the universe should be awash with electromagnetic radiation having a characteristic temperature of 5 K. Unfortunately, that initial work went largely unnoticed by the astronomical community at the time. Some sixteen years later, a Soviet astrophysicist, Yakov Borisovich Zel'dovich (1914–1987), and two Princeton University astrophysicists and cosmologists, Robert Dicke [Fig. 19.8 (*Top*)] and Jim Peebles [Fig. 19.8 (*Bottom*)], independently proposed that a background of blackbody microwave radiation (Section 6.6) should exist as relic radiation from the Big Bang. To test the theory, two colleagues of Dicke and Peebles at Princeton began to build a radio telescope that should be capable of detecting the proposed radiation.

Meanwhile, at Bell Telephone Laboratories at Crawford Hill, Holmdel, New Jersey, only 50 km (30 mi) from Princeton, physicists Arno Penzias and Robert Wilson, both seen in Fig. 19.9, were working to understand background radio noise in a horn antenna that was being used for a NASA communications project that involved bouncing signals off giant aluminum-covered balloons. The two had considered and rejected a number of possible causes, including interference from signals originating in New York City, and even the dung left behind by bats and pigeons. Af-

Fig. 19.9 Robert Woodrow Wilson (b. 1936) (left) and Arno Allan Penzias (b. 1933) standing at the giant horn antenna in Crawford Hill, Holmdel, New Jersey. (Public domain)

Fig. 19.8 *Top:* Robert Henry Dicke (1916–1997). (Courtesy of Princeton University Library) *Bottom:* Jim Peebles (b. 1935). (Juan Diego Soler; CC BY 2.0)

ter clearing out the dung, the pigeons returned and contributed to the surface coating of the antenna again. Sadly, in the name of science, the continually returning pigeons had to be destroyed. Having eliminated all other possibilities, and realizing that the background radio noise was coming from every direction in the sky, Penzias and Wilson concluded that the noise was probably astronomical in origin. Based on their sparse data, it seemed that the signal was associated with a temperature of about 3.5 K. Penzias and Wilson then called the physicists at Princeton University and spoke with Dicke, who immediately told his colleagues, "Boys, we've been scooped!" Penzias, Wilson, and the Princeton group agreed to publish two papers simultaneously, Penzias and Wilson would publish their observations and the Princeton group would publish the supporting theoretical foundation.

Penzias and Wilson had accidentally discovered what is known today as the cosmic microwave background (CMB). Modern high-precision observations of the cosmic microwave background obtained by the Planck satellite have confirmed

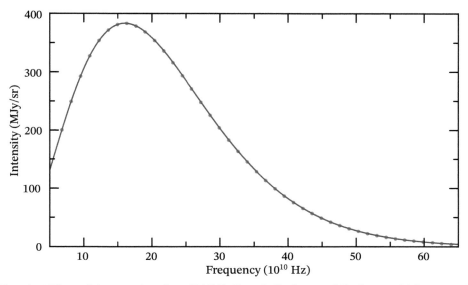

Fig. 19.10 The red dots are data from NASA's Cosmic Background Explorer, which operated from 1989 to 1993. The blue curve is that of a perfect blackbody with a temperature $T = 2.725\,48$ K. The error bars in the data are significantly smaller than the size of the red dots. (Data courtesy of NASA/GSFC)

that the radiation is an almost perfect blackbody radiation curve with an associated temperature of $2.725\,48$ K $\pm\ 0.000\,57$ K, or approximately 2.7 K. Figure 19.10 shows the observational data from the Cosmic Background Explorer (COBE) microwave satellite, a predecessor of Planck, that revealed the nearly perfect blackbody spectrum for the CMB. Although the steady-state theory could argue for a background radiation due to scattered light from galaxies, it could not produce the observed blackbody spectrum with the Big Bang theory's temperature prediction and so had to be discarded in favor of the Big Bang theory. In fact, the CMB measurements spectacularly verified, but of course do not prove, the theory (recall again the characteristics of a scientific theory from Section 1.2).

Key Point
The cosmic microwave background is the remnant fireball of the universe-wide Big Bang.

The Copernican Principle Revisited (Again)

By now you may be wondering, if all of the galaxies are moving away from us (except for those close to us) and their recessional velocities are directly proportional to their distances from us as the Hubble–Lemaître law claims, does this mean that we are the center of the universe after all? The answer to that question is no. This is basically the Copernican principle again, which argues that Earth does not occupy a special place in the universe. It isn't true in the Solar System, it's not true in the Milky Way Galaxy, and it's not true in the universe as a whole.

To see how this could be the case, remember first that the Big Bang did not occur at a particular location in infinite space, it occurred everywhere at the same time; it is just that when the Big Bang occurred, everything in the universe was essentially infinitesimally close together in a singularity of infinite mass–energy density. As the universe expanded, everything in it started moving away from everything else with no one point having a special place of prominence.

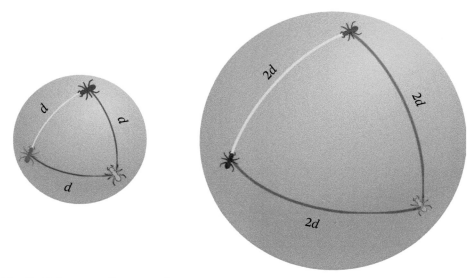

Fig. 19.11 An expanding ant universe.

We can think about this in terms of an analogy: Imagine three ants living on the surface of the expanding balloon pictured in Fig. 19.11. As is true of ants (but not the flying kind!), they are only able to explore the surface and have no idea that there is a "universe" beyond the surface of the balloon or inside it. Their sophisticated sensors can only detect what is going on at their two-dimensional surface of left-right–forward-backward. Initially, each ant is a distance d away from the other two, but after the radius of the balloon has doubled, the distances between each of the three ants has increased to a distance of $2d$. Which ant is at the "center" of the expansion? None of them are, since there is no center on the balloon's surface. The scientifically minded black ant measures distances and recessional velocities as the balloon inflates, discovering the balloon's Hubble–Lemaître law. The green ant is envious that it didn't discover the law first, and the blue ant is depressed to learn that it isn't at the center of their universe after all.

19.3   The Accelerating Universe

The Distance Ladder Revisited

Calibrating the Hubble–Lemaître law by accurately determining the Hubble constant is extremely difficult. Although it is fairly straightforward to measure the recessional velocities of galaxies from their redshifted spectral lines, measuring distances to very remote galaxies is much more challenging. Recall that you were introduced to the idea of the distance ladder on page 605, whereby determining distances to increasingly distant objects depends on methods that were first calibrated using methods for less distant objects.

The first "rung" on the distance ladder is based on light-travel times and geometrical methods within the Solar System. The next rung involves the geometrical method of parallax to relatively nearby stars, although the unprecedented precision of the measurements made by the Gaia spacecraft has pushed the limits of what "relatively nearby" means to much greater distances than ever before. Other geometrical methods, referred to as standard rulers, involve measuring *angular* diameters while also knowing actual diameters by other means; this method is based on the fact that angular diameters (α) are inversely proportional to distance d, or $\alpha \propto 1/d$ (remember that $\propto$ means "is proportional to"); see Fig. 19.12 for an example. Most of the other rungs of the distance ladder make use of standard candles, which are classes of objects with brightnesses that can be determined by properties of the objects themselves, such as the period–luminosity relationship for Cepheid variable stars. Knowing a standard candle's absolute magnitude and measuring its apparent magnitude means that the distance can be calculated from the inverse square law for light; see Section 6.7. Climbing the rungs of the distance ladder using standard candles requires relying on ever intrinsically brighter objects.

An incomplete list of the rungs of the distance ladder from closest to most distant include:

1. Accurate determination of distances across the Solar System (geometry or light-travel times; page 188);
2. Parallax measurements of stars and other objects based on knowing the value of the astronomical unit (geometry; page 605);
3. RR Lyra stars (standard candles; page 678);
4. The light-echo ring around SN 1987A shown in Fig. 17.17 (standard ruler; light-travel time gave the angular diameter);
5. The period–luminosity relationship for Cepheid variable stars (standard candles; page 666);
6. Novas (standard candles; page 690);
7. The brightest stars at the tip of the red giant branch of the Hertzsprung–Russell diagram (standard candles; Fig. 16.19);
8. The brightest galaxies in large clusters of galaxies (standard candles; Section 18.4);
9. White dwarf supernovas (standard candles; page 694);
10. The Hubble–Lemaître law (from redshifts).

The final rung of the distance ladder is the Hubble–Lemaître law itself. Once the Hubble–Lemaître law has been carefully calibrated it can be used to determine extreme distances by measuring spectrum redshifts. Much of the remainder of this chapter depends heavily on the calibrated Hubble–Lemaître law and white dwarf supernovas that are the brightest and therefore the most distant observable standard candles used in its calibration.

Another method of measuring enormous distances makes use of the direct detection of gravitational waves (page 708) and is referred to as standard sirens because it relies on the amplitudes of gravitational waves, just as the intensity of sound waves depends on their amplitudes. Standard sirens don't rely on any intermediate distances from the distance ladder. When combined with multimessenger redshift data, standard sirens give a completely independent method of determining Hubble's constant.

The lengths that astronomers will go to in order to determine distances indicates just how important that information is to understanding our universe.

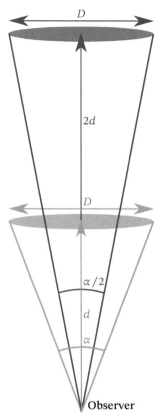

Fig. 19.12 If two objects have the same diameter D, but one object (blue) is twice as far away, the angular diameter of the more distant one is one-half the angular diameter of the closer one.

Fig. 19.13 *Top:* Saul Perlmutter (b. 1959) led the Supernova Cosmology Project. (Holger Motzkau; CC BY-SA 3.0) *Center:* Adam Guy Riess (b. 1969) was co-leader of the High-z Supernova Search Team. (Holger Motzkau; CC BY-SA 3.0) *Bottom:* Brian Schmidt (b. 1967) was co-leader of the High-z Supernova Search Team. (Markus Pössel; CC BY-SA 3.0)

Weirdness at Cosmological Distances

In the late 1990s, two teams of astronomers (Fig. 19.13) were pushing our observational limits for detecting white dwarf supernovas farther out into space than ever before. At these enormous distances, referred to as cosmological distances, they expected to find indications that the expansion of the universe would be slowing down. After all, because of gravity, if you threw a baseball into the air, it would immediately start to slow down, eventually reaching a maximum height, and then fall back to Earth, picking up speed as it fell (the acceleration due to gravity). In the same way, the collective gravitational pull of all of the mass–energy in the universe should cause the expansion to decelerate (don't forget that mass and energy can be thought of as being equivalent; $E = mc^2$). What those teams of researchers discovered instead was that the expansion of the universe isn't slowing down, it is actually speeding up! It is as if that ball you threw in the air slowed for a bit, but then suddenly shot up into the sky at an ever-increasing speed without any obvious force applied to oppose gravity. From our limited human intuition the universe can be a weird place, sometimes even for hardened astronomers.

Figure 19.14 shows a set of raw data for the apparent magnitudes of white dwarf supernovas over a range of redshift parameters [Equation (19.4)] out to $z = 1.3$.[4] The data have not been corrected for a host of complications, including how apparent magnitudes are affected by the light-scattering ability of dust, differing galaxy masses, variations in peak absolute magnitudes that require corrections for how individual white dwarf supernova brightnesses change over time, and other subtleties. When those adjustments are included, the scatter is reduced. Nevertheless, the graph illustrates an important point. Although the apparent magnitudes follow a straight line for small values of the redshift parameter, as z goes beyond about 0.4 the supernovas start getting dimmer more rapidly with z. This means that the distances to those higher redshift supernovas at earlier times are increasing more rapidly with z than they are for smaller-z galaxies in the more recent universe; from about $z = 0.4$ until today ($z = 0$), the expansion of the universe has been accelerating! A redshift parameter of $z = 0.4$ corresponds to looking about 4.4 billion years into the past at galaxies that are today 5.2 billion light-years (1600 Mpc) away from Earth. (That baseball you threw into the air? It is starting to pick up speed instead of slowing down.)

Dark Energy

As strange as it might seem, about 4.4 billion years ago the expansion of the universe started to accelerate instead of slow down. Before that time the mass–energy of the universe was slowing the universe's expansion, just as expected for gravitation behaving according to our intuition — but then it wasn't. Remember Einstein's "greatest blunder," the introduction of a cosmological constant (Λ) from page 779? Before universal expansion was discovered, Einstein had introduced the constant into his equation to force a static universe by tuning the constant to balance gravity. However, Einstein's balancing act couldn't actually work because it is unstable, like trying to balance one ball on top of another one; if the ball on top moves, even just

---

[4]Because it is easy to measure, and because it is directly related to the expansion of the universe, the redshift parameter z is used in cosmology to represent distance and light-travel time.

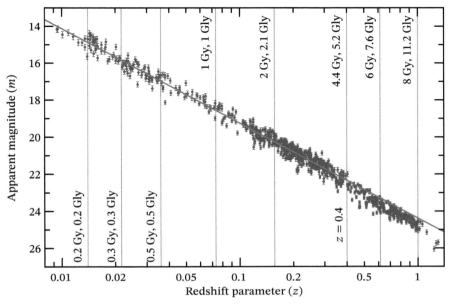

Fig. 19.14 The uncorrected apparent magnitudes (Section 15.2) of white dwarf supernovas with apparent brightness increasing vertically as a function of the logarithm of the redshift parameter z [Equation (19.4)]. Uncertainties in the apparent magnitude measurements are included. Earlier in the universe and farther from Earth increases to the right with increasing z. The solid red line shows the trend for $z < 0.4$, but the data bend below the line for $z > 0.4$. The vertical green lines indicate how long ago the light from those galaxies was emitted (in Gy) and how far away from Earth they are today (in Gly). Although it isn't shown because of the logarithmic nature of the horizontal axis, $z = 0$ corresponds to today ($z = 0$ would be infinitely far to the left). (Raw data courtesy of Betoule, et al., 2014,[5] and VizieR.)

the tiniest bit one direction or the other, it will accelerate away from the balance point and roll off (this is termed an "unstable equilibrium"). If the universe compressed even a tiny amount, it would collapse, but if it expanded infinitesimally, it would expand forever.

Ironically, Einstein's cosmological constant, Λ, seems to describe our accelerating universe with a properly selected value and so was not a blunder after all. But a constant having a particular value isn't the same as understanding what that constant is describing as a physical process. Today, the mysterious something associated with the cosmological constant is termed dark energy, paralleling the naming of the unknown dark matter. But giving it a name still isn't the same as understanding what dark energy is.

To be honest, at this point scientists don't really know what dark energy is, although there is certainly speculation. One possibility is that dark energy may be associated with the vacuum of space. Oddly, the vacuum of space isn't truly empty, it is more like a frothy foam where particles constantly pop into and out of existence. According to Heisenberg's uncertainty principle if a time interval is short enough, a matter–antimatter pair of particles such an electron and a positron

Key Point
Dark energy may be vacuum energy, which is a property of space itself.

[5]Betoule, M., Kessler, R., Guy, J., Mosher, J., et al. (2014). "Improved cosmological constraints from a joint analysis of the SDSS-II and SNLS supernova samples," *Astronomy & Astrophysics*, 568:A22–A53.

can suddenly appear, temporarily violating conservation of energy, as long as they disappear quickly enough not to get caught in the act. This particular version of quantum weirdness is actually a part of quantum theory and has an impact on measurements, allowing scientists to confirm its reality. These extremely temporary particles are called virtual particles. The implication is that "empty" space has a vacuum energy associated with it, along with a vacuum pressure. The hypothesis is that this vacuum pressure is a property of space itself, causing its expansion.

Key Point
4.4 billion years ago, at $z = 0.4$, space had expanded so much that dark energy and vacuum pressure began to dominate over the gravitational attraction of ordinary and dark matter.

But if dark energy is vacuum energy, why did it only begin to influence the expansion of the universe about 4.4 billion years ago? It seems that dark energy is always present in the vacuum of space, but in the distant past there was less space and less vacuum, implying less dark energy. As a result, the total gravitational attraction of all of the mass–energy in the universe dominated over dark energy, causing the expansion of the universe to slow, just like you expected your ball to do when you threw it into the air. But 4.4 billion years ago, when space had expanded enough, vacuum energy and its associated vacuum pressure began to dominate, pushing the expansion at an accelerating rate.

By the way, this acceleration still doesn't mean that the Local Group of galaxies will fly apart. We remain on a collision course with the Andromeda Galaxy. This is because within the Local Group the amount of mass–energy still overwhelms the repulsive nature of dark energy. But overall, we now live in a dark-energy dominated, accelerating universe.

In the spirit of transparency, there is *a slight problem* with the suggestion that the cosmological constant and dark energy are the result of vacuum energy. When theory is used to calculate the cosmological constant from vacuum energy, the result turns out to be off by about 122 orders of magnitude, or 10^{122}: that's a one with 122 zeros after it! At the time this text was written, dark energy continues to baffle, and science still has plenty of work to do.

19.4 The Early Universe

In this section we will be exploring the evolution of the universe from its earliest moments to the formation of its first stars, while in later sections we will consider the possibility that other universes might exist, and contemplate the distant future of our own universe. Once again, it is important to remember though that scientific theories are always based on sets of assumptions, and, as such, theories can never be proven, they can only be disproven (Chapter 1). Despite that caveat, constant testing and verification of the predictions generated from theories can lead to a sense of confidence that our ideas have at least some level of validity.

The Cosmological Principle

When Nicolaus Copernicus developed his heliocentric model of the Solar System that moved Earth into an orbit around the Sun (Section 4.2), rather than having the Sun traveling around Earth, he started us down a path of assuming that Earth, and therefore humanity, does not occupy a unique location in the universe. This concept became known as the Copernican principle.

Today, when scientists develop ideas about how the universe itself should have evolved to this point, and how it may continue to evolve into the distant future, a sort of grand version of the Copernican principle is employed that is called the cosmological principle. The idea is that no matter where in the universe an observer is, over sufficiently large distances the universe should look the same (a concept referred to as homogeneity), and no matter which direction an observer looks, the universe should look the same over sufficiently large fields of view (isotropy).

Certainly, the Solar System doesn't look the same from anywhere inside the Solar System, and looking in one direction within the Solar System doesn't give the same view as looking in another direction (think about looking toward the Sun versus looking perpendicular to the plane of the ecliptic), so the cosmological principle clearly doesn't apply on the scale of the Solar System. Neither does it apply on the scale of the Milky Way Galaxy or even on the scale of the Local Group of galaxies, but perhaps it does apply on much larger scales of hundreds of millions of light-years.

The cosmological principle also implies something else that is very important; namely that the same physical laws that apply in one location in the universe must apply everywhere in the universe. If that isn't true, and the universal gravitational constant, G, Planck's constant, h, the speed of light, c, the elementary charge, e, and other fundamental constants were different in different places, then stars, exoplanets, galaxies, and even atoms and spectral lines would differ throughout the universe (it may not even be possible to form stars with significantly different constants). It would therefore be impossible for homogeneity and isotropy to hold in that case. As has been assumed throughout this textbook, physical laws are believed to be both universal across space and constant in time.

In recent years, there has been some question about whether homogeneity and isotropy strictly hold, even on the largest scales, but there isn't any evidence that physical laws vary over space or time.

The Planck Limits

Today there is very strong evidence that the universe is 13.8 billion years old, and that it likely started out with essentially infinite density and infinite temperature in a space that is infinite in extent. Despite those remarkable conditions at the beginning of the universe, we know a great deal about how the universe evolved after the Big Bang (Section 19.2), but there are limits. We have talked about a number of fundamental constants of nature, such as the elementary charge, e (page 256), the mass of the electron, m_e (page 256), and the constant in Coulomb's law that describes the strength of the electrostatic force, k_E [Equation (6.10), page 209]. These and other constants are associated with fundamental characteristics of nature's particles or physical processes, but four fundamental constants in particular relate to gravitation, the realm of the subatomic world, heat and entropy, and nature's speed limit; all quantities that one would expect to be important in the exotic environment immediately following the Big Bang:

- G, the universal gravitational constant, is the fundamental constant of gravitation [e.g., Equation (5.7) on page 168], including in the general theory of relativity,

- h, Planck's constant, is the fundamental constant of quantum mechanics and the smallest features of nature [e.g., Equation (8.1), page 265],
- k_B, Boltzmann's constant, is the fundamental constant associated with heat and entropy [e.g., Equation (6.20)], and
- c, the speed of light, is the upper limit for how fast anything can travel through space, including information [Equation (6.1), page 190].

So how far back in time can we hope to explore the evolution of the universe? The answer to that question presumably depends on limitations imposed on us by nature, as described by its fundamental constants. Although not a true, rigorous derivation, it is possible to combine and arrange G, h, and c to give a result that has units of time:

$$t_P = \sqrt{\frac{Gh}{c^5}} \approx 10^{-43} \text{ s}. \tag{19.5}$$

Planck time

The numerical result for the so-called Planck time comes from plugging in each of the three constants and rounding off to the nearest power of 10. (The symbol $\approx$ means "is approximately equal to", and when only powers of 10 are given, the number is said to be an "order of magnitude estimate;" in other words, a very crude estimate of what the answer might actually be.) To put this extremely tiny number into perspective, 10^{-43} s is one ten-millionth of a trillionth of a trillionth of a trillionth of a second!

The maximum distance that information can travel in one Planck time is calculated by multiplying the speed limit of the universe, c, by the Planck time to give a Planck length of about

$$\ell_P = ct_P = \sqrt{\frac{Gh}{c^3}} \approx 10^{-35} \text{ m}, \tag{19.6}$$

Planck length

or about ten one-billionths of a trillionth of the size of the nucleus of an atom. As another way to think about how tiny this size is, the period at the end of this sentence is about 0.4 mm in diameter. If the Planck length was scaled up to the size of the period, the length of a one-meter long stick would become the distance that light would travel if it had 70,000 times the age of the universe to do so.

These Planck values may give us some idea of the farthest we can look back in time, or the smallest size that we can explore theoretically if our theories are up to the task, but there are other limitations as well, such as what matter is like in such an incredibly tiny space. By adding Boltzmann's constant, k_B, to our crude set of estimates, we find the Planck temperature given by

$$T_P = \sqrt{\frac{hc^5}{Gk_B^2}} \approx 10^{33} \text{ K}, \tag{19.7}$$

Planck temperature

which is roughly one hundred trillion trillion times the temperature at the center of the Sun. Obviously, scientists haven't explored the nature of matter at anywhere near that temperature, let alone at densities where all of the mass–energy of the universe is compressed into a volume having a diameter of about one Planck length.

One important realm in which we know our theories are not complete is when gravitation (described by the general theory of relativity) that governs the behavior of the universe on large scales reaches down into the realm of the very small (quantum mechanics). To truly explore nature at anywhere near the time scale of the Planck time or the length scale of the Planck length, a robust theory of quantum gravity would be necessary, but it doesn't yet exist. Nevertheless, our current theories do allow us to push our understanding of the early universe back to an extremely short time after the Big Bang, when its density was still incomprehensibly great and its temperature was unfathomably high.

Key Point
A theory of quantum gravity is a major gap in our current understanding at the intersection of gravitation and quantum mechanics.

Problems With the Standard Big Bang Theory

Revisiting the Cosmic Microwave Background

Regardless of whatever happened to trigger the Big Bang or events that happened within about one Planck time after the beginning, we know that we have an expanding and accelerating universe today. An important key to understanding the aftermath of the universe's creation is the 2.7 K blackbody spectrum of the cosmic microwave background (CMB); the afterglow of the Big Bang (recall Fig. 19.10).

The CMB Equilibrium-Temperature Problem

Although the standard Big Bang theory is able to explain the existence of the cosmic microwave background, subtle but very important problems exist. For example, why is the average temperature of the CMB so remarkably constant across the entire sky? This equilibrium-temperature problem[6] exists because even though the universe is 13.8 billion years old, there has not been enough time for light to travel from one side of the universe to the other. The only way for an equilibrium temperature to be established would be if information (and heat) could be exchanged for a sufficiently long period of time that an equilibrium environment would develop. Since information or anything else can't travel faster than the speed of light through space, information about temperature from one side of the universe could not have made it to the other of the universe even once, let alone enough times to reach equilibrium.

Even when the universe was only a tiny fraction of a second old, universal expansion means that light couldn't traverse the extremely short distance from one side of the universe to the other. According to the Hubble–Lemaître law, the farther an object is away from the observer, the faster that object is moving away from the observer. Even in the very early universe, beyond some distance from the "observer" objects are moving away from that "observer" at faster than the speed of light. This means that information could not have traveled to the "observer," let alone back and forth across the entire universe in order to reach an equilibrium temperature, even when the entire universe was only a bit larger than a Planck length.

Key Point
How did the temperature of the CMB get to be so close to being exactly the same across the entire observable universe?

[6]Astronomers refer to this as either the horizon problem or the homogeneity problem.

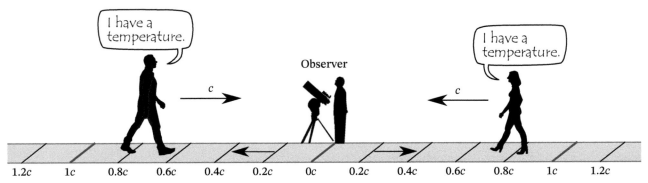

Fig. 19.15 Two people walking toward one another on a sidewalk that is stretching out in both directions. An observer is standing on a blue line that is always at the midpoint between them. The two started walking at just inside the red lines on the sidewalk.

If all of this seems a bit confusing, consider the analogy in Fig. 19.15. Two people are walking toward one another at some speed, c, to share information about their temperatures with an observer. But the sidewalk is no ordinary, stationary sidewalk. Instead, it is stretching in both directions away from its center, marked by the blue line where the observer is located. Because of the stretching, expansion cracks are moving away from the center, with the most distant cracks moving away from the center the fastest (sound familiar?). The two information carriers started walking toward one another from just inside the red-colored cracks, which are moving away from the center at the speed c. It takes the same amount of time for the two people to reach the middle and communicate their temperatures with the observer.

In the analogy, it should be apparent that the "people" moving toward one another are actually photons that are arriving at the observer at a time that is very slightly less than the age of the universe. Of course, photons that started their journeys closer to the middle wouldn't take the age of the universe to arrive in the middle, but the farther away they started, the longer it would take. The expanding sidewalk represents the expanding universe and is lengthening according to the Hubble–Lemaître law. The cracks in the sidewalk are points *in* space that are moving farther and farther apart due to the expansion *of* space (the "Big Stretch").

In reality, the actual situation is even more restrictive than the analogy implies. Rather than meeting in the middle, they need to share their temperature information at each other's starting location enough times to negotiate an equilibrium temperature. Only then will they arrive in the middle with knowledge of the same temperature. So how did they end up having the same temperature?

The CMB Temperature-Variations Problem

The cosmic microwave background isn't perfectly smooth across the sky. The European Space Agency's Planck satellite recorded the most precise temperature measurements of the CMB ever made and produced the map of its temperature variations that is shown in Fig. 19.16.

Figure 19.16 presents another challenge; why are there apparently random, albeit tiny, fluctuations in the temperature of the CMB of only about 0.0001 K or

Fig. 19.16 Tiny variations in temperature across the entire sky in the otherwise smooth cosmic microwave background of 2.725 48 K. Blue regions are cooler than average and red regions are warmer than average. The largest variations are no more than about 0.0001 K. The midplane of the Milky Way Galaxy is located horizontally across the middle of the temperature map, but its emission has been deleted so that the microwave background behind it is visible. (© ESA/Planck Collaboration; CC BY 4.0)

smaller? The Heisenberg uncertainty principle certainly allows for extremely tiny and very localized quantum fluctuations, such as the appearance and disappearance of virtual particles, but the physical sizes of the temperature variations apparent in Fig. 19.16 are thousands to millions of light-years across instead of being subatomic in size. Where did those variations come from? A complete explanation of the beginning of the universe needs to explain this temperature-variations problem. In science, saying that an explanation is very close and therefore "good enough" is never "good enough."

Geometry and the Flatness Problem

Yet another issue with the standard Big Bang theory has been known for a long time: why does the universe appear to be almost perfectly balanced between infinite expansion and ultimate collapse?

Go back to the thought experiment of throwing a baseball up into the air. Unless you are superhuman, if you throw a ball into the air it will fall back down again. This is because you didn't give it enough kinetic energy to escape Earth's gravitational potential energy well. On the other hand, if you threw the ball straight up faster than 11.2 km/s, the escape speed from the surface of Earth (see page 199), you would have given the ball enough kinetic energy to overcome the pull of gravity. The borderline case is if you threw the ball at exactly the escape speed. Although it would continue forever (ignoring everything else in the universe except an airless Earth), its speed would approach zero infinitely far from our planet and infinitely far into the future.

The flatness problem is equivalent to throwing the baseball upward with exactly the escape speed. Neglecting the accelerating expansion of the universe over

Key Point
Why does the universe appear to be so perfectly "tuned" for a flat geometry?

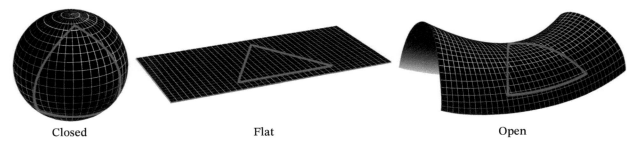

Closed   Flat   Open

Fig. 19.17 Three possible geometries of the universe. The three interior angles for a closed universe, a flat universe, and an open universe add up to more than 180°, exactly 180°, and less than 180°, respectively. (Adapted from NASA/WMAP Science Team)

Key Point
The universe can have one of three possible geometries: closed, flat, or open.

the past 4 1/2 billion years, the expansion rate appears to be almost perfectly tuned to the borderline case. That balance seems like an impossible coincidence.

The problem gets its name from the three geometries of the universe associated with not enough energy (closed), just the right amount of energy (flat), and too much energy (open). Analogies for the three geometries are shown in Fig. 19.17. You probably learned at one point in a geometry class that two parallel lines never cross and they never get farther apart, even when the lines extend infinitely far into space. That characteristic of parallel lines only applies to a flat universe. In a flat universe the three interior angles of a triangle will always add up to exactly 180 degrees as well. In a closed universe, if you draw two parallel lines somewhere in that universe, like longitude lines on Earth at its equator, they will end up crossing (Earth's longitude lines intersect at its poles). The three angles of a triangle in a closed universe also add up to more than 180°. Conversely, an open universe can be described as a saddle shape; two lines drawn parallel to one another on the saddle will end up diverging. The three angles of a triangle in an open universe will also add up to less than 180°. In thinking about the problem this way, why does the universe seem to be perfectly tuned to be flat, at least to the precision of our best measurements?

The Characteristics of a Scientific Theory Revisited

At this point it might be useful to once again return to Chapter 1, and Section 1.2 in particular. Recall that science doesn't have the luxury of simply ignoring inconvenient disagreements with a theory. We have seen that several types of observation pose problems for the standard Big Bang theory with its universe-wide, yet incredibly tiny, beginning. (There are a couple of other problems as well, that we won't concern ourselves with here.)

Is it possible that a more sophisticated version of the Big Bang theory can incorporate and resolve those problems? As always, any new or expanded theory must be consistent with all of the relevant information and must not conflict with known physical laws, unless one or more of the physical "laws" is found to be flawed or ultimately wrong. Borrowing from Plato's question regarding planetary motions, what can be done to "save the [Big Bang] phenomena?"

The Force(s) of Nature and Unification

A "Holy Grail" of theoretical physics is the effort to unify fundamental theories and their associated forces. It is almost an article of faith in physics that there should exist a single overarching theory that encompasses all of the forces.

It was James Clerk Maxwell (page 209) who combined the then-separate phenomena of electricity and magnetism through his four equations, and in so doing was able to show that electromagnetic waves also explained a third phenomenon of nature, namely light. Einstein's primary goal for much of his later life was to take that effort one step farther and unify electromagnetism with gravitation; his quest was ultimately unsuccessful.

Today, four fundamental forces of nature are known to exist:

- the gravitational force that was described classically by Newton's universal law of gravitation [Equation (5.7), page 168] and in modern form by Einstein's general theory of relativity (page 269),
- the strong nuclear force, a very short-range force that holds nuclei together (page 348),
- the weak nuclear force, a very short-range force of only about the diameter of a proton, that governs radioactive decay (page 254) and participates in nuclear reactions (e.g., Sections 8.8, 9.6, and 16.3), and
- the electromagnetic force, described as waves by Maxwell's equations (page 209) and in a modern view by photons (Section 8.5).

Our modern understanding of the last three forces is quantum mechanical in nature, but so far the gravitational force has steadfastly resisted quantum mechanics, as did the author of the general theory of relativity, Albert Einstein.

Today, physicists have been able to combine the electromagnetic force and the weak nuclear force into a very successful electroweak theory that functions at high energies. As required by science, the theory is testable in high-energy particle accelerators such as the Large Hadron Collider at CERN, the European Organization for Nuclear Research, located near Geneva, Switzerland.[7] At lower energies, the electroweak force separates into the electromagnetic and weak nuclear forces that we see in typical environments today. This separation happened just 10^{-11} s or 10 trillionths of a second following the Big Bang, leaving us with the four fundamental forces ever since. In other words, science has been able to test the nature of the universe to times earlier than 10^{-11} second following the birth of the universe.

An effort to combine the electroweak force with the strong nuclear force at even higher energies has had some success, although there are several competing models that still need to be resolved. These grand unified theories (GUTs) make predictions for very exotic particles of extremely high mass and energy that have not yet been observed; nor are they likely to be, given that the energies involved are far greater than anything currently achievable using Earth's most powerful particle accelerators, or by particle accelerators envisioned in the foreseeable future. However, it may be possible to deduce the existence of GUTs through subtle effects on known particles like neutrinos or the radioactive decay of protons, which have an estimated half-life of at least 10^{34} years, or roughly one trillion trillion times the current age of the universe.

Key Point
The four fundamental forces.

Key Point
The electroweak force separated into the electromagnetic force and the weak nuclear force 10^{-11} seconds after the Big Bang.

[7]CERN is the birthplace of the World Wide Web, originally created by Tim Berners-Lee (b. 1955) in 1989 for the purpose of sharing information among its researchers.

At the peak of the unification goal lies Einstein's original, but now extended, effort to marry gravitation with the three quantum-based fundamental forces. Doing so would also necessarily involve quantum gravity. Such a theory of complete unification is termed a theory of everything, or ToE. If such a unification of all of nature's fundamental forces did once exist in nature, it must have existed at unimaginably high energies (and temperatures) that are far beyond even those needed for GUTs.

Current efforts to develop a ToE involve an area of research referred to as string theory. The general idea is that particles can be described as vibrating strings and loops, where the characteristic sizes of these features are of the order of the Planck length. One consequence of these theories is the appearance of particles that carry information about gravity, known as gravitons, in the same way that photons carry information about the electromagnetic force. This means that string theory is a theory of quantum gravity. ToEs also typically include many more dimensions than the three dimensions of space and one dimension of time in the spacetime of the special and general theories of relativity. One version of a ToE postulates the presence of 11 dimensions, where the other seven dimensions are either "curled up" into immeasurably small size or our four-dimensional universe is embedded in dimensions that we cannot detect. This latter case would be similar to our ants on the two-dimensional surface of the balloon in Fig. 19.11 not knowing that the balloon's surface is inside a third spatial dimension that exists above and below the surface. Our developing understanding of the universe is getting very weird.

You might say that a ToE is the ultimate foundation upon which everything else stands, including GUTs.

After the Beginning

A Universe of ToEs and GUTs

Key Point
ToE separated into gravitation and GUTs at 10^{-43} s (the Planck time) and GUTs separated into the strong nuclear force and the electroweak force at 10^{-36} s.

The Big Bang produced a universe flooded with photons of enormous energy, consistent with the Planck temperature of 10^{33} K at a Planck time of 10^{-43} seconds. With photons of such energy, ordinary matter had no hope of existing, and fundamental forces were coupled together in a single unification. At about the Planck time, it is thought that gravitation split from the GUTs force, ending the ToE era. This means that gravity and GUTs ruled the universe until about 10^{-36} s, or one trillionth of a trillionth of a trillionth of a second after the Big Bang, at which point the strong nuclear force separated from the electroweak force. Meanwhile, as the universe expanded, the photons flooding the universe started to decrease in energy as their wavelengths were being stretched with the expansion. The increasing photon wavelengths also means that the temperature of the universe was dropping with universal expansion.

Cosmic Inflation

In the late 1970s and early 1980s, Alan Guth (b. 1947), Andrei Linde (b. 1948), and others developed a model for the very early universe that at once solved the CMB equilibrium-temperature problem, the CMB temperature-variations problem, the flatness problem, and others. As illustrated on the timeline in Fig. 19.18, starting

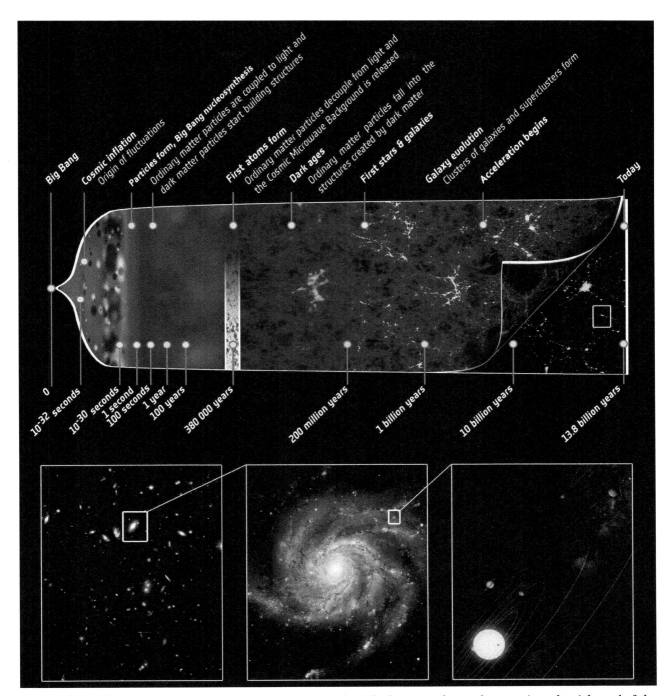

Fig. 19.18 *Top:* The history of the universe. The present day, and the location of our telescopes is at the right end of the diagram. Looking to the left we are seeing farther out into space and deeper in time. *Bottom row:* An enlargement of the white box near the lower right corner of the top figure. *Left:* A cluster of galaxies. *Center:* A large spiral galaxy indicating a location similar to where our Solar System would be in our Milky Way Galaxy. *Right:* Our Solar System. (Adaptation of a figure by NASA)

at about 10^{-36} second after the Big Bang, when the strong nuclear force separated from the electroweak force, until about 10^{-32} second or so, the universe experienced an incredibly fast period of expansion. Although this period of tremendous growth only lasted for about 10^{-33} second, the characteristic size of the universe increased by a factor of some 10^{26}, or one hundred trillion trillion. When this period of cosmic inflation began, the universe would have been about 100 billion times smaller than the size of a proton, but by the time it ended, the universe had grown to the size of a beach ball. After the inflationary period, universal expansion returned to a more normal rate.

To understand how inflation solves the CMB equilibrium-temperature problem, consider leaves changing color in the late fall, like those shown in Fig. 19.19. On a large scale, a leaf shows multiple colors, but if you zoom in on just one tiny section of the leaf you can still see the dominant green that it had during the summer months. In the extremely early universe it was impossible for information to travel from one side of the universe to the other, but very small sections (relatively speaking) could reach local equilibrium temperatures. As inflation took hold, those tiny sections of equilibrium were expanded dramatically and eventually grew into the uniform-temperature observable universe we see today.

Fig. 19.19 Fall leaves. (Dale A. Ostlie)

The CMB temperature-variations problem is solved when the extremely tiny quantum fluctuations due to Heisenberg's uncertainty principle were amplified and enlarged by 26 orders of magnitude because of inflation. Those quantum fluctuations at the beginning of the universe appear today as differences of only 0.0001 kelvins in the 2.7255 K cosmic microwave background.

Finally, inflation solves the flatness problem because space expanded so dramatically that any curvature was stretched out beyond any ability to measure it. Imagine our tiny ant on the balloon as the balloon expanded by a factor of one hundred trillion trillion under its feet. The ant would have no hope of telling if it lived on a balloon, on an infinitely flat sheet, or on an incredibly big saddle.

Scientists haven't yet been able to explain what drove inflation in the first place, and why it ended, but of course science doesn't yet have a satisfactory understanding of dark energy either, but it is encouraging that inflation does seem to explain a host of puzzles. At least particle physicists and cosmologists have the GUTs to explore these questions while dipping their ToEs in the primeval soup that immediately followed the Big Bang.

Quarks and Gluons

It was when the electroweak force separated into the weak nuclear force and the electromagnetic force at about the 10^{-11} s (or 10 picoseconds) mark that particles started to appear out of the cooling Big Bang fireball; in particular, exotic particles known as quarks and gluons (recall the brief mention on page 301). Unlike electrons, which are not made of anything else other than perhaps cosmic strings, protons and neutrons are composed of quarks. The quarks themselves are "glued" together inside protons and neutrons by gluons that are associated with the strong nuclear force that had just separated from the electroweak force. (Each proton and neutron is composed of three quarks.)

Today you won't find any free-roaming quarks, because gluons are really good at their job. To pull a pair of quarks apart you would have to add so much energy that you would form a second pair of quarks out of the energy that was put into the effort (remember that mass is just one form of energy; $E = mc^2$). But that wasn't the case in the still-extreme temperature of the universe following inflation; quarks and gluons were flying around unbound.

It was also about this time that the universe decided to give an advantage to matter over antimatter, although we don't yet understand why. Matter dominating over antimatter is a good thing for us so that we don't get hit by antimatter particles that would instantly annihilate particles in our matter selves to produce pairs of gamma-ray photons.

The Appearance of Protons, Neutrons, Electrons, and Neutrinos

By the time the universe reached 10 microseconds (10 μs) of age, its temperature had dropped to a cool 1 trillion kelvins. The drop in temperature and decreasing density were (and still are) associated with universal expansion. The temperature itself is a measure of the energy of the photons. Remember that as the universe continues to expand, the wavelengths of photons are stretched (the cosmological redshift), which corresponds to lowering energies and decreasing temperature.

At 1 trillion K, the universe was now cool enough for gluons to do their job of combining quarks together into more massive particles, including protons and neutrons. Shortly after protons and neutrons formed, electrons and positrons began to appear, along with neutrinos and antineutrinos. The universe was beginning to contain particles we are more familiar with.

When protons and neutrons first appeared, they did so in about equal numbers, but reactions with positrons and neutrinos began to skew the neutron/proton ratio toward more protons because protons are slightly less massive, and therefore have slightly less energy ($E = mc^2$) than neutrons:

$$n + e^+ \rightleftharpoons p^+ + \bar{\nu}_e$$
$$n + \nu_e \rightleftharpoons p^+ + e^-.$$

(The double arrows mean that the reactions can go both directions.) Isolated neutrons not contained in nuclei (that didn't yet exist) can undergo radioactive decay (page 257) with a half-life of 611 seconds, which produces additional protons and removes neutrons from the mix:

$$n \rightarrow p^+ + e^- + \bar{\nu}_e.$$

When the universe was one second old and the temperature had dropped to 10 billion kelvins, these reactions stopped, leaving a neutron/proton ratio of 1/7. These massive particles were still overwhelmed by massless photons, however, with photons outnumbering protons and neutrons by 2 billion to one. With so many high-energy photons around, the universe is said to have been radiation dominated.

It is important to note that we are now in a situation where the physics is actually very well understood, given that the temperature and density of the universe have decreased to levels that can be studied experimentally today.

Key Point
Normal matter started to form when the temperature of the expanding universe dropped below 1 trillion kelvins, 10 μs after the Big Bang.

Big Bang Nucleosynthesis

Key Point

When the temperature of
the universe had dropped
below 1 billion kelvins, 10
seconds after the Big Bang,
helium nuclei started to
form, with the reactions
lasting for 20 minutes.

When the temperature of the universe had dropped to about 1 billion kelvins, 10 seconds after it began, the strong nuclear force could finally hold nuclei together against the bombardment of now less-energetic photons. Because single protons are also $^1_1$H nuclei, there was already a universe full of hydrogen nuclei and free-roaming neutrons, but it wasn't until the 10-second mark that more massive nuclei could be built, which were almost exclusively nuclei of $^4_2$He. As with the proton–proton chain of hydrogen-burning in the Sun (Fig. 9.33), a series of steps were needed during Big Bang nucleosynthesis to produce helium-4 out of hydrogen-1:

$$^1_1\text{H} + n \rightarrow {}^2_1\text{H} + \gamma \qquad \text{(deuterium and a photon)}$$
$$^1_1\text{H} + {}^2_1\text{H} \rightarrow {}^3_2\text{He} + \gamma$$
$$^2_1\text{H} + {}^2_1\text{H} \rightarrow {}^3_2\text{He} + n$$
$$^2_1\text{H} + {}^2_1\text{H} \rightarrow {}^3_1\text{H} + {}^1_1\text{H} \qquad \text{(tritium and hydrogen)}$$
$$^3_2\text{He} + {}^2_1\text{H} \rightarrow {}^4_2\text{He} + {}^1_1\text{H}$$
$$^3_1\text{H} + {}^2_1\text{H} \rightarrow {}^4_2\text{He} + n.$$

Very tiny amounts of lithium-7 were also produced when tritium reacted with helium-4:

$$^3_1\text{H} + {}^4_2\text{He} \rightarrow {}^7_3\text{Li} + \gamma.$$

Key Point

After nucleosynthesis (by
mass): 75% hydrogen, 25%
helium-4, 0.01% deuterium
and helium-3, and a tiny bit
of lithium-7.

Twenty minutes after the Big Bang, nucleosynthesis stopped because the temperature and density of the expanding universe had dropped too low. When it was all over, for every 12 hydrogen nuclei, 1 helium-4 nucleus had been created. This corresponds to hydrogen-1 making up about 75% of the mass of all nuclei and helium-4 making up almost all of the remaining mass (25%), although a tiny amount (about 0.01% of the mass) was a mix of deuterium and helium-3, and 0.000 000 0001% of the mass was lithium-7.

Key Point

Some aspects of the theory
of Big Bang nucleosynthesis
have been tested
experimentally.

In 2020, at the Laboratory for Underground Astrophysics (LUNA) at the Gran Sasso National Laboratories in Italy, scientists were successfully able to measure the reaction rate for $^1_1$H + $^2_1$H by simulating the environment during Big Bang nucleosynthesis. The LUNA experiment fired 100 trillion protons (hydrogen nuclei) at a deuterium target every second, and found that several fusion reactions took place every day. From those extremely rare reactions, it was possible to extrapolate to determine the expected abundances of hydrogen, deuterium, helium-3, and helium-4, based on the experimentally measured hydrogen–deuterium reaction rate. The results agreed very well with theoretical expectations. All of the hydrogen-1 in your body was produced when quarks combined 10 μs after the Big Bang (remember that hydrogen is contained in water and organic molecules). A few grams of deuterium from Big Bang nucleosynthesis also exists inside your body today.

The situation isn't quite as positive for the abundance of lithium-7. As we understand it today, the only way that lithium-7 could have been produced was during Big Bang nucleosynthesis. The measured abundance of lithium-7 exceeds the theoretically expected value by a factor of three. Even though the abundance of lithium-7 is extremely low, the discrepancy is something that a complete understanding of the first 20 minutes following the Big Bang must still explain. As always, science must explain all of the data, convenient or otherwise.

The Formation of Neutral Atoms and the Appearance of the CMB

Forming atoms by binding electrons to nuclei is governed by the electromagnetic force. This couldn't happen during or immediately following Big Bang nucleosynthesis because the electromagnetic force is much weaker than the strong nuclear force that governs the formation of nuclei; the universe was still far too hot, and photons had far too much energy, to allow atoms to form.

It wasn't until the temperature of the universe dropped below about 100,000 K that photon energies had decreased enough so that electrons could finally begin attaching to nuclei to form atoms without the photons immediately destroying them again. This atom-building period didn't start until 300,000 years after Big Bang nucleosynthesis had ended and it lasted for 100,000 years. We will refer to this period as the era of the first atoms.

Big Bang photons had been constantly bouncing around off free electrons and charged nuclei in a dense mix before atoms started to form. But with the formation of atoms, the universe finally became transparent to those photons. This means that about 380,000 years after the Big Bang, photons could finally move around unimpeded. What we see today as the cosmic microwave background (page 784) are the photons that decoupled from matter during the era of the first atoms.

Looking deep into space and back in time we encounter the surface of last scattering. When we study CMB photons, we are studying the environment of the universe at the surface of last scattering where photons decoupled from matter. That environment depended on all of the processes that led up to that point in time, including Big Bang nucleosynthesis. The surface of last scattering also marks the boundary where the universe transitioned from being radiation-dominated to being matter-dominated.

As shown in Fig. 19.18, observing the early universe deeper into space and time beyond the surface of last scattering isn't currently possible, because it is filled with a fog of scattered photons. The problem is similar to trying to look deeper into the Sun than the photosphere: we must probe the Sun's interior using indirect methods based on our understanding of stellar astrophysics. What we have described about the early universe beyond the surface of last scattering is based on our understanding of quantum mechanics and gravitation, combined with some informed speculation regarding GUTs and ToEs, and what the universe looks like on our side of the surface of last scattering. The value of the redshift parameter at the surface of last scattering that is the boundary beyond which we cannot look deeper into space and farther back in time is $z = 1089$.

It is important to understand that when light decoupled from matter, the liberated photons didn't head off somewhere where matter wasn't present because there wasn't anywhere where matter wasn't present; both light and matter filled our observable universe at that time, as it does now. The decoupling was happening everywhere across the observable universe at the same time after expansion had lowered photon energies enough for atoms to form. Decoupling simply means that photons weren't scattering off charged particles anymore. Don't forget that when we look into deep space, we are looking back into deep time; we are seeing our observable universe at the surface of last scattering as it was 380,000 years after the Big Bang. Today's universe is still bathed in photons that were released from the surface of last scattering, along with photons that were created from physical

Key Point
The first atoms of hydrogen and helium formed between 300,000 and 400,000 years after the Big Bang.

Key Point
Photons of the cosmic microwave background began to decouple from matter when the first atoms formed.

Key Point
The surface of last scattering is the boundary beyond which we can't see with our telescopes. It is located at $z = 1089$.

processes today and at every period of time going back to 380,000 years after the Big Bang. When we look back 4.4 billion years we are seeing photons that were produced then and are finally reaching us now, just as the CMB photons are reaching us now from nearly 13.8 billion years ago.

As research continues, science should be able to refine our understanding of the earliest moments of our observable universe. Science is always a work in progress.

Forming The Cosmic Web

Our observable universe wasn't completely inactive between the end of Big Bang nucleosynthesis and the surface of last scattering. Along with all of the high-energy photons that appeared with the Big Bang, dark matter (page 747) was produced as well, making up 84% of all of the mass in the universe. Dark matter reacts gravitationally with other dark matter and ordinary matter, but it isn't influenced by the electromagnetic force, the strong nuclear force, and probably not the weak nuclear force. This means that dark matter can be gravitationally attracted to other dark matter particles but any gravitational clumping of dark matter can't be stopped by gas and radiation pressure, as happens in stars, for example. This is because both gas pressure and radiation pressure depend on the electromagnetic force.

Think back to Fig. 19.16 depicting the extremely tiny temperature variations in the cosmic microwave background that are the result of quantum fluctuations before inflation. Those tiny temperature fluctuations also correspond to very small density fluctuations in the early universe. The regions that are slightly cooler than average (the blue regions) are regions of slightly greater densities and stronger gravitational attraction, while regions slightly warmer than average have slightly lower densities and weaker gravitational attraction. Being unaffected by the very high-energy photons of the radiation-dominated universe, dark matter particles began to collect within the higher density regions. In this way, structures began to appear in the very early universe prior to the release of the cosmic microwave background photons at the surface of last scattering.

Following the decoupling of photons from matter and the transition to the matter-dominated universe, atoms throughout the universe began to be pulled gravitationally to the structures formed by dark matter. The result was an intricate and vast web of matter, both ordinary and dark, intertwined throughout the universe. This was also a period referred to as the dark ages, where the only light in the universe was the remnant CMB fireball of the Big Bang; after all, there still weren't any stars to generate new light.

Figure 19.20 shows a computer simulation of the cosmic web that was first mentioned on page 757. The simulation required one month of supercomputer time at the Max Planck Institute for Astrophysics and is based on 10 billion different point masses representing masses in the early universe. The actual cosmic web, similar to the simulation, has been observed directly as voids, walls of galaxies, clusters and superclusters of galaxies around the voids, and filaments composed of strings of galaxies, galaxy clusters, and superclusters. Observations have also revealed that up to one-half of all of the mass of ordinary matter in the universe may be contained in hot ionized gas along filaments within the cosmic web that has not been pulled

Key Point
Dark matter makes up about 84% of the mass of the universe and ordinary matter makes up the other 16%.

Key Point
The cosmic web formed when dark matter began collecting in regions of space where the density was slightly greater than other regions. The regions are the result of the quantum fluctuations that produced the tiny variations in temperature of the CMB.

Key Point
Galaxies and galaxy clusters and superclusters formed along filament and wall structures in the cosmic web. One-half of all ordinary matter remains as hot ionized gas in intergalactic space within filaments.

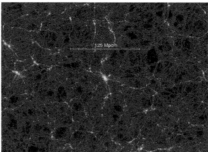

Fig. 19.20 The Millennium simulation of the cosmic web based on 10 billion test particles. The three frames are zooming in on a small portion of the simulation; The white scale lines correspond in length to from left-to-right: 1600 Mly (500 Mpc), 400 Mly (125 Mpc), and 100 Mly (31.25 Mpc). The closest view shows a structure in the center that is similar to the size of the Virgo supercluster. A movie of the simulation is available on the textbook's website. (Springel et al., 2005[8])

into galaxies and galaxy clusters. Note in the right-hand frame of Fig. 19.20 that mass also tends to accumulate at nodes in the cosmic web. As science demands, these theoretical simulations and others like them are strongly supported by observational evidence.

The First Stars and Galaxies

As gas fell into the gravitational wells of dark matter that collected at nodes and along filaments in the cosmic web, hundreds of thousands to perhaps one million solar masses of gas collected in regions only 100 light-years across.

This situation is not unlike the interstellar nurseries of dust and gas where star formation is occurring today, as discussed in Section 16.1. There are two important differences, however: the gas that produced the first stars didn't contain any elements heavier than helium (and a tiny amount of lithium), and the cloud temperatures were much higher (typically 1000 K instead of 10 K). Both of these conditions meant that the first stars to form would have been much more massive than today's stars. Computer models suggest that many of the first stars likely had masses between 100 M_{Sun} and 1000 M_{Sun}, much greater than today's most massive stars.

When the first stars ignited due to the collapsing gas clouds, suddenly the dark ages ended with spectacular bursts of new light about 200 million years after the Big Bang, corresponding to a redshift parameter of $z = 20$ and a distance from Earth of roughly 36 Gly. This means that, in order to see the very first stars and galaxies, we need to look 36 Gly away and 13.6 billion years into the past. Because these first stars were so massive, their surface temperatures were much higher than today's stars and their radiated light was in the ultraviolet portion of the electromagnetic spectrum (see Fig. 6.43). This intense, high-energy light ionized the hydrogen and helium gas in the neighborhoods of the newly formed stars, creating ionized-gas bubbles surrounding the stars. The intense radiation then blew the ionized gas outward, filling the eventual intergalactic medium with the ionized gas seen today.[9]

Key Point
The first stars ignited about 13.6 billion years ago, or 200 million years following the Big Bang. Those stars, and the galaxies in which they live, are 36 Gly away at $z = 20$.

[8]Springel, Volker, White, Simon D. M., Jenkins, Adrian, Frenk, Carlos S. (2005). "Simulations of the formation, evolution and clustering of galaxies and quasars," *Nature*, 435(7042):629–636.

[9]Astronomers refer to this time when the first stars formed as reionization, because all matter had already been ionized before the decoupling of photons with matter.

Key Point
The first stars with masses greater than 250 M_{Sun} created black holes that merged into supermassive black holes.

Because of their high masses, these first stars had short lifetimes. Stars with masses greater than 250 M_{Sun} simply collapsed at the end of their lifetime to form black holes. Over time, these unusually massive black holes would merge with others in their dark matter halos to form the supermassive black holes found in all of today's large galaxies (Section 18.5). Often the sources of active galactic nuclei, in the early universe these adolescent monsters devoured surrounding gases and stars, appearing today as the quasars we see when looking into the distant past.

Key Point
The first stars with masses between 100 and 250 M_{Sun} created the first heavy elements that enriched later star formation.

First-generation stars between 100 and 250 M_{Sun} exploded at the ends of their lives as core-collapse supernovas (page 697). However, instead of forming neutron stars or black holes, as core-collapse supernovas do today, those stars blasted all of their processed heavy elements into the cosmos, allowing star formation to proceed like it does now. This also means that planetary systems could form from those heavy elements, including one a few stellar generations later that formed around a 1 M_{Sun} G2 star with a small rocky planet in the star's Goldilocks zone. That planet, with an iron–nickel core and a large moon, would develop life on it, perhaps around a hot thermal vent or in a shallow pond, that would evolve enough for students to study astronomy and better understand their common origin as a species.

19.5 The Λ-Cold Dark Matter Model

To have a complete model for the expansion and evolution of the universe, all of the observational data must be included. From what you have learned so far in this chapter, this means that the model must incorporate:

- the Hubble–Lemaître law (page 777) and Hubble's constant (H_0),
- cosmic inflation (page 798),
- the cosmic microwave background (page 784) and its subtleties (page 793),
- 84% of all of the mass in the universe is cold dark matter (CDM) and the remaining 16% is ordinary matter (page 804),
- the relative amounts of hydrogen-1, helium-4, deuterium, helium-3, and lithium-7 present in the early universe before the formation of the first stars (page 802),
- the cosmic web (page 804),
- the formation of heavy elements from stars that originally only contained hydrogen and helium (page 806),
- the formation of supermassive black holes (page 806), and
- the accelerating universe and dark energy (the cosmological constant, Λ; page 788).

All of these features are consequences of the Λ-cold dark matter (Λ-CDM) model of cosmology, sometimes referred to as the standard model of cosmology.

Key Point
Of the total energy of the universe, in excess of 69% is dark energy, more than 26% is dark matter, and less than 5% is ordinary matter.

As the name suggests, the Λ-CDM model explicitly includes the dominant forms of energy in the universe, dark energy (symbolized by Λ), cold dark matter (CDM), and ordinary matter (even though it isn't in the title); see Table 19.1. Using $E = mc^2$ to write the mass of cold dark matter and ordinary matter in terms of energy, observational data from the Planck satellite shows that dark energy accounts for more than 69% of the energy of the universe, cold dark matter accounts for more than 26%, and ordinary matter is only about 5%. In other words, less than 5% of the energy contained in the universe is composed of stuff we really understand like

Table 19.1 The components of the Λ-CDM model. The first row indicates the percentage of mass–energy in the universe that each component accounts for. The remaining rows indicate whether or not the component is influenced by each fundamental force.

	Dark energy	Cold dark matter	Ordinary matter
Energy content	69%	26%	5%
Gravitational	no	yes	yes
Electromagnetic	no	no	yes
Strong nuclear	no	no	yes
Weak nuclear	no	probably not[a]	yes

[a]Experiments appear to rule out the weak nuclear force, but research is still ongoing.

protons and neutrons (other particles, such as electrons, neutrinos, and photons, make up a very small percentage of the energy of the present-day observable universe)! Until fairly recently, astronomy had focused exclusively on what we can see and measure through telescopes: less than 5% of the energy content of the universe.

Despite our lack of knowledge about the details of the non-luminous ("dark") components of the universe that make up 95% of its energy, the Λ-CDM model does a remarkably good job of describing the evolution of the observable universe from the Planck time to today. The model includes cosmic inflation in the first moments following the Big Bang, the era when radiation dominated the universe, the matter-dominated era, and now universal acceleration that began 4 1/2 billion years ago when the universe transitioned to being dominated by dark energy. Going forward, dark energy will continue to dominate the evolution of the universe, with all of the stuff that we can see being relegated to a minor support role in the continuing pageant of the universe.

Key Point
The universe has transitioned from being radiation-dominated to matter-dominated, and now to dark-energy-dominated.

Precision Cosmology

The cosmic microwave background provides a tremendous amount of information about the early universe. We have already seen that it tells us that our observable universe was able to reach a common equilibrium temperature before the surface of last scattering, as evidenced by the almost perfect blackbody spectrum of the CMB displayed in Fig. 19.10. We also know that quantum fluctuations in the moments after the Big Bang are seen today as the tiny temperature and density differences in the CMB (Fig. 19.16), which in turn led to structure in the universe in the form of the cosmic web. But even more information can be be mined from the data.

A consequence of the tiny temperature and density fluctuations in the early universe that show up in the CMB means that there should be evidence of sound waves that bounced around within those fluctuations. Since sound waves are just pressure changes that move from one place to another in solids, liquids, or gases, it is possible for sound waves to exist in space with extremely low gas densities, but those sound waves would have extremely long wavelengths! Back on page 352, when you studied helioseismology, you learned that wavelengths in guitar strings depend on the length of the strings. In the Sun and pulsating stars (Section 16.4), wavelengths

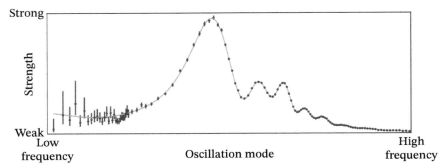

Fig. 19.21 The strengths of different modes of sound oscillations in the early universe. The dots (and error bars at low frequencies) are the observational data and the blue line is the theoretical fit to the data. (Adapted from ESA and Planck Collaboration, 2020[10])

depend on the size and characteristics of stars; the same is true in space. With density differences in the early universe, standing waves can be set up that depend on the density of the gas and the characteristic sizes of density fluctuations. Figure 19.21 shows measurements of different harmonics (or modes) imprinted on the CMB. In the same way that helioseismology, asteroseismology, and the study of earthquakes, moonquakes, and vibrations in other bodies can allow us to probe interiors that we can't directly measure, the harmonics in the CMB give precise information about the environment of our observable universe before the decoupling of radiation from matter. Remarkably, there is evidence that the influence of harmonics in the early universe persists in the large-scale structure of the observable universe today, through the distribution of galaxies in the cosmic web.

The CMB also provides information through the amount of polarized light (page 211) that is measured. Because photons were bouncing off electrons an enormous number of times before decoupling, the directions of those scatterings would have been completely random, with no preferred direction and therefore no light polarization. This changed when the first massive stars were born, because their intense ultraviolet light reionized gas surrounding the stars, and electrons were free to roam around again. When the CMB photons started scattering off the newly freed electrons, some light polarization occurred again. The amount of light polarization seen today gives information about when the first stars formed.

In order to completely understand the details of the cosmic background radiation, scientists must also account for the amount of energy lost by photons climbing out of the cooler and denser fluctuations (gravitational redshift). Gravitational lensing effects need to be included in the analysis as well.

When every influence on the cosmic microwave background is accounted for, and when the data are included in the Λ-cold dark matter model, a very detailed picture of the early observable universe emerges. It is from this work that we now know that less than 5% of the mass–energy of the observable universe is due to ordinary matter, a little more than 26% is due to cold dark matter, and the rest is from dark energy. The application of the Λ-CDM model also predicts the amount of hydrogen-1, helium-4, deuterium (hydrogen-2), helium-3, and lithium-7 that should have been present before the first stars; a prediction that generally agrees

[10]Planck Collaboration (2020). "Planck 2018 Results VI. Cosmological Parameters," *Astronomy & Astrophysics*, 641:A6–A74.

very well with observations, although some disagreement between theory and observations still exists regarding the abundance of lithium-7. In addition, the model data show that the geometry of the observable universe should be flat to a high degree of precision, and that the universe began accelerating nearly 4 1/2 billion years ago.

As science demands, a complete theory of nature must include all known observational and experimental data in a self-consistent way. The theory must also make testable predictions that may one day lead to falsification of the theory when predictions can no longer be reconciled with new results.

What about the Hubble Constant?

One of the products of the Λ-CDM model is the ability to determine the Hubble constant from data in the early universe. When the Planck satellite was launched in 2009, it conducted the most precise observations of the cosmic microwave background ever made. Following the completion of its data-gathering mission, another five years were spent analyzing the tremendous amount of data that was collected. Those data and the subsequent analysis gave a value for the Hubble constant of

$$H_0 = 20.7 \text{ km/s per Mly} = 67.7 \text{ km/s per Mpc}. \tag{19.8}$$

The Hubble constant (from measurements of the early universe)

If you compare this result with the result from distance-ladder measurements on page 780, you will notice that the two results differ by almost 10%. They actually differ so much that their error bars (discussed on page 177) don't overlap, implying that the difference between the two results is statistically significant. A tremendous amount of effort has gone into checking measurements, looking for mistakes, and seeing if detectors are giving systematically high or low results, but the discrepancy has persisted.

The smaller value of the Hubble constant from the Planck satellite data also implies an older age for the universe relative to the distance-ladder results. The Hubble time from the Planck data is

$$t_H = \frac{1}{H_0} = 4.561 \times 10^{17} \text{ s} = 14.5 \text{ Gy}. \tag{19.9}$$

The Hubble time (from measurements of the early universe)

Remember that the Hubble time assumes that the expansion of the universe has been constant throughout its history, so its value doesn't actually represent the true age of the universe. When all the details of an expanding universe are taken into account, including cosmic inflation, followed by a slowing expansion and then an acceleration due to dark energy that began almost 4 1/2 billion years ago, the current age of the universe is calculated to be 13.8 billion years, the value that has been used throughout this textbook.

At least at the time your textbook was written, astronomers have not been able to resolve the difference between the two sets of data. Are there still some undetected errors in one or both measurements? Given that the distance-ladder data are

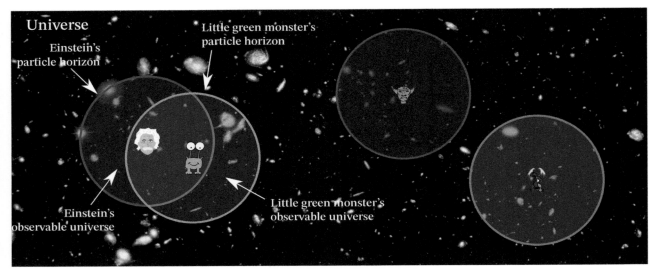

Fig. 19.22 Four different observable universes. The blue and green observable universes have some galaxies in common, but the red and teal ones do not overlap with any of the others shown. In reality, all of the objects seen in the background image are a small part of our, and Einstein's, observable universe because the image is a portion of the Extreme Deep Field image captured by the Hubble Space Telescope. (Adapted from NASA; ESA; G. Illingworth, D. Magee, and P. Oesch, University of California, Santa Cruz; R. Bouwens, Leiden University; and the HUDF09 Team)

sampling the relatively nearby universe, and therefore the relatively recent past, while the Planck satellite data are sampling the observable universe only 380,000 years after the Big Bang, are the results saying that there is something fundamental that is missing in our understanding of how the universe evolves? Is it dust ... again?! (page 731).

Science is a constant human quest for understanding.

19.6   The Meaning of "The Universe"

Can We See the Entire Universe?

Key Point
The particle horizon is a spherical "surface" from which light could theoretically reach the observer from the earliest moments of the universe. The particle horizon contains the observer's observable universe.

The answer to that question is no. What we do see is a part of the universe that is centered on us. Actually, your author sees a portion of the universe that is centered on him, someone else at a different location sees a part of the universe centered on her, and someone or something living somewhere in the Virgo cluster sees a part of the universe centered on his/hers/its location. Each of us sees out to a distance that corresponds to a sphere centered on us where photons (light particles) that left the "surface" of that sphere had time equal to the age of the universe to reach us. Since we are in different locations, each of us has our own personal maximum spherical "surface." That personal spherical surface is our particle horizon and the portion of the universe that each of us sees within our particle horizon is our personal observable universe.

Figure 19.22 depicts four different observable universes within the likely infinite universe. Einstein's observable universe overlaps the observable universe of the little green monster, meaning that they can both see some of the same galaxies.

Given that both Einstein and the little green monster are located barely within each others' observable universe, they can also see each others' galaxy, but in each others' distant past. Einstein would not yet exist when the little green monster studies the Milky Way and the little green monster wouldn't have been born (hatched?) when astronomers on Earth study its galaxy. In fact, given how close each being is to the particle horizon of the other's observable universe, they would likely see each other's galaxy while it was in its infancy.

It is often the case that someone (including an astronomer) might use the term or phrase "universe," "the universe," or "our universe" when the implication is our "observable universe," as opposed to the probably infinite universe of which our observable universe is only a portion. It is important to pay attention to context.

How Far Away is the Particle Horizon?

It may seem as though the distance to the particle horizon ought to be the speed of light times the age of the universe, giving 13.8 billion light-years, but that turns out not to be the case.[11] The culprit that messes up that simple argument is again the expansion of the universe. Refer back to the expanding sidewalk analogy of Fig. 19.15 for some additional insight. When the information carriers started walking toward one another just inside the red cracks, those cracks were moving away from the middle of the sidewalk at the speed of light. During the time it took for light to travel from just inside the red cracks to the middle, those red cracks kept moving farther and farther away. When the information carriers meet in the middle, their starting positions are now much farther away than when the information carriers started walking, but the information carriers are still carrying the information they had when they started their journeys. If you refer back to Fig. 19.14, this is why how far we look back in time to see receding galaxies doesn't correspond to a distance determined by simply multiplying the speed of light by the light-travel time; for example, if a galaxy is seen 8 billion years in the past, that galaxy is now 11.2 billion light-years from Earth. Today, the source of the 2.7 K CMB is more than 45 Gly away (almost 13.9 Gpc) and the particle horizon is 46.8 Gly (14.3 Gpc) distant, with a redshift parameter z that is infinity large.

Now imagine that the information carriers started toward the middle while located exactly on the red expansion cracks. Since those cracks are moving away from the observer at the speed of light, and the information carriers are moving toward the observer at the speed of light, the information carriers never will reach the middle, according to the observer. If the information carriers started their journeys beyond the red cracks, even though they are traveling at the speed of light relative to that local space (an expansion crack), the expansion of space will still carry them farther and farther away, relative to the observer.

At this point it is important to remember the postulates of Einstein's special theory of relativity on page 276: all observers must agree on the results of all experiments or observations, and all observers always measure the speed of light in a vacuum to be $c = 299{,}792{,}458$ m/s. The postulates remain true regardless of how fast the universe is expanding. Once again, this is because the speed of light is measured with respect to locations or objects *in* space, say between a distant galaxy and Earth, even though the distance between those locations or objects is increasing due

---

[11]This is a common mistake in news articles.

to the expansion *of* space (the distance between cracks in the expanding sidewalk is increasing). The effect on light is not to change its speed as measured by those objects, it is an expansion (or stretching) of the wavelength of the light, which is the cosmological redshift.

In reality, the situation is a bit more complicated than that. During cosmic inflation following the Big Bang, the stretching of space was carrying every photon away from our future location at faster than the speed of light. But as time went on, universal expansion slowed, eventually allowing photons to start moving toward our location. Imagine walking on the expanding sidewalk, but the entire sidewalk was stretching faster than you can walk. Even though you are walking toward the observer, you are being carried away from the observer. However, as the expansion slowed, the movement of the sidewalk decreased below walking speed and you were finally able to get closer to the observer. The particle horizon is that location where photons were emitted in the direction of our future position but were initially being carried farther away for a time before universal expansion slowed prior to 4.4 billion years ago, or $z > 0.4$, thus allowing the photons to begin making progress in closing the distance and finally reach us today.

The Hubble Horizon

Given the complex nature of the particle horizon, how far away is the location where galaxies are being carried away from us at faster than the speed of light today? The answer is closer than you may think. This location, known as the Hubble horizon, is given by the speed of light multiplied by the Hubble time. From the value for the Hubble time based on Planck data (page 809), the distance to the Hubble horizon turns out to be about

$$d_H = ct_H = \frac{c}{H_0} = 14.5 \, \text{Gly} = 4.4 \, \text{Gpc}. \tag{19.10}$$

Distance to the Hubble horizon (from measurements of the early universe)

which is much closer than the particle horizon. This distance to the Hubble horizon is called the Hubble distance. The Hubble distance corresponds to a redshift parameter of $z \approx 1.5$, not much farther out than the farthest white dwarf supernovas included in the Hubble–Lemaître law shown in Fig. 19.14.

The Hubble horizon isn't really a horizon in the sense that you can't see beyond it; the Hubble horizon is just that location in space where objects are moving away from us at faster than the speed of light.

Key Point
Expanding the expanding
sidewalk analogy.

To confuse the issue of the expansion of the universe a bit more, let's return to the expanding sidewalk analogy one last time to complete the story. All information carriers anywhere along the expanding sidewalk actually started out being carried away from the observer at faster than their walking speed, because the expansion of the universe started out faster than the speed of light. But as the speed of the expanding sidewalk began to slow down, more and more of the sidewalk slowed to sub-walking speed relative to the observer and the information carriers could start moving closer to the observer, with the ones that started at the sidewalk's "particle horizon" finally reaching the observer today. But at some point in time the speed of expansion of the sidewalk started to increase again, even before the information

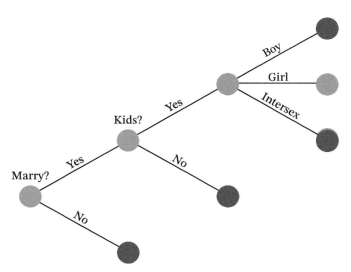

Fig. 19.23 An example of a many-worlds multiverse based on possible outcomes of events or decisions. A universe in which a couple has a girl after a proposal is indicated by the green universe sequence. The red universes exist in the many-worlds multiverse but are not universes in which a girl is born to the couple.

carriers from the sidewalk's "particle horizon" made it to the observer. Today, the acceleration has progressed enough that if information carriers beyond the sidewalk's "Hubble horizon" started their journeys now, they would never be able to reach the observer in the future, simply because they are already being carried away by the sidewalk faster than they can walk relative to the sidewalk. You could always test this yourself assuming that you have a very long, variable-speed treadmill handy! Start out by walking in the opposite direction of the motion of the treadmill, with the speed of the treadmill being faster than you can walk. Next, start slowing the treadmill down enough to make progress toward the observer at the end of the treadmill. If you have enough time you will reach the observer. However, as the treadmill starts speeding up again if you don't reach the observer soon, you never will because the constantly accelerating treadmill will eventually be moving faster than you can walk again, but this time it won't slow down and you will never reach the observer.

Are There Other Universes?

The likely infinite universe that we have been discussing throughout this textbook is the one in which we live (obviously), including our observable universe and any other observable universes within it. It came into existence at the time of the Big Bang and has been evolving ever since, but could there be other universes as well? The definitive answer is a resounding who knows? A hypothetical collection of universes is known as a multiverse and the universes within the multiverse go by a variety of names, including parallel universes, alternate universes, other universes, and many-worlds.

There are many proposed multiverse theories. One of those derives from what is known as the many-worlds interpretation of quantum mechanics. Think back to Schrödinger's cat in Fig. 8.37. That thought experiment raised the question of what happens to a cat if it is isolated in a box with a device that has a 50–50 chance of killing the cat? When the observer opens the box, is the cat dead or alive? In the many-worlds interpretation of quantum mechanics, the answer is yes! The idea is that during the event two universes appear, one in which the cat is dead and another in which the cat lives. Every time a decision point occurs in the probabilistic nature of quantum mechanics, all possible outcomes occur, just in different universes. If you asked someone to marry you, there is a universe in which the love of your life said yes and another in which that person turned you down. One example, as depicted in Fig. 19.23, shows a universe in which a couple agrees to marry and they eventually have a girl. Other universes exist, where they don't marry, or don't have kids, and so on. Besides those branches, an enormous number of other branches exist with associated universes that develop from all possible outcomes of events: get a haircut, turn right or left, go up the stairs now or later, hit a home run or fly out to deep center field, choose to skip the rest of this chapter (not a good choice by the way), ... However, it is important to note that there is no observational or experimental evidence for the many-worlds multiverse.

String theory, cosmic inflation, and quantum mechanics suggest that perhaps there are independent regions across an infinite universe of eternal cosmic inflation where that inflation stops. These regions in such a multiverse are sometimes referred to as **bubble universes**. For an infinite eternal-inflation universe there would naturally be an infinite number of infinite bubble universes. It is possible that these bubble universes could even have differing physical laws or fundamental constants. In this way, the universe that we live in with its highly tuned constants that are needed for us to exist is just one of an infinite number of possibilities. At this time there is no evidence that disconnected, separate bubble universes exist.

Yet another multiverse theory also derives from string theory and the attempt to develop a theory of everything (ToE). Gravity is by far the weakest of the four fundamental forces, being a trillion trillion times weaker than the weak nuclear force. From everyday experience, it may not be apparent that gravity is so weak, but think about what happens when you pick up your cellphone off a table. In that simple, act you are easily overcoming the gravitational pull of the entire planet on your cellphone. One of the great challenges of developing an all-encompassing theory of everything that includes quantum gravity is to explain why gravity is so weak.

Brane theory (think *membrane*) proposes that our universe exists on the "surface" of a brane, in a higher dimensional space known as the "bulk." A lower dimensional analogy for branes and the bulk are illustrated in Fig. 19.24. Branes within the bulk contain universes that could have physical laws and constants different from our own. It has been suggested that if two branes collide in the bulk, their universes are destroyed and new big bangs occur, creating new universes that expand and evolve. The feeble nature of gravity is explained in the theory because, unlike the three fundamental forces in GUTs (the electromagnetic force, the strong nuclear force, and the weak nuclear force) that only exist on our universe's brane, the long-range gravitational force can "leak" through the small fourth spatial dimension of our brane and into the bulk, interacting with mass in other brane universes. The gravity that we experience is only that portion that is contained within our brane. For sufficiently advanced civilizations, perhaps it would be possible to

Key Point
Gravity is by far the weakest of the four fundamental forces.

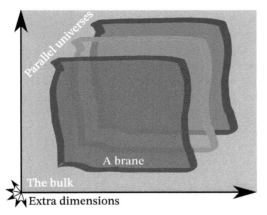

Fig. 19.24 Hypothetical branes. For each brane, two of three standard spatial dimensions are represented, although many others may exist as well. In addition to a time dimension, a small fourth spatial dimension (a "thickness") is also associated with each brane. Perhaps an infinite number of branes could be floating and moving around in the multidimensional bulk. Parallel universes exist on the "surfaces" of branes.

communicate between branes, and therefore between universes, by using gravitational waves to transmit information. Once again, no observational or experimental evidence exists to support brane theory, let alone communication from one universe to another.

Other multiverse theories also exist, but we won't spend any more time on those very theoretical ideas here.

Scientists and philosophers have argued for decades over whether or not theories like string theory and multiverse theories are actually scientific. One of the characteristics of a scientific theory, discussed in Section 1.2, is that it "must be able to make testable predictions that have the potential to be falsified." It is not yet clear that these theories can make testable predictions. After all, the GUTs forces cannot interact with matter or energy on other branes. Thus far, evidence of a gravitational interaction is lacking and, of course, we have not witnessed repeating big bangs, nor has there been any residual evidence of previous big bangs (at least so far).

Key Point
It has been argued that string theory and multiverse theories may not be truly scientific in nature.

19.7   The Scientist's Crystal Ball: Looking to the Future

In the final section of the final chapter it is fitting to consider the end of the universe. It is always a challenge forecasting the future of course, even by scientists using empirical evidence and falsifiable, if well-supported, theories of how the physical universe operates. Any errors in data about the universe, even very tiny ones, can lead to large errors over time, such as if the universe is closed, flat, or open (recall the butterfly effect discussed on page 445 in relation to weather forecasting). More important is our incomplete understanding of the relationship between gravitation and quantum mechanics; does the ability to forecast the future of the universe depend on a quantum theory of gravity and an ultimate theory of everything? What is dark matter? Is it a hint that our standard model of particle physics is incomplete because there are particles other than quarks, electrons, neutrinos, the

Higgs boson, and so on that we haven't discovered yet? And what is the dark energy that has been causing the expansion of our universe to accelerate over the past 4 1/2 billion years? Is it due to vacuum energy or something completely different?

Despite these questions and others, we can still use our best current understanding of the universe to make some educated conjectures about what might happen going forward. After all, there won't be any consequences to humanity if we are wrong, since it is extremely unlikely that humanity will be around at the end of the universe to go back and check the predictions!

The Big Freeze

What the majority of cosmologists see as the most likely outcome for the universe, based on what we know today, might be termed the Big Freeze. From the data collected by the Planck satellite and other satellites and observatories, the universe appears to be flat, and dark energy seems to behave like the cosmological constant that Einstein inserted into his general relativity theory and later considered to be his greatest blunder. If those assumptions are true, then our universe seems destined to expand forever.

Based on our knowledge of stellar evolution we know that all stars will eventually die simply because they only have a finite amount of mass available to burn. Our own Sun will puff up as it moves off the main sequence and onto the subgiant branch, having consumed all of the hydrogen in its core. It will then cool and expand dramatically as it evolves up the red giant branch of the Hertzsprung–Russell (H–R) diagram. If humanity or our evolutionary descendants are still around and haven't yet colonized the moons of the giant planets, they will be doomed. The expanding Sun will boil the Earth's oceans and cause the planet's atmosphere to escape. All that will be left is a charred, lifeless rock. But even if future Earthlings have moved out to the outer Solar System, they won't find a permanent reprieve. After the Sun burns all of the helium in its core along with some carbon, it will become so blotted that its outer envelope will escape out into space, forming a planetary nebula. What remains is a carbon–oxygen core having about six-tenths of the mass of the present Sun. The white dwarf star is destined to cool off forever, leaving a burned out relic of our once life-supporting home star. In addition, because the mass of the cooling white dwarf is only about 0.6 M_{Sun}, its gravitational strength will have decreased, causing the remaining Solar System objects to move farther away from the star, where it will be dark and very cold. To avoid this fate, Earthlings will need to find a different planetary system to call home.

Although gas-rich spiral galaxies may be able to continue making new stars for billions of years, eventually all of the dust and gas will be collected into future stars or become so dissipated that any additional star formation will come to an end, even after considering replenishment of the interstellar medium by stellar winds and supernovas. After many billions of years, all that will remain across the universe is dark matter, dead planets and moons, the burned-out husks of white dwarfs, neutron stars and black holes produced by supernovas, supermassive black holes at the centers of fading galaxies, and widely dispersed gas.

Then there is also the ongoing and accelerating stretching of spacetime caused by dark energy. Think about what happens to galaxies that are currently closer to us than the Hubble horizon. Because those galaxies are now moving away from us at ever-increasing speeds, at some point they will reach our future Hubble horizon, and they will move beyond it with speeds greater than the speed of light. We can see the light that was given off by those galaxies today, but after they move to the other side of the Hubble horizon, the light they give off will never reach us. If we could live long enough to observe those galaxies, they will simply get redder and redder, and fainter and fainter, until they fade from view altogether, and forever. As this continues, we will see fewer and fewer galaxies until all that is left are the galaxies in our Local Group, that are too tightly bound together by ordinary and dark matter to succumb to the accelerating nature of dark energy.

Even neutron stars, white dwarfs, planets, asteroids, comets, and all other matter may not exist forever in our Local Group. Science already knows that if left alone, neutrons decay into protons, electrons, and antineutrinos with a half-life of 611 seconds. Theory suggests that protons may decay as well. If they do, current experimental evidence indicates that their half-life must be greater than 10^{34} years (10 billion trillion trillion years). But given an infinite amount of time, how long it takes is ultimately irrelevant. Assuming that protons do decay, then they will be replaced by other particles in the standard model of particle physics. As a result, atomic nuclei will disintegrate, atoms will cease to exist as their electrons escape, and all matter will fall apart. Gone will be stellar corpses, planets, and everything else, all replaced by particles that are destined to be spread farther and farther apart in the ever-expanding universe.

Even black holes are destined to disappear eventually, including the supermassive black holes in the centers of galaxies. As mentioned on page 719, black holes evaporate very slowly via Hawking radiation. This happens through the loss of mass (and therefore energy), when one particle in a particle–antiparticle pair of virtual particles falls into a black hole from just outside the black hole's event horizon while the other one carries mass–energy away from the black hole. For a 1 M_{Sun} black hole, evaporation would take some 10^{64} years (ten thousand trillion trillion trillion trillion trillion years), and for a 100 billion M_{Sun} black hole evaporation wouldn't be completed for 2×10^{100} years, but with infinite time in an infinitely expanding universe, there is plenty of time. The end of the evaporation of black holes comes with the finality of bursts of light going off like extremely powerful flash bulbs across the universe.

With no remaining sources of new photons, the wavelengths of existing photons will continue to stretch and the photons' energies will continue to decrease. Our observable universe is destined to expand and cool forever — the Big Freeze.

Fade to black.

The Big Crunch

If the universe is closed rather than flat or open, at some point in the distant future, expansion will slow to a stop, turn around, and accelerate toward a Big Crunch. As was the case with the Big Bang, but in reverse, the universe won't fall toward a specific point in space; instead, the likely infinite universe will get denser and hotter everywhere, but will remain infinite in extent.

Rather than spacetime expanding, it will be contracting, carrying everything with it, including photons. From an observer's perspective, galaxies will start to move together, and toward the observer, with the most distant galaxies moving the fastest. In this way, the closest galaxies and those farthest away will all come together at the same time. As spacetime shrinks, the wavelengths of photons also shrink, becoming progressively more blueshifted. The shortening wavelengths also imply that photon frequencies increase and the photons increase in energy. The cosmic microwave background's characteristic blackbody temperature increases, slowly at first but then faster and faster. As the CMB wavelengths shorten, the background microwave radiation will move into the infrared portion of the electromagnetic spectrum, and then into the visible range, followed by the ultraviolet, x-rays, and ultimately gamma-rays. Over time, astronomers will be forced to rename the background radiation, again and again over billions of years; the cosmic infrared background, the cosmic visible background, … .

But it isn't just the CMB that will get hotter and more energetic over time. All of the light produced by the interstellar medium, stars, galaxies, active galactic nuclei, and so on will as well. Already being at higher energies than CMB photons, the shrinking of spacetime will drive them to much higher energies. As the universe continues to compress, this light will reach energies high enough to ignite stellar atmospheres, causing nuclear reactions in their atmospheres that will ultimate destroy them.

At the end, the universe will approach infinite density and temperature. Just before that happens, the environment will become so extreme that the electroweak force will replace the separate forces of electromagnetism and the weak nuclear force. This will be followed by GUTs subsuming the electroweak force and the strong nuclear force. And finally, assuming a unified gravity and GUTs exist, the universe will be governed by the ToE.

It has been suggested that the Big Crunch could be followed by a new Big Bang and a new universe, although there are some problems with that idea because of the second law of thermodynamics and increasing entropy. It could be that the scenario is accounted for by two banes bouncing off one another. In any case, the Big Crunch would be just as devastating as the Big Freeze, only faster and with a greater sense of paranoia, knowing that the universe is coming after you.

The Big Rip

If the cosmological constant doesn't describe dark energy, but another type of expansion is happening or will happen in the future, an even stranger end awaits the universe. Solutions to Einstein's general theory of relativity provide for the possibility of not only an accelerating universe, but also a sort of exponentially increasing hyper-accelerating universe powered by what has been termed phantom energy. The outcome of this hyper-acceleration is the Big Rip. The data don't exactly support the idea, but they don't completely rule it out either; in fact, there is an infinitesimally small difference between the two drastically different possibilities.

Initially the universe will appear to be progressing as though it is destined for the Big Freeze, with galaxies getting farther and farther apart. But phantom energy implies that even gravity can't hold objects together locally. At first, clusters of galaxies will begin to separate into their individual galaxies, then the galaxies themselves will begin to evaporate, with stars, dust, and gas in their outer regions drifting off into intergalactic space. Eventually the Local Group of galaxies will move away from one another. If there are still intelligent species in the merged Milky Way/Andromeda Galaxy, they will see its stars drifting away and the brightness of the galaxy will dim. As our merged galaxy and other galaxies become unbound due to the phantom energy hyper-expansion of spacetime, observers will start to see their planetary systems come apart as well, with Oort clouds dissipating first, followed by their Kuiper belts if they exist. As spacetime continues to stretch, first the orbits of the outer planets increase, moving farther and farther from their parent stars, followed by planets farther in. The planets are destined to wander the interstellar medium without the heat and security of their parents.

Phantom energy isn't done yet though. As time goes on, the energy overwhelms the gravitational forces holding stars and planets themselves together, causing them to be torn apart and dispersed into space as separate atoms and molecules. Phantom energy then overpowers the electromagnetic force holding atoms and molecules together, leaving electrons stripped from nuclei. Phantom energy eventually overcomes the strong nuclear force, tearing nuclei apart. There is nowhere to hide, as phantom energy ultimately tears everything apart, even ripping apart the fabric of spacetime itself.

The Bubble of Death

Finally, earning an honorable mention, is death by vacuum decay. With the discovery of the Higgs boson came another way for the universe, as we know it, to die. This unique and unsettling, although extremely remote, scenario could happen at any time in any of the previously discussed fates of the universe, and if the universe expands forever, will happen eventually, perhaps even before you finish reading the end of this sentence!

The discovery of the Higgs boson was the crowning achievement of the standard model of particle physics. It was the last particle predicted by the theory that hadn't been detected yet. With it came the Higgs field, akin to the electromagnetic field associated with photons or the gravitational field associated with gravitons (if quantum gravity is a reality). It is actually the Higgs field that gives particles their mass. Before the electroweak force separated into the electromagnetic force and the weak nuclear force, particles of all stripes could travel through space at the speed of light; but that all changed with the breakup of the electroweak force. Suddenly, most particles were getting slowed down by interacting with the Higgs field, with the exceptions of the photon and the gluon (and potentially the graviton). The stronger the particle's interaction with the Higgs field, the more massive the particle.

The problem is that when the electroweak force separated into the electromagnetic and weak nuclear forces, 0.1 nanosecond after the Big Bang, the Higgs field settled into what is termed a false vacuum, not unlike water caught in a high mountain lake near Salt Lake City, Utah (see Fig. 19.25). If something happened, like an earthquake disrupting the walls of the lake, the water would immediately flow down to a lower level with a lower potential energy. In this analogy, the high mountain lake would be the false vacuum, and the Great Salt Lake, where the water finally ends up, would be the true vacuum.[12] The properties of our universe are tied

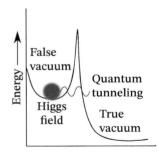

Fig. 19.25 The Higgs field is currently in a false vacuum state, but it could be "bumped" or quantum tunnel to a lower energy or true vacuum state.

[12]There is no outlet from the Great Salt Lake. The only way water leaves is through evaporation.

to the nature of the Higgs field and the false vacuum. If the Higgs field had ended up with a different value we wouldn't exist.

If the Higgs field somehow found its way to the true vacuum, even at one infinitesimal point in space, the properties of the universe in that location would change instantaneously, including all physical laws and constants. Even worse, that change in the nature of the universe would propagate outward in every direction, at very near the speed of light, because the true vacuum is the lowest energy state where nature always wants to find itself if given the chance (if a region of false vacuum is adjacent to a region of true vacuum, the false vacuum would instantly "cascade" down to the true vacuum). If this event happened beyond our Hubble horizon, the disruption would forever be kept from us because spacetime is expanding away from us faster than the speed of light in that region. On the other hand, if it happened inside our Hubble horizon, the expanding bubble of new universe would eventually get to us given enough time.

It's not that you could see it coming though. One of the things that many science fiction movies and shows get wrong is watching characters jumping out of the way of an oncoming laser beam or some other light-speed weapon. Since nothing travels faster than the speed of light, including information, you would only learn that the beam is coming at you when it reaches you. (Sorry about that tidbit of knowledge; your author hopes that it doesn't ruin sci-fi for you forever.) If the bubble is coming, you would never know what hit you. You would simply cease to exist; no pain, no fuss. You can't use it as an excuse for not doing your homework or preparing for your final exam, because as long as you exist in this observable universe, the deadlines are coming. On the other hand, if the bubble of death does arrive, consider the exam canceled; you just won't know ahead of time so you will still have to study anyway.

Fire and Ice

Some say the world will end in fire,
   Some say in ice.
From what I've tasted of desire
I hold with those who favor fire.
   But if it had to perish twice,
I think I know enough of hate
   To say that for destruction ice
Is also great
   And would suffice.

Robert Frost (1874–1963)

? Exercises

All of the exercises are designed to further develop *your understanding* of the material by thinking carefully and critically about what you have read in this chapter. Answers to selected exercises are available in the back of the book.

True/False

1. Cosmology is the science of understanding the evolution of stars.
2. (*Answer*) Hubble's original estimate of what is today known as the Hubble constant was within 10% of the currently accepted value.
3. Einstein added the cosmological constant to his equation for the evolution of the universe in order to force the universe to be static.
4. (*Answer*) According to the original Hubble–Lemaître law, if Galaxy B is 10 times farther from Earth than Galaxy A, then it is moving away from Earth at ten times of speed of Galaxy A.
5. There is no observational evidence that can distinguish between the Big Bang theory and the steady-state theory of the universe.
6. (*Answer*) The cosmological redshift can always be expressed by $z = v/c$.
7. The steady-state theory required that matter must be created in empty space to fill the void left behind by the cosmological expansion of the universe.
8. (*Answer*) Determining the distances to galaxies with the largest redshifts depends on accurate knowledge of the distances to nearby stars and galaxies.
9. The expansion of the universe began to accelerate about 8 billion years ago.
10. (*Answer*) Cosmologists understand what dark energy is because it seems to be described by the cosmological constant.
11. Dark energy seems to be a property of space itself.
12. (*Answer*) The vacuum of space is completely devoid of any particles or energy.
13. Dark energy only began to dominate the evolution of the universe when the volume of space had increased sufficiently due to earlier universal expansion.
14. (*Answer*) Virtual particles are a consequence of Heisenberg's uncertainty principle.
15. The cosmological principle is a fundamental physical law that all models of the early universe must be able to explain.
16. (*Answer*) The Planck limits are rough estimates of the limits of our physical understanding in extreme conditions.
17. The Planck length is roughly the size of a proton.
18. (*Answer*) Because the Big Bang happened everywhere in the universe at the same time, the cosmic microwave background must have automatically had the same temperature everywhere.
19. The tiny variations in the temperature of the cosmic microwave background are due to quantum fluctuations in the very early universe.
20. (*Answer*) The geometry of the universe appears to be open to a high degree of precision.
21. The cosmic inflation model of the very early universe explains several problems with the original Big Bang theory.
22. (*Answer*) It is thought that the universe increased in size by a factor of one hundred trillion trillion over a period of only 10^{-33} s.

23. The electroweak theory has yet to be tested in laboratories because the energies required are too high.
24. (*Answer*) A successful theory of everything must be able to describe how gravity works with quantum mechanics.
25. Gravitons are the hypothetical particles that theories of everything suggest could carry information about quantum gravity in the same way that photons carry information about the electromagnetic force.
26. (*Answer*) Quarks are particles that combine to form electrons and positrons.
27. The universe had cooled enough 10 μs after the Big Bang for quarks to combine to form protons and neutrons.
28. (*Answer*) Before Big Bang nucleosynthesis started, the universe contained more neutrons than protons.
29. The amount of lithium-7 in the universe agrees almost exactly with the amount predicted from calculations of Big Bang nucleosynthesis.
30. (*Answer*) The first atoms formed when the temperature of the universe dropped to 1 million kelvins.
31. Dark matter is not affected by gas and radiation pressure.
32. (*Answer*) The dark ages of cosmology was the period before decoupling.
33. The ultimate origin of the cosmic web is due to quantum fluctuations that occurred within 10^{-36} s after the Big Bang.
34. (*Answer*) Voids in the universe were created by enormous supernovas, that swept out regions of space that measured millions of light-years across.
35. Up to 50% of all of the ordinary mass in the universe is in the form of hot ionized gas found along filaments in the cosmic web.
36. (*Answer*) The first heavy elements were formed from stars with masses of between 100 M_{Sun} and 250 M_{Sun}.
37. The observable universe is all that exists.
38. (*Answer*) The distance to our particle horizon has always been what it is today and always will be.
39. Particles (photons, protons, electrons, et cetera) that are in our observable universe today have always been in our observable universe, and always will be.
40. (*Answer*) A galaxy that is located 15 billion light-years from Earth today emits light from a supernova explosion. That light will never reach Earth, even in an infinitely long period of time.
41. All observers, anywhere in the entire universe, share the same observable universe.
42. (*Answer*) All multiverse theories have made predictions that are, at least in principle, falsifiable.
43. In brane theory, universes could potentially interact with one another through collisions or gravitational waves.
44. (*Answer*) The strength of the gravitational force is greater than either the weak nuclear force or the electromagnetic force, as evidenced by the fact that it leads to planet formation, star formation, and even the formation of galaxies and superclusters of galaxies.

45. Two different values for the Hubble constant are given in the text. At the time the text was written cosmologists didn't yet understand why the two results are different.

46. (*Answer*) The Λ-cold dark matter model has been very successful in describing the evolution of the universe, despite not knowing what 95% of the mass–energy of the universe actually is.

47. It is possible that protons undergo radioactive decay in greater than 10^{34} years.

48. (*Answer*) The Big Freeze model implies that the universe will ultimately contain iced-over worlds, white dwarfs, neutron stars, and black holes of all sizes.

49. Supermassive black holes can evaporate in several trillion years.

50. (*Answer*) In the Big Rip scenario, all matter will be torn apart, even atomic nuclei.

51. If the Higgs field transitions from a false vacuum to the true vacuum anywhere within a Hubble distance from us, we will be able to watch our doom coming at us at the speed of light.

Multiple Choice

52. (*Answer*) A galaxy has a recessional radial velocity of 5000 km/s. Using the Hubble constant determined from the Planck satellite data, estimate the galaxy's distance from the Milky Way Galaxy.
    (a) 24 Mly (b) 50 Mly (c) 120 Mly
    (d) 240 Mly (e) 1070 Mly (f) 5000 Mly

53. The Big Bang is a particularly poor choice to describe the cosmological theory of the beginning and evolution of the universe because
    (a) there was no central location in the universe where an explosion occurred.
    (b) time didn't exist before the universe began.
    (c) galaxies are not moving out through space.
    (d) all of the above
    (e) (a) and (b) only

54. (*Answer*) As of June, 2021, the most distant galaxy ever detected was found in the constellation of Ursa Major. The Lyman-α line of hydrogen has a wavelength of 121.6 nm in a laboratory, but its wavelength as observed coming from the galaxy is 1470 nm. The name of the galaxy is GN-z__, where the blank is the approximate value of its redshift parameter. The galaxy is known as
    (a) GN-z0 (b) GN-z1 (c) GN-z2
    (d) GN-z11 (e) GN-z20 (f) GN-z1089

55. What is the value of the redshift parameter, z, for the Local Group (our local universe today)?
    (a) 0 (b) 0.4 (c) 1.5
    (d) 20 (e) 1089 (f) infinite

56. (*Answer*) What type of redshift(s) would be measured for light produced from just outside of the event horizon of a black hole in our Galaxy that is 5000 ly away and moving away from the Sun at 50 km/s?
    (a) Doppler redshift (b) gravitational redshift
    (c) cosmological redshift (d) (a), (b), and (c)
    (e) (a) and (b) only (f) (b) and (c) only

57. The steady-state theory of cosmology was ultimately rejected because
    (a) it required that matter be created in empty space for all eternity.

(b) the universe simply could not have existed for ever.
(c) it couldn't explain the existence of background radiation throughout the universe.
(d) it couldn't explain the blackbody radiation spectrum of the cosmic microwave background.
(e) all of the above
(f) (a) and (b) only
(g) (c) and (d) only

58. (*Answer*) An object has an angular diameter of 1 arcsecond ($1'' = 1/3600$ of one degree). If another, identical object is located five times farther away, what would its angular diameter be?
    (a) $1''$ (b) $2''$ (c) $5''$
    (d) $0.5''$ (e) $0.2''$ (f) $0.1''$

59. The temperature of the cosmic microwave background that is measured today is closest to
    (a) 2.7 K (b) 3.5 K (c) 5 K
    (d) 27 K (e) 1089 K (f) infinite

60. (*Answer*) What is the value of the redshift parameter, z, where the universe transitioned from slowing down to accelerating?
    (a) 0 (b) 0.4 (c) 1.5
    (d) 20 (e) 1089 (f) infinite

61. From Fig. 19.14, if a galaxy has a measured redshift parameter of 1, it is _____ from Earth and its light took _____ to reach us.
    (a) 0.2 Gly, 0.2 Gy (b) 0.3 Gly, 0.3 Gy
    (c) 0.5 Gly, 0.5 Gy (d) 1 Gly, 1 Gy
    (e) 2.1 Gly, 2 Gy (f) 5.2 Gly, 4.4 Gy
    (g) 7.6 Gly, 6 Gy (h) 11.2 Gly, 8 Gy

62. (*Answer*) Dark energy began to dominate the universe _____ billion years ago. Looking back in time, that moment corresponds to a redshift parameter of about

    _____ .
    (a) 0.2 Gy, 0.014 (b) 0.3 Gy, 0.021
    (c) 0.5 Gy, 0.035 (d) 1 Gy, 0.073
    (e) 2 Gy, 0.15 (f) 4.4 Gy, 0.4
    (g) 6 Gy, 0.62 (h) 8 Gy, 1.0

63. Dark matter influences the expansion of the universe
    (a) by contributing to the overall mass of the universe.
    (b) through its gravitational attraction.
    (c) by accelerating the expansion.
    (d) all of the above
    (e) none of the above
    (f) (a) and (b) only
    (g) (b) and (c) only

64. (*Answer*) Dark energy influences the expansion of the universe
    (a) by contributing to the overall mass of the universe.
    (b) through its gravitational attraction.
    (c) by driving the acceleration of the expansion.
    (d) all of the above
    (e) none of the above
    (f) (a) and (b) only
    (g) (b) and (c) only

65. The cosmological principle argues that
    (a) the universe should be homogeneous.
    (b) regardless of where the observer is located in the universe, it should look essentially the same over sufficiently large distances.

(c) the universe should be isotropic.
(d) the universe should look essentially the same in whatever direction an observer looks.
(e) our location in the universe is not unique.
(f) the universe has always looked about the same and it always will.
(g) all of the above
(h) (a)–(e) only.

66. (*Answer*) If we lived in a closed universe, the sum of the three interior angles of a triangle drawn on a sheet of paper would always add up to
(a) less than 180°.
(b) exactly 180°.
(c) more than 180°.
(d) the answer depends on how the triangle was drawn

67. The cosmic inflation model of the very early universe was developed to explain problems with the standard Big Bang theory, including
(a) why the universe appears to be almost perfectly flat.
(b) how the cosmic microwave background has such a remarkably constant temperature, regardless of the direction we look in the sky.
(c) how very tiny fluctuations in the temperature of cosmic microwave background developed that now cover distances of thousands to millions of light-years.
(d) all of the above
(e) (b) and (c) only

68. (*Answer*) Which of the following are fundamental forces known to exist in the universe today?
(a) gravitational force   (b) electromagnetic force
(c) weak nuclear force   (d) strong nuclear force
(e) all of the above

69. The fundamental forces that are combined into one overarching grand unified theory (GUT) are the
(a) gravitational force.   (b) electromagnetic force.
(c) weak nuclear force.   (d) strong nuclear force.
(e) all of the above   (f) (a) and (b) only
(g) (a), (b), and (c)   (h) (b), (c), and (d)

70. (*Answer*) Big Bang nucleosynthesis began _____ after the Big Bang when the temperature of the universe dropped below _____ kelvins.
(a) 10 ps, 1 trillion   (b) 10 μs, 1 trillion
(c) 10 ns, 1 billion   (d) 10 s, 1 billion
(e) 20 min, 1 million   (f) 300,000 y, 100,000
(g) (a), (b), and (c)   (h) (b), (c), and (d)

71. Which of the following was *not* produced by Big Bang nucleosynthesis?
(a) $^{2}_{1}$H   (b) $^{3}_{2}$He   (c) $^{4}_{2}$He   (d) $^{7}_{3}$Li   (e) $^{12}_{6}$C
(f) All of the above were produced by Big Bang nucleosynthesis.
(g) (d) and (e)

72. (*Answer*) Following Big Bang nucleosynthesis, the universe contained about _____ hydrogen-1 and _____ helium-4 by mass.
(a) 75%, 25%   (b) 75%, 0.01%   (c) 25%, 75%
(d) 25%, 0.01%   (e) 0.01%, 75%   (f) 0.01%, 25%

73. Dark matter makes up _____ of all of the mass in the universe.
(a) 0%   (b) 10%   (c) 16%   (d) 26%   (e) 69%
(f) 84%

74. (*Answer*) Dark matter is known to interact with which force(s)?
(a) gravitational force   (b) electromagnetic force
(c) weak nuclear force   (d) strong nuclear force
(e) all of the above   (f) (a) and (b) only
(g) (a), (b), and (c)   (h) (b), (c), and (d)

75. What is the value of the redshift parameter, z, for the cosmic microwave background?
(a) 0   (b) 0.4   (c) 1.5   (d) 20   (e) 1089
(f) infinite

76. (*Answer*) The distance to the surface of last scattering is currently _____ Gly.
(a) 0   (b) 13.8   (c) 36   (d) 45   (e) 46.2

77. What is the value of the redshift parameter, z, at the particle horizon?
(a) 0   (b) 0.4   (c) 1.5   (d) 20   (e) 1089
(f) infinite

78. (*Answer*) The distance to the particle horizon is currently _____ Gly.
(a) 0   (b) 13.8   (c) 36   (d) 45   (e) 46.8

79. What is the value of the redshift parameter, z, at the Hubble horizon?
(a) 0   (b) 0.4   (c) 1.5   (d) 20   (e) 1089
(f) infinite

80. (*Answer*) According to Planck satellite data, if galaxies at a distance of _____ Gly or greater had supernovas occur in them today, the light from those explosions would never reach Earth, even over an infinite period of time.
(a) 13.2   (b) 13.8   (c) 14.5   (d) 45   (e) 46.8

81. Looking into deep space, what is the value of the redshift parameter, z, where the dark ages ended?
(a) 0   (b) 0.4   (c) 1.5   (d) 20   (e) 1089
(f) infinite

82. (*Answer*) The first stars that formed in the universe are estimated to be _____ Gly from Earth.
(a) 0   (b) 13.8   (c) 36   (d) 45   (e) 46.2

83. The first stars in the universe
(a) had masses between 100 M_{Sun} and 1000 M_{Sun}.
(b) formed out of hydrogen, helium, and very tiny amounts of lithium-7.
(c) had very short lifetimes.
(d) were all responsible for producing the first generation of heavy elements that enriched the interstellar medium.
(e) all ended their lives in supernova explosions.
(f) all ultimately produced black holes.
(g) all of the above
(h) (a), (b), and (c) only
(i) (a)–(d) only
(j) (a)–(e) only

84. (*Answer*) Within the cosmic web of the universe, about 50% of ordinary matter in the universe is hot, ionized gas that is found in _____ while the remaining ordinary mass is found in _____.
(a) filaments; voids
(b) voids; filaments and nodes
(c) filaments; filaments, walls, and nodes
(d) voids; voids
(e) filaments; walls and voids
(f) walls and nodes; filaments, walls, and voids

85. Which of the following multiverse theories have been verified through testable predictions?
    (a) many-worlds interpretation of quantum mechanics
    (b) bubble universes
    (c) brane theory
    (d) all of the above
    (e) none of the above
    (f) (a) and (b) only

86. (*Answer*) Data from the cosmic microwave background have provided information about
    (a) the temperature of the blackbody spectrum of Big Bang photons.
    (b) quantum fluctuations in space during the very early universe.
    (c) the presence of sound waves in the early universe and the strengths of various harmonics.
    (d) the end of the dark ages by studying the polarization of CMB light.
    (e) all of the above
    (f) (a) and (b) only
    (g) (a), (b), and (c) only

87. Assuming that protons eventually decay, consequence(s) of the Big Freeze scenario for the end of the universe include
    (a) those galaxies not contained in our Local Group will eventually disappear from view beyond the Hubble horizon.
    (b) all stars will eventually die.
    (c) the inevitability that galaxies will eventually stop making new stars.
    (d) the temperature of the universe approaching absolute zero.
    (e) the universe ultimately containing only the corpses of dead planets, white dwarfs that cool to dead cinders, neutron stars, and black holes.
    (f) all of the above
    (g) (a), (b), and (c) only
    (h) (a), (b), (c), and (d) only

88. (*Answer*) If the universe is closed, the ultimate fate of the universe is the _____, or possibly the _____.
    (a) Big Freeze, Big Crunch
    (b) Big Crunch, Bubble of Death
    (c) Big Freeze, Big Rip
    (d) Big Rip, Bubble of Death
    (e) none of the above

89. Among the consequences of the universe ending in a Big Crunch,
    (a) the atmospheres of stars would ignite in nuclear explosions shortly before the final crunch, because of extremely blueshifted, energetic photons in the universe.
    (b) all galaxies beyond the Local Group would remain redshifted.
    (c) the entire universe would collapse to a single geometric point in space.
    (d) clocks would start running backward.
    (e) more than one of the above
    (f) none of the above

90. (*Answer*) If the accelerating expansion of the universe is governed by phantom energy instead of a cosmological constant-like dark energy
    (a) the universe will experience exponentially increasing hyper-acceleration.

(b) the Local Group of galaxies won't be able to stay together.
(c) the merged Milky Way/Andromeda Galaxy will be torn apart.
(d) the planets in our Solar System will be stripped from their orbits around the Sun.
(e) Earth will be destined to float on its own across the emptiness of space forever.
(f) the nuclei of atoms will be torn apart.
(g) all of the above
(h) (a), (b), and (c) only
(i) (a), (b), (c), and (d) only
(j) all except (d)
(k) all except (e)
(l) all except (f)

Short Answer

91. Explain why the Hubble time doesn't accurately represent the age of the universe.

92. (a) Using the approximate expression for the redshift parameter at recessional velocities much less than the speed of light, $z = v/c$, calculate its value for the fastest of the three galaxies in Example 19.1.
    (b) Is that galaxy's recessional velocity reasonably represented by the Hubble–Lemaître law on page 778? Why or why not?

93. (*Answer*)
    (a) Hubble's original value for the constant in the Hubble–Lemaître law was about 150 km/s per Mly. Based on that value, estimate the value of the Hubble time.
    (b) Based on what you have learned earlier in this text, why is that value clearly wrong?

94. Explain why the steady-state theory of cosmology was rejected in favor of the Big Bang theory.

95. Consider a raisin bread analogy of an expanding universe. Suppose that inside the bread there are five raisins in a line, each 1 cm from its neighbor(s). As the yeast causes the bread to expand (rise) each raisin moves away from its neighbor by the same amount. After one hour each raisin is now 2 cm apart, after two hours their separation has increased to 3 cm and so on.
    (a) Make a sketch of the locations of the five raisins when the (i) bread starts rising, (ii) after one hour, (iii) and again after two hours. Label each raisin starting from one end as A, B, C, D, and E.
    (b) Make a table of the distance of each raisin from raisin A before the bread starts to rise and then after one hour and after two hours as shown (i.e., complete columns 2, 3, and 4).

Raisin	d (0 h)	d (1 h)	d (2 h)	v (cm/h)
A	0 cm	0 cm	0 cm	0 cm/h
B	1 cm	2 cm	3 cm	
C				
D				
E				

    (c) Complete the last column by using the fact that speed equals distance divided by elapsed time, or $v = d/t$.

(d) Make a graph of speed (vertical axis) as a function of the original distances of each raisin from raisin A (horizontal axis). Include raisin A as well and be sure to label each point with the name of the raisin. If needed, a standard (Cartesian) sheet of graph paper can be downloaded from the textbook's website at `Graph Templates`.

(e) What is the slope of the straight line connecting the five points? The slope is found by dividing the difference in the speed of raisin E relative to raisin A by the distance of raisin E from raisin A. Make sure that you also include the correct units of cm/h per cm.

(f) The Hubble–Lemaître law is given by Equation (19.1), where the Hubble constant is the slope of the graph of galaxy velocity as a function of galaxy distance. What is the "Hubble–Lemaître law" for the raisin bread universe?

96. (a) Using Wien's blackbody color law given in Equation (6.12), calculate the observed peak wavelength of the cosmic microwave background in nanometers.

(b) The wavelength of the emitted photon corresponding to a given value of the redshift parameter, z, can be found by solving Equation (19.4) for λ_{emit}, giving $\lambda_{emit} = \lambda_{obs}/(1 + z)$. Using the value of z for the cosmic microwave background, calculate the peak wavelength of the blackbody spectrum of the CMB at the time the light was emitted.

(c) According to Table 6.1, what part of the electromagnetic spectrum does your answer to part (b) belong to?

(d) Use your result in part (b) and Equation (6.13) to make an estimate of the temperature of the CMB when its light was emitted.

(e) Compare your answer in part (d) to the surface temperature of the Sun.

(f) Explain why the temperature and peak wavelength of the CMB that we measure today are so different from your estimated values when the light was emitted?

97. Explain the general idea behind the construction of the distance ladder.

98. It was pointed out on page 788 that light emitted from a galaxy 4.4 billion years ago at a redshift parameter of $z = 0.4$ is just reaching us today, but that the galaxy is actually 5.2 billion light-years from us now. Explain how that could be true.

99. Why is the universe dark-energy-dominated today, but it wasn't before 4.4 billion years ago?

100. How does the cosmological principle imply that physical constants should be constant everywhere in the universe, and over all time?

101. The perfect cosmological principle is an extension of the cosmological principle. In addition to arguing that the universe should be homogeneous and isotropic, the perfect cosmological principle assumes that the universe is also constant throughout time. Does the Big Bang theory satisfy the perfect cosmological principle? What about the steady-state theory? Explain your answers.

102. Describe the flatness problem of cosmology, and how the cosmic inflation model of the very early universe is able to provide a solution.

103. The general theory of relativity has resisted being combined with grand unified theories to produce a theory of everything. What is fundamentally different about general relativity relative to the other three fundamental forces of nature?

104. (*Answer*) Why did it take until 10 seconds after the Big Bang before helium nuclei could form from protons and neutrons?

105. Helium-4 nuclei couldn't form through Big Bang nucleosynthesis until it became possible for the first step of the sequence on page 802 to start.

(a) A deuterium nucleus can be broken apart if it is struck by a photon having an energy of $E_{eV} = 2.2$ MeV. The equation for the energy of a photon [Equation (8.3)] can be solved for a photon's wavelength if its energy is known (in eV), giving

$$\lambda_{nm} = \frac{1240 \text{ nm eV}}{E_{eV}}.$$

Calculate the wavelength (in nanometers) of a photon having an energy of 2.2 MeV. (Don't forget to convert MeV to eV before carrying out the calculation.)

(b) Assuming that the wavelength of the photon calculated in part (a) corresponds to the peak of the blackbody radiation curve, the temperature from the blackbody peak wavelength equation [Equation (6.13)] can be used to calculate the temperature associated with the blackbody radiation. What was the temperature of the universe when deuterium started to form through Big Bang nucleosynthesis?

(c) How does your result in part (b) compare to the order-of-magnitude value given for the temperature when Big Bang nucleosynthesis began? Are they similar to within a factor of 10?

(d) All other nuclei require higher energy photons to break them apart. Could the rest of the Big Bang nucleosynthesis sequence occur once deuterium was able to form? Explain. (Congratulations: using what you have learned in this course you have just calculated the temperature when the first nuclei heavier than protons formed.)

106. How does the formation of the first atoms relate to the cosmic microwave background?

107. Repeat your calculation in Exercise 105 to calculate at what temperature atoms could start forming by capturing electrons. Recall that neutral helium has two electrons, but singly ionized helium only has one electron. To form singly ionized helium means that photons trying to break the ion apart must have energies less than 54.4 eV.

(a) What is the wavelength of a 54.4 eV photon?

(b) What is the temperature of blackbody radiation where the peak wavelength corresponds to your answer in part (a).

(c) How does your answer in part (b) compare to the order-of-magnitude temperature given in the text when the first atoms started to form?

(d) Why did it take longer before neutral hydrogen could form?

108. (a) Explain why lithium-7 is an important test of our understanding of Big Bang nucleosynthesis.

(b) Does the abundance of lithium-7 agree with theoretical predictions?

(c) What does your answer to part (b) suggest about our complete understanding of Big Bang nucleosynthesis?

109. Why did the universe transition from being dominated by radiation to being dominated by matter (both dark and regular) about 380,000 years after the Big Bang?

110. Explain how supermassive black holes likely formed.

111. Why is it that we can't see the entire universe?
112. Explain how it is possible for galaxies beyond the Hubble horizon to be moving away from us faster than the speed of light without violating the postulates of the special theory of relativity.
113. It was pointed out on page 803 that we see photons from the cosmic microwave background arriving at Earth today at the same time that photons from Jupiter, the Andromeda Galaxy, and distant quasars are arriving. Use the expanding sidewalk analogy from Fig. 19.15 to explain how this can happen in an expanding universe.
114. Until 4.4 billion years ago the universe was slowing down, but since that time it has been accelerating. What does that mean for galaxies that are within our Hubble horizon today? Will they always be within our Hubble horizon? Explain.
115. The text points out that when all of the mass of the universe is accounted for, cold dark matter makes up 84% of the total and ordinary matter is 16% of the total. But the text also states that dark energy comprises 69% of all of the energy in the universe. Explain how to reconcile these data.
116. Create a table that lists the following events in their appropriate order in time or distance since the Big Bang and/or the value for the redshift parameter, z:
    - Big Bang
    - today
    - time when quarks and gluons formed
    - time and the value of z at the end of the dark ages
    - time and the value of z when cosmic microwave background photons were released (decoupling)
    - time at the start of Big Bang nucleosynthesis
    - time when the first atoms formed
    - time of the appearance of the first protons and electrons
    - time when cosmic inflation started
    - time when the gravitational force likely separated from the GUTs forces
    - time when the strong nuclear force separated from the electroweak force
    - time when the weak nuclear force and the electromagnetic force separated.
    - time and the value of z at the start of universal acceleration
    - distance to the Hubble horizon and the corresponding value of z
    - distance to the particle horizon and the corresponding value of z
    - distance to the surface of last scattering and the corresponding value of z

Event	Time	Distance	z
Big Bang	0 s	⋯	∞
⋯	⋯	⋯	⋯
Today	13.8 Gy	⋯	0

117. Of the various ways in which the universe might end as described in Section 19.7, which one would you prefer? There is no right answer, but you should compare your choice with the others and explain why you selected it.

There is no end to education. It is not that you read a book, pass an examination, and finish with education. The whole of life, from the moment you are born to the moment you die, is a process of learning.

Jiddu Krishnamurti (1895–1986)

Appendix A
The 88 constellations

Name/ pronunciation[1]	Abbr.	Genitive/ pronunciation	English name	Brightest star
Andromeda an-DRAH-mih-duh	And	Andromedae an-DRAH-mih-dee	Chained Maiden	Alpheratz
Antlia ANT-lee-uh	Ant	Antliae ANT-lee-ee	Air Pump	α Antliae
Apus APP-us	Aps	Apodis APP-oh-diss	Bird of Paradise	α Apodis
Aquarius uh-QUAIR-ee-us	Aqr	Aquarii uh-QUAIR-ee-eye	Water Bearer	Sadalsuud
Aquila uh-QUILL-uh	Aql	Aquilae uh-QUILL-ee	Eagle	Altair
Ara AIR-uh	Ara	Arae AIR-ee	Altar	β Arae
Aries AIR-eez	Ari	Arietis uh-RYE-ih-tiss	Ram	Hamal
Auriga aw-RYE-guh	Aur	Aurigae aw-RYE-ghee	Charioteer	Capella
Boötes bo-OH-teez	Boo	Boötis bo-OH-tiss	Herdsman	Arcturus
Caelum SEE-lum	Cae	Caeli SEE-lye	Engraving Tool	α Caeli
Camelopardalis cuh-MEL-oh- PAR-duh-liss	Cam	Camelopardalis cuh-MEL-oh- PAR-duh-liss	Giraffe	β Camelopardalis
Cancer CAN-ser	Cnc	Cancri CANG-cry	Crab	Tarf
Canes Venatici CANE-eez ve-NAT-iss-eye	CVn	Canum Venaticorum CANE-um ve-nat-ih-COR-um	Hunting Dogs	Cor Caroli
Canis Major CANE-iss MAY-jer	CMa	Canis Majoris CANE-iss muh-JOR-iss	Great Dog	Sirius
Canis Minor CANE-iss MY-ner	CMi	Canis Minoris CANE-iss mih-NOR-iss	Lesser Dog	Procyon
Capricornus CAP-rih-CORN-us	Cap	Capricorni CAP-rih-CORN-eye	Sea Goat	Deneb Algedi
Carina cuh-REE-nuh	Car	Carinae cuh-REE-nee	Keel	Canopus
Cassiopeia CASS-ee-uh-PEE-uh	Cas	Cassiopeiae CASS-ee-uh-PEE-ye	Seated Queen	Schedar
Centaurus sen-TOR-us	Cen	Centauri sen-TOR-eye	Centaur	Rigil Kentaurus
Cepheus SEE-fee-us	Cep	Cephei SEE-fee-eye	King	Alderamin

[1] Finding charts are available at https://www.iau.org/public/themes/constellations/.

Name/ pronunciation	Abbr.	Genitive/ pronunciation	English name	Brightest star
Cetus SEE-tus	Cet	Ceti SEE-tie	Sea Monster	Diphda
Chamaeleon cuh-MEAL-yun	Cha	Chamaeleontis cuh-MEAL-ee-ON-tiss	Chameleon	α Chamaeleontis
Circinus SER-sin-us	Cir	Circini SER-sin-eye	Compass	α Circini
Columba cuh-LUM-buh	Col	Columbae cuh-LUM-bee	Dove	Phact
Coma Berenices COE-muh BER-uh-NICE-eez	Com	Comae Berenices COE-mee BER-uh-NICE-eez	Bernice's Hair	β Comae Berenices
Corona Australis cuh-ROE-nuh aw-STRAL-iss	CrA	Coronae Australis cuh-ROE-nee aw-STRAL-iss	Southern Crown	Meridiana
Corona Borealis cuh-ROE-nuh bor-ee-AL-iss	CrB	Coronae Borealis cuh-ROE-nee bor-ee-AL-iss	Northern Crown	Alphecca
Corvus COR-vus	Crv	Corvi COR-vye	Crow	Gienah
Crater CRAY-ter	Crt	Crateris cruh-TEE-riss	Cup	δ Crateris
Crux CRUCKS	Cru	Crucis CROO-siss	Southern Cross	Acrux
Cygnus SIG-nus	Cyg	Cygni SIG-nye	Swan	Deneb
Delphinus del-FIN-us	Del	Delphini del-FIN-eye	Dolphin	Rotanev
Dorado duh-RAH-do	Dor	Doradus duh-RAH-dus	Swordfish	α Doradus
Draco DRAY-co	Dra	Draconis druh-CONE-iss	Dragon	Eltanin
Equuleus eh-QUOO-lee-us	Equ	Equule eh-QUOO-lee-eye	Pony	Kitalpha
Eridanus ih-RID-un-us	Eri	Eridani ih-RID-un-eye	River Eridanus	Achernar
Fornax FOR-naks	For	Fornacis for-NAY-siss	Chemical Furnace	Dalim
Gemini JEM-uh-nye	Gem	Geminorum JEM-uh-NOR-um	Twins	Pollux
Grus GRUSS	Gru	Gruis GROO-iss	Crane	Alnair
Hercules HER-kyuh-leez	Her	Herculis HER-kyuh-liss	Hercules	Kornephoros
Horologium hor-uh-LOE-jee-um	Hor	Horologii hor-uh-LOE-jee-eye	Pendulum Clock	α Horologii
Hydra HIGH-druh	Hya	Hydrae HIGH-dree	Female Water Snake	Alphard
Hydrus HIGH-drus	Hyi	Hydri HIGH-dry	Male Water Snake	β Hydri
Indus IN-dus	Ind	Indi IN-dye	Indian	α Indi
Lacerta luh-SER-tuh	Lac	Lacertae uh-SER-tee	Lizard	α Lacertae
Leo LEE-oh	Leo	Leonis lee-OH-niss	Lion	Regulus

Name/ pronunciation	Abbr.	Genitive/ pronunciation	English name	Brightest star
Leo Minor LEE-oh MY-ner	LMi	Leonis Minoris lee-OH-niss mih-NOR-iss	Lesser Lion	Praecipua
Lepus LEP-us	Lep	Leporis LEP-or-iss	Hare	Arneb
Libra LEE-bruh	Lib	Librae LEE-bree	Weight Scales	Zubeneschamali
Lupus LOOP-us	Lup	Lupi LOOP-eye	Wolf	α Lupi
Lynx LINKS	Lyn	Lyncis LIN-siss	Lynx	α Lyncis
Lyra LYE-ruh	Lyr	Lyrae LYE-ree	Lyre	Vega
Mensa MEN-suh	Men	Mensae MEN-see	Table Mountain	α Mensae
Microscopium my-cruh-SCOPE-ee-um	Mic	Microscopii my-cruh-SCOPE-ee-eye	Microscope	γ Microscopii
Monoceros muh-NAH-ser-us	Mon	Monocerotis muh-NAH-ser-OH-tiss	Unicorn	β Monocerotis
Musca MUSS-cuh	Mus	Muscae MUSS-kee	Fly	α Muscae
Norma NOR-muh	Nor	Normae NOR-mee	Carpenter's Level	γ² Normae
Octans OCK-tanz	Oct	Octantis ock-TAN-tiss	Octant	ν Octantis
Ophiuchus OAF-ee-YOO-kus	Oph	Ophiuchi OAF-ee-YOO-kye	Serpent Bearer	Rasalhague
Orion oh-RYE-un	Ori	Orionis or-eye-OH-niss	Hunter	Rigel
Pavo PAY-vo	Pav	Pavonis puh-VOE-niss	Peacock	Peacock
Pegasus PEG-us-us	Peg	Pegasi PEG-us-eye	Winged Horse	Enif
Perseus PER-see-us	Per	Persei PER-see-eye	Hero	Mirfak
Phoenix FEE-nix	Phe	Phoenicis fuh-NICE-iss	Phoenix	Ankaa
Pictor PICK-ter	Pic	Pictoris pick-TOR-iss	Painter's Easel	α Pictoris
Pisces PICE-eez	Psc	Piscium PICE-ee-um	Fishes	Alpherg
Piscis Austrinus PICE-iss aw-STRY-nus	PsA	Piscis Austrini PICE-iss aw-STRY-nye	Southern Fish	Fomalhaut
Puppis PUP-iss	Pup	Puppis PUP-iss	Stern	Naos
Pyxis PIX-iss	Pyx	Pyxidis PIX-ih-diss	Compass	α Pyxidis
Reticulum rih-TICK-yuh-lum	Ret	Reticuli rih-TICK-yuh-lye	Eyepiece Graticule	α Reticuli
Sagitta suh-JIT-uh	Sge	Sagittae suh-JIT-ee	Arrow	γ Sagittae
Sagittarius SAJ-ih-TARE-ee-us	Sgr	Sagittarii SAJ-ih-TARE-ee-eye	Archer	Kaus Australis

Name/ pronunciation	Abbr.	Genitive/ pronunciation	English name	Brightest star
Scorpius SCOR-pee-us	Sco	Scorpii SCOR-pee-eye	Scorpion	Antares
Sculptor SCULP-ter	Scl	Sculptoris sculp-TOR-iss	Sculptor	α Sculptoris
Scutum SCOOT-um	Sct	Scuti SCOOT-eye	Shield	α Scuti
Serpens SER-punz	Ser	Serpentis ser-PEN-tiss	Serpent	Unukalhai
Sextans SEX-tunz	Sex	Sextantis sex-TAN-tiss	Sextant	α Sextantis
Taurus TOR-us	Tau	Tauri TOR-eye	Bull	Aldebaran
Telescopium tel-ih-SCOPE-ee-um	Tel	Telescopii tel-ih-SCOPE-ee-eye	Telescope	α Telescopii
Triangulum try-ANG-gyuh-lum	Tri	Trianguli try-ANG-gyuh-lye	Triangle	β Trianguli
Triangulum Australe try-ANG-gyuh-lum aw-STRAL-ee	TrA	Trianguli Australis try-ANG-gyuh-lye aw-STRAL-iss	Southern Triangle	Atria
Tucana too-KAY-nuh	Tuc	Tucanae too-KAY-nee	Toucan	α Tucanae
Ursa Major ER-suh MAY-jur	UMa	Ursae Majoris ER-suh muh-JOR-iss	Great Bear	Alioth
Ursa Minor ER-suh MY-ner	UMi	Ursae Minoris ER-suh mih-NOR-iss	Little Bear	Polaris
Vela VEE-luh	Vel	Velorum vee-LOR-um	Sails	γ^2 Velorum
Virgo VER-go	Vir	Virginis VER-jin-iss	Virgin	Spica
Volans VOH-lanz	Vol	Volantis vo-LAN-tiss	Flying Fish	β Volantis
Vulpecula vul-PECK-yuh-luh	Vul	Vulpeculae vul-PECK-yuh-lee	Fox	Anser

Appendix B
The 25 brightest stars

Name	App. mag. (m_V)	Abs. mag. (M_V)	Spectral type[1]	Distance (ly)	RA (hms)[2]	Dec (° ′ ″)[2]	Const.
Sirius	−1.46	1.42	A1V	8.60	06 45 08.917	−16 42 58.02	CMa
Canopus	−0.74	−5.60	A9II	309	06 23 57.110	−52 41 44.38	Car
Arcturus	−0.05	−0.30	K1.5III	36.7	14 15 39.672	+19 10 56.67	Boo
Rigel Kentaurus	0.01	4.38	G2V	4.39	14 39 36.494	−60 50 02.37	Cen
Vega	0.03	0.58	A0V	25.0	18 36 56.336	+38 47 01.28	Lyr
Capella A (a/b)[3]	0.07 (0.89/0.76)	0.30/0.17	G8III/G1III	42.8	05 16 41.359	+45 59 52.77	Aur
Rigel	0.13	−7.84	B8Ia	860	05 14 32.272	−08 12 05.90	Ori
Procyon	0.37	2.66	F5IV-V	11.46	07 39 18.119	+05 13 29.96	CMi
Achernar	0.46	−2.70	B6V	139	01 37 42.845	−57 14 12.31	Eri
Betelgeuse	0.50	−5.63	M1-M2Ia-Iab	548	05 55 10.305	+07 24 25.43	Ori
Hadar	0.58	−4.90	B1III	392	14 03 49.405	−60 22 22.93	Cen
Altair	0.76	2.20	A7V	16.72	19 50 46.999	+08 52 05.96	Aql
Acrux A/B[4]	0.76 (1.33/1.75)	−3.77 (−2.2/−2.7)	B0.5IV/B1V	98.7	12 26 35.895	−63 05 56.73	Cru
Aldebaran	0.86	−0.69	K5III	66.6	04 35 55.239	+16 30 33.49	Tau
Antares	0.91	−5.24	M1.5Iab-Ib	553	16 29 24.460	−26 25 55.21	Sco
Spica	0.97	−3.45	B1V	250	13 25 11.579	−11 09 40.75	Vir
Pollux	1.14	1.06	K0IIIb	33.8	07 45 18.950	+28 01 34.32	Gem
Fomalhaut	1.16	1.73	A4V	25.11	22 57 39.046	−29 37 20.05	PsA
Deneb	1.25	−6.93	A2Ia	1410	20 41 25.915	+45 16 49.22	Cyg
Mimosa	1.25	−3.41	B1IV	278	12 47 43.269	−59 41 19.58	Cru
Regulus	1.40	−0.53	B8IV	79.3	10 08 22.311	+11 58 01.95	Leo
Adhara	1.50	−4.07	B1.5II	405	06 58 37.549	−28 58 19.51	CMa
Castor	1.58	0.61	A1V	50.8	07 34 35.873	+31 53 17.82	Gem
Shaula	1.62	−4.60	B2IV	571	17 33 36.520	−37 06 13.76	Sco
Bellatrix	1.64	−2.80	B2V	252	05 25 07.863	+06 20 58.93	Ori

[1]The first letter and number are the Harvard Spectral Classification Scheme and the following Roman numeral is the luminosity class: I = supergiant, II = bright giant, III = giant, IV = subgiant, V = main sequence.

[2]The coordinates are for the start of the year 2000. Precession constantly changes coordinates.

[3]Capella is a combination of two binary star systems orbiting each other. The much brighter binary (Capella A) is composed of two giant stars (Aa and Ab), only separated by 0.74 au with an orbital period of 104 days. Capella Aa and Ab combine for an apparent visual magnitude of 0.07 and can't be resolved separately with the naked eye.

[4]Acrux A/B is a binary star system composed of two main sequence stars, but that binary is only one part of a six-star system. Acrux A and B together have a combined apparent visual magnitude of 0.76, an orbital period of 75.8 days, and are separated by 1.1 au.

Appendix C
The Messier catalog

Cat. num.	Common name	Object type	App. mag.	Distance (kly)	Angular size (′)	RA (h m)	Dec (° ′)	Const.
M1	Crab Nebula	supernova remnant	8.4	6.3	6 × 4	05 34.5	+22 01	Tau
M2		globular cluster	6.5	37.5	16.0	21 33.5	−00 49	Aqr
M3		globular cluster	6.2	33.9	18.0	13 42.2	+28 23	CVn
M4		globular cluster	5.6	7.2	36.0	16 23.6	−26 32	Sco
M5		globular cluster	5.6	24.5	23.0	15 18.6	+02 05	Ser
M6	Butterfly Cluster	open cluster	4.2	1.6	25.0	17 40.1	−32 13	Sco
M7	Ptolemy's Cluster	open cluster	3.3	0.8	80.0	17 53.9	−34 49	Sco
M8	Lagoon Nebula	star-forming nebula	6.0	5.2	90 × 40	18 03.8	−24 23	Sgr
M9		globular cluster	7.7	25.8	12.0	17 19.2	−18 31	Oph
M10		globular cluster	6.6	14.3	20.0	16 57.1	−04 06	Oph
M11	Wild Duck Cluster	open cluster	6.3	6.0	14.0	18 51.1	−06 16	Sct
M12		globular cluster	6.7	16.0	16.0	16 47.2	−01 57	Oph
M13	Great Hercules Cluster	globular cluster	5.8	25.1	20.0	16 41.7	+36 28	Her
M14		globular cluster	7.6	30.3	11.0	17 37.6	−03 15	Oph
M15		globular cluster	6.2	33.6	18.0	21 30.0	+12 10	Peg
M16	Eagle Nebula[1]	star-forming nebula	6.4	7.0	7.0	18 18.8	−13 47	Ser
M17	Swan Nebula[2]	star-forming nebula	6.0	5.0	11.0	18 20.8	−16 11	Sgr
M18		open cluster	7.5	4.9	9.0	18 19.9	−17 08	Sgr
M19		globular cluster	6.8	28.0	17.0	17 02.3	−26 16	Oph
M20	Trifid Nebula	star-forming nebula	9.0	5.2	28.0	18 02.6	−23 02	Sgr
M21		open cluster	6.5	4.3	13.0	18 04.6	−22 30	Sgr
M22		globular cluster	5.1	10.4	32.0	18 36.4	−23 54	Sgr
M23		open cluster	6.9	2.2	27.0	17 56.8	−19 01	Sgr
M24	Milky Way Patch	open cluster & star cloud	4.6	10.0	90	18 16.9	−18 29	Sgr
M25		open cluster	4.6	2.0	32.0	18 31.6	−19 15	Sgr
M26		open cluster	8.0	5.0	15.0	18 45.2	−09 24	Sct
M27	Dumbbell Nebula	planetary nebula	7.4	1.3	8.0 × 5.7	19 59.6	+22 43	Vul
M28		globular cluster	6.8	18.3	11.2	18 24.5	−24 52	Sgr
M29		open cluster	7.1	4.0	7.0	20 23.9	+38 32	Cyg
M30		globular cluster	7.2	26.1	12.0	21 40.4	−23 11	Cap
M31	Andromeda Galaxy	spiral galaxy	3.4	2500	178 × 63	00 42.7	+41 16	And
M32	Satellite galaxy of M31	elliptical galaxy	8.1	2500	8 × 6	00 42.7	+40 52	And
M33	Triangulum Galaxy[3]	spiral galaxy	5.7	2700	73 × 45	01 33.9	+30 39	Tri
M34		open cluster	5.5	1.4	35.0	02 42.0	+42 47	Per
M35		open cluster	5.3	2.8	28.0	06 08.9	+24 20	Gem
M36		open cluster	6.3	4.1	12.0	05 36.1	+34 08	Aur

[1] Also known as the Star Queen Nebula.
[2] Also known alternatively as the Omega Nebula, the Horseshoe Nebula, or the Lobster Nebula.
[3] Also known as the Pinwheel Galaxy.

Cat. num.	Common name	Object type	App. mag.	Distance (kly)	Angular size (')	RA (h m)	Dec (° ')	Const.
M37		open cluster	6.2	4.4	24.0	05 52.4	+32 33	Aur
M38		open cluster	7.4	4.2	21.0	05 28.4	+35 50	Aur
M39		open cluster	4.6	0.8	32.0	21 32.2	+48 26	Cyg
M40	Winnecke 4	double star[4]	8.4	1.1/0.4	0.8	12 22.4	+58 05	UMa
M41		open cluster	4.5	2.3	2.3	06 46.0	−20 44	CMa
M42	Great Orion Nebula	star-forming nebula	4.0	1.4	85 × 60	05 35.4	−05 24	Ori
M43	de Mairan's Nebula[5]	star-forming nebula	9.0	1.4	20 × 15	05 35.6	−05 16	Ori
M44	Praesepe[6]	open cluster	3.7	0.6	95.0	08 40.1	+19 59	Cnc
M45	Pleiades	open cluster	1.6	0.4	110.0	03 47.0	+24 07	Tau
M46		open cluster	6.0	5.4	27.0	07 41.8	−14 49	Pup
M47		open cluster	5.2	1.6	30.0	07 36.6	−14 30	Pup
M48		open cluster	5.5	1.5	54.0	08 13.8	−05 45	Hya
M49		elliptical galaxy	8.4	55,900	9 × 7.5	12 29.8	+08 00	Vir
M50		open cluster	5.9	3.2	16.0	07 03.2	−08 20	Mon
M51	Whirlpool Galaxy	spiral galaxy	8.4	37,000	11 × 7	13 29.9	+47 12	CVn
M52		open cluster	7.3	5.0	13.0	23 24.2	+61 35	Cas
M53		globular cluster	7.6	58.0	13.0	13 12.9	+18 10	Com
M54		globular cluster	7.6	87.4	12.0	18 55.1	−30 29	Sgr
M55		globular cluster	6.3	17.3	19.0	19 40.0	−30 58	Sgr
M56		globular cluster	8.3	32.9	8.8	19 16.6	+30 11	Lyr
M57	Ring Nebula	planetary nebula	8.8	2.3	1.4 × 1.0	18 53.6	+33 02	Lyr
M58		spiral galaxy	9.7	62,300	5.5 × 4.5	12 37.7	+11 49	Vir
M59		elliptical galaxy	9.6	48,700	5 × 3.5	12 42.0	+11 39	Vir
M60		elliptical galaxy	8.8	56,400	7 × 6	12 43.7	+11 33	Vir
M61		spiral galaxy	9.7	52,500	6 × 5.5	12 21.9	+04 28	Vir
M62		globular cluster	6.5	22.5	15.0	17 01.2	−30 07	Oph
M63	Sunflower Galaxy	spiral galaxy	8.6	27,000	10 × 6	13 15.8	+42 02	CVn
M64	Black Eye Galaxy	spiral galaxy	8.5	24,000	9.3 × 5.4	12 56.7	+21 41	Com
M65		spiral galaxy	9.3	35,000	8 × 1.5	11 18.9	+13 05	Leo
M66		spiral galaxy	8.9	36,000	8 × 2.5	11 20.2	+12 59	Leo
M67		open cluster	6.1	2.7	30.0	08 51.3	+11 48	Cnc
M68		globular cluster	7.8	33.3	11.0	12 39.5	−26 45	Hya
M69		globular cluster	7.6	29.7	9.8	18 31.4	−32 21	Sgr
M70		globular cluster	7.9	29.3	8.0	18 43.2	−32 18	Sgr
M71		globular cluster	8.2	13.0	7.2	19 53.8	+18 47	Sge
M72		globular cluster	6.6	55.4	6.6	20 53.5	−12 32	Aqr
M73		grouping of four stars[7]	9.0	2.5	2.8	20 58.9	−12 38	Aqr
M74		spiral galaxy	9.4	30,000	10.2 × 9.5	01 36.7	+15 47	Psc
M75		globular cluster	8.5	67.5	6.8	20 06.1	−21 55	Sgr
M76	Little Dumbbell Nebula[8]	planetary nebula	10.1	2.5	2.7 × 1.8	01 42.4	+51 34	Per
M77		spiral galaxy	8.9	47,000	7 × 6	02 42.7	−00 00	Cet
M78		star-forming nebula	8.3	1.35	8 × 6	05 46.7	+00 03	Ori

[4]Two stars that are at different distances but just happen to aligned in the sky; the two stars are not orbiting each other.
[5]Part of the Orion Nebula.
[6]Also known as the Beehive Cluster.
[7]A grouping of stars that are visually located near one another but are not gravitationally bound.
[8]Also known alternatively as the Cork Nebula the Butterfly Nebula, or the Barbell Nebula.

Cat. num.	Common name	Object type	App. mag.	Distance (kly)	Angular size (′)	RA (h m)	Dec (° ′)	Const.
M79		globular cluster	7.7	42.1	9.6	05 24.2	−24 31	Lup
M80		globular cluster	7.3	32.6	10.0	16 17.0	−22 59	Sco
M81	Bode's Galaxy	spiral galaxy	6.9	11,700	21 × 10	09 55.6	+69 04	UMa
M82	Cigar Galaxy	irregular galaxy	8.4	11,400	9 × 4	09 55.8	+69 41	UMa
M83	Southern Pinwheel Galaxy	barred spiral galaxy	7.6	27,800	11 × 10	13 37.0	−29 52	Hya
M84		lenticular galaxy	9.1	60,100	5.0	12 25.1	+12 53	Vir
M85		lenticular galaxy	9.1	58,200	7.1 × 5.2	12 25.4	+18 11	Com
M86		lenticular galaxy	8.9	54,900	7.5 × 5.5	12 26.2	+12 57	Vir
M87	Virgo A	elliptical galaxy	8.6	56,100	7.0	12 30.8	+12 24	Vir
M88		spiral galaxy	9.6	54,000	7 × 4	12 32.0	+14 25	Com
M89		elliptical galaxy	9.8	51,700	4.0	12 35.7	+12 33	Vir
M90		spiral galaxy	9.5	58,700	9.5 × 4.5	12 36.8	+13 10	Vir
M91		barred spiral galaxy	10.2	63,000	5.4 × 4.4	12 35.4	+14 30	Com
M92		globular cluster	6.4	26.7	14.0	17 17.1	+43 08	Her
M93		open cluster	6.0	3.6	22.0	07 44.6	−23 52	Pup
M94		spiral galaxy	8.2	14,500	7 × 3	12 50.9	+41 07	CVn
M95		barred spiral galaxy	9.7	32,600	4.4 × 3.3	10 44.0	+11 42	Leo
M96		spiral galaxy	9.2	31,300	6 × 4	10 46.8	+11 49	Leo
M97	Owl Nebula	planetary nebula	9.9	2.6	3.4 × 3.3	11 14.8	+55 01	UMa
M98		spiral galaxy	10.1	44,400	9.5 × 3.2	12 13.8	+14 54	Com
M99		spiral galaxy	9.9	45,200	5.4 × 4.8	12 18.8	+14 25	Com
M100		spiral galaxy	9.3	55,000	7 × 6	12 22.9	+15 49	Com
M101	Pinwheel Galaxy	spiral galaxy	7.9	20,900	22.0	14 03.2	+54 21	UMa
M102	Spindle Galaxy	lenticular galaxy	9.9	50,000	5.2 × 2.3	15 06.5	+55 46	Dra
M103		open cluster	7.4	8.5	6.0	01 33.2	+60 42	Cas
M104	Sombrero Galaxy	spiral galaxy	8.0	31,100	9 × 4	12 40.0	−11 37	Vir
M105		elliptical galaxy	9.3	32,000	2.0	10 47.8	+12 35	Leo
M106		spiral galaxy	8.4	23,700	19 × 8	12 19.0	+47 18	CVn
M107		globular cluster	7.9	20.9	13.0	16 32.5	−13 03	Oph
M108		spiral galaxy	10.0	46,000	8 × 1	11 11.5	+55 40	UMa
M109		barred spiral galaxy	9.8	83,500	7 × 4	11 57.6	+53 23	UMa
M110	satellite galaxy of M31	elliptical galaxy	8.5	2700	17 × 10	00 40.4	+41 41	And

Selected Answers

Chapter 1
2	True
4	False
6	False
8	True
10	(e)
12	(e)

Chapter 2
2	False
4	False
6	True
8	False
10	True
12	False
14	False
16	True
18	True
20	True
22	False
24	False
26	(d)
28	(f)
30	(b)
32	(a)
34	(b)
36	(c)
38	(e)
40	(e)
42	(a)
44	(d)
46	(b)
48	(b)
50	(e)
52	(d)
57	1,872,000 days, approximately 5125 1/4 years
59	18,812

Chapter 3
2	True
4	True
6	False
8	True
10	False
12	True
14	True
16	True
18	True
20	False
22	False
24	True
26	True
28	(d)
30	(e)
32	(c)
34	(c)
36	(d)
38	(d)
40	(b)
42	(e)
44	(d)

Chapter 4
2	True
4	False
6	True
8	False
10	False
12	True
14	True
16	False
18	False
20	False
22	False
24	True
26	True
28	True
30	False
32	True
34	(e)
36	(b)
38	(d)
40	(c)
42	(g)
44	(e)
46	(d)
48	(e)
50	(e)
52	(f)
63	3.0×10^{-6}. It is 1.0 for Earth's orbit.

Chapter 5
2	True
4	False
6	True
8	False
10	False
12	False
14	False
16	True
18	True
20	True
22	(f)
24	(f)
26	(c)
28	(c)
30	(a)
32	(b)
34	(d)
41	drawing, 22 km/s
48	you will feel significantly heavier (56%) than on Earth
52	1.1×10^{11} M_{Sun}

Chapter 6
2	True
4	True
6	False
8	False
10	False
12	False
14	True
16	True
18	False
20	True
22	False
24	False
26	True
28	True
30	(b)
32	(f)
34	(c)
36	(c)
38	(c)
40	(b)
42	(d)
44	(e)

Chapter 6 (cont.)

46	(b)
48	(c)
50	(d)
52	(b)
54	(d)
56	(b)
58	(d)
60	(a)
62	(c)
67	53 μs
70	It would move farther away because the force of gravitational attraction to Earth would decrease
85	(a) Two times hotter, (b) 820 nm
92	0.0032 W/m^2
95	(a) $R = 2GM/c^2$, (c) 4.2×10^{-6} R$_{\text{Sun}}$, (d) 2.9 km (1.8 mi), (e) a small town

Chapter 7

2	True
4	False
6	True
8	True
10	False
12	(d)
14	(f)
16	(a)
18	(f)
20	(g)
22	(e)
24	(b)
26	(c)

Chapter 8

2	False
4	False
6	True
8	True
10	False
12	False
14	False
16	True
18	False
20	True
22	(f)
24	(g)
26	(b)
28	(c)
30	(c)

32	(c)
34	(d)
36	(a)
38	(b)
40	(a)
42	(d)
44	(a)
46	(h)

Chapter 9

2	True
4	True
6	True
8	False
10	True
12	False
14	True
16	False
18	True
20	False
22	True
24	True
26	False
28	False
30	(j)
32	(e)
34	(e)
36	(c)
38	(b)
40	(c)
42	(d)
44	(e)
46	(a)
48	(b)
50	(e)
52	(d)
54	(c)
56	(d)
58	(e)

Chapter 10

2	True
4	False
6	True
8	False
10	True
12	False
14	True
16	False
18	False
20	True

22	False
24	(c)
26	(a)
28	(f)
30	(k)
32	(f)
34	(d)
36	(c)
38	(d)
40	(b)
42	(b)

Chapter 11

2	False
4	True
6	False
8	True
10	True
12	False
14	True
16	True
18	True
20	False
22	True
24	True
26	True
28	False
30	False
32	True
34	False
36	True
38	True
40	False
42	True
44	True
46	False
48	False
50	True
52	True
54	(d)
56	(f)
58	(c)
60	(g)
62	(c)
64	(e)
66	(f)
68	(b)
70	(g)
72	(b)
74	(f)
76	(b)

Chapter 11 (cont.)

78	(e)
80	(d)
82	(f)
84	(d)
86	(a)
88	(d)
90	(a)
96	0.44 eV, 2820 nm, infrared
103	The maria exhibit far less cratering

Chapter 12

2	False
4	True
6	False
8	True
10	False
12	True
14	True
16	False
18	False
20	True
22	False
24	True
26	(e)
28	(a)
30	(c)
32	(e)
34	(e)
36	(b)
38	(d)
40	(a)
42	(b)
44	(a)
46	(d)
48	(e)
50	(b)
52	(c)
54	(c)
56	(e)
58	(c)

Chapter 13

2	False
4	True
6	True
8	False
10	True
12	True
14	True
16	True

18	True
20	False
22	False
24	True
26	True
28	(e)
30	(b)
32	(g)
34	(g)
36	(c)
38	(g)
40	(b)
42	(e)
44	(b)
46	(c)
48	(f)
50	(a)
52	(e)
54	(c)
56	(b)
58	(f)
60	(a)
62	(e)
64	(b)
66	(a)
68	(e)
80	(a) Earth and the Moon both pull on the spaceship equally, (b) closer to the Moon because its lesser mass requires the spacecraft be closer to experience the same pull as applied by Earth, (c) it would "fall" toward the Moon, (d) L_1

Chapter 14

2	True
4	True
6	False
8	False
10	True
12	True
14	False
16	False
18	True
20	False
22	(c)
24	(d)
26	(c)
28	(d)

30	(d)
32	(a)
34	(g)
36	(c)
38	(d)
40	(f)
42	(b)
44	(c)
46	(c)
48	(d)
50	(b)
52	(d)
54	(d)
56	(g)
59	(a) 0.000 002 335 (b) 0.001 533 nm (c) blueshifted (d) 656.468 076 nm

Chapter 15

2	True
4	False
6	True
8	False
10	False
12	True
14	True
16	False
18	False
20	True
22	False
24	(f)
26	(a)
28	(g)
30	(e)
32	(e)
34	(a)
36	(d)
38	(e)
40	(a)
42	(b)
44	(c)
56	Pollux
58	No, during a part of its cycle Mira's apparent magnitude is fainter than 6, the detection limit for the human eye.
61	Shedir; 17
63	(a) 168 pc; 548 ly, (b) If this is 2022, we would learn about it in 22570. (c) Yes, but the light hasn't reached us yet.

Chapter 15 (cont.)

69   (a) approximately 6.5 au for Sirius A and 13.5 au for Sirius B. (b) 20 au, which is very close to the 19.7 au value given in Example 15.6. The value for Sirius A is about one-half of the value for Sirius B, which also makes sense since Sirius A is about twice the mass of Sirius B.

73   (a) 20 au, (b) 120 au

76   (a) 4400 K, (b) 660 nm, (c) red

81   174 R_{Sun}

Chapter 16

2   True
4   True
6   False
8   False
10   False
12   True
14   False
16   True
18   False
20   True
22   True
24   False
26   True
28   True
30   True
32   True
34   True
36   False
38   False
40   False
42   True
44   True
46   (c)
48   (d)
50   (f)
52   (a)
54   (d)
56   (b)
58   (f)
60   (a)
62   (e)
64   (f)
66   (f)
68   (d)
70   (a)
72   (d)
74   (d)

76   (f)
78   (f)
80   (d)
82   (c)
84   (a) $P = nk_B T$,
       (b) 1.38×10^{-14} N/m^2,
       (c) 7×10^{19} or 70 million trillion

Chapter 17

2   True
4   False
6   False
8   True
10   False
12   False
14   True
16   False
18   True
20   True
22   False
24   True
26   True
28   False
30   False
32   True
34   True
36   False
38   False
40   False
42   False
44   False
46   False
48   False
50   True
52   (c)
54   (a)
56   (e)
58   (b)
60   (d)
62   (d)
64   (c)
66   (a)
68   (e)
70   (c)
72   (c)
74   (g)
76   (e)
78   (d)
80   (d)
82   (b)
84   (e)

86   (b)
88   (e)
90   (e)
92   (b)
94   (a)
96   (f)

Chapter 18

2   False
4   False
6   True
8   False
10   False
12   True
14   True
16   False
18   True
20   False
22   True
24   True
26   False
28   False
30   True
32   False
34   True
36   True
38   False
40   True
42   (b)
44   (f)
46   (h)
48   (c)
50   (d)
52   (d)
54   (f)
56   (g)
58   (d)
60   (f)
62   (k)
64   (d)
66   (c)
68   (a)
70   (b)
72   (a)
74   (b)
76   (d)
79   (a) 0.16 M_{Sun}, (b) M
83   2.1
91   (a) 230, (b) mostly dark matter

Chapter 19

2	False	36	True	72	(a)
4	True	38	False	74	(a)
6	False	40	True	76	(d)
8	True	42	False	78	(e)
10	False	44	False	80	(c)
12	False	46	True	82	(c)
14	True	48	False	84	(c)
16	True	50	True	86	(e)
18	False	52	(d)	88	(b)
20	False	54	(d)	90	(k)
22	True	56	(e)	93	(a) 2 Gy, (b) younger than Earth
24	True	58	(e)	104	The universe had to cool off enough first, meaning that photon wavelengths had to stretch so that the photons lost enough energy not to blow the helium nuclei apart.
26	False	60	(b)		
28	False	62	(f)		
30	False	64	(c)		
32	False	66	(c)		
34	False	68	(e)		
		70	(d)		

Credits

Many of the images are made available through Creative Commons licenses as a part of the open access movement. Some of the figures in this text are derivative works of Creative Commons licensed works and therefore have been uploaded to Wikimedia Commons (https://commons.wikimedia.org) under the username: Astroskiandhike. Further information about each license can be found on the associated websites:

- CC0 1.0: Universal Public Domain Dedication; https://creativecommons.org/publicdomain/zero/1.0/deed.en
- CC BY 2.0: Attribution 2.0 Generic; https://creativecommons.org/licenses/by/2.0/
- CC BY-SA 2.5: Attribution-ShareAlike 2.5 Generic; https://creativecommons.org/licenses/by-sa/2.5/
- CC BY-SA 3.0: Attribution-ShareAlike 3.0 Unported; https://creativecommons.org/licenses/by-sa/3.0/
- CC BY-SA 3.0 IGO: Attribution-ShareAlike 3.0 IGO; https://creativecommons.org/licenses/by-sa/3.0/igo/
- CC BY-SA 3.0 NL: Attribution-ShareAlike 3.0 Netherlands; https://creativecommons.org/licenses/by-sa/3.0/nl/deed.en
- CC BY-SA 4.0: Attribution-ShareAlike 4.0 International; https://creativecommons.org/licenses/by-sa/4.0/
- CC BY 4.0: Attribution 4.0 International; https://creativecommons.org/licenses/by/4.0/
- The GNU Free Documentation License (copyleft) is also used, indicated as GFDL: https://www.gnu.org/copyleft/fdl.html. The documentation for the license can also be found on the individual chapter pages of textbook's website: https://www.AstronomyHumanQuest.com.

Images from NASA, USGS, and other agencies of the United States government are typically public domain.

Links to all publicly available images and line art are available on the individual chapter pages of textbook's website at https://AstronomyHumanQuest.com. In addition, further information about particular images or line art may be available through their associated web links.

Glossary

A | B | C | D | E | F | G | H | I | J | K | L | M | N | O | P | Q | R | S | T | U | V | W | X | Y | Z

A

absolute magnitude (M) The apparent magnitude that an object would have if it were located 10 pc (32.6 ly) from Earth. Absolute magnitudes provide a way to compare brightnesses of objects directly without concern for their differing distances from Earth.

absolute zero Theoretically, the coldest temperature that can be achieved in nature (0 K, −273.15°C, −459.67°F). At absolute zero all motion either stops or is reduced to the lowest level allowed by quantum mechanics.

absorption Light from a continuous spectrum that is absorbed by atoms in a cool gas when electrons in the atoms make transitions from lower energy levels to a higher energy levels. The absorbed photons must have energies that correspond to available energy transitions for the elections in the atoms. The photons that have been absorbed cause dark lines to appear on the brighter continuous spectrum at wavelengths corresponding to the energy transitions. See also emission and continuous.

acceleration ($\vec{a}$) The rate of change of an object's velocity over time. Acceleration is a vector quantity that points in the direction of the *change* in the velocity vector.

acceleration due to gravity (g) The rate at which a falling object accelerates toward a mass in the absence of any resistive forces such as air resistance.

accretion disk An orbiting disk of material that forms from gas and dust. The material in the disk can spiral onto the central object, increasing the central object's mass. Objects may also form within the disk, such as for protoplanetary disks.

active galactic nucleus (AGN) A supermassive black hole at the center of a galaxy, combined with an accretion disk and/or a dusty torus that feeds the supermassive black hole. The AGN can produce intense jets of radiation perpendicular to the accretion disk, with the jets extending thousands and even millions of light-years.

alchemy A medieval "science" with the goal of discovering the so-called philosopher's stone, which was believed capable of turning other metals, such as mercury, into gold.

Aldrin, Buzz (b. 1930) An American astronaut who became the second human, after Neil Armstrong, to set foot on the Moon during the Apollo 11 mission in 1969. Michael Collins, the third man on the mission team, orbited the Moon in the command module.

Alexander the Great (356–323 BCE), also known as Alexander III of Macedon. King of the ancient Greek kingdom of Macedon, who conquered Babylon and vast areas from Greece to Egypt, and also including northwest India. He opened the region up to the Greek language and influence.

al-Khwārizmī, Moḥammed ibn-Mūsā (c. 780–850) An Arab astronomer and mathematician who made important advances in algebra. The term "algebra" derives from an operation in the title of one of his books and the term "algorithm" comes from the Latinized version of his name.

ALMA (Atacama Large Millimeter/submillimeter Array) An array of 50 12-m diameter dish antennas, combined with an array of 12 7-m diameter dishes and 4 additional 12-m dishes at an elevation of 5060 m (16,600 ft) above sea level in the Atacama Desert on the Chajnantor plateau in northern Chili. The array operates in the electromagnetic wavelength range between 0.3 and 10 mm. The larger array can be moved into different configurations.

alpha particle (α) The name that Rutherford gave to the least penetrating radiation coming from uranium, polonium, and radium. Alpha particles are the nuclei of helium atoms, $^4_2$He.

Alpher, Ralph Asher (1921–2007) An American astrophysicist and cosmologist who worked on Big Bang nucleosynthesis with Robert Herman. They proposed that the universe should contain a uniform microwave background due to the Big Bang.

al-Shātir, Ibn (1304–1375) An Arab astronomer who made improvements to the original planetary models of Ptolemy.

amino acid A molecule having a NH_2 group, a COOH group, and a "side chain," a molecular group that defines each type of amino acid. Amino acids are building blocks of proteins necessary for life on Earth.

angular momentum ($\vec{L}$) A conserved quantity associated with rotation. Not to be confused with linear momentum. Technically, angular momentum is a vector quantity.

anode The positively charged terminal where electrons leave an electrical device, such as a cathode ray tube. See also cathode.

antarctic circle 66 1/2° S latitude. The southern latitude on Earth where the Sun never sets for one day each year (approximately December 21) and never rises for one day each year (approximately June 21).

Anticthon A hypothetical counter-Earth proposed by Philolaus.

antimatter For every type of particle of normal matter there exists an antimatter partner that is identical except for certain opposite quantum numbers such as electric charge.

antimuon (μ^+) A member of the class of particles known as leptons. The antimuon is the antimatter counterpart is the muon (μ^-).

antineutrino ($\bar{\nu}$) An elementary particle with no charge and an extremely small mass that is a member of the class of particles known as leptons. Neutrinos only interact very rarely with other matter and have a good chance of traveling unimpeded through a light-year's length of lead. There are three different types of neutrinos associated with the elementary particles, the electron (e^-), the muon (μ^-), and the tau (τ^-).

aphelion The point when a planet, dwarf planet, comet, or asteroid is farthest from the Sun in its orbit.

Apollonius of Perga (c. 262–190 BCE) Proposed the idea that orbits may not be perfectly geocentric circular motion, but could be eccentric or composed of deferents and epicycles.

apparent magnitude (m) A measure of the brightness of an object. In the original scheme by Hipparchus, the brightest stars were given a magnitude of 1, somewhat dimmer stars were classified as magnitude 2, and so on, with the dimmest stars visible to the naked eye being classified as magnitude 6. Apparent magnitude does not account for the effects of differing distances of objects from Earth.

archaeoastronomy The study of the intersection of archeology and astronomy that investigates the relationships between a society's observations of the heavens and its cultural significance.

Archimedes of Syracuse (c. 287–212 BCE) A physicist, astronomer, and mathematician of ancient Greece. Considered to be the greatest mathematician of antiquity, his many accomplishments include the principle of buoyancy that bears his name, the development of the concept of infinitesimals as a prelude to the calculus, and a procedure for estimating π (pi). He also authored *The Sand Reckoner* which includes a discussion of the heliocentric model of Aristarchus.

arcsecond ($''$) A unit of measurement for an angle. There are $60''$ in $1'$ or $3600''$ in $1°$.

arctic circle 66 1/2° N latitude. The latitude on Earth where the Sun never sets for one day each year (approximately June 21) and never rises for one day each year (approximately December 21).

Aristarchus of Samos (c. 310–230 BCE) A mathematician and astronomer of ancient Greece who first proposed a heliocentric model of the universe. Pythagoras also called Samos home.

Aristotle (384–322 BCE) Ancient Greek philosopher and scientist. He joined Plato's Academy at age 18 and went on to become what some scholars refer to as the first scientist. His work strongly influenced the development of science into the Renaissance (14th to the 17th century CE), and wasn't fully superseded until the Enlightenment, beginning in the latter half of the 17th century, with the work of Newton.

Armstrong, Neil (1930–2012) An American astronaut who was the first human to set foot on the Moon during the Apollo 11 mission in 1969. Buzz Aldrin also walked on the Moon during the mission, while Michael Collins, the third man on the mission team, orbited the Moon in the command module.

Asimov, Isaac (1920–1992) An American biochemist and author. He was a prolific writer of science for public consumption as well as science fiction. In his robot series of novels, Asimov developed what are often referred to as the three laws of robotics, specifying how robots are to interact with humans.

asteroid Small rocky or metallic objects ranging in size from pebbles to Ceres, a dwarf planet with a diameter of 1000 km.

asteroid belt The region between Mars and Jupiter where millions of small asteroids orbit the Sun. The dwarf planet, Ceres, is also located in the asteroid belt.

astrobiology The combination of astronomy, geology, planetary sciences, and biology with the goal of understanding the origin(s) of life on Earth, the conditions necessary for the development of life, and the search for extraterrestrial life. See also exobiology.

astrology The pseudoscientific belief that the Sun, Moon, planets, and stars, together with their relative positions at the time of your birth can influence your life.

astronomical unit (au) Originally defined as the average distance from Earth to the Sun. Today it is a defined to be an exact length in meters (1 au $\equiv$ 1.495 978 707 00 × 10^{11} m), while the average distance from Earth to the Sun is very nearly 1 au.

atom A nucleus surrounded by an associated cloud of electrons.

atomic mass (A) The total number of nucleons contained in an atomic nucleus.

atomic number (Z) The number of protons in the nucleus of an atom. The atomic number also corresponds to a particular element and its location in the periodic table.

aurora A light display produced by ions trapped in a planet's magnetic field that bounce back and forth between the planet's north and south magnetic poles. When ions collide with atoms in the atmosphere near the poles, the atoms are excited, causing emission lines to be generated.

autumnal equinox The point on the celestial sphere when the center of the Sun crosses the celestial equator moving from north to south. The coordinates of the autumnal equinox are RA = 12^h, dec = $0°$. The start of autumn is defined to occur when the center of the Sun crosses the autumnal equinox.

B

Baade, Walter (1893–1960) A German astonomer who, with Fritz Zwicky, first identified supernovas as a new class of astronomical object, and suggested supernovas may be transition types to neutron stars.

Bacon, Francis (1561–1626) An English philosopher, scientist, jurist, and orator who argued that empirical observation and measurement are necessary for scientific knowledge. He is often considered to be the father of the scientific method.

Balmer, Johan Jackob (1825–1898) A Swiss mathematician who found an integer-based equation that gave the wavelengths of the four visible spectral lines of hydrogen (Hα, Hβ, Hγ, and Hδ).

barred spiral galaxy (SB) A barred spiral galaxy shares the same characteristics as a spiral galaxy except that its spiral arms begin at the ends of a bar structure that passes through the galaxy's center.

Bartlett, Albert Allen (1923–2013) An American physicist and university professor who worked to help people understand the implications of exponential population growth.

baryon A class of particles with significant mass. Baryons include protons and neutrons.

Becquerel, Antoine Henri (1852–1908) A French physicist who discovered the first radioactive material, uranium, by noticing that it exposed a photographic plate despite the plate being wrapped in black paper.

Bell Burnell, Dame Susan Jocelyn (b. 1943) An astronomer from Northern Ireland who first detected pulsars.

Bessel, Friedrich (1784–1846) A German astronomer and mathematician who first measured the distance to a star by measuring its annual parallax.

beta particle (β) The name that Rutherford gave to the second most penetrating radiation coming from uranium (U), polonium (Po), and radium (Ra). Beta particles that possess negative charge are electrons.

Big Bang The event believed to have been the beginning of spacetime and the universe.

binary A system in which two objects are in orbit around one another.

biochemistry The subdiscipline of organic chemistry related to the chemical processes that are associated with life.

black hole An object where the force of gravity is so strong that all mass collapses to a singularity. Within the event horizon of a black hole, nothing, including light, can escape. The radius of a black hole is the Schwarzschild radius. Black holes are theorized to exist based on Einstein's general theory of relativity, although current evidence overwhelmingly supports their reality.

blackbody An object (a hot, dense gas or a solid) that produces a continuous electromagnetic spectrum given off by hot, dense gases or solids.

blackbody spectrum The continuous electromagnetic spectrum produced by an ideal blackbody. The intensity of the light varies with wavelength in a well-defined way. The wavelength that corresponds to the peak of the intensity distribution and the level of intensity at every wavelength depends only on the temperature of the hot gas or solid.

blueshift The shortening of an electromagnetic wave due to the decreasing distance of seperation between the source and the observer (Doppler effect), or light traveling into a gravitational well.

Bohr, Niels Henrik David (1885–1962) A Danish physicist who made fundamental contributions to the early development of our knowledge of the atom and quantum mechanics.

Bok, Bart (1906–1983) A Dutch-American astronomer who studied star-formation regions now known as Bok globules, and he also investigated the structure of the Milky Way Galaxy. Bok was the first to observe the globules in the 1940s and he theorized that they could be "cocoons" for star formation.

bolide A very bright meteor, appearing as a fireball, that explodes in the atmosphere.

Boltzmann, Ludwig (1844–1906) An Austrian physicist and one of Stefan's doctoral students who made numerous important contributions to the field of thermodynamics, particularly his development of statistical mechanics. Boltzmann was the first person to successfully derive the Stefan–Boltzmann blackbody luminosity law from basic physics.

Bonacci, Leonardo (c. 1170–1250), an Italian mathematician, also referred to as Fibonacci, who introduced the Hindu-Arabic base-10 decimal numerical system to western Europe. He is best known for the numerical series that bears his name.

Bondi, Hermann (1919–2005) An Austrian-born astrophysicist and cosmologist who, with Thomas Gold and Sir Fred Hoyle, proposed a steady-state theory of the universe in 1948.

boundary condition Constraints placed on a system at its ends or edges.

Boyle, Robert (1627–1691) Born in Waterford, Ireland and lived in London, England. Generally considered to be the first modern chemist, although he also believed in alchemy. He is best known for the relationship between gas pressure and volume known as Boyle's law.

brightness (b) The amount of light energy per square meter (m^2) arriving at a detector (the human eye, a CCD, etc.) each second from a source; the power per m^2.

Bruno, Giordano (1548–1600) A Dominican friar who was burned at the stake for heresy by the Roman Inquisition. His "crimes" included denial of several Catholic doctrines. He also believed in an infinite version of the heliocentric Copernican universe, with stars surrounded by planets that were inhabited by people.

Bunsen, Robert (1822–1899) A German scientist who worked with Kirchhoff investigating the emission spectra of various elements.

C

calcium and aluminum-rich inclusion (CAI) Small, round deposits found in carbonaceous chondrites. They are believed to be the first solids that formed out of the solar protoplanetary disk, making them the oldest material in the Solar System to be dated radiometrically.

calculus The branch of mathematics that deals with continuous change in numerical quantities. Differential calculus deals with how one quantity changes due to changes in another quantity; for example, acceleration is the change in the velocity of an object over a change in time. Integral calculus involves adding quantites that may continuously change. For example, a vehicle that is accelerating begins at one velocity and finishes accelerating at a different velocity; integral calculus adds together all of the infinitesimal changes in acceleration multiplied by infinitesimal intervals of time to determine how much the velocity has changed. As a second example, if the density of a body is known, even if it changes throughout the body, the mass of the body can be determined by adding together all of the infinitesimally small volumes multiplied by the densities in those volumes. Development of the calculus was absolutely fundamental in advancing science, starting with terrestrial and planetary motion.

Canadian Space Agency (CSA) The space agency of the Government of Canada.

Cannon, Annie Jump (1863–1941) An American astronomer who developed the modern Harvard spectral classification scheme for stellar spectra, based on previous contributions by Antonia Maury and Williamina Fleming. She was one of the "computers" under the direction of Edward Pickering at the Harvard College Observatory. Cannon classified more than 425,000 stellar objects during her career, earning her the nickname, "Census Taker of the Sky." Her work was published in the Henry Draper catalog, which remains an important stellar catalog today. In 1925 Cannon became the first recipient of the honorary doctorate from Oxford University and was awarded a professorship at Harvard in 1938. The Annie Jump Cannon Award is given annually by the American Astronomical Society to a woman astronomer in North America.

carbonaceous A body containing carbon and carbon-based compounds.

carbonaceous chondrite The oldest class of meteorites that have fallen to Earth. They contain melted or partially melted calcium and aluminum-rich inclusions that date to 4.57 billion years ago, making the inclusions older than anything else in the Solar System except for the Sun itself. These meteorites are used to determine when the Sun was born. They typically contain water, organic compounds, and other compounds that contain silicon, oxygen, or sulfur.

Cassini, Giovanni Domenico (1625–1712) An Italian–French astronomer, mathematician, and engineer. He was the first person to observe four of Saturn's moons. Along with Jean Richer, he also estimated the distance to Mars in 1672 using parallax, and thus made a reasonably accurate determination of the value of the astronomical unit.

cathode The negatively charged terminal where electrons enter an electrical device, such as a cathode ray tube. See also anode.

cathode ray A stream of negatively charged electrons that travel through a vacuum tube from a negatively charged terminal called a cathode toward a positively charged anode. Cathode ray tubes were used in old-style, bulky televisions and computer monitors.

Cavendish, Henry (1731–1810) A British physicist and chemist who first determined the density of Earth and discovered the element hydrogen.

celestial equator (CE) A hypothetical great circle that is located on the celestial sphere and is directly over Earth's equator.

celestial mechanics The portion of astrophysics that deals with how and why astronomical bodies move through space as they do. The discipline uses Kepler's laws, Newton's laws, and Newton's universal law of gravitation or Albert Einstein general theory of relativity to calculate motions.

celestial sphere (CS) An enormous imaginary sphere, centered on Earth, that ancient astronomers believed contained all of the stars. Although fictitious, the celestial sphere is a convenience that is still used today to mark the directions (coordinates) of objects in the sky.

centaur Small Solar System bodies larger than 1 km in diameter that are located between the orbits of Jupiter and Neptune.

center of mass The location where all of the mass of an object or a system of objects can be considered to be positioned for the purpose of determining bulk motions of the object or system of objects. The center of mass of a teeter-totter is where it would balance on a fulcrum, and the center of mass of a binary system is the location about which both stars orbit.

centripetal An acceleration or force directed toward the center of the curve that an object is tracing at that moment.

Cepheid A class of pulsating variable stars. Cepheids are very bright giant or subgiant stars within the instability strip of the Hertzsprung–Russell diagram, and they obey a period–luminosity relationship. These properties make Cepheids excellent determiners of distances out to millions of light-years, including to nearby galaxies. They serve as a critical "rung" of the distance ladder. δ Cephei is the namesake of the class.

Chandrasekhar, Subrahmanyan (1910–1995) An Indian–American theoretical astrophysicist who made numerous contributions to our understanding of the physical processes that govern the structure and evolution of stars. He received the Nobel Prize in Physics in 1983.

Chandrasekhar limit The upper mass limit of a white dwarf star ($1.4\,M_{Sun}$) that can be supported by degenerate gas (electron) pressure. Neutron stars also have a Chandrasehkar limit due to degenerate neutron pressure, but the strong nuclear force sets their upper mass limit of about $2.7\,M_{Sun}$.

charge-coupled device (CCD) An electronic detector that measures the number of photons that strike its surface in a given period of time. Along with their scientific uses, CCDs are found in modern digital cameras, including cell phone cameras.

chemical symbol The one or two letter symbol of an element in the periodic table, such as H for hydrogen, He for helium, etc.

chemically differentiate The process whereby gravity causes the heaviest elements to sink toward the center of a fluid or partially fluid astronomical body, leaving lighter elements closer to the surface.

chromosphere The layer of a star's atmosphere lying just above the photosphere.

cold dark matter (CDM) The general term for dark matter that is slowly moving relative to the speed of light and therefore can collect gravitationally into clumps rather than being sread evenly throughout the universe. The clumps of CDM attract gas, which leads to galaxy formation and new stars.

Collins, Michael (1930–2021) An American astronaut who piloted the command module orbiting the Moon during the Apollo 11 mission in 1969 when Neil Armstrong and Buzz Aldrin first walked on the Moon.

Columbus, Christopher (1451–1506) An Italian explorer, sailing with Spanish support, sought to find a trade route to the Far East by way of the Atlantic Ocean. The "Old World" at the time did not know of the existence of North and South America, which blocked Columbus's route to Asia. Columbus went on to lead three more voyages to the Americas.

coma Essentially an extremely thin atmosphere around the nucleus of a comet that is produced when the comet enters the inner Solar System. Heat from the Sun causes gases and dust in the nucleus to escape, thus forming the coma.

comet A frozen mixture of ices (primarily water and carbon dioxide), rock, and dust, that is left over from the formation of the Solar System. A comet can develop a coma, along with a dust tail and a gas tail (also referred to as an ion tail), while in the inner portion of the Solar System. The tails are the product of the sublimating ices caused by increased heating from the Sun.

Comte, Auguste (1798–1857) A French philosopher of science.

conduction The transport of heat by vibrations of atoms and molecules in a solid.

continental drift The early theory, proposed by , that the continents move across Earth's surface. The theory was later replaced by a more general theory of plate tectonics.

continuous electromagnetic radiation from a source that exists at all wavelengths, although the intensity may depend on wavelength. Blackbody radiation is an example of a continuous spectrum. See also absorption and emission.

convection The transport of energy in a fluid by mass motions of hot material "rising" from regions of high temperature to regions of lower temperature where heat energy is transferred to the surrounding material, at which time the now-cooler material "sinks" back down again to be reheated.

Copernican principle The philosophical argument that humanity, and by extension, Earth, should not be located in a preferred location in the cosmos such as the center of the Solar System, the center of the Milky Way Galaxy, or the center of the universe. Humanity, and Earth, do not hold any special status within the universe.

Copernican revolution The scientific revolution, begun by Nicolaus Copernicus, that led to a new, empirical and scientifically-based, way of understanding nature.

Copernicus, Nicolaus (1473–1543) A Polish mathematician who developed a simple heliocentric model of the universe that ultimately replaced the complex geocentric model that had previously been believed for nearly two thousand years. A Copernican was someone who accepted the heliocentric model.

core-collapse supernova The explosion of a massive star ($M \gtrsim 9\,M_{Sun}$) at the end of its life most often due to the destruction of an iron–nickel core.

corona The very low-density outer atmosphere of a star. It can change shape over time due to the solar cycle and extend for several stellar radii.

coronal hole A cooler, lower density region of the Sun's corona where magnetic field lines enter or leave and, because of the magnetic field lines' tremendous extent out into space, the lines are effectively open-ended. One form of the solar wind emanates from coronal holes.

coronal mass ejection (CME) Massive ejection of ionized coronal gas. CMEs can reach Earth in two to three of days and potentially damage communications satellites, ground-based power grids, and produce spectacular aurora.

Cortés, Hernán (1485–1547) A Spanish explorer who led an expedition to Mesoamerica in 1519 that resulted in the nearly complete destruction of the great cultures of that region through warfare and disease.

cosmic microwave background (CMB) The 2.725 48 K blackbody radiation that exists throughout the universe as the relic radiation from the Big Bang after the ongoing universal expansion has cooled the original fireball to today's level.

cosmic ray A charged particle traveling through space at near the speed of light. Cosmic rays are believed to be produced by supernovas and active galactic nuclei.

cosmic web The overall structure of dark matter, gas, voids, walls, and filaments of superclusters, across the universe that formed following the Big Bang.

cosmological constant (Λ) A parameter in the general relativistic equations of Einstein that controls the rate of acceleration of universal expansion.

cosmological redshift The lengthening of a photon's wavelength caused by the expansion of the universe.

cosmology In pre-modern societies it refers to the cultural mythology of the origin of the universe and its peoples. Today, cosmology is the scientific study of the beginning, evolution, and possible futures of the universe.

Coulomb, Charles-Augustin de (1736–1806) A French scientist who developed the mathematical law that describes the force of attraction or repulsion between two charged particles. See Coulomb's law.

Coulomb's law The force of attraction or repulsion between two charged particles. The law is virtually identical in mathematical form to Newton's universal law of gravitation.

cryovolcano Analogous to volcanoes on Earth except that molten rock magma is replaced by magma composed of substances such as water (H_2O), carbon dioxide (CO_2), methane (CH_4), and ammonia (NH_3).

cuneiform The oldest known form a writing, which was developed in Mesopotamia sometime in the latter half of the fourth millennium BCE by the Sumerians. Cuneiform writing was recorded on baked clay tablets.

Curie, Marie Sklodowska (1867–1934) A Polish-born and French-naturalized scientist who discovered several radioactive elements. Curie was the first woman to win a Nobel prize and the first person to win two Nobel prizes. She shared the first prize with her husband, Pierre Curie. She was also the first woman to hold a chair in the Physics Department at the University of Paris. The Curies' daughter, Irène Joliot-Curie, and son-in-law, Frédéric Joliot-Curie, were also Nobel prize winners.

Curtis, Heber Doust (1872–1942) An American astronomer who was the first to discover a jet coming from the center of the supergiant elliptical galaxy, M87 (a cD galaxy). Curtis also participated in the Great Debate of April 26, 1920 with Harlow Shapley concerning "spiral nebulae" and island universes. The debate was held at the National Academy of Sciences in Washington, D.C.

Cyrano de Bergerac (1619–1655) A French playwright and novelist.

D

Dalton, John (1766–1844) An English chemist and physicist who proposed the modern theory that matter is composed of a variety of atoms, and that all compounds are combinations of integer ratios of various types of atoms.

dark matter A form of matter that doesn't emit any light at any wavelength. The existence of dark matter is known because of its gravitational influence alone. To date the actual nature of dark matter is unknown and yet it makes up the vast majority of the mass of the universe.

Darwin, Charles (1809–1882) An English biologist, naturalist, and geologist who proposed that all life evolved from common ancestors.

Davis, Raymond Jr. (1914–2006) An American chemist who won the Nobel Prize in Physics for his work in the detection of solar neutrinos.

de Broglie, Louis-Victor-Pierre-Raymons (1892–1987) A physicist of French nobility who fist postulated that particles with mass can have wave-like properties just like photons do.

declination (dec) Coordinate lines on the celestial sphere that parallel Earth's lines of constant latitude. The origin of declination is Earth's equator. Declination is sometimes symbolized by δ. See also right ascension.

decoupling The end of scattering of photons off charged particles in the early universe when atoms first formed about 380,000 years after the Big Bang. Decoupling is associated with the surface of last scattering.

deferent The largest circle of a geometrical model for celestial motions surrounding an astronomical object that was believed to be the center of motion, usually Earth. In some models the Sun was inside a deferent circle.

degenerate gas An extremely dense gas with unusual properties. While under normal conditions the pressure of the gas is due to its particles moving around rapidly and running into each other, but the pressure of a degenerate gas occurs because the particles in the gas (either electrons or nuclei) have been squeezed so tightly that they cannot be compressed any further. The compression limit exists because all of the lowest energy levels have been filled. Until the degenerate nature of the gas is lifted, heating the gas doesn't cause the gas to expand as it normally would.

deGrasse Tyson, Neil (b. 1958) An American astrophysicist, author, and science communicator.

Democritus (c. 460–370 BCE) One of the ancient Greek philosophers who believed that matter could be broken down into tiny pieces called atoms. He also suggested that the universe contains multiple worlds and suns that could be destroyed through collisions.

density (ρ) The amount of mass or the number of particles in a specific volume of material. Density may also refer to the amount of something in a specified surface area or along a line.

Descartes, René (1596–1650) A French philosopher, physicist, and mathematician who is known for proposing a mechanical model of the Solar System and the greater universe that contains swirling vortices that carried the planets in their orbits. The common *xyz* rectangular coordinates bear his name as Cartesian coordinates.

deuterium An isotope of hydrogen (^2_1H) that contains one neutron in addition to its proton, making it twice as massive as ordinary hydrogen which only has one proton in the nucleus.

Dicke, Robert Henry (1916–1997) An American astrophysicist who calculated the temperature of the cosmic microwave background in 1964, although his work came 16 years after a previous prediction by Alpher and Herman. Zel'dovich had also rediscovered the prediction of Alpher and Herman in 1964.

Diggs, Thomas (1546–1595) An English astronomer who proposed that the celestial sphere should be replaced by an infinite universe containing an infinite number of stars.

Dirac, Paul (1902–1984) An English theoretical physicist who incorporated the special theory of relativity into quantum mechanics which produced the spin quantum number, m_s.

distance ladder The succession of methods for determining increasingly greater astronomical distances. Each "rung" on the ladder is based on the calibration of methods used to measure shorter distances.

DNA A nucleic acid with a double helix structure that is present in all living cells on Earth. The backbones are composed of alternating groups of deoxyribose (a sugar formed from hydrogen, carbon, and oxygen) and a phosphate group that contains phosphorus. Attached to the backbones are the nucleobases thymine, adenine, cytosine, and guanine, with thymine and adenine forming one base pair and guanine and thymine forming the second base pair. The base pairs connect the two helix structures.

Doppler effect The lengthening of a sound or light wave when the source and the observer are moving apart (lower pitch or redshift), or the shortening of the wave when the source and the observer are moving toward one another (higher pitch or blueshift).

draconic month The amount of time between consecutive crossings of the ecliptic plane by the Moon: 27.21 days. The term draconic comes from the ancient Chinese belief that a dragon would devour the Sun or the Moon during an eclipse.

Drake, Frank (b. 1930) An American astrophysicist who has been active in the search for extraterrestrial intelligence. He developed the Drake equation in 1961 that is a probabilistic equation identifying the various factors that would be necessary for humanity to detect extraterrestrial intelligence.

Draper, Henry (1837–1882) An American doctor and amateur astronomer who was one of the early developers of astrophotography and the uncle of Antonia Maury. His estate funded the famous Henry Draper catalog as a memorial. Stars are still often referred to by their HD numbers.

du Châtelet, Gabrielle Émilie Le Tonnelier de Breteuil, marquise (1706–1749) A French mathematician, physicist, and author who developed an early concept of the conservation of energy and had a major impact on the acceptance of Newton's *Principia* by translating it into French with included commentaries.

dust tail Dust that is liberated from a comet or asteroid that is pushed away from the Sun by the radiation pressure of sunlight. See also gas tail.

dwarf planet An object with sufficient mass to obtain a nearly spherical shape, and that directly orbits the Sun or another star rather than orbiting a planet, but without sufficient gravitational influence to sweep clean the region near its orbit.

E

eccentricity (*e*) Describes the "flatness" of an ellipse. $e = 0$ is a perfect circle and, as *e* increases toward 1, the ellipse approaches a straight line. Values of *e* equal to or greater than one are open ended; technically, a parabola or a hyperbola, respectively. *ae* is the distance of either of the two focal points of the ellipse from the center of the ellipse, where *a* is the semimajor axis of the ellipse.

eclipse An event when one celestial object passes in front of another. A solar eclipse occurs when the Moon partially or fully blocks our view of the Sun, which means that the Moon casts a shadow on Earth. A lunar eclipse occurs when the Moon passes partially or completely through Earth's shadow.

ecliptic The apparent path of the Sun across the stars over the course of one year. The ecliptic is also the plane of Earth's orbit around the Sun. The term, ecliptic, comes from "where eclipses occur." The ecliptic passes through the twelve zodiac constellations plus the constellation of Ophiuchus.

Eddington, Sir Arthur Stanley (1882–1944) A British astronomer, physicist, and mathematician who made important contributions to the study of stars and stellar evolution. He also led the 1919 expedition to the island of Principe (off the west coast of Africa) to record the May 19 solar eclipse. That expedition resulted in a major demonstration of the validity of Einstein's general theory of relativity.

Einstein, Albert (1879–1955) A German-born theoretical physicist who revolutionized our understanding of space and time. He was the architect of both the special and general theories of relativity. In addition, Einstein made many contributions to our understanding of light and atoms.

electromagnetic force One of the four fundamental forces of nature that includes the phenomena of electricity and magnetism.

electromagnetic radiation The energy contained within electromagnetic waves.

electromagnetic spectrum The wavelengths and frequencies of electromagnetic radiation. They range from long-wavelength and low-frequency radio waves, through microwave, infrared, visible, ultraviolet, x-rays, and very short-wavelength and high-frequency gamma rays. Table 6.1 lists the wavelength ranges of all parts of the electromagnetic spectrum.

electromagnetic wave Self-propagating waves of electric and magnetic fields that do not require any medium to pass through. The speed of an electromagnetic wave in a vacuum is the speed of light.

electron (e^-) An elementary particle that has a negative electric charge and is a member of the class of particles known as leptons. Electrons can exist in isolation or they can be located in orbitals around a nucleus. Electrons are approximately 1800 times less massive than either protons or neutrons. The electric charge of an electron is equal in magnitude but opposite in sign to the electric charge of a proton. Electrons are associated with electron neutrinos and their antimatter counterparts are positrons (e^+).

electron antineutrino ($\overline{\nu}_e$) An elementary particle that is a member of the class of particles known as leptons and the antimatter counterpart to the electron neutrino (ν_e).

electron neutrino (ν_e) An elementary particle that is a member of the class of particles known as leptons and one of three types of neutrinos. The electron neutrino is associated with the electron (e^-).

electron volt (eV) A unit of energy typically used in discussing the energy levels in atoms. $1\ \text{eV} = 1.60 \times 10^{-19}$ J. The definition comes from the amount of kinetic energy an electron would have after it is accelerated through 1 volt (1 V).

element An atom with a specific number of protons in its nucleus.

elementary charge (*e*) The smallest unit of charge that any particle, except quarks, can possess; $e = 1.60 \times 10^{-19}$ C.

ellipse A geometric shape that can be drawn by attaching the two ends of a string to two focal points and tracing out the path created while keeping the string taut. A circle is a special form on an ellipse in which the two focal points are located at the same position, the center of the circle.

emission Light at specific wavelengths produced when electrons "jump" from higher energy levels to lower energy levels in atoms, molecules, and ions. The transitions produce bright lines superimposed on a darker background. Emission can also be continuous in nature, such as when charged particles lose energy, generally in the presence of a magnetic field, or through vibrations or collisions in a hot, dense environment. See also absorption and continuous.

Emperor Julian of the Roman Empire (336–363 CE) had a deep respect for the ancient Greeks, including being a great supporter of the heliocentric model of Aristarchus.

energy (*E*) The capacity to do work. Total energy is a conserved quantity (the conservation of energy), meaning that it can be transformed from one form to another, but the total amount of energy cannot be created or destroyed. Various forms of energy include kinetic, potential, thermal, light, electric, magnetic, gravitational, nuclear, and mass.

Enlightenment An intellectual movement in Europe and America from the mid 1600s to roughly 1800 that embraced the idea that human reason can lead to understanding without invoking previous forms of authority, including monarchs or the Church. The Enlightenment is also referred to as the Age of Reason.

epicycle A smaller circle, centered on a deferent, that carries a celestial object in a geometrical model of heavenly motions. The epicycle/deferent scheme was used to mimic retrograde motion.

equinox A general term representing either the vernal equinox or the autumnal equinox.

equivalence principle The apparent equivalence of the effect of specific amoung of mass on motion and on the force of gravity. This equivalence led Einstein to realize that when a person falls she does not feel her own weight, which in turn led to the development of the general theory of relativity.

Eratosthenes of Cyrene (c. 276–195 BCE) successfully calculated the circumference of Earth using shadows. He also measured the tilt of Earth's axis and created the first world map that incorporated meridians and parallels. He may have also been able to determine the distances to the Moon and the Sun. Eratosthenes served as the head librarian of the Library of Alexandria. Cyrene is in present-day Libya, west of Alexandria, Egypt.

error bar A range of possible values that can be valid results of a measurement.

escape speed (v_{escape}) The minimum speed reqired for an object to escape from the gravitational force of a mass.

ether The ancient Greek philosopher scientists believed the ether to be a perfect fifth element that permeated the heavens (the four ancient Greek terrestrial elements being water, earth, air, and fire). In the 19th and the early 20th centuries, scientists assumed incorrectly that the ether must exist so that electromagnetic waves had a substance to propagate through. Also spelled æther.

Euclid of Alexandria (c. mid-4th–mid-3rd century BCE) A Greek mathematician often considered the "father of geometry."

Eudoxus of Cnidus (408–355 BCE) A student of Plato's who is considered as the greatest mathematician in classical Greece, and second only to Archimedes throughout all of antiquity. He developed insights into numbers as continuous quantities and not simply as integers. Eudoxus developed the first geometrical models of the universe based on Plato's conditions of perfect celestial motions centered on Earth. Eudoxus also created his own school in Cyzicus (near Erdek, in modern-day Turkey).

Euler, Leonhard (1707–1783) A Swiss mathematician who made crucial contributions to many areas of mathematics, astronomy, and physics.

European Space Agency (ESA) An organization composed of 20 member nations that has built and launched many important astronomical satellites and space probes. ESA has also collaborated with NASA on many projects, including the Hubble Space Telescope.

event horizon The "surface" of a black hole where the escape speed equals the speed of light.

evolution In its most general form, evolution means change over time. In biology, evolution generally refers to ongoing change in a species over many generations rather than an individual, although individuals can certainly evolve over their lifetimes. In astronomy and astrophysics, evolution represents changes in a specific object, or the universe itself, over its lifetime. For example, stellar evolution refers to how an individual star is born, changes internally and externally over time, and then dies.

evolutionary track The changing temperature and luminosity of a star as it ages, as plotted on a Hertzsprung–Russell diagram.

exobiology A subdiscipline of astrobiology focused on the conditions necessary for the development of life and the search for extraterrestrial life.

exoplanet A planet orbiting a star other than our Sun, or a "rogue" planet not associated with a particular star.

expansion of the universe The ongoing expansion of space resulting from the Big Bang. It causes the distance between geometrical points in space to increase and can be thought of as stretching the "fabric" of space.

F

Faber, Sandra Moore (b. 1944) An American astronomer who has made critical contributions to the study of dark matter, galaxies, and their evolution.

Fabricius, David (1564–1617) A German pastor who discovered the first variable star, Mira, in August 1596. Together with his son, Johannes Fabricius (1587–1615), he was also the first person to observe sunspots through a telescope in 1611.

falsifiable The inherent nature of a scientific theory that requires that predictions of the theory have the potential be proven to be incorrect, thus invalidating the theory itself.

Faraday, Michael (1791–1867) An English physicist and chemist who studied electromagnetic phenomena and electrochemical processes. Among his discoveries is the phenomenon that changing the electrical current in a wire can induce a current in a nearby coiled wire.

Fermi, Enrico (1901–1954) An Italian-born physicist who emigrated to the United States. Fermi made numerous contributions to physics in the areas of quantum mechanics, nuclear physics, particle physics, and statistical mechanics.

Fibonacci (c. 1170–1250), the nickname of Leonardo Bonacci.

field A mathematical representation of a scalar or a vector at every point in space. For example, a vector field could describe what the acceleration due to gravity is at every point in space surrounding a particular mass.

filament A "string" of superclusters, dark matter, and gas within the cosmic web.

fireball A very bright meteor falling through the atmosphere. See also bolide.

first quarter One phase of the Moon. One-half of the illuminated portion of the Moon is visible from Earth, with the amount of illuminated lunar surface *increasing* with time.

fission The breaking apart of a nucleus, producing two or more different nuclei.

Fizeau, Hippolyte (1819–1896) A French physicist who made the first reasonably accurate measurement of the speed of light.

Fleming, Williamina Paton Stevens (1857–1922) A Scottish-born astronomer who emigrated to the United States and worked as one of the "computers" under Edward Pickering at the Harvard College Observatory. Fleming developed a classification system for stars based on the strength of hydrogen lines in their spectra with class A having the strongest lines. The system was later modified by Annie Jump Cannon. Flemming also cataloged more than 10,000 stars that were published in the Henry Draper catalog.

fluid A gas or a liquid.

focal point One of two defining points of an ellipse, together with the sum of the distances from both foci to the same point on the ellipse. If the two focal points occupy the same location in space, the ellipse becomes a circle.

force ($\overrightarrow{F}$) The amount of push or pull applied on an object according to Newton's second law, $\overrightarrow{F} = m\overrightarrow{a}$. The SI units unit of force is the newton.

force of gravity The attraction of two bodies to one another as a consequence of their masses and distance of seperation. See Newton's universal law of gravitation or Albert Einstein's general theory of relativity.

Franklin, Benjamin (1706–1790) One of the Founding Fathers of the United States of America. He was not only a politician and statesman, but he was also a scientist and inventor who conducted extensive experiments in electricity. He proposed that electricity is a fluid and he is also credited with the invention of the lightning rod.

Fraunhofer, Joseph von (1787–1826) A German optician who discovered dark absorption lines in the spectrum of the Sun.

frequency (f) The number of times a traveling wave passes a point in one second or the number of oscillations completed in one second. Frequency is typically expressed in units of hertz (Hz).

Friedmann, Alexander (1888–1925) A Russian and Soviet mathematician and physicist who used Albert Einstein's general theory of relativity to propose that the universe could be expanding. His work was carried out independently of Georges Lemaître's.

full moon One phase of the Moon. The entire illuminated surface of the Moon is visible from Earth.

fusion The merging of nuclei to form a new type of nucleus.

G

Gaia A mission of the European Space Agency designed to measure the distances of billions of stars within the Milky Way Galaxy out to about 200,000 pc (200 kpc), or 650,000 ly (650 kly). The spacecraft can also measure motions, brightnesses, and compositions.

galaxy An enormous collection of stars, dust, and gas with its components orbiting its center of mass. Galaxies can have masses of many millions to trillions of solar masses.

Galileo Galilei (1564–1642) An Italian experimental physicist and observational astronomer who conducted some of the first true experiments in scientific research. Galileo, as he is usually referred to, was also the first person to develop a telescope specifically for astronomical observations. His observations helped to solidify the heliocentric model of our Solar System.

gamma ray The portion of the electromagnetic spectrum having the shortest wavelengths. Table 6.1 lists the wavelength ranges of all parts of the electromagnetic spectrum.

gas giant A giant planet that is primarily composed of gases in its interior. Within our Solar System, Jupiter and Saturn are gas giants.

gas tail Also known as an ion tail, it is produced by the solar wind pushing ions in the escaped gases of a comet directly away from the Sun. These charged particles are trapped along the Sun's magnetic field lines. See also dust tail.

general theory of relativity (GR) Einstein's theory that gravity is a consequence of curved spacetime around massive objects. General relativity extends the special theory of relativity by allowing for accelerating motions, including those due to "falling."

geocentric Earth-centered. The geocentric model of our universe had been the standard view until the Copernican revolution.

Gerard of Cremona (c. 1114–1187) An Italian who translated the astronomical work of Ptolemy, *Almagest*, from Arabic to Latin. While living in Toledo (in present-day Spain), Gerard translated 87 science and mathematics books from Arabic to Latin, and was a member of the Toledo School of Translators.

giant A star that is large and cool, and is burning helium in its core. A giant star has evolved off the main sequence, beyond the subgiant branch, and is located above the main sequence toward the upper right-hand corner of the Hertzsprung–Russell diagram, but it is not as luminous as a supergiant.

Gilbert, William (1544–1603) An English physicist, physician, and natural philosopher best known for his treatise on magnetism, *De Magnete, Magneticisque Corporibus, et de Magno Magnete Tellure* (*On the Magnet and Magnetic Bodies, and on the Great Magnet the Earth*), published in 1600. He was also one of the first to use the term "electricity" and distinguish between the two phenomena.

Global Positioning System (GPS) A system of United States satellites that are used to determine the latitude, longitude, and elevation of any point on or above the surface of Earth.

globular cluster An immense collection of tens of thousands to one million or more stars in a spheroidal distribution. The members of a globular cluster tend to be among the oldest stars in a galaxy. Globular clusters orbit around the center of mass of their parent galaxies and can be found far above or below the disks of spiral galaxies.

Gold, Thomas (1920–2004) An Austrian-born astrophysicist and cosmologist who helped develop a steady-state theory of the universe with Sir Fred Hoyle and Hermann Bondi in 1948.

Goldilocks zone A region around a star where life like we know it on Earth could potentially develop. The zone requires that liquid water could exist on a planet's surface. The Goldilocks zone is also known as the habitable zone.

Goodricke, John (1764–1786) An English astronomer who discovered the variability of δ Cephei and Algol (β Persei). He was inducted into the Royal Society in 1786 but never knew of the award, having died four days later of pneumonia at the age of 21.

grand unified theory (GUT) A theory that combines the electromagnetic force, the weak nuclear force, and the strong nuclear force.

granulation The mottled appearance of bright, hot rising gas and darker, cooler sinking gas at the top of the Sun's convection zone, which is visible at the base of the photosphere.

gravitational lensing The bending of light around massive objects due to the warping of spacetime as described by the general theory of relativity. Very small effects, such as produced by light bending around an exoplanet, is sometimes referred to as microlensing.

gravitational wave Propogating oscillations in spacetime that are generated by accelerating masses, as predicted by the general theory of relativity.

gravitational well A term that is used to describe the amount of gravitational potential energy associated with an object of mass attracting another massive object. Being "deeper" in the gravitational well of an object means that a second object requires more energy to climb its way out of the well and escape. The imagery of a well is simply a visualization tool; it does not represent any tangable object.

greenhouse effect The process by which visible light is able pass through atmospheric greenhouse gases, and the light energy then heats the surface of the planet. When the planet's surface reradiates the energy as infrared blackbody radiation, the greenhouse gases effectively absorbs the energy, hindering the energy from escaping back out into space. This process leads to heating, resulting in rising temperatures at the planet's surface and in its atmosphere. The greenhouse effect gets its name from the use of glass walls and ceilings to capture the energy from the Sun to heat the interior of the greenhouse. Glass behaves the same way that greenhouse gases do.

greenhouse gas A molecule that efficiently absorbs infrared light, but is largely transparent to visible light. Major greenhouse gases include carbon dioxide (CO_2), methane (CH_4), nitrous oxide (N_2O), and water vapor (H_2O).

Gutenberg, Johannes (1398–1468) Inventor of the mechanical printing press with movable type. His invention helped to make the recording and dissemination of new knowledge widespread, thus helping to fuel the Renaissance.

Guth, Alan (b. 1947) An American physicist and cosmologist who, with Andrei Linde and others, proposed a time of extremely rapid universal inflation between 10^{-36} and 10^{-32} seconds following the Big Bang.

H

habitable zone The region of space around a star where liquid water could exist on the surface of an Earth-like exoplanet. See also Goldilocks zone.

half-life The average amount of time required for one-half of the atoms of a particular radioisotope of an element to spontaneously emit particles and change into another type of element.

Halley, Edmond (1656–1742) An English astronomer and mathematician best known for his prediction of the return of the comet that bears his name.

harmonic A wave in a system that is constrained by certain boundary conditions, such as end nodes on a guitar string.

Harriot, Thomas (1560–1621) An English astronomer who drew a sketch of the surface of the Moon four months before Galileo did.

Harrison, John (1693–1776) An English clockmaker who won a £20,000 prize from British Parliament for the first successful design of a seafaring clock that allowed for the accurate determination of longitude while on a ship.

Harvard spectral classification scheme The modern classification of stellar spectra according to surface temperature, from hot to cool stars. The traditional sequence is OBAFGKM with additional classifications added in recent years for stars cooler than M, and for stars with unusual spectra.

Hawking, Stephen (1942–2018) An English physicist who was appointed the Lucasian Professor of Mathematics at the University of Cambridge. Hawking made major contributions to our understanding of black holes and cosmology.

Hawking radiation The process, first proposed by Stephen Hawking, where all black holes are eventually destined to evaporate due to the emission of particles from their event horizons. This occurs when virtual particles just beyond the event horizon appear and one falls into the black hole. The other escapes, carrying energy, and therefore mass, with it.

heat The amount of internal energy within a material due to the kinetic energy and potential energy associated with ions, atoms, and molecules. Heating is the process of transferring energy to a substance.

heavy element In astronomy, the term refers to any element more massive than helium. This is because hydrogen and helium make up the vast majority of all atoms and ions in the universe. Astronomers often refer to heavy elements as metals even though they may not have the characteristics of metals (for example, oxygen, is classified as a metal in astronomy.)

heavy-element poor Stars that have a low abundance of heavy elements, typically less than 1% by mass near their surfaces. The most heavy-element poor stars can have close to 0% of their surface gases comprised of heavy elements. Because heavy elements are formed during the evolution of stars, the more heavy-element poor a star is the fewer the number of generations of stars that preceeded its formation. Astronomers continue to search for the very first generation of stars that would have contained only hydrogen, helium, and a very tiny amount of lithium when they formed.

heavy-element rich Stars that have a relatively high abundance of heavy elements, typically greater than 1% by mass near their surfaces. The most heavy-element rich stars can have up to 3% of their surface gases comprised of heavy elements. Because heavy elements are formed during the evolution of stars, the more heavy-element rich a star is the greater the number of generations of stars that preceeded its formation.

Heezen, Bruce Charles (1924–1977) An American geologist who worked with Marie Tharp to produce the first topographic map of the entire ocean floor.

Heisenberg, Werner Karl (1901–1976) A German physicist who made pioneering contributions to the development of quantum mechanics, including his uncertainty principle.

Heisenberg uncertainty principle The quantum mechanical principle that the position and linear momentum, or the energy and time interval, of a particle cannot both be known to infinitesimal precision at the same time.

heliacal The first appearance of a rising planet or star after emerging from the glow of the Sun, or the last appearance of a planet or star before disappearing into the Sun's glow.

heliocentric Sun-centered. The heliocentric model, proposed by Nicolaus Copernicus, suggested that the Sun is the center of our Solar System.

heliocentrist Someone who subscribes to the heliocentric model of our Solar System.

helioseismology The study of the Sun's interior by analyzing the oscillations visible on the surface caused by chaotic motions of its convection zone. The convection zone causes the Sun to "ring" like a struck bell.

Herman, Robert (1914–1997) An American physicist who worked with Ralph Alpher on the prediction of a uniform microwave background due to the Big Bang.

Herschel, Frederick William (1738–1822) A renowned British astronomer and composer, born in Germany, who designed and built large telescopes that allowed him to make many important discoveries. Herschel was the first to realize that Uranus was not a star, and he discovered two of Saturn's moons, along with moons of Uranus. He also published major catalogs of nebulas and created the first map of the Milky Way Galaxy. William Herschel's sister, **Caroline Lucretia Herschel** (1750–1848), was a German astronomer who discovered several comets, and the first woman to be paid for her work. His son, **Sir John Frederick William Herschel** (1792–1871), was a British polymath who conducted important work in astronomy, botany, chemistry, and mathematics. John Herschel was also an inventor and photographer.

Hertz, Heinrich (1857–1894) A German physicist who first proved that electromagnetic waves exist.

hertz (Hz) The SI unit of frequency. The hertz is defined as the number of cycles per second, and has units of $1/s$.

Hertzsprung, Ejnar (1873–1967) A Danish chemist turned astronomer who, independently from Henry Norris Russell but at essentially the same time, developed an important diagram displaying stars according to their luminosities and surface temperatures. The diagram is still used today and is referred to as the Hertzsprung–Russell diagram or simply as the H–R diagram.

Hertzsprung–Russell diagram (H–R) An important method for displaying the luminosities and surface temperatures of stars that was developed independently by Ejnar Hertzsprung and Henry Norris Russell.

Hewish, Antony (1924–2021) A British radio astronomer who co-discovered pulsars with his graduate student, Jocelyn Bell Burnell.

Hilda asteroid A member of a group of asteroids named for 153 Hilda, the first asteroid found to be orbiting the Sun trapped in a 3:2 orbital resonance with Jupiter. The Hildas are three groups of asteroids that find themselves in one of the three Lagrange points, L_3, L_4, or L_5 every time they return to aphelion in their orbits. Each group cycles through the three points on three successive orbits at the same time that Jupiter completes two orbits.

Hipparchus of Nicaea (c. 190–120 BCE) An ancient Greece astronomer, mathematician, and geographer. He is credited with the discovery of the precession of the equinoxes, and being the founder of trigonometry. He also created a catalog of stars and developed the magnitude scale for stellar brightnesses.

Hoyle, Sir Fred (1915–2001) An English astrophysicist and cosmologist who popularized the term "Big Bang." Hoyle, together with Hermann Bondi and Thomas Gold, proposed a steady-state theory of the universe in 1948. Hoyle was also an early developer of the theory stellar nucleosynthesis, and promoted the idea of panspermia. In addition, Hoyle was a popularizer of science on radio and television, wrote science fiction and mysteries, wrote operas, and was a mountain climber.

Hubble , Edwin Powell (1889–1953) An American astronomer who made multiple critical contributions to our understanding of galaxies, galaxy distances, and the expansion of the universe, including the Hubble–Lemaître law.

Hubble constant (H_0) A fundamental constant of cosmology that describes how fast the universe is expanding. The Hubble constant also relates to the age of the universe since the Big Bang.

Hubble Space Telescope (HST) A space-based telescope operating at visible and ultraviolet wavelengths. The HST telescope was launched in 1990 and may last until 2030–2040. The HST is a joint project of NASA and ESA.

Hubble–Lemaître law The relationship between the observed recessional speeds of galaxies and their distances from the observer.

Huygens, Christiaan (1629–1695) A Dutch mathematician and scientist who made telescopic observations of the rings of Saturn and discovered its giant moon, Titan. He also made an estimate of the speed of light based on the observations of Rømer and measurements of the astronomical unit.

hydrocarbon An organic molecule that is composed entirely of hydrogen and carbon.

I

inertia The tendency of an object to resist change in its motion due to its mass.

inferior planet The planets orbiting the Sun that are closer to the Sun than Earth. See also superior planet.

inflation The idea that the universe underwent an extreme rate of growth between 10^{-36} s and 10^{-32} s after the Big Bang, increasing the size of the universe by a factor of 10^{26}.

infrared A region of the electromagnetic spectrum having wavelengths longer than visible light but shorter than microwaves. See Table 6.1 for a list of the wavelength ranges for all parts of the electromagnetic spectrum.

inorganic Molecules that don't contain carbon atoms. There are also a small number of molecules that do contain carbon but are generally not considered to be organic, such as carbon dioxide, cyanides, and carbonates.

instability strip A diagonal strip across the Hertzsprung–Russell diagram from bright and cool down to dim and hot where many of the pulsating variable stars are located, most notably Cepheids and RR Lyras. Stars passing through the instability strip as they evolve pulsate due to the "just right" depth of the partial helium ionization zone (He^+/He^{++}) in their atmospheres that temporarily traps and then releases photons from the star's interior.

International Astronomical Union (IAU) The official governing board of astronomy, responsible for such things as formally classifying and naming celestial objects, and defining the astronomical unit. It was the IAU that officially "demoted" Pluto from planetary to dwarf planet status in 2006.

interstellar medium (ISM) The gas and dust that resides between the stars, often in distinct clouds.

ion An atom or molecule with fewer (or occasionally with a greater number of) electrons than the total number of protons in its nucleus or nuclei. An ion has a positive (or occasionally negative) electric charge.

ionization The process of removing (or occassionally adding) electrons from an atom, molecule, or ion to alter its net electric charge. For example, stripping an electron from a He^+ (singly-ionized) ion produces a He^{++} (doubly-ionized) ion.

isotope An atom of a specific element that contains a particular number of neutrons in its nucleus. For example, hydrogen has three isotopes: ^1_1H (hydrogen's most common isotope), ^2_1H (deuterium), and ^3_1H (tritium).

J

James Web Space Telescope (JWST) A space-based infrared telescope that was launched in late December, 2021, to compliment the Hubble Space Telescope. The space telescope's primary mirror measures 6.5 m across. JWST's mission is to study star formation, exoplanets, galaxy formation, and the origin of the universe. JWST is a joint project of NASA, ESA, and the Canadian Space Agency.

Jansky, Karl Guthe (1905–1950) An American physicist who, in 1933, made the first radio wavelength observations of a celestial object. He worked for Bell Telephone Laboratories and set out to find sources of static in communications. Considered the father of radio astronomy, Jansky would have likely won the Nobel Prize for his work had he not died at the age of 44 from a massive stroke (Nobel Prize winners must be alive to receive the award).

Joliot-Curie, Irène (1897–1956) A French chemist and daughter of Marie and Pierre Curie. She shared the Nobel prize with her husband Frédéric for their discovery of artificial radioactivity.

Joule, James Prescott (1818–1889) An English physicist, mathematician, and beer maker who studied the relationship between heat and work.

joule (J) The SI unit of energy. The joule is defined as one newton times one meter (force times distance); $1 \, \text{N m} = 1 \, \text{kg m/s}^2 \, \text{m} = 1 \, \text{kg m}^2/\text{s}^2$.

K

Kaku, Michio (b. 1947) An American theoretical physicist, writer, and popularizer of science.

Kapteyn, Jacobus Cornelius (1851–1922) A Dutch astronomer who conducted extensive studies of the Milky Way Galaxy and created an early map of the Galaxy, known today as "Kapteyn's universe."

Kepler, Johannes (1571–1630) A German mathematician and heliocentrist who developed three laws of planetary motion, and first made the revolutionary proposal that planets orbit in ellipses, rather than in perfect circles. Kepler was also the first person to assume that the same physical processes that apply to the motions of the heavens also apply to Earth.

Kepler's laws Three laws of orbital motion, first partially developed by Johannes Kepler, and derived later in their complete form by Sir Isaac Newton using Newton's laws and his universal law of gravitation. Newton's derivations led to important modifications of Kepler's original "laws" that provide a method to determine masses of objects orbiting one another.

kilonova When two mutually-orbiting neutron stars spiral into one another due to the radiation of gravitational waves, they can produce an explosion that is about 1000 times more energetic than a nova but still 10 to 100 times less energetic than a core-collapse supernova. The product of the explosion is a black hole.

kinetic energy (KE) The energy associated with the motion of a mass through space. At speeds much less than the speed of light, kinetic energy is described by $\text{KE} = mv^2/2$. Kinetic energy is also symbolized as K. The SI unit of energy is the joule (J).

Kirchhoff, Gustav Robert (1824–1887) A Prussian/German physicist who made numerous important contributions in electrical circuits, optics, and thermal (blackbody) radiation.

Krishnamurti, Jiddu (1895–1986) An India-born philosopher and founder of an international educational center. His thoughts on educational goals included global outlook, concern for humanity intertwined with nature and the environment, and a scientific perspective.

Kuiper, Gerard (1905–1973) A Dutch–American astronomer who is sometimes referred to as the father of planetary science. The Kuiper belt was named in his honor.

Kuiper belt A doughnut-shaped collection of objects orbiting the Sun beyond Neptune. The Kuiper belt ranges from 30 au to 50 au and likely contains hundreds of thousands of ice-covered objects larger than 100 km across and possibly trillions of comets. The Kuiper belt is sometimes referred to as the Edgeworth–Kuiper belt.

L

Lagrange, Joseph-Louis (1736–1813) An Italian mathematician who made contributions to mathematics and our understanding of motion.

Lagrange point One of five locations where the total gravitational force produced by two large bodies while in orbit about their mutual center of mass (such as the Sun and Jupiter or Earth and the Moon) results in equilibrium points where much smaller objects such as asteroids or spacecraft can orbit in synchronization with the larger bodies.

Large Magellanic Cloud (LMC) A satellite galaxy of the Milky Way Galaxy that is visible to the naked eye in the southern hemisphere. The LMC appears near the Small Magellanic Cloud in the sky. The LMC is classified as a SBm, a barred spiral of the magellanic type.

late heavy bombardment A period of intense bombardment of the rocky planets and the Moon by asteroids and comets between 4 and 3.5 billion years ago.

Leavitt, Henrietta Swan (1868–1921) An American astronomer who discovered the period–luminosity relationship for Cepheid variable stars. Her discovery established Cepheids as the first standard candles, a critical "rung" in the distance ladder. Leavitt was one of the "computers" working under the direction of Edward Pickering at the Harvard College Observatory.

Leibniz, Gottfried Willhelm (1646–1716) A German mathematician and philosopher who discovered the calculus after, and independent of, Sir Isaac Newton.

Lemaître, Georges (1894–1966) A Belgian Catholic priest, astronomer, and mathematician who first showed mathematically that the expansion of the universe naturally follows from Einstein's general theory of relativity. Lemaître also provided data in his 1927 paper supporting his theoretical result, including a first estimate of the Hubble constant.

lepton A class of particles with very little mass. Leptons include electrons, positrons, neutrinos, and antineutrinos.

Leucippus An ancient Greek philosopher who was apparently the first to propose that every element is composed of a corresponding small, indivisible, and imperishable unit called an atom.

Levy, David (b. 1948) A Canadian astronomer, science writer, and popularizer of science who, along with Carolyn Shoemaker and Eugene Shoemaker, co-discovered the Shoemaker–Levy 9 comet.

light Electromagnetic radiation. A particle of light is referred to as a photon.

light curve The change in the amount of light coming from an object over time, such as due to an eclipse or from a variable star.

light-year (ly) A unit of distance that is typical of the distances between nearby stars. One light-year is defined as the distance that light travels in one year. Note that a light-year is *not* a measure of time. One light-year is related to one parsec by $1 \, \text{pc} = 3.261\,563\,777\,16 \, \text{ly}$. (One light-year is approximately 1/3 of a parsec.)

Linde , Andrei (b. 1948) A Russian–American physicist and cosmologist who, with Alan Guth and others, proposed the idea of inflation following the Big Bang.

line of nodes The imaginary line that is formed by the intersection of two orbital planes; e.g., the ecliptic and the plane of the Moon's orbit. The "nodes" are the two points where the orbit of one object intersects the orbit of the other (for example, Earth and our Moon).

linear momentum ($\vec{p}$) A vector quantity equal to the mass of a moving object times its velocity vector; $\vec{p} = m\vec{v}$. The total linear momentum of a system is conserved, but linear momentum can be exchanged between particles though interactions. Linear momentum is often simply referred to as momentum, but this can lead to some confusion with angular momentum, which is a different quantity.

Lippershey, Hans (1570–1619) A German–Dutch lensmaker who is believed to be the first inventor of the telescope.

Little Ice Age An extended period of unusually cool weather estimated to have occurred between about 950 CE and 1250 CE, although the start and end dates are debated. The Maunder minimum occurred during the deepest part of the Little Ice Age.

Lockyer, Sir Norman (1836–1920) An English astronomer who first discovered helium by identifying previously unknown spectral lines in the corona of the Sun. He also started, and was the first editor of the important scientific journal *Nature*.

logarithm A mathematical operation that returns an exponent for a specified base. For example, for a base-10 logarithm, the logarithm of $1 = 10^0$ is 0 (the exponent of the base, which is 10 in this case). Similarly, the base-10 logarithm of $10 = 10^1$ is 1, of $100 = 10^2$ is 2, and so on. The same applies to numbers that lie between powers of ten. For example, since $3.141\,59 = 10^{0.4971}$ the base-10 $\log(3.141\,59) = 0.4971$.

logarithmic The mathematical behavior of a quantity described by the exponent, x, when the numerical value of that quantity is expressed as 10^x, rather than as the actual value of that quantity.

long-period comet A comet with an orbital period of more than 200 years, and may have an orbital period of millions of years. Long-period comets are believed to originate in the Oort cloud.

Lowell, Percival (1855–1916) A wealthy businessman and American astronomer who suggested that there may be canals on Mars. He also erroneously calculated a possible location for Pluto based on faulty data. Lowell founded an observatory in Flagstaff, Arizona that bears his name.

luminosity (L) The rate at which energy is being emitted by an object. The luminosity of an object is the same as the object's power output. The unit of luminosity is the watt (W).

M

magma Molten rock.

magnetic dynamo The generation of a magnetic field through movement by rotation of an ionized gas or an electrically–conducting fluid, combined with convection. Magnetic dynamos are known to operate in many planets and stars, including Earth and the Sun.

magnitude scale A system, developed by Hipparchus of Nicaea, for describing the brightnesses of celestial objects. The first use of his system was in the star catalog he created. See also absolute magnitude and apparent magnitude.

main sequence The diagonal line of stars from upper left to lower right in the Hertzsprung–Russell diagram. Main-sequence stars are burning hydrogen into helium in their cores.

Marius, Simon (1573–1625) A German astronomer who independently discovered the four Galilean moons of Jupiter. He may have discovered them shortly before Galileo but didn't record his discovery until one day after Galileo did.

mass A characteristic of an object that describes its degree of resistance to acceleration resulting from an applied force. Mass also describes how strong the gravitational force associated with an object is. The apparent reality that mass describes both of those behaviors of matter is the basis for the equivalence principle. See also inertia.

Maunder, Annie Scott Dill Russell (1868–1947) An Irish astronomer who studied the solar corona and the periodicity of sunspot latitudes with her husband Walter Maunder. Annie Maunder was elected to the Royal Astronomical Society in 1916, the first year when women were selected for the honor. The Maunder minimum was named for the Maunders.

Maunder minimum A period of very low sunspot activity and minimal solar activity between about 1645 CE and 1715 CE. The Maunder minimum occurred during the deepest part of the Little Ice Age.

Maury, Antonia Coetana de Paiva Pereira (1866–1952) was an American astronomer who conducted important early work on the spectra of stars and calculated the first orbits of two spectroscopic binaries (the first system was discovered by Edward Pickering but Maury discovered the second one, β Aurigae). She was a niece of Henry Draper and her work was published in a catalog of stars in 1897 in a memorial to her uncle using a spectral classification scheme that she developed. She was one of the "computers" under Pickering at the Harvard College Observatory and she received the Annie Jump Cannon Award in 1943.

Maxwell, James Clerk (1831–1879) A Scottish mathematician and physicist who unified the forces of electricity and magnetism, and, at the same time, demonstrated mathematically the existence of electromagnetic waves. Maxwell also made important contributions to our understanding of the behavior of gases.

mechanical energy The combination of the kinetic energy and the potential energy of an object. Work alters mechanical energy.

Mendeleev, Dmitri (1834–1907) A Russian chemist who discovered the periodic chemical properties of elements when arranged in order of increasing atomic mass. His periodic table of the elements played a major role in organizing the field of chemistry and led to the prediction and discovery of other, previously unknown, elements.

meridian The arc across the sky that passes from the horizon point due north of the observer, through the observer's zenith, to the horizon point due south of the observer.

Mesopotamia The region between the Tigris and Euphrates rivers, also known as the Fertile Crescent, in what is present-day Iraq.

Messier, Charles (1730–1817) A French astronomer who searched for comets and compiled the catalog of star clusters, nebulas, and galaxies that bear his name.

meteor A streak in the sky produced by a meteoroid heating up due to friction, causing it to glow as it passes through the atmosphere at high speed.

meteorite A meteoroid that survived the passage through Earth's atmosphere and hit the ground.

meteoroid A rock, sometimes kilometers across and sometimes as small as a grain of sand, that is moving through space and could enter Earth's atmosphere. Most meteoroids are relics left over from the formation of the Solar System, although they may also originate from ejected material from other objects such as Mars, the Moon, or asteroids.

Michell, John (1724–1793) An English scientist and clergyman who proposed and built the original apparatus that was later modeled by Cavendish to measure the density of Earth, and thereby determining the value of the universal gravitational constant, G. He also first proposed the possible existence of black holes and that earthquakes travel in waves.

Michelson, Albert A. (1852–1931) An American physicist who is best known for the Michelson–Morley experiment and his measurements of the speed of light.

Michelson interferometer Designed to measure extremely short length differences, the apparatus uses a mirror to split a light beam, sending the separated beams in two different directions and then bringing them back together again to produce interference patterns. The apparatus was designed for the Michelson–Morley experiment but is used today for extremely precise length measurements, such aligning telescope mirror surfaces to a fraction of a wavelength of laser light.

Michelson–Morley experiment A series of experiments carried out by Albert A. Michelson and Edward W. Morley in an attempt to detect evidence of the ether that scientists of the late 1800s and early 1900s believed must permeate all of space. The ether was believed to be a medium that electromagnetic waves (light) needed to travel through. The device used in the experiments was a Michelson interferometer. A detection of the ether was never successful.

microwave The portion of the electromagnetic spectrum between radio and infrared wavelengths. Refer to Table 6.1.

Milky Way Galaxy The barred spiral galaxy (SBbc) composed of 200–400 billion stars, of which our Sun with its Solar System is a member.

millennium A period of 1000 years.

Miller, Stanley (1930–2007) An American biochemist who carried out a ground-breaking experiment in 1952 with Harold Urey in which amino acids were formed out of elements believed to have been present on early Earth.

Millikan, Robert (1868–1953) An American experimental physicist who successfully demonstrated Albert Einstein's law of the photoelectric effect. Millikan also first measured the elementary charge possessed by the electron, the proton, and other particles.

molecule A specific combination of atoms that are bound together as a single unit. A molecule may be composed of atoms of differing elements.

moon A body in orbit around a planet. Earth's moon is named with a capital M; e.g. Moon.

Morley, Edward W. (1838–1923) An American physicist who, with Albert A. Michelson, searched for the ether through what became known as the Michelson–Morley experiments. Morley also made precise measurements of the mass of the oxygen atom.

Muhammad ibn Abdullāh (c. 570–632) The last Prophet of Islam, to whom Muslims believe that God revealed the sacred text, the Quran.

Müller, Johannes (1436–1476) A German astronomer, also known by his adopted name, Regiomontanus, who wrote an important text, *Epitome of the Almagest*, that built on Ptolemy's work and included developments from medieval Hindu and Islamic astronomers and mathematicians.

multimessenger Information from an astronomical source that arrives through more than one type of "communication," such as via electromagnetic waves (photons), gravitational waves, or neutrinos.

muon (μ^-) An elementary particle that is a member of the class of particles known as leptons. The muon particle is associated with the muon neutrino and its antimatter counterpart is the antimuon (μ^+).

muon neutrino (ν_μ) An elementary particle that is a member of the class of particles known as leptons and one of three types of neutrinos. The muon neutrino is associated with the muon.

N

National Aeronautics and Space Administration (NASA) An independent agency of the United States of America, established in 1958. NASA's mission: *To reach for new heights and reveal the unknown so that what we do and learn will benefit all humankind.* In addition to its other activities, NASA conducts research in atmospheric and space flight and space exploration.

nebula An interstellar cloud of dust and gas. The plural of nebula is nebulas, or alternatively, nebulae.

neuromythology A commonly believed, but false, understanding of a particular aspect of how the human brain functions.

neutrino (ν) An elementary particle with no charge and a very small mass that is a member of the class of particles known as leptons. Neutrinos only interact very rarely with other matter and have a good chance of traveling unimpeded through a light-year's length of lead. There are three different types of neutrinos associated with the elementary particles, the electron (e^-), the muon (μ^-), and the tau (τ^-).

neutron (n) A subatomic particle that has no electric charge. Neutrons and protons combine to form the nuclei of atoms. Although neutrons can exist in isolation, they have finite lifetimes in that situation. A neutron is approximately 0.1% more massive than a proton. A neutron is composed of three quarks.

neutron star The remnant of a massive star of between 9 M_{Sun} and 25 M_{Sun} after it undergoes a core-collapse supernova explosion. A neutron star is essentially a giant nucleus without protons that has a characteristic size of a small city.

new moon One phase of the Moon. The entire portion of the Moon that is in shadow is facing Earth.

Newton, Sir Issac (1642–1727) An English physicist and mathematician who made numerous fundamental contributions to science, including the development of the calculus, the three laws of motion that bear his name, the universal law of gravitation, and work in the field of optics. His derivation of Kepler's laws of planetary motion represented a culmination of the Copernican revolution.

newton (N) The SI unit of force. The newton is equivalent to one kilogram times one meter per second squared ($kg\,m/s^2$).

Newton's laws Three general expressions regarding how an object moves under the influence of the vector sum of all forces acting on that object.

node Location on a standing wave where there is no motion.

north celestial pole (NCP) The point on the celestial sphere that is located directly above Earth's North Pole.

nova A thermonuclear explosion on the surface of a white dwarf star caused by material from a binary companion being deposited on the surface through an accretion disk.

nuclear reaction The conversion of one type of atomic nucleus into another, such as through nuclear fission or nuclear fusion.

nucleic acid Complex molecules found in the cells of all life on Earth. RNA and DNA are both nucleic acids.

nucleobase Molecular groups that attach to the backbones of RNA and DNA. The arrangement of nucleobases on the nucleic acid backbones code all of the genetic information of the organism.

nucleon A nucleon is a particle, either a proton or a neutron, that exists in an atomic nucleus.

nucleosynthesis The production of nuclei through nuclear reactions.

nucleus A dense collection of protons and neutron that form the center of an atom.

O

Occam's Razor The philosophical idea that if there are various ways of explaining a particular phenomenon, the simplest explanation is usually the most correct one. Attributed to William of Ockham (c. 1287–1347), although many others, including Pythagoras, Aristotle, and Ptolemy before him espoused similar ideas.

occult When one celestial object passes in front of another, blocking it from view. Occult is a verb and occultation is the associated noun.

Ørsted, Hans Christian (1777–1851) A Dutch physicist and scientist who discovered that a current passing through a wire produces a magnetic field.

olivine A mineral that is prevalent in Earth's mantle and in many types of meteorites. It is composed of varying amounts of the silicates Mg_2SiO_4 and Fe_2SiO_4.

Oort, Jan (1900–1992) A Dutch astronomer who made numerous contributions to astronomy, ranging from discoveries about the structure and motions of the Milky Way Galaxy to the statistical determination of the existence of the Oort cloud.

Oort cloud A roughly spherically symmetric distribution of icy bodies that are the source of the long-period comets.

open cluster A small cluster of stars, typically with a few hundred members, that are usually found near the galactic plane. Stars in open clusters tend to be relatively young in age.

optics The study of light.

orbital The quantum mechanical locations of electrons in atoms based on their four quantum numbers.

orbital resonance When two or more bodies, such as asteroids, moons, dwarf planets, or planets, have orbital periods that are simple integer ratios, meaning that the bodies align periodically. These periodic alignments are produced by tidal forces over time.

organic A molecule containing carbon atoms, often in the shapes of rings or strings. Some carbon-containing molecules, such as carbon dioxide (CO_2) are considered to be inorganic. A modern use of the term pertaining to how foods (or ideas or movements) are grown or developed is fundamentally different from the scientific meaning.

organic chemistry The subdiscipline of chemistry related to the unique chemistry of carbon-based molecules. See also biochemistry.

Orphic A cult of followers of the teachings of the mythical poet and musician, Orpheus, dating back to perhaps the 18th century, BCE.

P

panspermia The idea that primordial life could have developed elsewhere and was then transported by a meteoroid or asteroid to a different planet, where it served as the seed for evolution of life on the new world.

paradigm A long-held view, belief, or explanation that has become so entrenched in our way of thinking that it is inconceivable to most that it could be incorrect. A classic example is the paradigm of a geocentric universe.

parallax The shift in the alignment of an object with respect to more distant objects when the position of the observer changes.

parallax angle One-half of the full shift in angular position of an object due to the motion of the observer, such as the annual motion of Earth around the Sun.

parsec (pc) A unit of distance that is characteristic of the distances between nearby stars. The term derives from how it is measured; *par*allax arc*sec*ond.

Parsons, William, third Earl of Rosse (1800–1867) An Irish and English astronomer who built the Leviathan with a 72 in diameter mirror that was then the largest telescope in the world. Parsons also gave the Crab Nebula (M1) its name.

Pascal, Blaise (1623–1662) A French physicist, mathematician, and Catholic theologian. As a child prodigy, he published an important work in geometry at age 16. He also studied fluids and aspects of pressure.

pascal (Pa) The unit of pressure in the SI system of units. One pascal is defined as one newton per meter squared, or $1\,Pa \equiv 1\,N/m^2$. The pascal was named for Blaise Pascal.

Pauli, Wolfgang (1900–1958) An Austrian born theoretical physicist who demonstrated that no two electrons in an atom can have the same combination of four quantum numbers.

Pauli exclusion principle No two electrons in an atom can have the same combination the four quantum numbers, n, ℓ, m_ℓ, and m_s. The Pauli exclusion principle, combined with filling the electron energy levels of atoms from the lowest levels to higher levels, give rise to the chemical properties of atoms and the structure of the periodic table.

Payne-Gaposchkin, Cecilia Helena (1900–1979) A British-born astronomer and astrophysicist who emigrated to the United States in 1923, and became a United States citizen in 1931 (she changed her last name from Payne to Payne-Gaposchkin when she was married in 1934). With Harlow Shapley's encouragement, she earned her Ph.D. in 1925 from Radcliffe College (now a part of Harvard). In her doctoral dissertation she was the first person to determine the compositions of the stars and the dominant role that hydrogen and helium play in the universe. She also studied the structure of the Milky Way Galaxy and variable stars in both the Milky Way and the Magellanic Clouds. Payne-Gaposchkin spent her entire career at Harvard, being named "Astronomer" in 1938, and she became the first female full professor promoted from within the faculty in 1956. She later became the first woman to head a department at Harvard.

Peebles, Phillip James (Jim) Edwin (b. 1935) A Canadian astrophysicist and theoretical cosmologist who made many significant contributions to the field of cosmology.

peer review The process, commonly used in scientific publishing, of examining the results of research by an independent, and usually anonymous, peer or group of peers. A form of peer review can involve trying to reproduce the results of a scientific study by other peers or teams of peers.

Penzias, Arno Allan (b. 1933) An American physicist and radio astronomer who, with Robert Wilson, made the first detection of the cosmic microwave background in 1964.

percent (%) The term percent is a concatenated word meaning "per cent," or per one hundred; for example, there are 100 cents in one dollar, and 100 years in one century. When you see the symbol % following a number, the meaning is that you multiply the number by $1/100 = 0.01$.

perihelion The point of closest approach for a planet, comet, or asteroid in its orbit around the Sun.

periodic table An organizational scheme that arranges all of the elements in the universe based on atomic number and their chemical properties. The rows correspond to the principal quantum number (n) and the columns correspond to the angular momentum quantum number (ℓ). The energy levels fill in each row from lowest energy to highest in accordance with the Pauli exclusion principle.

period–luminosity relationship A relationship between the pulsation periods of certain classes of pulsating stars and their intrinsic luminosities or absolute magnitudes, especially heavy-element rich and heavy-element poor Cepheids, RR Lyras (generally heavy-element poor), and Miras. The P–L relationship is used to help evaluate our understanding of stellar structure and evolution as well as playing a critical role in determining distances up to millions of light-years or more.

phases of the Moon Although the Sun always illuminates the side of the Moon facing the Sun, the portion of the Moon's illuminated surface facing Earth changes as the Moon orbits Earth. Because a specific phase requires a particular alignment between the Sun, Earth, and the Moon, the lunar phase cycle takes approximately 29 1/2 days (the synodic period). The order of phases is new moon, waxing crescent, first quarter, waxing gibbous, full moon, waning gibbous, third quarter, waning crescent, and back to new moon.

Philolaus of Croton (c. 470–385 BCE) A follower of Pythagoras who suggested a universe with a central fire, with Earth, a counter-Earth (Anticthon), the Moon, the Sun, the planets, and the stars going around it in circles.

photon (γ) A particle of light that carries a quantum of energy $(E = hf = hc/\lambda)$ and a quantum of linear momentum $(p = E/c = h/\lambda)$.

photosphere The deepest region of the atmosphere of a star where most of the light energy originates.

Piazzi, Giuseppe (1746–1826) An Italian Catholic priest who was also a mathematician and astronomer. Piazzi is credited with the discovery of the first dwarf planet, Ceres, in 1801.

Pickering, Edward Charles (1846–1919) An American physicist and astronomer who discovered the first spectroscopic binary star system and was director of the Harvard College Observatory from 1877 until his death in 1919. He made important contributions to the study of stellar spectra through photography. Pickering recruited numerous women to assist him with data analysis, including Antonia Maury, Williamina Fleming, Annie Jump Cannon, and Henrietta Leavitt. The group of women was variously referred to as the "Harvard Computers," "Pickering's Women," or sometimes as "Pickering's Harem" by the scientific community at the time.

Planck, Max (1858–1947) A German theoretical physicist who first derived the curve that represents blackbody radiation as a function of wavelength for any specific temperature and, in the process, first introduced the concept of quantized energy.

Planck's constant (h) A fundamental constant of nature specific to quantum mechanics that is always involved in quantities at the smallest scales such as energy, linear momentum, and angular momentum $(h = 6.626\,070\,15 \times 10^{-34}\ \mathrm{J\,s})$.

planet An object with sufficient mass to obtain a nearly spherical shape, that directly orbits the Sun or another star rather than orbiting a planet, and has sufficient gravitational influence to sweep clean the region near its orbit. Rogue planets also appear to be roaming the Milky Way Galaxy without orbiting a parent star.

planetarium A theater where the positions and motions of the stars and planets is projected onto the ceiling.

planetary embryo A pre-planetary body that formed from much smaller objects known as planetesimals. Most ultimately merged to form planets or were destroyed through impacts. Planetary embryos can range in size from 1000 km up to the sizes of the Moon or Mars.

planetary nebula An expanding cloud of dust and gas that is the ejected envelope of a dying low-mass star with an original mass less than 9 M_{Sun}. Following the planetary nebula ejection, the final stage of evolution is a white dwarf star.

planetary system One or more planets orbiting a single star or a multiple star system.

planetesimal A small, rocky body that forms out of dust in an early protoplanetary disk around a forming star as the first step in planet formation. Most planetesimals merged to build planetary embryos and planets, became small Solar System bodies, or were ejected through close encounters with the giant planets. Planetesimals range in size from 1 km across to 1000 km across.

plate tectonics Earth's crust is broken up into segments known as plates. Plates move relative to one another, producing earthquakes, volcanoes, trenches, mountain ranges, and other geologic features when they collide with, or slide past, other plates.

Plato (c. 428–348 BCE) A student of Socrates and the teacher of Aristotle and Eudoxus, Plato founded the School of Athens, considered to be the first university. Plato's natural philosophy was strongly influenced by Pythagoras. Plato argued that the motions of celestial objects should follow perfect spheres or circles at constant speeds, centered on Earth.

Plutarch (c. 46–120 CE) A well-known ancient Greek biographer and historian.

polarized light Electromagnetic waves with the electric field vectors aligned.

polymath A person who has accomplishments in multiple intellectual or artistic areas.

positron (e^+) An elementary particle that is a member of the class of particles known as leptons and the antimatter partner to an electron. A positron is identical to an electron except that it has a positive charge instead of a negative charge.

potential energy (PE) The energy stored in a system that can be converted into other forms of energy such as kinetic energy, heat, or photons. Potential energy is also symbolized as U. The SI unit of energy is the joule (J).

power (P) The rate at which energy is used or absorbed. The standard unit of measure is the watt (W).

precession The wobble of Earth's axis with a period of approximately 26,000 years. This effect causes the north celestial pole and the south celestial pole to trace 23 1/2° radius circles on the celestial sphere. The equinoxes also move along the celestial equator with the same period, requiring astronomers to adjust the coordinates, right ascension and declination, relative to a specific reference date when making observations on any given night.

pre-modern A society that existed prior to the introduction of the modern scientific process of studying nature.

pressure Mathematically, pressure is the perpendicular force acting on a surface divided by the area of the surface, or $P = F_\perp/A$. The unit of pressure is the pascal (Pa). Physically, pressure is the result of particles bouncing off a surface or other particles, exerting a force during the collisions.

principal focus (F) One of two focal points or foci of an ellipse. The principal focus is the location of the center of mass of a system of two masses orbiting one another.

Principia (*Philosophiæ Naturalis Principia Mathematica*) The monumental treatise of Sir Isaac Newton, first published on July 5, 1687. The *Principia* contains Newton's three laws of motion, his universal law of gravitation, his derivation of Kepler's laws of planetary motion, the first viable theory of tides, and makes use of his development of the calculus.

prograde The usual direction of motion or rotation of a planet or other celestial object. In our Solar System that direction is counterclockwise when viewed from above Earth's North Pole, or the Sun's north pole.

prominence Hot gas rising above the surface of the Sun that follow magnetic field lines.

protein A molecule composed of long chains of amino acids.

proton (p^+) A subatomic particle with a positive electric charge. Protons may exist in isolation, or they may combine with other protons and neutrons to form the nucleus of an atom. A proton is approximately 0.1% less massive than a neutron (n). The electric charge of a proton is equal in magnitude but opposite in sign to the electric charge on an electron (e^-). A proton is composed of three quarks.

protoplanet A large planetary embryo where melting occurred in its core, leading to a chemically differentiated interior.

protoplanetary disk The disk that forms from the gravitational collapse of a nebula as a direct result of conservation of angular momentum.

pseudoscience A set of ideas or beliefs that may appear superficially to be scientific in nature but that do not make testable and verifiable predictions, and are therefore not fundamentally falsifiable. Pseudoscientific ideas are also often based on ad hoc assumptions that are not themselves testable, that may be based on supernatural causes, or are inconsistent with factual evidence. See also the scientific method.

Ptolemy, Claudius (c. 90–168 CE) The last of the great natural philosophers of the Greek tradition, he had Greek and Roman citizenship, wrote in Greek, and lived in Alexandria (his Roman name was Claudius Ptolemæus). Ptolemy was a mathematician, astronomer, astrologer, and geographer. His *Almagest* became the definitive work in astronomy until the Copernican revolution. His world maps influenced exploration, particularly sailing, until the 1500s.

pulsar A rapidly spinning neutron star possessing a strong magnetic field that beams electromagnetic radiation into space along the star's magnetic field axis.

pulsating star A star with variable luminosity, temperature, and radius due to internal processes.

Pythagoras of Samos (c. 570–495 BCE) An ancient Greek philosopher who believed that numbers were a true representation of nature. He also believed in the validity of abstract thinking and had a significant influence on Plato. A religious cult, known as the Pythagoreans, was founded on the teaching of Pythagoras.

Q

quantum A discrete quantity, usually associated with quantum mechanics. The term may also refer to a discontinuous change.

quantum gravity An as yet to be fully-developed theory of gravity that, unlike the general theory of relativity, would be fundamentally quantum mechanical.

quantum mechanical tunneling The ability of particles to penetrate a barrier without having enough kinetic energy to overcome the barrier's potential energy. This can occur under the right physical conditions because of the wavelike characteristics of particles and the Heisenberg uncertainty principle.

quantum mechanics A theory developed collectively by numerous scientists that describes how nature behaves at atomic and subatomic size scales.

quantum number A number that describes a discrete quantity in quantum mechanics, such as for electron orbitals in atoms; energy (n), angular momentum (ℓ), magnetic (m_ℓ), or spin (m_s). These quantum numbers govern the locations of the various elements on the periodic table.

quark The component particles that make up baryons, such as protons and neutrons.

quasar Active galactic nuclei in the early universe that produced extremely energetic and bright jets that are easily observed today.

R

radial velocity The component of an object's velocity vector that is directed toward or away from the observer. Radial velocities can be determined by measuring wavlength shifts (the Doppler effect).

radian (rad) The natural unit of angular measure: 2π rad $= 360°$ (a complete circle), 1 rad $\approx 57.3°$ and $1° \approx 0.0175$ rad.

radiation The emission of particles, including photons, from a source. The particles carry energy away from the source, reducing the amount of internal energy (heat) the source contains.

radiation pressure The pressure due to photons reflecting off, or being absorbed by, particles such as electrons, atoms, ions, or dust, or even larger objects, such as asteroids or solar sails.

radio The longest wavelengths of the electromagnetic spectrum; refer to Table 6.1.

radioactive The spontaneous emission of particles from a source, such as the atomic nucleus of an isotope. Each type of radioactive decay from a particular isotope has a characteristic half-life.

radioisotope An isotope of an element that is radioactive.

radiometric dating The process of analyzing the radioisotopes of a rock or other sample to determine the sample's age based on half-life information.

redshift The lengthening of an electromagnetic wave due to the motion of the source away from the observer (Doppler effect), light traveling out of a gravitational well, or expansion of the universe.

regolith A fine, dusty layer that can be found covering solid rock in some locations on the rocky planets, as well as across the surfaces of our Moon, other moons, and many asteroids.

relativity theory Developed by Albert Einstein, relativity theory (special and general) describe how space and time are linked and forever considered collectively as spacetime.

Renaissance, European An intellectual movement that existed between the 14th and 17th centuries. The Renaissance produced great innovation in art, music, literature, poetry, philosophy, religion, politics, science, and mathematics. An earlier Renaissance of the 12th century resulted in the rediscovery of the science and mathematics of the ancient Greeks.

retrograde The apparent motion of a planet or other celestial object when it is traveling or rotating in the opposite direction from its typical prograde motion.

Richer, Jean (1630–1696) A French astronomer who worked with Cassini to determine the distance to Mars in 1672, thereby making a reasonably accurate determination of the value of the astronomical unit.

right ascension (RA) Coordinate lines on the celestial sphere that parallel Earth's lines of constant longitude. The point of origin of right ascension is the vernal equinox, and the angle east of the vernal equinox is measured in hours associated with Earth's rotation, where $1^h = 15°$. Right ascension is sometimes symbolized by α. See also declination.

right triangle A triangle that contains one 90° angle. The sum of the angles of any flat triangle equals 180°. A right triangle provides straight-forward ways of determining relationships between its angles and the lengths of its sides through the mathematics of trigonometry.

RNA A nucleic acid that most often has a single helix structure that is present in all living cells on Earth. The backbone is composed of alternating groups of ribose (a sugar formed from hydrogen, carbon, and oxygen) and a phosphate group that contains phosphorus. Attached to the backbone are the nucleobases uracil, adenine, cytosine, and guanine. Various forms of RNA serve a variety of purposes in cell processes. For example, mRNA (messenger RNA) carries information about the genetic sequence of a gene, and aids in synthesizing proteins. mRNAs played a crucial role in the rapid development of vacines for the coronavirus pandemic.

Roche, Édouard (1820–1883) A French mathematician and astronomer who studied celestial mechanics. He is most well-known for showing that small objects that get too close to a massive object can be pulled apart by tidal forces.

rocky planet The class of planets that are Earth-like: rocky, small, and close to the Sun. The rocky planets are Mercury, Venus, Earth, and Mars. The Moon is also often thought of as a rocky planet. The rocky planets are also referred to as terrestrial planets.

Rømer, Ole Christensen (1644–1710) A Danish astronomer who first estimated the amount of time required for light to travel across Earth's orbit.

Röntgen, Wilhelm Conrad (1845–1923) A German physicist and mechanical engineer who discovered x-rays in 1895.

RR Lyra A class of pulsating variable stars that obey a nearly flat period–luminosity relationship. RR Lyras are subgiant stars passing through the instability strip on the H-R diagram. Although not as bright as Cepheids, they are still bright enough to be used as standard candles for determining distances to globular clusters.

Rubin, Vera Cooper (1928–2016) An American astronomer who, in the 1970s, confirmed the existence of dark matter through observations of flat rotation curves of spiral galaxies.

Rumford, Sir Benjamin Thompson, Count (1753–1814) Born in America and later emigrating to England, Rumford carried out early experiments on the transfer of kinetic energy into heat via the boring of cannon barrels.

Russell, Henry Norris (1877–1957) An American astronomer who was one of two scientists to develop the Hertzsprung–Russell diagram. Russell conducted his work independent of, but at about the same time as, Ejnar Hertzsprung. Russell also made a number of other important contributions to the field regarding the spectra of light and stellar evolution. Russell served as director of the Princeton University Observatory from 1912 to 1947.

Rutherford, Ernest (1871–1937) A New Zealand-born British physicist and the 1st Baron Rutherford of Nelson, who conducted pioneering research in nuclear physics, and radioactivity, and the existence of the atomic nucleus.

S

Sagan, Carl Edward (1934–1996) An American astronomer and planetary scientist. Along with being a prolific research scientist, Sagan was an author of science and science fiction books, narrator of a popular television series, and the foremost popularizer of astronomy during his lifetime.

Saha, Meghnad (1893–1956) An Indian astrophysicist who conducted theoretical work in understanding the ionization of atoms. His famous Saha equation is one of the foundations in understanding the atmospheres of stars.

saros cycle An 18.03 year cycle of lunar and solar eclipses corresponding to 223 synodic lunar months and 242 draconic months.

scalar A quantity that is described by a single number and its unit, such as mass, which is given in kilograms.

scattered disk A vast collection of small Solar System bodies that exist beyond the Kuiper belt. Scattered disk objects tend to have highly eccentric orbits.

Schiaparelli, Giovanni Virginio (1835–1910) An Italian astronomer who produced an early map of Mars that depicted what he referred to as canali (channels). The term "canali" was misinterpreted by some, including Percival Lowell, to mean canals, presumably built by intelligent beings.

Schrödinger, Erwin Rudolf Alexander (1887–1961) An Austrian physicist who made significant contributions to the development of quantum mechanics. Particularly important is the Schrödinger equation that describes the wave-like behavior of particles, first postulated by de Broglie.

Schwarzschild, Karl (1973–1916) A German physicist and astronomer. Among his many accomplishments was applying Einstein's general theory of relativity to the development of our understanding of black holes. He carried out his calculations on black holes while serving in the German army during World War I. He died as a result of a disease he contracted while on the Russian front.

Schwarzschild radius The radius of a black hole from its center to the event horizon.

scientific method A process of analysis of nature that leads to further advances based on the predictions of a theory or that can potentially prove that a scientific theory is invalid. Although the scientific method is often described as very formulaic, the process of science is rarely so well-defined or executed.

SETI The Institute's mission is in part to "explore, understand, and explain the origin and nature of life in the universe." A portion of its efforts has been to scan the heavens for signs of intelligent life through signals in the electromagnetic spectrum. SETI is an acronym for "Search for ExtraTerrestrial Intelligence."

Seleucus of Seleucia (c. 190–c. 150 BCE) Supported the heliocentric model of Aristarchus. The peak of his work was in the time around 150 BCE. The year of his death is unknown. Seleucia was located near on the Tigris River, near present-day Madain, Iraq.

semimajor axis (a) One-half of the length of the long (major) axis of an ellipse.

Shapley, Harlow (1885–1972) An American astronomer who used globular clusters to estimate the size of the Milky Way Galaxy. Shapley served as head of the Harvard College Observatory and president of the American Association for the Advancement of Science, helped to establish and fund the National Science Foundation, and was a political activist. Shapley also participated in the Great Debate about whether or not "spiral nebulas" are within or external to the Milky Way Galaxy ("island universes").

Shelley, Mary Wollstonecraft (1797–1851) An English writer who authored the Gothic novel, *Frankenstein*, considered by some to be the first science fiction novel.

Shelton, Ian (b. 1957) A Canadian astronomer who discovered the naked-eye core-collapse supernova, SN 1987A, in the Large Magellanic Cloud, a satellite galaxy of the Milky Way Galaxy.

Shoemaker–Levy 9 A comet that had been orbiting Jupiter for 20 to 30 years before it broke apart during a close encounter with the gas giant in 1992. The multiple fragments impacted Jupiter on the next orbit over the course of six days between July 16 and July 24, 1994. The collisions were watched worldwide by astronomers and the general public alike, providing significant information about Jupiter. Shoemaker–Levy 9 was discovered by Carolyn Shoemaker, Eugene Shoemaker, and David Levy.

Shoemaker, Carolyn (1929–2021) An American astronomer who was a co-discoverer of the Shoemaker–Levy 9 comet, together with her husband, Eugene Shoemaker, and their colleague David Levy. She once held the record for the most comets discovered by a single individual.

Shoemaker, Eugene (1928–1997) An American planetary geologist who studied terrestrial craters formed by meteor impacts. He also trained Apollo astronauts in geology for their lunar missions. In addition, he was a co-discoverer of the Shoemaker–Levy 9 comet, together with his wife Carolyn Shoemaker and their colleague David Levy. Gene Shoemaker had been very disappointed to not have had the opportunity to go to the Moon and "[bang] on it with my own hammer". In 1999 his ashes were deposited on the Moon when the Lunar Prospector spacecraft crashed into it, making him the first person to have his remains left on another world.

SI units (SI) The international system of units that uses kilograms (kg), meters (m), seconds (s), and amperes (A). The system abbreviation comes from the French: *Le Système international d'unités*.

sidereal With respect to the distant stars. A sidereal period is one complete cycle with respect to the stars.

Sidereus Nuncius The 24-page booklet, written by Galileo in 1610, in which he describes some of his first observations of the heavens with his small telescope.

silicate The primary building material for rock. A silicate is any compound containing combinations of silicon and oxygen atoms (for example, SiO_4 or SiO_5).

Simplicius of Cilicia (c. 490–560 CE) Considered to be one of the last of the Neoplatonists and antiquity's last great philosopher. He wrote extensively on Aristotle. Cilicia was along the southeastern coast of modern Turkey.

singularity A mathematical term meaning a point where a value becomes infinite. A singularity is also used in astronomy to describe the point at the center of a non-spinning black hole where mass density is infinite.

Sitting Bull (1831–1890) A Native American holy man and leader of the Hunkpapa Lakota tribe who led their fight against policies of the United States government. His leadership led to the defeat of Lt. Col. George Armstrong Custer in the Battle of the Little Big-

horn on June 25, 1876. Sitting Bull was killed by Indian agency police on the Standing Rock Indian Reservation when they tried to arrest him on December 15, 1890.

Small Magellanic Cloud (SMC) A small irregular satellite galaxy of the Milky Way Galaxy that is visible to the naked eye in the southern hemisphere. The SMC appears near the LMC in the sky and is being destroyed by the LMC. The SMC is classified as Irr.

small Solar System body Objects in the Solar System smaller than dwarf planets.

SN1987A A naked-eye core-collapse supernova that appeared in the Large Magellanic Cloud on February 14, 1987.

Socrates (c. 470–399 BCE) The teacher of Plato. Socrates proposed that a problem should be solved by breaking it down into a series of questions that yield to logical hypotheses. He is also credited with developing the teaching style known as the Socratic method that is still in use today.

solar cycle The 11-year cycle of solar activity, including the number and latitude of sunspots, and the frequency and strength of solar flares, prominences, and coronal mass ejections. However, the orientation of the Sun's magnetic field flips every 11 years as well. This means that the full solar cycle (the time it takes for the Sun's magnetic polarity to return to its starting configuration) is actually 22 years.

solar flare An energetic explosion on the surface of the Sun capable of releasing energy ranging from the equivalent of a ten megaton bomb to over one billion megatons. They also send particles through space that can travel 1 au in as little as thirty minutes. Flares can affect radio communications on Earth.

solar mass (M_{Sun}) A unit of mass equaling the mass of the Sun ($1\ M_{Sun} = 1.988\ 47 \times 10^{30}$ kg).

solar nebula The initial cloud of dust and gas that the Sun and our Solar System formed out of.

Solar System The system of planets, dwarf planets, moons, asteroids, meteoroids, comets, and dust that are gravitationally bound to the Sun.

solar wind Ionized gas coming from the Sun's atmosphere that travels through the Solar System. The temperature of the solar wind is close to one million kelvins.

sound Oscillating pressure waves that pass through a substance. The pressure waves oscillate in the same direction as their motion.

south celestial pole (SCP) The point on the celestial sphere that is located directly above Earth's South Pole.

spacetime The three dimensions of space and the dimension of time combined into a four-dimensional coordinate system where all four dimensions are linked according to relativity theory.

special theory of relativity Describes the affect on spacetime due to the motion of the observer at a constant velocity relative to an observed object or event. At low speeds, special relativity corresponds to our everyday experiences as understood by Galileo and Newton. At speeds approaching the speed of light, the affect on spacetime differs dramatically from what our intuition would suggest: moving clocks appear to run more slowly and lengths in the direction of motion appear to contract. Special relativity also produced Einstein's famous mass–energy equation: $E = mc^2$.

spectral line An absorption or emission line produced by atoms, molecules, or ions.

spectroscope An optical device that seperates visible light into an observable spectrum.

spectroscopy The study of the absorption and emission lines (the spectra) of atoms, molecules, and ions.

spectrum The intensity of light (or the absence of light) at wavelengths across electromagnetic radiation.

speed of light (*c*) The speed at which an electromagnetic wave travels in a vacuum. The speed of light is the maximum speed that anything can travel, including information; $c \equiv 299{,}792{,}458$ m/s, or approximately 300,000 km/s.

spin–orbit coupling When the orbital and rotation periods of a body (a moon, a planet, or a star) are a small integer ratio. This ratio results from tidal forces produced by each body on the other.

spiral galaxy (S) A galaxy with a visible disk where young, bright, massive stars highlight spiral arm patterns. Spiral galaxies also have spheroidal and/or boxy/peanut bulges, stellar halos, hot gas halos, and dark matter halos. Spiral galaxies tend to be among the more massive galaxies in the universe.

standard candle A celestial object whose luminosity (or absolute magnitude) is known due to some intrinsic property without needing to know its distance first. Standard candles can be used to determine distances by measuring an object's apparent magnitude and then comparing it with the object's absolute magnitude.

star An object with sufficient mass to have, or have had, fusion reactions taking place in its deep interior.

statistical mechanics The study of thermodynamics by means of a statistical analysis of the motions and energies of very large numbers of particles, such as atoms, molecules, ions, electrons, nuclei, or photons.

steady-state theory A 1948 theory that proposed that the universe is eternal, having neither a beginning or an end. The theory included the idea of extremely slow but continuous creation of matter to replace matter that is spread out due to the expansion of the universe.

Stefan, Jožef (1835–1893) A Slovene Austrian physicist who was able to show that the luminosity of a blackbody is directly related to its surface area and its surface temperature.

Stefan–Boltzmann blackbody luminosity law , which is typically referred to as the Stefan–Boltzmann law. The relationship between the luminosity of a blackbody radiator (a hot, dense gas or a solid) and its surface area and surface temperature.

Strömgren, Bengt Georg Daniel (1908–1987) A Danish astronomer and astrophysicist. Strömgren was a bit of a child prodigy, publishing articles in astronomy as a boy and earning his doctorate before the age of 21.

strong nuclear force The fundamental force of nature that holds protons and neutrons together within the nucleus of an atom.

Struve, Otto (1897–1963) A Ukrainian–American astronomer who served as director of several observatories, published more than 900 articles and books, and served as Editor in Chief of a major research journal, *The Astrophysical Journal*.

subatomic Refers to a component of an atom or a size scale smaller than an atom.

subgiant The evolutionary stage of a star that places it on the horizontal branch of the Hertzsprung–Russell diagram, immediately following the main sequence. A subgiant star has depleted hydrogen in its core.

sublimate The change of a solid directly into a gas. Water ice and carbon dioxide ice (dry ice) are well-known examples.

Sumer An ancient civilization in southern Mesopotamia, that may have been established sometime between c. 5500 and 4000 BCE. It is believed that the Sumerians developed the first primitive form of writing known as cuneiform.

Sumerian A member of the ancient civilization of Sumer. The term also refers to the language spoken by the Sumer civilization.

summer solstice The point on the celestial sphere when the center of the Sun reaches its greatest northerly declination. The coordinates of the summer solstice are RA = 6^h, dec = 23 1/2°. The start of summer is defined to occur when the center of the Sun reaches the summer solstice.

Sun A main sequence G2 star that is also our parent star. The Sun is also referred to as Sol from which the term "solar" derives.

sunspot Dark, cooler, regions on the surface of the Sun that are produced by the emergence of localized strong magnetic field lines.

supercluster A cluster of galaxy clusters and groups.

supergiant A star that is burning hydrogen, helium, and perhaps progressively heavier elements in shells around a condensed core. A supergiant star, which is near the end of its lifetime, is very luminous and cool, and so is located in the upper right-hand corner of the Hertzsprung–Russell diagram. A supergiant star can have a radius nearly as large as the orbit of Jupiter.

superior planet The planets orbiting the Sun that are farther from the Sun than Earth. See also inferior planet.

supermassive black hole (SMBH) A black hole that has a mass of millions or billions of solar masses. SMBHs are found at the centers of all large galaxies.

supernova (SN) The extremely energetic explosion of a massive star or a white dwarf star. The plural is supernovas, or alternatively, supernovae.

surface of last scattering Looking back into deep space and deep time, it is the spherical surface around the observer where decoupling occurred. It is also where photons of the cosmic microwave background were finally able to travel throughout the universe without significant scattering off free electrons and protons.

surface temperature ($T_{surface}$) The temperature at the surface of a planet or star. A star doesn't have a definite "surface" so a specific location in the photosphere of the star is chosen.

synchronous rotation When the rotation period of a body is equal to its orbital period as a consequence of tidal forces.

synodic The alignment of an object (e.g., the Moon or a planet) with the Sun and Earth. A synodic period is the time interval between two such alignments. Note that a synodic period for a planet or the Moon is not a 360° cycle, because of Earth's orbital motion.

system 1. A collection of objects, including their energies, angular momenta, and linear momenta. A closed system prohibits the exchange of particles, energy, angular momentum, or linear momentum with anything not part of the system. 2. An organizational scheme or types of measurements.

T

tau (τ^-) An elementary particle that is a member of the class of particles known as leptons. The tau particle is associated with the tau neutrino.

tau neutrino (ν_τ) An elementary particle that is a member of the class of particles known as leptons and one of three types of neutrinos. The tau neutrino is associated with the tau (τ^-).

Taylor, Jane (1783–1824) An English poet and novelist who is credited with the children's poem, "The Star," better known as "Twinkle, twinkle little star."

tectonic plate See plate tectonics.

temperature (*T*) A quantity that is proportional to the amount of heat energy contained in a substance. Note that temperature is not the same thing as heat itself.

Thales of Miletus (c. 624–546 BCE) A pre-Socratic Greek philosopher known as one of the Seven Sages of Greece. Aristotle referred to Thales as the first philosopher of the Greek tradition. Ruins of Thales's home region, Miletus, are just north of the present-day village of Balat, Turkey. Miletus is approximately 40 km (25 mi) southwest of Samos, home of Pythagoras.

Tharp, Marie (1920–2006) An American oceanographer who, together with Bruce C. Heezen, produced the first topographic map of the entire ocean floor.

theory A basic set of assumptions upon which a major aspect of nature is described. A scientific theory must be self-consistent, without ad-hoc components to explain observations or experimental results. A scientific theory must also be able to make predictions (hypotheses) concerning the results of observations or experiments that can potentially prove that the theory is invalid; in other words, the theory must be falsifiable.

theory of everything (ToE) A theory that combines all four fundamental forces of nature into one overarching theory. A ToE must necessarily include a theory of quantum gravity.

thermal Relating to heat.

thermodynamics The study of heat and how heat energy is transferred from one substance to another, as well as how heat is converted into other forms of energy. See also statistical mechanics.

third quarter One phase of the Moon. One-half of the illuminated portion of the Moon is visible from Earth, with the amount of illuminated lunar surface *decreasing* with time.

tholin A dark, reddish, tar-like material containing organic molecules, found on bodies in the outer Solar System.

Thomson, Sir William, Lord Kelvin (1824–1907) An Irish mathematical physicist who made important contributions to the study of thermodynamics.

tide The difference in force vectors on opposite sides of a body caused by one side being closer to a massive object than the other side is. Tidal forces can cause an object to be stretched along the direction toward the center of the massive object or even pulled apart if the tidal force is great enough.

Tombaugh, Clyde (1906–1997) An American astronomer who discovered Pluto in 1930 by using a blink comparator to compare the same regions of the sky taken at different times in order to detect the dwarf planet's motion across the background stars.

tonne A unit of mass associated with the International System of Units (SI). One tonne is defined as being equal to 1000 kg, and is equivalent to 2204.62 lb_m, or 1.10231 ton. The official SI unit is megagram (Mg). Tonne is also referred to as metric ton.

torque Torque is the application of a force perpendicular to a line connecting the location of the applied force to a rotation axis, multiplied by the distance between the applied force and the rotation axis ($\tau = F_\perp r$). The use of a wrench to tighten a bolt is an applied example; the longer the wrench and/or the greater the applied perpendicular force, the greater the torque.

Transiting Exoplanet Survey Satellite (TESS) Launched in April, 2018, TESS is a NASA exoplanet-hunting mission to study nearly 85% of the sky and more than 200,000 stars.

transition region The layer of a star's atmosphere between the chromosphere and the corona. It is within the transition region that the temperature climbs very rapidly with height over a short distance until coronal temperatures are reached.

trans-Neptunian object (TNO) An object orbiting the Sun near the ecliptic and beyond the orbit of Neptune. A TNO may be a dwarf planet.

trigonometry The area of mathematics that deals with relationships between angles and distances related to triangles. The triangles need not be defined on flat surfaces; for example, they could be drawn on the surfaces of spheres like the celestial sphere or Earth.

tritium An isotope of hydrogen that contains one proton (which is what makes it a hydrogen isotope) as well as two neutrons: $^3_1$H.

trojan A small Solar System body that is trapped in a 1:1 orbital resonance with a planet. Trojans lead or trail the planet by 60° near the L_4 and L_5 Lagrange points.

tropic of Cancer 23 1/2° N latitude. The latitude on Earth north of the equator where the Sun is directly overhead at midday when the Sun is on the summer solstice. This happens on about June 21.

tropic of Capricorn 23 1/2° S latitude. The southern latitude on Earth where the Sun is directly overhead at midday on approximately December 21 when the Sun is on the winter solstice.

Twain, Mark (1835–1910), An American writer and humorist. Mark Twain was the pen name of Samuel Langhorne Clemens.

Tycho Brahe (1546–1601) A Danish astronomer who carried out precise observations of the locations of the naked-eye planets relative to the stars on the celestial sphere. Tycho Brahe is typically referred by his first name, Tycho. Tycho hired Johannes Kepler to be his assistant.

U

ultraviolet A portion of the electromagnetic spectrum with wavelengths shorter than visible; see Table 6.1.

universal gravitational constant (G) A fundamental constant of nature that describes the strength of the gravitational force. $G = 6.674\,30 \times 10^{-11}\,\mathrm{N\,m^2/kg^2}$

universal law of gravitation Newton's equation describing how two objects exert equal and opposite forces of attraction on one another due solely to the product of their masses divided by the square of the distance between their centers of mass. The direction of the force vectors are directly toward each other's center of mass.

Urey, Harold (1893–1981) An American chemist who, together with Stanley Miller, conducted a ground-breaking experiment in 1952 that produced amino acids out of molecules believed to have been present on early Earth.

V

Van Allen, James (1914–2006) An American space scientist who flew sounding rockets with radiation detectors, discovering what are known today as the Van Allen radiation belts.

variable star A stars that varies in luminosity by either internal or external processes. Examples include pulsating stars, eclipsing binary star systems, novas, kilonovas, and supernovas.

vector A quantity that is described by both a magnitude (a number and its unit) and direction. For example, velocity is given by how fast something is moving (its speed in m/s, which is a scalar) and a direction, such as traveling straight north. A vector is distinguished from a scalar symbolically by being typeset in bold face with an arrow above the symbol ($\vec{v}$ for a vector and v for a scalar).

velocity The rate of change of an object's position over time. Velocity is a vector quantity.

vernal equinox (Υ) The point on the celestial sphere when the center of the Sun crosses the celestial equator moving from south to north. The coordinates of the vernal equinox are RA = 0^h, dec = $0°$. The start of spring is defined to occur when the center of the Sun crosses the vernal equinox. The vernal equinox is also referred to as the First Point of Aries.

Verne, Jules Gabriel (1828–1905) A French writer who penned what are considered today early science fiction works; *Journey to the Center of the Earth*, and *Twenty Thousand Leagues Under the Sea*.

virtual particle A particle that pops into and out of existence too quickly to be observed; a consequence of Heisenberg's uncertainty principle. Virtual particles appear as matter–antimatter pairs.

visible The portion of the electromagnetic spectrum that the human eye is sensitive to, with wavelengths from 400 nm to 700 nm; see Table 6.1.

void A region of space where very few galaxies exist and that is surrounded by bubble-like structures of superclusters and filaments.

W

wall Structures within the cosmic web that form the boundaries of voids.

waning crescent One phase of the Moon. Less than one-half of the illuminated portion of the Moon is visible from Earth, with the amount of illuminated lunar surface *decreasing* with time.

waning gibbous One phase of the Moon. More than one-half, but less than the entire illuminated surface of the Moon is visible from Earth, with the amount of illuminated surface *decreasing* with time.

watt (W) The SI unit of power. The watt is the amount of energy given off, absorbed, or used every second. One watt is equivalent to one joule per second (J/s).

wavelength (λ) The length of a repeating wave, as measured from peak to peak, or from trough to trough.

waxing crescent One phase of the Moon. Less than one-half of the illuminated portion of the Moon is visible from Earth, with the amount of illuminated lunar surface *increasing* with time.

waxing gibbous One phase of the Moon. More than one-half, but less than the entire illuminated surface of the Moon is visible from Earth, with the amount of illuminated surface *increasing* with time.

weak nuclear force The fundmanetal force of nature that controls certain types of radioactive decay.

Wedgwood, Josiah (1730–1795) An English potter and scientist who developed a pyrometer capable of measuring the temperature associated with particular glowing colors in his kilns.

Wegener, Alfred (1880–1930) A German geologist and physicist who proposed the first theory of continental drift. He was also a polar researcher and meteorologist.

weight The force due to gravity acting on a mass. Weight = mg, which is just a specific form of Newton's second law, $\vec{F} = m\vec{a}$.

Welles, George Orson (1915–1985) An American actor, writer, and producer. Among his many successful productions was the radio adaptation of *The War of the Worlds* by H. G. Wells.

Wells, Herbert George (H. G.) (1866–1946) An English author who published several books of science fiction, including *The War of the Worlds*. He is considered by some to be the "father of science fiction."

Whipple, Fred (1906–2004) An American astronomer who first used the phrase "dirty snowball" to describe a comet. Whipple carried out his work at the Harvard College Observatory for more than 70 years.

white dwarf star A very dim and hot star located in the lower left-hand corner of the Hertzsprung–Russell diagram. A white dwarf is the end product of the evolution of a star with an initial mass of less than about 9 M_{Sun}. White dwarf stars always have masses of less than 1.4 M_{Sun} due to the Chandrasekhar limit.

white dwarf supernova The detonation of a white dwarf star in a binary star system when it accretes so much mass that it exceeds the Chandrasekhar limit.

Wien, Wilhelm (1864–1928) A German physicist who received the Nobel Prize in 1911 for his theoretical work on understanding the blackbody spectrum.

Wilson, Robert Woodrow (b. 1936) An American physicist and radio astronomer who, with Arno Penzias, made the first detection of the cosmic microwave background in 1964.

winter solstice The point on the celestial sphere when the center of the Sun reaches its greatest southerly declination. The coordinates of the winter solstice are RA = 18^h, dec = $-23 1/2°$. The start of winter is defined to occur when the center of the Sun reaches the winter solstice.

work The amount of energy required to move an object a distance, d, in the direction of an applied force vector; $W = Fd$.

X

x-ray A portion of the electromagnetic spectrum with wavelengths shorter than ultraviolet light but longer than gamma rays. All of the wavelength regions are listed Table 6.1.

Y

year (y) The time required for Earth to complete one full orbit around the Sun (1 y = 365.2564 d = 3.1558×10^7 s).

Young, Thomas (1733–1829) An English physician who also did work in light, physiology, musical harmony, and Egyptology. He was the first person to measure the wavelengths of visible light with his famous double-slit experiment.

Z

Zel'dovich, Yakov Borisovich (1914–1987) A Soviet astrophysicist who proposed the existence of the cosmic microwave background in the early 1960's. Also known by YaB.

zenith The position in the sky directly above the observer.

zodiac A series of twelve constellations spread long the ecliptic, separated by $30°$ intervals. The zodiac was useful in ancient cultures for referencing locations in the heavens. Although not used in modern astronomy, the zodiac has remained an important part of the pseudoscience of astrology. There is a thirteenth constellation that lies along the ecliptic, Ophiuchus, that is not included in the astrological zodiac.

Zwicky, Fritz (1898–1974) A Bulgarian-born Swiss astronomer. He was the first to propose the existence of dark matter in 1933. Zwicky also worked with colleague Walter Baade to suggest that supernovas are a new class of astronomical object, and that supernovas may be transition types to neutron stars.

Index

absolute zero, 214, 229, 230, 643
acceleration ($\vec{a}$), 158–162
  centripetal (a_c), 160–162, 166, 167, 173, 313
  definition, 158
  due to gravity (g), 134, 166, 579, 710
accretion, 557, 647
accretion disks
  active galactic nuclei (AGN)
    see galaxies
  black holes (supermassive)
    see black holes, supermassive, accretion disks
  protoplanetary, 649, 650
  stellar, 690, 691
    white dwarf supernovas, 693
active galactic nuclei (AGN)
  see galaxies
Adams, John Couch, 176
Africa, 56
Age of Reason
  see Enlightenment
al-Khwārizmī, Mohammed ibn-Mūsā, 73
al-Shātir, Ibn, 74, 94
Aladin, 674
alchemy, 153, 248
Alexander the Great, 34, 59, 63
algorithm, 73
alpha (α) particles, 255, 257, 288, 289, 298–300, 342, 658, 670, 696
  see also helium (He)
Alpher, Ralph Asher, 784
aluminum (Al), 194, 254, 515, 520, 556, 557, 580
  aluminum-26 (radioactive), 700
American Academy of Arts and Sciences, 224
American Association for the Advancement of Science (AAAS), 729
American Astronomical Society (AAS), 620
American Revolution, 224
amino acids, 520, 526, 557, 569–571
  glycine (CH_2NH_2COOH), 526, 642
  spectrum, 570
  see also astrobiology; nucleobases
amino radical (NH_2), 570
ammonia (NH_3), 465, 468, 483, 486, 487, 557, 569, 571, 642, 653, 696
  ice, 468, 476, 494, 526, 557

ammonium hydrosulfide (NH_4SH), 468
  ice, 468
ancient Greece
  see Greece (ancient)
Andereson, Thomas David, 690
angular momentum ($\vec{L}$), 336, 359, 371–372, 374, 554–556, 560, 561, 649, 694, 701–703, 705, 707, 718
  definition (magnitude), 194
  Earth–Moon system, 372–375
  protoplanetary disk formation, 554–556
  quantized, 294, 295
  see also conservation laws; Bohr, Niels, semi-classical atomic model
Antarctica, 745
Anthropocene, 435
anthropology, 5, 27
Antichthon, 71
antimatter, 345–346, 352, 531, 719, 789, 801
Apollonius of Perga, 66–67, 70
Applewhite, Marshall, 521
Aquinas, Saint Thomas, 64
Arab astronomy and mathematics, 73–74, 92
archaeoastronomy, 26–51
archaeology, 27
Archimedes of Syracuse, 61, 71, 72
  The Sand Reckoner, 71
arcminute (definition), 56
arcsecond (definition), 56
arctic circle, 450
Arecibo message
  see extraterrestrial intelligence, reaching out
Arecibo Observatory (radio), 385, 579, 580, 584, 596
argon (Ar), 248, 351, 386, 412, 696
Aristarchus of Samos, 71
  heliocentric universe, 71–72, 94, 604
  distance to the Sun, 187
Aristotelian–Ptolemaic worldview, 117, 119–121, 132, 133
  see also Aristotle; Ptolemy
Aristotle, 55, 59, 60, 63–64, 89, 94, 116, 117, 119–121, 140, 147, 208
  celestial motions, 64
  elements, 64, 70, 134, 247
  geocentric crystalline spheres, 64
  "nature abhors a vacuum", 64, 70

observations and empirical information, 63–64
physics, 64, 133, 134, 162
prime mover, 64
Arizona, 33
armillary sphere, 66
arrow of time
  see thermodynamics
arXiv.org, 152
Asia, western, 56
Asimov, Isaac, 567
asterism, 15
asteroids, 366, 380, 398, 405, 508, 510–520, 522, 528, 534, 536, 552, 557, 558, 570
  active, 528
  asteroid belt, 220, 362–363, 502, 508–510, 516, 519, 552, 559
    Kirkwood gaps, 511–512
  carbonaceous, 519, 552, 562
  Hildas, 513, 552
  Hungarias, 514
  ice-covered, 528
  metallic, 519, 529, 535, 557
  orbital resonances, 512–514
  rubble-pile, 519, 535
  stony, 517–519, 529, 535, 552, 557, 562
  stony–iron, 557
  trojans
    see Solar System
asteroids (specific)
  Astraea, 514
  Bennu, 518–519
    regolith, 518
  Ceres, 130, 502, 510, 514, 552
  Eros, 517–519
    regolith, 517
  Gaspra, 363, 518
  Ida, 363, 518
    Dactyl, 363, 518
    escape speed, 201
  Juno, 514
  Pallas, 514, 516, 552
  Psyche, 519
  Vesta, 514–517, 552, 557
    Rheasilvia impact crater, 515
  see also dwarf planets
astrobiology, 567
  amino acids and proteins, 569–570
  exoplanets
    biosignatures, 573
  first life on Earth, 568–572
  ingredients, 569

RNA and DNA, 570–571
panspermia, 572, 783
astrology
  see pseudoscience
astronomical unit (au), 113, 188–189
  formally defined, 189
  relationship to parsecs (pc) and
    light-years (ly), 607
astrophysics, 310
Atacama Large Millimeter/
  submillimeter Array (ALMA), 594,
  596, 597, 645, 731
atomic bomb
  see also Einstein, Albert; Szilárd, Leó
atomic bombs, 25, 199, 273, 343, 697
atoms, 213, 247–249, 336, 639, 643
  atomic mass (*A*), 300, 345
  atomic number (*Z*), 249, 299, 300,
    345
  isotopes, 299–300, 345
  nucleus, 291, 298–301, 336, 627, 707
    discovery, 288–290
    energy released per nucleon, 695
  see also quantum mechanics
Aubrey, John, 47
auroras, 326, 471
autumnal equinox
  see celestial sphere
Avogadro's number, 249, 269
Aztecs
  see Mesoamerica

B612 Foundation, 530
Baade, Walter, 747
Babylon
  see Mesopotamia, Babylonians
Bacon, Francis, 179
bacteria, 213
Balmer, Johann Jakob, 290, 291
bar, 475
  millibar, 475
bar magnet, 208, 322, 323, 330, 332, 401,
  402, 404
Bartlett, Albert Allen, 431
basalts, 391, 406, 407
Battle of Hastings, 502
Bay of Fundy, New Brunswick,
  Canada, 372, 373
Bayeux tapestry, 502
Becquerel, Henri, 254
Bell Burnell, Jocelyn, 702–703
Beringia land bridge, 27
Berners-lee, Tim, 797
Bernoulli, Johann, 152
beryllium (Be), 346, 351, 424, 657
Bessel, Friedrich, 604
beta (*β*) particles, 255, 256, 288
  see also electrons (*e*⁻); positrons
    (*e*⁺); radioactivity
Biden, President Joseph R. Jr., 452

Big Bang, 819
Big Bang nucleosynthesis
  see cosmology
Big Bang theory, xiii, 7, 9, 301, 641, 669,
  674, 696, 709, 711, 767, 781, 791,
  798, 804
  problems with, 793–796
  see also cosmology
Big Bear
  see constellations, Ursa Major
Big Dipper
  see constellations, Ursa Major
binary systems, 175
  asteroids, 536
    contact binary, 510
    Ida and Dactyl, 199, 363
  Neptune–Triton capture hypothesis,
    490
  stars, 542
    black holes, 708, 712, 717
    eclipsing, 666
    η Carina, 645
    mass determinations, 611–616, 626
    neutron stars, 707
    planetary nebulas, 687
    Procyon, 623
    Sirius, 687
    white dwarf stars, 690, 693, 707
    see also black holes, neutron stars,
      white dwarf stars
biochemistry
  see chemistry
biology, 5, 7, 334, 370, 640
  evolution, 5
  genetics, 5
  microbiology, 310
  plant, 443
black holes, 235, 263, 531, 596, 599,
  707–720, 749, 806
  black holes have no hair theorem,
    718
  event horizon, 710, 720, 745, 746
  falling-in thought experiment,
    719–720
  Hawking radiation, 719, 817
  magnetic fields, 763
  photon spheres, 716
  singularity, 710, 719
  sizes, 760
  smallest stellar black hole, 711
  spacetime curvature, 714
  supermassive, 744, 753, 756, 760–764,
    806
    accretion disks, 745, 760, 762, 763
    jets, 760–761
    mergers, 767, 768
  see also accretion disks; binary
    systems, stars; relativity theory,
    gravitational

blackbody radiation, 213–220, 228, 253,
  259, 336, 620, 625, 639, 785, 807
  intensity–color spectrum curves
    see blackbody spectrum
  Stefan's law (flux), 218–219, 321
  Stefan–Boltzmann law (luminosity),
    219–220, 315, 464, 623, 624, 633,
    649, 650
  temperature, 217, 464, 623, 626
  Wien's blackbody color law, 215–217,
    265, 628
  see also blackbody spectrum;
    electromagnetic theory
blackbody spectrum, 214–215, 244–246
blueshift
  see Doppler effect
Bohr, Niels, 290–292, 295, 627
  Einstein–Bohr debate, 298
  semi-classical atomic model,
    290–292, 294, 296
    quantized angular momentum
      (*L*), 291
Bok globules, 646, 685
  NGC 281, 646
Bok, Bart, 646
bolides
  see meteors
Boltzmann, Ludwig, 220, 224, 229, 264,
  265, 339
Bondi, Sir Hermann, 783
Bonn Observatory, 619
boron (B), 299, 351
Boyle, Robert, 147
Brahe, Tycho, 26, 119–121, 123–125,
  133, 173–174, 310
  Uraniborg observatory, 120
Breyvichko, Andrey, 532
brightest stars, 833
brightness (*b*), 222, 599–604
  brightness ratio–magnitude
    difference, 602, 610
  inverse square law, 222, 505, 666
  relationship to luminosity (*L*), 223,
    604
British Parliament, 83
bronze, 119, 247
Bronze Age, 247
brown dwarfs, 544, 652–653, 749
Brownian motion, 269
Bruno, Giordano, 118–119, 133, 566
Bunsen, Robert, 246, 251–253, 288
Burney, Venetia, 363
Byzantine Empire, 73, 92
Byzantium, 72

calcium (Ca), 253, 520, 556, 619, 622,
  696
calculus, 199
Calvin, John, 93

Cannon, Annie Jump, 589, 618–620, 622, 653
carbon (C), 27, 248, 249, 251, 253, 258, 300, 301, 316, 349, 417, 475, 483, 506, 520, 556, 562, 569, 570, 580, 620, 642, 651, 655, 657–662, 685, 687, 689, 694, 696, 766
  diamond, 475
  spectrum, 251, 252
carbon dating
  see radiometric dating
carbon dioxide (CO_2), 248, 380, 386–389, 395, 412–417, 438–441, 443, 444, 450, 465, 483, 486, 530, 551, 557, 571, 573, 594, 598, 642, 696
  ice, 414, 415, 486, 494, 526, 557
carbon monoxide (CO), 386, 448, 486, 491, 530, 642
  ice, 504, 506
carboxylic acid (COOH), 570
Carter, President Jimmy, 461, 581
cartography, 728
Cassander (King), 63
Cassini mission, 394, 461, 465, 467, 473, 484, 485, 487, 488, 492
  Huygens probe, 461, 484
Cassini, Giovanni Domenico, 188, 491
cataclysmic variable stars, 690
catalogs
  Bonner Durchmusterung catalog (BD), 619
  Gaia, 631, 687
  Henry Draper catalog (HD), 618, 619
  Messier catalog, 674, 691, 739, 749, 834–836
  New General Catalog of Nebulae and Clusters of Stars (NGC), 674
  Smithsonian Astronomical Observatory Star catalog (SAO), 619
  Yale University Star catalog, 619
catalysts, 655
Cavendish, Henry, 191, 711
celestial sphere, 50, 61–64, 66, 68–70, 74–86, 95, 97–100, 121, 134, 310, 734
  autumnal equinox, 20, 31, 33, 43, 68, 79–81, 86, 102, 103
  celestial equator, 62, 75–82, 84, 85, 100, 103, 734
  equatorial coordinate system, 84–86
    declination (dec), 84, 85
    right ascension (RA), 84, 85
  north celestial pole, 16, 17, 49, 50, 61, 62, 75–78, 80, 82, 85, 86, 100, 103
  south celestial pole, 16, 61, 62, 75, 76, 78, 82, 86, 100, 103
  summer solstice, 20, 31, 33, 43, 48, 65, 79, 80, 86, 100, 103

vernal equinox, 20, 31, 33, 43, 68, 79–81, 84–86, 102, 103, 508
  First Point in Aries, 86
winter solstice, 20, 31, 33, 43, 48, 79, 85, 86, 102, 103
centaurs, 528, 552, 570
  Chariklo, 528
  Chiron, 528
  Hidalgo, 528
center of mass, 173, 175, 507, 544, 545, 612–615, 626, 673, 730
  equation, 613, 614
centripetal acceleration
  see acceleration ($\vec{a}$)
Cepheids
  see pulsating stars
Ceres
  see asteroids; dwarf planets
CERN, 797
Chaco Canyon, New Mexico, 33–34
  Fajada Butte, 33–34
    Sun Dagger, 33
  SN 1054, 691
Chandra X-Ray Observatory, 761
Chandrasekhar limit, 693–694, 706
Chandrasekhar, Subrahmanyan (Chandra), 631, 693
charge-coupled device (CCD), 288, 600, 601
Chelyabinsk fireball
  see meteors, bolides
chemical reactions, 299
chemical symbol, 300
chemistry, 153, 191, 224, 246–251, 255, 264, 310, 334, 370, 443, 488, 655
  organic, 569
    biochemistry, 569
    see also organic compounds and molecules
  see also molecules
Chile, 731
  Andes Mountains, 594
    Cerro Pachón, 749
  Chajnantor plateau, 731
China, 56
  astronomy (ancient), 49–51, 56, 58
    calendar, 50
    cosmology, 50–51
    eclipses, 50
    equinoxes, 49
    prominences, 49
    solstices, 49
    sunspots, 49
    supernovas (SN 1006 and SN 1054), 691
chlorine (Cl), 249, 251, 351, 642
chondrules, 520
chromosomes, 571
circumpolar, 16, 75, 76, 78, 82
Civil War, American, 209, 214

climate
  see Earth, climate change
Clinton, Presidet William (Bill), 421
clocks
  atomic, 83, 280, 281, 372
  grandfather, 83
  pendulum, 83
  seafaring, 83
Collins, Michael, 379
color index, 675, 676
color–magnitude diagrams, 674–678
  main-sequence turn-off points, 675–677, 765
  tests of stellar evolution, 676
  see also stellar evolution
Colorado, 33
Colorado Plateau, 398
Colorado River, 398, 399
Columbus, Christopher, 27, 34, 68, 93, 728
comets, 24–25, 365–366, 380, 398, 405, 510, 520–530, 534, 535, 552, 558, 560, 570, 572
  coma, 366, 521
  composition, 525
  Encke-type, 524
  formation, 564
  Halley-type, 524
  Jupiter-family, 524
  long-period, 366, 523–524
  main-belt, 528
  nucleus, 366, 521
  periodic, 522
  short-period, 366, 524, 564
  sungrazer, 523
  tails, 24, 521
    dust, 366, 522
    gas, 366, 521
    radiation pressure, 522
comets (*specific*)
  Hartley 2, 522
comets (*specific*)
  1I/Oumuamua, 528–530
  2I/Borisov, 529–530
  Borrelly, 526
  Churyumov–Gerasimenko, 366, 521, 525
  Comet McNaught, 24
  Encke, 524
  Great Comet of 1577, 120, 121
  Hale–Bopp, 520, 521, 524
  Halley, 194, 365, 522, 524, 526
    retrograde orbit, 524
  Hartley 2, 522
  Kohoutek, 523
  Lovejoy, 521
  Shoemaker–Levy 9, 517, 524–525
  Temple 1, 525
  Wild 2, 525

computer modeling, 350, 443
  butterfly effect, 445
  Solar System formation, 561
  supercomputers, 334–335, 443
  supernovas, 699
  uncertainty, 444–445
  see also science, models
Comte, Auguste, 251
conduction, 425
conservation laws, 192–201
  angular momentum ($\vec{L}$), 194–196
    see also angular momentum
  closed system, 192
  conserved quantities, 192, 194, 224
  electric charge, 300, 345, 346
  energy (E)
    see also energy
  energy (E), 196–201, 224, 229
  lepton number, 346
  linear momentum ($\vec{p}$), 193–194
    see also linear momentum
  nucleon number, 300, 345, 346
Constantine the Great, 72
Constantinople, 72, 92
constellations, 14–17, 98, 99, 829–832
  Andromeda, 169
  Aquarius, 23, 86, 550
  Aquila, 686, 734
  Aries, 23, 86, 599
  Auriga, 15, 734
  Boötes, 235
  Boötes, 15
  Camelopardalis, 15
  Cancer, 23, 86, 98
  CanesVenatici, 15
  Canis Major, 49, 85, 600, 601, 603, 612
  Canis Minor, 603, 623, 734
  Capricorn, 86
  Capricornus, 23, 98
  Carina, 641
  Cassiopeia, 28, 119, 646
  Centaurus, 761, 762
  Cepheus, 103, 542
  Cetus, 636, 665, 764
  Circinus, 691
  Coma Berenices, 15, 756
  Corona Borealis, 15, 29, 32
  Crux, 636
  Cygnus, 217, 542, 712
  Draco, 15, 49, 86, 103, 684, 685
  Eridanus, 17
  Gemini, 15, 23, 526, 636
  Hercules, 15, 103, 579, 584
  Hydra, 708, 741
  Leo, 23
  Leo Minor, 15
  Libra, 23
  Lupus, 691, 692
  Lynx, 15
  Lyra, 103, 542, 678, 712

Monoceros, 734
Musca, 694
Ophiuchus, 23, 625
Orion, 17, 49, 76, 85, 216, 526, 553,
  599, 600, 603, 641, 642, 644, 664,
  665, 729, 734
Pegasus, 542, 686, 690, 764
Phoenix, 768
Pisces, 23, 86
Piscis Austrinus, 636
Pyxis, 686
Sagittarius, 23, 86, 577, 640, 729, 731,
  733, 734, 743
Scorpius, 23, 686
Sculptor, 756
Serpens, 15, 643
Taurus, 17, 23, 31, 44, 86, 133, 239,
  241, 526, 628, 636, 673, 691, 692
Ursa Major, 14–17, 29, 49, 767
  pointing stars, 14
Ursa Minor, 14, 15, 29, 86
Vela, 642, 686
Virgo, 15, 23, 86, 542, 694, 741, 756
Contact, 577, 596
continental drift
  see Earth, plate tectonics
coordinated universal time (UTC), 83
Copernican principle, 549, 567–568,
  576, 733, 740, 785, 790
Copernican revolution, 6, 94–97, 132,
  400
Copernicus, Nicolaus, 7, 74, 93–94, 114,
  119, 127, 136, 146, 310, 566, 604,
  732
  Commentariolus, 94, 95, 97, 100, 115,
    127
  De revolutionibus orbium coelestium,
    95–97, 116, 118, 119, 136, 141
  heliocentric universe, 93–97, 121,
    188, 604
copper (Cu), 119, 247, 254
Cortés, Hernán
  see Mesoamerica, Aztecs, Spanish
Cosmic Background Explorer (COBE),
  785
cosmic microwave background (CMB),
  784, 802–804, 807
  harmonics, 807–808
  polarized light, 808
  problems with, 793–795
cosmic rays, 696, 746, 747
cosmological constant, 779, 788, 816
cosmological distances, 788
cosmological principle, 790–791
  perfect, 825
cosmology, 27, 28, 50, 750, 777, 780
  Big Bang nucleosynthesis, 802, 804
  cosmic web
    see universe, structures
  dark ages, 804

end of the story
  Big Crunch, 817–818
  Big Freeze, 816–817
  Big Rip, 818–819
  Bubble of Death, 819–820
geometries, 795–796
  flat, 816
  Hubble horizon, 812–813, 816, 820
  inflation, 798–800, 804
  Λ-cold dark matter model (Λ-CMD),
    806–810
  observable universe, 810–813
  particle horizon, 810–812
  radiation-dominated universe, 801
  sound waves, 807
  surface of last scattering, 803–804,
    807–808
  the first atoms, 802–803
  see also Big Bang theory
Coulomb, Charles-Augustin de, 209,
  291
Crab Supernova Remnant
  see supernovas, SN 1054
critical thinking, 179–180
  see science
Curie, Marie, 254–255, 295, 627
Curie, Pierre, 254, 627
Curtis, Heber Doust, 749, 760
cyanobacteria, 395

da Vinci, Leonardo, 93
Dalton, John, 248, 288
Daniel K. Inouye Solar Telescope, 320
dark energy, 788–790, 816
dark matter, 747–749, 753, 804–805
  cold (CDM), 748, 769
  hot (HDM), 748
Darwin, Charles, 7, 19
Davis, Raymond Jr., 350
Davy, Sir Humphry, 248
Dawn mission, 503, 514
day
  sidereal, 17, 74, 99, 100, 106, 595, 702
  solar, 17, 99, 100
days of the week, 22
de Bergerac, Savinien de Cyrano, 460
de Broglie, Louis, 293–295, 627
  particle waves
    see quantum mechanics
Dead Sea, 25, 398, 399
decibel scale, 600
Deep Impact mission, 525
degenerate electron gas, 661
deGrasse Tyson, Neil, 358, 726
deities
  see mythology
Democritus, 248
Denmark, Hveen (island), 120
density (ρ), 282, 338–340, 349
  definition, 338

density waves
  see Milky Way Galaxy, spiral arms
deoxyribonucleic acid (DNA), 520,
    570–571, 580, 633
  archaeological analysis, 27, 48
  deoxyribose (sugar), 570
Descartes, René, 147, 178, 179
  Geometry, 147
determinism, 179
deuterium ($^2_1$H)
  see hydrogen (H)
Dicke, Robert Henry, 784
differential rotation, 317
diffraction gratings, 241–243
  reflection, 243
  transmission, 261
Diggs, Thomas, 118, 133
digital camera, 288
dinosaurs, 435, 442
  extinction, 199, 442, 533
Dirac, Paul Adrien Maurice, 294, 627
distance ladder, 605, 668, 694, 733, 786
distance modulus ($m − M$), 682
DNA
  see deoxyribonucleic acid (DNA)
Doppler effect, 473, 645, 717, 733, 757,
    763
  light, 281–284, 321, 544
    blueshift, 283, 284, 321, 390, 544,
      545, 717, 739, 757
    redshift, 283, 284, 321, 390, 544,
      545, 717, 720, 739, 783
  sound, 281–282
Double Asteroid Redirect Test mission
    (DART), 536
Drake, Frank, 574, 579
Draper, Henry, 618
du Châtelet, Émilie, 178–179, 197–198,
    200
dust, 239, 366, 444, 554–556, 558, 646,
    649, 653, 685
  formation, 556
  interstellar, 170, 213, 239, 241, 553,
    570, 640, 641, 643, 731–733,
    736–738, 756
dwarf planets, 177, 363–365, 489,
    502–510, 520, 552, 558, 570
dwarf planets (*specific*)
  Ceres, 363, 364, 502–504, 509, 510,
    514, 516, 557
    impact craters, 533
    see also asteroids
  Eris, 365
  Gonggong, 365
  Haumea, 364
    moons, 509
  Makemake, 364
  Pluto, 177, 363, 364, 461, 489, 502,
    504–510
    Charon, 506–508

discovery, 177
    orbital resonance with Neptune,
      508
    organic molecules, 506
    retrograde rotation, 508
  Quaoar, 364
  Sedna, 365, 509, 510, 744

Earth, 6, 7, 13, 113, 122, 169, 176, 220,
    223, 359, 366, 380, 394–405, 441,
    461, 464, 471, 514, 581, 603, 716
  acceleration due to gravity (g), 134,
    166
  age, 5, 342, 551
    formation, 560
    oceans, 568
    oldest life, 5, 568
    see pseudoscience, young Earth
      "theory"
  antarctic circle, 62, 79, 101, 102
    latitude, 20, 100
  arctic circle, 19, 61, 79, 101
    latitude, 19, 100
    motion of the Sun, 19
  atmosphere, 25, 394–397, 468, 736
  auroras, 326, 404, 639
  carbonaceous rock, 438
  climate change, 333, 437–453, 567
    computer modeling, 443–447
    consequences, 447–450
    evidence, 447
    fossil fuels, 439, 444
    global warming, 437–443
    international agreements, 450–453
    permafrost, 444, 450–451
    weather and climate, 437
  coordinate system, 81–83
  crust, 426
    continental, 426
    oceanic, 426
  density, 191, 359, 421
  earthquakes
    epicenter, 424
    hypocenter, 423
    P and S waves, 422–424
    Richter scale, 530
  eastern hemisphere, 20
  equator, 18, 19, 62, 75–82, 84, 101, 113
    latitude, 19
  escape speed, 199, 200
  flat Earth "theory"
    see pseudoscience
  geology, 380
  Himalyas, Mount Everest, 417
  hurricanes, 395, 445–447
  hydrothermal vents, 568
  ice ages, 27
    see also Earth, Little Ice Age
  impact craters, 533
  interior, 421–428

innermost inner core (solid), 424
    lithosphere, 426
    mantle, 424
    outer core (fluid), 424
    outermost inner core (solid), 424
  jet streams, 396
  latitude, 19, 61
    effects on star positions, 16–20,
      75–76
  Little Ice Age, 328, 333, 440
  magnetic field, 401–405, 422, 424,
    426, 442, 470, 471, 746
    reversals, 404
  mantle convection, 425
  map, 734
  mass, 191, 359
  Medieval Warm Period, 440
  mid-Atlantic Ridge, 400
  mythology, 362
  North Pole, 19, 61, 75, 77–81, 97, 98,
    101, 208
  northern hemisphere, 14, 15, 18, 20,
    82
  ocean currents, 397, 468
  orbit, 41, 98, 99, 109, 441, 510
    eccentricity, 441
  orbital plane
    see ecliptic
  pale blue dot
    see Sagan, Carl, *A Pale Blue Dot*
  plate tectonics, 81, 399–400, 426
    mid-ocean ridges, 426
    seafloor spreading, 400, 426
  population growth, 431–436
    projections, 432
  prime meridian (Royal
    Observatory), 81, 84
    see also coordinated universal time
  radius (size), 359, 360
  Ring of Fire, 400, 426
  rotation, 98, 99, 102
  rotation (retrograde), 468
  rotation axis, 97, 100, 103, 470
    tilt, 100–103, 113, 412
    see also ecliptic, tilt
  sea levels, 27
  seasons, 17–20, 100–102
  South Pole, 61, 75, 81, 97
  southern hemisphere, 20, 82
  surface, 397–400
  trenches, 426
  tropic of Cancer, 19, 34, 61, 65, 79,
    101
    latitude, 19, 86, 100
  tropic of Capricorn, 61, 79, 101
    latitude, 20, 86, 100
  Van Allen radiation belts, 402
  volcanoes, 394, 426–428
    Hawai'ian Islands, 419, 426–427
    shield volcanoes, 393, 419, 426

water (H_2O), 398, 486, 552
  see also water (H_2O)
weather, 395–397, 437, 468
  air pressure, 397
weathering, 380, 398–399
  see also meteorites
eclipses, 21–22, 48, 50, 58
  cycles, 34, 40–41
    saros, 41, 50, 110
    tritos, 50
  line of nodes, 109
  lunar, 21, 22, 40, 41, 50, 56, 107–110
    partial, 65
  penumbra, 108
  solar, 21, 40, 41, 50, 56, 110–111, 116,
    253, 318
    2017 August 21 (Great American
      Eclipse), 21, 111
    2024 April 8, 110
    annular, 110
    partial, 110
    total, 110
  umbra, 108
ecliptic, 18, 21, 23, 24, 61, 66, 78–79,
  98–103, 109, 114, 470, 508, 542,
  551, 556, 566
  tilt, 61–62, 100
  see also Earth, orbit, rotation axis
Eddington, Sir Arthur, 629, 777
Edison, Thomas, 267
Egypt, 35, 59, 61
  Alexandria, 59, 65, 68, 72
    Library of, 66
  astronomy, 59
  Nile River, 49, 65, 207
  Syene (Aswan), 65
Einstein cross, 764
Einstein, Albert, xx, 6, 7, 10, 136, 172,
  263, 266–274, 288, 290, 293, 295,
  341, 351, 375, 543, 625, 627, 629,
  694, 695, 707, 708, 711–714, 717,
  718, 749, 764, 777, 779, 797
  annus mirabilis, 269, 285, 344
  atomic bomb, 273–274
  bicycle (angular momentum, $\vec{L}$), 371
  music, 267
  photoelectric effect, 285–288
  quantum mechanics, 272
    Einstein–Bohr debate, 298
  relativity theory
    see relativity theory
  search for a unified field theory, 272
  theory of light emission, lasers, 272
Eisenhower, President Dwight D., 273
electric fields, 209–211, 255, 256
  see also electromagnetic theory
electricity
  see also electromagnetic theory

electromagnetic force, 797, 800, 803
  see electromagnetic theory;
    unification
electromagnetic radiation
  see electromagnetic waves (light);
    photons
electromagnetic spectrum
  see electromagnetic waves (light),
    spectra; specific elements
electromagnetic theory, 207–211, 269,
  276, 279
  blackbody radiation
    crisis in classical physics, 264–266,
      285
  electricity, 207–209
    cathode rays, 253, 255
    Coulomb's electrostatic force law,
      209, 348, 657
    current, 202, 209, 276, 323, 331
    effect on charged particles, 253
  fields, 209–211, 255–256, 266
  magnetism, 124, 207–209
    bar magnets, 208, 210, 322, 331
    compass, 209
    effect on charged particles, 256,
      322
  Maxwell's equations, 209, 264, 272,
    287, 290
electromagnetic waves (light),
  201–214, 274, 276, 285, 294, 595
  electromagnetic radiation, 212, 228,
    322, 336
  frequency–wavelength–speed
    relationship, 207
  Kirchhoff's rules, 246, 292
  polarization, 211–212
  radio
    amplitude modulation (AM), 597
    frequency modulation (FM), 597
    very high frequency (VHF),
      television, 597
  scattering
    wavelength–frequency
      dependence, 239–240
  spectra, 238, 286, 590, 595, 598
    absorption lines, 244, 246, 253,
      283, 292, 619, 620, 626, 639, 738
    colors, 204–207, 213, 239, 243, 281,
      283, 468, 620, 625, 673, 675, 685
    continuous, 214, 215, 246
    emission lines, 246, 253, 292, 626,
      639, 738
    Fraunhofer's lines, 241–244, 622
    frequencies and wavelengths,
      205–207
    gamma (γ) rays, 212
    infrared light, 212, 642–644
    microwaves, 212
    radio waves, 212
    ultraviolet light, 212, 570, 642

  visible light, 212
  x-rays, 212, 255
  see also blackbody radiation;
    quantum mechanics, energy
    level; specific elements
speed of light (c)
  see fundamental constants
x-rays
  discovery of, 253–254
  see also photons; waves
electron degeneracy, 689, 693
electron spin-flip, 738
electron volt (eV), 286
electrons (e^-), 259, 276, 287–296,
  298–301, 325, 327, 331, 334, 337,
  342, 345–347, 351, 386, 404, 422,
  554, 597, 599, 620, 621, 633, 640,
  653, 661, 668, 670–673, 689, 693,
  695, 698, 706, 738, 763, 801, 815
  charge, 256
  discovery, 255–256
  mass, 256
electroweak theory, 797, 800
  see also unification
elementary particles, 301, 346, 720
  see also specific particles
elements, 248–251, 345
  abundances and production,
    695–697
  see also atoms; periodic table of the
    elements; by name
Elliot, T. S., 776
ellipses, 124–126
  eccentricity (e), 126
  focal points, 125
  semimajor (a) and semiminor (b)
    axes, 125
energy (E), 217, 223, 291, 300, 336, 338,
  531, 656, 694
  heat
    see thermodynamics
  kinetic, 197, 223, 348, 466, 490, 554,
    657, 662
  mechanical, 223
  potential, 197, 223, 348, 466
  work, 223
    definition, 197
  see also conservation laws; ideal gas
    law
energy levels
  see quantum mechanics
Enlightenment, 178–180, 208
entropy
  see thermodynamcis
equivalence principle
  see relativity theory, general
Eratosthenes of Cyrene, 65–66, 68, 89,
  169, 187
  circumference of Earth, 65–66

Eris
  see dwarf planets
escape speed, 198–201, 366, 465, 490,
    528, 533, 710, 720, 747, 766
ethane (C_2H_6), 486, 569
  ice, 557
ether, 64, 147, 178, 276, 282
  Greeks (ancient), 64, 70, 247
  search for, 274–276, 279
  wind, 274–276, 592
  see also Aristotle, elements
Euclid of Alexandria, 147
Eudoxus of Cnidus, 60–64, 66, 70, 78, 88
Euler, Leonhard, 513
Euphrates River, 56
Europe, 34, 58, 73
European Renaissance, 93
European Southern Observatory, 593
European Space Agency (ESA), 363,
    599
evening star, 24
Event Horizon Telescope (EHT), 745,
    760, 761
evolution (biological), 7, 9, 19, 567, 571
  human, 580
exobiology, 568
exoplanets, 284, 542–551, 560, 574, 749,
    764
  atmospheres, 551
  detection methods, 543–546
    direct imaging, 543
    gravitational microlensing,
      543–544, 749
    radial velocity, 544–545, 548
    transits, 545–546
  hot Jupiters, 545, 548, 552, 560
  masses, 547–548
  mini Neptunes, 548
  orbits, 548–549, 552
  parent stars, 552
  rogue planets, 544, 552, 563
    escape, 561
  super Earths, 548
exponential function, 433
  global pandemics, 436
  population growth, 433–436
extinctions
  see mass extinctions
extraterrestrial intelligence, 574–581
  Drake equation, 574–576
  Fermi paradox, 578–579
  reaching out
    Arecibo message, 579–580
    radio and television signals, 579
    spacecraft, 580–581
  Search for Extraterrestrial
    Intelligence (SETI), 529, 574
  Wow signal, 577–578

Faber, Sandra, 769
Fabricius, David, 328, 665
fall (first day), 18
  see also celestial sphere, autumnal
    equinox
false vacuum, 819
falsifiable, 5, 6, 8, 71, 279, 815
Faraday, Michael, 209
Fermi, Enrico, 578
Fertile Crescent, 56, 57
Fibonacci, 73
filament, 757
finding charts, 665
Fire and Ice, 820
First Peoples of the Americas, 27
  see also Native Americans
First Point in Aries
  see vernal equinox
Five-hundred-meter Aperture
  Spherical Telescope (FAST), 596, 597
Fizeau, Hippolyte, 190
Flat Earth Society
  see pseudoscience, flat Earth
    "theory"
Fleming, Williamina Paton Stevens,
    617–618, 620
flux, 218
forces ($\vec{F}$)
  centripetal, 313, 703
  net, 163
  resistive, 134, 162
  see also electromagnetic theory,
    electricity; gravitation;
    Newton's laws
formaldehyde (CH_2O), 526
Four-Corners region, southwest
  United States, 33
frame dragging, 718
France, Paris, 92, 179, 188, 189
Frankland, Edward, 253
Franklin, Benjamin, 180, 208
Fraunhofer, Joseph von, 241, 252, 288,
    622
Frederick II, King, 178
French Academy of Sciences, 188
Friedmann, Alexander, 777
Frost, Robert, 820
Fuller, Richard Buckmister, 642
fullerenes
  buckyballs (C_{60}), 642
fundamental constants, 186
  Boltzmann's constant (k_B), 229, 256,
    339, 792
  electricity, derived (k_E), 210, 791
  elementary charge (e), 256, 259, 288,
    291, 295, 298–300, 348, 625, 791
  magnetism (μ_m), 210
  mass of the electron (m_e), 256, 791
  Planck's constant (h), 265, 285,
    291–293, 297, 341, 791, 792

speed of light (c), 6, 190, 256, 607,
    791, 792
  defined, 190
  measurements, 135, 189–190
  wavelength/frequency
    dependence, 238
  see also relativity theory
Stefan–Boltzmann constant, derived
  (σ), 218
universal gravitational constant (G),
    168, 170, 186, 191–192, 222, 256,
    791
  see also gravitation
fundamental forces of nature, 797

Gaia mission, 608, 609, 631, 636, 651,
    653, 733, 755, 758, 787
galactic cannibalism, 767
galactic clusters
  see star clusters, open (galactic)
Galapagos Islands, 19
galaxies, 726–769
  active galactic nuclei (AGN), 806
    accretion disks, 760, 762, 763
  dark matter halos, 752, 766
  first stars, 766, 805–806
  flat rotation curves, 747
  mergers, 766
  most distant seen, 767
  number in the universe, 750
  primeval, 766
  tuning fork diagram, 750–752
  types
    active galaxies, 761–763
    diffuse, supergiants (cD), 753
    dwarf irregulars (dIrr), 752
    dwarf spheroidals (dSph), 752, 767
    ellipticals (E), 750, 753
    grand-design spirals, 754
    interacting, 752
    lenticulars (S0 and SB0), 750
    magellanic, barred (SBm), 752
    spirals (S and SB), 747, 750,
      753–754, 816
    starbursts, 752
    ultra-diffuse galaxies (UDG), 752,
      767
    ultra-faint dwarfs (UFD), 752, 766,
      767
galaxies (*specific*)
  Andromeda (M31), 169, 749, 750,
    754, 756
    merger with the Milky Way
      Galaxy, 757–759, 767
  Antennae, 759
  Centaurus A
    jet, 761
    supermassive black hole, 761
  GN-z11, 766–767

Large Magellanic Cloud (LMC), 25, 666, 668, 699, 733, 752, 754, 758
Leo A, 755
Leo T, 755
M61, 739
M87 (Virgo A), 750, 756, 760
   globular clusters, 753
   jet, 760–761
   supermassive black hole (M87*), 761
M110, 750
Milky Way Galaxy
   see Milky Way Galaxy
NGC 4565, 736
NGC 5866, 750
Phoenix dwarf, 755
Rose, 759
Sculptor dwarf, 755
Small Magellanic Cloud (SMC), 25, 666, 733, 752, 754, 758
Southern Ellipse, 759
Southern Pinwheel (M83), 739
Tadpole, 759
Triangulum (M33), 754
Ursa Minor dwarf, 755
Whirlpool (M51), 754
galaxy clusters, 804
   intracluster gas, 756
   superclusters, 756–757, 804
galaxy clusters (specific)
   Abell 370, 764
   Abell 2744 (Pandora's Cluster), 756
   Coma, 747
   Local Group, 279, 754–757, 763
   SPT2349-56 (protocluster), 767
   Stephan's Quintuplet, 759
   superclusters
      Local Supercluster (Virgo Supercluster), 757
   Virgo, 756
Galaxy Zoo, 769
Galileo Galilei, 131–141, 146, 166, 176, 189, 196, 198, 310, 328, 362, 419, 432, 476, 491, 727
   Assayer, 136
   acceleration due to gravity, 134
   Dialogue Concerning the Two Chief World Systems, Ptolemaic and Copernican, 139–141, 147
   experimental testing, 136
   Galilean relativity, 134
   language of mathematics, 136
   law of inertia, 134, 162
   Letter to the Grand Duchess Christian of Tuscany, 138
   microscope invention, 134
   pendulum and grandfather clocks, 134
   Roman Catholic Church
      see Roman Catholic Church

Sidereus Nuncius, 132–134, 136, 137, 362, 476
   telescope invention (refracting), 131–132
   The Two Sciences, 140
   trial, 136–141
Galileo mission, 363, 461, 473, 483, 488, 525
Galle, Johann Gottfried, 176
gamma (γ) rays, 219, 238, 255, 288
   see also electromagnetic waves (light), spectra; radioactivity
Geiger, Hans, 289
general theory of relativity
   see relativity theory, general
genes, 571
genetics, 27
geocentric universe, 7, 60, 68, 70, 74, 98, 112, 120–121, 123, 127, 133
geology, 5, 310, 370, 443
   fossil record, 5, 6
geomagnetic storm, 404
George III, King, 224
Gerard of Cremona, 92
Giant Magellan Telescope, 593
Giese, Tiedemann (Bishop of Chelmno), 95
Gilbert, William, 124, 208
   De Magnete, 208
Giotto mission, 194
global navigation satellite systems (GNSS), 83
Global Positioning System (GPS), 83, 280
global warming
   see Earth, climate change
globular clusters
   see star clusters
gluons, 301, 800–801
gold (Au), 119, 153, 248, 289, 299, 580, 709
Gold, Thomas, 783
Goldilocks zone
   see habitable zone
Goodricke, John, 666
grand unified theories (GUTs), 797, 798, 803
   see also unification
gravitation, 8, 102, 336, 803
   general theory of relativity
      see relativity theory, general
   Newtonian, 148, 150, 154, 165–172, 336, 495, 540, 545, 611, 710, 718
      acceleration due to gravity, 168, 169
      law of gravity, 125, 168–171, 176, 187, 191, 199, 209, 270, 280, 797
   potential energy, 646, 654, 656, 698
   quantum mechanics
      see quantum mechanics

universal gravitational constant
   see fundamental constants
gravitational force
   see gravitation; unification
gravitational lensing, 764
   see relativity theory, general
gravitational waves
   see relativity theory, general
gravitons
   see quantum mechanics, gravitation
gravity assist, 461
Gravity Probe B mission, 718
Gravity Recovery and Interior Laboratory mission, 428
Great Debate, 749–750
Great Plains of North America, 27
Greece (ancient), 59–70, 92, 97, 120, 127, 186, 310
   astronomy, 15, 59, 92
   Athens, 60, 61, 63
   flat Earth myth, 65
   see also specific persons
greenhouse effect, 388–389, 394–395, 437, 441, 464
   greenhouse gases, 388, 486
   see also Earth, climate change
Greenwich mean time (GMT), 83
Greenwich, London, United Kingdom, 81
Gregorian calendar, 37, 118, 136
Gutenberg, Johannes, 93

habitable zone, 398, 549–550, 569, 574
half-life
   see radioactivity, radioactive decay
Hall, Asaph, 420
Halley, Sir Edmond, 188, 365
Hamlet, 228
harmonics
   see cosmic microwave background (CMB); sound
Harold, King, 502
Harriot, Thomas, 132
Harvard Astronomical Observatory, 616, 618, 622, 627
Harvard spectral classification scheme, 616–623, 628–631, 666
   extended, 652–653
   see also stars, stellar spectra
Haumea
   see dwarf planets
Hawai'i, Mauna Kea volcano, 593, 597
Hawking, Stephen, 149, 719
Hawkins, Gerald, 48
heat
   see thermodynamics
Heaven's Gate, 520
heavy elements, 316, 551, 557, 651, 668–669
Heezen, Bruce Charles, 399

Heisenberg's uncertainty principle
  see quantum mechanics
Heisenberg, Werner, 294, 295, 627
heliacal rising, 40, 44, 58
heliocentric universe, 7, 70–72, 94–100, 102, 112–118, 121, 132, 146, 187
  see also Aristarchus of Samos; Copernicus, Nicolaus
helium (He), 251, 255, 257, 288, 289, 295, 298, 299, 301, 316, 319, 325, 340, 342, 345–347, 349, 350, 422, 465, 466, 468, 475, 551, 556, 570, 619, 620, 640, 641, 643, 651, 653, 655, 658, 659, 662, 669, 670, 685, 695, 696, 766, 802
  discovery, 251–253
  spectrum, 251–253, 255
  see also alpha (α) particles; radioactivity
helium hydride (HeH$^+$), 641
Herbig–Haro objects, 760
  HH 30, 649
  HH 212, 645, 649
Herman, Robert, 784
Herschel, Caroline Lucretia, 176
Herschel, Sir Frederick William, 176, 362, 492
Herschel, Sir John Frederick William, 176, 618, 728
Hertz, Heinrich Rudolf, 211
Hertzsprung, Ejnar, 618, 627–629, 631, 777
Hertzsprung–Russell (H–R) diagram, 627–631, 633, 673, 675, 765, 816
  giant stars, 630, 631, 633, 653
  hypergiant stars, 664
  luminosity classes, 630
  main-sequence stars, 630, 631, 633, 639, 652–655, 658, 676, 677, 688, 749
    masses, 630
    radii, 630
  spectral classes, 628, 630
  supergiant stars, 630, 631, 633, 653, 700, 712
  white dwarf stars, 630, 631, 633, 653
    masses, 630
    see also white dwarf stars
  see also color–magnitude diagrams; stellar evolution
Hewish, Antony, 702
Higgs boson, 301, 816, 819
Higgs field, 819
Himalayan mountains, 59
Hipparchus of Nicaea, 67–70, 72, 87, 102, 127, 176, 187, 599–601, 604, 718
  star catalog, 68, 69
  trigonometry, 68

Hitler, Adolf, 266, 267
  Nazi Germany, 272–274, 579, 598
Homer, 34
homogeneity, 791
hoodoos, 642
Hooke, Robert, 150, 151, 154
Hoover, J. Edgar, 274
hot dark matter (HDM), 748
House Committee on Un-American Activities, 730
Hoyle, Sir Fred, 783
Hubble constant, 779, 780, 809
Hubble diagram, 778
Hubble sequence
  see galaxies, tuning fork diagram
Hubble Space Telescope (HST), 239, 243, 408, 481, 484, 491, 525, 542, 592, 599, 642, 750, 760, 761, 773
  spherical aberration, 769
Hubble time, 780
Hubble, Edwin Powell, 750, 778
Hubble–Lemaître law, 778
Hume, David, 180
Huygens, Christiaan, 189, 491
Hyades
  see star clusters, open (galactic)
hydrocarbons, 485, 486, 488, 491, 493, 494
hydrochloric acid (HCl), 386
Hydrogen (H)
  see also quantum mechanics, energy levels
hydrogen (H), 247–249, 251, 259, 290–292, 294–296, 299–301, 316, 325, 340, 343, 345–347, 349, 350, 422, 465, 475, 506, 551, 569–571, 580, 619, 620, 640, 641, 643, 651, 653, 654, 669, 695, 696, 766, 802
  deuterium ($_1^2$H), 300, 347–349, 390, 416, 552, 802
  discovery of, 191
  molecular (H$_2$), 340, 465, 468, 475, 556, 569
  spectrum, 246, 251–253, 283
    21-cm (spin-flip), 580, 738–739
    Balmer series, 290–291, 619
    Lyman series, 763
  tritium ($_1^3$H), 300, 802
hydrogen cyanide (HCN), 526
hydrostatic equilibrium, 336–337
hypothesis, 5

Iberian Peninsula, 73, 74
ideal gas law, 339
  particle kinetic energy, 338–339
  particle speed, 339
  pressure (P), 338–340
Imperial system of units, 154
inclination, 113
India, 59, 73

astronomy, 59, 92
inductive reasoning, 93
industrial revolution, 432
inertia, 134
  see Newton's laws, first law
inflation
  see cosmology
infrared light
  see electromagnetic waves (light), spectra
Inhofe, Senator James (Oklahoma), 437
InSight mission, 430
Institute of Science and Astronomical Science (ISAS), 363
interference
  see waves
interferometry, 592, 708
  see also Michelson interferometer
intergalactic medium, 763
International Asteroid Day, 530
International Astronomical Union (IAU), 15, 177, 189, 312, 363, 364, 487, 502, 510, 529, 665
  Minor Planet Center, 508, 511, 512
International Celestial Reference System (ICRAS), 763
International Space Station, 704
International System of Units (SI), 155
interstellar medium, 640–645, 669, 738
  magnetic field, 640
  nebulas (clouds), 170, 239, 359, 641–645, 738
inverse square law, 167
  electrostatic, 209
  gravitation, 168, 192, 199, 209, 222, 495
  light, 220–223, 505, 603–604, 623, 626, 633, 787
    brightness (b), 223
Io torus, 478
ionization, 554, 650, 670
  energies, 620–622
    see also quantum mechanics, energy levels
ions, 255, 331, 380, 389, 404, 597, 620, 621, 671
Iraq, 56
iron (Fe), 208, 210, 253, 316, 319, 322, 323, 334, 366, 404, 407, 422, 424, 428–430, 478, 480, 483, 487, 515, 516, 519, 529, 531, 552, 556, 557, 569, 619, 642, 662, 672, 693, 695–698
  spectrum, 251
iron oxide (FeO), 407
iron sulfide (FeS), 429
isotopes
  see atoms
isotropy, 791

Israel, 398
    Jerusalem, 59
Ivins, Marsha, 713

James Webb Space Telescope (JWST),
    514, 542, 543, 551, 599, 642
    see also Lagrange points
Jansky, Karl Guthe, 595, 743, 746
Japan Aerospace Exploration Agency,
    599
Jefferson, President Thomas, 180
Joliot-Curie, Frédéric, 255
Joliot-Curie, Iréne, 255
Jordan, 398
Joule, James Prescott, 197
jovian planets
    see giant planets
Juno mission, 461, 467, 471, 473
Jupiter, 22, 112, 123, 130, 137, 169, 174,
    176, 220, 359, 361–363, 368, 419,
    441, 460–468, 475, 492, 508, 509,
    514, 544, 547, 552, 630
    atmosphere, 465–468
        composition, 465–466
        Great Red Spot, 466, 468, 470
        pressure, 461
    aurora, 471, 472
    density, 466
    distance, 113
    formation, 558
        migration, 561
    Galilean moons, 130, 133, 134, 362,
        472
        water, 486
    interior, 473–476
        energy output, 464, 466, 468
    Io torus, 472
    magnetic field, 470–472, 483, 596
    moons, 565
        Amalthea, 484
        Callisto, 480, 483–484
        Europa, 480–482, 488, 550
        Galilean moons, 133, 471, 476–484
        Ganymede, 480, 482–483, 550
        Io, 189, 478–480, 550
    mythology, 361, 362
    orbit, 363, 502, 509–512
    rotation, 464, 466
        axis, 470
    superior planet, 23, 113
    trojans, 511, 513, 552, 563

Kaku, Michio, 541, 542
Kant, Immanuel, 180
Kapteyn, Jacobus Cornelius, 729, 730
    Kapteyn's universe, 729
Karl G. Jansky Very Large Array
    (VLA), 596
Kelvin, Lord
    see Thomson, Sir William

Kepler mission, 542, 545, 549
Kepler's laws (complete), 420, 611, 747
    derived, 173–175
        first law (law of ellipses), 175
        second law (law of equal areas), 175
        third law (harmonic law), 174, 175,
            196, 611, 737, 744
    see also Kepler, Johannes, laws of
        orbits (original form)
Kepler, Johannes, 6, 8, 26, 91, 121–125,
    131, 134, 136, 146, 166, 188, 208,
    310, 413, 419, 567
    Astronomia Nova, 124
    fall of circular motion, 123–124
    Harmonices Mundi, 124
    joins Tycho Brahe, 123
    laws of orbits (original form),
        124–130
        first law (law of ellipses), 125–127
        second law (law of equal areas),
            127–128
        third law (harmonic law), 128–130
    Mysterium Cosmographicum, 122, 132
Keplerian orbits
    see Kepler's laws (complete)
kilonovas, 707–708, 717
King, Martin Luther, Jr, 638
Kirchhoff's rules
    see electromagnetic waves (light),
        spectra
Kirchhoff, Gustav Robert, 214, 246,
    252, 264, 288, 292
Krishnamurti, Jiddu, 827
Kuiper belt
    see Solar System
Kuiper, Gerard, 365
Kulik, Leonid Alekseyevich, 530–531
Kyoto Protocol, 451, 452

La Silla Observatory, Chile, 26
laboratory sciences, 334
Lagrange points, 513–514, 563, 564, 567
    James Webb Space Telescope
        (JWST), 514
Lagrange, Joseph-Louis, 513
languages
    Akkadian (Mesopotamia), 56
    Arabic, 92
    Aramaic, 56
    Greek, 72, 92
    Hebrew, 56
    Latin, 72, 92
Las Campanas Observatory, 593, 699
Laser Interferometry Gravitational-
    wave Observatory (LIGO), 708
lasers, 221, 272, 277, 281, 372, 592
late heavy bombardment
    see Solar System, formation
Lauterbach, Anton, 117
lava tubes, 406

Le Gentil, Guillaume, 188
Le Verrier, Urbain, 176
lead (Pb), 191, 253, 254, 346
Leaning Tower of Pisa, 134
learning outcomes, xiv–xv
Leavitt, Henrietta Swan, 666, 732
Leibniz, Gottfried, 151–152, 154
Leif Erikson, 27
Lemaître, Georges, 777
    primeval atom hypothesis, 779, 783
leptons, 346–347
    lepton number
        see conservation laws
Leucippus, 248
Levy, David, 517, 524
Lexel, Anders Johan, 176
Lick Observatory, 749
light
    see eelectromagnetic waves (light);
        photons
light curve
    Cepheid variable star, 666, 667
        see also pulsating stars
    transit (exoplanet), 546
        see also exoplanets, detection
            methods
light quanta, 265, 269, 285
    Einstein's explanation, 285–288
        see also photons
light-year (ly), 607
    relationship to parsecs (pc) and
        astronomical units (au), 607
linear momentum ($\vec{p}$), 300, 336, 338,
    341, 342, 349, 371, 471, 531
    definition, 193
        see also conservation laws
linguistics, 27
Lippershey, Hans, 131
liquid metal, 519
    hydrogen (H), 475
    mercury (Hg), 475
        see also Solar System, giant planets,
            interiors
lithium (Li), 249, 295, 299, 301, 347,
    669, 696, 766
Locke, John, 180
Lockyer, Sir Norman, 253
logarithms, 600
Lorentz, Hendric, 295
Lowell Observatory, 177
Lowell, Percival, 177, 408
Lucas, George, 488
luminosity (L), 222, 609, 623, 624, 626,
    627, 633, 646, 649, 655, 656, 658,
    660, 661, 666, 668, 672, 735, 743,
    761, 762
    blackbody radiation, 217–220
    relationship to brightness (b), 223,
        604
        see also magnitude scale, absolute

Lunar Prospector mission, 517
Luther, Martin, 93, 117–118, 122
Lyman alpha forest (hydrogen), 763
  see also electromagnetic waves
    (light), spectra, Lyman series

Macedonia, 59, 63
Machiavelli, Niccolò, 93
Magellanic Stream, 758, 759, 765, 767
magnesium (Mg), 253, 316, 424, 556,
    569, 696, 700
magnesium hydride (MgH), 619
magnetic dynamo, 331, 386, 422, 424,
    429, 430, 470, 475, 639
magnetic fields, 209–211, 256, 276, 331,
    352, 461, 471, 483, 492, 522, 560,
    590, 599, 639, 735, 738
  see also electromagnetic theory,
    fields; specific object
magnetism
  see electromagnetic theory
magnitude scale, 68, 599–603
  absolute, 626, 627, 629, 666
    definition, 609
    relationship to luminosity (L),
      609–611, 626
  apparent, 602–604, 626–629, 666, 787
  relationship to brightness ratios,
    600–604
  see also Hipparchus of Nicaea
main sequence
  see Hertzsprung–Russell (H–R)
    diagram
Makemake
  see dwarf planets
Mariner missions
  Mariner 4, 411
  Mariner 10, 383, 385
Marius, Simon, 133, 477
Mars, 6, 22, 28, 32, 112, 116, 121, 123,
    124, 128, 130, 176, 220, 359, 380,
    408–421, 514, 519, 552, 557, 558,
    560, 630, 716
  acceleration due to gravity, 412
  age, 560
  atmosphere, 412–414, 417
    ancient, 415–417
  atmospheric pressure, 412
  carbon dioxide (CO$_2$), 413, 486
    ice, 414, 415
    see also carbon dioxide (CO$_2$)
  color, 24, 430
  density, 359, 551
  distance, 113
  dust storms, 413
  escape speed, 199
  formation, 560
    size, 562
  greenhouse effect, 412
  ice caps, 414

impact craters, 533
  Intrepid, 533
  see also meteors
interior, 430–431
life, 420–421
magnetic field, 419, 430, 470
marsquakes, 430
Martians, 408–411, 488
mass, 359, 551
moons, 419–420
  Deimos, 419, 510
  Phobos, 419, 493, 510
motion in the sky
  retrograde 2022–2023, 22, 23, 61,
    63, 114, 115
mythology, 361, 362
orbit, 188, 363, 502, 509–512
  eccentricity, 126, 127, 413
parallax, 188
radius (size), 359, 360
rotation axis
  tilt, 412
shield volcanoes, 380, 412, 413
  Olympus Mons, 414, 417–419, 431
spectrum, 244
superior planet, 23, 113
surface map, 418, 419
temperature, 412
The War of the Worlds, 409
Valles Marineris, 412, 417, 418
water (H$_2$O), 398, 413–415, 438
  ice, 414–416, 486, 552
  see also water (H$_2$O)
Mars Express mission, 415, 416
Mars Global Surveyor mission, 410
Mars Reconnaissance Orbiter mission,
    410, 415
Marsden, Ernest, 289
mass, 154, 169
  relationship to weight, 169
mass extinctions, 395, 435
massive compact halo objects
    (MACHOs), 749
matter-dominated universe, 803, 804
Maunder minimum, 440
  see Sun, solar cycle
Maunder, Annie Scott Dill Russell, 328
Maunder, Edward Walter, 328
Maury, Antonia, 618, 620, 628–630
MAVEN mission, 417
Max Planck Society for the
    Advancement of Science, 266
Maxwell's equations
  see electromagnetic theory
Maxwell, James Clerk, 209–211, 237,
    249, 264–266, 269, 272, 274, 276,
    279, 287, 290, 292, 294, 595, 797
May, Brian, 530
  see also Queen (rock band)

Maya
  see Mesoamerica
measurement
  error bars, 177
  uncertainty, 177
Mediterranean Sea, 48, 59, 72
Melanchthon, Philipp, 118, 122
Mendeleev, Dmitri, 249, 264, 288, 295
Mercury, 22, 112, 123, 126, 128, 132,
    176, 220, 359, 375, 380–386, 473,
    514
  advance of perihelion, 270, 285, 375
  atmosphere, 380–382, 398
  cooling stresses, 383
  density, 359
  impact craters, 382–383, 392, 533
    Caloris Basin, 383
    ejecta rays, 383
    see also meteors
  inferior planet, 23, 70, 113
  interior, 429
  lava flows, 384
  magnetic field, 380, 381, 385, 394,
    401, 429, 470
  mass, 359
  micrometeorites, 380
  mythology, 361, 362
  orbit, 114, 548, 549, 659, 739, 745,
    747, 760
    advance of perihelion, 6, 270, 280,
      708, 718
    eccentricity, 413, 549
  radius (size), 359, 360, 405
  regolith, 382
  rotation, 394
  spin–orbit coupling (resonance),
    375–376
  surface, 382, 386
    age, 384
    temperature, 380
  water (H$_2$O), 398
    ice, 407
    see also water (H$_2$O)
  water (H$_2$O)ice, 385
mercury (Hg), 248, 475
Mesoamerica, 34–44, 56, 73
  Aztecs
    Moctezuma II, 44
    Spanish, 44
  base-20 number system
    see number systems
  Kukulkán
    see Mesoamerica, Quetzalcóatl
  Maya, 19, 36–44, 50, 58
    2012 CE, 37
    astronomy, 40–44, 110
    calendars, 37–40, 44
    Chichén Itzá, Yucatán peninsula,
      42–44
    Classic Period, 36

codexes, 36, 44
mythology, 37
observatory (Caracol), 43
pyramid (Castillo), 43
Venus, 44
Olmec, 19, 34, 35, 37, 41, 208
Quetzalcóatl (Kukulcán), 36, 41–44
Teotihuacán, 19, 41, 42
Toltec, 42
Mesopotamia, 56–60
Assyria, 58
Babylonians, 35, 59, 92
astronomy, 58–59
calendar, 56
cuneiform, 58
eclipses, 58
Sumerians, 92
astronomy, 15, 56–58
creation myth, 58
cuneiform, 56, 58
MUL.APIN, 58
number system (base 60), *see*
number systems, base 60
MESSENGER Orbiter mission, 380,
385, 386, 429
Messier, Charles, 674, 749
metals
see heavy elements
meteorites, 28, 29, 366, 380, 398, 405,
534, 552, 571, 572
age, 342
Allan Hills 84001, 421, 572
Allende, 520
calcium–aluminum inclusions
(CAI), 520, 556
carbonaceous, 552, 557
carbonaceous chondrites, 520, 552,
556, 570, 571
Chelyabinsk, 532–533
Church of the Chelyabinsk
Meteorite, 532
fusion crust, 532
impact craters
Barringer's Meteor Crater,
Arizona, 25, 533
Chicxulub crater, 199, 435, 442,
518, 533, 534
Maniitsoq, Greenland, 533, 534
see also dinosaurs, extinction
impact frequency and energy,
533–536
Mars, 366, 421, 560, 572
metallic, 516, 552, 557
Moon, 366, 572
Murchison, 556, 557, 570, 571
radiometric dating, 520
solar nebula, 557
stony, 517, 552, 557
stony–iron, 366, 519, 552, 557
meteoroids, 366, 557, 559

meteorology, 443
meteors, 24–25, 366, 487, 494, 557
airbursts, 25
bolides, 25, 28, 520
Chelyabinsk, 25, 532–534
Tunguska, 534
fireballs, 25
meteor showers, 25, 526–528
Perseids, 25
radiant, 526, 528
methane (CH_4), 388, 438, 439, 444, 450,
465, 468, 475, 484–487, 491, 551,
557, 569, 571, 573, 642, 653, 696
ice, 468, 476, 504, 506, 526, 557
methane hydrate, 468
methanol (CH_3OH), 526, 571
Mexico, 19
Yucatán peninsula, 36
Chicxulub crater, 199, 435
see also meteorites, impact craters
Michelangelo, 93
Michell, Reverend John, 191, 235, 711
Michelson interferometer, 276, 597
Michelson–Morley experiment,
274–276, 279, 592, 708
Michelson, Albert A., 274
microgravity, 713
microphysics, 673
microwaves
see electromagnetic waves (light),
spectra
Middle East, 34, 59, 73
midnight Sun, 19
Milky Way Galaxy, 12, 25, 26, 29, 32,
132, 133, 142, 169, 193, 199, 279,
305, 398, 523, 529, 542, 544, 549,
552, 567, 568, 574, 576, 578, 581,
596, 597, 608, 622, 662, 666, 671,
673, 674, 691, 693, 706, 726, 727,
733–750, 754, 756, 758, 763
ages, 765–766
boxy/peanut bulge, 731, 743, 747,
754, 765
collision with Andromeda, 757
disks, 736–738, 747
thick disk, 737, 765
thin disk, 737, 743, 765
face-on depiction, 739
Fermi bubbles, 746
galactic poles, 734
globular clusters, 753, 765
see also star clusters, globular
clusters
halos, 736
dark matter halo, 747–749
hot corona halo, 736, 747
inner stellar halo, 736, 747, 765
outer stellar halo, 736, 765
long bar, 739, 743, 747, 765
luminosity, 735–738

Magellanic Stream, 767
magnetic fields, 733
general, 746
map
Gaia, 733
Herschel, 728
Kapteyn, 729, 749
Shapley, 729–730
mass, 735–738
mergers, 757–758
Andromeda Galaxy, 757–759, 767
midplane, 758
tilt to celestial equator, 734
oldest stars, 765, 766
rotation curve, 747–748
flat, 747
Keplerian, 747
rigid-body, 747
Sagittarius A, 743
Sagittarius A*, 743–746, 753, 760
magnetic field, 746
satellite galaxies, 747, 752
spiral arms, 739–743
density waves, 739–742
winding problem, 740, 741
zone of avoidance, 730–733
Miller, Stanley, 569
Miller–Urey experiment, 569–570
Millikan, Robert, 256, 287
molecules, 213, 248–249, 299, 336, 643
see also
amino acids;
amino radical (NH_2);
ammonia (NH_3);
ammonium hydrosulfide
(NH_4SH);
carbon dioxide (CO_2);
carbon monoxide (CO);
carboxylic acid ($COOH$);
deoxyribonucleic acid (DNA);
ethane (C_2H_6);
formaldehyde (CH_2O);
fullerenes [buckyballs (C_{60})];
hydrogen (H), deuterium (^2_1H);
hydrogen (H), molecular (H_2);
hydrogen (H), tritium (^3_1H);
hydrogen cyanide (HCN);
iron oxide (FeO);
iron sulfide (FeS);
magnesium hydride (MgH);
methane (CH_4);
methane hydrate;
methanol (CH_3OH);
nitrogen, molecular (N_2);
nitrous oxide (N_2O);
nucleobases;
oxygen, molecular [O_2 and O_3
(ozone)];
ribonucleic acid (RNA);
silicates;

sulfur dioxide (SO_2);
  sulfuric acid (H_2SO_4);
  tetracholorethylene (C_2Cl_4);
  titanium oxide (TiO);
  water (H_2O);
  zirconium silicate ($ZrSiO_4$)
month
  draconic, 110
  sidereal, 40, 106, 110
  synodic, 40, 41, 106, 110
Moon, 22, 28, 64, 65, 114, 133, 405–408,
    517
  age, 5, 259, 342, 551, 560
  Apollo missions, 183, 407, 428, 517,
    520
    rocks, 342
  crust, 408
  distance from Earth, 186–187, 372
  escape speed, 199, 200
  formation, 565
  highlands, 406
  impact craters, 405, 533
    see also meteorites
  interior, 428
  magnetic field, 405, 407
  maria, 406
  moonquakes, 428
  motion in the sky, 21, 60, 62, 66, 67,
    69, 70
  orbit, 41, 109
    conservation of angular
      momentum, 195
    orbital period, 106
    orbital plane, 109
    tilt, 565
  phases, 20–22, 57, 104–107
  radiometric dating, 406
    see also radiometric dating
  radius (size), 405
  regolith, 405
  spectrum, 244
  water (H_2O), 407
    ice, 407
    see also water (H_2O)
More, Sir Thomas, 93
Morley, Edward W., 274
morning star, 24
Mount Wilson Observatory, 749
Müller, Johannes
  see Regiomontanus
multimessenger astronomy, 708
muons (μ), 346
mythology, 8, 13, 15, 26, 81, 252, 365
  Aphrodite, 361, 362
  Ares, 361, 362
  centaurs, 528
  Ceres, 363
  creation, 27
    Skidi Pawnee, 27–32
    see also Pawnee, Skidi

Cronus, 362
deities, 8, 13, 15, 26, 28, 40–42, 56, 64,
    81, 92, 146, 572
Demeter, 363
Egyptian
  Bennu, 519
  Isis, 49
  Osiris, 49, 519
Gaia, 362
Greek, 49
Greek and Roman, 361–362
Helios, 253, 362
Hermes, 361, 362
Jupiter, 361, 362, 477
Mars, 361, 362
Mayan, 37
Mercury, 361, 362
Neptune, 362
Olympians, 362, 363
Pluto, 363
Poseidon, 362, 363
Roman, 49
Saturn, 362
Selene, 565
Sol, 362
supernatural, 6, 56, 64, 70, 92, 146,
    180, 553
Theia, 565
Titans, 70, 253, 362, 565
Trojans, 528
Urania, 120
Uranus, 362
Venus, 361, 362
Vesta, 514
Zeus, 361–363, 477
  Callisto, 477
  Europa, 477
  Ganymede, 477
  Io, 477

National Academy of Sciences (NAS),
    749
National Aerodynamics and Space
    Administration (NASA), 7, 363, 598
National Institutes of Health (NIH), 7
national parks (United States)
  Arches National Park, 12, 16
  Bryce Canyon National Park, 239,
    642, 643
  Chaco Culture National Historical
    Park, 33
  Crater Lake National Park, 212
  Grand Canyon National Park, 398,
    399
  Grand Teton National Park, 14, 397,
    398, 727
  Hawai'i Volcanoes National Park,
    398, 406, 427
  Joshua Tree National Park, 16

Yellowstone National Park, 248, 398,
    427
  Zion National Park, 398, 399
National Science Foundation (NSF), 7
Native Americans, 18, 32
  Ancestral Puebloan Peoples
    (Anasazi), 33–34
  Inuit, 18, 30
  Pawnee, 27–32
    Europeans arrive, 27
    Skidi, 27–32
    treaties, United States of America,
      27
  petroglyphs
    supernovas (SN 1006 and
      SN 1054), 691
Near Earth Asteroid Rendezvous
    (NEAR)–Shoemaker mission, 517
nebulas, 239, 553–554
  Carina Nebula, 641, 644–646, 673
  Crab Nebula, 705
  Eagle Nebula, 643
    Pillars of Creation, 642, 643
  Gum Nebula, 642
  Horsehead Nebula, 642
  NGC 281, 646
  Orion Molecular Complex, 641
  Orion Nebula, 553, 643–646, 673, 674
  reflection, 240, 673
  Tarantula Nebula, 699
  see also planetary nebulas
neon (Ne), 316, 386, 570, 572, 658,
    660–662, 685, 687, 694, 696
Neptune, 134, 223, 359, 361–363, 365,
    461–464, 468–470, 492, 508, 514,
    552, 563, 564, 604
  atmosphere, 468–470
  discovery, 176–177
  formation, 558
  Great Dark Spot, 470
  interior, 473–476
    energy output, 464, 468
  magnetic field, 470, 471
  moons, 565
  mythology, 362
  orbit, 502, 739, 747
  rotation, 464
    axis, 469, 470
  Triton, 489–491, 510
    cryovolcanoes, 491
    orbit (retrograde), 489, 551, 558
    planetary embryo, 565
    rotation (retrograde), 551
  trojans, 366, 511, 524, 528, 552, 564
Nettles, Bonnie, 521
neuromythologies, xviii–xix
neutrinos (ν), 300, 342, 639, 640, 748,
    797, 815
  electron neutrinos (ν_e), 345–347,
    350–352, 671, 698, 801

antineutrinos ($\overline{\nu}_e$), 671, 801
muon neutrinos (ν_μ), 352
observatories
　Homestake Mine chlorine
　　experiment, 350–351
　Super-Kamiokande water
　　experiment, 351
oscillations, 352
solar, 350–352, 673, 698
supernovas, 698, 700
tau neutrinos (ν_τ), 352
neutron degeneracy, 707
neutron stars, 171, 700–709, 712, 714,
　717, 747, 749, 806
mass limit, 707
maximum mass, 707, 711
neutrons (n), 300, 301, 345–347, 390,
　662, 695, 698, 707, 710, 711, 749,
　801
radioactive decay, 801
New Horizons mission, 173, 461, 504
New Mexico, 33
Santa Fe, 33
Newton's laws, 151, 154–165, 192, 336,
　400, 523
first law (law of inertia), 162–163,
　165, 193, 196
second law (force, $\overrightarrow{F} = m\overrightarrow{a}$),
　163–165, 167, 256, 290, 313, 336,
　338, 341
third law (action/reaction pairs),
　164–165, 167, 192, 337, 374
Newton, Sir Isaac, 6, 96, 125, 145–154,
　165, 166, 169, 175, 176, 179, 180,
　186, 187, 191, 198, 248, 264, 265,
　310, 370, 432, 513, 567, 611, 625,
　710, 711, 740, 747
annus mirabilis, 148, 165, 167, 269
calculus, 151–152, 154, 199
cannon thought experiment, 173,
　713
force, 148
gravitation
　see gravitation, Newtonian
Kepler's laws
　see Kepler's laws
light, 149–150, 201, 238, 285
　color, 149
　Newton's rings, 201, 203, 205
Newtonian physics, 7, 179, 365
　absolute space and time, 276
*Philosophiæ Naturalis Principia
　Mathematica*, 150–151, 155, 173,
　178–180, 196, 365
telescope invention (reflecting), 149,
　591
nickel (Ni), 422, 424, 428–430, 478, 515,
　516, 519, 552, 556, 557, 693,
　695–697

Nile River
　see Egypt
nitrogen (N), 248, 249, 251, 253, 299,
　316, 349, 444, 491, 506, 556,
　569–571, 580, 642, 651, 655, 696
　ice, 491
　spectrum, 252
nitrogen, molecular (N_2), 248, 299, 380,
　386, 387, 389, 395, 412, 465, 486,
　494, 505, 642, 696
　ice, 504, 506, 557
nitrous oxide (N_2O), 248, 438
noble gas, 641
Normans, 92
north celestial pole
　see celestial sphere
North Star
　see stars, Polaris
Novara da Ferrara, Domenico Maria,
　94
novas, 25, 28, 119, 690–691, 707, 749
　GK Persei, 690
　recurrent, 691
nuclear physics, 336
nuclear reactions, 299, 300, 346, 347,
　349–352, 554, 639, 650, 651, 653,
　659–662, 671, 673, 685, 691, 693,
　697, 698
　carbon–carbon burning, 694
　conservation laws, 345, 346
　deuterium burning, 649, 650, 652
　fission, 273, 345, 697
　fusion, 345, 697
　heavy elements, 662
　helium burning, 658, 662
　　triple alpha process, 657, 658
　hydrogen burning, 653
　　CNO cycle, 654–655, 696
　　proton–proton chain, 345–349,
　　　351, 654, 695, 802
　shell burning, 658
　see also stellar evolution
nucleic acids, 570
nucleobases, 570, 571
　adenine, 570, 571
　base pairs, 570
　cytosine, 570, 571
　quanine, 570, 571
　thymine, 570, 571
　uracil, 570, 571
nucleons, 300, 345–347, 695, 707
　energy released per, 695
　see also protons (p^+); neutrons (n)
nucleosynthesis
　Big Bang
　　see cosmology
　see also nuclear reactions
nucleus
　see atoms
number density, 681

number systems
　base 10 (decimal), 35, 73
　base 20 (vigesimal), 34–36, 73
　base 60 (sexagesimal), 56
　Hindu–Arabic, 73
　Roman, 73

Obama, President Barack, 452, 769
Occam's razor, 115–116, 310
Odoacer, King Flavius, 73
olivine, 425, 515
Olmec
　see Mesoamerica
Olympic Games
　Summer 1936, 579, 598
　　Owens, Jesse, 579
　Summer 2012, 197
　Winter 2014, 532
Oort, Jan, 366, 523
open clusters
　see star clusters, open (galactic)
orbital geometry
　aphelion, 126
　ellipse, 126
　　principal focus, 126
　perihelion, 126
　see also ellipse; Kepler's laws
orbital resonance, 480
　see also asteroids
organic chemistry
　see chemistry, organic
organic compounds and molecules,
　395, 483, 488, 504, 506, 519, 520,
　526, 557, 569, 570, 642
　see also molecules
Orpheus, 71
Orphic cult, 70
Ørsted, Hans Christian, 209
Osiander, Andreas, 96, 116
OSIRIS-REx mission, 519
Ottoman Empire, 73
Ottoman Turks, 92
oxygen (O), 248, 249, 251, 253, 299, 316,
　349, 390, 424, 444, 506, 556, 569,
　570, 580, 620, 621, 642, 651, 655,
　658, 660–662, 685, 687, 689, 694,
　696
　spectrum, 252
oxygen, molecular
　O_2, 248, 380, 387, 395, 465, 642, 696
　O_3 (ozone), 248, 395, 443, 669

pair production
　see virtual particles
Palestine, 398
pandemics
　Covid-19 2019, 436, 567
　influenza 1918, 432, 436
Pangaea, 400

Panoramic Survey Telescope and Rapid
    Response System Observatory
    (Pan-STARRS), 529, 755
panspermia
    see astrobiology
parallax, 71–72, 119, 120, 186–188,
    604–608, 611, 626, 627, 629, 668,
    678, 733, 786
    angle (*p*), 606, 608
    spectroscopic
        see spectroscopic parallax
Paranal Observatory, 593
pareidolia, 411
Paris Academy, 179
Paris Climate Agreement, 452–453
parsecs (pc), 605–608
    relationship to light-years (ly) and
        astronomical units (au), 607
Parsons, Lawrence, fourth Earl of
    Rosse, 591
Parsons, William, third Earl of Rosse,
    591
pascal (Pa), 336, 475
Pauli exclusion principle, 295, 641, 661,
    738
Pauli, Wolfgang, 295, 627
Payne-Gaposchkin, Cecilia, 620–622,
    627, 729
peer review
    see science
penicillin, 432
Penzias, Arno Allan, 784
percentage (%), 66, 857
period–luminosity relationship
    see pulsating stars
periodic table of the elements,
    249–251, 264, 299, 316, 641, 709,
    738
    quantum mechanics, 295
Persia, 73
phantom energy, 818
Philip II, King, 63
Philolaus of Croton, 71
phosphates, 570
phosphorus (P), 569, 570, 580, 642
photons (γ), 285–288, 290–293, 298,
    327, 341, 342, 345–347, 349, 352,
    388, 389, 404, 412, 443, 590, 592,
    593, 600, 601, 620, 621, 627, 639,
    640, 646, 649, 654, 671, 686, 698,
    700, 704, 705, 716, 717, 720, 738,
    739, 763
    energy (*E*), 265, 285, 291, 341, 598
        in electron volts (eV), 286
    linear momentum ($\vec{p}$), 341
    photoelectric effect, 287, 293
    see also light (electromagnetic
        waves)
photosynthesis, 395
photovoltaic cell, 288

physics
    classical, 264
    modern, 7, 265, 266
pi-day, 266
Piazzi, Giuseppe, 362
Pickering, Edward Charles, 616–618,
    627, 628, 666, 729
    Pickering's Harvard "computers",
        617–620, 622, 627, 666
Pioneer missionS
    Pioneer 10, 461
Pioneer missions
    attached plaque, 580
    Pioneer 10, 706
    Pioneer 11, 461, 484, 488, 706
plague, 269
    Black Death, 432
    Great Plague of Europe, 17th
        century, 148
Planck limits, 791–793
    length, 792
    temperature, 792, 798
    time, 792, 798
Planck satellite, 784, 794, 806, 809
Planck, Max, 264–266, 269, 285, 288,
    292, 294, 295, 297, 627
planetarium, 75
planetary defense system, 536
planetary nebulas, 685–687, 816
    Butterfly Nebula (NGC 6302), 686
    Cat's Eye Nebula, 684, 685, 687
    Eight-Burst Nebula (NGC 3132), 686
    magnetic fields, 687
    multiple star systems, 687
    multiple stars, 686
    NGC 2818, 686
    NGC 6751, 686
planetary planetary nebulas
    LL Pegasi, 686
planetary science, 310, 370
planets, 552
    definition, 653
    distances, 112–113
    formation, 359, 551–566
        accretion disks, 649, 650
        hot Jupiters, 560
        magnetic fields, 649
        pebble accretion, 557
        planet migration, 560–561
        planetary embryos, 516, 552, 557,
            558
        planetesimals, 520, 557
        protoplanetary disks, 530,
            553–556, 558, 560, 649, 650
        protoplanets, 516, 519, 552, 557
        snow line, 562
        see also accretion disks; Solar
            System, formation
    inferior, 23, 70, 112

    motions in the sky, 22–24, 60, 66, 69,
        97, 112–115
        brightness changes, 24
        retrograde motion, 22, 63, 67
        see also retrograde motion
    naked-eye, 23, 28, 51, 63, 112
    orbits
        inclinations, 113–114
        line of nodes, 114
        resonance, 552
    ring systems, 491–496
        discoveries, 491–492
        gaps, 494–495
        lifetimes, 496
        shepherd moons, 496
        sources, 492–494
        waves, 496
    superior, 23, 24, 63, 66, 69, 113
    see also specific planets; Solar
        System; exoplanets
Plato, 12, 60–64, 66, 71, 74, 94, 96, 97,
    114, 122, 141, 142
    ideas and observations, 61
    "save the planetary phenomena", 61,
        63, 70
    uniform and circular motion, 60–61
platonic solids, 122, 124, 142
Pleiades
    see star clusters, open (galactic)
Pluto
    see dwarf planets
Polaris
    see stars
polonium (Po), 254, 255
polymath, 93
Pope, Alexander, 489
positrons (e^+), 345–347, 351, 801
potassium (K), 249, 551, 642
pound
    force (pound, lb_f), 154
    mass (lb_m), 154
power, 9, 217–219, 223, 315, 326, 328,
    342–344
    see also luminosity
pre-Columbian, 34
pre-modern society, 14, 16, 17, 21, 22,
    24–26, 32, 40, 92, 104, 107, 108,
    110, 359, 502
precession
    line of nodes, 109, 114
    of the equinoxes, 68, 69, 86, 97, 100,
        102–103, 109
pressure (*P*), 207, 322, 336, 337
    definition, 336
    radiation pressure, 341–342, 366, 522
    sonic booms, 282, 322
    see also hydrostatic equilibrium;
        ideal gas law
prime numbers, 579
printing press, 93

prisms, 238
proteins, 570, 571
  see also astrobiology
protons (p^+), 299–301, 345–349, 351,
    390, 654, 657, 662, 672, 694, 695,
    698, 707–709, 738, 749, 801
  radioactive decay, 797
  see also hydogen (H)
protoplanetary disks
  see planets, formation
protostars, 650
pseudosciece
  young Earth "theory", 567
pseudoscience, 8–9
  astrology, 8–9, 13, 23, 58, 117
    zodiac, 23, 58, 66, 70, 98
  flat Earth "theory", 65, 567
  young Earth "theory", 6
Ptolemæus, Claudius
  see Ptolemy
Ptolemy, 68–70, 72–74, 78, 94–96, 112,
    116, 117, 119–121, 124, 127, 128,
    133, 147, 187, 728
  Almagest, 69, 73, 74, 92, 93, 114, 176
  Geography, 68, 93
Ptolemy III Euergetes, pharaoh, 66
Ptolemy, King, 63
pulsars, 581, 701–708, 712, 746
  little green men, 702–703
  magnetic fields, 704, 705
  orbiting planets, 542
  precession, 718
pulsating stars, 665–673, 685, 702, 730,
    733
  asteroseismology, 673
  Cepheids, 666–671, 673, 678, 694,
    730, 731, 750, 778, 787
    heavy-element poor (W Virginis
      stars or Type II), 668, 732
    heavy-element rich (classical,
      Type I), 668, 731
  instability strip, 671, 672
  mechanism, 669–672
  partial ionization zone, 670
  period–luminosity relationship,
    666–668, 673, 732, 733, 750, 778,
    787
  RR Lyras, 671, 678, 730
    heavy-element poor, 671
Puzio, Michael, 519
pyramids
  Egyptian, 34, 59, 89
    Great Pyramid, Giza, 48, 103, 567
  Mayan, 43
Pythagoras of Samos, 60–61, 71, 94,
    116, 122, 124
  abstract thinking, 60
  numbers, 60

quantum gravity, 793, 798
quantum mechanics, 7, 9, 269, 272,
    288, 293–298, 316, 335, 336, 341,
    473, 590, 627, 710, 718, 803
  atoms
    electron orbitals, 294, 331, 671, 738
  chemistry, 334
  de Broglie's wavelength for massive
    particles, 293, 294, 349
  electron double-slit experiment, 293,
    298
  energy
    see electron volt (eV)
  energy levels, 296, 316, 350, 388, 404,
    620, 621, 633, 641, 661, 668, 673,
    689, 738
    greenhouse gas molecules, 388
    ground state, 292
    hydrogen (H), 292, 343
    ionization energies, 621, 627
    iron (Fe), 334, 672
    water (H_2O), 388
  gravitation, 718–719
    gravitons, 798, 819
  Heisenberg's uncertainty principle,
    297, 349, 719, 789, 795, 800
  linear momentum ($\vec{p}$), 293, 294, 297
  periodic table of the elements, 295
  Planck's constant (h)
    see fundamental constants
  probabilities, 295–298
    "God does not play dice", 298
  quantum (quanta, pl.), 265
  quantum numbers, 291, 296, 661
    magnetic (m_ℓ), 294
    orbital angular momentum (ℓ),
      294
    principal (n), 294
    spin (m_s), 294, 738
  Schrödinger's cat, 297, 298
  Schrödinger's equation, 294–295
  tunneling, 305, 349, 650, 658, 660
  wave–particle duality, 285, 293–294,
    349
Quaoar
  see dwarf planets
quarks, 301, 800–801, 815
quasars, 702, 761–764, 806
  see also galaxies, types, active
    galaxies
Queen (rock band), 530

radial velocity, 544–545, 548, 733, 763
radiation, 212, 255
  see also alpha; beta; gamma; light
    (electromagnetic waves)
radio waves
  see electromagnetic waves (light),
    spectra

radioactivity, 254–255, 627
  radioactive decay, 257–259
    half-life, 257–259, 280, 281, 297,
      515, 556, 700
  radiometric dating, 27, 258–259, 300,
    384
    carbon dating, 258
    meteorites, 342, 520, 673
    Moon rocks, 342, 673
  radium (Ra), 255
  radon (Rn), 257
Reagan, President Ronald, 730
recombination
  see cosmology, the first atoms
redshift, 777
  cosmological, 763, 781–783, 812
    redshift parameter (z), 781–782
  gravitational, 717, 720, 783, 808
  radial velocity
    see Doppler effect, light
Regiomontanus, 74, 94
  Epitome of the Almagest, 74
Reinhold, Erasmus, 118, 123
relativity theory
  constancy of the speed of light, 277,
    717
  Galilean, 134
  general, 7, 9, 172, 269–271, 285, 335,
    543, 597, 629, 711, 712, 717, 718,
    749, 764, 777, 779
    accelerating-laboratory thought
      experiment, 713–714
    advance of perihelion of Mercury,
      6, 270–271, 280, 285, 375, 708,
      718
    equivalence principle, 713
    falling-man thought experiment,
      712–713
    gravitation, 270
    gravitation and quantum
      mechanics, 718–719
    gravitational lensing, 544, 749,
      764, 808
    gravitational time dilation, 716,
      720
    gravitational waves, 694, 707–708,
      717
    solar eclipse of 1919, 271
    spacetime curvature, 270–271,
      708, 712–716
    tests, 270–271, 715, 717–718
  spacetime, 7, 269
  special, 270, 277–281, 285, 351, 627,
    713
    $E = mc^2$, 269, 344–345, 695
    length contraction, 281
    postulates of, 277
    tests, 279–281
    time dilation, 280, 720

religions, 7, 13
  Christian, 72, 92, 118
    Bible, 25, 117, 121, 138, 153, 180
    Fourth Crusade, 92
    Protestant Reformation, 93, 117, 122, 136
    see also Roman Catholic Church
  Hindu, 73, 92
  Islam, 92
    astronomy, 73–74
    Mecca, 73
    Medina, 73
    month, 73
    Muhammad, 73
    Ramadan, 74
  Mayan, 37
  polytheism, 73
Remy, Danica, 530
Renaissance, European, 74, 92–93
resonances, 480
  music, 480
  orbital, 519, 552
    Enceladus–Dione system, 488
    Galilean moons, 480
    Kirkwood gaps, 528
    migration, 561–564
    Moon and TESS, 542
    Neptune and plutinos, 509, 510
    Neptune and Pluto, 509
    ring systems, 494–496
    TRAPPIST-1 system, 550
  spin–orbit coupling, 375–376, 550, 552
  synchronous rotation, 372, 374, 479, 484, 488, 489, 507, 549, 550, 552, 565
  Tacoma Narrows Bridge, 480
  see also Lagrange points; Solar System, trojans; tides
retrograde motion, 22–24, 51, 61, 63, 66, 67, 69, 72, 95, 97, 114–115
  orbits
    moons, 484
    Triton, 558
  rotation, 359, 360, 381, 463, 469, 470, 478, 551
Revolutionary War, American, 83
Rheticus, Georg Joachim, 95, 96, 118
ribonucleic acid (RNA), 520, 570–571
  ribose (sugar), 570
Richer, Jean, 188
Richters, Grigorij, 530
ringwoodite, 425
RNA
  see ribonucleic acid (RNA)
Roche distance limit, 493
Roche, Édouard, 493
rogue asteroids and comets, 528–530, 552

rogue planets
  see exoplanets
Roman Catholic Church
  Congregation of the Holy Office of the Inquisition, 136
    Index Librorum Prohibitorium, 136, 139–141
  popes and cardinals, 95, 118, 136, 138, 139, 141
  see also Galileo Galilei, trial
Roman Empire, 72–73, 224
  emperors, 72, 73
Rømer, Ole Christensen, 189
Roosevelt, President Franklin Delano, 729
  New Deal, 729
Rosetta mission, 525
Royal Observatory, 81, 84
Royal Society, 150, 152, 224
RR Lyra variable stars
  see pulsating stars
Rubin, Vera C., 747, 749
Rumford, Count Benjamin Thompson, 224
Russell, Henry Norris, 622, 629, 631
Rutherford, Ernest, 288, 298, 627, 696
  early atomic model, 289, 290
Röntgen, Wilhelm Conrad, 253, 255

Sagan, Carl, 3, 453, 461, 567, 569, 577–579, 581, 596, 684, 697
  A Pale Blue Dot, 453
Saha's equation, 621
Saha, Meghnad, 621, 627
Saint-Exupéry, Antoine de
  The Little Prince, 530
Sandia National Laboratories, 531
Sargon of Akkad, 56
SARS-CoV-2, 436
  see also pandemics, Covid-19 2019
Saturn, 112, 123, 128, 133, 176, 359, 361, 460–468, 491, 508, 514, 563
  atmosphere, 465–468
    composition, 465–466
    pressure, 461
  density, 466
  distance, 113
  formation, 558
    migration, 561
  interior, 473–476
    energy output, 466, 468
  magnetic field, 470, 471
  moons
    Daphnis, 495
    Enceladus, 487–488, 494
    Mimas, 488, 494
    Pan, 495
    Pandora, 496
    Prometheus, 496
    Titan, 484–487, 569

mythology, 362
  orbit, 114
  ring system, 133, 495, 496
    Cassini division, 494
    E ring, 494
    see also planets, ring systems
  rotation, 464
    axis, 470
  superior planet, 23, 113
scalars, 155–158
scattered disk
  see Solar System
Schiaparelli, Giovanni Virginio, 408
Schrödinger, Erwin, 294, 295, 627
Schrödinger's equation
  see quantum mechanics
Schwarzschild, Karl, 631
Schweickart, Rusty, 530
science
  central question, 13, 590, 633
  creativity, 13
  critical thinking, 9, 92, 437, 633
  evidence, 5–7, 9, 25, 310, 311, 390, 411, 551–553, 567, 578, 604, 699, 701, 708
  human enterprise, 4, 731
  models, 334–335, 350, 443, 553, 560
    see also computer modeling
  paradigm, 6–7, 310
  patterns, 633
  peer review, 7–8, 152, 440, 447
  scientific method, 4, 310, 350, 731
  scientific revolution, 6–7, 93, 97
  scientific theories, 5–6, 9, 92, 334, 350, 553
    self-consistency, 6, 310, 553, 560, 566, 796
  society, 9–10
  universality of physical law, 185–230, 310, 336, 408, 551–553, 560, 566, 572, 590, 624–625
Sedna
  see dwarf planets
Seleucus of Seleucia, 72
SETI
  see extraterrestrial intelligence
Shakespeare, William, 93, 228, 489
Shapley, Harlow, 729, 749, 777
Shapley, Martha Betz, 730
Shelley, Mary Wollstonecraft, 567
Shelton, Ian, 699
Shepard, Alan, 183
Shoemaker, Carolyn, 517, 524
Shoemaker, Eugene, 517, 524
Shovell, Sir Clowdesley, 83
sidereal time, 17
silica, 424
silicates, 424, 696
silicon (Si), 316, 424, 556, 569, 620, 642, 695, 696, 766

silver (Ag), 119
SIMBAD, 674
Simplicius of Cilicia, 60, 140
singularity
   see black holes
Sitting Bull, 309
Sky & Telescope magazine, 665
SkyMapper telescope, 669
Sloan Digital Sky Survey (SDSS), 755
Socrates, 34, 60
sodium (Na), 249, 251, 253, 551, 592,
   619
   spectrum, 252
Sofaer, Anna, 33
Solar and Heliospheric Observatory
   (SoHO), 319, 323
Solar Dynamics Observatory (SDO),
   325, 333
solar panel, 288
Solar System, 6, 489, 552, 581, 816
   age, 342, 520, 556, 673
   asteroids
      see asteroids
   comets
      see comets
   cryogeysers, 481, 488
   cryovolcanoes, 487, 491, 494, 504, 506
   dwarf planets
      see dwarf planets
   formation
      aluminum-26 radioactive decay,
         556, 557, 700
      asteroid belt, 562
      ice giants, 558, 563
      ices, 557
      Kuiper belt, 564
      late heavy bombardment, 384, 385,
         406, 408, 551, 560, 564
      magnetic field, 560
      Oort cloud, 562
      orbital resonances, 561–562
      orbits, 558
      planet migration, 560–565
      planetary embryos, 557, 564, 565
      planetesimals, 564
      protoplanetary disk, 556–560, 565,
         566
      protoplanets, 557
      rocky planets, 559
      scattered disk, 564
      scattering small bodies, 562
      snow line, 557
      solar nebula, 359, 366, 387, 390,
         466, 476, 520
      solar wind, 558
      Theia, 565
      third-ice-giant scenario, 563
      see also planets, formation;
         resonances
   gas giants, 361, 551, 565

   atmospheres, 465–468
      see also Jupiter; Saturn
   giant planets, 361–362, 460–496, 551
      auroras, 471
      general characteristics, 462–463
      interiors, 473–476
      magnetic fields, 470
      moons, 476–491, 570
      rotation, 464
      rotation axes, 470
      temperatures and heat, 463–464
      see also Jupiter; Saturn; Uranus;
         Neptune
   ice giants, 362, 551, 564, 565
      atmospheres, 468–470
      see also Uranus; Neptune
   Kuiper belt, 365, 366, 461, 489, 490,
      504, 505, 508–510, 524, 552, 560,
      570
      Arrokoth, 510
      moons, 362
      ices, 552
   near-Earth objects, 510
   Oort cloud, 366, 523, 524, 552, 560
   plutinos, 509, 510
   ring systems, 362
      see also planets
   rocky planets, 359–361, 379–431, 461,
      462, 551
      see also Mercury; Venus; Earth;
         Mars
   scattered disk, 365, 366, 509, 510,
      524, 552, 560
   size, 186–189
      scale model, 367–370
   small Solar System bodies, 510
   trans-Neptunian objects, 363, 502,
      504, 508, 509, 552
   trojans, 511
      Jupiter's, 511, 513, 552, 563
      Neptune's, 366, 511, 524, 528, 552
   uniqueness, 560
sound, 207, 211, 353, 807
   harmonics, 808
   speed of, 282
   see also Doppler effect
south celestial pole
   see celestial sphere
southern hemisphere, 15
Soviet Academy of Sciences, 531
Soviet Union, 730
Space Shuttle Atlantis, 713
space weather
   see Sun
spacetime
   see relativity theory
Spain, 27, 745
   Ferdinand, King of, 68
   Isabella, Queen of, 68
   Toledo, 92

Spanish flu
   see pandemics, influenza 1918
special theory of relativity
   see relativity theory, special
spectroscopic parallax, 677–678
spectroscopy, 243, 251, 316, 627, 733
   spectrographs, spectrometers,
      spectroscopes, 243, 246
spectrum
   see electromagnetic waves (light),
      spectra
speed of light (*c*)
   see fundamental constants
Spinoza, Baruch, 180
spiral nebulas, 730
   see also galaxies, spiral (S and SB)
Spitzer Space Telescope, 550
spring (first day), 18
   see also celestial sphere, vernal
      equinox
standard candles, 678, 682, 787
   Cepheids, 666, 778, 780
      see also pulsating stars
   white dwarf supernovas, 694, 780
standard rulers, 787
star charts, 665
star clusters
   ages, 676–677
   galactic
      see star clusters, open (galactic)
   globular, 673, 678, 730–732, 747, 765
      Hercules Cluster, 579, 584
      NGC 6293, 673–675
      ω Centauri, 673, 674
   open (galactic), 674, 677, 730
      Hyades, 29, 44, 526, 628–630
      M67, 673–677
      Pleiades, 29, 31, 32, 44, 133, 239,
         241, 526, 628–630, 673–675, 677
star quake, 705
star trails
   see stars, motion in the sky
Star Trek, 531
   Vulcan, 270
Star Wars, 488
star-tar
   see tholin
Stardust mission, 525, 526
stars
   classification
      see Harvard spectral classification
         scheme
   mass determinations
      see binary systems, stars
   mass loss, 686
   motions in the sky, 14–17, 60, 61, 74,
      76–78, 97–100
      annual, 98–100
      daily, 97–98
      star trails, 16

radii, 623–624, 627
spectra, 616, 621–623, 626
  compositions, 622, 626
  see also electromagnetic waves
    (light), spectra
surface temperatures, 622, 623
twinkling, 593
stars (*specific*)
38 Aries, 599
61 Cygni, 604
51 Pegasi, 542
134 Tau, 636
Acrux, 636
Aldebaran, 44, 94, 526, 628, 636
Alioth, 14
Alkaid, 14
Alnitak, 642
Antares, 28
Arcturus, 15, 235
Bellatrix, 17, 665
Betelgeuse, 17, 85, 216, 220, 526, 599,
    603, 606, 609, 623, 630, 664
  spectrum, 244
Capella, 28
  spectrum, 244
Castor, 526, 636
  spectrum, 244
Cygnus X-1, 712
δ Cephei, 666
Deneb, 217
Dubhe, 14
Errai, 103
η Carina, 645
Fomalhaut, 636
γ Cephei, 542
GK Persei, 690
Lalande 21185, 220, 630
LL Pegasi, 685
Megrez, 14
Merak, 14
Mimosa, 636
Mira (*o* Ceti), 636, 665, 672, 680
Mizar, 14, 49
PDS 70
  planet formation, 558
Phecda (Phad), 14, 16
Polaris, 14, 16, 28, 32, 49, 51, 76, 82,
    86, 100, 103, 208, 666
Pollux, 526, 636
  spectrum, 244
Procyon A, 526, 599, 623
  spectrum, 244
Procyon B, 623, 630
Proxima Centauri, 603, 606, 608, 609,
    625
  exoplanets, 550
R Virgo, 15
Rigel, 17, 216, 729
S2, 744
Saiph, 17

Sanduleak −69 202, 700
Sirius, 220
Sirius A, 28, 49, 85, 526, 599, 601, 612,
    615, 618, 623, 687
  spectrum, 244
Sirius B, 612, 615, 630, 687
Thuban, 49, 86, 103
TRAPPIST-1
  exoplanets, 550, 585
  orbital resonances, 550
TW Hydrae
  protoplanetary disk, 558
UY Scuti, 664
V1154 Cygni, 666
V883 Orionis
  protoplanetary disk, 557
Vega, 28, 103, 603
statistical mechanics, 224, 228, 264
  see also thermodynamics
steady-state theory, 785
Stefan, Jožef, 218
Stefan–Boltzmann blackbody
    luminosity law
  see blackbody radiation
stellar evolution, 553–554, 639–678, 685
  asymptotic giant branch, 656, 659,
    675, 676, 685
  evolutionary tracks, 646, 650, 655,
    658, 659, 662–664, 676, 687, 688,
    765
  gravity versus nucler energy sources,
    654
  helium core flash, 661–662
  horizontal branch, 675
    blueward, 656, 658, 671, 675
    redward, 656, 658, 671
  main sequence, 650–652, 656, 658,
    675
    lifetimes, 654
    zero-age main sequence, 651
  massive stars, 662–663, 712
    size, 664
  nebular free-fall, 646–647
  post-main sequence, 654–664
  pre-main sequence, 646–650, 653,
    657
  protostars, 649
  red giant branch, 656, 658, 675
  shell burning, 658, 659
    hydrogen, 655
  subgiant branch, 655–657, 671, 675
  superwind, 686
  see also Hertzsprung–Russell (H–R)
    diagram
Stellarium, 15, 17, 23, 99, 103, 115, 553,
    600
Stonehenge monument, 34, 44–48
Stoney, George Johnstone, 256
Strasbourg astronomical Data Center
    (CDS), 674

Stratospheric Observatory for Infrared
    Astronomy (SOFIA), 598
string theory, 798
  see also unification
Strömgren, Bengt, 631
strong nuclear force, 348, 706–707, 710,
    797, 802, 803
  see also unification
Struve, Otto, 622
study aids, xvi–xviii
study skills, xix–xxiii
Stukeley, William, 165
sulfur (S), 316, 478, 556, 569, 642, 696
sulfur dioxide (SO_2), 386
sulfuric acid (H_2SO_4), 386, 387,
    389–391
summer (first day), 18–20
  see also celestial sphere, summer
    solstice
summer solstice
  see celestial sphere
Sun, 22, 28, 113, 169, 309–354, 544, 551,
    552, 639, 664
  age, 5, 342, 349, 520, 551, 673
    energy sources, 342–343
  apparent magnitude, 604
  atmosphere, 253, 317–328, 622, 641
    chromosphere, 317, 318, 320, 322,
    333, 622, 639
    composition, 316, 465
    corona, 318, 320, 622, 639, 716
    photosphere, 317, 320, 322, 349,
    622, 639
    transition region, 317, 318, 320,
    321, 622, 639
  brightness, 223
  characteristics, 310–317
    basic data, 311
  composition, 349, 620, 673
  computer modeling, 316, 336, 349,
    350, 352, 353
  convection zone, 320–322, 331, 349,
    353, 422, 639
  core
    conditions, 349
    nuclear, 349, 353
  coronal holes, 327–328, 639
  coronal mass ejections, 326–328, 639
  daily motion, 21
  differential rotation, 317, 331–332,
    353, 422
  distance to, 187
  flares, 324–325, 328, 403, 639
  formation
    protoplanetary disk, 359
    see also planets, formation
  granulation, 320–322, 639, 673
  helioseismology, 352–354, 639, 673
    oscillations, 353
  interior, 349

composition, 349, 353
density, 352, 353
temperature, 352
luminosity, 223, 315, 343, 344, 349, 352, 387, 388, 441, 673
magnetic field, 322–328, 330–333, 354, 386, 422, 470, 473, 639, 702
magnetic dynamo, 331
see also electromagnetic theory
mass, 313, 349
motion in the sky, 17–20, 60, 61, 66, 67, 69, 78–81
mythology, 362
nuclear reactions, 351, 352
see also nuclear reactions
prominences, 49, 325–326, 328, 639
eruptive, 325
radiation zone, 349, 639
radius, 311, 349, 639, 673, 760
rotation axis, 556, 566
shock waves, 322, 333, 639
solar cycle, 317, 328–333, 639
butterfly diagram, 329
Maunder minimum, 328–329
solar maximum, 328
solar minimum, 328
solar flares, 318
solar neutrino problem, 350–352
solar oscillations, 639
solar wind, 327–328, 366, 402, 403, 470, 471, 522, 639
space weather, 328
spectrum, 241, 244, 245
blackbody, 244, 253
Fraunhofer lines, 241–244, 622
sunspots, 49, 133, 322–324, 328, 639
magnetic fields, 470
penumbra, 322
umbra, 322
surface temperature, 315
temperature, 349, 639, 673
supermassive black holes
see black holes
supernatural
see mythology
supernovas, 25, 26, 33, 691–700, 747
core-collapse supernovas, 697–700, 708, 709, 711, 766, 806
neutrinos, 700
neutron stars, 700
photons, 700
shock waves, 700, 742
supernova remnants, 766
testing theory, 699–700
white dwarf supernovas, 693–694
accretion, 693
white dwarf mergers, 694
supernovas (*specific*)
SN 185, 691
SN 1006, 26, 691–692

SN 1054 (Crab supernova), 26, 33, 50, 691, 705
SN 1181, 26
SN 1572 (Tycho's supernova), 26, 119, 124, 691, 694
SN 1604 (Kepler's supernova), 26, 124, 691, 694, 699
SN 1987A, 26, 691, 699–700
neutrinos, 700
superstition, 8, 13, 25, 26, 58
comets, 44
eclipses, 58
Sweden, Hveen (island), 120
Swift, Jonathan, 420
Gulliver's Travels, 420
symmetry, 168
synchronous rotation
see resonance
Système international d'unités
see International System of Units (SI)
Szilárd, Leó, 273
atomic bomb, 273

tardigrades, 572
taus (τ), 346
Taylor, Jane, 689
telescopes
combining multiple telescopes, 593
ground-based
atmospheric windows, 594–598
guide stars, 592
infrared, 598
radio and microwave, 594–598, 743
interferometry
see interferometry
invention, 131–132, 149, 590
see also Galileo Galilei; Newton, Sir Isaac
reflecting, 590
advantages, 591–592
challenges, 592–593
refracting, 590
space-based, 598–599
see also specific telescopes and observatories
Teller, Edward, 273
temperature (T), 225, 282, 338, 339, 349
see also thermodynamics, heat
temperature scales
Celsius, 213
Fahrenheit, 213
Kelvin, 213–214
terrestrial planets
see Solar System, rocky planets
tetrachloroethylene (C_2Cl_4), 351
Thales of Miletus, 59–60, 64, 89, 207, 208
Tharp, Marie, 399

theory of everything (ToEs), 797–798, 803
see also unification
thermodynamics, 224, 264
arrow of time, 230
entropy, 228–230
heat, 223–228, 336, 466, 553, 646
temperature, 225, 338
heat (energy) transport, 226, 342
conduction, 227
convection, 226–227, 321, 655
radiation, 227–228
laws of, 229–230
see also statistical mechanics
Thirty Meter Telescope International Observatory, 593
tholin, 569, 570
Thomson, Sir Joseph John, 255, 288
Thomson, Sir William (Lord Kelvin), 185, 214
tides, 196, 370–376, 493
neap, 375
spring, 374
see also resonances
Tigris River, 56
titanium oxide (TiO), 619
Toledo School of Translators, 92
Tolkien, J. R. R., 487
Tombaugh, Clyde, 177, 504
tonne, 45
torque ($\vec{\tau}$), 195, 371–372
see also conservation laws, angular momentum ($\vec{L}$)
Transiting Exoplanet Survey Satellite (TESS), 542, 545
tritium ($^3_1$H)
see hydrogen (H)
Trojan war, 34
trojans
see Solar System
true vacuum, 819
Truman, President Harry S., 273
Trump, President Donald J., 436, 452
Tunguska event, 530–531
shock wave, 530
Twain, Mark, 437, 501
Twinkle, twinkle little star, 689
Tycho Brahe
see Brahe, Tycho

ultraviolet light
see electromagnetic waves (light), spectra
unidentified aerial phenomena, 567
unidentified flying objects (UFOs), 520, 567, 578
unification, 796–798
uniform and circular motion, 61, 66
see also Plato

United Nations, 530
    United Nations Education, Science,
        and Cultural Organization
        (UNESCO)
        World Heritage Sites, 33, 41, 44,
            49, 568, 729
United States (US) customary units,
    154
universe
    age, 5, 6, 780
    expanding, 6
    steady-state theory, 783
    structures
        cosmic web, 757, 804–805
        filaments, 757, 804
        voids, 757, 804
        walls, 757, 804
    structures (specific)
        Great Attractor, 757
        Laniakea, 757
    see also cosmology
universities, earliest European
    curriculum (trivium and
        quadrivium), 92
    University of Bologna (Italy), 92, 94
    University of Oxford (England), 92
    University of Paris (France), 92
uranium (U), 253–255, 257, 259, 273,
    288, 697
Uranus, 176, 177, 359, 361, 362,
    461–464, 468–470, 492, 508, 514,
    563, 564
    atmosphere, 468–470
    discovery, 176
    formation, 558
    interior, 473–476
        energy output, 468
    magnetic field, 470, 471
    moons
        Miranda, 489
    mythology, 362
    rotation
        axis, 469, 470
        retrograde, 464, 469, 470, 551
Urey, Harold, 569
Utah, 33, 398, 417
    Great Salt Lake, 17, 398, 819
    Salt Lake City, 77, 819

vacuum decay, 819
vacuum energy, 790
vacuum pressure, 790
Van Allen, James, 402
variable stars (pulsating)
    see pulsating stars
vectors, 155–158, 209, 554
velocity ($\vec{v}$), 158–163
    definition, 158
    radial
        see radial velocity

Venera 13 lander, 391
Venus, 22, 24, 28, 32, 36, 41, 42, 58, 112,
    123, 132, 176, 220, 359, 380,
    386–394, 441, 514, 691
    atmosphere, 386–390, 398
        pressure, 386–388
    carbon dioxide (CO_2), 386
        see also carbon dioxide (CO_2)
    density, 359
    distance, 113
    greenhouse effect, 388–389, 412, 438
    impact craters, 391–392
    inferior planet, 23, 70, 113
    interior, 429–430
    magnetic field, 394, 470
    mass, 359
    mythology, 361, 362
    oceans, 380
    orbit, 114
    phases, 133
    radius (size), 359, 360
    rotation (retrograde), 390–391, 468,
        551
        axis, 469
    shield volcanoes, 380, 393–394, 419,
        430
    spectrum, 244
    surface, 391–394
        age, 392
    temperature, 412
    transit, 188
    water (H_2O), 398, 438, 552
        see also water (H_2O)
Vera C. Rubin Observatory, 749
vernal equinox
    see celestial sphere
Verne, Jules Gabriel, 422, 567
Very Large Telescope (VLT), 593, 597
Very Long Baseline Array (VLBA), 597
very long baseline interferometry
    (VLBI), 597, 746, 763
Vesta
    see asteroids
Viking missions
    Viking 1, 410, 419, 716
    Viking 2, 716
Vikings, Norse, 27
Virgo graviational wave observatory,
    708
virtual particles, 790
visible light
    see electromagnetic waves (light),
        spectra
visual, auditory, kinaesthetic (VAK),
    xix
VizieR, 674, 788
volcanism
    hotspots, 431
        shield volcanoes, 426
        Yellowstone National Park, 427

Moon, 406
    see also specific planets or moons
Voltaire (François-Marie Arouet), 178,
    180
Voyager missions
    golden records, 461, 581
    Voyager 1, 173, 453, 461, 477, 478,
        488, 492, 624, 706
    Voyager 2, 173, 461, 469, 470, 473,
        477, 484, 488, 489, 491, 492, 706

W Virginis stars
    see pulsating stars, Chepheids
wadsleyite, 425
Waldheim, Kurt (United Nations
    Secretary-General), 461, 581
wandering stars
    see planets, motions in the sky
water (H_2O), 28, 40, 50, 51, 64, 132, 134,
    147, 203, 207, 211, 213, 214,
    224–228, 230, 238, 239, 247, 248,
    316, 331, 335–338, 349, 359, 361,
    365, 372–374, 380, 382, 383,
    386–391, 395–398, 407, 408, 412,
    413, 415, 417, 438, 443, 444, 446,
    447, 449, 464–466, 468, 478,
    480–483, 486–488, 504, 518, 525,
    530, 549, 551, 552, 557, 558, 560,
    562, 565, 569, 594, 596, 598, 642,
    653, 657, 689, 696, 741, 802, 819
    ice, 380, 385, 386, 395, 407, 414, 416,
        468, 476, 480, 481, 483, 486–489,
        491–494, 503, 504, 506, 508, 509,
        515, 526, 528, 557, 571, 686
watt (W), 218
Watt, James, 218
Watzenrode the Younger, Lucas
    (Prince-Bishop of Warmia), 93
waves
    characteristics, 201–203
    harmonics, 352
        nodes, 352
    interference, 201, 203–205, 243, 274,
        293
    light
        see electromagnetic theory
    music, 203
    speed, 352
weak nuclear force, 797, 800
    see also unification
weakly interacting massive particles
    (WIMPs), 748
Wedgwood, Josiah, 213
Wegener, Alfred, 399
weight, 154, 169, 170
    relationship to mass, 169
Welles, George Orson, 410
Wells, H.G., 409, 567
western hemisphere, 27

white dwarf stars, 687–690, 702, 706, 709, 714, 749, 816
   mass limit, 693
Wien, Wilhelm, 215, 264
William of Ockham, 116
William the Conqueror, 502
Wilson, Robert Woodrow, 784
winter (first day), 18–20
   see also celestial sphere, winter solstice
winter solstice
   see celestial sphere
winter triangle, 526, 600, 623

Wollaston, William, 252
World Health Organization, 436
World War I, 255, 432, 531, 750, 777
World War II, 266, 267, 273–274, 531, 579, 730, 747
   Manhattan Project, 273
   radar, 595
World Wide Web, 797
Wyoming, 487

x-rays
   see electromagnetic waves (light), spectra

Young, Thomas, 205
   double-slit experiment, 206, 241

Zel'dovich, Yakov Borisovich, 784
zinc (Zn), 569
zirconium (Zr), 424
zirconium silicate ($ZrSiO_4$), 568
zodiac
   see pseudoscience, astrology
Zwicky, Fritz, 747, 764